Straight Line
$y = m x + b$ where $m = $ slope $= \Delta y / \Delta x$. Δy is the amount y changes when x changes by Δx.

Quadratic Equation
Roots of $ax^2 + bx + c = 0$ are

$$x_{1,2} = \frac{-b \pm \sqrt{b^2 - 4ac}}{2a}$$

If $(b^2 - 4ac) \geq 0$, roots are real; if $(b^2 - 4ac) < 0$, roots are complex conjugates.

Trigonometric Identities
$\sin(-\alpha) = -\sin \alpha$ $\cos(-\alpha) = \cos \alpha$

$\sin(\omega t \pm 90°) = \pm \cos \omega t$ $\sin(\omega t \pm 180°) = -\sin \omega t$

$\cos(\omega t \pm 90°) = \mp \sin \omega t$ $\cos(\omega t \pm 180°) = -\cos \omega t$

$$\sin^2 \omega t = \frac{1}{2}(1 - \cos 2\omega t)$$

Natural Logarithms
Given $e^x = y$. x may be found by noting that $\ln e^x = x$.
For example, if $e^{-5t} = 0.7788$, then

$\ln e^{-5t} = \ln 0.7788$

$\quad -5t = -0.25$

$\qquad t = 0.05$ s

Calculus
$\dfrac{d}{dx}(ax) = a$ $\dfrac{d}{dx}(\sin ax) = a \cos ax$

$\dfrac{d}{dx}(e^{ax}) = ae^{ax}$ $\dfrac{d}{dx}(\cos ax) = -a \sin ax$

$\dfrac{d}{dx}(uv) = u\dfrac{dv}{dx} + v\dfrac{du}{dx}$

$\displaystyle\int x dx = \frac{x^2}{2}$ $\displaystyle\int \sin ax dx = -\frac{1}{a}\cos ax$

$\displaystyle\int \frac{dx}{x} = \ln x$ $\displaystyle\int \cos ax dx = \frac{1}{a}\sin ax$

Other
$$\frac{1}{1 + x} \approx 1 - x \quad (x \ll 1)$$

CIRCUIT ANALYSIS WITH DEVICES

Theory and Practice

CIRCUIT ANALYSIS WITH DEVICES

Theory and Practice

Allan H. Robbins
Red River College, Manitoba

Wilhelm C. Miller
Red River College, Manitoba

THOMSON

DELMAR LEARNING

Australia ■ Canada ■ Mexico ■ Singapore ■ Spain ■ United Kingdom ■ United States

Circuit Analysis with Devices: Theory and Practice

Allan H. Robbins and Wilhelm C. Miller

Vice President Technology and Trades SBU:
Alar Elken

Editorial Director:
Sandy Clark

Senior Acquisitions Editors:
Greg Clayton/Dave Garza

Senior Development Editor:
Michelle Ruelos Cannistraci

Marketing Director:
Maura Theriault

Channel Manager:
Fair Huntoon

Marketing Coordinator:
Brian McGrath

Production Director:
Mary Ellen Black

Production Manager:
Larry Main

Production Coordinator:
Dawn Jacobson

Senior Project Editor:
Christopher Chien

Art/Design Coordinator:
Francis Hogan

Technology Project Manager:
David Porush

Technology Project Specialist:
Kevin Smith

Senior Editorial Assistant:
Dawn Daugherty

ISBN: 1–4018–7984–5

NOTICE TO THE READER

Contents

III
Capacitance and Inductance 335

Preface

We would like to welcome you to the first edition of *Circuit Analysis with Devices: Theory and Practice*. If you are a student, we hope it will make your journey into learning circuit theory and electronics easier and more rewarding; if you are an instructor, we hope that it will better assist you in your role as an educator.

The Book and Who It Is For

Circuit Analysis with Devices: Theory and Practice was developed specifically for use in introductory circuit analysis/basic electronics courses. Written primarily for electronics technology students at higher education colleges, universities, and career schools, as well as industry training programs, it covers fundamentals of dc and ac circuits, methods of analysis, capacitance, inductance, magnetic circuits, basic transients, introductory semiconductor theory, and basic electronic devices and their application (including amplifiers, oscillators, etc.). When students successfully complete a course using this book, they will have a good working knowledge of basic circuit principles and a demonstrated ability to solve a variety of circuit- and electronic-related problems.

Text Organization

The book contains 31 chapters and is divided into six main parts: Foundation DC Concepts, Basic DC Analysis, Capacitance and Inductance, Foundation AC Concepts, Impedance Networks, and Foundation Electronic Concepts. Chapters 1 through 4 are introductory. They cover the foundation concepts of voltage, current, resistance, Ohm's Law, and power. Chapters 5 through 9 focus on dc analysis methods. Included are Kirchhoff's Laws, series and parallel circuits, mesh and nodal analysis, Y and Δ transformations, source transformations, Thévenin's and Norton's theorems, the maximum power transfer theorem, and so on. Chapters 10 through 14 cover capacitance, magnetism, and inductance, plus magnetic circuits and simple dc transients. Chapters 15 through 17 cover foundation concepts of ac, ac voltage generation, the basic ideas of frequency, period, phase, and so on. Phasors and the impedance concept are introduced and used to solve simple problems. Power in ac circuits is investigated and the concept of power factor and the power triangle are introduced. Chapters 18 through 23 then apply these ideas. Topics include ac versions of earlier dc techniques such as mesh and nodal analysis, Thévenin's theorem, and so on, as well as new ideas such as resonance, filters, Bode techniques, and transformers. Chapters 24 to 31 cover basic electronic devices and their application, including diodes, rectifier circuits, power supply filtering, basic transistor theory, transistor amplifiers, operational amplifiers, op-amp application, oscillators, thyristers, and optical semiconductor devices.

Several appendices round out the book: Appendix A provides supplementary information on PSpice and MultiSIM; Appendix B reviews determinants and the solution of simultaneous equations; Appendix C provides additional material on the maximum power transfer theorem; and finally, Appendix D contains answers to odd-numbered end-of-chapter problems.

Required Background

Students need a working knowledge of basic algebra and trigonometry and the ability to solve second-order linear equations such as those found in mesh analysis. They should be familiar with the SI metric system and the atomic nature of matter. In terms of higher math, calculus is introduced gradually in later chapters to aid in the development of ideas. However, optional derivations and problems using calculus (which are provided for enrichment purposes) are marked by an ∫ icon and may be omitted in those programs that do not require the use of calculus.

Features of the Book

Since the most important attribute of a book is its value to the user, we have created a textbook and companion learning package with features to help students and staff in a variety of ways. These features include

- **A clearly written, easy-to-understand style** that emphasizes principles and concepts.
- **Over 1600 diagrams and photos.** Color and 3D visual effects are used to illustrate and clarify ideas and to aid visual learners.
- **Examples.** Hundreds of examples worked out in step-by-step detail help promote understanding and guide the student in problem solving.

Each chapter opens with Key Terms, Outline, Objectives, Chapter Preview, and Putting It in Perspective.

Series Circuits

5

■ **KEY TERMS**

Electric Circuit
Ground
Kirchhoff's Voltage Law
Loading Effect (Ammeter)
Ohmmeter Design
Point Sources
Series Connection
Total Equivalent Resistance
Voltage Divider Rule
Voltage Subscripts

■ **OUTLINE**

Series Circuits
Kirchhoff's Voltage Law
Resistors in Series
Voltage Sources in Series
Interchanging Series Components
The Voltage Divider Rule
Circuit Ground
Voltage Subscripts
Internal Resistance of Voltage
  Sources
Ammeter Loading Effects
Circuit Analysis Using Computers

■ **OBJECTIVES**

After studying this chapter, you will be able to

- determine the total resistance in a series circuit and calculate circuit current,
- use Ohm's law and the voltage divider rule to solve for the voltage across all resistors in the circuit,
- express Kirchhoff's voltage law and use it to analyze a given circuit,
- solve for the power dissipated by any resistor in a series circuit and show that the total power dissipated is exactly equal to the power delivered by the voltage source,
- solve for the voltage between any two points in a series or parallel circuit,
- calculate the loading effect of an ammeter in a circuit,
- use computers to assist in the analysis of simple series circuits.

CHAPTER PREVIEW

In the previous chapter we examined the interrelation of current, voltage, resistance, and power in a single resistor circuit. In this chapter we will expand on these basic concepts to examine the behavior of circuits having several resistors in series.

We will use Ohm's law to derive the voltage divider rule and to verify Kirchhoff's voltage law. A good understanding of these important principles provides an important base upon which further circuit analysis techniques are built. Kirchhoff's voltage law and Kirchhoff's current law, which will be covered in the next chapter, are fundamental in understanding *all* electrical and electronic circuits.

After developing the basic framework of series circuit analysis, we will apply the ideas to analyze and design simple voltmeters and ohmmeters. While meters are usually covered in a separate instruments or measurements course, we examine these circuits merely as an application of the concepts of circuit analysis.

Similarly, we will observe how circuit principles are used to explain the loading effect of an ammeter placed in series with a circuit. ■

Gustav Robert Kirchhoff

KIRCHHOFF WAS A GERMAN PHYSICIST born on March 12, 1824, in Königsberg, Prussia. His first research was on the conduction of electricity, which led to his presentation of the laws of closed electric circuits in 1845. Kirchhoff's current law and Kirchhoff's voltage law apply to all electrical circuits and therefore are fundamentally important in understanding circuit operation. Kirchhoff was the first to verify that an electrical impulse travelled at the speed of light.

Although these discoveries have immortalized Kirchhoff's name in electrical science, he is better known for his work with R. W. Bunsen in which he made major contributions to the study of spectroscopy and advanced the research into blackbody radiation.

Kirchhoff died in Berlin on October 17, 1887. ■

PUTTING IT IN PERSPECTIVE

- **Over 2000 end-of-chapter problems, Practice Problems, and In-Process Learning Check problems.**

- **Practice Problems.** These follow closely to where key ideas have been introduced and encourage the student to practice the skills just learned.

- **In-Process Learning Checks.** These are short, self-test quizzes that provide a quick review of the material just learned and help identify learning gaps.

- **Putting It into Practice.** These are tasks (one per chapter) that present users with a group of more challenging, project-like problems that require them to reason their way through realistic situations similar to what may be encountered in practice.

- **Putting It in Perspective.** These are short vignettes that provide interesting background on people, events, and ideas that led to major advances or contributions in electrical science.

- **Chapter Previews** to provide a context and a brief overview for the upcoming chapter and to answer the question "Why am I learning this?"

- **Competency-based objectives** to define the knowledge or skill that the student is expected to gain from each chapter.

- **Key terms** at the beginning of each chapter to identify new terms to be introduced.

Hundreds of Examples with detailed solutions are found throughout each chapter.

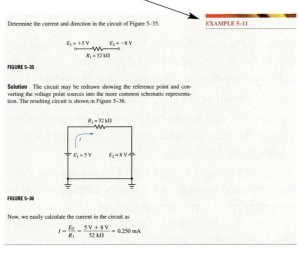

MultiSIM and PSpice are used to illustrate circuit simulations. End-of-chapter problems can be solved by using these computer simulation programs.

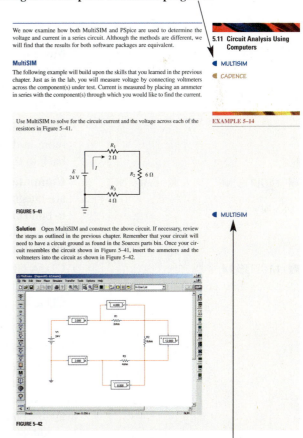

In-Process Learning Checks provide a quick review of each section.

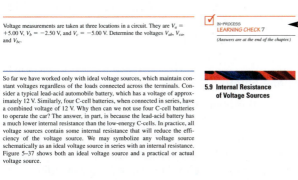

MultiSIM circuits are available on the accompanying CD. A MultiSIM icon is placed beside those selected circuits.

Practice Problems build problem-solving skills and test student's understanding.

Putting It into Practice boxes are found at the end of the chapters and describe a work-related problem.

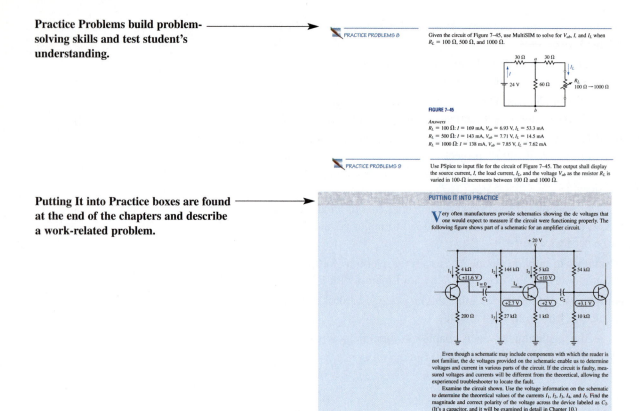

PRACTICE PROBLEMS 8

Given the circuit of Figure 7–45, use MultiSIM to solve for V_{ab}, I, and I_L when $R_L = 100\ \Omega$, $500\ \Omega$, and $1000\ \Omega$.

FIGURE 7–45

Answers
$R_L = 100\ \Omega$: $I = 169$ mA, $V_{ab} = 6.93$ V, $I_L = 53.3$ mA
$R_L = 500\ \Omega$: $I = 143$ mA, $V_{ab} = 7.71$ V, $I_L = 14.5$ mA
$R_L = 1000\ \Omega$: $I = 138$ mA, $V_{ab} = 7.85$ V, $I_L = 7.62$ mA

PRACTICE PROBLEMS 9

Use PSpice to input file for the circuit of Figure 7–45. The output shall display the source current, I, the load current, I_L, and the voltage V_{ab} as the resistor R_L is varied in 100-Ω increments between 100 Ω and 1000 Ω.

PUTTING IT INTO PRACTICE

Very often manufacturers provide schematics showing the dc voltages that one would expect to measure if the circuit were functioning properly. The following figure shows part of a schematic for an amplifier circuit.

Even though a schematic may include components with which the reader is not familiar, the dc voltages provided on the schematic enable us to determine voltages and current in various parts of the circuit. If the circuit is faulty, measured voltages and currents will be different from the theoretical, allowing the experienced troubleshooter to locate the fault.

Examine the circuit shown. Use the voltage information on the schematic to determine the theoretical values of the currents I_1, I_2, I_3, I_4, and I_5. Find the magnitude and correct polarity of the voltage across the device labeled as C_2. (It's a capacitor, and it will be examined in detail in Chapter 10.)

- **Margin Notes.** These include Practical Notes (which provide practical information, e.g., tips on how to use meters) plus general notes that provide additional information or add a perspective to the material being studied.

- **Answers to odd-numbered problems** in an appendix.

- **RealAudio clips.** These clips (found on the textbook's web site) present a more in-depth discussion of the most difficult topic for each chapter and tie back to the text via an icon in the text margin.

◀ MULTISIM   ◀ CADENCE
- **Computer simulations.** MultiSIM and Cadence PSpice simulations (which are integrated throughout the text) provide step-by-step instructions on how to build circuits on your screen, plus actual screen captures to show you what you should see when you run the simulations. Specific simulation problems are indicated by MultiSim and Cadence symbols.

◀ MULTISIM
- **CD.** A CD in back of the book includes MultiSIM Circuit files. You will see a MultiSIM icon placed beside those figures that are available on circuit files. CD also includes Textbook Edition of MultiSIM.

About PSPice Versions and Editors

The version of PSpice used in this book (the version current at the time of writing) is Orcad PSpice Student Version 9.1 from Cadence Design Systems Inc. (To download a free copy, see Appendix A. An alternative choice is PSpice Lite available on CD-ROM.) With Orcad PSpice, you get a choice of schematic capture editors—at run time, you can select either the *Orcad Capture* editor or the *Schematics* editor. Since *Capture* is Orcad's preferred editor, we have chosen to use it in this book. However, for those who prefer the user-friendly interface that PSpice Version 8 users are familiar with, we have provided duplicate coverage using *Schematics* on the book's web site.

More on the Learning Package

The complete ancillary package was developed to achieve two goals:

1. To assist students in learning the essential information needed to prepare for the field of electronics.
2. To assist instructors in planning and implementing their instructional programs for the most efficient use of time and other resources.

The *Circuit Analysis with Devices: Theory and Practice* package was created as an integrated whole. Supplements are linked to and integrated with the text to create a comprehensive supplement package that supports students and instructors. The package includes:

Laboratory Manual

The lab manual contains over 40 hands-on labs, most with integrated computer simulation exercises plus a comprehensive guide to equipment and laboratory measurements. Solutions to the lab manual may be found in the Instructor's Guide. ISBN: 1401811582.

Instructor's Guide

Contains step-by-step solutions to all end-of-chapter (even and odd) problems, including waveforms, circuit diagrams and more. ISBN: 1401812384.

e.resource™

Available all on one CD-ROM are tools and instructional resources to enrich your classroom. The elements of e.resource link directly to the text and tie together to provide a unified instructional system. ISBN: 1401811574.

Features contained in the e.resource include:

PowerPoint® Presentation Slides. Provides customizable presentations for classroom use. Slides are prepared for every chapter of the book that help you present key points and concepts. Graphics from the Image Library or your own images can be imported to create individualized classroom presentations.

Image Library. Includes selected full-color images from the textbook, providing the instructor with another means of promoting student understanding. The Image Library allows the instructor to display or print images for a classroom presentation.

Computerized Testbank. Over 1000 questions for use in creating tests of varying levels so you can assess student comprehension.

Electronics Technology Homepage. You can link directly to the Delmar Electronics Technology web site and to the textbook's Online Companion for additional resources.

Online Companion™

◀ **Online Companion**

This companion web site is intended for use by both educators and students. It provides ongoing assistance in the form of Real Audio Sound Files, technology updates, additional problems, circuit schematics, and general information.

Please visit our web site at *www.electronictech.com* for more details.

To the Student

Learning circuit theory and electronics should be challenging, interesting, and (hopefully) fun. However, it is also hard work, since the knowledge and skills that you seek can only be gained through practice. We offer a few guidelines.

1. As you go through the material, try to gain an appreciation of where the theory comes from—i.e., the basic experimental laws on which it is based. This will help you better understand the foundation ideas on which the theory is built.

2. Learn the terminology and definitions. Important new terms are introduced frequently. Learn what they mean and where they are used.

3. Study each new section carefully and be sure that you understand the basic ideas and how they are put together. Work your way through the examples with your calculator. Try the practice problems, then the end-of-chapter problems. Not every concept will be clear immediately and most likely many will require several readings before you gain an adequate understanding.

4. When you are ready, test your understanding using the In-Process Learning Checks (self-quizzes) located in each chapter.

5. When you have mastered the material, move on to the next block. For those concepts that you are having difficulty with, consult your instructor or some other authoritative source.

Calculators for Circuit Analysis and Electronics

You will need a good scientific calculator. A good calculator will permit you to more easily master the numerical aspects of problem solving, thereby leaving you more time to concentrate on the theory itself. This is especially true for ac, where complex number work dominates. There are some inexpensive calculators on the market that handle complex-number arithmetic almost as easily as real-number arithmetic. Such calculators save an enormous amount of time. You should acquire such a calculator (after consulting with your instructor), and learn to use it proficiently.

Acknowledgments

Many people have contributed to the development of *Circuit Analysis with Devices: Theory and Practice,* and we would like to thank them. First, the reviewers and accuracy checkers: no textbook can be successful without the dedication and commitment of such people. We thank the following:

Reviewers

Sami Antoun, DeVry University, Columbus, OH

G. Thomas Bellarmine, Florida A & M University

Harold Broberg, Purdue University

William Conrad, IUPUI—Indiana University, Purdue University

David Delker, Kansas State University

Timothy Haynes, Haywood Community College

Bruce Johnson, University of Nevada

Jim Pannell, DeVry University, Irving, TX

Alan Price, DeVry University, Pomona, CA

Philip Regalbuto, Trident Technical College

Carlo Sapijaszko, DeVry University, Orlando, FL

Jeffrey Schwartz, DeVry University, Long Island City, NY

John Sebeson, DeVry University, Addison, IL

Parker Sproul, DeVry University, Phoenix, AZ

Lloyd E. Stallkamp, Montana State University

Roman Stemprok, University of Texas

Richard Sturtevant, Springfield Tech Community College

Technical Accuracy Reviewers

Chia-chi Tsui, DeVry University, Long Island City, NY

Rudy Hofer, Conestoga College, Kitchener, Ontario, Canada

Marie Sichler, Red River College, Winnipeg, Manitoba, Canada

The following firms and individuals supplied photographs, diagrams, and other useful information:

Allen-Bradley	Illinois Capacitor Inc.
AT & T	Interactive Images Technologies
AVX Corporation	JBL Professional
B + K Precision	John Fluke Mfg. Co. Inc.
Bourns Inc.	Siemens Solar Industries
Butterworth & Co. Ltd.	Tektronix
Cadence Design Systems Inc.	Transformers Manufacturers Inc.
Condor DC Power Supplies Inc.	Vansco Electronics

We express our deep appreciation to the staff at Delmar Learning for their tireless efforts in putting this book together: To Greg Clayton, our Electronics Editor, for direction and encouragement; Michelle Ruelos Cannistraci, our Development Editor, for encouragement, advice, and making sure we got things done on time; Christopher Chien, Senior Project Editor, for his skill in editing and pulling the final project together; Francis Hogan, Art & Design Coordinator for his guidance in preparing the art; Larry Main, Production Manager, and their staffs for making the project work on such an impossibly short deadline. We also wish to thank Larry Goldberg and his colleagues at Shepherd Incorporated for guiding the book through the editing, design, and production stages. A special thanks to all of you.

Lastly, we thank our wives and families for their support and perseverance during the preparation of this book.

Allan H. Robbins
Wilhelm C. Miller
July, 2003

About the Authors

Allan H. Robbins graduated from engineering with a Bachelor's degree and a Master's degree in Electrical Engineering. In graduate school, he specialized in circuit theory. Allan, who was formerly head of the Department of Electrical and Computer Technology at Red River College, has been an instructor for over 30 years. In addition to his academic career, he has been a consultant and a small business partner. He began writing as a contributing author for Osborne-McGraw-Hill in the computer field and is also joint author of one other textbook. He has served as Section Chairman for the IEEE and as a member of the board for the Electronics Industry Association of Manitoba.

Wilhelm (Will) C. Miller obtained a diploma in Electronic Engineering Technology from Red River Community College (now RRC) and later graduated from the University of Winnipeg with a degree in Physics and Mathematics. He worked in the communications field for 10 years, including a one-year assignment with Saudi PTT in Jeddah, Saudi Arabia. Will was an instructor in the Electronics and Computer Engineering Technologies for 20 years, having taught at Red River College and College of The Bahamas (Nassau, Bahamas). He currently serves as Chair of the EET programs at Red River College. Will is the Chair of the Panel of Examiners for CTTAM (the Certified Technicians and Technologists Association of Manitoba) and is past president of the board of directors of CTTAM.

Foundation dc Concepts

I

Circuit theory provides the tools and concepts needed to understand and analyze electrical and electronic circuits. The foundations of this theory were laid down over the past several hundred years by a number of pioneer researchers. In 1780, Alessandro Volta of Italy developed an electric cell (battery) that provided the first source of what we now call dc voltage. Around the same time, the concept of current was evolved (even though nothing was known about the atomic structure of matter until much later). In 1826, Georg Simon Ohm of Germany brought the two ideas together and experimentally determined the relationship between voltage and current in a resistive circuit. This result, known as Ohm's law, set the stage for the development of modern-day circuit theory.

In Part I, we examine the foundation of this theory. We look at voltage, current, power, energy, and the relationships between them. The ideas developed here are the fundamental ideas upon which all circuit theory is built. ■

■ KEY TERMS

Application Packages
Base
Block Diagram
Circuit Theory
Conversion Factor
Engineering Notation
Exponent
Horsepower
Joule
Newton
Pictorial Diagram
Power of Ten Notation
Prefixes
Programming Language
Resistance
Schematic Diagram
Scientific Notation
SI System
SPICE
Watt

■ OUTLINE

Introduction
The SI System of Units
Converting Units
Power of Ten Notation
Prefixes
Circuit Diagrams
Circuit Analysis Using Computers

■ OBJECTIVES

After studying this chapter, you will be able to

- describe the SI system of measurement,
- convert between various sets of units,
- use power of ten notation to simplify handling of large and small numbers,
- express electrical units using standard prefix notation such as μA, kV, mW, etc.,
- use a sensible number of significant digits in calculations,
- describe what block diagrams are and why they are used,
- convert a simple pictorial circuit to its schematic representation,
- describe generally how computers fit in the electrical and electronic circuit analysis picture.

Introduction

1

An electrical circuit is a system of interconnected components such as resistors, capacitors, inductors, voltage sources, and so on. The electrical behavior of these components is described by a few basic experimental laws. These laws and the principles, concepts, mathematical relationships, and methods of analysis that have evolved from them are known as **circuit theory.**

Much of circuit theory deals with problem solving and numerical analysis. When you analyze a problem or design a circuit, for example, you are typically required to compute values for voltage, current, and power. In addition to a numerical value, your answer must include a unit. The system of units used for this purpose is the SI system (Systéme International). The SI system is a unified system of metric measurement; it encompasses not only the familiar MKS (meters, kilograms, seconds) units for length, mass, and time, but also units for electrical and magnetic quantities as well.

Quite frequently, however, the SI units yield numbers that are either too large or too small for convenient use. To handle these, engineering notation and a set of standard prefixes have been developed. Their use in representation and computation is described and illustrated.

Since circuit theory is somewhat abstract, diagrams are used to help present ideas. We look at several types—schematic, pictorial, and block diagrams—and show how to use them to represent circuits and systems.

We conclude the chapter with a brief look at computer usage in circuit analysis and design. Several popular prepackaged software products are described, with special emphasis on Orcad PSpice® from Cadence Design Systems Inc. and Electronic Workbench's MultiSIM® from Interactive Image Technologies Ltd., the two software tools used throughout the book. ■

PUTTING IT IN PERSPECTIVE

Hints on Problem Solving

DURING THE ANALYSIS OF ELECTRIC circuits, you will find yourself solving quite a few problems. An organized approach helps. Listed below are some useful guidelines:

1. Make a sketch (e.g., a circuit diagram), mark on it what you know, then identify what it is that you are trying to determine. Watch for "implied data" such as the phrase "the capacitor is initially uncharged". (As you will find out later, this means that the initial voltage on the capacitor is zero.) Be sure to convert all implied data to explicit data.
2. Think through the problem to identify the principles involved, then look for relationships that tie together the unknown and known quantities.
3. Substitute the known information into the selected equation(s) and solve for the unknown. (For complex problems, the solution may require a series of steps involving several concepts. If you cannot identify the complete set of steps before you start, start anyway. As each piece of the solution emerges, you are one step closer to the answer. You may make false starts. However, even experienced people do not get it right on the first try every time. Note also that there is seldom one "right" way to solve a problem. You may therefore come up with an entirely different correct solution method than the authors do.)
4. Check the answer to see that it is sensible—that is, is it in the "right ballpark"? Does it have the correct sign? Do the units match? ■

1.1 Introduction

◀ **Online Companion**

Technology has dramatically changed the way we do things; we now have computers in our homes, electronic control systems in our cars, cellular phones that can be used just about anywhere, robots that assemble products on production lines, and so on.

A first step to understanding these technologies is electric circuit theory. Circuit theory provides you with the knowledge of basic principles that you need to understand the behavior of electric and electronic devices, circuits, and systems. In this book, we develop and explore its basic ideas.

Before We Begin

Before we begin, let us look at a few examples of the technology at work. (As you go through these, you will see devices, components, and ideas that have not yet been discussed. You will learn about these later. For the moment, just concentrate on the general ideas.)

As a first example, consider Figure 1–1, which shows a home theater system. This system relies on electrical and electronic circuits, magnetic circuits, and laser technology for its operation. For example, resistors, capacitors, and integrated circuits are used to control the voltages and currents that operate its motors and to amplify its audio and video signals, while laser circuitry is used to read data from the disks. The speaker system (which includes subwoofers) relies on magnetic circuits for its operation, while another magnetic circuit (the power transformer) drops the ac voltage from the 120-volt wall outlet voltage to the lower levels required to power the system.

Figure 1–2 shows another example. Here, a computer-generated screen shot of the magnetic flux pattern for an electric motor illustrates the use of computers in research and design. Programmed to apply basic magnetic circuit fundamentals to complex shapes, software packages such as that used here help make it possible to develop more efficient and better performing motors, computer disk drives, audio speaker systems, and the like.

FIGURE 1–1   A home theater system.

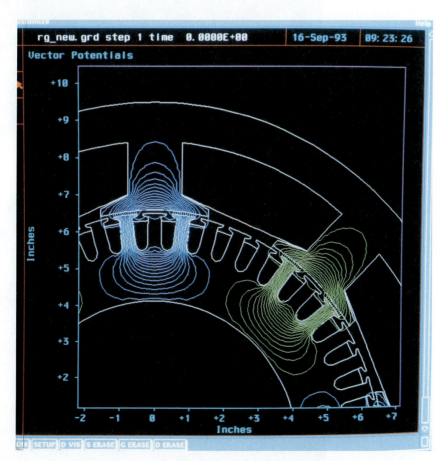

FIGURE 1–2   Computer-generated magnetic flux pattern for a dc motor, armature only excited. (*Courtesy GE Research and Development Center*)

Figure 1–3 shows another application, a manufacturing facility where fine pitch surface-mount (SMT) components are placed on printed circuit boards at high speed using laser centering and optical verification. The bottom row of Figure 1–4 shows how small these components are. Computer control provides the high precision needed to accurately position parts as tiny as these.

FIGURE 1–3   Laser centering and optical verification in a manufacturing process. *(Courtesy Vansco Electronics Ltd.)*

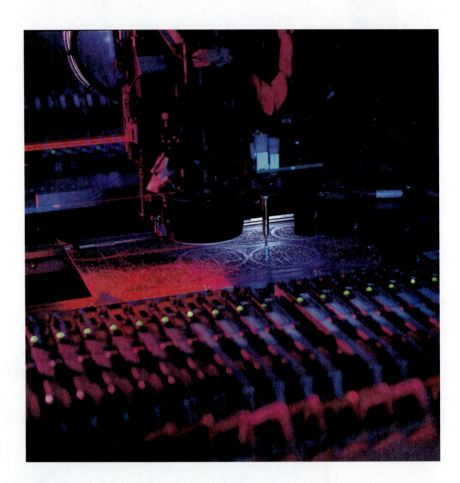

FIGURE 1–4   Some typical electronic components. The small components at the bottom are surface mount parts that are installed on printed circuit boards by the machine shown in Figure 1–3.

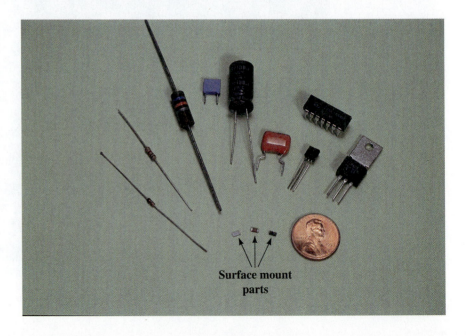

Surface mount
parts

Before We Move On

Before we move on, we should note that, as diverse as these applications are, they all have one thing in common: all are rooted in the principles of circuit theory.

The solution of technical problems requires the use of units. At present, two major systems—the English (US Customary) and the metric—are in everyday use. For scientific and technical purposes, however, the English system has been almost totally superseded. In its place the **SI system** is used. Table 1–1 shows a few frequently encountered quantities with units expressed in both systems.

The SI system combines the MKS metric units and the electrical units into one unified system: See Tables 1–2 and 1–3. (Do not worry about the electrical units yet. We define them later, starting in Chapter 2.) The units in Table 1–2 are defined units, while the units in Table 1–3 are derived units, obtained by combining units from Table 1–2. Note that some symbols and abbreviations use capital letters while others use lowercase letters.

A few non-SI units are still in use. For example, electric motors are commonly rated in horsepower, and wires are frequently specified in AWG sizes (American Wire Gage, Section 3.2). On occasion, you will need to convert non-SI units to SI units. Table 1–4 may be used for this purpose.

1.2 The SI System of Units

TABLE 1–1 Common Quantities
1 meter = 100 centimeters
= 39.37 inches
1 millimeter = 39.37 mils
1 inch = 2.54 centimeters
1 foot = 0.3048 meter
1 yard = 0.9144 meter
1 mile = 1.609 kilometers
1 kilogram = 1000 grams
= 2.2 pounds
1 gallon (US) = 3.785 liters

Definition of Units

When the metric system came into being in 1792, the meter was defined as one tenmillionth of the distance from the north pole to the equator and the second as $\frac{1}{60} \times \frac{1}{60} \times \frac{1}{24}$ of the mean solar day. Later, more accurate definitions based on physical laws of nature were adopted. The meter is now defined as the distance travelled

TABLE 1–2 Some SI Base Units			
Quantity	**Symbol**	**Unit**	**Abbreviation**
Length	l	meter	m
Mass	m	kilogram	kg
Time	t	second	s
Electric current	I, i	ampere	A
Temperature	T	kelvin	K

TABLE 1–3 Some SI Derived Units*			
Quantity	**Symbol**	**Unit**	**Abbreviation**
Force	F	newton	N
Energy	W	joule	J
Power	P, p	watt	W
Voltage	V, v, E, e	volt	V
Charge	Q, q	coulomb	C
Resistance	R	ohm	Ω
Capacitance	C	farad	F
Inductance	L	henry	H
Frequency	f	hertz	Hz
Magnetic flux	Φ	weber	Wb
Magnetic flux density	B	tesla	T

*Electrical and magnetic quantities will be explained as you progress through the book. As in Table 1–2, the distinction between capitalized and lowercase letters is important.

TABLE 1–4 Conversions

	When You Know	Multiply By	To Find
Length	inches (in)	0.0254	meters (m)
	feet (ft)	0.3048	meters (m)
	miles (mi)	1.609	kilometers (km)
Force	pounds (lb)	4.448	newtons (N)
Power	horsepower (hp)	746	watts (W)
Energy	kilowatthour (kWh)	3.6×10^6	joules† (J)
	foot-pound (ft-lb)	1.356	joules† (J)

† 1 joule $=$ 1 newton-meter.

by light in a vacuum in 1/299 792 458 of a second, while the second is defined in terms of the period of a cesium-based atomic clock. The definition of the kilogram is the mass of a specific platinum-iridium cylinder (the international prototype), preserved at the International Bureau of Weights and Measures in France.

Relative Size of the Units*

To gain a feel for the SI units and their relative size, refer to Tables 1–1 and 1–4. Note that 1 meter is equal to 39.37 inches; thus, 1 inch equals 1/39.37 = 0.0254 meter or 2.54 centimeters. A force of one pound is equal to 4.448 newtons; thus, 1 **newton** is equal to 1/4.448 = 0.225 pound of force, which is about the force required to lift a ¼-pound weight. One **joule** is the work done in moving a distance of one meter against a force of one newton. This is about equal to the work required to raise a quarter-pound weight one meter. Raising the weight one meter in one second requires about one **watt** of power.

The watt is also the SI unit for electrical power. A typical electric lamp, for example, dissipates power at the rate of 60 watts, and a toaster at a rate of about 1000 watts.

The link between electrical and mechanical units can be easily established. Consider an electrical generator. Mechanical power input produces electrical power output. If the generator were 100% efficient, then one watt of mechanical power input would yield one watt of electrical power output. This clearly ties the electrical and mechanical systems of units together.

However, just how big is a watt? While the above examples suggest that the watt is quite small, in terms of the rate at which a human can work it is actually quite large. For example, a person can do manual labor at a rate of about 60 watts when averaged over an 8-hour day—just enough to power a standard 60-watt electric lamp continuously over this time! A horse can do considerably better. Based on experiment, James Watt determined that a strong dray horse could average 746 watts. From this, he defined the **horsepower** (hp) as 1 horsepower = 746 watts. This is the figure that we still use today.

1.3 Converting Units

Sometimes quantities expressed in one unit must be converted to another. For example, suppose you want to determine how many kilometers there are in ten miles. Given that 1 mile is equal to 1.609 kilometers, Table 1–1, you can write 1 mi = 1.609 km, using the abbreviations in Table 1–4. Now multiply both sides by 10. Thus, 10 mi = 16.09 km.

---

*Paraphrased from Edward C. Jordan and Keith Balmain, *Electromagnetic Waves and Radiating Systems,* Second Edition. (Englewood Cliffs, New Jersey: Prentice-Hall, Inc, 1968).

This procedure is quite adequate for simple conversions. However, for complex conversions, it may be difficult to keep track of units. The procedure outlined next helps. It involves writing units into the conversion sequence, cancelling where applicable, then gathering up the remaining units to ensure that the final result has the correct units.

To get at the idea, suppose you want to convert 12 centimeters to inches. From Table 1–1, 2.54 cm = 1 in. From this, you can write

$$\frac{2.54 \text{ cm}}{1 \text{ in}} = 1 \quad \text{or} \quad \frac{1 \text{ in}}{2.54 \text{ cm}} = 1 \qquad (1–1)$$

The quantities in Equation 1–1 are called **conversion factors.** As you can see, conversion factors have a value of 1 and thus you can multiply them times any expression without changing the value of that expression. For example, to complete the conversion of 12 cm to inches, choose the second ratio (so that units cancel), then multiply. Thus,

$$12 \text{ cm} = 12 \text{ cm} \times \frac{1 \text{ in}}{2.54 \text{ cm}} = 4.72 \text{ in}$$

When you have a chain of conversions, select factors so that all unwanted units cancel. This provides an automatic check on the final result as illustrated in part (b) of Example 1–1.

EXAMPLE 1–1

Given a speed of 60 miles per hour (mph),

a. convert it to kilometers per hour,

b. convert it to meters per second.

Solution

a. Recall, 1 mi = 1.609 km. Thus,

$$1 = \frac{1.609 \text{ km}}{1 \text{ mi}}$$

Now multiply both sides by 60 mi/h and cancel units:

$$60 \text{ mi/h} = \frac{60 \text{ mi}}{h} \times \frac{1.609 \text{ km}}{1 \text{ mi}} = 96.54 \text{ km/h}$$

b. Given that 1 mi = 1.609 km, 1 km = 1000 m, 1 h = 60 min, and 1 min = 60 s, choose conversion factors as follows:

$$1 = \frac{1.609 \text{ km}}{1 \text{ mi}}, \quad 1 = \frac{1000 \text{ m}}{1 \text{ km}}, \quad 1 = \frac{1 \text{ h}}{60 \text{ min}}, \quad \text{and } 1 = \frac{1 \text{ min}}{60 \text{ s}}$$

Thus,

$$\frac{60 \text{ mi}}{h} = \frac{60 \text{ mi}}{h} \times \frac{1.609 \text{ km}}{1 \text{ mi}} \times \frac{1000 \text{ m}}{1 \text{ km}} \times \frac{1 \text{ h}}{60 \text{ min}} \times \frac{1 \text{ min}}{60 \text{ s}} = 26.8 \text{ m/s}$$

You can also solve this problem by treating the numerator and denominator separately. For example, you can convert miles to meters and hours to seconds, then divide (see Example 1–2). In the final analysis, both methods are equivalent.

EXAMPLE 1–2

Do Example 1–1(b) by expanding the top and bottom separately.

Solution

$$60 \text{ mi} = 60 \text{ mi} \times \frac{1.609 \text{ km}}{1 \text{ mi}} \times \frac{1000 \text{ m}}{1 \text{ km}} = 96\,540 \text{ m}$$

$$1 \text{ h} = 1 \text{ h} \times \frac{60 \text{ min}}{1 \text{ h}} \times \frac{60 \text{ s}}{1 \text{ min}} = 3600 \text{ s}$$

Thus, velocity = 96 540 m/3600 s = 26.8 m/s as above.

PRACTICE PROBLEMS 1

1. Area = πr^2. Given $r = 8$ inches, determine area in square meters (m²).
2. A car travels 60 feet in 2 seconds. Determine

   a. its speed in meters per second,

   b. its speed in kilometers per hour.

For part (b), use the method of Example 1–1, then check using the method of Example 1–2.

Answers
1. 0.130 m²; 2. a. 9.14 m/s, b. 32.9 km/h

1.4 Power of Ten Notation

Electrical values vary tremendously in size. In electronic systems, for example, voltages may range from a few millionths of a volt to several thousand volts, while in power systems, voltages of up to several hundred thousand are common. To handle this large range, the **power of ten notation** (Table 1–5) is used.

To express a number in power of ten notation, move the decimal point to where you want it, then multiply the result by the power of ten needed to restore the number to its original value. Thus, 247 000 = 2.47 × 10⁵. (The number 10 is called the **base,** and its power is called the **exponent.**) An easy way to determine the exponent is to count the number of places (right or left) that you moved the decimal point. Thus,

$$247\,000 = 2\,4\,7\,0\,0\,0 = 2.47 \times 10^5$$

$$5\,4\,3\,2\,1$$

Similarly, the number 0.003 69 may be expressed as 3.69 × 10⁻³ as illustrated below.

$$0.003\,69 = 0.0\,0\,3\,6\,9 = 3.69 \times 10^{-3}$$

$$1\,2\,3$$

TABLE 1–5 Common Power of Ten Multipliers	
$1\,000\,000 = 10^6$	$0.000001 = 10^{-6}$
$100\,000 = 10^5$	$0.00001 = 10^{-5}$
$10\,000 = 10^4$	$0.0001 = 10^{-4}$
$1\,000 = 10^3$	$0.001 = 10^{-3}$
$100 = 10^2$	$0.01 = 10^{-2}$
$10 = 10^1$	$0.1 = 10^{-1}$
$1 = 10^0$	$1 = 10^0$

Multiplication and Division Using Powers of Ten

To multiply numbers in power of ten notation, multiply their base numbers, then add their exponents. Thus,

$$(1.2 \times 10^3)(1.5 \times 10^4) = (1.2)(1.5) \times 10^{(3+4)} = 1.8 \times 10^7$$

For division, subtract the exponents in the denominator from those in the numerator. Thus,

$$\frac{4.5 \times 10^2}{3 \times 10^{-2}} = \frac{4.5}{3} \times 10^{2-(-2)} = 1.5 \times 10^4$$

EXAMPLE 1–3

Convert the following numbers to power of ten notation, then perform the operation indicated:

a. 276×0.009,

b. $98\ 200/20$.

Solution

a. $276 \times 0.009 = (2.76 \times 10^2)(9 \times 10^{-3}) = 24.8 \times 10^{-1} = 2.48$

b. $\dfrac{98\ 200}{20} = \dfrac{9.82 \times 10^4}{2 \times 10^1} = 4.91 \times 10^3$

Addition and Subtraction Using Powers of Ten

To add or subtract, first adjust all numbers to the same power of ten. It does not matter what exponent you choose, as long as all are the same.

EXAMPLE 1–4

Add 3.25×10^2 and 5×10^3

a. using 10^2 representation,

b. using 10^3 representation.

Solution

a. $5 \times 10^3 = 50 \times 10^2$. Thus, $3.25 \times 10^2 + 50 \times 10^2 = 53.25 \times 10^2$

b. $3.25 \times 10^2 = 0.325 \times 10^3$. Thus, $0.325 \times 10^3 + 5 \times 10^3 = 5.325 \times 10^3$, which is the same as 53.25×10^2 as found in part a.

Powers

Raising a number to a power is a form of multiplication (or division if the exponent is negative). For example,

$$(2 \times 10^3)^2 = (2 \times 10^3)(2 \times 10^3) = 4 \times 10^6$$

In general, $(N \times 10^n)^m = N^m \times 10^{nm}$. In this notation, $(2 \times 10^3)^2 = 2^2 \times 10^{3 \times 2} = 4 \times 10^6$ as before.

Integer fractional powers represent roots. Thus, $4^{1/2} = \sqrt{4} = 2$ and $27^{1/3} = \sqrt[3]{27} = 3$.

NOTES . . .

Use common sense when handling numbers. With calculators, for example, it is often easier to work directly with numbers in their original form than to convert them to power of ten notation. (As an example, it is more sensible to multiply 276×0.009 directly than to convert to power of ten notation as we did in Example 1–3(a).) If the final result is needed as a power of ten, you can convert as a last step.

EXAMPLE 1–5

Expand the following:

a. $(250)^3$ b. $(0.0056)^2$ c. $(141)^{-2}$ d. $(60)^{1/3}$

Solution

a. $(250)^3 = (2.5 \times 10^2)^3 = (2.5)^3 \times 10^{2 \times 3} = 15.625 \times 10^6$

b. $(0.0056)^2 = (5.6 \times 10^{-3})^2 = (5.6)^2 \times 10^{-6} = 31.36 \times 10^{-6}$

c. $(141)^{-2} = (1.41 \times 10^2)^{-2} = (1.41)^{-2} \times (10^2)^{-2} = 0.503 \times 10^{-4}$

d. $(60)^{1/3} = \sqrt[3]{60} = 3.915$

PRACTICE PROBLEMS 2

Determine the following:

a. $(6.9 \times 10^5)(0.392 \times 10^{-2})$

b. $(23.9 \times 10^{11})/(8.15 \times 10^5)$

c. $14.6 \times 10^2 + 11.2 \times 10^1$ (Express in 10^2 and 10^1 notation.)

d. $(29.6)^3$

e. $(0.385)^{-2}$

Answers
a. 2.70×10^3; b. 2.93×10^6; c. $15.72 \times 10^2 = 157.2 \times 10^1$; d. 25.9×10^3; e. 6.75

1.5 Prefixes

Scientific and Engineering Notation

In scientific and engineering work, a very large or a very small number is usually expressed as a base number times a power of 10. If such a number is written with one digit to the left of the decimal place, it is said to be in **scientific notation**, e.g., 2.47×10^5 is in scientific notation, but 24.7×10^4 is not. In circuit theory, however, we are more interested in **engineering notation.** In this notation, only certain powers of 10 as in Table 1–6 are used. These specific powers have **prefix** abbreviations and symbols as shown, and it is preferable to use these rather than powers of 10. Thus, although a current of 0.0045 A (amperes) can be expressed as 4.5×10^{-3} A, it is preferable to express it as 4.5 mA (or 4.5 milliamps). Usually we select a prefix that results in a base number between 0.1 and 999. However, there are often several equally acceptable ways to express a quantity. For example, 15×10^{-5} s is commonly expressed as either 150 μs or 0.15 ms.

TABLE 1–6   Engineering Prefixes

Power of 10	Prefix	Symbol
10^{12}	tera	T
10^9	giga	G
10^6	mega	M
10^3	kilo	k
10^{-3}	milli	m
10^{-6}	micro	μ
10^{-9}	nano	n
10^{-12}	pico	p

EXAMPLE 1–6

Express the following in engineering notation:

a. 10×10^4 volts   b. 0.1×10^{-3} watts   c. 250×10^{-7} seconds

Solution

a. 10×10^4 V $= 100 \times 10^3$ V $= 100$ kilovolts $= 100$ kV

b. 0.1×10^{-3} W $= 0.1$ milliwatts $= 0.1$ mW

c. 250×10^{-7} s $= 25 \times 10^{-6}$ s $= 25$ microseconds $= 25$ μs

Convert 0.1 MV to kilovolts (kV).

EXAMPLE 1–7

Solution

$$0.1 \text{ MV} = 0.1 \times 10^6 \text{ V} = (0.1 \times 10^3) \times 10^3 \text{ V} = 100 \text{ kV}$$

Remember that a prefix represents a power of ten and thus the rules for power of ten computation apply. For example, when adding or subtracting, adjust to a common base, as illustrated in Example 1–8.

Compute the sum of 1 ampere (amp) and 100 milliamperes.

EXAMPLE 1–8

Solution   Adjust to a common base, either amps (A) or milliamps (mA). Thus,

$$1 \text{ A} + 100 \text{ mA} = 1 \text{ A} + 100 \times 10^{-3} \text{ A} = 1 \text{ A} + 0.1 \text{ A} = 1.1 \text{ A}$$

Alternatively, $1 \text{ A} + 100 \text{ mA} = 1000 \text{ mA} + 100 \text{ mA} = 1100 \text{ mA}$.

1. Convert 1800 kV to megavolts (MV).

PRACTICE PROBLEMS 3

2. In Chapter 4, we show that voltage is the product of current times resistance—that is, $V = I \times R$, where V is in volts, I is in amperes, and R is in ohms. Given $I = 25$ mA and $R = 4$ kΩ, convert these to power of ten notation, then determine V.

3. If $I_1 = 520 \text{ } \mu\text{A}$, $I_2 = 0.157 \text{ mA}$, and $I_3 = 2.75 \times 10^{-4} \text{ A}$, what is $I_1 + I_2 + I_3$ in mA? In microamps?

Answers
1. 1.8 MV; 2. 100 V; 3. 0.952 mA, 952 μA

1. All conversion factors have a value of what?

IN-PROCESS
LEARNING CHECK 1

(Answers are at the end of the chapter.)

2. Convert 14 yards to centimeters.

3. What units does the following reduce to?

$$\frac{\text{km}}{\text{h}} \times \frac{\text{m}}{\text{km}} \times \frac{\text{h}}{\text{min}} \times \frac{\text{min}}{\text{s}}$$

4. Express the following in engineering notation:
   a. 4270 ms   b. 0.001 53 V   c. 12.3×10^{-4} s

5. Express the result of each of the following computations as a number times 10 to the power indicated:
   a. 150×120 as a value times 10^4; as a value times 10^3.
   b. $300 \times 6/0.005$ as a value times 10^4; as a value times 10^5; as a value times 10^6.
   c. $430 + 15$ as a value times 10^2; as a value times 10^1.
   d. $(3 \times 10^{-2})^3$ as a value times 10^{-6}, as a value times 10^{-5}.

6. Express each of the following as indicated.
   a. 752 μA in mA.
   b. 0.98 mV in μV.
   c. 270 μs + 0.13 ms in μs and in ms.

NOTES . . .

Most calculations that you do in circuit theory are done using a hand calculator. While calculators often display 8 digits or more, the question is how many should you retain in your answer? While it is tempting to show all the digits that you see on your display, a little thought will show that this is not sensible, since you seldom know answers accurate to so many digits. In engineering practice, answers are frequently rounded to 3 digits, and that is the number that we usually use in this book. (There are exceptions where it is sometimes necessary to include more digits to adequately illustrate the theory.) Thus, as a useful guideline, you should show only 3 digits in your final answer unless it makes sense to show more.

1.6 Circuit Diagrams

Electrical and electronic circuits are constructed using components such as batteries, switches, resistors, capacitors, transistors, interconnecting wires, etc. To represent these circuits on paper, diagrams are used. In this book, we use three types: block diagrams, schematic diagrams, and pictorials.

Block Diagrams

Block diagrams describe a circuit or system in simplified form. The overall problem is broken into blocks, each representing a portion of the system or circuit. Blocks are labelled to indicate what they do or what they contain, then interconnected to show their relationship to each other. General signal flow is usually from left to right and top to bottom. Figure 1–5, for example, represents an audio amplifier. Although you have not covered any of its circuits yet, you should be able to follow the general idea quite easily—sound is picked up by the microphone, converted to an electrical signal, amplified by a pair of amplifiers, then output to the speaker, where it is converted back to sound. A power supply energizes the system. The advantage of a block diagram is that it gives you the overall picture and helps you understand the general nature of a problem. However, it does not provide detail.

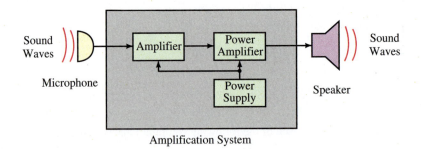

FIGURE 1–5　An example block diagram. Pictured is a simplified representation of an audio amplification system.

Pictorial Diagrams

Pictorial diagrams are one of the types of diagrams that provide detail. They help you visualize circuits and their operation by showing components as they actually appear. For example, the circuit of Figure 1–6 consists of a battery, a switch, and an electric lamp, all interconnected by wire. Operation is easy to visualize—when the switch is closed, the battery causes current in the circuit, which lights the lamp. The battery is referred to as the source and the lamp as the load.

Schematic Diagrams

While pictorial diagrams help you visualize circuits, they are cumbersome to draw. **Schematic diagrams** get around this by using simplified, standard symbols to represent components; see Table 1–7. (The meaning of these symbols will be made clear as you progress through the book.) In Figure 1–7(a), for example, we have used some of these symbols to create a schematic for the circuit of Figure 1–6. Each component has been replaced by its corresponding circuit symbol.

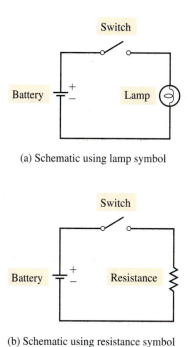

(a) Schematic using lamp symbol

(b) Schematic using resistance symbol

FIGURE 1–7 Schematic representation of Figure 1–6. The lamp has a circuit property called resistance (discussed in Chapter 3).

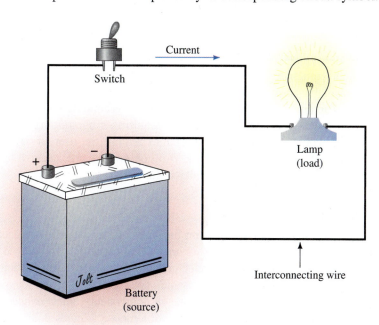

FIGURE 1–6 A pictorial diagram. The battery is referred to as a *source* while the lamp is referred to as a *load*. (The + and − on the battery are discussed in Chapter 2.)

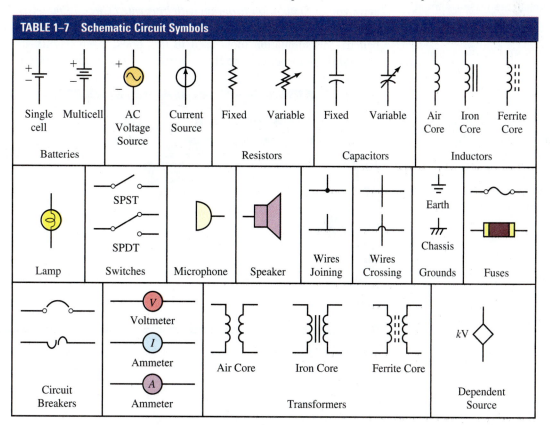

TABLE 1–7	Schematic Circuit Symbols

Single cell	Multicell	AC Voltage Source	Current Source	Fixed	Variable	Fixed	Variable	Air Core	Iron Core	Ferrite Core
Batteries				Resistors		Capacitors		Inductors		

Lamp	Switches (SPST, SPDT)	Microphone	Speaker	Wires Joining	Wires Crossing	Grounds (Earth, Chassis)	Fuses

Circuit Breakers	Voltmeter / Ammeter / Ammeter	Air Core	Iron Core	Ferrite Core	Dependent Source
		Transformers			

When choosing symbols, choose those that are appropriate to the occasion. Consider the lamp of Figure 1–7(a). As we will show later, the lamp possesses a property called **resistance** that causes it to resist the passage of charge. When you wish to emphasize this property, use the resistance symbol rather than the lamp symbol, as in Figure 1–7(b).

When we draw schematic diagrams, we usually draw them with horizontal and vertical lines joined at right angles as in Figure 1–7. This is standard practice. (At this point you should glance through some later chapters, e.g., Chapter 7, and study additional examples.)

1.7 Circuit Analysis Using Computers

Personal computers are used extensively for analysis and design. Software tools available for such tasks fall into two broad categories: prepackaged application programs (application packages) and programming languages. **Application packages** solve problems without requiring programming on the part of the user, while **programming languages** require the user to write code for each type of problem to be solved. We look only at prepackaged application software in this book. First, consider simulation software.

Circuit Simulation Software

Simulation software solves problems by simulating the behavior of electrical and electronic circuits rather than by solving sets of equations. To analyze a circuit, you "build" it on your screen by selecting components (resistors, capacitors, transistors, etc.) from a library of parts, which you then position and interconnect to form the desired circuit. You can change component values, connections, and analysis options instantly with the click of a mouse. Figures 1–8 and 1–9 show two examples. Software products such as these permit you to set up and test your circuit on the computer screen without the need to build a hardware prototype.

Most simulation packages use a software engine called **SPICE**, an acronym for *Simulation Program with Integrated Circuit Emphasis.* Two of the most popular products are Orcad PSpice and MultiSIM (Electronic Workbench), the simulation tools used in this book. Each has its strong points. MultiSIM, for instance, more closely models an actual workbench (complete with

NOTES . . .

1. The version of Electronics Workbench used in this book is MultiSIM 2001, V6 Educational, the version current at the time of writing. (There is also a student version available from MultiSIM that is suitable for use with this book.)

2. The version of PSpice used here is the Orcad Student Editon Release 9.1 from Cadence Design Systems Inc., the version current at the time of writing.

3. If you have different versions of these software products, your screens may look slightly different than the photos included in this book.

4. Should any significant changes occur in these software packages during the lifetime of this edition, updated material will be posted on our web site—see the Preface for our web site address.

FIGURE 1–8 Computer screen showing circuit analysis using MultiSIM.

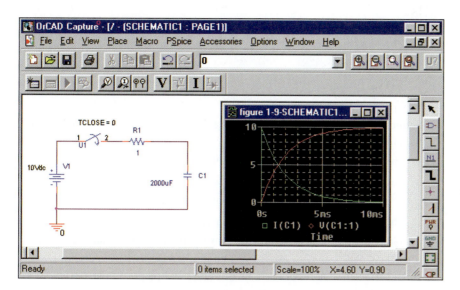

FIGURE 1–9   Computer screen showing circuit analysis using Orcad PSpice.

realistic meters) than does PSpice, but PSpice has a more complete analysis capability. For example, it determines and displays important information (such as phase angles in ac analyses and current waveforms in transient analysis) that MultiSIM, as of this writing, does not.

Prepackaged Math Software

Another useful category of software includes mathematics packages such as MATLAB and MathCAD. In contrast to the previous products (which use simulation to determine voltages and currents), these programs use numerical analysis techniques to solve equation and plot data. For example, they are great for solving sets of simultaneous equations as are encountered during mesh or nodal analysis (Chapters 8 and 19) where you enter data and let the computer do the numerical work. In addition, they may provide you with built-in electronic handbooks or toolkits that contain many useful formulas and circuit diagrams that can save you a great deal of time and effort.

Programming Languages

Many problems can also be solved using programming languages such as C++ and Java, however, we do not consider programming languages in this book.

PROBLEMS

1.3 Converting Units

1. Perform the following conversions:

   a. 27 minutes to seconds

   b. 0.8 hours to seconds

   c. 2 h 3 min 47 s to s

   d. 35 horsepower to watts

   e. 1827 W to hp

   f. 23 revolutions to degrees

2. Perform the following conversions:
   a. 27 feet to meters
   b. 2.3 yd to cm
   c. 36° F to degrees C
   d. 18 (US) gallons to liters
   e. 100 sq. ft to m²
   f. 124 sq. in. to m²
   g. 47-pound force to newtons

3. Set up conversion factors, compute the following, and express the answer in the units indicated.
   a. The area of a plate 1.2 m by 70 cm in m².
   b. The area of a triangle with base 25 cm, height 0.5 m in m².
   c. The volume of a box 10 cm by 25 cm by 80 cm in m³.
   d. The volume of a sphere with 10 in. radius in m³.

4. An electric fan rotates at 300 revolutions per minute. How many degrees is this per second?

5. If the surface mount robot machine of Figure 1–3 places 15 parts every 12 s, what is its placement rate per hour?

6. If your laser printer can print 8 pages per minute, how many pages can it print in one tenth of an hour?

7. A car gets 27 miles per US gallon. What is this in kilometers per liter?

8. The equatorial radius of the earth is 3963 miles. What is the earth's circumference in kilometers at the equator?

9. A wheel rotates 18° in 0.02 s. How many revolutions per minute is this?

10. The height of horses is sometimes measured in "hands," where 1 hand = 4 inches. How many meters tall is a 16-hand horse? How many centimeters?

11. Suppose $s = vt$ is given, where s is distance travelled, v is velocity, and t is time. If you travel at $v = 60$ mph for 500 seconds, you get upon unthinking substitution $s = vt = (60)(500) = 30,000$ miles. What is wrong with this calculation? What is the correct answer?

12. A round pizza has a circumference of 47 inches. How long does it take for a pizza cutter traveling at 0.12 m/s to cut diagonally across it?

13. Joe S. was asked to convert 2000 yd/h to meters per second. Here is Joe's work: velocity = 2000 × 0.9144 × 60/60 = 1828.8 m/s. Determine conversion factors, write units into the conversion, and find the correct answer.

14. The mean distance from the earth to the moon is 238 857 miles. Radio signals travel at 299 792 458 m/s. How long does it take a radio signal to reach the moon?

15. If you walk at a rate of 3 km/h for 8 minutes, 5 km/h for 1.25 h, then continue your walk at a rate of 4 km/h for 12 minutes, how far will you have walked in total?

16. Suppose you walk at a rate of 2 mph for 12 minutes, 4 mph for 0.75 h, then finish off at 5 mph for 15 minutes. How far have you walked in total?

17. You walk for 15 minutes at a rate of 2 km/h, then 18 minutes at 5 km/h, and for the remainder, your speed is 2.5 km/h. If the total distance covered is 2.85 km, how many minutes did you spend walking at 2.5 km/h?

18. You walk for 16 minutes at a rate of 1.5 mph, speed up to 3.5 mph for awhile, then slow down to 3 mph for the last 12 minutes. If the total distance covered is 1.7 miles, how long did you spend walking at 3.5 mph?

19. Your plant manager asks you to investigate two machines. The cost of electricity for operating machine #1 is 43 cents/minute, while that for machine #2 is $200.00 per 8-hour shift. The purchase price and production capacity

for both machines are identical. Based on this information, which machine should you purchase and why?

20. Given that 1 hp = 550 ft-lb/s, 1 ft = 0.3048 m, 1 lb = 4.448 N, 1 J = 1 N-m, and 1 W = 1 J/s, show that 1 hp = 746 W.

1.4   Power of Ten Notation

21. Express each of the following in power of ten notation with one nonzero digit to the left of the decimal point:

    a. 8675
    b. 0.008 72
    c. 12.4×10^2
    d. 37.2×10^{-2}
    e. $0.003\ 48 \times 10^5$
    f. $0.000\ 215 \times 10^{-3}$
    g. 14.7×10^0

22. Express the answer for each of the following in power of ten notation with one nonzero digit to the left of the decimal point.

    a. $(17.6)(100)$
    b. $(1400)(27 \times 10^{-3})$
    c. $(0.15 \times 10^6)(14 \times 10^{-4})$
    d. $1 \times 10^{-7} \times 10^{-4} \times 10.65$
    e. $(12.5)(1000)(0.01)$
    f. $(18.4 \times 10^0)(100)(1.5 \times 10^{-5})(0.001)$

23. Repeat the directions in Question 22 for each of the following.

    a. $\dfrac{125}{1000}$
    b. $\dfrac{8 \times 10^4}{(0.001)}$
    c. $\dfrac{3 \times 10^4}{(1.5 \times 10^6)}$
    d. $\dfrac{(16 \times 10^{-7})(21.8 \times 10^6)}{(14.2)(12 \times 10^{-5})}$

24. Determine answers for the following

    a. $123.7 + 0.05 + 1259 \times 10^{-3}$
    b. $72.3 \times 10^{-2} + 1 \times 10^{-3}$
    c. $86.95 \times 10^2 - 383$
    d. $452 \times 10^{-2} + (697)(0.01)$

25. Convert the following to power of 10 notation and, without using your calculator, determine the answers.

    a. $(4 \times 10^3)(0.05)^2$
    b. $(4 \times 10^3)(-0.05)^2$
    c. $\dfrac{(3 \times 2 \times 10)^2}{(2 \times 5 \times 10^{-1})}$
    d. $\dfrac{(30 + 20)^{-2}(2.5 \times 10^6)(6000)}{(1 \times 10^3)(2 \times 10^{-1})^2}$
    e. $\dfrac{(-0.027)^{1/3}(-0.2)^2}{(23 + 1)^0 \times 10^{-3}}$

26. For each of the following, convert the numbers to power of ten notation, then perform the indicated computations. Round your answer to four digits:

    a. $(452)(6.73 \times 10^4)$
    b. $(0.009\ 85)(4700)$
    c. $(0.0892)/(0.000\ 067\ 3)$
    d. $12.40 - 236 \times 10^{-2}$
    e. $(1.27)^3 + 47.9/(0.8)^2$
    f. $(-643 \times 10^{-3})^3$
    g. $[(0.0025)^{1/2}][1.6 \times 10^4]$
    h. $[(-0.027)^{1/3}]/[1.5 \times 10^{-4}]$
    i. $\dfrac{(3.5 \times 10^4)^{-2} \times (0.0045)^2 \times (729)^{1/3}}{[(0.008\ 72) \times (47)^3] - 356}$

27. For the following,
    a. convert numbers to power of ten notation, then perform the indicated computation,
    b. perform the operation directly on your calculator without conversion. What is your conclusion?

$$\text{i. } 842 \times 0.0014 \qquad \text{ii. } \frac{0.0352}{0.007\ 91}$$

28. Express each of the following in conventional notation:
    a. 34.9×10^4 b. 15.1×10^0
    c. 234.6×10^{-4} d. 6.97×10^{-2}
    e. $45\ 786.97 \times 10^{-1}$ f. 6.97×10^{-5}

29. One coulomb (Chapter 2) is the amount of charge represented by 6 240 000 000 000 000 000 electrons. Express this quantity in power of ten notation.

30. The mass of an electron is 0.000 000 000 000 000 000 000 000 000 000 899 9 kg. Express as a power of 10 with one non-zero digit to the left of the decimal point.

31. If 6.24×10^{18} electrons pass through a wire in 1 s, how many pass through it during a time interval of 2 hr, 47 min and 10 s?

32. Compute the distance traveled in meters by light in a vacuum in 1.2×10^{-8} second.

33. How long does it take light to travel 3.47×10^5 km in a vacuum?

34. How far in km does light travel in one light-year?

35. While investigating a site for a hydroelectric project, you determine that the flow of water is 3.73×10^4 m³/s. How much is this in liters/hour?

36. The gravitational force between two bodies is $F = 6.6726 \times 10^{-11} \dfrac{m_1 m_2}{r^2}$ N, where masses m_1 and m_2 are in kilograms and the distance r between gravitational centers is in meters. If body 1 is a sphere of radius 5000 miles and density of 25 kg/m³, and body 2 is a sphere of diameter 20 000 km and density of 12 kg/m³, and the distance between centers is 100 000 miles, what is the gravitational force between them?

1.5 Prefixes

37. What is the appropriate prefix and its abbreviation for each of the following multipliers?
    a. 1000   b. 1 000 000
    c. 10^9 d. 0.000 001
    e. 10^{-3} f. 10^{-12}

38. Express the following in terms of their abbreviations, e.g., microwatts as μW. Pay particular attention to capitalization (e.g., V, not v, for volts).
    a. milliamperes   b. kilovolts
    c. megawatts   d. microseconds
    e. micrometers   f. milliseconds
    g. nanoamps

39. Express the following in the most sensible engineering notation (e.g., 1270 μs = 1.27 ms).
    a. 0.0015 s   b. 0.000 027 s   c. 0.000 35 ms

40. Convert the following:

    a. 156 mV to volts   b. 0.15 mV to microvolts

    c. 47 kW to watts   d. 0.057 MW to kilowatts

    e. 3.5×10^4 volts to kilovolts   f. 0.000 035 7 amps to microamps

41. Determine the values to be inserted in the blanks.

    a. 150 kV = ___ $\times 10^3$ V = ___ $\times 10^6$ V

    b. 330 μW = ___ $\times 10^{-3}$ W = ___ $\times 10^{-5}$ W

42. Perform the indicated operations and express the answers in the units indicated.

    a. 700 μA − 0.4 mA = ___ μA = ___ mA

    b. 600 MW + 300 $\times 10^4$ W = ___ MW

43. Perform the indicated operations and express the answers in the units indicated.

    a. 330 V + 0.15 kV + 0.2 $\times 10^3$ V = ___ V

    b. 60 W + 100 W + 2700 mW = ___ W

44. The voltage of a high voltage transmission line is 1.15×10^5 V. What is its voltage in kV?

45. You purchase a 1500 W electric heater to heat your room. How many kW is this?

46. Consider Figure 1–10. As you will learn in Chapter 6, $I_4 = I_1 + I_2 + I_3$. If $I_1 = 1.25$ mA, $I_2 = 350$ μA and $I_3 = 250 \times 10^{-5}$ A, what is I_4?

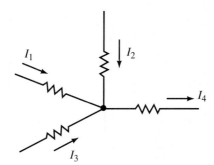

FIGURE 1–10

47. For Figure 1–11, $I_1 + I_2 - I_3 + I_4 = 0$. If $I_1 = 12$ A, $I_2 = 0.150$ kA and $I_4 = 250 \times 10^{-1}$ A, what is I_3?

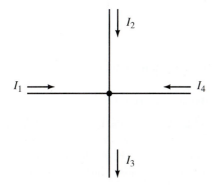

FIGURE 1–11

48. While repairing an antique radio, you come across a faulty capacitor designated 39 mmfd. After a bit of research, you find that "mmfd" is an obsolete unit meaning "micromicrofarads". You need a replacement capacitor of equal value. Consulting Table 1–6, what would 39 "micromicrofarads" be equivalent to?

49. A radio signal travels at 299 792.458 km/s and a telephone signal at 150 m/μs. If they originate at the same point, which arrives first at a destination 5000 km away? By how much?

50. a. If 0.045 coulomb of charge (Question 29) passes through a wire in 15 ms, how many electrons is this?

    b. At the rate of 9.36×10^{19} electrons per second, how many coulombs pass a point in a wire in 20 μs?

1.6 Circuit Diagrams

51. Consider the pictorial diagram of Figure 1–12. Using the appropriate symbols from Table 1–7, draw this in schematic form. Hint: In later chapters, there are many schematic circuits containing resistors, inductors, and capacitors. Use these as aids.

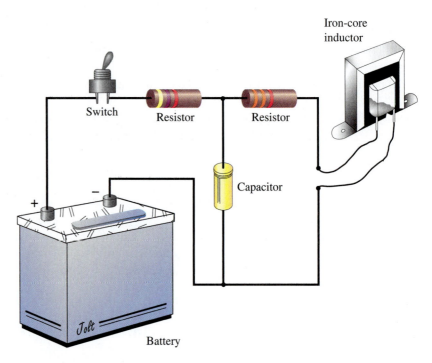

FIGURE 1–12

52. Draw the schematic diagram for a simple flashlight.

1.7 Circuit Analysis Using Computers

53. Many electronic and computer magazines carry advertisements for computer software tools such as PSpice, Mathcad, Matlab, plus others. Investigate a few of these magazines in your school's library; by studying such advertisements, you can gain valuable insight into what modern software packages are able to do.

✓ **ANSWERS TO IN-PROCESS LEARNING CHECKS**

In-Process Learning Check 1

1. One

2. 1280 cm

3. m/s

4. a. 4.27 s

   b. 1.53 mV

   c. 1.23 ms

5. a. $1.8 \times 10^4 = 18 \times 10^3$

   b. $36 \times 10^4 = 3.6 \times 10^5 = 0.36 \times 10^6$

   c. $4.45 \times 10^2 = 44.5 \times 10^1$

   d. $27 \times 10^{-6} = 2.7 \times 10^{-5}$

6. a. 0.752 mA

   b. 980 μV

   c. 400 μs = 0.4 ms

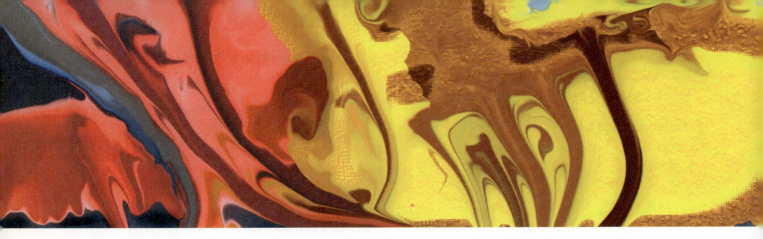

■ KEY TERMS

Ampere

Ampere-hour

Atom

Battery

Capacity

Cell

Circuit Breaker

Conductor

Coulomb

Coulomb's Law

Current

Electric Charge

Electron

Free Electrons

Fuse

Insulator

Ion

Loosely Bound

Neutron

Polarity

Potential Difference

Proton

Semiconductor

Shell

Switch

Tightly Bound

Valence

Volt

■ OUTLINE

Atomic Theory Review

The Unit of Electrical Charge: The
  Coulomb

Voltage

Current

Practical DC Voltage Sources

Measuring Voltage and Current

Switches, Fuses, and Circuit Breakers

■ OBJECTIVES

After studying this chapter, you will be able to

- describe the basic makeup of an atom,

- explain the relationships between valence shells, free electrons, and conduction,

- describe the fundamental (coulomb) force within an atom and the energy required to create free electrons,

- describe what ions are and how they are created,

- describe the characteristics of conductors, insulators, and semiconductors,

- describe the coulomb as a measure of charge,

- define voltage,

- describe how a battery "creates" voltage,

- explain current as a movement of charge and how voltage causes current in a conductor,

- describe important battery types and their characteristics,

- describe how to measure voltage and current.

Voltage and Current

2

A basic electric circuit consisting of a source of electrical energy, a switch, a load, and interconnecting wire is shown in Figure 2–1. When the switch is closed, current in the circuit causes the light to come on. This circuit is representative of many common circuits found in practice, including those of flashlights and automobile headlight systems. We will use it to help develop an understanding of voltage and current.

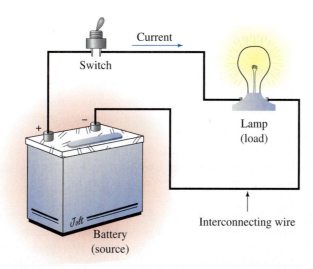

FIGURE 2–1 A basic electric circuit.

Elementary atomic theory shows that the current in Figure 2–1 is actually a flow of charges. The cause of their movement is the "voltage" of the source. While in Figure 2–1 this source is a battery, in practice it may be any one of a number of practical sources including generators, power supplies, solar cells, and so on.

In this chapter we look at the basic ideas of voltage and current. We begin with a discussion of atomic theory. This leads us to free electrons and the idea of current as a movement of charge. The fundamental definitions of voltage and current are then developed. Following this, we look at a number of common voltage sources. The chapter concludes with a discussion of voltmeters and ammeters and the measurement of voltage and current in practice. ∎

PUTTING IT IN PERSPECTIVE

The Equations of Circuit Theory

IN THIS CHAPTER you meet the first of the equations and formulas that we use to describe the relationships of circuit theory. Remembering formulas is made easier if you clearly understand the principles and concepts on which they are based. As you may recall from high school physics, formulas can come about in only one of three ways: through experiment, by definition, or by mathematical manipulation.

Experimental Formulas

Circuit theory rests on a few basic experimental results. These are results that can be proven in no other way; they are valid solely because experiment has shown them to be true. The most fundamental of these are called "laws." Four examples are Ohm's law, Kirchhoff's current law, Kirchhoff's voltage law, and Faraday's law. (These laws will be met in various chapters throughout the book.) When you see a formula referred to as a law or an experimental result, remember that it is based on experiment and cannot be obtained in any other way.

Defined Formulas

Some formulas are created by definition, i.e., we make them up. For example, there are 60 seconds in a minute because we define the second as 1/60 of a minute. From this we get the formula $t_{sec} = 60 \times t_{min}$.

Derived Formulas

This type of formula or equation is created mathematically by combining or manipulating other formulas. In contrast to the other two types of formulas, the only way that a derived relationship can be obtained is by mathematics.

An awareness of where circuit theory formulas come from is important to you. This awareness not only helps you understand and remember formulas, it helps you understand the very foundations of the theory—the basic experimental premises upon which it rests, the important definitions that have been made, and the methods by which these foundation ideas have been put together. This can help enormously in understanding and remembering concepts. ■

2.1 Atomic Theory Review

The basic structure of an atom is shown symbolically in Figure 2–2. It consists of a nucleus of protons and neutrons surrounded by a group of orbiting electrons. As you learned in physics, the electrons are negatively charged (−), while the protons are positively charged (+). Each atom (in its normal state) has an equal number of electrons and protons, and since their charges are equal and opposite, they cancel, leaving the atom electrically neutral, i.e., with zero net charge. The nucleus, however, has a net positive charge, since it consists of positively charged protons and uncharged neutrons.

The basic structure of Figure 2–2 applies to all elements, but each element has its own unique combination of electrons, protons, and neutrons. For example, the hydrogen atom, the simplest of all atoms, has 1 proton and 1 electron, while the copper atom has 29 electrons, 29 protons, and 35 neutrons. Silicon, which is important because of its use in transistors and other electronic devices, has 14 electrons, 14 protons, and 14 neutrons.

In the model of Figure 2–2, electrons that have approximately the same orbital radiuses may be thought of as forming shells. This gives us the simplified picture of Figure 2–3, where we have grouped closely spaced orbits into shells designated *K, L, M, N*, etc. Only certain numbers of electrons can exist within each shell, and no electrons can exist in the space between shells. The maximum that any shell can hold is $2n^2$ where n is the shell number. Thus, there can be up to 2 electrons in the *K* shell, up to 8 in the *L* shell, up to 18 in the *M* shell, and up to 32 in the *N* shell. The number in any shell depends on

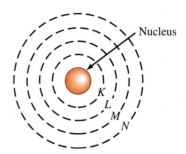

FIGURE 2–3 Simplified representation of the atom. Electrons travel in roughly spherical orbits called "shells."

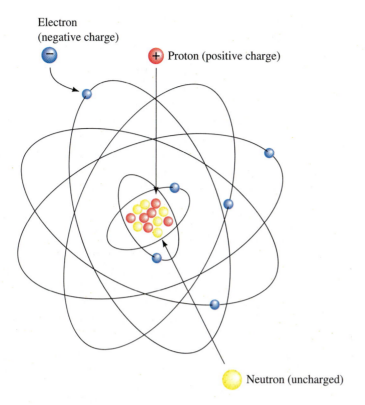

Electron (negative charge)

+ Proton (positive charge)

Neutron (uncharged)

FIGURE 2–2 Bohr model of the atom. Electrons travel around the nucleus at incredible speeds, making billions of trips in a fraction of a second. The force of attraction between the electrons and the protons in the nucleus keeps them in orbit.

the element. For instance, the copper atom, which has 29 electrons, has all 3 of its inner shells completely filled but its outer shell (shell *N*) has only 1 electron, Figure 2–4. This outermost shell is called its **valence shell,** and the electron in it is called its **valence electron.**

No element can have more than 8 valence electrons; when a valence shell has 8 electrons, it is filled. As we shall see, the number of valence electrons that an element has directly affects its electrical properties.

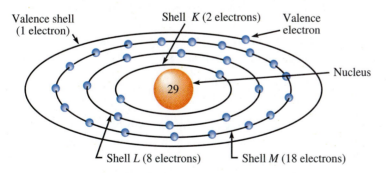

Valence shell (1 electron)

Shell *K* (2 electrons)

Valence electron

Nucleus

Shell *L* (8 electrons)

Shell *M* (18 electrons)

FIGURE 2–4 Copper atom. Since the valence electron is only weakly attracted to the nucleus, it is said to be "loosely bound".

Electrical Charge

In the previous paragraphs, we mentioned the word "charge". However, we need to look at its meaning in more detail. First, we should note that electrical charge is an intrinsic property of matter that manifests itself in the form of forces—electrons repel other electrons but attract protons, while protons repel

◀ **Online Companion**

each other but attract electrons. It was through studying these forces that scientists determined that the charge on the electron is negative while that on the proton is positive.

However, the way in which we use the term "charge" extends beyond this. To illustrate, consider again the basic atom of Figure 2–2. It has equal numbers of electrons and protons, and since their charges are equal and opposite, they cancel, leaving the atom as a whole uncharged. However, if the atom acquires additional electrons (leaving it with more electrons than protons), we say that it (the atom) is negatively charged; conversely, if it loses electrons and is left with fewer electrons than protons, we say that it is positively charged. The term "charge" in this sense denotes an imbalance between the number of electrons and protons present in the atom.

Now move up to the macroscopic level. Here, substances in their normal state are also generally uncharged; that is, they have equal numbers of electrons and protons. However, this balance is easily disturbed—electrons can be stripped from their parent atoms by simple actions such as walking across a carpet, sliding off a chair, or spinning clothes in a dryer. (Recall "static cling".) Consider two additional examples from physics. Suppose you rub an ebonite (hard rubber) rod with fur. This action causes a transfer of electrons from the fur to the rod. The rod therefore acquires an excess of electrons and is thus negatively charged. Similarly, when a glass rod is rubbed with silk, electrons are transferred from the glass rod to the silk, leaving the rod with a deficiency and, consequently, a positive charge. Here again, charge refers to an imbalance of electrons and protons.

As the above examples illustrate, "charge" can refer to the charge on an individual electron or to the charge associated with a whole group of electrons. In either case, this charge is denoted by the letter Q, and its unit of measurement in the SI system is the coulomb. (The definition of the coulomb is considered shortly.) In general, the charge Q associated with a group of electrons is equal to the product of the number of electrons times the charge on each individual electron. Since charge manifests itself in the form of forces, charge is defined in terms of these forces. This is discussed next.

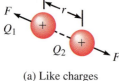

(a) Like charges repel

(b) Unlike charges attract

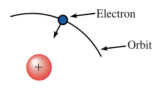

(c) The force of attraction keeps electrons in orbit

FIGURE 2–5 Coulomb law forces.

Coulomb's Law

The force between charges was studied by the French scientist Charles Coulomb (1736–1806). Coulomb determined experimentally that the force between two charges Q_1 and Q_2 (Figure 2–5) is directly proportional to the product of their charges and inversely proportional to the square of the distance between them. Mathematically, Coulomb's law states

$$F = k\frac{Q_1 Q_2}{r^2} \quad \text{[newtons, N]} \qquad (2\text{–}1)$$

where Q_1 and Q_2 are the charges in coulombs (to be defined in Section 2.2), r is the center-to-center spacing between them in meters, and $k = 9 \times 10^9$. Coulomb's law applies to aggregates of charges as in Figure 2–5(a) and (b), as well as to individual electrons within the atom as in (c).

As Coulomb's law indicates, force decreases inversely as the square of distance; thus, if the distance between two charges is doubled, the force decreases to $(1/2)^2 = 1/4$ (i.e., one quarter) of its original value. Because of this relationship, electrons in outer orbits are less strongly attracted to the nucleus than those in inner orbits; that is, they are less **tightly bound** to the nucleus than those close by. Valence electrons are the least tightly bound and will, if they acquire sufficient energy, escape from their parent atoms.

Free Electrons

The amount of energy required to escape depends on the number of electrons in the valence shell. If an atom has only a few valence electrons, there will be a relatively weak attraction between these electrons and the nucleus and only a small amount of additional energy is needed. For example, for a metal like copper, valence electrons can gain sufficient energy from heat alone (thermal energy), even at room temperature, to escape from their parent atoms and wander from atom to atom throughout the material as depicted in Figure 2–6. (Note that these electrons do not leave the substance, they simply wander from the valence shell of one atom to the valence shell of another. The material therefore remains electrically neutral.) Such electrons are called **free electrons.** In copper, there are of the order of 10^{23} free electrons per cubic centimeter at room temperature. As we shall see, it is the presence of this large number of free electrons that makes copper such a good conductor of electric current. On the other hand, if the valence shell is full (or nearly full), valence electrons are much more tightly bound. Such materials have few (if any) free electrons.

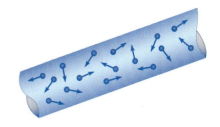

FIGURE 2–6   Random motion of free electrons in a conductor.

Ions

As noted earlier, when a previously neutral atom gains or loses an electron, it acquires a net electrical charge. The charged atom is referred to as an **ion.** If the atom loses an electron, it is called a **positive ion;** if it gains an electron, it is called a **negative ion.**

Conductors, Insulators, and Semiconductors

The atomic structure of matter affects how easily charges, i.e., electrons, move through a substance and hence how it is used electrically. Electrically, materials are classified as conductors, insulators, or semiconductors.

Conductors

Materials through which charges move easily are termed **conductors.** The most familiar examples are metals. Good metal conductors have large numbers of free electrons that are able to move about easily. In particular, silver, copper, gold, and aluminum are excellent conductors. Of these, copper is the most widely used. Not only is it an excellent conductor, it is inexpensive and easily formed into wire, making it suitable for a broad spectrum of applications ranging from common house wiring to sophisticated electronic equipment. Aluminum, although it is only about 60% as good a conductor as copper, is also used, mainly in applications where light weight is important, such as in overhead power transmission lines. Silver and gold are too expensive for general use. However, gold, because it oxidizes less than other materials, is used in specialized applications; for example, critical electrical connectors in electronic equipment use it because it makes a more reliable connection than other materials.

Insulators

Materials that do not conduct (e.g., glass, porcelain, plastic, rubber, and so on) are termed **insulators.** The covering on electric lamp cords, for example, is an insulator. It is used to prevent the wires from touching and to protect us from electric shock.

Insulators do not conduct because they have full or nearly full valence shells and thus their electrons are tightly bound. However, when high enough voltage is applied, the force is so great that electrons are literally torn from their parent atoms, causing the insulation to break down and conduction to occur. In air, you see this as an arc or flashover. In solids, charred insulation usually results.

Semiconductors

Silicon and germanium (plus a few other materials) have half-filled valence shells and are thus neither good conductors nor good insulators. Known as **semiconductors,** they have unique electrical properties that make them important to the electronics industry. The most important material is silicon. It is used to make transistors, diodes, integrated circuits, and other electronic devices. Semiconductors have made possible personal computers, DVD systems, cell phones, calculators, and a host of other electronic products. You will study them in great detail in your electronics courses.

IN-PROCESS
LEARNING CHECK 1

(Answers are at the end of the chapter.)

1. Describe the basic structure of the atom in terms of its constituent particles: electrons, protons, and neutrons. Why is the nucleus positively charged? Why is the atom as a whole electrically neutral?

2. What are valence shells? What does the valence shell contain?

3. Describe Coulomb's law and use it to help explain why electrons far from the nucleus are loosely bound.

4. What are free electrons? Describe how they are created, using copper as an example. Explain what role thermal energy plays in the process.

5. Briefly distinguish between a normal (i.e., uncharged) atom, a positive ion, and a negative ion.

2.2 The Unit of Electrical Charge: The Coulomb

As noted in the previous section, the unit of electrical charge in the SI system is the coulomb (C). The **coulomb** is defined as the charge carried by 6.24×10^{18} electrons. Thus, if an electrically neutral (i.e., uncharged) body has 6.24×10^{18} electrons removed, it will be left with a net positive charge of 1 coulomb, i.e., $Q = 1$ C. Conversely, if an uncharged body has 6.24×10^{18} electrons added, it will have a net negative charge of 1 coulomb, i.e., $Q = -1$ C. Usually, however, we are more interested in the charge moving through a wire. In this regard, if 6.24×10^{18} electrons pass through a wire, we say that the charge that passed through the wire is 1 C.

We can now determine the charge on 1 electron. It is $Q_e = 1/(6.24 \times 10^{18}) = 1.602 \times 10^{-19}$ C.

EXAMPLE 2–1

An initially neutral body has 1.7 μC of negative charge removed. Later, 18.7×10^{11} electrons are added. What is the body's final charge?

Solution Initially the body is neutral, i.e., $Q_{initial} = 0$ C. When 1.7 μC of electrons is removed, the body is left with a positive charge of 1.7 μC. Now, 18.7×10^{11} electrons are added back. This is equivalent to

$$18.7 \times 10^{11} \text{ electrons} \times \frac{1 \text{ coulomb}}{6.24 \times 10^{18} \text{ electrons}} = 0.3 \text{ μC}$$

of negative charge. The final charge on the body is therefore $Q_f = 1.7$ μC − 0.3 μC = +1.4 μC.

To get an idea of how large a coulomb is, we can use Coulomb's law. If it were possible to place 2 charges of 1 coulomb each 1 meter apart, the force between them would be

$$F = (9 \times 10^9) \frac{(1 \text{ C})(1 \text{ C})}{(1 \text{ m})^2} = 9 \times 10^9 \text{ N, i.e., about 1 million tons!}$$

PRACTICE PROBLEMS 1

1. Positive charges $Q_1 = 2$ μC and $Q_2 = 12$ μC are separated center to center by 10 mm. Compute the force between them. Is it attractive or repulsive?

2. Two equal charges are separated by 1 cm. If the force of repulsion between them is 9.7×10^{-2} N, what is their charge? What may the charges be, both positive, both negative, or 1 positive and 1 negative?

3. After 10.61×10^{13} electrons are added to a metal plate, it has a negative charge of 3 μC. What was its initial charge in coulombs?

Answers
1. 2160 N, repulsive; 2. 32.8 nC, both (+) or both (−); 3. 14 μC (+)

When charges are detached from one body and transferred to another, a *potential difference* or *voltage* results between them. A familiar example is the voltage that develops when you walk across a carpet. Voltages in excess of ten thousand volts can be created in this way. (We will define the volt rigorously very shortly.) This voltage is due entirely to the separation of positive and negative charges, i.e., charges that have been pulled apart.

Figure 2–7 illustrates another example. During electrical storms, electrons in thunderclouds are stripped from their parent atoms by the forces of turbulence and carried to the bottom of the cloud, leaving a deficiency of electrons (positive charge) at the top and an excess (negative charge) at the bottom. The force of repulsion then drives electrons away beneath the cloud, leaving the ground positively charged. Hundreds of millions of volts are created in this way. (This is what causes the air to break down and a lightning discharge to occur.)

Practical Voltage Sources

As the preceding examples show, voltage is created solely by the separation of positive and negative charges. However, static discharges and lightning strikes are not practical sources of electricity. We now look at practical sources. A common example is the battery. In a battery, charges are separated by chemical action. An ordinary alkaline flashlight battery, Figure 2–8, illustrates the concept. The alkaline material (a mixture of manganese dioxide, graphite, and electrolytic) and a gelled zinc powder mixture, separated by a paper barrier soaked in electrolyte, are placed in a steel can. The can is connected to the top cap to form one electrode and the zinc mixture, by means of the brass pin, is connected to the bottom to form the other electrode. (The bottom is insulated from the rest of the can). Chemical reactions result in an excess of electrons in the zinc mix (making it negative) and a deficiency in the manganese dioxide mix (making it positive). This separation of charges creates a voltage of approximately 1.5 V, with the top end cap + and the bottom of the can −. The battery is useful as a source since its chemical action creates a continuous supply of energy that is able to do useful work, such as light a lamp or run a motor.

2.3 Voltage

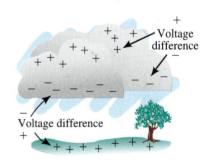

FIGURE 2–7 Voltages created by separation of charges in a thundercloud. The force of repulsion drives electrons away beneath the cloud, creating a voltage between the cloud and earth as well. If voltage becomes large enough, the air breaks down and a lightning discharge occurs.

NOTES . . .

The source of Figure 2–8 is more properly called a cell than a battery, since "cell" refers to a single cell while "battery" refers to a group of cells. However, through common usage, such cells are referred to as batteries. In what follows, we will also call them batteries.

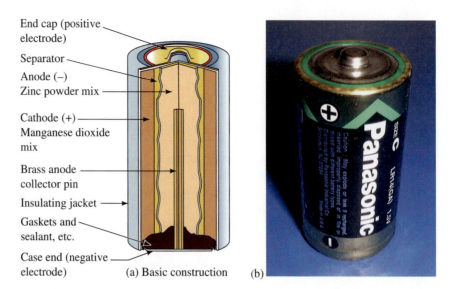

End cap (positive electrode)
Separator
Anode (–)
Zinc powder mix
Cathode (+)
Manganese dioxide mix
Brass anode collector pin
Insulating jacket
Gaskets and sealant, etc.
Case end (negative electrode)

(a) Basic construction   (b)

FIGURE 2–8   Alkaline cell. Voltage is created by the separation of charges due to chemical action. Nominal cell voltage is 1.5 V.

Potential Energy

The concept of voltage is tied into the concept of potential energy. We therefore look briefly at energy.

In mechanics, potential energy is the energy that a body possesses because of its position. For example, a bag of sand hoisted by a rope over a pulley has the potential to do work when it is released. The amount of work that went into giving it this potential energy is equal to the product of force times the distance through which the bag was lifted (i.e., work equals force times distance). In the SI system, force is measured in newtons and distance in meters. Thus, work has the unit newton-meters (which we call **joules,** Table 1–4, Chapter 1).

In a similar fashion, work is required to move positive and negative charges apart. This gives them potential energy. To understand why, consider again the cloud of Figure 2–7, redrawn in Figure 2–9. Assume the cloud is ini-

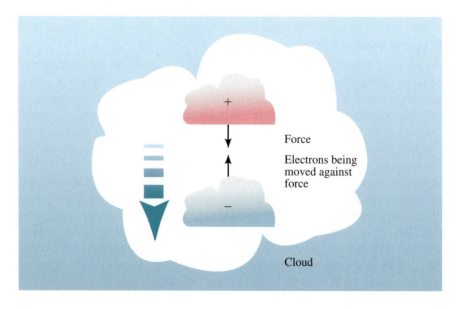

Force

Electrons being moved against force

Cloud

FIGURE 2–9   Work (force × distance) is required to move the charges apart.

tially uncharged. Now assume a charge of Q electrons is moved from the top of the cloud to the bottom. The positive charge left at the top of the cloud exerts a force on the electrons that tries to pull them back as they are being moved away. Since the electrons are being moved against this force, work (force times distance) is required. Since the separated charges experience a force to return to the top of the cloud, they have the potential to do work if released, i.e., they possess potential energy.

NOTES . . .

The discussion found here may seem a bit abstract and somewhat detached from our ordinary experience that suggests that voltage is the "force or push" that moves electric current through a circuit. While both viewpoints are correct (we look at the latter in great detail, starting in Chapter 4), in order to establish a coherent analytic theory, we need a rigorous definition. Equation 2–2 provides that definition. Although it is a bit abstract, it gives us the foundation upon which rests many of the important circuit relationships that you will encounter shortly.

Definition of Voltage: The Volt

In electrical terms, a difference in potential energy is defined as **voltage.** In general, the amount of energy required to separate charges depends on the voltage developed and the amount of charge moved. By definition, *the voltage between two points is one volt if it requires one joule of energy to move one coulomb of charge from one point to the other.* In equation form,

$$V = \frac{W}{Q} \quad \text{[volts, V]} \qquad (2\text{–}2)$$

where W is energy in joules, Q is charge in coulombs, and V is the resulting voltage in volts.

Note carefully that voltage is defined between points. For the case of the battery, for example, voltage appears between its terminals. Thus, voltage does not exist at a point by itself; it is always determined with respect to some other point. (For this reason, voltage is also called **potential difference.** We often use the terms interchangeably.) Note also that, although we considered static electricity in developing the energy argument, the same conclusion results regardless of how you separate the charges; this may be by chemical means as in a battery, by mechanical means as in a generator, by photoelectric means as in a solar cell, and so on.

Alternate arrangements of Equation 2–2 are useful:

$$W = QV \quad \text{[joules, J]} \qquad (2\text{–}3)$$

$$Q = \frac{W}{V} \quad \text{[coulombs, C]} \qquad (2\text{–}4)$$

If it takes 35 J of energy to move a charge of 5 C from one point to another, what is the voltage between the two points?

EXAMPLE 2–2

Solution

$$V = \frac{W}{Q} = \frac{35 \text{ J}}{5 \text{ C}} = 7 \text{ J/C} = 7 \text{ V}$$

---

1. The voltage between two points is 19 V. How much energy is required to move 67×10^{18} electrons from one point to the other?

2. The potential difference between two points is 140 mV. If 280 μJ of work are required to move a charge Q from one point to the other, what is Q?

PRACTICE PROBLEMS 2

Answers
1. 204 J; 2. 2 mC

Symbol for DC Voltage Sources

Consider again Figure 2–1. The battery is the source of electrical energy that moves charges around the circuit. This movement of charges, as we will soon see, is called an electric current. Because one of the battery's terminals is always positive and the other is always negative, current is always in the same direction. Such a unidirectional current is called **dc** or **direct current,** and the battery is called a **dc source.** Symbols for dc sources are shown in Figure 2–10. The long bar denotes the positive terminal. On actual batteries, the positive terminal is usually marked POS (+) and the negative terminal NEG (−).

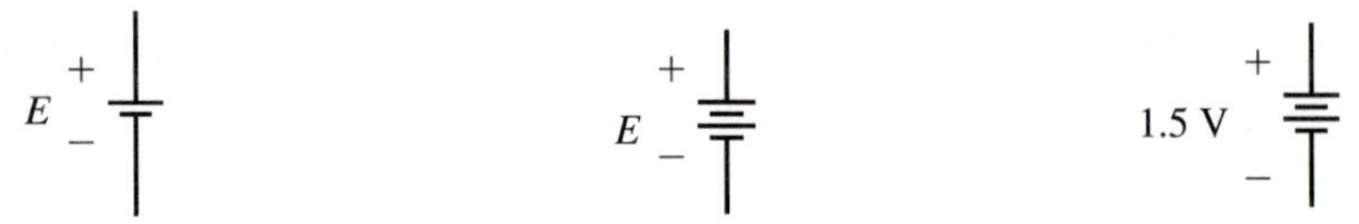

(a) Symbol for a cell   (b) Symbol for a battery   (c) A 1.5 volt battery

FIGURE 2–10   Battery symbol. The long bar denotes the positive terminal and the short bar the negative terminal. Thus, it is not necessary to put + and − signs on the diagram. For simplicity, we use the symbol shown in (a) throughout this book.

2.4 Current

Earlier, you learned that there are large numbers of free electrons in metals like copper. These electrons move randomly throughout the material (Figure 2–6), but their net movement in any given direction is zero.

Assume now that a battery is connected as in Figure 2–11. Since electrons are attracted by the positive pole of the battery and repelled by the negative pole, they move around the circuit, passing through the wire, the lamp, and the battery. This movement of charge is called an **electric current.** The more electrons per second that pass through the circuit, the greater is the current. Thus, current is the *rate of flow* (or *rate of movement*) of charge.

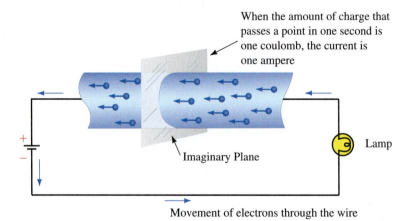

When the amount of charge that passes a point in one second is one coulomb, the current is one ampere

Imaginary Plane

Lamp

Movement of electrons through the wire

FIGURE 2–11   Electron flow in a conductor. Electrons (−) are attracted to the positive (+) pole of the battery. As electrons move around the circuit, they are replenished at the negative pole of the battery. This flow of charge is called an electric current.

The Ampere

Since charge is measured in coulombs, its rate of flow is coulombs per second. In the SI system, one coulomb per second is defined as 1 **ampere** (commonly abbreviated A). From this, we get that *1 ampere is the current in a circuit when*

1 coulomb of charge passes a given point in 1 second (Figure 2–11). The symbol for current is *I*. Expressed mathematically,

$$I = \frac{Q}{t} \quad \text{[amperes, A]} \qquad (2\text{–}5)$$

where *Q* is the charge (in coulombs) and *t* is the time interval (in seconds) over which it is measured. *In Equation 2–5, it is important to note that t does not represent a discrete point in time but is the interval of time during which the transfer of charge occurs.* Alternate forms of Equation 2–5 are

$$Q = It \text{ [coulombs, C]} \qquad (2\text{–}6)$$

and

$$t = \frac{Q}{I} \quad \text{[seconds, s]} \qquad (2\text{–}7)$$

EXAMPLE 2–3

If 840 coulombs of charge pass through the imaginary plane of Figure 2–11 during a time interval of 2 minutes, what is the current?

Solution   Convert *t* to seconds. Thus,

$$I = \frac{Q}{t} = \frac{840 \text{ C}}{(2 \times 60)\text{s}} = 7 \text{ C/s} = 7 \text{ A}$$

PRACTICE PROBLEMS 3

1. Between *t* = 1 ms and *t* = 14 ms, 8 μC of charge pass through a wire. What is the current?

2. After the switch of Figure 2–1 is closed, current *I* = 4 A. How much charge passes through the lamp between the time the switch is closed and the time that it is opened 3 minutes later?

Answers
1. 0.615 mA; 2. 720 C

Although Equation 2–5 is the theoretical definition of current, we never actually use it to measure current. In practice, we use an instrument called an ammeter (Section 2.6). However, it is an extremely important equation that we will soon use to develop other relationships.

Current Direction

In the early days of electricity, it was believed that current was a movement of positive charge and that these charges moved around the circuit from the positive terminal of the battery to the negative as depicted in Figure 2–12(a). Based on this, all the laws, formulas, and symbols of circuit theory were developed. (We now refer to this direction as the **conventional current direction.**) After the discovery of the atomic nature of matter, it was learned that what actually moves in metallic conductors are electrons and that they move through the circuit as in Figure 2–12(b). This direction is called the **electron flow direction.** We thus have two possible representations for current direction and a choice has to be made. *In this book, we use the conventional direction (see Notes).*

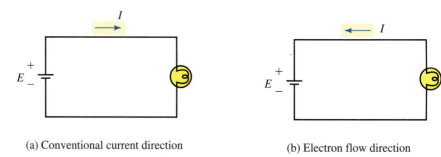

(a) Conventional current direction

(b) Electron flow direction

FIGURE 2–12   Conventional current versus electron flow. In this book, we use conventional current.

NOTES . . .

A perfectly coherent theory can be built around either of the directions of Figure 2–12, and many circuit analysis and electronics textbooks have been written from each viewpoint. There are, however, compelling reasons for choosing the conventional direction. Among these are (1) Standard electronic symbols used in circuit diagrams and found in manufacturer's data books are based on it, (2) Computer software such as PSpice and MultiSIM utilizes it, and (3) All Engineering level and virtually all Technology level (EET) college programs teach it.

Alternating Current (AC)

So far, we have considered only dc. Before we move on, we will briefly mention ac or alternating current. **Alternating current** is current that changes direction cyclically, i.e., charges alternately flow in one direction, then in the other in a circuit. The most common ac source is the commercial ac power system that supplies energy to your home. We mention it here because you will encounter it briefly in Section 2.5. It is covered in detail in Chapter 15.

IN-PROCESS
LEARNING CHECK 2

(Answers are at the end of the chapter.)

1. Body A has a negative charge of 0.2 μC and body B has a charge of 0.37 μC (positive). If 87×10^{12} electrons are transferred from A to B, what are the charges in coulombs on A and on B after the transfer?

2. Briefly describe the mechanism of voltage creation using the alkaline cell of Figure 2–8 to illustrate.

3. When the switch in Figure 2–1 is open, the current is zero, yet free electrons in the copper wire are moving about. Describe their motion. Why does their movement not constitute an electric current?

4. If 12.48×10^{20} electrons pass a certain point in a circuit in 2.5 s, what is the current in amperes?

5. For Figure 2–1, assume a 12-V battery. The switch is closed for a short interval, then opened. If $I = 6$ A and the battery expends 230 040 J moving charge through the circuit, how long was the switch closed?

2.5 Practical DC Voltage Sources

Batteries

Batteries are the most common dc source. They are made in a variety of shapes, sizes, and ratings, from miniaturized button batteries capable of delivering only a few microamps to large automotive batteries capable of delivering hundreds of amps. Common sizes are the AAA, AA, C, and D as illustrated in the various photos of this chapter. All batteries use unlike conductive electrodes immersed in an electrolyte. Chemical interaction between the electrodes and the electrolyte creates the voltage of the battery.

Primary and Secondary Batteries

Batteries eventually become "discharged." Some types of batteries, however, can be "recharged." Such batteries are called **secondary** batteries. Other types, called **primary** batteries, cannot be recharged. A familiar example of a secondary battery is the automobile battery. It can be recharged by passing current through it opposite to its discharge direction. A familiar example of a primary cell is the flashlight battery.

FIGURE 2–13 Alkaline batteries. From left to right, a 9–V rectangular battery, an AAA cell, a D cell, an AA cell, and a C cell.

Types of Batteries and Their Applications

The voltage of a battery, its service life, and other characteristics depend on the material from which it is made.

Alkaline

This is the most widely used, general-purpose primary cell available. Alkaline batteries are used in flashlights, portable radios, TV remote controllers, cassette players, cameras, toys, and so on. They come in various sizes as depicted in Figure 2–13. Their nominal cell voltage is 1.5 V.

Carbon-Zinc

Also called a **dry cell,** the carbon-zinc battery was for many years the most widely used primary cell, but it has now given way to other types such as the alkaline battery. Its nominal cell voltage is 1.5 volts.

Lithium

Lithium batteries (Figure 2–14) feature small size and long life (e.g., shelf lives of 10 to 20 years). Applications include watches, pacemakers, cameras, and battery backup of computer memories. Several types of lithium cells are available, with voltages from of 2 V to 3.5 V and current ratings from the microampere to the ampere range.

Nickel-Cadmium

Commonly called "Ni-Cads," these are the most popular, general-purpose rechargeable batteries available. They have long service lives, operate over wide temperature ranges, and are manufactured in many styles and sizes, including C, D, AAA, and AA. Inexpensive chargers make it economically feasible to use nickel-cadmium batteries for home entertainment equipment.

Lead-Acid

This is the familiar automotive battery. Its basic cell voltage is about 2 volts, but typically, six cells are connected internally to provide 12 volts at its terminals. Lead-acid batteries are capable of delivering large current (in excess of 100 A) for short periods as required, for example, to start an automobile.

FIGURE 2–14 An assortment of lithium batteries. The battery on the computer motherboard is for memory backup.

Battery Capacity

Batteries run down under use. However, an estimate of their useful life can be determined from their **capacity**, i.e., their **ampere-hour** rating. (The ampere-hour rating of a battery is equal to the product of its current drain times the length of time that you can expect to draw the specified current before the battery becomes unusable.) For example, a battery rated at 200 Ah can theoretically supply 20 A for 10 h, or 5 A for 40 h, etc. The relationship between capacity, life, and current drain is

$$\text{life} = \frac{\text{capacity}}{\text{current drain}} \qquad (2\text{–}8)$$

The capacity of batteries is not a fixed value as suggested above but is affected by discharge rates, operating schedules, temperature, and other factors. At best, therefore, capacity is an estimate of expected life under certain conditions. Table 2–1 illustrates approximate service capacities for several sizes of carbon-zinc batteries at three values of current drain at 20° C. Under the conditions listed, the AA cell has a capacity of (3 mA)(450 h) = 1350 mAh at a drain of 3 mA, but its capacity decreases to (30 mA)(32 h) = 960 mAh at a drain of 30 mA. Figure 2–15 shows a typical variation of capacity of a Ni-Cad battery with changes in temperature.

TABLE 2–1 Capacity-Current Drain of Selected Carbon-Zinc Cells

Cell	Starting Drain (mA)	Service Life (h)
AA	3.0	450
	15.0	80
	30.0	32
C	5.0	520
	25.0	115
	50.0	53
D	10.0	525
	50.0	125
	100.0	57

Courtesy T. R. Crompton, *Battery Reference Book*, Butterworths & Co. (Publishers) Ltd.

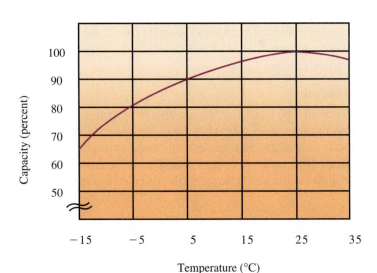

FIGURE 2–15 Typical variation of capacity versus temperature for a Ni-Cad battery.

Other Characteristics

Because batteries are not perfect, their terminal voltage drops as the amount of current drawn from them increases. (This issue is considered in Chapter 5.) In addition, battery voltage is affected by temperature and other factors that affect their chemical activity. However, these factors are not considered in this book.

Assume the battery of Figure 2–15 has a capacity of 240 Ah at 25° C. What is its capacity at −15° C?

Solution From the graph, capacity at –15° C is down to 65%. Thus, capacity = 0.65 × 240 = 156 Ah.

EXAMPLE 2–4

Cells in Series and Parallel

Cells may be connected as in Figures 2–16 and 2–17 to increase their voltage and current capabilities. This is discussed in later chapters.

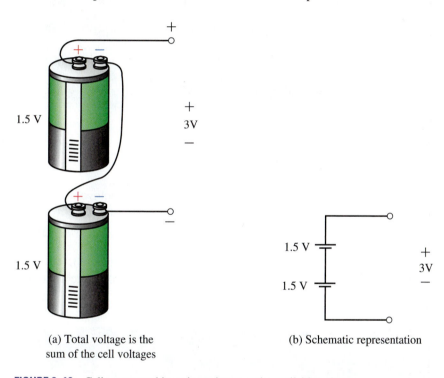

(a) Total voltage is the sum of the cell voltages

(b) Schematic representation

FIGURE 2–16 Cells connected in series to increase the available voltage.

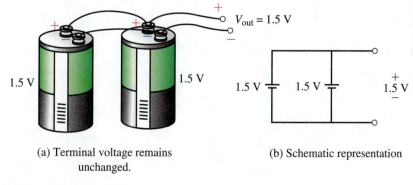

(a) Terminal voltage remains unchanged.

(b) Schematic representation

FIGURE 2–17 Cells connected in parallel to increase the available current. (Both must have the same voltage.) Do not do this for extended periods of time.

Electronic Power Supplies

Electronic systems such as TV sets, VCRs, computers, and so on, require dc for their operation. Except for portable units, which use batteries, they obtain their power from the commercial ac power lines by means of built-in power supplies (Figure 2–18). Such supplies convert the incoming ac to the dc voltages required by the equipment. Power supplies are also used in electronic laboratories. These are usually variable to provide the range of voltages needed for prototype development and circuit testing. Figure 2–19 shows a variable supply.

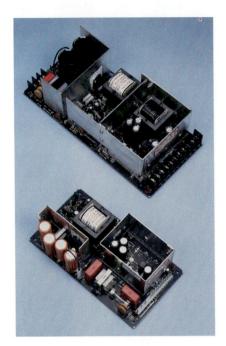

FIGURE 2–18 Fixed power supplies. *(Courtesy of Condor PC Power Supplies Inc.)*

FIGURE 2–19 A variable laboratory power supply.

Solar Cells

Solar cells convert light energy to electrical energy using photovoltaic means. The basic cell consists of two layers of semiconductor material. When light strikes the cell, many electrons gain enough energy to cross from one layer to the other to create a dc voltage.

Solar energy has a number of practical applications. Figure 2–20, for example, shows an array of solar panels supplying power to a commercial ac net-

FIGURE 2–20 Solar panels, Davis California Pacific Gas & Electric PVUSA (Photovoltaic for Utility Scale Applications). Solar panels produce dc, which must be converted to ac before being fed into the ac system. This plant is rated 174 kilowatts. *(Courtesy Siemens Solar Industries, Camarillo, California.)*

work. In remote areas, solar panels are used to power communications systems and irrigation pumps. In space, they are used to power satellites. In everyday life, they are used to power hand-held calculators.

DC Generators

Direct current (dc) generators, which convert mechanical energy to electrical energy, are another source of dc. They create voltage by means of a coil of wire rotated through a magnetic field. Their principle of operation is similar to that of ac generators (discussed in Chapter 15).

2.6 Measuring Voltage and Current

Voltage and current are measured in practice using instruments called **voltmeters** and **ammeters.** While voltmeters and ammeters are available as individual instruments, they are more commonly combined into a multipurpose instrument called a **multimeter** (Figure 2–21). (Although both digital and analog instruments are illustrated in the photo, we will consider only digital multimeters (DMMs) and leave the consideration of analog instruments to your lab course.)

Terminal Designations

Multimeters typically have a set of terminals marked $V\Omega$, A, and COM as can be seen in Figure 2–21. Terminal $V\Omega$ is the terminal to use to measure voltage and resistance, while terminal A is used for current measurement. The terminal marked COM is the common terminal for all measurements. (Some multimeters combine the $V\Omega$ and A terminals into one terminal marked $V\Omega$A.) On

(a) Hand-held digital multimeter (DMM).
(*Reproduced with permission from the John Fluke Mfg. Co., Inc.*)

(b) Analog multimeter.

FIGURE 2–21   Multimeters. These are multipurpose test instruments that you can use to measure voltage, current, and resistance. Some meters use terminal markings of + and −, others use $V\Omega$ and COM, and so on. Color coded test leads (red and black) are industry standard.

NOTES . . .

DMMs as Learning Tools

Voltage and current as presented earlier in this chapter are rather abstract concepts involving energy, charge, and charge movement. Voltmeters and ammeters are introduced at this point to help present the ideas in more physically meaningful terms. In particular, we concentrate on DMMs. Experience has shown them to be powerful learning tools. For example, when dealing with the sometimes difficult topics of voltage polarity conventions, current direction conventions, and so on (as in later chapters), the use of DMMs showing readings complete with signs for voltage polarity and current direction provides clarity and aids understanding in a way that simply drawing arrows and putting numbers on diagrams does not. You will find that in the first few chapters of this book DMMs are used for this purpose quite frequently.

some instruments the VΩ terminal is called the + terminal and the COM terminal is called the − terminal (Figure 2–22).

Function Selection

DMMs generally include a function selector switch (or alternatively, a set of push buttons) that permit you to select the quantity to be measured—e.g., dc voltage, ac voltage, resistance, dc current, and ac current—and you must set the meter to the desired function before you make a measurement, Figure 2–22. Note the symbols on the dial. The symbol $\overline{\overline{\text{V}}}$ denotes dc voltage, $\tilde{\text{A}}$ denotes ac voltage, Ω denotes resistance, and so on. When set to dc volts, the meter measures and displays the voltage between its VΩ (or +) and COM (or −) terminals as indicated in Figure 2–22(a); when set to dc current, it measures the current passing through it, i.e., the current entering its A (or +) terminal and leaving its COM (or −) terminal. Be sure to note the sign of the measured quantity. (DMMs generally have an **autopolarity** feature that automatically determines the sign for you.) Thus, if the meter is connected with its + lead connected to the + terminal of the source, the display will show 47.2 V as indicated, while if the leads are reversed, the display will show −47.2 V. Similarly, if the leads are reversed for current measurement (so that current enters the COM terminal), the display will show −3.6 A. Be sure to observe the standard color convention for lead connection (Practical Notes).

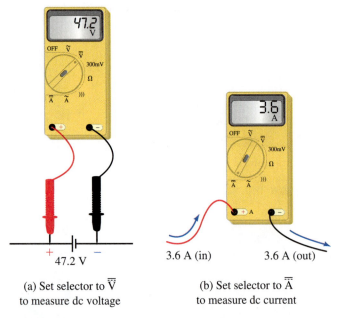

(a) Set selector to $\overline{\overline{\text{V}}}$ to measure dc voltage

(b) Set selector to $\overline{\overline{\text{A}}}$ to measure dc current

FIGURE 2–22   Measuring voltage and current with a multimeter. Be sure to set the selector switch to the correct function before you energize the circuit and connect the red lead to the VΩ (+) terminal and the black lead to the COM (−) terminal.

NOTES . . .

Most DMMs have internal circuitry that automatically selects the correct range for voltage measurement. Such instruments are called "autoranging" or "autoscaling" devices.

How to Measure Voltage

Since voltage is the potential difference between two points, you measure voltage by placing the voltmeter leads *across* the component whose voltage you wish to determine as we saw in Figure 2–22(a). Figure 2–23 shows another example. To measure the voltage across the lamp, place one lead on each side of the lamp as shown. If the meter is not autoscale and you have no idea how large the voltage is, set the meter to its highest range, then work your way down to avoid damage to the instrument.

PRACTICAL NOTES . . .

By convention, DMMs and VOMs have one red lead and one black lead, with the red lead connected to the (+) or VΩA terminal of the meter and the black connected to the (−) or COM terminal. If, when so connected, the voltmeter indicates a positive value, the point where the red lead is touching is positive with respect to the point where the black lead is touching; inversely, if the meter indicates negative, the point where the red lead is touching is negative with respect to the point where the black lead is connected. For current measurements, if an ammeter indicates a positive value, this means that the direction of current is into its (+) or VΩA terminal and out of its (−) or COM terminal; conversely, if the reading is negative, this means that the direction of current is into the meter's COM terminal and out of its (+) or VΩA terminal.

How to Measure Current

As indicated by Figure 2–22(b), the current that you wish to measure must pass *through* the meter. Consider Figure 2–24(a). To measure this current, open the circuit as in (b) and insert the ammeter. The sign of the reading will be positive if current enters the A or (+) terminal or negative if it enters the COM (or −) terminal as described in the Practical Notes.

Meter Symbols

So far, we have shown meters pictorially. Usually, however, they are shown schematically. The schematic symbol for a voltmeter is a circle with the letter *V*, while the symbol for an ammeter is a circle with the letter *I*. The circuits of Figures 2–23 and 2–24 have been redrawn (Figure 2–25) to indicate this.

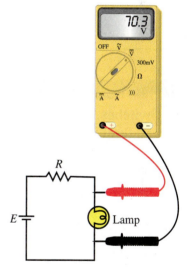

FIGURE 2–23 To measure voltage, place the voltmeter leads across the component whose voltage you wish to determine. If the voltmeter reading is positive, the point where the red lead is connected is positive with respect to the point where the black lead is connected.

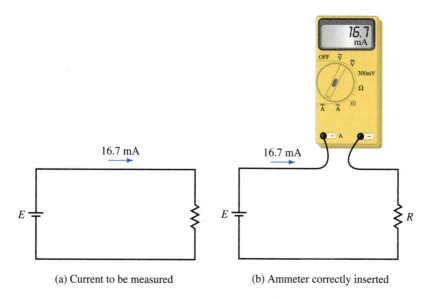

(a) Current to be measured

(b) Ammeter correctly inserted

FIGURE 2–24 To measure current, insert the ammeter into the circuit so that the current you wish to measure passes through the instrument. The reading is positive here because current enters the + (A) terminal.

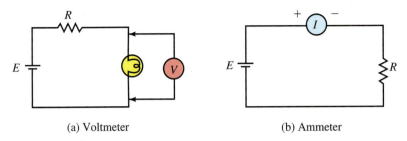

(a) Voltmeter   (b) Ammeter

FIGURE 2–25   Schematic symbols for voltmeter and ammeter.

2.7 Switches, Fuses, and Circuit Breakers

Switches

The most basic switch is a single-pole, single-throw (SPST) switch as shown in Figure 2–26. With the switch open, the current path is broken and the lamp is off; with it closed, the lamp is on. This type of switch is used, for example, for light switches in homes.

Figure 2–27(a) shows a single-pole, double-throw (SPDT) switch. Two of these switches may be used as in (b) for two-way control of a light. This type of arrangement is sometimes used for stairway lights; you can turn the light on or off from either the bottom or the top of the stairs.

Many other configurations of switches exist in practice. However, we will leave the topic at this point.

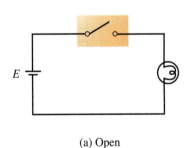

(a) Open

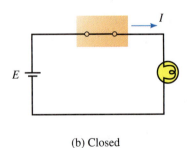

(b) Closed

FIGURE 2–26   Single-pole, single-throw (SPST) switch.

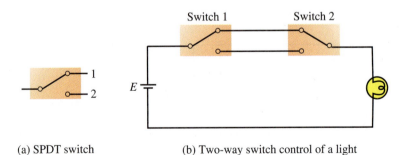

(a) SPDT switch   (b) Two-way switch control of a light

FIGURE 2–27   Single-pole, single-throw (SPDT) switch.

Fuses and Circuit Breakers

Fuses and circuit breakers are wired into a circuit between the source and the load as illustrated in Figure 2–28 to protect equipment or wiring against excessive current. For example, in your home, if you connect too many appliances to an outlet, the fuse or circuit breaker in your electrical panel "blows." This opens the circuit to protect against overloading and possible fire. Fuses and circuit breakers may also be installed in equipment such as your automobile to protect against internal faults. Figure 2–29 shows a variety of fuses and breakers.

Fuses use a metallic element that melts when current exceeds a preset value. Thus, if a fuse is rated at 3 A, it will "blow" if more than 3 amps passes through it. Fuses are made as fast-blow and slow-blow types. Fast-blow fuses are very fast; typically, they blow in a fraction of a second. Slow-blow fuses, on the other hand, react more slowly so that they do not blow on small, momentary overloads.

Circuit breakers work on a different principle. When the current exceeds the rated value of a breaker, the magnetic field produced by the excessive current operates a mechanism that trips open a switch. After the fault or overload condition has been cleared, the breaker can be reset and used again. Since they are mechanical devices, their operation is slower than that of a fuse; thus, they do not "pop" on momentary overloads as, for example, when you start a motor.

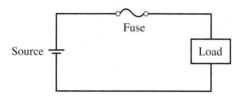

FIGURE 2–28 Using a fuse to protect a circuit.

(a) A variety of fuses and circuit breakers.

FIGURE 2–29 Fuses and circuit breakers.

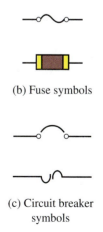

(b) Fuse symbols

(c) Circuit breaker symbols

PROBLEMS

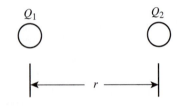

Q_1 Q_2

$\leftarrow\!\!\!-\!\!-\!\!- r \!\!-\!\!-\!\!-\!\!\!\rightarrow$

FIGURE 2–30

2.1 Atomic Theory Review

1. How many free electrons are there in the following at room temperature?

   a. 1 cubic meter of copper

   b. a 5 m length of copper wire whose diameter is 0.163 cm

2. Two charges are separated by a certain distance, Figure 2–30. How is the force between them affected if

   a. the magnitudes of both charges are doubled?

   b. the distance between the charges is tripled?

3. Two charges are separated by a certain distance. If the magnitude of one charge is doubled and the other tripled and the distance between them halved, how is the force affected?

4. A certain material has 4 electrons in its valence shell and a second material has 1. Which is the better conductor?

5. a. What makes a material a good conductor? (In your answer, consider valence shells and free electrons.)

   b. Besides being a good conductor, list two other reasons why copper is so widely used.

   c. What makes a material a good insulator?

   d. Normally air is an insulator. However, during lightning discharges, conduction occurs. Briefly discuss the mechanism of charge flow in this discharge.

6. a. Although gold is very expensive, it is sometimes used in electronics as a plating on contacts. Why?

   b. Why is aluminum sometimes used when its conductivity is only about 60% as good as that of copper?

2.2 The Unit of Electrical Charge: The Coulomb

7. What do we mean when we say that a body is "charged"?

8. Compute the electrical force between the following charges and state whether it is attractive or repulsive.

   a. A + 1 μC charge and a +7 μC charge, separated 10 mm

   b. $Q_1 = 8$ μC and $Q_2 = -4$ μC, separated 12 cm

   c. Two electrons separated by 12×10^{-8} m

   d. An electron and a proton separated by 5.3×10^{-11} m

   e. An electron and a neutron separated by 5.7×10^{-11} m

9. The force between a positive charge and a negative charge that are 2 cm apart is 180 N. If $Q_1 = 4$ μC, what is Q_2? Is the force attraction or repulsion?

10. If you could place a charge of 1 C on each of two bodies separated 25 cm center to center, what would be the force between them in newtons? In tons?

11. The force of repulsion between two charges separated by 50 cm is 0.02 N. If $Q_2 = 5Q_1$, determine the charges and their possible signs.

12. How many electrons does a charge of 1.63 μC represent?

13. Determine the charge possessed by 19×10^{13} electrons.

14. An electrically neutral metal plate acquires a negative charge of 47 μC. How many electrons were added to it?

15. A metal plate has 14.6×10^{13} electrons added. Later, 1.3 μC of charge is added. If the final charge on the plate is 5.6 μC, what was its initial charge?

2.3   Voltage

16. Sliding off a chair and touching someone can result in a shock. Explain why.

17. If 360 joules of energy are required to transfer 15 C of charge through the lamp of Figure 2–1, what is the voltage of the battery?

18. If 600 J of energy are required to move 9.36×10^{19} electrons from one point to the other, what is the potential difference between the two points?

19. If 1.2 kJ of energy are required to move 500 mC from one point to another, what is the voltage between the two points?

20. How much energy is required to move 20 mC of charge through the lamp of Figure 2–23?

21. How much energy is gained by a charge of 0.5 µC as it moves through a potential difference of 8.5 kV?

22. If the voltage between two points is 100 V, how much energy is required to move an electron between the two points?

23. Given a voltage of 12 V for the battery in Figure 2–1, how much charge is moved through the lamp if it takes 57 J of energy to move it?

2.4   Current

24. For the circuit of Figure 2–1, if 27 C pass through the lamp in 9 seconds, what is the current in amperes?

25. If 250 µC pass through the ammeter of Figure 2–31 in 5 ms, what will the meter read?

26. If the current $I = 4$ A in Figure 2–1, how many coulombs pass through the lamp in 7 ms?

27. How much charge passes through the circuit of Figure 2–24 in 20 ms?

28. How long does it take for 100 µC to pass a point if the current is 25 mA?

29. If 93.6×10^{12} electrons pass through a lamp in 5 ms, what is the current?

30. The charge passing through a wire is given by $q = 10t + 4$, where q is in coulombs and t in seconds,
    a. How much charge has passed at $t = 5$ s?
    b. How much charge has passed at $t = 8$ s?
    c. What is the current in amps?

31. The charge passing through a wire is $q = (80t + 20)$ C. What is the current? Hint: Choose two arbitrary values of time and proceed as in Question 30.

32. How long does it take 312×10^{19} electrons to pass through the circuit of Figure 2–31 if the ammeter reads 8 A?

33. If 1353.6 J are required to move 47×10^{19} electrons through the lamp of Figure 2–31 in 1.3 min, what are E and I?

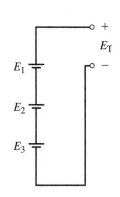

FIGURE 2–31

2.5   Practical DC Voltage Sources

34. What do we mean by dc? By ac?

35. Consider three batteries connected as in Figure 2–32.
    a. If $E_1 = 1.47$ V, $E_2 = 1.61$ V and $E_3 = 1.58$ V, what is E_T?
    b. If the connection to source 3 is reversed, what is E_T?

36. How do you charge a secondary battery? Make a sketch. Can you charge a primary battery?

37. A battery rated 1400 mAh supplies 28 mA to a load. How long can it be expected to last?

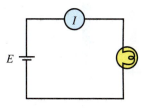

FIGURE 2–32

38. What is the approximate service life of the D cell of Table 2–1 at a current drain of 10 mA? At 50 mA? At 100 mA? What conclusion do you draw from these results?

39. The battery of Figure 2–15 is rated at 81 Ah at 5° C. What is the expected life (in hours) at a current draw of 5 A at −15° C?

40. The battery of Figure 2–15 is expected to last 17 h at a current drain of 1.5 A at 25° C. How long do you expect it to last at 5° C at a current drain of 0.8 A?

41. In the engineering workplace, you sometimes have to make estimations based on the information you have available. In this vein, assume you have a battery-operated device that uses the C cell of Table 2–1. If the device draws 10 mA, what is the estimated time (in hours) that you will be able to use it?

2.6   Measuring Voltage and Current

42. The digital voltmeter of Figure 2–33 has autopolarity. For each case, determine its reading.

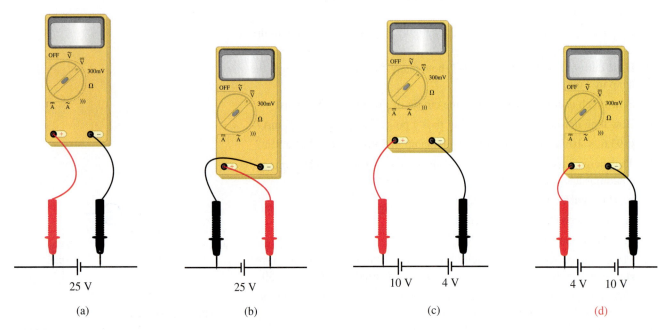

(a)   (b)   (c)   (d)

FIGURE 2–33

43. The current in the circuit of Figure 2–34 is 9.17 mA. Which ammeter correctly indicates the current? (a) Meter 1, (b) Meter 2, (c) both.

44. What is wrong with the statement that the voltage through the lamp of Figure 2–23 is 70.3 V?

45. What is wrong with the metering scheme shown in Figure 2–35? Fix it.

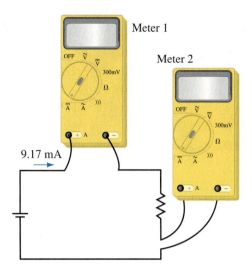

Meter 1

Meter 2

9.17 mA

FIGURE 2–34

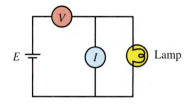

FIGURE 2–35 What is wrong here?

2.7 Switches, Fuses, and Circuit Breakers

46. It is desired to control a light using two switches as indicated in Table 2–2. Draw the required circuit.

47. Fuses have a current rating so that you can select the proper size to protect a circuit against overcurrent. They also have a voltage rating. Why? Hint: Read the section on insulators, i.e., Section 2.1.

TABLE 2–2

Switch 1	Switch 2	Lamp
Open	Open	Off
Open	Closed	On
Closed	Open	On
Closed	Closed	On

 ANSWERS TO IN-PROCESS LEARNING CHECKS

In-Process Learning Check 1

1. An atom consists of a nucleus of protons and neutrons orbited by electrons. The nucleus is positive because protons are positive, but the atom is neutral because it contains the same number of electrons as protons, and their charges cancel.

2. The valence shell is the outermost shell. It contains the atom's valence electrons. The number of electrons in this shell determines the properties of the material with regard whether it is a conductor, insulator, or semiconductor.

3. The force between charged particles is proportional to the product of their charges and inversely proportional to the square of their spacing. Since force decreases as the square of the spacing, electrons far from the nucleus experience little force of attraction.

4. If a loosely bound electron gains sufficient energy, it may break free from its parent atom and wander throughout the material. Such an electron is called a free electron. For materials like copper, heat (thermal energy) can give an electron enough energy to dislodge it from its parent atom.

5. A normal atom is neutral because it has the same number of electrons as protons and their charges cancel. An atom that has lost an electron is called a positive ion, while an atom that has gained an electron is called a negative ion.

In-Process Learning Check 2

1. $Q_A = 13.7 \ \mu C$ (pos.), $Q_B = 13.6 \ \mu C$ (neg.)

2. Chemical action creates an excess of electrons in the zinc mixture and a deficiency in the manganese dioxide mix. This separation of charges results in a voltage of 1.5 V. The zinc mixture is connected to the top end cap by the steel can (making it the positive electrode) and the manganese dioxide mixture is connected to the bottom of the can by the brass pin (making it the negative electrode).

3. Motion is random. Since the net movement in all directions is zero, current is zero.

4. 80 A

5. 3195 s

■ KEY TERMS

Color Codes
Conductance
Diode
Ohmmeter
Open Circuit
Photocell
Resistance
Resistivity
Short Circuit
Superconductance
Temperature Coefficient
Thermistor
Varistor
Wire Gauge

■ OUTLINE

Resistance of Conductors
Electrical Wire Tables
Resistance of Wires—Circular Mils
Temperature Effects
Types of Resistors
Color Coding of Resistors
Measuring Resistance—The Ohmmeter
Thermistors
Photoconductive Cells
Nonlinear Resistance
Conductance
Superconductors

■ OBJECTIVES

After studying this chapter, you will be able to

• calculate the resistance of a section of conductor, given its cross-sectional area and length,

• convert between areas measured in square mils, square meters, and circular mils,

• use tables of wire data to obtain the cross-sectional dimensions of various gauges of wire and predict the allowable current for a particular gauge of wire,

• use the temperature coefficient of a material to calculate the change in resistance as the temperature of the sample changes,

• use resistor color codes to determine the resistance and tolerance of a given fixed-composition resistor,

• demonstrate the procedure for using an ohmmeter to determine circuit continuity and to measure the resistance of both an isolated component and one that is located in a circuit,

• develop an understanding of various ohmic devices such as thermistors and photocells,

• develop an understanding of the resistance of nonlinear devices such as varistors and diodes,

• calculate the conductance of any resistive component.

Resistance

3

You have been introduced to the concepts of voltage and current in previous chapters and have found that current involves the movement of charge. In a conductor, the charge carriers are the free electrons that are moved due to the voltage of an externally applied source. As these electrons move through the material, they constantly collide with atoms and other electrons within the conductor. In a process similar to friction, the moving electrons give up some of their energy in the form of heat. These collisions represent an opposition to charge movement that is called **resistance.** The greater the opposition (i.e., the greater the resistance), the smaller will be the current for a given applied voltage.

Circuit components (called **resistors**) are specifically designed to possess resistance and are used in almost all electronic and electrical circuits. Although the resistor is the most simple component in any circuit, its effect is very important in determining the operation of a circuit.

Resistance is represented by the symbol R (Figure 3–1) and is measured in units of ohms (after Georg Simon Ohm). The symbol for ohms is the capital Greek letter omega (Ω).

In this chapter, we examine resistance in its various forms. Beginning with metallic conductors, we study the factors that affect resistance in conductors. Following this, we look at commercial resistors, including both fixed and variable types. We then discuss important nonlinear resistance devices and conclude with an overview of superconductivity and its potential impact and use. ∎

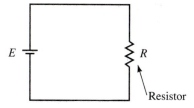

FIGURE 3–1 Basic resistive circuit.

PUTTING IT IN PERSPECTIVE

3.1 Resistance of Conductors

◀ **Online Companion**

Georg Simon Ohm and Resistance

ONE OF THE FUNDAMENTAL RELATIONSHIPS of circuit theory is that between voltage, current, and resistance. This relationship and the properties of resistance were investigated by the German physicist Georg Simon Ohm (1787–1854) using a circuit similar to that of Figure 3–1. Working with Volta's recently developed battery and wires of different materials, lengths, and thicknesses, Ohm found that current depended on both voltage and resistance. For example, for a fixed resistance, he found that doubling the voltage doubled the current, tripling the voltage tripled the current, and so on. Also, for a fixed voltage, Ohm found that the opposition to current was directly proportional to the length of the wire and inversely proportional to its cross-sectional area. From this, he was able to define the resistance of a wire and show that current was inversely proportional to this resistance; e.g., when he doubled the resistance, he found that the current decreased to half of its former value.

These two results when combined form what is known as Ohm's law. (You will study Ohm's law in great detail in Chapter 4.) Ohm's results are of such fundamental importance that they represent the real beginnings of what we now call electrical circuit analysis. ∎

As mentioned in the chapter preview, conductors are materials that permit the flow of charge. However, conductors do not all behave the same way. Rather, we find that the resistance of a material is dependent upon several factors:

- Type of material
- Length of the conductor
- Cross-sectional area
- Temperature

If a certain length of wire is subjected to a current, the moving electrons will collide with other electrons within the material. Differences at the atomic level of various materials cause variation in how the collisions affect resistance. For example, silver has more free electrons than copper, and so the resistance of a silver wire will be less than the resistance of a copper wire having the identical dimensions. We may therefore conclude the following:

The resistance of a conductor is dependent upon the type of material.

If we were to double the length of the wire, we can expect that the number of collisions over the length of the wire would double, thereby causing the resistance to also double. This effect may be summarized as follows:

The resistance of a metallic conductor is directly proportional to the length of the conductor.

A somewhat less intuitive property of a conductor is the effect of cross-sectional area on the resistance. As the cross-sectional area is increased, the moving electrons are able to move more freely through the conductor, just as water moves more freely through a large-diameter pipe than a small-diameter pipe. If the cross-sectional area is doubled, the electrons would be involved in half as many collisions over the length of the wire. We may summarize this effect as follows:

The resistance of a metallic conductor is inversely proportional to the cross-sectional area of the conductor.

The factors governing the resistance of a conductor at a given temperature may be summarized mathematically as follows:

$$R = \frac{\rho\ell}{A} \quad \text{[ohms, } \Omega\text{]} \tag{3–1}$$

where

ρ = resistivity, in ohm-meters (Ω-m)
ℓ = length, in meters (m)
A = cross-sectional area, in square meters (m^2).

In the previous equation the lowercase Greek letter rho (ρ) is the constant of proportionality and is called the **resistivity** of the material. Resistivity is a physical property of a material and is measured in ohm-meters (Ω-m) in the SI system. Table 3–1 lists the resistivities of various materials at a temperature of 20° C. The effects on resistance due to changes in temperature will be examined in Section 3.4.

Since most conductors are circular, as shown in Figure 3–2, we may determine the cross-sectional area from either the radius or the diameter as follows:

$$A = \pi r^2 = \pi \left(\frac{d}{2}\right)^2 = \frac{\pi d^2}{4} \qquad (3\text{–}2)$$

TABLE 3–1	**Resistivity of Materials, ρ**
Material	**Resistivity, ρ, at 20° C (Ω-m)**
Silver	1.645×10^{-8}
Copper	1.723×10^{-8}
Gold	2.443×10^{-8}
Aluminum	2.825×10^{-8}
Tungsten	5.485×10^{-8}
Iron	12.30×10^{-8}
Lead	22×10^{-8}
Mercury	95.8×10^{-8}
Nichrome	99.72×10^{-8}
Carbon	3500×10^{-8}
Germanium	20–2300*
Silicon	$\cong$500*
Wood	10^8–10^{14}
Glass	10^{10}–10^{14}
Mica	10^{11}–10^{15}
Hard rubber	10^{13}–10^{16}
Amber	5×10^{14}
Sulphur	1×10^{15}
Teflon	1×10^{16}

*The resistivities of these materials are dependent upon the impurities within the materials.

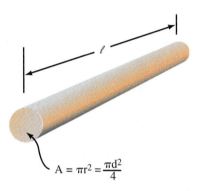

$$A = \pi r^2 = \frac{\pi d^2}{4}$$

FIGURE 3–2   Conductor with a circular cross-section.

Most homes use solid copper wire having a diameter of 1.63 mm to provide electrical distribution to outlets and light sockets. Determine the resistance of 75 meters of a solid copper wire having the above diameter.

EXAMPLE 3–1

Solution   We will first calculate the cross-sectional area of the wire using equation 3–2.

$$A = \frac{\pi d^2}{4}$$

$$= \frac{\pi (1.63 \times 10^{-3} \text{ m})^2}{4}$$

$$= 2.09 \times 10^{-6} \text{ m}^2$$

Now, using Table 3–1, the resistance of the length of wire is found as

$$R = \frac{\rho \ell}{A}$$

$$= \frac{(1.723 \times 10^{-8} \ \Omega\text{-m})(75 \text{ m})}{2.09 \times 10^{-6} \text{ m}^2}$$

$$= 0.619 \ \Omega$$

PRACTICE PROBLEMS 1

Find the resistance of a 100-m long tungsten wire that has a circular cross-section with a diameter of 0.1 mm ($T = 20°$ C).

Answer
698 Ω

EXAMPLE 3–2

Bus bars are bare solid conductors (usually rectangular) used to carry large currents within buildings such as power generating stations, telephone exchanges, and large factories. Given a piece of aluminum bus bar as shown in Figure 3–3, determine the resistance between the ends of this bar at a temperature of 20° C.

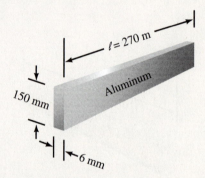

FIGURE 3–3   Conductor with a rectangular cross-section.

Solution   The cross-sectional area is

$$A = (150 \text{ mm})(6 \text{ mm})$$
$$= (0.15 \text{ m})(0.006 \text{ m})$$
$$= 0.0009 \text{ m}^2$$
$$= 9.00 \times 10^{-4} \text{ m}^2$$

The resistance between the ends of the bus bar is determined as

$$R = \frac{\rho\ell}{A}$$
$$= \frac{(2.825 \times 10^{-8} \ \Omega\text{-m})(270 \text{ m})}{9.00 \times 10^{-4} \text{ m}^2}$$
$$= 8.48 \times 10^{-3} \ \Omega = 8.48 \text{ m}\Omega$$

IN-PROCESS
LEARNING CHECK 1

(Answers are at the end of the chapter.)

1. Given two lengths of wire having identical dimensions. If one wire is made of copper and the other is made of iron, which wire will have the greater resistance? How much greater will the resistance be?

2. Given two pieces of copper wire that have the same cross-sectional area, determine the relative resistance of the one that is twice as long as the other.

3. Given two pieces of copper wire that have the same length, determine the relative resistance of the one that has twice the diameter of the other.

Although the SI system is the standard measurement for electrical and other physical quantities, the English system is still used extensively in the United States and to a lesser degree throughout the rest of the English-speaking world. One area that has been slow to convert to the SI system is the designation of cables and wires, where the American Wire Gauge (AWG) is the primary system used to denote wire diameters. In this system, each wire diameter is assigned a gauge number. As Figure 3–4 shows, the higher the AWG number, the smaller the diameter of the cable or wire, e.g., AWG 22 gauge wire is a smaller diameter than AWG 14 gauge. Since cross-sectional area is inversely proportional to the square of the diameter, a given length of 22-gauge wire will have more resistance than an equal length of 14-gauge wire. Because of the difference in resistance, we can intuitively deduce that large-diameter cables will be able to handle more current than smaller-diameter cables. Table 3–2 provides a listing of data for standard bare copper wire.

Even though Table 3–2 provides data for solid conductors up to AWG 4/0, most applications do not use solid conductor sizes beyond AWG 10. Solid conductors are difficult to bend and are easily damaged by mechanical flexing. For this reason, large-diameter cables are nearly always stranded rather than solid. Stranded wires and cables use anywhere from seven strands, as shown in Figure 3–5, to in excess of a hundred strands.

As one might expect, stranded wire uses the same AWG notation as solid wire. Consequently, AWG 10 stranded wire will have the same cross-sectional conductor area as AWG 10 solid wire. However, due to the additional space lost between the conductors, the stranded wire will have a larger overall diameter than the solid wire. Also, because the individual strands are coiled as a helix, the overall strand length will be slightly longer than the cable length.

Wire tables similar to Table 3–2 are available for stranded copper cables and for cables constructed of other materials (notably aluminum).

3.2 Electrical Wire Tables

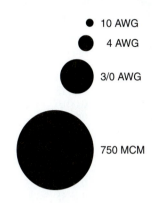

FIGURE 3–4   Typical conductor cross-sections (actual size).

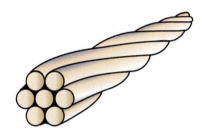

FIGURE 3–5   Stranded wire (7 strands).

Calculate the resistance of 200 feet of AWG 16 solid copper wire at 20° C.

EXAMPLE 3–3

Solution   From Table 3–2, we see that AWG 16 wire has a resistance of 4.02 Ω per 1000 feet. Since we are given a length of only 200 feet, the resistance will be determined as

$$R = \left(\frac{4.02\ \Omega}{1000\ \text{ft}}\right)(200\ \text{ft}) = 0.804\ \Omega$$

By examining Table 3–2, several important points may be observed:

- If the wire size increases by three gauge sizes, the cross-sectional area will approximately double. Since resistance is inversely proportional to cross-sectional area, a given length of larger-diameter cable will have a resistance that is approximately half as large as the resistance of a similar length of the smaller-diameter cable.

- If there is a difference of three gauge sizes between cables, then the larger-diameter cable will be able to handle approximately twice as much current as the smaller-diameter cable. The amount of current that a conductor can safely handle is directly proportional to the cross-sectional area.

- If the wire size increases by ten gauge sizes, the cross-sectional area will increase by a factor of about ten. Due to the inverse relationship between resistance and cross-sectional area, the larger-diameter cable will have about one tenth the resistance of a similar length of the smaller-diameter cable.

- For a 10-gauge difference in cable sizes, the larger-diameter cable will have ten times the cross-sectional area of the smaller-diameter cable and so it will be able to handle approximately ten times more current.

TABLE 3–2   Standard Solid Copper Wire at 20° C

Size (AWG)	Diameter (inches)	(mm)	Area (CM)	(mm²)	Resistance (Ω/1000 ft)	Current Capacity (A)
56	0.0005	0.012	0.240	0.000122	43 200	
54	0.0006	0.016	0.384	0.000195	27 000	
52	0.0008	0.020	0.608	0.000308	17 000	
50	0.0010	0.025	0.980	0.000497	10 600	
48	0.0013	0.032	1.54	0.000779	6 750	
46	0.0016	0.040	2.46	0.00125	4 210	
45	0.0019	0.047	3.10	0.00157	3 350	
44	0.0020	0.051	4.00	0.00243	2 590	
43	0.0022	0.056	4.84	0.00245	2 140	
42	0.0025	0.064	6.25	0.00317	1 660	
41	0.0028	0.071	7.84	0.00397	1 320	
40	0.0031	0.079	9.61	0.00487	1 080	
39	0.0035	0.089	12.2	0.00621	847	
38	0.0040	0.102	16.0	0.00811	648	
37	0.0045	0.114	20.2	0.0103	521	
36	0.0050	0.127	25.0	0.0127	415	
35	0.0056	0.142	31.4	0.0159	331	
34	0.0063	0.160	39.7	0.0201	261	
33	0.0071	0.180	50.4	0.0255	206	
32	0.0080	0.203	64.0	0.0324	162	
31	0.0089	0.226	79.2	0.0401	131	
30	0.0100	0.254	100	0.0507	104	
29	0.0113	0.287	128	0.0647	81.2	
28	0.0126	0.320	159	0.0804	65.3	
27	0.0142	0.361	202	0.102	51.4	
26	0.0159	0.404	253	0.128	41.0	0.75*
25	0.0179	0.455	320	0.162	32.4	
24	0.0201	0.511	404	0.205	25.7	1.3*
23	0.0226	0.574	511	0.259	20.3	
22	0.0253	0.643	640	0.324	16.2	2.0*
21	0.0285	0.724	812	0.412	12.8	
20	0.0320	0.813	1 020	0.519	10.1	3.0*
19	0.0359	0.912	1 290	0.653	8.05	
18	0.0403	1.02	1 620	0.823	6.39	5.0†
17	0.0453	1.15	2 050	1.04	5.05	
16	0.0508	1.29	2 580	1.31	4.02	10.0†
15	0.0571	1.45	3 260	1.65	3.18	
14	0.0641	1.63	4 110	2.08	2.52	15.0†
13	0.0720	1.83	5 180	2.63	2.00	
12	0.0808	2.05	6 530	3.31	1.59	20.0†
11	0.0907	2.30	8 230	4.17	1.26	
10	0.1019	2.588	10 380	5.261	0.998 8	30.0†
9	0.1144	2.906	13 090	6.632	0.792 5	
8	0.1285	3.264	16 510	8.367	0.628 1	
7	0.1443	3.665	20 820	10.55	0.498 1	
6	0.1620	4.115	26 240	13.30	0.395 2	
5	0.1819	4.620	33 090	16.77	0.313 4	
4	0.2043	5.189	41 740	21.15	0.248 5	
3	0.2294	5.827	52 620	26.67	0.197 1	
2	0.2576	6.543	66 360	33.62	0.156 3	
1	0.2893	7.348	83 690	42.41	0.123 9	
1/0	0.3249	8.252	105 600	53.49	0.098 25	
2/0	0.3648	9.266	133 100	67.43	0.077 93	
3/0	0.4096	10.40	167 800	85.01	0.061 82	
4/0	0.4600	11.68	211 600	107.2	0.049 01	

*This current is suitable for single conductors and surface or loose wiring.

†This current may be accommodated in up to three wires in a sheathed cable. For four to six wires, the current in each wire must be reduced to 80% of the indicated value. For seven to nine wires, the current in each wire must be reduced to 70% of the indicated value.

If AWG 14 solid copper wire is able to handle 15 A of current, determine the expected current capacity of AWG 24 and AWG 8 copper wire at 20° C.

EXAMPLE 3–4

Solution　Since AWG 24 is ten sizes smaller than AWG 14, the smaller cable will be able to handle about one tenth the capacity of the larger-diameter cable.
　　AWG 24 will be able to handle approximately 1.5 A of current.
　　AWG 8 is six sizes larger than AWG 14. Since current capacity doubles for an increase of three sizes, AWG 11 would be able to handle 30 A and AWG 8 will be able to handle 60 A.

---

1. From Table 3–2 find the diameters in millimeters and the cross-sectional areas in square millimeters of AWG 19 and AWG 30 solid wire.

2. By using the cross-sectional areas for AWG 19 and AWG 30, approximate the areas that AWG 16 and AWG 40 should have.

3. Compare the actual cross-sectional areas as listed in Table 3–2 to the areas found in Problems 1 and 2 above. (You will find a slight variation between your calculated values and the actual areas. This is because the actual diameters of the wires have been adjusted to provide optimum sizes for manufacturing.)

PRACTICE PROBLEMS 2

Answers
1. $d_{AWG19} = 0.912$ mm　　$A_{AWG19} = 0.653$ mm^2
　 $d_{AWG30} = 0.254$ mm　　$A_{AWG30} = 0.0507$ mm^2
2. $A_{AWG16} \cong 1.31$ mm^2　　$A_{AWG40} \cong 0.0051$ mm^2
3. $A_{AWG16} = 1.31$ mm^2　　$A_{AWG40} = 0.00487$ mm^2

---

1. AWG 12-gauge wire is able to safely handle 20 amps of current. How much current should an AWG 2-gauge cable be able to handle?

2. The electrical code actually permits up to 120 A for the above cable. How does the actual value compare to your theoretical value? Why do you think there is a difference?

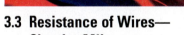

IN-PROCESS
LEARNING CHECK 2

(Answers are at the end of the chapter.)

The American Wire Gauge system for specifying wire diameters was developed using a unit called the **circular mil** (CM), which is defined as the area contained within a circle having a diameter of 1 mil (1 mil = 0.001 inch). A **square mil** is defined as the area contained in a square having side dimensions of 1 mil. By referring to Figure 3–6, it is apparent that the area of a circular mil is smaller than the area of a square mil.

3.3 Resistance of Wires— Circular Mils

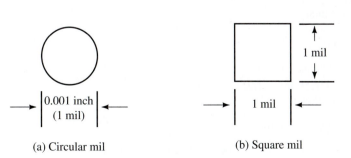

(a) Circular mil　　　　(b) Square mil

FIGURE 3–6

Because not all conductors have circular cross-sections, it is occasionally necessary to convert areas expressed in square mils into circular mils. We will now determine the relationship between the circular mil and the square mil.

Suppose that a wire has the circular cross section shown in Figure 3–6(a). By applying Equation 3–2, the area, in square mils, of the circular cross section is determined as follows:

$$A = \frac{\pi d^2}{4}$$

$$= \frac{\pi (1 \text{ mil})^2}{4}$$

$$= \frac{\pi}{4} \text{ sq. mil}$$

From the above derivation the following relations must apply:

$$1 \text{ CM} = \frac{\pi}{4} \text{ sq. mil} \tag{3–3}$$

$$1 \text{ sq. mil} = \frac{4}{\pi} \text{ CM} \tag{3–4}$$

The greatest advantage of using the circular mil to express areas of wires is the simplicity with which calculations may be made. Unlike previous area calculations which involved the use of π, area calculations may be reduced to simply finding the square of the diameter.

If we are given a circular cross section with a diameter, d (in mils) the area of this cross-section is determined as

$$A = \frac{\pi d^2}{4} \quad \text{[square mils]}$$

Using Equation 3–4, we convert the area from square mils to circular mils. Consequently, if the diameter of a circular conductor is given in mils, we determine the area in circular mils as

$$A_{CM} = d_{mil}^2 \quad \text{[circular mils, CM]} \tag{3–5}$$

EXAMPLE 3–5

Determine the cross-sectional area in circular mils of a wire having the following diameters:

a. 0.0159 inch (AWG 26 wire)

b. 0.500 inch

Solution

a. $d = 0.0159$ inch

$\quad = (0.0159 \text{ inch})(1000 \text{ mils/inch})$

$\quad = 15.9$ mils

Now, using Equation 3–5, we obtain

$$A_{CM} = (15.9)^2 = 253 \text{ CM}$$

From Table 3–2, we see that the previous result is precisely the area given for AWG 26 wire.

b. $d = 0.500$ inch
   $= (0.500 \text{ inch})(1000 \text{ mils/inch})$
   $= 500$ mils

$$A_{CM} = (500)^2 = 250\ 000 \text{ CM}$$

In Example 3–5(b) we see that the cross-sectional area of a cable may be a large number when it is expressed in circular mils. In order to simplify the units for area, the Roman numeral M is often used to represent 1000. If a wire has a cross-sectional area of 250 000 CM, it is more easily written as 250 MCM.

Clearly, this is a departure from the SI system, where M is used to represent one million. Since there is no simple way to overcome this conflict, the student working with cable areas expressed in MCM will need to remember that the M stands for one thousand and not for one million.

a. Determine the cross-sectional area in square mils and in circular mils of a copper bus bar having cross-sectional dimensions of 0.250 inch × 6.00 inch.

EXAMPLE 3–6

b. If this copper bus bar were to be replaced by AWG 2/0 cables, how many cables would be required?

Solution

a. $A_{\text{sq. mil}} = (250 \text{ mils})(6000 \text{ mils})$
   $= 1\ 500\ 000$ sq. mils

The area in circular mils is found by applying Equation 3–4, and this will be

$$A_{CM} = (250 \text{ mils})(6000 \text{ mils})$$

$$= (1\ 500\ 000 \text{ sq. mils})\left(\frac{4}{\pi} \text{ CM/sq. mil}\right)$$

$$= 1\ 910\ 000 \text{ CM}$$
$$= 1910 \text{ MCM}$$

b. From Table 3–2, we see that AWG 2/0 cable has a cross-sectional area of 133.1 MCM (133 100 CM), and so the bus bar is equivalent to the following number of cables:

$$n = \frac{1910 \text{ MCM}}{133.1 \text{ MCM}} = 14.4$$

This example illustrates that 15 cables would need to be installed to be equivalent to a single 6-inch by 0.25-inch bus bar. Due to the expense and awkwardness of using this many cables, we see the economy of using solid bus bar. The main disadvantage of using bus bar is that the conductor is not covered with an insulation, and so the bus bar does not offer the same protection as cable. However, since bus bar is generally used in locations where only experienced technicians are permitted access, this disadvantage is a minor one.

TABLE 3–3 **Resistivity of Conductors, ρ**	
Material	**Resistivity, ρ, at 20° C (CM-Ω/ft)**
Silver	9.90
Copper	10.36
Gold	14.7
Aluminum	17.0
Tungsten	33.0
Iron	74.0
Lead	132.
Mercury	576.
Nichrome	600.

As we have seen in Section 3.1, the resistance of a conductor was determined to be

$$R = \frac{\rho\ell}{A}$$

Although the original equation used SI units, the equation will also apply if the units are expressed in any other convenient system. If cable length is generally expressed in feet and the area in circular mils, then the resistivity must be expressed in the appropriate units. Table 3–3 gives the resistivities of some conductors represented in circular mil-ohms per foot.

The following example illustrates how Table 3–3 may be used to determine the resistance of a given section of wire.

EXAMPLE 3–7

Determine the resistance of an AWG 16 copper wire at 20° C if the wire has a diameter of 0.0508 inch and a length of 400 feet.

Solution   The diameter in mils is found as

$$d = 0.0508 \text{ inch} = 50.8 \text{ mils}$$

Therefore the cross-sectional area (in circular mils) of AWG 16 is

$$A_{CM} = 50.8^2 = 2580 \text{ CM}$$

Now, by applying Equation 3–1 and using the appropriate units, we obtain the following:

$$R = \frac{\rho\ell}{A_{CM}}$$

$$= \frac{\left(10.36 \dfrac{\text{CM-}\Omega}{\text{ft}}\right)(400 \text{ ft})}{2580 \text{ CM}}$$

$$= 1.61 \ \Omega$$

PRACTICE PROBLEMS 3

1. Determine the resistance of 1 mile (5280 feet) of AWG 19 copper wire at 20° C, if the cross-sectional area is 1290 CM.

2. Compare the above result with the value that would be obtained by using the resistance (in ohms per thousand feet) given in Table 3–2.

3. An aluminum conductor having a cross-sectional area of 1843 MCM is used to transmit power from a high-voltage dc (HVDC) generating station to a large urban center. If the city is 900 km from the generating station, determine the resistance of the conductor at a temperature of 20° C. (Use 1 ft ≡ 0.3048 m.)

Answers
1. 42.4 Ω; 2. 42.5 Ω; 3. 27.2 Ω

IN-PROCESS
LEARNING CHECK 3

(Answers are at the end of the chapter.)

A conductor has a cross-sectional area of 50 square mils. Determine the cross-sectional area in circular mils, square meters, and square millimeters.

Section 3.1 indicated that the resistance of a conductor will not be constant at all temperatures. As temperature increases, more electrons will escape their orbits, causing additional collisions within the conductor. For most conducting materials, the increase in the number of collisions translates into a relatively linear increase in resistance, as shown in Figure 3–7.

3.4 Temperature Effects

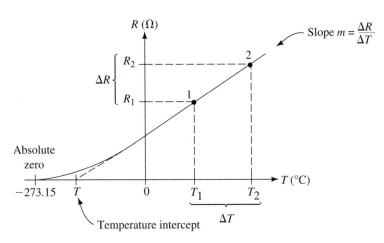

FIGURE 3–7 Temperature effects on resistance of a conductor.

The rate at which the resistance of a material changes with a variation in temperature is dependent on the **temperature coefficient** of the material, which is assigned the Greek letter alpha (α). Some materials have only very slight changes in resistance, while other materials demonstrate dramatic changes in resistance with a change in temperature.

Any material for which resistance increases as temperature increases is said to have a **positive temperature coefficient.**

For semiconductor materials such as carbon, germanium, and silicon, increases in temperature allow electrons to escape their usually stable orbits and become free to move within the material. Although additional collisions do occur within the semiconductor, the effect of the collisions is minimal when compared with the contribution of the extra electrons to the overall flow of charge. As the temperature increases, the number of charge electrons increases, resulting in more current. Therefore, an increase in temperature results in a decrease in resistance. Consequently, these materials are referred to as having **negative temperature coefficients.**

Table 3–4 gives the temperature coefficients, α per degree Celsius, of various materials at 20° C and at 0° C.

If we consider that Figure 3–7 illustrates how the resistance of copper changes with temperature, we observe an almost linear increase in resistance as the temperature increases. Further, we see that as the temperature is decreased to **absolute zero** ($T = -273.15°$ C), the resistance approaches zero.

In Figure 3–7, the point at which the linear portion of the line is extrapolated to cross the abscissa (temperature axis) is referred to as the **temperature intercept** or the **inferred absolute temperature** T of the material.

By examining the straight-line portion of the graph, we see that we have two similar triangles, one with the apex at point 1 and the other with the apex at point 2. The following relationship applies for these similar triangles.

$$\frac{R_2}{T_2 - T} = \frac{R_1}{T_1 - T}$$

TABLE 3–4   Temperature Intercepts and Coefficients for Common Materials

	T (°C)	α (°C)^1 at 20° C	α (°C)^1 at 0° C
Silver	−243	0.003 8	0.004 12
Copper	−234.5	0.003 93	0.004 27
Aluminum	−236	0.003 91	0.004 24
Tungsten	−202	0.004 50	0.004 95
Iron	−162	0.005 5	0.006 18
Lead	−224	0.004 26	0.004 66
Nichrome	−2270	0.000 44	0.000 44
Brass	−480	0.002 00	0.002 08
Platinum	−310	0.003 03	0.003 23
Carbon		−0.000 5	
Germanium		−0.048	
Silicon		−0.075	

This expression may be rewritten to solve for the resistance, R_2 at any temperature, T_2 as follows:

$$R_2 = \frac{T_2 - T}{T_1 - T} R_1 \qquad (3\text{–}6)$$

An alternate method of determining the resistance, R_2 of a conductor at a temperature, T_2 is to use the temperature coefficient, α of the material. Examining Table 3–4, we see that the temperature coefficient is not a constant for all temperatures, but rather is dependent upon the temperature of the material. The temperature coefficient for any material is defined as

$$\alpha = \frac{m}{R_1} \qquad (3\text{–}7)$$

The value of α is typically given in chemical handbooks. In the above expression, α is measured in $(°C)^{-1}$, R_1 is the resistance in ohms at a temperature, T_1, and m is the slope of the linear portion of the curve ($m = \Delta R/\Delta T$). It is left as an end-of-chapter problem for the student to use Equations 3–6 and 3–7 to derive the following expression from Figure 3–6.

$$R_2 = R_1[1 + \alpha_1(T_2 - T_1)] \qquad (3\text{–}8)$$

EXAMPLE 3–8

An aluminum wire has a resistance of 20 Ω at room temperature (20° C). Calculate the resistance of the same wire at temperatures of −40° C, 100° C, and 200° C.

Solution   From Table 3–4, we see that aluminum has a temperature intercept of –236° C.

At $T = -40°$ C:
The resistance at −40° C is determined using Equation 3–6.

$$R_{-40°\,C} = \left[\frac{-40°\,C - (-236°\,C)}{20°\,C - (-236°\,C)} \right] 20\,\Omega = \left(\frac{196°\,C}{256°\,C} \right) 20\,\Omega = 15.3\,\Omega$$

At $T = 100°$ C:

$$R_{100°\,C} = \left[\frac{100°\,C - (-236°\,C)}{20°\,C - (-236°\,C)} \right] 20\,\Omega = \left(\frac{336°\,C}{256°\,C} \right) 20\,\Omega = 26.3\,\Omega$$

At $T = 200°$ C:

$$R_{200° \text{ C}} = \left[\frac{200° \text{ C} - (-236° \text{ C})}{20° \text{ C} - (-236° \text{ C})}\right] 20 \; \Omega = \left(\frac{436° \text{ C}}{256° \text{ C}}\right) 20 \; \Omega = 34.1 \; \Omega$$

The above phenomenon indicates that the resistance of conductors changes quite dramatically with changes in temperature. For this reason manufacturers generally specify the range of temperatures over which a conductor may operate safely.

Tungsten wire is used as filaments in incandescent light bulbs. Current in the wire causes the wire to reach extremely high temperatures. Determine the temperature of the filament of a 100-W light bulb if the resistance at room temperature is measured to be 11.7 Ω and when the light is on, the resistance is determined to be 144 Ω.

EXAMPLE 3–9

Solution　If we rewrite Equation 3–6, we are able to solve for the temperature T_2 as follows

$$T_2 = (T_1 - T)\frac{R_2}{R_1} + T$$

$$= [20° \text{ C} - (-202° \text{ C})]\frac{144 \; \Omega}{11.7 \; \Omega} + (-202° \text{ C})$$

$$= 2530° \text{ C}$$

A HVDC (high-voltage dc) transmission line must be able to operate over a wide temperature range. Calculate the resistance of 900 km of 1843 MCM aluminum conductor at temperatures of $-40°$ C and $+40°$ C.

PRACTICE PROBLEMS 4

Answers
20.8 Ω, 29.3 Ω

Explain what is meant by the terms *positive temperature coefficient* and *negative temperature coefficient*. To which category does aluminum belong?

IN-PROCESS
LEARNING CHECK 4

(Answers are at the end of the chapter.)

Virtually all electric and electronic circuits involve the control of voltage and/or current. The best way to provide such control is by inserting appropriate values of resistance into the circuit. Although various types and sizes of resistors are used in electrical and electronic applications, all resistors fall into two main categories: fixed resistors and variable resistors.

3.5　Types of Resistors

Fixed Resistors

As the name implies, **fixed resistors** are resistors having resistance values that are essentially constant. There are numerous types of fixed resistors, ranging in size from almost microscopic (as in integrated circuits) to high-power resistors that are capable of dissipating many watts of power. Figure 3–8 illustrates the basic structure of a **molded carbon composition** resistor.

As shown in Figure 3–8, the molded carbon composition resistor consists of a carbon core mixed with an insulating filler. The ratio of carbon to filler determines the resistance value of the component: the higher the proportion of carbon, the lower the resistance. Metal leads are inserted into the carbon core, and then the entire resistor is encapsulated with an insulated coating. Carbon composition resistors are available in resistances from less than 1 Ω to 100 MΩ and typically have power ratings from ⅛ W to 2 W. Figure 3–9 shows various sizes of resistors, with the larger resistors being able to dissipate more power than the smaller resistors.

Although carbon-core resistors have the advantages of being inexpensive and easy to produce, they tend to have wide tolerances and are susceptible to large

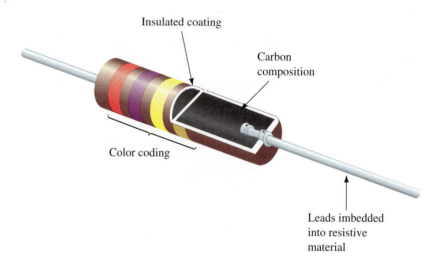

FIGURE 3–8   Structure of a molded carbon composition resistor.

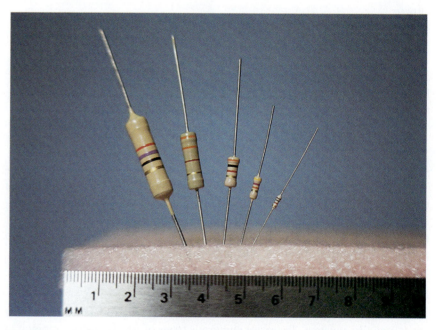

FIGURE 3–9   Actual size of carbon resistors (2W, 1W, ½W, ¼W, ⅛W).

changes in resistance due to temperature variation. As shown in Figure 3–10, the resistance of a carbon composition resistor may change by as much as 5% when temperature is changed by 100° C.

Other types of fixed resistors include **carbon film, metal film, metal oxide, wire-wound,** and **integrated circuit packages.**

If fixed resistors are required in applications where precision is an important factor, then film resistors are usually employed. These resistors consist of either carbon, metal, or metal-oxide film deposited onto a ceramic cylinder. The desired resistance is obtained by removing part of the resistive material, resulting in a helical pattern around the ceramic core. If variation of resistance due to temperature is not a major concern, then low-cost carbon is used. However, if close tolerances are required over a wide temperature range, then the resistors are made of films consisting of alloys such as nickel chromium, constantum, or manganin, which have very small temperature coefficients.

Occasionally a circuit requires a resistor to be able to dissipate large quantities of heat. In such cases, wire-wound resistors may be used. These resistors are constructed of a metal alloy wound around a hollow porcelain core, which is then covered with a thin layer of porcelain to seal it in place. The porcelain is able to quickly dissipate heat generated due to current through the wire. Figure 3–11 shows a few of the various types of power resistors available.

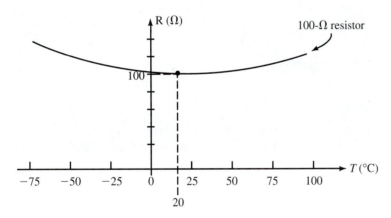

FIGURE 3–10 Variation in resistance of a carbon composition fixed resistor.

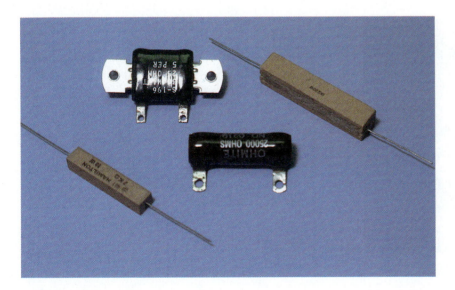

FIGURE 3–11 Power resistors.

In circuits where the dissipation of heat is not a major design consideration, fixed resistances may be constructed in miniature packages (called integrated circuits or ICs) capable of containing many individual resistors. The obvious advantage of such packages is their ability to conserve space on a circuit board. Figure 3–12 illustrates a typical resistor IC package.

Variable Resistors

Variable resistors provide indispensable functions that we use in one form or another almost daily. These components are used to adjust the volume of our radios, set the level of lighting in our homes, and adjust the heat of our stoves and furnaces. Figure 3–13 shows the internal and the external view of typical variable resistors.

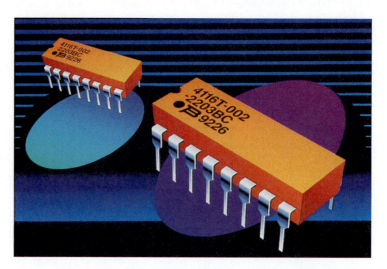

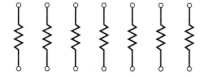

(a) Internal resistor arrangement

(b) Integrated resistor network. (*Courtesy of Bourns, Inc.*)

FIGURE 3–12

(a) External view of variable resistors

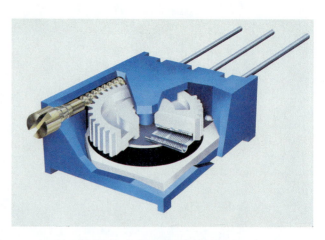

(b) Internal view of variable resistor

FIGURE 3–13   Variable resistors. (*Courtesy of Bourns, Inc.*)

In Figure 3–14, we see that variable resistors have three terminals, two of which are fixed to the ends of the resistive material. The central terminal is connected to a wiper that moves over the resistive material when the shaft is rotated with either a knob or a screwdriver. The resistance between the two outermost terminals will remain constant, while the resistance between the central terminal and either terminal will change according to the position of the wiper.

If we examine the schematic of a variable resistor as shown in Figure 3–14(b), we see that the following relationship must apply:

$$R_{ac} = R_{ab} + R_{bc} \tag{3–9}$$

Variable resistors are used for two principal functions. **Potentiometers,** shown in Figure 3–14(c), are used to adjust the amount of potential (voltage) provided to a circuit. **Rheostats,** the connections and schematic of which are shown in Figure 3–15, are used to adjust the amount of current within a circuit. Applications of potentiometers and rheostats will be covered in later chapters.

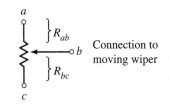

(b) Terminals of a variable resistor

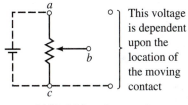

(c) Variable resistor used as a potentiometer

FIGURE 3–14 (a) Variable resistors. (*Courtesy of Bourns, Inc.*)

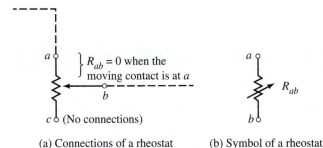

(a) Connections of a rheostat   (b) Symbol of a rheostat

FIGURE 3–15

3.6 Color Coding of Resistors

Large resistors such as the wire-wound resistors or the ceramic-encased power resistors have their resistor values and tolerances printed on their cases. Smaller resistors, whether constructed of a molded carbon composition or a metal film, may be too small to have their values printed on the component. Instead, these smaller resistors are usually covered by an epoxy or similar insulating coating over which several colored bands are printed radially as shown in Figure 3–16.

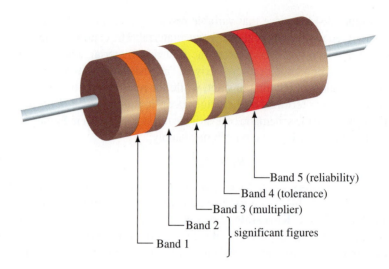

Band 5 (reliability)

Band 4 (tolerance)

Band 3 (multiplier)

Band 2

Band 1

significant figures

FIGURE 3–16   Resistor color codes.

The colored bands provide a quickly recognizable code for determining the value of resistance, the tolerance (in percentage), and occasionally the expected reliability of the resistor. The colored bands are always read from left to right, left being defined as the side of the resistor with the band nearest to it.

The first two bands represent the first and second digits of the resistance value. The third band is called the multiplier band and represents the number of zeros following the first two digits; it is usually given as a power of ten. The fourth band indicates the tolerance of the resistor, and the fifth band (if present) is an indication of the expected reliability of the component. The reliability is a statistical indication of the expected number of components that will no longer have the indicated resistance value after 1000 hours of use. For example, if a particular resistor has a reliability of 1%, it is expected that after 1000 hours of use, no more than 1 resistor in 100 is likely to be outside the specified range of resistance as indicated in the first four bands of the color codes. Table 3–5 shows the colors of the various bands and the corresponding values.

TABLE 3–5 **Resistor Color Codes**					
Color	Band 1 Sig. Fig.	Band 2 Sig. Fig.	Band 3 Multiplier	Band 4 Tolerance	Band 5 Reliability
Black		0	$10^0 = 1$		
Brown	1	1	$10^1 = 10$		1%
Red	2	2	$10^2 = 100$		0.1%
Orange	3	3	$10^3 = 1\ 000$		0.01%
Yellow	4	4	$10^4 = 10\ 000$		0.001%
Green	5	5	$10^5 = 100\ 000$		
Blue	6	6	$10^6 = 1\ 000\ 000$		
Violet	7	7	$10^7 = 10\ 000\ 000$		
Gray	8	8			
White	9	9			
Gold			0.1	5%	
Silver			0.01	10%	
No color				20%	

Determine the resistance of a carbon film resistor having the color codes shown in Figure 3–17.

EXAMPLE 3–10

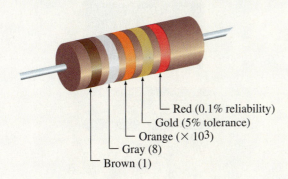

Red (0.1% reliability)
Gold (5% tolerance)
Orange ($\times 10^3$)
Gray (8)
Brown (1)

FIGURE 3–17

Solution   From Table 3–5, we see that the resistor will have a value determined as

$$R = 18 \times 10^3 \ \Omega \pm 5\%$$
$$= 18 \ \text{k}\Omega \pm 0.9 \ \text{k}\Omega \text{ with a reliability of } 0.1\%$$

This specification indicates that the resistance will fall between 17.1 kΩ and 18.9 kΩ. After 1000 hours, we would expect that no more than 1 resistor in 1000 would fall outside the specified range.

A resistor manufacturer produces carbon composition resistors of 100 MΩ, with a tolerance of $\pm 5\%$. What will be the color codes on the resistor? (Left to right)

PRACTICE PROBLEMS 5

Answer
Brown, Black, Violet, Gold

The **ohmmeter** is an instrument that is generally part of a multimeter (usually including a voltmeter and an ammeter) and is used to measure the resistance of a component. Although it has limitations, the ohmmeter is used almost daily in service shops and laboratories to measure resistance of components and also to determine whether a circuit is faulty. In addition, the ohmmeter may also be used to determine the condition of semiconductor devices such as diodes and transistors. Figure 3–18 shows a typical digital ohmmeter.

In order to measure the resistance of an isolated component or circuit, the ohmmeter is placed across the component under test, as shown in Figure 3–19. The resistance is then simply read from the meter display.

When using an ohmmeter to measure the resistance of a component that is located in an operating circuit, the following steps should be observed:

1. As shown in Figure 3–20(a), remove all power supplies from the circuit or component to be tested. If this step is not followed, the ohmmeter reading will, at best, be meaningless, and the ohmmeter may be severely damaged.

2. If you wish to measure the resistance of a particular component, it is necessary to isolate the component from the rest of the circuit. This is done by disconnecting at least one terminal of the component from the balance of the circuit as shown in Figure 3–20(b). If this step is not followed, in all

3.7 Measuring Resistance— The Ohmmeter

FIGURE 3–18 Digital ohmmeter *(Reproduced with permission from the John Fluke Mfg. Co., Inc.)*

FIGURE 3–19 Ohmmeter used to measure an isolated component.

Voltage source

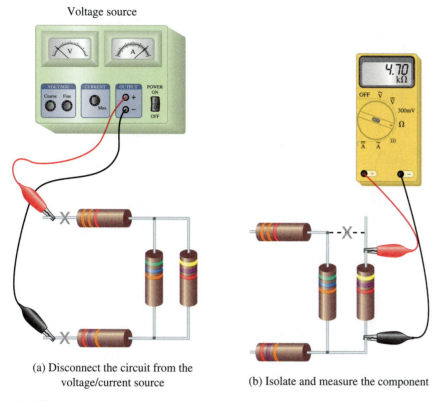

(a) Disconnect the circuit from the
voltage/current source

(b) Isolate and measure the component

FIGURE 3–20   Using an ohmmeter to measure resistance in a circuit.

likelihood the resistance reading indicated by the ohmmeter will not be the resistance of the desired resistor, but rather the resistance of the combination.

3. As shown in Figure 3–20(b), connect the two probes of the ohmmeter across the component to be measured. The black and red leads of the ohmmeter may be interchanged when measuring resistors. When measuring resistance of other components, however, the measured resistance will be dependent upon the direction of the sensing current. Such devices are covered briefly in a later section of this chapter.

4. Ensure that the ohmmeter is on the correct range to provide the most accurate reading. For example, although a digital multimeter (DMM) can measure a reading for a 1.2-kΩ resistor on the 2-MΩ range, the same ohmmeter will provide additional significant digits (hence more precision) when it is switched to the 2-kΩ range. For analog meters, the best accuracy is obtained when the needle is approximately in the center of the scale.

5. When you are finished, turn the ohmmeter off. Because the ohmmeter uses an internal battery to provide a small sensing current, it is possible to drain the battery if the probes accidentally connect together for an extended period.

In addition to measuring resistance, the ohmmeter may also be used to indicate the continuity of a circuit. Many modern digital ohmmeters have an audible tone to indicate that a circuit is unbroken from one point to another point. As demonstrated in Figure 3–21(a), the audible tone of a digital ohmmeter allows the user to determine continuity without having to look away from the circuit under test.

Ohmmeters are particularly useful instruments in determining whether a given circuit has been short circuited or open circuited.

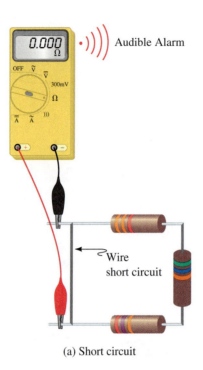

Audible Alarm

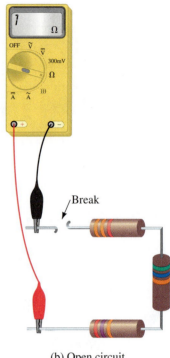

Break

Wire
short circuit

(a) Short circuit

(b) Open circuit

FIGURE 3–21

PRACTICAL NOTES . . .

When a digital ohmmeter measures an open circuit, the display on the meter will usually be the digit 1 at the left-hand side, with no following digits. This reading should not be confused with a reading of 1 Ω, 1 kΩ, or 1 MΩ, which would appear on the right-hand side of the display.

A **short circuit** occurs when a low-resistance conductor such as a piece of wire or any other conductor is connected between two points in a circuit. Due to the very low resistance of the short circuit, current will bypass the rest of the circuit and go through the short. An ohmmeter will indicate a very low (theoretically zero) resistance when used to measure across a short circuit.

An **open circuit** occurs when a conductor is broken between the points under test. An ohmmeter will indicate infinite resistance when used to measure the resistance of a circuit having an open circuit.

Figure 3–21 illustrates circuits having a short circuit and an open circuit.

 PRACTICE PROBLEMS 6

An ohmmeter is used to measure across the terminals of a switch.

a. What will the ohmmeter indicate when the switch is closed?

b. What will the ohmmeter indicate when the switch is opened?

Answers
a. 0 Ω (short circuit)
b. ∞ (open circuit)

3.8 Thermistors

In Section 3.4 we saw how resistance changes with changes in temperature. While this effect is generally undesirable in resistors, there are many applications that use electronic components having characteristics that vary according to changes in temperature. Any device or component that causes an electrical change due to a physical change is referred to as a **transducer.**

A **thermistor** is a two-terminal transducer in which resistance changes significantly with changes in temperature (hence a thermistor is a "thermal resistor"). The resistance of thermistors may be changed either by external tempera-

(a) Photograph

(b) Symbol

FIGURE 3–22 Thermistors.

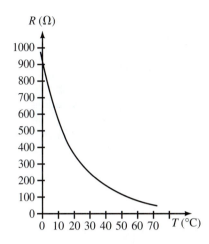

FIGURE 3–23 Thermistor resistance as a function of temperature.

ture changes or by changes in temperature caused by current through the component. By applying this principle, thermistors may be used in circuits to control current and to measure or control temperature. Typical applications include electronic thermometers and thermostatic control circuits for furnaces. Figure 3–22 shows a typical thermistor and its electrical symbol.

Thermistors are constructed of oxides of various materials such as cobalt, manganese, nickel, and strontium. As the temperature of the thermistor is increased, the outermost (valence) electrons in the atoms of the material become more active and break away from the atom. These extra electrons are now free to move within the circuit, thereby causing a reduction in the resistance of the component (negative temperature coefficient). Figure 3–23 shows how resistance of a thermistor varies with temperature effects.

Referring to Figure 3–22, determine the approximate resistance of a thermistor at each of the following temperatures:

PRACTICE PROBLEMS 7

a. 10° C.

b. 30° C.

c. 50° C.

Answers
a. 550 Ω; b. 250 Ω; c. 120 Ω

Photoconductive cells or **photocells** are two-terminal transducers that have a resistance determined by the amount of light falling on the cell. Most photocells are constructed of either cadmium sulfide (CdS) or cadmium selenide (CdSe) and are sensitive to light having wavelengths between 4000 Å (blue light) and 10 000 Å (infrared). The angstrom (Å) is a unit commonly used to measure the wavelength of light and has a dimension given as 1 Å = 1 $\times$ 10^{-10} m. Light, which is a form of energy, strikes the material of the photocell and causes the release of valence electrons, thereby reducing the resistance of the component. Figure 3–24 shows the structure, symbol, and resistance characteristics of a typical photocell.

Photocells may be used to measure light intensity and/or to control lighting. They are typically used as part of a security system.

3.9 Photoconductive Cells

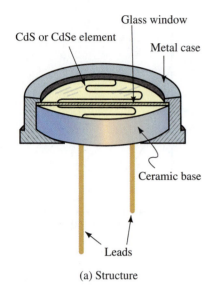

Glass window
CdS or CdSe element
Metal case

Ceramic base

Leads

(a) Structure

(b) Symbol of a photocell

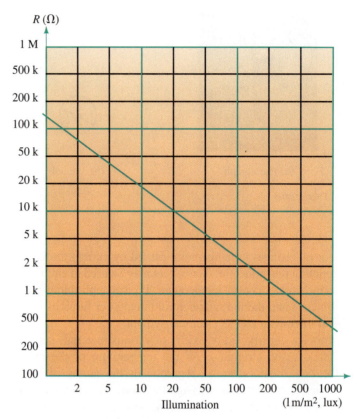

(c) Resistance versus illumination

FIGURE 3–24 Photocell.

3.10 Nonlinear Resistance

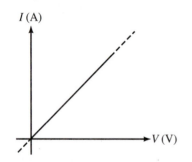

FIGURE 3–25 Linear current-voltage relationship.

Up to this point, the components we have examined have had values of resistance that were essentially constant for a given temperature (or, in the case of a photocell, for a given amount of light). If we were to examine the current versus voltage relationship for these components, we would find that the relationship is linear, as shown in Figure 3–25.

If a device has a linear (straight-line) current-voltage relation then it is referred to as an **ohmic device.** (The linear current-voltage relationship will be will be covered in greater detail in the next chapter.) Often in electronics, we use components that do not have a linear current-voltage relationship; these devices are referred to as **nonohmic devices.** On the other hand, some components, such as the thermistor, can be shown to have both an ohmic region and a nonohmic region. For large current through the thermistor, the component will get hotter. This increase in temperature will result in a decrease of resistance. Consequently, for large currents, the thermistor is a nonohmic device.

We will now briefly examine two common nonohmic devices.

Diodes

The diode is a semiconductor device that permits charge to flow in only one direction. Figure 3–26 illustrates the appearance and the symbol of a typical diode.

Conventional current through a diode is in the direction from the anode toward the cathode (the end with the line around the circumference). When current is in this direction, the diode is said to be **forward biased** and operating in its **forward region.** Since a diode has very little resistance in its forward region, it is often approximated as a short circuit.

If the circuit is connected such that the direction of current is from the cathode to the anode (against the arrow in Figure 3–26), the diode is **reverse**

biased and operating in its **reverse region.** Due to the high resistance of a reverse-biased diode, it is often approximated as an open circuit.

Although this textbook does not attempt to provide an in-depth study of diode theory, Figure 3–27 shows the basics of diode operation both when forward biased and when reverse biased.

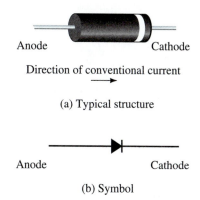

(a) Typical structure

(b) Symbol

FIGURE 3–26   Diode.

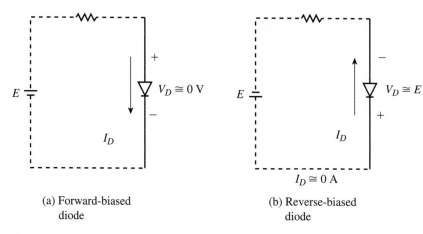

(a) Forward-biased diode

(b) Reverse-biased diode

FIGURE 3–27   Current-voltage relation for a silicon diode.

Because an ohmmeter uses an internal voltage source to generate a small sensing current, the instrument may easily be used to determine the terminals (and hence the direction of conventional flow) of a diode. (See Figure 3–28.)

PRACTICAL NOTES . . .

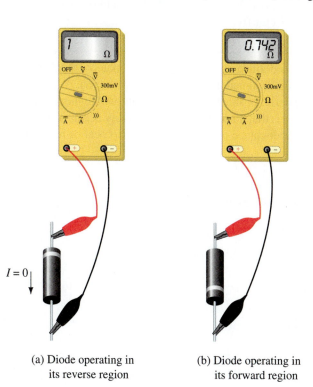

(a) Diode operating in its reverse region

(b) Diode operating in its forward region

FIGURE 3–28   Determining diode terminals with an ohmmeter.

If we measure the resistance of the diode in both directions, we will find that the resistance will be low when the positive terminal of the ohmmeter is connected to the anode of the diode. When the positive terminal is connected to the cathode, virtually no current will occur in the diode and so the indication on the ohmmeter will be infinite resistance normally shown by the numeral 1 on the left.

(a) Photograph

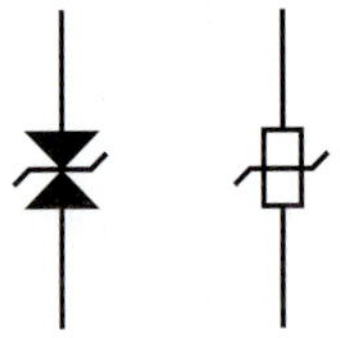

(b) Varistor symbols

FIGURE 3–29   Varistors.

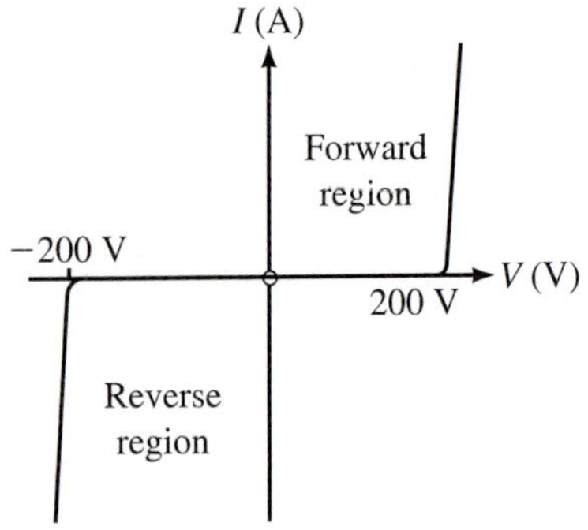

Varistors

Varistors, as shown in Figure 3–29, are semiconductor devices that have very high resistances when the voltage across the varistors is below the breakdown value. However, when the voltage across a varistor (either polarity) exceeds the rated value, the resistance of the device suddenly becomes very small, allowing charge to flow. Figure 3–30 shows the current-voltage relation for varistors.

Varistors are used in sensitive circuits, such as those in computers, to ensure that if the voltage suddenly exceeds a predetermined value, the varistor will effectively become a short circuit to the unwanted signal, thereby protecting the rest of the circuit from excessive voltage.

FIGURE 3–30   Current-voltage relation of a 200-V (peak) varistor.

3.11 Conductance

Conductance, G, is defined as the measure of a material's ability to allow the flow of charge and is assigned the SI unit the siemens (S). A large conductance indicates that a material is able to conduct current well, whereas a low value of conductance indicates that a material does not readily permit the flow of charge. Mathematically, conductance is defined as the reciprocal of resistance. Thus

$$G = \frac{1}{R} \quad \text{[siemens, S]} \tag{3–10}$$

where R is resistance, in ohms (Ω).

EXAMPLE 3–11

Determine the conductance of the following resistors:

    a. 5 Ω

    b. 100 kΩ

    c. 50 mΩ

Solution

a. $G = \dfrac{1}{5\ \Omega} = 0.2\ \text{S} = 200\ \text{mS}$

b. $G = \dfrac{1}{100\ \text{k}\Omega} = 0.01\ \text{mS} = 10\ \mu\text{S}$

c. $G = \dfrac{1}{50\ \text{m}\Omega} = 20\ \text{S}$

PRACTICE PROBLEMS 8

1. A given cable has a conductance given as 5.0 mS. Determine the value of the resistance, in ohms.

2. If the conductance is doubled, what happens to the resistance?

Answers
1. 200 Ω; 2. It halves

Although the SI unit of conductance (siemens) is almost universally accepted, older books and data sheets list conductance in the unit given as the mho (ohm spelled backwards) and having an upside-down omega, ℧, as the symbol. In such a case, the following relationship holds:

$$1 \, ℧ = 1 \, S \qquad \text{(3–11)}$$

PRACTICE PROBLEMS 9

A specification sheet for a radar transmitter indicates that one of the components has a conductance of 5 $\mu\mu$℧.

a. Express the conductance in the proper SI prefix and unit.

b. Determine the resistance of the component, in ohms.

Answers
a. 5 pS; b. $2 \times 10^{11} \, \Omega$

As you have seen, all power lines and distribution networks have internal resistance, which results in energy loss due to heat as charge flows through the conductor. If there was some way of eliminating the resistance of the conductors, electricity could be transmitted farther and more economically. The idea that energy could be transmitted without losses along a "superconductor" transmission line was formerly a distant goal. However, recent discoveries in high-temperature superconductivity promise the almost magical ability to transmit and store energy with no loss in energy.

In 1911, the Dutch physicist Heike Kamerlingh Onnes discovered the phenomenon of superconductivity. Studies of mercury, tin, and lead verified that the resistance of these materials decreases to no more than one ten-billionth of the room temperature resistance when subjected to temperatures of 4.6 K, 3.7 K, and 6 K respectively. Recall that the relationship between kelvins and degrees Celsius is as follows:

$$T_K = T_{(° C)} + 273.15° \qquad \text{(3–12)}$$

The temperature at which a material becomes a superconductor is referred to as the **critical temperature**, T_C, of the material. Figure 3–31 shows how the resistance of a sample of mercury changes with temperature. Notice how the resistance suddenly drops to zero at a temperature of 4.6 K.

Experiments with currents in supercooled loops of superconducting wire have determined that the induced currents will remain undiminished for many years within the conductor provided that the temperature is maintained below the critical temperature of the conductor.

A peculiar, seemingly magical property of superconductors occurs when a permanent magnet is placed above the superconductor. The magnet will float

3.12 Superconductors

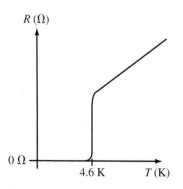

FIGURE 3–31 Critical temperature of mercury.

FIGURE 3–32 The Meissner effect: A magnetic cube hovers above a disk of ceramic superconductor. The disk is kept below its critical temperature in a bath of liquid nitrogen. *(Courtesy of AT&T Bell Laboratories/AT&T Archives)*

PUTTING IT INTO PRACTICE

You are a troubleshooting specialist working for a small telephone company. One day, word comes in that an entire subdivision is without telephone service. Everyone suspects that a cable was cut by one of several backhoe operators working on a waterline project near the subdivision. However, no one is certain exactly where the cut occurred. You remember that the resistance of a length of wire is determined by several factors, including the length. This gives you an idea for determining the distance between the telephone central office and location of the cut.

First you go to the telephone cable records, which show that the subdivision is served by 26-gauge copper wire. Then, since each customer's telephone is connected to the central office with a pair of wires, you measure the resistance of several loops from the central office. As expected, some of the measurements indicate open circuits. However, several pairs of the wire were shorted by the backhoe, and each of these pairs indicates a total resistance of 338 Ω. How far from the central office did the cut occur?

above the surface of the conductor as if it is defying the law of gravity, as shown in Figure 3–32.

This principle, which is referred to as the *Meissner effect* (named after Walther Meissner), may be simply stated as follows:

When a superconductor is cooled below its critical temperature, magnetic fields may surround but not enter the superconductor.

The principle of superconductivity is explained in the behavior of electrons within the superconductor. Unlike conductors, which have electrons moving randomly through the conductor and colliding with other electrons [Figure 3–33(a)], the electrons in superconductors form pairs which move through the material in a manner similar to a band marching in a parade. The orderly motion of electrons in a superconductor, shown in Figure 3–33(b), results in an ideal conductor, since the electrons no longer collide.

The economy of having a high critical temperature has led to the search for high-temperature superconductors. In recent years, research at the IBM Zurich Research Laboratory in Switzerland and the University of Houston in Texas has yielded superconducting materials that are able to operate at temperatures as high as 98 K (−175° C). While this temperature is still very low, it means that superconductivity can now be achieved by using the readily available liquid nitrogen rather than the much more expensive and rarer liquid helium.

Superconductivity has been found in such seemingly unlikely materials as ceramics consisting of barium, lanthanum, copper, and oxygen. Research is now centered on developing new materials that become superconductors at ever higher temperatures and that are able to overcome the disadvantages of the early ceramic superconductors.

Very expensive, low-temperature superconductivity is currently used in some giant particle accelerators and, to a limited degree, in electronic components (such as superfast Josephson junctions and SQUIDs, i.e., superconducting quantum interference devices, which are used to detect very small magnetic fields). Once research produces commercially viable, high-temperature superconductors, however, the possibilities of the applications will be virtually limitless. High-temperature superconductivity promises to yield improvements in transportation, energy storage and transmission, computers, and medical treatment and research.

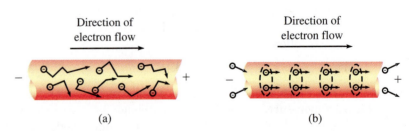

FIGURE 3–33 (a) In conductors, electrons are free to move in any direction through the conductor. Energy is lost due to collisions with atoms and other electrons, giving rise to the resistance of the conductor. (b) In superconductors, electrons are bound in pairs and travel through the conductor in step, avoiding all collisions. Since there is no energy loss, the conductor has no resistance.

PROBLEMS

3.1   Resistance of Conductors

1. Determine the resistance, at 20° C, of 100 m of solid aluminum wire having the following radii:

   a. 0.5 mm

   b. 1.0 mm

   c. 0.005 mm

   d. 0.5 cm

2. Determine the resistance, at 20° C, of 200 feet of iron conductors having the following cross sections:

   a. 0.25 inch by 0.25 inch square

   b. 0.125 inch diameter round

   c. 0.125 inch by 4.0 inch rectangle

3. A 250-foot length of solid copper bus bar, shown in Figure 3–34, is used to connect a voltage source to a distribution panel. If the bar is to have a resistance of 0.02 Ω at 20° C, calculate the required height of the bus bar (in inches).

4. Nichrome wire is used to construct heating elements. Determine the length of 1.0-mm-diameter Nichrome wire needed to produce a heating element that has a resistance of 2.0 Ω at a temperature of 20° C.

5. A copper wire having a diameter of 0.80 mm is measured to have a resistance of 10.3 Ω at 20° C. How long is this wire in meters? How long is the wire in feet?

6. A piece of aluminum wire has a resistance, at 20° C, of 20 Ω. If this wire is melted down and used to produce a second wire having a length four times the original length, what will be the resistance of the new wire at 20° C? (Hint: The volume of the wire has not changed.)

7. Determine the resistivity (in ohm-meters) of a carbon-based graphite cylinder having a length of 6.00 cm, a diameter of 0.50 mm, and a measured resistance of 3.0 Ω at 20° C. How does this value compare with the resistivity given for carbon?

8. A solid circular wire of length 200 m and diameter of 0.4 mm has a resistance measured to be 357 Ω at 20° C. Of what material is the wire constructed?

9. A 2500-m section of alloy wire has a resistance of 32 Ω. If the wire has a diameter of 1.5 mm, determine the resistivity of the material in ohm-meters. Is this alloy a better conductor than copper?

10. A section of iron wire having a diameter of 0.030 inch is measured to have a resistance of 2500 Ω (at a temperature of 20° C).

    a. Determine the cross-sectional area in square meters and in square millimeters. (Note: 1 inch = 2.54 cm = 25.4 mm)

    b. Calculate the length of the wire in meters.

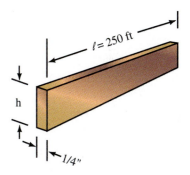

FIGURE 3–34

3.2   Electrical Wire Tables

11. Use Table 3–2 to determine the resistance of 300 feet of AWG 22 and AWG 19 solid copper conductors. Compare the diameters and the resistances of the wires.

12. Use Table 3–2 to find the resistance of 250 m of AWG 8 and AWG 2 solid copper conductors. Compare the diameters and the resistances of the wires.

13. Determine the maximum current that could be handled by AWG 19 wire and by AWG 30 wire.

14. If AWG 8 is rated at a maximum of 40 A, how much current could AWG 2 handle safely?

15. A spool of AWG 36 copper transformer wire is measured to have a resistance of 550 Ω at a temperature of 20° C. How long is this wire in meters?

16. How much current should AWG 36 copper wire be able to handle?

3.3   Resistance of Wires—Circular Mils

17. Determine the area in circular mils of the following conductors ($T = 20°$ C):

    a. Circular wire having a diameter of 0.016 inch

    b. Circular wire having a diameter of 2.0 mm

    c. Rectangular bus bar having dimensions 0.25 inch by 6.0 inch

18. Express the cross-sectional areas of the conductors of Problem 17 in square mils and in square millimeters.

19. Calculate the resistance, at 20° C, of 400 feet of copper conductors having the cross-sectional areas given in Problem 17.

20. Determine the diameter in inches and in millimeters of circular cables having cross-sectional areas as given below: (Assume the cables to be solid conductors.)

    a. 250 CM

    b. 1000 CM

    c. 250 MCM

    d. 750 MCM

21. A 200-foot length of solid copper wire is measured to have a resistance of 0.500 Ω.

    a. Determine the cross-sectional area of the wire in both square mils and circular mils.

    b. Determine the diameter of the wire in mils and in inches.

22. Repeat Problem 21 if the wire had been made of Nichrome.

23. A spool of solid copper wire having a diameter of 0.040 inch is measured to have a resistance of 12.5 Ω (at a temperature of 20° C).

    a. Determine the cross-sectional area in both square mils and circular mils.

    b. Calculate the length of the wire in feet.

24. Iron wire having a diameter of 30 mils was occasionally used for telegraph transmission. A technician measures a section of telegraph line to have a resistance of 2500 Ω (at a temperature of 20° C).

    a. Determine the cross-sectional area in both square mils and circular mils.

    b. Calculate the length of the wire in feet and in meters. (Note: 1 ft = 0.3048 m.) Compare your answer to the answer obtained in Problem 10.

3.4   Temperature Effects

25. An aluminum conductor has a resistance of 50 Ω at room temperature. Find the resistance of the same conductor at $-30°$ C, 0° C, and at 200° C.

26. AWG 14 solid copper house wire is designed to operate within a temperature range of $-40°$ C to $+90°$ C. Calculate the resistance of 200 circuit feet of wire at both temperatures. Note: A circuit foot is the length of cable needed for a current to travel to and from a load.

27. A given material has a resistance of 20 Ω at room temperature (20° C) and 25 Ω at a temperature of 85° C.

    a. Does the material have a positive or a negative temperature coefficient? Explain briefly.

    b. Determine the value of the temperature coefficient, α, at 20° C.

    c. Assuming the resistance versus temperature function to be linear, determine the expected resistance of the material at 0° C (the freezing point of water) and at 100° C (the boiling point of water).

28. A given material has a resistance of 100 Ω at room temperature (20° C) and 150 Ω at a temperature of −25° C.

    a. Does the material have a positive or a negative temperature coefficient? Explain briefly.

    b. Determine the value of the temperature coefficient, α, at 20° C.

    c. Assuming the resistance versus temperature function to be linear, determine the expected resistance of the material at 0° C (the freezing point of water) and at −40° C.

29. An electric heater is made of Nichrome wire. The wire has a resistance of 15.2 Ω at a temperature of 20° C. Determine the resistance of the Nichrome wire when the temperature of the wire is increased to 260° C.

30. A silicon diode is measured to have a resistance of 500 Ω at 20° C. Determine the resistance of the diode if the temperature of the component is increased with a soldering iron to 30° C. (Assume that the resistance versus temperature function is linear.)

31. An electrical device has a linear temperature response. The device has a resistance of 120 Ω at a temperature of −20° C and a resistance of 190 Ω at a temperature of 120° C.

    a. Calculate the resistance at a temperature of 0° C.

    b. Calculate the resistance at a temperature of 80° C.

    c. Determine the temperature intercept of the material.

32. Derive the expression of Equation 3–8.

3.5   Types of Resistors

33. A 10-kΩ variable resistor has its wiper (movable terminal b) initially at the bottom terminal, c. Determine the resistance R_{ab} between terminals a and b and the resistance R_{bc} between terminals b and c under the following conditions:

    a. The wiper is at c.

    b. The wiper is one-fifth of the way around the resistive surface.

    c. The wiper is four-fifths of the way around the resistive surface.

    d. The wiper is at a.

34. The resistance between wiper terminal b and bottom terminal c of a 200-kΩ variable resistor is measured to be 50 kΩ. Determine the resistance that would be measured between the top terminal, a and the wiper terminal, b.

3.6   Color Coding of Resistors

35. Given resistors having the following color codes (as read from left to right), determine the resistance, tolerance and reliability of each component. Express the uncertainty in both percentage and ohms.

   a. Brown   Green   Yellow   Silver

   b. Red   Gray   Gold   Gold   Yellow

   c. Yellow   Violet   Blue   Gold

   d. Orange   White   Black   Gold   Red

36. Determine the color codes required if you need the following resistors for a project:

   a. 33 kΩ $\pm$ 5%, 0.1% reliability

   b. 820 Ω $\pm$ 10%

   c. 15 Ω $\pm$ 20%

   d. 2.7 MΩ $\pm$ 5%

3.7   Measuring Resistance—The Ohmmeter

37. Explain how an ohmmeter may be used to determine whether a light bulb is burned out.

38. If an ohmmeter were placed across the terminal of a switch, what resistance would you expect to measure when the contacts of the switch are closed? What resistance would you expect to measure when the contacts are opened?

39. Explain how you could use an ohmmeter to determine approximately how much wire is left on a spool of AWG 24 copper wire.

40. An analog ohmmeter is used to measure the resistance of a two-terminal component. The ohmmeter indicates a resistance of 1.5 kΩ. When the leads of the ohmmeter are reversed, the meter indicates that the resistance of the component is an open circuit. Is the component faulty? If not, what kind of component is being tested?

3.8   Thermistors

41. A thermistor has the characteristics shown in Figure 3–23.

   a. Determine the resistance of the device at room temperature, 20° C.

   b. Determine the resistance of the device at a temperature of 40° C.

   c. Does the thermistor have a positive or a negative temperature coefficient? Explain.

3.9   Photoconductive Cells

42. For the photocell having the characteristics shown in Figure 3–24(c), determine the resistance

   a. in a dimly lit basement having an illuminance of 10 lux

   b. in a home having an illuminance of 50 lux

   c. in a classroom having an illuminance of 500 lux

3.11 Conductance

43. Calculate the conductance of the following resistances:

    a. $0.25 \ \Omega$

    b. $500 \ \Omega$

    c. $250 \ k\Omega$

    d. $12.5 \ M\Omega$

44. Determine the resistance of components having the following conductances:

    a. $62.5 \ \mu S$

    b. $2500 \ mS$

    c. $5.75 \ mS$.

    d. $25.0 \ S$

45. Determine the conductance of 1000 m of AWG 30 solid copper wire at a temperature of $20° \ C$.

46. Determine the conductance of 200 feet of aluminum bus bar (at a temperature of $20° \ C$) which has a cross-sectional dimension of 4.0 inches by 0.25 inch. If the temperature were to increase, what would happen to the conductance of the bus bar?

 ANSWERS TO IN-PROCESS LEARNING CHECKS

In-Process Learning Check 1

1. Iron wire will have approximately seven times more resistance than copper.

2. The longer wire will have twice the resistance of the shorter wire.

3. The wire having the greater diameter will have one quarter the resistance of the small-diameter wire.

In-Process Learning Check 2

1. 200 A

2. The actual value is less than the theoretical value. Since only the surface of the cable is able to dissipate heat, the current must be decreased to prevent heat build-up.

In-Process Learning Check 3

$A = 63.7 \ CM$
$A = 3.23 \times 10^{-8} \ m^2 = 0.0323 \ mm^2$

In-Process Learning Check 4

Positive temperature coefficient means that resistance of a material increases as temperature increases. Negative temperature coefficient means that the resistance of a material decreases as the temperature increases. Aluminum has a positive temperature coefficient.

■ **OBJECTIVES**

After studying this chapter, you will be able to

- compute voltage, current, and resistance in simple circuits using Ohm's law,
- use the voltage reference convention to determine polarity,
- describe how voltage, current, and power are related in a resistive circuit,
- compute power in dc circuits,
- use the power reference convention to describe the direction of power transfer,
- compute energy used by electrical loads,
- determine energy costs,
- determine the efficiency of machines and systems,
- use PSpice and MultiSIM to solve Ohm's law problems.

Ohm's Law, Power, and Energy

4

CHAPTER PREVIEW

In the previous two chapters, you studied voltage, current, and resistance separately. In this chapter, we consider them together. Beginning with Ohm's law, you will study the relationship between voltage and current in a resistive circuit, reference conventions, power, energy and efficiency. Also in this chapter, we begin our study of computer methods. Two application packages are considered here; they are Orcad PSpice (from Cadence Design Systems Inc.) and MultiSIM (the successor to Electronics Workbench V5). ∎

PUTTING IT IN PERSPECTIVE

Georg Simon Ohm

IN CHAPTER 3, WE LOOKED BRIEFLY at Ohm's experiments. We now take a look at Ohm the person.

Georg Simon Ohm was born in Erlangen, Bavaria, on March 16, 1787. His father was a master mechanic who determined that his son should obtain an education in science. Although Ohm became a teacher in a high school, he had aspirations to receive a university appointment. The only way that such an appointment could be realized would be if Ohm could produce important results through scientific research. Since the science of electricity was in its infancy, and because the electric cell had recently been invented by the Italian Conte Alessandro Volta, Ohm decided to study the behavior of current in resistive circuits. Because equipment was expensive and hard to come by, Ohm made much of his own, thanks, in large part, to his father's training. Using this equipment, Ohm determined experimentally that the amount of current transmitted along a wire was directly proportional to its cross-sectional area and inversely proportional to its length. From these results, Ohm was able to define resistance and show that there was a simple relationship between voltage, resistance, and current. This result, now known as Ohm's law, is probably the most fundamental relationship in circuit theory. However, when published in 1827, Ohm's results were met with ridicule. As a result, not only did Ohm miss out on a university appointment, he was forced to resign from his high school teaching position. While Ohm was living in poverty and shame, his work became known and appreciated outside Germany. In 1842, Ohm was appointed a member of the Royal Society. Finally, in 1849, he was appointed as a professor at the University of Munich, where he was at last recognized for his important contributions. ∎

4.1 Ohm's Law

Consider the circuit of Figure 4–1. Using a circuit similar in concept to this, Ohm determined experimentally that *current in a resistive circuit is directly proportional to its applied voltage and inversely proportional to its resistance.* In equation form, Ohm's law states

$$I = \frac{E}{R} \quad [\text{amps, A}] \tag{4–1}$$

where

E is the voltage in volts,

R is the resistance in ohms,

I is the current in amperes.

From this you can see that the larger the applied voltage, the larger the current, while the larger the resistance, the smaller the current.

The proportional relationship between voltage and current described by Equation 4–1 may be demonstrated by direct substitution as indicated in Figure 4–2. For a fixed resistance, doubling the voltage as shown in (b) doubles the current, while tripling the voltage as shown in (c) triples the current, and so on.

The inverse relationship between resistance and current is demonstrated in Figure 4–3. For a fixed voltage, doubling the resistance as shown in (b) halves the current, while tripling the resistance as shown in (c) reduces the current to one third of its original value, and so on.

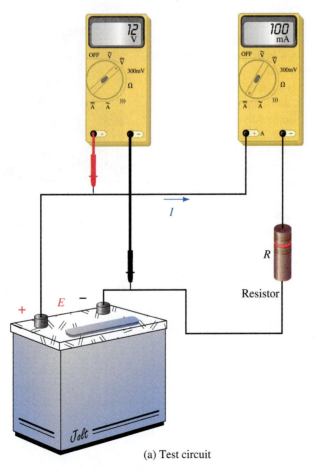

(a) Test circuit

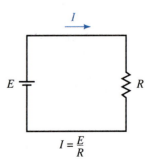

$$I = \frac{E}{R}$$

(b) Schematic, meters not shown

FIGURE 4–1 Circuit for illustrating Ohm's law.

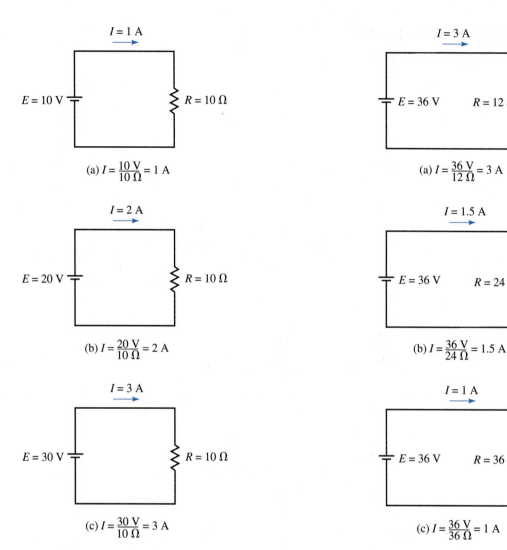

$$I = 1\ A$$

$$E = 10\ V \qquad R = 10\ \Omega$$

$$\text{(a) } I = \frac{10\ V}{10\ \Omega} = 1\ A$$

$$I = 2\ A$$

$$E = 20\ V \qquad R = 10\ \Omega$$

$$\text{(b) } I = \frac{20\ V}{10\ \Omega} = 2\ A$$

$$I = 3\ A$$

$$E = 30\ V \qquad R = 10\ \Omega$$

$$\text{(c) } I = \frac{30\ V}{10\ \Omega} = 3\ A$$

$$I = 3\ A$$

$$E = 36\ V \qquad R = 12\ \Omega$$

$$\text{(a) } I = \frac{36\ V}{12\ \Omega} = 3\ A$$

$$I = 1.5\ A$$

$$E = 36\ V \qquad R = 24\ \Omega$$

$$\text{(b) } I = \frac{36\ V}{24\ \Omega} = 1.5\ A$$

$$I = 1\ A$$

$$E = 36\ V \qquad R = 36\ \Omega$$

$$\text{(c) } I = \frac{36\ V}{36\ \Omega} = 1\ A$$

FIGURE 4–2 For a fixed resistance, current is directly proportional to voltage; thus, doubling the voltage as in (b) doubles the current, while tripling the voltage as in (c) triples the current and so on.

FIGURE 4–3 For a fixed voltage, current is inversely proportional to resistance; thus, doubling the resistance as in (b) halves the current, while tripling the resistance as in (c) results in one third the current and so on.

Ohm's law may also be expressed in the following forms by rearrangement of Equation 4–1:

$$E = IR \text{ [volts, V]} \qquad \textbf{(4–2)}$$

and

$$R = \frac{E}{I} \quad \text{[ohms, } \Omega\text{]} \qquad \textbf{(4–3)}$$

When using Ohm's law, be sure to express all quantities in base units of volts, ohms, and amps as in Examples 4–1 to 4–3, or utilize the relationships between prefixes as in Example 4–4.

A 27-Ω resistor is connected to a 12-V battery. What is the current?

EXAMPLE 4–1

Solution   Substituting the resistance and voltage values into Ohm's law yields

$$I = \frac{E}{R} = \frac{12\ V}{27\ \Omega} = 0.444\ A$$

EXAMPLE 4–2

The lamp of Figure 4–4 draws 25 mA when connected to a 6-V battery. What is its resistance?

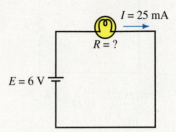

FIGURE 4–4

Solution   Using Equation 4–3,

$$R = \frac{E}{I} = \frac{6\ V}{25 \times 10^{-3}\ A} = 240\ \Omega$$

EXAMPLE 4–3

If 125 µA is the current in a resistor with color bands red, red, yellow, what is the voltage across the resistor?

Solution   Using the color code of Chapter 3, $R = 220\ k\Omega$. From Ohm's law, $E = IR = (125 \times 10^{-6}\ A)(220 \times 10^3\ \Omega) = 27.5\ V$.

EXAMPLE 4–4

A resistor with the color code brown, red, yellow is connected to a 30-V source. What is *I*?

Solution   When *E* is in volts and *R* is in kΩ, the answer comes out directly in mA. From the color code, $R = 120\ k\Omega$. Thus,

$$I = \frac{E}{R} = \frac{30\ V}{120\ k\Omega} = 0.25\ mA$$

Traditionally, circuits are drawn with the source on the left and the load on the right as indicated in Figures 4–1 to 4–3. However, you will also encounter circuits with other orientations. For these, the same principles apply; as you saw in Figure 4–4, simply draw the current arrow pointing out from the positive end of the source and apply Ohm's law in the usual manner. More examples are shown in Figure 4–5.

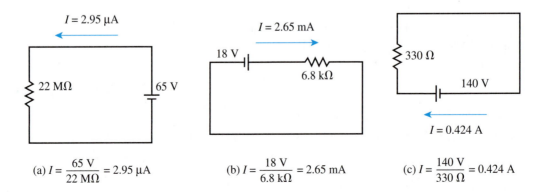

(a) $I = \dfrac{65\ V}{22\ M\Omega} = 2.95\ µA$ (b) $I = \dfrac{18\ V}{6.8\ k\Omega} = 2.65\ mA$ (c) $I = \dfrac{140\ V}{330\ \Omega} = 0.424\ A$

FIGURE 4–5

1. a. For the circuit of Figure 4–2(a), show that halving the voltage halves the current.

   b. For the circuit of Figure 4–3(a), show that halving the resistance doubles the current.

   c. Are these results consistent with the verbal statement of Ohm's law?

2. For each of the following, draw the circuit with values marked, then solve for the unknown.

   a. A 10 000-milliohm resistor is connected to a 24-V battery. What is the resistor current?

   b. How many volts are required to establish a current of 20 μA in a 100-kΩ resistor?

   c. If 125 V is applied to a resistor and 5 mA results, what is the resistance?

3. For each circuit of Figure 4–6 determine the current, including its direction (i.e., the direction that the current arrow should point).

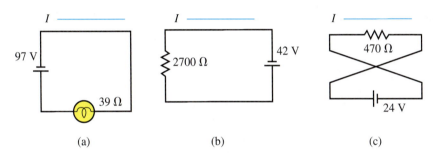

FIGURE 4–6   What is the current direction and value of *I* for each case?

Answers

1. a. 5 V/10 Ω = 0.5 A   b. 36 V/6 Ω = 6 A   c. Yes

2. a. 2.4 A   b. 2.0 V   c. 25 kΩ

3. a. 2.49 A, left   b. 15.6 mA, right   c. 51.1 mA, left

Ohm's Law in Graphical Form

The relationship between current and voltage described by Equation 4-1 may be shown graphically as in Figure 4–7. The graphs, which are straight lines, show clearly that the relationship between voltage and current is linear, i.e., that current is directly proportional to voltage.

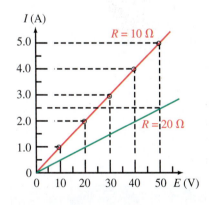

FIGURE 4–7   Graphical representation of Ohm's law. The red plot is for a 10-Ω resistor while the green plot is for a 20-Ω resistor.

Open Circuits

Current can only exist where there is a conductive path (e.g., a length of wire). For the circuit of Figure 4–8, *I* equals zero since there is no conductor between points *a* and *b*. We refer to this as an *open circuit*. Since *I* = 0, substitution of this into Equation 4–3 yields

$$R = \frac{E}{I} = \frac{E}{0} \Rightarrow \infty \text{ ohms}$$

Thus, an open circuit has infinite resistance.

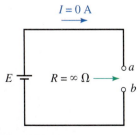

FIGURE 4–8   An open circuit has infinite resistance.

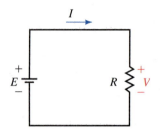

FIGURE 4–9 Symbols used to represent voltages. *E* is used for source voltages while *V* is used for voltage across circuit components such as resistors.

Voltage Symbols

Two different symbols are used to represent voltage. For sources, use uppercase *E;* for loads (and other components), use uppercase *V.* This is illustrated in Figure 4–9.

Using the symbol *V,* Ohm's law may be rewritten in its several forms as

$$I = \frac{V}{R} \quad [\text{amps}] \qquad \qquad (4\text{–}4)$$

$$V = IR \quad [\text{volts}] \qquad \qquad (4\text{–}5)$$

$$R = \frac{V}{I} \quad [\text{ohms}] \qquad \qquad (4\text{–}6)$$

These relationships hold for every resistor in a circuit, no matter how complex the circuit. Since $V = IR$, these voltages are often referred to as *IR drops.*

EXAMPLE 4–5

NOTE . . .

In the interest of brevity (as in Figure 4–10), we sometimes draw only a portion of a circuit with the rest of the circuit implied rather than shown explicitly.

Thus, this circuit contains a source and connecting wire, even though they are not shown here.

The current through each resistor of Figure 4–10 is $I = 0.5$ A (see Note). Compute V_1 and V_2.

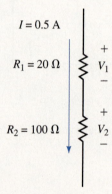

FIGURE 4–10 Ohm's law applies to each resistor.

Solution $V_1 = IR_1 = (0.5 \text{ A})(20 \text{ }\Omega) = 10$ V. Note, *I* is also the current through R_2. Thus, $V_2 = IR_2 = (0.5 \text{ A})(100 \text{ }\Omega) = 50$ V.

4.2 Voltage Polarity and Current Direction

◀ **Online Companion**

So far, we have paid little attention to the polarity of voltages across resistors—that is, at which end should we place the + sign for voltage and at which end should we place the − sign? However, the issue is of extreme importance. Fortunately, there is a simple relationship between current direction and voltage polarity. To get at the idea, consider Figure 4–11(a). Here the polarity of *V* is obvious since the resistor is connected directly to the source. This makes the top end of the resistor positive with respect to the bottom end, and $V = E = 12$ V as indicated by the meters.

Now consider current. The direction of *I* is from top to bottom through the resistor as indicated by the current direction arrow. Examining voltage polarity, we see that the plus sign for *V* is at the tail of this arrow. This observation turns out to be true in general and gives us a convention for marking voltage polarity on circuit diagrams. *For voltage across a resistor, always place the plus sign at the tail of the current reference arrow.* Two additional examples are shown in Figure 4–11(b).

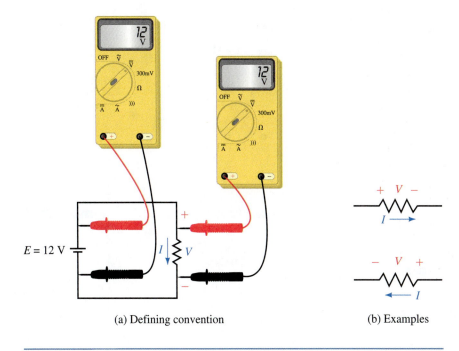

(a) Defining convention

(b) Examples

FIGURE 4–11   Convention for voltage polarity. Place the plus sign for *V* at the tail of the current direction arrow.

For each resistor of Figure 4–12, compute *V* and show its polarity.

PRACTICE PROBLEMS 2

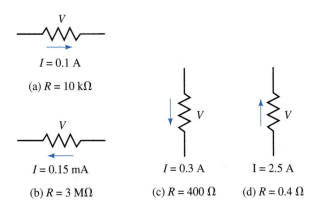

V

I = 0.1 A

(a) *R* = 10 kΩ

V

I = 0.15 mA

(b) *R* = 3 MΩ

V

I = 0.3 A

(c) *R* = 400 Ω

V

I = 2.5 A

(d) *R* = 0.4 Ω

FIGURE 4–12

Answers
a. 1000 V, + at left; b. 450 V, + at right; c. 120 V, + at top; d. 1 V, + at bottom

1. A resistor has color bands brown, black, and red and a current of 25 mA. Determine the voltage across it.

2. For a resistive circuit, what is *I* if *E* = 500 V and *R* is open-circuited? Will the current change if the voltage is doubled?

3. A certain resistive circuit has voltage *E* and resistance *R*. If *I* = 2.5 A, what will be the current if:

   a. *E* remains unchanged but *R* is doubled?

   b. *E* remains unchanged but *R* is quadrupled?

   c. *E* remains unchanged but *R* is reduced to 20% of its original value?

   d. *R* is doubled and *E* is quadrupled?

4. The voltmeters of Figure 4–13 have autopolarity. Determine the reading of each meter, its magnitude and sign.

IN-PROCESS
LEARNING CHECK 1

(Answers are at the end of the chapter.)

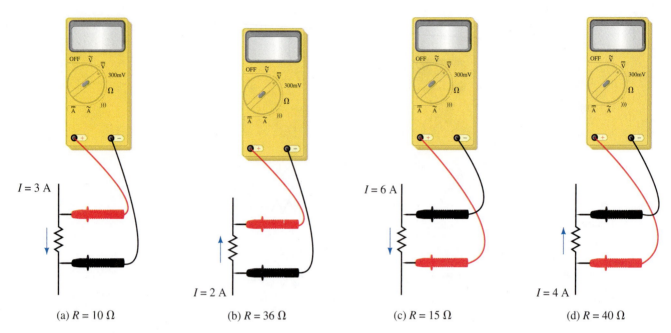

(a) $R = 10\ \Omega$ (b) $R = 36\ \Omega$ (c) $R = 15\ \Omega$ (d) $R = 40\ \Omega$

FIGURE 4–13

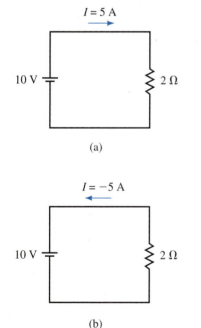

(a)

(b)

FIGURE 4–14   Two representations of the same current.

Before We Move On

Before we move on, we will comment on one more aspect of current representation. First, note that to completely specify current, you must include both its value and its direction. (This is why we show current direction reference arrows on circuit diagrams.) Normally we show the current coming out of the plus (+) terminal of the source as in Figure 4–14(a). (Here, $I = E/R = 5$ A in the direction shown. This is the actual direction of the current.) As you can see from this (and all preceding examples in this chapter), determining the actual current direction in single-source networks is easy. However, when analyzing complex circuits (such as those with multiple sources as in later chapters), it is not always easy to tell in advance in what direction all currents will be. As a result, when you solve such problems, you may find that some currents have negative values. What does this mean?

To get at the answer, consider both parts of Figure 4–14. In (a), current is shown in the usual direction, while in (b), it is shown in the opposite direction. To compensate for the reversed direction, we have changed the sign of I. The interpretation placed on this is that a positive current in one direction is the same as a negative current in the opposite direction. Therefore, (a) and (b) are two representations of the same current. Thus, if during the solution of a problem you obtain a positive value for current, this means that its actual direction is the same as the reference arrow; if you obtain a negative value, its direction is opposite to the reference arrow. This is an important idea and one that you will use many times in later chapters. It was introduced at this point to help explain power flow in the electric car of upcoming Example 4–9. However, apart from the electric car example, we will leave its consideration and use to later chapters. That is, we will continue to use the representation of Figure 4–14(a).

4.3 Power

Power is familiar to all of us, at least in a general sort of way. We know, for example, that electric heaters and light bulbs are rated in watts (W) and that motors are rated in horsepower (or watts), both being units of power as discussed in Chapter 1. We also know that the higher the watt rating of a device, the more energy we can get out of it per unit time. Figure 4–15 illustrates the

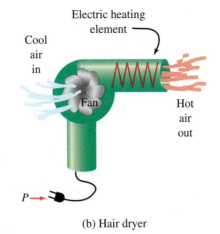

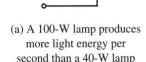

(a) A 100-W lamp produces
    more light energy per
    second than a 40-W lamp

(b) Hair dryer

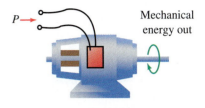

(c) A 10-hp motor can do more work in
    a given time than a ½-hp motor

FIGURE 4–15   Energy conversion. Power *P* is a measure of the rate of energy conversion.

idea. In (a), the greater the power rating of the light, the more light energy that it can produce per second. In (b), the greater the power rating of the heater, the more heat energy it can produce per second. In (c), the larger the power rating of the motor, the more mechanical work that it can do per second.

As you can see, power is related to energy, which is the capacity to do work. Formally, **power** is defined as the rate of doing work or, equivalently, as the rate of transfer of energy. The symbol for power is *P*. By definition,

$$P = \frac{W}{t} \quad \text{[watts, W]} \qquad \textbf{(4–7)}$$

where *W* is the work (or energy) in joules and *t* is the corresponding time interval of *t* seconds (see Notes).

The SI unit of power is the watt. From Equation 4–7, we see that *P* also has units of joules per second. If you substitute $W = 1$ J and $t = 1$ s you get $P = 1$ J/1 s $= 1$ W. From this, you can see that *one watt equals one joule per second*. Occasionally, you also need power in horsepower. To convert, recall that 1 hp = 746 watts.

Power in Electrical and Electronic Systems

Since our interest is in electrical power, we need expressions for *P* in terms of electrical quantities. Recall from Chapter 2 that voltage is defined as work per unit charge and current as the rate of transfer of charge, i.e.,

$$V = \frac{W}{Q} \qquad \textbf{(4–8)}$$

and

$$I = \frac{Q}{t} \qquad \textbf{(4–9)}$$

From Equation 4–8, $W = QV$. Substituting this into Equation 4–7 yields $P = W/t = (QV)/t = V(Q/t)$. Replacing Q/t with *I*, we get

$$P = VI \quad \text{[watts, W]} \qquad \textbf{(4–10)}$$

and, for a source,

$$P = EI \quad \text{[watts, W]} \qquad \textbf{(4–11)}$$

NOTES . . .

1. Time *t* in Equation 4–7 is a time interval, not an instantaneous point in time.

2. The symbol for energy is *W* and the abbreviation for watts is W. Multiple use of symbols is common in technology. In such cases, you have to look at the context in which a symbol is used to determine its meaning.

Additional relationships are obtained by substituting $V = IR$ and $I = V/R$ into Equation 4–10:

$$P = I^2R \quad [\text{watts, W}] \tag{4–12}$$

and

$$P = \frac{V^2}{R} \quad [\text{watts, W}] \tag{4–13}$$

EXAMPLE 4–6

Compute the power supplied to the electric heater of Figure 4–16 using all three electrical power formulas.

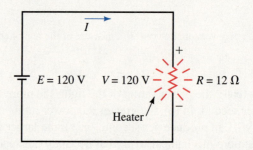

FIGURE 4–16 Power to the load (i.e., the heater) can be computed from any of the power formulas.

Solution $I = V/R = 120 \text{ V}/12 \text{ }\Omega = 10 \text{ A}$. Thus, the power may be calculated as follows:

a. $P = VI = (120 \text{ V})(10 \text{ A}) = 1200 \text{ W}$

b. $P = I^2R = (10 \text{ A})^2(12 \text{ }\Omega) = 1200 \text{ W}$

c. $P = V^2/R = (120 \text{ V})^2/12 \text{ }\Omega = 1200 \text{ W}$

Note that all give the same answer, as they must.

EXAMPLE 4–7

Compute the power to each resistor in Figure 4–17 using Equation 4–13.

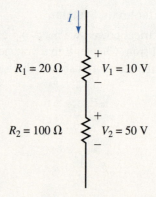

FIGURE 4–17

Solution You must use the appropriate voltage in the power equation. For resistor R_1, use V_1; for resistor R_2, use V_2.

a. $P_1 = V_1^2/R_1 = (10 \text{ V})^2/20 \text{ }\Omega = 5 \text{ W}$

b. $P_2 = V_2^2/R_2 = (50 \text{ V})^2/100 \text{ }\Omega = 25 \text{ W}$

If the dc motor of Figure 4–15(c) draws 6 A from a 120-V source,

EXAMPLE 4–8

a. Compute its power input in watts.

b. Assuming the motor is 100% efficient (i.e., that all electrical power supplied to it is output as mechanical power), compute its power output in horsepower.

Solution

a. $P_{in} = VI = (120 \text{ V})(6 \text{ A}) = 720 \text{ W}$

b. $P_{out} = P_{in} = 720$ W. Converting to horsepower, $P_{out} = (720 \text{ W})/(746 \text{ W/hp}) = 0.965$ hp.

PRACTICE PROBLEMS 3

a. Show that $I = \sqrt{\dfrac{P}{R}}$

b. Show that $V = \sqrt{PR}$

c. A 100-Ω resistor dissipates 169 W. What is its current?

d. A 3-Ω resistor dissipates 243 W. What is the voltage across it?

e. For Figure 4–17, $I = 0.5$ A. Use Equations 4-10 and 4-12 to compute power to each resistor. Compare your answers to the answers of Example 4–7.

Answers
c. 1.3 A;   d. 27 V;   e. $P_1 = 5$ W, $P_2 = 25$ W (same)

Power Rating of Resistors

Resistors must be able to safely dissipate their heat without damage. For this reason, resistors are rated in watts. (For example, composition resistors of the type used in electronics are made with standard ratings of ⅛, ¼, ½, 1, and 2 W as you saw in Figure 3–8.) To provide a safety margin, it is customary to select a resistor that is capable of dissipating two or more times its computed power. By overrating a resistor, it will run a little cooler.

PRACTICAL NOTES . . .

A properly chosen resistor is able to dissipate its heat safely without becoming excessively hot. However, if through bad design or subsequent component failure its current becomes excessive, it will overheat and damage may result, as shown in Figure 4–18. One of the symptoms of overheating is that the resistor becomes noticeably hotter than other resistors in the circuit. (Be careful, however, as you might get burned if you try to check by touch.) Component failure may also be detected by smell. Burned components have a characteristic odor that you will soon come to recognize. If you detect any of these symptoms, turn the equipment off and look for the source of the problem. Note, however, an overheated component is often the symptom of a problem, rather than its cause.

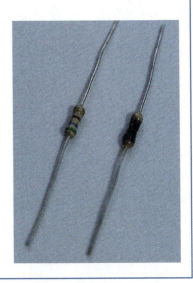

FIGURE 4–18 The resistor on the right has been damaged by overheating.

Measuring Power

Power can be measured using a device called a wattmeter. However, since wattmeters are used primarily for ac power measurement, we will hold off their consideration until Chapter 17. (You seldom need a wattmeter for dc circuits since you can determine power directly as the product of voltage times current and V and I are easy to measure.)

4.4 Power Direction Convention

For circuits with one source and one load, energy flows from the source to the load and the direction of power transfer is obvious. For circuits with multiple sources and loads, however, the direction of energy flow in some parts of the network may not be at all apparent. We therefore need to establish a clearly defined power transfer direction convention.

A resistive load (Figure 4–19) may be used to illustrate the idea. Since the direction of power flow can only be into a resistor, never out of it (since resistors do not produce energy), we define the positive direction of power transfer as from the source to the load as in (a) and indicate this by means of an arrow: $P{\rightarrow}$. We then adopt the convention that, *for the relative voltage polarities and current and power directions shown in Figure 4–19(a), when power transfer is in the direction of the arrow, it is positive, whereas when it is in the direction opposite to the arrow, it is negative.*

To help interpret the convention, consider Figure 4–19(b), which highlights the source end. From this we see that *power out of a source is positive when both the current and power arrows point out from the source, both I and P have positive values, and the source voltage has the polarity indicated.*

Now consider Figure 4–19(c), which highlights the load end. Note the relative polarity of the load voltage and the direction of the current and power arrows. From this, we see that *power to a load is positive when both the current and power direction arrows point into the load, both have positive values, and the load voltage has the polarity indicated.*

In Figure 4–19(d), we have generalized the concept. The box may contain either a source or a load. If P has a positive value, power transfer is into the box; if P has a negative value, its direction is out.

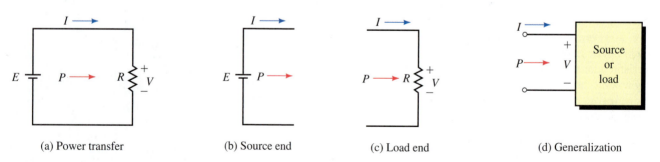

(a) Power transfer (b) Source end (c) Load end (d) Generalization

FIGURE 4–19 Reference convention for power.

Use the previous convention to describe power transfer for the electric vehicle of Figure 4–20.

EXAMPLE 4–9

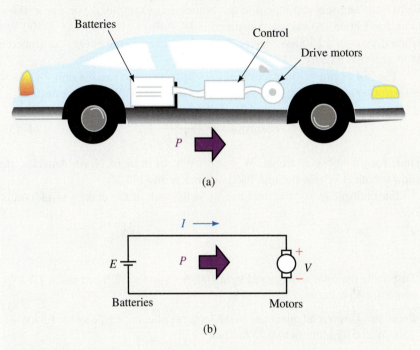

(a)

(b)

FIGURE 4–20

Solution   During normal operation, the batteries supply power to the motors, and current and power are both positive, Fig. 4–20(b). However, when the vehicle is going downhill, its motors are driven by the weight of the car and they act as generators. Since the motors now act as the source and the batteries as the load, the actual current is opposite in direction to the reference arrow shown and is thus negative (recall Figure 4–14). Thus, $P = VI$ is negative. The interpretation is, therefore, that power transfer is in the direction opposite to the power reference arrow. For example, if $V = 48$ volts and $I = -10$ A, then $P = VI = (48\text{ V})(-10\text{ A}) = -480$ W. This is consistent with what is happening, since minus 480 W *into* the motors is the same as plus 480 W *out*. This 480 W of power flows from the motors to the batteries, helping to charge them as the car goes downhill.

Earlier (Equation 4–7), we defined power as the rate of doing work. When you transpose this equation, you get the formula for **energy:**

4.5 Energy

$$W = Pt \qquad\qquad \textbf{(4–14)}$$

If t is measured in seconds, W has units of watt-seconds (i.e., joules, J), while if t is measured in hours, W has units of watthours (Wh). Note that in Equation 4–14, P must be constant over the time interval under consideration. If it is not, apply Equation 4–14 to each interval over which P is constant as described later in this section. (For the more general case, you need calculus, which we don't consider here.)

The most familiar example of energy usage is the energy that we use in our homes and pay for on our utility bills. This energy is the energy used by the lights and electrical appliances in our homes. For example, if you run a 100-W lamp for 1 hour, the energy consumed is $W = Pt = (100 \text{ W})(1\text{h}) = 100$ Wh, while if you run a 1500-W electric heater for 12 hours, the energy consumed is $W = (1500 \text{ W})(12 \text{ h}) = 18\,000$ Wh.

The last example illustrates that the watthour is too small a unit for practical purposes. For this reason, we use **kilowatthours** (kWh). By definition,

$$\text{energy}_{(\text{kWh})} = \frac{\text{energy}_{(\text{Wh})}}{1000} \qquad \textbf{(4–15)}$$

Thus, for the above example, $W = 18$ kWh. In most of North America, the kilowatthour (kWh) is the unit used on your utility bill.

For multiple loads, the total energy is the sum of the energy of individual loads.

EXAMPLE 4–10

Determine the total energy used by a 100-W lamp for 12 hours and a 1.5-kW heater for 45 minutes.

Solution   Convert all quantities to the same set of units, e.g., convert 1.5 kW to 1500 W and 45 minutes to 0.75 h. Then,

$$W = (100 \text{ W})(12 \text{ h}) + (1500 \text{ W})(0.75 \text{ h}) = 2325 \text{ Wh} = 2.325 \text{ kWh}$$

Alternatively, convert all power to kilowatts first. Thus,

$$W = (0.1 \text{ kW})(12 \text{ h}) + (1.5 \text{ kW})(0.75 \text{ h}) = 2.325 \text{ kWh}$$

EXAMPLE 4–11

Suppose you use the following electrical appliances: a 1.5-kW heater for 7½ hours; a 3.6-kW broiler for 17 minutes; three 100-W lamps for 4 hours; a 900-W toaster for 6 minutes. At $0.09 per kilowatthour, how much will this cost you?

Solution   Convert time in minutes to hours. Thus,

$$W = (1500)(7\tfrac{1}{2}) + (3600)\left(\frac{17}{60}\right) + (3)(100)(4) + (900)\left(\frac{6}{60}\right)$$

$$= 13\,560 \text{ Wh} = 13.56 \text{ kWh}$$

$$\text{cost} = (13.56 \text{ kWh})(\$0.09/\text{kWh}) = \$1.22$$

In those areas of the world where the SI system dominates, the **megajoule** (MJ) is sometimes used instead of the kWh (as the kWh is not an SI unit). The relationship is 1 kWh = 3.6 MJ.

Watthour Meter

In practice, energy is measured by watthour meters, many of which are electro-mechanical devices that incorporate a small electric motor whose speed is proportional to power to the load. This motor drives a set of dials through a gear train (Figure 4–21). Since the angle through which the dials rotate depends on

FIGURE 4–21 Watthour meter. This meter uses a gear train to drive the dials. Newer meters have digital displays.

the speed of rotation (i.e., power consumed) and the length of time that this power flows, the dial position indicates energy used. Note however, that electromechanical devices are starting to give way to electronic meters, which perform this function electronically and display the result on digital readouts.

Law of Conservation of Energy

Before leaving this section, we consider the law of conservation of energy. It states that energy can neither be created nor destroyed, but is instead converted from one form to another. You saw examples of this above—for example, the conversion of electrical energy into heat energy by a resistor, and the conversion of electrical energy into mechanical energy by a motor. In fact, several types of energy may be produced simultaneously. For example, electrical energy is converted to mechanical energy by a motor, but some heat is also produced. This results in a lowering of efficiency, a topic we consider next.

Poor efficiency results in wasted energy and higher costs. For example, an inefficient motor costs more to run than an efficient one for the same output. An inefficient piece of electronic gear generates more heat than an efficient one, and this heat must be removed, resulting in increased costs for fans, heat sinks, and the like.

Efficiency can be expressed in terms of either energy or power. Power is generally easier to measure, so we usually use power. The efficiency of a device or system (Figure 4–22) is defined as the ratio of power output P_{out} to power input P_{in}, and it is usually expressed in percent and denoted by the Greek letter η (eta). Thus,

$$\eta = \frac{P_{out}}{P_{in}} \times 100\% \qquad \text{(4–16)}$$

In terms of energy,

$$\eta = \frac{W_{out}}{W_{in}} \times 100\% \qquad \text{(4–17)}$$

4.6 Efficiency

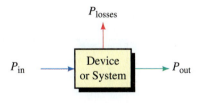

FIGURE 4–22 Input power equals output power plus losses.

Since $P_{in} = P_{out} + P_{losses}$, efficiency can also be expressed as

$$\eta = \frac{P_{out}}{P_{out} + P_{losses}} \times 100\% = \frac{1}{1 + \dfrac{P_{losses}}{P_{out}}} \times 100\% \qquad (4\text{--}18)$$

The efficiency of equipment and machines varies greatly. Large power transformers, for example, have efficiencies of 98% or better, while many electronic amplifiers have efficiencies lower than 50%. Note that efficiency will always be less than 100%.

EXAMPLE 4–12

A 120-V dc motor draws 12 A and develops an output power of 1.6 hp.

a. What is its efficiency?

b. How much power is wasted?

Solution

a. $P_{in} = EI = (120\text{ V})(12\text{ A}) = 1440\text{ W}$, and $P_{out} = 1.6\text{ hp} \times 746\text{ W/hp} = 1194\text{ W}$. Thus,

$$\eta = \frac{P_{out}}{P_{in}} = \frac{1194\text{ W}}{1440\text{ W}} \times 100 = 82.9\%$$

b. $P_{losses} = P_{in} - P_{out} = 1440 - 1194 = 246\text{ W}$

EXAMPLE 4–13

The efficiency of a power amplifier is the ratio of the power delivered to the load (e.g., speakers) to the power drawn from the power supply. Generally, this efficiency is not very high. For example, suppose a power amplifier delivers 400 W to its speaker system. If the power loss is 509 W, what is its efficiency?

Solution

$$P_{in} = P_{out} + P_{losses} = 400\text{ W} + 509\text{ W} = 909\text{ W}$$

$$\eta = \frac{P_{out}}{P_{in}} \times 100\% = \frac{400\text{ W}}{909\text{ W}} \times 100\% = 44\%$$

For systems with subsystems or components in cascade (Figure 4–23), overall efficiency is the product of the efficiencies of each individual part, where efficiencies are expressed in decimal form. Thus,

$$\eta_T = \eta_1 \times \eta_2 \times \eta_3 \times \cdots \times \eta_n \qquad (4\text{--}19)$$

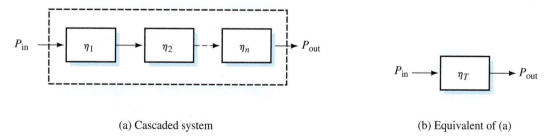

(a) Cascaded system   (b) Equivalent of (a)

FIGURE 4–23   For systems in cascade, the resultant efficiency is the product of the efficiencies of the individual stages.

EXAMPLE 4–14

a. For a certain system, $\eta_1 = 95\%$, $\eta_2 = 85\%$, and $\eta_3 = 75\%$. What is η_T?

b. If $\eta_T = 65\%$, $\eta_2 = 80\%$, and $\eta_3 = 90\%$, what is η_1?

Solution

a. Convert all efficiencies to a decimal value, then multiply. Thus, $\eta_T = \eta_1\eta_2\eta_3 = (0.95)(0.85)(0.75) = 0.61$ or 61%.

b. $\eta_1 = \eta_T/(\eta_2\eta_3) = (0.65)/(0.80 \times 0.90) = 0.903$ or 90.3%

EXAMPLE 4–15

A motor drives a pump through a gearbox (Figure 4–24). Power input to the motor is 1200 W. How many horsepower are delivered to the pump?

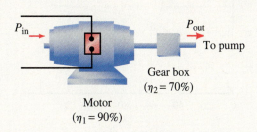

(a) Physical system

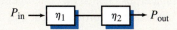

(b) Block diagram.

FIGURE 4–24   Motor driving pump through a gear box.

Solution   The efficiency of the motor-gearbox combination is $\eta_T = (0.90)(0.70) = 0.63$. The output of the gearbox (and hence the input to the pump) is $P_{out} = \eta_T \times P_{in} = (0.63)(1200 \text{ W}) = 756 \text{ W}$. Converting to horsepower, $P_{out} = (756 \text{ W})/(746 \text{ W/hp}) = 1.01 \text{ hp}$.

EXAMPLE 4–16

The motor of Figure 4–24 is operated from 9:00 A.M. to 12:00 noon and from 1:00 P.M. to 5:00 P.M. each day, for 5 days a week, outputting 7 hp to a load. At $0.085/kWh, it costs $22.19 per week for electricity. What is the efficiency of the motor/gearbox combination?

Solution

$$W_{in} = \frac{\$22.19/\text{wk}}{\$0.085/\text{kWh}} = 261.1 \text{ kWh/wk}$$

The motor operates 35 h/wk. Thus,

$$P_{in} = \frac{W_{in}}{t} = \frac{261.1 \text{ kWh/wk}}{35 \text{ h/wk}} = 7460 \text{ W}$$

$$\eta_T = \frac{P_{out}}{P_{in}} = \frac{(7 \text{ hp} \times 746 \text{ W/hp})}{7460 \text{ W}} = 0.7$$

Thus, the motor/gearbox efficiency is 70%.

4.7 Nonlinear and Dynamic Resistances

All resistors considered so far have constant values that do not change with voltage or current. Such resistors are termed **linear** or **ohmic** since their current-voltage (*I-V*) plot is a straight line. However, the resistance of some materials changes with voltage or current as you saw in Chapter 3, Section 3.10. These materials are termed **nonlinear** because their *I-V* plot is curved (Figure 4–25).

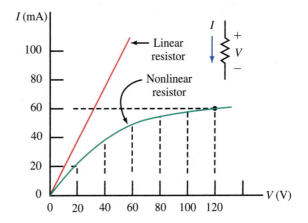

FIGURE 4–25   Linear and non-linear resistance characteristics.

Since the resistance of all materials changes with temperature, all resistors are to some extent nonlinear, since they all produce heat and this heat changes their resistance. For most resistors, however, this effect is small over their normal operating range, and such resistors are considered to be linear. (The commercial resistors shown in Figure 3–8 and most others that you will encounter in this book are linear.)

Since an *I-V* plot is a graph of Ohm's law, resistance can be computed from the ratio *V/I*. First, consider the linear plot of Figure 4–25. Because the slope is constant, the resistance is constant and you can compute *R* at any point. For example, at $V = 10$ V, $I = 20$ mA, and $R = 10$ V/20 mA $= 500\ \Omega$. Similarly, at $V = 20$ V, $I = 40$ mA, and $R = 20$ V/40 mA $= 500\ \Omega$, which is the same as before. This is true at all points on this linear curve. The resulting resistance is referred to as dc resistance, R_{dc}. Thus, $R_{dc} = 500\ \Omega$.

An alternate way to compute resistance is illustrated in Figure 4–26. At point 1, $V_1 = I_1 R$. At point 2, $V_2 = I_2 R$. Subtracting voltages and solving for *R* yields

$$R = \frac{V_2 - V_1}{I_2 - I_1} = \frac{\Delta V}{\Delta I} \quad [\text{ohms}, \Omega] \qquad (4\text{--}20)$$

where $\Delta V/\Delta I$ is the inverse of the slope of the line. (Here, Δ is the Greek letter delta. It is used to represent a change or increment in value.) To illustrate, if you select ΔV to be 20 V, you find that the corresponding ΔI from Figure 4–26 is 40 mA. Thus, $R = \Delta V/\Delta I = 20$ V/40 mA $= 500\ \Omega$ as before. Resistance calculated as in Figure 4–26 is called **ac** or **dynamic resistance.** For linear resistors, $R_{ac} = R_{dc}$.

Now consider the nonlinear resistance plot of Figure 4–25. At $V = 20$ V, $I = 20$ mA. Therefore, $R_{dc} = 20$ V/20 mA $= 1.0$ kΩ; at $V = 120$ V, $I = 60$ mA, and $R_{dc} = 120$ V/60 mA $= 2.0$ kΩ. This resistance therefore increases with applied voltage. However, for small variations about a fixed point on the curve, the ac resistance will be constant. This is an important concept that you will utilize later in your study of electronics.

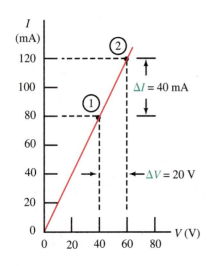

FIGURE 4–26 $R = \Delta V/\Delta I =$ 20 V/40 mA $= 500\ \Omega$.

We end our introduction to Ohm's law by solving several simple problems using Electronics Workbench's MultiSIM and Cadence Design System's Orcad PSpice. As noted in Chapter 1, these are application packages that work from a circuit schematic that you build on your screen. (Since the details are different, we will consider the two products separately, using the circuit of Figure 4–27 to get started.) Because this is our first look at circuit simulation, considerable detail is included. (Although the procedures may seem complex, they become quite intuitive with a little practice.) When setting up a problem for analysis, there are generally several ways that you can proceed, and with experience you will learn shortcuts. In the meantime, to get started, use the methods suggested here.

4.8 Computer-Aided Circuit Analysis

◀ MULTISIM

◀ CADENCE

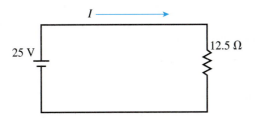

FIGURE 4–27 Simple circuit to illustrate computer analysis.

MultiSIM

Figure 4–28 shows a MultiSIM user interface screen. Along the top and sides are menu items and icons. When you position your mouse pointer over an item, a drop-down box opens to indicate the purpose of your selection. (Do this—i.e., with your mouse identify various icons, including the Indicators icon.) If you left click an item, you activate it—for example, if you click a parts bin icon, the corresponding parts bin opens. Now create and simulate the circuit of Figure 4–27. First read the MSM (MultiSIM) Operational Notes; then proceed as follows:

• Click the New icon (Figure 4–28) or menu item File/New to create a new circuit file

• Click the Sources icon to open the sources bin, click the battery symbol (to select it), move your pointer into the white area of the screen where you want the battery, then click to place it. Similarly, add the ground symbol. Close the Sources bin to avoid screen clutter.

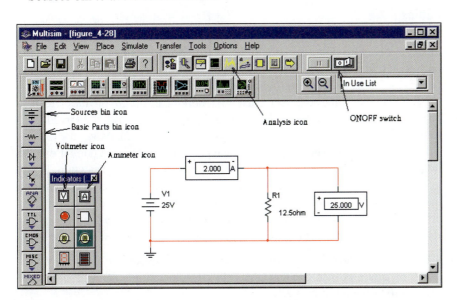

FIGURE 4–28 MultiSIM (Electronics Workbench) simulation of the circuit of Figure 4–27.

NOTES . . .

MSM Operational Notes

1. Unless directed otherwise, use the left button for all operations.

2. To select a component previously placed on the screen, click it. Four black square dots appear, indicating selection.

3. To deselect a selected component, move the cursor to an empty spot on the screen and click the left button.

4. To drag a component, place the pointer over it, press and hold the left button, drag the mouse to position the component, then release.

5. To delete a component, select it, then press the delete key.

6. To rotate a component, select it, then while holding the Ctrl key down, press the R key. (We denote this operation as Ctrl/R. Each Ctrl/R rotates the component by 90°.)

7. Sometimes you need a connector dot. To get a connector dot, place the pointer on a white space on the screen, click the right button, then click Place Junction. Position as needed.

8. A maximum of 4 wires may be joined at any point (junction).

9. Sometimes the order of wiring a circuit has an effect. If you have trouble getting wires to join properly, try changing the order in which you wire the circuit.

10. All MultiSIM circuits require a ground.

- Click the Basic parts bin icon; select a virtual resistor and temporarily place it on the screen. Now click the placed resistor, rotate it by 90° then reposition (see MSM Operational Notes 2, 4, and 6).

- Open the Indicators parts bin and click the ammeter symbol. A Component Browser dialog box opens. Double click AMMETER_H and position it as in Figure 4–28. Similarly, place the voltmeter using VOLTMETER_V.

- "Wire" the circuit. To do this, place the mouse pointer tip at the top battery terminal and when a crosshair cursor appears, click. This attaches a wire to the terminal. Now move the cursor to the ammeter pulling the wire with you. Click again and it attaches. Now add wires from the bottom of the source to the ground, from the ammeter to the top of the resistor, and from the bottom of the resistor to the wire connecting the source to the ground. Finally, wire in the voltmeter. (Usually the order of the wiring is not important, but see MSM Operational Notes 8 and 9.)

- Reposition any wire that needs to be moved by dragging it with the mouse to its new position.

- To change the battery voltage, double click the battery symbol and when the dialog box opens, click the Value tab (if not already selected), type **25,** select V, then click OK.

- Similarly, double click the resistor symbol, click the Value tab, type **12.5,** select Ω, then click OK.

- Save the circuit as **figure 4-28** (or any suitable file name you wish).

- Activate the circuit by clicking the ON/OFF power switch at the top right corner of the MSM window.

After a bit of run time, the voltmeter should show 25 V and the ammeter 2 A as indicated in Figure 4–28.

PRACTICE PROBLEMS 4

Repeat the above example, except reverse the source voltage symbol so that the circuit is driven by a −25 V source. Note the voltmeter and ammeter readings and compare to the above solution. Reconcile these results with the voltage polarity and current direction conventions discussed in this chapter and in Chapter 2.

PSpice

PSpice provides you with a variety of analysis options. In this chapter, we will illustrate two. Our first example, Figure 4–29, shows bias point analysis, while our second, Figure 4–30, illustrates DC Sweep. In later chapters, we will study others, including AC and transient analyses.

To begin an analysis, you must first build the circuit on the screen, a process referred to as **schematic capture.** Figure 4–29 shows the Orcad Capture user interface screen (see PSpice Operational Note 12). Along the top and sides are menu items and icons. When you position your mouse pointer over an item, a drop-down box opens to indicate the purpose of your selection. If you left click the item, you activate it—for example, if you click the Place Part tool, the Place Part dialog box opens. Now create and simulate the circuit of Figure 4–27. First read the PSpice Operational Notes, study the location of key icons and menu items on Figure 4–29, and proceed as follows:

- Double click the Capture icon on your Windows screen (or click Start, Programs, select PSpice Student, then click Capture Student). This opens the Capture Session log frame.

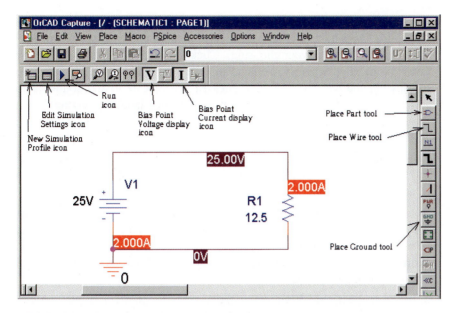

FIGURE 4–29   PSpice simulation of Figure 4–27 using the bias point analysis technique.

NOTES . . .

PSpice Operational Notes

1. Unless directed otherwise, use the left mouse button for all operations.

2. To select a particular component on your schematic, place the pointer on it and click. The selected component changes color.

3. To deselect a component, move the pointer to an empty place on the screen and click.

4. To delete a component from your schematic, select it, then press the Delete key.

5. To drag a component, place the pointer over it, press and hold the left button, drag the component to where you want it, and then release.

6. To change a default value, place the cursor over the numerical value (not the component symbol) and double click. Change the appropriate value.

7. To rotate a component, select it, click the right button, then click Rotate. Alternately, hold the Ctrl key and press the R key. (This is denoted as Ctrl/R).

8. There must be no space between a value and its unit. Thus, use 25V, not 25 V, etc.

9. All PSpice circuits must have a ground.

10. As you build a circuit, you should frequently click the Save Document icon to save your work in case something goes wrong.

11. Since PSpice creates a lot of intermediate files, it is a good idea to use a separate folder for each problem. For the example of Figure 4–29, a typical path would be C:\PSpice\figure 4-29, and in folder figure 4-29 you would save file figure 4-29.

12. In its original form, PSpice used a simple schematic capture editor called *Schematics*. When Orcad purchased PSpice a few years ago, they redesigned it to use their own *Capture* editor. This means PSpice users now have a choice between two editors. In this book, we use *Capture*. However, our website includes *Schematic* versions of all PSpice examples from this book for those who prefer to use Schematics.

- Click the Create document menu item (or File/New/Project). A New Project dialog box (Appendix A, Figure A-1) opens. In the Name box, type **figure 4-29** (or a suitable name of your choosing). Click the Analog or Mixed A/D. Now enter the location where you intend to save your PSpice file. (If you have already created the desired folder, click Browse, locate it, and double click it. This causes the directory path to appear in the box labeled Location. If you have not, click Browse/Create Dir. . . and create your new folder (see PSPice Operational Note 11). Click OK.

- Dialog box Create PSpice Project opens. Click Create a Blank Project, then OK. A screen labeled Schematic 1 Page 1 appears. Click the screen anywhere to activate the toolbar.

- You are now ready to create the circuit on your screen. Click the Place Part tool. The Place Part dialog box opens. (If this is the first time that PSpice has been accessed, no libraries will be present. To add these, click Add Library. From the Browse File dialog box, select the eval, source, analog, and special libraries; then click Open.)

- A list of parts appears. If you know the name of the part you want, type it in the Part box window. For example, the dc source is called VDC. Thus, type **VDC,** then click OK. (If you don't know the name, scan the list until you find it, click it, and then click OK.) A battery symbol appears. Position it as in Figure 4–29, click the left button to place it, then press the Esc key to end placement (or right click the mouse and left click End Mode).

- Click the Place Part tool, type **R,** and click OK. Note that the resistor that appears is horizontal. Rotate it three times (see PSpice Operational Note 7) for reasons described in Appendix A. Place as shown using the left button, then press Esc.

- Click the Place Ground tool, select 0/SOURCE, click OK, place the symbol, then press Esc.

- To wire the circuit, click the Place Wire tool, position the cursor in the little box at the top end of the battery symbol, and click. This attaches a wire. Now move the cursor to the top of the resistor, pulling the wire with you. Click again and it attaches. Wire the rest of the circuit in this manner, then press Esc to end wire placement. To improve the looks of the circuit, reposition wires or components as desired by dragging them.

- You now need to update component values. To change the resistor value, double click its 1k default value (not its symbol), type **12.5** into the Value box, then click OK. Similarly, change the source voltage to 25V (not 25 V— see PSpice Operational Note 8).

- Click the New Profile Simulation icon, enter a name (e.g., **figure 4-29**) in the Name box, leave None in the Inherit From box, then click Create. A Simulations Settings box opens. Click the Analysis tab, select Analysis type, Bias Point, then click OK.

- Click the Save document icon to save your work, then the Run icon. After a short execution time, an inactive Output window opens. Close it.

- On the menu bar, click the bias point voltage and current display icons V and I. The computed values will appear on the screen. Note that $I = 2$ A as expected. (To cancel the display of answers, click the V and I icons again.)

- To exit from PSpice, click the $\times$ in the box in the upper right-hand corner of the screen, and click yes to save changes.

A bias point analysis is useful for simple dc problems. However, for more complex problems, you may need to use meters. Let us illustrate the technique by

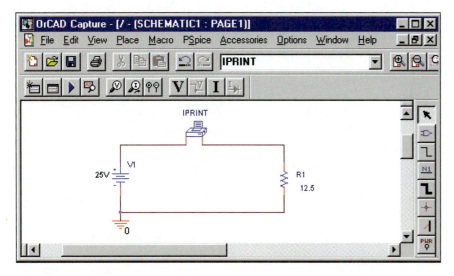

FIGURE 4–30 PSpice simulation of Figure 4–27 using an ammeter to measure current.

installing an ammeter as in Figure 4–30. In PSpice, the ammeter is designated IPRINT.

- Click the Create Document icon and create the circuit of Figure 4–30 as you did Figure 4–29, except add the ammeter. To add the ammeter, click the Place Part tool, type **IPRINT** in the Part box, click OK, then place the ammeter and wire it in.

- IPRINT is a general purpose ammeter, and you need to configure it for dc operation. Double click its symbol and a Properties editor opens. Click the Parts tab (bottom of screen), locate a cell labeled DC, type **yes** into the cell, click Apply, then close the editor by clicking the × box in its upper right hand corner. (Caution: Be sure to click the × for the Properties editor, not the × for PSpice, as you don't want to exit PSpice.)

- For this problem, you need to use DC Sweep. First, click the New Simulation Profile icon, for Name, type in **figure 4-30,** then click Create. Click the Analysis tab and under Analysis type, select DC Sweep, then from the Options list select Primary Sweep. Under Sweep Variable, choose Voltage Source. On your schematic, check the name of the voltage source—it is likely V1. If so, enter V1 in the Name box, click the Value list button, type **25V** into the box, then click OK. (Unfortunately, although this correctly sets the voltage of your source to 25 V, it does not update your screen display.) Click the Save Document icon to save your work.

- Click the Run icon. On the Output window that appears after simulation, click View and Output File on the menu bar, then scroll to the bottom of the file where you will find the answers.

V_V1   I(V_PRINT1)

2.500E+01   2.000E+00

The symbol V_V1 represents the voltage of source V_1 and 2.500E+01 represents its value (in exponent form). Thus, $V_1 = 2.500E+01 = 2.5 \times 10^1 = 25$ (which is the value you entered earlier). Similarly, I(V_PRINT1) represents the current measured by component IPRINT. Thus the current in the circuit is $I = 2.000E+00 = 2.0 \times 10^0 = 2.0$ (which, as you can see, is correct).

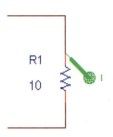

FIGURE 4–31   Using a current marker to display current.

Plotting Ohm's Law

PSpice can be used to compute and graph results. By varying the source voltage of Figure 4–27 and plotting *I* for example, you can obtain a plot of Ohm's law. Proceed as follows.

- Ensure the session log is on your screen, then click menu item File and reopen file figure 4-29. Change the resistor value to 10 ohms.
- Click the Edit Simulation Profile icon, and select DC Sweep from the Analysis tab and Primary Sweep from the Options list. For Sweep Variable, choose Voltage Source, and in the Name box type in the name of your voltage source. (It is probably V1.) Choose Linear as the sweep type, type **0V** into the Start Value box, **100V** into the End Value box, **5V** into the Increment box, and click OK. (This will sweep the voltage from 0 to 100 V in 5-V steps.)
- Click the Current Marker icon and position the marker as shown in Figure 4–31. Click the Save Document icon to save your work. Click the Run icon. The Ohm's law plot of Figure 4–32 appears on your screen. Using the cursor (see Appendix A), read values from the graph and verify with your calculator.

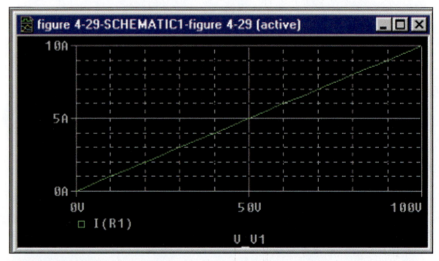

FIGURE 4–32   Ohm's law plot for the circuit of Figure 4–29 with $R = 10\ \Omega$.

PUTTING IT INTO PRACTICE

You have been assigned to perform a design review and cost analysis of an existing product. The product uses a 12-V ±5% power supply. After a catalog search, you find a new power supply with a specification of 12 V ±2%, which, because of its newer technology, costs only five dollars more than the one you are currently using. Although it provides improved performance for the product, your supervisor won't approve it because of the extra cost. However, the supervisor agrees that if you can bring the cost differential down to $3.00 or less, you can use the new supply. You then look at the schematic and discover that a number of precision 1% resistors are used because of the wide tolerance of the old supply. (As a specific example, a ±1% tolerance 220-kΩ resistor is used instead of a standard 5% tolerance resistor because the loose tolerance of the original power supply results in current through the resistor that is out of spec. Its current must be between 50 mA to 60 mA when the 12-V supply is connected across it. Several other such instances are present in the product, a total of 15 in all.) You begin to wonder whether you could replace the precision resistors (which cost $0.24 each) with standard 5% resistors (which cost $0.03 each) and save enough money to satisfy your supervisor. Perform an analysis to determine if your hunch is correct.

PROBLEMS

4.1 Ohm's Law

1. For the circuit of Figure 4–33, determine the current I for each of the following. Express your answer in the most appropriate unit: amps, milliamps, microamps, etc.

   a. $E = 40$ V, $R = 20$ Ω
   b. $E = 35$ mV, $R = 5$ mΩ
   c. $V = 200$ V, $R = 40$ kΩ
   d. $E = 10$ V, $R = 2.5$ MΩ
   e. $E = 7.5$ V, $R = 2.5 \times 10^3$ Ω
   f. $V = 12$ kV, $R = 2$ MΩ

2. Determine R for each of the following. Express your answer in the most appropriate unit: ohms, kilohms, megohms, etc.

   a. $E = 50$ V, $I = 2.5$ A
   b. $E = 37.5$ V, $I = 1$ mA
   c. $E = 2$ kV, $I = 0.1$ kA
   d. $E = 4$ kV, $I = 8 \times 10^{-4}$ A

3. For the circuit of Figure 4–33, compute V for each of the following:

   a. 1 mA, 40 kΩ
   b. 10 μA, 30 kΩ
   c. 10 mA, 4×10^4 Ω
   d. 12 A, 3×10^{-2} Ω

4. A 48-Ω hot water heater is connected to a 120-V source. What is the current drawn?

5. When plugged into a 120-V wall outlet, an electric lamp draws 1.25 A. What is its resistance?

6. What is the potential difference between the two ends of a 20-kΩ resistor when its current is 3×10^{-3} A?

7. How much voltage can be applied to a 560-Ω resistor if its current must not exceed 50 mA?

8. A relay with a coil resistance of 240 Ω requires a minimum of 50 mA to operate. What is the minimum voltage that will cause it to operate?

9. For Figure 4–33, if $E = 30$ V and the conductance of the resistor is 0.2 S, what is I? Hint: See Section 3.11, Chapter 3.

10. If $I = 36$ mA when $E = 12$ V, what is I if the 12-V source is

    a. replaced by an 18-V source?

    b. replaced by a 4-V source?

11. Current through a resistor is 15 mA. If the voltage drop across the resistor is 33 V, what is its color code?

12. For the circuit of Figure 4–34,

    a. If $E = 28$ V, what does the meter indicate?

    b. If $E = 312$ V, what does the meter indicate?

13. For the circuit of Figure 4–34, if the resistor is replaced with one with red, red, black color bands, at what voltage do you expect the fuse to blow?

14. A 20-V source is applied to a resistor with color bands brown, black, red, and silver

    a. Compute the nominal current in the circuit.

    b. Compute the minimum and maximum currents based on the tolerance of the resistor.

15. An electromagnet is wound with AWG 30 copper wire. The coil has 800 turns and the average length of each turn is 3 inches. When connected to a 48-V dc source, what is the current

    a. at 20° C?
    b. at 40° C?

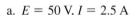

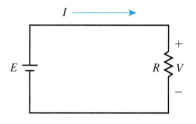

FIGURE 4–33

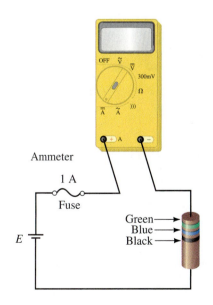

Ammeter

1 A

Fuse

Green
Blue
Black

E

FIGURE 4–34

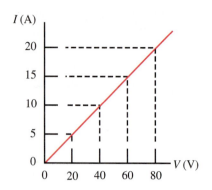

FIGURE 4–35

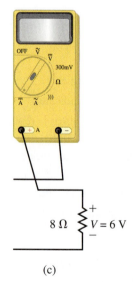

(a) $I = 3$ A

(b) $V = 60$ V

(c) $I = 6$ A

(d) $V = 105$ V

FIGURE 4–36   All resistors are 15 Ω.

16. You are to build an electromagnet with 0.643-mm-diameter copper wire. To create the required magnetic field, the current in the coil must be 1.75 A at 20° C. The electromagnet is powered from a 9.6-V dc source. How many meters of wire do you need to wind the coil?

17. A resistive circuit element is made from 100 m of 0.5-mm-diameter aluminum wire. If the current at 20° C is 200 mA, what is the applied voltage?

18. Prepare an Ohm's law graph similar to Figure 4–7 for a 2.5-kΩ and a 5-kΩ resistor. Compute and plot points every 5 V from $E = 0$ V to 25 V. Reading values from the graph, find current at $E = 14$ V.

19. Figure 4–35 represents the I-V graph for the circuit of Figure 4–33. What is R?

20. For a resistive circuit, E is quadrupled and R is halved. If the new current is 24 A, what was the original current?

21. For a resistive circuit, $E = 100$ V. If R is doubled and E is changed so that the new current is double the original current, what is the new value of E?

22. You need to measure the resistance of an electric heater element, but have only a 12-V battery and an ammeter. Describe how you would determine its resistance. Include a sketch.

23. If 25 m of 0.1-mm-diameter Nichrome wire is connected to a 12-V battery, what is the current at 20° C?

24. If the current is 0.5 A when a length of AWG 40 copper wire is connected to 48 V, what is the length of the wire in meters? Assume 20° C.

4.2 Voltage Polarity and Current Direction

25. For each resistor of Figure 4–36, determine voltage V and its polarity or current I and its direction as applicable.

26. The ammeters of Figure 4–37 have autopolarity. Determine their readings, magnitude, and polarity.

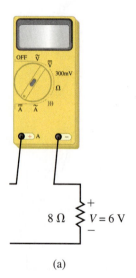

(a)

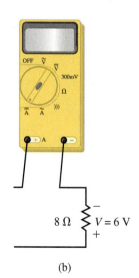

(b)

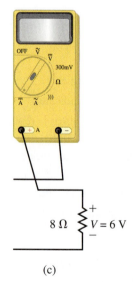

(c)

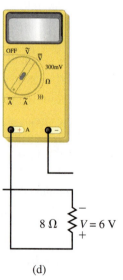

(d)

FIGURE 4–37

4.3 Power

27. A resistor dissipates 723 joules of energy in 3 minutes and 47 seconds. Calculate the rate at which energy is being transferred to this resistor in joules per second. What is its power dissipation in watts?

28. How long does it take for a 100-W soldering iron to dissipate 1470 J?

29. A resistor draws 3 A from a 12-V battery. How much power does the battery deliver to the resistor?

30. A 120-V electric coffee maker is rated 960 W. Determine its resistance and rated current.

31. A 1.2-kW electric heater has a resistance of 6 Ω. How much current does it draw?

32. A warning light draws 125 mA when dissipating 15 W. What is its resistance?

33. How many volts must be applied to a 3-Ω resistor to result in a power dissipation of 752 W?

34. What IR drop occurs when 90 W is dissipated by a 10-Ω resistor?

35. A resistor with color bands brown, black, and orange dissipates 0.25 W. Compute its voltage and current.

36. A 2.2-kΩ resistor with a tolerance of ±5% is connected to a 12-V dc source. What is the possible range of power dissipated by the resistor?

37. A portable radio transmitter has an input power of 0.455 kW. How much current does it draw from a 12-V battery?

38. For a resistive circuit, $E = 12$ V.

    a. If the load dissipates 8 W, what is the current in the circuit?

    b. If the load dissipates 36 W, what is the load resistance?

39. A motor delivers 3.56 hp to a load. How many watts is this?

40. The load on a 120-V circuit consists of six 100-W lamps, a 1.2-kW electric heater, and an electric motor drawing 1500 W. If the circuit is fused at 30 A, what happens when a 900-W toaster is plugged in? Justify your answer.

41. A 0.27-kΩ resistor is rated 2 W. Compute the maximum voltage that can be applied and the maximum current that it can carry without exceeding its rating.

42. Determine which, if any, of the following resistors may have been damaged by overheating. Justify your answer.

    a. 560 Ω, ½ W, with 75 V across it.

    b. 3 Ω, 20 W, with 4 A through it.

    c. ¼ W, with 0.25 mA through it and 40 V across it.

43. A 25-Ω resistor is connected to a power supply whose voltage is 100 V ±5%. What is the possible range of power dissipated by the resistor?

44. A load resistance made from copper wire is connected to a 24-V dc source. The power dissipated by the load when the wire temperature is 20° C is 192 W. What will be the power dissipated when the temperature of the wire drops to −10° C? (Assume the voltage remains constant.)

4.4 Power Direction Convention

45. Each block of Figure 4–38 may be a source or a load. For each, determine power and its direction.

46. The 12-V battery of Figure 4–39 is being "charged" by a battery charger. The current is 4.5 A as indicated.

    a. What is the direction of current?

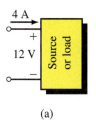

(a)

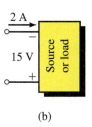

(b)

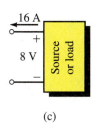

(c)

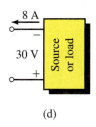

(d)

FIGURE 4–38

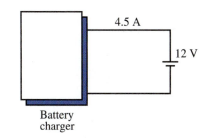

FIGURE 4–39

b. What is the direction of power flow?

c. What is the power to the battery?

4.5 Energy

47. A 40-W night safety light burns for 9 hours.

    a. Determine the energy used in joules.

    b. Determine the energy used in watthours.

    c. At $0.08/kWh, how much does it cost to run this light for 9 hours?

48. An indicator light on a control panel operates continuously, drawing 20 mA from a 120-V supply. At $0.09 per kilowatthour, how much does it cost per year to operate the light?

49. Determine the total cost of using the following at $0.11 per kWh:

    a. a 900-W toaster for 5 minutes,

    b. a 120-V, 8-A heater for 1.7 hours,

    c. an 1100-W dishwasher for 36 minutes,

    d. a 120-V, 288-Ω soldering iron for 24 minutes.

50. An electric device with a cycle time of 1 hour operates at full power (400 W) for 15 minutes, at half power for 30 minutes, then cuts off for the remainder of the hour. The cycle repeats continuously. At $0.10/kWh, determine the yearly cost of operating this device.

51. While the device of Problem 50 is operating, two other loads as follows are also operating:

    a. a 4-kW heater, continuously,

    b. a 3.6-kW heater, on for 12 hours per day.

    Calculate the yearly cost of running all loads.

52. At $0.08 per kilowatthour, it costs $1.20 to run a heater for 50 hours from a 120-V source. How much current does the heater draw?

53. If there are 24 slices in a loaf of bread and you have a two-slice, 1100-W toaster that takes 1 minute and 45 seconds to toast a pair of slices, at $0.13/kWh, how much does it cost to toast a loaf?

4.6 Efficiency

54. The power input to a motor with an efficiency of 85% is 690 W. What is its power output?

    a. in watts   b. in hp

55. The power output of a transformer with $\eta = 97\%$ is 50 kW. What is its power in?

56. For a certain device, $\eta = 94\%$. If losses are 18 W, what are P_{in} and P_{out}?

57. The power input to a device is 1100 W. If the power lost due to various inefficiencies is 190 W, what is the efficiency of the device?

58. A 240-V, 4.5-A water heater produces heat energy at the rate of 3.6 MJ per hour. Compute

    a. the efficiency of the water heater,

    b. the annual cost of operation at $0.09/kWh if the heater is on for 6 h/day.

59. A 120-V dc motor with an efficiency of 89% draws 15 A from the source. What is its output horsepower?

60. A 120-V dc motor develops an output of 3.8 hp. If its efficiency is 87%, how much current does it draw?

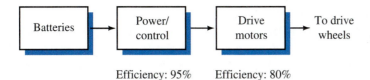

Efficiency: 95%   Efficiency: 80%

FIGURE 4–40

61. The power/control system of an electric car consists of a 48-V onboard battery pack, electronic control/drive unit, and motor (Figure 4–40). If 180 A are drawn from the batteries, how many horsepower are delivered to the drive wheels?

62. Show that the efficiency of n devices or systems in cascade is the product of their individual efficiencies, i.e., that $\eta_T = \eta_1 \times \eta_2 \times \cdots \times \eta_n$.

63. A 120-V dc motor drives a pump through a gearbox (Figure 4–24). If the power input to the pump is 1100 W, the gearbox has an efficiency of 75%, and the power input to the motor is 1600 W, determine the horsepower output of the motor.

64. If the motor of Problem 63 is protected by a 15-A circuit breaker, will it open? Compute the current to find out.

65. If the overall efficiency of a radio transmitting station is 55% and it transmits at 35 kW for 24 h/day, compute the cost of energy used per day at $0.09/kWh.

66. In a factory, two machines, each delivering 27 kW are used on an average for 8.7 h/day, 320 days/year. If the efficiency of the newer machine is 87% and that of the older machine is 72%, compute the difference in cost per year of operating them at $0.10 per kilowatthour.

4.7 Nonlinear and Dynamic Resistances

67. A voltage-dependent resistor has the *I-V* characteristic of Figure 4–41.
   a. At $V = 25$ V, what is I? What is R_{dc}?
   b. At $V = 60$ V, what is I? What is R_{dc}?
   c. Why are the two values different?

68. For the resistor of Figure 4–41:
   a. Determine $R_{dynamic}$ for V between 0 and 40 V.
   b. Determine $R_{dynamic}$ for V greater than 40 V.
   c. If V changes from 20 V to 30 V, how much does I change?
   d. If V changes from 50 V to 70 V, how much does I change?

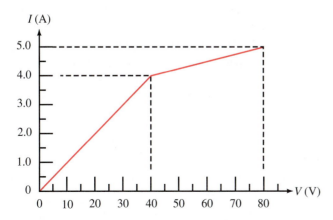

FIGURE 4–41

4.8 Computer-Aided Circuit Analysis

◀ MULTISIM

69. Set up the circuit of Figure 4–33 and solve for currents for the voltage/resistance pairs of Problem 1a, 1c, 1d, and 1e.

◀ MULTISIM

70. A battery charger with a voltage of 12.9 V is used to charge a battery, Figure 4–42(a). The internal resistance of the charger is 0.12 Ω and the voltage of the partially run-down battery is 11.6 V. The equivalent circuit for the charger/battery combination is shown in (b). You reason that, since the two voltages are in opposition, net voltage for the circuit will be 12.9 V − 11.6 V = 1.3 V and, thus, the charging current I will be 1.3 V/0.12 Ω = 10.8 A. Set up the circuit of (b) and use MultiSIM to verify your conclusion.

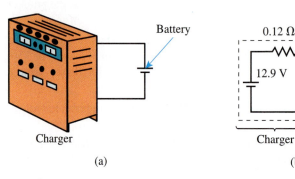

(a)   (b) Equivalent

FIGURE 4–42

◀ MULTISIM

71. Open the Basic parts bin, click the switch icon, and select SPDT. Place on the screen and double click its symbol. When the dialog box opens, select the Value tab, type the letter A, and click OK. [This relabels the switch as (A). Press the A key on the keyboard several times and note that the switch opens and closes.] Select a second switch and label it (B), then add a 12-V dc source, so as to set up the two-way light control circuit of Figure 2–27. Operate the switches and determine whether you have successfully achieved the desired control.

◀ CADENCE

72. Repeat Problem 69 using PSpice.

◀ CADENCE

73. Repeat Problem 70 using PSpice.

◀ CADENCE

74. Repeat the Ohm's law analysis (Figure 4–32) except use $R = 25$ Ω and sweep the source voltage from −10 V to +10 V in 1-V increments. On this graph, what does the negative value for current mean?

◀ CADENCE

75. The cursor may be used to read values from PSpice graphs. Get the graph from Problem 74 on the screen and:

a. Click Trace on the menu bar, select Cursor, click Display, then position the cursor on the graph, and click again. The cursor reading is indicated in the box in the lower right hand corner of the screen.

b. The cursor may be positioned using the mouse or the left and right arrow keys. Position the cursor at 2 V and read the current. Verify by Ohm's law. Repeat at a number of other points, both positive and negative.

✓ **ANSWERS TO IN-PROCESS LEARNING CHECKS**

In-Process Learning Check 1

1. 25 V

2. 0 A, No

3. a. 1.25 A; b. 0.625 A; c. 12.5 A; d. 5 A

4. a. 30 V; b. −72 V; c. −90 V; d. 160 V

Basic dc Analysis

II

Part I provided the foundation upon which proficient circuit analysis is built. The terms, units, and definitions used in the previous chapters have developed the necessary vocabulary that is used throughout electrical and electronics technology.

In Part II, we build on this foundation, applying the concepts developed in Part I to initially analyze series dc circuits. Several very important rules, laws, and theorems will be developed in the following chapters, providing additional tools needed to extend circuit theory to the analysis of parallel and series-parallel circuits. Circuit theorems will be used to simplify the most complex circuit into an equivalent circuit that is represented as a single source and a single resistor. Although most circuits are much more complex than the circuits used in this text, all circuits, even the most complex, follow the same laws.

Several different techniques are often available to analyze a given circuit. In order to give a broad overview, this textbook uses the various methods for illustrative purposes. You are encouraged to use different techniques where possible, so that you develop the skills to become proficient in circuit analysis. ∎

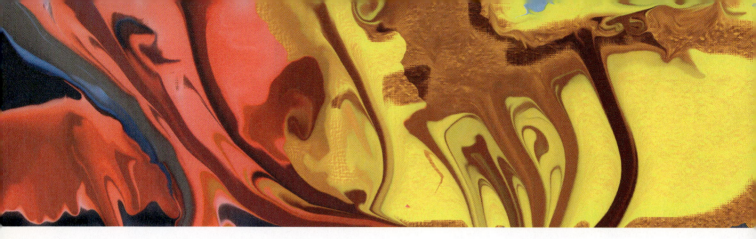

KEY TERMS

Electric Circuit
Ground
Kirchhoff's Voltage Law
Loading Effect (Ammeter)
Ohmmeter Design
Point Sources
Series Connection
Total Equivalent Resistance
Voltage Divider Rule
Voltage Subscripts

OUTLINE

Series Circuits
Kirchhoff's Voltage Law
Resistors in Series
Voltage Sources in Series
Interchanging Series Components
The Voltage Divider Rule
Circuit Ground
Voltage Subscripts
Internal Resistance of Voltage
   Sources
Ammeter Loading Effects
Circuit Analysis Using Computers

OBJECTIVES

After studying this chapter, you will be able to

• determine the total resistance in a series circuit and calculate circuit current,

• use Ohm's law and the voltage divider rule to solve for the voltage across all resistors in the circuit,

• express Kirchhoff's voltage law and use it to analyze a given circuit,

• solve for the power dissipated by any resistor in a series circuit and show that the total power dissipated is exactly equal to the power delivered by the voltage source,

• solve for the voltage between any two points in a series or parallel circuit,

• calculate the loading effect of an ammeter in a circuit,

• use computers to assist in the analysis of simple series circuits.

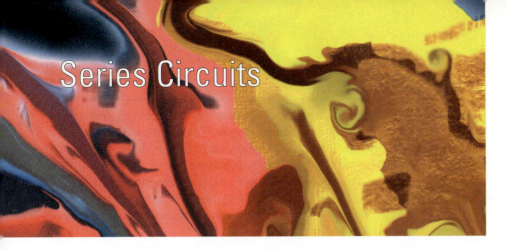

Series Circuits

5

In the previous chapter we examined the interrelation of current, voltage, resistance, and power in a single resistor circuit. In this chapter we will expand on these basic concepts to examine the behavior of circuits having several resistors in series.

We will use Ohm's law to derive the voltage divider rule and to verify Kirchhoff's voltage law. A good understanding of these important principles provides an important base upon which further circuit analysis techniques are built. Kirchhoff's voltage law and Kirchhoff's current law, which will be covered in the next chapter, are fundamental in understanding *all* electrical and electronic circuits.

After developing the basic framework of series circuit analysis, we will apply the ideas to analyze and design simple voltmeters and ohmmeters. While meters are usually covered in a separate instruments or measurements course, we examine these circuits merely as an application of the concepts of circuit analysis.

Similarly, we will observe how circuit principles are used to explain the loading effect of an ammeter placed in series with a circuit. ■

CHAPTER PREVIEW

Gustav Robert Kirchhoff

KIRCHHOFF WAS A GERMAN PHYSICIST born on March 12, 1824, in Königsberg, Prussia. His first research was on the conduction of electricity, which led to his presentation of the laws of closed electric circuits in 1845. Kirchhoff's current law and Kirchhoff's voltage law apply to all electrical circuits and therefore are fundamentally important in understanding circuit operation. Kirchhoff was the first to verify that an electrical impulse travelled at the speed of light.

Although these discoveries have immortalized Kirchhoff's name in electrical science, he is better known for his work with R. W. Bunsen in which he made major contributions in the study of spectroscopy and advanced the research into blackbody radiation.

Kirchhoff died in Berlin on October 17, 1887. ■

PUTTING IT IN PERSPECTIVE

5.1 Series Circuits

Conventional flow

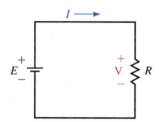

FIGURE 5–1

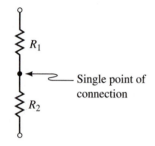

FIGURE 5–2   Resistors in series.

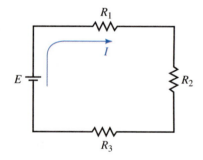

FIGURE 5–3   Series circuit.

An **electric circuit** is the combination of any number of sources and loads connected in any manner that allows charge to flow. The electric circuit may be simple, such as a circuit consisting of a battery and a light bulb. Or the circuit may be very complex, such as the circuits contained within a television set, microwave oven, or computer. However, no matter how complicated, each circuit follows fairly simple rules in a predictable manner. Once these rules are understood, any circuit may be analyzed to determine the operation under various conditions.

All electric circuits obtain their energy either from a direct current (dc) source or from an alternating current (ac) source. In the next few chapters, we examine the operation of circuits supplied by dc sources. Although ac circuits have fundamental differences when compared with dc circuits, the laws, theorems, and rules that you learn in dc circuits apply directly to ac circuits as well.

In the previous chapter, you were introduced to a simple dc circuit consisting of a single voltage source (such as a chemical battery) and a single load resistance. The schematic representation of such a simple circuit was covered in Chapter 4 and is shown again in Figure 5–1.

While the circuit of Figure 5–1 is useful in deriving some important concepts, very few practical circuits are this simple. However, we will find that even the most complicated dc circuits can generally be simplified to the circuit shown.

We begin by examining the most simple connection, the **series connection.** In Figure 5–2, we have two resistors, R_1 and R_2, connected at a single point in what is said to be a series connection.

Two elements are said to be in series if they are connected at a single point and if there are no other current-carrying connections at this point.

A **series circuit** is constructed by combining various elements in series, as shown in Figure 5–3. Current will leave the positive terminal of the voltage source, move through the resistors, and return to the negative terminal of the source.

In the circuit of Figure 5–3, we see that the voltage source, E, is in series with R_1, R_1 is in series with R_2, and R_2 is in series with E. By examining this circuit, another important characteristic of a series circuit becomes evident. In an analogy similar to water flowing in a pipe, current entering an element must be the same as the current leaving the element. Now, since current does not leave at any of the connections, we conclude that the following must be true.

The current is the same everywhere in a series circuit.

While the above statement seems self-evident, we will find that this will help to explain many of the other characteristics of a series circuit.

✓ IN-PROCESS
LEARNING CHECK 1

(Answers are at the end of the chapter.)

What two conditions determine whether two elements are connected in series?

The Voltage Polarity Convention Revisited

The $+$, $-$ sign convention of Figure 5–4(a) has a deeper meaning than so far considered. *Voltage exists between points, and when we place a $+$ at one point and a $-$ at another point, we define this to mean that we are looking at the voltage at the point marked $+$ with respect to the point marked $-$.* Thus, in Figure 5–4(b), $V = 6$ volts means that point a is 6 V positive with respect to point b. Since the red lead of the meter is placed at point a and the black at point b, the meter will indicate $+6$ V.

Now consider Figure 5–5(b). [Part (a) has been repeated from Figure 5–4(b) for reference.] Here, we have placed the plus sign at point *b,* meaning that you are looking at the voltage at *b* with respect to *a.* Since point *b* is 6 V negative with respect to point *a, V* will have a value of minus 6 volts, i.e., *V* = −6 volts. Note also that the meter indicates −6 volts since we have reversed its leads and the red lead is now at point *b* and the black at point *a. It is important to realize here that the voltage across R has not changed. What has changed is how we are looking at it and how we have connected the meter to measure it.* Thus, since the actual voltage is the same in both cases, (a) and (b) are equivalent representations.

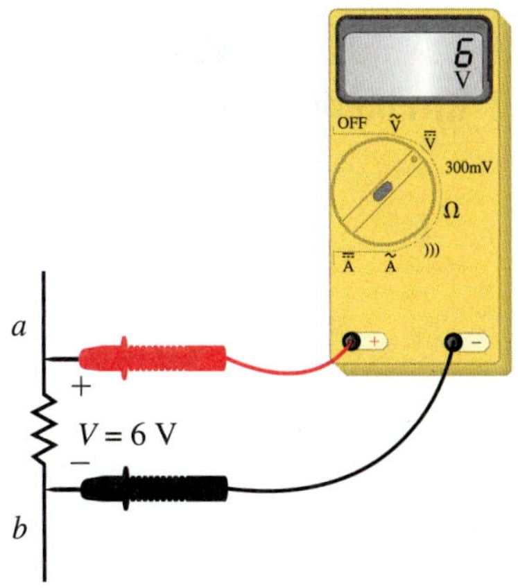

FIGURE 5–4 The +, − symbology means that you are looking at the voltage at the point marked + with respect to the point marked −.

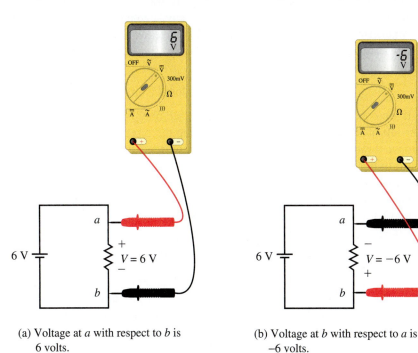

(a) Voltage at *a* with respect to *b* is 6 volts.

(b) Voltage at *b* with respect to *a* is −6 volts.

FIGURE 5–5 Two representations of the same voltage.

Consider Figure 5–6. In (a), *I* = 3 A, while in (b), *I* = −3 A. Using the voltage polarity convention, determine the voltages across the two resistors and show that they are equal.

EXAMPLE 5–1

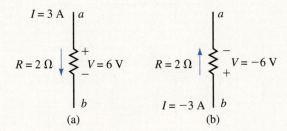

FIGURE 5–6

Solution   In each case, place the plus sign at the tail of the current direction arrow. Then, for (a), you are looking at the polarity of *a* with respect to *b* and you get *V* = *IR* = (3 A)(2 Ω) = 6 V as expected. Now consider (b). The polarity markings mean that you are looking at the polarity of *b* with respect to *a* and you get *V* = *IR* = (−3 A)(2 Ω) = −6 V. This means that point *b* is 6 V negative with respect to *a,* or equivalently, *a* is 6 V positive with respect to *b.* Thus, the two voltages are equal.

5.2 Kirchhoff's Voltage Law

◀ **Online Companion**

Next to Ohm's law, one of the most important laws of electricity is Kirchhoff's voltage law (KVL) which states the following:

The summation of voltage rises and voltage drops around a closed loop is equal to zero. Symbolically, this may be stated as follows:

$$\Sigma V = 0 \quad \text{for a closed loop} \qquad (5\text{–}1)$$

In the above symbolic representation, the uppercase Greek letter sigma (Σ) stands for summation and V stands for voltage rises and drops. A **closed loop** is defined as any path that originates at a point, travels around a circuit, and returns to the original point without retracing any segments.

An alternate way of stating Kirchhoff's voltage law is as follows:

The summation of voltage rises is equal to the summation of voltage drops around a closed loop.

$$\Sigma E_{\text{rises}} = \Sigma V_{\text{drops}} \quad \text{for a closed loop} \qquad (5\text{–}2)$$

If we consider the circuit of Figure 5–7, we may begin at point a in the lower left-hand corner. By arbitrarily following the direction of the current, I, we move through the voltage source, which represents a rise in potential from point a to point b. Next, in moving from point b to point c, we pass through resistor R_1, which presents a potential drop of V_1. Continuing through resistors R_2 and R_3, we have additional drops of V_2 and V_3 respectively. By applying Kirchhoff's voltage law around the closed loop, we arrive at the following mathematical statement for the given circuit:

$$E - V_1 - V_2 - V_3 = 0$$

Although we chose to follow the direction of current in writing Kirchhoff's voltage law equation, it would be just as correct to move around the circuit in the opposite direction. In this case the equation would appear as follows:

$$V_3 + V_2 + V_1 - E = 0$$

By simple manipulation, it is quite easy to show that the two equations are identical.

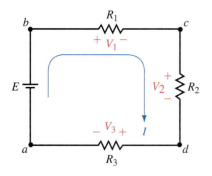

FIGURE 5–7   Kirchoff's voltage law.

EXAMPLE 5–2

Verify Kirchhoff's voltage law for the circuit of Figure 5–8.

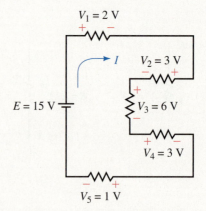

FIGURE 5–8

Solution   If we follow the direction of the current, we write the loop equation as

$$15\,\text{V} - 2\,\text{V} - 3\,\text{V} - 6\,\text{V} - 3\,\text{V} - 1\,\text{V} = 0$$

Verify Kirchhoff's voltage law for the circuit of Figure 5–9.

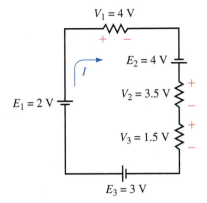

FIGURE 5–9

Answer
$2\,V - 4\,V + 4\,V - 3.5\,V - 1.5\,V + 3\,V = 0$

Define Kirchhoff's voltage law.

IN-PROCESS
LEARNING CHECK 2

(Answers are at the end of the chapter.)

Almost all complicated circuits can be simplified. We will now examine how to simplify a circuit consisting of a voltage source in series with several resistors. Consider the circuit shown in Figure 5–10.

5.3 Resistors in Series

Since the circuit is a closed loop, the voltage source will cause a current I in the circuit. This current in turn produces a voltage drop across each resistor, where

$$V_x = IR_x$$

Applying Kirchhoff's voltage law to the closed loop gives

$$E = V_1 + V_2 + \cdots + V_n$$
$$= IR_1 + IR_2 + \cdots + IR_n$$
$$= I(R_1 + R_2 + \cdots + R_n)$$

If we were to replace all the resistors with an equivalent total resistance, R_T, then the circuit would appear as shown in Figure 5–11.

However, applying Ohm's law to the circuit of Figure 5–11 gives

$$E = IR_T \qquad\qquad \textbf{(5–3)}$$

Since the circuit of Figure 5–11 is equivalent to the circuit of Figure 5–10, we conclude that this can only occur if the total resistance of the n series resistors is given as

$$R_T = R_1 + R_2 + \cdots + R_n \quad [\text{ohms}, \Omega] \qquad \textbf{(5–4)}$$

If each of the n resistors has the same value, then the total resistance is determined as

$$R_T = nR \quad [\text{ohms}, \Omega] \qquad\qquad \textbf{(5–5)}$$

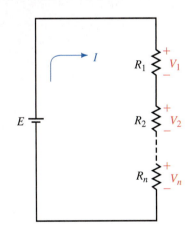

FIGURE 5–10

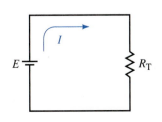

FIGURE 5–11

EXAMPLE 5–3

Determine the total resistance for each of the networks shown in Figure 5–12.

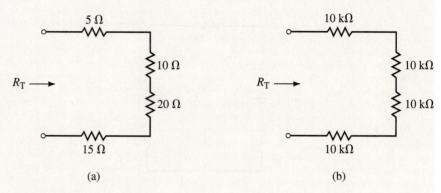

(a)   (b)

FIGURE 5–12

Solution

a. $R_T = 5\ \Omega + 10\ \Omega + 20\ \Omega + 15\ \Omega = 50.0\ \Omega$

b. $R_T = 4(10\ k\Omega) = 40.0\ k\Omega$

Any voltage source connected to the terminals of a network of series resistors will provide the same current as if a single resistance, having a value of R_T, were connected between the open terminals. From Ohm's law we get

$$I = \frac{E}{R_T} \quad [\text{amps, A}] \qquad (5\text{–}6)$$

The power dissipated by each resistor is determined as

$$P_1 = V_1 I = \frac{V_1{}^2}{R_1} = I^2 R_1 \quad [\text{watts, W}]$$

$$P_2 = V_2 I = \frac{V_2{}^2}{R_2} = I^2 R_2 \quad [\text{watts, W}] \qquad (5\text{–}7)$$

$$\vdots$$

$$P_n = V_n I = \frac{V_n{}^2}{R_n} = I^2 R_n \quad [\text{watts, W}]$$

In Chapter 4, we showed that the power delivered by a voltage source to a circuit is given as

$$P_T = EI \quad [\text{watts, W}] \qquad (5\text{–}8)$$

Since energy must be conserved, the power delivered by the voltage source is equal to the total power dissipated by all the resistors. Hence

$$P_T = P_1 + P_2 + \cdots + P_n \quad [\text{watts, W}] \qquad (5\text{–}9)$$

For the series circuit shown in Figure 5–13, find the following quantities:

a. Total resistance, R_T.

b. Circuit current, I.

c. Voltage across each resistor.

d. Power dissipated by each resistor.

e. Power delivered to the circuit by the voltage source.

f. Verify that the power dissipated by the resistors is equal to the power delivered to the circuit by the voltage source.

EXAMPLE 5–4

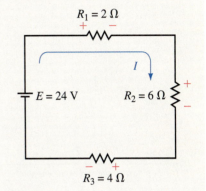

FIGURE 5–13

Solution

a. $R_T = 2\ \Omega + 6\ \Omega + 4\ \Omega = 12.0\ \Omega$

b. $I = (24\ \text{V})/(12\ \Omega) = 2.00\ \text{A}$

c. $V_1 = (2\ \text{A})(2\ \Omega) = 4.00\ \text{V}$

  $V_2 = (2\ \text{A})(6\ \Omega) = 12.0\ \text{V}$

  $V_3 = (2\ \text{A})(4\ \Omega) = 8.00\ \text{V}$

d. $P_1 = (2\ \text{A})^2(2\ \Omega) = 8.00\ \text{W}$

  $P_2 = (2\ \text{A})^2(6\ \Omega) = 24.0\ \text{W}$

  $P_3 = (2\ \text{A})^2(4\ \Omega) = 16.0\ \text{W}$

e. $P_T = (24\ \text{V})(2\ \text{A}) = 48.0\ \text{W}$

f. $P_T = 8\ \text{W} + 24\ \text{W} + 16\ \text{W} = 48.0\ \text{W}$

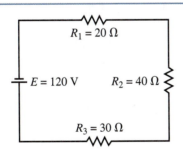

FIGURE 5–14

PRACTICE PROBLEMS 2

For the series circuit shown in Figure 5–14, find the following quantities:

a. Total resistance, R_T.

b. The direction and magnitude of the current, I.

c. Polarity and magnitude of the voltage across each resistor.

d. Power dissipated by each resistor.

e. Power delivered to the circuit by the voltage source.

f. Show that the power dissipated is equal to the power delivered.

IN-PROCESS
LEARNING CHECK 3

(Answers are at the end of the chapter.)

Three resistors, R_1, R_2, and R_3, are in series. Determine the value of each resistor if $R_T = 42$ kΩ, $R_2 = 3R_1$, and $R_3 = 2R_2$.

5.4 Voltage Sources in Series

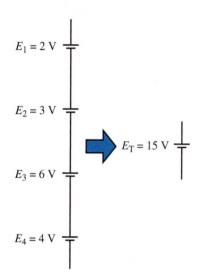

FIGURE 5–15

If a circuit has more than one voltage source in series, then the voltage sources may effectively be replaced by a single source having a value that is the sum or difference of the individual sources. Since the sources may have different polarities, it is necessary to consider polarities in determining the resulting magnitude and polarity of the equivalent voltage source.

If the polarities of all the voltage sources are such that the sources appear as voltage rises in given direction, then the resultant source is determined by simple addition, as shown in Figure 5–15.

If the polarities of the voltage sources do not result in voltage rises in the same direction, then we must compare the rises in one direction to the rises in the other direction. The magnitude of the resultant source will be the sum of the rises in one direction minus the sum of the rises in the opposite direction. The polarity of the equivalent voltage source will be the same as the polarity of whichever direction has the greater rise. Consider the voltage sources shown in Figure 5–16.

If the rises in one direction were equal to the rises in the opposite direction, then the resultant voltage source would be equal to zero.

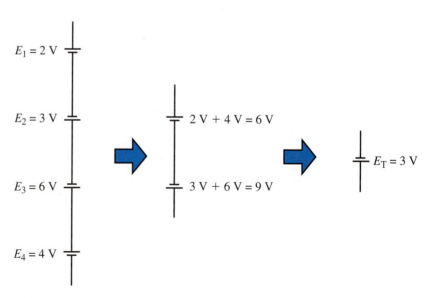

FIGURE 5–16

A typical lead-acid automobile battery consists of six cells connected in series. If the voltage between the battery terminals is measured to be 13.06 V, what is the average voltage of each cell within the battery?

IN-PROCESS
LEARNING CHECK 4

(Answers are at the end of the chapter.)

5.5 Interchanging Series Components

The order of series components may be changed without affecting the operation of the circuit.

The two circuits in Figure 5–17 are equivalent.

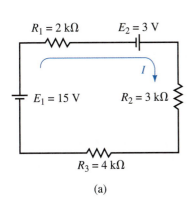

(a)

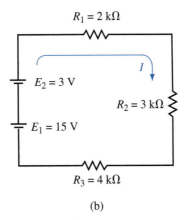

(b)

FIGURE 5–17

Very often, once the circuits have been redrawn it becomes easier to visualize the circuit operation. Therefore, we will regularly use the technique of interchanging components to simplify circuits before we analyze them.

Simplify the circuit of Figure 5–18 into a single source in series with the four resistors. Determine the direction and magnitude of the current in the resulting circuit.

EXAMPLE 5–5

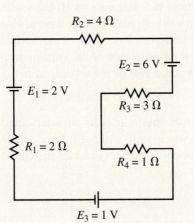

FIGURE 5–18

Solution   We may redraw the circuit by the two steps shown in Figure 5–19. It is necessary to ensure that the voltage sources are correctly moved since it is quite easy to assign the wrong polarity. Perhaps the easiest way is to imagine that we slide the voltage source around the circuit to the new location.

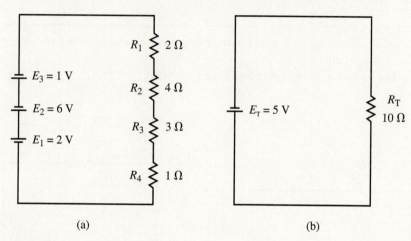

(a)   (b)

FIGURE 5–19

The current in the resulting circuit will be in a counterclockwise direction around the circuit and will have a magnitude determined as

$$I = \frac{E_T}{R_T} = \frac{6\ \text{V} + 1\ \text{V} - 2\ \text{V}}{2\ \Omega + 4\ \Omega + 3\ \Omega + 1\ \Omega} = \frac{5\ \text{V}}{10\ \Omega} = 0.500\ \text{A}$$

Because the circuits are in fact equivalent, the current direction determined in Figure 5–19 also represents the direction for the current in the circuit of Figure 5–18.

5.6 The Voltage Divider Rule

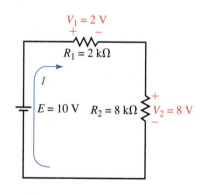

FIGURE 5–20

The voltage dropped across any series resistor is proportional to the magnitude of the resistor. The total voltage dropped across all resistors must equal the applied voltage source(s) by KVL.

Consider the circuit of Figure 5–20.

We see that the total resistance $R_T = 10\ \text{k}\Omega$ results in a circuit current of $I = 1\ \text{mA}$. From Ohm's law, R_1 has a voltage drop of $V_1 = 2.0\ \text{V}$, while R_2, which is four times as large as R_1, has four times as much voltage drop, $V_2 = 8.0\ \text{V}$.

We also see that the summation of the voltage drops across the resistors is exactly equal to the voltage rise of the source, namely,

$$E = 10\ \text{V} = 2\ \text{V} + 8\ \text{V}$$

The voltage divider rule allows us to determine the voltage across any series resistance in a single step, without first calculating the current. We have seen that for any number of resistors in series the current in the circuit is determined by Ohm's law as

$$I = \frac{E}{R_T} \quad \text{[Amps, A]} \tag{5–10}$$

where the two resistors in Figure 5–20 result in a total resistance of

$$R_T = R_1 + R_2$$

By again applying Ohm's law, the voltage drop across any resistor in the series circuit is calculated as

$$V_x = IR_x$$

Now, by substituting Equation 5–4 into the above equation we write the **voltage divider rule** for two resistors as a simple equation:

$$V_x = \frac{R_x}{R_T} E = \frac{R_x}{R_1 + R_2} E$$

In general, for any number of resistors the voltage drop across any resistor may be found as

$$V_x = \frac{R_x}{R_T} E \qquad\qquad \textbf{(5–11)}$$

Use the voltage divider rule to determine the voltage across each of the resistors in the circuit shown in Figure 5–21. Show that the summation of voltage drops is equal to the applied voltage rise in the circuit.

EXAMPLE 5–6

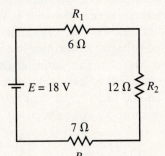

FIGURE 5–21

Solution

$$R_T = 6\ \Omega + 12\ \Omega + 7\ \Omega = 25.0\ \Omega$$

$$V_1 = \left(\frac{6\ \Omega}{25\ \Omega}\right)(18\ \text{V}) = 4.32\ \text{V}$$

$$V_2 = \left(\frac{12\ \Omega}{25\ \Omega}\right)(18\ \text{V}) = 8.64\ \text{V}$$

$$V_3 = \left(\frac{7\ \Omega}{25\ \Omega}\right)(18\ \text{V}) = 5.04\ \text{V}$$

The total voltage drop is the summation

$$V_T = 4.32\ \text{V} + 8.64\ \text{V} + 5.04\ \text{V} = 18.0\ \text{V} = E$$

EXAMPLE 5–7

Using the voltage divider rule, determine the voltage across each of the resistors of the circuit shown in Figure 5–22.

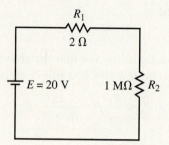

FIGURE 5–22

Solution

$$R_T = 2\ \Omega + 1\ 000\ 000\ \Omega = 1\ 000\ 002\ \Omega$$

$$V_1 = \left(\frac{2\ \Omega}{1\ 000\ 002\ \Omega}\right)(20\ V) \approx 40\ \mu V$$

$$V_2 = \left(\frac{1.0\ M\Omega}{1.000\ 002\ M\Omega}\right)(20\ V) = 19.999\ 86\ V$$

$$\approx 20.0\ V$$

The previous example illustrates two important points that occur regularly in electronic circuits. If a single series resistance is very large in comparison with the other series resistances, then the voltage across that resistor will be essentially the total applied voltage. On the other hand, if a single resistance is very small in comparison with the other series resistances, then the voltage drop across the small resistor will be essentially zero. As a general rule, if a series resistor is more than 100 times larger than another series resistor, then the effect of the smaller resistor(s) may be effectively neglected.

PRACTICE PROBLEMS 3

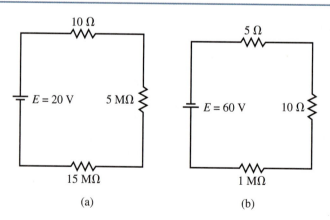

(a)   (b)

FIGURE 5–23

For the circuits shown in Figure 5–23, determine the approximate voltage drop across each resistor without using a calculator. Compare your approximations to the actual values obtained with a calculator.

Answers

a. $V_{10\text{-}\Omega} \cong 0$ $V_{5\text{-}M\Omega} \cong 5.00$ V $V_{10\text{-}M\Omega} \cong 15.0$ V

     $V_{10\text{-}\Omega} = 10.0\ \mu$V $V_{5\text{-}M\Omega} = 5.00$ V $V_{10\text{-}M\Omega} = 15.0$ V

b. $V_{5\text{-}\Omega} \cong 0$ $V_{10\text{-}\Omega} \cong 0$ $V_{1\text{-}M\Omega} \cong 60$ V

     $V_{5\text{-}\Omega} = 0.300$ mV $V_{10\text{-}\Omega} = 0.600$ mV $V_{1\text{-}M\Omega} = 60.0$ V

The voltage drops across three resistors are measured to be 10.0 V, 15.0 V, and 25.0 V. If the largest resistor is 47.0 kΩ, determine the sizes of the other two resistors.

IN-PROCESS
LEARNING CHECK 5

(Answers are at the end of the chapter.)

5.7 Circuit Ground

(a) Circuit ground or reference

(b) Chassis ground

FIGURE 5–24

Perhaps one of the most misunderstood concepts in electronics is that of ground. This misunderstanding leads to many problems when circuits are designed and analyzed. The standard symbol for circuit ground is shown in Figure 5–24(a), while the symbol for chassis ground is shown in Figure 5–24(b).

In its most simple definition, **ground** is simply an "arbitrary electrical point of reference" or "common point" in a circuit. Using the ground symbol in this manner usually allows the circuit to be sketched more simply. When the ground symbol is used arbitrarily to designate a point of reference, it would be just as correct to redraw the circuit schematic showing all the ground points connected together or indeed to redraw the circuit using an entirely different point of reference. The circuits shown in Figure 5–25 are exactly equivalent circuits even though the circuits of Figure 5–25(a) and 5–25(c) use different points of reference.

While the ground symbol is used to designate a common point of reference within a circuit, it usually has a greater meaning to the technologist or engineer. Very often, the metal chassis of an appliance is connected to the circuit ground. Such a connection is referred to as a **chassis ground** and is usually designated as shown in Figure 5–24(b).

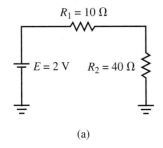

(a)

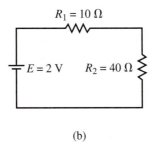

(b)

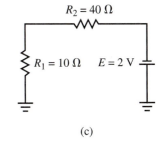

(c)

FIGURE 5–25

In order to help prevent electrocution, the chassis ground is usually further connected to the **earth ground** through a connection provided at the electrical outlet box. In the event of a failure within the circuit, the chassis would redirect current to ground (tripping a breaker or fuse), rather than presenting a hazard to an unsuspecting operator.

As the name implies, the earth ground is a connection that is bonded to the earth, either through water pipes or by a connection to ground rods. Everyone is familiar with the typical 120-Vac electrical outlet shown in Figure 5–26. The rounded terminal of the outlet is always the ground terminal and is used not only in ac circuits by may also be used to provide a common point for dc circuits. When a circuit is bonded to the earth through the ground terminal the ground symbol no longer represents an arbitrary connection, but rather represents a very specific type of connection.

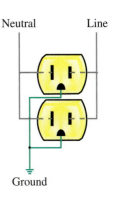

FIGURE 5–26 Ground connection in a typical 120-Vac outlet.

(Answers are at the end of the chapter.)

5.8 Voltage Subscripts

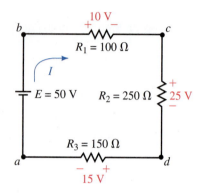

FIGURE 5–27

If you measure the resistance between the ground terminal of the 120-Vac plug and the metal chassis of a microwave oven to be zero ohms, what does this tell you about the chassis of the oven?

Double Subscripts

As you have already seen, voltages are always expressed as the potential difference between two points. In a 9-V battery, there is a 9-volt rise in potential from the negative terminal to the positive terminal. Current through a resistor results in a voltage drop across the resistor such that the terminal from which charge leaves is at a lower potential than the terminal into which charge enters. We now examine how voltages within any circuit may be easily described as the voltage between two points. If we wish to express the voltage between two points (say points a and b in a circuit), then we express such a voltage in a subscripted form (e.g., V_{ab}), where the first term in the subscript is the point of interest and the second term is the point of reference.

Consider the series circuit of Figure 5–27.

If we label the points within the circuit a, b, c, and d, we see that point b is at a higher potential than point a by an amount equal to the supply voltage. We may write this mathematically as $V_{ba} = +50$ V. Although the plus sign is redundant, we show it here to indicate that point b is at a higher potential than point a. If we examine the voltage at point a with respect to point b, we see that a is at a lower potential than b. This may be written mathematically as $V_{ab} = -50$ V.

From the above illustration, we make the following general statement:

$$V_{ab} = -V_{ba}$$

for any two points a and b within a circuit.

Current through the circuit results in voltage drops across the resistors as shown in Figure 5–27. If we determine the voltage drops on all resistors and show the correct polarities, then we see that the following must also apply:

$$V_{bc} = +10 \text{ V} \qquad V_{cb} = -10 \text{ V}$$
$$V_{cd} = +25 \text{ V} \qquad V_{dc} = -25 \text{ V}$$
$$V_{da} = +15 \text{ V} \qquad V_{ad} = -15 \text{ V}$$

If we wish to determine the voltage between any other two points within the circuit, it is a simple matter of adding all the voltages between the two points, taking into account the polarities of the voltages. The voltage between points b and d would be determined as follows:

$$V_{bd} = V_{bc} + V_{cd} = 10 \text{ V} + 25 \text{ V} = +35 \text{ V}$$

Similarly, the voltage between points b and a could be determined by using the voltage drops of the resistors:

$$V_{ba} = V_{bc} + V_{cd} + V_{da} = 10 \text{ V} + 25 \text{ V} + 15 \text{ V} = +50 \text{ V}$$

Notice that the above result is precisely the same as when we determined V_{ba} using only the voltage source. This result indicates that the voltage between two points is not dependent upon the path taken.

For the circuit of Figure 5–28, find the voltages V_{ac}, V_{ad}, V_{cf}, and V_{eb}.

EXAMPLE 5–8

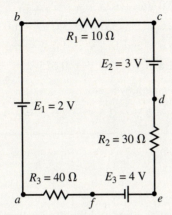

FIGURE 5–28

Solution   First, we determine that the equivalent supply voltage for the circuit is

$$E_T = 3\,V + 4\,V - 2\,V = 5.0\,V$$

with a polarity such that current will move in a counterclockwise direction within the circuit.

Next, we determine the voltages on all resistors by using the voltage divider rule and assigning polarities based on the direction of the current.

$$V_1 = \frac{R_1}{R_T}E_T$$

$$= \left(\frac{10\,\Omega}{10\,\Omega + 30\,\Omega + 40\,\Omega}\right)(5.0\,V) = 0.625\,V$$

$$V_2 = \frac{R_2}{R_T}E_T$$

$$= \left(\frac{30\,\Omega}{10\,\Omega + 30\,\Omega + 40\,\Omega}\right)(5.0\,V) = 1.875\,V$$

$$V_3 = \frac{R_3}{R_T}E_T$$

$$= \left(\frac{40\,\Omega}{10\,\Omega + 30\,\Omega + 40\,\Omega}\right)(5.0\,V) = 2.50\,V$$

The voltages appearing across the resistors are as shown in Figure 5–29.

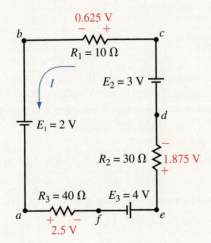

FIGURE 5–29

Finally, we solve for the voltages between the indicated points:

$$V_{ac} = -2.0\,\text{V} - 0.625\,\text{V} = -2.625\,\text{V}$$
$$V_{ad} = -2.0\,\text{V} - 0.625\,\text{V} + 3.0\,\text{V} = +0.375\,\text{V}$$
$$V_{cf} = +3.0\,\text{V} - 1.875\,\text{V} + 4.0\,\text{V} = +5.125\,\text{V}$$
$$V_{eb} = +1.875\,\text{V} - 3.0\,\text{V} + 0.625\,\text{V} = -0.500\,\text{V}$$

Or, selecting the opposite path, we get

$$V_{eb} = +4.0\,\text{V} - 2.5\,\text{V} - 2.0\,\text{V} = -0.500\,\text{V}$$

PRACTICAL NOTES . . .

Since most students initially find it difficult to determine the correct polarity for the voltage between two points, we present a simplified method to correctly determine the polarity and voltage between any two points within a circuit.

1. Determine the circuit current. Calculate the voltage drop across all components.

2. Polarize all resistors based upon the direction of the current. The terminal at which the current enters is assigned to be positive, while the terminal at which the current leaves is assigned to be negative.

3. In order to determine the voltage at point a with respect to point b, start at point b. Refer to Figure 5–30. Now, imagine that you walk around the circuit to point a.

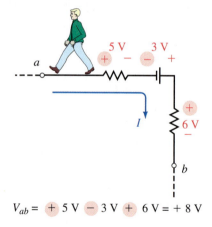

$$V_{ab} = \boxed{+}\ 5\,\text{V}\ \boxed{-}\ 3\,\text{V}\ \boxed{+}\ 6\,\text{V} = +\,8\,\text{V}$$

FIGURE 5–30

4. As you "walk" around the circuit, add the voltage drops and rises as you get to them. The assigned polarity of the voltage at any component (whether it is a source or a resistor) is positive if the voltage rises as you "walk" through the component and is negative if the voltage decreases as you pass through the component.

5. The resulting voltage, V_{ab}, is the numerical sum of all the voltages between a and b.

For Figure 5–30, the voltage V_{ab} is determined as

$$V_{ab} = 6\,\text{V} - 3\,\text{V} + 5\,\text{V} = 8\,\text{V}$$

Find the voltage V_{ab} in the circuit of Figure 5–31.

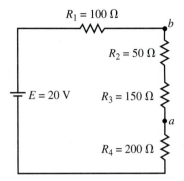

FIGURE 5–31

Answer
$V_{ab} = -8.00$ V

Single Subscripts

In a circuit which has a reference point (or ground point), most voltages will be expressed with respect to the reference point. In such a case it is no longer necessary to express a voltage using a dual subscript. Rather if we wish to express the voltage at point *a* with respect to ground, we simply refer to this as V_a. Similarly, the voltage at point *b* would be referred to as V_b. Therefore, any voltage that has only a single subscript is always referenced to the ground point of the circuit.

For the circuit of Figure 5–32, determine the voltages V_a, V_b, V_c, and V_d.

EXAMPLE 5–9

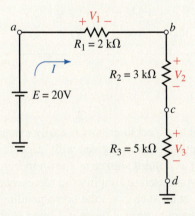

FIGURE 5–32

Solution Applying the voltage divider rule, we determine the voltage across each resistor as follows:

$$V_1 = \frac{2\ \text{k}\Omega}{2\ \text{k}\Omega + 3\ \text{k}\Omega + 5\ \text{k}\Omega}(20\ \text{V}) = 4.00\ \text{V}$$

$$V_2 = \frac{3\ \text{k}\Omega}{2\ \text{k}\Omega + 3\ \text{k}\Omega + 5\ \text{k}\Omega}(20\ \text{V}) = 6.00\ \text{V}$$

$$V_3 = \frac{5\ \text{k}\Omega}{2\ \text{k}\Omega + 3\ \text{k}\Omega + 5\ \text{k}\Omega}(20\ \text{V}) = 10.00\ \text{V}$$

Now we solve for the voltage at each of the points as follows:

$$V_a = 4\text{ V} + 6\text{ V} + 10\text{ V} = +20\text{ V} = E$$
$$V_b = 6\text{ V} + 10\text{ V} = +16.0\text{ V}$$
$$V_c = +10.0\text{ V}$$
$$V_d = 0\text{ V}$$

If the voltage at various points in a circuit is known with respect to ground, then the voltage between the points may be easily determined as follows:

$$V_{ab} = V_a - V_b \quad [\text{volts, V}] \tag{5-12}$$

EXAMPLE 5–10

For the circuit of Figure 5–33, determine the voltages V_{ab} and V_{cb} given that $V_a = +5$ V, $V_b = +3$ V, and $V_c = -8$ V.

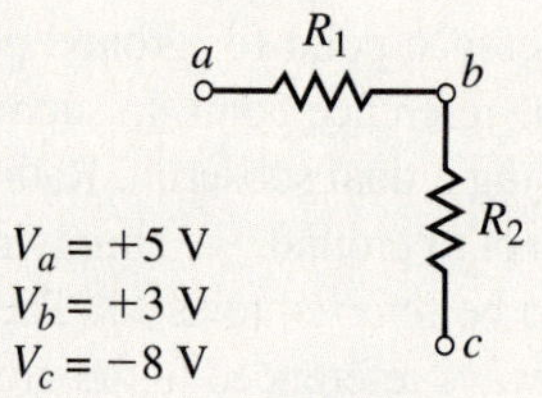

$$V_a = +5\text{ V}$$
$$V_b = +3\text{ V}$$
$$V_c = -8\text{ V}$$

FIGURE 5–33

Solution

$$V_{ab} = +5\text{ V} - (+3\text{ V}) = +2\text{ V}$$
$$V_{cb} = -8\text{ V} - (+3\text{ V}) = -11\text{ V}$$

Point Sources

The idea of voltages with respect to ground is easily extended to include voltage sources. When a voltage source is given with respect to ground, it may be simplified in the circuit as a **point source** as shown in Figure 5–34.

Point sources are often used to simplify the representation of circuits. We need to remember that in all such cases the corresponding points always represent voltages with respect to ground (even if ground is not shown).

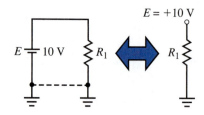

FIGURE 5–34

Determine the current and direction in the circuit of Figure 5–35.

EXAMPLE 5–11

$E_1 = +5 \text{ V}$ $E_2 = -8 \text{ V}$

$R_1 = 52 \text{ k}\Omega$

FIGURE 5–35

Solution   The circuit may be redrawn showing the reference point and converting the voltage point sources into the more common schematic representation. The resulting circuit is shown in Figure 5–36.

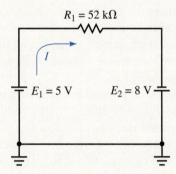

$R_1 = 52 \text{ k}\Omega$

I

$E_1 = 5 \text{ V}$ $E_2 = 8 \text{ V}$

FIGURE 5–36

Now, we easily calculate the current in the circuit as

$$I = \frac{E_T}{R_1} = \frac{5 \text{ V} + 8 \text{ V}}{52 \text{ k}\Omega} = 0.250 \text{ mA}$$

Voltage measurements are taken at three locations in a circuit. They are $V_a = +5.00$ V, $V_b = -2.50$ V, and $V_c = -5.00$ V. Determine the voltages V_{ab}, V_{ca}, and V_{bc}.

IN-PROCESS
LEARNING CHECK 7

(Answers are at the end of the chapter.)

5.9 Internal Resistance of Voltage Sources

So far we have worked only with ideal voltage sources, which maintain constant voltages regardless of the loads connected across the terminals. Consider a typical lead-acid automobile battery, which has a voltage of approximately 12 V. Similarly, four C-cell batteries, when connected in series, have a combined voltage of 12 V. Why then can we not use four C-cell batteries to operate the car? The answer, in part, is because the lead-acid battery has a much lower internal resistance than the low-energy C-cells. In practice, all voltage sources contain some internal resistance that will reduce the efficiency of the voltage source. We may symbolize any voltage source schematically as an ideal voltage source in series with an internal resistance. Figure 5–37 shows both an ideal voltage source and a practical or actual voltage source.

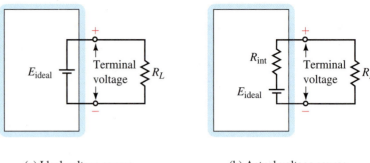

(a) Ideal voltage source   (b) Actual voltage source

FIGURE 5–37

The voltage that appears between the positive and negative terminals is called the **terminal voltage.** In an ideal voltage source, the terminal voltage will remain constant regardless of the load connected. An ideal voltage source will be able to provide as much current as the circuit demands. However, in a practical voltage source, the terminal voltage is dependent upon the value of the load connected across the voltage source. As expected, the practical voltage source sometimes is not able to provide as much current as the load demands. Rather the current in the circuit is limited by the combination of the internal resistance and the load resistance.

Under a no-load condition ($R_L = \infty\ \Omega$), there is no current in the circuit and so the terminal voltage will be equal to the voltage appearing across the ideal voltage source. If the output terminals are shorted together ($R_L = 0\ \Omega$), the current in the circuit will be a maximum and the terminal voltage will be equal to approximately zero. In such a situation, the voltage dropped across the internal resistance will be equal to the voltage of the ideal source.

The following example helps to illustrate the above principles.

EXAMPLE 5–12

Two batteries having an open-terminal voltage of 12 V are used to provide current to the starter of a car having a resistance of 0.10 Ω. If one battery has an internal resistance of 0.02 Ω and the second battery has an internal resistance of 100 Ω, calculate the current through the load and the resulting terminal voltage for each of the batteries.

Solution   The circuit for each of the batteries is shown in Figure 5–38.

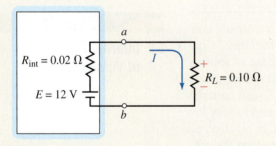

(a) Low internal resistance

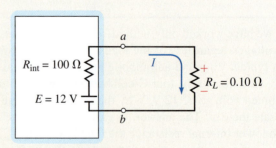

(b) High internal resistance

FIGURE 5–38

$R_{int} = 0.02\ \Omega$:

$$I = \frac{12\ V}{0.02\ \Omega + 0.10\ \Omega} = 100.\ A$$

$$V_{ab} = (100\ A)(0.10\ \Omega) = 10.0\ V$$

$R_{int} = 100\ \Omega$:

$$I = \frac{12\ V}{100\ \Omega + 0.10\ \Omega} = 0.120\ A$$

$$V_{ab} = (0.120\ A)(0.10\ \Omega) = 0.0120\ V$$

This simple example helps to illustrate why a 12-V automobile battery (which is actually 14.4 V) is able to start a car while eight 1.5-V flashlight batteries connected in series will have virtually no measurable effect when connected to the same circuit.

5.10 Ammeter Loading Effects

As you have already learned, ammeters are instruments which measure the current in a circuit. In order to use an ammeter, the circuit must be disconnected and the ammeter placed in series with the branch for which the current is to be determined. Since an ammeter uses the current in the circuit to provide a reading, it will affect the circuit under measurement. This effect is referred to as **meter loading.** All instruments, regardless of type, will load the circuit to some degree. The amount of loading is dependent upon both the instrument and the circuit being measured. For any meter, we define the loading effect as follows:

$$\text{loading effect} = \frac{\text{theoretical value} - \text{measured value}}{\text{theoretical value}} \times 100\% \quad \textbf{(5–13)}$$

For the series circuits of Figure 5–39, determine the current in each circuit. If an ammeter having an internal resistance of 250 Ω is used to measure the current in the circuits, determine the current through the ammeter and calculate the loading effect for each circuit.

EXAMPLE 5–13

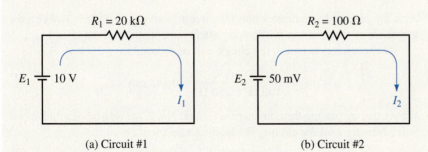

$R_1 = 20\ k\Omega$ $R_2 = 100\ \Omega$

$E_1 \quad 10\ V$ $E_2 \quad 50\ mV$

I_1 I_2

(a) Circuit #1   (b) Circuit #2

FIGURE 5–39

Solution   Circuit No. 1: The current in the circuit is

$$I_1 = \frac{10\ V}{20\ k\Omega} = 0.500\ mA$$

Now, by placing the ammeter into the circuit as shown in Figure 5–40(a) the resistance of the ammeter will slightly affect the operation of the circuit.

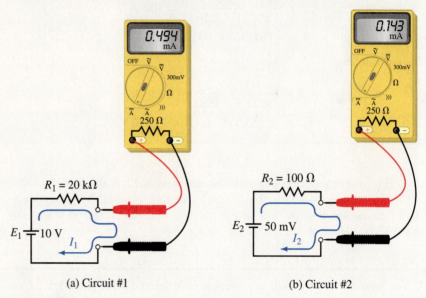

(a) Circuit #1   (b) Circuit #2

FIGURE 5–40

The resulting current in the circuit will be reduced to

$$I_1 = \frac{10 \text{ V}}{20 \text{ k}\Omega + 0.25 \text{ k}\Omega} = 0.494 \text{ mA}$$

Circuit No. 1: We see that by placing the ammeter into circuit No. 1, the resistance of the meter slightly affects the operation of the circuit. Applying Equation 5–16 gives the loading effect as

$$\text{loading effect} = \frac{0.500 \text{ mA} - 0.494 \text{ mA}}{0.500 \text{ mA}} \times 100\%$$

$$= 1.23\%$$

Circuit No. 2: The current in the circuit is also found as

$$I_2 = \frac{50 \text{ mV}}{100 \text{ }\Omega} = 0.500 \text{ mA}$$

Now, by placing the ammeter into the circuit as shown in Figure 5–40(b), the resistance of the ammeter will greatly affect the operation of the circuit.

The resulting current in the circuit will be reduced to

$$I_2 = \frac{50 \text{ mV}}{100 \text{ }\Omega + 250 \text{ }\Omega} = 0.143 \text{ mA}$$

We see that by placing the ammeter into circuit No. 2, the resistance of the meter will adversely load the circuit. The loading effect will be

$$\text{loading effect} = \frac{0.500 \text{ mA} - 0.143 \text{ mA}}{0.500 \text{ mA}} \times 100\%$$

$$= 71.4\%$$

The results of this example indicate that an ammeter, which usually has fairly low resistance, will not significantly load a circuit having a resistance of several thousand ohms. However, if the same meter is used to measure current in a circuit having low values of resistance, then the loading effect will be substantial.

We now examine how both MultiSIM and PSpice are used to determine the voltage and current in a series circuit. Although the methods are different, we will find that the results for both software packages are equivalent.

MultiSIM

The following example will build upon the skills that you learned in the previous chapter. Just as in the lab, you will measure voltage by connecting voltmeters across the component(s) under test. Current is measured by placing an ammeter in series with the component(s) through which you would like to find the current.

Use MultiSIM to solve for the circuit current and the voltage across each of the resistors in Figure 5–41.

5.11 Circuit Analysis Using Computers

◀ MULTISIM

◀ CADENCE

EXAMPLE 5–14

◀ MULTISIM

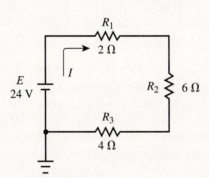

FIGURE 5–41

Solution   Open MultiSIM and construct the above circuit. If necessary, review the steps as outlined in the previous chapter. Remember that your circuit will need to have a circuit ground as found in the Sources parts bin. Once your circuit resembles the circuit shown in Figure 5–41, insert the ammeters and the voltmeters into the circuit as shown in Figure 5–42.

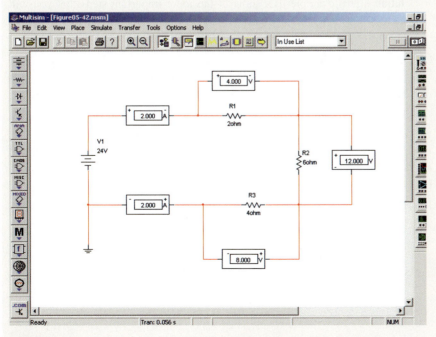

FIGURE 5–42

Notice that an extra ammeter is placed into the circuit. The only reason for this is to show that the current is the same everywhere in a series circuit.

Once all ammeters and voltmeters are inserted with the correct polarities, you may run the simulator by moving the toggle switch to the ON position. Your indicators should show the same readings as the values shown in Figure 5–42. If any of the values indicated by the meters are negative, you will need to disconnect the meter(s) and reverse the terminals by using the Ctrl R function.

Although this example is very simple, it illustrates some very important points that you will find useful when simulating circuit operation.

1. All voltmeters are connected across the components for which we are trying to measure the voltage drop.

2. All ammeters are connected in series with the components through which we are trying to find the current.

3. A ground symbol (or reference point) is required by all circuits that are to be simulated by MultiSIM.

PSpice

While PSpice has some differences compared to MultiSIM, we find that there are also many similarities. The following example shows how to use PSpice to analyze the previous circuit. In this example, you will use the Voltage Differential tool (one of three markers) to find the voltage across various components in a circuit. If necessary, refer to Appendix A to find the Voltage Differential tool. You may wish to experiment with other markers, namely the Voltage Level marker (which indicates voltage with respect to ground) and the Current Into Pin marker.

EXAMPLE 5–15

Use PSpice to solve for the circuit current and the voltage across each of the resistors in Figure 5–41.

Solution This example lists some of the more important steps that you will need to follow. For more detail, refer to Appendix A and the PSpice example in Chapter 4.

- Open the CIS Demo software.
- Once you are in the Capture session frame, click on the menu item File, select New, and then click on Project.
- In the New Project box, type **Ch 5 PSpice 1** in the Name text box. Ensure that the Analog or Mixed-Signal Circuit Wizard is activated.
- You will need to add libraries for your project. Select the breakout.olb, and eval.olb libraries. Click Finish.

- You should now be in the Capture schematic editor page. Click anywhere to activate it. Build the circuit as shown in Figure 5–43. Remember to rotate the components to provide for the correct node assignments. Change the component values as required.

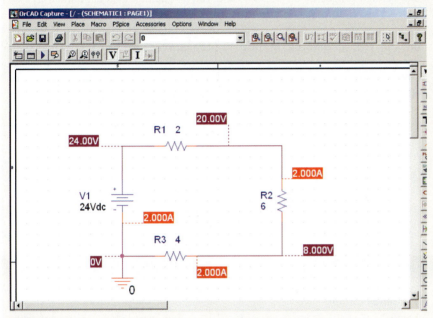

FIGURE 5–43

- Click the New Simulation Profile icon and enter a name (e.g., **Figure 5–43**) in the Name text box. You will need to enter the appropriate settings for this project in the Simulation Setting box. Click the Analysis tab, and select Bias Point from the Analysis type list. Click OK and save the document.

- Click on the run icon. You will observe the A/D Demo screen. When you close this screen, you will observe the bias voltages and currents. From these results, we have the following:

$$V_1 = 24 \text{ V} - 20 \text{ V} = 4 \text{ V}$$
$$V_2 = 20 \text{ V} - 8 \text{ V} = 12 \text{ V}$$
$$V_3 = 8 \text{ V}$$

For the supply voltage of 24 V, the current is 2.00 A. Clearly, these results are consistent with the theoretical calculations and the results obtained using Electronics Workbench.

- Save your project and exit from PSpice.

You are part of a research team in the electrical metering department of a chemical processing plant. As part of your work, you regularly measure voltages between 200 V and 600 V. The only voltmeter available to you today has voltage ranges of 20 V, 50 V, and 100 V. Clearly, you cannot safely use the voltmeter to measure the expected voltages. However, you recall from your electrical course that you can use a voltage divider network to predictably reduce voltages. In order to keep current levels to safe values, you decide to use resistors in the megohm range. Without changing any internal circuitry of the voltmeter, show how you can use large-value resistors to change the 100-V range of the voltmeter to effectively measure a maximum of 1000 V. (Naturally, you would take extra precautions when measuring these voltages.) Show the schematic of the design, including the location of your voltmeter.

PROBLEMS

5.1   Series Circuits

1. The voltmeters of Figure 5–44 have autopolarity. Determine the reading of each meter, giving the correct magnitude and sign.

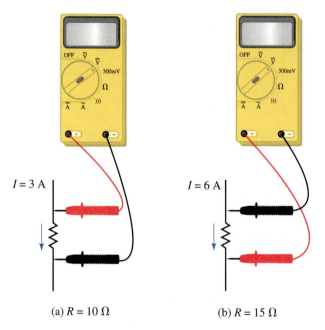

(a) $R = 10\ \Omega$ (b) $R = 15\ \Omega$

FIGURE 5–44

2. The voltmeters of Figure 5–45 have autopolarity. Determine the reading of each meter, giving the correct magnitude and sign.

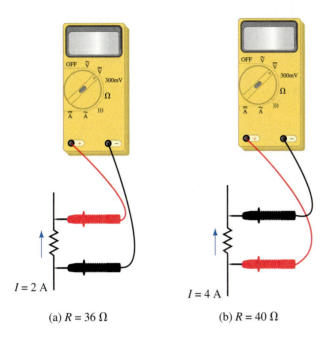

(a) $R = 36\ \Omega$ (b) $R = 40\ \Omega$

FIGURE 5–45

3. All resistors in Figure 5–46 are 15 Ω. For each case, determine the magnitude and polarity of voltage V.

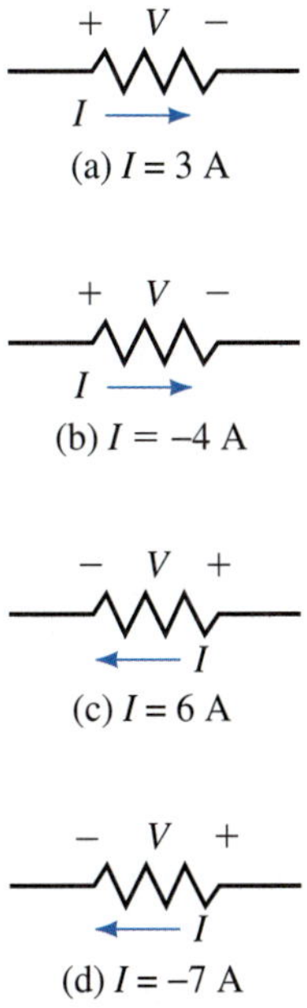

(a) $I = 3$ A

(b) $I = -4$ A

(c) $I = 6$ A

(d) $I = -7$ A

FIGURE 5–46   All resistors are 15 Ω.

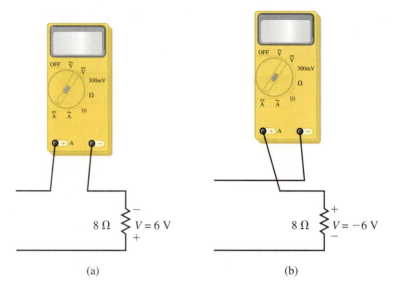

FIGURE 5–47

4. The ammeters of Figure 5–47 have autopolarity. Determine their readings, giving the correct magnitude and sign.

5.2 Kirchhoff's Voltage Law

5. Determine the unknown voltages in the networks of Figure 5–48.

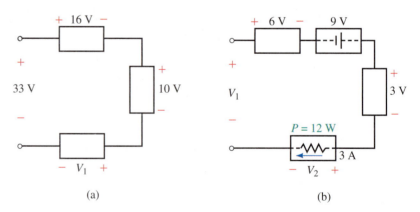

FIGURE 5–48

6. Determine the unknown voltages in the networks of Figure 5–49.
7. Solve for the unknown voltages in the circuit of Figure 5–50.
8. Solve for the unknown voltages in the circuit of Figure 5–51.

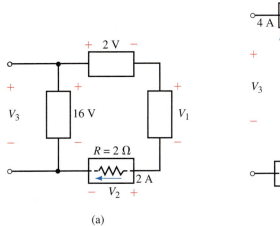

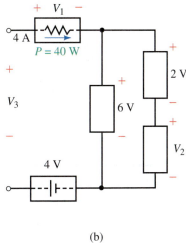

(a)   (b)

FIGURE 5–49

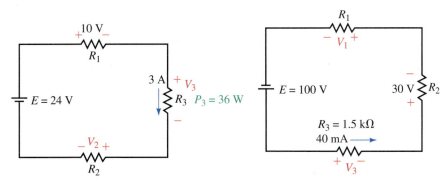

FIGURE 5–50   **FIGURE 5–51**

5.3 Resistors in Series

9. Determine the total resistance of the networks shown in Figure 5–52.

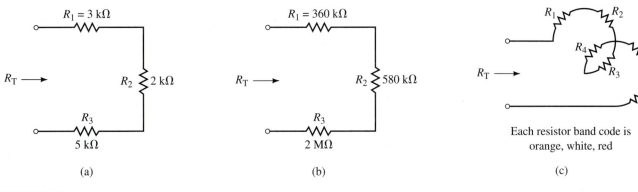

(a)   (b)   (c)

Each resistor band code is
orange, white, red

FIGURE 5–52

10. Determine the unknown resistance in each of the networks in Figure 5–53.

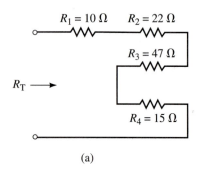

(a)

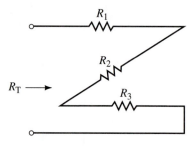

(b) Each resistor band code is
brown, red, orange

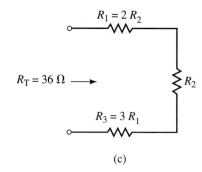

(c)

FIGURE 5–53

11. For the circuits shown in Figure 5–54, determine the total resistance, R_T, and the current, I.

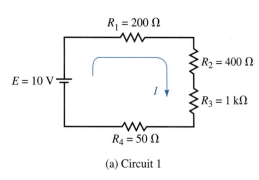

(a) Circuit 1

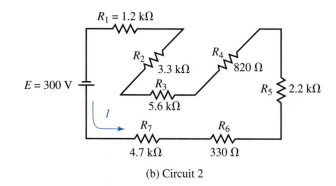

(b) Circuit 2

FIGURE 5–54

12. The circuits of Figure 5–55 have the total resistance, R_T, as shown. For each of the circuits find the following:

a. The magnitude of current in the circuit.

b. The total power delivered by the voltage source.

c. The direction of current through each resistor in the circuit.

d. The value of the unknown resistance, R.

e. The voltage drop across each resistor.

f. The power dissipated by each resistor. Verify that the summation of powers dissipated by the resistors is equal to the power delivered by the voltage source.

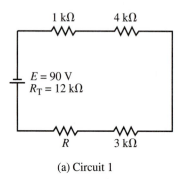

(a) Circuit 1

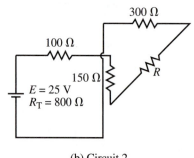

(b) Circuit 2

FIGURE 5–55

13. For the circuit of Figure 5–56, find the following quantities:

   a. The circuit current.

   b. The total resistance of the circuit.

   c. The value of the unknown resistance, R.

   d. The voltage drop across all resistors in the circuit.

   e. The power dissipated by all resistors.

14. The circuit of Figure 5–57 has a current of 2.5 mA. Find the following quantities:

   a. The total resistance of the circuit.

   b. The value of the unknown resistance, R_2.

   c. The voltage drop across each resistor in the circuit.

   d. The power dissipated by each resistor in the circuit.

15. For the circuit of Figure 5–58, find the following quantities:

   a. The current, I.

   b. The voltage drop across each resistor.

   c. The voltage across the open terminals a and b.

16. Refer to the circuit of Figure 5–59:

   a. Use Kirchhoff's voltage law to find the voltage drops across R_2 and R_3.

   b. Determine the magnitude of the current, I.

   c. Solve for the unknown resistance, R_1.

17. Repeat Problem 16 for the circuit of Figure 5–60.

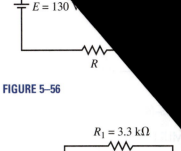

FIGURE 5–56

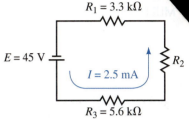

FIGURE 5–57

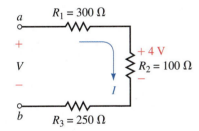

FIGURE 5–58

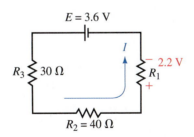

FIGURE 5–59

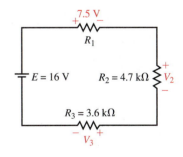

FIGURE 5–60

18. Refer to the circuit of Figure 5–61:

   a. Find R_T.

   b. Solve for the current, I.

   c. Determine the voltage drop across each resistor.

   d. Verify Kirchhoff's voltage law around the closed loop.

   e. Find the power dissipated by each resistor.

   f. Determine the minimum power rating of each resistor, if resistors are available with the following power ratings: ⅛ W, ¼ W, ½ W, 1 W, and 2 W.

   g. Show that the power delivered by the voltage source is equal to the summation of the powers dissipated by the resistors.

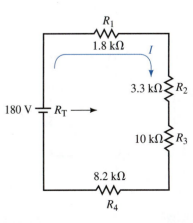

FIGURE 5–61

at Problem 18 for the circuit of Figure 5–62.

to the circuit of Figure 5–63.

lculate the voltage across each resistor.

termine the values of the resistors R_1 and R_2.

ve for the power dissipated by each of the resistors.

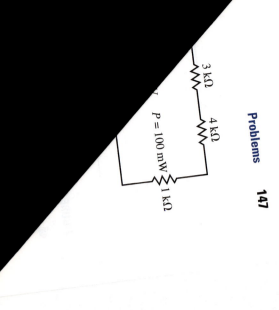

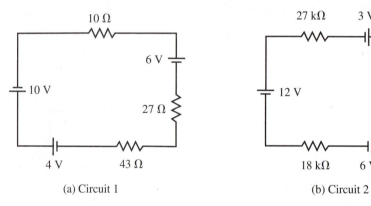

FIGURE 5–63

5.5 Interchanging Series Components

21. Redraw the circuits of Figure 5–64, showing a single voltage source for each circuit. Solve for the current in each circuit.

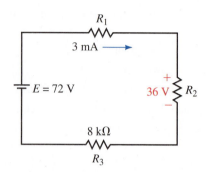

(a) Circuit 1   (b) Circuit 2

FIGURE 5–64

22. Use the information given to determine the polarity and magnitude of the unknown voltage source in each of the circuits of Figure 5–65.

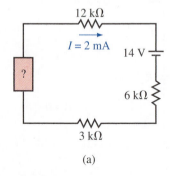

(a)

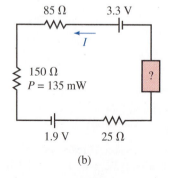

(b)

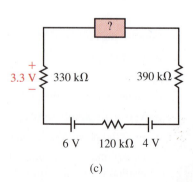

(c)

FIGURE 5–65

5.6   Voltage Divider Rule

23. Use the voltage divider rule to determine the voltage across each resistor in the circuits of Figure 5–66. Use your results to verify Kirchhoff's voltage law for each circuit.

24. Repeat Problem 23 for the circuits of Figure 5–67.

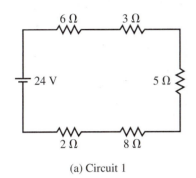

(a) Circuit 1

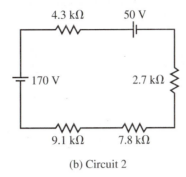

(b) Circuit 2

FIGURE 5–66

◀ MULTISIM

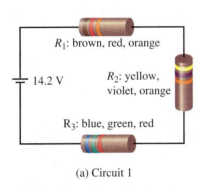

(a) Circuit 1

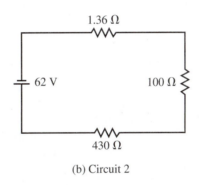

(b) Circuit 2

FIGURE 5–67

25. Refer to the circuits of Figure 5–68:
    a. Find the values of the unknown resistors.
    b. Calculate the voltage across each resistor.
    c. Determine the power dissipated by each resistor.

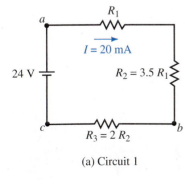

(a) Circuit 1

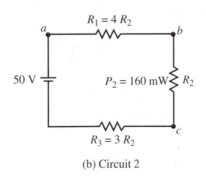

(b) Circuit 2

FIGURE 5–68

26. Refer to the circuits of Figure 5–69:

    a. Find the values of the unknown resistors using the voltage divider rule.

    b. Calculate the voltage across R_1 and R_3.

    c. Determine the power dissipated by each resistor.

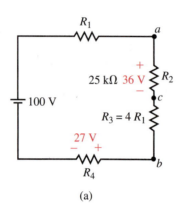
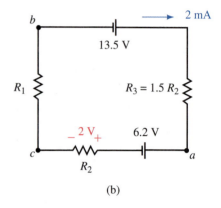

(a)   (b)

FIGURE 5–69

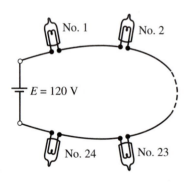

$R = 25\ \Omega$/light bulb

FIGURE 5–70

27. A string of 24 series light bulbs is connnected to a 120-V supply as shown in Figure 5–70.

    a. Solve for the current in the circuit.

    b. Use the voltage divider rule to find the voltage across each light bulb.

    c. Calculate the power dissipated by each bulb.

    d. If a single light bulb were to become an open circuit, the entire string would stop working. To prevent this from occurring, each light bulb has a small metal strip which shorts the light bulb when the filament fails. If two bulbs in the string were to burn out, repeat Steps (a) through (c).

    e. Based on your calculations of Step (d), what do you think would happen to the life expectancy of the remaining light bulbs if the two faulty bulbs were not replaced?

28. Repeat Problem 27 for a string consisting of 36 light bulbs.

5.8 Voltage Subscripts

29. Solve for the voltages V_{ab} and V_{bc} in the circuits of Figure 5–68.

30. Repeat Problem 29 for the circuits of Figure 5–69.

31. For the circuits of Figure 5–71, determine the voltage across each resistor and calculate the voltage V_a.

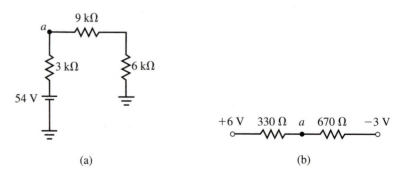

(a)   (b)

FIGURE 5–71

32. Given the circuits of Figure 5–72:

    a. Determine the voltage across each resistor.

    b. Find the magnitude and direction of the current in the 180-kΩ resistor.

    c. Solve for the voltage V_a.

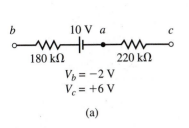

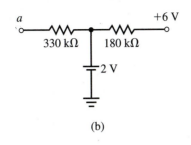

FIGURE 5–72

(a)   (b)

5.9 Internal Resistance of Voltage Sources

33. A battery is measured to have an open-terminal voltage of 14.2 V. When this voltage is connected to a 100-Ω load, the voltage measured between the terminals of the battery drops to 6.8 V.

    a. Determine the internal resistance of the battery.

    b. If the 100-Ω load were replaced with a 200-Ω load, what voltage would be measured across the terminals of the battery?

34. The voltage source shown in Figure 5–73 is measured to have an open-circuit voltage of 24 V. When a 10-Ω load is connected across the terminals, the voltage measured with a voltmeter drops to 22.8 V.

    a. Determine the internal resistance of the voltage source.

    b. If the source had only half the resistance determined in (a), what voltage would be measured across the terminals with the 10-Ω resistor connected?

FIGURE 5–73

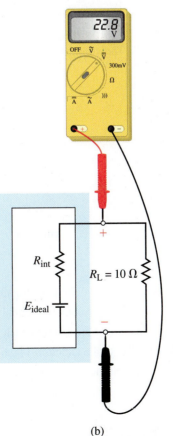

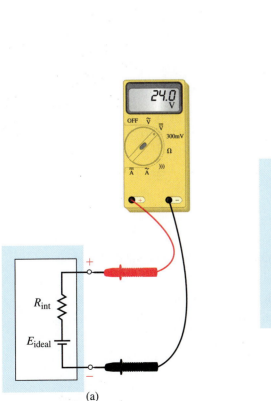

(a)   (b)

5.10 Ammeter Loading Effects

35. For the series circuits of Figure 5–74, determine the current in each circuit. If an ammeter having an internal resistance of 50 Ω is used to measure the current in the circuits, determine the current through the ammeter and calculate the loading effect for each circuit.

36. Repeat Problem 39 if the ammeter has a resistance of 10 Ω.

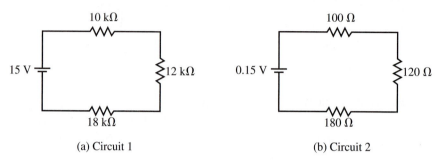

(a) Circuit 1   (b) Circuit 2

FIGURE 5–74

5.11 Circuit Analysis Using Computers

◀ MULTISIM

37. Refer to the circuits of Figure 5–66. Use MultiSIM to find the following:
    a. The current in each circuit.
    b. The voltage across each resistor in the circuit.

◀ MULTISIM

38. Given the circuit of Figure 5–75, use MultiSIM to determine the following:
    a. The current through the voltage source, *I.*
    b. The voltage across each resistor.
    c. The voltage between terminals *a* and *b.*
    d. The voltage, with respect to ground, at terminal *c.*

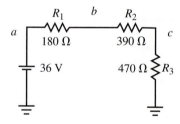

◀ MULTISIM

FIGURE 5–75

◀ CADENCE

39. Refer to the circuit of Figure 5–62. Use PSpice to find the following:
    a. The current in the circuit.
    b. The voltage across each resistor in the circuit.

◀ CADENCE

40. Refer to the circuit of Figure 5–61. Use PSpice to find the following:
    a. The current in the circuit.
    b. The voltage across each resistor in the circuit.

ANSWERS TO IN-PROCESS LEARNING CHECKS

In-Process Learning Check 1

1. Two elements are connected at only one node.
2. No current-carrying element is connected to the common node.

In-Process Learning Check 2

The summation of voltage drops and rises around any closed loop is equal to zero; or the summation of voltage rises is equal to the summation of voltage drops around a closed loop.

In-Process Learning Check 3

$R_1 = 4.2 \text{ k}\Omega$
$R_2 = 12.6 \text{ k}\Omega$
$R_3 = 25.2 \text{ k}\Omega$

In-Process Learning Check 4

$E_{CELL} = 2.18 \text{ V}$

In-Process Learning Check 5

$R_1 = 18.8 \text{ k}\Omega$
$R_2 = 28.21 \text{ k}\Omega$

In-Process Learning Check 6

The chassis of the oven is grounded when it is connected to the electrical outlet.

In-Process Learning Check 7

$V_{ab} = 7.50 \text{ V}$
$V_{ca} = -10.0 \text{ V}$
$V_{bc} = 2.5 \text{ V}$

■ **OBJECTIVES**

After studying this chapter you will be able to

- recognize which elements and branches in a given circuit are connected in parallel and which are connected in series,

- calculate the total resistance and conductance of a network of parallel resistances,

- determine the current in any resistor in a parallel circuit,

- solve for the voltage across any parallel combinations of resistors,

- apply Kirchhoff's current law to solve for unknown currents in a circuit,

- explain why voltage sources of different magnitudes must never be connected in parallel,

- use the current divider rule to solve for the current through any resistor of a parallel combination,

- identify and calculate the loading effects of a voltmeter connected into a circuit,

- use MultiSIM to observe loading effects of a voltmeter,

- use PSpice to evaluate voltage and current in a parallel circuit.

Parallel Circuits

6

Two fundamental circuits form the basis of all electrical circuits. They are the series circuit and the parallel circuit. The previous chapter examined the principles and rules which applies to series circuits. In this chapter we study the **parallel** (or **shunt**) circuit and examine the rules governing the operation of these circuits.

Figure 6–1 illustrates a simple example of several light bulbs connected in parallel with one another and a battery supplying voltage to the bulbs.

This illustration shows one of the important differences between the series circuit and the parallel circuit. The parallel circuit will continue to operate even though one of the light bulbs may have a defective (open) filament. Only the defective light bulb will no longer glow. If a circuit were made up of several light bulbs in series, however, the defective light bulb would prevent any current in the circuit, and so all the light bulbs would be off. ■

CHAPTER PREVIEW

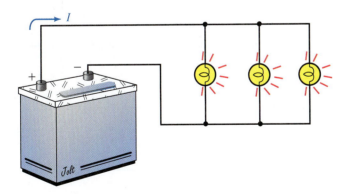

FIGURE 6–1 Simple parallel circuit.

PUTTING IT IN PERSPECTIVE

Luigi Galvani and the Discovery of Nerve Excitation

LUIGI GALVANI WAS BORN IN BOLOGNA, Italy, on September 9, 1737.

Galvani's main expertise was in anatomy, a subject in which he was appointed lecturer at the university in Bologna.

Galvani discovered that when the nerves of frogs were connected to sources of electricity, the muscles twitched. Although he was unable to determine where the electrical pulses originated within the animal, Galvani's work was significant and helped to open further discoveries in nerve impulses.

Galvani's name has been adopted for the instrument called the **galvanometer,** which is used for detecting very small currents.

Luigi Galvani died in Bologna on December 4, 1798. Although he made many contributions to science, Galvani died poor, shrouded in controversy due to his refusal to swear allegiance to Napolean. ∎

6.1 Parallel Circuits

The illustration of Figure 6–1 shows that one terminal of each light bulb is connected to the positive terminal of the battery and that the other terminal of the light bulb is connected to the negative terminal of the battery. These points of connection are often referred to as **nodes.**

Elements or branches are said to be in a parallel connection when they have exactly two nodes in common.

Figure 6–2 shows several different ways of sketching parallel elements. The elements between the nodes may be any two-terminal devices such as voltage sources, resistors, light bulbs, and the like.

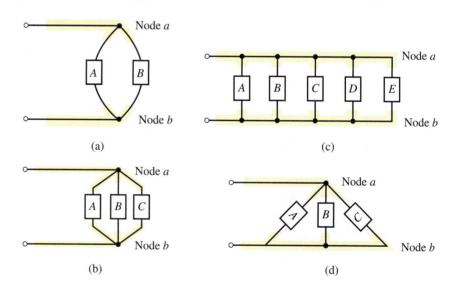

(a)   (c)

(b)   (d)

FIGURE 6–2   Parallel elements.

In the illustrations of Figure 6–2, notice that every element has two terminals and that each of the terminals is connected to one of the two nodes.

Very often, circuits contain a combination of series and parallel components. Although we will study these circuits in greater depth in later chapters, it is important at this point to be able to recognize the various connections in a given network. Consider the networks shown in Figure 6–3.

When analyzing a particular circuit, it is usually easiest to first designate the nodes (we will use lowercase letters) and then to identify the types of connections. Figure 6–4 shows the nodes for the networks of Figure 6–3.

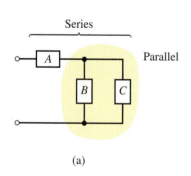

(a)

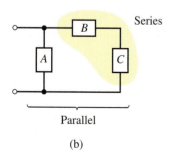

(b)

FIGURE 6–3   Series-parallel combinations.

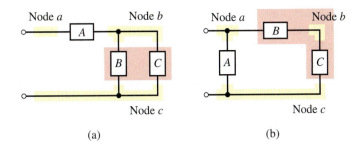

FIGURE 6–4   (a)   (b)

In the circuit of Figure 6–4(a), we see that element B is in parallel with element C since they each have nodes b and c in common. This parallel combination is now seen to be in series with element A.

In the circuit of Figure 6–4(b), element B is in series with element C since these elements have a single common node: node b. The branch consisting of the series combination of elements B and C is then determined to be in parallel with element A.

Recall that Kirchhoff's voltage law was extremely useful in understanding the operation of the series circuit. In a similar manner, Kirchhoff's current law is the underlying principle that is used to explain the operation of a parallel circuit. Kirchhoff's current law states the following:

The summation of currents entering a node is equal to the summation of currents leaving the node.

An analogy that helps us understand the principle of Kirchhoff's current law is the flow of water. When water flows in a closed pipe, the amount of water entering a particular point in the pipe is exactly equal to the amount of water leaving, since there is no loss. In mathematical form, Kirchhoff's current law is stated as follows:

$$\Sigma\, I_{\text{entering node}} = \Sigma\, I_{\text{leaving node}} \qquad \textbf{(6–1)}$$

Figure 6–5 is an illustration of Kirchhoff's current law. Here we see that the node has two currents entering, $I_1 = 5$ A and $I_5 = 3$ A, and three currents leaving, $I_2 = 2$ A, $I_3 = 4$ A, and $I_4 = 2$ A. Now we can see that Equation 6–1 applies in the illustration, namely.

$$\Sigma\, I_{\text{in}} = \Sigma\, I_{\text{out}}$$
$$5\,\text{A} + 3\,\text{A} = 2\,\text{A} + 4\,\text{A} + 8\,\text{A}$$
$$8\,\text{A} = 8\,\text{A}\ (\text{checks!})$$

6.2 Kirchhoff's Current Law

◀ **Online Companion**

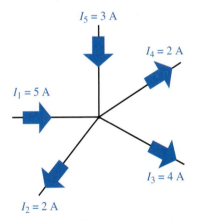

FIGURE 6–5   Kirchhoff's current law.

Verify that Kirchhoff's current law applies at the node shown in Figure 6–6.

PRACTICE PROBLEMS 1

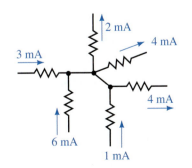

FIGURE 6–6

Answer
$3\,\text{mA} + 6\,\text{mA} + 1\,\text{mA} = 2\,\text{mA} + 4\,\text{mA} + 4\,\text{mA}$

Quite often, when we analyze a given circuit, we are unsure of the direction of current through a particular element within the circuit. In such cases, we assume a reference direction and base further calculations on this assumption. If our assumption is incorrect, calculations will show that the current has a negative sign. The negative sign simply indicates that the current is in fact opposite to the direction selected as the reference. The following example illustrates this very important concept.

EXAMPLE 6–1

Determine the magnitude and correct direction of the currents I_3 and I_5 for the network of Figure 6–7.

Solution   Although points a and b are in fact the same node, we treat the points as two separate nodes with 0 Ω resistance between them.

Since Kirchhoff's current law must be valid at point a, we have the following expression for this node:

$$I_1 = I_2 + I_3$$

and so

$$I_3 = I_1 - I_2$$
$$= 2\,A - 3\,A = -1\,A$$

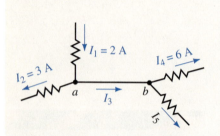

FIGURE 6–7

Notice that the reference direction of current I_3 was taken to be from a to b, while the negative sign indicates that the current is in fact from b to a.

Similarly, using Kirchhoff's current law at point b gives

$$I_3 = I_4 + I_5$$

which gives current I_5 as

$$I_5 = I_3 - I_4$$
$$= -1\,A - 6\,A = -7\,A$$

The negative sign indicates that the current I_5 is actually towards node b rather than away from the node. The actual directions and magnitudes of the currents are illustrated in Figure 6–8.

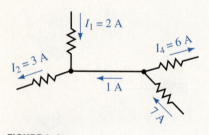

FIGURE 6–8

EXAMPLE 6–2

Find the magnitudes of the unknown currents for the circuit of Figure 6–9.

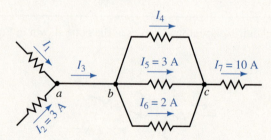

FIGURE 6–9

Solution   If we consider point a, we see that there are two unknown currents, I_1 and I_3. Since there is no way to solve for these values, we examine the currents at point b, where we again have two unknown currents, I_3 and I_4. Finally we observe that at point c there is only one unknown, I_4. Using Kirchhoff's current law we solve for the unknown current as follows:

$$I_4 + 3\,\text{A} + 2\,\text{A} = 10\,\text{A}$$

Therefore,

$$I_4 = 10\,\text{A} - 3\,\text{A} - 2\,\text{A} = 5\,\text{A}$$

Now we can see that at point b the current entering is

$$I_3 = 5\,\text{A} + 3\,\text{A} + 2\,\text{A} = 10\,\text{A}$$

And finally, by applying Kirchhoff's current law at point a, we determine that the current I_1 is

$$I_1 = 10\,\text{A} - 3\,\text{A} = 7\,\text{A}$$

Determine the unknown currents in the network of Figure 6–10.

EXAMPLE 6–3

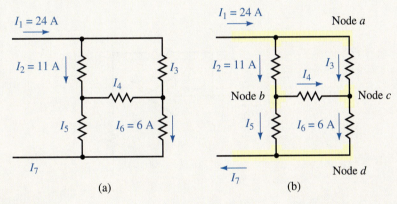

FIGURE 6–10

Solution   We first assume reference directions for the unknown currents in the network.

Since we may use the analogy of water moving through conduits, we can easily assign directions for the currents I_3, I_5, and I_7. However, the direction for the current I_4 is not as easily determined, so we arbitrarily assume that its direction is to the right. Figure 6–10(b) shows the various nodes and the assumed current directions.

By examining the network, we see that there is only a single source of current $I_1 = 24$ A. Using the analogy of water pipes, we conclude that the current leaving the network is $I_7 = I_1 = 24$ A.

Now, applying Kirchhoff's current law to node a, we calculate the current I_3 as follows:

$$I_1 = I_2 + I_3$$

Therefore,

$$I_3 = I_1 - I_2 = 24\,\text{A} - 11\,\text{A} = 13\,\text{A}$$

Similarly, at node c, we have

$$I_3 + I_4 = I_6$$

Therefore,

$$I_4 = I_6 - I_3 = 6\,\text{A} - 13\,\text{A} = -7\,\text{A}$$

Although the current I_4 is opposite to the assumed reference direction, we do not change its direction for further calculations. We use the original direction together with the negative sign; otherwise the calculations would be needlessly complicated.

Applying Kirchhoff's current law at node b, we get

$$I_2 = I_4 + I_5$$

which gives

$$I_5 = I_2 - I_4 = 11\ \text{A} - (-7\ \text{A}) = 18\ \text{A}$$

Finally, applying Kirchhoff's current law at node d gives

$$I_5 + I_6 = I_7$$

resulting in

$$I_7 = I_5 + I_6 = 18\ \text{A} + 6\ \text{A} = 24\ \text{A}$$

PRACTICE PROBLEMS 2

Determine the unknown currents in the network of Figure 6–11.

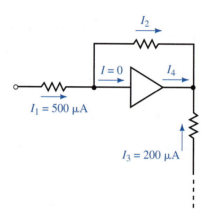

FIGURE 6–11

Answers
$I_2 = 500\ \mu\text{A},\ I_4 = -700\ \mu\text{A}$

6.3 Resistors in Parallel

A simple parallel circuit is constructed by combining a voltage source with several resistors as shown in Figure 6–12.

The voltage source will result in current from the positive terminal of the source toward node a. At this point the current will split between the various resistors and then recombine at node b before continuing to the negative terminal of the voltage source.

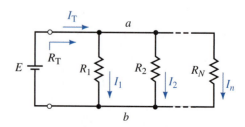

FIGURE 6–12

This circuit illustrates a very important concept of parallel circuits. If we were to apply Kirchhoff's voltage law around each closed loop in the parallel circuit of Figure 6–12, we would find that the voltage across all parallel resistors is exactly equal, namely $V_{R_1} = V_{R_2} = V_{R_3} = E$. Therefore, by applying Kirchhoff's voltage law, we make the following statement:

The voltage across all parallel elements in a circuit will be the same.

The above principle allows us to determine the equivalent resistance, R_T, of any number of resistors connected in parallel. The equivalent resistance, R_T, is the effective resistance "seen" by the source and determines the total current, I_T, provided to the circuit. Applying Kirchhoff's current law to the circuit of Figure 6–11, we have the following expression:

$$I_T = I_1 + I_2 + \cdots + I_n$$

However, since Kirchhoff's voltage law also applies to the parallel circuit, the voltage across each resistor must be equal to the supply voltage, E. The total current in the circuit, which is determined by the supply voltage and the equivalent resistance, may now be written as

$$\frac{E}{R_T} = \frac{E}{R_1} + \frac{E}{R_2} + \cdots + \frac{E}{R_n}$$

Simplifying the above expression gives us the general expression for total resistance of a parallel circuit as

$$\frac{1}{R_T} = \frac{1}{R_1} + \frac{1}{R_2} + \cdots + \frac{1}{R_n} \quad \text{(siemens, S)} \qquad \text{(6–2)}$$

Since conductance was defined as the reciprocal of resistance, we may write the above equation in terms of conductance, namely,

$$G_T = G_1 + G_2 + \cdots + G_n \quad \text{(S)} \qquad \text{(6–3)}$$

Whereas series resistors had a total resistance determined by the summation of the particular resistances, we see that any number of parallel resistors have a total conductance determined by the summation of the individual conductances.

The equivalent resistance of n parallel resistors may be determined in one step as follows:

$$R_T = \frac{1}{\frac{1}{R_1} + \frac{1}{R_2} + \cdots + \frac{1}{R_n}} \quad (\Omega) \qquad \text{(6–4)}$$

An important effect of combining parallel resistors is that the resultant resistance will always be smaller than the smallest resistor in the combination.

EXAMPLE 6–4

Solve for the total conductance and total equivalent resistance of the circuit shown in Figure 6–13.

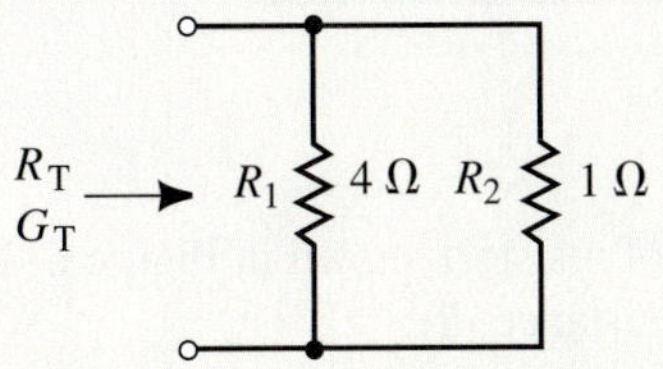

FIGURE 6–13

Solution   The total conductance is

$$G_T = G_1 + G_2 = \frac{1}{4\ \Omega} + \frac{1}{1\ \Omega} = 1.25\ \text{S}$$

The total equivalent resistance of the circuit is

$$R_T = \frac{1}{G_T} = \frac{1}{1.25\ \text{S}} = 0.800\ \Omega$$

Notice that the equivalent resistance of the parallel resistors is indeed less than the value of each resistor.

EXAMPLE 6–5

Determine the conductance and resistance of the network of Figure 6–14.

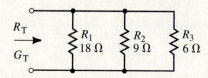

FIGURE 6–14

Solution   The total conductance is

$$G_T = G_1 + G_2 + G_3$$

$$= \frac{1}{18\ \Omega} + \frac{1}{9\ \Omega} + \frac{1}{6\ \Omega}$$

$$= 0.0\overline{5}\ \text{S} + 0.1\overline{1}\ \text{S} + 0.1\overline{6}\ \text{S}$$

$$= 0.3\overline{3}\ \text{S}$$

where the overbar indicates that the number under it is repeated infinitely to the right.

The total resistance is

$$R_T = \frac{1}{0.3\overline{3}\ \text{S}} = 3.00\ \Omega$$

PRACTICE PROBLEMS 3

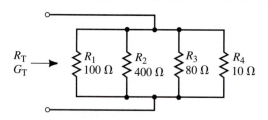

FIGURE 6–15

For the parallel network of resistors shown in Figure 6–15, find the total conductance, G_T and the total resistance, R_T.

Answers
$G_T = 0.125\ \text{S},\ R_T = 8.00\ \Omega$

n Equal Resistors in Parallel

If we have *n* equal resistors in parallel, each resistor, *R*, has the same conductance, *G*. By applying Equation 6–3, the total conductance is found:

$$G_T = nG$$

The total resistance is now easily determined as

$$R_T = \frac{1}{G_T} = \frac{1}{nG} = \frac{R}{n}$$ **(6–5)**

For the networks of Figure 6–16, calculate the total resistance.   **EXAMPLE 6–6**

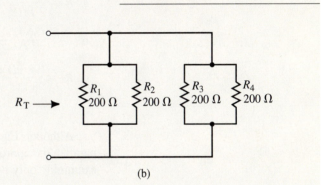

(a)   (b)

FIGURE 6–16

Solution

a. $R_T = \dfrac{18\ k\Omega}{3} = 6\ k\Omega$

b. $R_T = \dfrac{200\ \Omega}{4} = 50\ \Omega$

Two Resistors in Parallel

Very often circuits have only two resistors in parallel. In such a case, the total resistance of the combination may be determined without the necessity of determining the conductance.

For two resistors, Equation 6–4 is written

$$R_T = \frac{1}{\dfrac{1}{R_1} + \dfrac{1}{R_2}}$$

By cross multiplying the terms in the denominator, the expression becomes

$$R_T = \frac{1}{\dfrac{R_1 + R_2}{R_1 R_2}}$$

Thus, for two resistors in parallel we have the following expression:

$$R_T = \frac{R_1 R_2}{R_1 + R_2}$$ **(6–6)**

For two resistors connected in parallel, the equivalent resistance is found by the product of the two values divided by the sum.

EXAMPLE 6–7

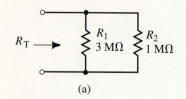

Determine the total resistance of the resistor combinations of Figure 6–17.

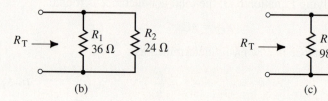

(a)   (b)   (c)

FIGURE 6–17

Solution

a. $R_T = \dfrac{(3\ M\Omega)(1\ M\Omega)}{3\ M\Omega + 1\ M\Omega} = 0.75\ M\Omega = 750\ k\Omega$

b. $R_T = \dfrac{(36\ \Omega)(24\ \Omega)}{36\ \Omega + 24\ \Omega} = 14.4\ \Omega$

c. $R_T = \dfrac{(98\ k\Omega)(2\ k\Omega)}{98\ k\Omega + 2\ k\Omega} = 1.96\ k\Omega$

Although Equation 6–6 is intended primarily to solve for two resistors in parallel, the approach may also be used to solve for any number of resistors by examining only two resistors at a time.

EXAMPLE 6–8

Calculate the total resistance of the resistor combination of Figure 6–18.

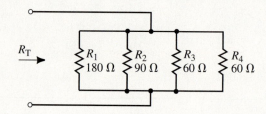

FIGURE 6–18

Solution   By grouping the resistors into combinations of two, the circuit may be simplified as shown in Figure 6–19.

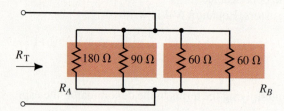

FIGURE 6–19

The equivalent resistance of each of the indicated combinations is determined as follows:

$$R_A = \frac{(180\ \Omega)(90\ \Omega)}{180\ \Omega + 90\ \Omega} = 60\ \Omega$$

$$R_B = \frac{(60\ \Omega)(60\ \Omega)}{60\ \Omega + 60\ \Omega} = 30\ \Omega$$

The circuit can be further simplified as a combination of two resistors shown in Figure 6–20.

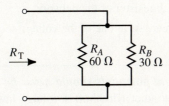

FIGURE 6–20

The resultant equivalent resistance is

$$R_T = \frac{(60\ \Omega)(30\ \Omega)}{60\ \Omega + 30\ \Omega} = 20\ \Omega$$

Three Resistors in Parallel

Using an approach similar to the derivation of Equation 6–6, we may arrive at an equation that solves for three resistors in parallel. Indeed, it is possible to write a general equation to solve for four resistors, five resistors, etc. Although such an equation is certainly useful, students are discouraged from memorizing such lengthy expressions. You will generally find that it is much more efficient to remember the principles upon which the equation is constructed. Consequently, the derivation of Equation 6–7 is left up to the student.

$$R_T = \frac{R_1 R_2 R_3}{R_1 R_2 + R_1 R_3 + R_2 R_3} \qquad \textbf{(6–7)}$$

PRACTICE PROBLEMS 4

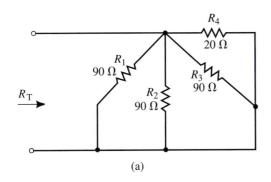

(a)

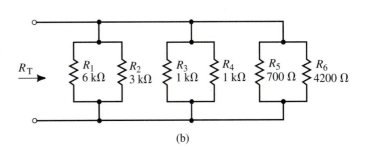

(b)

FIGURE 6–21

Find the total equivalent resistance for each network in Figure 6–21.

Answers
a. 12 Ω; b. 240 Ω

IN-PROCESS
LEARNING CHECK 1

(Answers are at the end of the chapter.)

If the circuit of Figure 6–21(a) is connected to a 24-V voltage source, determine the following quantities:

a. The total current provided by the voltage source.

b. The current through each resistor of the network.

c. Verify Kirchhoff's current law at one of the voltage source terminals.

6.4 Voltage Sources in Parallel

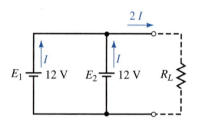

FIGURE 6–22 Voltage sources in parallel.

Voltage sources of different potentials should never be connected in parallel, since to do so would contradict Kirchhoff's voltage law. However, when two equal potential sources are connected in parallel, each source will deliver half the required circuit current. For this reason automobile batteries are sometimes connected in parallel to assist in starting a car with a "weak" battery. Figure 6–22 illustrates this principle.

Figure 6–23 shows that if voltage sources of two different potentials are placed in parallel, Kirchhoff's voltage law would be violated around the closed loop. In practice, if voltage sources of different potentials are placed in parallel, the resulting closed loop can have a very large current. The current will occur even though there may not be a load connected across the sources. Example 6–9 illustrates the large currents that can occur when two parallel batteries of different potential are connected.

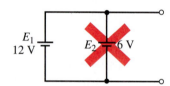

FIGURE 6–23   Voltage sources of different voltages must never be placed in parallel.

EXAMPLE 6–9

A 12-V battery and a 6-V battery (each having an internal resistance of 0.05 Ω) are inadvertently placed in parallel as shown in Figure 6–24. Determine the current through the batteries.

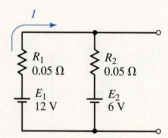

FIGURE 6–24

Solution   From Ohm's law,

$$I = \frac{E_T}{R_T} = \frac{12\,V - 6\,V}{0.05\,\Omega + 0.05\,\Omega} = 60\,A$$

This example illustrates why batteries of different potential must never be connected in parallel. Tremendous currents will occur within the sources resulting in the possibility of a fire or explosion.

When we examined series circuits we determined that the current in the series circuit was the same everywhere in the circuit, whereas the voltages across the series elements were typically different. The voltage divider rule (VDR) was used to determine the voltage across all resistors within a series network.

In parallel networks, the voltage across all parallel elements is the same. However, the currents through the various elements are typically different. The current divider rule (CDR) is used to determine how current entering a node is split between the various parallel resistors connected to the node.

Consider the network of parallel resistors shown in Figure 6–25.

6.5 Current Divider Rule

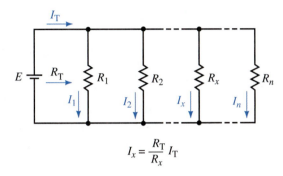

$$I_x = \frac{R_T}{R_x} I_T$$

FIGURE 6–25   Current divider rule.

If this network of resistors is supplied by a voltage source, the total current in the circuit is

$$I_T = \frac{E}{R_T} \qquad (6\text{--}8)$$

Since each of the n parallel resistors has the same voltage, E, across its terminals, the current through any resistor in the network is given as

$$I_x = \frac{E}{R_x} \qquad (6\text{--}9)$$

By rewriting Equation 6–8 as $E = I_T R_T$ and then substituting this into Equation 6–9, we obtain the current divider rule as follows:

$$I_x = \frac{R_T}{R_x} I_T \qquad (6\text{--}10)$$

An alternate way of writing the current divider rule is to express it in terms of conductance. Equation 6–10 may be modified as follows:

$$I_x = \frac{G_x}{G_T} I_T \qquad (6\text{--}11)$$

The current divider rule allows us to calculate the current in any resistor of a parallel network if we know the total current entering the network. Notice the similarity between the voltage divider rule (for series components) and the current divider rule (for parallel components). The main difference is that the current divider rule of Equation 6–11 uses circuit conductance rather than resistance. While this equation is useful, it is generally easier to use resistance to calculate current.

If the network consists of only two parallel resistors, then the current through each resistor may be found in a slightly different way. Recall that for two resistors in parallel, the total parallel resistance is given as

$$R_T = \frac{R_1 R_2}{R_1 + R_2}$$

Now, by substituting this expression for total resistance into Equation 6–10, we obtain

$$I_1 = \frac{I_T R_T}{R_1}$$

$$= \frac{I_T\left(\dfrac{R_1 R_2}{R_1 + R_2}\right)}{R_1}$$

which simplifies to

$$I_1 = \frac{R_2}{R_1 + R_2} I_T \qquad (6\text{--}12)$$

Similarly,

$$I_2 = \frac{R_1}{R_1 + R_2} I_T \qquad (6\text{--}13)$$

Several other important characteristics of parallel networks become evident.

If current enters a parallel network consisting of any number of equal resistors, then the current entering the network will split equally between all of the resistors.

If current enters a parallel network consisting of several values of resistance, then the smallest value of resistor in the network will have the largest amount of current. Inversely, the largest value of resistance will have the smallest amount of current.

This characteristic may be simplified by saying that *most of the current will follow the path of least resistance.*

EXAMPLE 6–10

For the network of Figure 6–26, determine the currents I_1, I_2, and I_3.

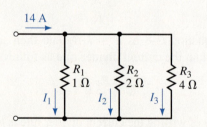

FIGURE 6–26

Solution   First, we calculate the total conductance of the network.

$$G_T = \frac{1}{1\,\Omega} + \frac{1}{2\,\Omega} + \frac{1}{4\,\Omega} = 1.75\text{ S}$$

Now the currents may be evaluated as follows:

$$I_1 = \frac{G_1}{G_T} I_T = \left(\frac{1\text{ S}}{1.75\text{ S}}\right) 14\text{ A} = 8.00\text{ A}$$

$$I_2 = \frac{G_2}{G_T} I_T = \left(\frac{0.5\text{ S}}{1.75\text{ S}}\right) 14\text{ A} = 4.00\text{ A}$$

$$I_3 = \frac{G_3}{G_T} I_T = \left(\frac{0.25\text{ S}}{1.75\text{ S}}\right) 14\text{ A} = 2.00\text{ A}$$

An alternate approach is to use circuit resistance, rather than conductance.

$$R_T = \frac{1}{G_T} = \frac{1}{1.75 \text{ S}} = 0.571 \text{ } \Omega$$

$$I_1 = \frac{R_T}{R_1} I_T = \left(\frac{0.571 \text{ } \Omega}{1 \text{ } \Omega}\right) 14 \text{ A} = 8.00 \text{ A}$$

$$I_2 = \frac{R_T}{R_2} I_T = \left(\frac{0.571 \text{ } \Omega}{2 \text{ } \Omega}\right) 14 \text{ A} = 4.00 \text{ A}$$

$$I_3 = \frac{R_T}{R_3} I_T = \left(\frac{0.571 \text{ } \Omega}{5 \text{ } \Omega}\right) 14 \text{ A} = 2.00 \text{ A}$$

For the network of Figure 6–27, determine the currents I_1, I_2, and I_3.

EXAMPLE 6–11

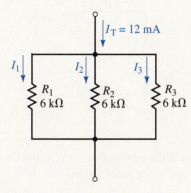

FIGURE 6–27

Solution   Since all the resistors have the same value, the incoming current will split equally between the resistances. Therefore,

$$I_1 = I_2 = I_3 = \frac{12 \text{ mA}}{3} = 4.00 \text{ mA}$$

Determine the currents I_1 and I_2 in the network of Figure 6–28.

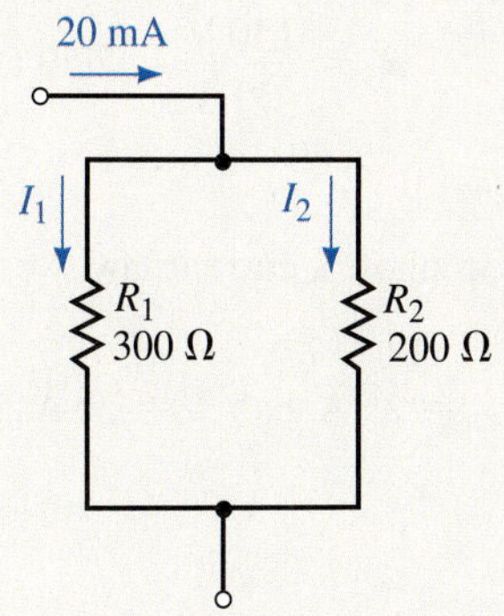

EXAMPLE 6–12

FIGURE 6–28

Solution Because we have only two resistors in the given network, we use Equations 6–12 and 6–13:

$$I_1 = \frac{R_2}{R_1 + R_2} I_T = \left(\frac{200 \ \Omega}{300 \ \Omega + 200 \ \Omega}\right)(20 \text{ mA}) = 8.00 \text{ mA}$$

$$I_2 = \frac{R_1}{R_1 + R_2} I_T = \left(\frac{300 \ \Omega}{300 \ \Omega + 200 \ \Omega}\right)(20 \text{ mA}) = 12.0 \text{ mA}$$

EXAMPLE 6–13

Determine the resistance R_1 so that current will divide as shown in the network of Figure 6–29.

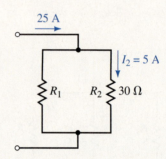

FIGURE 6–29

Solution There are several methods that may be used to solve this problem. We will examine only two of the possibilities.

Method I: Since we have two resistors in parallel, we may use Equation 6–13 to solve for the unknown resistor:

$$I_2 = \frac{R_1}{R_1 + R_2} I_T$$

$$5 \text{ A} = \left(\frac{R_1}{R_1 + 30 \ \Omega}\right)(25 \text{ A})$$

Using algebra, we get

$$(5 \text{ A})R_1 + (5 \text{ A})(30 \ \Omega) = (25 \text{ A})R_1$$

$$(20 \text{ A})R_1 = 150 \text{ V}$$

$$R_1 = \frac{150 \text{ V}}{20 \text{ A}} = 7.50 \ \Omega$$

Method II: By applying Kirchhoff's current law, we see that the current in R_1 must be

$$I_1 = 25 \text{ A} - 5 \text{ A} = 20 \text{ A}$$

Now, since elements in parallel must have the same voltage across their terminals, the voltage across R_1 must be exactly the same as the voltage across R_2. By Ohm's law, the voltage across R_2 is

$$V_2 = (5\ A)(30\ \Omega) = 150\ V$$

And so

$$R_1 = \frac{150\ V}{20\ A} = 7.50\ \Omega$$

As expected, the results are identical. This example illustrates that there is usually more than one method for solving a given problem. Although the methods are equally correct, we see that the second method in this example is less involved.

Use the current divider rule to calculate the unknown currents for the networks of Figure 6–30.

PRACTICE PROBLEMS 5

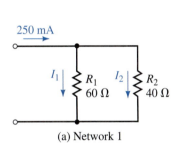

(a) Network 1

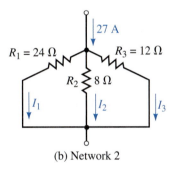

(b) Network 2

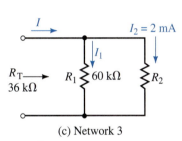

(c) Network 3

FIGURE 6–30

Answers
Network 1: $I_1 = 100$ mA, $I_2 = 150$ mA
Network 2: $I_1 = 4.50$ A, $I_2 = 13.5$ A $I_3 = 9.00$ A
Network 3: $I_1 = 3.00$ mA, $I = 5.00$ mA

Four resistors are connected in parallel. The values of the resistors are 1 Ω, 3 Ω, 4 Ω, and 5 Ω.

a. Using only a pencil and a piece of paper (no calculator), determine the current through each resistor if the current through the 5-Ω resistor is 6 A.

b. Again, without a calculator, solve for the total current applied to the parallel combination.

c. Use a calculator to determine the total parallel resistance of the four resistors. Use the current divider rule and the total current obtained in part (b) to calculate the current through each resistor.

IN-PROCESS
LEARNING CHECK 2

(Answers are at the end of the chapter.)

6.6 Analysis of Parallel Circuits

We will now examine how to use the principles developed in this chapter when analyzing parallel circuits. In the examples to follow, we find that the laws of conservation of energy apply equally well to parallel circuits as to series circuits. Although we choose to analyze circuits a certain way, remember that there is usually more than one way to arrive at the correct answer. As you become more proficient at circuit analysis you will generally use the most efficient method. For now, however, use the method with which you feel most comfortable.

EXAMPLE 6–14

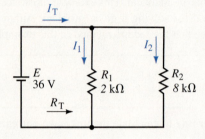

FIGURE 6–31

For the circuit of Figure 6–31, determine the following quantities:

a. R_T

b. I_T

c. Power delivered by the voltage source

d. I_1 and I_2 using the current divider rule

e. Power dissipated by the resistors

Solution

a. $R_T = \dfrac{R_1 R_2}{R_1 + R_2} = \dfrac{(2\text{ k}\Omega)(8\text{ k}\Omega)}{2\text{ k}\Omega + 8\text{ k}\Omega} = 1.6\text{ k}\Omega$

b. $I_T = \dfrac{E}{R_T} = \dfrac{36\text{ V}}{1.6\text{ k}\Omega} = 22.5\text{ mA}$

c. $P_T = EI_T = (36\text{ V})(22.5\text{ mA}) = 810\text{ mW}$

d. $I_2 = \dfrac{R_1}{R_1 + R_2} I_T = \left(\dfrac{2\text{ k}\Omega}{2\text{ k}\Omega + 8\text{ k}\Omega}\right)(22.5\text{ mA}) = 4.5\text{ mA}$

   $I_1 = \dfrac{R_2}{R_1 + R_2} I_T = \left(\dfrac{8\text{ k}\Omega}{2\text{ k}\Omega + 8\text{ k}\Omega}\right)(22.5\text{ mA}) = 18.0\text{ mA}$

e. Since we know the voltage across each of the parallel resistors must be 36 V, we use this voltage to determine the power dissipated by each resistor. It would be equally correct to use the current through each resistor to calculate the power. However, it is generally best to use given information rather than calculated values to perform further calculations since it is then less likely that an error is carried through.

$$P_1 = \frac{E^2}{R_1} = \frac{(36\text{ V})^2}{2\text{ k}\Omega} = 648\text{ mW}$$

$$P_2 = \frac{E^2}{R_2} = \frac{(36\text{ V})^2}{8\text{ k}\Omega} = 162\text{ mW}$$

Notice that the power delivered by the voltage source is exactly equal to the total power dissipated by the resistors, namely $P_T = P_1 + P_2$.

Refer to the circuit of Figure 6–32:

EXAMPLE 6–15

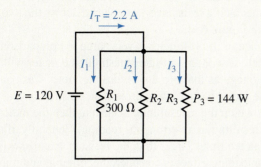

FIGURE 6–32

a. Solve for the total power delivered by the voltage source.

b. Find the currents I_1, I_2, and I_3.

c. Determine the values of the unknown resistors R_2 and R_3.

d. Calculate the power dissipated by each resistor.

e. Verify that the power dissipated is equal to the power delivered by the voltage source.

Solution

a. $P_T = EI_T = (120 \text{ V})(2.2 \text{ A}) = 264 \text{ W}$

b. Since the three resistors of the circuit are in parallel, we know that the voltage across all resistors must be equal to $E = 120$ V.

$$I_1 = \frac{V_1}{R_1} = \frac{120 \text{ V}}{300 \text{ }\Omega} = 0.4 \text{ A}$$

$$I_3 = \frac{P_3}{V_3} = \frac{144 \text{ W}}{120 \text{ V}} = 1.2 \text{ A}$$

Because KCL must be maintained at each node, we determine the current I_2 as

$$I_2 = I_T - I_1 - I_3$$
$$= 2.2 \text{ A} - 0.4 \text{ A} - 1.2 \text{ A} = 0.6 \text{ A}$$

c. $R_2 = \dfrac{V_2}{I_2} = \dfrac{120 \text{ V}}{0.6 \text{ A}} = 200 \text{ }\Omega$

Although we could use the calculated current I_3 to determine the resistance, it is best to use the given data in calculations rather than calculated values.

$$R_3 = \frac{V_3{}^2}{P_3} = \frac{(120 \text{ V})^2}{144 \text{ W}} = 100 \text{ }\Omega$$

d. $P_1 = \dfrac{V_1{}^2}{R_1} = \dfrac{(120 \text{ V})^2}{300 \text{ }\Omega} = 48 \text{ W}$

$P_2 = I_2 E_2 = (0.6 \text{ A})(120 \text{ V}) = 72 \text{ W}$

e. $P_{in} = P_{out}$

$264 \text{ W} = P_1 + P_2 + P_3$

$264 \text{ W} = 48 \text{ W} + 72 \text{ W} + 144 \text{ W}$

$264 \text{ W} = 264 \text{ W}$ (checks!)

6.7 Voltmeter Loading Effects

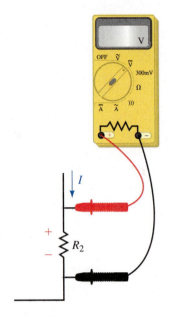

FIGURE 6–33

In the previous chapter, we observed that a voltmeter is essentially a meter movement in series with a current-limiting resistance. When a voltmeter is placed across two terminals to provide a voltage reading, the circuit is affected in the same manner as if a resistance were placed across the two terminals. The effect is shown in Figure 6–33.

If the resistance of the voltmeter is very large in comparison with the resistance across which the voltage is to be measured, the meter will indicate essentially the same voltage as that present before the meter was connected. On the other hand, if the meter has an internal resistance which is near in value to the resistance across which the measurement is taken, then the meter will adversely load the circuit, resulting in an erroneous reading. Generally, if the meter resistance is more than ten times larger than the resistance across which the voltage is taken, then the loading effect is considered negligible and may be ignored.

In the circuit of Figure 6–34, there is no current in the circuit since the terminals a and b are open circuited. The voltage appearing between the open terminals must be $V_{ab} = 10$ V. Now, if we place a voltmeter having an internal resistance of 200 kΩ between the terminals, the circuit is closed, resulting in a small current. The complete circuit appears as shown in Figure 6–35.

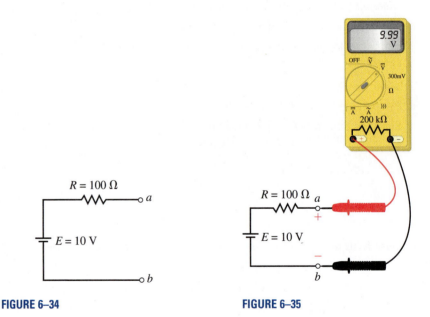

FIGURE 6–34   **FIGURE 6–35**

The reading indicated on the face of the meter is the voltage that occurs across the internal resistance of the meter. Applying Kirchhoff's voltage law to the circuit, this voltage is

$$V_{ab} = \frac{200 \text{ k}\Omega}{200 \text{ k}\Omega + 100 \Omega}(10 \text{ V}) = 9.995 \text{ V}$$

Clearly, the reading on the face of the meter is essentially equal to the expected value of 10 V. Recall from the previous chapter that we defined the loading effect of a meter as follows:

$$\text{loading effect} = \frac{\text{actual value} - \text{reading}}{\text{actual value}} \times 100\%$$

For the circuit of Figure 6–35, the voltmeter has a loading effect of

$$\text{loading effect} = \frac{10 \text{ V} - 9.995 \text{ V}}{10 \text{ V}} \times 100\% = 0.05\%$$

This loading error is virtually undetectable for the circuit given. The same would not be true if we had a circuit as shown in Figure 6–36 and used the same voltmeter to provide a reading.

Again, if the circuit were left open circuited, we would expect that $V_{ab} = 10$ V.

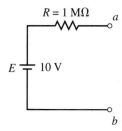

FIGURE 6–36

By connecting the 200-kΩ voltmeter between the terminals, as shown in Figure 6–37, we see that the voltage detected between terminals a and b will no longer be the desired voltage; rather,

$$V_{ab} = \frac{200 \text{ k}\Omega}{200 \text{ k}\Omega + 1 \text{ M}\Omega}(10 \text{ V}) = 1.667 \text{ V}$$

The loading effect of the meter in this circuit is

$$\text{loading effect} = \frac{10 \text{ V} - 1.667 \text{ V}}{10 \text{ V}} \times 100\% = 83.33\%$$

The previous illustration is an example of a problem that can occur when taking measurements in electronic circuits. When an inexperienced technician or technologist obtains an unforeseen result, he or she assumes that something is wrong with either the circuit or the instrument. In fact, both the circuit and the instrument are behaving in a perfectly predictable manner. The tech merely forgot to take into account the meter's loading effect. All instruments have limitations and we must always be aware of these limitations.

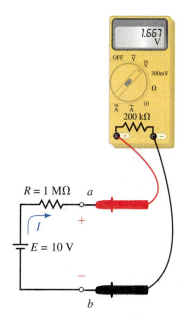

FIGURE 6–37

A digital voltmeter having an internal resistance of 5 MΩ is used to measure the voltage across terminals a and b in the circuit of Figure 6–37.

a. Determine the reading on the meter.

b. Calculate the loading effect of the meter.

EXAMPLE 6–16

Solution

a. The voltage applied to the meter terminals is

$$V_{ab} = \left(\frac{5 \text{ M}\Omega}{1 \text{ M}\Omega + 5 \text{ M}\Omega}\right)(10 \text{ V}) = 8.33 \text{ V}$$

b. The loading effect is

$$\text{loading error} = \frac{10 \text{ V} - 8.33 \text{ V}}{10 \text{ V}} \times 100\% = 16.7\%$$

All instruments have a loading effect on the circuit in which a measurement is taken. If you were given two voltmeters, one with an internal resistance of 200 kΩ and another with an internal resistance of 1 MΩ, which meter would load a circuit more? Explain.

IN-PROCESS
LEARNING CHECK 3

(Answers are at the end of the chapter.)

6.8 Computer Analysis

◀ MULTISIM

◀ CADENCE

As you have already seen, computer simulation is useful in providing a visualization of the skills you have learned. We will use both MultiSIM and PSpice to "measure" voltage and current in parallel circuits. One of the most useful features of MultiSIM is the program's ability to accurately simulate the operation of a real circuit. In this section, you will learn how to change the settings of the multimeter to observe meter loading in a circuit.

MultiSIM

EXAMPLE 6–17

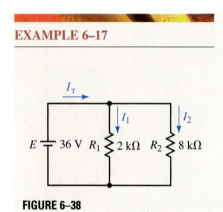

FIGURE 6–38

Use MultiSIM to determine the currents I_T, I_1, and I_2 in the circuit of Figure 6–38. This circuit was analyzed previously in Example 6–14.

Solution After opening the Circuit window:

- Select the components for the circuit from the Parts bin toolbars. You need to select the battery and ground symbol from the Sources toolbar. The resistors are obtained from the Basic toolbar.

- Once the circuit is completely wired, you may select the ammeters from the Indicators toolbar. Make sure that the ammeters are correctly placed into the circuit. Remember that the solid bar on the ammeter is connected to the lower-potential side of the circuit or branch.

- Simulate the circuit by clicking on the power switch. You should see the same results as shown in Figure 6–39.

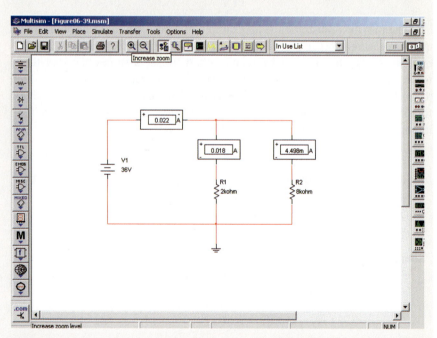

◀ MULTISIM

FIGURE 6–39

Notice that these results are consistent with those found in Example 6–14.

EXAMPLE 6–18

Use MultiSIM to demonstrate the loading effect of the voltmeter used in Figure 6–37. The voltmeter is to have internal resistance of 200 kΩ.

Solution   After opening the Circuit window:

• Construct the circuit by placing the battery, resistor, and ground as shown in Figure 6–37.

• Select the multimeter from the Instruments toolbar.

• Enlarge the multimeter by double clicking on the symbol.

• Click on the Settings button on the multimeter face.

• Change the voltmeter resistance to 200 kΩ. Accept the new value by clicking on OK.

• Run the simulation by clicking on the power switch. The resulting display is shown in Figure 6–40.

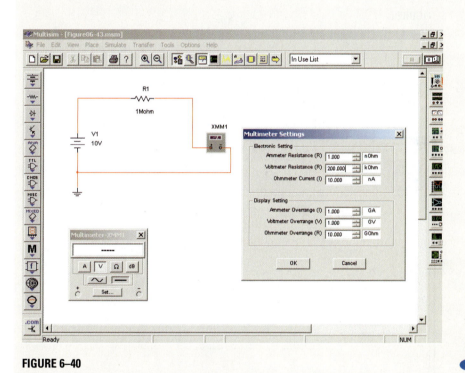

FIGURE 6–40

◀ **MULTISIM**

PSpice

In previous PSpice examples, we used the bias point analysis to obtain the dc current in a circuit. In this chapter we will again use the same analysis technique to examine parallel circuits.

EXAMPLE 6–19

Use PSpice to determine the currents in the circuit of Figure 6–41.

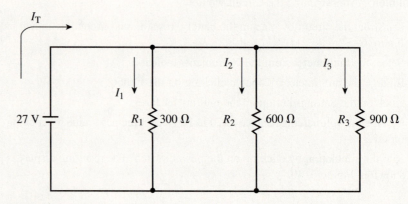

FIGURE 6–41

Solution

• Open the CIS Demo software and construct the circuit as illustrated.

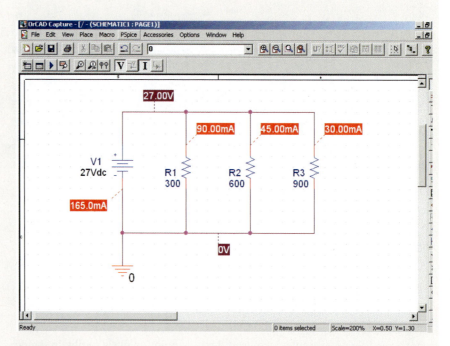

FIGURE 6–42

• Click on the New Simulation Profile and select Bias Point analysis.

• After running the project, you will observe a display of the circuit currents and voltages. The currents are $I(R1) = 90$ mA, $I(R2) = 45$ mA, $I(R3) = 30$ mA, and $I(V1) = 165$ mA.

PRACTICE PROBLEMS 6

Use MultiSIM to determine the current in each resistor of the circuit of Figure 6–21(a) if a 24-V voltage source is connected across the terminals of the resistor network.

Answers
$I_1 = I_2 = I_3 = 0.267$ A, $I_4 = 1.20$ A

PRACTICE PROBLEMS 7

Use PSpice to determine the current in each resistor of the circuit shown in Figure 6–43.

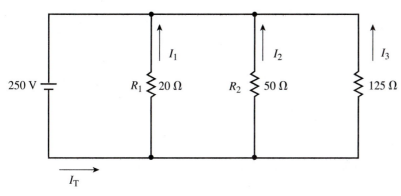

FIGURE 6–43

Answers
$I_1 = 12.5$ A, $I_2 = 5.00$ A, $I_3 = 2.00$ A, $I_T = 19.5$ A

PUTTING IT INTO PRACTICE

You have been hired as a consultant to a heating company. One of your jobs is to determine the number of 1000-W heaters that can be safely handled by an electrical circuit. All of the heaters in any circuit are connected in parallel. Each circuit operates at a voltage of 240 V and is rated for a maximum of 20 A. The normal operating current of the circuit should not exceed 80% of the maximum rated current. How many heaters can be safely installed in each circuit? If a room requires 5000 W of heaters to provide adequate heat during the coldest weather, how many circuits must be installed in this room?

PROBLEMS

6.1 Parallel Circuits

1. Indicate which of the elements in Figure 6–44 are connected in parallel and which elements are connected in series.

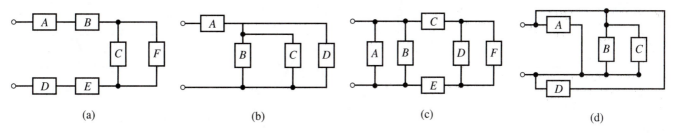

(a)   (b)   (c)   (d)

FIGURE 6–44

2. For the networks of Figure 6–45, indicate which resistors are connected in series and which resistors are connected in parallel.

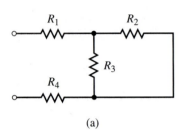

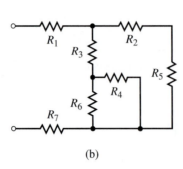

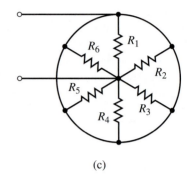

(a)

(b)

(c)

FIGURE 6–45

3. Without changing the component positions, show at least one way of connecting all the elements of Figure 6–46 in parallel.

4. Repeat Problem 3 for the elements shown in Figure 6–47.

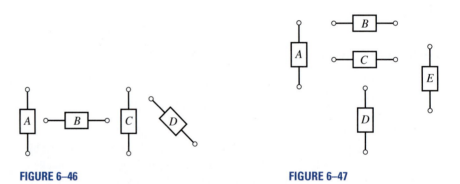

FIGURE 6–46

FIGURE 6–47

6.2   Kirchhoff's Current Law

5. Use Kirchhoff's current law to determine the magnitudes and directions of the indicated currents in each of the networks shown in Figure 6–48.

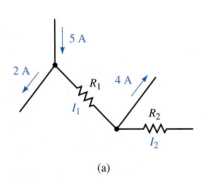

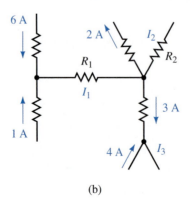

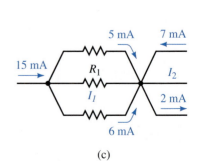

(a)

(b)

(c)

FIGURE 6–48

6. For the circuit of Figure 6–49, determine the magnitude and direction of each of the indicated currents.

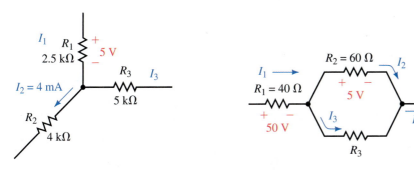

FIGURE 6–49   **FIGURE 6–50**

7. Consider the network of Figure 6–50:
   a. Calculate the currents I_1, I_2, I_3, and I_4.
   b. Determine the value of the resistance R_3.
8. Find each of the unknown currents in the networks of Figure 6–51.

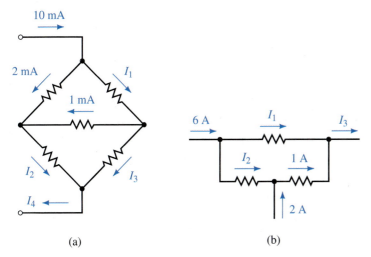

(a)   (b)

FIGURE 6–51

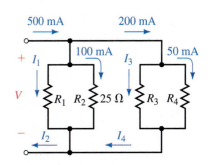

FIGURE 6–52

9. Refer to the network of Figure 6–52:
   a. Use Kirchhoff's current law to solve for the unknown currents, I_1, I_2, I_3, and I_4.
   b. Calculate the voltage, V, across the network.
   c. Determine the values of the unknown resistors, R_1, R_3, and R_4.
10. Refer to the network of Figure 6–53:
   a. Use Kirchhoff's current law to solve for the unknown currents.
   b. Calculate the voltage, V, across the network.
   c. Determine the required value of the voltage source, E. (Hint: Use Kirchhoff's voltage law.)

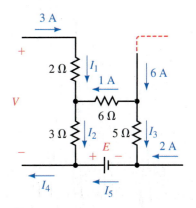

FIGURE 6–53

6.3 Resistors in Parallel

11. Calculate the total conductance and total resistance of each of the networks shown in Figure 6–54.

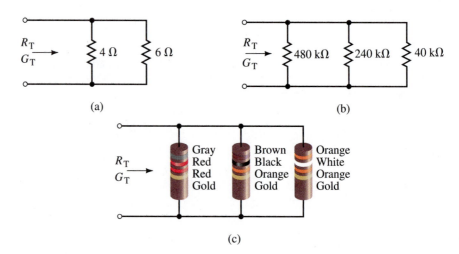

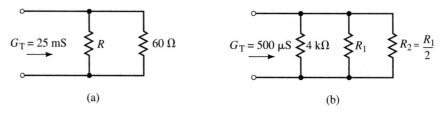

FIGURE 6–54

12. For the networks of Figure 6–55, determine the value of the unknown resistance(s) to result in the given total conductance.

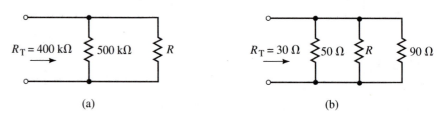

FIGURE 6–55

13. For the networks of Figure 6–56, determine the value of the unknown resistance(s) to result in the total resistances given.

FIGURE 6–56

14. Determine the value of each unknown resistor in the network of Figure 6–57, so that the total resistance is 100 kΩ.

15. Refer to the network of Figure 6–58:

    a. Calculate the values of R_1, R_2, and R_3 so that the total resistance of the network is 200 Ω.

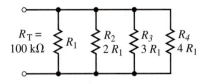

FIGURE 6–57

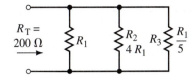

FIGURE 6–58

b. If R_3 has a current of 2 A, determine the current through each of the other resistors.

c. How much current must be applied to the entire network?

16. Refer to the network of Figure 6–59:

a. Calculate the values of R_1, R_2, R_3, and R_4 so that the total resistance of the network is 100 kΩ.

b. If R_4 has a current of 2 mA, determine the current through each of the other resistors.

c. How much current must be applied to the entire network?

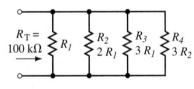

FIGURE 6–59

17. Refer to the network of Figure 6–60:

a. Find the voltages across R_1 and R_2.

b. Determine the current I_2.

18. Refer to the network of Figure 6–61:

a. Find the voltages across R_1, R_2 and R_3.

b. Calculate the current I_2.

c. Calculate the current I_3.

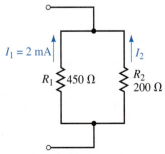

FIGURE 6–60

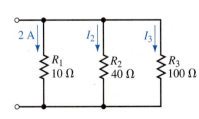

FIGURE 6–61

19. Determine the total resistance of each network of Figure 6–62.

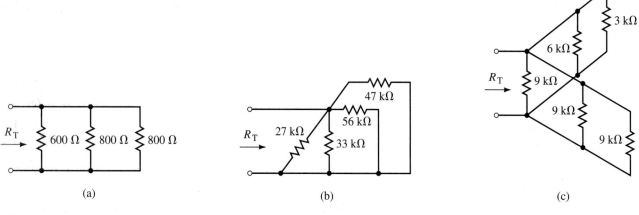

(a)   (b)   (c)

FIGURE 6–62

20. Determine the total resistance of each network of Figure 6–63.

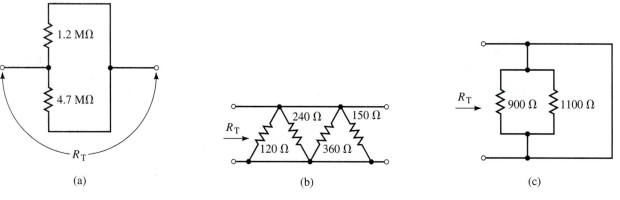

(a)   (b)   (c)

FIGURE 6–63

21. Determine the values of the resistors in the circuit of Figure 6–64, given the indicated conditions.

22. Given the indicated conditions, calculate all currents and determine all resistor values for the circuit of Figure 6–65.

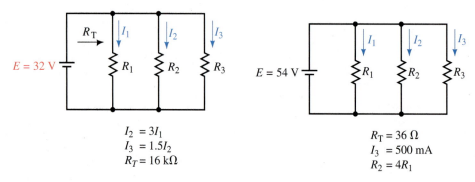

$I_2 = 3I_1$
$I_3 = 1.5I_2$
$R_T = 16\ \text{k}\Omega$

$R_T = 36\ \Omega$
$I_3 = 500\ \text{mA}$
$R_2 = 4R_1$

FIGURE 6–64   **FIGURE 6–65**

23. Without using a pencil, paper, or a calculator, determine the resistance of each network of Figure 6–66.

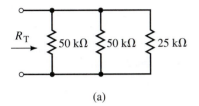

(a)

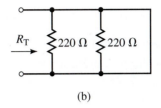

(b)

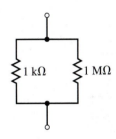

(c)

FIGURE 6–66

24. Without using a pencil, paper, or calculator, determine the approximate resistance of the network of Figure 6–67.

25. Without using a pencil, paper, or a calculator, approximate the total resistance of the network of Figure 6–68.

26. Derive Equation 6–7, which is used to calculate the total resistance of three parallel resistors.

FIGURE 6–67

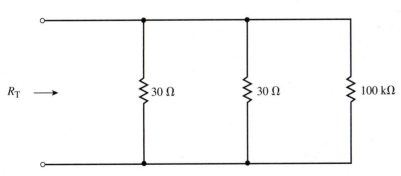

FIGURE 6–68

6.4   Voltage Sources in Parallel

27. Two 20-V batteries are connected in parallel to provide current to a 100-V load as shown in Figure 6–69. Determine the current in the load and the current in each battery.

28. Two lead-acid automobile batteries are connected in parallel, as shown in Figure 6–70, to provide additional starting current. One of the batteries is fully charged at 14.2 V and the other battery has discharged to 9 V. If the internal resistance of each battery is 0.01 Ω, determine the current in the batteries. If each battery is intended to provide a maximum current of 150 A, should this method be used to start a car?

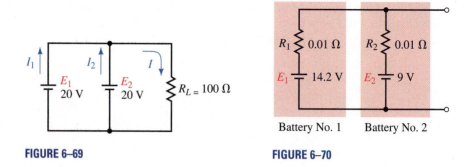

FIGURE 6–69

FIGURE 6–70

6.5   Current Divider Rule

29. Use the current divider rule to find the currents I_1 and I_2 in the networks of Figure 6–71.

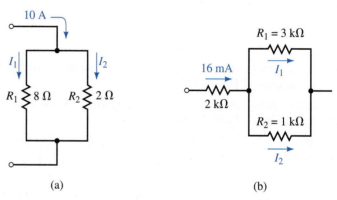

(a)

(b)

FIGURE 6–71

30. Repeat Problem 29 for the networks of Figure 6–72.

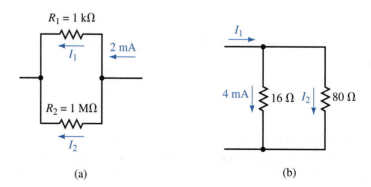

(a)

(b)

FIGURE 6–72

31. Use the current divider rule to determine all unknown currents for the networks of Figure 6–73.

32. Repeat Problem 31 for the networks of Figure 6–74.

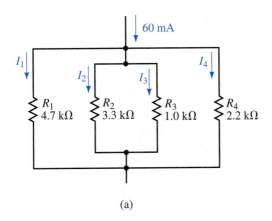

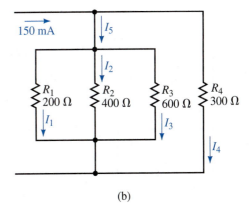

(a)   (b)

FIGURE 6–73

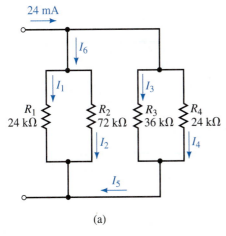

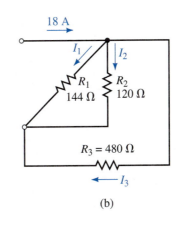

(a)   (b)

FIGURE 6–74

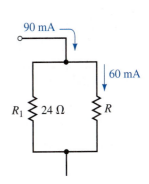

33. Use the current divider rule to determine the unknown resistance in the network of Figure 6–75.

FIGURE 6–75

34. Use the current divider rule to determine the unknown resistance in the network of Figure 6–76.

35. Refer to the circuit of Figure 6–77:

   a. Determine the equivalent resistance, R_T, of the circuit.

   b. Solve for the current I.

   c. Use the current divider rule to determine the current in each resistor.

   d. Verify Kirchhoff's current law at node a.

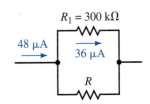

FIGURE 6–76

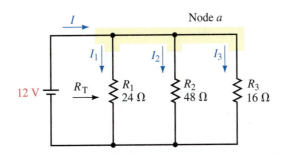

FIGURE 6–77

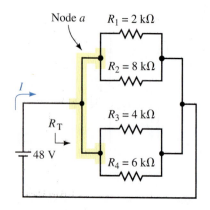

Node *a* $R_1 = 2\text{ k}\Omega$

$R_2 = 8\text{ k}\Omega$

$R_3 = 4\text{ k}\Omega$

R_T

48 V

$R_4 = 6\text{ k}\Omega$

FIGURE 6–78

◀ MULTISIM

36. Repeat Problem 35 for the circuit of Figure 6–78.

6.6   Analysis of Parallel Circuits

37. Refer to the circuit of Figure 6–79:

   a. Find the total resistance, R_T, and solve for the current, I, through the voltage source.

   b. Find all of the unknown currents in the circuit.

   c. Verify Kirchhoff's current law at node *a*.

   d. Determine the power dissipated by each resistor. Verify that the total power dissipated by the resistors is equal to the power delivered by the voltage source.

38. Repeat Problem 37 for the circuit of Figure 6–80.

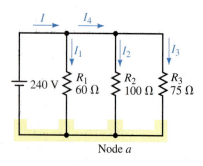

I I_4

I_1 I_2 I_3

240 V R_1 60 Ω R_2 100 Ω R_3 75 Ω

Node *a*

FIGURE 6–79

◀ MULTISIM

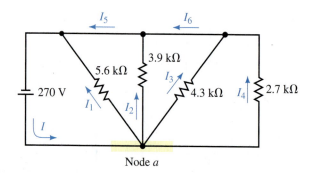

I_5 I_6

3.9 kΩ

5.6 kΩ

I_3

270 V

I_1 I_2 4.3 kΩ I_4 2.7 kΩ

I

Node *a*

FIGURE 6–80

◀ MULTISIM

39. Refer to the circuit of Figure 6–81:

   a. Calculate the current through each resistor in the circuit.

   b. Determine the total current supplied by the voltage source.

   c. Find the power dissipated by each resistor.

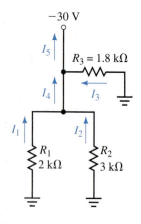

−30 V

I_5

$R_3 = 1.8\text{ k}\Omega$

I_4 I_3

I_1 I_2

R_1 2 kΩ R_2 3 kΩ

FIGURE 6–82

◀ MULTISIM

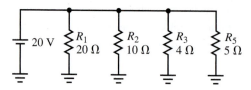

20 V R_1 20 Ω R_2 10 Ω R_3 4 Ω R_5 5 Ω

FIGURE 6–81

40. Refer to the circuit of Figure 6–82.

   a. Solve for the indicated currents.

   b. Find the power dissipated by each resistor.

   c. Verify that the power delivered by the voltage source is equal to the total power dissipated by the resistors.

41. Given the circuit of Figure 6–83:

    a. Determine the values of all resistors.

    b. Calculate the currents through R_1, R_2, and R_4.

    c. Find the currents I_1 and I_2.

    d. Find the power dissipated by resistors R_2, R_3, and R_4.

42. A circuit consists of four resistors connected in parallel and connected to a 20-V source as shown in Figure 6–84. Determine the minimum power rating of each resistor if resistors are available with the following power ratings: ⅛ W, ¼ W, ½ W, 1 W, and 2 W.

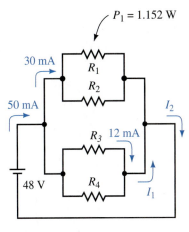

FIGURE 6–83

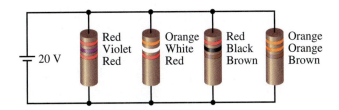

FIGURE 6–84

43. For the circuit of Figure 6–85, determine each of the indicated currents. If the circuit has a 15-A fuse as shown, is the current enough to cause the fuse to open?

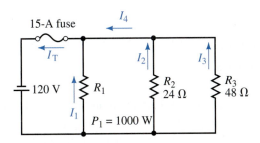

FIGURE 6–85

44. a. For the circuit of Figure 6–85, calculate the value of R_3 that will result in a circuit current of exactly $I_T = 15$ A.

    b. If the value of R_3 is increased above the value found in part (a), what will happen to the circuit current, I_T?

6.7   Voltmeter Loading Effects

45. A voltmeter having a 1-MΩ internal resistance is used to measure the indicated voltage in the circuit shown in Figure 6–86.

    a. Determine the voltage reading which will be indicated by the meter.

    b. Calculate the voltmeter's loading effect when used to measure the indicated voltage.

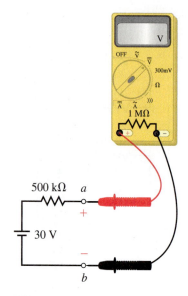

FIGURE 6–86

◀ MULTISIM

46. Repeat Problem 45 if the 500-kΩ resistor of Figure 6–86 is replaced with a 2-MΩ resistor.

47. An inexpensive analog voltmeter is used to measure the voltage across terminals a and b of the circuit shown in Figure 6–87. If the voltmeter indicates that the voltage $V_{ab} = 1.2$ V, what is the actual voltage of the source if the resistance of the meter is 50 kΩ?

48. What would be the reading if a digital meter having an internal resistance of 10 MΩ is used instead of the analog meter of Problem 47?

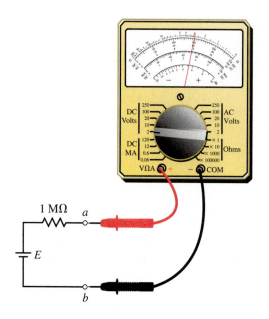

FIGURE 6–87

6.8 Computer Analysis

◀ MULTISIM

49. Use MultiSIM to solve for the current through each resistor in the circuit of Figure 6–79.

◀ MULTISIM

50. Use MultiSIM to solve for the current through each resistor in the circuit of Figure 6–80.

◀ MULTISIM

51. Use MultiSIM to simulate a voltmeter with an internal resistance of 1 MΩ used to measure voltage as shown in Figure 6–86.

◀ MULTISIM

52. Use MultiSIM to simulate a voltmeter with an internal resistance of 500 kΩ used to measure voltage as shown in Figure 6–86.

◀ CADENCE

53. Use PSpice to solve for the current through each resistor in the circuit of Figure 6–79.

◀ CADENCE

54. Use PSpice to solve for the current through each resistor in the circuit of Figure 6–80.

✓ ANSWERS TO IN-PROCESS LEARNING CHECKS

In-Process Learning Check 1

a. $I = 2.00$ A

b. $I_1 = I_2 = I_3 = 0.267$ A, $I_4 = 1.200$ A

c. $3(0.267$ A$) + 1.200$ A $= 2.00$ A (as required)

In-Process Learning Check 2

a. $I_{1\Omega} = 30.0$ A, $I_{3\Omega} = 10.0$ A, $I_{4\Omega} = 7.50$ A

b. $I_T = 53.5$ A

c. $R_T = 0.561$ Ω. (The currents are the same as those determined in part a.)

In-Process Learning Check 3

The voltmeter with the smaller internal resistance would load the circuit more, since more of the circuit current would enter the instrument.

■ OBJECTIVES

After studying this chapter, you will be able to

- find the total resistance of a network consisting of resistors connected in various series-parallel configurations,

- solve for the current through any branch or component of a series-parallel circuit,

- determine the difference in potential between any two points in a series-parallel circuit,

- calculate the voltage drop across a resistor connected to a potentiometer,

- analyze how the size of a load resistor connected to a potentiometer affects the output voltage,

- calculate the loading effects of a voltmeter or ammeter when used to measure the voltage or current in any circuit,

- use PSpice to solve for voltages and currents in series-parallel circuits,

- use MultiSIM to solve for voltages and currents in series-parallel circuits.

Series-Parallel Circuits

7

Most circuits encountered in electronics are neither simple series circuits nor simple parallel circuits, but rather a combination of the two. Although series-parallel circuits appear to be more complicated than either of the previous types of circuits analyzed to this point, we find that the same principles apply.

This chapter examines how Kirchhoff's voltage and current laws are applied to the analysis of series-parallel circuits. We will also observe that voltage and current divider rules apply to the more complex circuits. In the analysis of series-parallel circuits, we often simplify the given circuit to enable us to more clearly see how the rules and laws of circuit analysis apply. Students are encouraged to redraw circuits whenever the solution of a problem is not immediately apparent. This technique is used by even the most experienced engineers, technologists, and technicians.

In this chapter, we begin by examining simple resistor circuits. The principles of analysis are then applied to more practical circuits such as those containing zener diodes and transistors. The same principles are then applied to determine the loading effects of voltmeters and ammeters in more complex circuits.

After analyzing a complex circuit, we want to know whether the solutions are in fact correct. As you have already seen, electrical circuits usually may be studied in more than one way to arrive at a solution. Once currents and voltages for a circuit have been found, it is very easy to determine whether the resultant solution verifies the law of conservation of energy, Kirchhoff's current law, and Kirchhoff's voltage law. If there is any discrepancy (other than rounding error), there is an error in the calculation! ■

Benjamin Franklin

BENJAMIN FRANKLIN WAS BORN IN BOSTON, Massachusetts, in 1706. Although Franklin is best known as a great statesman and diplomat, he also furthered the cause of science with his experiments in electricity. This particularly includes his work with the Leyden jar, which was used to store electric charge. In his famous experiment of 1752, he used a kite to demonstrate that lightning is an electrical event. It was Franklin who postulated that positive and negative electricity are in fact a single "fluid."

Although Franklin's major accomplishments came as a result of his work in achieving independence of the Thirteen Colonies, he was nonetheless a notable scientist.

Benjamin Franklin died in his Philadelphia home on February 12, 1790, at the age of eighty-four. ■

7.1 The Series-Parallel Network

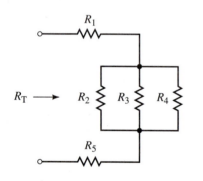

FIGURE 7–1

In electric circuits, we define a **branch** as any portion of a circuit that can be simplified as having two terminals. The components between the two terminals may be any combination of resistors, voltage sources, or other elements. Many complex circuits may be separated into a combination of both series and/or parallel elements, while other circuits consist of even more elaborate combinations that are neither series nor parallel.

In order to analyze a complicated circuit, it is important to be able to recognize which elements are in series and which elements or branches are in parallel. Consider the network of resistors shown in Figure 7–1.

We immediately recognize that the resistors R_2, R_3, and R_4 are in parallel. This parallel combination is in series with the resistors R_1 and R_5. The total resistance may now be written as follows:

$$R_T = R_1 + (R_2 \| R_3 \| R_4) + R_5$$

EXAMPLE 7–1

For the network of Figure 7–2, determine which resistors and branches are in series and which are in parallel. Write an expression for the total equivalent resistance, R_T.

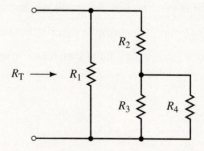

FIGURE 7–2

Solution First, we recognize that the resistors R_3 and R_4 are in parallel: $(R_3 \| R_4)$.

Next, we see that this combination is in series with the resistor R_2: $[R_2 + (R_3 \| R_4)]$.

Finally, the entire combination is in parallel with the resistor R_1. The total resistance of the circuit may now be written as follows:

$$R_T = R_1 \| [R_2 + (R_3 \| R_4)]$$

For the network of Figure 7–3, determine which resistors and branches are in series and which are in parallel. Write an expression for the total resistance, R_T.

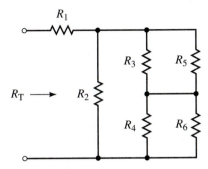

FIGURE 7–3

Answer
$R_T = R_1 + R_2 \parallel [(R_3 \parallel R_5) + (R_4 \parallel R_6)]$

7.2 Analysis of Series-Parallel Circuits

Series-parallel networks are often difficult to analyze because they initially appear confusing. However, the analysis of even the most complex circuit is simplified by following some fairly basic steps. By practicing (not memorizing) the techniques outlined in this section, you will find that most circuits can be reduced to groupings of series and parallel combinations. In analyzing such circuits, it is imperative to remember that the rules for analyzing series and parallel elements still apply.

The same current occurs through all series elements.
The same voltage occurs across all parallel elements.

In addition, remember that Kirchhoff's voltage law and Kirchhoff's current law apply for all circuits regardless of whether the circuits are series, parallel, or series-parallel. The following steps will help to simplify the analysis of series-parallel circuits:

1. Whenever necessary, redraw complicated circuits showing the source connection at the left-hand side. All nodes should be labelled to ensure that the new circuit is equivalent to the original circuit. You will find that as you become more experienced at analyzing circuits, this step will no longer be as important and may therefore be omitted.

2. Examine the circuit to determine the strategy that will work best in analyzing the circuit for the required quantities. You will usually find it best to begin the analysis of the circuit at the components most distant to the source.

3. Simplify recognizable combinations of components wherever possible, redrawing the resulting circuit as often as necessary. Keep the same labels for corresponding nodes.

4. Determine the equivalent circuit resistance, R_T.

5. Solve for the total circuit current. Indicate the directions of all currents and label the correct polarities of the voltage drops on all components.

6. Calculate how currents and voltages split between the elements of the circuit.

7. Since there are usually several possible ways at arriving at solutions, verify the answers by using a different approach. The extra time taken in this step will usually ensure that the correct answer has been found.

EXAMPLE 7–2

◀ **Online Companion**

Consider the circuit of Figure 7–4.

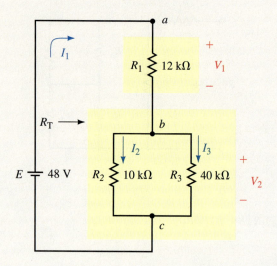

FIGURE 7–4

a. Find R_T.

b. Calculate I_1, I_2, and I_3.

c. Determine the voltages V_1 and V_2.

Solution   By examining the circuit of Figure 7–4, we see that resistors R_2 and R_3 are in parallel. This parallel combination is in series with the resistor R_1.

The combination of resistors may be represented by a simple series network shown in Figure 7–5. Notice that the nodes have been labelled using the same notation.

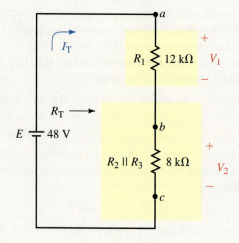

FIGURE 7–5

a. The total resistance of the circuit may be determined from the combination

$$R_T = R_1 + R_2 \| R_3$$

$$R_T = 12\ k\Omega + \frac{(10\ k\Omega)(40\ k\Omega)}{10\ k\Omega + 40\ k\Omega}$$

$$= 12\ k\Omega + 8\ k\Omega = 20\ k\Omega$$

b. From Ohm's law, the total current is

$$I_T = I_1 = \frac{48\ V}{20\ k\Omega} = 2.4\ mA$$

The current I_1 will enter node b and then split between the two resistors R_2 and R_3. This current divider may be simplified as shown in the partial circuit of Figure 7–6.

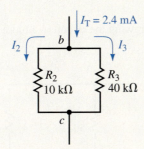

FIGURE 7–6

Applying the current divider rule to these two resistors gives

$$I_2 = \frac{(40\ k\Omega)(2.4\ mA)}{10\ k\Omega + 40\ k\Omega} = 1.92\ mA$$

$$I_3 = \frac{(10\ k\Omega)(2.4\ mA)}{10\ k\Omega + 40\ k\Omega} = 0.48\ A$$

c. Using the above currents and Ohm's law, we determine the voltages:

$$V_1 = (2.4\ mA)(12\ k\Omega) = 28.8\ V$$

$$V_3 = (0.48\ mA)(40\ k\Omega) = 19.2\ V = V_2$$

In order to check the answers, we may simply apply Kirchhoff's voltage law around any closed loop which includes the voltage source:

$$\sum V = E - V_1 - V_3$$
$$= 48\ V - 28.8\ V - 19.2\ V$$
$$= 0\ V\ (\text{checks!})$$

The solution may be verified by ensuring that the power delivered by the voltage source is equal to the summation of powers dissipated by the resistors.

Use the results of Example 7–2 to verify that the law of conservation of energy applies to the circuit of Figure 7–4 by showing that the voltage source delivers the same power as the total power dissipated by all resistors.

IN-PROCESS
LEARNING CHECK 1

(Answers are at the end of the chapter.)

EXAMPLE 7–3

Find the voltage V_{ab} for the circuit of Figure 7–7.

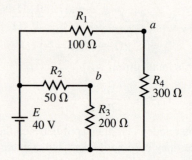

FIGURE 7–7

Solution   We begin by redrawing the circuit in a more simple representation as shown in Figure 7–8.

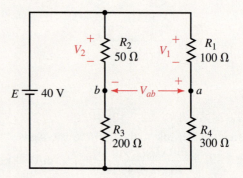

FIGURE 7–8

From Figure 7–8, we see that the original circuit consists of two parallel branches, where each branch is a series combination of two resistors.

If we take a moment to examine the circuit, we see that the voltage V_{ab} may be determined from the combination of voltages across R_1 and R_2. Alternatively, the voltage may be found from the combination of voltages across R_3 and R_4.

As usual, several methods of analysis are possible. Because the two branches are in parallel, the voltage across each branch must be 40 V. Using the voltage divider rules allows us to quickly calculate the voltage across each resistor. Although equally correct, other methods of calculating the voltages would be more lengthy.

$$V_2 = \frac{R_2}{R_2 + R_3}E$$

$$= \left(\frac{50\ \Omega}{50\ \Omega + 200\ \Omega}\right)(40\ \text{V}) = 8.0\ \text{V}$$

$$V_1 = \frac{R_1}{R_1 + R_4}E$$

$$= \left(\frac{100\ \Omega}{100\ \Omega + 300\ \Omega}\right)(40\ \text{V}) = 10.0\ \text{V}$$

As shown in Figure 7–9, we apply Kirchhoff's voltage law to determine the voltage between terminals a and b.

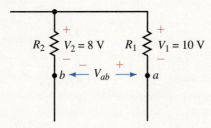

FIGURE 7–9

$$V_{ab} = -10.0 \text{ V} + 8.0 \text{ V} = -2.0 \text{ V}$$

Consider the circuit of Figure 7–10:

a. Find the total resistance R_T "seen" by the source E.

b. Calculate I_T, I_1, and I_2.

c. Determine the voltages V_2 and V_4.

Solution We begin the analysis by redrawing the circuit. Since we generally like to see the source on the left-hand side, one possible way of redrawing the resultant circuit is shown in Figure 7–11. Notice that the polarities of voltages across all resistors have been shown.

EXAMPLE 7–4

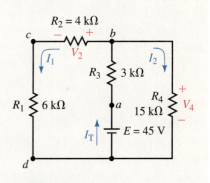

FIGURE 7–10

◀ **MULTISIM**

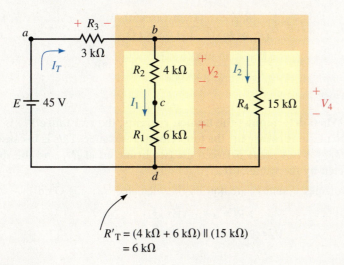

$$R'_T = (4 \text{ k}\Omega + 6 \text{ k}\Omega) \parallel (15 \text{ k}\Omega)$$
$$= 6 \text{ k}\Omega$$

FIGURE 7–11

a. From the redrawn circuit, the total resistance of the circuit is

$$R_T = R_3 + [(R_1 + R_2)\|R_4]$$

$$= 3\text{ k}\Omega + \frac{(4\text{ k}\Omega + 6\text{ k}\Omega)(15\text{ k}\Omega)}{(4\text{ k}\Omega + 6\text{ k}\Omega) + 15\text{ k}\Omega}$$

$$= 3\text{ k}\Omega + 6\text{ k}\Omega = 9.00\text{ k}\Omega$$

b. The current supplied by the voltage source is

$$I_T = \frac{E}{R_T} = \frac{45\text{ V}}{9\text{ k}\Omega} = 5.00\text{ mA}$$

We see that the supply current divides between the parallel branches as shown in Figure 7–12.

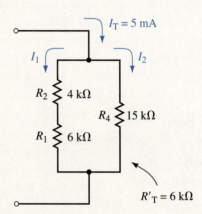

FIGURE 7–12

Applying the current divider rule, we calculate the branch currents as

$$I_1 = I_T\frac{R'_T}{(R_1 + R_2)} = \frac{(5\text{ mA})(6\text{ k}\Omega)}{4\text{ k}\Omega + 6\text{ k}\Omega} = 3.00\text{ mA}$$

$$I_2 = I_T\frac{R'_T}{R_4} = \frac{(5\text{ mA})(6\text{ k}\Omega)}{15\text{ k}\Omega} = 2.00\text{ mA}$$

Notice: When determining the branch currents, the resistance R'_T is used in the calculations rather the the total circuit resistance. This is because the current $I_T = 5$ mA splits between the two branches of R'_T and the split is not affected by the value of R_3.

c. The voltages V_2 and V_4 are now easily calculated by using Ohm's law:

$$V_2 = I_1R_2 = (3\text{ mA})(4\text{ k}\Omega) = 12.0\text{ V}$$

$$V_4 = I_2R_4 = (2\text{ mA})(15\text{ k}\Omega) = 30.0\text{ V}$$

For the circuit of Figure 7–13, find the indicated currents and voltages.

EXAMPLE 7–5

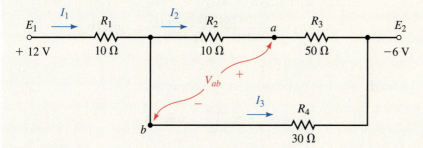

FIGURE 7–13

◀ MULTISIM

Solution   Because the above circuit contains voltage point sources, it is easier to analyze if we redraw the circuit to help visualize the operation.

The point sources are voltages with respect to ground, and so we begin by drawing a circuit with the reference point as shown in Figure 7–14.

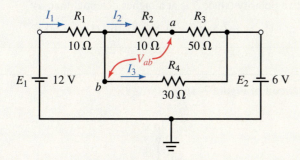

FIGURE 7–14

Now, we can see that the circuit may be further simplified by combining the voltage sources ($E = E_1 + E_2$) and by showing the resistors in a more suitable location. The simplified circuit is shown in Figure 7–15.

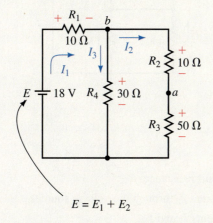

FIGURE 7–15

The total resistance "seen" by the equivalent voltage source is

$$R_T = R_1 + [R_4\|(R_2 + R_3)]$$
$$= 10\ \Omega + \frac{(30\ \Omega)(10\ \Omega + 50\ \Omega)}{30\ \Omega + (10\ \Omega + 50\ \Omega)} = 30.0\ \Omega$$

And so the total current provided into the circuit is

$$I_1 = \frac{E}{R_T} = \frac{18\ \text{V}}{30\ \Omega} = 0.600\ \text{A}$$

At node b this current divides between the two branches as follows:

$$I_3 = \frac{(R_2 + R_3)I_1}{R_4 + R_2 + R_3} = \frac{(60\ \Omega)(0.600\ \text{A})}{30\ \Omega + 10\ \Omega + 50\ \Omega} = 0.400\ \text{A}$$
$$I_2 = \frac{R_4 I_1}{R_4 + R_2 + R_3} = \frac{(30\ \Omega)(0.600\ \text{A})}{30\ \Omega + 10\ \Omega + 50\ \Omega} = 0.200\ \text{A}$$

The voltage V_{ab} has the same magnitude as the voltage across the resistor R_2, but with a negative polarity (since b is at a higher potential than a):

$$V_{ab} = -I_2 R_2 = -(0.200\ \text{A})(10\ \Omega) = -2.0\ \text{V}$$

PRACTICE PROBLEMS 2

Consider the circuit of Figure 7–16:

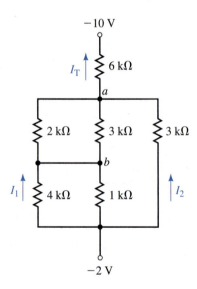

FIGURE 7–16

a. Find the total circuit resistance, R_T.
b. Determine the current I_T through the voltage sources.
c. Solve for the currents I_1 and I_2.
d. Calculate the voltage V_{ab}.

Answers
a. $R_T = 7.20\ \text{k}\Omega$; b. $I_T = 1.11\ \text{mA}$; c. $I_1 = 0.133\ \text{mA}$, $I_2 = 0.444\ \text{mA}$;
d. $V_{ab} = -0.800\ \text{V}$

We now examine how the methods developed in the first two sections of this chapter are applied when analyzing practical circuits. You may find that some of the circuits introduce you to unfamiliar devices. For now, you do not need to know precisely how these devices operate, simply that the voltages and currents in the circuits follow the same rules and laws that you have used up to now.

7.3 Applications of Series-Parallel Circuits

The circuit of Figure 7–17 is referred to as a *bridge circuit* and is used extensively in electronic and scientific instruments.

EXAMPLE 7–6

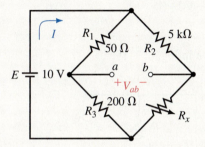

FIGURE 7–17

Calculate the current I and the voltage V_{ab} when
a. $R_x = 0\ \Omega$ (short circuit)
b. $R_x = 15\ k\Omega$
c. $R_x = \infty$ (open circuit)

Solution
a. $R_x = 0\ \Omega$:
The circuit is redrawn as shown in Figure 7–18.

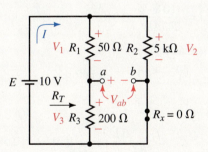

FIGURE 7–18

The voltage source "sees" a total resistance of

$$R_T = (R_1 + R_3)\|R_2 = 250\ \Omega\|5000\ \Omega = 238\ \Omega$$

resulting in a source current of

$$I = \frac{10\ V}{238\ \Omega} = 0.042\ A = 42.2\ mA$$

The voltage V_{ab} may be determined by solving for voltage across R_1 and R_2.

The voltage across R_1 will be constant regardless of the value of the variable resistor R_x. Hence

$$V_1 = \left(\frac{50\ \Omega}{50\ \Omega + 200\ \Omega}\right)(10\ \text{V}) = 2.00\ \text{V}$$

Now, since the variable resistor is a short circuit, the entire source voltage will appear across the resistor R_2, giving

$$V_2 = 10.0\ \text{V}$$

And so

$$V_{ab} = -V_1 + V_2 = -2.00\ \text{V} + 10.0\ \text{V} = +8.00\ \text{V}$$

b. $R_x = 15\ \text{k}\Omega$:

The circuit is redrawn in Figure 7–19.

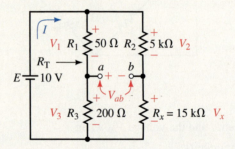

FIGURE 7–19

The voltage source "sees" a circuit resistance of

$$R_T = (R_1 + R_3)\|(R_2 + R_x)$$
$$= 250\ \Omega\|20\ \text{k}\Omega = 247\ \Omega$$

which results in a source current of

$$I = \frac{10\ \text{V}}{247\ \Omega} = 0.0405\ \text{A} = 40.5\ \text{mA}$$

The voltages across R_1 and R_2 are

$$V_1 = 2.00\ \text{V} \quad \text{(as before)}$$

$$V_2 = \frac{R_2}{R_2 + R_x}E$$

$$= \left(\frac{5\ \text{k}\Omega}{5\ \text{k}\Omega + 15\ \text{k}\Omega}\right)(10\ \text{V}) = 2.50\ \text{V}$$

Now the voltage between terminals a and b is found as

$$V_{ab} = -V_1 + V_2$$
$$= -2.0\ \text{V} + 2.5\ \text{V} = +0.500\ \text{V}$$

c. $R_x = \infty$:

The circuit is redrawn in Figure 7–20.

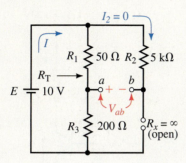

FIGURE 7–20

Because the second branch is an open circuit due to the resistor R_x, the total resistance "seen" by the source is

$$R_T = R_1 + R_3 = 250 \ \Omega$$

resulting in a source current of

$$I = \frac{10 \ V}{250 \ \Omega} = 0.040 \ A = 40.0 \ mA$$

The voltages across R_1 and R_2 are

$$V_1 = 2.00 \ V \quad \text{(as before)}$$
$$V_2 = 0 \ V \quad \text{(since the branch is open)}$$

And so the resulting voltage between terminals a and b is

$$V_{ab} = -V_1 + V_2$$
$$= -2.0 \ V + 0 \ V = -2.00 \ V$$

The previous example illustrates how voltages and currents within a circuit are affected by changes elsewhere in the circuit. In the example, we saw that the voltage V_{ab} varied from -2 V to $+8$ V, while the total circuit current varied from a minimum value of 40 mA to a maximum value of 42 mA. These changes occurred even though the resistor R_x varied from 0 Ω to ∞.

A **transistor** is a three-terminal device that may be used to amplify small signals. In order for the transistor to operate as an amplifier, however, certain dc conditions must be met. These conditions set the "bias point" of the transistor. The bias current of a transistor circuit is determined by a dc voltage source and several resistors. Although the operation of the transistor is outside the scope of this chapter, we can analyze the bias circuit of a transistor using elementary circuit theory.

EXAMPLE 7–7

Use the given conditions to determine I_C, I_E, V_{CE}, and V_B for the transistor circuit of Figure 7–21.

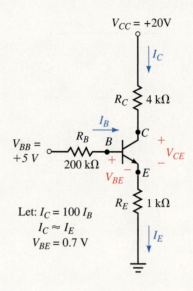

Let: $I_C = 100\, I_B$
$I_C \approx I_E$
$V_{BE} = 0.7$ V

FIGURE 7–21

Solution   In order to simplify the work, the circuit of Figure 7–21 is separated into two circuits: one circuit containing the known voltage V_{BE} and the other containing the unknown voltage V_{CE}.

Since we always start will the given information, we redraw the circuit containing the known voltage V_{BE} as illustrated in Figure 7–22.

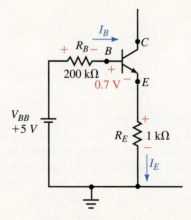

FIGURE 7–22

Although the circuit of Figure 7–22 initially appears to be a series circuit, we see that this cannot be the case, since we are given that $I_E \cong I_C = 100I_B$. We know that the current everywhere in a series circuit must be the same. However, Kirchhoff's voltage law still applies around the closed loop, resulting in the following:

$$V_{BB} = R_B I_B + V_{BE} + R_E I_E$$

The previous expression contains two unknowns, I_B and I_E (V_{BE} is given). From the given information we have the current $I_E \cong 100I_B$, which allows us to write

$$V_{BB} = R_B I_B + V_{BE} + R_E(100I_B)$$

Solving for the unknown current I_B, we have

$$5.0\text{ V} = (200\text{ k}\Omega)I_B + 0.7\text{ V} + (1\text{ k}\Omega)(100I_B)$$

$$(300\text{ k}\Omega)I_B = 5.0\text{ V} - 0.7\text{ V} = 4.3\text{ V}$$

$$I_B = \frac{4.3\text{ V}}{300\text{ k}\Omega} = 14.3\text{ }\mu\text{A}$$

The current $I_E \cong I_C = 100I_B = 1.43$ mA.

As mentioned previously, the circuit can be redrawn as two separate circuits. The circuit containing the unknown voltage V_{CE} is illustrated in Figure 7–23. Notice that the resistor R_E appears in both Figure 7–22 and Figure 7–23.

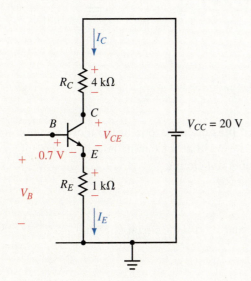

FIGURE 7–23

Applying Kirchhoff's voltage law around the closed loop of Figure 7–23, we have the following:

$$V_{R_C} + V_{CE} + V_{R_E} = V_{CC}$$

The voltage V_{CE} is found as

$$
\begin{aligned}
V_{CE} &= V_{CC} - V_{R_C} - V_{R_E} \\
&= V_{CC} - R_C I_C - R_E I_E \\
&= 20.0\text{ V} - (4\text{ k}\Omega)(1.43\text{ mA}) - (1\text{ k}\Omega)(1.43\text{ mA}) \\
&= 20.0\text{ V} - 5.73\text{ V} - 1.43\text{ V} = 12.8\text{ V}
\end{aligned}
$$

Finally, applying Kirchhoff's voltage law from B to ground, we have

$$
\begin{aligned}
V_B &= V_{BE} + V_{R_E} \\
&= 0.7\text{ V} + 1.43\text{ V} \\
&= 2.13\text{ V}
\end{aligned}
$$

Use the given information to find V_G, I_D, and V_{DS} for the circuit of Figure 7–24.

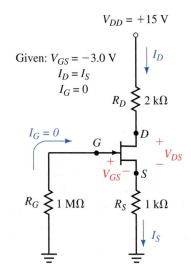

$V_{DD} = +15$ V

Given: $V_{GS} = -3.0$ V
$I_D = I_S$
$I_G = 0$

R_D 2 kΩ

$I_G = 0$

R_G 1 MΩ R_S 1 kΩ

FIGURE 7–24

Answers
$V_G = 0$, $I_D = 3.00$ mA, $V_{DS} = 6.00$ V

The **universal bias** circuit is one of the most common transistor circuits used in amplifiers. We will now examine how to use circuit analysis principles to analyze this important circuit.

EXAMPLE 7–8

Determine the I_C and V_{CE} for the circuit of Figure 7–25.

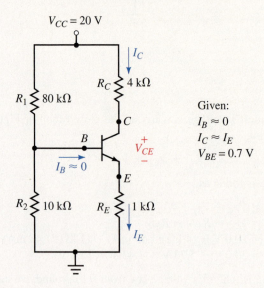

$V_{CC} = 20$ V

R_C 4 kΩ

R_1 80 kΩ

Given:
$I_B \approx 0$
$I_C \approx I_E$
$V_{BE} = 0.7$ V

$I_B \approx 0$

R_2 10 kΩ R_E 1 kΩ

FIGURE 7–25

Solution   If we examine the above circuit, we see that since $I_B \approx 0$, we may assume that R_1 and R_2 are effectively in series. This assumption would be incorrect if the current I_B was not very small compared to the currents through R_1 and R_2. We use the voltage divider rule to solve for the voltage, V_B. (For this reason, the universal bias circuit is often referred to as **voltage divider bias**.)

$$V_B = \frac{R_2}{R_1 + R_2} V_{CC}$$

$$= \left[\frac{10 \text{ k}\Omega}{80 \text{ k}\Omega + 10 \text{ k}\Omega}\right](20 \text{ V})$$

$$= 2.22 \text{ V}$$

Next, we use the value for V_B and Kirchhoff's voltage law to determine the voltage across R_E.

$$V_{RE} = 2.22 \text{ V} - 0.7 \text{ V} = 1.52 \text{ V}$$

Applying Ohm's law, we now determine the current I_E.

$$I_E = \frac{1.52 \text{ V}}{1 \text{ k}\Omega} = 1.52 \text{ mA} \cong I_C$$

Finally, applying Kirchhoff's voltage law and Ohm's law, we determine V_{CE} as follows:

$$V_{CC} = V_{RC} + V_{CE} + V_{RE}$$
$$V_{CE} = V_{CC} - V_{RC} - V_{RE}$$
$$= 20 \text{ V} - (1.52 \text{ mA})(4 \text{ k}\Omega) - (1.52 \text{ mA})(1 \text{ k}\Omega)$$
$$= 12.4 \text{ V}$$

A **zener diode** is a two-terminal device similar to a varistor (refer to Chapter 3). When the voltage across the zener diode attempts to go above the rated voltage for the device, the zener diode provides a low-resistance path for the extra current. Due to this action, a relatively constant voltage, V_Z, is maintained across the zener diode. This characteristic is referred to as **voltage regulation** and has many applications in electronic and electrical circuits. Once again, although the theory of operation of the zener diode is outside the scope of this chapter, we are able to apply simple circuit theory to examine how the circuit operates.

For the voltage regulator circuit of Figure 7–26, calculate I_Z, I_1, I_2, and P_Z.

EXAMPLE 7–9

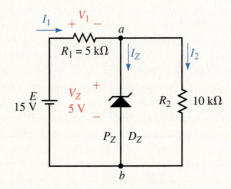

FIGURE 7–26

Solution If we take a moment to examine the circuit, we see that the zener diode is placed in parallel with the resistor R_2. This parallel combination is in series with the resistor R_1 and the voltage source, E.

In order for the zener diode to operate as a regulator, the voltage across the diode would have to be above the zener voltage without the diode present. If we remove the zener diode, the circuit would appear as shown in Figure 7–27.

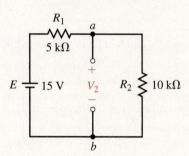

FIGURE 7–27

From Figure 7–27, we may determine the voltage V_2 which would be present without the zener diode in the circuit. Since the circuit is a simple series circuit, the voltage divider rule may be used to determine V_2:

$$V_2 = \frac{R_2}{R_1 + R_2}E = \left(\frac{10\ \text{k}\Omega}{5\ \text{k}\Omega + 10\ \text{k}\Omega}\right)(15\ \text{V}) = 10.0\ \text{V}$$

When the zener diode is placed across the resistor R_2 the device will operate to limit the voltage to $V_Z = 5\ \text{V}$.

Because the zener diode is operating as a voltage regulator, the voltage across both the diode and the resistor R_2 must be the same, namely 5 V. The parallel combination of D_Z and R_2 is in series with the resistor R_1, and so the voltage across R_1 is easily determined from Kirchhoff's voltage law as

$$V_1 = E - V_Z = 15\ \text{V} - 5\text{V} = 10\ \text{V}$$

Now, from Ohm's law, the currents I_1 and I_2 are easily found to be

$$I_2 = \frac{V_2}{R_2} = \frac{5\ \text{V}}{10\ \text{k}\Omega} = 0.5\ \text{mA}$$

$$I_1 = \frac{V_1}{R_1} = \frac{10\ \text{V}}{5\ \text{k}\Omega} = 2.0\ \text{mA}$$

Applying Kirchoff's current law at node a, we get the zener diode current as

$$I_Z = I_1 - I_2 = 2.0\ \text{mA} - 0.5\ \text{mA} = 1.5\ \text{mA}$$

Finally, the power dissipated by the zener diode must be

$$P_Z = V_Z I_Z = (5\ \text{V})(1.5\ \text{mA}) = 7.5\ \text{mW}$$

IN-PROCESS
LEARNING CHECK 2

(Answers are at the end of the chapter.)

Use the results of Example 7–9 to show that the power delivered to the circuit by the voltage source of Figure 7–26, is equal to the total power dissipated by the resistors and the zener diode.

1. Determine I_1, I_Z, and I_2 for the circuit of Figure 7–26 if the resistor R_1 is increased to 10 kΩ.

2. Repeat Problem 1 if R_1 is increased to 30 kΩ.

Answers

1. $I_1 = 1.00$ mA, $I_2 = 0.500$ mA, $I_Z = 0.500$ mA

2. $I_1 = I_2 = 0.375$ mA, $I_Z = 0$ mA. (The voltage across the zener diode is not sufficient for the device to come on.)

7.4 Potentiometers

As mentioned in Chapter 3, variable resistors may be used as potentiometers as shown in Figure 7–28 to control voltage into another circuit.

The volume control on a receiver or amplifier is an example of a variable resistor used as a potentiometer. When the movable terminal is at the uppermost position, the voltage appearing between terminals b and c is simply calculated by using the voltage divider rule as

$$V_{bc} = \left(\frac{50 \text{ k}\Omega}{50 \text{ k}\Omega + 50 \text{ k}\Omega}\right)(120 \text{ V}) = 60 \text{ V}$$

Alternatively, when the movable terminal is at the lowermost position, the voltage between terminals b and c is $V_{bc} = 0$ V, since the two terminals are effectively shorted and the voltage across a short circuit is always zero.

The circuit of Figure 7–28 represents a potentiometer having an output voltage which is adjustable between 0 and 60 V. This output is referred to as the **unloaded output,** since there is no load resistance connected between the terminals b and c. If a load resistance were connected between these terminals, the output voltage, called the **loaded output,** would no longer be the same. The following example is an illustration of circuit loading.

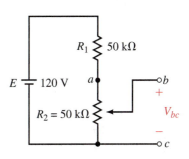

FIGURE 7–28

EXAMPLE 7–10

For the circuit of Figure 7–29, determine the range of the voltage V_{bc} as the potentiometer varies between its minimum and maximum values.

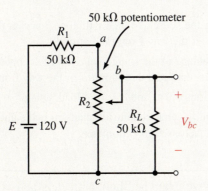

FIGURE 7–29

Solution The minimum voltage between terminals b and c will occur when the movable contact is at the lowermost contact of the variable resistor. In this position, the voltage $V_{bc} = 0$ V, since the terminals b and c are shorted.

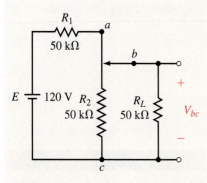

FIGURE 7–30

The maximum voltage V_{bc} occurs when the movable contact is at the uppermost contact of the variable resistor. In this position, the circuit may be represented as shown in Figure 7–30.

In Figure 7–30, we see that the resistance R_2 is in parallel with the load resistor R_L. The voltage between terminals b and c is easily determined from the voltage divider rule, as follows:

$$V_{bc} = \frac{R_2\|R_L}{(R_2\|R_L) + R_1}E$$

$$= \left(\frac{25 \text{ k}\Omega}{25 \text{ k}\Omega + 50 \text{ k}\Omega}\right)(120 \text{ V}) = 40 \text{ V}$$

We conclude that the voltage at the output of the potentiometer is adjustable from 0 V to 40 V for a load resistance of $R_L = 50$ kΩ.

By inspection, we see that an unloaded potentiometer in the circuit of Figure 7–29 would have an output voltage of 0 V to 60 V.

PRACTICE PROBLEMS 5

Refer to the circuit of Figure 7–29.

a. Determine the output voltage range of the potentiometer if the load resistor is $R_L = 5$ kΩ.

b. Repeat (a) if the load resistor is $R_L = 500$ kΩ.

c. What conclusion may be made about the output voltage of a potentiometer when the load resistance is large in comparison with the potentiometer resistance?

Answers
a. 0 to 10 V

b. 0 to 57.1 V

c. When R_L is large in comparison to the potentiometer resistance, the output voltage will better approximate the unloaded voltage. (In this example, the unloaded voltage is 0 to 60 V.)

IN-PROCESS
LEARNING CHECK 3

(Answers are at the end of the chapter.)

A 20-kΩ potentiometer is connected across a voltage source with a 2-kΩ load resistor connected between the wiper (center terminal) and the negative terminal of the voltage source.

a. What percentage of the source voltage will appear across the load when the wiper is one-fourth of the way from the bottom?

b. Determine the percentage of the source voltage which appears across the load when the wiper is one-half and three-fourths of the way from the bottom.

c. Repeat the calculations of (a) and (b) for a load resistor of 200 kΩ.

d. From the above results, what conclusion can you make about the effect of placing a large load across a potentiometer?

7.5 Loading Effects of Instruments

In Chapters 5 and 6 we examined how ammeters and voltmeters affect the operation of simple series circuits. The degree to which the circuits are affected is called the **loading effect** of the instrument. Recall that, in order for an instrument to provide an accurate indication of how a circuit operates, the loading effect should ideally be zero. In practice, it is impossible for any instrument to have zero loading effect, since all instruments absorb some energy from the circuit under test, thereby affecting circuit operation.

In this section, we will determine how instrument loading affects more complex circuits.

Calculate the loading effects if a digital multimeter, having an internal resistance of 10 MΩ, is used to measure V_1 and V_2 in the circuit of Figure 7–31.

EXAMPLE 7–11

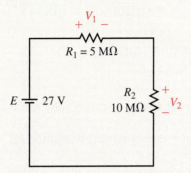

FIGURE 7–31

Solution   In order to determine the loading effect for a particular reading, we need to calculate both the unloaded voltage and the loaded voltage.

For the circuit given in Figure 7–31, the unloaded voltage across each resistor is

$$V_1 = \left(\frac{5\ \text{M}\Omega}{5\ \text{M}\Omega + 10\ \text{M}\Omega}\right)(27\ \text{V}) = 9.0\ \text{V}$$

$$V_2 = \left(\frac{10\ \text{M}\Omega}{5\ \text{M}\Omega + 10\ \text{M}\Omega}\right)(27\ \text{V}) = 18.0\ \text{V}$$

When the voltmeter is used to measure V_1, the result is equivalent to connecting a 10-MΩ resistor across resistor R_1, as shown in Figure 7–32.

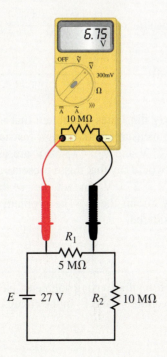

FIGURE 7–32

The voltage appearing across the parallel combination of R_1 and resistance of the voltmeter is calculated as

$$V_1 = \left(\frac{5\ \text{M}\Omega\|10\ \text{M}\Omega}{(5\ \text{M}\Omega\|10\ \text{M}\Omega) + 10\ \text{M}\Omega}\right)(27\ \text{V})$$

$$= \left(\frac{3.33\ \text{M}\Omega}{13.3\ \text{M}\Omega}\right)(27\ \text{V})$$

$$= 6.75\ \text{V}$$

Notice that the measured voltage is significantly less than the 9 V that we had expected to measure.

With the voltmeter connected across resistor R_2, the circuit appears as shown in Figure 7–33.

The voltage appearing across the parallel combination of R_2 and resistance of the voltmeter is calculated as

$$V_2 = \left(\frac{10\ \text{M}\Omega\|10\ \text{M}\Omega}{5\ \text{M}\Omega + (10\ \text{M}\Omega\|10\ \text{M}\Omega)}\right)(27\ \text{V})$$

$$= \left(\frac{5.0\ \text{M}\Omega}{10.0\ \text{M}\Omega}\right)(27\ \text{V})$$

$$= 13.5\ \text{V}$$

Again, we notice that the measured voltage is quite a bit less than the 18 V that we had expected.

Now the loading effects are calculated as follows.

When measuring V_1:

$$\text{loading effect} = \frac{9.0\ \text{V} - 6.75\ \text{V}}{9.0\ \text{V}} \times 100\%$$

$$= 25\%$$

When measuring V_2:

$$\text{loading effect} = \frac{18.0\ \text{V} - 13.5\ \text{V}}{18.0\ \text{V}} \times 100\%$$

$$= 25\%$$

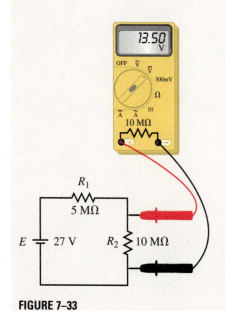

FIGURE 7–33

PRACTICAL NOTES . . .

Whenever an instrument is used to measure a quantity, the operator must always consider the loading effects of the instrument.

This example clearly illustrates a problem that novices often make when they are taking voltage measurements in high-resistance circuits. If the measured voltages $V_1 = 6.75$ V and $V_2 = 13.50$ V are used to verify Kirchhoff's voltage law, the novice would say that this represents a contradiction of the law (since 6.75 V + 13.50 V ≠ 27.0 V). In fact, we see that the circuit is behaving exactly as predicted by circuit theory. The problem occurs when instrument limitations are not considered.

 PRACTICE PROBLEMS 6

Calculate the loading effects if an analog voltmeter, having an internal resistance of 200 kΩ, is used to measure V_1 and V_2 in the circuit of Figure 7–31.

Answers
V_1: loading effect = 94.3%
V_2: loading effect = 94.3%

For the circuit of Figure 7–34, calculate the loading effect if a 5.00-Ω ammeter is used to measure the currents I_T, I_1, and I_2.

EXAMPLE 7–12

Solution We begin by determining the unloaded currents in the circuit. Using Ohm's law, we solve for the currents I_1 and I_2:

$$I_1 = \frac{100 \text{ mV}}{25 \text{ } \Omega} = 4.0 \text{ mA}$$

$$I_2 = \frac{100 \text{ mV}}{5 \text{ } \Omega} = 20.0 \text{ mA}$$

Now, by Kirchhoff's current law,

$$I_T = 4.0 \text{ mA} + 20.0 \text{ mA} = 24.0 \text{ mA}$$

If we were to insert the ammeter into the branch with resistor R_1, the circuit would appear as shown in Figure 7–35.

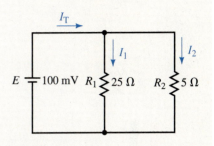

FIGURE 7–34

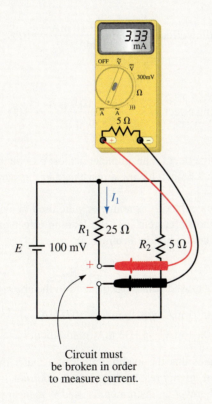

FIGURE 7–35

The current through the ammeter would be

$$I_1 = \frac{100 \text{ mV}}{25 \text{ } \Omega + 5 \text{ } \Omega} = 3.33 \text{ mA}$$

If we were to insert the ammeter into the branch with resistor R_2, the circuit would appear as shown in Figure 7–36.
The current through the ammeter would be

$$I_2 = \frac{100 \text{ mV}}{5 \text{ } \Omega + 5 \text{ } \Omega} = 10.0 \text{ mA}$$

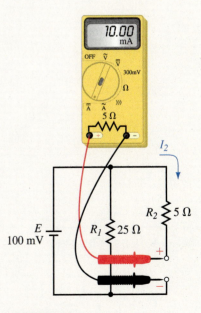

FIGURE 7–36

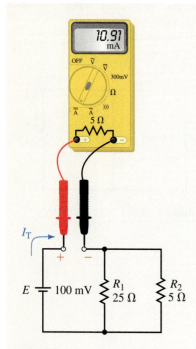

FIGURE 7–37

If the ammeter were inserted into the circuit to measure the current I_T, the equivalent circuit would appear as shown in Figure 7–37.

The total resistance of the circuit would be

$$R_T = 5\ \Omega + 25\ \Omega \| 5\ \Omega = 9.17\ \Omega$$

This would result in a current I_T determined by Ohm's law as

$$I_T = \frac{100\ \text{mV}}{9.17\ \Omega}$$

$$= 10.9\ \text{mA}$$

The loading effects for the various current measurements are as follows: When measuring I_1,

$$\text{loading effect} = \frac{4.0\ \text{mA} - 3.33\ \text{mA}}{4.0\ \text{mA}} \times 100\%$$

$$= 16.7\%$$

When measuring I_2,

$$\text{loading effect} = \frac{20\ \text{mA} - 10\ \text{mA}}{20\ \text{mA}} \times 100\%$$

$$= 50\%$$

When measuring I_T,

$$\text{loading effect} = \frac{24\ \text{mA} - 10.9\ \text{mA}}{24\ \text{mA}} \times 100\%$$

$$= 54.5\%$$

Notice that the loading effect for an ammeter is most pronounced when it is used to measure current in a branch having a resistance in the same order of magnitude as the meter.

You will also notice that if this ammeter were used in a circuit to verify the correctness of Kirchhoff's current law, the loading effect of the meter would produce an apparent contradiction. From KCL,

$$I_T = I_1 + I_2$$

By substituting the measured values of current into the above equation, we have

$$10.91\ \text{mA} = 3.33\ \text{mA} + 10.0\ \text{mA}$$

$$10.91\ \text{mA} \neq 13.33\ \text{mA (contradiction)}$$

This example illustrates that the loading effect of a meter may severely affect the current in a circuit, giving results that seem to contradict the laws of circuit theory. Therefore, wherever an instrument is used to measure a particular quantity, we must always take into account the limitations of the instrument and question the validity of a resulting reading.

PRACTICE PROBLEMS 7

Calculate the readings and the loading error if an ammeter having an internal resistance of 1 Ω is used to measure the currents in the circuit of Figure 7–34.

Answers

$I_{T(RDG)} = 19.4\ \text{mA}$; loading error $= 19.4\%$

$I_{1(RDG)} = 3.85\ \text{mA}$; loading error $= 3.85\%$

$I_{T(RDG)} = 16.7\ \text{mA}$; loading error $= 16.7\%$

MultiSIM

The analysis of series-parallel circuits using MultiSIM is almost identical to the methods used in analyzing series and parallel circuits in previous chapters. The following example illustrates that MultiSIM results in the same solutions as those obtained in Example 7–4.

7.6 Circuit Analysis Using Computers

◀ MULTISIM   ◀ CADENCE

Given the circuit of Figure 7–38, use MultiSIM to find the following quantities:

a. Total resistance, R_T

b. Voltages V_2 and V_4

c. Currents I_T, I_1, and I_2

EXAMPLE 7–13

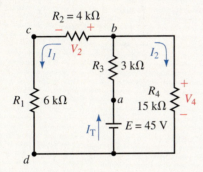

FIGURE 7–38

Solution

a. We begin by constructing the circuit as shown in Figure 7–39. This circuit is identical to that shown in Figure 7–38, except that the voltage source has been omitted and a multimeter (from the Instruments button on the Parts bin toolbar) inserted in its place. The ohmmeter function is then selected and the power switch turned on. The resistance is found to be $R_T = 9.00$ kΩ.

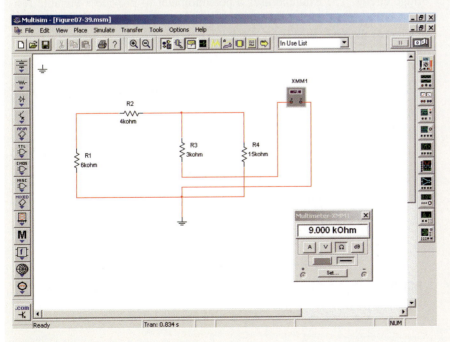

FIGURE 7–39

b. Next, we remove the multimeter and insert the 45-V source. Ammeters and voltmeters are inserted, as shown in Figure 7–40.

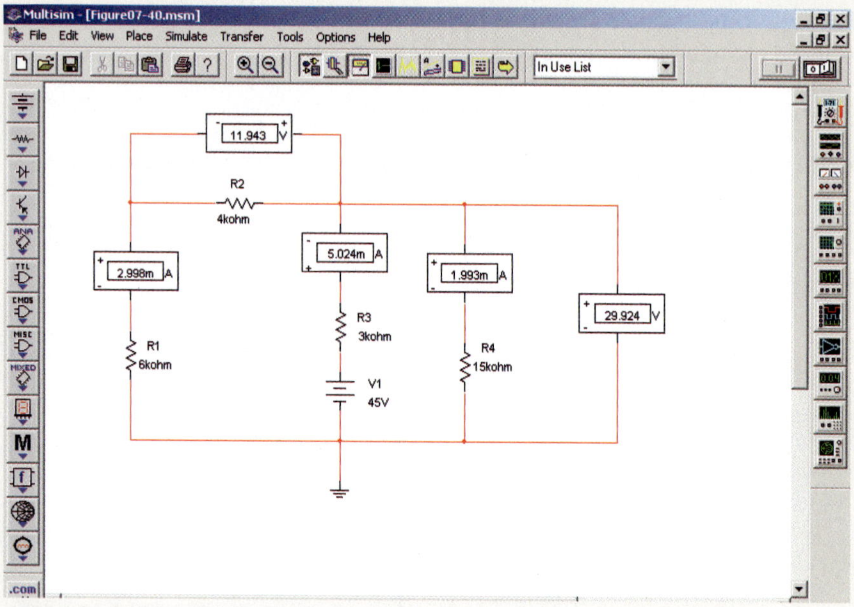

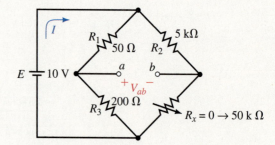

MULTISIM

FIGURE 7–40

From the results we have $I_T = 5.00$ mA, $I_1 = 3.00$ mA, $I_2 = 2.00$ mA.

c. The required voltages are $V_2 = 12.0$ V and $V_4 = 30.0$ V. These results are consistent with those obtained in Example 7–4.

The following example uses MultiSIM to determine the voltage across a bridge circuit. The example uses a potentiometer to provide a variable resistance in the circuit. MultiSIM is able to provide a display of voltage (on a multimeter) as the resistance is changed.

EXAMPLE 7–14

Given the circuit of Figure 7–41, use MultiSIM to determine the values of I and V_{ab} when $R_x = 0$ Ω, 15 kΩ, and 50 kΩ.

FIGURE 7–41

Solution

1. We begin by constructing the circuit as shown in Figure 7–42. The potentiometer is selected from the Basic button on the Parts bin toolbar. Ensure that the potentiometer is inserted as illustrated.

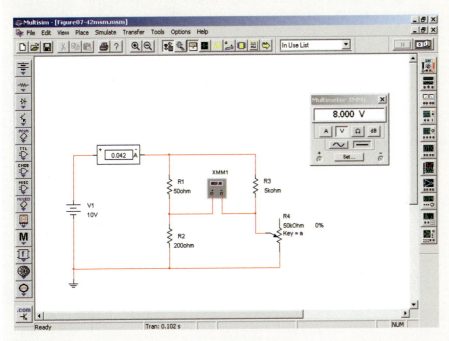

FIGURE 7–42

◀ MULTISIM

2. Double click on the potentiometer symbol and change its value to 50 kΩ. Notice that the increment value is set for 5%. Adjust this value so that it is at 10%. We will use this in a following step.

3. Once the circuit is completely built, the power switch is turned on. Notice that the resistor value is at 50%. This means that the potentiometer is adjusted so that its value is 25 kΩ. The following step will change the value of the potentiometer to 0 Ω.

4. Double click on the potentiometer symbol. The value of the resistor is now easily changed by either Shift **A** (to incrementally increase the value) or **A** (to incrementally decrease the value). As you decrease the value of the potentiometer, you should observe that the voltage displayed on the multimeter is also changing. It takes a few seconds for the display to stabilize. As shown in Figure 7–42, you should observe that $I = 42.0$ mA and $V_{ab} = 8.00$ V when $R_x = 0$ Ω.

5. Finally, after adjusting the value of the potentiometer, we obtain the following readings:

$$R_x = 15 \text{ k}\Omega \text{ (30\%):} \quad I = 40.5 \text{ mA and } V_{ab} = 0.500 \text{ V}$$
$$R_x = 50 \text{ k}\Omega \text{ (100\%):} \quad I = 40.2 \text{ mA and } V_{ab} = -0.192 \text{ V}$$

PSpice

PSpice is somewhat different from MultiSIM in the way it handles potentiometers. In order to place a variable resistor into a circuit, it is necessary to set the parameters of the circuit to sweep through a range of values. The following example illustrates the method used to provide a graphical display of output voltage and source current for a range of resistance. Although the method is different, the results are consistent with those of the previous example.

EXAMPLE 7–15

Use PSpice to provide a graphical display of voltage V_{ab} and current I as R_x is varied from 0 to 50 kΩ in the circuit of Figure 7–41.

Solution   PSpice uses global parameters to represent numeric values by name. This will permit us to set up an analysis that sweeps a variable (in this case a resistor) through a range of values.

- Open the CIS Demo software and move to the Capture schematic as outlined in previous PSpice examples. You may wish to name your project Ch 7 PSpice 1.
- Build the circuit as shown in Figure 7–43. Remember to rotate the components to provide for the correct node assignments. Change all component values (except R4) as required.

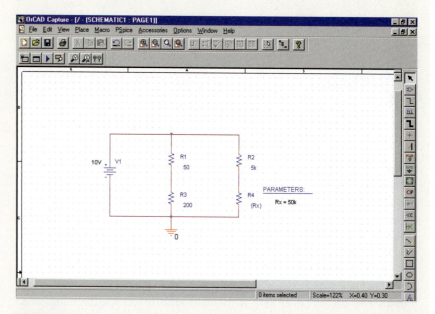

FIGURE 7–43

- Double click on the component value for R4. Enter **{Rx}** in the V$\underline{a}$lue text box of the Display Properties for this resistor. The curly braces tell PSpice to evaluate the parameter and use its value.
- Click on the Place part tool. Select the Special library and click on the PARAM part. Place the PARAM part adjacent to resistor R4.
- Double click on PARAMETERS. Click on New. Type **Rx** in the Property Name text box and click OK. Click in the cell below the Rx column and enter **50k** as the default value for this resistor. Click on Apply. Click on Display and select Name and Value from the Display Format. Exit the Property Editor.

- Click on the New Simulation icon and give the simulation a name such as Figure 7-43.
- In the Simulation Settings, select the Analysis Tab. The Analysis type is DC Sweep. Select Primary Sweep from the Options listing. In the Sweep variable box select Global Parameter. Type **Rx** in the Parameter name text box. Select the Sweep type to be linear and set the limits as follows:

Start value: 100

End value: 50k

Increment: 100.

These settings will change the resistor value from 100 Ω to 50 kΩ in 100-Ω increments. Click OK.

- Click on the Run icon. You will see a blank screen with the abscissa (horizontal axis) showing R_x scaled from 0 to 50 kΩ.
- PSpice is able to plot most circuit variables as a function of R_x. In order to request a plot of V_{ab}, click on Trace and Add Trace. Enter **V(R3:1) – V(R4:1)** in the Trace Expression text box. The voltage V_{ab} is the voltage between node 1 of R_3 and node 1 of R_4. Click OK.
- Finally, to obtain a plot of the circuit current I (current through the voltage source), we need to first add an extra axis. Click on Plot and then click Add Y Axis. To obtain a plot of the current, click on Trace and Add Trace. Select **I(V1)**. The resulting display on the monitor is shown in Figure 7–44.

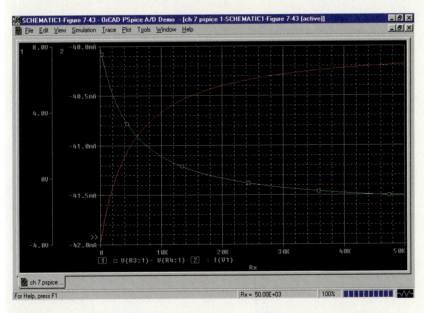

FIGURE 7–44

Notice that the current shown is negative. This is because PSpice sets the reference direction through a voltage source from the positive terminal to the negative terminal. One way of removing the negative sign is to request the current as $-I(V1)$.

PRACTICE PROBLEMS 8

Given the circuit of Figure 7–45, use MultiSIM to solve for V_{ab}, I, and I_L when $R_L = 100 \ \Omega$, 500 Ω, and 1000 Ω.

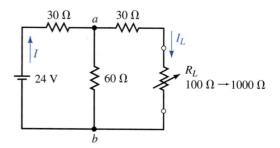

FIGURE 7–45

Answers
$R_L = 100 \ \Omega$: $I = 169$ mA, $V_{ab} = 6.93$ V, $I_L = 53.3$ mA
$R_L = 500 \ \Omega$: $I = 143$ mA, $V_{ab} = 7.71$ V, $I_L = 14.5$ mA
$R_L = 1000 \ \Omega$: $I = 138$ mA, $V_{ab} = 7.85$ V, $I_L = 7.62$ mA

PRACTICE PROBLEMS 9

Use PSpice to input file for the circuit of Figure 7–45. The output shall display the source current, I, the load current, I_L, and the voltage V_{ab} as the resistor R_L is varied in 100-Ω increments between 100 Ω and 1000 Ω.

PUTTING IT INTO PRACTICE

Very often manufacturers provide schematics showing the dc voltages that one would expect to measure if the circuit were functioning properly. The following figure shows part of a schematic for an amplifier circuit.

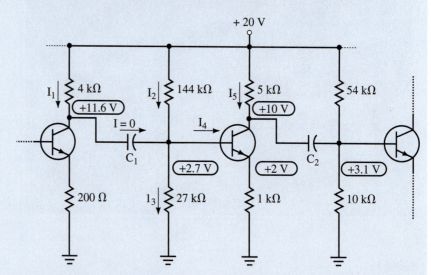

Even though a schematic may include components with which the reader is not familiar, the dc voltages provided on the schematic enable us to determine voltages and current in various parts of the circuit. If the circuit is faulty, measured voltages and currents will be different from the theoretical, allowing the experienced troubleshooter to locate the fault.

Examine the circuit shown. Use the voltage information on the schematic to determine the theoretical values of the currents I_1, I_2, I_3, I_4, and I_5. Find the magnitude and correct polarity of the voltage across the device labeled as C_2. (It's a capacitor, and it will be examined in detail in Chapter 10.)

PROBLEMS

7.1 The Series-Parallel Network

1. For the networks of Figure 7–46, determine which resistors and branches are in series and which are in parallel. Write an expression for the total resistance, R_T.

2. For each of the networks of Figure 7–47, write an expression for the total resistance, R_T.

3. Write an expression for both R_{T_1} and R_{T_2} for the networks of Figure 7–48.

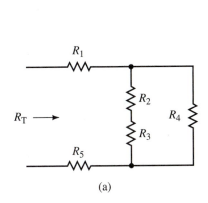
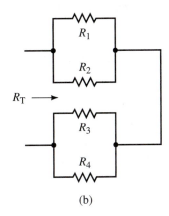

(a) (b)

FIGURE 7–46

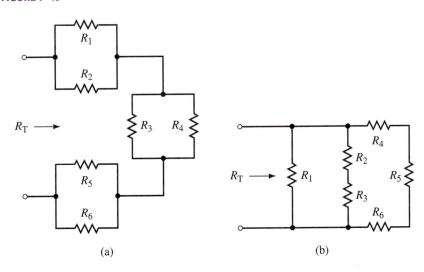

(a) (b)

FIGURE 7–47

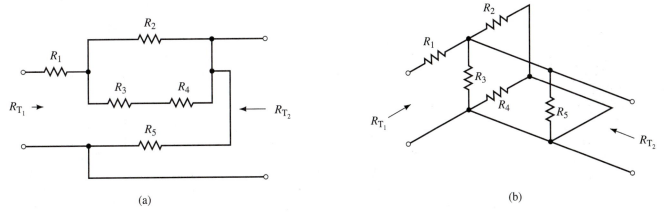

(a) (b)

FIGURE 7–48

4. Write an expression for both R_{T_1} and R_{T_2} for the networks of Figure 7–49.

5. Resistor networks have total resistances as given below. Sketch a circuit that corresponds to each expression.

   a. $R_T = (R_1\|R_2\|R_3) + (R_4\|R_5)$

   b. $R_T = R_1 + (R_2\|R_3) + [R_4\|(R_5 + R_6)]$

6. Resistor networks have total resistances as given below. Sketch a circuit that corresponds to each expression.

   a. $R_T = [(R_1\|R_2) + (R_3\|R_4)]\|R_5$

   b. $R_T = (R_1\|R_2) + R_3 + [(R_4 + R_5)\|R_6$

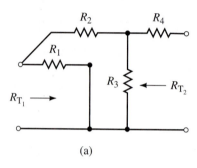
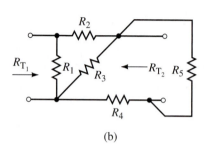

(a) (b)

FIGURE 7–49

7.2 Analysis of Series-Parallel Circuits

7. Determine the total resistance of each network in Figure 7–50.

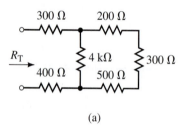
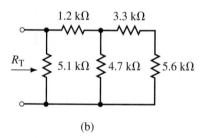

(a) (b)

FIGURE 7–50

8. Determine the total resistance of each network in Figure 7–51.

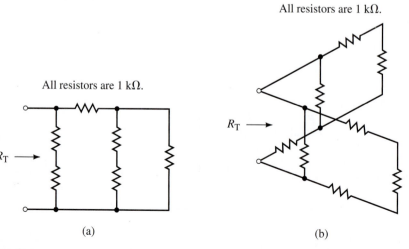

(a) (b)

FIGURE 7–51

9. Calculate the resistances R_{ab} and R_{cd} in the circuit of Figure 7–52.

10. Calculate the resistances R_{ab} and R_{bc} in the circuit of Figure 7–53.

11. Refer to the circuit of Figure 7–54:

   Find the following quantities:

   a. R_T

   b. I_T, I_1, I_2, I_3, I_4

   c. V_{ab}, V_{bc}.

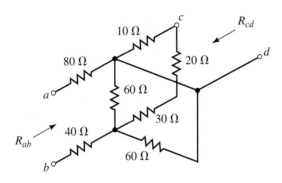

FIGURE 7–52

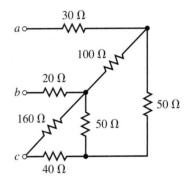

FIGURE 7–53

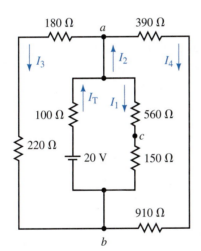

FIGURE 7–54

◀ MULTISIM

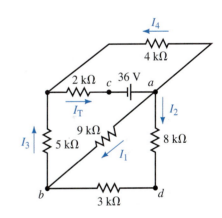

FIGURE 7–55

12. Refer to the circuit of Figure 7–55:

   Find the following quantities:

   a. R_T (equivalent resistance "seen" by the voltage source).

   b. I_T, I_1, I_2, I_3, I_4

   c. V_{ab}, V_{bc}, V_{cd}.

13. Refer to the circuit of Figure 7–56:

   a. Find the currents $I_1, I_2, I_3, I_4, I_5,$ and I_6.

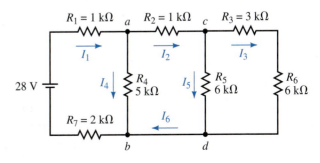

FIGURE 7–56

◀ MULTISIM

b. Solve for the voltages V_{ab} and V_{cd}.

c. Verify that the power delivered to the circuit is equal to the summation of powers dissipated by the resistors.

14. Refer to the circuit of Figure 7–57:

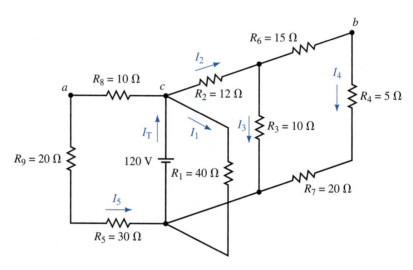

FIGURE 7–57

a. Find the currents I_1, I_2, I_3, I_4, and I_5.

b. Solve for the voltages V_{ab} and V_{bc}.

c. Verify that the power delivered to the circuit is equal to the summation of powers dissipated by the resistors.

15. Refer to the circuits of Figure 7–58:

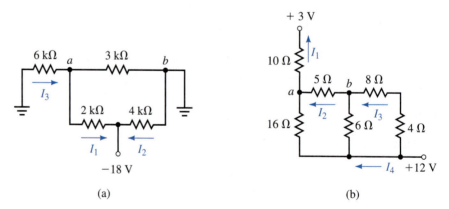

(a)   (b)

FIGURE 7–58

a. Find the indicated currents.

b. Solve for the voltage V_{ab}.

c. Verify that the power delivered to the circuit is equal to the summation of powers dissipated by the resistors.

16. Refer to the circuit of Figure 7–59:

a. Solve for the currents I_1, I_2, and I_3 when $R_x = 0\ \Omega$ and when $R_x = 5\ \text{k}\Omega$.

b. Calculate the voltage V_{ab} when $R_x = 0\ \Omega$ and when $R_x = 5\ \text{k}\Omega$.

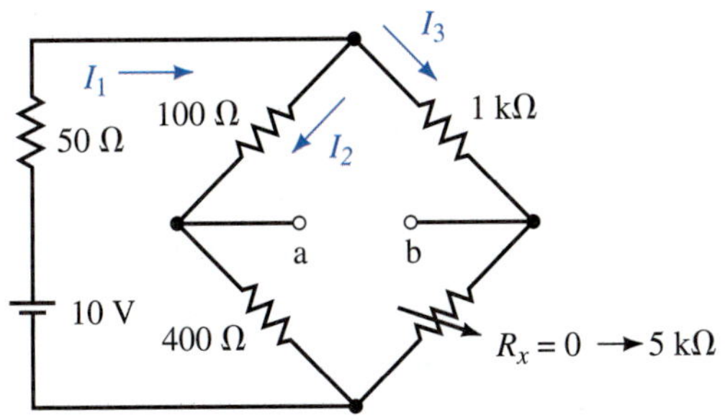

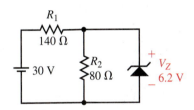

FIGURE 7–59

7.3 Applications of Series-Parallel Circuits

17. Solve for all currents and voltage drops in the circuit of Figure 7–60. Verify that the power delivered by the voltage source is equal to the power dissipated by the resistors and the zener diode.

18. Refer to the circuit of Figure 7–61:
    a. Determine the power dissipated by the 6.2-V zener diode. If the zener diode is rated for a maximum power of ¼ W, is it likely to be destroyed?
    b. Repeat Part (a) if the resistance R_1 is doubled.

19. Given the circuit of Figure 7–62, determine the range of R (maximum and minimum values) that will ensure that the output voltage $V_L = 5.6$ V while the maximum power rating of the zener diode is not exceeded.

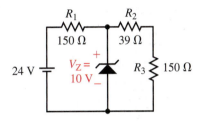

FIGURE 7–60

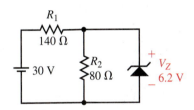

FIGURE 7–61

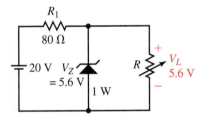

FIGURE 7–62

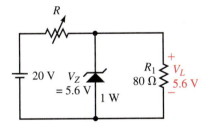

FIGURE 7–63

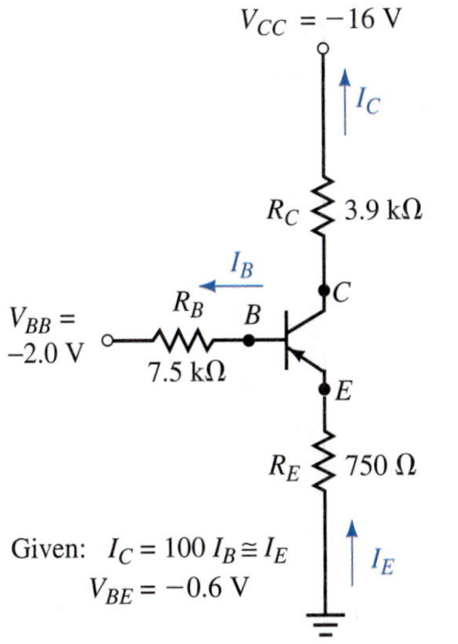

20. Given the circuit of Figure 7–63, determine the range of R (maximum and minimum values) that will ensure that the output voltage $V_L = 5.6$ V while the maximum power rating of the zener diode is not exceeded.

21. Given the circuit of Figure 7–64, determine V_B, I_C, and V_{CE}.

22. Repeat Problem 21 if R_B is increased to 10 kΩ. (All other quantities remain unchanged.)

Given: $I_C = 100 \, I_B \cong I_E$
$V_{BE} = -0.6$ V

FIGURE 7–64

23. Consider the circuit of Figure 7–65 and the indicated values:

    a. Determine I_D.

    b. Calculate the required value for R_S.

    c. Solve for V_{DS}.

24. Consider the circuit of Figure 7–66 and the indicated values:

    a. Determine I_D and V_G.

    b. Design the required values for R_S and R_D.

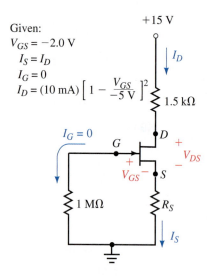

Given:
$V_{GS} = -2.0$ V
$I_S = I_D$
$I_G = 0$
$I_D = (10 \text{ mA}) \left[1 - \dfrac{V_{GS}}{-5 \text{ V}} \right]^2$

FIGURE 7–65

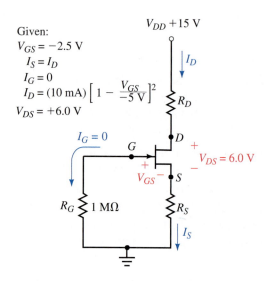

Given:
$V_{GS} = -2.5$ V
$I_S = I_D$
$I_G = 0$
$I_D = (10 \text{ mA}) \left[1 - \dfrac{V_{GS}}{-5 \text{ V}} \right]^2$
$V_{DS} = +6.0$ V

FIGURE 7–66

25. Calculate I_C and V_{CE} for the circuit of Figure 7–67.

26. Calculate I_C and V_{CE} for the circuit of Figure 7–68.

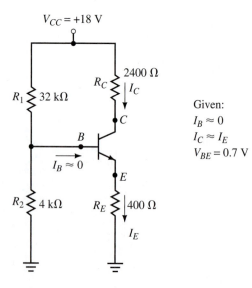

Given:
$I_B \approx 0$
$I_C \approx I_E$
$V_{BE} = 0.7$ V

FIGURE 7–67

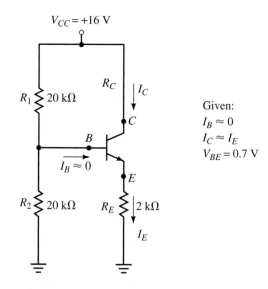

Given:
$I_B \approx 0$
$I_C \approx I_E$
$V_{BE} = 0.7$ V

FIGURE 7–68

7.4   Potentiometers

27. Refer to the circuit of Figure 7–69:

 a. Determine the range of voltages that will appear across R_L as the potentiometer is varied between its minimum and maximum values.

 b. If R_2 is adjusted to be 2.5 kΩ, what will be the voltage V_L? If the load resistor is now removed, what voltage would appear between terminals a and b?

28. Repeat Problem 27 using the load resistor $R_L = 30$ kΩ.

29. If the potentiometer of Figure 7–70 is adjusted so that $R_2 = 200$ Ω, determine the voltages V_{ab} and V_{bc}.

30. Calculate the values of R_1 and R_2 required in the potentiometer of Figure 7–70 if the voltage V_L across the 50-Ω load resistor is to be 6.0 V.

31. Refer to the circuit of Figure 7–71:

 a. Determine the range of output voltage (minimum to maximum) that can be expected as the potentiometer is adjusted from minimum to maximum.

 b. Calculate R_2 when $V_{out} = 20$ V.

32. In the circuit of Figure 7–71, what value of R_2 results in an output voltage of 40 V?

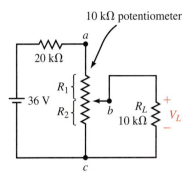

FIGURE 7–69

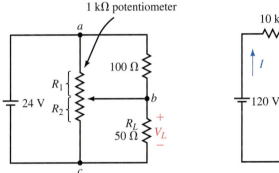

FIGURE 7–70

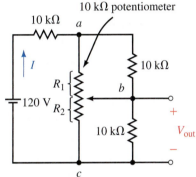

FIGURE 7–71

33. Given the circuit of Figure 7–72, calculate the output voltage V_{out} when $R_L = 0$ Ω, 250 Ω, and 500 Ω.

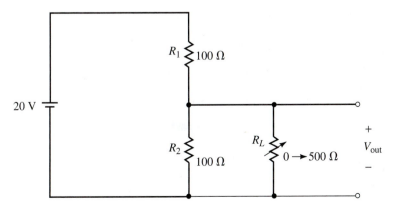

FIGURE 7–72

34. Given the circuit of Figure 7–73, calculate the output voltage V_{out} when $R_L = 0\ \Omega$, 500 Ω, and 1000 Ω.

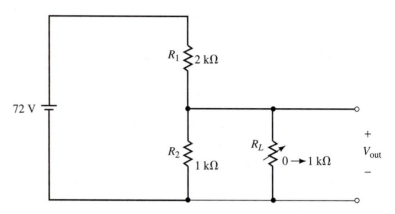

FIGURE 7–73

7.5   Loading Effects of Instruments

35. A voltmeter having a sensitivity $S = 20\ k\Omega/V$ is used on the 10-V range (200-$k\Omega$ total internal resistance) to measure voltage across the 750-$k\Omega$ resistor of Figure 7–74. The voltage indicated by the meter is 5.00 V.

    a. Determine the value of the supply voltage, E.

    b. What voltage will be present across the 750-$k\Omega$ resistor when the voltmeter is removed from the circuit?

    c. Calculate the loading effect of the meter when used as shown.

    d. If the same voltmeter is used to measure the voltage across the 200-$k\Omega$ resistor, what voltage would be indicated?

36. The voltmeter of Figure 7–75 has a sensitivity $S = 2\ k\Omega/V$.

    a. If the meter is used on its 50-V range ($R = 100\ k\Omega$) to measure the voltage across R_2, what will be the meter reading and the loading error?

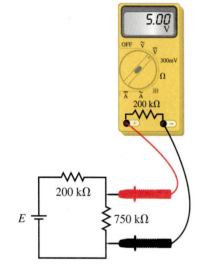

FIGURE 7–74

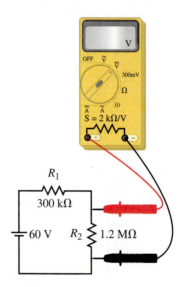

FIGURE 7–75

◀ MULTISIM

b. If the meter is changed to its 20-V range ($R = 40$ kΩ) determine the reading on this range and the loading error. Will the meter be damaged on this range? Will the loading error be less or more than the error in Part (a)?

37. An ammeter is used to measure current in the circuit shown in Figure 7–76.

   a. Explain how to correctly connect the ammeter to measure the current I_1.

   b. Determine the values indicated when the ammeter is used to measure each of the indicated currents in the circuit.

   c. Calculate the loading effect of the meter when measuring each of the currents.

38. Suppose the ammeter in Figure 7–76 has an internal resistance of 0.5 Ω:

   a. Determine the values indicated when the ammeter is used to measure the indicated currents in the circuit.

   b. Calculate the loading effect of the meter when measuring each of the currents.

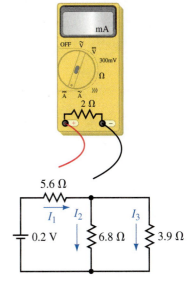

FIGURE 7–76

7.6 Circuit Analysis Using Computers

39. Use MultiSIM to solve for V_2, V_4, I_T, I_1, and I_2 in the circuit of Figure 7–10.   ◀ MULTISIM

40. Use MultiSIM to solve for V_{ab}, I_1, I_2, and I_3 in the circuit of Figure 7–13.   ◀ MULTISIM

41. Use to MultiSIM solve for the meter reading in the circuit of Figure 7–75 if the meter is used on its 50-V range.   ◀ MULTISIM

42. Repeat Problem 41 if the meter is used on its 20-V range.   ◀ MULTISIM

43. Use PSpice to solve for V_2, V_4, I_T, I_1, and I_2 in the circuit of Figure 7–10.   ◀ CADENCE

44. Use PSpice to solve for V_{ab}, I_1, I_2, and I_3 in the circuit of Figure 7–13.

45. Use PSpice to obtain a display of V_{ab} and I_1 in the circuit of Figure 7–59. Let R_x change from 500 Ω to 5 kΩ using 100-Ω increments.   ◀ CADENCE   ◀ CADENCE

✓ **ANSWERS TO IN-PROCESS LEARNING CHECKS**

In-Process Learning Check 1

$P_T = 115.2$ mW, $P_1 = 69.1$ mW, $P_2 = 36.9$ mW, $P_3 = 9.2$ mW
$P_1 + P_2 + P_3 = 115.2$ mW as required.

In-Process Learning Check 2

$P_T = 30.0$ mW, $P_{R1} = 20.0$ mW, $P_{R2} = 2.50$ mW, $P_Z = 7.5$ mW
$P_{R1} + P_{R2} + P_Z = 30$ mW as required.

In-Process Learning Check 3

$R_L = 2$ kΩ:
 a. $N = ¼$: $V_L = 8.7\%$ of V_{in}

 b. $N = ½$: $V_L = 14.3\%$ of V_{in}
  $N = ¾$: $V_L = 26.1\%$ of V_{in}
$R_L = 200$ kΩ:
 c. $N = ¼$: $V_L = 24.5\%$ of V_{in}
  $N = ½$: $V_L = 48.8\%$ of V_{in}
  $N = ¾$: $V_L = 73.6\%$ of V_{in}
 d. If $RL \gg R_1$, the loading effect is minimal.

■ **OBJECTIVES**

After studying this chapter you will be able to

- convert a voltage source into an equivalent current source,
- convert a current source into an equivalent voltage source,
- analyze circuits having two or more current sources in parallel,
- write and solve branch equations for a network,
- write and solve mesh equations for a network,
- write and solve nodal equations for a network,
- convert a resistive delta to an equivalent wye circuit or a wye to its equivalent delta circuit and solve the resulting simplified circuit,
- determine the voltage across or current through any portion of a bridge network,
- use PSpice to analyze multiloop circuits;
- use MultiSIM to analyze multiloop circuits.

Methods of Analysis

8

The networks you have worked with so far have generally had a single voltage source and could be easily analyzed using techniques such as Kirchhoff's voltage law and Kirchhoff's current law. In this chapter, you will examine circuits that have more than one voltage source or that cannot be easily analyzed using techniques studied in previous chapters.

The methods used in determining the operation of complex networks will include branch-current analysis, mesh (or loop) analysis, and nodal analysis. Although any of the above methods may be used, you will find that certain circuits are more easily analyzed using one particular approach. The advantages of each method will be discussed in the appropriate section.

In using the techniques outlined above, it is assumed that the networks are **linear bilateral networks.** The term **linear** indicates that the components used in the circuit have voltage-current characteristics that follow a straight line. Refer to Figure 8–1.

The term **bilateral** indicates that the components in the network will have characteristics that are independent of the direction of the current through the element or the voltage across the element. A resistor is an example of a linear bilateral component since the voltage across a resistor is directly proportional to the current through it and the operation of the resistor is the same regardless of the direction of the current.

In this chapter you will be introduced to the conversion of a network from a delta (Δ) configuration to an equivalent wye (Y) configuration. Conversely, we will examine the transformation from a Y configuration to an equivalent Δ configuration. You will use these conversions to examine the operation of an unbalanced bridge network. ■

CHAPTER PREVIEW

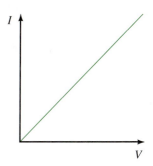

(a) Linear V-I characteristics

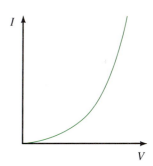

(b) Non-linear V-I characteristics

FIGURE 8–1

Sir Charles Wheatstone

CHARLES WHEATSTONE WAS BORN IN GLOUCESTER, England, on February 6, 1802. Wheatstone's original interest was in the study of acoustics and musical instruments. However, he gained fame and a knighthood as a result of inventing the telegraph and improving the electric generator.

Although he did not invent the bridge circuit, Wheatstone used one for measuring resistance very precisely. He found that when the currents in the Wheatstone bridge are exactly balanced, the unknown resistance can be compared to a known standard.

Sir Charles died in Paris, France, on October 19, 1875. ∎

8.1 Constant-Current Sources

FIGURE 8–2 Ideal constant current source.

All the circuits presented so far have used voltage sources as the means of providing power. However, the analysis of certain circuits is easier if you work with current rather than with voltage. Unlike a voltage source, a **constant-current source** maintains the same current in its branch of the circuit regardless of how components are connected external to the source. The symbol for a constant-current source is shown in Figure 8–2.

The direction of the current source arrow indicates the direction of conventional current in the branch. In previous chapters you learned that the magnitude and the direction of current through a voltage source varies according to the size of the circuit resistances and how other voltage sources are connected in the circuit. For current sources, the voltage across the current source depends on how the other components are connected.

EXAMPLE 8–1

Refer to the circuit of Figure 8–3:

a. Calculate the voltage V_S across the current source if the resistor is 100 Ω.

b. Calculate the voltage if the resistor is 2 kΩ.

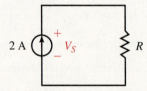

FIGURE 8–3

Solution The current source maintains a constant current of 2 A through the circuit. Therefore,

a. $V_S = V_R = (2\text{ A})(100\text{ }\Omega) = 200\text{ V}$.

b. $V_S = V_R = (2\text{ A})(2\text{ k}\Omega) = 4000\text{ V}$.

If the current source is the only source in the circuit, then the polarity of voltage across the source will be as shown in Figure 8–3. This, however, may not be the case if there is more than one source. The following example illustrates this principle.

Determine the voltages V_1, V_2, and V_S and the current I_S for the circuit of Figure 8–4.

EXAMPLE 8–2

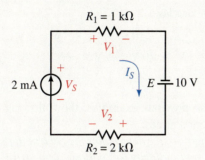

FIGURE 8–4

Solution Since the given circuit is a series circuit, the current everywhere in the circuit must be the same, namely

$$I_S = 2 \text{ mA}$$

Using Ohm's law,

$$V_1 = (2 \text{ mA})(1 \text{ k}\Omega) = 2.00 \text{ V}$$
$$V_2 = (2 \text{ mA})(2 \text{ k}\Omega) = 4.00 \text{ V}$$

Applying Kirchhoff's voltage law around the closed loop,

$$\sum V = V_S - V_1 - V_2 + E = 0$$
$$V_S = V_1 + V_2 - E$$
$$= 2 \text{ V} + 4 \text{ V} - 10 \text{ V} = -4.00 \text{ V}$$

From the above result, you see that the actual polarity of V_S is opposite to that assumed.

Calculate the currents I_1 and I_2 and the voltage V_S for the circuit of Figure 8–5.

EXAMPLE 8–3

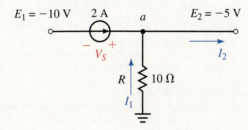

FIGURE 8–5

Solution Because the 5-V supply is effectively across the load resistor,

$$I_1 = \frac{5 \text{ V}}{10 \text{ }\Omega} = 0.5 \text{ A} \quad \text{(in the direction assumed)}$$

Applying Kirchhoff's current law at point a,

$$I_2 = 0.5\text{ A} + 2.0\text{ A} = 2.5\text{ A}$$

From Kirchhoff's voltage law,

$$\sum V = -10\text{ V} + V_S + 5\text{ V} = 0\text{ V}$$
$$V_S = 10\text{ V} - 5\text{ V} = +5\text{ V}$$

By examining the previous examples, the following conclusions may be made regarding current sources:

The constant-current source determines the current in its branch of the circuit.

The magnitude and polarity of voltage appearing across a constant-current source are dependent upon the network in which the source is connected.

8.2 Source Conversions

◀ **Online Companion**

In the previous section you were introduced to the ideal constant-current source. This is a source that has no internal resistance included as part of the circuit. As you recall, voltage sources always have some series resistance, although in some cases this resistance is so small in comparison with other circuit resistance that it may effectively be ignored when determining the operation of the circuit. Similarly, a constant-current source will always have some shunt (or parallel) resistance. If this resistance is very large in comparison with the other circuit resistance, the internal resistance of the source may once again be ignored. **An ideal current source has an infinite shunt resistance.**

Figure 8–6 shows equivalent voltage and current sources.

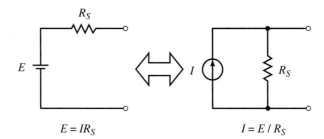

$$E = IR_S \qquad\qquad\qquad\qquad I = E/R_S$$

FIGURE 8–6

If the internal resistance of a source is considered, the source, whether it is a voltage source or a current source, is easily converted to the other type. The current source of Figure 8–6 is equivalent to the voltage source if

$$I = \frac{E}{R_S} \qquad\qquad (8\text{–}1)$$

and the resistance in both sources is R_S.

Similarly, a current source may be converted to an equivalent voltage source by letting

$$E = IR_S \qquad\qquad (8\text{–}2)$$

These results may be easily verified by connecting an external resistance, R_L, across each source. The sources can be equivalent only if the voltage across R_L is the same for both sources. Similarly, the sources are equivalent only if the current through R_L is the same when connected to either source.

Consider the circuit shown in Figure 8–7. The voltage across the load resistor is given as

$$V_L = \frac{R_L}{R_L + R_S}E \tag{8–3}$$

The current through the resistor R_L is given as

$$I_L = \frac{E}{R_L + R_S} \tag{8–4}$$

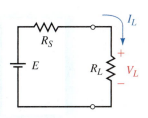

FIGURE 8–7

Next, consider an equivalent current source connected to the same load as shown in Figure 8–8. The current through the resistor R_L is given by

$$I_L = \frac{R_S}{R_S + R_L}I$$

But, when converting the source, we get

$$I = \frac{E}{R_S}$$

And so

$$I_L = \left(\frac{R_S}{R_S + R_L}\right)\left(\frac{E}{R_S}\right)$$

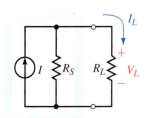

FIGURE 8–8

This result is equivalent to the current obtained in Equation 8–4. The voltage across the resistor is given as

$$V_L = I_L R_L$$
$$= \left(\frac{E}{R_S + R_L}\right)R_L$$

The voltage across the resistor is precisely the same as the result obtained in Equation 8–3. We therefore conclude that the load current and voltage drop are the same whether the source is a voltage source or an equivalent current source.

NOTES . . .

Although the sources are equivalent, currents and voltages within the sources may no longer be the same. The sources are only equivalent with respect to elements connected external to the terminals.

Convert the voltage source of Figure 8–9(a) into a current source and verify that the current, I_L, through the load is the same for each source.

EXAMPLE 8–4

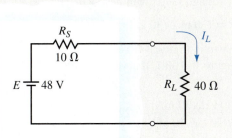

(a)

$$I = \frac{48\ \text{V}}{10\ \Omega} = 4.8\ \text{A}$$

(b)

FIGURE 8–9

Solution   The equivalent current source will have a current magnitude given as

$$I = \frac{48 \text{ V}}{10 \text{ }\Omega} = 4.8 \text{ A}$$

The resulting circuit is shown in Figure 8–9(b).

For the circuit of Figure 8–9(a), the current through the load is found as

$$I_L = \frac{48 \text{ V}}{10 \text{ }\Omega + 40 \text{ }\Omega} = 0.96 \text{ A}$$

For the equivalent circuit of Figure 8–9(b), the current through the load is

$$I_L = \frac{(4.8 \text{ A})(10 \text{ }\Omega)}{10 \text{ }\Omega + 40 \text{ }\Omega} = 0.96 \text{ A}$$

Clearly the results are the same.

EXAMPLE 8–5

Convert the current source of Figure 8–10(a) into a voltage source and verify that the voltage, V_L, across the load is the same for each source.

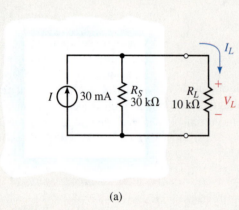
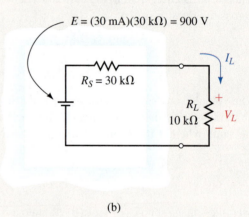

(a)

(b)

FIGURE 8–10

Solution   The equivalent voltage source will have a magnitude given as

$$E = (30 \text{ mA})(30 \text{ k}\Omega) = 900 \text{ V}$$

The resulting circuit is shown in Figure 8–10(b).

For the circuit of Figure 8–10(a), the voltage across the load is determined as

$$I_L = \frac{(30 \text{ k}\Omega)(30 \text{ mA})}{30 \text{ k}\Omega + 10 \text{ k}\Omega} = 22.5 \text{ mA}$$

$$V_L = I_L R_L = (22.5 \text{ mA})(10 \text{ k}\Omega) = 225 \text{ V}$$

For the equivalent circuit of Figure 8–10(b), the voltage across the load is

$$V_L = \frac{10 \text{ k}\Omega}{10 \text{ k}\Omega + 30 \text{ k}\Omega}(900 \text{ V}) = 225 \text{ V}$$

Once again, we see that the circuits are equivalent.

1. Convert the voltage sources of Figure 8–11 into equivalent current sources.

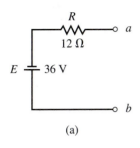
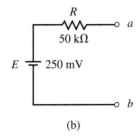

(a)   (b)

FIGURE 8–11

2. Convert the current sources of Figure 8–12 into equivalent voltage sources.

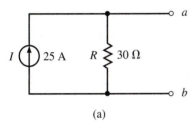
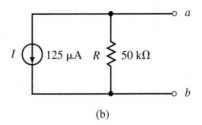

(a)   (b)

FIGURE 8–12

Answers
1. a. $I = 3.00$ A (downward) in parallel with $R = 12 \ \Omega$
   b. $I = 5.00 \ \mu$A (upward) in parallel with $R = 50 \ k\Omega$
2. a. $E = V_{ab} = 750$ V in series with $R = 30 \ \Omega$
   b. $E = V_{ab} = -6.25$ V in series with $R = 50 \ k\Omega$

When several current sources are placed in parallel, the circuit may be simplified by combining the current sources into a single current source. The magnitude and direction of this resultant source is determined by adding the currents in one direction and then subtracting the currents in the opposite direction.

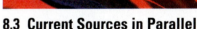

8.3 Current Sources in Parallel and Series

Simplify the circuit of Figure 8–13 and determine the voltage V_{ab}.

EXAMPLE 8–6

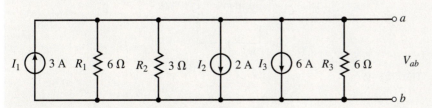

FIGURE 8–13

Solution   Since all of the current sources are in parallel, they can be replaced by a single current source. The equivalent current source will have a direction that is the same as both I_2 and I_3, since the magnitude of current in the downward direction is greater than the current in the upward direction. The equivalent current source has a magnitude of

$$I = 2\,\text{A} + 6\,\text{A} - 3\,\text{A} = 5\,\text{A}$$

as shown in Figure 8–14(a).

The circuit is further simplified by combining the resistors into a single value:

$$R_T = 6\,\Omega\|3\,\Omega\|6\,\Omega = 1.5\,\Omega$$

The equivalent circuit is shown in Figure 8–14(b).

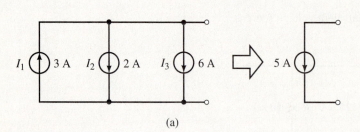

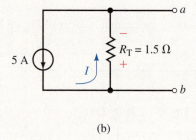

(a)   (b)

FIGURE 8–14

The voltage V_{ab} is found as

$$V_{ab} = -(5\,\text{A})(1.5\,\Omega) = -7.5\,\text{V}$$

EXAMPLE 8–7

Reduce the circuit of Figure 8–15 into a single current source and solve for the current through the resistor R_L.

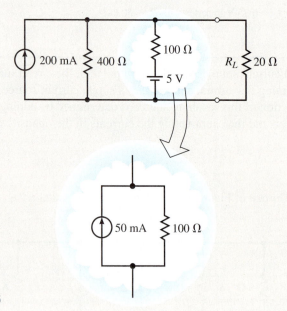

FIGURE 8–15

Solution   The voltage source in this circuit is converted to an equivalent current source as shown. The resulting circuit may then be simpified to a single current source where

$$I_S = 200 \text{ mA} + 50 \text{ mA} = 250 \text{ mA}$$

and

$$R_S = 400 \text{ }\Omega \| 100 \text{ }\Omega = 80 \text{ }\Omega$$

The simplified circuit is shown in Figure 8–16.

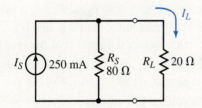

FIGURE 8–16

The current through R_L is now easily calculated as

$$I_L = \left(\frac{80 \text{ }\Omega}{80 \text{ }\Omega + 20 \text{ }\Omega} \right)(250 \text{ mA}) = 200 \text{ mA}$$

Current sources should never be placed in series. If a node is chosen between the current sources, it becomes immediately apparent that the current entering the node is not the same as the current leaving the node. Clearly, this cannot occur since there would then be a violation of Kirchhoff's current law (see Figure 8–17).

NOTES . . .

Current sources of different values are never placed in series.

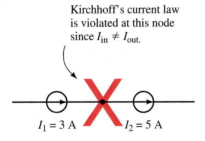

Kirchhoff's current law
is violated at this node
since $I_{in} \neq I_{out}$.

$I_1 = 3 \text{ A}$ $I_2 = 5 \text{ A}$

FIGURE 8–17

1. Briefly explain the procedure for converting a voltage source into an equivalent current source.

2. What is the most important rule determining how current sources are connected into a circuit?

IN-PROCESS
LEARNING CHECK 1

(Answers are at the end of the chapter.)

8.4 Branch-Current Analysis

In previous chapters we used Kirchhoff's circuit law and Kirchhoff's voltage law to solve equations for circuits having a single voltage source. In this section, you will use these powerful tools to analyze circuits having more than one source.

Branch-current analysis allows us to directly calculate the current in each branch of a circuit. Since the method involves the analysis of several simultaneous linear equations, you may find that a review of determinants is in order. Appendix B has been included to provide a review of the mechanics of solving simultaneous linear equations.

When applying branch-current analysis, you will find the technique listed below useful.

1. Arbitrarily assign current directions to each branch in the network. If a particular branch has a current source, then this step is not necessary since you already know the magnitude and direction of the current in this branch.

2. Using the assigned currents, label the polarities of the voltage drops across all resistors in the circuit.

3. Apply Kirchhoff's voltage law around each of the closed loops. Write just enough equations to include all branches in the loop equations. If a branch has only a current source and no series resistance, it is not necessary to include it in the KVL equations.

4. Apply Kirchhoff's current law at enough nodes to ensure that all branch currents have been included. In the event that a branch has only a current source, it will need to be included in this step.

5. Solve the resulting simultaneous linear equations.

EXAMPLE 8–8

Find the current in each branch in the circuit of Figure 8–18.

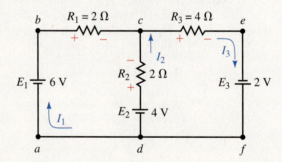

FIGURE 8–18

Solution

Step 1: Assign currents as shown in Figure 8–18.

Step 2: Indicate the polarities of the voltage drops on all resistors in the circuit, using the assumed current directions.

Step 3: Write the Kirchhoff voltage law equations.

Loop *abcda:* $6\text{ V} - (2\ \Omega)I_1 + (2\ \Omega)I_2 - 4\text{ V} = 0\text{ V}$

Notice that the circuit still has one branch which has not been included in the KVL equations, namely the branch *cefd.* This branch would be included if a loop equation for *cefdc* or for *abcefda* were written. There is no reason for choosing one loop over another, since the overall result will remain unchanged even though the intermediate steps will not give the same results.

Loop *cefdc:* $4\text{ V} - (2\ \Omega)I_2 - (4\ \Omega)I_3 + 2\text{ V} = 0\text{ V}$

Now that all branches have been included in the loop equations, there is no need to write any more. Although more loops exist, writing more loop equations would needlessly complicate the calculations.

Step 4: Write the Kirchhoff current law equation(s).

By applying KCL at node c, all branch currents in the network are included.

Node c: $\qquad\qquad I_3 = I_1 + I_2$

To simplify the solution of the simultaneous linear equations we write them as follows:

$$2I_1 - 2I_2 + 0I_3 = 2$$
$$0I_1 - 2I_2 - 4I_3 = -6$$
$$1I_1 + 1I_2 - 1I_3 = 0$$

The principles of linear algebra (Appendix B) allow us to solve for the determinant of the denominator as follows:

$$D = \begin{vmatrix} 2 & -2 & 0 \\ 0 & -2 & -4 \\ 1 & 1 & -1 \end{vmatrix}$$

$$= 2\begin{vmatrix} -2 & -4 \\ 1 & -1 \end{vmatrix} - 0\begin{vmatrix} -2 & 0 \\ 1 & -1 \end{vmatrix} + 1\begin{vmatrix} -2 & 0 \\ -2 & -4 \end{vmatrix}$$

$$= 2(2 + 4) - 0 + 1(8) = 20$$

Now, solving for the currents, we have the following:

$$I_1 = \frac{\begin{vmatrix} 2 & -2 & 0 \\ -6 & -2 & -4 \\ 0 & 1 & -1 \end{vmatrix}}{D}$$

$$= \frac{2\begin{vmatrix} -2 & -4 \\ 1 & -1 \end{vmatrix} - (-6)\begin{vmatrix} -2 & 0 \\ 1 & -1 \end{vmatrix} + 0\begin{vmatrix} -2 & 0 \\ -2 & -4 \end{vmatrix}}{20}$$

$$= \frac{2(2 + 4) + 6(2) + 0}{20} = \frac{24}{20} = 1.200 \text{ A}$$

$$I_2 = \frac{\begin{vmatrix} 2 & 2 & 0 \\ 0 & -6 & -4 \\ 1 & 0 & -1 \end{vmatrix}}{D}$$

$$= \frac{2\begin{vmatrix} -6 & -4 \\ 0 & -1 \end{vmatrix} - 0\begin{vmatrix} 2 & 0 \\ 0 & -1 \end{vmatrix} + 1\begin{vmatrix} 2 & 0 \\ -6 & -4 \end{vmatrix}}{20}$$

$$= \frac{2(6) + 0 + 1(-8)}{20} = \frac{4}{20} = 0.200 \text{ A}$$

$$I_3 = \frac{\begin{vmatrix} 2 & -2 & 2 \\ 0 & -2 & -6 \\ 1 & 1 & 0 \end{vmatrix}}{D}$$

$$= \frac{2\begin{vmatrix} -2 & -6 \\ 1 & 0 \end{vmatrix} - 0\begin{vmatrix} -2 & 2 \\ 1 & 0 \end{vmatrix} + 1\begin{vmatrix} -2 & 2 \\ -2 & -6 \end{vmatrix}}{20}$$

$$= \frac{2(6) - 0 + 1(12 + 4)}{20} = \frac{28}{20} = 1.400 \text{ A}$$

EXAMPLE 8–9

Find the currents in each branch of the circuit shown in Figure 8–19. Solve for the voltage V_{ab}.

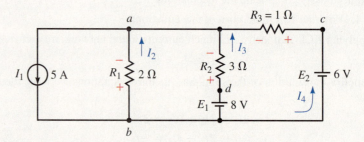

FIGURE 8–19

Solution   Notice that although the above circuit has four currents, there are only three **unknown** currents: I_2, I_3, and I_4. The current I_1 is given by the value of the constant-current source. In order to solve this network we will need three linear equations. As before, the equations are determined by Kirchhoff's voltage and current laws.

Step 1: The currents are indicated in the given circuit.

Step 2: The polarities of the voltages across all resistors are shown.

Step 3: Kirchhoff's voltage law is applied at the indicated loops:

Loop *badb*: $-(2\ \Omega)(I_2) + (3\ \Omega)(I_3) - 8\ \text{V} = 0\ \text{V}$
Loop *bacb*: $-(2\ \Omega)(I_2) + (1\ \Omega)(I_4) - 6\ \text{V} = 0\ \text{V}$

Step 4: Kirchhoff's current law is applied as follows:

Node *a*: $I_2 + I_3 + I_4 = 5\ \text{A}$

Rewriting the linear equations,

$$-2I_2 + 3I_3 + 0I_4 = 8$$
$$-2I_2 + 0I_3 + 1I_4 = 6$$
$$1I_2 + 1I_3 + 1I_4 = 5$$

The determinant of the denominator is evaluated as

$$D = \begin{vmatrix} -2 & 3 & 0 \\ -2 & 0 & 1 \\ 1 & 1 & 1 \end{vmatrix} = 11$$

Now solving for the currents, we have

$$I_2 = \frac{\begin{vmatrix} 8 & 3 & 0 \\ 6 & 0 & 1 \\ 5 & 1 & 1 \end{vmatrix}}{D} = \frac{11}{11} = -1.00\ \text{A}$$

$$I_3 = \frac{\begin{vmatrix} -2 & 8 & 0 \\ -2 & 6 & 1 \\ 1 & 5 & 1 \end{vmatrix}}{D} = \frac{22}{11} = 2.00\ \text{A}$$

$$I_4 = \frac{\begin{vmatrix} -2 & 3 & 8 \\ -2 & 0 & 6 \\ 1 & 1 & 5 \end{vmatrix}}{D} = \frac{44}{11} = 4.00\ \text{A}$$

The current I_2 is negative, which simply means that the actual direction of the current is opposite to the chosen direction.

Although the network may be further analyzed using the assumed current directions, it is easier to understand the circuit operation by showing the actual current directions as in Figure 8–20.

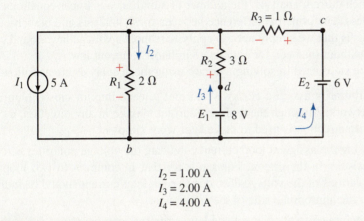

$$I_2 = 1.00 \text{ A}$$
$$I_3 = 2.00 \text{ A}$$
$$I_4 = 4.00 \text{ A}$$

FIGURE 8–20

Using the actual direction for I_2,

$$V_{ab} = +(2 \text{ } \Omega)(1 \text{ A}) = +2.00\text{V}$$

 **PRACTICE PROBLEMS 2**

Use branch-current analysis to solve for the indicated currents in the circuit of Figure 8–21.

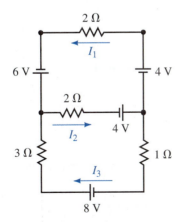

FIGURE 8–21

Answers
$I_1 = 3.00$ A, $I_2 = 4.00$ A, $I_3 = 1.00$ A

8.5 Mesh (Loop) Analysis

In the previous section you used Kirchhoff's laws to solve for the current in each branch of a given network. While the methods used were relatively simple, branch-current analysis is awkward to use because it generally involves solving several simultaneous linear equations. It is not difficult to see that the number of equations may be prohibitively large even for a relatively simple circuit.

A better approach and one that is used extensively in analyzing linear bilateral networks is called **mesh (or loop) analysis.** While the technique is similar to branch-current analysis, the number of simultaneous linear equations tends to be less. The principal difference between mesh analysis and branch-current analysis is that we simply need to apply Kirchhoff's voltage law around closed loops without the need for applying Kirchhoff's current law.

The steps used in solving a circuit using mesh analysis are as follows:

1. Arbitrarily assign a clockwise current to each interior closed loop in the network. Although the assigned current may be in any direction, a clockwise direction is used to make later work simpler.

2. Using the assigned loop currents, indicate the voltage polarities across all resistors in the circuit. For a resistor that is common to two loops, the polarities of the voltage drop due to each loop current should be indicated on the appropriate side of the component.

3. Applying Kirchhoff's voltage law, write the loop equations for each loop in the network. Do not forget that resistors that are common to two loops will have two voltage drops, one due to each loop.

4. Solve the resultant simultaneous linear equations.

5. Branch currents are determined by algebraically combining the loop currents that are common to the branch.

EXAMPLE 8–10

Find the current in each branch for the circuit of Figure 8–22.

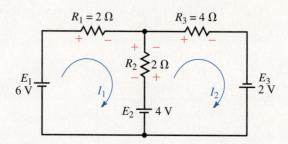

FIGURE 8–22

Solution

Step 1: Loop currents are assigned as shown in Figure 8–22. These currents are designated I_1 and I_2.

Step 2: Voltage polarities are assigned according to the loop currents. Notice that the resistor R_2 has two different voltage polarities due to the different loop currents.

Step 3: The loop equations are written by applying Kirchhoff's voltage law in each of the loops. The equations are as follows:

Loop 1: $6\text{ V} - (2\text{ }\Omega)I_1 - (2\text{ }\Omega)I_1 + (2\text{ }\Omega)I_2 - 4\text{ V} = 0$

Loop 2: $4\text{ V} - (2\text{ }\Omega)I_2 + (2\text{ }\Omega)I_1 - (4\text{ }\Omega)I_2 + 2\text{ V} = 0$

Note that the voltage across R_2 due to the currents I_1 and I_2 is indicated as two separated terms, where one term represents a voltage drop in the direction of I_1 and the other term represents a voltage rise in the same direction. The magnitude and polarity of the voltage across R_2 is determined by the actual size and directions of the loop currents. The above loop equations may be simplified as follows:

Loop 1: $(4\text{ }\Omega)I_1 - (2\text{ }\Omega)I_2 = 2\text{ V}$

Loop 2: $-(2\text{ }\Omega)I_1 + (6\text{ }\Omega)I_2 = 6\text{ V}$

Using determinants, the loop equations are easily solved as

$$I_1 = \frac{\begin{vmatrix} 2 & -2 \\ 6 & 6 \end{vmatrix}}{\begin{vmatrix} 4 & -2 \\ -2 & 6 \end{vmatrix}} = \frac{12 + 12}{24 - 4} = \frac{24}{20} = 1.20\text{ A}$$

and

$$I_2 = \frac{\begin{vmatrix} 4 & 2 \\ -2 & 6 \end{vmatrix}}{\begin{vmatrix} 4 & -2 \\ -2 & 6 \end{vmatrix}} = \frac{24 + 4}{24 - 4} = \frac{28}{20} = 1.40\text{ A}$$

From the above results, we see that the currents through resistors R_1 and R_3 are I_1 and I_2 respectively.

The branch current for R_2 is found by combining the loop currents through this resistor:

$$I_{R_2} = 1.40\text{ A} - 1.20\text{ A} = 0.20\text{ A (upward)}$$

The results obtained by using mesh analysis are exactly the same as those obtained by branch-current analysis. Whereas branch-current analysis required three equations, this approach requires the solution of only two simultaneous linear equations. Mesh analysis also requires that only Kirchhoff's voltage law be applied and clearly illustrates why mesh analysis is preferred to branch-current analysis.

If the circuit being analyzed contains current sources, the procedure is a bit more complicated. The circuit may be simplified by converting the current source(s) to voltage sources and then solving the resulting network using the procedure shown in the previous example. Alternatively, you may not wish to alter the circuit, in which case the current source will provide one of the loop currents.

EXAMPLE 8–11

Determine the current through the 8-V battery for the circuit shown in Figure 8–23.

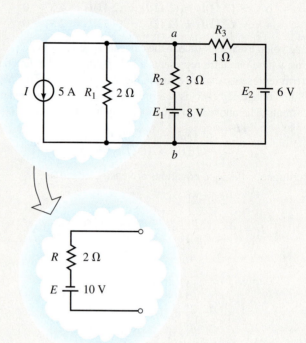

FIGURE 8–23

Solution   Convert the current source into an equivalent voltage source. The equivalent circuit may now be analyzed by using the loop currents shown in Figure 8–24.

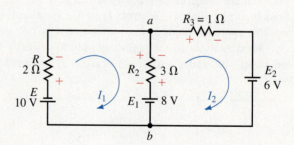

FIGURE 8–24

Loop 1: $-10 \text{ V} - (2\ \Omega)I_1 - (3\ \Omega)I_1 + (3\ \Omega)I_2 - 8 \text{ V} = 0$
Loop 2: $8 \text{ V} - (3\ \Omega)I_2 + (3\ \Omega)I_1 - (1\ \Omega)I_2 - 6 \text{ V} = 0$

Rewriting the linear equations, you get the following:

Loop 1: $(5\ \Omega)I_1 - (3\ \Omega)I_2 = -18 \text{ V}$
Loop 2: $-(3\ \Omega)I_1 + (4\ \Omega)I_2 = 2 \text{ V}$

Solving the equations using determinants, we have the following:

$$I_1 = \frac{\begin{vmatrix} -18 & -3 \\ 2 & 4 \end{vmatrix}}{\begin{vmatrix} 5 & -3 \\ -3 & 4 \end{vmatrix}} = -\frac{66}{11} = -6.00 \text{ A}$$

$$I_2 = \frac{\begin{vmatrix} 5 & -18 \\ -3 & 2 \end{vmatrix}}{\begin{vmatrix} 5 & -3 \\ -3 & 4 \end{vmatrix}} = -\frac{44}{11} = -4.00 \text{ A}$$

If the assumed direction of current in the 8-V battery is taken to be I_2, then

$$I = I_2 - I_1 = -4.00 \text{ A} - (-6.00 \text{ A}) = 2.00 \text{ A}$$

The direction of the resultant current is the same as I_2 (upward).

The circuit of Figure 8–23 may also be analyzed without converting the current source to a voltage source. Although the approach is generally not used, the following example illustrates the technique.

Determine the current through R_1 for the circuit shown in Figure 8–25.

EXAMPLE 8–12

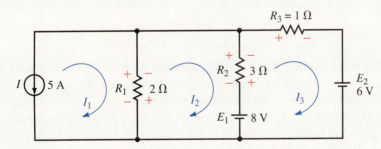

FIGURE 8–25

Solution   By inspection, we see that the loop current $I_1 = -5$ A. The mesh equations for the other two loops are as follows:

Loop 2: $-(2 \text{ }\Omega)I_2 + (2 \text{ }\Omega)I_1 - (3 \text{ }\Omega)I_2 + (3 \text{ }\Omega)I_3 - 8 \text{ V} = 0$
Loop 3: $8 \text{ V} - (3 \text{ }\Omega)I_3 + (3 \text{ }\Omega)I_2 - (1 \text{ }\Omega)I_3 - 6 \text{ V} = 0$

Although it is possible to analyze the circuit by solving three linear equations, it is easier to substitute the known value $I_1 = -5$A into the mesh equation for loop 2, which may now be written as

Loop 2: $-(2 \text{ }\Omega)I_2 - 10 \text{ V} - (3 \text{ }\Omega)I_2 + (3 \text{ }\Omega)I_3 - 8 \text{ V} = 0$

The loop equations may now be simplified as

Loop 2: $(5 \text{ }\Omega)I_2 - (3 \text{ }\Omega)I_3 = -18 \text{ V}$
Loop 3: $-(3 \text{ }\Omega)I_2 + (4 \text{ }\Omega)I_3 = 2 \text{ V}$

The simultaneous linear equations are solved as follows:

$$I_2 = \frac{\begin{vmatrix} -18 & -3 \\ 2 & 4 \end{vmatrix}}{\begin{vmatrix} 5 & -3 \\ -3 & 4 \end{vmatrix}} = -\frac{66}{11} = -6.00 \text{ A}$$

$$I_3 = \frac{\begin{vmatrix} 5 & -18 \\ -3 & 2 \end{vmatrix}}{\begin{vmatrix} 5 & -3 \\ -3 & 4 \end{vmatrix}} = -\frac{44}{11} = -4.00 \text{ A}$$

The calculated values of the assumed reference currents allow us to determine the actual current through the various resistors as follows:

$$I_{R_1} = I_1 - I_2 = -5 \text{ A} - (-6 \text{ A}) = 1.00 \text{ A} \quad \text{downward}$$
$$I_{R_2} = I_3 - I_2 = -4 \text{ A} - (-6 \text{ A}) = 2.00 \text{ A} \quad \text{upward}$$
$$I_{R_3} = -I_3 = 4.00 \text{ A} \quad \text{left}$$

These results are consistent with those obtained in Example 8–9.

Format Approach for Mesh Analysis

A very simple technique may be used to write the mesh equations for any linear bilateral network. When this format approach is used, the simultaneous linear equations for a network having n independent loops will appear as follows:

$$R_{11}I_1 - R_{12}I_2 - R_{13}I_3 - \cdots - R_{1n}I_n = E_1$$
$$-R_{21}I_1 + R_{22}I_2 - R_{23}I_3 - \cdots - R_{2n}I_n = E_2$$
$$\vdots$$
$$-R_{n1}I_1 - R_{n2}I_2 - R_{n3}I_3 - \cdots + R_{nn}I_n = E_n$$

The terms $R_{11}, R_{22}, R_{33}, \ldots, R_{nn}$ represent the total resistance in each loop and are found by simply adding all the resistances in a particular loop. The remaining resistance terms are called the **mutual resistance** terms. These resistances represent resistance which is shared between two loops. For example, the mutual resistance R_{12} is the resistance in loop 1, which is located in the branch between loop 1 and loop 2. If there is no resistance between two loops, this term will be zero.

The terms containing $R_{11}, R_{22}, R_{33}, \ldots, R_{nn}$ are positive, and all of the mutual resistance terms are negative. This characteristic occurs because all currents are assumed to be clockwise.

If the linear equations are correctly written, you will find that the coefficients along the principal diagonal ($R_{11}, R_{22}, R_{33}, \ldots, R_{nn}$) will be positive. All other coefficients will be negative. Also, if the equations are correctly written, the terms will be symmetrical about the principal diagonal, e.g., $R_{12} = R_{21}$.

The terms $E_1, E_2, E_3, \ldots, E_n$ are the summation of the voltage rises in the direction of the loop currents. If a voltage source appears in the branch shared by two loops, it will be included in the calculation of the voltage rise for each loop.

The method used in applying the format approach of mesh analysis is as follows:

1. Convert current sources into equivalent voltage sources.

2. Assign clockwise currents to each independent closed loop in the network.

3. Write the simultaneous linear equations in the format outlined.

4. Solve the resulting simultaneous linear equations.

Solve for the currents through R_2 and R_3 in the circuit of Figure 8–26.

EXAMPLE 8–13

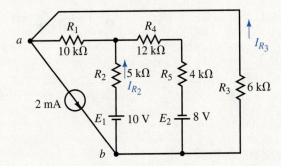

FIGURE 8–26

Solution

Step 1: Although we see that the circuit has a current source, it may not be immediately evident how the source can be converted into an equivalent voltage source. Redrawing the circuit into a more recognizable form, as shown in Figure 8–27, we see that the 2-mA current source is in parallel with a 6-kΩ resistor. The source conversion is also illustrated in Figure 8–27.

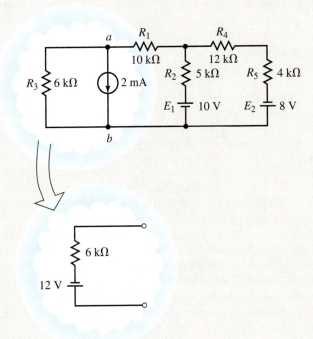

FIGURE 8–27

Step 2: Redrawing the circuit is further simplified by labelling some of the nodes, in this case a and b. After performing a source conversion, we have the two-loop circuit shown in Figure 8–28. The current directions for I_1 and I_2 are also illustrated.

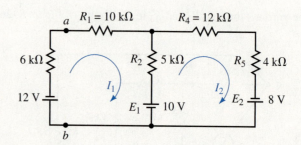

FIGURE 8–28

Step 3: The loop equations are

Loop 1: $(6 \text{ k}\Omega + 10 \text{ k}\Omega + 5 \text{ k}\Omega)I_1 - (5 \text{ k}\Omega)I_2 = -12 \text{ V} - 10 \text{ V}$
Loop 2: $-(5 \text{ k}\Omega)I_1 + (5 \text{ k}\Omega + 12 \text{ k}\Omega + 4 \text{ k}\Omega)I_2 = 10 \text{ V} + 8 \text{ V}$

In loop 1, both voltages are negative since they appear as voltage drops when following the direction of the loop current.
These equations are rewritten as

$$(21 \text{ k}\Omega)I_1 - (5 \text{ k}\Omega)I_2 = -22 \text{ V}$$
$$-(5 \text{ k}\Omega)I_1 + (21 \text{ k}\Omega)I_2 = 18 \text{ V}$$

Step 4: In order to simplify the solution of the previous linear equations, we may eliminate the units (kΩ and V) from our calculations. By inspection, we see that the units for current must be in milliamps. Using determinants, we solve for the currents I_1 and I_2 as follows:

$$I_1 = \frac{\begin{vmatrix} -22 & -5 \\ 18 & 21 \end{vmatrix}}{\begin{vmatrix} 21 & -5 \\ -5 & 21 \end{vmatrix}} = \frac{-462 + 90}{441 - 25} = \frac{-372}{416} = -0.894 \text{ mA}$$

$$I_2 = \frac{\begin{vmatrix} 21 & -22 \\ -5 & 18 \end{vmatrix}}{\begin{vmatrix} 21 & -5 \\ -5 & 21 \end{vmatrix}} = \frac{378 - 110}{441 - 25} = \frac{268}{416} = 0.644 \text{ mA}$$

The current through resistor R_2 is easily determined to be

$$I_2 - I_1 = 0.644 \text{ mA} - (-0.894 \text{ mA}) = 1.54 \text{ mA}   \text{(upward)}$$

The current through R_3 is not found as easily. A common mistake is to say that the current in R_3 is the same as the current through the 6-kΩ resistor of the circuit in Figure 8–28. **This is not the case.** Since this resistor was part of the source conversion it is no longer in the same location as in the original circuit.
Although there are several ways of finding the required current, the method used here is the application of Ohm's law. If we examine Figure 8–26, we see that the voltage across R_3 is equal to V_{ab}. From Figure 8–28, we see that we determine V_{ab} by using the calculated value of I_1.

$$V_{ab} = -(6 \text{ k}\Omega)I_1 - 12 \text{ V} = -(6 \text{ k}\Omega)(-0.894 \text{ mA}) - 12 \text{ V} = -6.64 \text{ V}$$

The above calculation indicates that the current through R_3 is upward (since point a is negative with respect to point b). The current has a value of

$$I_{R_3} = \frac{6.64 \text{ V}}{6 \text{ k}\Omega} = 1.11 \text{ mA}$$

Use mesh analysis to find the loop currents in the circuit of Figure 8–29.

Answers
$I_1 = 3.00$ A, $I_2 = 2.00$ A, $I_3 = 5.00$ A

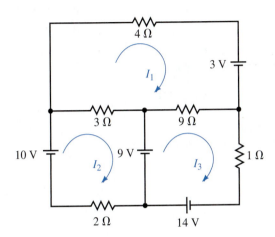

FIGURE 8–29

8.6 Nodal Analysis

In the previous section we applied Kirchhoff's voltage law to arrive at loop currents in a network. In this section we will apply Kirchhoff's current law to determine the potential difference (voltage) at any node with respect to some arbitrary reference point in a network. Once the potentials of all nodes are known, it is a simple matter to determine other quantities such as current and power within the network.

The steps used in solving a circuit using **nodal analysis** are as follows:

1. Arbitrarily assign a reference node within the circuit and indicate this node as **ground.** The reference node is usually located at the bottom of the circuit, although it may be located anywhere.

2. Convert each voltage source in the network to its equivalent current source. This step, although not absolutely necessary, makes further calculations easier to understand.

3. Arbitrarily assign voltages ($V_1, V_2, \ldots, V_n$) to the remaining nodes in the circuit. (Remember that you have already assigned a reference node, so these voltages will all be with respect to the chosen reference.)

4. Arbitrarily assign a current direction to each branch in which there is no current source. Using the assigned current directions, indicate the corresponding polarities of the voltage drops on all resistors.

5. With the exception of the reference node (ground), apply Kirchhoff's current law at each of the nodes. If a circuit has a total of $n + 1$ nodes (including the reference node), there will be n simultaneous linear equations.

6. Rewrite each of the arbitrarily assigned currents in terms of the potential difference across a known resistance.

7. Solve the resulting simultaneous linear equations for the voltages ($V_1, V_2, \ldots, V_n$).

EXAMPLE 8–14

Given the circuit of Figure 8–30, use nodal analysis to solve for the voltage V_{ab}.

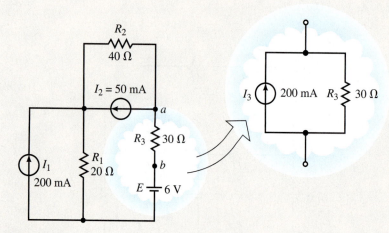

FIGURE 8–30

Solution

Step 1: Select a convenient reference node.

Step 2: Convert the voltage sources into equivalent current sources. The equivalent circuit is shown in Figure 8–31.

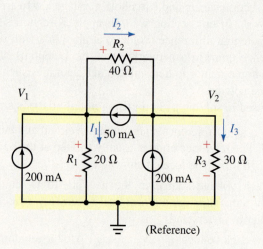

FIGURE 8–31

Steps 3 and 4: Arbitrarily assign node voltages and branch currents. Indicate the voltage polarities across all resistors according to the assumed current directions.

Step 5: We now apply Kirchhoff's current law at the nodes labelled as V_1 and V_2:

Node V_1:
$$\sum I_{\text{entering}} = \sum I_{\text{leaving}}$$
$$200 \text{ mA} + 50 \text{ mA} = I_1 + I_2$$

Node V_2:
$$\sum I_{\text{entering}} = \sum I_{\text{leaving}}$$
$$200 \text{ mA} + I_2 = 50 \text{ mA} + I_3$$

Step 6: The currents are rewritten in terms of the voltages across the resistors as follows:

$$I_1 = \frac{V_1}{20\ \Omega}$$

$$I_2 = \frac{V_1 - V_2}{40\ \Omega}$$

$$I_3 = \frac{V_2}{30\ \Omega}$$

The nodal equations become

$$200\text{ mA} + 50\text{ mA} = \frac{V_1}{20\ \Omega} + \frac{V_1 - V_2}{40\ \Omega}$$

$$200\text{ mA} + \frac{V_1 - V_2}{40\ \Omega} = 50\text{ mA} + \frac{V_2}{30\ \Omega}$$

Substituting the voltage expressions into the original nodal equations, we have the following simultaneous linear equations:

$$\left(\frac{1}{20\ \Omega} + \frac{1}{40\ \Omega}\right)V_1 - \left(\frac{1}{40\ \Omega}\right)V_2 = 250\text{ mA}$$

$$-\left(\frac{1}{40\ \Omega}\right)V_1 + \left(\frac{1}{30\ \Omega} + \frac{1}{40\ \Omega}\right)V_2 = 150\text{ mA}$$

These may be further simplified as

$$(0.075\text{ S})V_1 - (0.025\text{ S})V_2 = 250\text{ mA}$$

$$-(0.025\text{ S})V_1 + (0.058\overline{3})V_2 = 150\text{ mA}$$

Step 7: Use determinants to solve for the nodal voltages as

$$V_1 = \frac{\begin{vmatrix} 0.250 & -0.025 \\ 0.150 & 0.058\overline{3} \end{vmatrix}}{\begin{vmatrix} 0.075 & -0.025 \\ 0.025 & 0.058\overline{3} \end{vmatrix}}$$

$$= \frac{(0.250)(0.058\overline{3}) - (0.150)(-0.025)}{(0.075)(0.058\overline{3}) - (-0.025)(-0.025)}$$

$$= \frac{0.018\overline{3}}{0.00375} = 4.89\text{ V}$$

and

$$V_2 = \frac{\begin{vmatrix} 0.075 & 0.250 \\ -0.025 & 0.150 \end{vmatrix}}{\begin{vmatrix} 0.075 & 0.025 \\ -0.025 & 0.058\overline{3} \end{vmatrix}}$$

$$= \frac{(0.075)(0.150) - (-0.025)(0.250)}{0.00375}$$

$$= \frac{0.0175}{0.00375} = 4.67\text{ V}$$

If we go back to the original circuit of Figure 8–30, we see that the voltage V_2 is the same as the voltage V_a, namely

$$V_a = 4.67 \text{ V} = 6.0 \text{ V} + V_{ab}$$

Therefore, the voltage V_{ab} is simply found as

$$V_{ab} = 4.67 \text{ V} - 6.0 \text{ V} = -1.33 \text{ V}$$

EXAMPLE 8–15

Determine the nodal voltages for the circuit shown in Figure 8–32.

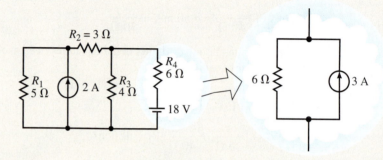

FIGURE 8–32

Solution   By following the steps outlined, the circuit may be redrawn as shown in Figure 8–33.

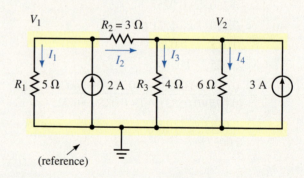

(reference)

FIGURE 8–33

Applying Kirchhoff's current law to the nodes corresponding to V_1 and V_2, the following nodal equations are obtained:

$$\sum I_{\text{leaving}} = \sum I_{\text{entering}}$$

Node V_1: $I_1 + I_2 = 2 \text{ A}$

Node V_2: $I_3 + I_4 = I_2 + 3 \text{ A}$

The currents may once again be written in terms of the voltages across the resistors:

$$I_1 = \frac{V_1}{5\,\Omega}$$

$$I_2 = \frac{V_1 - V_2}{3\,\Omega}$$

$$I_3 = \frac{V_2}{4\,\Omega}$$

$$I_4 = \frac{V_2}{6\,\Omega}$$

The nodal equations become

Node V_1: $\dfrac{V_1}{5\,\Omega} + \dfrac{(V_1 - V_2)}{3\,\Omega} = 2\text{ A}$

Node V_2: $\dfrac{V_2}{4\,\Omega} + \dfrac{V_2}{6\,\Omega} = \dfrac{(V_1 - V_2)}{3\,\Omega} + 3\text{ A}$

These equations may now be simplified as

Node V_1: $\left(\dfrac{1}{5\,\Omega} + \dfrac{1}{3\,\Omega}\right)V_1 - \left(\dfrac{1}{3\,\Omega}\right)V_2 = 2\text{ A}$

Node V_2: $-\left(\dfrac{1}{3\,\Omega}\right)V_1 + \left(\dfrac{1}{4\,\Omega} + \dfrac{1}{6\,\Omega} + \dfrac{1}{3\,\Omega}\right)V_2 = 3\text{ A}$

The solutions for V_1 and V_2 are found using determinants:

$$V_1 = \frac{\begin{vmatrix} 2 & -0.333 \\ 3 & 0.750 \end{vmatrix}}{\begin{vmatrix} 0.533 & -0.333 \\ -0.333 & 0.750 \end{vmatrix}} = \frac{2.500}{0.289} = 8.65\text{ V}$$

$$V_2 = \frac{\begin{vmatrix} 0.533 & 2 \\ -0.333 & 3 \end{vmatrix}}{\begin{vmatrix} 0.533 & -0.333 \\ -0.333 & 0.750 \end{vmatrix}} = \frac{2.267}{0.289} = 7.85\text{ V}$$

In the previous two examples, you may have noticed that the simultaneous linear equations have a format similar to that developed for mesh analysis. When we wrote the nodal equation for node V_1 the coefficient for the variable V_1 was positive, and it had a magnitude given by the summation of the conductance attached to this node. The coefficient for the variable V_2 was negative and had a magnitude given by the mutual conductance between nodes V_1 and V_2.

Format Approach

A simple format approach may be used to write the nodal equations for any network having $n + 1$ nodes. Where one of these nodes is denoted as the reference node, there will be n simultaneous linear equations, which will appear as follows:

$$G_{11}V_1 - G_{12}V_2 - G_{13}V_3 - \cdots - R_{1n}V_n = I_1$$
$$-G_{21}V_1 + G_{22}V_2 - G_{23}V_3 - \cdots - R_{2n}V_n = I_2$$
$$\vdots$$
$$-G_{n1}V_1 - G_{n2}V_2 - G_{n3}V_3 - \cdots + R_{nn}V_n = I_n$$

The coefficients (constants) G_{11}, G_{22}, G_{33}, . . . , G_{nn} represent the summation of the conductances attached to the particular node. The remaining coefficients are called the **mutual conductance** terms. For example, the mutual conductance G_{23} is the conductance attached to node V_2, which is common to node V_3. If there is no conductance that is common to two nodes, then this term would be zero. Notice that the terms G_{11}, G_{22}, G_{33}, . . . , G_{nn} are positive and that the mutual conductance terms are negative. Further, if the equations are written correctly, then the terms will be symmetrical about the principal diagonal, e.g., $G_{23} = G_{32}$.

The terms V_1, V_2, . . . , V_n are the unknown node voltages. Each voltage represents the potential difference between the node in question and the reference node.

The terms I_1, I_2, . . . , I_n are the summation of current sources entering the node. If a current source has a current such that it is leaving the node, then the current is simply assigned as negative. If a particular current source is shared between two nodes, then this current must be included in both nodal equations.

The method used in applying the format approach of nodal analysis is as follows:

1. Convert voltage sources into equivalent current sources.
2. Label the reference node as $\perp$. Label the remaining nodes as V_1, V_2, . . . , V_n.
3. Write the linear equation for each node using the format outlined.
4. Solve the resulting simultaneous linear equations for V_1, V_2, . . . , V_n.

The next examples illustrate how the format approach is used to solve circuit problems.

EXAMPLE 8–16

Determine the nodal voltages for the circuit shown in Figure 8–34.

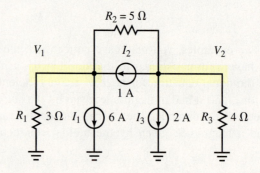

FIGURE 8–34

Solution   The circuit has a total of three nodes: the reference node (at a potential of zero volts) and two other nodes, V_1 and V_2.

By applying the format approach for writing the nodal equations, we get two equations:

Node V_1: $\left(\dfrac{1}{3\ \Omega}+\dfrac{1}{5\ \Omega}\right)V_1-\left(\dfrac{1}{5\ \Omega}\right)V_2=-6\ \text{A}+1\ \text{A}$

Node V_2: $-\left(\dfrac{1}{5\ \Omega}\right)V_1+\left(\dfrac{1}{5\ \Omega}+\dfrac{1}{4\ \Omega}\right)V_2=-1\ \text{A}-2\ \text{A}$

On the right-hand sides of the above, those currents that are leaving the nodes are given a negative sign.

These equations may be rewritten as

Node V_1: $\quad\quad\quad (0.533\ \text{S})V_1-(0.200\ \text{S})V_2=-5\ \text{A}$

Node V_2: $\quad\quad\quad -(0.200\ \text{S})V_1+(0.450\ \text{S})V_2=-3\ \text{A}$

Using determinants to solve these equations, we have

$$V_1=\dfrac{\begin{vmatrix}-5 & -0.200 \\ -3 & 0.450\end{vmatrix}}{\begin{vmatrix}0.533 & -0.200 \\ -0.200 & 0.450\end{vmatrix}}=\dfrac{-2.85}{0.200}=-14.3\ \text{V}$$

$$V_2=\dfrac{\begin{vmatrix}0.533 & -5 \\ -0.200 & -3\end{vmatrix}}{\begin{vmatrix}0.533 & -0.200 \\ -0.200 & 0.450\end{vmatrix}}=\dfrac{-2.60}{0.200}=-13.0\ \text{V}$$

Use nodal analysis to find the nodal voltages for the circuit of Figure 8–35. Use the answers to solve for the current through R_1.

EXAMPLE 8–17

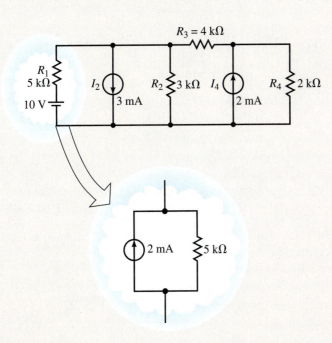

FIGURE 8–35

Solution In order to apply nodal analysis, we must first convert the voltage source into its equivalent current source. The resulting circuit is shown in Figure 8–36.

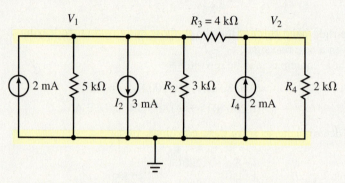

FIGURE 8–36

Labelling the nodes and writing the nodal equations, we obtain the following:

Node V_1: $\left(\dfrac{1}{5\text{ k}\Omega} + \dfrac{1}{3\text{ k}\Omega} + \dfrac{1}{4\text{ k}\Omega}\right)V_1 - \left(\dfrac{1}{4\text{ k}\Omega}\right)V_2 = 2\text{ mA} - 3\text{ mA}$

Node V_2: $\qquad -\left(\dfrac{1}{4\text{ k}\Omega}\right)V_1 + \left(\dfrac{1}{4\text{ k}\Omega} + \dfrac{1}{2\text{ k}\Omega}\right)V_2 = 2\text{ mA}$

Because it is inconvenient to use kilohms and milliamps throughout our calculations, we may eliminate these units in our calculations. You have already seen that any voltage obtained by using these quantities will result in the units being "volts." Therefore the nodal equations may be simplified as

Node V_1: $\qquad\qquad (0.7833)V_1 - (0.2500)V_2 = -1$

Node V_2: $\qquad\qquad -(0.2500)V_1 + (0.750)V_2 = 2$

The solutions are as follows:

$$V_1 = \frac{\begin{vmatrix} -1 & -0.250 \\ 2 & 0.750 \end{vmatrix}}{\begin{vmatrix} 0.7833 & -0.250 \\ -0.250 & 0.750 \end{vmatrix}} = \frac{-0.250}{0.525} = -0.476\text{ V}$$

$$V_2 = \frac{\begin{vmatrix} 0.7833 & -1 \\ -0.250 & 2 \end{vmatrix}}{\begin{vmatrix} 0.7833 & -0.250 \\ -0.250 & 0.750 \end{vmatrix}} = \frac{1.3167}{0.525} = 2.51\text{ V}$$

Using the values derived for the nodal voltages, it is now possible to solve for any other quantities in the circuit. To determine the current through resistor $R_1 = 5\text{ k}\Omega$, we first reassemble the circuit as it appeared originally. Since the node voltage V_1 is the same in both circuits, we use it in determining the desired current. The resistor may be isolated as shown in Figure 8–37.

The current is easily found as

$$I = 10\text{ V} - \frac{(-0.476\text{ V})}{5\text{ k}\Omega} = 2.10\text{ mA} \quad (\text{upward})$$

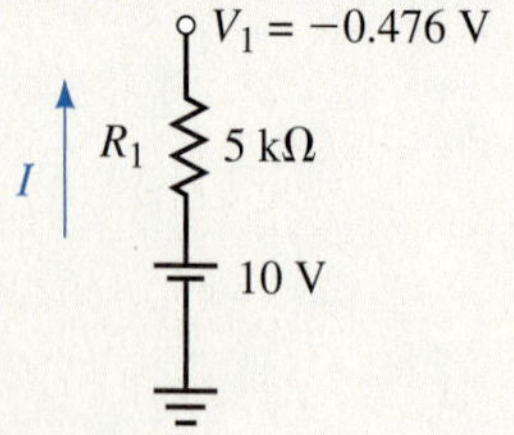

NOTES . . .

A common mistake is that the current is determined by using the equivalent circuit rather than the original circuit. You must remember that the circuits are only equivalent external to the conversion.

$V_1 = -0.476$ V

$R_1 \lessgtr 5\text{ k}\Omega$

I

10 V

FIGURE 8–37

Use nodal analysis to determine the node voltages for the circuit of Figure 8–38.

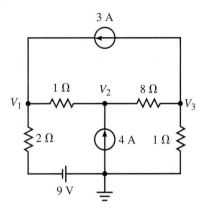

FIGURE 8–38

Answers
$V_1 = 3.00$ V, $V_2 = 6.00$ V, $V_3 = -2.00$ V

Delta-Wye Conversion

You have previously examined resistor networks involving series, parallel, and series-parallel combinations. We will next examine networks that cannot be placed into any of the above categories. While these circuits may be analyzed using techniques developed earlier in this chapter, there is an easier approach. For example, consider the circuit shown in Figure 8–39.

This circuit could be analyzed using mesh analysis. However, you see that the analysis would involve solving four simultaneous linear equations, since there are four separate loops in the circuit. If we were to use nodal analysis, the solution would require determining three node voltages, since there are three nodes in addition to a reference node. Unless a computer is used, both techniques are very time-consuming and prone to error.

As you have already seen, it is occasionally easier to examine a circuit after it has been converted to some equivalent form. We will now develop a technique for converting a circuit from a **delta** (or pi) into an equivalent **wye** (or **tee**) **circuit.** Consider the circuits shown in Figure 8–40. We start by making the assumption that the networks shown in Figure 8–40(a) are equivalent to those shown in Figure 8–40(b). Then, using this assumption, we will determine the mathematical relationships between the various resistors in the equivalent circuits.

The circuit of Figure 8–40(a) can be equivalent to the circuit of Figure 8–40(b) only if the resistance "seen" between any two terminals is exactly the same. If we were to connect a source between terminals *a* and *b* of the "Y," the resistance between the terminals would be

$$R_{ab} = R_1 + R_2 \tag{8–5}$$

But the resistance between terminals *a* and *b* of the "Δ" is

$$R_{ab} = R_C \| (R_A + R_B) \tag{8–6}$$

8.7 Delta-Wye (Pi-Tee) Conversion

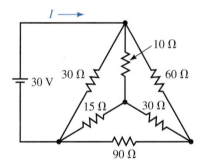

FIGURE 8–39

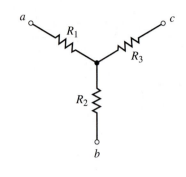

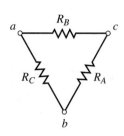

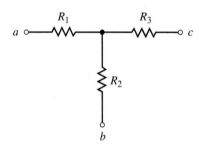

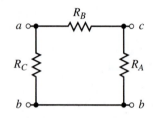

(a) Wye ("Y") or Tee ("T") network   (b) Delta ("Δ") or Pi ("Π") network

FIGURE 8–40

Combining Equations 8–5 and 8–6, we get

$$R_1 + R_2 = \frac{R_C(R_A + R_B)}{R_A + R_B + R_C}$$

$$R_1 + R_2 = \frac{R_A R_C + R_B R_C}{R_A + R_B + R_C}$$

(8–7)

Using a similar approach between terminals b and c, we get

$$R_2 + R_3 = \frac{R_A R_B + R_A R_C}{R_A + R_B + R_C}$$

(8–8)

and between terminals c and a we get

$$R_1 + R_3 = \frac{R_A R_B + R_B R_C}{R_A + R_B + R_C}$$

(8–9)

If Equation 8–8 is subtracted from Equation 8–7, then

$$R_1 + R_2 - (R_2 + R_3) = \frac{R_A R_C + R_B R_C}{R_A + R_B + R_C} - \frac{R_A R_B + R_A R_C}{R_A + R_B + R_C}$$

$$R_1 - R_3 = \frac{R_B R_C - R_A R_B}{R_A + R_B + R_C}$$

(8–10)

Adding Equations 8–9 and 8–10, we get

$$R_1 + R_3 + R_1 - R_3 = \frac{R_A R_B + R_B R_C}{R_A + R_B + R_C} + \frac{R_B R_C - R_A R_B}{R_A + R_B + R_C}$$

$$2R_1 = \frac{2R_B R_C}{R_A + R_B + R_C}$$

(8–11)

$$R_1 = \frac{R_B R_C}{R_A + R_B + R_C}$$

Using a similar approach, we obtain

$$R_2 = \frac{R_A R_C}{R_A + R_B + R_C} \qquad \text{(8–12)}$$

$$R_3 = \frac{R_A R_B}{R_A + R_B + R_C} \qquad \text{(8–13)}$$

Notice that any resistor connected to a point of the "Y" is obtained by finding the product of the resistors connected to the same point in the "Δ" and then dividing by the sum of all the "Δ" resistances.

If all the resistors in a Δ circuit have the same value, R_Δ, then the resulting resistors in the equivalent Y network will also be equal and have a value given as

$$R_Y = \frac{R_\Delta}{3} \qquad \text{(8–14)}$$

Find the equivalent Y circuit for the Δ circuit shown in Figure 8–41.

EXAMPLE 8–18

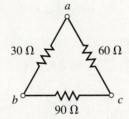

FIGURE 8–41

Solution From the circuit of Figure 8–41, we see that we have the following resistor values:

$$R_A = 90 \ \Omega$$
$$R_B = 60 \ \Omega$$
$$R_C = 30 \ \Omega$$

Applying Equations 8–11 through 8–13 we have the following equivalent "Y" resistor values:

$$R_1 = \frac{(30 \ \Omega)(60 \ \Omega)}{30 \ \Omega + 60 \ \Omega + 90 \ \Omega}$$
$$= \frac{1800 \ \Omega}{180} = 10 \ \Omega$$

$$R_2 = \frac{(30 \ \Omega)(90 \ \Omega)}{30 \ \Omega + 60 \ \Omega + 90 \ \Omega}$$
$$= \frac{2700 \ \Omega}{180} = 15 \ \Omega$$

$$R_3 = \frac{(60 \ \Omega)(90 \ \Omega)}{30 \ \Omega + 60 \ \Omega + 90 \ \Omega}$$
$$= \frac{5400 \ \Omega}{180} = 30 \ \Omega$$

The resulting circuit is shown in Figure 8–42.

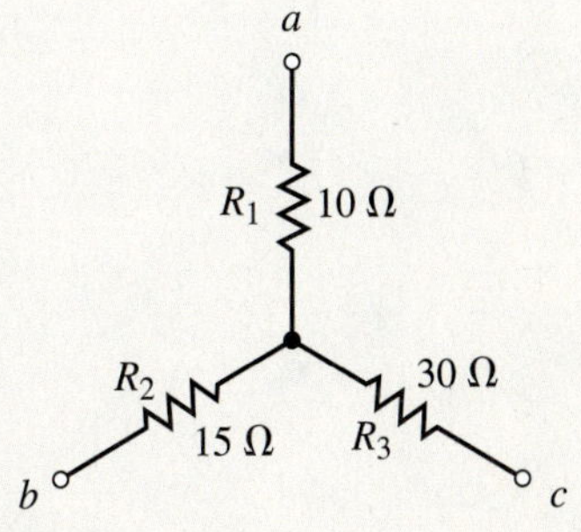

FIGURE 8–42

Wye-Delta Conversion

By using Equations 8–11 to 8–13, it is possible to derive another set of equations that allow the conversion from a "Y" into an equivalent "Δ." Examining Equations 8–11 through 8–13, we see that the following must be true:

$$R_A + R_B + R_C = \frac{R_A R_B}{R_3} = \frac{R_A R_C}{R_2} = \frac{R_B R_C}{R_1}$$

From the above expression we may write the following two equations:

$$R_B = \frac{R_A R_1}{R_2} \tag{8-15}$$

$$R_C = \frac{R_A R_1}{R_3} \tag{8-16}$$

Now, substituting Equations 8–15 and 8–16 into Equation 8–11, we have the following:

$$R_1 = \frac{\left(\dfrac{R_A R_1}{R_2}\right)\left(\dfrac{R_A R_1}{R_3}\right)}{R_A + \left(\dfrac{R_A R_1}{R_2}\right) + \left(\dfrac{R_A R_1}{R_3}\right)}$$

By factoring R_A out of each term in the denominator, we are able to arrive at

$$R_1 = \frac{\left(\dfrac{R_A R_1}{R_2}\right)\left(\dfrac{R_A R_1}{R_3}\right)}{R_A\left[1 + \left(\dfrac{R_1}{R_2}\right) + \left(\dfrac{R_1}{R_3}\right)\right]}$$

$$R_1 = \frac{\left(\dfrac{R_A R_1 R_1}{R_2 R_3}\right)}{\left[1 + \left(\dfrac{R_1}{R_2}\right) + \left(\dfrac{R_1}{R_3}\right)\right]}$$

$$= \frac{\left(\dfrac{R_A R_1 R_1}{R_2 R_3}\right)}{\left(\dfrac{R_1 R_2 + R_1 R_3 + R_2 R_3}{R_2 R_3}\right)}$$

$$= \frac{R_A R_1 R_1}{R_1 R_2 + R_1 R_3 + R_2 R_3}$$

Rewriting the above expression gives

$$R_A = \frac{R_1 R_2 + R_1 R_3 + R_2 R_3}{R_1} \tag{8-17}$$

Similarly,

$$R_B = \frac{R_1 R_2 + R_1 R_3 + R_2 R_3}{R_2} \tag{8-18}$$

and

$$R_C = \frac{R_1 R_2 + R_1 R_3 + R_2 R_3}{R_3} \tag{8-19}$$

In general, we see that the resistor in any side of a "Δ" is found by taking the sum of all two-product combinations of "Y" resistor values and then dividing by the resistance in the "Y," which is located directly opposite to the resistor being calculated.

If the resistors in a Y network are all equal, then the resultant resistors in the equivalent Δ circuit will also be equal and given as

$$R_\Delta = 3R_Y \qquad\qquad \textbf{(8–20)}$$

Find the Δ network equivalent of the Y network shown in Figure 8–43.

EXAMPLE 8–19

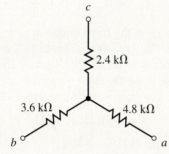

FIGURE 8–43

Solution   The equivalent Δ network is shown in Figure 8–44.

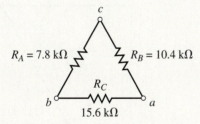

FIGURE 8–44

The values of the resistors are determined as follows:

$$R_A = \frac{(4.8\ \text{k}\Omega)(2.4\ \text{k}\Omega) + (4.8\ \text{k}\Omega)(3.6\ \text{k}\Omega) + (2.4\ \text{k}\Omega)(3.6\ \text{k}\Omega)}{4.8\ \text{k}\Omega}$$

$$= 7.8\ \text{k}\Omega$$

$$R_B = \frac{(4.8\ \text{k}\Omega)(2.4\ \text{k}\Omega) + (4.8\ \text{k}\Omega)(3.6\ \text{k}\Omega) + (2.4\ \text{k}\Omega)(3.6\ \text{k}\Omega)}{3.6\ \text{k}\Omega}$$

$$= 10.4\ \text{k}\Omega$$

$$R_C = \frac{(4.8\ \text{k}\Omega)(2.4\ \text{k}\Omega) + (4.8\ \text{k}\Omega)(3.6\ \text{k}\Omega) + (2.4\ \text{k}\Omega)(3.6\ \text{k}\Omega)}{2.4\ \text{k}\Omega}$$

$$= 15.6\ \text{k}\Omega$$

EXAMPLE 8–20

Given the circuit of Figure 8–45, find the total resistance, R_T, and the total current, I.

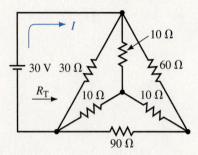

FIGURE 8–45

Solution As is often the case, the given circuit may be solved in one of two ways. We may convert the "Δ" into its equivalent "Y," and solve the circuit by placing the resultant branches in parallel, or we may convert the "Y" into its equivalent "Δ." We choose to use the latter conversion since the resistors in the "Y" have the same value. The equivalent "Δ" will have all resistors given as

$$R_\Delta = 3(10 \; \Omega) = 30 \; \Omega$$

The resulting circuit is shown in Figure 8–46(a).

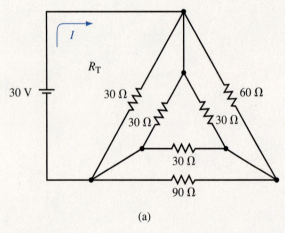

(a)

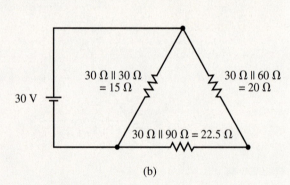

(b)

FIGURE 8–46

We see that the sides of the resulting "Δ" are in parallel, which allows us to simplify the circuit even further as shown in Figure 8–46(b). The total resistance of the circuit is now easily determined as

$$R_T = 15 \; \Omega \| (20 \; \Omega + 22.5 \; \Omega)$$
$$= 11.09 \; \Omega$$

This results in a circuit current of

$$I = \frac{30 \; V}{11.09 \; \Omega} = 2.706 \; A$$

PRACTICE PROBLEMS 5

Convert the Δ network of Figure 8–44 into an equivalent Y network. Verify that the result you obtain is the same as that found in Figure 8–43.

Answers
$R_1 = 4.8 \text{ k}\Omega, \ R_2 = 3.6 \text{ k}\Omega, \ R_3 = 2.4 \text{ k}\Omega$

8.8 Bridge Networks

In this section you will be introduced to the **bridge network.** Bridge networks are used in electronic measuring equipment to precisely measure resistance in dc circuits and similar quantities in ac circuits. The bridge circuit was originally used by Sir Charles Wheatstone in the mid-nineteenth century to measure resistance by balancing small currents. The Wheatstone bridge circuit is still used to measure resistance very precisely. The digital bridge, shown in Figure 8–47, is an example of one such instrument.

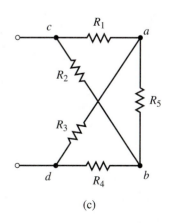

FIGURE 8–47 Digital bridge used for precisely measuring resistance, inductance, and capacitance.

You will use the techniques developed earlier in the chapter to analyze the operation of these networks. Bridge circuits may be shown in various configurations as seen in Figure 8–48.

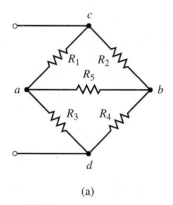

(a)

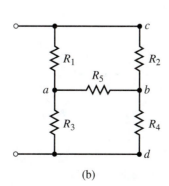

(b)

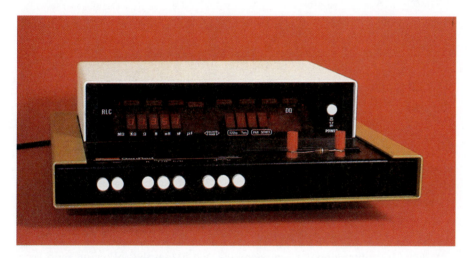

(c)

FIGURE 8–48

Although a bridge circuit may appear in one of three forms, you can see that they are equivalent. There are, however, two different states of bridges: the balanced bridge and the unbalanced bridge.

A **balanced bridge** is one in which the current through the resistance R_5 is equal to zero. In practical circuits, R_5 is generally a variable resistor in series with a sensitive galvanometer. When the current through R_5 is zero, then it follows that

$$V_{ab} = (R_5)(0 \text{ A}) = 0 \text{ V}$$

$$I_{R_1} = I_{R_3} = \frac{V_{cd}}{R_1 + R_3}$$

$$I_{R_2} = I_{R_4} = \frac{V_{cd}}{R_2 + R_4}$$

But the voltage V_{ab} is found as

$$V_{ab} = V_{ad} - V_{bd} = 0$$

Therefore, $V_{ad} = V_{bd}$ and

$$R_3 I_{R_3} = R_4 I_{R_4}$$

$$R_3\left(\frac{V_{cd}}{R_1 + R_3}\right) = R_4\left(\frac{V_{cd}}{R_2 + R_4}\right),$$

which simplifies to

$$\frac{R_3}{R_1 + R_3} = \frac{R_4}{R_2 + R_4}$$

Now, if we invert both sides of the equation and simplify, we get the following:

$$\frac{R_1 + R_3}{R_3} = \frac{R_2 + R_4}{R_4}$$

$$\frac{R_1}{R_3} + 1 = \frac{R_2}{R_4} + 1$$

Finally, by subtracting 1 from each side, we obtain the following ratio for a balanced bridge:

$$\frac{R_1}{R_3} = \frac{R_2}{R_4} \tag{8–21}$$

From Equation 8–21, we notice that a bridge network is balanced whenever the ratios of the resistors in the two arms are the same.

An **unbalanced bridge** is one in which the current through R_5 is not zero, and so the above ratio does not apply to an unbalanced bridge network. Figure 8–49 illustrates each condition of a bridge network.

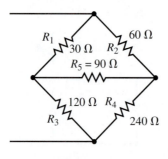

(a) Balanced bridge

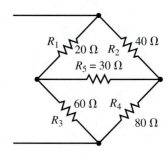

(b) Unbalanced bridge

FIGURE 8–49

If a balanced bridge appears as part of a complete circuit, its analysis is very simple since the resistor R_5 may be removed and replaced with either a short (since $V_{R_5} = 0$) or an open (since $I_{R_5} = 0$).

However, if a circuit contains an unbalanced bridge, the analysis is more complicated. In such cases, it is possible to determine currents and voltages by using mesh analysis, nodal analysis, or by using Δ to Y conversion. The following examples illustrate how bridges may be analyzed.

Solve for the currents through R_1 and R_4 in the circuit of Figure 8–50.

EXAMPLE 8–21

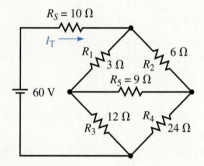

FIGURE 8–50

Solution We see that the bridge of the above circuit is balanced (since $R_1/R_3 = R_2/R_4$). Because the circuit is balanced, we may remove R_5 and replace it with either a short circuit (since the voltage across a short circuit is zero) or an open circuit (since the current through an open circuit is zero). The remaining circuit is then solved by one of the methods developed in previous chapters. Both methods will be illustrated to show that the results are exactly the same.

Method 1: If R_5 is replaced by an open, the result is the circuit shown in Figure 8–51.

The total circuit resistance is found as

$$R_T = 10\ \Omega + (3\ \Omega + 12\ \Omega)\|(6\ \Omega + 24\ \Omega)$$
$$= 10\ \Omega + 15\ \Omega\|30\ \Omega$$
$$= 20\ \Omega$$

The circuit current is

$$I_T = \frac{60\ \text{V}}{20\ \Omega} = 3.0\ \text{A}$$

The current in each branch is then found by using the current divider rule:

$$I_{R_1} = \left(\frac{30\ \Omega}{30\ \Omega + 15\ \Omega}\right)(3.0\ \text{A}) = 2.0\ \text{A}$$

$$I_{R_4} = \frac{10\ \Omega}{24\ \Omega + 6\ \Omega}(3.0\ \text{A}) = 1.0\ \text{A}$$

Method 2: If R_5 is replaced with a short circuit, the result is the circuit shown in Figure 8–52.

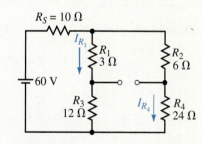

FIGURE 8–51

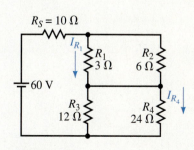

FIGURE 8–52

The total circuit resistance is found as

$$R_T = 10\,\Omega + (3\,\Omega \| 6\,\Omega) + (12\,\Omega \| 24\,\Omega)$$
$$= 10\,\Omega + 2\,\Omega + 8\,\Omega$$
$$= 20\,\Omega$$

The above result is precisely the same as that found using Method 1. Therefore the circuit current will remain as $I_T = 3.0$ A.

The currents through R_1 and R_4 may be found by the current divider rule as

$$I_{R_1} = \left(\frac{6\,\Omega}{6\,\Omega + 3\,\Omega}\right)(3.0\text{ A}) = 2.0\text{ A}$$

and

$$I_{R_4} = \left(\frac{12\,\Omega}{12\,\Omega + 24\,\Omega}\right)(3.0\text{ A}) = 1.0\text{ A}$$

Clearly, these results are precisely those obtained in Method 1, illustrating that the methods are equivalent. Remember, though, R_5 can be replaced with a short circuit or an open circuit only when the bridge is balanced.

EXAMPLE 8–22

$R_S = 6\,\Omega$

R_1 $6\,\Omega$ R_2 $12\,\Omega$
R_5 $18\,\Omega$
30 V I_2
R_3 $3\,\Omega$ I_3 R_4 $3\,\Omega$
I_1

FIGURE 8–53

Use mesh analysis to find the currents through R_1 and R_5 in the unbalanced bridge circuit of Figure 8–53.

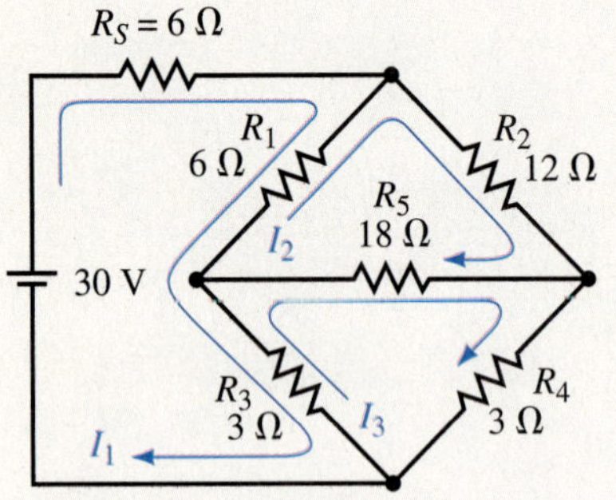

Solution   After assigning loop currents as shown, we write the loop equations as

Loop 1: $(15\,\Omega)I_1 - (6\,\Omega)I_2 - (3\,\Omega)I_3 = 30$ V

Loop 2: $-(6\,\Omega)I_1 + (36\,\Omega)I_2 - (18\,\Omega)I_3 = 0$

Loop 3: $-(3\,\Omega)I_1 - (18\,\Omega)I_2 + (24\,\Omega)I_3 = 0$

The determinant for the denominator will is

$$D = \begin{vmatrix} 15 & -6 & -3 \\ -6 & 36 & -18 \\ -3 & -18 & 24 \end{vmatrix} = 6264$$

Notice that, as expected, the elements in the principal diagonal are positive and that the determinant is symmetrical around the principal diagonal.

The loop currents are now evaluated as

$$I_1 = \frac{\begin{vmatrix} 30 & -6 & -3 \\ 0 & 36 & -18 \\ 0 & -18 & 24 \end{vmatrix}}{D} = \frac{16\,200}{6264} = 2.586\text{ A}$$

$$I_2 = \frac{\begin{vmatrix} 15 & 30 & -3 \\ -6 & 0 & -18 \\ -3 & 0 & 24 \end{vmatrix}}{D} = \frac{5940}{6264} = 0.948\text{ A}$$

$$I_3 = \frac{\begin{vmatrix} 15 & -6 & 30 \\ -6 & 36 & 0 \\ -3 & -18 & 0 \end{vmatrix}}{D} = \frac{6480}{6264} = 1.034\text{ A}$$

The current through R_1 is found as

$$I_{R_1} = I_1 - I_2 = 2.586 \text{ A} - 0.948 \text{ A} = 1.638 \text{ A}$$

The current through R_5 is found as

$$I_{R_5} = I_3 - I_2 = 1.034 \text{ A} - 0.948 \text{ A}$$
$$= 0.086 \text{ A} \quad \text{to the right}$$

The previous example illustrates that if the bridge is not balanced, there will always be some current through resistor R_5. The unbalanced circuit may also be easily analyzed using nodal analysis, as in the following example.

Determine the node voltages and the voltage V_{R_5} for the circuit of Figure 8–54.   **EXAMPLE 8–23**

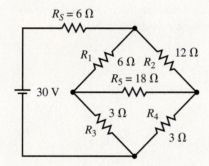

FIGURE 8–54

Solution By converting the voltage source into an equivalent current source, we obtain the circuit shown in Figure 8–55.

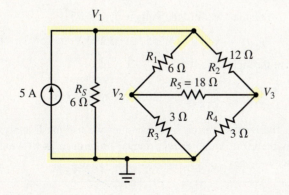

FIGURE 8–55

The nodal equations for the circuit are as follows:

Node 1: $\left(\dfrac{1}{6\ \Omega} + \dfrac{1}{6\ \Omega} + \dfrac{1}{12\ \Omega}\right)V_1 - \left(\dfrac{1}{6\ \Omega}\right)V_2 - \left(\dfrac{1}{12\ \Omega}\right)V_3 = 5\ \text{A}$

Node 2: $-\left(\dfrac{1}{6\ \Omega}\right)V_1 + \left(\dfrac{1}{6\ \Omega} + \dfrac{1}{3\ \Omega} + \dfrac{1}{18\ \Omega}\right)V_2 - \left(\dfrac{1}{18\ \Omega}\right)V_3 = 0\ \text{A}$

Node 3: $-\left(\dfrac{1}{12\ \Omega}\right)V_1 - \left(\dfrac{1}{18\ \Omega}\right)V_2 + \left(\dfrac{1}{3\ \Omega} + \dfrac{1}{12\ \Omega} + \dfrac{1}{18\ \Omega}\right)V_3 = 0\ \text{A}$

The linear equations are

Node 1: $\quad 0.4167V_1 - 0.1667V_2 - 0.0833V_3 = 5$

Node 2: $\quad -0.1667V_1 + 0.5556V_2 - 0.0556V_3 = 0$

Node 3: $\quad -0.0833V_1 - 0.0556V_2 + 0.4722V_3 = 0$

The determinant of the denominator is

$$D = \begin{vmatrix} 0.4167 & -0.1667 & -0.0833 \\ -0.1667 & 0.5556 & -0.0556 \\ -0.0833 & -0.0556 & 0.4722 \end{vmatrix} = 0.08951\ \text{A}$$

Again notice that the elements on the principal diagonal are positive and that the determinant is symmetrical about the principal diagonal.

The node voltages are calculated to be

$$V_1 = \dfrac{\begin{vmatrix} 5 & -0.1667 & -0.0833 \\ 0 & 0.5556 & -0.0556 \\ 0 & -0.0556 & 0.4722 \end{vmatrix}}{D} = \dfrac{1.2963}{0.08951} = 14.48\ \text{A}$$

$$V_2 = \dfrac{\begin{vmatrix} 0.4167 & 5 & -0.0833 \\ -0.1667 & 0 & -0.0556 \\ -0.0833 & 0 & 0.4722 \end{vmatrix}}{D} = \dfrac{0.41667}{0.08951} = 4.66\ \text{V}$$

$$V_3 = \dfrac{\begin{vmatrix} 0.4167 & -0.1667 & 5 \\ -0.1667 & 0.0556 & 0 \\ -0.0833 & -0.0556 & 0 \end{vmatrix}}{D} = \dfrac{0.2778}{0.08951} = 3.10\ \text{V}$$

Using the above results, we find the voltage across $R5$:

$$V_{R_5} = V_2 - V_3 = 4.655\ \text{V} - 3.103\ \text{V} = 1.55\ \text{V}$$

and the current through R_5 is

$$I_{R_5} = \dfrac{1.55\ \text{V}}{18\ \Omega} = 0.086\ \text{A} \quad \text{to the right}$$

As expected, the results are the same whether we use mesh analysis or nodal analysis. It is therefore a matter of personal preference as to which approach should be used.

A final method for analyzing bridge networks involves the use of Δ to Y conversion. The following example illustrates the method used.

Find the current through R_5 for the circuit shown in Figure 8–56.

EXAMPLE 8–24

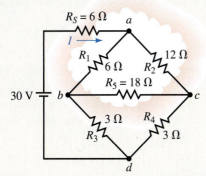

FIGURE 8–56

Solution By inspection we see that this circuit is not balanced, since

$$\frac{R_1}{R_3} \neq \frac{R_2}{R_4}$$

Therefore, the current through R_5 cannot be zero. Notice, also, that the circuit contains two possible Δ configurations. If we choose to convert the top Δ to its equivalent Y, we get the circuit shown in Figure 8–57.

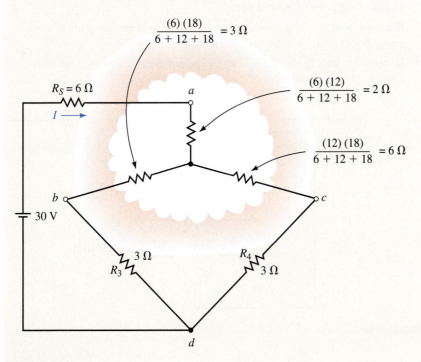

FIGURE 8–57

By combining resistors, it is possible to reduce the complicated circuit to the simple series circuit shown in Figure 8–58.

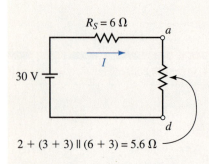

$R_S = 6 \ \Omega$

30 V

a

I

d

$2 + (3 + 3) \parallel (6 + 3) = 5.6 \ \Omega$

FIGURE 8–58

The circuit of Figure 8–58 is easily analyzed to give a total circuit current of

$$I = \frac{30 \ \text{V}}{6 \ \Omega + 2 \ \Omega + 3.6 \ \Omega} = 2.59 \ \text{A}$$

Using the calculated current, it is possible to work back to the original circuit. The currents in the resistors R_3 and R_4 are found by using the current divider rule for the corresponding resistor branches, as shown in Figure 8–57.

$$I_{R_3} = \frac{(6 \ \Omega + 3 \ \Omega)}{(6 \ \Omega + 3 \ \Omega) + (3 \ \Omega + 3 \ \Omega)}(2.59 \ \text{A}) = 1.55 \ \text{A}$$

$$I_{R_4} = \frac{(3 \ \Omega + 3 \ \Omega)}{(6 \ \Omega + 3 \ \Omega) + (3 \ \Omega + 3 \ \Omega)}(2.59 \ \text{A}) = 1.03 \ \text{A}$$

These results are exactly the same as those found in Examples 8–21 and 8–22. Using these currents, it is now possible to determine the voltage V_{bc} as

$$V_{bc} = -(3 \ \Omega)I_{R_4} + (3 \ \Omega)I_{R_3}$$
$$= (-3 \ \Omega)(1.034 \ \text{A}) + (3 \ \Omega)(3.103 \ \text{A})$$
$$= 1.55 \ \text{V}$$

The current through R_5 is determined to be

$$I_{R_5} = \frac{1.55 \ \text{V}}{18 \ \Omega} = 0.086 \ \text{A} \quad \text{to the right}$$

IN-PROCESS
LEARNING CHECK 2

(Answers are at the end of the chapter.)

1. For a balanced bridge, what will be the value of voltage between the midpoints of the arms of the bridge?

2. If a resistor or sensitive galvanometer is placed between the arms of a balanced bridge, what will be the current through the resistor?

3. In order to simplify the analysis of a balanced bridge, how may the resistance R_5, between the arms of the bridge, be replaced?

PRACTICE PROBLEMS 6

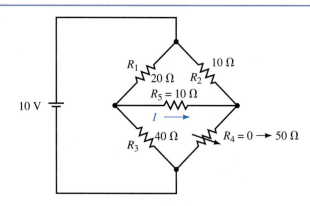

R_1
$20 \ \Omega$
$10 \ \Omega$
R_2
$R_5 = 10 \ \Omega$
10 V
I
$40 \ \Omega$
R_3
$R_4 = 0 \rightarrow 50 \ \Omega$

FIGURE 8–59

1. For the circuit shown in Figure 8–59, what value of R_4 will ensure that the bridge is balanced?

2. Determine the current I through R_5 in Figure 8–59 when $R_4 = 0 \ \Omega$ and when $R_4 = 50 \ \Omega$.

Answers
1. 20 V; 2. 286 mA, -52.6 mA

MultiSIM and PSpice are able to analyze a circuit without the need to convert between voltage and current sources or having to write lengthy linear equations. It is possible to have the program output the value of voltage across or current through any element in a given circuit. The following examples were previously analyzed using several other methods throughout this chapter.

8.9 Circuit Analysis Using Computers

◀ MULTISIM

◀ CADENCE

Given the circuit of Figure 8–60, use MultiSIM to find the voltage V_{ab} and the current through each resistor.

EXAMPLE 8–25

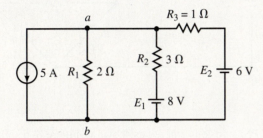

FIGURE 8–60

Solution   The circuit is entered as shown in Figure 8–61. The current source is obtained by clicking on the Sources button in the Parts bin toolbar. As before, it is necessary to include a ground symbol in the schematic although the original circuit of Figure 8–60 did not have one. Make sure that all values are changed from the default values to the required circuit values.

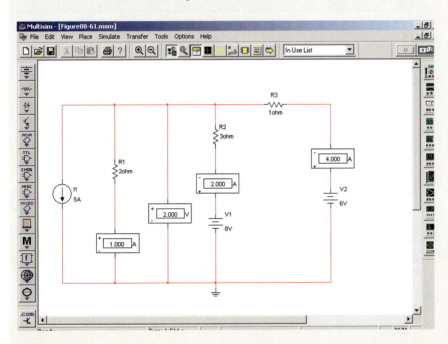

FIGURE 8–61

◀ MULTISIM

From the above results, we have the following values:

$$V_{ab} = 2.00 \text{ V}$$

$$I_{R_1} = 1.00 \text{ A} \quad \text{(downward)}$$

$$I_{R_2} = 2.00 \text{ A} \quad \text{(upward)}$$

$$I_{R_3} = 4.00 \text{ A} \quad \text{(to the left)}$$

PSpice

EXAMPLE 8–26

Use PSpice to find currents through R_1 and R_5 in the circuit of Figure 8–62.

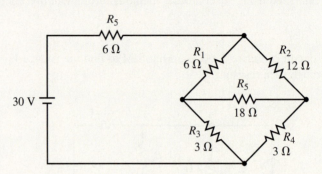

FIGURE 8–62

◀ CADENCE

Solution   The PSpice file is entered as shown in Figure 8–63. Ensure that you enter **Yes** in the DC cell of the Properties Editor for each IPRINT part.

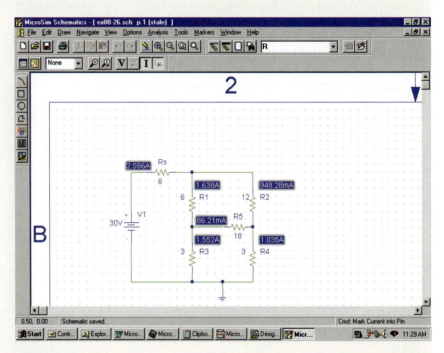

FIGURE 8–63

Once you have selected a New Simulation Profile, click on the Run icon. Select View and Output File to see the results of the simulation. The currents are $I_{R_1} = 1.64$ A and $I_{R_5} = 86.2$ mA. These results are consistent with those obtained in Example 8–22.

Use MultiSIM to determine the currents, I_T, I_{R_1} and I_{R_4} in the circuit of Figure 8–50. Compare your results to those obtained in Example 8–21.

PRACTICE PROBLEMS 7

Answers
$I_T = 3.00$ A, $I_{R_1} = 2.00$ A and $I_{R_4} = 1.00$ A

Use PSpice to input the circuit of Figure 8–50. Determine the value of I_{R_5} when $R_4 = 0$ Ω and when $R_4 = 48$ Ω.
Note: Since PSpice cannot let $R_4 = 0$ Ω, you will need to let it equal some very small value such as 1 μΩ (1e-6).

PRACTICE PROBLEMS 8

Answers
$I_{R_5} = 1.08$ A when $R_4 = 0$ Ω and $I_{R_5} = 0.172$ A when $R_4 = 48$ Ω

Use PSpice to input the circuit of Figure 8–54, so that the output file will provide the currents through R_1, R_2, and R_5. Compare your results to those obtained in Example 8–23.

PRACTICE PROBLEMS 9

PUTTING IT INTO PRACTICE

S train gauges are manufactured from very fine wire mounted on insulated surfaces that are then glued to large metal structures. These instruments are used by civil engineers to measure the movement and mass of large objects such as bridges and buildings. When the very fine wire of a strain gauge is subjected to stress, its effective length is increased (due to stretching) or decreased (due to compression). This change in length results in a corresponding minute change in resistance. By placing one or more strain gauges into a bridge circuit, it is possible to detect variation in resistance, ΔR. This change in resistance can be calibrated to correspond to an applied force. Consequently, it is possible to use such a bridge as a means of measuring very large masses. Consider that you have two strain gauges mounted in a bridge as shown in the accompanying figure.

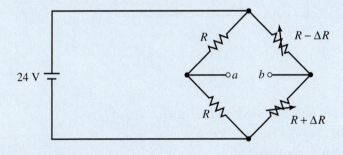

Strain gauge bridge. $R = 100$ Ω

The variable resistors, R_2 and R_4 are strain gauges that are mounted on opposite sides of a steel girder used to measure very large masses. When a mass is applied to the girder, the strain gauge on one side of the girder will compress, reducing the resistance. The strain gauge on the other side of the girder will stretch, increasing the resistance. When no mass is applied, there will be neither compression nor stretching and so the bridge will be balanced, resulting in a voltage $V_{ab} = 0$ V.

Write an expression for ΔR as a function of V_{ab}. Assume that the scale is calibrated so that resistance variation of $\Delta R = 0.02$ Ω corresponds to a mass of 5000 kg. Determine the measured mass if $V_{ab} = -4.20$ mV.

8.1 Constant-Current Sources

1. Find the voltage V_S for the circuit shown in Figure 8–64.
2. Find the voltage V_S for the circuit shown in Figure 8–65.

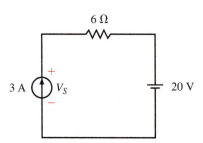

FIGURE 8–64

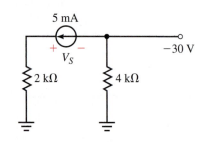

FIGURE 8–65

3. Refer to the circuit of Figure 8–66:
   a. Find the current I_3.
   b. Determine the voltages V_S and V_1.
4. Consider the circuit of Figure 8–67:
   a. Calculate the voltages V_2 and V_S.
   b. Find the currents I and I_3.

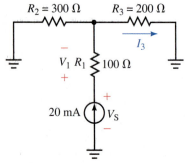

FIGURE 8–66

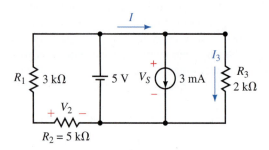

FIGURE 8–67

5. For the circuit of Figure 8–68, find the currents I_1 and I_2.
6. Refer to the circuit of Figure 8–69:
   a. Find the voltages V_S and V_2.
   b. Determine the current I_4.

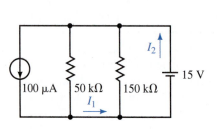

FIGURE 8–68

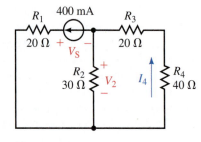

FIGURE 8–69

7. Verify that the power supplied by the sources is equal to the summation of the powers dissipated by the resistors in the circuit of Figure 8–68.

8. Verify that the power supplied by the source in the circuit of Figure 8–69 is equal to the summation of the powers dissipated by the resistors.

8.2   Source Conversions

9. Convert each of the voltage sources of Figure 8–70 into its equivalent current source.

10. Convert each of the current sources of Figure 8–71 into its equivalent voltage source.

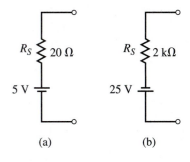

(a)   (b)

FIGURE 8–70

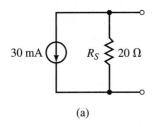 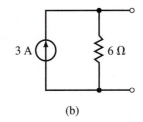

(a)   (b)

FIGURE 8–71

11. Refer to the circuit of Figure 8–72:

   a. Solve for the current through the load resistor using the current divider rule.

   b. Convert the current source into its equivalent voltage source and again determine the current through the load.

12. Find V_{ab} and I_2 for the network of Figure 8–73.

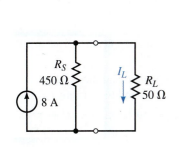

FIGURE 8–72

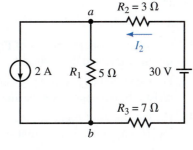

FIGURE 8–73

13. Refer to the circuit of Figure 8–74:

   a. Convert the current source and the 330-Ω resistor into an equivalent voltage source.

   b. Solve for the current I through R_L.

   c. Determine the voltage V_{ab}.

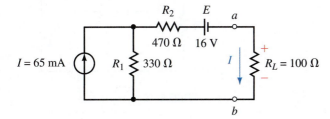

FIGURE 8–74

14. Refer to the circuit of Figure 8–75:

   a. Convert the voltage source and the 36-Ω resistor into an equivalent current source.

   b. Solve for the current I through R_L.

   c. Determine the voltage V_{ab}.

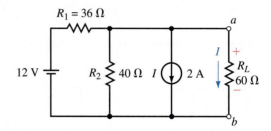

FIGURE 8–75

8.3   Current Sources in Parallel and Series

15. Find the voltage V_2 and the current I_1 for the circuit of Figure 8–76.

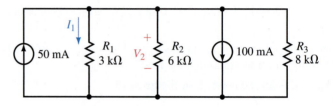

◄ MULTISIM   **FIGURE 8–76**

16. Convert the voltage sources of Figure 8–77 into current sources and solve for the current I_1 and the voltage V_{ab}.

17. For the circuit of Figure 8–78 convert the current source and the 2.4-kΩ resistor into a voltage source and find the voltage V_{ab} and the current I_3.

18. For the circuit of Figure 8–78, convert the voltage source and the series resistors into an equivalent current source.

   a. Determine the current I_2.

   b. Solve for the voltage V_{ab}.

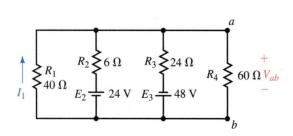

FIGURE 8–77

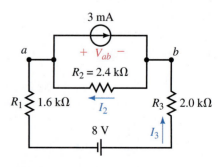

FIGURE 8–78

8.4 Branch-Current Analysis

19. Write the branch-current equations for the circuit shown in Figure 8–79 and solve for the branch currents using determinants.

20. Refer to the circuit of Figure 8–80:
    a. Solve for the current I_1 using branch-current analysis.
    b. Determine the voltage V_{ab}.

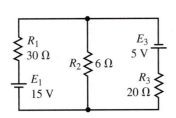

FIGURE 8–79

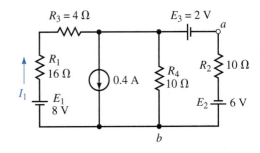

FIGURE 8–80

21. Write the branch-current equations for the circuit shown in Figure 8–81 and solve for the current I_2.

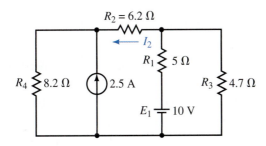

FIGURE 8–81

22. Refer to the circuit shown in Figure 8–82:
    a. Write the branch-current equations.
    b. Solve for the currents I_1 and I_2.
    c. Determine the voltage V_{ab}.

23. Refer to the circuit shown in Figure 8–83:
    a. Write the branch-current equations.
    b. Solve for the current I_2.
    c. Determine the voltage V_{ab}.

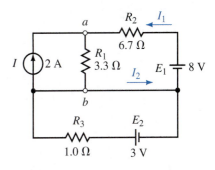

FIGURE 8–82

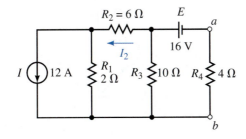

FIGURE 8–83

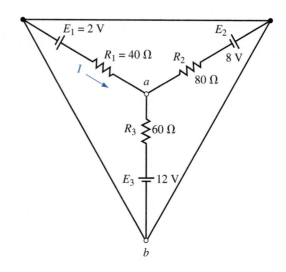

FIGURE 8–84

24. Refer to the circuit shown in Figure 8–84:
    a. Write the branch-current equations.
    b. Solve for the current I.
    c. Determine the voltage V_{ab}.

8.5   Mesh (Loop) Analysis

25. Write the mesh equations for the circuit shown in Figure 8–79 and solve for the loop currents.

26. Use mesh analysis for the circuit of Figure 8–80 to solve for the current I_1.

27. Use mesh analysis to solve for the current I_2 in the circuit of Figure 8–81.

28. Use mesh analysis to solve for the loop currents in the circuit of Figure 8–83. Use your results to determine I_2 and V_{ab}.

29. Use mesh analysis to solve for the loop currents in the circuit of Figure 8–84. Use your results to determine I and V_{ab}.

30. Using mesh analysis, determine the current through the 6-Ω resistor in the circuit of Figure 8–85.

31. Write the mesh equations for the network in Figure 8–86. Solve for the loop currents using determinants.

32. Repeat Problem 31 for the network in Figure 8–87.

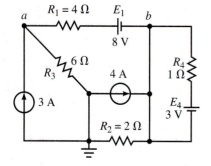

FIGURE 8–85

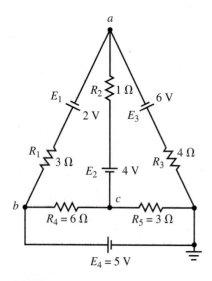

FIGURE 8–86

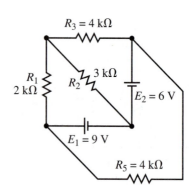

FIGURE 8–87

8.6   Nodal Analysis

33. Write the nodal equations for the circuit of Figure 8–88 and solve for the nodal voltages.

34. Write the nodal equations for the circuit of Figure 8–89 and determine the voltage V_{ab}.

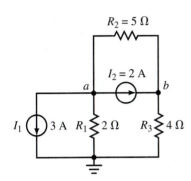

FIGURE 8–88

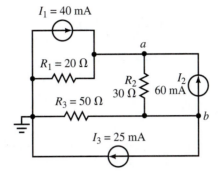

FIGURE 8–89

35. Repeat Problem 33 for the circuit of Figure 8–90.

36. Repeat Problem 34 for the circuit of Figure 8–91.

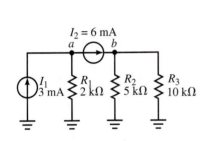

FIGURE 8–90

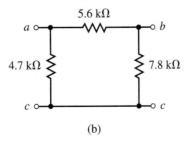

FIGURE 8–91

◀ MULTISIM

37. Write the nodal equations for the circuit of Figure 8–86 and solve for $V_{6\,\Omega}$.

38. Write the nodal equations for the circuit of Figure 8–85 and solve for $V_{6\,\Omega}$.

8.7   Delta-Wye (Pi-Tee) Conversion

39. Convert each of the Δ networks of Figure 8–92 into its equivalent Y configuration.

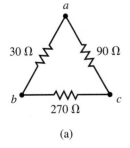

(a)

5.6 kΩ

4.7 kΩ   7.8 kΩ

(b)

FIGURE 8–92

40. Convert each of the Δ networks of Figure 8–93 into its equivalent Y configuration.

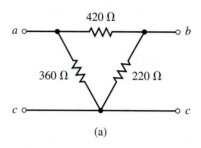

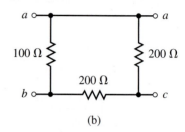

(a)

(b)

FIGURE 8–93

41. Convert each of the Y networks of Figure 8–94 into its equivalent Δ configuration.

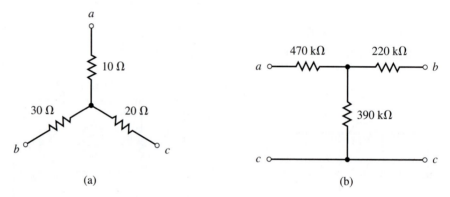

(a)

(b)

FIGURE 8–94

42. Convert each of the Y networks of Figure 8–95 into its equivalent Δ configuration.

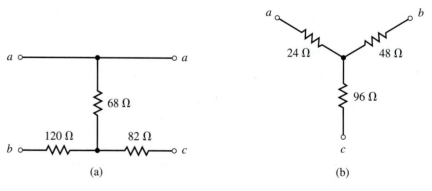

(a)

(b)

FIGURE 8–95

43. Using Δ-Y or Y-Δ conversion, find the current *I* for the circuit of Figure 8–96.

44. Using Δ-Y or Y-Δ conversion, find the current *I* and the voltage V_{ab} for the circuit of Figure 8–97.

45. Repeat Problem 43 for the circuit of Figure 8–98.

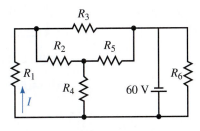

All resistors are 4.5 kΩ

FIGURE 8–96

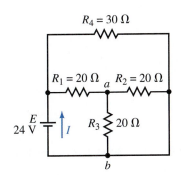

FIGURE 8–97

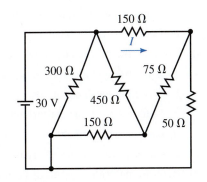

FIGURE 8–98

46. Repeat Problem 44 for the circuit of Figure 8–99.

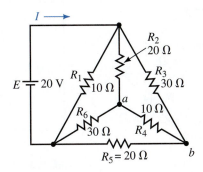

FIGURE 8–99

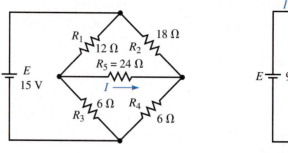

FIGURE 8–100

◀ MULTISIM

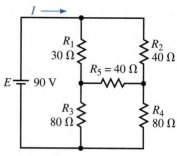

FIGURE 8–101

◀ MULTISIM

8.8 Bridge Networks

47. Refer to the bridge circuit of Figure 8–100:
    a. Is the bridge balanced? Explain.
    b. Write the mesh equations.
    c. Calculate the current through R_5.
    d. Determine the voltage across R_5.

48. Consider the bridge circuit of Figure 8–101:
    a. Is the bridge balanced? Explain.
    b. Write the mesh equations.
    c. Determine the current through R_5.
    d. Calculate the voltage across R_5.

49. Given the bridge circuit of Figure 8–102, find the current through each resistor.

50. Refer to the bridge circuit of Figure 8–103:
    a. Determine the value of resistance R_x such that the bridge is balanced.
    b. Calculate the current through R_5 when $R_x = 0\ \Omega$ and when $R_x = 10\ k\Omega$.

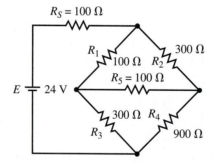

FIGURE 8–102

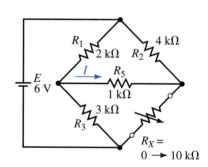

FIGURE 8–103

8.9 Circuit Analysis Using Computers

51. Use MultiSIM to solve for the currents through all resistors of the circuit shown in Figure 8–86.

 MULTISIM

52. Use MultiSIM to solve for the voltage across the 5-kΩ resistor in the circuit of Figure 8–87.

 MULTISIM

53. Use PSpice to solve for the currents through all resistors in the circuit of Figure 8–96.

 CADENCE

54. Use PSpice to solve for the currents through all resistors in the circuit of Figure 8–97.

 CADENCE

✓ **ANSWERS TO IN-PROCESS LEARNING CHECKS**

In-Process Learning Check 1

1. A voltage source E in series with a resistor R is equivalent to a current source having an ideal current source $I = E/R$ in parallel with the same resistance, R.

2. Current sources are never connected in series.

In-Process Learning Check 2

1. Voltage is zero.

2. Current is zero.

3. R_5 can be replaced with either a short circuit or an open circuit.

■ **OBJECTIVES**

After studying this chapter you will be able to

- apply the superposition theorem to determine the current through or voltage across any resistance in a given network,

- state Thévenin's theorem and determine the Thévenin equivalent circuit of any resistive network,

- state Norton's theorem and determine the Norton equivalent circuit of any resistive network,

- determine the required load resistance of any circuit to ensure that the load receives maximum power from the circuit,

- apply Millman's theorem to determine the current through or voltage across any resistor supplied by any number of sources in parallel,

- state the reciprocity theorem and demonstrate that it applies for a given single-source circuit,

- state the substitution theorem and apply the theorem in simplifying the operation of a given circuit.

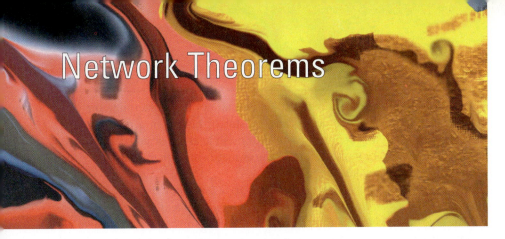

Network Theorems

9

I n this chapter you will learn how to use some basic theorems which will allow the analysis of even the most complex resistive networks. The theorems that are most useful in analyzing networks are the superposition, Thévenin, Norton, and maximum power transfer theorems.

You will also be introduced to other theorems which, while useful for providing a well-rounded appreciation of circuit analysis, have limited use in the analysis of circuits. These theorems, which apply to specific types of circuits, are the substitution, reciprocity, and Millman theorems. Your instructor may choose to omit the latter theorems without any loss in continuity. ■

André Marie Ampère

ANDRÉ MARIE AMPÈRE WAS BORN in Polémieux, Rhône, near Lyon, France on January 22, 1775. As a youth, Ampère was a brilliant mathematician who was able to master advanced mathematics by the age of twelve. However, the French Revolution, and the ensuing anarchy that swept through France from 1789 to 1799, did not exclude the Ampère family. Ampère's father, who was a prominent merchant and city official in Lyon, was executed under the guillotine in 1793. Young André suffered a nervous breakdown from which he never fully recovered. His suffering was further compounded in 1804, when after only five years of marriage, Ampère's wife died.

Even so, Ampère was able to make profound contributions to the field of mathematics, chemistry, and physics. As a young man, Ampère was appointed as professor of chemistry and physics in Bourg. Napoleon was a great supporter of Ampère's work, though Ampère had a reputation as an "absent-minded professor." Later he moved to Paris, where he taught mathematics.

Ampère showed that two current-carrying wires were attracted to one another when the current in the wires was in the same direction. When the current in the wires was in the opposite direction, the wires repelled. This work set the stage for the discovery of the principles of electric and magnetic field theory. Ampère was the first scientist to use electromagnetic principles to measure current in a wire. In recognition of his contribution to the study of electricity, current is measured in the unit of amperes.

Despite his personal suffering, Ampère remained a popular, friendly human being. He died of pneumonia in Marseille on June 10, 1836 after a brief illness. ■

9.1 Superposition Theorem

The **superposition theorem** is a method that allows us to determine the current through or the voltage across any resistor or branch in a network. The advantage of using this approach instead of mesh analysis or nodal analysis is that it is not necessary to use determinants or matrix algebra to analyze a given circuit. The theorem states the following:

The total current through or voltage across a resistor or branch may be determined by summing the effects due to each independent source.

In order to apply the superposition theorem it is necessary to remove all sources other than the one being examined. In order to "zero" a voltage source, we **replace it with a short circuit,** since the voltage across a short circuit is zero volts. A current source is zeroed by **replacing it with an open circuit,** since the current through an open circuit is zero amps.

If we wish to determine the power dissipated by any resistor, we must first find either the voltage across the resistor or the current through the resistor:

$$P = I^2R = \frac{V^2}{R}$$

EXAMPLE 9–1

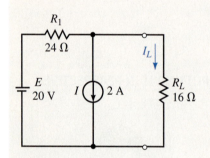

FIGURE 9–1

Consider the circuit of Figure 9–1:

a. Determine the current in the load resistor, R_L.

b. Verify that the superposition theorem does not apply to power.

Solution

a. We first determine the current through R_L due to the voltage source by removing the current source and replacing in with an open circuit (zero amps) as shown in Figure 9–2.

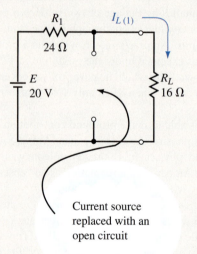

Current source replaced with an open circuit

FIGURE 9–2

The resulting current through R_L is determined from Ohm's law as

$$I_{L(1)} = \frac{20\text{V}}{16\ \Omega + 24\ \Omega} = 0.500\text{ A}$$

Next, we determine the current through R_L due to the current source by removing the voltage source and replacing it with a short circuit (zero volts) as shown in Figure 9–3.

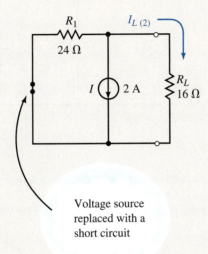

Voltage source
replaced with a
short circuit

FIGURE 9–3

The resulting current through R_L is found with the current divider rule as

$$I_{L(2)} = -\left(\frac{24\ \Omega}{24\ \Omega + 16\ \Omega}\right)(2\ \text{A}) = -1.20\ \text{A}$$

The resultant current through R_L is found by applying the superposition theorem:

$$I_L = 0.5\ \text{A} - 1.2\ \text{A} = -0.700\ \text{A}$$

The negative sign indicates that the current through R_L is opposite to the assumed reference direction. Consequently, the current through R_L will, in fact, be upward with a magnitude of 0.7 A.

b. If we assume (incorrectly) that the superposition theorem applies for power, we would have the power due the first source given as

$$P_1 = I_{L(1)}^2 R_L = (0.5\ \text{A})^2(16\ \Omega) = 4.0\ \text{W}$$

and the power due the second source as

$$P_2 = I_{L(2)}^2 R_L = (1.2\ \text{A})^2(16\ \Omega) = 23.04\ \text{W}$$

The total power, if superposition applies, would be

$$P_T = P_1 + P_2 = 4.0\ \text{W} + 23.04\ \text{W} = 27.04\ \text{W}$$

Clearly, this result is wrong, since the actual power dissipated by the load resistor is correctly given as

$$P_L = I_L^2 R_L = (0.7\ \text{A})^2(16\ \Omega) = 7.84\ \text{W}$$

The superposition theorem may also be used to determine the voltage across any component or branch within the circuit.

EXAMPLE 9–2

Determine the voltage drop across the resistor R_2 of the circuit shown in Figure 9–4.

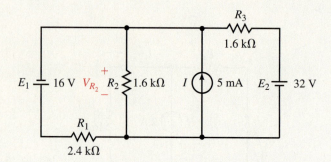

FIGURE 9–4

Solution   Since this circuit has three separate sources, it is necessary to determine the voltage across R_2 due to each individual source.

First, we consider the voltage across R_2 due to the 16-V source as shown in Figure 9–5.

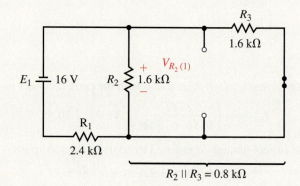

FIGURE 9–5

The voltage across R_2 will be the same as the voltage across the parallel combination of $R_2 \| R_3 = 0.8$ kΩ. Therefore,

$$V_{R2(1)} = -\left(\frac{0.8 \text{ k}\Omega}{0.8 \text{ k}\Omega + 2.4 \text{ k}\Omega}\right)(16 \text{ V}) = -4.00 \text{ V}$$

The negative sign in the above calculation simply indicates that the voltage across the resistor due to the first source is opposite to the assumed reference polarity.

Next, we consider the current source. The resulting circuit is shown in Figure 9–6.

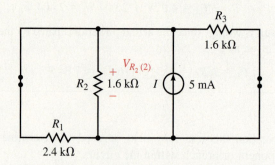

FIGURE 9–6

From this circuit, you can observe that the total resistance "seen" by the current source is

$$R_T = R_1 \| R_2 \| R_3 = 0.6 \text{ k}\Omega$$

The resulting voltage across R_2 is

$$V_{R_2(2)} = (0.6 \text{ k}\Omega)(5 \text{ mA}) = 3.00 \text{ V}$$

Finally, the voltage due to the 32-V source is found by analyzing the circuit of Figure 9–7.

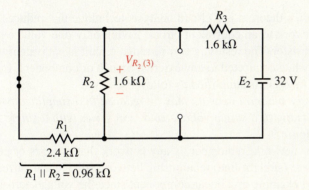

FIGURE 9–7

The voltage across R_2 is

$$V_{R_2(3)} = \left(\frac{0.96\text{k}\Omega}{0.96 \text{ k}\Omega + 1.6 \text{ k}\Omega} \right)(32 \text{ V}) = 12.0 \text{ V}$$

By superposition, the resulting voltage is

$$V_{R_2} = -4.0 \text{ V} + 3.0 \text{ V} + 12.0 \text{ V} = 11.0 \text{ V}$$

Use the superposition theorem to determine the voltage across R_1 and R_3 in the circuit of Figure 9–4.

Answers
$V_{R_1} = 27.0 \text{ V}, \ V_{R_3} = 21.0 \text{ V}$

PRACTICE PROBLEMS 1

Use the final results of Example 9–2 and Practice Problem 1 to determine the power dissipated by the resistors in the circuit of Figure 9–4. Verify that the superposition theorem does not apply to power.

IN-PROCESS
LEARNING CHECK 1

(Answers are at the end of the chapter.)

In this section, we will apply one of the most important theorems of electric circuits. **Thévenin's theorem** allows even the most complicated circuit to be reduced to a single voltage source and a single resistance. The importance of such a theorem becomes evident when we try to analyze a circuit as shown in Figure 9–8.

   If we wanted to find the current through the variable load resistor when $R_L = 0$, $R_L = 2 \text{ k}\Omega$, and $R_L = 5 \text{ k}\Omega$ using existing methods, we would need to analyze the entire circuit three separate times. However, if we could reduce the entire circuit external to the load resistor to a single voltage source in series with a resistor, the solution becomes very easy.

9.2 Thévenin's Theorem

◀ **Online Companion**

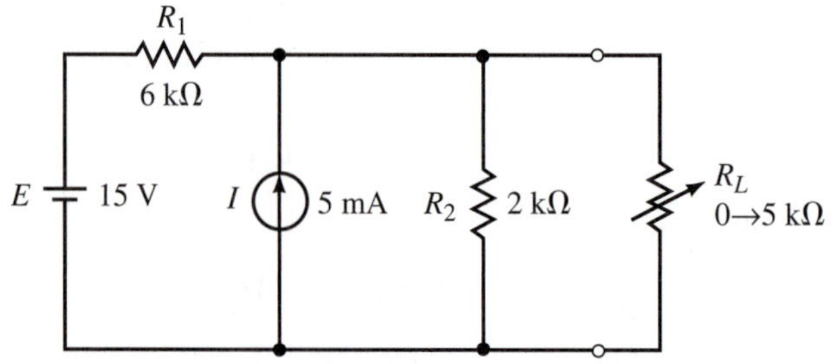

R_1
6 kΩ

E — 15 V I ⊙ 5 mA R_2 ⧚ 2 kΩ R_L 0→5 kΩ

FIGURE 9–8

Thévenin's theorem is a circuit analysis technique that reduces any linear bilateral network to an equivalent circuit having only one voltage source and one series resistor. The resulting two-terminal circuit is equivalent to the original circuit when connected to any external branch or component. In summary, Thévenin's theorem is simplified as follows:

Any linear bilateral network may be reduced to a simplified two-terminal circuit consisting of a single voltage source in series with a single resistor as shown in Figure 9–9.

A linear network, remember, is any network that consists of components having a linear (straight-line) relationship between voltage and current. A resistor is a good example of a linear component since the voltage across a resistor increases proportionally to an increase in current through the resistor. Voltage and current sources are also linear components. In the case of a voltage source, the voltage remains constant although current through the source may change.

A bilateral network is any network that operates in the same manner regardless of the direction of current in the network. Again, a resistor is a good example of a bilateral component, since the magnitude of current through the resistor is not dependent upon the polarity of voltage across the component. (A diode is not a bilateral component, since the magnitude of current through the device is dependent upon the polarity of the voltage applied across the diode.)

The following steps provide a technique that converts any circuit into its Thévenin equivalent:

1. Remove the load from the circuit.

2. Label the resulting two terminals. We will label them as *a* and *b,* although any notation may be used.

3. Set all sources in the circuit to zero.

   Voltage sources are set to zero by replacing them with short circuits (zero volts).

   Current sources are set to zero by replacing them with open circuits (zero amps).

4. Determine the Thévenin equivalent resistance, R_{Th}, by calculating the resistance "seen" between terminals *a* and *b.* It may be necessary to redraw the circuit to simplify this step.

5. Replace the sources removed in Step 3, and determine the open-circuit voltage between the terminals. If the circuit has more than one source, it may be necessary to use the superposition theorem. In that case, it will be necessary to determine the open-circuit voltage due to each source separately and then determine the combined effect. The resulting open-circuit voltage will be the value of the Thévenin voltage, E_{Th}.

6. Draw the Thévenin equivalent circuit using the resistance determined in Step 4 and the voltage calculated in Step 5. As part of the resulting circuit, include that portion of the network removed in Step 1.

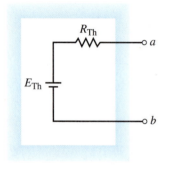

R_{Th} ∘ *a*

E_{Th}

∘ *b*

FIGURE 9–9   Thévenin equivalent circuit.

Determine the Thévenin equivalent circuit external to the resistor R_L for the circuit of Figure 9–10. Use the Thévenin equivalent circuit to calculate the current through R_L.

EXAMPLE 9–3

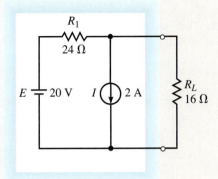

FIGURE 9–10

Solution
Steps 1 and 2: Removing the load resistor from the circuit and labelling the remaining terminals, we obtain the circuit shown in Figure 9–11.

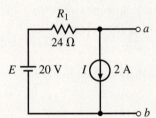

FIGURE 9–11

Step 3: Setting the sources to zero, we have the circuit shown in Figure 9–12.

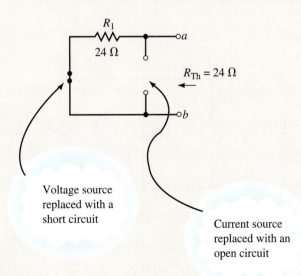

Voltage source replaced with a short circuit

Current source replaced with an open circuit

FIGURE 9–12

Step 4: The Thévenin resistance between the terminals is $R_{Th} = 24\ \Omega$.

Step 5: From Figure 9–11, the open-circuit voltage between terminals a and b is found as

$$V_{ab} = 20\ V - (24\ \Omega)(2\ A) = -28.0\ V$$

Step 6: The resulting Thévenin equivalent circuit is shown in Figure 9–13.

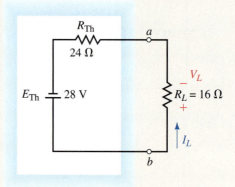

FIGURE 9–13

Using this Thévenin equivalent circuit, we easily find the current through R_L as

$$I_L = \left(\frac{28\ V}{24\ \Omega + 16\ \Omega} \right) = 0.700\ A \quad \text{(upward)}$$

This result is the same as that obtained by using the superposition theorem in Example 9–1.

EXAMPLE 9–4

Find the Thévenin equivalent circuit of the indicated area in Figure 9–14. Using the equivalent circuit, determine the current through the load resistor when $R_L = 0$, $R_L = 2\ k\Omega$, and $R_L = 5\ k\Omega$.

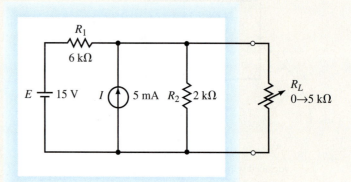

FIGURE 9–14

Solution
Steps 1, 2, and 3: After removing the load, labelling the terminals, and setting the sources to zero, we have the circuit shown in Figure 9–15.

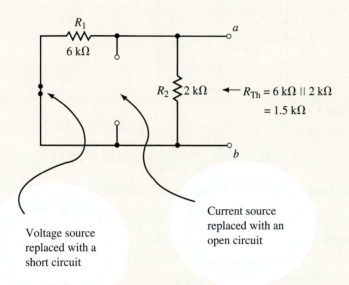

R_1
6 kΩ
a
R_2 2 kΩ $\leftarrow R_{Th} = 6\ k\Omega\ ||\ 2\ k\Omega$
$= 1.5\ k\Omega$
b

Current source replaced with an open circuit

Voltage source replaced with a short circuit

FIGURE 9–15

Step 4: The Thévenin resistance of the circuit is

$$R_{Th} = 6\ k\Omega || 2\ k\Omega = 1.5\ k\Omega$$

Step 5: Although several methods are possible, we will use the superposition theorem to find the open-circuit voltage V_{ab}. Figure 9–16 shows the circuit for determining the contribution due to the 15-V source.

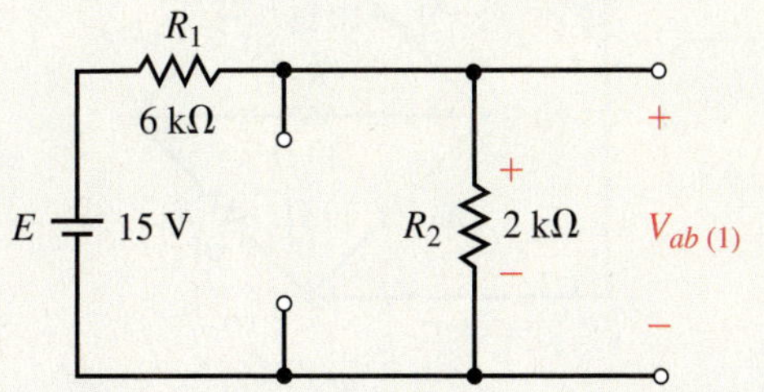

R_1
6 kΩ

E 15 V R_2 2 kΩ $V_{ab\ (1)}$

FIGURE 9–16

$$V_{ab(1)} = \left(\frac{2\ k\Omega}{2\ k\Omega + 6\ k\Omega} \right)(15\ V) = +3.75\ V$$

Figure 9–17 shows the circuit for determining the contribution due to the 5-mA source.

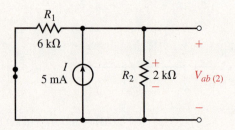

FIGURE 9–17

$$V_{ab(2)} = \left(\frac{(2 \text{ k}\Omega)(6 \text{ k}\Omega)}{2 \text{ k}\Omega + 6 \text{ k}\Omega}\right)(5 \text{ mA}) = +7.5 \text{ V}$$

The Thévenin equivalent voltage is

$$E_{Th} = V_{ab(1)} + V_{ab(2)} = +3.75 \text{ V} + 7.5 \text{ V} = 11.25 \text{ V}$$

Step 6: The resulting Thévenin equivalent circuit is shown in Figure 9–18.

From this circuit, it is now an easy matter to determine the current for any value of load resistor:

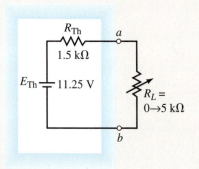

FIGURE 9–18

$R_L = 0 \ \Omega$: $I_L = \dfrac{11.25 \text{ V}}{1.5 \text{ k}\Omega} = 7.5 \text{ mA}$

$R_L = 2 \ k\Omega$: $I_L = \dfrac{11.25 \text{ V}}{1.5 \text{ k}\Omega + 2 \text{ k}\Omega} = 3.21 \text{ mA}$

$R_L = 5 \ k\Omega$: $I_L = \dfrac{11.25 \text{ V}}{1.5 \text{ k}\Omega + 5 \text{ k}\Omega} = 1.73 \text{ mA}$

EXAMPLE 9–5

Find the Thévenin equivalent circuit external to R_5 in the circuit in Figure 9–19. Use the equivalent circuit to determine the current through the resistor.

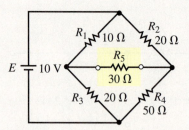

FIGURE 9–19

Solution Notice that the circuit is an unbalanced bridge circuit. If the techniques of the previous chapter had to be used, we would need to solve either three mesh equations or three nodal equations.

Steps 1 and 2: Removing the resistor R_5 from the circuit and labelling the two terminals a and b, we obtain the circuit shown in Figure 9–20.

By examining the circuit shown in Figure 9–20, we see that it is no simple task to determine the equivalent circuit between terminals a and b. The process is simplified by redrawing the circuit as illustrated in Figure 9–21.

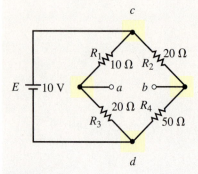

FIGURE 9–20

Notice that the circuit of Figure 9–21 has nodes *a* and *b* conveniently shown at the top and bottom of the circuit. Additional nodes (node *c* and node *d*) are added to simplify the task of correctly placing resistors between the nodes.

After simplifying a circuit, it is always a good idea to ensure that the resulting circuit is indeed an equivalent circuit. You may verify the equivalence of the two circuits by confirming that each component is connected between the same nodes for each circuit.

Now that we have a circuit that is easier to analyze, we find the Thévenin equivalent of the resultant.

Step 3: Setting the voltage source to zero by replacing it with a short, we obtain the circuit shown in Figure 9–22.

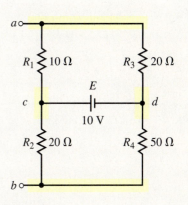

FIGURE 9–21

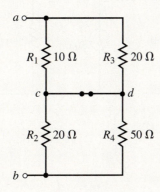

FIGURE 9–22

Step 4: The resulting Thévenin resistance is

$$R_{Th} = 10 \ \Omega \| 20 \ \Omega + 20 \ \Omega \| 50 \ \Omega$$
$$= 6.67 \ \Omega + 14.29 \ \Omega = 20.95 \ \Omega$$

Step 5: The open-circuit voltage between terminals *a* and *b* is found by first indicating the loop currents I_1 and I_2 in the circuit of Figure 9–23.

Because the voltage source, *E*, provides a constant voltage across the resistor combinations R_1-R_3 and R_2-R_4, we simply use the voltage divider rule to determine the voltage across the various components:

$$V_{ab} = -V_{R_1} + V_{R_2}$$
$$= -\frac{(10 \ \Omega)(10 \ V)}{30 \ \Omega} + \frac{(20 \ \Omega)(10 \ V)}{70 \ \Omega}$$
$$= -0.476 \ V$$

Note: The above technique could not be used if the source had some series resistance, since then the voltage provided to resistor combinations R_1-R_3 and R_2-R_4 would no longer be the entire supply voltage but rather would be dependent upon the value of the series resistance of the source.

Step 6: The resulting Thévenin circuit is shown in Figure 9–24.

From the circuit of Figure 9–24, it is now possible to calculate the current through the resistor R_5 as

$$I = \frac{0.476 \ V}{20.95 \ \Omega + 30 \ \Omega} = 9.34 \ mA \quad \text{(from } b \text{ to } a\text{)}$$

This example illustrates the importance of labelling the terminals that remain after a component or branch is removed. If we had not labelled the terminals and drawn an equivalent circuit, the current through R_5 would not have been found as easily.

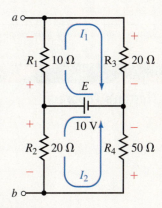

FIGURE 9–23

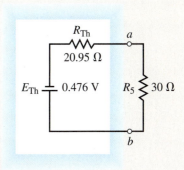

FIGURE 9–24

Find the Thévenin equivalent circuit external to resistor R_1 in the circuit of Figure 9–1.

Answer
$R_{Th} = 16\ \Omega,\ E_{Th} = 52\ V$

Use Thévenin's theorem to determine the current through load resistor R_L for the circuit of Figure 9–25.

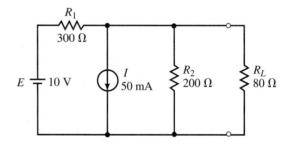

FIGURE 9–25

Answer
$I_L = 10.0\ mA$ upward

(Answers are at the end of the chapter.)

In the circuit of Figure 9–25, what would the value of R_1 need to be in order that the Thévenin resistance is equal to $R_L = 80\ \Omega$?

9.3 Norton's Theorem

Norton's theorem is a circuit analysis technique that is similar to Thévenin's theorem. By using this theorem the circuit is reduced to a single current source and one parallel resistor. As with the Thévenin equivalent circuit, the resulting two-terminal circuit is equivalent to the original circuit when connected to any external branch or component. In summary, **Norton's theorem** may be simplified as follows:

Any linear bilateral network may be reduced to a simplified two-terminal circuit consisting of a single current source and a single shunt resistor as shown in Figure 9–26.

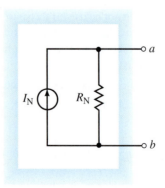

FIGURE 9–26   Norton equivalent circuit.

The following steps provide a technique that allows the conversion of any circuit into its Norton equivalent:

1. Remove the load from the circuit.

2. Label the resulting two terminals. We will label them as *a* and *b,* although any notation may be used.

3. Set all sources to zero. As before, voltage sources are set to zero by replacing them with short circuits and current sources are set to zero by replacing them with open circuits.

4. Determine the Norton equivalent resistance, R_N, by calculating the resistance seen between terminals *a* and *b.* It may be necessary to redraw the circuit to simplify this step.

5. Replace the sources removed in Step 3, and determine the current that would occur in a short if the short were connected between terminals *a* and *b.* If the original circuit has more than one source, it may be necessary to use the superposition theorem. In this case, it will be necessary to determine the short-circuit current due to each source separately and then determine the combined effect. The resulting short-circuit current will be the value of the Norton current I_N.

6. Sketch the Norton equivalent circuit using the resistance determined in Step 4 and the current calculated in Step 5. As part of the resulting circuit, include that portion of the network removed in Step 1.

The Norton equivalent circuit may also be determined directly from the Thévenin equivalent circuit by using the source conversion technique developed in Chapter 8. As a result, the Thévenin and Norton circuits shown in Figure 9–27 are equivalent.

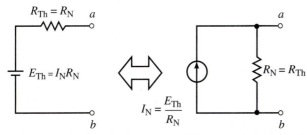

Thévenin equivalent circuit   Norton equivalent circuit

FIGURE 9–27

From Figure 9–27 we see that the relationship between the circuits is as follows:

$$E_{Th} = I_N R_N \qquad\qquad \textbf{(9–1)}$$

$$I_N = \frac{E_{Th}}{R_{Th}} \qquad\qquad \textbf{(9–2)}$$

EXAMPLE 9–6

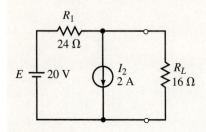

FIGURE 9–28

Determine the Norton equivalent circuit external to the resistor R_L for the circuit of Figure 9–28. Use the Norton equivalent circuit to calculate the current through R_L. Compare the results to those obtained using Thévenin's theorem in Example 9–3.

Solution

Steps 1 and 2: Remove load resistor R_L from the circuit and label the remaining terminals as a and b. The resulting circuit is shown in Figure 9–29.

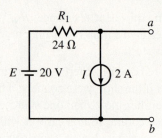

FIGURE 9–29

Step 3: Zero the voltage and current sources as shown in the circuit of Figure 9–30.

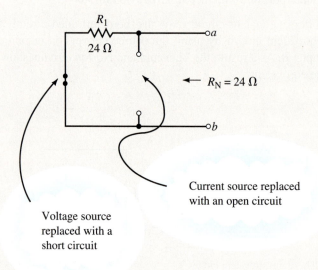

Current source replaced with an open circuit

Voltage source replaced with a short circuit

FIGURE 9–30

Step 4: The resulting Norton resistance between the terminals is

$$R_N = R_{ab} = 24 \ \Omega$$

Step 5: The short-circuit current is determined by first calculating the current through the short due to each source. The circuit for each calculation is illustrated in Figure 9–31.

Voltage Source, E: The current in the short between terminals a and b [Figure 9–31(a)] is found from Ohm's law as

$$I_{ab(1)} = \frac{20 \ \text{V}}{24 \ \Omega} = 0.833 \ \text{A}$$

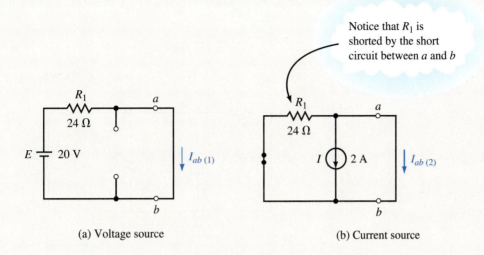

Notice that R_1 is shorted by the short circuit between a and b

(a) Voltage source

(b) Current source

FIGURE 9–31

Current Source, I: By examining the circuit for the current source [Figure 9–31(b)] we see that the short circuit between terminals a and b effectively removes R_1 from the circuit. Therefore, the current through the short will be

$$I_{ab(2)} = -2.00 \text{ A}$$

Notice that the current I_{ab} is indicated as being a negative quantity. As we have seen before, this result merely indicates that the actual current is opposite to the assumed reference direction.

Now, applying the superposition theorem, we find the Norton current as

$$I_N = I_{ab(1)} + I_{ab(2)} = 0.833 \text{ A} - 2.0 \text{ A} = -1.167 \text{ A}$$

As before, the negative sign indicates that the short-circuit current is actually from terminal b toward terminal a.

Step 6: The resultant Norton equivalent circuit is shown in Figure 9–32. Now we can easily find the current through load resistor R_L by using the current divider rule:

$$I_L = \left(\frac{24 \text{ }\Omega}{24 \text{ }\Omega + 16 \text{ }\Omega}\right)(1.167 \text{ A}) = 0.700 \text{ A} \quad \text{(upward)}$$

By referring to Example 9–3, we see that the same result was obtained by finding the Thévenin equivalent circuit. An alternate method of finding the Norton equivalent circuit is to convert the Thévenin circuit found in Example 9–3 into its equivalent Norton circuit shown in Figure 9–33.

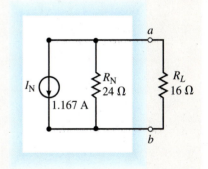

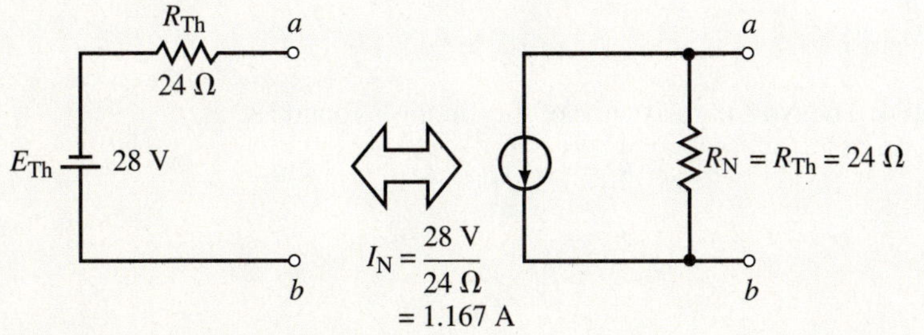

FIGURE 9–32

R_{Th}
24 Ω
E_{Th} 28 V

$I_N = \dfrac{28 \text{ V}}{24 \text{ }\Omega}$
$= 1.167 \text{ A}$

$R_N = R_{Th} = 24 \text{ }\Omega$

FIGURE 9–33

EXAMPLE 9–7

Find the Norton equivalent of the circuit external to resistor R_L in the circuit in Figure 9–34. Use the equivalent circuit to determine the load current I_L when $R_L = 0, 2$ kΩ, and 5 kΩ.

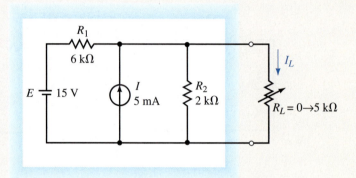

FIGURE 9–34

Solution

Steps 1, 2, and 3: After removing the load resistor, labelling the remaining two terminals a and b, and setting the sources to zero, we have the circuit of Figure 9–35.

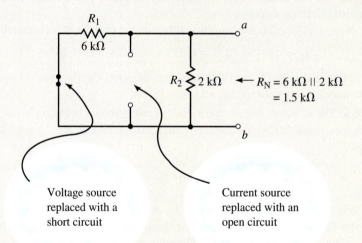

FIGURE 9–35

Step 4: The Norton resistance of the circuit is found as

$$R_N = 6 \text{ k}\Omega \| 2 \text{ k}\Omega = 1.5 \text{ k}\Omega$$

Step 5: The value of the Norton constant-current source is found by determining the current effects due to each independent source acting on a short circuit between terminals a and b.

Voltage Source, E: Referring to Figure 9–36(a), a short circuit between terminals a and b eliminates resistor R_2 from the circuit. The short-circuit current due to the voltage source is

$$I_{ab(1)} = \frac{15 \text{ V}}{6 \text{ k}\Omega} = 2.50 \text{ mA}$$

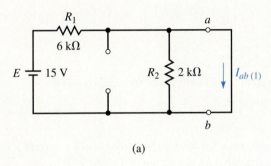

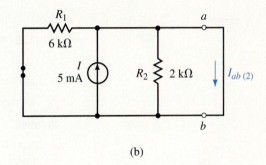

(a)   (b)

FIGURE 9–36

Current Source, I: Referring to Figure 9–36(b), the short circuit between terminals a and b eliminates both resistors R_1 and R_2. The short-circuit current due to the current source is therefore

$$I_{ab(2)} = 5.00 \text{ mA}$$

The resultant Norton current is found from superposition as

$$I_N = I_{ab(1)} + I_{ab(2)} = 2.50 \text{ mA} + 5.00 \text{ mA} = 7.50 \text{ mA}$$

Step 6: The Norton equivalent circuit is shown in Figure 9–37.

Let $R_L = 0$: The current I_L must equal the source current, and so

$$I_L = 7.50 \text{ mA}$$

Let $R_L = 2 \text{ k}\Omega$: The current I_L is found from the current divider rule as

$$I_L = \left(\frac{1.5 \text{ k}\Omega}{1.5 \text{ k}\Omega + 2 \text{ k}\Omega}\right)(7.50 \text{ mA}) = 3.21 \text{ mA}$$

Let $R_L = 5 \text{ k}\Omega$: Using the current divider rule again, the current I_L is found as

$$I_L = \left(\frac{1.5 \text{ k}\Omega}{1.5 \text{ k}\Omega + 5 \text{ k}\Omega}\right)(7.50 \text{ mA}) = 1.73 \text{ mA}$$

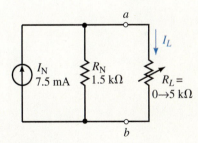

FIGURE 9–37

Comparing the above results to those obtained in Example 9–4, we see that they are precisely the same.

EXAMPLE 9–8

Consider the circuit of Figure 9–38:

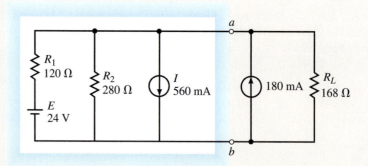

FIGURE 9–38

a. Find the Norton equivalent circuit external to terminals a and b.
b. Determine the current through R_L.

Solution

a. **Steps 1 and 2:** After removing the load, (which consists of a current source in parallel with a resistor), we have the circuit of Figure 9–39.

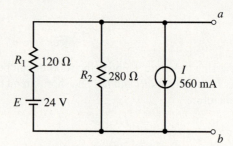

FIGURE 9–39

Step 3: After zeroing the sources, we have the network shown in Figure 9–40.

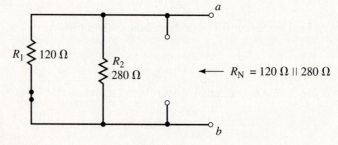

FIGURE 9–40

Step 4: The Norton equivalent resistance is found as

$$R_N = 120 \ \Omega \| 280 \ \Omega = 84 \ \Omega$$

Step 5: In order to determine the Norton current we must again determine the short-circuit current due to each source separately and then combine the results using the superposition theorem.

Voltage Source, E: Referring to Figure 9–41(a), notice that the resistor R_2 is shorted by the short circuit between terminals a and b and so the current in the short circuit is

$$I_{ab(1)} = \frac{24 \text{ V}}{120 \text{ }\Omega} = 0.2 \text{ A} = 200 \text{ mA}$$

Current Source, I: Referring to Figure 9–41(b), the short circuit between terminals a and b will now eliminate both resistors. The current through the short will simply be the source current. However, since the current will not be from a to b but rather in the opposite direction, we write

$$I_{ab(2)} = -560 \text{ mA}$$

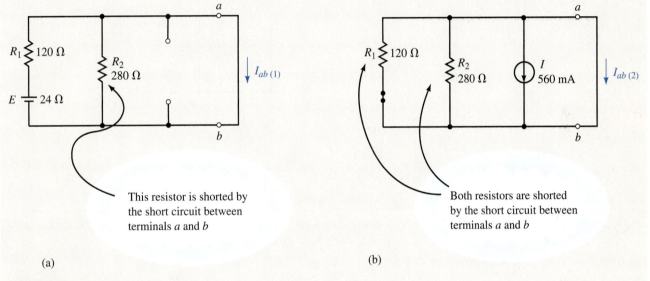

This resistor is shorted by the short circuit between terminals a and b

Both resistors are shorted by the short circuit between terminals a and b

(a)   (b)

FIGURE 9–41

Now the Norton current is found as the summation of the short-circuit currents due to each source:

$$I_N = I_{ab(1)} + I_{ab(2)} = 200 \text{ mA} + (-560 \text{ mA}) = -360 \text{ mA}$$

The negative sign in the above calculation for current indicates that if a short circuit were placed between terminals a and b, current would actually be in the direction from b to a. The Norton equivalent circuit is shown in Figure 9–42.

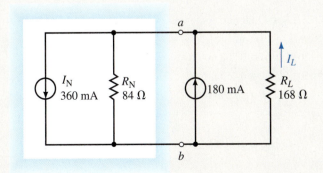

FIGURE 9–42

b. The current through the load resistor is found by applying the current divider rule:

$$I_L = \left(\frac{84\ \Omega}{84\ \Omega + 168\ \Omega}\right)(360\ \text{mA} - 180\ \text{mA}) = 60\ \text{mA} \quad \text{(upward)}$$

PRACTICE PROBLEMS 4

Find the Norton equivalent of the circuit shown in Figure 9–43. Use the source conversion technique to determine the Thévenin equivalent of the circuit between points *a* and *b*.

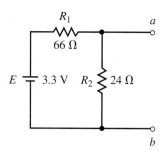

FIGURE 9–43

Answers
$R_\text{N} = R_\text{Th} = 17.6\ \Omega$, $I_\text{N} = 0.05\ \text{A}$, $E_\text{Th} = 0.88\ \text{V}$

PRACTICE PROBLEMS 5

Find the Norton equivalent external to R_L in the circuit of Figure 9–44. Solve for the current I_L when $R_L = 0$, 10 kΩ, 50 kΩ, and 100 kΩ.

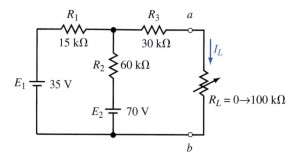

FIGURE 9–44

Answers
$R_\text{N} = 42\ \text{k}\Omega$, $I_\text{N} = 1.00\ \text{mA}$
For $R_L = 0$: $I_L = 1.00\ \text{mA}$
For $R_L = 10\ \text{k}\Omega$: $I_L = 0.808\ \text{mA}$
For $R_L = 50\ \text{k}\Omega$: $I_L = 0.457\ \text{mA}$
For $R_L = 100\ \text{k}\Omega$: $I_L = 0.296\ \text{mA}$

1. Show the relationship between the Thévenin equivalent circuit and the Norton equivalent circuit. Sketch each circuit.
2. If a Thévenin equivalent circuit has $E_{Th} = 100$ mV and $R_{Th} = 500$ Ω, draw the corresponding Norton equivalent circuit.
3. If a Norton equivalent circuit has $I_N = 10$ μA and $R_N = 20$ kΩ, draw the corresponding Thévenin equivalent circuit.

IN-PROCESS
LEARNING CHECK 3

(Answers are at the end of the chapter.)

In amplifiers and in most communication circuits such as radio receivers and transmitters, it is often desired that the load receive the maximum amount of power from a source.

The **maximum power transfer theorem** states the following:

A load resistance will receive maximum power from a circuit when the resistance of the load is exactly the same as the Thévenin (Norton) resistance looking back at the circuit.

The proof for the maximum power transfer theorem is determined from the Thévenin equivalent circuit and involves the use of calculus. This theorem is proved in Appendix C.

From Figure 9–45 we see that once the network has been simplified using either Thévenin's or Norton's theorem, maximum power will occur when

$$R_L = R_{Th} = R_N \qquad (9\text{–}3)$$

Examining the equivalent circuits of Figure 9–45, shows that the following equations determine the power delivered to the load:

$$P_L = \frac{\left(\dfrac{R_L}{R_L + R_{Th}} \times E_{Th}\right)^2}{R_L}$$

which gives

$$P_L = \frac{E_{Th}^2 R_L}{(R_L + R_{Th})^2} \qquad (9\text{–}4)$$

Similarly,

$$P_L = \left(\frac{I_N R_N}{R_L + R_N}\right)^2 \times R_L \qquad (9\text{–}5)$$

Under maximum power conditions ($R_L = R_{Th} = R_N$), the above equations may be used to determine the maximum power delivered to the load and may therefore be written as

$$P_{max} = \frac{E_{Th}^2}{4R_{Th}} \qquad (9\text{–}6)$$

$$P_{max} = \frac{I_N^2 R_N}{4} \qquad (9\text{–}7)$$

9.4 Maximum Power Transfer Theorem

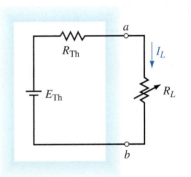

(a)

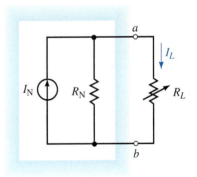

(b)

FIGURE 9–45

EXAMPLE 9–9

For the circuit of Figure 9–46, sketch graphs of V_L, I_L, and P_L as functions of R_L.

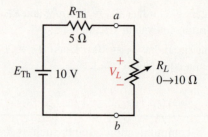

FIGURE 9–46

TABLE 9–1

R_L (Ω)	V_L (V)	I_L (A)	P_L (W)
0	0	2.000	0
1	1.667	1.667	2.778
2	2.857	1.429	4.082
3	3.750	1.250	4.688
4	4.444	1.111	4.938
5	5.000	1.000	5.000
6	5.455	0.909	4.959
7	5.833	0.833	4.861
8	6.154	0.769	4.734
9	6.429	0.714	4.592
10	6.667	0.667	4.444

Solution   We may first set up a table of data for various values of resistance, R_L. See Table 9–1. Voltage and current values are determined by using the voltage divider rule and Ohm's law respectively. The power P_L for each value of resistance is determined by finding the product $P_L = V_L I_L$, or by using Equation 9–4.

If the data from Table 9–1 are plotted on linear graphs, the graphs will appear as shown in Figures 9–47, 9–48, and 9–49.

Notice in the graphs that although voltage across the load increases as R_L increases, the power delivered to the load will be a maximum when $R_L = R_{Th} = 5\ \Omega$. The reason for this apparent contradiction is because, as R_L increases, the reduction in current more than offsets the corresponding increase in voltage.

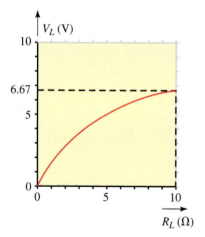

FIGURE 9–47   Voltage versus R_L.

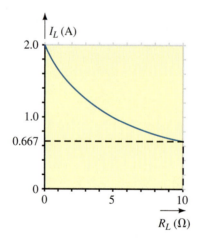

FIGURE 9–48   Current versus R_L.

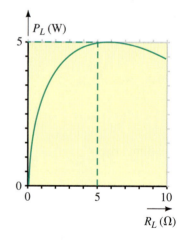

FIGURE 9–49   Power versus R_L.

EXAMPLE 9–10

Consider the circuit of Figure 9–50:

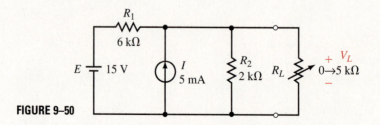

FIGURE 9–50

a. Determine the value of load resistance required to ensure that maximum power is transferred to the load.

b. Find V_L, I_L, and P_L when maximum power is delivered to the load.

Solution

a. In order to determine the conditions for maximum power transfer, it is first necessary to determine the equivalent circuit external to the load. We may determine either the Thévenin equivalent circuit or the Norton equivalent circuit. This circuit was analyzed in Example 9–4 using Thévenin's theorem, and we determined the equivalent circuit to be as shown in Figure 9–51.

Maximum power will be transferred to the load when $R_L = 1.5 \text{ k}\Omega$.

b. Letting $R_L = 1.5 \text{ k}\Omega$, we see that half of the Thévenin voltage will appear across the load resistor and half will appear across the Thévenin resistance. So, at maximum power,

$$V_L = \frac{E_{Th}}{2} = \frac{11.25 \text{ V}}{2} = 5.625 \text{ V}$$

$$I_L = \frac{5.625 \text{ V}}{1.5 \text{ k}\Omega} = 3.750 \text{ mA}$$

The power delivered to the load is found as

$$P_L = \frac{V_L^2}{R_L} = \frac{(5.625 \text{ V})^2}{1.5 \text{ k}\Omega} = 21.1 \text{ mW}$$

Or, alternatively using current, we calculate the power as

$$P_L = I_L^2 R_L = (3.75 \text{ mA})^2 (1.5 \text{ k}\Omega) = 21.1 \text{ mW}$$

In solving this problem, we could just as easily have used the Norton equivalent circuit to determine required values.

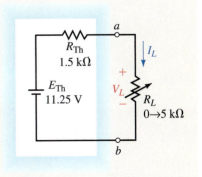

FIGURE 9–51

Recall that efficiency was defined as the ratio of output power to input power:

$$\eta = \frac{P_{out}}{P_{in}}$$

or as a percentage:

$$\eta = \frac{P_{out}}{P_{in}} \times 100\%$$

By using the maximum power transfer theorem, we see that under the condition of maximum power the efficiency of the circuit is

$$\eta = \frac{P_{out}}{P_{in}} \times 100\%$$
$$= \frac{\dfrac{E_{Th}^2}{4R_{Th}}}{\dfrac{E_{Th}^2}{2R_{Th}}} \times 100\% = 0.500 \times 100\% = 50\% \qquad (9\text{–}8)$$

For communication circuits and for many amplifier circuits, 50% represents the maximum possible efficiency. At this efficiency level, the voltage presented to the following stage would only be half of the maximum terminal voltage.

In power transmission such as the 115-Vac, 60-Hz power in your home, the condition of maximum power is not a requirement. Under the condition of maximum power transfer, the voltage across the load will be reduced to half of the maximum available terminal voltage. Clearly, if we are working with power supplies, we would like to ensure that efficiency is brought as close to 100% as possible. In such cases, load resistance R_L is kept much larger than the internal resistance of the voltage source (typically $R_L \geq 10R_{\text{int}}$), ensuring that the voltage appearing across the load will be very nearly equal to the maximum terminal voltage of the voltage source.

EXAMPLE 9–11

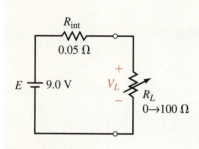

FIGURE 9–52

Refer to the circuit of Figure 9–52, which represents a typical dc power supply.

a. Determine the value of R_L needed for maximum power transfer.

b. Determine terminal voltage V_L and the efficiency when the value of the load resistor is $R_L = 50\ \Omega$.

c. Determine terminal voltage V_L and the efficiency when the value of the load resistor is $R_L = 100\ \Omega$.

Solution

a. For maximum power transfer, the load resistor will be given as $R_L = 0.05\ \Omega$. At this value of load resistance, the efficiency will be only 50%.

b. For $R_L = 50\ \Omega$, the voltage appearing across the output terminals of the voltage source is

$$V_L = \left(\frac{50\ \Omega}{50\ \Omega + 0.05\ \Omega}\right)(9.0\ \text{V}) = 8.99\ \text{V}$$

The efficiency is

$$\eta = \frac{P_{\text{out}}}{P_{\text{in}}} \times 100\%$$

$$= \frac{\dfrac{(8.99\ \text{V})^2}{50\ \Omega}}{\dfrac{(9.0\ \text{V})^2}{50.05\ \Omega}} \times 100\%$$

$$= \frac{1.6168\ \text{W}}{1.6184\ \text{W}} \times 100\% = 99.90\%$$

c. For $R_L = 100\ \Omega$, the voltage appearing across the output terminals of the voltage source is

$$V_L = \left(\frac{100\ \Omega}{100\ \Omega + 0.05\ \Omega}\right)(9.0\ \text{V}) = 8.995\ 50\ \text{V}$$

The efficiency is

$$\eta = \frac{P_{\text{out}}}{P_{\text{in}}} \times 100\%$$

$$= \frac{\dfrac{(8.9955\ \text{V})^2}{100\ \Omega}}{\dfrac{(9.0\ \text{V})^2}{100.05\ \Omega}} \times 100\%$$

$$= \frac{1.6168\ \text{W}}{1.6184\ \text{W}} \times 100\% = 99.95\%$$

From this example, we see that if efficiency is important, as it is in power transmission, then the load resistance should be much larger than the resistance of the source (typically $R_L \geq 10R_{\text{int}}$). If, on the other hand, it is more important to ensure maximum power transfer, then the load resistance should equal the source resistance ($R_L = R_{\text{int}}$).

PRACTICE PROBLEMS 6

Refer to the circuit of Figure 9–44. For what value of R_L will the load receive maximum power? Determine the power when $R_L = R_N$, when $R_L = 25$ kΩ, and when $R_L = 50$ kΩ.

Answers
$R_L = 42$ kΩ: $P_L = 10.5$ mW
$R_L = 25$ kΩ: $P_L = 9.82$ mW
$R_L = 50$ kΩ: $P_L = 10.42$ mW

IN-PROCESS
LEARNING CHECK 4

(Answers are at the end of the chapter.)

A Thévenin equivalent circuit consists of $E_{Th} = 10$ V and $R_{Th} = 2$ kΩ. Determine the efficiency of the circuit when
a. $R_L = R_{Th}$
b. $R_L = 0.5R_{Th}$
c. $R_L = 2R_{Th}$

IN-PROCESS
LEARNING CHECK 5

(Answers are at the end of the chapter.)

1. In what instances is maximum power transfer a desirable characteristic of a circuit?
2. In what instances is maximum power transfer an undesirable characteristic of a circuit?

The **substitution theorem** states the following:

Any branch within a circuit may be replaced by an equivalent branch, provided the replacement branch has the same current through it and voltage across it as the original branch.

This theorem is best illustrated by examining the operation of a circuit. Consider the circuit of Figure 9–53.

The voltage V_{ab} and the current I in the circuit of Figure 9–53 are given as

$$V_{ab} = \left(\frac{6\text{ k}\Omega}{4\text{ k}\Omega + 6\text{ k}\Omega}\right)(10\text{ V}) = +6.0\text{ V}$$

and

$$I = \frac{10\text{ V}}{4\text{ k}\Omega + 6\text{ k}\Omega} = 1\text{ mA}$$

9.5 Substitution Theorem

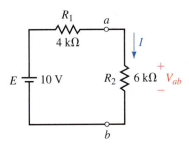

FIGURE 9–53

The resistor R_2 may be replaced with any combination of components, provided that the resulting components maintain the above conditions. We see that the branches of Figure 9–54 are each equivalent to the original branch between terminals a and b of the circuit in Figure 9–53.

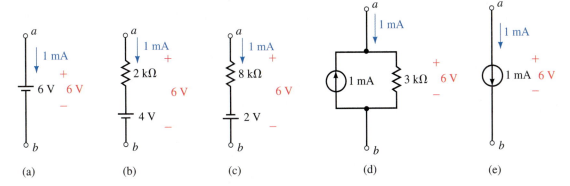

(a)   (b)   (c)   (d)   (e)

FIGURE 9–54

Although each of the branches in Figure 9–54 is different, the current entering or leaving each branch will be same as that in the original branch. Similarly, the voltage across each branch will be the same. If any of these branches is substituted into the original circuit, the balance of the circuit will operate in the same way as the original. It is left as an exercise for the student to verify that each circuit behaves the same as the original.

This theorem allows us to replace any branch within a given circuit with an equivalent branch, thereby simplifying the analysis of the remaining circuit.

EXAMPLE 9–12

If the indicated portion in the circuit of Figure 9–55 is to be replaced with a current source and a 240-Ω shunt resistor, determine the magnitude and direction of the required current source.

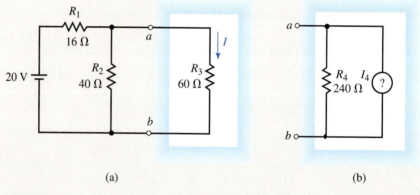

(a)   (b)

FIGURE 9–55

Solution   The voltage across the branch in the original circuit is

$$V_{ab} = \left(\frac{40\ \Omega \| 60\ \Omega}{16\ \Omega + (40\ \Omega \| 60\ \Omega)}\right)(20\ \text{V}) = \left(\frac{24\ \Omega}{16\ \Omega + 24\ \Omega}\right)(20\ \text{V}) = 12.0\ \text{V}$$

which results in a current of

$$I = \frac{12.0\ \text{V}}{60\ \Omega} = 0.200\ \text{A} = 200\ \text{mA}$$

In order to maintain the same terminal voltage, $V_{ab} = 12.0$ V, the current through resistor $R_4 = 240\ \Omega$ must be

$$I_{R_4} = \frac{12.0\ \text{V}}{240\ \Omega} = 0.050\ \text{A} = 50\ \text{mA}$$

Finally, we know that the current entering terminal a is $I = 200$ mA. In order for Kirchhoff's current law to be satisfied at this node, the current source must have a magnitude of 150 mA and the direction must be downward, as shown in Figure 9–56.

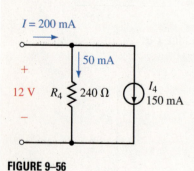

FIGURE 9–56

Millman's theorem is used to simplify circuits having several parallel voltage sources as illustrated in Figure 9–57. Although any of the other theorems developed in this chapter will work in this case, Millman's theorem provides a much simpler and more direct equivalent.

In circuits of the type shown in Figure 9–57, the voltage sources may be replaced with a single equivalent source as shown in Figure 9–58.

To find the values of the equivalent voltage source E_{eq} and series resistance R_{eq}, we need to convert each of the voltage sources of Figure 9–57 into its equivalent current source using the technique developed in Chapter 8. The value of each current source would be determined by using Ohm's law (i.e., $I_1 = E_1/R_1$, $I_2 = E_2/R_2$, etc.). After the source conversions are completed, the circuit appears as shown in Figure 9–59.

From the circuit of Figure 9–59 we see that all of the current sources have the same direction. Clearly, this will not always be the case since the direction of each current source will be determined by the initial polarity of the corresponding voltage source.

9.6 Millman's Theorem

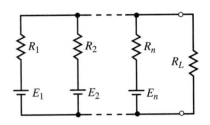

FIGURE 9–57

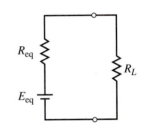

FIGURE 9–58

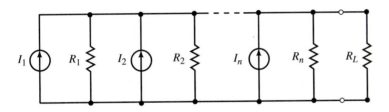

FIGURE 9–59

It is now possible to replace the n current sources with a single current source having a magnitude given as

$$I_{eq} = \sum_{x=0}^{n} I_x = I_1 + I_2 + I_3 + \cdots + I_n \qquad (9\text{--}9)$$

which may be written as

$$I_{eq} = \frac{E_1}{R_1} + \frac{E_2}{R_2} + \frac{E_3}{R_3} + \cdots + \frac{E_n}{R_n} \qquad (9\text{--}10)$$

If the direction of any current source is opposite to the direction shown, then the corresponding magnitude would be subtracted, rather than added. From Figure 9–59, we see that removing the current sources results in an equivalent resistance given as

$$R_{eq} = R_1 \| R_2 \| R_3 \| \cdots \| R_n \qquad (9\text{--}11)$$

which may be determined as

$$R_{eq} = \frac{1}{G_{eq}} = \frac{1}{\dfrac{1}{R_1} + \dfrac{1}{R_2} + \dfrac{1}{R_3} + \cdots + \dfrac{1}{R_n}} \qquad (9\text{--}12)$$

The general expression for the equivalent voltage is

$$E_{eq} = I_{eq}R_{eq} = \frac{\dfrac{E_1}{R_1} + \dfrac{E_2}{R_2} + \dfrac{E_3}{R_3} + \cdots + \dfrac{E_n}{R_n}}{\dfrac{1}{R_1} + \dfrac{1}{R_2} + \dfrac{1}{R_3} + \cdots + \dfrac{1}{R_n}} \qquad (9\text{--}13)$$

EXAMPLE 9–13

Use Millman's Theorem to simplify the circuit of Figure 9–60 so that it has only a single source. Use the simplified circuit to find the current in the load resistor, R_L.

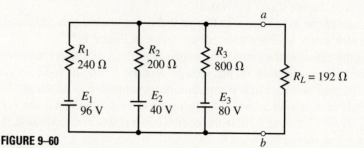

FIGURE 9–60

Solution From Equation 9–13, we express the equivalent voltage source as

$$V_{ab} = E_{eq} = \frac{\dfrac{-96\ V}{240\ \Omega} + \dfrac{40\ V}{200\ \Omega} + \dfrac{-80\ V}{800\ \Omega}}{\dfrac{1}{240\ \Omega} + \dfrac{1}{200\ \Omega} + \dfrac{1}{800\ \Omega}}$$

$$V_{ab} = \frac{-0.300}{10.42\ mS} = -28.8\ V$$

The equivalent resistance is

$$R_{eq} = \frac{1}{\dfrac{1}{240\ \Omega} + \dfrac{1}{200\ \Omega} + \dfrac{1}{800\ \Omega}} = \frac{1}{10.42\ mS} = 96\ \Omega$$

The equivalent circuit using Millman's theorem is shown in Figure 9–61. Notice that the equivalent voltage source has a polarity that is opposite to the originally assumed polarity. This is because the voltage sources E_1 and E_3 have magnitudes which overcome the polarity and magnitude of the source E_2.

From the equivalent circuit of Figure 9–61, it is a simple matter to determine the current through the load resistor:

$$I_L = \frac{28.8\ V}{96\ \Omega + 192\ \Omega} = 0.100\ A = 100\ mA \quad (upward)$$

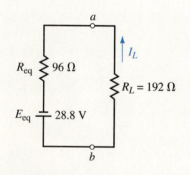

FIGURE 9–61

9.7 Reciprocity Theorem

The **reciprocity theorem** is a theorem that can only be used with single-source circuits. This theorem, however, may be applied to either voltage sources or current sources. The theorem states the following:

Voltage Sources

A voltage source causing a current I *in any branch of a circuit may be removed from the original location and placed into that branch having the current* I. *The voltage source in the new location will produce a current in the original source location that is exactly equal to the originally calculated current,* I.

When applying the reciprocity theorem for a voltage source, the following steps must be followed:

1. The voltage source is replaced by a short circuit in the original location.

2. The polarity of the source in the new location is such that the current direction in that branch remains unchanged.

Current Sources

A current source causing a voltage V *at any node of a circuit may be removed from the original location and connected to that node. The current source in the new location will produce a voltage in the original source location that is exactly equal to the originally calculated voltage,* V.

When applying the reciprocity theorem for a current source, the following conditions must be met:

1. The current source is replaced by an open circuit in the original location.
2. The direction of the source in the new location is such that the polarity of the voltage at the node to which the current source is now connected remains unchanged.

The following examples illustrate how the reciprocity theorem is used within a circuit.

Consider the circuit of Figure 9–62:

a. Calculate the current *I*.

b. Remove voltage source *E* and place it into the branch with R_3. Show that the current through the branch which formerly had *E* is now the same as the current *I*.

Solution

a. $V_{12\,\Omega} = \left(\dfrac{8\,\Omega \| 12\,\Omega}{4\,\Omega + (8\,\Omega \| 12\,\Omega)}\right)(22\text{ V}) = \left(\dfrac{4.8}{8.8}\right)(22\text{ V}) = 12.0\text{ V}$

$I = \dfrac{V_{12\,\Omega}}{12\,\Omega} = \dfrac{12.0\text{ V}}{12\,\Omega} = 1.00\text{ A}$

b. Now removing the voltage source from its original location and moving it into the branch containing the current *I*, we obtain the circuit shown in Figure 9–63.

EXAMPLE 9–14

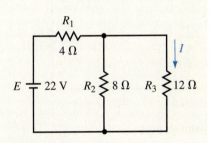

FIGURE 9–62

FIGURE 9–63

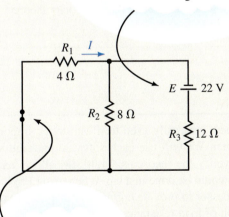

Polarity of the source is such that the current direction remains unchanged.

When *E* is removed it is replaced by a short circuit.

For the circuit of Figure 9–63, we determine the current I as follows:

$$V_{4\,\Omega} = \left(\frac{4\;\Omega\|8\;\Omega}{12\;\Omega + (4\;\Omega\|8\;\Omega)}\right)(22\text{ V}) = \left(\frac{2.\overline{6}}{14.\overline{6}}\right)(22\text{ V}) = 4.00\text{ V}$$

$$I = \frac{V_{4\,\Omega}}{4\;\Omega} = \frac{4.00\text{ V}}{4\;\Omega} = 1.00\text{ A}$$

From this example, we see that the reciprocity theorem does indeed apply.

EXAMPLE 9–15

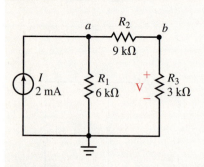

FIGURE 9–64

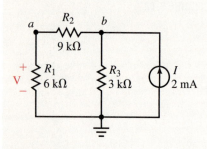

FIGURE 9–65

Consider the circuit shown in Figure 9–64:

a. Determine the voltage V across resistor R_3.

b. Remove the current source I and place it between node b and the reference node. Show that the voltage across the former location of the current source (node a) is now the same as the voltage V.

Solution

a. The node voltages for the circuit of Figure 9–64 are determined as follows:

$$R_T = 6\text{ k}\Omega\|(9\text{ k}\Omega + 3\text{ k}\Omega) = 4\text{ k}\Omega$$

$$V_a = (2\text{ mA})(4\text{ k}\Omega) = 8.00\text{ V}$$

$$V_b = \left(\frac{3\text{ k}\Omega}{3\text{ k}\Omega + 9\text{ k}\Omega}\right)(8.0\text{ V}) = 2.00\text{ V}$$

b. After relocating the current source from the original location, and connecting it between node b and ground, we obtain the circuit shown in Figure 9–65.

The resulting node voltages are now found as follows:

$$R_T = 3\text{ k}\Omega\|(6\text{ k}\Omega + 9\text{ k}\Omega) = 2.50\text{ k}\Omega$$

$$V_b = (2\text{ mA})(2.5\text{ k}\Omega) = 5.00\text{ V}$$

$$V_a = \left(\frac{6\text{ k}\Omega}{6\text{ k}\Omega + 9\text{ k}\Omega}\right)(5.0\text{ V}) = 2.00\text{ V}$$

From the above results, we conclude that the reciprocity theorem again applies for the given circuit.

9.8 Circuit Analysis Using Computers

 MULTISIM   CADENCE

MultiSIM® and PSpice® are easily used to illustrate the important theorems developed in this chapter. We will use each software package in a somewhat different approach to verify the theorems. MultiSIM allows us to "build and test" a circuit just as it would be done in a lab. When using PSpice we will activate the Probe postprocessor to provide a graphical display of voltage, current, and power as a function of load resistance.

MultiSIM

Use MultiSIM to find both the Thévenin and the Norton equivalent circuits external to the load resistor in the circuit of Figure 9–66:

EXAMPLE 9–16

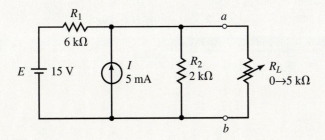

FIGURE 9–66

Solution

1. Using MultiSIM, construct the circuit as shown in Figure 9–67.

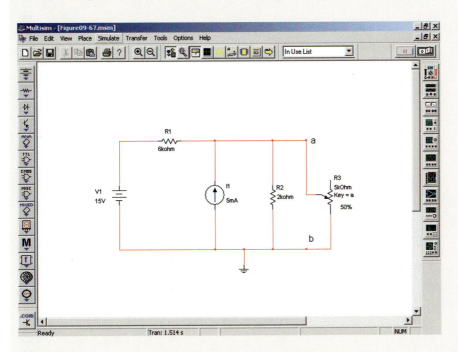

FIGURE 9–67

◀ MULTISIM

2. Just as in a lab, we will use a multimeter to find the open circuit (Thévenin) voltage and the short circuit (Norton) current. As well, the multimeter is used to measure the Thévenin (Norton) resistance. The steps in these measurements are essentially the same as those used to theoretically determine the equivalent circuits.

a. We begin by removing the load resistor, R_L from the circuit (using the Edit menu and the Cut/Paste menu items). The remaining terminals are labeled as *a* and *b*.

b. The Thévenin voltage is measured by simply connecting the multimeter between terminals *a* and *b*. After clicking on the power switch, we obtain a reading of $E_{Th} = 11.25$ V as shown in Figure 9–68.

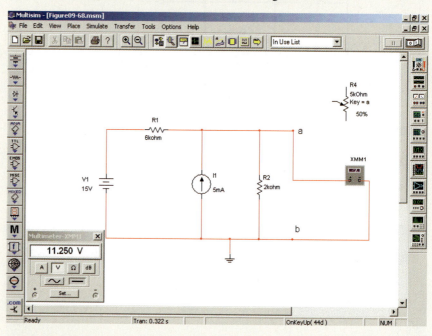

FIGURE 9–68

c. With the multimeter between terminals *a* and *b*, the Norton current is easily measured by switching the multimeter to its ammeter range. After clicking on the power switch, we obtain a reading of $I_N = 7.50$ mA as shown in Figure 9–69.

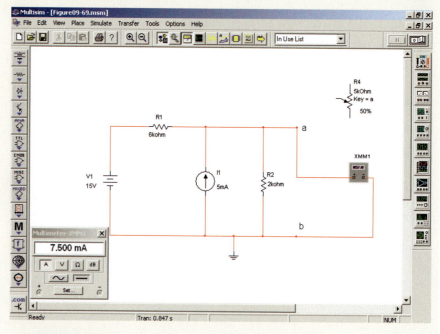

FIGURE 9–69

d. In order to measure the Thévenin resistance, the voltage source is removed and replaced by a short circuit (wire) and the current source is removed and replaced by an open circuit. Now, with the multimeter connected between terminals *a* and *b* it is set to measure resistance (by clicking on the Ω button). After clicking on the power switch, we have the display shown in Figure 9–70. The multimeter provides the Thévenin resistance as $R_{Th} = 1.5 \text{ k}\Omega$.

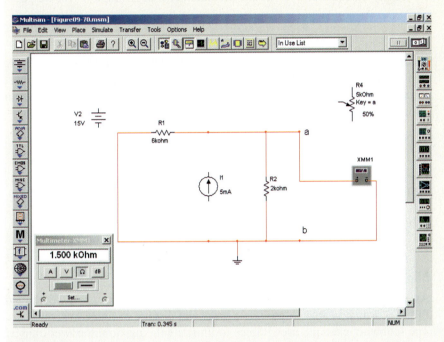

FIGURE 9–70

3. Using the measured results, we are able to sketch both the Thévenin equivalent and the Norton equivalent circuit as illustrated in Figure 9–71.

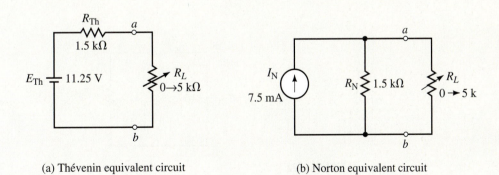

(a) Thévenin equivalent circuit (b) Norton equivalent circuit

FIGURE 9–71

These results are consistent with those obtained in Example 9–4.

Note: From the previous example we see that it is not necessary to directly measure the Thévenin (Norton) resistance since the value can be easily calculated from the Thévenin voltage and the Norton current. The following equation is an application of Ohm's law and always applies when finding an equivalent circuit.

$$R_{Th} = R_N = \frac{E_{Th}}{I_N} \tag{9–14}$$

Applying equation 9–14 to the measurements of Example 9–16, we get

$$R_{Th} = R_N = \frac{11.25 \text{ V}}{7.50 \text{ mA}} = 1.50 \text{ k}\Omega$$

Clearly, this is the same result as that obtained when we went through the extra step of removing the voltage and current sources. This approach is the most practical method and is commonly used when actually measuring the Thévenin (Norton) resistance of a circuit.

PRACTICE PROBLEMS 7

Use MultiSIM to find both the Thévenin and the Norton equivalent circuits external to the load resistor in the circuit of Figure 9–25.

Answers
$E_{Th} = 2.00$ V, $I_N = 16.67$ mA, $R_{Th} = R_N = 120 \ \Omega$

PSpice

◀ CADENCE

As we have already seen, PSpice has an additional postprocessor called PROBE, which is able to provide a graphical display of numerous variables. The following example uses PSpice to illustrate the maximum power transfer theorem.

EXAMPLE 9–17

Use PSpice to input the circuit of Figure 9–72 and use the Probe postprocessor to display output voltage, current, and power as a function of load resistance.

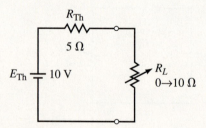

◀ CADENCE

FIGURE 9–72

Solution   The circuit is constructed as shown in Figure 9–73.

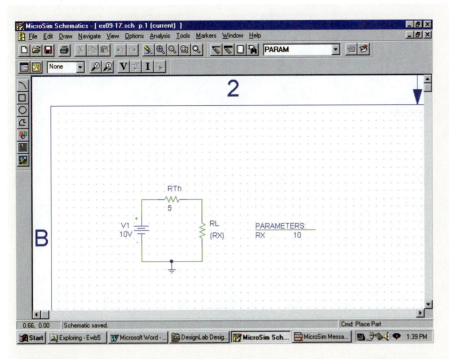

FIGURE 9–73

- Double click on each resistor in the circuit and change the Reference cells to RTH and RL. Click on Apply to accept the changes.

- Double click on the value for RL and enter **{Rx}.** Place the PARAM part adjacent to RL. Use the Property Editor to assign a default value of 10 Ω to Rx. Click on Apply. Have the display show the name and value and then exit the Property Editor.

- Adjust the Simulation Settings to result in a DC sweep of the load resistor from 0.1 Ω to 10 Ω in 0.1 Ω increments. (Refer to Example 7–15 for the complete procedure.)

- Click on the Run icon once the circuit is complete.

- Once the design is simulated, you will see a blank screen with the abscissa (horizontal axis) showing RX scaled from 0 to 10 Ω.

- Since we would like to have a simultaneous display of voltage, current, and power, it is necessary to do the following:

To display V_L: Click Trace and then Add Trace. Select **V(RL:1).** Click OK and the load voltage will appear as a function of load resistance.

To display I_L: First add another axis by clicking on Plot and Add Y Axis. Next, click Trace and then Add Trace. Select **I(RL).** Click OK and the load current will appear as a function of load resistance.

To display P_L: Add another Y axis. Click Trace and Add Trace. Now, since power is not one of the options that can be automatically selected, it is necessary to enter the power into the Trace Expression box. One method of doing this is to enter **I(RL)*V(RL:1)** and then click OK. Adjust the limits of the Y axis by clicking on Plot and Axis Settings. Click on the Y Axis tab and select the User Defined Data Range. Set the limits from 0W to 5W. The display will appear as shown in Figure 9–74.

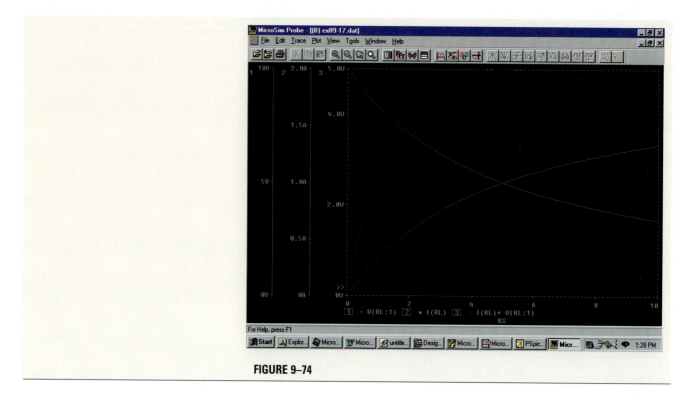

FIGURE 9–74

PRACTICE PROBLEMS 8

Use PSpice to input the circuit of Figure 9–66. Use the Probe postprocessor to obtain voltage, current, and power for the load resistor as it is varied from 0 to 5 kΩ.

PUTTING IT INTO PRACTICE

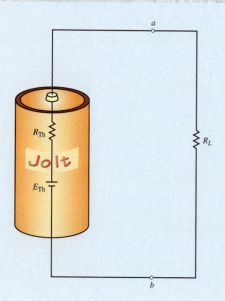

A simple battery cell (such as a "D" cell) can be represented as a Thévenin equivalent circuit as shown in the accompanying figure.

The Thévenin voltage represents the open-circuit (or unloaded) voltage of the battery cell, while the Thévenin resistance is the internal resistance of the battery. When a load resistance is connected across the terminals of the battery, the voltage V_{ab} will decrease due to the voltage drop across the internal resistor. By taking two measurements, it is possible to find the Thévenin equivalent circuit of the battery.

When no load is connected between the terminals of the battery, the terminal voltage is found to be $V_{ab} = 1.493$ V. When a resistance of $R_L = 10.6$ Ω is connected across the terminals, the voltage is measured to be $V_{ab} = 1.430$ V. Determine the Thévenin equivalent circuit of the battery. Use the measurements to determine the efficiency of the battery for the given load.

PROBLEMS

9.1 Superposition Theorem

1. Given the circuit of Figure 9–75, use superposition to calculate the current through each of the resistors.

2. Use superposition to determine the voltage drop across each of the resistors of the circuit in Figure 9–76.

3. Use superposition to solve for the voltage V_a and the current I in the circuit of Figure 9–77.

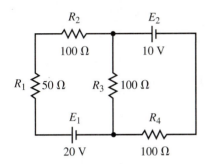

FIGURE 9–75

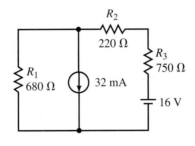

FIGURE 9–76

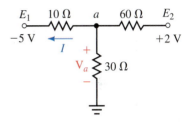

FIGURE 9–77

4. Using superposition, find the current through the 480-Ω resistor in the circuit of Figure 9–78:

5. Given the circuit of Figure 9–79, what must be the value of the unknown voltage source to ensure that the current through the load is $I_L = 5$ mA as shown. Verify the results using superposition.

6. If the load resistor in the circuit of Figure 9–80 is to dissipate 120 W, determine the value of the unknown voltage source. Verify the results using superposition.

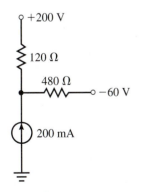

FIGURE 9–78

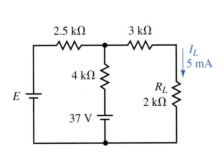

FIGURE 9–79

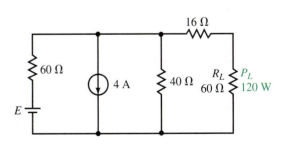

FIGURE 9–80

9.2 Thévenin's Theorem

7. Find the Thévenin equivalent external to R_L in circuit of Figure 9–81. Use the equivalent circuit to find V_{ab}.

8. Repeat Problem 7 for the circuit of Figure 9–82.

9. Repeat Problem 7 for the circuit of Figure 9–83.

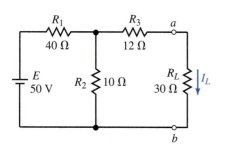

FIGURE 9–81

◀ MULTISIM

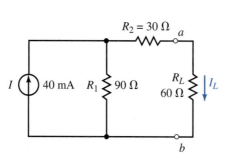

FIGURE 9–82

◀ MULTISIM

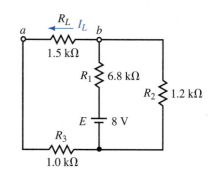

FIGURE 9–83

10. Repeat Problem 7 for the circuit of Figure 9–84.

11. Refer to the circuit of Figure 9–85:

    a. Find the Thévenin equivalent circuit external to R_L.

    b. Use the equivalent circuit to determine V_{ab} when $R_L = 20\ \Omega$ and when $R_L = 50\ \Omega$.

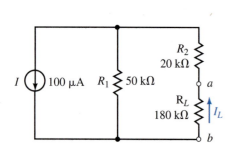

FIGURE 9–84

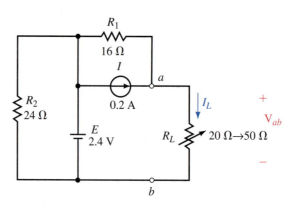

FIGURE 9–85

12. Refer to the circuit of Figure 9–86:

    a. Find the Thévenin equivalent circuit external to R_L.

    b. Use the equivalent circuit to determine V_{ab} when $R_L = 10\ k\Omega$ and when $R_L = 20\ k\Omega$.

13. Refer to the circuit of Figure 9–87:

    a. Find the Thévenin equivalent circuit external to the indicated terminals.

    b. Use the Thévenin equivalent circuit to determine the current through the indicated branch.

14. Refer to the circuit of Figure 9–88:

    a. Find the Thévenin equivalent circuit external to R_L.

    b. Use the Thévenin equivalent circuit to find V_L.

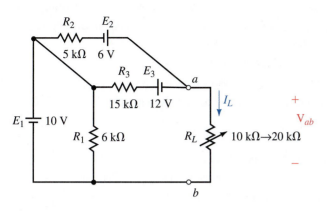

FIGURE 9–86

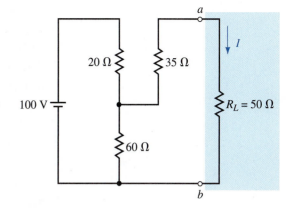

FIGURE 9–87

◀ **MULTISIM**

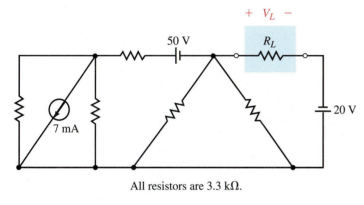

All resistors are 3.3 kΩ.

FIGURE 9–88

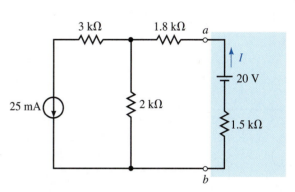

FIGURE 9–89

◀ **MULTISIM**

15. Refer to the circuit of Figure 9–89:

    a. Find the Thévenin equivalent circuit external to the indicated terminals.

    b. Use the Thévenin equivalent circuit to determine the current through the indicated branch.

16. Refer to the circuit of Figure 9–90:

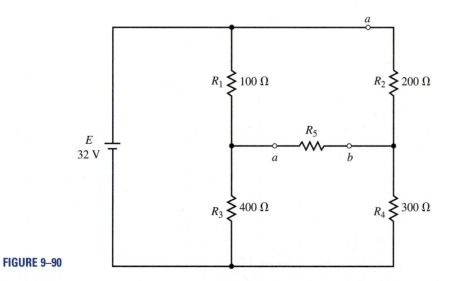

FIGURE 9–90

    a. Find the Thévenin equivalent circuit external to the indicated terminals.

    b. If $R_5 = 1$ kΩ, use the Thévenin equivalent circuit to determine the voltage V_{ab} and the current through this resistor.

17. Refer to the circuit of Figure 9–91:

    a. Find the Thévenin equivalent circuit external to R_L.

    b. Use the Thévenin equivalent circuit to find the current I when $R_L = 0$, 10 kΩ, and 50 kΩ.

18. Refer to the circuit of Figure 9–92:

    a. Find the Thévenin equivalent circuit external to R_L.

    b. Use the Thévenin equivalent circuit to find the power dissipated by R_L.

19. Repeat Problem 17 for the circuit of Figure 9–93.

20. Repeat Problem 17 for the circuit of Figure 9–94.

21. Find the Thévenin equivalent circuit of the network external to the indicated branch as shown in Figure 9–95.

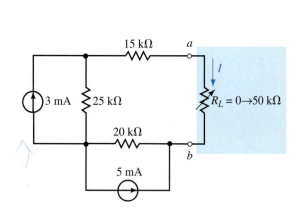

FIGURE 9–91

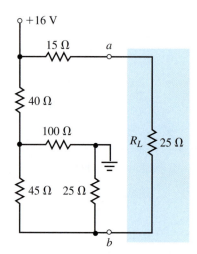

FIGURE 9–92

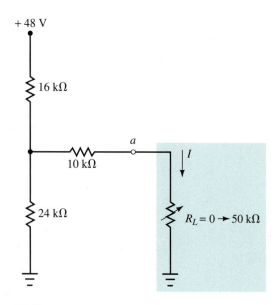

FIGURE 9–93

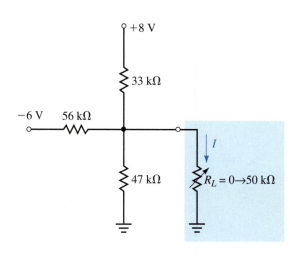

FIGURE 9–94

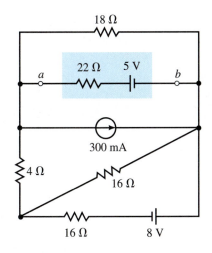

FIGURE 9–95

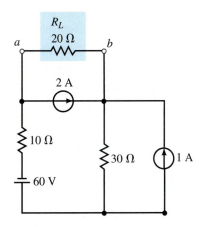

FIGURE 9–96

22. Refer to the circuit of Figure 9–96.

    a. Find the Thévenin equivalent circuit external to the indicated terminals.

    b. Use the Thévenin equivalent circuit to determine the current through the indicated branch.

23. Repeat Problem 22 for the circuit of Figure 9–97.

24. Repeat Problem 22 for the circuit of Figure 9–98.

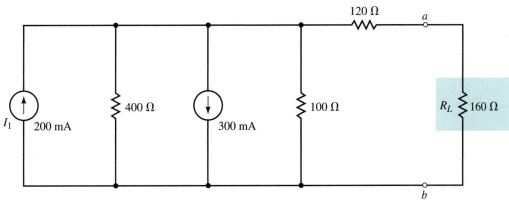

FIGURE 9–97

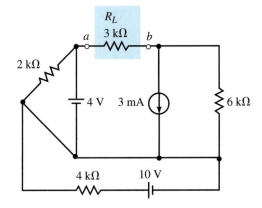

FIGURE 9–98

9.3   Norton's Theorem

25. Find the Norton equivalent circuit external to R_L in the circuit of Figure 9–81. Use the equivalent circuit to find I_L for the circuit.

26. Repeat Problem 25 for the circuit of Figure 9–82.

27. Repeat Problem 25 for the circuit of Figure 9–83.

28. Repeat Problem 25 for the circuit of Figure 9–84.

29. Refer to the circuit of Figure 9–85:

    a. Find the Norton equivalent circuit external to R_L.

    b. Use the equivalent circuit to determine I_L when $R_L = 20\ \Omega$ and when $R_L = 50\ \Omega$.

30. Refer to the circuit of Figure 9–86:

    a. Find the Norton equivalent circuit external to R_L.

    b. Use the equivalent circuit to determine I_L when $R_L = 10\ k\Omega$ and when $R_L = 20\ k\Omega$.

31. a. Find the Norton equivalent circuit external to the indicated terminals of Figure 9–87.

    b. Convert the Thévenin equivalent circuit of Problem 13 to its Norton equivalent.

32. a. Find the Norton equivalent circuit external to R_L in the circuit of Figure 9–88.

    b. Convert the Thévenin equivalent circuit of Problem 14 to its Norton equivalent.

33. Repeat Problem 31 for the circuit of Figure 9–91.

34. Repeat Problem 31 for the circuit of Figure 9–92.

35. Repeat Problem 31 for the circuit of Figure 9–95.

36. Repeat Problem 31 for the circuit of Figure 9–96.

9.4   Maximum Power Transfer Theorem

37. a. For the circuit of Figure 9–91, determine the value of R_L so that maximum power is delivered to the load.

    b. Calculate the value of the maximum power which can be delivered to the load.

    c. Sketch the curve of power versus resistance as R_L is adjusted from $0\ \Omega$ to $50\ k\Omega$ in increments of $5\ k\Omega$.

38. Repeat Problem 37 for the circuit of Figure 9–94.

39. a. For the circuit of Figure 9–99, find the value of R so that $R_L = R_{Th}$.

    b. Calculate the maximum power dissipated by R_L.

40. Repeat Problem 39 for the circuit of Figure 9–100.

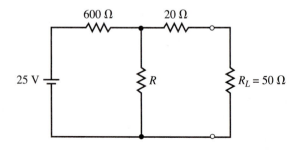

FIGURE 9–99

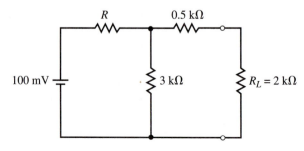

FIGURE 9–100

41. a. For the circuit of Figure 9–101, determine the values of R_1 and R_2 so that the 32-kΩ load receives maximum power.

    b. Calculate the maximum power delivered to R_L.

42. Repeat Problem 41 if the load resistor has a value of $R_L = 25$ kΩ.

9.5 Substitution Theorem

43. If the indicated portion of the circuit in Figure 9–102 is to be replaced with a voltage source and a 50-Ω series resistor, determine the magnitude and polarity of the resulting voltage source.

44. If the indicated portion of the circuit in Figure 9–102 is to be replaced with a current source and a 200-Ω shunt resistor, determine the magnitude and direction of the resulting current source.

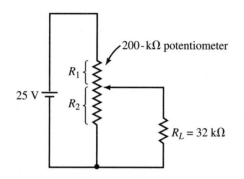

FIGURE 9–101

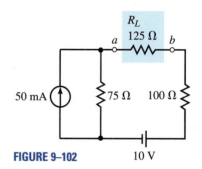

FIGURE 9–102

9.6 Millman's Theorem

45. Use Millman's theorem to find the current through and the power dissipated by R_L in the circuit of Figure 9–103.

46. Repeat Problem 45 for the circuit of Figure 9–104.

47. Repeat Problem 45 for the circuit of Figure 9–105.

48. Repeat Problem 45 for the circuit of Figure 9–106.

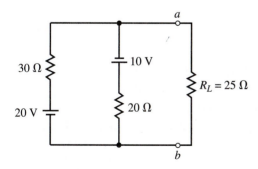

FIGURE 9–103

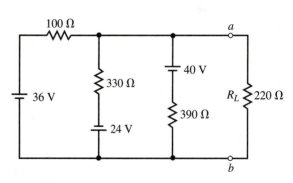

FIGURE 9–104

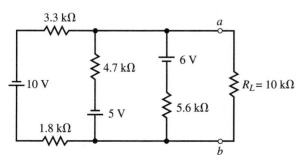

FIGURE 9–105

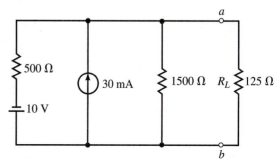

FIGURE 9–106

9.7   Reciprocity Theorem

49. a. Determine the current *I* in the circuit of Figure 9–107.

    b. Show that reciprocity applies for the given circuit.

50. Repeat Problem 49 for the circuit of Figure 9–108.

51. a. Determine the voltage *V* in the circuit of Figure 9–109.

    b. Show that reciprocity applies for the given circuit.

52. Repeat Problem 51 for the circuit of Figure 9–110.

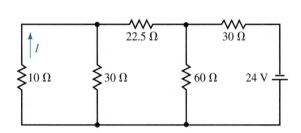

FIGURE 9–107

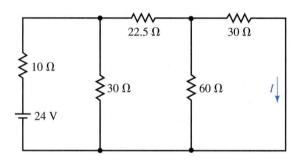

FIGURE 9–108

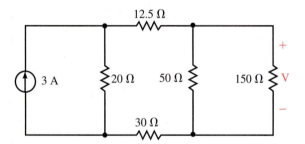

FIGURE 9–109

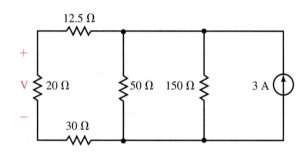

FIGURE 9–110

9.8   Circuit Analysis Using Computers

◀ MULTISIM

53. Use MultiSIM to find both the Thévenin and the Norton equivalent circuits external to the load resistor in the circuit of Figure 9–81.

◀ MULTISIM

54. Repeat Problem 53 for the circuit of Figure 9–82.

◀ CADENCE

55. Use the schematic editor of PSpice to input the circuit of Figure 9–83 and use the PROBE postprocessor to display output voltage, current, and power as a function of load resistance. Use the cursor in the PROBE postprocessor to determine the value of load resistance for which the load will receive maximum power. Let the load resistance vary from 100 Ω to 4000 Ω in increments of 100 Ω.

◀ CADENCE

56. Repeat Problem 55 for the circuit of Figure 9–84. Let the load resistance vary from 1 kΩ to 100 kΩ in increments of 1 kΩ.

✓ **ANSWERS TO IN-PROCESS LEARNING CHECKS**

In-Process Learning Check 1

$P_{R_1} = 304$ mW, $P_{R_2} = 76$ mW, $P_{R_3} = 276$ mW
Assuming that superposition applies for power:
$P_{R_1(1)} = 60$ mW, $P_{R_1(2)} = 3.75$ mW, $P_{R_1(3)} = 60$ mW
But $P_{R_1} = 304$ mW · 123.75 mW

In-Process Learning Check 2

$R_1 = 133 \ \Omega$

In-Process Learning Check 3

1. Refer to Figure 9–27.
2. $I_N = 200 \ \mu$A, $R_N = 500 \ \Omega$
3. $E_{Th} = 0.2$ V, $R_{Th} = 20$ kΩ

In-Process Learning Check 4

a. $\eta = 50\%$.
b. $\eta = 33.3\%$.
c. $\eta = 66.7\%$.

In-Process Learning Check 5

1. In communication circuits and some amplifiers, maximum power transfer is a desirable characteristic.

2. In power transmission and dc voltage sources maximum power transfer is not desirable.

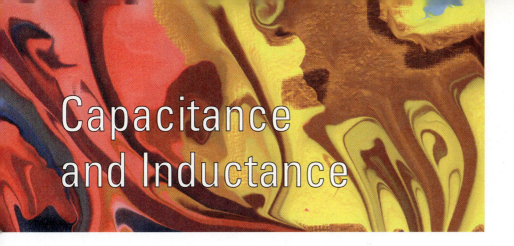

Capacitance and Inductance

III

Resistance, inductance, and capacitance are the three basic circuit properties that we use to control voltages and currents in electrical and electronic circuits. However, each behaves in a fundamentally different way. Resistance, for example (as you learned in earlier chapters), opposes current, while (as you will soon see) inductance opposes any change in current, and capacitance opposes any change in voltage. In addition, resistance dissipates energy, while inductance and capacitance both store energy—inductance in its magnetic field and capacitance in its electric field.

Circuit elements that are built to possess capacitance are called capacitors, while elements built to possess inductance are called inductors. In Part III of this book, we explore these elements, their properties, and their behavior in electric circuits.

Note

Part III of this book deals with the basics of capacitors, inductors, magnetic circuits, and simple transients. However, some colleges and universities cover some of these topics in an order different than presented here, for example, by teaching ac before transients. To provide flexibility, Part III has been organized so that all or parts of Chapters 11, 12, and 14 may be delayed if desired (without loss of continuity) to fit individual curriculums. ■

■ OBJECTIVES

After studying this chapter, you will be able to

• describe the basic construction of capacitors,

• explain how capacitors store charge,

• define capacitance,

• describe what factors affect capacitance and in what way,

• describe the electric field of a capacitor,

• compute the breakdown voltages of various materials,

• describe various types of commercial capacitors,

• compute the capacitance of capacitors in series and in parallel combinations,

• compute capacitor voltage and current for simple time-varying waveforms,

• determine stored energy,

• describe capacitor faults and the basic troubleshooting of capacitors.

Capacitors and Capacitance

10

A capacitor is a circuit component designed to store electrical charge. If you connect a dc voltage source to a capacitor, for example, the capacitor will "charge" to the voltage of the source. If you then disconnect the source, the capacitor will remain charged, i.e., its voltage will remain constant at the value that it attained while connected to the source (assuming no leakage). Because of this tendency to hold voltage, *a capacitor opposes changes in voltage.* It is this characteristic that gives capacitors their unique properties.

Capacitors are widely used in electrical and electronic applications. They are used in radio and TV systems, for example, to tune in signals, in cameras to store the charge that fires the photoflash, on pump and refrigeration motors to increase starting torque, in electric power systems to increase operating efficiency, and so on. A photo of some typical capacitors is shown at right.

Capacitance is the electrical property of capacitors: it is a measure of how much charge a capacitor can hold. In this chapter, we look at capacitance and its basic properties. In Chapter 11, we look at capacitors in dc and pulse circuits; in later chapters, we look at capacitors in ac applications. ■

CHAPTER PREVIEW

Typical capacitors.

Michael Faraday and the Field Concept

PUTTING IT IN PERSPECTIVE

THE UNIT OF CAPACITANCE, the farad, is named after Michael Faraday (1791–1867). Born in England to a working class family, Faraday received limited education. Nonetheless, he was responsible for many of the fundamental discoveries of electricity and magnetism. Lacking mathematical skills, he used his intuitive ability rather than mathematical models to develop conceptual pictures of basic phenomena. It was his development of the field concept that made it possible to map out the fields that exist around magnetic poles and electrical charges.

To get at this idea, recall from Chapter 2 that unlike charges attract and like charges repel, i.e., a force exists between electrical charges. We call the region where this force acts an electric field. To visualize this field, we use Faraday's field concept and draw lines of force (or flux lines) that show at every point in space the magnitude and direction of the force. Now, rather than supposing that one charge exerts a force on another, we instead visualize that the original charges create a field in space and that other charges introduced into this field experience a force due to the field. This concept is helpful in studying certain aspects of capacitors, as you will see in this chapter.

The development of the field concept had a significant impact on science. We now picture several important phenomena in terms of fields, including electric fields, gravitation, and magnetism. When Faraday published his theory in 1844, however, it was not taken seriously, much like Ohm's work two decades earlier. It is also interesting to note that the development of the field concept grew out of Faraday's research into magnetism, not electric charge. ■

10.1 Capacitance

A **capacitor** consists of two conductors separated by an insulator. One of its basic forms is the parallel-plate capacitor shown in Figure 10–1. It consists of two metal plates separated by a nonconducting material (i.e., an insulator) called a **dielectric.** The dielectric may be air, oil, mica, plastic, ceramic, or other suitable insulating material.

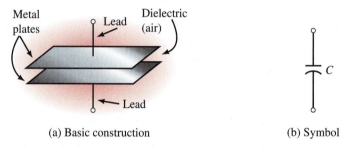

(a) Basic construction   (b) Symbol

FIGURE 10–1 Parallel-plate capacitor.

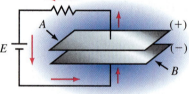

FIGURE 10–2 Capacitor during charging. At the instant that you connect the source, there is a momentary surge of current as electrons are pulled from Plate A and an equal number deposited on plate B. This leaves the top plate positively charged and the bottom plate negatively charged. When charging is complete, there is no further movement of electrons, and the current is thus zero.

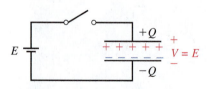

FIGURE 10–3 Capacitor after charging. When you disconnect the source, electrons are trapped on the bottom plate and cannot return—thus, charge is stored.

Since the plates of the capacitor are metal, they contain huge numbers of free electrons. In their normal state, however, they are uncharged, that is, there is no excess or deficiency of electrons on either plate. If a dc source is now connected (Figure 10–2), electrons are pulled from the top plate by the positive potential of the battery and the same number deposited on the bottom plate. This leaves the top plate with a deficiency of electrons (i.e., positive charge) and the bottom plate with an excess (i.e., negative charge). In this state, the capacitor is said to be **charged.** (Note that no current can pass through the dielectric between the plates—thus, the movement of electrons illustrated in Figure 10–2 will cease when the capacitor reaches full charge.)

If Q coulombs of electrons are moved during the charging process (leaving the top plate with a deficiency of Q electrons and the bottom with an excess of Q), we say that the capacitor has a charge of Q.

If we now disconnect the source (Figure 10–3), the excess electrons that were moved to the bottom plate remain trapped as they have no way to return to the top plate. The capacitor therefore remains charged even though no source is present. Because of this, we say that *a capacitor can store charge.* Capacitors with little leakage (Section 10.5) can hold their charge for a considerable time.

Large capacitors charged to high voltages contain a great deal of energy and can give you a bad shock. Always **discharge** such capacitors after power has been removed if you intend to handle them. You can do this by shorting a wire across their leads. Electrons then return to the top plate, restoring the charge balance and reducing the capacitor voltage to zero. (However, you also need to be concerned about residual voltage due to dielectric absorption. This is discussed in Section 10.5.)

Definition of Capacitance

The amount of charge Q that a capacitor can store depends on the applied voltage. Experiments show that for a given capacitor, Q is proportional to voltage. Let the constant of proportionality be C. Then,

$$Q = CV \qquad\qquad (10\text{–}1)$$

Rearranging terms yields

$$C = \frac{Q}{V} \quad \text{(farads, F)} \qquad\qquad (10\text{–}2)$$

The term C is defined as the capacitance of the capacitor. As indicated, its unit is the **farad.** By definition, *the **capacitance** of a capacitor is one farad if it stores one coulomb of charge when the voltage across its terminals is one volt.* The farad, however, is a very large unit. Most practical capacitors range in size from picofarads (pF or 10^{-12} F) to microfarads (μF or 10^{-6} F). The larger the value of C, the more charge that the capacitor can hold for a given voltage.

EXAMPLE 10–1

a. How much charge is stored on a 10-μF capacitor when it is connected to a 24-volt source?

b. The charge on a 20-nF capacitor is 1.7 μC. What is its voltage?

Solution

a. From Equation 10–1, $Q = CV$. Thus, $Q = (10 \times 10^{-6}\ \text{F})(24\ \text{V}) = 240\ \mu\text{C}$.

b. Rearranging Equation 10–1, $V = Q/C = (1.7 \times 10^{-6}\ \text{C})/(20 \times 10^{-9}\ \text{F}) = 85$ V.

10.2 Factors Affecting Capacitance

Effect of Area

As shown by Equation 10–2, capacitance is directly proportional to charge. This means that the more charge you can put on a capacitor's plates for a given voltage, the greater will be its capacitance. Consider Figure 10–4. The capacitor of (b) has four times the area of (a). Since it has the same number of free electrons per unit area, it has four times the total charge and hence four times the capacitance. This turns out to be true in general, that is, *capacitance is directly proportional to plate area.*

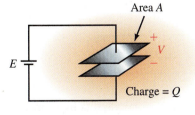

(a) Capacitor with area A
and charge Q

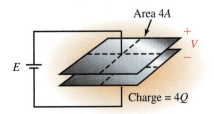

(b) Plates with four times the area
have four times the charge and
therefore, four times the capacitance

FIGURE 10–4 For a fixed separation, capacitance is proportional to plate area.

Effect of Spacing

Now consider Figure 10–5. Since the top plate has a deficiency of electrons and the bottom plate an excess, a force of attraction exists across the gap. For a fixed spacing as in (a), the charges are in equilibrium. Now move the plates closer together as in (b). As spacing decreases, the force of attraction increases, pulling more electrons from within the material of plate *B* to its top surface. This creates a deficiency of electrons in the lower levels of *B*. To replenish these, the source moves additional electrons around the circuit, leaving *A* with an even greater deficiency and *B* with an even greater excess. The charge on the plates therefore increases and hence, according to Equation 10–2, so does the capacitance. We therefore conclude that decreasing spacing increases capacitance, and vice versa. In fact, as we will show later, *capacitance is inversely proportional to plate spacing.*

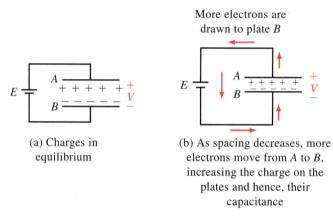

(a) Charges in equilibrium

(b) As spacing decreases, more electrons move from *A* to *B*, increasing the charge on the plates and hence, their capacitance

FIGURE 10–5 Decreasing spacing increases capacitance.

TABLE 10–1 Relative Dielectric Constants (also called Relative Permittivities)

Material	ϵ_r (Nominal Values)
Vacuum	1
Air	1.0006
Ceramic	30–7500
Mica	5.5
Mylar	3
Oil	4
Paper (dry)	2.2
Polystyrene	2.6
Teflon	2.1

Effect of Dielectric

Capacitance also depends on the dielectric. Consider Figure 10–6(a), which shows an air-dielectric capacitor. If you substitute different materials for air, the capacitance increases. Table 10–1 shows the factor by which capacitance increases for a number of different materials. For example, if Teflon® is used instead of air, capacitance is increased by a factor of 2.1. This factor is called the **relative dielectric constant** or **relative permittivity** of the material. (Permittivity is a measure of how easy it is to establish electric flux in a material.) Note that high-permittivity ceramic increases capacitance by as much as 7500, as indicated in Figure 10–6(b).

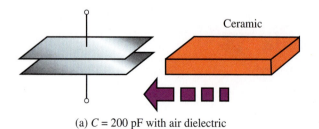

(a) $C = 200$ pF with air dielectric

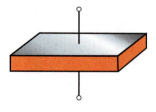

(b) $C = 1.5$ μF with high permittivity ceramic dielectric

FIGURE 10–6 The factor by which a dielectric causes capacitance to increase is termed its *relative dielectric constant*. The ceramic used here has a value of 7500.

Capacitance of a Parallel-Plate Capacitor

From the above observations, we see that capacitance is directly proportional to plate area, inversely proportional to plate separation, and dependent on the dielectric. In equation form,

$$C = \in \frac{A}{d} \quad \text{(F)} \qquad \text{(10–3)}$$

where area A is in square meters and spacing d is in meters.

Dielectric Constant

The constant $\in$ in Equation 10–3 is the **absolute dielectric constant** of the insulating material. Its units are farads per meter (F/m). For air or vacuum, $\in$ has a value of $\in_o = 8.854 \times 10^{-12}$ F/m. For other materials, $\in$ is expressed as the product of the relative dielectric constant, $\in_r$ (shown in Table 10–1), times $\in_o$. That is,

$$\in = \in_r \in_o \qquad \text{(10–4)}$$

Consider, again, Equation 10–3: $C = \in A/d = \in_r \in_o A/d$. Note that $\in_o A/d$ is the capacitance of a vacuum- (or air-) dielectric capacitor. Denote it by C_o. Then, for any other dielectric,

$$C = \in_r C_o \qquad \text{(10–5)}$$

EXAMPLE 10–2

Compute the capacitance of a parallel-plate capacitor with plates 10 cm by 20 cm, separation of 5 mm, and

a. an air dielectric,

b. a ceramic dielectric with a relative permittivity of 7500.

Solution   Convert all dimensions to meters. Thus, A = (0.1 m)(0.2 m) = 0.02 m², and $d = 5 \times 10^{-3}$ m.

a. For air, $C = \in_o A/d = (8.854 \times 10^{-12})(2 \times 10^{-2})/(5 \times 10^{-3}) = 35.4 \times 10^{-12}$ F = 35.4 pF.

b. For ceramic with $\in_r = 7500$, $C = 7500(35.4 \text{ pF}) = 0.266 \text{ μF}$.

EXAMPLE 10–3

A parallel-plate capacitor with air dielectric has a value of $C = 12$ pF. What is the capacitance of a capacitor that has the following:

a. The same separation and dielectric but five times the plate area?

b. The same dielectric but four times the area and one-fifth the plate spacing?

c. A dry paper dielectric, six times the plate area, and twice the plate spacing?

Solution

a. Since the plate area has increased by a factor of five and everything else remains the same, C increases by a factor of five. Thus, $C = 5(12 \text{ pF}) = 60 \text{ pF}$.

b. With four times the plate area, C increases by a factor of four. With one-fifth the plate spacing, C increases by a factor of five. Thus, $C = (4)(5)(12 \text{ pF}) = 240 \text{ pF}$.

c. Dry paper increases C by a factor of 2.2. The increase in plate area increases C by a factor of six. Doubling the plate spacing reduces C by one-half. Thus, $C = (2.2)(6)(\frac{1}{2})(12 \text{ pF}) = 79.2 \text{ pF}$.

1. A capacitor with plates 7.5 cm × 8 cm and plate separation of 0.1 mm has an oil dielectric:

   a. Compute its capacitance;

   b. If the charge on this capacitor is 0.424 μC, what is the voltage across its plates?

2. For a parallel-plate capacitor, if you triple the plate area and halve the plate spacing, how does capacitance change?

3. For the capacitor of Figure 10–6, if you use mica instead of ceramic, what will be the capacitance?

4. What is the dielectric for the capacitor of Figure 10–7(b)?

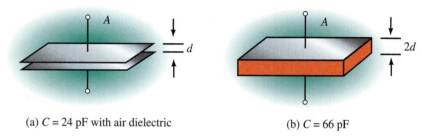

(a) $C = 24$ pF with air dielectric   (b) $C = 66$ pF

FIGURE 10–7

10.3 Electric Fields

Electric Flux

Electric fields are force fields that exist in the region surrounding charged bodies. Some familiarity with electric fields is necessary to understand dielectrics and their effect on capacitance. We now look briefly at the key ideas.

Consider Figure 10–8(a). As noted in Chapter 2, unlike charges attract and like charges repel, i.e., a force exists between them. The region where this force exists is called an **electric field.** To visualize this field, we use Faraday's field concept. The direction of the field is defined as the direction of force on a positive charge. It is therefore directed outward from the positive charge and inward toward the negative charge as shown. Field lines never cross, and the density of the lines indicates the strength of the field; i.e., the more dense the lines, the stronger the field. Figure 10–8(b) shows the field of a parallel-plate capacitor. In this case, the field is uniform across the gap, with some fringing near its edges. **Electric flux** lines are represented by the Greek letter ψ (psi).

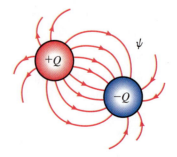 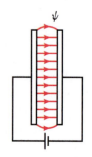

(a) Field about a pair of positive   (b) Field of parallel
and negative charges   plate capacitor

FIGURE 10–8   Some example electric fields.

Electric Field Intensity

The strength of an electric field, also called its **electric field intensity,** is the force per unit charge that the field exerts on a small, positive test charge, Q_t. Let the field strength be denoted by $\mathscr{E}$. Then, by definition,

$$\mathscr{E} = F/Q_t \quad \text{(newtons/coulomb, N/C)} \qquad \textbf{(10–6)}$$

To illustrate, let us determine the field about a point charge, Q. When the test charge is placed near Q, it experiences a force of $F = kQQ_t/r^2$ (Coulomb's law, Chapter 2). The constant in Coulomb's law is actually equal to $1/4\pi\in$. Thus, $F = QQ_t/4\pi\in r^2$, and from Equation 10–6,

$$\mathscr{E} = \frac{F}{Q_t} = \frac{Q}{4\pi\in r^2} \quad \text{(N/C)} \qquad \textbf{(10–7)}$$

Electric Flux Density

Because of the presence of $\in$ in Equation 10–7, the electric field intensity depends on the medium in which the charge is located. Let us define a new quantity, D, that is independent of the medium. Let

$$D = \in \mathscr{E} \qquad \textbf{(10–8)}$$

D is known as **electric flux density.** Although not apparent here, D represents the density of flux lines in space, that is,

$$D = \frac{\text{total flux}}{\text{area}} = \frac{\psi}{A} \qquad \textbf{(10–9)}$$

where ψ is the flux passing through area A.

Electric Flux (Revisited)

Consider Figure 10–9. Flux ψ is due to the charge Q. Although we will not prove it, in the SI system the number of flux lines emanating from a charge Q is equal to the charge itself, that is,

$$\psi = Q \quad \text{(C)} \qquad \textbf{(10–10)}$$

An easy way to visualize this is to think of one flux line as emanating from each positive charge on the body as shown in Figure 10–9. Then, as indicated, the total number of lines is equal to the total number of charges.

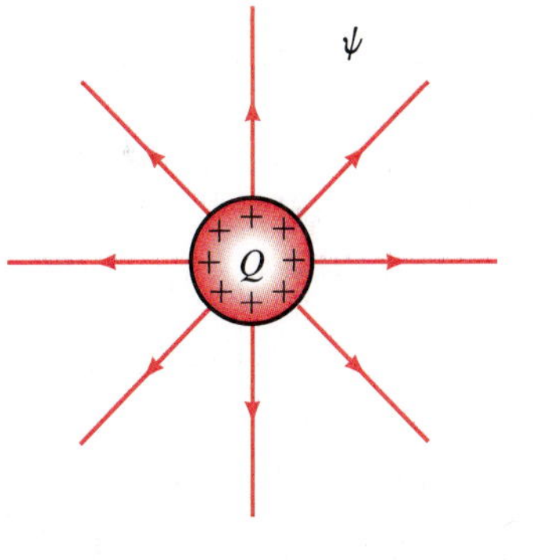

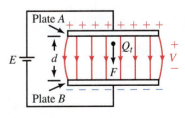

FIGURE 10–9 In the SI system, total flux ψ equals charge Q.

Field of a Parallel-Plate Capacitor

Now consider a parallel-plate capacitor (Figure 10–10). The field here is created by the charge distributed over its plates. Since plate A has a deficiency of electrons, it looks like a sheet of positive charge, while plate B looks like a sheet of negative charge. A positive test charge Q_t between these sheets is therefore repelled by the positive sheet and attracted by the negative sheet.

Now move the test charge from plate B to plate A. The work W required to move the charge against the force F is force times distance. Thus,

$$W = Fd \quad \text{(J)} \qquad \textbf{(10–11)}$$

FIGURE 10–10 Work moving test charge Q_t is force times distance.

In Chapter 2, we defined voltage as work divided by charge, i.e., $V = W/Q$. Since the charge here is the test charge, Q_t, the voltage between plates A and B is

$$V = W/Q_t = (Fd)/Q_t \quad \text{(V)} \qquad \textbf{(10–12)}$$

Now divide both sides by d. This yields $V/d = F/Q_t$. But $F/Q_t = \mathscr{E}$, from Equation 10–6. Thus,

$$\mathscr{E} = V/d \quad \text{(V/m)} \tag{10–13}$$

Equation 10–13 shows that the electric field strength between capacitor plates is equal to the voltage across the plates divided by the distance between them. Thus, if $V = 30$ V and $d = 10$ mm, $\mathscr{E} = 3000$ V/m.

EXAMPLE 10–4

Suppose that the electric field intensity between the plates of a capacitor is 50 000 V/m when 80 V is applied:

a. What is the plate spacing if the dielectric is air? If the dielectric is ceramic?

b. What is $\mathscr{E}$ if the plate spacing is halved?

Solution

a. $\mathscr{E} = V/d$, independent of dielectric. Thus,

$$d = \frac{V}{\mathscr{E}} = \frac{80 \text{ V}}{50 \times 10^3 \text{ V/m}} = 1.6 \times 10^{-3} \text{ m}$$

b. Since $\mathscr{E} = V/d$, $\mathscr{E}$ will double to 100 000 V/m.

PRACTICE PROBLEMS 1

1. What happens to the electric field intensity of a capacitor if you do the following:
   a. Double the applied voltage?
   b. Triple the applied voltage and double the plate spacing?
2. If the electric field intensity of a capacitor with polystyrene dielectric and plate size 2 cm by 4 cm is 100 kV/m when 50 V is applied, what is its capacitance?

Answers
1. a. Doubles, b. Increases by a factor of 1.5; 2. 36.8 pF

Capacitance (Revisited)

With the above background, we can examine **capacitance** a bit more rigorously. Recall, $C = Q/V$. Using the above relationships yields

$$C = \frac{Q}{V} = \frac{\psi}{V} = \frac{AD}{\mathscr{E}d} = \frac{D}{\mathscr{E}}\left(\frac{A}{d}\right) = \epsilon\frac{A}{d}$$

This confirms Equation 10–3 that we developed intuitively in Section 10.2.

10.4 Dielectrics

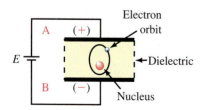

FIGURE 10–11 Effect of the capacitor's electric field on an atom of its dielectric.

As you saw in Figure 10–6, a dielectric increases capacitance. We now examine why. Consider Figure 10–11. For a charged capacitor, electron orbits (which are normally circular) become elliptical as electrons are attracted toward the positive (+) plate and repelled from the negative (−) plate. This makes the end of the atom nearest the positive plate appear negative while its other end appears positive. Such atoms are **polarized.** Throughout the bulk of the dielectric, the negative end of a polarized atom is adjacent to the positive end of another polarized atom, and the effects cancel. However, at the surfaces of the dielectric, there are no atoms to cancel, and the net effect is as if a layer of negative charge exists on the surface of the dielectric at the positive plate and a layer of positive charge at the negative plate. This makes the plates appear closer together, thus increasing capacitance. Materials for which the effect is largest result in the greatest increase in capacitance.

Voltage Breakdown

If the voltage of Figure 10–11 is increased beyond a critical value, the force on the electrons is so great that they are literally torn from orbit. This is called **dielectric breakdown** and the electric field intensity (Equation 10–13) at breakdown is called the **dielectric strength** of the material. For air, breakdown occurs when the voltage gradient reaches 3 kV/mm. The breakdown strengths for other materials are shown in Table 10–2. Since the quality of a dielectric depends on many factors, dielectric strength varies from sample to sample.

Breakdown is not limited to capacitors; it can occur with any type of electrical apparatus whose insulation is stressed beyond safe limits. (For example, air breaks down and flashovers occur on high-voltage transmission lines when they are struck by lightning.) The shape of conductors also affects breakdown voltage. Breakdown occurs at lower voltages at sharp points than at blunt points. This effect is made use of in lightning arresters.

TABLE 10–2 Dielectric Strength*	
Material	**kV/mm**
Air	3
Ceramic (high $\in_r$)	3
Mica	40
Mylar	16
Oil	15
Polystyrene	24
Rubber	18
Teflon®	60

*Values depend on the composition of the material. These are the values we use in this book.

EXAMPLE 10–5

A capacitor with plate dimensions of 2.5 cm by 2.5 cm and a ceramic dielectric with $\in_r = 7500$ experiences breakdown at 2400 V. What is C?

Solution   From Table 10–2 dielectric strength = 3 kV/mm. Thus, $d = 2400$ V/3000 V/mm = 0.8 mm = 8×10^{-4} m. So

$$C = \in_r \in_o A/d$$

$$= (7500)(8.854 \times 10^{-12})(0.025 \text{ m})^2/(8 \times 10^{-4} \text{ m})$$

$$= 51.9 \text{ nF}$$

PRACTICAL NOTES . . .

Breakdown in solid dielectrics is generally destructive. Typically, a carbonized pinhole results and since this is a conductive path, the dielectric is no longer a useful insulator. Air, vacuum, and insulating liquids like oil, on the other hand, recover from the flashover once it is extinguished.

 PRACTICE PROBLEMS 2

1. At what voltage will breakdown occur for a mylar dielectric capacitor with plate spacing of 0.25 cm?

2. An air-dielectric capacitor breaks down at 500 V. If the plate spacing is doubled and the capacitor is filled with oil, at what voltage will breakdown occur?

Answers
1. 40 kV; 2. 5 kV

Capacitor Voltage Rating

Because of dielectric breakdown, capacitors are rated for maximum operating voltage (called **working voltage**) by their manufacturer (indicated on the capacitor as WVDC or **working voltage dc**). If you operate a capacitor beyond its working voltage, you may damage it.

So far, we have assumed ideal capacitors. However, real capacitors have several nonideal characteristics.

10.5 Nonideal Effects

Leakage Current

When a charged capacitor is disconnected from its source, it will eventually discharge. This is because no insulator is perfect and a small amount of charge "leaks" through the dielectric. Similarly, a small leakage current will pass through its dielectric when a capacitor is connected to a source.

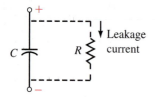

FIGURE 10–12 Leakage current.

The effect of **leakage** is modeled by a resistor in Figure 10–12. Except for electrolytic capacitors (see Sec. 10.6), leakage is very small and R is very large, typically hundreds of megohms. The larger R is, the longer a capacitor can hold its charge. For most applications, leakage can be neglected and you can treat the capacitor as ideal.

Dielectric Absorption

When a capacitor is discharged by temporarily shorting its leads, it should have zero volts when the short is removed. However, atoms sometimes remain partially polarized, and when the short is removed, they cause a residual voltage to appear across the capacitor. This effect is known as **dielectric absorption.** In electronic circuits, the voltage due to dielectric absorption can upset circuit voltage levels; in TV tubes and electrical power apparatus, it can result in large and potentially dangerous voltages. You may have to put the short back on to complete the discharge.

Temperature Coefficient

Because dielectrics are affected by temperature, capacitance may change with temperature. If capacitance increases with increasing temperature, the capacitor is said to have a **positive temperature coefficient;** if it decreases, the capacitor has a **negative temperature coefficient;** if it remains essentially constant, the capacitor has a **zero temperature coefficient.**

The temperature coefficient is specified as a change in capacitance in parts per million (ppm) per degree Celsius. Consider a 1-μF capacitor. Since 1 μF = 1 million pF, 1 ppm is 1 pF. Thus, a 1-μF capacitor with a temperature coefficient of 200 ppm/° C could change as much as 200 pF per degree Celsius. If the circuit you are designing is sensitive to a capacitor's value, you may have to use a capacitor with a small temperature coefficient.

10.6 Types of Capacitors

Since no single capacitor type suits all applications, capacitors are made in a variety of types and sizes. Among these are fixed and variable types with differing dielectrics and recommended areas of application.

Fixed Capacitors

Fixed capacitors are often identified by their dielectric. Common dielectric materials include ceramic, plastic, and mica, plus, for electrolytic capacitors, aluminum and tantalum oxide. Design variations include tubular and interleaved plates. The interleave design (Figure 10–13) uses multiple plates to increase effective plate area. A layer of insulation separates plates, and alternate plates are connected together. The tubular design (Figure 10–14) uses sheets of metal foil separated by an insulator such as plastic film. Fixed capacitors are encapsulated in plastic, epoxy resin, or other insulating material and identified with value, tolerance, and other appropriate data either via body markings or color coding. Electrical characteristics and physical size depend on the dielectric used.

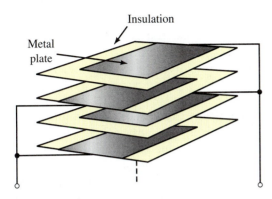

FIGURE 10–13   Stacked capacitor construction. The stack is compressed, leads attached, and the unit coated with epoxy resin or other insulating material.

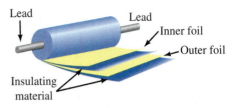

FIGURE 10–14   Tubular capacitor with axial leads.

Ceramic Capacitors

First, consider ceramic. The permittivity of ceramic varies widely (as indicated in Table 10–1). At one end are ceramics with extremely high permittivity. These permit packaging a great deal of capacitance in a small space, but yield capacitors whose characteristics vary widely with temperature and operating voltage. However, they are popular in limited temperature applications where small size and cost are important. At the other end are ceramics with highly stable characteristics. They yield capacitors whose values change little with temperature, voltage, or aging. However, since their dielectric constants are relatively low (typically 30 to 80), these capacitors are physically larger than those made using high-permittivity ceramic. Many surface mount capacitors (considered later in this section) use ceramic dielectrics.

Plastic Film Capacitors

Plastic-film capacitors are of two basic types: film/foil or metalized film. **Film/foil** capacitors use metal foil separated by plastic film as in Figure 10–14, while **metallized-film** capacitors have their foil material vacuum-deposited directly onto plastic film. Film/foil capacitors are generally larger than metallized-foil units, but have better capacitance stability and higher insulation resistance. Typical film materials are polyester, Mylar, polypropylene, and polycarbonate. Figure 10–15 shows a selection of plastic-film capacitors.

Metalized-film capacitors are **self-healing.** Thus, if voltage stress at an imperfection exceeds breakdown, an arc occurs that evaporates the metallized area around the fault, isolating the defect. (Film/foil capacitors are not self-healing.)

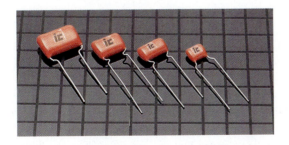

FIGURE 10–15   Radial lead film capacitors. *(Courtesy Illinois Capacitor Inc.)*

Mica Capacitors

Mica capacitors are low in cost with low leakage and good stability. Available values range from a few picofarads to about 0.1 μF.

Electrolytic Capacitors

Electrolytic capacitors provide large capacitance (i.e., up to several hundred thousand microfarads) at a relatively low cost. (Their capacitance is large because they have a very thin layer of oxide as their dielectric.) However, their leakage is relatively high and breakdown voltage relatively low. Electrolytics have either aluminum or tantalum as their plate material. Tantalum devices are smaller than aluminum devices, have less leakage, and are more stable.

The basic aluminum electrolytic capacitor construction is similar to that of Figure 10–14, with strips of aluminum foil separated by gauze saturated with an electrolyte. During manufacture, chemical action creates a thin oxide layer that acts as the dielectric. This layer must be maintained during use. For this reason, electrolytic capacitors are polarized (marked with a + and − sign), and the plus (+) terminal must always be kept positive with respect to the minus (−) terminal. Electrolytic capacitors have a **shelf life;** that is, if they are not used for an extended period, they may fail when powered up again. Figure 10–16 shows the symbol for an electrolytic capacitor, while Figure 10–17 shows a selection of actual devices.

Tantalum capacitors come in two basic types: wet slug and solid dielectric. Figure 10–18 shows a cutaway view of a solid tantalum unit. The slug, made from powdered tantalum, is highly porous and provides a large internal surface area that is coated with an oxide to form the dielectric. Tantalum capacitors are polarized and must be inserted into a circuit properly.

FIGURE 10–16 Symbol for an electrolytic capacitor.

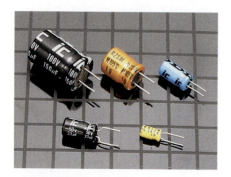

FIGURE 10–17 Radial lead aluminium electrolytic capacitors. *(Courtesy Illinois Capacitor Inc.)*

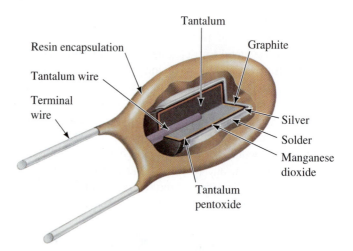

FIGURE 10–18 Cutaway view of a solid tantalum capacitor. *(Courtesy AVX Corporation.)*

Surface Mount Capacitors

Many electronic products now use **surface mount devices** (SMDs). (SMDs do not have connection leads, but are soldered directly onto printed circuit boards.) Figure 10–19 shows a surface mount, ceramic chip capacitor. Such devices are extremely small and provide high packaging density—recall Figure 1–4, Chapter 1.

Silver, Nickel, Solder
┌End Terminations┐

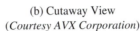

Margin   Precious
Metal
Electrodes

(b) Cutaway View
(*Courtesy AVX Corporation*)

Terminations

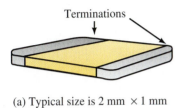

(a) Typical size is 2 mm × 1 mm
(See Figure 1–4)

FIGURE 10–19   Surface mount ceramic chip capacitor.

Variable Capacitors

The most common variable capacitor is that used in radio tuning circuits (Figure 10–20). It has a set of stationary plates and a set of movable plates that are ganged together and mounted on a shaft. As the shaft is rotated, the movable plates mesh with the stationary plates, changing the effective surface area (and hence the capacitance).

Another adjustable type is the **trimmer** or **padder** capacitor, which is used for fine adjustments, usually over a very small range. In contrast to the variable capacitor (which is frequently varied by the user), a trimmer is usually set to its required value, then never touched again.

(a) Variable capacitor of type used in radios

(b) Symbol

FIGURE 10–20

PRACTICE PROBLEMS 3

A 2.5-μF capacitor has a tolerance of +80% and −20%. Determine what its maximum and minimum values could be.

Answer
4.5 μF and 2 μF

10.7 Capacitors in Parallel and Series

Capacitors in Parallel

For capacitors in parallel, the effective plate area is the sum of the individual plate areas; thus, the total capacitance is the sum of the individual capacitances. This is easily shown. Consider Figure 10–21. The charge on each capacitor is given by Equation 10–1, i.e., $Q_1 = C_1V$ and $Q_2 = C_2V$. Thus $Q_T = Q_1 + Q_2 = C_1V + C_2V = (C_1 + C_2)\,V$. But $Q_T = C_TV$. Thus, $C_T = C_1 + C_2$. For more than two capacitors,

$$C_T = C_1 + C_2 + \cdots + C_N \qquad (10\text{–}14)$$

That is, *the total capacitance of capacitors in parallel is the sum of their individual capacitances.*

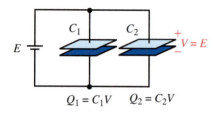

$Q_1 = C_1V \qquad Q_2 = C_2V$

(a) Parallel capacitors

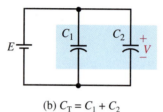

(b) $C_T = C_1 + C_2$

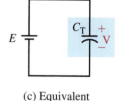

(c) Equivalent

FIGURE 10–21   Capacitors in parallel. Total capacitance is the sum of the individual capacitances.

EXAMPLE 10–6

A 10-μF, a 15-μF, and a 100-μF capacitor are connected in parallel across a 50-V source. Determine the following:

a. Total capacitance.

b. Total charge stored.

c. Charge on each capacitor.

Solution

a. $C_T = C_1 + C_2 + C_3 = 10\ \mu\text{F} + 15\ \mu\text{F} + 100\ \mu\text{F} = 125\ \mu\text{F}$

b. $Q_T = C_TV = (125\ \mu\text{F})(50\ \text{V}) = 6.25\ \text{mC}$

c. $Q_1 = C_1V = (10\ \mu\text{F})(50\ \text{V}) = 0.5\ \text{mC}$

   $Q_2 = C_2V = (15\ \mu\text{F})(50\ \text{V}) = 0.75\ \text{mC}$

   $Q_3 = C_3V = (100\ \mu\text{F})(50\ \text{V}) = 5.0\ \text{mC}$

Check: $Q_T = Q_1 + Q_2 + Q_3 = (0.5 + 0.75 + 5.0)\ \text{mC} = 6.25\ \text{mC}$.

PRACTICE PROBLEMS 4

1. Three capacitors are connected in parallel. If $C_1 = 20\ \mu$F, $C_2 = 10\ \mu$F and $C_T = 32.2\ \mu$F, what is C_3?

2. Three capacitors are paralleled across an 80-V source, with $Q_T = 0.12$ C. If $C_1 = 200\ \mu$F and $C_2 = 300\ \mu$F, what is C_3?

3. Three capacitors are paralleled. If the value of the second capacitor is twice that of the first and the value of the third is one quarter that of the second and the total capacitance is 70 μF, what are the values of each capacitor?

Answers

1. 2.2 μF; 2. 1000 μF; 3. 20 μF, 40 μF and 10 μF

Capacitors in Series

For capacitors in series (Figure 10–22), the same charge appears on each. Thus, $Q = C_1 V_1$, $Q = C_2 V_2$, etc. Solving for voltages yields $V_1 = Q/C_1$, $V_2 = Q/C_2$, and so on. Applying KVL, we get $V = V_1 + V_2 + \ldots + V_N$. Therefore,

$$V = \frac{Q}{C_1} + \frac{Q}{C_2} + \cdots + \frac{Q}{C_N} = Q\left(\frac{1}{C_1} + \frac{1}{C_2} + \cdots + \frac{1}{C_N}\right)$$

But $V = Q/C_T$. Equating this with the right side and cancelling Q yields

$$\frac{1}{C_T} = \frac{1}{C_1} + \frac{1}{C_2} + \cdots + \frac{1}{C_N} \qquad \textbf{(10–15)}$$

For two capacitors in series, this reduces to

$$C_T = \frac{C_1 C_2}{C_1 + C_2} \qquad \textbf{(10–16)}$$

For N equal capacitors in series, Equation 10–15 yields $C_T = C/N$.

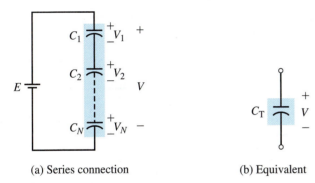

(a) Series connection   (b) Equivalent

FIGURE 10–22   Capacitors in series: $\dfrac{1}{C_T} = \dfrac{1}{C_1} + \dfrac{1}{C_2} + \cdots + \dfrac{1}{C_N}$.

EXAMPLE 10–7

NOTES . . .

1. For capacitors in parallel, total capacitance is always larger than the largest capacitance, while for capacitors in series, total capacitance is always smaller than the smallest capacitance.

2. The formula for capacitors in parallel is similar to the formula for resistors in series, while the formula for capacitors in series is similar to the formula for resistors in parallel.

Refer to Figure 10–23(a):

a. Determine C_T.

b. If 50 V is applied across the capacitors, determine Q.

c. Determine the voltage on each capacitor.

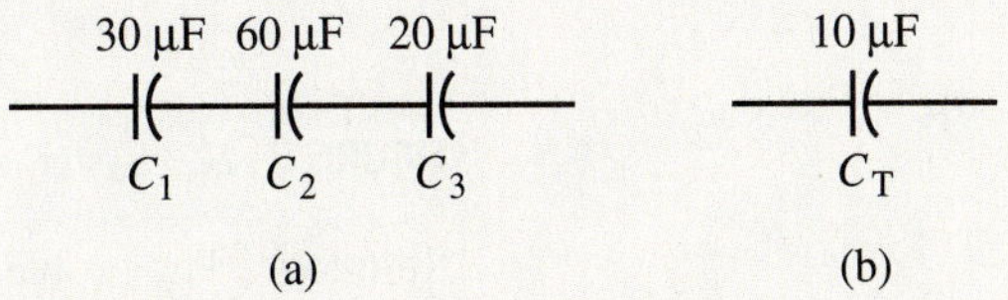

30 μF 60 μF 20 μF   10 μF

C_1 C_2 C_3 C_T

(a)   (b)

FIGURE 10–23

Solution

a. $\dfrac{1}{C_T} = \dfrac{1}{C_1} + \dfrac{1}{C_2} + \dfrac{1}{C_3} = \dfrac{1}{30\ \mu F} + \dfrac{1}{60\ \mu F} + \dfrac{1}{20\ \mu F}$

$= 0.0333 \times 10^6 + 0.0167 \times 10^6 + 0.05 \times 10^6 = 0.1 \times 10^6$

Therefore as indicated in (b),

$$C_T = \frac{1}{0.1 \times 10^6} = 10 \ \mu F$$

b. $Q = C_T V = (10 \times 10^{-6} \ F)(50 \ V) = 0.5 \ mC$

c. $V_1 = Q/C_1 = (0.5 \times 10^{-3} \ C)/(30 \times 10^{-6} \ F) = 16.7 \ V$

$\quad V_2 = Q/C_2 = (0.5 \times 10^{-3} \ C)/(60 \times 10^{-6} \ F) = 8.3 \ V$

$\quad V_3 = Q/C_3 = (0.5 \times 10^{-3} \ C)/(20 \times 10^{-6} \ F) = 25.0 \ V$

Check: $V_1 + V_2 + V_3 = 16.7 + 8.3 + 25 = 50 \ V$.

EXAMPLE 10–8

For the circuit of Figure 10–24(a), determine C_T.

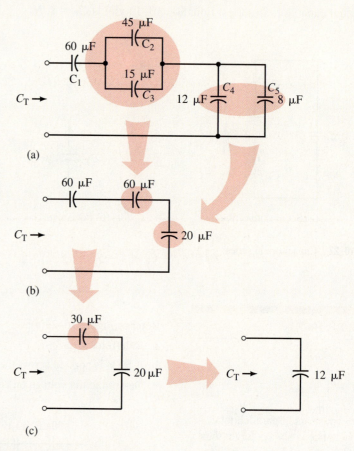

(a)

(b)

(c)

FIGURE 10–24   Systematic reduction.

Solution   The problem is easily solved through step-by-step reduction. C_2 and C_3 in parallel yield 45 μF + 15 μF = 60 μF. C_4 and C_5 in parallel total 20 μF. The reduced circuit is shown in (b). The two 60-μF capacitances in series reduce to 30 μF. The series combination of 30 μF and 20 μF can be found from Equation 10–16. Thus,

$$C_T = \frac{30 \ \mu F \times 20 \ \mu F}{30 \ \mu F + 20 \ \mu F} = 12 \ \mu F$$

Alternately, you can reduce (b) directly using Equation 10–15. Try it.

Voltage Divider Rule for Series Capacitors

For capacitors in series (Figure 10–25) a simple voltage divider rule can be developed. Recall, for individual capacitors, $Q_1 = C_1V_1$, $Q_2 = C_2V_2$, etc., and for the complete string, $Q_T = C_TV_T$. As noted earlier, $Q_1 = Q_2 = \ldots = Q_T$. Thus, $C_1V_1 = C_TV_T$. Solving for V_1 yields

$$V_1 = \left(\frac{C_T}{C_1}\right)V_T$$

This type of relationship holds for all capacitors. Thus,

$$V_x = \left(\frac{C_T}{C_x}\right)V_T \qquad\qquad (10\text{--}17)$$

From this, you can see that the voltage across a capacitor is inversely proportional to its capacitance, that is, the smaller the capacitance, the larger the voltage, and vice versa. Other useful variations are

$$V_1 = \left(\frac{C_2}{C_1}\right)V_2, \qquad V_1 = \left(\frac{C_3}{C_1}\right)V_3, \qquad V_2 = \left(\frac{C_3}{C_2}\right)V_3, \qquad \text{etc.}$$

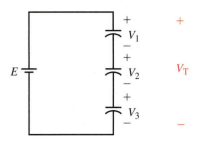

FIGURE 10–25 Capacitive voltage divider.

---

1. Verify the voltages of Example 10–7 using the voltage divider rule for capacitors.

2. Determine the voltage across each capacitor of Figure 10–24 if the voltage across C_5 is 30 V.

Answers
1. $V_1 = 16.7$ V, $V_2 = 8.3$ V, $V_3 = 25.0$ V; 2. $V_1 = 10$ V, $V_2 = V_3 = 10$ V, $V_4 = V_5 = 30$ V

PRACTICE PROBLEMS 5

As noted earlier (Figure 10–2), during charging, electrons are moved from one plate of a capacitor to the other plate. Several points should be noted.

1. This movement of electrons constitutes a current.

2. This current lasts only long enough for the capacitor to charge. When the capacitor is fully charged, current is zero.

3. Current in the circuit during charging is due solely to the movement of electrons from one plate to the other around the external circuit; no current passes through the dielectric between the plates.

4. As charge is deposited on the plates, the capacitor voltage builds. However, this voltage does not jump to full value immediately since it takes time to move electrons from one plate to the other. (Billions of electrons must be moved.)

5. Since voltage builds up as charging progresses, the difference in voltage between the source and the capacitor decreases and hence the rate of movement of electrons (i.e., the current) decreases as the capacitor approaches full charge.

Figure 10–26 shows what the voltage and current look like during the charging process. As indicated, the current starts out with an initial surge, then decays to zero while the capacitor voltage gradually climbs from zero to full voltage. The charging time typically ranges from nanoseconds to milliseconds, depending on the resistance and capacitance of the circuit. (We study these relationships in detail in Chapter 11.) A similar surge (but in the opposite direction) occurs during discharge.

10.8 Capacitor Current and Voltage

◀ **Online Companion**

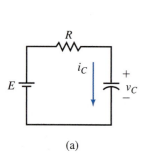

(a)

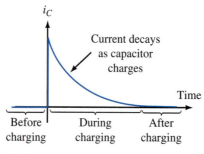

(b) Current surge during charging. Current is zero when fully charged.

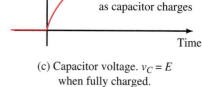

(c) Capacitor voltage. $v_C = E$ when fully charged.

FIGURE 10–26 The capacitor does not charge instantaneously, as a finite amount of time is required to move electrons around the circuit.

As Figure 10–26 indicates, current exists only while the capacitor voltage is changing. This observation turns out to be true in general, that is, *current in a capacitor exists only while capacitor voltage is changing.* The reason is not hard to understand. As you saw before, a capacitor's dielectric is an insulator and consequently no current can pass through it (assuming zero leakage). The only charges that can move, therefore, are the free electrons that exist on the capacitor's plates. When capacitor voltage is constant, these charges are in equilibrium, no net movement of charge occurs, and the current is thus zero. However, if the source voltage is increased, additional electrons are pulled from the positive plate; inversely, if the source voltage is decreased, excess electrons on the negative plate are returned to the positive plate. Thus, in both cases, capacitor current results when capacitor voltage is changed. As we show next, this current is proportional to the rate of change of voltage. Before we do this, however, we need to look at symbols.

Symbols for Time-Varying Voltages and Currents

Quantities that vary with time are called **instantaneous** quantities. *Standard industry practice requires that we use lowercase letters for time-varying quantities, rather than capital letters as for dc.* Thus, we use v_C and i_C to represent changing capacitor voltage and current rather than V_C and I_C. (Often we drop the subscripts and just use v and i.) Since these quantities are functions of time, they may also be shown as $v_C(t)$ and $i_C(t)$.

Capacitor *v-i* Relationship

The relationship between charge and voltage for a capacitor is given by Equation 10–1. For the time-varying case, it is

$$q = Cv_C \tag{10–18}$$

But current is the rate of movement of charge. In calculus notation, this is $i_C = dq/dt$. Differentiating Equation 10–18 yields

$$i_C = \frac{dq}{dt} = \frac{d}{dt}(Cv_C) \tag{10–19}$$

Since C is constant, we get

$$i_C = C\frac{dv_C}{dt} \quad \text{(A)} \tag{10–20}$$

Equation 10–20 shows that *current through a capacitor is equal to C times the rate of change of voltage across it.* This means that the faster the voltage changes, the larger the current, and vice versa. It also means that if the voltage is constant, the current is zero (as we noted earlier).

Reference conventions for voltage and current are shown in Figure 10–27. As usual, the plus sign goes at the tail of the current arrow. If the voltage is increasing, dv_C/dt is positive and the current is in the direction of the reference arrow; if the voltage is decreasing, dv_C/dt is negative and the current is opposite to the arrow.

The derivative dv_C/dt of Equation 10–20 is the slope of the capacitor voltage versus time curve. When capacitor voltage varies linearly with time (i.e., the relationship is a straight line as in Figure 10–28), Equation 10–20 reduces to

$$i_C = C\frac{\Delta v_C}{\Delta t} = C\frac{\text{rise}}{\text{run}} = C \times \text{slope of the line} \qquad \textbf{(10–21)}$$

FIGURE 10–27 The $+$ sign for v_c goes at the tail of the current arrow.

A signal generator applies voltage to a 5-μF capacitor with a waveform as in Figure 10–28(a). The voltage rises linearly from 0 to 10 V in 1 ms, falls linearly to -10 V at $t = 3$ ms, remains constant until $t = 4$ ms, rises to 10 V at $t = 5$ ms, and remains constant thereafter.

EXAMPLE 10–9

a. Determine the slope of v_C in each time interval.

b. Determine the current and sketch its graph.

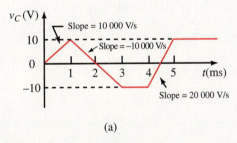
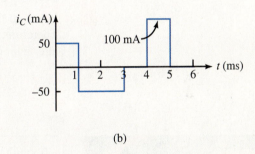

(a)   (b)

FIGURE 10–28

Solution

a. We need the slope of v_C during each time interval where slope = rise/run = $\Delta v/\Delta t$.

0 ms to 1 ms: $\Delta v = 10$ V; $\Delta t = 1$ ms; Therefore, slope = 10 V/1 ms = 10 000 V/s.

1 ms to 3 ms: Slope = -20 V/2 ms = $-10\,000$ V/s.

3 ms to 4 ms: Slope = 0 V/s.

4 ms to 5 ms: Slope = 20 V/1 ms = 20 000 V/s.

b. $i_C = Cdv_C/dt = C$ times slope. Thus,

0 ms to 1 ms: $i = (5 \times 10^{-6}$ F$)(10\,000$ V/s$) = 50$ mA.

1 ms to 3 ms: $i = -(5 \times 10^{-6}$F$)(10\,000$ V/s$) = -50$ mA.

3 ms to 4 ms: $i = (5 \times 10^{-6}$ F$)(0$ V/s$) = 0$ A.

4 ms to 5 ms: $i = (5 \times 10^{-6}$ F$)(20\,000$ V/s$) = 100$ mA.

The current is plotted in Figure 10–28(b).

EXAMPLE 10–10

The voltage across a 20-μF capacitor is $v_C = 100\, t\, e^{-t}$ V. Determine current i_C.

Solution   Differentiation by parts using $\dfrac{d(uv)}{dt} = u\dfrac{dv}{dt} + v\dfrac{du}{dt}$ with $u = 100\, t$ and $v = e^{-t}$ yields

$$i_C = C\frac{d}{dt}(100\ t\,e^{-t}) = 100\ C\frac{d}{dt}(t\,e^{-t}) = 100\ C\left(t\frac{d}{dt}(e^{-t}) + e^{-t}\frac{dt}{dt}\right)$$

$$= 2000 \times 10^{-6}(-t\,e^{-t} + e^{-t})\ \text{A} = 2.0\,(1 - t)e^{-t}\,\text{mA}$$

10.9 Energy Stored by a Capacitor

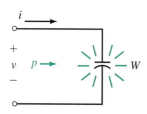

FIGURE 10–29   Storing energy in a capacitor.

An ideal capacitor does not dissipate power. When power is transferred to a capacitor, all of it is stored as energy in the capacitor's electric field. When the capacitor is discharged, this stored energy is returned to the circuit.

To determine the stored energy, consider Figure 10–29. Power is given by $p = vi$ watts. Using calculus (see), it can be shown that the stored energy is given by

$$W = \frac{1}{2}CV^2 \quad \text{(J)} \tag{10–22}$$

where V is the voltage across the capacitor. This means that the energy at any time depends on the value of the capacitor's voltage at that time.

Deriving Equation 10–22

Power to the capacitor (Figure 10–29) is given by $p = vi$, where $i = Cdv/dt$. Therefore, $p = Cvdv/dt$. However, $p = dW/dt$. Equate the two values of p, then after a bit of manipulation, you can integrate. Thus,

$$W = \int_0^t p\,dt = C\int_0^t v\frac{dv}{dt}\,dt = C\int_0^V v\,dv = \frac{1}{2}CV^2$$

10.10 Capacitor Failures and Troubleshooting

Although capacitors are quite reliable, they may fail because of misapplication, excessive voltage, current, temperature, or simply because they age. They can short internally, leads may become open, dielectrics may become excessively leaky, and they may fail catastrophically due to incorrect use. (If an electrolytic capacitor is connected with its polarity reversed, for example, it may explode.) Capacitors should be used well within their rating limits. Excessive voltage can lead to dielectric puncture creating pinholes that short the plates together. High temperatures may cause an increase in leakage and/or a permanent shift in capacitance. High temperatures may be caused by inadequate heat removal, excessive current, lossy dielectrics, or an operating frequency beyond the capacitor's rated limit. Generally capacitors are so inexpensive that you simply replace them if you suspect they are faulty. To help locate faulty capacitors, you can sometimes use an ohmmeter as described next.

Basic Testing with an Ohmmeter

Some basic (out-of-circuit) tests can be made with an analog ohmmeter. The ohmmeter can detect opens and shorts and, to a certain extent, leaky dielectrics. First, ensure that the capacitor is discharged, then set the ohmmeter to its highest range and connect it to the capacitor. (For electrolytic devices, ensure that the plus (+) side of the ohmmeter is connected to the plus (+) side of the capacitor.)

Initially, the ohmmeter reading should be low, then for a good capacitor gradually increase to infinity as the capacitor charges through the ohmmeter circuit. (Or at least a very high value, since most good capacitors, except electrolytics, have a resistance of hundreds of megohms.) For small capacitors, however, the time to charge may be too short to yield useful results.

Faulty capacitors respond differently. If a capacitor is shorted, the meter resistance reading will stay low. If it is leaky, the reading will be lower than normal. If it is open circuited, the meter will indicate infinity immediately, without dipping to zero when first connected.

Capacitor Testers

Ohmmeter testing of capacitors has its limitations; other tools may be needed. Figure 10–30 shows two of them. The DMM in (a) can measure capacitance and display it directly on its readout. The LCR (inductance, capacitance, resistance) analyzer in (b) can determine capacitance as well as detect opens and shorts. More sophisticated testers are available that determine capacitance value, leakage at rated voltage, dielectric absorption, and so on.

(a) Measuring C with a DMM. (Not all DMMs can measure capacitance)

(b) Capacitor/inductor analyzer. (*Courtesy B + K Precision*)

FIGURE 10–30   Capacitor testing.

FIGURE 10–31

10.1   Capacitance

1. For Figure 10–31, determine the charge on the capacitor, its capacitance, or the voltage across it as applicable for each of the following.
   a. $E = 40$ V, $C = 20$ μF
   b. $V = 500$ V, $Q = 1000$ μC
   c. $V = 200$ V, $C = 500$ nF
   d. $Q = 3 \times 10^{-4}$ C, $C = 10 \times 10^{-6}$ F
   e. $Q = 6$ mC, $C = 40$ μF
   f. $V = 1200$ V, $Q = 1.8$ mC

2. Repeat Question 1 for the following:
   a. $V = 2.5$ kV, $Q = 375$ μC
   b. $V = 1.5$ kV, $C = 0.04 \times 10^{-4}$ F
   c. $V = 150$ V, $Q = 6 \times 10^{-5}$ C
   d. $Q = 10$ μC, $C = 400$ nF
   e. $V = 150$ V, $C = 40 \times 10^{-5}$ F
   f. $Q = 6 \times 10^{-9}$ C, $C = 800$ pF

3. The charge on a 50–μF capacitor is 10×10^{-3} C. What is the potential difference between its terminals?

4. When 10 μC of charge is placed on a capacitor, its voltage is 25 V. What is the capacitance?

5. You charge a 5-μF capacitor to 150 V. Your lab partner then momentarily places a resistor across its terminals and bleeds off enough charge that its voltage falls to 84 V. What is the final charge on the capacitor?

10.2   Factors Affecting Capacitance

6. A capacitor with circular plates 0.1 m in diameter and an air dielectric has 0.1 mm spacing between its plates. What is its capacitance?

7. A parallel-plate capacitor with a mica dielectric has dimensions of 1 cm × 1.5 cm and separation of 0.1 mm. What is its capacitance?

8. For the capacitor of Problem 7, if the mica is removed, what is its new capacitance?

9. The capacitance of an oil-filled capacitor is 200 pF. If the separation between its plates is 0.1 mm, what is the area of its plates?

10. A 0.01-μF capacitor has ceramic with a dielectric constant of 7500. If the ceramic is removed, the plate separation doubled, and the spacing between plates filled with oil, what is the new value for *C?*

11. A capacitor with a Teflon dielectric has a capacitance of 33 μF. A second capacitor with identical physical dimensions but with a Mylar dielectric carries a charge of 55×10^{-4} C. What is its voltage?

12. The plate area of a capacitor is 4.5 in.^{2} and the plate separation is 5 mils. If the relative permittivity of the dielectric is 80, what is *C?*

10.3   Electric Fields

13. a. What is the electric field strength $\mathscr{E}$ at a distance of 1 cm from a 100-mC charge in transformer oil?
    b. What is $\mathscr{E}$ at twice the distance?

14. Suppose that 150 V is applied across a 100-pF parallel-plate capacitor whose plates are separated by 1 mm. What is the electric field intensity $\mathscr{E}$ between the plates?

10.4   Dielectrics

15. An air-dielectric capacitor has plate spacing of 1.5 mm. How much voltage can be applied before breakdown occurs?

16. Repeat Problem 15 if the dielectric is mica and the spacing is 2 mils.

17. A mica-dielectric capacitor breaks down when E volts is applied. The mica is removed and the spacing between plates doubled. If breakdown now occurs at 500 V, what is E?

18. Determine at what voltage the dielectric of a 200 nF Mylar capacitor with a plate area of 0.625 m^2 will break down.

19. Figure 10–32 shows several gaps, including a parallel-plate capacitor, a set of small spherical points, and a pair of sharp points. The spacing is the same for each. As the voltage is increased, which gap breaks down for each case?

20. If you continue to increase the source voltage of Figures 10–32(a), (b), and (c) after a gap breaks down, will the second gap also break down? Justify your answer.

10.5 Nonideal Effects

21. A 25-μF capacitor has a negative temperature coefficient of 175 ppm/$^\circ$ C. By how much and in what direction might it vary if the temperature rises by 50° C? What would be its new value?

22. If a 4.7-μF capacitor changes to 4.8 μF when the temperature rises 40° C, what is its temperature coefficient?

10.7 Capacitors in Parallel and Series

23. What is the equivalent capacitance of 10 μF, 12 μF, 22 μF, and 33 μF connected in parallel?

24. What is the equivalent capacitance of 0.10 μF, 220 nF, and 4.7×10^{-7} F connected in parallel?

25. Repeat Problem 23 if the capacitors are connected in series.

26. Repeat Problem 24 if the capacitors are connected in series.

27. Determine C_T for each circuit of Figure 10–33.

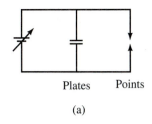

(a)

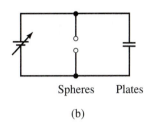

(b)

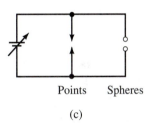

(c)

FIGURE 10–32 Source voltage is increased until one of the gaps breaks down. (The source has high internal resistance to limit current following breakdown.)

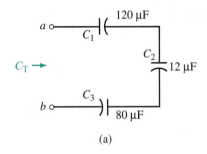

(a)

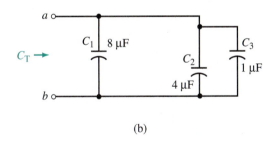

(b)

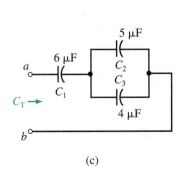

(c)

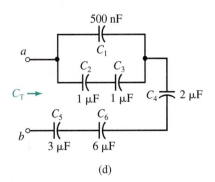

(d)

FIGURE 10–33

28. Determine total capacitance looking in at the terminals for each circuit of Figure 10–34.

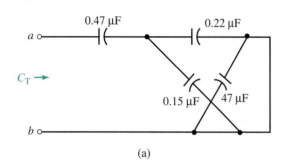

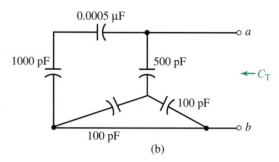

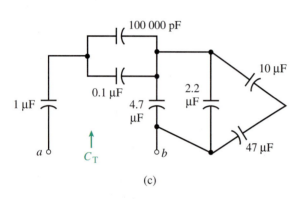

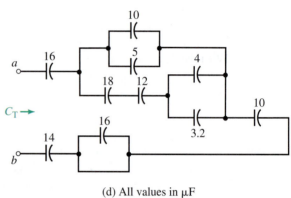

(d) All values in μF

FIGURE 10–34

29. A 30-μF capacitor is connected in parallel with a 60-μF capacitor, and a 10-μF capacitor is connected in series with the parallel combination. What is C_T?

30. For Figure 10–35, determine C_x.

31. For Figure 10–36, determine C_3 and C_4.

32. For Figure 10–37, determine C_T.

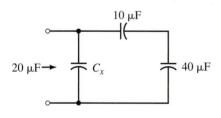

FIGURE 10–35

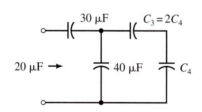

FIGURE 10–36

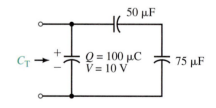

FIGURE 10–37

33. You have capacitors of 22 μF, 47 μF, 2.2 μF and 10 μF. Connecting these any way you want, what is the largest equivalent capacitance you can get? The smallest?

34. A 10-μF and a 4.7-μF capacitor are connnected in parallel. After a third capacitor is added to the circuit, C_T = 2.695 μF. What is the value of the third capacitor? How is it connected?

35. Consider capacitors of 1 μF, 1.5 μF, and 10 μF. If C_T = 10.6 μF, how are the capacitors connected?

36. For the capacitors of Problem 35, if C_T = 2.304 μF, how are the capacitors connected?

37. For Figures 10–33(c) and (d), find the voltage on each capacitor if 100 V is applied to terminals *a–b*.

38. Use the voltage divider rule to find the voltage across each capacitor of Figure 10–38.

39. Repeat Problem 38 for the circuit of Figure 10–39.

40. For Figure 10–40, $V_x = 50$ V. Determine C_x and C_T.

41. For Figure 10–41, determine C_x.

42. A dc source is connected to terminals *a–b* of Figure 10–35. If C_x is 12 μF and the voltage across the 40-μF capacitor is 80 V,

   a. What is the source voltage?

   b. What is the total charge on the capacitors?

   c. What is the charge on each individual capacitor?

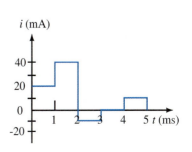

FIGURE 10–38

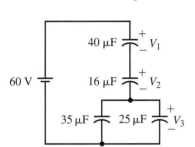

FIGURE 10–39

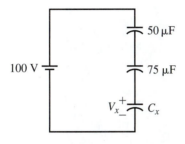

FIGURE 10–40

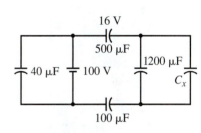

FIGURE 10–41

10.8   Capacitor Current and Voltage

43. The voltage across the capacitor of Figure 10–42(a) is shown in (b). Sketch current i_C scaled with numerical values.

44. The current through a 1-μF capacitor is shown in Figure 10–43. Sketch voltage v_C scaled with numerical values. Voltage at $t = 0$ s is 0 V.

45. If the voltage across a 4.7-μF capacitor is $v_C = 100e^{-0.05t}$ V, what is i_C?

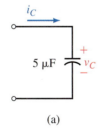

(a)

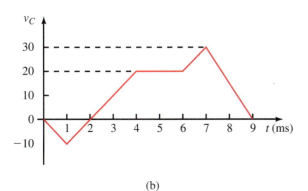

(b)

FIGURE 10–42

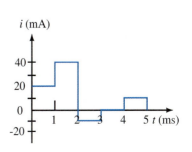

FIGURE 10–43

10.9   Energy Stored by a Capacitor

46. For the circuit of Figure 10–38, determine the energy stored in each capacitor.

47. For Figure 10–42, determine the capacitor's energy at each of the following times: $t = 0$, 1 ms, 4 ms, 5 ms, 7 ms, and 9 ms.

10.10   Capacitor Failures and Troubleshooting

48. For each case shown in Figure 10–44, what is the likely fault?

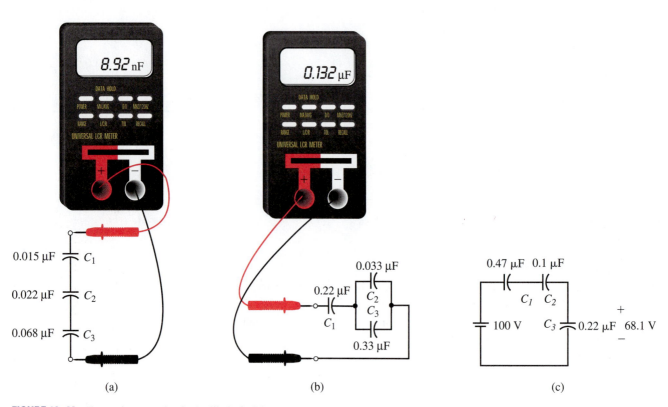

(a)   (b)   (c)

FIGURE 10–44   For each case, what is the likely fault?

ANSWERS TO IN-PROCESS LEARNING CHECKS

In-Process Learning Check 1

1. a. 2.12 nF
    b. 200 V

2. It becomes 6 times larger.
3. 1.1 nF
4. mica

■ OBJECTIVES

After studying this chapter, you will be able to

- explain why transients occur in *RC* circuits,
- explain why an uncharged capacitor looks like a short circuit when first energized,
- describe why a capacitor looks like an open circuit to steady state dc,
- describe charging and discharging of simple *RC* circuits with dc excitation,
- determine voltages and currents in simple *RC* circuits during charging and discharging,
- plot voltage and current transients,
- understand the part that time constants play in determining the duration of transients,
- compute time constants,
- describe the use of charging and discharging waveforms in simple timing applications,
- calculate the pulse response of simple *RC* circuits,
- solve simple *RC* transient problems using PSpice and MultiSIM.

Capacitor Charging, Discharging, and Simple Waveshaping Circuits

11

As you saw in Chapter 10, a capacitor does not charge instantaneously; instead, voltages and currents take time to reach their new values. This time (depicted in Figure 10–26) depends on the capacitance of the circuit and the resistance through which it charges—the larger the resistance and capacitance, the longer it takes. (Similar comments hold for discharge.) Since the voltages and currents that exist during charging and discharging are transitory in nature, they are called **transients.** Transients do not last very long, typically only a fraction of a second. However, they are important to us for a number of reasons, some of which you will learn in this chapter.

Transients occur in both capacitive and inductive circuits. In capacitive circuits, they occur because capacitor voltage cannot change instantaneously; in inductive circuits, they occur because inductor current cannot change instantaneously. In this chapter, we look at capacitive transients; in Chapter 14, we look at inductive transients. As you will see, many of the basic principles are the same. ∎

Desirable and Undesirable Transients

TRANSIENTS OCCUR IN CAPACITIVE AND inductive circuits whenever circuit conditions are changed, for example, by the sudden application of a voltage, the switching in or out of a circuit element, or the malfunctioning of a circuit component. Some transients are desirable and useful; others occur under abnormal conditions and are potentially destructive in nature.

An example of the latter is the transient that results when lightning strikes a power line. Following a strike, the line voltage, which may have been only a few thousand volts before the strike, momentarily rises to many hundreds of thousands of volts or higher, then rapidly decays, while the current, which may have been only a few hundred amps, suddenly rises to many times its normal value. Although these transients do not last very long, they can cause serious damage.

Some transient effects, on the other hand, are useful. For example, many electronic devices and circuits (such as oscillators and timers) utilize transient effects due to capacitor charging and discharging as the basis for their operation. ∎

11.1 Introduction

◀ **Online Companion**

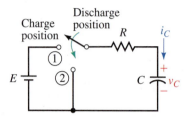

FIGURE 11–1 Circuit for studying capacitor charging and discharging. Transient voltages and currents result when the circuit is switched.

Capacitor Charging

Capacitor charging and discharging may be studied using the simple circuit of Figure 11–1. We will begin with charging. First, assume the capacitor is uncharged and that the switch is open. Now move the switch to the charge position, Figure 11–2(a). At the instant the switch is closed the current jumps to E/R amps, then decays to zero, while the voltage, which is zero at the instant the switch is closed, gradually climbs to E volts. This is shown in (b) and (c). The shapes of these curves are easily explained.

First, consider voltage. In order to change capacitor voltage, electrons must be moved from one plate to the other. Even for a relatively small capacitor, billions of electrons must be moved. This takes time. Consequently, *capacitor voltage cannot change instantaneously, i.e., it cannot jump abruptly from one value to another.* Instead, it climbs gradually and smoothly as illustrated in Figure 11–2(b). Stated another way, capacitor voltage must be **continuous** at all times.

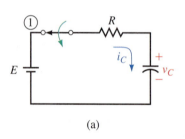

(a)

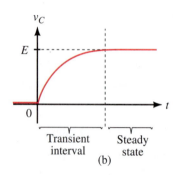

(b)

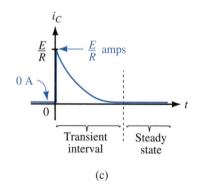

(c)

FIGURE 11–2 Capacitor voltage and current during charging. Time $t = 0$s is defined as the instant the switch is moved to the charge position. The capacitor is initially uncharged.

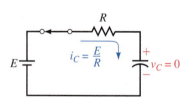

(a) Circuit as it looks just after the switch is moved to the charge position; v_C is still zero

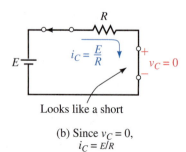

(b) Since $v_C = 0$,
$i_C = E/R$

FIGURE 11–3 An uncharged capacitor looks like a short circuit.

Now consider current. The movement of electrons noted above is a current. As indicated in Figure 11–2(c), this current jumps abruptly from 0 to E/R amps, i.e., the current is **discontinuous.** To understand why, consider Figure 11–3(a). Since capacitor voltage cannot change instantaneously, its value just after the switch is closed will be the same as it was just before the switch is closed, namely 0 V. Since the voltage across the capacitor just after the switch is closed is zero (even though there is current through it), *the capacitor looks momentarily like a short circuit.* This is indicated in (b). This is an important observation and is true in general, that is, *an uncharged capacitor looks like a short circuit at the instant of switching.* Applying Ohm's law yields $i_C = E/R$ amps. This agrees with what we indicated in Figure 11–2(c).

Finally, note the trailing end of the current curve Figure 11–2(c). Since the dielectric between the capacitor plates is an insulator, no current can pass through it. This means that the current in the circuit, which is due entirely to the movement of electrons from one plate to the other through the battery, must decay to zero as the capacitor charges.

Steady State Conditions

When the capacitor voltage and current reach their final values and stop changing (Figure 11–2(b) and (c)), the circuit is said to be in **steady state.** Figure 11–4(a) shows the circuit after it has reached steady state. Note that $v_C = E$ and $i_C = 0$. Since the capacitor has voltage across it but no current through

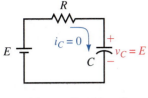

(a) $v_C = E$ and $i_C = 0$

(b) Equivalent circuit for the capacitor

FIGURE 11–4   Charging circuit after it has reached steady state. Since the capacitor has voltage across it but no current, it looks like an open circuit in steady state dc.

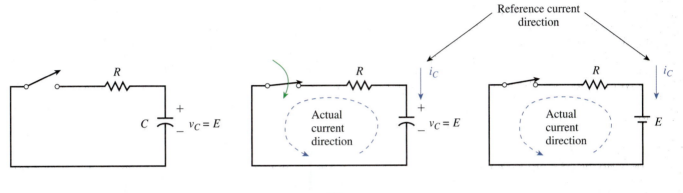

(a) Voltage $v_C = E$ just before the switch is closed

(b) Immediately after the switch is closed, v_C still equals E

(c) Capacitor therefore momentarily looks like a voltage source. Ohm's law yields $i_C = -E/R$

FIGURE 11–5   A charged capacitor looks like a voltage source at the instant of switching. Current is negative since it is opposite in direction to the current reference arrow.

it, it looks like an open circuit as indicated in (b). This is also an important observation and one that is true in general, that is, *a capacitor looks like an open circuit to steady state dc.*

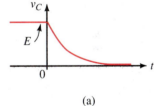

(a)

Capacitor Discharging

Now consider the discharge case, Figures 11–5 and 11–6. First, assume the capacitor is charged to E volts and that the switch is open, Figure 11–5(a). Now close the switch. Since the capacitor has E volts across it just before the switch is closed, and since its voltage cannot change instantaneously, it will still have E volts across it just after as well. This is indicated in (b). The capacitor therefore looks momentarily like a voltage source, (c) and the current thus jumps immediately to $-E/R$ amps. (Note that the current is negative since it is opposite in direction to the reference arrow.) The voltage and current then decay to zero as indicated in Figure 11–6.

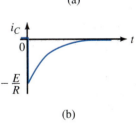

(b)

FIGURE 11–6   Voltage and current during discharge. Time $t = 0$ s is defined as the instant the switch is moved to the discharge position.

EXAMPLE 11–1

For Figure 11–1, $E = 40$ V, $R = 10\ \Omega$, and the capacitor is initially uncharged. The switch is moved to the charge position and the capacitor allowed to charge fully. Then the switch is moved to the discharge position and the capacitor allowed to discharge fully. Sketch the voltages and currents and determine the values at switching and in steady state.

Solution   The current and voltage curves are shown in Figure 11–7. Initially, $i = 0$ A since the switch is open. Immediately after it is moved to the charge position, the current jumps to $E/R = 40$ V/10 $\Omega = 4$ A; then it decays to zero. At the same time, v_C starts at 0 V and climbs to 40 V. When the switch is moved to the discharge position, the capacitor looks momentarily like a 40-V source and the current jumps to –40 V/10 $\Omega = -4$ A; then it decays to zero. At the same time, v_C also decays to zero.

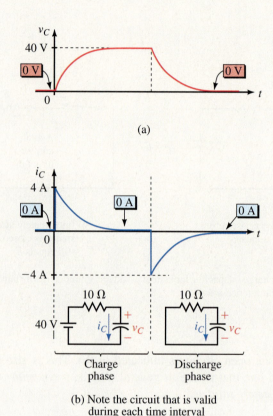

(a)

(b) Note the circuit that is valid during each time interval

FIGURE 11–7   A charge/discharge example.

The Meaning of Time in Transient Analysis

The time t used in transient analysis is measured from the instant of switching. Thus, $t = 0$ in Figure 11–2 is defined as the instant the switch is moved to charge, while in Figure 11–6, it is defined as the instant the switch is moved to discharge. Voltages and currents are then represented in terms of this time as $v_C(t)$ and $i_C(t)$. For example, the voltage across a capacitor at $t = 0$ s is denoted as $v_C(0)$, while the voltage at $t = 10$ ms is denoted as $v_C(10$ ms$)$, and so on.

A problem arises when a quantity is discontinuous as is the current of Figure 11–2(c). Since its value is changing at $t = 0$ s, $i_C(0)$ cannot be defined. To get around this problem, we define two values for 0 s. We define $t = 0^-$ s as $t = 0$ s just prior to switching and $t = 0^+$ s as $t = 0$ s just after switching. In Figure 11–2(c), therefore, $i_C(0^-) = 0$ A while $i_C(0^+) = E/R$ amps. For Figure 11–6, $i_C(0^-) = 0$ A and $i_C(0^+) = -E/R$ amps. Note that $v_c(0^+) = v_c(0^-)$.

Exponential Functions

As we will soon show, the waveforms of Figures 11–2 and 11–6 are exponential and vary according to e^{-x} or $(1 - e^{-x})$, where e is the base of the natural logarithm. Fortunately, exponential functions are easy to evaluate with modern calculators using their e^x function. You will need to be able to evaluate both e^{-x} and $(1 - e^{-x})$ for any value of x. Table 11–1 shows a tabulation of values for both cases. Note that as x gets larger, e^{-x} gets smaller and approaches zero, while $(1 - e^{-x})$ gets larger and approaches 1. These observations will be important to you in what follows.

TABLE 11–1   Table of Exponentials

x	e^{-x}	$1 - e^{-x}$
0	1	0
1	0.3679	0.6321
2	0.1353	0.8647
3	0.0498	0.9502
4	0.0183	0.9817
5	0.0067	0.9933

PRACTICE PROBLEMS 1

1. Use your calculator and verify the entries in Table 11–1. Be sure to change the sign of x before using the e^x function. Note that $e^{-0} = e^0 = 1$ since any quantity raised to the zeroth power is one.

2. Plot the computed values on graph paper and verify that they yield curves that look like those shown in Figure 11–2(b) and (c).

11.2 Capacitor Charging Equations

We will now develop equations for voltages and current during charging. Consider Figure 11–8. KVL yields

$$v_R + v_C = E \qquad (11\text{--}1)$$

But $v_R = Ri_C$ and $i_C = Cdv_C/dt$ (Equation 10-20). Thus, $v_R = RCdv_C/dt$. Substituting this into Equation 11–1 yields

$$RC\frac{dv_C}{dt} + v_C = E \qquad (11\text{--}2)$$

Equation 11–2 can be solved for v_C using basic calculus (see) The result is

$$v_C = E(1 - e^{-t/RC}) \qquad (11\text{--}3)$$

where R is in ohms, C is in farads, t is in seconds and $e^{-t/RC}$ is the exponential function discussed earlier. The product RC has units of seconds. (This is left as an exercise for the student to show.)

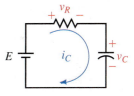

FIGURE 11–8   Circuit for the charging case. Capacitor is initially uncharged.

Solving Equation 11–2 (Optional Derivation—see Notes)

First, rearrange Equation 11–2:

$$\frac{dv_C}{dt} = \frac{1}{RC}(E - v_C)$$

Rearrange again:

$$\frac{dv_C}{E - v_C} = \frac{dt}{RC}$$

Now multiply both sides by -1 and integrate.

$$\int_0^{v_C} \frac{dv_C}{v_C - E} = -\frac{1}{RC}\int_0^t dt$$

$$\Big[\ln(v_C - E)\Big]_0^{v_C} = \Big[-\frac{t}{RC}\Big]_0^t$$

NOTES . . .

Optional problems and derivations are marked by a ∫ icon. These may be omitted without loss of continuity by those who do not require calculus.

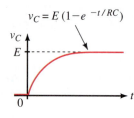

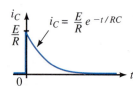

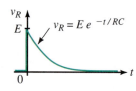

FIGURE 11–9   Curves for the circuit of Figure 11–8.

Next, substitute integration limits,

$$\ln(v_C - E) - \ln(-E) = -\frac{t}{RC}$$

$$\ln\left(\frac{v_C - E}{-E}\right) = -\frac{t}{RC}$$

Finally, take the inverse log of both sides. Thus,

$$\frac{v_C - E}{-E} = e^{-t/RC}$$

When you rearrange this, you get Equation 11–3. That is,

$$v_C = E(1 - e^{-t/RC})$$

Now consider the resistor voltage. From Equation 11–1, $v_R = E - v_C$. Substituting v_C from Equation 11–3 yields $v_R = E - E(1 - e^{-t/RC}) = E - E + Ee^{-t/RC}$. After cancellation, you get

$$v_R = Ee^{-t/RC} \tag{11–4}$$

Now divide both sides by R. Since $i_C = i_R = v_R/R$, this yields

$$i_C = \frac{E}{R}e^{-t/RC} \tag{11–5}$$

The waveforms are shown in Figure 11–9. Values at any time may be determined by substitution.

EXAMPLE 11–2

Suppose $E = 100$ V, $R = 10$ kΩ, and $C = 10$ μF:
a. Determine the expression for v_C.
b. Determine the expression for i_C.
c. Compute the capacitor voltage at $t = 150$ ms.
d. Compute the capacitor current at $t = 150$ ms.
e. Locate the computed points on the curves.

Solution
a. $RC = (10 \times 10^3\ \Omega)(10 \times 10^{-6}\ F) = 0.1$ s. From Equation 11–3,
   $v_C = E(1 - e^{-t/RC}) = 100(1 - e^{-t/0.1}) = 100(1 - e^{-10t})$ V.
b. From Equation 11–5, $i_C = (E/R)e^{-t/RC} = (100\ V/10\ k\Omega)e^{-10t} = 10e^{-10t}$ mA.
c. At $t = 0.15$ s, $v_C = 100(1 - e^{-10t}) = 100(1 - e^{-10(0.15)}) = 100(1 - e^{-1.5})$
   $= 100(1 - 0.223) = 77.7$ V.
d. $i_C = 10e^{-10t}$ mA $= 10e^{-10(0.15)}$ mA $= 10e^{-1.5}$ mA $= 2.23$ mA.
e. The corresponding points are shown in Figure 11–10.

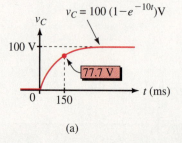

(a)

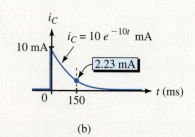

(b)

FIGURE 11–10   The computed points plotted on the v_C and i_C curves.

◀ MULTISIM

In the above example, we expressed voltage as $v_C = 100(1 - e^{-t/0.1})$ and as $100(1 - e^{-10t})$ V. Similarly, current can be expressed as $i_C = 10e^{-t/0.1}$ or as $10e^{-10t}$ mA. Although some people prefer one notation over the other, both are correct and you can use them interchangeably.

PRACTICE PROBLEMS 2

1. Determine additional voltage and current points for Figure 11–10 by computing values of v_C and i_C at values of time from $t = 0$ s to $t = 500$ ms at 100-ms intervals. Plot the results.

2. The switch of Figure 11–11 is closed at $t = 0$ s. If $E = 80$ V, $R = 4$ kΩ, and $C = 5$ μF, determine expressions for v_C and i_C. Plot the results from $t = 0$ s to $t = 100$ ms at 20-ms intervals. Note that charging takes less time here than for Problem 1.

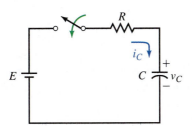

FIGURE 11–11

Answers

1.

t(ms)	v_C(V)	i_c(mA)
0	0	10
100	63.2	3.68
200	86.5	1.35
300	95.0	0.498
400	98.2	0.183
500	99.3	0.067

2. $80(1 - e^{-50t})$V $20e^{-50t}$ mA

t(ms)	v_C(V)	i_c(mA)
0	0	20
20	50.6	7.36
40	69.2	2.70
60	76.0	0.996
80	78.6	0.366
100	79.4	0.135

EXAMPLE 11–3

For the circuit of Figure 11–11, $E = 60$ V, $R = 2$ kΩ, and $C = 25$ μF. The switch is closed at $t = 0$ s, opened 40 ms later and left open. Determine equations for capacitor voltage and current and plot.

Solution $RC = (2 \text{ k}\Omega)(25 \text{ μF}) = 50$ ms. As long as the switch is closed (i.e., from $t = 0$ s to 40 ms), the following equations hold:

$$v_C = E(1 - e^{-t/RC}) = 60(1 - e^{-t/50 \text{ ms}}) \text{ V}$$

$$i_C = (E/R)e^{-t/RC} = 30e^{-t/50 \text{ ms}} \text{ mA}$$

Voltage starts at 0 V and rises exponentially. At $t = 40$ ms, the switch is opened, interrupting charging. At this instant, $v_C = 60(1 - e^{-(40/50)}) = 60(1 - e^{-0.8}) = 33.0$ V. Since the switch is left open, the voltage remains constant at 33 V thereafter as indicated in Figure 11–12. (The dotted curve shows how the voltage would have kept rising if the switch had remained closed.)

    Now consider current. The current starts at 30 mA and decays to $i_C = 30e^{-(40/50)}$ mA $= 13.5$ mA at $t = 40$ ms. At this point, the switch is opened, and the current drops instantly to zero. (The dotted line shows how the current would have decayed if the switch had not been opened.)

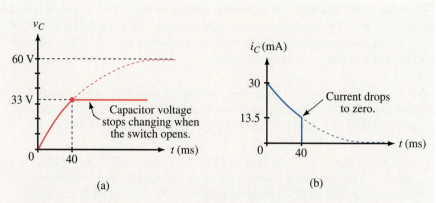

(a)   (b)

FIGURE 11–12   Incomplete charging. The switch of Figure 11–11 was opened at $t = $ 40 ms, causing charging to cease.

The Time Constant

The rate at which a capacitor charges depends on the product of R and C. This product is known as the **time constant** of the circuit and is given the symbol τ (the Greek letter tau). As noted earlier, RC has units of seconds. Thus,

$$\tau = RC \text{ (seconds, s)} \tag{11–6}$$

Using τ, Equations 11–3 to 11–5 can be written as

$$v_C = E(1 - e^{-t/\tau}) \tag{11–7}$$

$$i_C = \frac{E}{R}e^{-t/\tau} \tag{11–8}$$

and

$$v_R = Ee^{-t/\tau} \tag{11–9}$$

Duration of a Transient

The length of time that a transient lasts depends on the exponential function $e^{-t/\tau}$. As t increases, $e^{-t/\tau}$ decreases, and when it reaches zero, the transient is gone. Theoretically, this takes infinite time. In practice, however, over 99% of the transition takes place during the first five time constants (i.e., transients are within 1% of their final value at $t = 5\tau$). This can be verified by direct substitution. At $t = 5\tau$, $v_C = E(1 - e^{-t/\tau}) = E(1 - e^{-5}) = E(1 - 0.0067) = 0.993E$, meaning that the transient has achieved 99.3% of its final value. Similarly, the current falls to within 1% of its final value in five time constants. Thus, *for all practical purposes, transients can be considered to last for only five time constants* (Figure 11–13). Figure 11–14 summarizes how transient

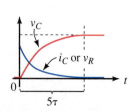

FIGURE 11–13   Transients last five time constants.

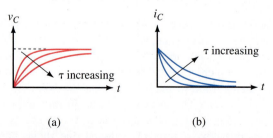

(a)   (b)

FIGURE 11–14   Illustrating how voltage and current in an *RC* circuit are affected by its time constant. The larger the time constant, the longer the capacitor takes to charge.

voltages and currents are affected by the time constant of a circuit—the larger the time constant, the longer the duration of the transient.

For the circuit of Figure 11–11, how long will it take for the capacitor to charge if $R = 2$ kΩ and $C = 10$ µF?

Solution $\tau = RC = (2$ k$\Omega)(10$ µF$) = 20$ ms. Therefore, the capacitor charges in 5 $\tau = 100$ ms.

The transient in a circuit with $C = 40$ µF lasts 0.5 s. What is R?

Solution   5 $\tau = 0.5$ s. Thus, $\tau = 0.1$ s and $R = \tau/C = 0.1$ s$/(40 \times 10^{-6}$ F$)$ $= 2.5$ kΩ.

Universal Time Constant Curves

Let us now plot capacitor voltage and current with their time axes scaled as multiples of τ and their vertical axes scaled in percent. The results, Figure 11–15, are **universal time constant curves.**

Points are computed from $v_C = 100(1 - e^{-t/\tau})$ and $i_C = 100e^{-t/\tau}$. For example, at $t = \tau$, $v_C = 100(1 - e^{-t/\tau}) = 100(1 - e^{-\tau/\tau}) = 100(1 - e^{-1}) = 63.2$ V, i.e., 63.2%, and $i_C = 100e^{-\tau/\tau} = 100e^{-1} = 36.8$ A, which is 36.8%, and so on.) These curves provide an easy method to determine voltages and currents with a minimum of computation.

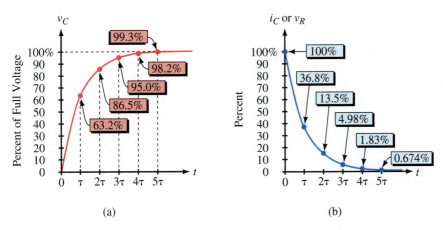

FIGURE 11–15   Universal voltage and current curves for *RC* circuits.

Using Figure 11–15, compute v_C and i_C at two time constants into charge for a circuit with $E = 25$ V, $R = 5$ kΩ, and $C = 4$ µF. What is the corresponding value of time?

Solution   At $t = 2 \tau$, v_C equals 86.5% of E or 0.865(25 V) = 21.6 V, Similarly, $i_C = 0.135I_0 = 0.135$ $(E/R) = 0.675$ mA. These values occur at $t = 2 \tau = 2RC = 40$ ms.

1. If the capacitor of Figure 11–16 is uncharged, what is the current immediately after closing the switch?

2. Given $i_C = 50e^{-20t}$ mA.
   a. What is τ?
   b. Compute the current at $t = 0^+$ s, 25 ms, 50 ms, 75 ms, 100 ms, and 250 ms and sketch it. Verify answers using the universal time constant curves.

3. Given $v_C = 100(1 - e^{-50t})$ V, compute v_C at the same time points as in Problem 2 and sketch.

4. For Figure 11–16, determine expressions for v_C and i_C. Compute capacitor voltage and current at $t = 0.6$ s. Verify answers using the universal time constant curves.

5. Refer to Figure 11–10:
   a. What are $v_C(0^-)$ and $v_C(0^+)$?
   b. What are $i_C(0^-)$ and $i_C(0^+)$?
   c. What are the steady state voltage and current?

6. For the circuit of Figure 11–11, the current just after the switch is closed is 2 mA. The transient lasts 40 ms and the capacitor charges to 80 V. Determine E, R, and C.

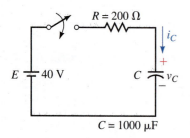

FIGURE 11–16

11.3 Capacitor with an Initial Voltage

Suppose a previously charged capacitor has not been discharged and thus still has voltage on it. Let this voltage be denoted as V_0. If the capacitor is now placed in a circuit like that in Figure 11–16, the voltage and current during charging will be affected by the initial voltage. In this case, Equations 11–7 and 11–8 become

$$v_C = E + (V_0 - E)e^{-t/\tau} \qquad (11\text{–}10)$$

$$i_C = \frac{E - V_0}{R}e^{-t/\tau} \qquad (11\text{–}11)$$

A few comments are in order about these equations. Consider Equation 11–10. When you set $t = 0$, you get $v_C = E + (V_0 - E) = V_0$. This agrees with our assertion that the capacitor was initially charged to V_0. If you now set $t = \infty$, you get $v_C = E$ which confirms that the capacitor charges to E volts as expected. Consider Equation 11–11. When you set $t = 0$, you get $i_C = (E - V_0)/R$. Recalling that an initially charged capacitor looks like a voltage source (see Figure 11–5 (c)), you can see that if you replace C in Figure 11–16 with a source V_0, the current at $t = 0$ will be $(E - V_0)/R$ as noted. Note also that these revert to their original forms when you set $V_0 = 0$ V.

EXAMPLE 11–7

Suppose the capacitor of Figure 11–16 has 25 volts on it with polarity shown at the time the switch is closed.
   a. Determine the expression for v_C.
   b. Determine the expression for i_C.
   c. Compute v_C and i_C at $t = 0.1$ s.
   d. Sketch v_C and i_C.

Solution $\tau = RC = (200 \ \Omega)(1000 \ \mu F) = 0.2$ s

a. From Equation 11–10,

$$v_C = E + (V_0 - E)e^{-t/\tau} = 40 + (25 - 40)e^{-t/0.2} = 40 - 15e^{-5t} \ V$$

b. From Equation 11–11,

$$i_C = \frac{E - V_0}{R}e^{-t/\tau} = \frac{40 - 25}{200}e^{-5t} = 75e^{-5t} \ mA$$

c. At $t = 0.1$ s,

$$v_C = 40 - 15e^{-5t} = 40 - 15e^{-0.5} = 30.9 \ V$$
$$i_C = 75e^{-5t} \ mA = 75e^{-0.5} \ mA = 45.5 \ mA$$

d. The waveforms are shown in Figure 11–17 with the above points plotted.

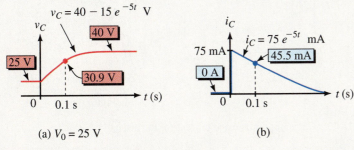

(a) $V_0 = 25$ V   (b)

FIGURE 11–17   Capacitor with an initial voltage.

◀ MULTISIM

PRACTICE PROBLEMS 3

Repeat Example 11–7 for the circuit of Figure 11–16 if $V_0 = -150$ V.

Answers

a. $40 - 190e^{-5t}$ V

b. $0.95e^{-5t}$ A

c. -75.2 V; 0.576 A

d. Curves are similar to Figure 11–17 except that v_C starts at -150 V and rises to 40 V while i_C starts at 0.95 A and decays to zero.

To determine the discharge equations, move the switch to the discharge position (Figure 11–18). KVL yields $v_R + v_C = 0$. Substituting $v_R = RCdv_C/dt$ from Section 11.2 yields

$$RC\frac{dv_C}{dt} + v_C = 0 \qquad (11\text{–}12)$$

This can be solved for v_C using basic calculus. The result is

$$v_C = V_0e^{t/RC} \qquad (11\text{–}13)$$

where V_0 is the voltage on the capacitor at the instant the switch is moved to discharge. Now consider the resistor voltage. Since $v_R + v_C = 0$, $v_R = -v_C$ and

$$v_R = -V_0e^{-t/RC} \qquad (11\text{–}14)$$

Now divide both sides by R. Since $i_C = i_R = v_R/R$,

$$i_C = -\frac{V_0}{R}e^{-t/RC} \qquad (11\text{–}15)$$

11.4 Capacitor Discharging Equations

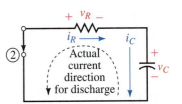

FIGURE 11–18   Discharge case. Initial capacitor voltage is V_0. Note the reference for i_C. (To conform to the standard voltage/current reference convention, i_C must be drawn in this direction so that the + sign for v_C is at the tail of the current arrow.) Since the actual current direction is opposite to the reference direction, i_C will be negative. This is indicated in Figure 11–19(b).

Note that this is negative, since, during discharge, the current is opposite in direction to the reference arrow of Figure 11–18. Voltage v_C and current i_C are shown in Figure 11–19. As in the charging case, *discharge transients last five time constants.* You can also write these equations in terms of τ, e.g., $v_C = V_0e^{-t/\tau}$, etc.

In Equations 11–13 to 11–15, V_0 represents the voltage on the capacitor at the instant the switch is moved to the discharge position. If the switch has been in the charge position long enough for the capacitor to fully charge, $V_0 = E$ and Equations 11–13 and 11–15 become $v_C = Ee^{-t/RC}$ and $i_C = -(E/R)e^{-t/RC}$ respectively.

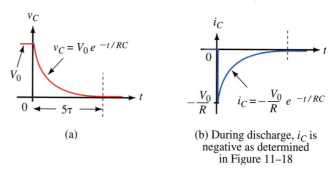

(a)

(b) During discharge, i_C is negative as determined in Figure 11–18

FIGURE 11–19   Capacitor voltage and current for the discharge case.

EXAMPLE 11–8

For the circuit of Figure 11–18, assume the capacitor is charged to 100 V before the switch is moved to the discharge position. Suppose $R = 5 \text{ k}\Omega$ and $C = 25$ µF. After the switch is moved to discharge,

a. Determine the expression for v_C.

b. Determine the expression for i_C.

c. Compute the voltage and current at 0.375 s.

Solution $RC = (5 \text{ k}\Omega)(25 \text{ µF}) = 0.125$ s and $V_0 = 100$ V. Therefore,

a. $v_C = V_0e^{-t/RC} = 100e^{-t/0.125} = 100e^{-8t}$ V.

b. $i_C = -(V_0/R)e^{-t/RC} = -20e^{-8t}$ mA.

c. At $t = 0.375$ s,

$$v_C = 100e^{-8t} = 100e^{-3} = 4.98 \text{ V}$$

$$i_C = -20e^{-8t} \text{ mA} = -20e^{-3} \text{ mA} = -0.996 \text{ mA}$$

Let us verify the answers of Example 11–8 using the appropriate universal time constant curve. As noted, $\tau = 0.125$ s, thus 0.375 s $= 3\tau$. From Figure 11–15(b), we see that capacitor voltage has fallen to 4.98% of E at 3τ. This is $(0.0498)(100 \text{ V}) = 4.98$ V as we computed earlier. Current can be verified similarly. I will leave this for you to do.

The charge and discharge equations and the universal time constant curves apply only to circuits of the forms shown in Figures 11–2 and 11–5. Fortunately, many circuits can be reduced to these forms using standard circuit reduction techniques such as series and parallel combinations, source conversions, Thévenin's theorem, and so on. Once a circuit has been reduced to its series equivalent, you can use any of the techniques that we have developed so far.

11.5 More Complex Circuits

For the circuit of Figure 11–20(a), determine expressions for v_C and i_C. Capacitors are initially uncharged.

EXAMPLE 11–9

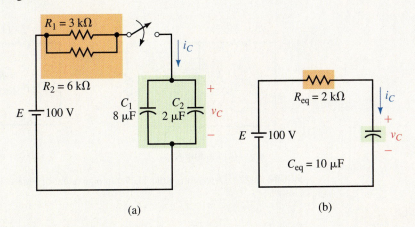

FIGURE 11–20

(a)

(b)

Solution Reduce circuit (a) to circuit (b).

$$R_{eq} = R_1 \| R_2 = 2.0 \text{ k}\Omega; \qquad C_{eq} = C_1 + C_2 = 10 \text{ }\mu\text{F}.$$
$$R_{eq}C_{eq} = (2 \text{ k}\Omega)(10 \times 10^{-6} \text{ F}) = 0.020 \text{ s}$$

Thus,

$$v_C = E(1 - e^{-t/R_{eq}C_{eq}}) = 100(1 - e^{-t/0.02}) = 100(1 - e^{-50t}) \text{ V}$$
$$i_C = \frac{E}{R_{eq}}e^{-t/R_{eq}C_{eq}} = \frac{100}{2000}e^{-t/0.02} = 50 \text{ } e^{-50t} \text{ mA}$$

The capacitor of Figure 11–21 is initially uncharged. Close the switch at $t = 0$ s.

EXAMPLE 11–10

a. Determine the expression for v_C.
b. Determine the expression for i_C.
c. Determine capacitor current and voltage at $t = 5$ ms and $t = 10$ ms.

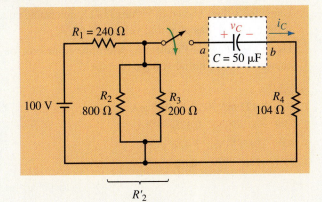

FIGURE 11–21

Solution   Reduce the circuit to its series equivalent using Thévenin's theorem:

$$R'_2 = R_2 \| R_3 = 160 \ \Omega$$

From Figure 11–22(a),

$$R_{Th} = R_1 \| R'_2 + R_4 = 240 \| 160 + 104 = 96 + 104 = 200 \ \Omega$$

From Figure 11–22(b),

$$V'_2 = \left(\frac{R'_2}{R_1 + R'_2} \right) E = \left(\frac{160}{240 + 160} \right) \times 100 \ V = 40 \ V$$

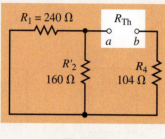

(a) Finding R_{Th} (b) Finding E_{Th}

FIGURE 11–22   Determining the Thévenin equivalent of Figure 11–21 following switch closure.

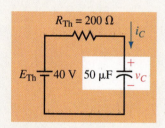

FIGURE 11–23   The Thévenin equivalent of Figure 11–21.

From KVL, $E_{Th} = V'_2 = 40$ V. The resultant equivalent circuit is shown in Figure 11–23.

$$\tau = R_{Th}C = (200 \ \Omega)(50 \ \mu F) = 10 \ ms$$

a. $v_C = E_{Th}(1 - e^{-t/\tau}) = 40(1 - e^{-100t}) \ V$

b. $i_C = \dfrac{E_{Th}}{R_{Th}} e^{-t/\tau} = \dfrac{40}{200} e^{-t/0.01} = 200e^{-100t} \ mA$

c. At t = 5 ms, $i_C = 200e^{-100(5 \ ms)} = 121$ mA. Similarly, $v_C = 15.7$ V. Similarly, at 10 ms, $i_C = 73.6$ mA and $v_C = 25.3$ V.

PRACTICE PROBLEMS 4

1. For Example 11–10, determine v_C and i_C at 5 ms and 10 ms using the universal time constant curves and compare to above. (You will have to estimate the point on the curves for $t = 5$ ms.)

2. For Figure 11–21, if $R_1 = 400 \ \Omega$, $R_2 = 1200 \ \Omega$, $R_3 = 300 \ \Omega$, $R_4 = 50 \ \Omega$, $C = 20 \ \mu F$, and $E = 200$ V, determine equations for v_C and i_C.

3. Using the values shown in Figure 11–21, determine equations for v_C and i_C if the capacitor has an initial voltage of 60 V.

4. Using the values of Problem 2, determine equations for v_C and i_C if the capacitor has an initial voltage of −50 V.

Answers

1. 1.57 V, 121 mA; 25.3 V, 73.6 mA

2. $75(1 - e^{-250t})$ V; $0.375e^{-250t}$ A

3. $40 + 20e^{-100t}$ V; $-0.1e^{-100t}$ A

4. $75 - 125e^{-250t}$ V; $0.625e^{-250t}$ A

Notes About Time References and Time Constants

1. So far, we have dealt with charging and discharging problems separately. For these, we define $t = 0$ s as the instant the switch is moved to the charge position for charging problems and to the discharge position for discharging problems.

2. When you have both charge and discharge cases in the same example, you need to establish clearly what you mean by "time." We use the following procedure:

   a. Define $t = 0$ s as the instant the switch is moved to the first position, then determine corresponding expressions for v_C and i_C. These expressions and the corresponding time scale are valid until the switch is moved to its new position.

   b. When the switch is moved to its new position, shift the time reference and make $t = 0$ s the time at which the switch is moved to its new position, then determine corresponding expressions for v_C and i_C. These new expressions are only valid from the new $t = 0$ s reference point. The old expressions are not valid on the new time scale.

   c. We now have two time scales for the same graph. However, we generally only show the first scale explicitly; the second scale is implied rather than shown. The new equations must use the new scale.

   d. Use τ_C to represent the time constant for charging and τ_d to represent the time constant for discharging. Since the equivalent resistance and capacitance for discharging may be different than that for charging, the time constants may be different for the two cases.

The capacitor of Figure 11–24(a) is uncharged. The switch is moved to position 1 for 10 ms, then to position 2, where it remains.

a. Determine v_C during charge.

b. Determine i_C during charge.

c. Determine v_C during discharge.

d. Determine i_C during discharge.

e. Sketch the charge and discharge waveforms.

EXAMPLE 11–11

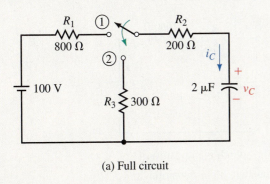

(a) Full circuit

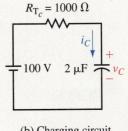

(b) Charging circuit
$R_{T_c} = R_1 + R_2$

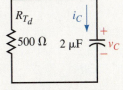

(c) Discharging circuit
$V_0 = 100$ V at $t = 0$ s

FIGURE 11–24 R_{T_c} is the total resistance of the charge circuit, while R_{T_d} is the total resistance of the discharge circuit.

Solution   Figure 11–24(b) shows the equivalent charging circuit. Here,

$$\tau_C = (R_1 + R_2)C = (1 \text{ k}\Omega)(2 \text{ μF}) = 2.0 \text{ ms}.$$

a. $v_C = E(1 - e^{-t/\tau_c}) = 100(1 - e^{-500t})$ V

b. $i_C = \dfrac{E}{R_{T_c}}e^{-t/\tau_c} = \dfrac{100}{1000}e^{-500t} = 100e^{-500t}$ mA

Since $5\tau_c = 10$ ms, charging is complete by the time the switch is moved to discharge. Thus, $V_0 = 100$ V when discharging begins.

c. Figure 11–24(c) shows the equivalent discharge circuit.

$$\tau_d = (500 \ \Omega)(2 \text{ μF}) = 1.0 \text{ ms}$$
$$v_C = V_0 e^{-t/\tau_d} = 100e^{-1000t} \text{ V}$$

where $t = 0$ s has been redefined for discharge as noted earlier.

d. $i_C = -\dfrac{V_0}{R_2 + R_3}e^{-t/\tau_d} = -\dfrac{100}{500}e^{-1000t} = -200e^{-1000t}$ mA

e. See Figure 11–25. Note that discharge is more rapid than charge since $\tau_d < \tau_c$.

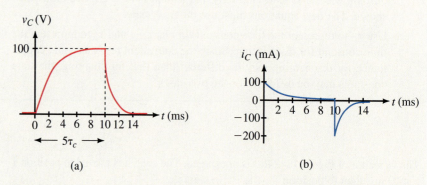

(a)   (b)

FIGURE 11–25   Waveforms for the circuit of Figure 11–24. Note that τ_d is shorter than τ_c.

EXAMPLE 11–12

The capacitor of Figure 11–26 is uncharged. The switch is moved to position 1 for 5 ms, then to position 2 and left there.

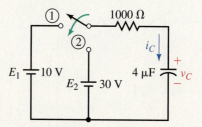

◀ MULTISIM

FIGURE 11–26

a. Determine v_C while the switch is in position 1.
b. Determine i_C while the switch is in position 1.
c. Compute v_C and i_C at $t = 5$ ms.
d. Determine v_C while the switch is in position 2.
e. Determine i_C while the switch is in position 2.
f. Sketch the voltage and current waveforms.
g. Determine v_C and i_C at $t = 10$ ms.

Solution

$$\tau_c = \tau_d = RC = (1 \text{ k}\Omega)(4 \text{ }\mu\text{F}) = 4 \text{ ms}$$

a. $v_C = E_1(1 - e^{-t/\tau_c}) = 10(1 - e^{-250t}) \text{ V}$

b. $i_C = \dfrac{E_1}{R}e^{-t/\tau_c} = \dfrac{10}{1000}e^{-250t} = 10e^{-250t} \text{ mA}$

c. At $t = 5$ ms,

$$v_C = 10(1 - e^{-250 \times 0.005}) = 7.14 \text{ V}$$
$$i_C = 10e^{-250 \times 0.005} \text{ mA} = 2.87 \text{ mA}$$

d. In position 2, $E_2 = 30$ V, and $V_0 = 7.14$ V. Use Equation 11–10:

$$v_C = E_2 + (V_0 - E_2)e^{-t/\tau_d} = 30 + (7.14 - 30)e^{-250t}$$
$$= 30 - 22.86e^{-250t} \text{ V}$$

where $t = 0$ s has been redefined for position 2.

e. $i_C = \dfrac{E_2 - V_0}{R}e^{-t/\tau_d} = \dfrac{30 - 7.14}{1000}e^{-250t} = 22.86e^{-250t} \text{ mA}$

f. See Figure 11–27.

g. $t = 10$ ms is 5 ms into the new time scale. Thus, $v_C = 30 - 22.86e^{-250(5 \text{ ms})} = 23.5$ V and $i_C = 22.86e^{-250(5 \text{ ms})} = 6.55$ mA. Values are plotted on the graph.

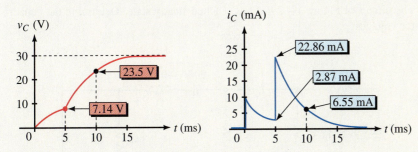

FIGURE 11–27   Capacitor voltage and current for the circuit of Figure 11–26.

In Figure 11–28(a), the capacitor is initially uncharged. The switch is moved to the charge position, then to the discharge position, yielding the current shown in (b). The capacitor takes 1.75 ms to discharge. Determine the following:

a. E.   b. R_1.   c. C.

EXAMPLE 11–13

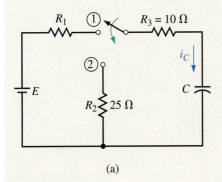

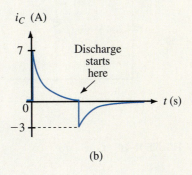

(a)   (b)

FIGURE 11–28

Solution

a. Since the capacitor charges fully, it has a value of E volts when switched to discharge. The discharge current spike is therefore

$$-\frac{E}{10\ \Omega + 25\ \Omega} = -3\ \text{A}$$

Thus, $E = 105$ V.

b. The charging current spike has a value of

$$\frac{E}{10\ \Omega + R_1} = 7\ \text{A}$$

Since $E = 105$ V, this yields $R_1 = 5\ \Omega$.

c. $5\tau_d = 1.75$ ms. Therefore $\tau_d = 350$ μs. But $\tau_d = (R_2 + R_3)C$. Thus, $C = 350\ \text{μs}/35\ \Omega = 10\ \text{μF}$.

RC Circuits in Steady State DC

When an *RC* circuit reaches steady state dc, its capacitors look like open circuits and a transient analysis is not needed—see box.

EXAMPLE 11–14

The circuit of Figure 11–29(a) has reached steady state. Determine the capacitor voltages.

◀ MULTISIM

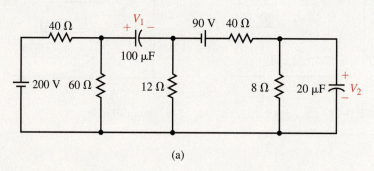

(a)

FIGURE 11–29 *Continues*

Solution Replace all capacitors with open circuits. Thus,

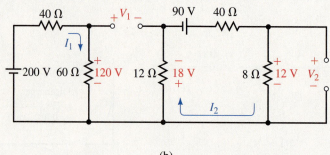

(b)

◀ MULTISIM

FIGURE 11–29 *Continued*

$$I_1 = \frac{200 \text{ V}}{40 \text{ }\Omega + 60 \text{ }\Omega} = 2 \text{ A}, \qquad I_2 = \frac{90 \text{ V}}{40 \text{ }\Omega + 8 \text{ }\Omega + 12 \text{ }\Omega} = 1.5 \text{ A}$$

KVL: $V_1 - 120 - 18 = 0$. Therefore, $V_1 = 138$ V. Further,

$$V_2 = (8 \text{ }\Omega)(1.5 \text{ A}) = 12 \text{ V}$$

PRACTICE PROBLEMS 5

1. The capacitor of Figure 11–30(a) is initially unchanged. At $t = 0$ s, the switch is moved to position 1 and 100 ms later, to position 2. Determine v_C and i_C for position 2.

2. Repeat for Figure 11–30(b). Hint: Use Thévenin's theorem.

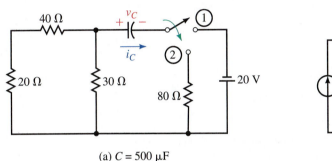

(a) $C = 500$ µF

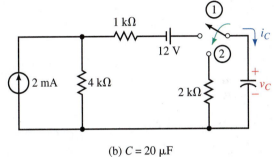

(b) $C = 20$ µF

FIGURE 11–30

3. The circuit of Figure 11–31 has reached steady state. Determine source currents I_1 and I_2.

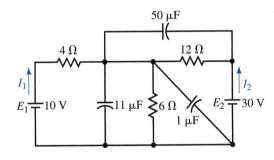

FIGURE 11–31

Answers

1. $20e^{-20t}$ V; $-0.2e^{-20t}$ A

2. $12.6e^{-25t}$ V; $-6.3e^{-25t}$ mA

3. 0 A; 1.67 A

RC circuits are used to create delays for alarm, motor control, and timing applications. Figure 11–32 shows an alarm application. The alarm unit contains a threshold detector, and when the input to this detector exceeds a preset value, the alarm is turned on.

11.6 An *RC* Timing Application

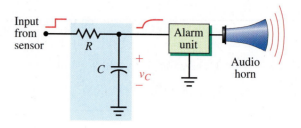

(a) Delay circuit

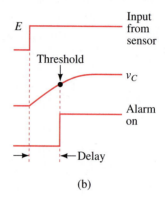

(b)

FIGURE 11–32 Creating a time delay with an RC circuit.

EXAMPLE 11–15

The circuit of Figure 11–32 is part of a building security system. When an armed door is opened, you have a specified number of seconds to disarm the system before the alarm goes off. If $E = 20$ V, $C = 40$ μF, the alarm is activated when v_C reaches 16 V, and you want a delay of at least 25 s, what values of R is needed?

Solution $v_C = E(1 - e^{-t/RC})$. After a bit of manipulation, you get

$$e^{-t/RC} = \frac{E - v_C}{E}$$

Taking the natural log of both sides yields

$$-\frac{t}{RC} = \ln\left(\frac{E - v_C}{E}\right)$$

At $t = 25$ s, $v_C = 16$ V. Thus,

$$-\frac{t}{RC} = \ln\left(\frac{20 - 16}{20}\right) = \ln 0.2 = -1.6094$$

Substituting $t = 25$ s and $C = 40$ μF yields

$$R = \frac{t}{1.6094C} = \frac{25 \text{ s}}{1.6094 \times 40 \times 10^{-6}} = 388 \text{ k}\Omega$$

Choose the next higher standard value, namely 390 kΩ.

PRACTICE PROBLEMS 6

1. Suppose you want to increase the disarm time of Example 11–15 to at least 35 s. Compute the new value of *R*.

2. If, in Example 11–15, the threshold is 15 V and $R = 1\ M\Omega$, what is the disarm time?

Answers
1. 544 kΩ. Use 560 kΩ.

2. 55.5 s

---

1. Refer to Figure 11–16:
   a. Determine the expression for v_C when $V_0 = 80$ V. Sketch v_C.
   b. Repeat (a) if $V_0 = 40$ V. Why is there no transient?
   c. Repeat (a) if $V_0 = -60$ V.

2. For Part (c) of Question 1, v_C starts at -60 V and climbs to $+40$ V. Determine at what time v_C passes through 0 V, using the technique of Example 11–15.

3. For the circuit of Figure 11–18, suppose $R = 10\ k\Omega$ and $C = 10\ \mu F$:
   a. Determine the expressions for v_C and i_C when $V_0 = 100$ V. Sketch v_C and i_C.
   b. Repeat (a) if $V_0 = -100$ V.

4. Repeat Example 11–12 if voltage source 2 is reversed, i.e., $E_2 = -30$ V.

5. The switch of Figure 11–33(a) is closed at $t = 0$ s. The Norton equivalent of the circuit in the box is shown in (b). Determine expressions for v_C and i_C. The capacitor is initially uncharged.

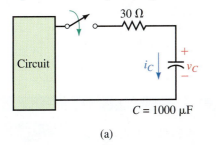
(a)

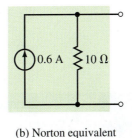
(b) Norton equivalent

FIGURE 11–33   Hint: Use a source transformation.

In previous sections, we looked at the response of *RC* circuits to switched dc inputs. In this section, we consider the effect that *RC* circuits have on pulse waveforms. Since many electronic devices and systems utilize pulse or rectangular waveforms, including computers, communications systems, and motor control circuits, these are important considerations.

11.7 Pulse Response of *RC* Circuits

Pulse Basics

A **pulse** is a voltage or current that changes from one level to the other and back again as in Figure 11–34(a) and (b). A **pulse train** is a repetitive stream of pulses, as in (c). If a waveform's high time equals its low time, as in (d), it is called a **square wave**.

(a) Positive pulse

(b) Negative pulse

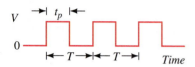

(c) Pulse train. T is referred to as the period of the pulse train

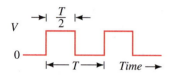

(d) Square wave

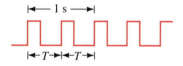

(e) PRR = 2 pulses/s

FIGURE 11–34 Ideal pulses and pulse waveforms.

The length of each cycle of a pulse train is termed its **period,** $T,$ and the number of pulses per second is defined as its **pulse repetition rate** (PRR) or **pulse repetition frequency** (PRF). For example, in (e), there are two complete cycles in one second; therefore, the PRR = 2 pulses/s. With two cycles every second, the time for one cycle is $T = \frac{1}{2}$ s. Note that this is 1/PRR. This is true in general. That is,

$$T = \frac{1}{PRR} \quad \text{s} \qquad (11\text{–}16)$$

The width, t_p, of a pulse relative to its period [Figure 11–24(c)] is its **duty cycle.** Thus,

$$\text{duty cycle} = \frac{t_p}{T} \times 100\% \qquad (11\text{–}17)$$

A square wave [Figure 11–24(d)] therefore has a 50% duty cycle, while a waveform with $t_p = 1.5$ μs and a period of 10 μs has a duty cycle of 15%.

In practice, waveforms are not ideal, that is, they do not change from low to high or high to low instantaneously. Instead, they have finite **rise** and **fall times.** Rise and fall times are denoted as t_r and t_f and are measured between the 10% and 90% points as indicated in Figure 11–35(a). **Pulse width** is measured at the 50% point. The difference between a real waveform and an ideal waveform is often slight. For example, rise and fall times of real pulses may be only a few nanoseconds and when viewed on an oscilloscope, as in Figure 11–35(b), appear to be ideal. In what follows, we will assume ideal waveforms.

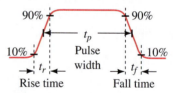

(a) Pulse definitions

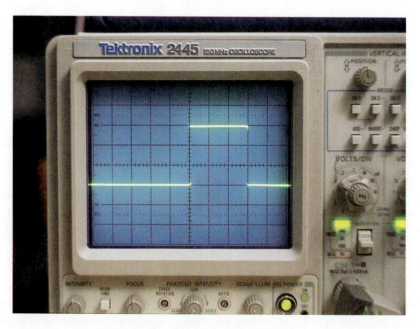

(b) Pulse waveform viewed on an oscilloscope

FIGURE 11–35 Practical pulse waveforms.

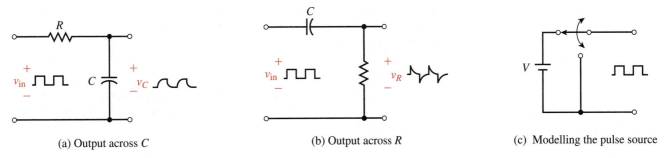

(a) Output across *C*   (b) Output across *R*   (c) Modelling the pulse source

FIGURE 11–36   *RC* circuits with pulse input. Although we have modelled the source here as a battery and a switch, in practice, pulses are usually created by electronic circuits.

The Effect of Pulse Width

The width of a pulse relative to a circuit's time constant determines how it is affected by an *RC* circuit. Consider Figure 11–36. In (a), the circuit has been drawn to focus on the voltage across *C;* in (b), it has been drawn to focus on the voltage across *R*. (Otherwise, the circuits are identical.) An easy way to visualize the operation of these circuits is to assume that the pulse is generated by a switch that is moved rapidly back and forth between *V* and common as in (c). This alternately creates a charge and discharge circuit, and thus all of the ideas developed in this chapter apply directly.

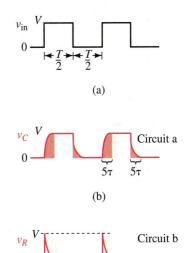

FIGURE 11–37   Pulse width much greater than 5 τ. Note that the shaded areas indicate where the capacitor is charging and discharging. Spikes occur on the input voltage transitions.

Pulse Width $t_p \gg 5\,\tau$

First, consider the ouput of circuit (a). When the pulse width and time between pulses are very long compared with the circuit time constant, the capacitor charges and discharges fully, Figure 11–37(b). (This case is similar to what we have already seen in this chapter.) Note, that charging and discharging occur at the transitions of the pulse. The transients therefore increase the rise and fall times of the output. In high-speed circuits, this may be a problem. (You will learn more about this in your digital electronics courses.)

EXAMPLE 11–16

A square wave is applied to the input of Figure 11–36(a). If $R = 1\ \text{k}\Omega$ and $C = 100\ \text{pF}$, estimate the rise and fall time of the output signal using the universal time constant curve of Figure 11–15(a).

Solution   Here, $\tau = RC = (1 \times 10^3)(100 \times 10^{-12}) = 100$ ns. From Figure 11–15(a), note that v_C reaches the 10% point at about 0.1 τ, which is (0.1)(100 ns) = 10 ns. The 90% point is reached at about 2.3 τ, which is (2.3)(100 ns) = 230 ns. The rise time is therefore approximately 230 ns – 10 ns = 220 ns. The fall time will be the same.

Now consider the circuit in Figure 11–36(b). Here, current i_C will be similar to that of Figure 11–7(b), except that the pulse widths will be narrower. Since voltage $v_R = R\,i_C$, the output will be a series of short, sharp spikes that occur at input transitions as in Figure 11–37(c). Under the conditions here (i.e., pulse width much greater than the circuit time constant), v_R is an

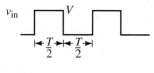

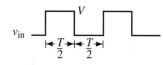

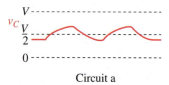

FIGURE 11–38   Pulse width equal 5 τ. These waveforms are the same as Figure 11–37, except that the transitions last relatively longer.

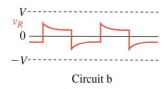

FIGURE 11–39   Pulse width much less than 5 τ. The circuit does not have time to charge or discharge substantially.

approximation to the derivative of v_{in} and the circuit is called a **differentiator circuit.** Such circuits have important practical uses.

Pulse Width $t_p = 5\,\tau$

These waveforms are shown in Figure 11–38. Since the pulse width is 5τ, the capacitor fully charges and discharges during each pulse. Thus waveforms here will be similar to what we have just seen.

Pulse Width $t_p \ll 5\,\tau$

This case differs from what we have seen so far in this chapter only in that the capacitor does not have time to charge and discharge significantly between pulses. The result is that switching occurs on the early (nearly straight line) part of the charging and discharging curves and thus, v_C is roughly triangular in shape, Figure 11–39(a). As shown below, it has an average value of $V/2$. Under the conditions here, v_C is the approximate integral of v_{in} and the circuit is called an **integrator circuit.**

It should be noted that v_C does not reach the steady state shown in Figure 11–39 immediately. Instead, it works its way up over a period of five time constants (Figure 11–40). To illustrate, assume an input square wave of 5 V with a pulse width of 0.1 s and $\tau = 0.1$ s.

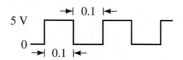

(a) Input waveform

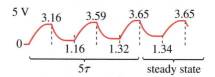

(b) Output voltage v_C

FIGURE 11–40   Circuit takes five time constants to reach a steady state.

◀ **MULTISIM**

Pulse 1 $v_C = E(1 - e^{-t/\tau})$. At the end of the first pulse ($t = 0.1$ s), v_C has climbed to $v_C = 5(1 - e^{-0.1/0.1}) = 5(1 - e^{-1}) = 3.16$ V. From the end of pulse 1 to the beginning of pulse 2 (i.e., over an interval of 0.1 s), v_C decays from 3.16 V to $3.16e^{-0.1/0.1} = 3.16e^{-1} = 1.16$ V.

Pulse 2 v_C starts at 1.16 V and 0.1 s later has a value of $v_C = E + (V_0 - E)e^{-t/\tau} = 5 + (1.16 - 5)e^{-0.1/0.1} = 5 - 3.84e^{-1} = 3.59$ V. It then decays to $3.59e^{-1} = 1.32$ V over the next 0.1 s.

Continuing in this manner, the remaining values for Figure 11–40(b) are determined. After $5\,\tau$, v_C cycles between 1.34 and 3.65 V, with an average of $(1.34 + 3.65)/2 = 2.5$ V, or half the input pulse amplitude.

Verify the remaining points of Figure 11–40(b).

PRACTICE PROBLEMS 7

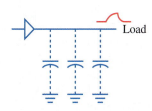

(a) Unloaded driver

(b) Distorted signal

FIGURE 11–41 Distortion caused by capacitive loading.

Capacitive Loading

Capacitance occurs whenever conductors are separated by insulating material. This means that capacitance exists between wires in cables, between traces on printed circuit boards, and so on. In general, this capacitance is undesirable but it cannot be avoided. It is called **stray capacitance.** Fortunately, stray capacitance is often so small that it can be neglected. However, in high-speed circuits, it may cause problems.

To illustrate, consider Figure 11–41. The electronic driver of (a) produces square pulses. However, when it drives a long line as in (b), stray capacitance loads it and increases the signal's rise and fall times (since capacitance takes time to charge and discharge). If the rise and fall times become excessively long, the signal reaching the load may be so degraded that the system malfunctions. (Capacitive loading is a serious issue but we will leave it for future courses to deal with.)

11.8 Transient Analysis Using Computers

◀ MULTISIM

◀ CADENCE

MultiSIM and PSpice are well suited for studying transients as they both incorporate easy to use graphing facilities that you can use to plot results directly on the screen. When plotting transients, you must specify the time duration for your plot—i.e., the length of time that you expect the transient to last. (This time is designated TSTOP by both MultiSIM and PSpice.) A good value to start with is 5 τ where τ is the time constant of the circuit. (For complex circuits, if you do not know τ, make an estimate, run a simulation, adjust the time scale, and repeat until you get an acceptable plot.)

MultiSIM

As a first example, consider the *RC* charging circuit of Figure 11–42. Let us plot the capacitor voltage waveform, then using the cursor, determine voltages

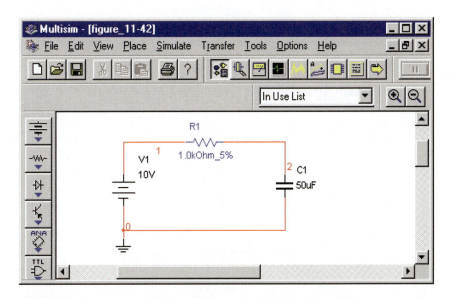

FIGURE 11–42 MultiSIM example. No switch is required as the transient solution is initiated by software. Since $\tau = 50$ ms, run the simulation to 0.25 s (i.e., 5τ).

MultiSIM Operational Notes

1. The version of MultiSIM at the time of this writing permits you to plot only voltage waveforms. If you want to graph a current waveform, plot the voltage across a resistor and use Ohm's law.

2. MultiSIM gives you the option of plotting waveforms via its standard graphing facility or via its oscilloscope. (In this chapter, we will use the standard facility; in Chapter 14 we introduce the oscilloscope.)

3. You can only plot voltages with respect to ground. However, you can put the ground wherever you want.

4. You don't need a switch to initiate a transient. Simply build the circuit without a switch and select transient analysis. MultiSIM then performs the transient simulation and plots the results.

5. MultiSIM generates time steps automatically when it plots waveforms. Sometimes, however, it does not generate enough and you get a jagged curve. If this happens, click *Minimum number of time points* in the transient analysis dialog box (the dialog box in which you enter the value for TSTOP), and type in a larger number (say 1000). Experiment until you get a suitably smooth curve.

6. Before you start, use your cursor to review and identify the icons on your screen, as you will be using some new ones here.

at *t* = 50 ms and *t* = 150 ms. Read the MultiSIM Operational Notes; then proceed as follows:

- Create the circuit of Figure 11–42 on your screen. (Use a virtual capacitor and rotate it once only so that its "1" end is at the top as described in Appendix A.)

- Click *Options/Preferences/Show node names* and determine the node number for the top end of the capacitor.

- Click the analysis icon (or *Simulate/Analysis*) and select *Transient Analysis*. Estimate the duration of the transient (it is 0.25 s) and enter this value as TSTOP. From the *Initial Conditions* box, choose *Set to zero*. Click the *Output Variables* tab, select the node determined in the previous step, click *Plot during simulation,* then *Simulate*.

- The waveform of Figure 11–43 should appear. Expand to full screen. Click the *Show/Hide Grid* icon on the Analysis Graphs menu bar, then the *Show/Hide cursor* icon. Drag the cursors to 50 ms and 150 ms and read the voltages.

Analysis of Results

As indicated in Figure 11–43, $v_C = 6.32$ V at $t = 50$ ms and 9.50 V at $t = 150$ ms. (Check by substitution into $v_C = 10(1 - e^{-20t})$. You will find that results agree exactly.)

Initial Conditions in MultiSIM

Let us change the above problem to include an initial voltage of 20 V on the capacitor. Get the circuit of Figure 11–42 back on the screen and double click the capacitor symbol; in the dialog box, click *Initial conditions,* type in **20,** ensure that units are set to V, and click *OK*. Click the *Analysis* icon and select *Transient Analysis;* from the *Initial Conditions* box, choose *User-defined*. Run the simulation and note that the transient starts at 20 V and decays to its steady state value of 10 V in five time constants as expected. With your cursor, measure several values from the graph and verify using Equation 11–10.

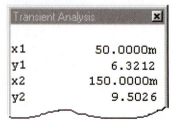

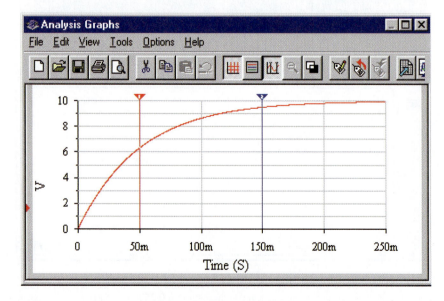

FIGURE 11–43   Solution for the circuit of Figure 11–42.

Another Example

Using the clock source from the Sources bin, build the circuit of Figure 11–44. (The clock, with its default settings, produces a square wave that cycles between 0 V and 5 V with a cycle length of $T = 1$ ms.) This means that its *on* time t_p is $T/2 = 500$ μs. Since the time constant of Figure 11–44 is $\tau = RC = 50$ μs, t_p is greater than 5τ and a waveform similar to that of Figure 11–37(c) should result. To verify, follow the procedure of the previous example, except set End Time (TSTOP) to 0.0025 in the *Analysis/Transient* dialog box. The waveform of Figure 11–45 appears. Note that output spikes occur on the transitions of the input waveform as predicted.

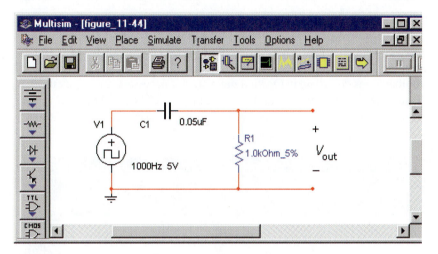

FIGURE 11–44   Applying a square wave source.

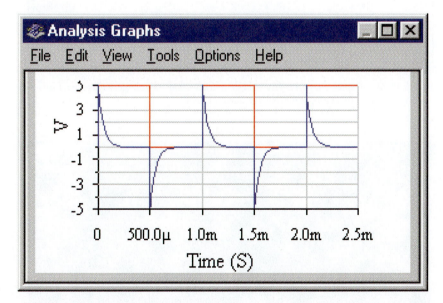

FIGURE 11–45   Waveform for the circuit of Figure 11–44. The red waveform is the input voltage and the blue is the output.

PSpice

As a first example, consider Figure 11–2 with $R = 200\ \Omega$, $C = 50\ \mu F$ and $E = 40$ V. Let the capacitor be initially uncharged (i.e., $V_0 = 0$ V). First, read the PSpice Operational Notes, then proceed as follows:

• Create the circuit on the screen as in Figure 11–46. (The switch can be found in the EVAL library as part Sw_tClose.) Remember to rotate the capacitor three times as discussed in Appendix A, then set its initial condition *(IC)* to zero. To do this, double click the capacitor symbol, type **0V** into the Property editor cell labeled IC, click Apply, then close the editor. Click the New Simulation Profile icon, enter a name (e.g., **Figure 11–47**) then click Create. In the Simulation Settings box, click the Analysis tab, select Time Domain (Transient) and in Options, select General Settings. Set the duration of the transient (TSTOP) to 50ms (i.e., five time constants). Find the voltage marker on the toolbar and place as shown.

• Click the Run icon. When simulation is complete, a trace of capacitor voltage versus time (the green trace of Figure 11–47) appears. Click Plot (on the toolbar) then Add Y Axis to create the second axis. Activate the additional toolbar icons described in Operational Note 5 if necessary, then click the Add Trace icon on the new toolbar. In the dialog box, click I(C1) (assuming your capacitor is designated C_1), then OK. This adds the current trace.

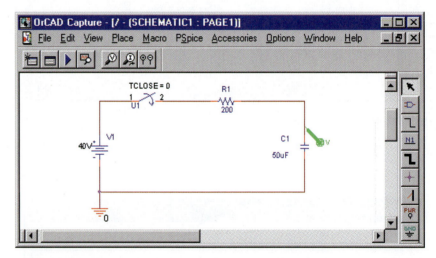

FIGURE 11–46 PSpice example. The voltage marker displays voltage with respect to ground, which, in this case, is the voltage across C_1.

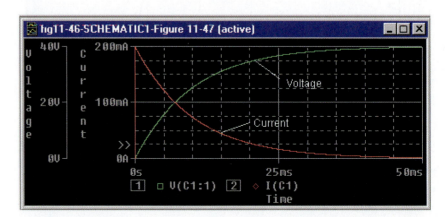

FIGURE 11–47 Waveforms for the circuits of Figures 11–46 and 11–48.

Analysis of Results

Click the Toggle cursor icon, then use the cursor to determine values from the screen. For example, at $t = 5$ ms, you should find $v_C = 15.7$ V and $i_C = 121$ mA. (An analytic solution for this circuit (which is Figure 11–23) may be found in Example 11–10, part (c). It agrees exactly with the PSpice solution.)

As a second example, consider the circuit of Figure 11–21 (shown as Figure 11–48). Create the circuit using the same general procedure as in the previous example, except do not rotate the capacitor. Again, be sure to set V_0 (the initial capacitor voltage) to zero. In the Simulation Profile box, set TSTOP to 50ms. Place differential voltage markers (found on the toolbar at the top of the screen) across C to graph the capacitor voltage. Run the analysis, create a second axis, then add the current plot. You should get the same graph (i.e., Figure 11–47) as you got for the previous example, since its circuit is the Thévenin equivalent of this one.

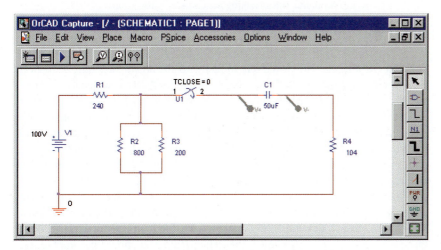

FIGURE 11–48   The differential markers display the voltage across C_1.

As a final example, consider Figure 11–49(a), which shows double switching action.

The capacitor of Figure 11–49(a) has an initial voltage of -10 V. The switch is moved to the charge position for 1 s, then to the discharge position where it remains. Determine curves for v_C and I_C.

EXAMPLE 11–17

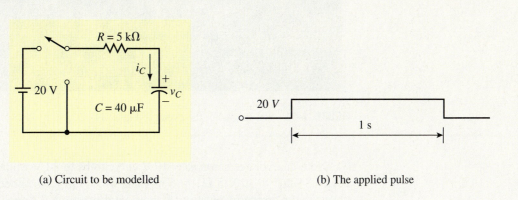

(a) Circuit to be modelled   (b) The applied pulse

FIGURE 11–49   Creating a charge/discharge waveform using PSpice. *Continues*

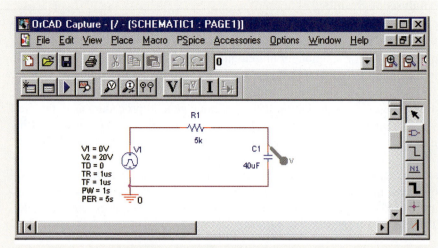

(c) Modeling the switching action using pulse source

FIGURE 11–49   *Continued*

Solution   PSpice has no switch that implements the above switching sequence. However, moving the switch first to charge then to discharge is equivalent to placing 20 V across the *RC* combination for the charge time, then 0 V thereafter as indicated in (b). You can do this with a pulse source (VPULSE) as indicated in (c). (VPULSE is found in the SOURCE library.) Create the circuit of Figure 11–49(c) on your screen. Note the parameters listed beside the VPULSE symbol. Click each in turn and set as indicated; e.g., click *V*1 and when its parameter box opens, type **0V**. (This defines a pulse with a period of 5 s, a width of 1 s, rise and fall times of 1 μs, amplitude of 20 V, and an initial value of 0 V.) Double click the capacitor symbol and set *IC* to –10V in the Properties Editor; click Apply then close. Click the New Simulation Profile icon and set TSTOP to 2s. Place a Voltage Marker as shown, then click Run. You should get the voltage trace of Figure 11–50 on the screen. Add the second axis and the current trace as described in the previous examples. The red current curve should appear.

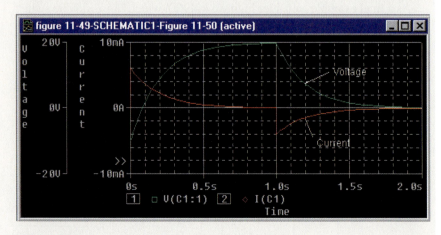

FIGURE 11–50   Waveforms for the circuit of Figure 11–49.

Note that voltage starts at − 10 V and climbs to 20 V while current starts at $(E - V_0)/R = 30$ V/5 kΩ = 6 mA and decays to zero. When the switch is turned to the discharge position, the current drops from 0 A to − 20 V/5 kΩ = −4 mA and then decays to zero while the voltage decays from 20 V to zero. Thus the solution checks.

PUTTING IT INTO PRACTICE

An electronic device employs a timer circuit of the kind shown in Figure 11–32(a), i.e., an *RC* charging circuit and a threshold detector. (Its timing waveforms are thus identical to those of Figure 11–32(b)). The input to the *RC* circuit is a 0 to 5 V ±4% step, $R = 680\,k\Omega \pm 10\%$, $C = 0.22\,\mu F \pm 10\%$, the threshold detector activates at $v_C = 1.8\,V \pm 0.05\,V$ and the required delay is 67 ms ±18 ms. You test a number of units as they come off the production line and find that some do not meet the timing spec. Perform a design review and determine the cause. Redesign the timing portion of the circuit in the most economical way possible.

PROBLEMS

11.1 Introduction

1. The capacitor of Figure 11–51 is uncharged.

   a. What are the capacitor voltage and current just after the switch is closed?

   b. What are the capacitor voltage and current after the capacitor is fully charged?

2. Repeat Problem 1 if the 20-V source is replaced by a −60-V source.

3. a. What does an uncharged capacitor look like at the instant of switching?

   b. What does a charged capacitor look like at the instant of switching?

   c. What does a capacitor look like to steady state dc?

   d. What do we mean by $i(0^-)$? By $i(0^+)$?

4. For a charging circuit, $E = 25\,V$, $R = 2.2\,k\Omega$, and the capacitor is initially uncharged. The switch is closed at $t = 0$. What is $i(0^+)$?

5. For a charging circuit, $R = 5.6\,k\Omega$ and $v_C(0^-) = 0\,V$. If $i(0^+) = 2.7\,mA$, what is E?

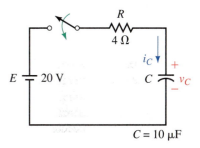

FIGURE 11–51

11.2 Capacitor Charging Equations

6. The switch of Figure 11–51 is closed at $t = 0$ s. The capacitor is initially uncharged.

   a. Determine the equation for charging voltage v_C.

   b. Determine the equation for charging current i_C.

   c. By direct substitution, compute v_C and i_C at $t = 0^+$ s, 40 μs, 80 μs, 120 μs, 160 μs, and 200 μs.

   d. Plot v_C and i_C on graph paper using the results of (c). Hint: See Example 11–2.

7. Repeat Problem 6 if $R = 500\,\Omega$, $C = 25\,\mu F$, and $E = 45\,V$, except compute and plot values at $t = 0^+$ s, 20 ms, 40 ms, 60 ms, 80 ms, and 100 ms.

8. The switch of Figure 11–52 is closed at $t = 0$ s. Determine the equations for capacitor voltage and current. Compute v_C and i_C at $t = 50$ ms.

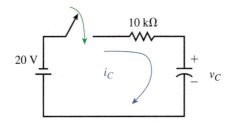

FIGURE 11–52 $V_0 = 0\,V$, $C = 10\,\mu F$.

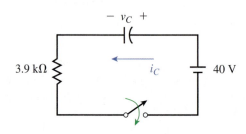

FIGURE 11–53 $C = 10\ \mu F$, $V_0 = 0$ V.

9. Repeat Problem 8 for the circuit of Figure 11–53.

10. The capacitor of Figure 11–2 is uncharged at the instant the switch is closed. If $E = 80$ V, $C = 10\ \mu F$, and $i_C(0^+) = 20$ mA, determine the equations for v_C and i_C.

11. Determine the time constant for the circuit of Figure 11–51. How long (in seconds) will it take for the capacitor to charge?

12. A capacitor takes 200 ms to charge. If $R = 5$ kΩ, what is C?

13. For Figure 11–51, the capacitor voltage with the switch open is 0 V. Close the switch at $t = 0$ and determine capacitor voltage and current at $t = 0^+$, 40 μs, 80 μs, 120 μs, 160 μs, and 200 μs using the universal time constant curves.

14. If $i_C = 25e^{-40t}$ A, what is the time constant τ and how long will the transient last?

15. For Figure 11–2, the current jumps to 3 mA when the switch is closed. The capacitor takes 1 s to charge. If $E = 75$ V, determine R and C.

16. For Figure 11–2, if $v_C = 100(1 - e^{-50t})$ V and $i_C = 25e^{-50t}$ mA, what are E, R, and C?

17. For Figure 11–2, determine E, R, and C if the capacitor takes 5 ms to charge, the current at 1 time constant after the switch is closed is 3.679 mA, and the capacitor charges to 45 volts steady state.

18. For Figure 11–2, $v_C(\tau) = 41.08$ V and $i_C(2\tau) = 219.4$ mA. Determine E and R.

11.3 Capacitor with an Initial Voltage

19. The capacitor of Figure 11–51 has an initial voltage. If $V_0 = 10$ V, what is the current just after the switch is closed?

20. Repeat Problem 19 if $V_0 = -10$ V.

21. For the capacitor of Figure 11–51, $V_0 = 30$ V.

   a. Determine the expression for charging voltage v_C.

   b. Determine the expression for current i_C.

   c. Sketch v_C and i_C.

22. Repeat Problem 21 if $V_0 = -5$ V.

11.4 Capacitor Discharging Equations

23. For the circuit of Figure 11–54, assume the capacitor is charged to 50 V before the switch is closed.

   a. Determine the equation for discharge voltage v_C.

   b. Determine the equation for discharge current i_C.

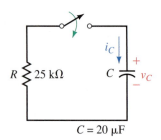

FIGURE 11–54

c. Determine the time constant of the circuit.

d. Compute v_C and i_C at $t = 0^+$ s, $t = \tau$, 2τ, 3τ, 4τ, and 5τ.

e. Plot the results of (d) with the time axis scaled in seconds and time constants.

24. The initial voltage on the capacitor of Figure 11–54 is 55 V. The switch is closed at $t = 0$. Determine capacitor voltage and current at $t = 0^+$, 0.5 s, 1 s, 1.5 s, 2 s, and 2.5 s using the universal time constant curves.

25. A 4.7-µF capacitor is charged to 43 volts. If a 39-kΩ resistor is then connected across the capacitor, what is its voltage 200 ms after the resistor is connected?

26. The initial voltage on the capacitor of Figure 11–54 is 55 V. The switch is closed at $t = 0$ s and opened 1 s later. Sketch v_C. What is the capacitor's voltage at $t = 3.25$ s?

27. For Figure 11–55, let $E = 200$ V, $R_2 = 1$ kΩ, and $C = 0.5$ µF. After the capacitor has fully charged in position 1, the switch is moved to position 2.

   a. What is the capacitor voltage immediately after the switch is moved to position 2? What is its current?

   b. What is the discharge time constant?

   c. Determine discharge equations for v_C and i_C.

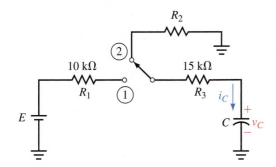

FIGURE 11–55

28. For Figure 11–55, C is fully charged before the switch is moved to discharge. Current just after it is moved is $i_C = -4$ mA and C takes 20 ms to discharge. If $E = 80$ V, what are R_2 and C?

11.5   More Complex Circuits

29. The capacitors of Figure 11–56 are uncharged. The switch is closed at $t = 0$. Determine the equation for v_C. Compute v_C at one time constant using the equation and the universal time constant curve. Compare answers.

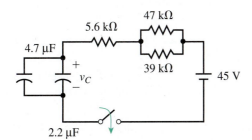

FIGURE 11–56

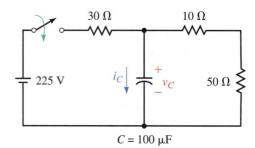

FIGURE 11–57

30. For Figure 11–57, the switch is closed at $t = 0$. Given $V_0 = 0$ V.
   a. Determine the equations for v_C and i_C.
   b. Compute the capacitor voltage at $t = 0^+$ 2, 4, 6, 8, 10, and 12 ms.
   c. Repeat (b) for the capacitor current.
   d. Why does 225 V/30 Ω also yield $i(0^+)$?

31. Repeat Problem 30, parts (a) to (c) for the circuit of Figure 11–58.

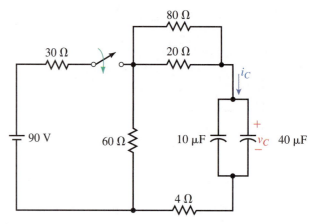

FIGURE 11–58

32. Consider again Figure 11–55. Suppose $E = 80$ V, $R_2 = 25$ kΩ, and $C = 0.5$ μF:
   a. What is the charge time constant?
   b. What is the discharge time constant?
   c. With the capacitor initially discharged, move the switch to position 1 and determine equations for v_C and i_C during charge.
   d. Move the switch to the discharge position. How long does it take for the capacitor to discharge?
   e. Sketch v_C and i_C from the time the switch is placed in charge to the time that the capacitor is fully discharged. Assume the switch is in the charge position for 80 ms.

33. For the circuit of Figure 11–55, the capacitor is initially uncharged. The switch is first moved to charge, then to discharge, yielding the current shown in Figure 11–59. The capacitor fully charges in 12.5 s. Determine E, R_2, and C.

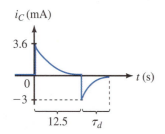

FIGURE 11–59

34. Refer to the circuit of Figure 11–60:
    a. What is the charge time constant?
    b. What is the discharge time constant?
    c. The switch is in position 2 and the capacitor is uncharged. Move the switch to position 1 and determine equations for v_C and i_C.
    d. After the capacitor has charged for two time constants, move the switch to position 2 and determine equations for v_C and i_C during discharge.
    e. Sketch v_C and i_C.

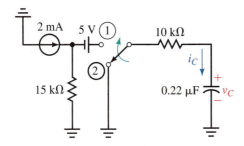

FIGURE 11–60

35. Determine the capacitor voltages and the source current for the circuit of Figure 11–61 after it has reached steady state.

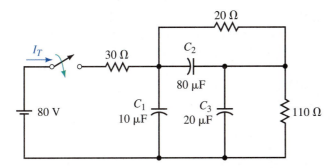

FIGURE 11–61

36. A black box containing dc sources and resistors has open-circuit voltage of 45 volts as in Figure 11–62(a). When the output is shorted as in (b), the short-circuit current is 1.5 mA. A switch and an uncharged 500-μF capacitor are connected as in (c). Determine the capacitor voltage and current 25 s after the switch is closed.

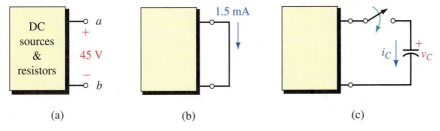

| (a) | (b) | (c) |

FIGURE 11–62

11.6 An *RC* Timing Application

37. For the alarm circuit of Figure 11–32, if the input from the sensor is 5 V, $R = 750$ kΩ, and the alarm is activated at 15 s when $v_C = 3.8$ V, what is C?

38. For the alarm circuit of Figure 11–32, the input from the sensor is 5 V, $C = 47$ μF, and the alarm is activated when $v_C = 4.2$ V. Choose the nearest standard resistor value to achieve a delay of at least 37 s.

11.7 Pulse Response of *RC* Circuits

39. Consider the waveform of Figure 11–63.

    a. What is the period?

    b. What is the duty cycle?

    c. What is the PRR?

40. Repeat Problem 39 for the waveform of Figure 11–64.

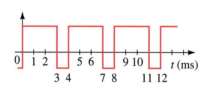

FIGURE 11–64

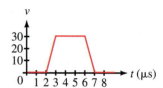

FIGURE 11–65

41. Determine the rise time, fall time, and pulse width for the pulse in Figure 11–65.

42. A single pulse is input to the circuit of Figure 11–66. Assuming that the capacitor is initially uncharged, sketch the output for each set of values below:

    a. $R = 2$ kΩ, $C = 1$ μF.

    b. $R = 2$ kΩ, $C = 0.1$ μF.

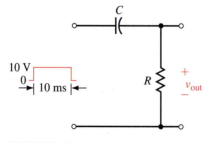

FIGURE 11–66

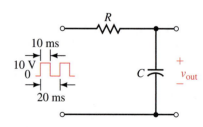

FIGURE 11–67

43. A step is applied to the circuit of Figure 11–67. If $R = 150$ Ω and $C = 20$ pF, estimate the rise time of the output voltage.

44. A pulse train is input to the circuit of Figure 11–67. Assuming that the capacitor is initially uncharged, sketch the output for each set of values below after the circuit has reached steady state:

    a. $R = 2$ kΩ, $C = 0.1$ μF.

    b. $R = 20$ kΩ, $C = 1.0$ μF.

FIGURE 11–63

(in left margin, with waveform graph)

t (μs) axis: 0 1 2 3 4 5 6 7 8 9 10 12 14 16

11.8 Transient Analysis Using Computers

45. Graph capacitor voltage for the circuit of Figure 11–2 with $E = -25$ V, $R = 40$ Ω, $V_0 = 0$ V, and $C = 400$ μF (see Note 1). Scale values from the plot at $t = 20$ ms using the cursor. Compare to the results you get using Equation 11–3 or the curve of Figure 11–15(a). ◀ MULTISIM

46. Obtain a plot of voltage versus time across R for the circuit of Figure 11–68 (see Note 1). Assume an initially uncharged capacitor. Use the cursor to read voltage $t = 50$ ms and use Ohm's law to compute current. Compare to the values determined analytically. Repeat if $V_0 = 100$ V. ◀ MULTISIM

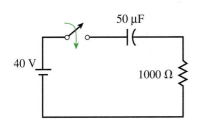

FIGURE 11–68

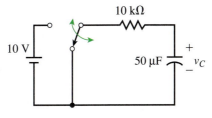

FIGURE 11–69

47. Consider Figure 11–58. Using MultiSIM and assuming zero initial conditions for both capacitors, do the following (see Note 3).
    a. Plot capacitor voltage for the circuit of Figure 11–58 and find v_C at $t = 4$ ms.
    b. Determine the current in the 4-Ω resistor at $t = 3.5$ ms.

48. The switch of Figure 11–69 is initially in the discharge position and the capacitor is uncharged. Move the switch to charge for 1 s, then to discharge where it remains. Solve for v_C (see Note 2). (Hint: Use MultiSIM's time delay (TD) switch. Double click it and set TON to 1 s and TOFF to 0. This will cause the switch to move to the charge position for 1 s, then return to discharge where it stays.) With the cursor, determine the peak voltage. Using Equation 11–2, compute v_C at $t = 1$ s and compare. ◀ MULTISIM

49. Using PSpice, graph capacitor voltage and current for a charging circuit with $E = -25$ V, $R = 40$ Ω, $V_0 = 0$ V, and $C = 400$ μF. Scale values from the plot using the cursor. Compare to the results you get using Equations 11–3 and 11–5 or the curves of Figure 11–15. ◀ CADENCE

50. Repeat the problem of Question 46 using PSpice. Plot both voltage and current. ◀ CADENCE

51. The switch of Figure 11–70 is closed at $t = 0$ s. Using PSpice, plot voltage and current waveforms. Use the cursor to determine v_C and i_C at $t = 10$ ms. ◀ CADENCE

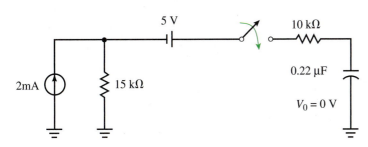

FIGURE 11–70

NOTES . . .

1. This problem (like most MultiSIM transient problems) does not require a switch since MultiSIM automatically initiates a transient solution when you click Simulate. However, some problems (such as Problem 48) do.

2. Omit this problem if you are using MultiSIM 2001 as its time delay switch may not work properly.

3. Since MultiSIM plots only voltages with respect to ground, you must place the ground in the position that you want to use as the reference point. This means that in some problems, you may have to move the ground and re-simulate to get some of the answers.

◀ CADENCE

52. Using PSpice, redo Example 11–17 with the switch in the charge position for 0.5 s and everything else the same. With your calculator, compute v_C and i_C at 0.5 s and compare them to the PSpice plots. Repeat for i_C just after moving the switch to the discharge position.

◀ CADENCE

53. Use PSpice to solve for voltages and currents in the circuit of Figure 11–61. From this, determine the final (steady state) voltages and currents and compare them to the answers of Problem 35.

✓ **ANSWERS TO IN-PROCESS LEARNING CHECKS**

In-Process Learning Check 1

1. 0.2 A
2. a. 50 ms
   b.

t(ms)	i_c(mA)
0	50
25	30.3
50	18.4
75	11.2
100	6.8
250	0.337

3.

t(ms)	v_c(V)
0	0
25	71.3
50	91.8
75	97.7
100	99.3
250	100

4. $40(1 - e^{-5t})$ V, $200e^{-5t}$ mA, 38.0 V, 9.96 mA
5. a. $v_C(0^+) = v_C(0^-) = 0$; b. $i_C(0^-) = 0$; $i_C(0^+) = 10$ mA; c. 100 V, 0 A
6. 80 V, 40 kΩ, 0.2 μF

In-Process Learning Check 2

1. a. $40 + 40e^{-5t}$ V. v_C starts at 80 V and decays exponentially to 40 V.

   b. There is no transient since initial value = final value.

   c. $40 - 100e^{-5t}$ V. v_C starts at -60 V and climbs exponentially to 40 V.

2. 0.1833 s

3. a. $100e^{-10t}$ V; $-10e^{-10t}$ mA; v_C starts at 100 V and decays to 0 in 0.5 s (i.e., 5 time constants); i_C starts at -10 mA and decays to 0 in 0.5 s

   b. $-100e^{-10t}$ V; $10e^{-10t}$ mA; v_C starts at -100 V and decays to 0 in 0.5 s (i.e., 5 time constants); i_C starts at 10 mA and decays to 0 in 0.5 s

4. a., b., and c. Same as Example 11–12

   d. $-30 + 37.14e^{-250t}$ V   e. $-37.14e^{-250t}$ mA

f. v_C (V)

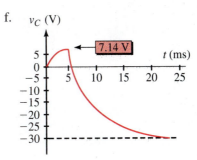

i_C (mA)

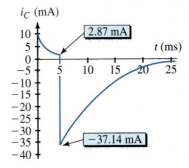

5. $6(1 - e^{-25t})$ V; $150e^{-25t}$ mA

■ OBJECTIVES

After studying this chapter, you will be able to

- represent magnetic fields using Faraday's flux concept,
- describe magnetic fields quantitatively in terms of flux and flux density,
- explain what magnetic circuits are and why they are used,
- determine magnetic field intensity or magnetic flux density from a *B-H* curve,
- solve series magnetic circuits,
- solve series-parallel magnetic circuits,
- compute the attractive force of an electromagnet,
- explain the domain theory of magnetism,
- describe the demagnetization process.

Magnetism and Magnetic Circuits

12

CHAPTER PREVIEW

Many common devices rely on magnetism. Familiar examples include computer disk drives, tape recorders, VCRs, transformers, motors, generators, and so on. To understand their operation, you need a knowledge of magnetism and magnetic circuit principles. In this chapter, we look at fundamentals of magnetism, relationships between electrical and magnetic quantities, magnetic circuit concepts, and methods of analysis. In Chapter 13, we look at electromagnetic induction and inductance, and in Chapter 24, we apply magnetic principles to the study of transformers. ■

PUTTING IT IN PERSPECTIVE

Magnetism and Electromagnetism

WHILE THE BASIC FACTS about magnetism have been known since ancient times, it was not until the early 1800s that the connection between electricity and magnetism was made and the foundations of modern electromagnetic theory laid down.

In 1819, Hans Christian Oersted, a Danish scientist, demonstrated that electricity and magnetism were related when he showed that a compass needle was deflected by a current-carrying conductor. The following year, Andre Ampere (1775–1836) showed that current-carrying conductors attract or repel each other just like magnets. However, it was Michael Faraday (recall Chapter 10) who developed our present concept of the magnetic field as a collection of flux lines in space that conceptually represent both the intensity and the direction of the field. It was this concept that led to an understanding of magnetism and the development of important practical devices such as the transformer and the electric generator.

In 1873, James Clerk Maxwell (see photo), a Scottish scientist, tied the then known theoretical and experimental concepts together and developed a unified theory of electromagnetism that predicted the existence of radio waves. Some 30 years later, Heinrich Hertz, a German physcist, showed experimentally that such waves existed, thus verifying Maxwell's theories and paving the way for modern radio and television. ■

12.1 The Nature of a Magnetic Field

Magnetism refers to the force that acts between magnets and magnetic materials. We know, for example, that magnets attract pieces of iron, deflect compass needles, attract or repel other magnets, and so on. This force acts at a distance and without the need for direct physical contact. The region where the force is felt is called the "field of the magnet" or simply, its **magnetic field.** Thus, *a magnetic field is a force field.*

Magnetic Flux

Faraday's flux concept (recall Putting It Into Perspective, Chapter 10) helps us visualize this field. Using Faraday's representation, magnetic fields are shown as lines in space. These lines, called **flux lines** or **lines of force,** show the direction and intensity of the field at all points. This is illustrated in Figure 12–1 for the field of a bar magnet. As indicated, the field is strongest at the **poles** of the magnet (where flux lines are most dense), its direction is from north (N) to south (S) external to the magnet, and flux lines never cross. The symbol for magnetic flux (Figure 12–1) is the Greek letter Φ (phi).

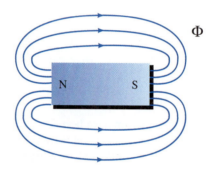

FIGURE 12–1   Field of a bar magnet. Magnetic flux is denoted by the symbol Φ.

Figure 12–2 shows what happens when two magnets are brought close together. In (a), unlike poles attract, and flux lines pass from one magnet to the other. In (b), like poles repel, and the flux lines are pushed back as indicated by the flattening of the field between the two magnets.

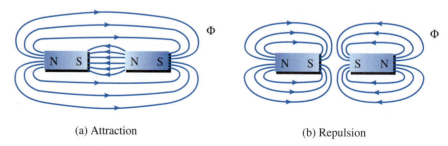

(a) Attraction   (b) Repulsion

FIGURE 12–2   Field patterns due to attraction and repulsion.

Ferromagnetic Materials

Magnetic materials (materials that are attracted by magnets such as iron, nickel, cobalt, and their alloys) are called **ferromagnetic** materials. Ferromagnetic materials provide an easy path for magnetic flux. This is illustrated in Figure 12–3 where the flux lines take the longer (but easier) path through the soft iron, rather than the shorter path that they would normally take (recall Figure 12–1). Note, however, that nonmagnetic materials (plastic, wood, glass, and so on) have no effect on the field.

Figure 12–4 shows an application of these principles. Part (a) shows a simplified representation of a loudspeaker, and part (b) shows expanded details of its magnetic field. The field is created by the permanent magnet, and the iron pole pieces guide the field and concentrate it in the gap where the speaker coil is placed. (For a description of how the speaker works, see Section 12.4.) Within the iron structure, the flux crowds together at sharp interior corners, spreads apart at exterior corners, and is essentially uniform elsewhere. This is characteristic of magnetic fields in iron.

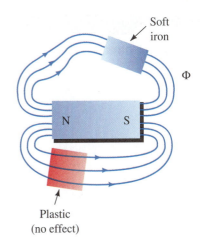

FIGURE 12–3 Magnetic field follows the longer (but easier) path through the iron. The plastic has no effect on the field.

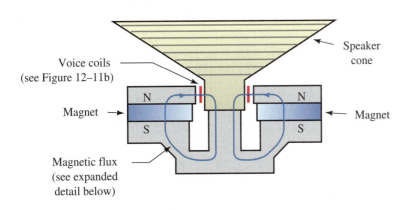

(a) Simplified representation of the magnetic field. Here, the complex field of (b) is represented symbolically by a single line

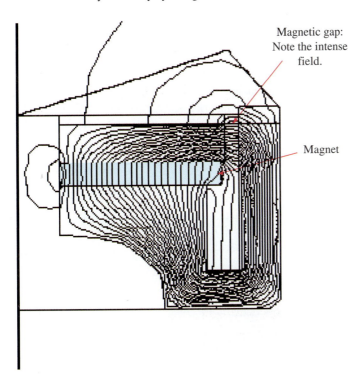

(b) Magnetic field pattern for the loud speaker. The field is symmetrical so only half the structure is shown. *(Courtesy JBL Professional)*

FIGURE 12–4 Magnetic circuit of a loudspeaker. The magnetic structure and voice coil are called a "speaker motor." The field is created by the permanent magnet.

12.2 Electromagnetism

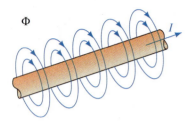

(a) Magnetic field produced
by current. Field is
proportional to *I*

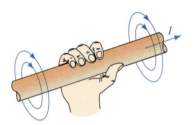

(b) Right-hand rule

FIGURE 12–5 Field about a current-carrying conductor. If current is reversed, the field remains concentric but the direction of the flux lines reverses.

Most applications of magnetism involve magnetic effects due to electric currents. We look first at some basic principles. Consider Figure 12–5. The current, *I*, creates a magnetic field that is concentric about the conductor, uniform along its length, and whose strength is directly proportional to *I*. Note the direction of the field. It may be remembered with the aid of the **right-hand rule.** As indicated in (b), imagine placing your right hand around the conductor with your thumb pointing in the direction of current. Your fingers then point in the direction of the field. If you reverse the direction of the current, the direction of the field reverses. If the conductor is wound into a coil, the fields of its individual turns combine, producing a resultant field as in Figure 12–6. The direction of the coil flux can also be remembered by means of a simple rule: curl the fingers of your right hand around the coil in the direction of the current and your thumb will point in the direction of the field. If the direction of the current is reversed, the field also reverses. Provided no ferromagnetic material is present, the strength of the coil's field is directly proportional to its current.

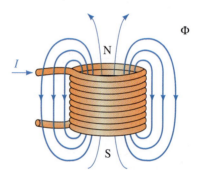

FIGURE 12–6 Field produced by a coil.

If the coil is wound on a ferromagnetic core as in Figure 12–7 (transformers are built this way), almost all flux is confined to the core, although a small amount (called stray or leakage flux) passes through the surrounding air. However, now that ferromagnetic material is present, the core flux is no longer proportional to current. The reason for this is discussed in Section 12.14.

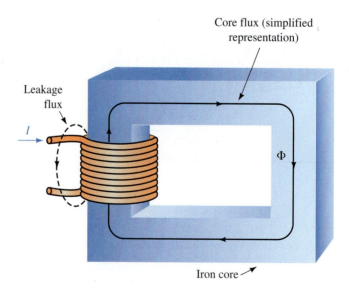

FIGURE 12–7 For ferromagnetic materials, most flux is confined to the core.

As noted in Figure 12–1, magnetic flux is represented by the symbol Φ. In the SI system, the unit of flux is the **weber** (Wb), in honor of pioneer researcher Wilhelm Eduard Weber, 1804–1891. However, we are often more interested in **flux density** B (i.e., flux per unit area) than in total flux Φ. Since flux Φ is measured in Wb and area A in m², flux density is measured as Wb/m². However, to honor Nikola Tesla (another early researcher, 1856–1943) the unit of flux density is called the **tesla** (T) where 1 T = 1 Wb/m². Flux density is found by dividing the total flux passing perpendicularly through an area by the size of the area, Figure 12–8. That is,

$$B = \frac{\Phi}{A} \quad \text{(tesla, T)} \qquad (12\text{–}1)$$

Thus, if $\Phi = 600\ \mu\text{Wb}$ of flux pass perpendicularly through an area $A = 20 \times 10^{-4}\ \text{m}^2$, the flux density is $B = (600 \times 10^{-6}\ \text{Wb})/(20 \times 10^{-4}\ \text{m}^2) = 0.3\ \text{T}$. The greater the flux density, the stronger the field.

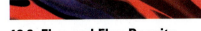

12.3 Flux and Flux Density

NOTES . . .

Although the weber appears at this point to be just an abstract quantity, it can in fact be linked to the familiar electrical system of units. For example if you pass a conductor through a magnetic field such that the conductor cuts the flux lines at the rate of 1 Wb per second, the voltage induced is 1 V.

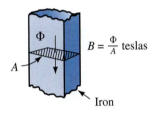

FIGURE 12–8   Concept of flux density. 1 T = 1 Wb/m².

For the magnetic core of Figure 12–9, the flux density at cross section 1 is $B_1 = 0.4$ T. Determine B_2.

EXAMPLE 12–1

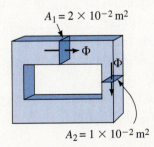

FIGURE 12–9

Solution $\Phi = B_1 \times A_1 = (0.4\ \text{T})(2 \times 10^{-2}\ \text{m}^2) = 0.8 \times 10^{-2}\ \text{Wb}$. Since all flux is confined to the core, the flux at cross section 2 is the same as at cross section 1. Therefore,

$$B_2 = \Phi/A_2 = (0.8 \times 10^{-2}\ \text{Wb})/(1 \times 10^{-2}\ \text{m}^2) = 0.8\ \text{T}$$

1. Refer to the core of Figure 12–8:
   a. If A is 2 cm $\times$ 2.5 cm and $B = 0.4$ T, compute Φ in webers.
   b. If A is 0.5 inch by 0.8 inch and $B = 0.35$ T, compute Φ in webers.
2. In Figure 12–9, if $\Phi = 100 \times 10^{-4}$ Wb, compute B_1 and B_2.

Answers
1. a. 2×10^{-4} Wb; b. 90.3 μWb; 2. 0.5 T; 1.0 T

To gain a feeling for the size of magnetic units, note that the strength of the earth's field is approximately 50 μT near the earth's surface, the field of a large generator or motor is on the order of 1 or 2 T, and the largest fields yet produced (using superconducting magnets) are on the order of 25 T.

Other systems of units (now largely superseded) are the CGS system and the English systems. In the CGS system, flux is measured in maxwells and flux density in gauss. In the English system, flux is measured in lines and flux density in lines per square inch. Conversion factors are given in Table 12–1. We use only the SI system in this book.

TABLE 12–1   Magnetic Units Conversion Table

System	Flux (Φ)	Flux Density (B)
SI	webers (Wb)	teslas (T) 1 T = 1 Wb/m^2
English	lines 1 Wb = 10^8 lines	lines/in^2 1 T = 6.452 $\times 10^4$ lines/in^2
CGS	maxwells 1 Wb = 10^8 maxwells	gauss 1 gauss = 1 maxwell/cm^2 1 T = 10^4 gauss

1. A magnetic field is a _____ field.
2. With Faraday's flux concept, the density of lines represents the _____ of the field and their direction represents the _____ of the field.
3. Three ferromagnetic materials are _____, _____, and _____.
4. The direction of a magnetic field is from _____ to _____ outside a magnet.
5. For Figures 12–5 and 12–6, if the direction of current is reversed, sketch what the fields look like.
6. If the core shown in Figure 12–7 is plastic, sketch what the field will look like.
7. Flux density B is defined as the ratio Φ/A, where A is the area (parallel, perpendicular) to Φ.
8. For Figure 12–9, if A_1 is 2 cm $\times$ 2.5 cm, B_1 is 0.5 T, and $B_2 = 0.25$ T, what is A_2?

12.4 Magnetic Circuits

Most practical applications of magnetism use magnetic structures to guide and shape magnetic fields by providing a well-defined path for flux. Such structures are called **magnetic circuits.** Magnetic circuits are found in motors, generators, computer disk drives, tape recorders, and so on. The speaker of Figure 12–4 illustrates the concept. It uses a powerful magnet to create flux and an iron circuit to guide the flux to the air gap to provide the intense field required by the voice coil. Note how effectively it does its job; almost the entire flux produced by the magnet is confined to the iron path with little leakage into the air.

A second example is shown in Figures 12–10 and 12–11. Tape recorders, VCRs, and computer disk drives all store information magnetically on iron oxide coated surfaces for later retrieval and use. The basic tape recorder scheme is shown symbolically in Figure 12–10. Sound picked up by a microphone is converted to an electrical signal, amplified, and the output applied to the record head. The record head is a small magnetic circuit. Current from the amplifier passes through its coil, creating a magnetic field that magnetizes the moving tape. The magnetized patterns on the tape correspond to the original sound input.

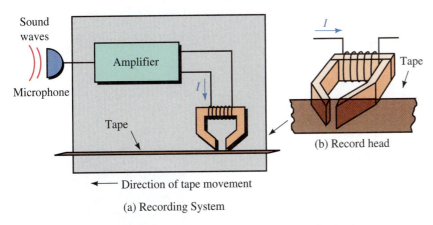

FIGURE 12–10   The recording head of a tape recorder is a magnetic circuit.

During playback the magnetized tape is passed by a playback head, as shown in Figure 12–11(a). Voltages induced in the playback coil are amplified and applied to a speaker. The speaker (b) utilizes a flexible cone to reproduce sound. A coil of fine wire attached at the apex of this cone is placed in the field of the speaker air gap. Current from the amplifier passes through this coil, creating a

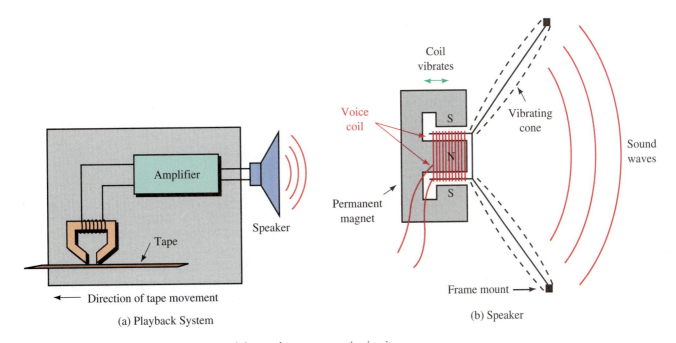

FIGURE 12–11   Both the playback system and the speaker use magnetic circuits.

varying field that interacts with the fixed field of the speaker magnet, creating force that causes the cone to vibrate. Since these vibrations correspond to the magnetized patterns on the tape, the original sound is reproduced. Computer disk drives use a similar record/playback scheme; in this case, binary logic patterns are stored and retrieved rather than music and voice.

12.5 Air Gaps, Fringing, and Laminated Cores

Consider again Figure 12–10. Note the gap in the record head. This is called an air gap. Most practical magnetic circuits have air gaps that are critical to the circuit's operation. At gaps, **fringing** occurs, causing a decrease in flux density in the gap as in Figure 12–12(a). For short gaps, fringing can usually be neglected. Alternatively, correction can be made by increasing each cross-sectional dimension of the gap by the size of the gap to approximate the decrease in flux density.

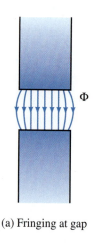

(a) Fringing at gap

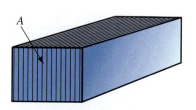

(b) Laminated section. Effective magnetic area is less than the physical area

FIGURE 12–12   Fringing and laminations.

EXAMPLE 12–2

A core with cross-sectional dimensions of 2.5 cm by 3 cm has a 0.1-mm gap. If flux density $B = 0.86$ T in the iron, what is the approximate uncorrected and corrected flux density in the gap?

Solution   Neglecting fringing, gap area is the same as the core area. Thus, $B_g \approx 0.86$ T. Correcting for fringing yields

$$\Phi = BA = (0.86 \text{ T})(2.5 \times 10^{-2} \text{ m})(3 \times 10^{-2} \text{ m}) = 0.645 \text{ mWb}$$

$$A_g \simeq (2.51 \times 10^{-2} \text{ m})(3.01 \times 10^{-2} \text{ m}) = 7.555 \times 10^{-4} \text{ m}^2$$

Thus,

$$B_g \simeq 0.645 \text{ mWb}/7.555 \times 10^{-4} \text{ m}^2 = 0.854 \text{ T}$$

Now consider laminations. Many practical magnetic circuits (such as transformers) use thin sheets of stacked iron or steel as in Figure 12–12(b). Since the core is not a solid block, its effective cross-sectional area (i.e., the actual area of iron) is less than its physical area. A **stacking factor,** defined as the ratio of the actual area of ferrous material to the physical area of the core, permits you to determine the core's effective area.

A laminated section of core has cross-sectional dimensions of 0.03 m by 0.05 m and a stacking factor of 0.9.

a. What is the effective area of the core?

b. Given $\Phi = 1.4 \times 10^{-3}$ Wb, what is the flux density, B?

Answers
 a. 1.35×10^{-3} m²; b. 1.04 T

Magnetic circuits may have sections of different materials. For example, the circuit of Figure 12–13 has sections of cast iron, sheet steel, and an air gap. For this circuit, flux Φ is the same in all sections. Such a circuit is called a **series magnetic circuit.** Although the flux is the same in all sections, the flux density in each section may vary, depending on its effective cross-sectional area as you saw earlier.

A circuit may also have elements in parallel (Figure 12–14). At each junction, the sum of fluxes entering is equal to the sum leaving. This is the counterpart of Kirchhoff's current law. Thus, for Figure 12–14, if $\Phi_1 = 25$ μWb and $\Phi_2 = 15$ μWb, then $\Phi_3 = 10$ μWb. For cores that are symmetrical about the center leg, $\Phi_2 = \Phi_3$.

12.6 Series Elements and Parallel Elements

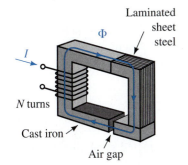

FIGURE 12–13   Series magnetic circuit. Flux Φ is the same throughout.

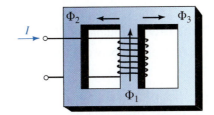

FIGURE 12–14   The sum of the flux entering a junction equals the sum leaving. Here, $\Phi_1 = \Phi_2 + \Phi_3$.

1. Why is the flux density in each section of Figure 12–13 different?

2. For Figure 12–13, $\Phi = 1.32$ mWb, the cross section of the core is 3 cm by 4 cm, the laminated section has a stacking factor of 0.8, and the gap is 1 mm. Determine the flux density in each section, taking fringing into account.

3. If the core of Figure 12–14 is symmetrical about its center leg, $B_1 = 0.4$ T, and the cross-sectional area of the center leg is 25 cm², what are Φ_2 and Φ_3?

 IN-PROCESS
LEARNING CHECK 2

(Answers are at the end of the chapter.)

12.7 Magnetic Circuits with DC Excitation

We now look at the analysis of magnetic circuits with dc excitation. There are two basic problems to consider: (1) given the flux, to determine the current required to produce it and (2) given the current, to compute the flux produced. To help visualize how to solve such problems, we first establish an analogy between magnetic circuits and electric circuits.

MMF: The Source of Magnetic Flux

Current through a coil creates magnetic flux. The greater the current or the greater the number of turns, the greater will be the flux. This flux-producing ability of a coil is called its **magnetomotive force** (mmf) and is measured in **ampere-turns.** Given the symbol $\mathscr{F}$, it is defined as

$$\mathscr{F} = NI \quad \text{(ampere-turns, At)} \tag{12–2}$$

Thus, a coil with 100 turns and 2.5 amps will have an mmf of 250 ampere-turns, while a coil with 500 turns and 4 amps will have an mmf of 2000 ampere-turns.

Reluctance, $\mathfrak{R}$: Opposition to Magnetic Flux

Flux in a magnetic circuit also depends on the opposition that the circuit presents to it. Termed **reluctance,** this opposition depends on the dimensions of the core and the material of which it is made. Like the resistance of a wire, reluctance is directly proportional to length and inversely proportional to cross-sectional area. In equation form,

$$\mathfrak{R} = \frac{\ell}{\mu A} \quad \text{(At/Wb)} \tag{12–3}$$

where μ is a property of the core material called its **permeability** (discussed in Section 12.8). Permeability is a measure of how easy it is to establish flux in a material. Ferromagnetic materials have high permeability and hence low $\mathfrak{R}$, while nonmagnetic materials have low permeability and high $\mathfrak{R}$.

Ohm's Law for Magnetic Circuits

The relationship between flux, mmf, and reluctance is

$$\Phi = \mathscr{F}/\mathfrak{R} \quad \text{(Wb)} \tag{12–4}$$

This relationship is similar to Ohm's law and is depicted symbolically in Figure 12–15. (Remember however that flux, unlike electric current, does not flow—see note in Section 12.1.)

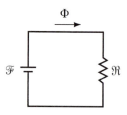

FIGURE 12–15 Electric circuit analogy of a magnetic circuit. $\Phi = \mathscr{F}/\mathfrak{R}$.

EXAMPLE 12–3

For Figure 12–16, if the reluctance of the magnetic circuit is $\mathfrak{R} = 12 \times 10^4$ At/Wb, what is the flux in the circuit?

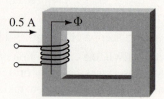

FIGURE 12–16   N = 300 turns

Solution

$$\mathscr{F} = NI = (300)(0.5 \text{ A}) = 150 \text{ At}$$
$$\Phi = \mathscr{F}/\mathfrak{R} = (150 \text{ At})/(12 \times 10^4 \text{ At/Wb}) = 12.5 \times 10^{-4} \text{ Wb}$$

In Example 12–3, we assumed that the reluctance of the core was constant. This is only approximately true under certain conditions. In general, it is not true, since $\mathfrak{R}$ is a function of flux density. Thus Equation 12–4 is not really very useful, since for ferromagnetic material, $\mathfrak{R}$ depends on flux, the very quantity that you are trying to find. The main use of Equations 12–3 and 12–4 is to provide an analogy between electric and magnetic circuit analysis.

We now look at a more practical approach to analyzing magnetic circuits. First, we require a quantity called **magnetic field intensity,** H (also known as **magnetizing force**). It is a measure of the mmf per unit length of a circuit.

To get at the idea, suppose you apply the same mmf (say 600 At) to two circuits with different path lengths (Figure 12–17). In (a), you have 600 ampere-turns of mmf to "drive" flux through 0.6 m of core; in (b), you have the same mmf but it is spread across only 0.15 m of path length. Thus the mmf per unit length in the second case is more intense. Based on this idea, one can define magnetic field intensity as the ratio of applied mmf to the length of path that it acts over. Thus,

$$H = \mathfrak{F}/\ell = NI/\ell \quad \text{(At/m)} \tag{12–5}$$

For the circuit of Figure 12–17(a), $H = 600 \text{ At}/0.6 \text{ m} = 1000 \text{ At/m}$, while for the circuit of (b), $H = 600 \text{ At}/0.15 \text{ m} = 4000 \text{ At/m}$. Thus, in (a) you have 1000 ampere-turns of "driving force" per meter of length to establish flux in the core, whereas in (b) you have four times as much. (However, you won't get four times as much flux, since the opposition to flux varies with the density of the flux.)

Rearranging Equation 12–5 yields an important result:

$$NI = H\ell \quad \text{(At)} \tag{12–6}$$

In an analogy with electric circuits (Figure 12–18), the NI product is an **mmf source,** while the $H\ell$ product is an **mmf drop.**

The Relationship between *B* and *H*

From Equation 12–5, you can see that magnetizing force, H, is a measure of the flux-producing ability of the coil (since it depends on NI). You also know that B is a measure of the resulting flux (since $B = \Phi/A$). Thus, B and H are related. The relationship is

$$B = \mu H \tag{12–7}$$

where μ is the permeability of the core (recall Equation 12–3).

It was stated earlier that permeability is a measure of how easy it is to establish flux in a material. To see why, note from Equation 12–7 that the larger the value of μ, the larger the flux density for a given H. However, H is proportional to current; therefore, the larger the value of μ, the larger the flux density for a given magnetizing current. From this, it follows that the larger the permeability, the more flux you get for a given magnetizing current.

In the SI system, μ has units of webers per ampere-turn-meter. The permeability of free space is $\mu_0 = 4\pi \times 10^{-7}$. For all practical purposes, the permeability of air and other nonmagnetic materials (e.g., plastic) is the same as for a vacuum. Thus, in air gaps,

$$B_g = \mu_0 H_g = 4\pi \times 10^{-7} \times H_g \tag{12–8}$$

Rearranging Equation 12–8 yields

$$H_g = \frac{B_g}{4\pi \times 10^{-7}} = 7.96 \times 10^5 B_g \quad \text{(At/m)} \tag{12–9}$$

12.8 Magnetic Field Intensity and Magnetization Curves

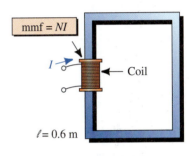

(a) A long path

(b) A short path

FIGURE 12–17 By definition, $H = $ mmf/length $= NI/\ell$.

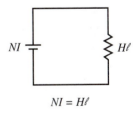

$$NI = H\ell$$

FIGURE 12–18 Circuit analogy, $H\ell$ model.

For Figure 12–16, the core cross section is 0.05 m × 0.08 m. If a gap is cut in the core and H in the gap is 3.6×10^5 At/m, what is the flux Φ in the core? Neglect fringing.

Answer
1.81 mWb

B-H Curves

For ferromagnetic materials, μ is not constant but varies with flux density and there is no easy way to compute it. In reality, however, it isn't μ that you are interested in: What you really want to know is, given B, what is H, and vice versa. A set of curves, called *B-H* or *magnetization* curves, provides this information. (These curves are obtained experimentally and are available in handbooks. A separate curve is required for each material.) Figure 12–19 shows typical curves for cast iron, cast steel, and sheet steel.

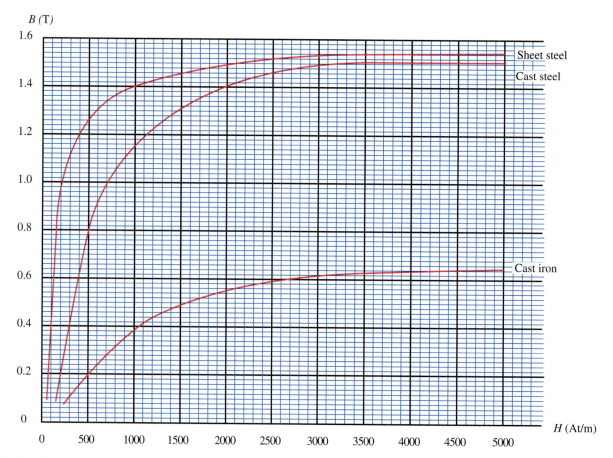

FIGURE 12–19   *B-H* curves for selected materials.

If $B = 1.4$ T for sheet steel, what is H?

EXAMPLE 12–4

Solution   Enter Figure 12–19 on the axis at $B = 1.4$ T, continue across until you encounter the curve for sheet steel, then read the corresponding value for H as indicated in Figure 12–20: $H = 1000$ At/m.

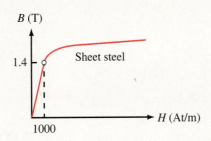

FIGURE 12–20   For sheet steel, H = 1000 At/m when $B = 1.4$ T.

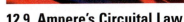

PRACTICE PROBLEMS 4

The cross section of a sheet steel core is 0.1 m $\times$ 0.1 m and its stacking factor is 0.93. If $\Phi = 13.5$ mWb, what is H?

Answer
1500 At/m

One of the key relationships in magnetic circuit theory is **Ampere's circuital law.** Ampere's law was determined experimentally and is a generalization of the relationship $\mathscr{F} = NI = H\ell$ that we developed earlier. Ampere showed that the algebraic sum of mmfs around a closed loop in a magnetic circuit is zero, regardless of the number of sections or coils. That is,

12.9 Ampere's Circuital Law

$$\sum_{\circlearrowleft} \mathscr{F} = 0 \qquad \text{(12–10)}$$

This can be rewrittten as

$$\sum_{\circlearrowleft} NI = \sum_{\circlearrowleft} H\ell \quad \text{At} \qquad \text{(12–11)}$$

which states that the sum of applied mmfs around a closed loop equals the sum of the mmf drops. The summation is algebraic and terms are additive or subtractive, depending on the direction of flux and how the coils are wound. To illustrate, consider again Figure 12–13. Here,

$$NI - H_{\text{iron}}\ell_{\text{iron}} - H_{\text{steel}}\ell\text{steel} - H_g\ell_g = 0$$

Thus,

$$NI = \underbrace{H_{\text{iron}}\ell_{\text{iron}} + H_{\text{steel}}\ell_{\text{steel}} + H_g\ell_g}$$

$\underbrace{NI}_{\text{Applied mmf}} \qquad \text{sum of mmf drops}$

The path to use for the $H\ell$ terms is the mean (average) path.

You now have two magnetic circuit models (Figure 12–21). While the reluctance model (a) is not very useful for solving problems, it helps relate magnetic circuit problems to familiar electrical circuit concepts. The Ampere's law model, on the other hand, permits us to solve practical problems. We look at how to do this in the next section.

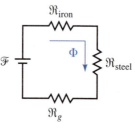

(a) Reluctance model

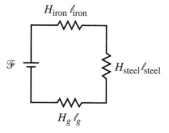

(b) Ampere's circuital law model

FIGURE 12–21   Two models for the magnetic circuit of Figure 12–13.

IN-PROCESS
LEARNING CHECK 3

(Answers are at the end of the chapter.)

1. If the mmf of a 200-turn coil is 700 At, the current in the coil is _____ amps.

2. For Figure 12–17, if $H = 3500$ At/m and $N = 1000$ turns, then for (a), I is _____ A, while for (b), I is _____ A.

3. For cast iron, if $B = 0.5$ T, then $H =$ _____ At/m.

4. A series circuit consists of one coil, a section of iron, a section of steel, and two air gaps (of different sizes). Draw the Ampere's law model.

5. Which is the correct answer for the circuit of Figure 12–22?

   a. Ampere's law around loop 1 yields ($NI = H_1\ell_1 + H_2\ell_2$, or $NI = H_1\ell_1 - H_2\ell_2$).

   b. Ampere's law around loop 2 yields ($0 = H_2\ell_2 + H_3\ell_3$, or $0 = H_2\ell_2 - H_3\ell_3$).

6. For the circuit of Figure 12–23, the length ℓ to use in Ampere's law is (0.36 m, 0.32 m, 0.28 m). Why?

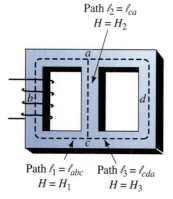

Path $\ell_2 = \ell_{ca}$
$H = H_2$

Path $\ell_1 = \ell_{abc}$ Path $\ell_3 = \ell_{cda}$
$H = H_1$ $H = H_3$

FIGURE 12–22

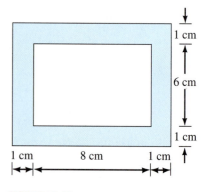

FIGURE 12–23

You now have the tools needed to solve basic magnetic circuit problems. We will begin with series circuits where Φ is known and we want to find the excitation to produce it. Problems of this type can be solved using four basic steps:

1. Compute *B* for each section using $B = \Phi/A$.
2. Determine *H* for each magnetic section from the *B-H* curves. Use $H_g = 7.96 \times 10^5 B_g$ for air gaps.
3. Compute *NI* using Ampere's circuital law.
4. Use the computed *NI* to determine coil current or turns as required. (Circuits with more than one coil are handled as in Example 12–6.)

Be sure to use the mean path through the circuit when applying Ampere's law. Unless directed otherwise, neglect fringing.

12.10 Series Magnetic Circuits: Given Φ, Find *NI*

◀ **Online Companion**

PRACTICAL NOTES . . .

Magnetic circuit analysis is not as precise as electric circuit analysis because (1) the assumption of uniform flux density breaks down at sharp corners as you saw in Figure 12–4, and (2) the *B-H* curve is a mean curve and has considerable uncertainty as discussed later (Section 12–14).

   Although the answers are approximate, they are adequate for most purposes.

If the core of Figure 12–24 is cast iron and $\Phi = 0.1 \times 10^{-3}$ Wb, what is the coil current?

EXAMPLE 12–5

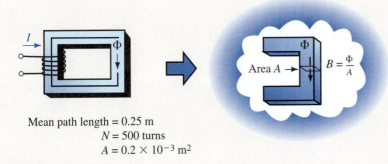

Mean path length = 0.25 m
N = 500 turns
A = 0.2 × 10⁻³ m²

FIGURE 12–24

Solution   Following the four steps outlined above:

1. The flux density is

$$B = \frac{\Phi}{A} = \frac{0.1 \times 10^{-3}}{0.2 \times 10^{-3}} = 0.5 \text{ T}$$

2. From the *B-H* curve (cast iron), Figure 12–19, H = 1550 At/m.

3. Apply Ampere's law. There is only one coil and one core section. Length = 0.25 m. Thus,

$$H\ell = 1550 \times 0.25 = 388 \text{ At} = NI$$

4. Divide by *N:*

$$I = 388/500 = 0.78 \text{ amps}$$

EXAMPLE 12–6

NOTES . . .

Since magnetic circuits are nonlinear, you cannot use superposition, that is, you cannot consider each coil of Figure 12–25 by itself, then sum the results. You must consider them simultaneously as we did in Example 12–6.

A second coil is added as shown in Figure 12–25. If $\Phi = 0.1 \times 10^{-3}$ Wb as before, but $I_1 = 1.5$ amps, what is I_2?

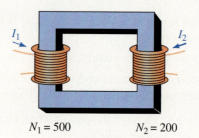

$N_1 = 500$ $N_2 = 200$

FIGURE 12–25

Solution From the previous example, you know that a current of 0.78 amps in coil 1 produces $\Phi = 0.1 \times 10^{-3}$ Wb. But you already have 1.5 amps in coil 1. Thus, coil 2 must be wound in opposition so that its mmf is subtractive. Applying Ampere's law yields $N_1I_1 - N_2I_2 = H\ell$. Hence,

$$(500)(1.5\,A) - 200I_2 = 388\ \text{At}$$

and so $I_2 = 1.8$ amps.

More Examples

If a magnetic circuit contains an air gap, add another element to the conceptual models (recall Figure 12–21). Since air represents a poor magnetic path, its reluctance will be high compared with that of iron. Recalling our analogy to electric circuits, this suggests that the mmf drop across the gap will be large compared with that of the iron. You can see this in the following example.

EXAMPLE 12–7

The core of Figure 12–24 has a 0.008-m gap cut as shown in Figure 12–26. Determine how much the current must increase to maintain the original core flux. Neglect fringing.

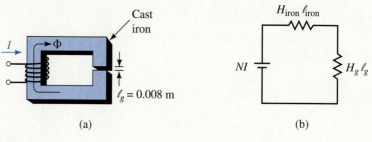

(a)   (b)

FIGURE 12–26

Solution

Iron

$\ell_{iron} = 0.25 - 0.008 = 0.242$ m. Since Φ does not change, *B* and *H* will be the same as before. Thus, $B_{iron} = 0.5$ T and $H_{iron} = 1550$ At/m.

Air Gap

B_g is the same as B_{iron}. Thus, $B_g = 0.5$ T and $H_g = 7.96 \times 10^5 B_g = 3.98 \times 10^5$ At/m.

Ampere's Law

$NI = H_{iron}\ell_{iron} + H_g\ell_g = (1550)(0.242) + (3.98 \times 10^5)(0.008) = 375 + 3184 = 3559$ At. Thus, $I = 3559/500 = 7.1$ amps. Note that the current had to increase from 0.78 amp to 7.1 amps in order to maintain the same flux, over a ninefold increase.

The laminated sheet steel section of Figure 12–27 has a stacking factor of 0.9. Compute the current required to establish a flux of $\Phi = 1.4 \times 10^{-4}$ Wb. Neglect fringing.

EXAMPLE 12–8

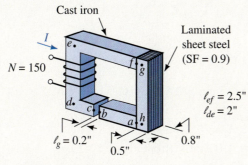

FIGURE 12–27

Cross section = 0.5" × 0.8" (all members)
$\Phi = 1.4 \times 10^{-4}$ Wb

Solution Convert all dimensions to metric.

Cast Iron

$$\ell_{iron} = \ell_{adef} - \ell_g = 2.5 + 2 + 2.5 - 0.2 = 6.8 \text{ in} = 0.173 \text{ m}$$
$$A_{iron} = (0.5 \text{ in})(0.8 \text{ in}) = 0.4 \text{ in}^2 = 0.258 \times 10^{-3} \text{ m}^2$$
$$B_{iron} = \Phi/A_{iron} = (1.4 \times 10^{-4})/(0.258 \times 10^{-3}) = 0.54 \text{ T}$$
$$H_{iron} = 1850 \text{ At/m} \quad \text{(from Figure 12–19)}$$

Sheet Steel

$$\ell_{steel} = \ell_{fg} + \ell_{gh} + \ell_{ha} = 0.25 + 2 + 0.25 = 2.5 \text{ in} = 6.35 \times 10^{-2} \text{ m}$$
$$A_{steel} = (0.9)(0.258 \times 10^{-3}) = 0.232 \times 10^{-3} \text{ m}^2$$
$$B_{steel} = \Phi/A_{steel} = (1.4 \times 10^{-4})/(0.232 \times 10^{-3}) = 0.60 \text{ T}$$
$$H_{steel} = 125 \text{ At/m} \quad \text{(from Figure 12–19)}$$

Air Gap

$$\ell_g = 0.2 \text{ in} = 5.08 \times 10^{-3} \text{ m}$$
$$B_g = B_{iron} = 0.54 \text{ T}$$
$$H_g = (7.96 \times 10^5)(0.54) = 4.3 \times 10^5 \text{ At/m}$$

Ampere's Law

$$NI = H_{iron}\ell_{iron} + H_{steel}\ell_{steel} + H_g\ell_g$$
$$= (1850)(0.173) + (125)(6.35 \times 10^{-2}) + (4.3 \times 10^5)(5.08 \times 10^{-3})$$
$$= 320 + 7.9 + 2184 = 2512 \text{ At}$$
$$I = 2512/N = 2512/150 = 16.7 \text{ amps}$$

EXAMPLE 12–9

Figure 12–28 shows a portion of a solenoid. Flux $\Phi = 4 \times 10^{-4}$ Wb when $I = 2.5$ amps. Find the number of turns on the coil.

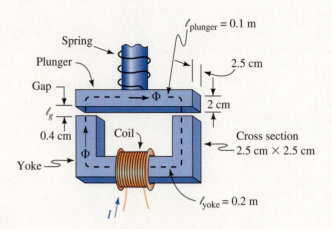

FIGURE 12–28   Solenoid. All parts are cast steel.

Solution

Yoke

$$A_{yoke} = 2.5 \text{ cm} \times 2.5 \text{ cm} = 6.25 \text{ cm}^2 = 6.25 \times 10^{-4} \text{ m}^2$$

$$B_{yoke} = \frac{\Phi}{A_{yoke}} = \frac{4 \times 10^{-4}}{6.25 \times 10^{-4}} = 0.64 \text{ T}$$

$$H_{yoke} = 410 \text{ At/m} \quad \text{(from Figure 12–19)}$$

Plunger

$$A_{plunger} = 2.0 \text{ cm} \times 2.5 \text{ cm} = 5.0 \text{ cm}^2 = 5.0 \times 10^{-4} \text{ m}^2$$

$$B_{plunger} = \frac{\Phi}{A_{plunger}} = \frac{4 \times 10^{-4}}{5.0 \times 10^{-4}} = 0.8 \text{ T}$$

$$H_{plunger} = 500 \text{ At/m} \quad \text{(from Figure 12–19)}$$

Air Gap

There are two identical gaps. For each,

$$B_g = B_{yoke} = 0.64 \text{ T}$$

Thus,

$$H_g = (7.96 \times 10^5)(0.64) = 5.09 \times 10^5 \text{ At/m}$$

The results are summarized in Table 12–2.

Ampere's Law

$$NI = H_{yoke}\ell_{yoke} + H_{plunger}\ell_{plunger} + 2H_g\ell_g = 82 + 50 + 2(2036) = 4204 \text{ At}$$

$$N = 4204/2.5 = 1682 \text{ turns}$$

TABLE 12–2

Material	Section	Length (m)	A (m²)	B (T)	H (At/m)	Hℓ (At)
Cast steel	yoke	0.2	6.25×10^{-4}	0.64	410	82
Cast steel	plunger	0.1	5×10^{-4}	0.8	500	50
Air	gap	0.4×10^{-2}	6.25×10^{-4}	0.64	5.09×10^5	2036

Series-parallel magnetic circuits are handled using the sum of fluxes principle (Figure 12–14) and Ampere's law.

The core of Figure 12–29 is cast steel. Determine the current to establish an air-gap flux $\Phi_g = 6 \times 10^{-3}$ Wb. Neglect fringing.

EXAMPLE 12–10

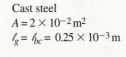

Cast steel
$A = 2 \times 10^{-2}\,\mathrm{m}^2$
$\ell_g = \ell_{bc} = 0.25 \times 10^{-3}\,\mathrm{m}$

$\Phi_g = \Phi_3$

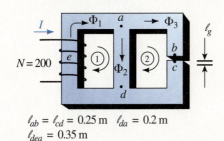

$\ell_{ab} = \ell_{cd} = 0.25\,\mathrm{m}$ $\quad \ell_{da} = 0.2\,\mathrm{m}$
$\ell_{dea} = 0.35\,\mathrm{m}$

FIGURE 12–29

Solution Consider each section in turn.

Air Gap

$$B_g = \Phi_g/A_g = (6 \times 10^{-3})/(2 \times 10^{-2}) = 0.3\,\mathrm{T}$$
$$H_g = (7.96 \times 10^5)(0.3) = 2.388 \times 10^5\,\mathrm{At/m}$$

Sections ab and cd

$$B_{ab} = B_{cd} = B_g = 0.3\,\mathrm{T}$$
$$H_{ab} = H_{cd} = 250\,\mathrm{At/m} \quad \text{(from Figure 12–19)}$$

Ampere's Law (Loop 2)

$\sum_{\circlearrowleft} NI = \sum_{\circlearrowleft} H\ell$. Since you are going opposite to flux in leg *da*, the corresponding term (i.e., $H_{da}\ell_{da}$) will be subtractive. Also, $NI = 0$ for loop 2. Thus,

$$0 = \sum_{\circlearrowleft\ \text{loop2}} H\ell$$
$$0 = H_{ab}\ell_{ab} + H_g\ell_g + H_{cd}\ell_{cd} - H_{da}\ell_{da}$$
$$= (250)(0.25) + (2.388 \times 10^5)(0.25 \times 10^{-3}) + (250)(0.25) - 0.2H_{da}$$
$$= 62.5 + 59.7 + 62.5 - 0.2H_{da} = 184.7 - 0.2H_{da}$$

Thus, $0.2H_{da} = 184.7$ and $H_{da} = 925\,\mathrm{At/m}$. From Figure 12–19, $B_{da} = 1.12$ T.

$$\Phi_2 = B_{da}A = 1.12 \times 0.02 = 2.24 \times 10^{-2}\,\mathrm{Wb}$$
$$\Phi_1 = \Phi_2 + \Phi_3 = 2.84 \times 10^{-2}\,\mathrm{Wb}.$$
$$B_{dea} = \Phi_1/A = (2.84 \times 10^{-2})/0.02 = 1.42\,\mathrm{T}$$
$$H_{dea} = 2125\,\mathrm{At/m} \quad \text{(from Figure 12–19)}$$

Ampere's Law (Loop 1)

$$NI = H_{dea}\ell_{dea} + H_{ad}\ell_{ad} = (2125)(0.35) + 184.7 = 929\,\mathrm{At}$$
$$I = 929/200 = 4.65\,\mathrm{A}$$

PRACTICE PROBLEMS 5

The cast-iron core of Figure 12–30 is symmetrical. Determine current *I*. Hint: To find *NI*, you can write Ampere's law around either loop. Be sure to make use of symmetry.

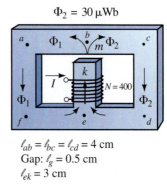

$\Phi_2 = 30\ \mu\text{Wb}$

$\ell_{ab} = \ell_{bc} = \ell_{cd} = 4$ cm
Gap: $\ell_g = 0.5$ cm
$\ell_{ek} = 3$ cm

Core dimensions: 1 cm × 1 cm

FIGURE 12–30

Answer
6.5 A

12.12 Series Magnetic Circuits: Given *NI*, Find Φ

In previous problems, you were given the flux and asked to find the current. We now look at the converse problem: given *NI*, find the resultant flux. For the special case of a core of one material and constant cross section (Example 12–11) this is straightforward. For all other cases, trial and error must be used.

EXAMPLE 12–11

For the circuit of Figure 12–31, *NI* = 250 At. Determine Φ.

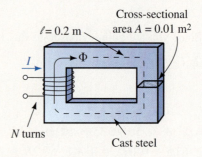

$\ell = 0.2$ m

Cross-sectional
area $A = 0.01$ m^2

N turns

Cast steel

FIGURE 12–31

Solution $H\ell = NI$. Thus, $H = NI/\ell = 250/0.2 = 1250$ At/m. From the *B-H* curve of Figure 12–19, $B = 1.24$ T. Therefore, $\Phi = BA = 1.24 \times 0.01 = 1.24 \times 10^{-2}$ Wb.

For circuits with two or more sections, the process is not so simple. Before you can find *H* in any section, for example, you need to know the flux density. However, in order to determine flux density, you need to know *H*. Thus, neither Φ nor *H* can be found without knowing the other first.

To get around this problem, use trial and error. First, take a guess at the value for flux, compute *NI* using the 4-step procedure of Section 12.10, then compare the computed *NI* against the given *NI*. If they agree, the problem is solved. If they don't, adjust your guess and try again. Repeat the procedure until you are within 5% of the given *NI*.

The problem is how to come up with a good first guess. For circuits of the type of Figure 12–32, note that $NI = H_{steel}\ell_{steel} + H_g\ell_g$. As a first guess, assume that the reluctance of the air gap is so high that the full mmf drop appears across the gap. Thus, $NI \simeq H_g\ell_g$, and

$$H_g \simeq NI/\ell_g \qquad\qquad (12\text{–}12)$$

You can now apply Ampere's law to see how close to the given *NI* your trial guess is (see Notes).

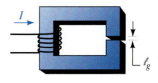

$$\mathcal{F} = H_{steel}\,\ell_{steel} + H_g\,\ell_g$$
$$\simeq H_g\,\ell_g \text{ if } H_g\,\ell_g \gg H_{steel}\,\ell_{steel}$$

FIGURE 12–32

The core of Figure 12–32 is cast steel, *NI* = 1100 At, the cross-sectional area everywhere is 0.0025 m², ℓ_g = 0.002 m, and ℓ_{steel} = 0.2 m. Determine the flux in the core.

EXAMPLE 12–12

Solution
Initial Guess
Assume that 90% of the mmf appears across the gap. The applied mmf is 1100 At. Ninety percent of this is 990 At. Thus, $H_g \simeq 0.9NI/\ell = 990/0.002 = 4.95 \times 10^5$ At/m and $B_g = \mu_0 H_g = (4\pi \times 10^{-7})(4.95 \times 10^5) = 0.62$ T.

Trial 1
Since the area of the steel is the same as that of the gap, the flux density is the same, neglecting fringing. Thus, $B_{steel} = B_g = 0.62$ T. From the *B-H* curve, $H_{steel} = 400$ At/m. Now apply Ampere's law:

$$NI = H_{steel}\ell_{steel} + H_g\ell_g = (400)(0.2) + (4.95 \times 10^5)(0.002)$$
$$= 80 + 990 = 1070 \text{ At}$$

This answer is 2.7% lower than the given *NI* of 1100 At and is therefore acceptable. Thus, Φ = *BA* = 0.62 × 0.0025 = 1.55 × 10⁻³ Wb.

NOTES . . .

Since you know that some of the mmf drop appears across the steel, you can start at less than 100% for the gap. Common sense and a bit of experience helps. The relative size of the mmf drops also depends on the core material. For cast iron, the percentage drop across the iron is larger than the percentage across a similar piece of sheet steel or cast steel. This is illustrated in Examples 12–12 and 12–13.

The initial guess in Example 12–12 yielded an acceptable answer on the first trial. (You are seldom this lucky.)

If the core of Example 12–12 is cast iron instead of steel, compute Φ.

EXAMPLE 12–13

Solution Because cast iron has a larger H for a given flux density (Figure 12–19), it will have a larger Hℓ drop and less will appear across the gap. Assume 75% across the gap.

Initial Guess

$$H_g \simeq 0.75\, NI/\ell = (0.75)(1100)/0.002 = 4.125 \times 10^5 \text{ At/m}.$$
$$B_g = \mu_0 H_g = (4\mu \times 10^{-7})(4.125 \times 10^5) = 0.52 \text{ T}.$$

Trial 1

$B_{iron} = B_g$. Thus, $B_{iron} = 0.52$ T. From the *B-H* curve, $H_{iron} = 1700$ At/m.

Ampere's Law

$$NI = H_{iron}\ell_{iron} + H_g\ell_g = (1700)(0.2) + (4.125 \times 10^5)(0.002)$$
$$= 340 + 825 = 1165 \text{ At} \quad \text{(high by 5.9\%)}$$

Trial 2
Reduce the guess by 5.9% to $B_{iron} = 0.49$ T. Thus, $H_{iron} = 1500$ At/m (from the *B-H* curve) and $H_g = 7.96 \times 10^5 B_g = 3.90 \times 10^5$ At/m.

Ampere's Law

$$NI = H_{iron}\ell_{iron} + H_g\ell_g = (1500)(0.2) + (3.90 \times 10^5)(0.002)$$
$$= 300 + 780 = 1080 \text{ At}$$

The error is now 1.82%, which is excellent. Thus, $\Phi = BA = (0.49)(2.5 \times 10^{-3}) = 1.23 \times 10^{-3}$ Wb. If the error had been larger than 5%, a third trial would have been needed.

12.13 Force Due to an Electromagnet

Electromagnets are used in relays, door bells, lifting magnets, and so on. For an electromagnetic relay as in Figure 12–33, it can be shown that the force created by the magnetic field is

$$F = \frac{B_g^2 A_g}{2\mu_0} \tag{12–13}$$

where B_g is flux density in the gap in teslas, A_g is gap area in square meters, and F is force in newtons.

EXAMPLE 12–14

Figure 12–33 shows a typical relay. The force due to the current-carrying coil pulls the pivoted arm against spring tension to close the contacts and energize the load. If the pole face is ¼ inch square and $\Phi = 0.5 \times 10^{-4}$ Wb, what is the pull on the armature in pounds?

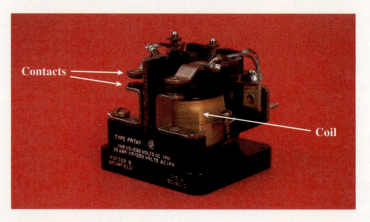

FIGURE 12–33 A typical relay.

Solution Convert to metric units.

$$A_g = (0.25 \text{ in})(0.25 \text{ in}) = 0.0625 \text{ in}^2 = 0.403 \times 10^{-4} \text{ m}^2$$

$$B_g = \Phi/A_g = (0.5 \times 10^{-4})/(0.403 \times 10^{-4}) = 1.24 \text{ T}$$

Thus,

$$F = \frac{B_g^2 A}{2\mu_0} = \frac{(1.24)^2(0.403 \times 10^{-4})}{2(4\pi \times 10^{-7})} = 24.66 \text{ N} = 5.54 \text{ lb}$$

Figure 12–34 shows how a relay is used in practice. When the switch is closed, the energized coil pulls the armature down. This closes the contacts and energizes the load. When the switch is opened, the spring pulls the contacts open again. Schemes like this use relatively small currents to control large loads. In addition, they permit remote control, as the relay and load may be a considerable distance from the actuating switch.

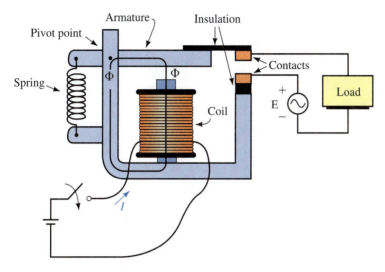

FIGURE 12–34   Controlling a load with a relay.

Magnetic properties are related to atomic structure. Each atom of a substance, for example, produces a tiny atomic-level magnetic field because its moving (i.e., orbiting) electrons constitute an atomic-level current and currents create magnetic fields. For nonmagnetic materials, these fields are randomly oriented and cancel. However, for ferromagnetic materials, the fields in small regions, called **domains** (Figure 12–35), do not cancel. (Domains are of microscopic size, but are large enough to hold from 10^{17} to 10^{21} atoms.) If the domain fields in a ferromagnetic material line up, the material is magnetized; if they are randomly oriented, the material is not magnetized.

Magnetizing a Specimen

A nonmagnetized specimen can be magnetized by making its domain fields line up. Figure 12–36 shows how this can be done. As current through the coil is increased, the field strength increases and more and more domains align themselves in the direction of the field. If the field is made strong enough, almost all domain fields line up and the material is said to be in **saturation** (the almost flat portion of the *B-H* curve). In saturation, the flux density increases slowly as magnetization intensity increases. This means that once the material is in saturation, you cannot magnetize it much further no matter how hard you try. Path 0-*a* traced from the nonmagnetized state to the saturated state is termed the **dc curve** or **normal magnetization curve.** (This is the *B-H* curve that you used earlier when you solved magnetic circuit problems.)

12.14 Properties of Magnetic Materials

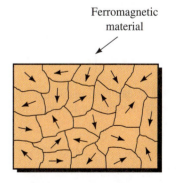

FIGURE 12–35   Random orientation of microscopic fields in a nonmagnetized ferromagnetic material. The small regions are called domains.

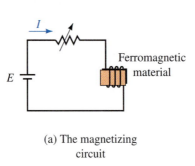

(a) The magnetizing
circuit

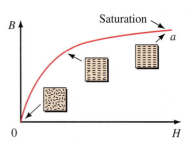

(b) Progressive change in the
domain orientations as the
field is increased. H is
proportional to current I

FIGURE 12–36   The magnetization process.

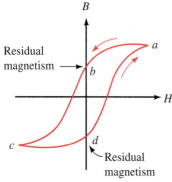

FIGURE 12–37   Hysteresis loop.

Hysteresis

If you now reduce the current to zero, you will find that the material still retains some magnetism, called **residual magnetism** (Figure 12–37, point b). If now you reverse the current, the flux reverses and the bottom part of the curve can be traced. By reversing the current again at d, the curve can be traced back to point a. The result is called a **hysteresis loop.** A major source of uncertainty in magnetic circuit behavior should now be apparent: As you can see, flux density depends not just on current, it also depends on which arm of the curve the sample is magnetized on, i.e., it depends on the circuit's past history. For this reason, B-H curves are the average of the two arms of the hysteresis loop, i.e., the dc curve of Figure 12–36.

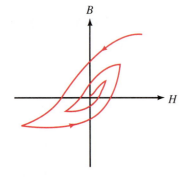

FIGURE 12–38   Demagnetization by successively shrinking the hysteresis loop.

The Demagnetization Process

As indicated above, simply turning the current off does not demagnetize ferromagnetic material. To demagnetize it, you must successively decrease its hysteresis loop to zero as in Figure 12–38. You can place the specimen inside a coil that is driven by a variable ac source and gradually decrease the coil current to zero, or you can use a fixed ac supply and gradually withdraw the specimen from the field. Such procedures are used by service personnel to "degauss" TV picture tubes.

12.15 Measuring Magnetic Fields

One way to measure magnetic field strength is to use the **Hall effect** (after E. H. Hall). The basic idea is illustrated in Figure 12–39. When a strip of semiconductor material such as indium arsenide is placed in a magnetic field, a small voltage, called the Hall voltage, V_H, appears across opposite edges. For a fixed current I, V_H is proportional to magnetic field strength B. Instruments using this principle are known as **Hall-effect gaussmeters.** To measure a magnetic field with such a meter, insert its probe into the field perpendicular to the field (Figure 12–40). The meter indicates flux density directly.

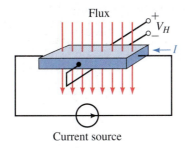

FIGURE 12–39   The Hall effect.

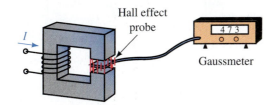

FIGURE 12–40   Magnetic field measurement.

PROBLEMS

12.3   Flux and Flux Density

1. Refer to Figure 12–41:
   a. Which area, A_1 or A_2, do you use to calculate flux density?
   b. If $\Phi = 28$ mWb, what is flux density in teslas?
2. For Figure 12–41, if $\Phi = 250$ μWb, $A_1 = 1.25$ in^2, and $A_2 = 2.0$ in^2, what is the flux density in the English system of units?
3. The toroid of Figure 12–42 has a circular cross section and $\Phi = 628$ μWb. If $r_1 = 8$ cm and $r_2 = 12$ cm, what is the flux density in teslas?
4. If r_1 of Figure 12–42 is 3.5 inches and r_2 is 4.5 inches, what is the flux density in the English system of units if $\Phi = 628$ μWb?

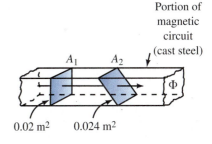

Portion of magnetic circuit (cast steel)

0.02 m^2 0.024 m^2

FIGURE 12–41

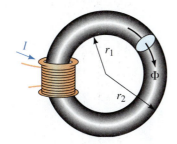

FIGURE 12–42

12.5   Air Gaps, Fringing, and Laminated Cores

5. If the section of core in Figure 12–43 is 0.025 m by 0.04 m, has a stacking factor of 0.85, and $B = 1.45$ T, what is Φ in webers?

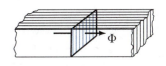

FIGURE 12–43

12.6 Series Elements and Parallel Elements

6. For the iron core of Figure 12–44, flux density $B_2 = 0.6$ T. Compute B_1 and B_3.

7. For the section of iron core of Figure 12–45, if $\Phi_1 = 12$ mWb and $\Phi_3 = 2$ mWb, what is B_2?

8. For the section of iron core of Figure 12–45, if $B_1 = 0.8$ T and $B_2 = 0.6$ T, what is B_3?

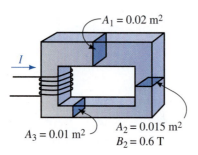

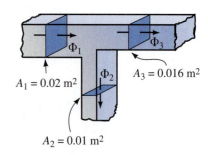

FIGURE 12–44   **FIGURE 12–45**

12.8 Magnetic Field Intensity and Magnetization Curves

9. A core with dimensions 2 cm $\times$ 3 cm has a magnetic intensity of 1200 At/m. What is Φ if the core is cast iron? If it is cast steel? If it is sheet steel with SF = 0.94?

10. Figure 12–46 shows the two electric circuit equivalents for magnetic circuits. Show that μ in $\mathfrak{R} = \ell/\mu A$ is the same as μ in $B = \mu H$.

11. Consider again Figure 12–42. If $I = 10$ A, $N = 40$ turns, $r_1 = 5$ cm, and $r_2 = 7$ cm, what is H in ampere-turns per meter?

12.9 Ampere's Circuital Law

12. Let H_1 and ℓ_1 be the magnetizing force and path length respectively, where flux Φ_1 exists in Figure 12–47 and similarly for Φ_2 and Φ_3. Write Ampere's law around each of the windows.

13. Assume that a coil N_2 carrying current I_2 is added on leg 3 of the core shown in Figure 12–47 and that it produces flux directed upward. Assume, however, that the net flux in leg 3 is still downward. Write the Ampere's law equations for this case.

14. Repeat Problem 13 if the net flux in leg 3 is upward but the directions of Φ_1 and Φ_2 remain as in Figure 12–47.

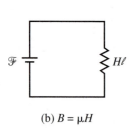

(a) $\mathfrak{R} = \dfrac{\ell}{\mu A}$

(b) $B = \mu H$

FIGURE 12–46 $\mathcal{F} = NI$.

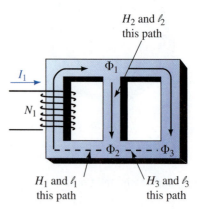

FIGURE 12–47

12.10   Series Magnetic Circuits: Given Φ, Find *NI*

15. Find the current *I* in Figure 12–48 if $\Phi = 0.16$ mWb.

16. Let everything be the same as in Problem 15 except that the cast steel portion is replaced with laminated sheet steel with a stacking factor of 0.85.

17. A gap of 0.5 mm is cut in the cast steel portion of the core in Figure 12–48. Find the current for $\Phi = 0.128$ mWb. Neglect fringing.

18. Two gaps, each 1 mm, are cut in the circuit of Figure 12–48, one in the cast steel portion and the other in the cast iron portion. Determine current for $\Phi = 0.128$ mWb. Neglect fringing.

19. The cast iron core of Figure 12–49 measures 1 cm × 1.5 cm, $\ell_g = 0.3$ mm, the air gap flux density is 0.426 T and $N = 600$ turns. The end pieces are half circles. Taking into account fringing, find current *I*.

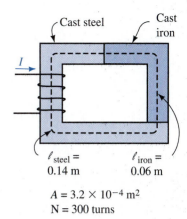

$A = 3.2 \times 10^{-4}$ m^2
N = 300 turns

FIGURE 12–48

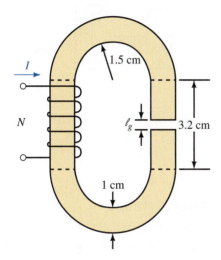

FIGURE 12–49

20. For the circuit of Figure 12–50, $\Phi = 141$ μWb and $N = 400$ turns. The bottom member is sheet steel with a stacking factor of 0.94, while the remainder is cast steel. All pieces are 1 cm × 1 cm. The length of the cast steel path is 16 cm. Find current *I*.

21. For the circuit of Figure 12–51, $\Phi = 30$ μWb and $N = 2000$ turns. Neglecting fringing, find current *I*.

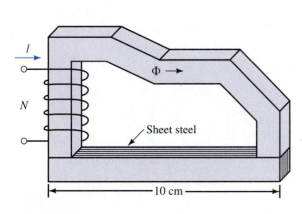

FIGURE 12–50

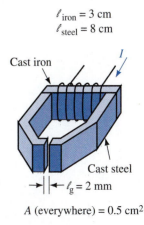

$\ell_{iron} = 3$ cm
$\ell_{steel} = 8$ cm

A (everywhere) = 0.5 cm^2

FIGURE 12–51

22. For the circuit of Figure 12–52, $\Phi = 25{,}000$ lines. The stacking factor for the sheet steel portion is 0.95. Find current I.

23. A second coil of 450 turns with $I_2 = 4$ amps is wound on the cast steel portion of Figure 12–52. Its flux is in opposition to the flux produced by the original coil. The resulting flux is 35 000 lines in the counterclockwise direction. Find the current I_1.

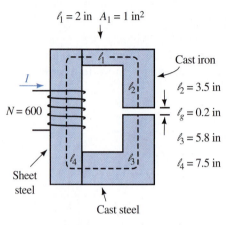

$\ell_1 = 2$ in $A_1 = 1$ in²

$N = 600$

Cast iron
$\ell_2 = 3.5$ in
$\ell_g = 0.2$ in
$\ell_3 = 5.8$ in
$\ell_4 = 7.5$ in

Sheet steel

Cast steel

Area of all sections (except A_1) = 2 in²

FIGURE 12–52

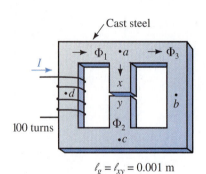

Cast steel

Φ_1 •a Φ_3

I

•d

x

y

•b

100 turns

Φ_2

•c

$\ell_g = \ell_{xy} = 0.001$ m
$\ell_{abc} = 0.14$ m
$\ell_{cda} = 0.16$ m
$\ell_{ax} = \ell_{cy} = 0.039$ m
$A = 4$ cm² everywhere

FIGURE 12–53

12.11   Series-Parallel Magnetic Circuits

24. For Figure 12–53, if $\Phi_g = 80$ μWb, find I.

25. If the circuit of Figure 12–53 has no gap and $\Phi_3 = 0.2$ mWb, find I.

12.12   Series Magnetic Circuits: Given *NI*, Find Φ

26. A cast steel magnetic circuit with $N = 2500$ turns, $I = 200$ mA, and a cross-sectional area of 0.02 m² has an air gap of 0.00254 m. Assuming 90% of the mmf appears across the gap, estimate the flux in the core.

27. If $NI = 644$ At for the cast steel core of Figure 12–54, find the flux, Φ.

28. A gap $\ell = 0.004$ m is cut in the core of Figure 12–54. Everything else remains the same. Find the flux, Φ.

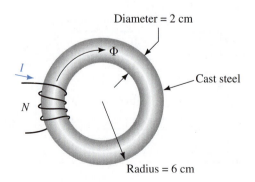

Diameter = 2 cm

Φ

Cast steel

I

N

Radius = 6 cm

FIGURE 12–54

12.13   Force Due to an Electromagnet

29. For the relay of Figure 12–34, if the pole face is 2 cm by 2.5 cm and a force of 2 pounds is required to close the gap, what flux (in webers) is needed?

30. For the solenoid of Figure 12–28, $\Phi = 4 \times 10^{-4}$ Wb. Find the force of attraction on the plunger in newtons and in pounds.

✓   ## ANSWERS TO IN-PROCESS LEARNING CHECKS

In-Process Learning Check 1

1. Force
2. Strength, direction
3. Iron, nickel, cobalt
4. North, south
5. Same except direction of flux reversed
6. Same as Figure 12–6, since plastic does not affect the field.
7. Perpendicular
8. 10 cm^2

In-Process Learning Check 2

1. While flux is the same throughout, the effective area of each section differs.

2. $B_{\text{iron}} = 1.1$ T; $B_{\text{steel}} = 1.38$ T; $B_g = 1.04$ T
3. $\Phi_2 = \Phi_3 = 0.5$ mWb

In-Process Learning Check 3

1. 3.5 A
2. a. 2.1 A;

   b. 0.525 A
3. 1550 At/m
4. Same as Figure 12–21(b) except add $H_{g_2}\ell_{g_2}$.
5. a. $NI = H_1\ell_1 + H_2\ell_2$;

   b. $0 = H_2\ell_2 - H_3\ell_3$
6. 0.32 m; use the mean path length.

■ KEY TERMS

Air-Core Coils

Back Voltage

Choke

Counter EMF

Faraday's Law

Flux Linkage

Henry

Induced Voltage

Inductance

Inductor

Iron-Core Coils

Lenz's Law

Reactors

Self-Inductance

Stray Inductance

Stray or Parasitic Capacitance

■ OUTLINE

Electromagnetic Induction

Induced Voltage and Induction

Self-Inductance

Computing Induced Voltage

Inductances in Series and Parallel

Practical Considerations

Inductance and Steady State DC

Energy Stored by an Inductance

Inductor Troubleshooting Hints

■ OBJECTIVES

After studying this chapter, you will be able to

- describe what an inductor is and what its effect on circuit operation is,
- explain Faraday's law and Lenz's law,
- compute induced voltage using Faraday's law,
- define inductance,
- compute voltage across an inductance,
- compute inductance for series and parallel configurations,
- compute inductor voltages and currents for steady state dc excitation,
- compute energy stored in an inductance,
- describe common inductor problems and how to test for them.

Inductance and Inductors

13

In this chapter, we look at self-inductance (usually just called inductance) and inductors. To get at the idea, note that when current flows through a conductor, it creates a magnetic field in space surrounding the conductor. As you will see, this magnetic field affects circuit operation. To describe this effect, we use the circuit parameter **inductance.** Inductance is due entirely to the magnetic field created by current, and its effect is to oppose any change in the current that created the field—thus, in a sense, inductance can be likened to inertia in a mechanical system. The advantage of using inductance is that you do not have to deal with the magnetic field directly—instead you can work entirely in terms of circuit quantities, voltage, current, and inductance.

A circuit element built to possess inductance is called an **inductor.** In its simplest form an inductor is simply a coil of wire, Figure 13–1(a). Ideally, inductors have only inductance. However, since they are made of wire, practical inductors also have some resistance. Initially, however, we assume that this resistance is negligible and treat inductors as ideal (i.e., we assume that they have no property other than inductance). (Coil resistance is considered in Sections 13.6 and 13.7.) In practice, inductors are also referred to as **chokes** (because they try to limit or "choke" current change) or as **reactors** (for reasons to be discussed in Chapter 16). In this chapter, we refer to them mainly as inductors.

On circuit diagrams and in equations, inductance is represented by the letter L. Its circuit symbol is a coil as shown in Figure 13–1(b). The unit of inductance is the **henry.**

Inductors are used in many places. In radios, they are part of the tuning circuit that you adjust when you select a station. In fluorescent lamps, they are part of the ballast circuit that limits current when the lamp is turned on; in power systems, they are part of the protection circuitry used to control short-circuit currents during fault conditions. ■

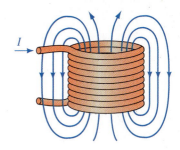

(a) A basic inductor

(b) Ideal inductor symbol

FIGURE 13–1   Inductance is due to the magnetic field created by an electric current.

PUTTING IT IN PERSPECTIVE

The Discovery of Electromagnetic Induction

MOST OF OUR IDEAS CONCERNING INDUCTANCE and induced voltages are due to Michael Faraday (recall Chapter 12) and Joseph Henry (1797–1878). Working independently (Faraday in England and Henry—shown at left—in the USA), they discovered, almost simultaneously, the fundamental laws governing electromagnetic induction.

While experimenting with magnetic fields, Faraday developed the transformer. He wound two coils on an iron ring and energized one of them from a battery. As he closed the switch energizing the first coil, Faraday noticed that a momentary voltage was induced in the second coil, and when he opened the switch, he found that a momentary voltage was again induced but with opposite polarity. When the current was steady, no voltage was produced at all.

Faraday explained this effect in terms of his magnetic lines of flux concept. When current was first turned on, he visualized the lines as springing outward into space; when it was turned off, he visualized the lines as collapsing inward. He then visualized that voltage was produced by these lines as they cut across circuit conductors. Companion experiments showed that voltage was also produced when a magnet was passed through a coil or when a conductor was moved through a magnetic field. Again, he visualized these voltages in terms of flux cutting a conductor.

Working independently in the United States, Henry discovered essentially the same results. In fact, Henry's work preceded Faraday's by a few months, but because he did not publish them first, credit was given to Faraday. However, Henry is credited with the discovery of self-induction, and in honor of his work the unit of inductance was named the henry. ■

13.1 Electromagnetic Induction

NOTES . . .

Since we work with time-varying flux linkages in this chapter, we use ϕ rather than Φ for flux (as we did in Chapter 12). This is in keeping with the standard practice of using lowercase symbols for time-varying quantities and uppercase symbols for dc quantities.

Inductance depends on **induced voltage.** Thus, we begin with a review of electromagnetic induction. First, we look at Faraday's and Henry's results. Consider Figure 13–2. In (a), a magnet is moved through a coil of wire, and this action induces a voltage in the coil. When the magnet is thrust into the coil, the meter deflects upscale; when it is withdrawn, the meter deflects downscale, indicating that polarity has changed. The voltage magnitude is proportional to how fast the magnet is moved. In (b), when the conductor is moved through the field, voltage is induced. If the conductor is moved to the right, its far end is positive; if it is moved to the left, the polarity reverses and its far end becomes negative. Again, the voltage magnitude is proportional to how fast the wire is moved. In (c), voltage is induced in coil 2 due to the magnetic field created by the current in coil 1. At the instant the switch is closed, the meter kicks upscale; at the instant it is opened, the meter kicks downscale. In (d) voltage is induced in a coil by its own current. At the instant the switch is closed, the top end of the coil becomes positive, while at the instant it is opened, the polarity reverses and the top end becomes negative.

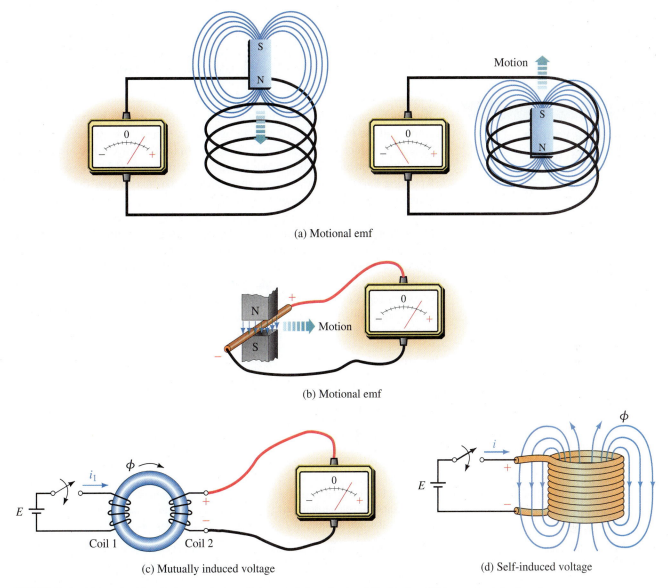

(a) Motional emf

(b) Motional emf

(c) Mutually induced voltage

(d) Self-induced voltage

FIGURE 13–2   Illustrating Faraday's experiments. Voltage is induced only while the flux linking a circuit is changing.

Faraday's Law

Based on these observations, Faraday concluded *that voltage is induced in a circuit whenever the flux linking* (i.e., passing through) *the circuit is changing and that the magnitude of the voltage is proportional to the rate of change of the flux linkages.* This result, known as **Faraday's law,** is also sometimes stated in terms of the rate of cutting flux lines. We look at this viewpoint in Chapter 15.

Lenz's Law

Heinrich Lenz (a Russian physicist, 1804–1865) determined a companion result. He showed that *the polarity of the induced voltage is such as to oppose the cause producing it.* This result is known as **Lenz's law.**

13.2 Induced Voltage and Induction

We now focus on inductors, Figure 13–2(d). As noted earlier, inductance is due entirely to the magnetic field created by current. Consider Figure 13–3 (which shows the inductor at three instants of time). In (a) the current is constant, and since the magnetic field is due to this current, the magnetic field is also constant. Applying Faraday's law, we note that, because the flux linking the coil is not changing, the induced voltage is zero. Now consider (b). Here, the current (and hence the field) is increasing. According to Faraday's law, a voltage is induced that is proportional to how fast the field is changing and according to Lenz's law, the polarity of this voltage must be such as to oppose the increase in current. Thus, the polarity of the voltage is as shown. Note that the faster the current increases, the larger the opposing voltage. Now consider (c). Since the current is decreasing, Lenz's law shows that the polarity of the induced voltage reverses, that is, the collapsing field produces a voltage that tries to keep the current going. Again, the faster the rate of change of current, the larger is this voltage.

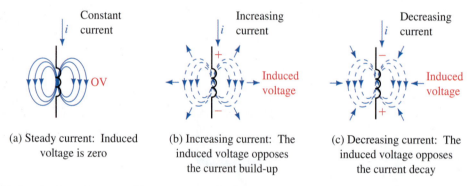

(a) Steady current: Induced voltage is zero

(b) Increasing current: The induced voltage opposes the current build-up

(c) Decreasing current: The induced voltage opposes the current decay

FIGURE 13–3 Self-induced voltage due to a coil's own current. The induced voltage opposes the current change. Note carefully the polarities in (b) and (c).

Counter EMF

Because the induced voltage in Figure 13–3 tries to counter (i.e., opposes) changes in current, it is called a **counter emf** or **back voltage.** Note carefully, however, that this voltage does not oppose current, it opposes only changes in current. It also does not prevent the current from changing; it only prevents it from changing abruptly. The result is that current in an inductor changes gradually and smoothly from one value to another as indicated in Figure 13–4(b). The effect of inductance is thus similar to the effect of inertia in a mechanical system. The flywheel used on an engine, for example, prevents abrupt changes in engine speed but does not prevent the engine from gradually changing from one speed to another.

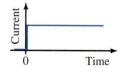

(a) Current cannot jump from one value to another like this

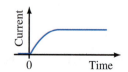

(b) Current must change smoothly with no abrupt jumps

FIGURE 13–4 Current in inductance.

Iron-Core and Air-Core Inductors

As Faraday discovered, the voltage induced in a coil depends on flux linkages, and flux linkages depend on core materials. Coils with ferromagnetic cores (called **iron-core coils**) have their flux almost entirely confined to their cores, while coils wound on nonferromagnetic materials do not. (The latter are sometimes called **air-core coils** because all nonmagnetic core materials have the same permeability as air and thus behave magnetically the same as air.)

First, consider the iron-core case, Figure 13–5. Ideally, all flux lines are confined to the core and hence pass through (link) all turns of the winding. The product of flux times the number of turns that it passes through is defined as the **flux linkage** of the coil. For Figure 13–5, ϕ lines pass through N turns yielding a flux linkage of $N\phi$. By Faraday's law, the induced voltage is proportional to the rate of change of $N\phi$. In the SI system, the constant of proportionality is one and Faraday's law for this case may therefore be stated as

$$e = N \times \text{the rate of change of } \phi \qquad \textbf{(13–1)}$$

In calculus notation,

$$e = N\frac{d\phi}{dt} \quad \text{(volts, V)} \qquad \textbf{(13–2)}$$

where ϕ is in webers, t in seconds, and e in volts. Thus if the flux changes at the rate of 1 Wb/s in a 1 turn coil, the voltage induced is 1 volt.

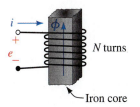

FIGURE 13–5   When flux ϕ passes through all N turns, the flux linking the coil is $N\phi$.

NOTES . . .

Equation 13–2 is sometimes shown with a minus sign. However, the minus sign is unnecessary. In circuit theory, we use Equation 13–2 to determine the magnitude of the induced voltage and Lenz's law to determine its polarity.

If the flux through a 200-turn coil changes steadily from 1 Wb to 4 Wb in one second, what is the voltage induced?

EXAMPLE 13–1

Solution   The flux changes by 3 Wb in one second. Thus, its rate of change is 3 Wb/s.

$$e = N \times \text{rate of change of flux}$$
$$= (200 \text{ turns})(3 \text{ Wb/s}) = 600 \text{ volts}$$

Now consider an air-core inductor (Figure 13–6). Since not all flux lines pass through all windings, it is difficult to determine flux linkages as above. However, (since no ferromagnetic material is present) flux is directly proportional to current. In this case, then, since induced voltage is proportional to the rate of change of flux, and since flux is proportional to current, induced voltage will be proportional to the rate of change of current. Let the constant of proportionality be L. Thus,

$$e = L \times \text{rate of change of current} \qquad \textbf{(13–3)}$$

In calculus notation, this can be written as

$$e = L\frac{di}{dt} \quad \text{(volts, V)} \qquad \textbf{(13–4)}$$

L is called the **self-inductance** of the coil, and in the SI system its unit is the henry. (This is discussed in more detail in Section 13.3.)

We now have two equations for coil voltage. Equation 13–4 is the more useful form for this chapter, while Equation 13–2 is the more useful form for the circuits of Chapter 24. We look at Equation 13–4 in the next section.

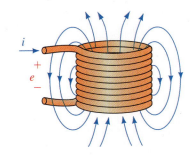

FIGURE 13–6   The flux linking the coil is proportional to current. Flux linkage is LI.

(Answers are at the end of the chapter.)

1. Which of the current graphs shown in Figure 13–7 cannot be the current in an inductor? Why?

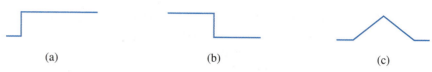

| (a) | (b) | (c) |

FIGURE 13–7

2. Compute the flux linkage for the coil of Figure 13–5, given $\phi = 500$ mWb and $N = 1200$ turns.

3. If the flux ϕ of Question 2 changes steadily from 500 mWb to 525 mWb in 1 s, what is the voltage induced in the coil?

4. If the flux ϕ of Question 2 changes steadily from 500 mWb to 475 mWb in 100 ms, what is the voltage induced?

13.3 Self-Inductance

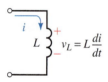

FIGURE 13–8 Voltage-current reference convention. As usual, the plus sign for voltage goes at the tail of the current arrow.

In the preceding section, we showed that the voltage induced in a coil is $e = Ldi/dt$, where L is the self-inductance of the coil (usually referred to simply as inductance) and di/dt is the rate of change of its current. In the SI system, L is measured in henries. As can be seen from Equation 13–4, it is the ratio of voltage induced in a coil to the rate of change of current producing it. From this, we get the definition of the henry. By definition, *the inductance of a coil is one henry if the voltage created by its changing current is one volt when its current changes at the rate of one ampere per second.*

In practice, the voltage across an inductance is denoted by v_L rather than by e (Figure 13–8). Thus,

$$v_L = L\frac{di}{dt} \quad \text{(V)} \qquad\qquad \textbf{(13–5)}$$

EXAMPLE 13–2

If the current through a 5-mH inductance changes at the rate of 1000 A/s, what is the voltage induced?

Solution

$$v_L = L \times \text{rate of change of current}$$
$$= (5 \times 10^{-3} \text{ H})(1000 \text{ A/s}) = 5 \text{ volts}$$

PRACTICE PROBLEMS 1

1. The voltage across an inductance is 250 V when its current changes at the rate of 10 mA/μs. What is L?

2. If the voltage across a 2-mH inductance is 50 volts, how fast is the current changing?

Answers
1. 25 mH; 2. 25×10^3 A/s

Inductance Formulas

Inductance for some simple shapes can be determined using the principles of Chapter 12. For example, the approximate inductance of the coil of Figure 13–9 can be shown to be

$$L = \frac{\mu N^2 A}{\ell} \quad \text{(H)} \qquad \text{(13–6)}$$

where ℓ is in meters, A is in square meters, N is the number of turns, and μ is the permeability of the core. (Details can be found in many physics books.)

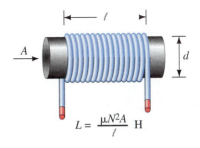

$$L = \frac{\mu N^2 A}{\ell} \ \text{H}$$

FIGURE 13–9   Approximate inductance formula for a single-layer coil.

A 0.15-m-long air-core coil has a radius of 0.006 m and 120 turns. Compute its inductance.

EXAMPLE 13–3

Solution

$$A = \pi r^2 = 1.131 \times 10^{-4} \ \text{m}^2$$
$$\mu = \mu_0 = 4\pi \times 10^{-7}$$

Thus,

$$L = 4\pi \times 10^{-7} (120)^2 (1.131 \times 10^{-4})/0.15 = 13.6 \ \mu\text{H}$$

The accuracy of Equation 13–6 breaks down for small ℓ/d ratios. (If ℓ/d is greater than 10, the error is less than 4%.) Improved formulas may be found in design handbooks, such as the *Radio Amateur's Handbook* published by the American Radio Relay League (ARRL).

To provide greater inductance in smaller spaces, iron cores are sometimes used. Unless the core flux is kept below saturation, however, permeability varies and inductance is not constant. To get constant inductance an air gap may be used (Figure 13–10). If the gap is wide enough to dominate, coil inductance is approximately

$$L = \frac{\mu_0 N^2 A_g}{\ell_g} \quad \text{(H)} \qquad \text{(13–7)}$$

where μ_0 is the permeability of air, A_g is the area of the air gap, and ℓ_g is its length. (See end-of-chapter Problem 11.) Another way to increase inductance is to use a ferrite core (Section 13.6).

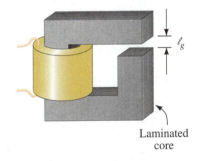

Laminated core

FIGURE 13–10   Iron-core coil with air gap. The gap keeps the core from going into saturation.

The inductor of Figure 13–10 has 1000 turns, a 5-mm gap, and a cross-sectional area at the gap of $5 \times 10^{-4} \ \text{m}^2$. What is its inductance?

EXAMPLE 13–4

Solution

$$L = (4\pi \times 10^{-7})(1000)^2(5 \times 10^{-4})/(5 \times 10^{-3}) = 0.126 \ \text{H}$$

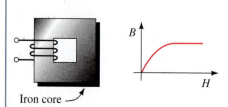

PRACTICAL NOTES . . .

Iron core

FIGURE 13–11 This coil does not have a fixed inductance because its flux is not proportional to its current.

1. Since inductance is due to a conductor's magnetic field, it depends on the same factors that the magnetic field depends on. The stronger the field for a given current, the greater the inductance. Thus, a coil of many turns will have more inductance than a coil of a few turns (L is proportional to N^2) and a coil wound on a magnetic core will have greater inductance than a coil wound on a nonmagnetic form.

2. However, if a coil is wound on a magnetic core, the core's permeability μ may change with flux density. Since flux density depends on current, L becomes a function of current. For example, the inductor of Figure 13–11 has a nonlinear inductance due to core saturation. All inductors encountered in this book are assumed to be linear, i.e., of constant value.

IN-PROCESS
LEARNING CHECK 2

(Answers are at the end of the chapter.)

1. The voltage across an inductance whose current changes uniformly by 10 mA in 4 μs is 70 volts. What is its inductance?

2. If you triple the number of turns in the inductor of Figure 13–10, but everything else remains the same, by what factor does the inductance increase?

13.4 Computing Induced Voltage

◀ **Online Companion**

Earlier, we determined that the voltage across an inductance is given by $v_L = L\,di/dt$, where the voltage and current references are shown in Figure 13–8. Note that the polarity of v_L depends on whether the current is increasing or decreasing. For example, if the current is increasing, di/dt is positive and so v_L is positive, while if the current is decreasing, di/dt is negative and v_L is negative.

To compute voltage, we need to determine di/dt. In general, this requires calculus. However, since di/dt is slope, you can determine voltage without calculus for currents that can be described by straight lines, as in Figure 13–12. For any Δt segment, slope $= \Delta i/\Delta t$, where Δi is the amount that the current changes during time interval Δt.

EXAMPLE 13–5

Figure 13–12 is the current through a 10-mH inductance. Determine voltage v_L and sketch it.

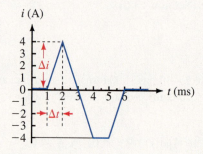

FIGURE 13–12

Solution Break the problem into intervals over which the slope is constant, determine the slope for each segment, then compute voltage using $v_L = L \times$ slope for that interval:

0 to 1 ms:	Slope = 0. Thus, $v_L = 0$ V.
1 ms to 2 ms:	Slope $= \Delta i/\Delta t = 4$ A/(1 $\times$ 10^{-3} s) = 4 $\times$ 10^3 A/s. Thus, $v_L = L\Delta i/\Delta t = (0.010$ H$)(4 \times 10^3$ A/s$) = 40$ V.
2 ms to 4 ms:	Slope $= \Delta i/\Delta t = -8$ A/(2 $\times$ 10^{-3} s) = -4×10^3 A/s. Thus, $v_L = L\Delta i/\Delta t = (0.010$ H$)(-4 \times 10^3$ A/s$) = -40$ V.
4 ms to 5 ms:	Slope = 0. Thus, $v_L = 0$ V.
5 ms to 6 ms:	Same slope as from 1 ms to 2 ms. Thus, $v_L = 40$ V.

The voltage waveform is shown in Figure 13–13.

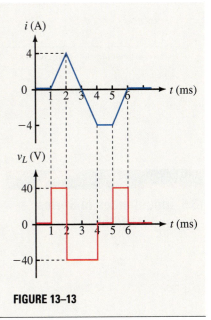

FIGURE 13–13

For currents that are not linear functions of time, you need to use calculus as illustrated in the following example.

What is the equation for the voltage across a 12.5 H inductance whose current is $i = te^{-t}$ amps?

EXAMPLE 13–6

Solution Differentiate by parts using

$$\frac{d(uv)}{dt} = u\frac{dv}{dt} + v\frac{du}{dt} \text{ with } u = t \text{ and } v = e^{-t}$$

Thus,

$$v_L = L\frac{di}{dt} = L\frac{d}{dt}(te^{-t}) = L[t(-e^{-t}) + e^{-t}] = 12.5e^{-t}(1 - t) \text{ volts}$$

1. Figure 13–14 shows the current through a 5-H inductance. Determine voltage v_L and sketch it.

2. If the current of Figure 13–12 is applied to an unknown inductance and the voltage from 1 ms to 2 ms is 28 volts, what is L?

3. ∫ The current in a 4-H inductance is $i = t^2e^{-5t}$ A. What is voltage v_L?

Answers

1. v_L is a square wave. Between 0 and 2 ms, its value is 15 V; between 2 ms and 4 ms, its value is -15 V, etc.

2. 7 mH; 3. $4e^{-5t}(2t - 5t^2)$ V

PRACTICE PROBLEMS 2

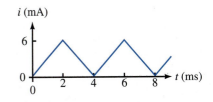

FIGURE 13–14

1. An inductance L_1 of 50 mH is in series with an inductance L_2 of 35 mH. If the voltage across L_1 at some instant is 125 volts, what is the voltage across L_2 at that instant? Hint: Since the same current passes through both inductances, the rate of change of current is the same for both.

2. Current through a 5-H inductance changes linearly from 10 A to 12 A in 0.5 s. Suppose now the current changes linearly from 2 mA to 6 mA in 1 ms. Although the currents are significantly different, the induced voltage is the same in both cases. Why? Compute the voltage.

13.5 Inductances in Series and Parallel

For inductances in series or parallel, the equivalent inductance is found by using the same rules that you used for resistance. For the series case (Figure 13–15) the total inductance is the sum of the individual inductances:

$$L_T = L_1 + L_2 + L_3 + \cdots + L_N \tag{13–8}$$

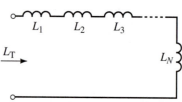

FIGURE 13–15 $L_T = L_1 + L_2 + \cdots + L_N$.

For the parallel case (Figure 13–16),

$$\frac{1}{L_T} = \frac{1}{L_1} + \frac{1}{L_2} + \frac{1}{L_3} + \cdots + \frac{1}{L_N} \tag{13–9}$$

For two inductances, Equation 13–9 reduces to

$$L_T = \frac{L_1 L_2}{L_1 + L_2} \tag{13–10}$$

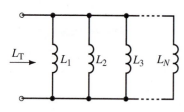

FIGURE 13–16 $\frac{1}{L_T} = \frac{1}{L_1} + \frac{1}{L_2} + \cdots + \frac{1}{L_N}$

EXAMPLE 13–7

Find L_T for the circuit of Figure 13–17.

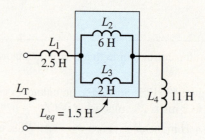

FIGURE 13–17

Solution   The parallel combination of L_2 and L_3 is

$$L_{eq} = \frac{L_2 L_3}{L_2 + L_3} = \frac{6 \times 2}{6 + 2} = 1.5 \text{ H}$$

This is in series with L_1 and L_4. Thus, $L_T = 2.5 + 1.5 + 11 = 15$ H.

1. For Figure 13–18, $L_T = 2.25$ H. Determine L_x.

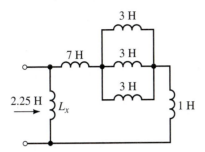

FIGURE 13–18

2. For Figure 13–15, the current is the same in each inductance and $v_1 = L_1 di/dt$, $v_2 = L_2 di/dt$, and so on. Apply KVL and show that $L_T = L_1 + L_2 + L_3 + \ldots + L_N$.

Answer
1. 3 H

Core Types

The type of core used in an inductor depends to a great extent on its intended use and frequency range. (Although you have not studied frequency yet, you can get a feel for frequency by noting that the electrical power system operates at low frequency [60 cycles per second, called 60 hertz], while radio and TV systems operate at high frequency [hundreds of megahertz].) Inductors used in audio or power supply applications generally have iron cores (because they need large inductance values), while inductors for radio-frequency circuits generally use air or ferrite cores. (Ferrite is a mixture of iron oxide in a ceramic binder. It has characteristics that make it suitable for high-frequency work.) Iron cannot be used, however, since it has large power losses at high frequencies (for reasons discussed in Chapter 17).

Variable Inductors

Inductors can be made so that their inductance is variable. In one approach, inductance is varied by changing coil spacing with a screwdriver adjustment. In another approach (Figure 13–19), a threaded ferrite slug is screwed in or out of the coil to vary its inductance. (Since ferrite contains ferromagnetic material, it increases the flux in the core and hence, its inductance.)

13.6 Practical Considerations

FIGURE 13–19 A variable inductor with its ferrite core removed for viewing.

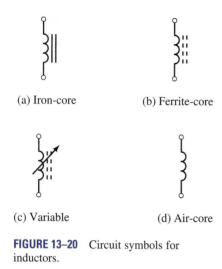

(a) Iron-core   (b) Ferrite-core

(c) Variable   (d) Air-core

FIGURE 13–20   Circuit symbols for inductors.

Circuit Symbols

Figure 13–20 shows inductor symbols. Iron-cores are identified by double solid lines, while dashed lines denote a ferrite core. (Air-core inductors have no core symbol.) An arrow indicates a variable inductor.

Coil Resistance

Ideally, inductors have only inductance. However, since inductors are made of imperfect conductors (e.g., copper wire), they also have resistance. (We can view this resistance as being in series with the coil's inductance as indicated in Figure 13–21(a). Also shown is stray capacitance, considered next.) Although coil resistance is generally small, it cannot always be ignored and thus, must sometimes be included in the analysis of a circuit. In Section 13.7, we show how this resistance is taken into account in dc analysis; in later chapters, you will learn how to take it into account in ac analysis.

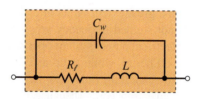

(a) Real inductors have stray capacitance and winding resistance

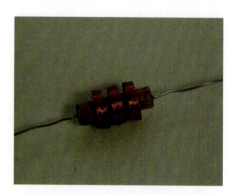

(b) Separating coil into sections helps reduce stray capacitance

FIGURE 13–21   A ferrite-core choke.

Stray Capacitance

Because the turns of an inductor are separated from each other by insulation, a small amount of capacitance exists from winding to winding. This capacitance is called **stray** or **parasitic capacitance.** Although this capacitance is distributed from turn to turn, its effect can be approximated by lumping it as in Figure 13–21(a). The effect of stray capacitance depends on frequency. At low frequencies, it can usually be neglected; at high frequencies, it may have to be taken into account as you will see in later courses. Some coils are wound in multiple sections, Figure 13–21(b) to reduce stray capacitance.

Stray Inductance

Because inductance is due entirely to the magnetic effects of electric current, all current-carrying conductors have inductance. This means that leads on circuit components such as resistors, capacitors, transistors, and so on, all have inductance, as do traces on printed circuit boards and wires in cables. We call this inductance **"stray inductance."** Fortunately, in many cases, the stray inductance is so small that it can be neglected (see Practical Notes).

We now look at inductive circuits with constant dc current. Consider Figure 13–22. The voltage across an ideal inductance with constant dc current is zero because the rate of change of current is zero. This is indicated in (a). Since the inductor has current through it but no voltage across it, it looks like a short circuit, (b). This is true in general, that is, *an ideal inductor looks like a short circuit in steady state dc.* (This should not be surprising since it is just a piece of wire to dc.) For a nonideal inductor, its dc equivalent is its coil resistance (Figure 13–23). For steady state dc, problems can be solved using simple dc analysis techniques.

13.7 Inductance and Steady State DC

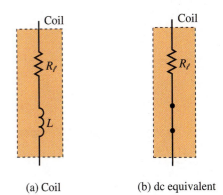

(a) Coil   (b) dc equivalent

FIGURE 13–23 Steady state dc equivalent of a coil with winding resistance.

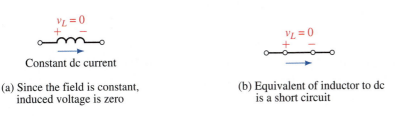

(a) Since the field is constant, induced voltage is zero

(b) Equivalent of inductor to dc is a short circuit

FIGURE 13–22 Inductance looks like a short circuit to steady state dc.

In Figure 13–24(a), the coil resistance is 14.4 Ω. What is the steady state current I?

EXAMPLE 13–8

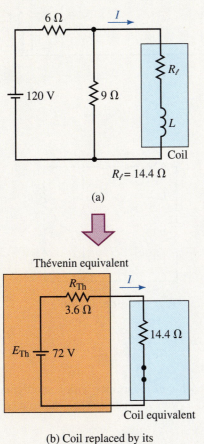

$R_\ell = 14.4\ \Omega$

(a)

Thévenin equivalent

(b) Coil replaced by its dc equivalent

FIGURE 13–24

Solution Reduce the circuit as in (b).

$$E_{Th} = (9/15)(120) = 72 \text{ V}$$
$$R_{Th} = 6\Omega\|9\Omega = 3.6 \ \Omega$$

Now replace the coil by its dc equivalent circuit as in (b). Thus,

$$I = E_{Th}/R_T = 72/(3.6 + 14.4) = 4\text{A}$$

EXAMPLE 13–9

The resistance of coil 1 in Figure 13–25(a) is 30 Ω and that of coil 2 is 15 Ω. Find the voltage across the capacitor assuming steady state dc.

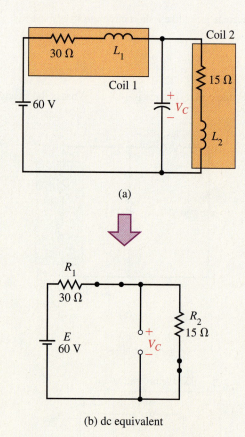

(a)

(b) dc equivalent

FIGURE 13–25

Solution Replace each coil inductance with a short circuit and the capacitor with an open circuit. As you can see from (b), the voltage across C is the same as the voltage across R_2. Thus,

$$V_C = \frac{R_2}{R_1 + R_2}E = \left(\frac{15 \ \Omega}{45 \ \Omega}\right)(60 \text{ V}) = 20 \text{ V}$$

For Figure 13–26, find I, V_{C_1}, and V_{C_2} in the steady state.

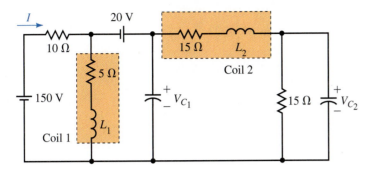

FIGURE 13–26

MULTISIM

Answers
10.7 A; 63 V; 31.5 V

When power flows into an inductor, energy is stored in its magnetic field. When the field collapses, this energy is returned to the circuit. For an ideal inductor, $R_\ell = 0$ ohm and hence no power is dissipated; thus, an ideal inductor has zero power loss.

To determine the energy stored by an ideal inductor, consider Figure 13–27. Power to the inductor is given by $p = v_L i$ watts, where $v_L = L di/dt$. By summing this power (see next ▌), the energy is found to be

$$W = \frac{1}{2}Li^2 \quad \text{(J)} \tag{13–11}$$

where i is the instantaneous value of current. When current reaches its steady state value I, $W = \frac{1}{2}LI^2$ J. This energy remains stored in the field as long as the current continues. When the current goes to zero, the field collapses and the energy is returned to the circuit.

13.8 Energy Stored by an Inductance

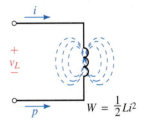

FIGURE 13–27 Energy is stored in the magnetic field of an inductor.

The coil of Figure 13–28(a) has a resistance of 15 Ω. When the current reaches its steady state value, the energy stored is 12 J. What is the inductance of the coil?

EXAMPLE 13–10

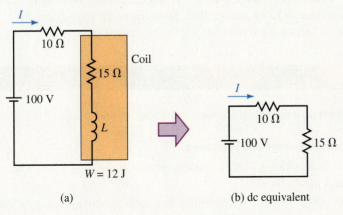

(a)

(b) dc equivalent

FIGURE 13–28

Solution From (b),

$$I = 100 \text{ V}/25 \text{ }\Omega = 4 \text{ A}$$

$$W = \frac{1}{2}LI^2 \text{ J}$$

$$12 \text{ J} = \frac{1}{2}L(4 \text{ A})^2$$

Thus,

$$L = 2(12)/4^2 = 1.5 \text{ H}$$

Deriving Equation 13–11

The power to the inductor in Figure 13–27 is given by $p = v_L i$, where $v_L = Ldi/dt$. Therefore, $p = Lidi/dt$. However, $p = dW/dt$. Integrating yields

$$W = \int_0^t p\,dt = \int_0^t Li\frac{di}{dt}dt = L\int_0^i i\,di = \frac{1}{2}Li^2$$

13.9 Inductor Troubleshooting Hints

Inductors may fail by either opening or shorting. Failures may be caused by misuse, defects in manufacturing, or faulty installation.

Open Coil

Opens can be the result of poor solder joints or broken connections. First, make a visual inspection. If nothing wrong is found, disconnect the inductor and check it with an ohmmeter. An open-circuited coil has infinite resistance.

Shorts

Shorts can occur between windings or between the coil and its core (for an iron-core unit). A short may result in excessive current and overheating. Again, check visually. Look for burned insulation, discolored components, an acrid odor, and other evidence of overheating. An ohmmeter can be used to check for shorts between windings and the core. However, checking coil resistance for shorted turns is often of little value, especially if only a few turns are shorted. This is because the shorting of a few windings may not change the overall resistance enough to be measurable. Sometimes the only conclusive test is to substitute a known good inductor for the suspected one.

PROBLEMS

Unless otherwise indicated, assume ideal inductors and coils.

13.2   Induced Voltage and Inductance

1. If the flux linking a 75-turn coil changes at the rate of 3 Wb/s, what is the voltage across the coil?

2. If 80 volts is induced when the flux linking a coil changes at a uniform rate from 3.5 mWb to 4.5 mWb in 0.5 ms, how many turns does the coil have?

3. Flux changing at a uniform rate for 1 ms induces 60 V in a coil. What is the induced voltage if the same flux change takes place in 0.01 s?

13.3 Self-Inductance

4. The current in a 0.4-H inductor is changing at the rate of 200 A/s. What is the voltage across it?

5. The current in a 75-mH inductor changes uniformly by 200 μA in 0.1 ms. What is the voltage across it?

6. The voltage across an inductance is 25 volts when the current changes at 5 A/s. What is L?

7. The voltage induced when current changes uniformly from 3 amps to 5 amps in a 10-H inductor is 180 volts. How long did it take for the current to change from 3 to 5 amps?

8. Current changing at a uniform rate for 1 ms induces 45 V in a coil. What is the induced voltage if the same current change takes place in 100 μs?

9. Compute the inductance of the air-core coil of Figure 13–29, given $\ell = 20$ cm, $N = 200$ turns, and $d = 2$ cm.

10. The iron-core inductor of Figure 13–30 has 2000 turns, a cross-section of 1.5×1.2 inches, and an air gap of 0.2 inch. Compute its inductance.

11. The iron-core inductor of Figure 13–30 has a high-permeability core. Therefore, by Ampere's law, $NI \simeq H_g\ell_g$. Because the air gap dominates, saturation does not occur and the core flux is proportional to the current, i.e., the flux linkage equals LI. In addition, since all flux passes through the coil, the flux linkage equals $N\Phi$. By equating the two values of flux linkage and using ideas from Chapter 12, show that the inductance of the coil is

$$L = \frac{\mu_0 N^2 A_g}{\ell_g}$$

13.4 Computing Induced Voltage

12. Figure 13–31 shows the current in a 0.75-H inductor. Determine v_L and plot its waveform.

13. Figure 13–32 shows the current in a coil. If the voltage from 0 to 2 ms is 100 volts, what is L?

14. Why is Figure 13–33 not a valid inductor current? Sketch the voltage across L to show why. Pay particular attention to $t = 10$ ms.

15. Figure 13–34 shows the graph of the voltage across an inductance. The current changes from 4 A to 5 A during the time interval from 4 s to 5 s.

    a. What is L?

    b. Determine the current waveform and plot it.

    c. What is the current at $t = 10$ s?

16. If the current in a 25-H inductance is $i_L = 20e^{-12t}$ mA, what is v_L?

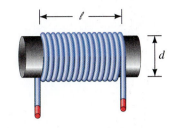

FIGURE 13–29

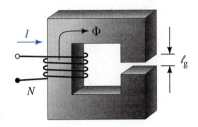

FIGURE 13–30

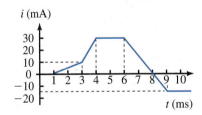

FIGURE 13–31

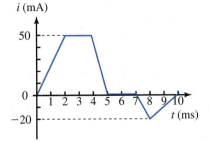

FIGURE 13–32

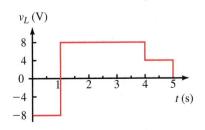

FIGURE 13–33

FIGURE 13–34

13.5 Inductances in Series and Parallel

17. What is the equivalent inductance of 12 mH, 14 mH, 22 mH, and 36 mH connected in series?

18. What is the equivalent inductance of 0.010 H, 22 mH, 86×10^{-3} H, and 12000 μH connected in series?

19. Repeat Problem 17 if the inductances are connected in parallel.

20. Repeat Problem 18 if the inductances are connected in parallel.

21. Determine L_T for the circuits of Figure 13–35.

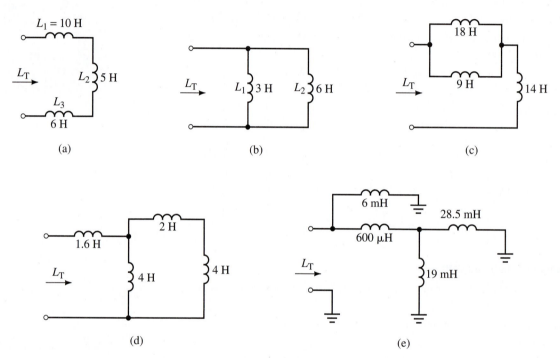

FIGURE 13–35

22. Determine L_T for the circuits of Figure 13–36.

23. A 30-μH inductance is connected in series with a 60-μH inductance, and a 10-μH inductance is connected in parallel with the series combination. What is L_T?

24. For Figure 13–37, determine L_x.

25. For the circuits of Figure 13–38, determine L_3 and L_4.

26. You have inductances of 24 mH, 36 mH, 22 mH and 10 mH. Connecting these any way you want, what is the largest equivalent inductance you can get? The smallest?

27. A 6-H and a 4-H inductance are connected in parallel. After a third inductance is added, $L_T = 4$ H. What is the value of the third inductance and how was it connected?

28. Inductances of 2 H, 4 H, and 9 H are connected in a circuit. If $L_T = 3.6$ H, how are the inductors connected?

29. Inductances of 8 H, 12 H, and 1.2 H are connected in a circuit. If $L_T = 6$ H, how are the inductors connected?

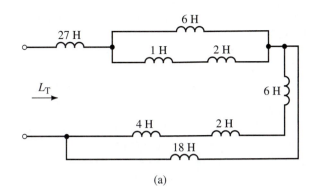

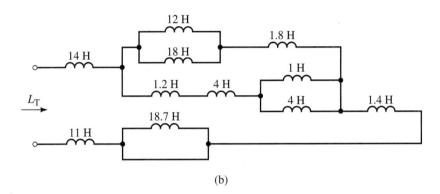

(b)

FIGURE 13–36

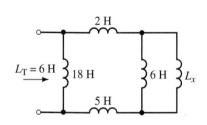

FIGURE 13–37

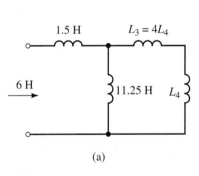

(a)

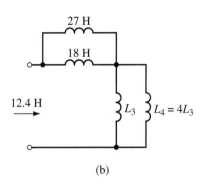

(b)

FIGURE 13–38

30. For inductors in parallel (Figure 13–39), the same voltage appears across each. Thus, $v = L_1 di_1/dt$, $v = L_2 di_2/dt$, etc. Apply KCL and show that $1/L_T = 1/L_1 + 1/L_2 + \cdots + 1/L_N$.

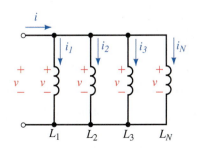

FIGURE 13–39

31. By combining elements, reduce each of the circuits of Figure 13–40 to their simplest form.

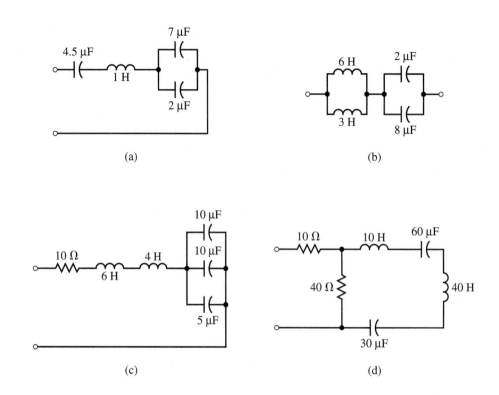

(a)

(b)

(c)

(d)

13.7 Inductance and Steady State DC

32. For each of the circuits of Figure 13–41, the voltages and currents have reached their final (steady state) values. Solve for the quantities indicated.

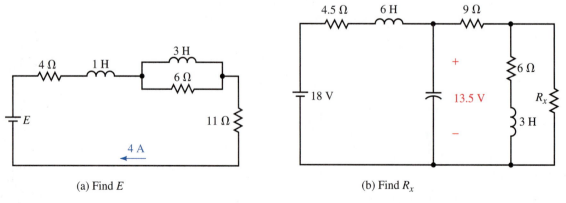

(a) Find E

(b) Find R_x

13.8   Energy Stored by an Inductance

33. Find the energy stored in the inductor of Figure 13–42.

34. In Figure 13–43, $L_1 = 2L_2$. The total energy stored is $W_T = 75$ J. Find L_1 and L_2.

13.9   Inductor Troubleshooting Hints

35. In Figure 13–44, an inductance meter measures 7 H. What is the likely fault?

36. Referring to Figure 13–45, an inductance meter measures $L_T = 8$ mH. What is the likely fault?

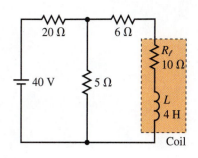

FIGURE 13–42

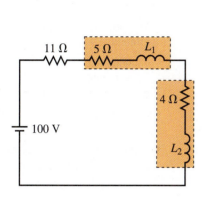

FIGURE 13–43

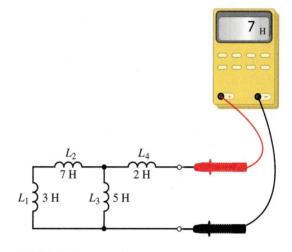

FIGURE 13–44

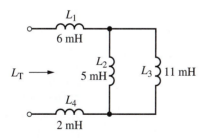

FIGURE 13–45

 ANSWERS TO IN-PROCESS LEARNING CHECKS

In-Process Learning Check 1

1. Both a and b. Current cannot change instantaneously.

2. 600 Wb-turns

3. 30 V

4. −300 V

In-Process Learning Check 2

1. 28 mH

2. 9 times

In-Process Learning Check 3

1. 87.5 V

2. Rate of change of current is the same; 20 V

■ **OBJECTIVES**

After studying this chapter, you will be able to

- explain why transients occur in *RL* circuits,

- explain why an inductor with zero initial conditions looks like an open circuit when first energized,

- compute time constants for *RL* circuits,

- compute voltage and current transients in *RL* circuits during the current buildup phase,

- compute voltage and current transients in *RL* circuits during the current decay phase,

- explain why an inductor with non-zero initial conditions looks like a current source when disturbed,

- solve moderately complex *RL* transient problems using circuit simplification techniques,

- solve *RL* transient problems using MultiSIM and PSpice.

Inductive Transients

14

I n Chapter 11, you learned that transients occur in capacitive circuits because capacitor voltage cannot change instantaneously. In this chapter, you will learn that transients occur in inductive circuits because inductor current cannot change instantaneously. Although the details differ, you will find that many of the basic ideas are the same.

Inductive transients result when circuits containing inductance are disturbed. More so than capacitive transients, inductive transients are potentially destructive and dangerous. For example, when you break the current in an inductive circuit, an extremely large and damaging voltage may result.

In this chapter, we study basic *RL* transients. We look at transients during current buildup and decay and learn how to calculate the voltages and currents that result. ■

CHAPTER PREVIEW

PUTTING IT IN PERSPECTIVE

Inductance, the Dual of Capacitance

INDUCTANCE IS THE DUAL of capacitance. This means that the effect that inductance has on circuit operation is identical with that of capacitance if you interchange the term current for voltage, open circuit for short circuit, and so on. For example, for simple dc transients, current in an *RL* circuit has the same form as voltage in an *RC* circuit: they both rise to their final value exponentially according to $1 - e^{-t/\tau}$. Similarly, voltage across inductance decays in the same manner as current through capacitance, i.e., according to $e^{-t/\tau}$. In fact, as you will see, there is complete duality between all the equations that describe transient voltage and current behavior in capacitive and inductive circuits.

Duality applies to steady state and initial condition representations as well. To steady state dc, for example, a capacitor looks like an open circuit, while an inductor looks like a short circuit. Similarly, the dual of a capacitor that looks like a short circuit at the instant of switching is an inductor that looks like an open circuit. Finally, the dual of a capacitor that has an initial condition of V_0 volts is an inductance with an initial condition of I_0 amps.

The principle of duality is helpful in circuit analysis as it lets you transfer the principles and concepts learned in one area directly into another. You will find, for example, that many of the ideas you learned in Chapter 11 reappear here in their dual form. ■

14.1 Introduction

◀ Online Companion

As you saw in Chapter 11, when a circuit containing capacitance is disturbed, voltages and currents do not change to their new values immediately, but instead pass through a transitional phase as the circuit capacitance charges or discharges. The voltages and currents during this transitional interval are called **transients.** In a dual fashion, transients occur when circuits containing inductances are disturbed. In this case, however, transients occur because current in inductance cannot change instantaneously.

To get at the idea, consider Figure 14–1. In (a), we see a purely resistive circuit. At the instant the switch is closed, current jumps from 0 to E/R as required by Ohm's law. Thus, no transient (i.e., transitional phase) occurs because current reaches its final value immediately. Now consider (b). Here, we have added inductance. At the instant the switch is closed, a counter emf appears across the inductance. This voltage attempts to stop the current from changing and consequently slows its rise. Current thus does not jump to E/R immediately as in (a), but instead climbs gradually and smoothly as in (b). The larger the inductance, the longer the transition takes.

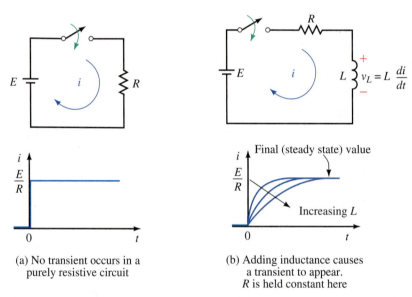

(a) No transient occurs in a purely resistive circuit

(b) Adding inductance causes a transient to appear. R is held constant here

FIGURE 14–1   Transient due to inductance. Adding inductance to a resistive circuit as in (b) slows the current rise and fall, thus creating a transient.

Continuity of Current

As Figure 14–1(b) illustates, *current through an inductance cannot change instantaneously, i.e., it cannot jump abruptly from one value to another, but must be continuous at all values of time.* This observation is known as the statement of **continuity of current for inductance** (see Notes). You will find this statement of great value when analyzing circuits containing inductance. We will use it many times in what follows.

Inductor Voltage

Now consider inductor voltage. With the switch open as in Figure 14–2(a), the current in the circuit and voltage across L are both zero. Now close the switch. Immediately after the switch is closed, the current is still zero, (since it cannot change instantaneously). Since $v_R = Ri$, the voltage across R is also zero and thus the full source voltage appears across L as shown in (b). The inductor volt-

NOTES . . .

The continuity statement for inductor current has a sound mathematical basis. Recall, induced voltage is proportional to the rate of change of current. In calculus notation,

$$v_L = L\frac{di}{dt}$$

This means that the faster that current changes, the larger the induced voltage. If inductor current could change from one value to another instantaneously as in Figure 14–1(a), its rate of change (i.e., di/dt) would be infinite and hence the induced voltage would be infinite. But infinite voltage is not possible. Thus we conclude that inductor current cannot change instantaneously.

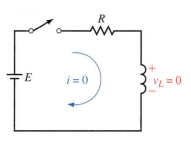

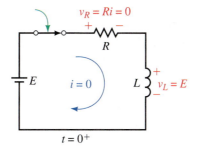

$$t = 0^+$$

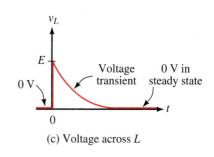

(a) Circuit before switch is closed. Current $i = 0$

(b) Circuit just after the switch has been closed. Current is still equal to zero. Thus, $v_L = E$

(c) Voltage across L

FIGURE 14–2 Voltage across L.

age therefore jumps from 0 V just before the switch is closed to E volts just after. It then decays to zero, since, as we saw in Chapter 13, the voltage across inductance is zero for steady state dc. This is indicated in (c).

Open-Circuit Equivalent of an Inductance

Consider again Figure 14–2(b). Note that just after the switch is closed, the inductor has voltage across it but no current through it. It therefore momentarily looks like an open circuit. This is indicated in Figure 14–3. This observation is true in general, that is, *an inductor with zero initial current looks like an open circuit at the instant of switching.* (Later, we extend this statement to include inductors with nonzero initial currents.)

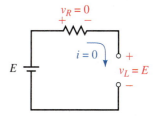

FIGURE 14–3 Inductor with zero initial current looks like an open circuit at the instant the switch is closed.

Initial Condition Circuits

Voltages and currents in circuits immediately after switching must sometimes be calculated. These can be determined with the aid of the open-circuit equivalent. By replacing inductances with open circuits, you can see what a circuit looks like just after switching. Such a circuit is called an **initial condition circuit.**

A coil and two resistors are connected to a 20-V source as in Figure 14–4(a). Determine source current i and inductor voltage v_L at the instant the switch is closed.

EXAMPLE 14–1

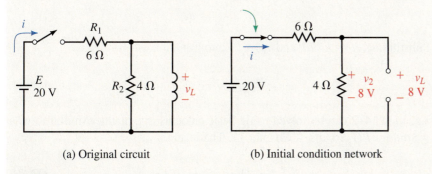

(a) Original circuit

(b) Initial condition network

FIGURE 14–4

Solution Replace the inductance with an open circuit. This yields the network shown in (b). Thus $i = E/R_T = 20 \text{ V}/10 \text{ }\Omega = 2$ A and the voltage across R_2 is $v_2 = (2 \text{ A})(4 \text{ }\Omega) = 8$ V. Since $v_L = v_2$, $v_L = 8$ volts as well.

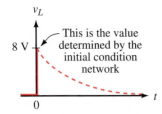

This is the value determined by the initial condition network

FIGURE 14–5 The initial condition network yields only the value at $t = 0^+$ s.

Initial condition networks yield voltages and currents only at the instant of switching, i.e., at $t = 0^+$ s. Thus, the value of 8 V calculated in Example 14–1 is only a momentary value as illustrated in Figure 14–5. Sometimes such an initial value is all that you need. In other cases, you need the complete solution. This is considered next, in Section 14.2.

PRACTICE PROBLEMS 1

Determine all voltages and currents in the circuit of Figure 14–6 immediately after the switch is closed and in steady state.

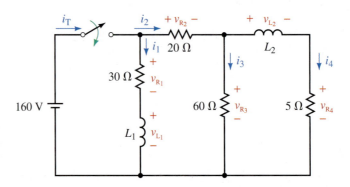

MULTISIM

FIGURE 14–6

Answers
Initial: $v_{R_1} = 0$ V; $v_{R_2} = 40$ V; $v_{R_3} = 120$ V; $v_{R_4} = 0$ V; $v_{L_1} = 160$ V; $v_{L_2} = 120$ V; $i_T = 2$ A; $i_1 = 0$ A; $i_2 = 2$ A; $i_3 = 2$ A; $i_4 = 0$ A.

Steady State: $v_{R_1} = 160$ V; $v_{R_2} = 130$ V; $v_{R_3} = v_{R_4} = 30$ V; $v_{L_1} = v_{L_2} = 0$ V; $i_T = 11.83$ A; $i_1 = 5.33$ A; $i_2 = 6.5$ A; $i_3 = 0.5$ A; $i_4 = 6.0$ A

14.2 Current Buildup Transients

FIGURE 14–7 KVL yields $v_L + v_R = E$.

Current

We will now develop equations to describe voltages and current during energization. Consider Figure 14–7. KVL yields

$$v_L + v_R = E \qquad \text{(14–1)}$$

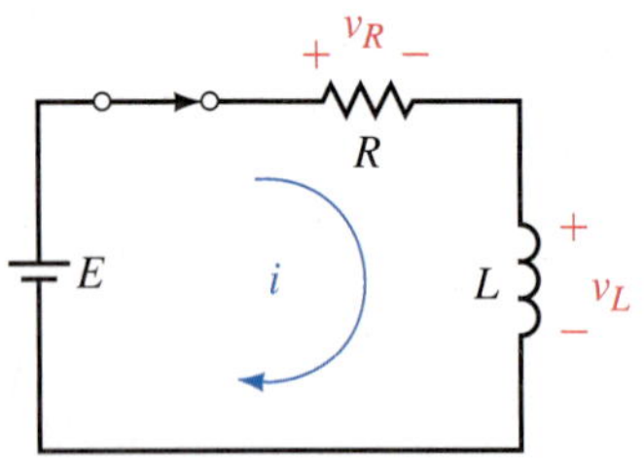

Substituting $v_L = L\,di/dt$ and $v_R = Ri$ into Equation 14–1 yields

$$L\frac{di}{dt} + Ri = E \qquad \text{(14–2)}$$

Equation 14–2 can be solved using basic calculus in a manner similar to what we did for *RC* circuits in Chapter 11. The result is

$$i = \frac{E}{R}\left(1 - e^{-Rt/L}\right) \quad \text{(A)} \qquad \text{(14–3)}$$

where R is in ohms, L is in henries, and t is in seconds. Equation 14–3 describes current buildup. Values of current at any point in time can be found by direct substitution as we illustrate next. Note that E/R is the final (steady state) current since the inductor looks like a short circuit to steady state dc (recall Sec. 13.7).

For the circuit of Figure 14–7, suppose $E = 50$ V, $R = 10\ \Omega$, and $L = 2$ H:

a. Determine the expression for i.

b. Compute and tabulate values of i at $t = 0^+$, 0.2, 0.4, 0.6, 0.8, and 1.0 s.

c. Using these values, plot the current.

d. What is the steady state current?

EXAMPLE 14–2

Solution

a. Substituting the values into Equation 14–3 yields

$$i = \frac{E}{R}(1 - e^{-Rt/L}) = \frac{50\ \text{V}}{10\ \Omega}(1 - e^{-10t/2}) = 5\,(1 - e^{-5t})\ \text{amps}$$

b. At $t = 0^+$ s, $i = 5(1 - e^{-5t}) = 5(1 - e^0) = 5(1 - 1) = 0$ A.

At $t = 0.2$ s, $i = 5(1 - e^{-5(0.2)}) = 5(1 - e^{-1}) = 3.16$ A.

At $t = 0.4$ s, $i = 5(1 - e^{-5(04)}) = 5(1 - e^{-2}) = 4.32$ A.

Continuing in this manner, you get Table 14–1.

c. Values are plotted in Figure 14–8. Note that this curve looks exactly like the curves we determined intuitively in Figure 14–1(b).

d. Steady state current is $E/R = 50$ V/10 $\Omega = 5$ A. This agrees with the curve of Figure 14–8.

TABLE 14–1

Time	Current
0	0
0.2	3.16
0.4	4.32
0.6	4.75
0.8	4.91
1.0	4.97

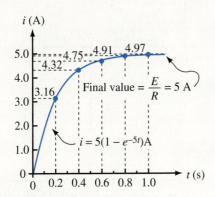

FIGURE 14–8 Current buildup transient.

◀ MULTISIM

Circuit Voltages

With i known, circuit voltages can be determined. Consider voltage v_R. Since $v_R = Ri$, when you multiply R times Equation 14–3, you get

$$v_R = E(1 - e^{-Rt/L}) \quad \text{(V)} \qquad (14\text{–}4)$$

Note that v_R has exactly the same shape as the current. Now consider v_L. Voltage v_L can be found by subtracting v_R from E as per Equation 14–1:

$$v_L = E - v_R = E - E(1 - e^{-Rt/L}) = E - E + Ee^{-Rt/L}$$

Thus,

$$v_L = Ee^{-Rt/L} \qquad (14\text{–}5)$$

An examination of Equation 14–5 shows that v_L has an initial value of E at $t = 0^+$ s and then decays exponentially to zero. This agrees with our earlier observation in Figure 14–2(c).

EXAMPLE 14–3

Repeat Example 14–2 for voltage v_L.

Solution

a. From equation 14–5,

$$v_L = Ee^{-Rt/L} = 50e^{-5t} \text{ volts}$$

b. At $t = 0^+$ s, $v_L = 50e^{-5t} = 50e^0 = 50(1) = 50$ V.

At $t = 0.2$ s, $v_L = 50e^{-5(0.2)} = 50e^{-1} = 18.4$ V.

At $t = 0.4$ s, $v_L = 50e^{-5(0.4)} = 50e^{-2} = 6.77$ V.

Continuing in this manner, you get Table 14–2.

c. The waveform is shown in Figure 14–9.

d. Steady state voltage is 0 V, as you can see in Figure 14–9.

TABLE 14–2

Time (s)	Voltage (V)
0	50.0
0.2	18.4
0.4	6.77
0.6	2.49
0.8	0.916
1.0	0.337

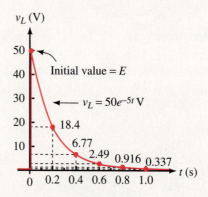

FIGURE 14–9   Inductor voltage transient.

◄ MULTISIM

PRACTICE PROBLEMS 2

For the circuit of Figure 14–7, with $E = 80$ V, $R = 5$ kΩ, and $L = 2.5$ mH:

a. Determine expressions for i, v_L, and v_R.

b. Compute and tabulate values at $t = 0^+$, 0.5, 1.0, 1.5, 2.0, and 2.5 μs.

c. At each point in time, does $v_L + v_R = E$?

d. Plot i, v_L, and v_R using the values computed in (b).

Answers

a. $i = 16(1 - e^{-2 \times 10^6 t})$ mA; $v_L = 80e^{-2 \times 10^6 t}$ V; $v_R = 80(1 - e^{-2 \times 10^6 t})$

b.

$t(\mu s)$	v_L (V)	i_L (mA)	v_R (V)
0	80	0	0
0.5	29.4	10.1	50.6
1.0	10.8	13.8	69.2
1.5	3.98	15.2	76.0
2.0	1.47	15.7	78.5
2.5	0.539	15.9	79.5

c. Yes

d. i and v_R have the shape shown in Figure 14–8, while v_L has the shape shown in Figure 14–9, with values according to the table shown in b.

Time Constant

In Equations 14–3 to 14–5 L/R is the time constant of the circuit.

$$\tau = \frac{L}{R} \quad \text{(s)} \qquad \text{(14–6)}$$

Note that τ has units of seconds. (This is left as an exercise for the student.) Equations 14–3, 14–4, and 14–5 may now be written as

$$i = \frac{E}{R}(1 - e^{-t/\tau}) \quad \text{(A)} \qquad \text{(14–7)}$$

$$v_L = Ee^{-t/\tau} \quad \text{(V)} \qquad \text{(14–8)}$$

$$v_R = E(1 - e^{-t/\tau}) \quad \text{(V)} \qquad \text{(14–9)}$$

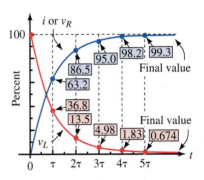

FIGURE 14–10   Universal time constant curves for the *RL* circuit.

Curves are plotted in Figure 14–10 versus time constant. As expected, transitions take approximately 5τ; thus, *for all practical purposes, inductive transients last five time constants.*

EXAMPLE 14–4

In a circuit where $L = 2$ mH, transients last 50 μs. What is R?

Solution   Transients last five time constants. Thus, $\tau = 50$ μs/5 = 10 μs. Now $\tau = L/R$. Therefore, $R = L/\tau = 2$ mH/10 μs = 200 Ω.

EXAMPLE 14–5

For an *RL* circuit, $i = 40(1 - e^{-5t})$ A and $v_L = 100e^{-5t}$ V.

a. What are E and τ?

b. What is R?

c. Determine L.

Solution

a. From Equation 14–8, $v_L = Ee^{-t/\tau} = 100e^{-5t}$. Therefore, $E = 100$ V and $\tau = \frac{1}{5} = 0.2$s.

b. From Equation 14–7,

$$i = \frac{E}{R}(1 - e^{-t/\tau}) = 40(1 - e^{-5t}).$$

Therefore, $E/R = 40$ A and $R = E/40$ A = 100 V/40 A = 2.5 Ω.

c. $\tau = L/R$. Therefore, $L = R\tau = (2.5)(0.2) = 0.5$ H.

It is sometimes easier to solve problems using the universal time constant curves than it is to solve the equations. (Be sure to convert curve percentages to a decimal value first, e.g., 63.2% to 0.632.) To illustrate, consider the problem of Examples 14–2 and 14–3. From Figure 14–10 at $t = \tau = 0.2$ s, $i = 0.632E/R$ and $v_L = 0.368E$. Thus, $i = 0.632(5$ A$) = 3.16$ A and $v_L = 0.368(50$ V$) = 18.4$ V as we found earlier.

The effect of inductance and resistance on transient duration is shown in Figure 14–11. The larger the inductance, the longer the transient for a given resistance. Resistance has the opposite effect: for a fixed inductance, the larger the resistance, the shorter the transient. [This is not hard to understand. As R increases, the circuit looks more and more resistive. If you get to a point where

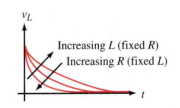

FIGURE 14–11   Effect of R and L on transient duration.

inductance is negligible compared with resistance, the circuit looks purely resistive, as in Figure 14–1(a), and no transient occurs.]

1. For the circuit of Figure 14–12, the switch is closed at $t = 0$ s.
   a. Determine expressions for v_L and i.
   b. Compute v_L and i at $t = 0^+$, 10 μs, 20 μs, 30 μs, 40 μs, and 50 μs.
   c. Plot curves for v_L and i.

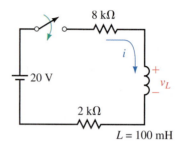

FIGURE 14–12

2. For the circuit of Figure 14–7, $E = 85$ V, $R = 50$ Ω, and $L = 0.5$ H. Use the universal time constant curves to determine v_L and i at $t = 20$ ms.

3. For a certain RL circuit, transients last 25 s. If $L = 10$ H and steady state current is 2 A, what is E?

4. An RL circuit has $E = 50$ V and $R = 10$ Ω. The switch is closed at $t = 0$ s. What is the current at the end of 1.5 time constants?

14.3 Interrupting Current in an Inductive Circuit

We now look at what happens when inductor current is interrupted. Consider Figure 14–13. At the instant the switch is opened, the field begins to collapse, which induces a voltage in the coil. If inductance is large and current is high, a great deal of energy is released in a very short time, creating a huge voltage that may damage equipment and create a shock hazard. (This induced voltage is referred to as an **inductive kick.**) For example, abruptly breaking the current

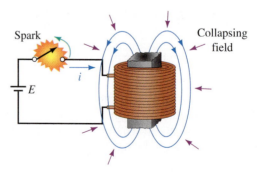

FIGURE 14–13 The sudden collapse of the magnetic field when the switch is opened causes a large induced voltage across the coil. (Several thousand volts may result.) The switch arcs over due to this voltage.

through a large inductor (such as a motor or generator field coil) can create voltage spikes up to several thousand volts, a value large enough to draw long arcs as indicated in Figure 14–13. Even moderate sized inductances in electronic systems can create enough voltage to cause damage if protective circuitry is not used.

The dynamics of the switch flashover are not hard to understand. When the field collapses, the voltage across the coil rises rapidly. Part of this voltage appears across the switch. As the switch voltage rises, it quickly exceeds the breakdown strength of air, causing a flashover between its contacts. Once struck, the arc is easily maintained, as it creates ionized gases that provide a relatively low resistance path for conduction. As the contacts spread apart, the arc elongates and eventually extinguishes as the coil energy is dissipated and coil voltage drops below that required to sustain the arc.

There are several important points to note here:

1. Flashovers, as in Figure 14–13, are generally undesirable. However, they can be controlled through proper engineering design. (One way is to use a discharge resistor, as in the next example; another way is to use a diode, as you will see in your electronics course.)

2. On the other hand, the large voltages created by breaking inductive currents have their uses. One is in the ignition system of automobiles, where current in the primary winding of a transformer coil is interrupted at the appropriate time by a control circuit to create the spark needed to fire the engine.

3. It is not possible for us to rigorously analyze the circuit of Figure 14–13 because the resistance of the arc changes as the switch opens. However, the main ideas can be established by studying circuits using fixed resistors. This we do next.

The Basic Ideas

We begin with the circuit of Figure 14–14. Assume the switch is closed and the circuit is in steady state. Since the inductance looks like a short circuit [Figure 14–15(a)], its current is $i_L = 120\text{ V}/30\ \Omega = 4\text{ A}$.

NOTES . . .

The intuitive explanation here has a sound mathematical basis. Recall, back emf (induced voltage) across a coil is given by

$$v_L = L\frac{di}{dt} \approx L\frac{\Delta i}{\Delta t}$$

where Δi is the change in current and Δt is the time interval over which the change takes place. When you open the switch, current begins to drop immediately toward zero. Since Δi is finite, and $\Delta t \to 0$ the ratio $\Delta i/\Delta t$ is very large and thus, the voltage across L rises to a very large value, causing a flashover to occur. After the flashover, current has a path through which to decay and thus Δt, although small, no longer approaches zero. The result is a large but finite voltage spike across L.

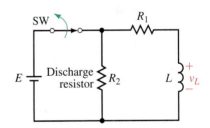

FIGURE 14–14   Discharge resistor R_2 helps limit the size of the induced voltage.

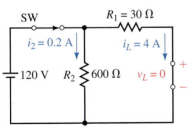

(a) Circuit just before the switch is opened

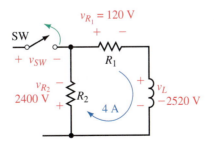

(b) Circuit just after SW is opened. Since coil voltage polarity is opposite to that shown, v_L is negative

FIGURE 14–15   Circuit of Figure 14–14 immediately before and after the switch is opened. Coil voltage jumps form 0 V to −2520 V for this example.

◀ MULTISIM

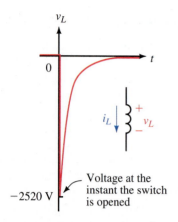

-2520 V

FIGURE 14–16   Voltage spike for the circuit of Figure 14–14. This voltage is more than 20 times larger than the source voltage.

Now open the switch. Just prior to opening the switch, $i_L = 4$ A; therefore, just after opening the switch, it must still be 4 A. As indicated in (b), this 4 A passes through resistances R_1 and R_2, creating voltages $v_{R_1} = 4$ A $\times$ 30 Ω = 120 V and $v_{R_2} = 4$ A $\times$ 600 Ω = 2400 V with the polarity shown. From KVL, $v_L + v_{R_1} + v_{R_2} = 0$. Therefore at the instant the switch is opened,

$$v_L = -(v_{R_1} + v_{R_2}) = -2520 \text{ volts}$$

appears across the coil, yielding a negative voltage spike as in Figure 14–16. Note that this spike is more than 20 times larger than the source voltage. As we see in the next section, the size of this spike depends on the ratio of R_2 to R_1; the larger the ratio, the larger the voltage.

Consider again Figure 14–15. Note that current i_2 changes abruptly from 0.2 A just prior to switching to -4 A just after. This is permissible, however, since i_2 does not pass through the inductor and only currents through inductance cannot change abruptly.

PRACTICE PROBLEMS 3

Figure 14–16 shows the voltage across the coil of Figure 14–14. Make a similar sketch for the voltage across the switch and across resistor R_2. Hint: Use KVL to find v_{SW} and v_{R_2}.

Answer
v_{SW}: With the switch closed, $v_{SW} = 0$ V; When the switch is opened, v_{SW} jumps to 2520 V, then decays to 120 V. v_{R_2}: Identical in shape to Figure 14–16 except that v_{R_2} begins at -2400 V instead of -2520 V.

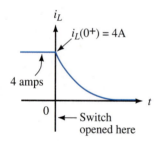

FIGURE 14–17   Inductor current for the circuit of Figure 14–15.

Inductor Equivalent at Switching

Figure 14–17 shows the current through L of Figure 14–15. Because the current is the same immediately after switching as it is immediately before, it is constant over the interval from $t = 0^-$ s to $t = 0^+$ s. Since this is true in general, we see that *an inductance with an initial current looks like a current source at the instant of switching*. Its value is the value of the current at switching, Figure 14–18. In any given problem, you can use either representation, but depending on what you are focusing on, you might choose one over the other.

(a) Current at switching   (b) Current source equivalent

FIGURE 14–18   An inductor carrying current looks like a current source at the instant of switching.

We now look at equations for the **de-energizing transient** voltages and currents described in the previous section.

Consider Figure 14–19(a). Let the initial current in the inductor be denoted as I_0 amps. Now open the switch as in (b). KVL yields $v_L + v_{R_1} + v_{R_2} = 0$. Substituting $v_L = Ldi/dt$, $v_{R_1} = R_1i$, and $v_{R_2} = R_2i$ yields $Ldi/dt + (R_1 + R_2)i = 0$. Now using calculus, it can be shown that

$$i = I_0e^{-t/\tau'} \quad \text{(A)} \tag{14–10}$$

where

$$\tau' = \frac{L}{R_T} = \frac{L}{R_1 + R_2} \quad \text{(s)} \tag{14–11}$$

is the time constant of the discharge circuit. If the circuit is in steady state before the switch is opened, initial current $I_0 = E/R_1$ and Equation 14–10 becomes

$$i = \frac{E}{R_1}e^{-t/\tau'} \quad \text{(A)} \tag{14–12}$$

14.4 De-energizing Transients

NOTES . . .

As we go through this material, you should focus on the basic principles involved, rather than just the resulting equations. You will probably forget formulas, but if you understand principles, you should be able to reason your way through many problems using just the basic current and voltage relationships.

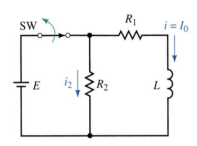

(a) Immediately before the switch is opened

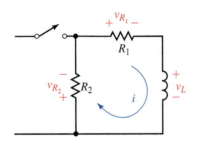

(b) Decay circuit

FIGURE 14–19   Circuit for studying decay transients.

For Figure 14–19(a), assume the current has reached steady state with the switch closed. Suppose that $E = 120$ V, $R_1 = 30$ Ω, $R_2 = 600$ Ω, and $L = 126$ mH:

a. Determine I_0.

b. Determine the decay time constant.

c. Determine the equation for the current decay.

d. Compute the current i at $t = 0.5$ ms.

EXAMPLE 14–6

Solution

a. Consider Figure 14–19(a). Since the circuit is in a steady state, the inductor looks like a short circuit to dc. Thus, $I_0 = E/R_1 = 4$ A.

b. Consider Figure 14–19(b). $\tau' = L/(R_1 + R_2) = 126$ mH/630 Ω $= 0.2$ ms.

c. $i = I_0e^{-t/\tau'} = 4e^{-t/0.2 \text{ ms}}$ A.

d. At $t = 0.5$ ms, $i = 4e^{-0.5 \text{ ms}/0.2 \text{ ms}} = 4e^{-2.5} = 0.328$ A.

Now consider voltage v_L. It can be shown to be

$$v_L = V_0 e^{-t/\tau'} \tag{14–13}$$

where V_0 is the voltage across L just after the switch is opened. Letting $i = I_0$ in Figure 14–19(b), you can see that $V_0 = -I_0(R_1 + R_2) = -I_0 R_T$. Thus Equation 14–13 can be written as

$$v_L = -I_0 R_T e^{-t/\tau'} \tag{14–14}$$

Finally, if the current has reached steady state before the switch is opened, $I_0 = E/R_1$, and Equation 14–14 becomes

$$v_L = -E\left(1 + \frac{R_2}{R_1}\right)e^{-t/\tau'} \tag{14–15}$$

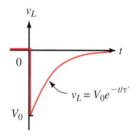

Note that v_L starts at V_0 volts (which is negative) and decays to zero as shown in Figure 14–20.

Now consider the resistor voltages. Each is the product of resistance times current (Equation 14–10). Thus,

$$v_{R_1} = R_1 I_0 e^{-t/\tau'} \tag{14–16}$$

and

$$v_{R_2} = R_2 I_0 e^{-t/\tau'} \tag{14–17}$$

FIGURE 14–20 Inductor voltage during decay phase. V_0 is negative.

If current has reached steady state before switching, these become

$$v_{R_1} = E e^{-t/\tau'} \tag{14–18}$$

and

$$v_{R_2} = \frac{R_2}{R_1} E e^{-t/\tau'} \tag{14–19}$$

Substituting the values of Example 14–6 into these equations, we get for the circuit of Figure 14–19 $v_L = -2520 e^{-t/0.2 \text{ ms}}$ V, $v_{R_1} = 120 e^{-t/0.2 \text{ ms}}$ V and $v_{R_2} = 2400 e^{-t/0.2 \text{ ms}}$ V. These can also be written as $v_L = -2520 e^{-5000t}$ V and so on if desired.

Decay problems can also be solved using the decay portion of the universal time constant curves shown in Figure 14–10.

EXAMPLE 14–7

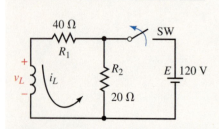

FIGURE 14–21

The circuit of Figure 14–21 is in steady state with the switch closed. Use Figure 14–10 to find i_L and v_L at $t = 2\,\tau$ after the switch is opened.

Solution $I_0 = E/R_1 = 3$ A. At $t = 2\,\tau$, current will have decayed to 13.5%. Therefore, $i_L = 0.135 I_0 = 0.405$ A and $v_L = -(R_1 + R_1)i = -(60\ \Omega)(0.405\ \text{A}) = -24.3$ V. (Alternately, $V_0 = -(3\ \text{A})(60\ \Omega) = -180$ V. At $t = 2\,\tau$, this has decayed to 13.5%. Therefore, $v_L = 0.135(-180\ \text{V}) = -24.3$ V as above.)

The equations developed so far apply only to circuits of the forms of Figures 14–7 or 14–19. Fortunately, many circuits can be reduced to these forms using circuit reduction techniques such as series and parallel combinations, source conversions, Thévenin's theorem, and so on.

14.5 More Complex Circuits

Determine i_L for the circuit of Figure 14–22(a) if $L = 5$ H.

EXAMPLE 14–8

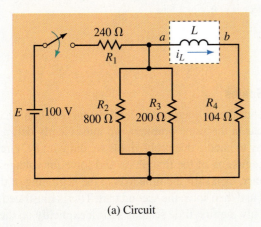

(a) Circuit

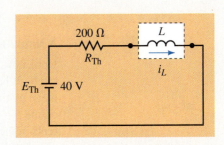

(b) Thévenin equivalent

FIGURE 14–22

Solution   The circuit can be reduced to its Thévenin equivalent (b) as you saw in Chapter 11 (Section 11.5). For this circuit, $\tau = L/R_{Th} = 5$ H/200 Ω = 25 ms. Now apply Equation 14–7. Thus,

$$i_L = \frac{E_{Th}}{R_{Th}} (1 - e^{-t/\tau}) = \frac{40}{200} (1 - e^{-t/25 \text{ ms}}) = 0.2 \, (1 - e^{-40t}) \quad \text{(A)}$$

For the circuit of Example 14–8, at what time does current reach 0.12 amps?

EXAMPLE 14–9

Solution

$$i_L = 0.2(1 - e^{-40t}) \text{ (A)}$$

Thus,

$$0.12 = 0.2(1 - e^{-40t}) \quad \text{(Figure 14–23)}$$
$$0.6 = 1 - e^{-40t}$$
$$e^{-40t} = 0.4$$

Taking the natural log of both sides,

$$\ln e^{-40t} = \ln 0.4$$
$$-40t = -0.916$$
$$t = 22.9 \text{ ms}$$

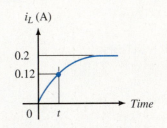

FIGURE 14–23

1. For the circuit of Figure 14–22, let $E = 120$ V, $R_1 = 600$ Ω, $R_2 = 3$ kΩ, $R_3 = 2$ kΩ, $R_4 = 100$ Ω, and $L = 0.25$ H:

   a. Determine i_L and sketch it.

   b. Determine v_L and sketch it.

2. Let everything be as in Problem 1 except L. If $i_L = 0.12$ A at $t = 20$ ms, what is L?

Answers

1. a. $160(1 - e^{-2000t})$ mA; b. $80e^{-2000t}$ V. i_L climbs from 0 to 160 mA with the waveshape of Figure 14–1(b), reaching steady state in 2.5 ms. v_L looks like Figure 14–2(c). It starts at 80 V and decays to 0 V in 2.5 ms.

2. 7.21 H

A Note About Time Scales

Until now, we have considered energization and de-energization phases separately. When both occur in the same problem, we must clearly define what we mean by time. One way to handle this problem (as we did with *RC* circuits) is to define $t = 0$ s as the beginning of the first phase and solve for voltages and currents in the usual manner, then shift the time axis to the beginning of the second phase, redefine $t = 0$ s and then solve the second part. This is illustrated in Example 14–10. Note that only the first time scale is shown explicitly on the graph.

EXAMPLE 14–10

Refer to the circuit of Figure 14–24:

a. Close the switch at $t = 0$ and determine equations for i_L and v_L.

b. At $t = 300$ ms, open the switch and determine equations for i_L and v_L during the decay phase.

c. Determine voltage and current at $t = 100$ ms and at $t = 350$ ms.

d. Sketch i_L and v_L. Mark the points from (c) on the sketch.

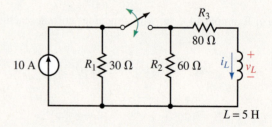

FIGURE 14–24

Solution

a. Convert the circuit to the left of L to its Thévenin equivalent. As indicated in Figure 14–25(a), $R_{Th} = 60\|30 + 80 = 100$ Ω. From (b), $E_{Th} = V_2$, where

$$V_2 = (10 \text{ A})(20 \text{ Ω}) = 200 \text{ V}$$

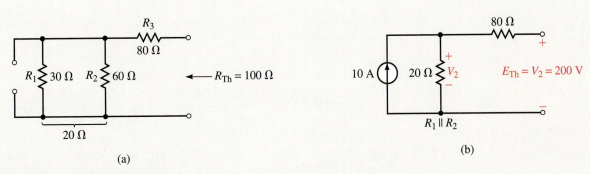

FIGURE 14–25

The Thévenin equivalent circuit is shown in Figure 14–26(a). $\tau = L/R_{Th} =$ 50 ms. Thus during current buildup,

$$i_L = \frac{E_{Th}}{R_{Th}}(1 - e^{-t/\tau}) = \frac{200}{100}(1 - e^{-t/50 \text{ ms}}) = 2(1 - e^{-20t}) \quad \text{A}$$

$$v_L = E_{Th}e^{-t/\tau} = 200\,e^{-20t} \quad \text{V}$$

b. Current build-up is sketched in Figure 14–26(b). Since $5\tau = 250$ ms, current is in steady state when the switch is opened at 300 ms. Thus $I_0 = 2$ A. When the switch is opened, current decays to zero through a resistance of $60 + 80 = 140\ \Omega$ as shown in Figure 14–27. Thus, $\tau' = 5\text{H}/140\ \Omega = 35.7$ ms. If $t = 0$ s is redefined as the instant the switch is opened, the equation for the decay is

$$i_L = I_0 e^{-t/\tau'} = 2e^{-t/35.7 \text{ ms}} = 2e^{-28t} \quad \text{A}$$

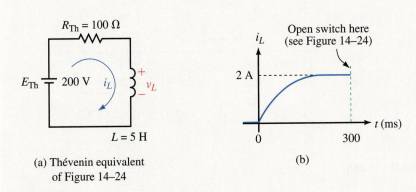

(a) Thévenin equivalent
    of Figure 14–24

(b)

FIGURE 14–26 Circuit and current during the buildup phase.

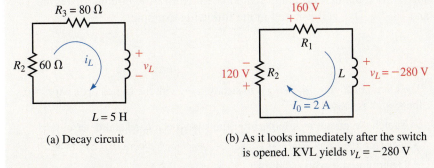

(a) Decay circuit

(b) As it looks immediately after the switch
    is opened. KVL yields $v_L = -280$ V

FIGURE 14–27 The circuit of Figure 14–24 as it looks during the decay phase.

Now consider voltage. As indicated in Figure 14–27(b), the voltage across L just after the switch is open is $V_0 = -280$ V. Thus

$$v_L = V_0 e^{-t/\tau'} = -280 e^{-28t} \text{ V}$$

c. You can use the universal time constant curves at $t = 100$ ms since 100 ms represents 2τ. At 2τ, current has reached 86.5% of its final value. Thus, $i_L = 0.865(2 \text{ A}) = 1.73$ A. Voltage has fallen to 13.5%. Thus $v_L = 0.135(200 \text{ V}) = 27.0$ V. Now consider $t = 350$ ms: Note that this is 50 ms into the decay portion of the curve. However, since 50 ms is not a multiple of τ', it is difficult to use the curves. Therefore, use the equations. Thus,

$$i_L = 2 \text{ A } e^{-28(50 \text{ ms})} = 2 \text{ A } e^{-1.4} = 0.493 \text{ A}$$
$$v_L = (-280 \text{ V})e^{-28(50 \text{ ms})} = (-280 \text{ V})e^{-1.4} = -69.0 \text{ V}$$

d. The above points are plotted on the waveforms of Figure 14–28.

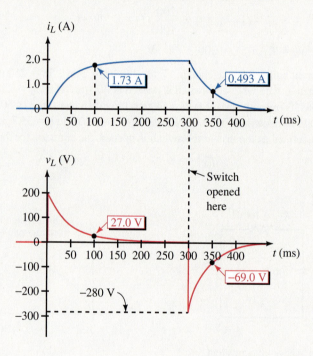

FIGURE 14–28

The basic principles that we have developed in this chapter permit us to solve problems that do not correspond exactly to the circuits of Figure 14–7 and 14–19. This is illustrated in the following example.

EXAMPLE 14–11

The circuit of Figure 14–29(a) is in steady state with the switch open. At $t = 0$ s, the switch is closed.

a. Sketch the circuit as it looks after the switch is closed and determine τ'.

b. Determine current i_L at $t = 0^+$ s.

c. Determine the expression for i_L.

d. Determine v_L at $t = 0^+$ s.

e. Determine the expression for v_L.

f. How long does the transient last?

g. Sketch i_L and v_L.

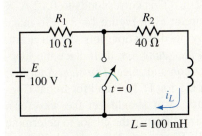

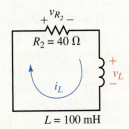

(a) Steady state current with the switch open is $\dfrac{100\text{ V}}{50\text{ }\Omega} = 2\text{ A}$

(b) Decay circuit $\tau' = \dfrac{L}{R_2} = 2.5$ ms

FIGURE 14–29

◀ MULTISIM

Solution

a. When you close the switch, you short out the E-R_1 branch, leaving the decay circuit of (b). Thus $\tau' = L/R_2 = 100$ mH/40 Ω = 2.5 ms.

b. In steady state with the switch open, $i_L = I_0 = 100$ V/50 Ω = 2 A. This is the current just before the switch is closed. Therefore, just after the switch is closed, i_L will still be 2 A (Figure 14–30).

c. i_L decays from 2 A to 0. From Equation 14–10, $i_L = I_0 e^{-t/\tau'} = 2e^{-t/2.5\text{ms}} = 2e^{-400t}$ A.

d. KVL yields $v_L = -v_{R_2} = -R_2 I_0 = -(40\text{ }\Omega)(2\text{A}) = -80$ V. Thus, $V_0 = -80$ V.

e. v_L decays from -80 V to 0. Thus, $v_L = V_0 e^{-t/\tau'} = -80e^{-400t}$ V.

f. Transients last $5\tau' = 5(2.5$ ms$) = 12.5$ ms.

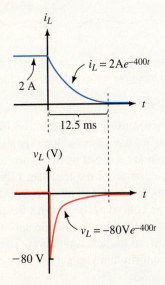

FIGURE 14–30

14.6 *RL* Transients Using Computers

NOTES . . .

1. Only basic steps are given for the following computer examples, as procedures here are similar to those of Chapter 11. If you need help, refer back to Chapter 11 or to Appendix A.

2. When scaling values from a computer plot, it is not always possible to place the cursor exactly where you want it (because of the nature of simulation programs). Consequently, you may have to set it as closely as you can get, then estimate the value you are trying to measure.

MultiSIM

MultiSIM can easily plot voltage, but it has no simple way to plot current. If you want to determine current, use Ohm's law and the applicable voltage waveform. To illustrate, consider Figure 14–22(a). Since the inductor current passes through R_4, we can use the voltage across R_4 to monitor current. Suppose we want to know at what time i_L reaches 0.12 A. (This corresponds to 0.12 A $\times$ 104 Ω = 12.48 V across R_4.) Create the circuit as in Figure 14–31. (To make connections to the R2/R3 branch, you may need junction dots. To get a dot, click your right mouse button and select *Place Junction,* then place the junction where you need it.) Click the Analysis icon and select *Transient Analysis;* in the Transient Analysis dialog box, enter 0.1 for TSTOP and for Initial Conditions, select *Set to zero.* After simulation, you should get the waveform shown in Figure 14–31(a). Expand to full size, then use the cursor to determine the time at which voltage equals 12.48 V. Figure 14–31(b) shows 22.9 ms (which agrees with the answer we obtained earlier in Example 14–9).

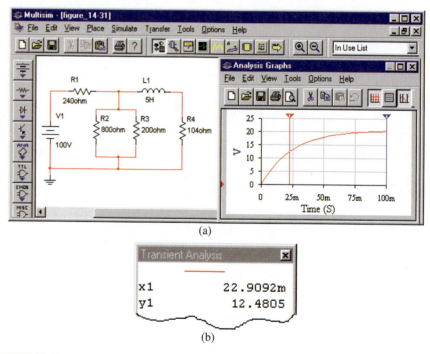

(a)

(b)

FIGURE 14–31 (a) MultiSIM representation of Figure 14–22. No switch is required as the transient solution is initiated by software. (b) Values scaled from the waveform of (a).

Using MultiSIM's Oscilloscope

Waveforms may be observed using MultiSIM's oscilloscope. Build the circuit of Figure 14–32(a). (Here we have used a clock source as a simple way to apply a step voltage to the circuit at $t = 0$. Set its frequency to 0.5 Hz so that its period is much longer than the duration of the transient.) Double click the scope and set its time base to 10 ms/div, set Channel A to 5 V/div and Y position to −3. (This starts the trace at the bottom of the screen.) Set the trigger level to a small positive value (say 1 V), select *Sing* (for single shot trace) and *Ext;* then click the ON/OFF power switch to initiate simulation. The waveform of (b) should result. With the cursor, confirm that you get the answer of Figure 14–31(b).

PSpice

RL transients are handled much like *RC* transients. As a first example, consider Figure 14–33. The circuit is in steady state with the switch closed. At $t = 0$, the

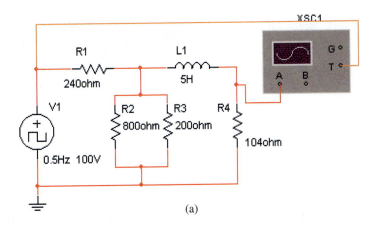

(a)

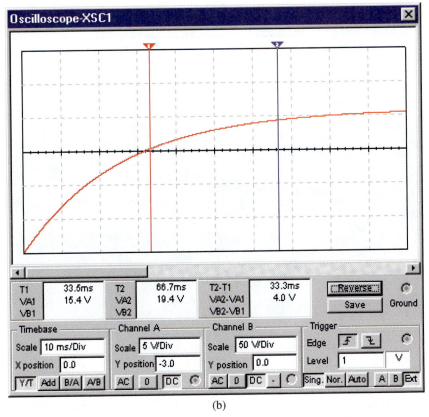

(b)

FIGURE 14–32   The scope needs no ground as MultiSIM grounds it automatically.

switch is opened. Use PSpice to plot inductor voltage and current, then use the cursor to determine values at $t = 100$ ms. Verify manually.

Preliminary   First, note that after the switch is opened, current builds up through R_1, R_2, and R_3 in series. Thus, the time constant of the circuit is $\tau = L_1/R_T = 3$ H/30 $\Omega = 0.1$ s. Now proceed as follows:

- Build the circuit on the screen (see Notes 1 and 2). Click the New Profile icon and name the file. In the Simulations Settings box, select transient analysis and set TSTOP to 0.5 (five time constants). Click OK.

- Click the Run icon. When simulation is complete, a trace of inductor voltage versus time appears. Create a second $\underline{Y}$ Axis, then add the current trace I(L1). You should now have the curves of Figure 14–34 on the screen. (The Y-axes can be labeled if desired as described in Appendix A.)

NOTES...

PSpice

1. Since the process here is similar to that of Chapter 11, only abbreviated instructions are given.

2. You have two choices: You can let PSpice automatically determine the initial condition (that is, the initial inductor current I_0) as we discussed in Appendix A, or you can compute it and enter it yourself. In this example, we let PSpice do it.

3. If you want to set the initial condition yourself, compute it as in Note 4 and double click the inductor symbol; in the Property Editor that opens, type **3A** into the cell labeled IC, click Apply, then close the editor. Run the simulation in the normal manner. Try it.

4. For this problem, it is easy to determine the initial inductor current. Note that before the switch is opened, the R_3/L_1 branch has 12 V across it. Since the circuit is in steady state dc, the inductor looks like a short circuit and $I_0 = 12$ V/4 $\Omega = 3$ A, confirming the result shown in Figure 14–34.

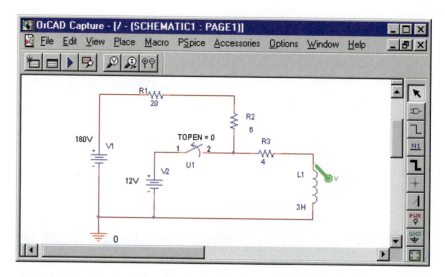

FIGURE 14–33 PSpice can easily determine both voltage and current transients. The voltage marker displays voltage across the inductance. The current display is created after the simulation is run.

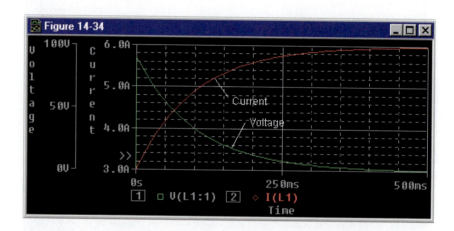

FIGURE 14–34 Inductor voltage and current for the circuit of Figure 14–33.

Results Consider Figure 14–33. With the switch open, steady state current is 180 V/30 Ω = 6 A and the initial current is 3 A. Thus, the current should start at 3 A and rise to 6 A in 5 time constants. (It does.) Inductor voltage should start at 180 V − (3 A)(30 Ω) = 90 V and decay to 0 V in 5 time constants. (It does). Thus, the solution checks. Now, with the cursor, scale voltage and current values at t = 100 ms. You should get 33.1 V for v_L and 4.9 A for i_L. (To check, note that the equations for inductor voltage and current are $v_L = 90\,e^{-10t}$ V and $i_L = 6 - 3\,e^{-10t}$ A respectively. Substitute t = 100 ms into these and verify results.)

EXAMPLE 14–12

Consider the circuit of Figure 14–24, Example 14–10. The switch is closed at t = 0 and opened 300 ms later. Prepare a PSpice analysis of this problem and determine v_L and i_L at t = 100 ms and at t = 350 ms.

Solution PSpice doesn't have a switch that both opens and closes. However, you can simulate such a switch by using two switches as in Figure 14–35. Begin by creating the circuit on the screen using IDC for the current source. Now double click TOPEN of switch U2 and set it to 300 ms. Click the New Profile icon and name the file. In the Simulation Settings box, select transient analysis then

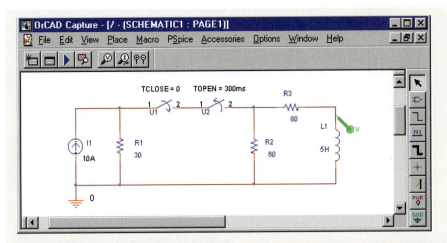

FIGURE 14–35   Simulating the circuit of Example 14–10. Two switches are used to model the closing and opening of the switch of Figure 14–24.

type in a value of 0.5 for TSTOP. Run the simulation, create a second Y-axis, then add the current trace I(L1). You should now have the curves of Figure 14–36 on the screen. (Compare to Figure 14–28.) Using the cursor, read values at $t = 100$ ms and 350 ms. You should get approximately 27 V and 1.73 A at $t = 100$ ms and –69 V and 490 mA at $t = 350$ ms. Note how well these agree with the results of Example 14–10.

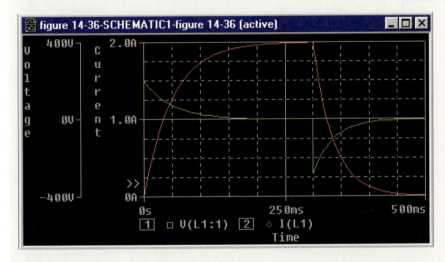

FIGURE 14–36   Inductor voltage and current for the circuit of Figure 14–35.

PUTTING IT INTO PRACTICE

The first sample of a new product that your company has designed has an indicator light that fails. (Symptom: When you turn a new unit on, the indicator light comes on as it should. However, when you turn the power off and back on, the lamp does not come on again.) You have been asked to investigate the problem and design a fix. You acquire a copy of the schematic and study the portion of the circuit where the indicator lamp is located. As shown in the accompanying figure, the lamp is used to indicate the status of the coil; the light

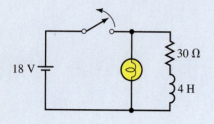

is to be on when the coil is energized and off when it is not. Immediately, you see the problem, solder in one component and the problem is fixed. Write a short note to your supervisor outlining the nature of the problem, explaining why the lamp burned out and why your design modification fixed the problem. Note also that your modification did not result in any substantial increase in power consumption (i.e., you did not use a resistor). Note: This problem requires a diode. If you have not had an introduction to electronics, you will need to obtain a basic electronics book and read about it.

PROBLEMS

14.1   Introduction

1. a. What does an inductor carrying no current look like at the instant of switching?

   b. For each circuit of Figure 14–37, determine i_S and v_L immediately after the switch is closed.

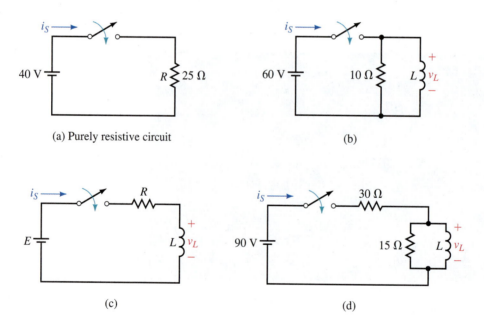

FIGURE 14–37   No value is needed for L here as it does not affect the solution.

2. Determine all voltages and currents in Figure 14–38 immediately after the switch is closed.

3. Repeat Problem 2 if L_1 is replaced with an uncharged capacitor.

14.2   Current Buildup Transients

4. a. If $i_L = 8(1 - e^{-500t})$ A, what is the current at $t = 6$ ms?

   b. If $v_L = 125e^{-500t}$ V, what is the voltage v_L at $t = 5$ ms?

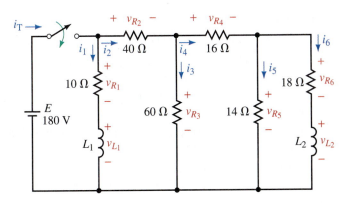

FIGURE 14-38

5. The switch of Figure 14–39 is closed at $t = 0$ s.
   a. What is the time constant of the circuit?
   b. How long is it until current reaches its steady value?
   c. Determine the equations for i_L and v_L.
   d. Compute values for i_L and v_L at intervals of one time constant from $t = 0$ to $5\,\tau$.
   e. Sketch i_L and v_L. Label the axis in τ and in seconds.

6. Close the switch at $t = 0$ s and determine equations for i_L and v_L for the circuit of Figure 14–40. Compute i_L and v_L at $t = 1.8$ ms.

7. Repeat Problem 5 for the circuit of Figure 14–41 with $L = 4$ H.

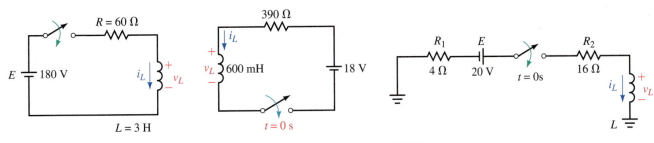

FIGURE 14-39   **FIGURE 14-40**   **FIGURE 14-41**

8. For the circuit of Figure 14–39, determine inductor voltage and current at $t = 50$ ms using the universal time constant curve of Figure 14–10.

9. Close the switch at $t = 0$ s and determine equations for i_L and v_L for the circuit of Figure 14–42. Compute i_L and v_L at $t = 3.4$ ms.

10. Using Figure 14–10, find v_L at one time constant for the circuit of Figure 14–42.

11. For the circuit of Figure 14–1(b), the voltage across the inductance at the instant the switch is closed is 80 V, the final steady state current is 4 A, and the transient lasts 0.5 s. Determine E, R, and L.

12. For an RL circuit, $i_L = 20(1 - e^{-t/\tau})$ mA and $v_L = 40e^{-t/\tau}$ V. If the transient lasts 0.625 ms, what are E, R, and L?

13. For Figure 14–1(b), if $v_L = 40e^{-2000t}$ V and the steady state current is 10 mA, what are E, R, and L?

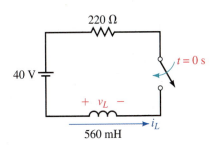

FIGURE 14-42

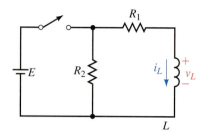

FIGURE 14–43

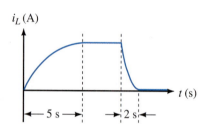

i_L (A)

FIGURE 14–44

5 s   2 s

t (s)

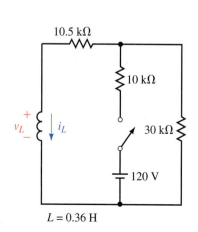

10.5 kΩ

10 kΩ

30 kΩ

v_L i_L

120 V

$L = 0.36$ H

FIGURE 14–46

14.4   De-energizing Transients

14. For Figure 14–43, $E = 80$ V, $R_1 = 200$ Ω, $R_2 = 300$ Ω, and $L = 0.5$ H.

a. When the switch is closed, how long does it take for i_L to reach steady state? What is its steady state value?

b. When the switch is opened, how long does it take for i_L to reach steady state? What is its steady state value?

c. After the circuit has reached steady state with the switch closed, it is opened. Determine equations for i_L and v_L.

15. For Figure 14–43, $R_1 = 20$ Ω, $R_2 = 230$ Ω, and $L = 0.5$ H, and the inductor current has reached a steady value of 5 A with the switch closed. At $t = 0$ s, the switch is opened.

a. What is the decay time constant?

b. Determine equations for i_L and v_L.

c. Compute values for i_L and v_L at intervals of one time constant from $t = 0$ to $5\,\tau$.

d. Sketch i_L and v_L. Label the axis in τ and in seconds.

16. Using the values from Problem 15, determine inductor voltage and current at $t = 3\tau$ using the universal time constant curves shown in Figure 14–10.

17. Given $v_L = -2700\,Ve^{-100t}$. Using the universal time constant curve, find v_L at $t = 20$ ms.

18. For Figure 14–43, the inductor voltage at the instant the switch is closed is 150 V and $i_L = 0$ A. After the circuit has reached steady state, the switch is opened. At the instant the switch is opened, $i_L = 3$ A and v_L jumps to -750 V. The decay transient lasts 5 ms. Determine E, R_1, R_2, and L.

19. For Figure 14–43, $L = 20$ H. The current during buildup and decay is shown in Figure 14–44. Determine R_1 and R_2.

20. For Figure 14–43, when the switch is moved to energization, $i_L = 2$ A $(1 - e^{-10t})$. Now open the switch after the circuit has reached steady state and redefine $t = 0$ s as the instant the switch is opened. For this case, $v_L = -400\,Ve^{-25t}$. Determine E, R_1, R_2, and L.

14.5   More Complex Circuits

21. For the coil of Figure 14–45 $R_\ell = 1.7$ Ω and $L = 150$ mH. Determine coil current at $t = 18.4$ ms.

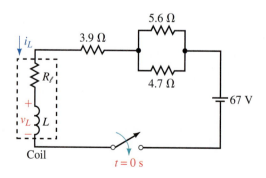

5.6 Ω

3.9 Ω

i_L

R_ℓ

4.7 Ω

67 V

v_L L

Coil

$t = 0$ s

FIGURE 14–45

22. Refer to Figure 14–46:

a. What is the energizing circuit time constant?

b. Close the switch and determine the equation for i_L and v_L during current buildup.

c. What is the voltage across the inductor and the current through it at $t = 20$ μs?

23. For Figure 14–46, the circuit has reached steady state with the switch closed. Now open the switch.

a. Determine the de-energizing circuit time constant.

b. Determine the equations for i_L and v_L.

c. Find the voltage across the inductor and current through it at $t = 17.8$ μs using the equations determined above.

24. Repeat Part (c) of Problem 23 using the universal time constant curves shown in Figure 14–10.

25. a. Repeat Problem 22, Parts (a) and (b) for the circuit of Figure 14–47.

b. What are i_L and v_L at $t = 25$ ms?

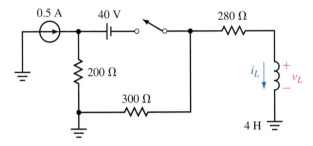

FIGURE 14–47

26. Repeat Problem 23 for the circuit of Figure 14–47, except find v_L and i_L at $t = 13.8$ ms.

27. An unknown circuit containing dc sources and resistors has an open-circuit voltage of 45 volts. When its output terminals are shorted, the short-circuit current is 0.15 A. A switch, resistor, and inductance are connected (Figure 14–48). Determine the inductor current and voltage 2.5 ms after the switch is closed.

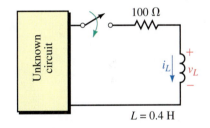

FIGURE 14–48

28. The circuit of Figure 14–49 is in steady state with the switch in position 1. At $t = 0$, it is moved to position 2, where it remains for 1.0 s. It is then moved to position 3, where it remains. Sketch curves for i_L and v_L from $t = 0^-$ until the circuit reaches steady state in position 3. Compute the inductor voltage and current at $t = 0.1$ s and at $t = 1.1$ s.

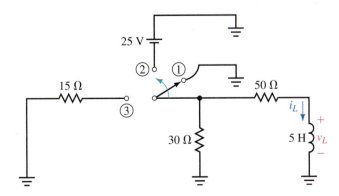

FIGURE 14–49

14.6   *RL* Transients Using Computers

29. The switch of Figure 14–46 is closed at $t = 0$ and remains closed. Graph the voltage across *L* and find v_L at 20 μs using the cursor.

MULTISIM CADENCE

30. For the circuit of Figure 14–47, close the switch at $t = 0$ and find v_L at $t = 10$ ms. (For PSpice, use current source IDC.)

MULTISIM CADENCE

31. For Figure 14–6, let $L_1 = 30$ mH and $L_2 = 90$ mH. Close the switch at $t = 0$ and find the current in the 30 Ω resistor at $t = 2$ ms. (Answer: 4.61 A) [Hint for MultiSIM users: Redraw the circuit with L_1 and the 30 Ω resistor interchanged. Finally, use Ohm's law.]

MULTISIM CADENCE

32. For Figure 14–41, let $L = 4$ H. Solve for v_L and, using the cursor, measure values at $t = 200$ ms and 500 ms. For PSpice users, also find current at these times.

CADENCE

33. We solved the circuit of Figure 14–22(a) by reducing it to its Thévenin equivalent. Using PSpice, analyze the circuit in its original form and plot the inductor current. Check a few points on the curve by computing values according to the solution of Example 14–8 and compare to values obtained from screen.

CADENCE

34. The circuit of Figure 14–46 is in steady state with the switch open. At $t = 0$, the switch is closed. It remains closed for 150 μs and is then opened and left open. Compute and plot i_L and v_L. With the cursor, determine values at $t = 60$ μs and at $t = 165$ μs.

✓ ANSWERS TO IN-PROCESS LEARNING CHECKS

In-Process Learning Check 1

1. a. $20e^{-100\,000t}$ V; $2(1 - e^{-100\,000t})$ mA

b.

$t(\mu s)$	v_L(V)	i_L(mA)
0	20	0
10	7.36	1.26
20	2.71	1.73
30	0.996	1.90
40	0.366	1.96
50	0.135	1.99

c.

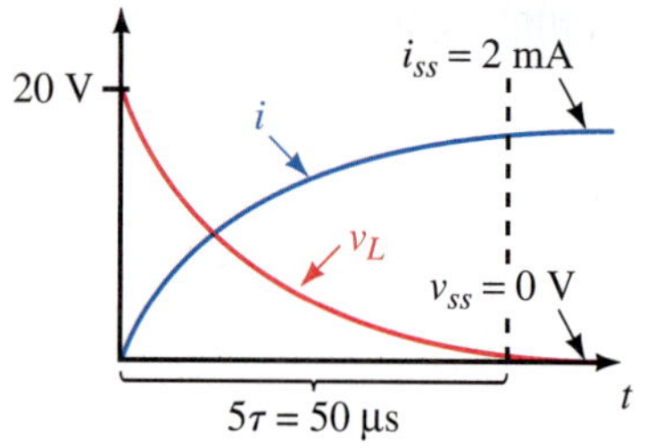

2. 11.5 V; 1.47 A
3. 4 V
4. 3.88 A

Foundation AC Concepts

IV

In previous chapters, we concentrated mostly on dc. We now turn our attention to ac (alternating current).

AC is important to us for a number of reasons. Firstly, it is the basis of the electrical power system that supplies our homes and businesses with electrical energy. AC is used instead of dc because it has several important advantages, the chief one being that ac power can be transmitted easily and efficiently over long distances. However, the importance of ac extends far beyond its use in the electrical power industry. The study of electronics, for example, deals to a large extent with ac, whether it be in the field of audio systems, communications systems, control systems, or any number of other areas. In fact, nearly every electrical and electronic device that we use in our daily lives operates from or involves the use of ac in some way.

We begin Part IV of this book with a look at fundamental ac concepts. We examine ways to generate ac voltages, methods used to represent ac voltages and currents, relationships between ac quantities in resistive, inductive, and capacitive circuits, and, finally, the meaning and representation of power in ac systems. This sets the stage for succeeding chapters that deal with ac circuit analysis techniques, including ac versions of the various methods that you have used for dc circuits throughout previous chapters of this book. ■

■ KEY TERMS

ac

Alternating Current

Alternating Voltage

Amplitude

Angular Velocity

Average Value

Cycle

Effective Value

Frequency

Hertz

Instantaneous Value

Lag

Lead

Oscilloscope

Peak Value

Period

Phase Shifts

Phasor

RMS

Sine Wave

Trapezoidal Rule

■ OUTLINE

Introduction

Generating AC Voltages

Voltage and Current Conventions for AC

Frequency, Period, Amplitude, and
  Peak Value

Angular and Graphic Relationships for
  Sine Waves

Voltages and Currents as Functions
  of Time

Introduction to Phasors

AC Waveforms and Average Value

Effective (RMS) Values

Rate of Change of a Sine Wave
  (Derivative)

AC Voltage and Current Measurement

Circuit Analysis Using Computers

■ OBJECTIVES

After studying this chapter, you will be
able to

• explain how ac voltages and currents differ from dc,

• draw waveforms for ac voltage and
  currents and explain what they
  mean,

• explain the voltage polarity and current direction conventions used for
  ac,

• describe the basic ac generator and
  explain how ac voltage is generated,

• define and compute frequency,
  period, amplitude, and peak-to-peak
  values,

• compute instantaneous sinusoidal
  voltage or current at any instant in
  time,

• define the relationships between ω,
  T, and f for a sine wave,

• define and compute phase differences between waveforms,

• use phasors to represent sinusoidal
  voltages and currents,

• determine phase relationships
  between waveforms using phasors,

• define and compute average values
  for time-varying waveforms,

• define and compute effective (rms)
  values for time-varying waveforms,

• use MultiSIM and PSpice to study
  ac waveforms.

AC Fundamentals

15

Alternating currents (ac) are currents that alternate in direction (usually many times per second), passing first in one direction, then in the other through a circuit. Such currents are produced by voltage sources whose polarities alternate between positive and negative (rather than being fixed as with dc sources). By convention, alternating currents are called *ac currents* and alternating voltages are called *ac voltages*.

The variation of an ac voltage or current versus time is called its waveform. Since waveforms vary with time, they are designated by lowercase letters *v(t), i(t), e(t),* and so on, rather than by uppercase letters *V, I,* and *E* as for dc. Often we drop the functional notation and simply use *v, i,* and *e.*

While many waveforms are important to us, the most fundamental is the sine wave (also called sinusoidal ac). In fact, the sine wave is of such importance that many people associate the term ac with sinusoidal, even though ac refers to any quantity that alternates with time.

In this chapter, we look at basic ac principles, including the generation of ac voltages and ways to represent and manipulate ac quantities. These ideas are then used throughout the remainder of the book to develop methods of analysis for ac circuits. ■

Thomas Alva Edison

NOWADAYS WE TAKE IT FOR GRANTED that our electrical power systems are ac. (This is reinforced every time you see a piece of equipment rated "60 hertz ac".) However, this was not always the case. In the late 1800s, a fierce battle—the so-called "war of the currents"—raged in the emerging electrical power industry. The forces favoring the use of dc were led by Thomas Alva Edison, and those favoring the use of ac were led by George Westinghouse (Chapter 23) and Nikola Tesla (Chapter 24).

Edison, a prolific inventor who gave us the electric light, the phonograph, and many other great inventions as well, fought vigorously for dc. He had spent a considerable amount of time and money on the development of dc power and had a lot at stake, in terms of both money and prestige. So determined was Edison in this battle that he first persuaded the state of New York to adopt ac for its newly devised electric chair, and then pointed at it with horror as an example of how deadly ac was. Ultimately, however, the combination of ac's advantages over dc and the stout opposition of Tesla and Westinghouse won the day for ac.

Edison was born in 1847 in Milan, Ohio. Most of his work was done at two sites in New Jersey—first at a laboratory in Menlo Park, and later at a much larger laboratory in West Orange, where his staff at one time numbered around 5,000. He received patents as inventor or co-inventor on nearly 1300 inventions—an astonishing feat that made him probably the greatest inventor of all time.

Thomas Edison died at the age of 84 on October 18, 1931. ∎

15.1 Introduction

Previously you learned that dc sources have fixed polarities and magnitudes and thus produce currents with constant value and unchanging direction, as illustrated in Figure 15–1. In contrast, the voltages of ac sources alternate in polarity and vary in magnitude and thus produce currents that vary in magnitude and alternate in direction.

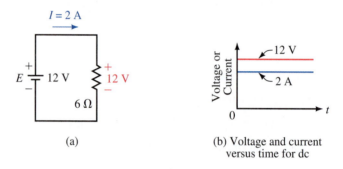

(a)

(b) Voltage and current versus time for dc

FIGURE 15–1   In a dc circuit, voltage polarities and current directions do not change.

Sinusoidal AC Voltage

To illustrate, consider the voltage at the wall outlet in your home. Called a **sine wave** or **sinusoidal ac waveform** (for reasons discussed in Section 15.5), this voltage has the shape shown in Figure 15–2. Starting at zero, the voltage increases to a positive maximum, decreases to zero, changes polarity, increases to a negative maximum, then returns again to zero. One complete variation is referred to as a **cycle.** Since the waveform repeats itself at regular intervals as in (b), it is called a **periodic** waveform.

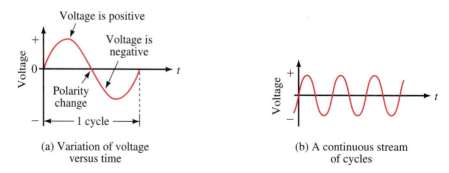

(a) Variation of voltage versus time

(b) A continuous stream of cycles

FIGURE 15–2   Sinusoidal ac waveforms. Values above the axis are positive while values below are negative.

Symbol for an AC Voltage Source

The symbol for a sinusoidal voltage source is shown in Figure 15–3. Note that a lowercase *e* is used to represent voltage rather than *E,* since it is a function of time. Polarity marks are also shown although, since the polarity of the source varies, their meaning has yet to be established.

FIGURE 15–3   Symbol for a sinusoidal voltage source. Lowercase letter *e* is used to indicate that the voltage varies with time.

Sinusoidal AC Current

Figure 15–4 shows a resistor connected to an ac source. During the first half-cycle, the source voltage is positive; therefore, the current is in the clockwise direction. During the second half-cycle, the voltage polarity reverses; therefore, the current is in the counterclockwise direction. Since current is proportional to voltage, its shape is also sinusoidal (Figure 15–5).

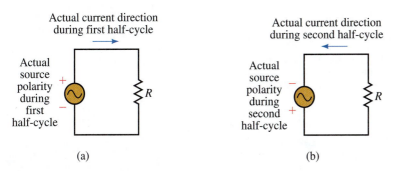

FIGURE 15–4   Current direction reverses when the source polarity reverses.

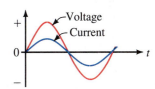

FIGURE 15–5   Current has the same wave shape as voltage.

15.2 Generating AC Voltages

One way to generate an ac voltage is to rotate a coil of wire at constant angular velocity in a fixed magnetic field, Figure 15–6. (Slip rings and brushes connect the coil to the load.) The magnitude of the resulting voltage is proportional to the rate at which flux lines are cut (Faraday's law, Chapter 13), and its polarity is dependent on the direction the coil sides move through the field. Since the rate of cutting flux varies with time, the resulting voltage will also vary with time. For example in (a), since the coil sides are moving parallel to the field, no flux lines are being cut and the induced voltage at this instant (and hence the current) is zero. (This is defined as the 0° position of the coil.) As the coil rotates from the 0° position, coil sides *AA'* and *BB'* cut across flux lines; hence, voltage builds, reaching a peak when flux is cut at the maximum rate in the 90° position as in (b). Note the polarity of the voltage and the direction of current. As the coil rotates further, voltage decreases, reaching zero at the 180° position when the coil sides again move parallel to the field as in (c). At this point, the coil has gone through a half-revolution.

During the second half-revolution, coil sides cut flux in directions opposite to that which they did in the first half revolution; hence, the polarity of the induced voltage reverses. As indicated in (d), voltage reaches a peak at the 270° point, and, since the polarity of the voltage has changed, so has the direction of current. When the coil reaches the 360° position, voltage is again zero and the cycle starts over. Figure 15–7 shows one cycle of the resulting waveform. Since the coil rotates continuously, the voltage produced will be a repetitive, periodic waveform as you saw in Figure 15–2(b). Current will be periodic also.

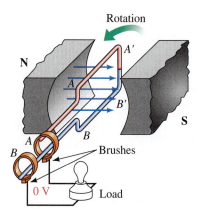

(a) 0° Position: Coil sides move parallel to flux lines. Since no flux is being cut, induced voltage is zero

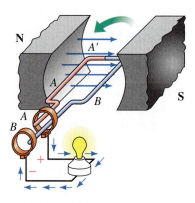

(b) 90° Position: Coil end A is positive with respect to B. Current direction is out of slip ring A

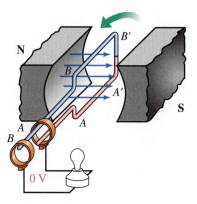

(c) 180° Position: Coil again cutting no flux. Induced voltage is zero

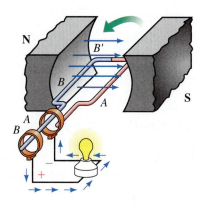

(d) 270° Position: Voltage polarity has reversed, therefore, current direction has also reversed

FIGURE 15–6   Generating an ac voltage. The 0° position of the coil is defined as in (a) where the coil sides move parallel to the flux lines.

PRACTICAL NOTES . . .

In practice, the coil of Figure 15–6 consists of many turns wound on an iron core. The coil, core, and slip rings rotate as a unit. In Figure 15–6, the magnetic field is fixed and the coil rotates. While small generators are built this way, large ac generators usually have the opposite construction, that is, their coils are fixed and the magnetic field is rotated instead. In addition, large ac generators are usually made as three-phase machines with three sets of coils instead of one. This is covered in Chapter 24. However, although its details are oversimplified, the generator of Figure 15–6 gives a true picture of the voltage produced by a real ac generator.

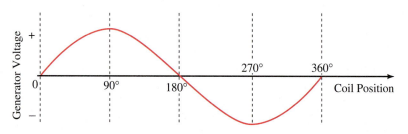

FIGURE 15–7   Coil voltage versus angular position.

Time Scales

The horizontal axis of Figure 15–7 is scaled in degrees. Often we need it scaled in time. The length of time required to generate one cycle depends on the velocity of rotation. To illustrate, assume that the coil rotates at 600 rpm (revolutions per minute). Six hundred revolutions in one minute equals 600 rev/60 s = 10 revolutions in one second. At ten revolutions per second, the time for one revolution is one tenth of a second, i.e., 100 ms. Since one cycle is 100 ms, a half-cycle is 50 ms, a quarter-cycle is 25 ms, and so on. Figure 15–8 shows the waveform rescaled in time.

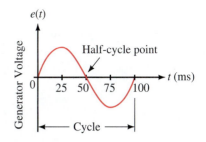

FIGURE 15–8   Cycle scaled in time. At 600 rpm, the cycle length is 100 ms.

Instantaneous Value

As Figure 15–8 shows, the coil voltage changes from instant to instant. The value of voltage at any point on the waveform is referred to as its **instantaneous value.** This is illustrated in Figure 15–9. Figure 15–9(a) shows a photograph of an actual waveform, and (b) shows it redrawn, with values scaled from the photo. For this example, the voltage has a peak value of 40 volts and a cycle time of 6 ms. From the graph, we see that at $t = 0$ ms, the voltage is zero. At $t = 0.5$ ms, it is 20 V. At $t = 2$ ms, it is 35 V. At $t = 3.5$ ms, it is -20 V, and so on.

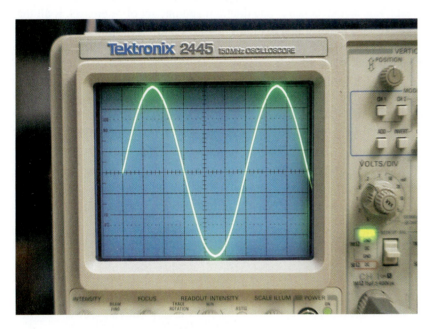

(a) Sinusoidal voltage

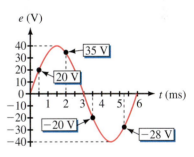

(b) Values scaled from the photograph

FIGURE 15–9   Instantaneous values.

Electronic Signal Generators

AC waveforms may also be created electronically using signal generators. In fact, with signal generators, you are not limited to sinusoidal ac. The general-purpose lab signal generator of Figure 15–10, for example, can produce a variety of variable-frequency waveforms, including sinusoidal, square wave, triangular, and so on. Waveforms such as these are commonly used to test electronic gear.

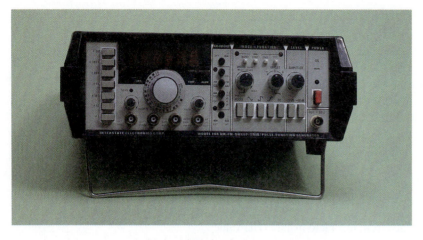

(a) A typical signal generator

Sine wave

Square wave

Triangle wave

(b) Sample waveforms

FIGURE 15–10 Electronic signal generators produce waveforms of different shapes.

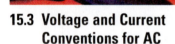

15.3 Voltage and Current Conventions for AC

In Section 15.1, we looked briefly at voltage polarities and current directions. At that time, we used separate diagrams for each half-cycle (Figure 15–4). However, this is unnecessary; one diagram and one set of references is all that is required. This is illustrated in Figure 15–11. First, we assign reference polarities for the source and a reference direction for the current. We then use the convention that, *when e has a positive value, its actual polarity is the same as the reference polarity, and when e has a negative value, its actual polarity is opposite to that of the reference.* For current, we use the convention that *when i has a positive value, its actual direction is the same as the reference arrow, and when i has a negative value, its actual direction is opposite to that of the reference.*

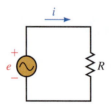

(a) References for voltage and current

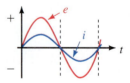

(b) During the first half-cycle, voltage polarity and current direction are as shown in (a). Therefore, *e* and *i* are positive. During the second half-cycle, voltage polarity and current direction are opposite to that shown in (a). Therefore, *e* and *i* are negative

FIGURE 15–11 AC voltage and current reference conventions.

To illustrate, consider Figure 15–12. (Parts (b) and (c) show snapshots at two instants of time.) At time t_1, *e* has a value of 10 volts. This means that at this instant, the voltage of the source is 10 V and its top end is positive with respect to its bottom end as indicated in (b). With a voltage of 10 V and a resistance of 5 Ω, the instantaneous value of current is $i = e/R = 10 \text{ V}/5 \text{ Ω} = 2 \text{ A}$. Since *i* is positive, the current is in the direction of the reference arrow.

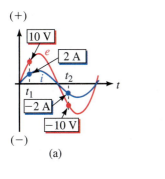

(a)

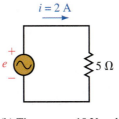

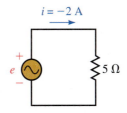

(b) Time t_1: $e = 10$ V and $i = 2$ A. Thus voltage and current have the polarity and direction indicated

(c) Time t_2: $e = -10$ V and $i = -2$ A. Thus, voltage polarity is opposite to that indicated and current direction is opposite to the arrow direction.

FIGURE 15–12 Illustrating the ac voltage and current convention.

Now consider time t_2. Here, $e = -10$ V. This means that source voltage is again 10 V, but now its top end is negative with respect to its bottom end. Again applying Ohm's law, you get $i = e/R = -10$ V/5 $\Omega = -2$ A. Since i is negative, current is actually opposite in direction to the reference arrow. This is indicated in (c).

The above concept is valid for any ac signal, regardless of waveshape.

Figure 15–13(b) shows one cycle of a triangular voltage wave. Determine the current and its direction at $t = 0, 1, 2, 3, 4, 5, 6, 7, 8, 9, 10, 11,$ and 12 μs and sketch.

EXAMPLE 15–1

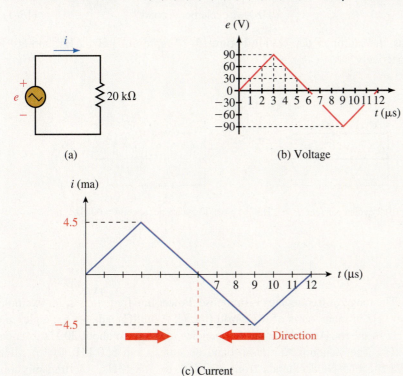

(a)

(b) Voltage

(c) Current

FIGURE 15–13

Solution   Apply Ohm's law at each point in time. At $t = 0$ μs, $e = 0$ V, so $i = e/R = 0$ V/20 k$\Omega = 0$ mA. At $t = 1$ μs, $e = 30$ V. Thus, $i = e/R = 30$ V/20 k$\Omega = 1.5$ mA. At $t = 2$ μs, $e = 60$ V. Thus, $i = e/R = 60$ V/20 k$\Omega = 3$ mA. Continuing in this manner, you get the values shown in Table 15–1. The waveform is plotted as Figure 15–13(c).

TABLE 15–1   Values for Example 15–1

t (μs)	e (V)	i (mA)
0	0	0
1	30	1.5
2	60	3.0
3	90	4.5
4	60	3.0
5	30	1.5
6	0	0
7	−30	−1.5
8	−60	−3.0
9	−90	−4.5
10	−60	−3.0
11	−30	−1.5
12	0	0

1. Let the source voltage of Figure 15–11 be the waveform of Figure 15–9. If $R = 2.5$ kΩ, determine the current at $t = 0, 0.5, 1, 1.5, 3, 4.5$, and 5.25 ms.

2. For Figure 15–13, if $R = 180$ Ω, determine the current at $t = 1.5, 3, 7.5$, and 9 μs.

Answers
1. 0, 8, 14, 16, 0, −16, −11.2 (all mA)
2. 0.25, 0.5, −0.25, −0.5 (all A)

15.4 Frequency, Period, Amplitude, and Peak Value

Periodic waveforms (i.e., waveforms that repeat at regular intervals), regardless of their waveshape, may be described by a group of attributes such as frequency, period, amplitude, peak value, and so on.

Frequency

The number of cycles per second of a waveform is defined as its **frequency.** In Figure 15–14(a), one cycle occurs in one second; thus its frequency is one cycle per second. Similarly, the frequency of (b) is two cycles per second and that of (c) is 60 cycles per second. Frequency is denoted by the lowercase letter f. In the SI system, its unit is the **hertz** (Hz, named in honor of pioneer researcher Heinrich Hertz, 1857–1894). By definition,

$$1 \text{ Hz} = 1 \text{ cycle per second} \qquad (15\text{–}1)$$

Thus, the examples depicted in Figure 15–14 represent 1 Hz, 2 Hz, and 60 Hz respectively.

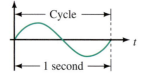
(a) 1 cycle per second = 1 Hz

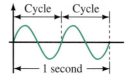
(b) 2 cycles per second = 2 Hz

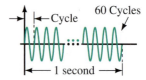
(c) 60 cycles per second = 60 Hz

FIGURE 15–14   Frequency is measured in hertz (Hz).

The range of frequencies is immense. Power line frequencies, for example, are 60 Hz in many parts of the world (the USA and Canada for example) and 50 Hz in others. Audible sound frequencies range from about 20 Hz to about 20 kHz. The standard AM radio band occupies from 550 kHz to 1.6 MHz, while the FM band extends from 88 MHz to 108 MHz. TV transmissions occupy several bands in the 54-MHz to 890-MHz range. Above 300 GHz are optical and X-ray frequencies.

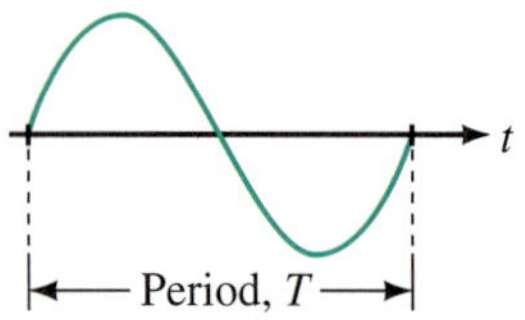

FIGURE 15–15   Period T is the duration of one cycle, measured in seconds.

Period

The **period,** T, of a waveform, (Figure 15–15) is the duration of one cycle. It is the inverse of frequency. To illustrate, consider again Figure 15–14. In (a), the frequency is 1 cycle per second; thus, the duration of each cycle is $T = 1$ s. In (b), the frequency is two cycles per second; thus, the duration of each cycle is $T = 1/2$ s, and so on. In general,

$$T = \frac{1}{f} \quad (s) \qquad\qquad \textbf{(15–2)}$$

and

$$f = \frac{1}{T} \quad (Hz) \qquad\qquad \textbf{(15–3)}$$

Note that these definitions are independent of wave shape.

a. What is the period of a 50-Hz voltage?

b. What is the period of a 1-MHz current?

Solution

$$(a) \quad T = \frac{1}{f} = \frac{1}{50 \text{ Hz}} = 20 \text{ ms}$$

$$(b) \quad T = \frac{1}{f} = \frac{1}{1 \times 10^6 \text{ Hz}} = 1 \text{ μs}$$

EXAMPLE 15–2

Figure 15–16 shows an oscilloscope trace of a square wave. Each horizontal division represents 50 μs. Determine the frequency.

EXAMPLE 15–3

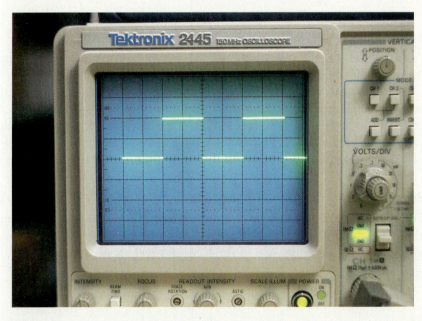

FIGURE 15–16 The concepts of frequency and period apply also to nonsinusoidal waveforms. Here, $T = 4$ div $\times$ 50 μs/div = 200 μs.

Solution Since the wave repeats itself every 200 μs, its period is 200 μs and

$$f = \frac{1}{200 \times 10^{-6} \text{ s}} = 5 \text{ kHz}$$

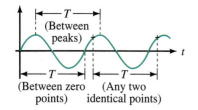

FIGURE 15–17 Period may be measured between any two corresponding points.

The period of a waveform can be measured between any two corresponding points (Figure 15–17). Often it is measured between zero points because they are easy to establish on an oscilloscope trace.

EXAMPLE 15–4

Determine the period and frequency of the waveform of Figure 15–18.

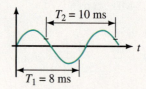

FIGURE 15–18

Solution   Time interval T_1 does not represent a period as it is not measured between corresponding points. Interval T_2, however, is. Thus, $T = 10$ ms and

$$f = \frac{1}{T} = \frac{1}{10 \times 10^{-3} \text{ s}} = 100 \text{ Hz}$$

Amplitude and Peak-to-Peak Value

The **amplitude** of a sine wave is the distance from its average to its peak. Thus, the amplitude of the voltage in Figures 15–19(a) and (b) is E_m.

   Peak-to-peak voltage is also indicated in Figure 15–19(a). It is measured between minimum and maximum peaks. Peak-to-peak voltages are denoted $E_{p\text{-}p}$ or $V_{p\text{-}p}$ in this book. (Some authors use $V_{pk\text{-}pk}$ or the like.) Similarly, peak-to-peak currents are denoted as $I_{p\text{-}p}$. To illustrate, consider again Figure 15–9. The amplitude of this voltage is $E_m = 40$ V, and its peak-to-peak voltage is $E_{p\text{-}p} = 80$ V.

Peak Value

The **peak value** of a voltage or current is its maximum value with respect to zero. Consider Figure 15–19(b). Here, a sine wave rides on top of a dc value, yielding a peak that is the sum of the dc voltage and the ac waveform amplitude. For the case indicated, the peak voltage is $E + E_m$.

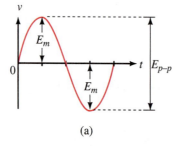

(a)

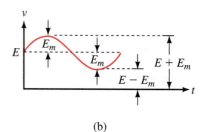

(b)

FIGURE 15–19 Definitions.

1. What is the period of an ac power system whose frequency is 60 Hz?

2. If you double the rotational speed of an ac generator, what happens to the frequency and period of the waveform?

3. If the generator of Figure 15–6 rotates at 3000 rpm, what is the period and frequency of the resulting voltage? Sketch four cycles and scale the horizontal axis in units of time.

4. For the waveform of Figure 15–9, list all values of time at which $e = 20$ V and $e = -35$ V. Hint: Sine waves are symmetrical.

5. Which of the waveform pairs of Figure 15–20 are valid combinations? Why?

✓ IN-PROCESS
LEARNING CHECK 1

(Answers are at the end of the chapter.)

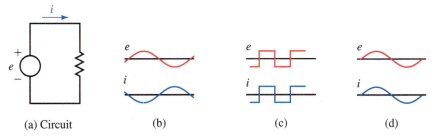

| (a) Circuit | (b) | (c) | (d) |

FIGURE 15–20   Which waveform pairs are valid?

6. For the waveform in Figure 15–21, determine the frequency.

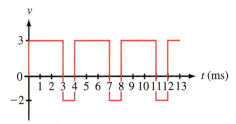

FIGURE 15–21

7. Two waveforms have periods of $T_1 = 10$ ms and $T_2 = 30$ ms respectively. Which has the higher frequency? Compute the frequencies of both waveforms.

8. Two sources have frequencies f_1 and f_2 respectively. If $f_2 = 20f_1$, and T_2 is 1 μs, what is f_1? What is f_2?

9. Consider Figure 15–22. What is the frequency of the waveform?

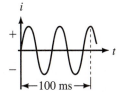

FIGURE 15–22

10. For Figure 15–11, if $f = 20$ Hz, what is the current direction at $t = 12$ ms, 37 ms, and 60 ms? Hint: Sketch the waveform and scale the horizontal axis in ms. The answers should be apparent.

11. A 10-Hz sinusoidal current has a value of 5 amps at $t = 25$ ms. What is its value at $t = 75$ ms? See hint in Problem 10.

15.5 Angular and Graphic Relationships for Sine Waves

The Basic Sine Wave Equation

Consider again the generator of Figure 15–6, reoriented and redrawn in end view as Figure 15–23. As the coil rotates, the voltage produced is

$$e = E_m \sin \alpha \quad (V) \qquad (15\text{–}4)$$

where E_m is the maximum coil voltage and α is the instantaneous angular position of the coil. (For a given generator and rotational velocity, E_m is constant.) Note that $\alpha = 0°$ represents the horizontal position of the coil and that one complete cycle corresponds to 360°. Equation 15–4 states that the voltage at any point on the sine wave may be found by multiplying E_m times the sine of the angle at that point.

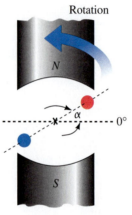

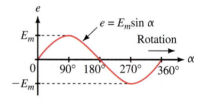

(a) End view showing coil position   (b) Voltage waveform

FIGURE 15–23   Coil voltage versus angular position.

EXAMPLE 15–5

If the amplitude of the waveform of Figure 15–23(b) is $E_m = 100$ V, determine the coil voltage at 30° and 330°.

Solution   At $\alpha = 30°$, $e = E_m \sin \alpha = 100 \sin 30° = 50$ V. At 330°, $e = 100 \sin 330° = -50$ V. These are shown on the graph of Figure 15–24.

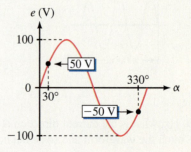

FIGURE 15–24

Table 15–2 is a tabulation of voltage versus angle computed from $e = 100 \sin \alpha$. Use your calculator to verify each value, then plot the result on graph paper. The resulting waveshape should look like Figure 15–24.

PRACTICE PROBLEMS 2

TABLE 15–2 Data for Plotting $e = 100 \sin \alpha$	
Angle α	Voltage e
0	0
30	50
60	86.6
90	100
120	86.6
150	50
180	0
210	−50
240	−86.6
270	−100
300	−86.6
330	−50
360	0

Angular Velocity, ω

The rate at which the generator coil rotates is called its **angular velocity.** If the coil rotates through an angle of 30° in one second, for example, its angular velocity is 30° per second. Angular velocity is denoted by the Greek letter ω (omega). For the case cited, $\omega = 30°/s$. (Normally angular velocity is expressed in radians per second instead of degrees per second. We will make this change shortly.) When you know the angular velocity of a coil and the length of time that it has rotated, you can compute the angle through which it has turned. For example, a coil rotating at 30°/s rotates through an angle of 30° in one second, 60° in two seconds, 90° in three seconds, and so on. In general,

$$\alpha = \omega t \qquad \text{(15–5)}$$

Expressions for t and ω can now be found. They are

$$t = \frac{\alpha}{\omega} \quad \text{(s)} \qquad \text{(15–6)}$$

$$\omega = \frac{\alpha}{t} \qquad \text{(15–7)}$$

If the coil of Figure 15–23 rotates at $\omega = 300°/s$, how long does it take to complete one revolution?

Solution One revolution is 360°. Thus,

$$t = \frac{\alpha}{\omega} = \frac{360 \text{ degrees}}{300 \dfrac{\text{degrees}}{\text{s}}} = 1.2 \text{ s}$$

Since this is one period, we should use the symbol T. Thus, $T = 1.2$ s, as in Figure 15–25.

EXAMPLE 15–6

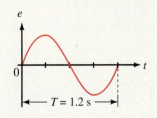

FIGURE 15–25

If the coil of Figure 15–23 rotates at 3600 rpm, determine its angular velocity, ω, in degrees per second.

Answer
21 600 deg/s

Radian Measure

In practice, ω is usually expressed in radians per second, where radians and degrees are related by the identity

$$2\pi \text{ radians} = 360° \qquad\qquad \textbf{(15–8)}$$

One radian therefore equals $360°/2\pi = 57.296°$. A full circle, as shown in Figure 15–26(a), can be designated as either 360° or 2π radians. Likewise, the cycle length of a sinusoid, shown in Figure 15–26(b), can be stated as either 360° or 2π radians; a half-cycle as 180° or π radians, and so on.

(a) 360° = 2π radians

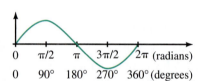

(b) Cycle length scaled in degrees
and radians

FIGURE 15–26   Radian measure.

TABLE 15–3 Selected Angles in Degrees and Radians	
Degrees	**Radians**
30	$\pi/6$
45	$\pi/4$
60	$\pi/3$
90	$\pi/2$
180	π
270	$3\pi/2$
360	2π

To convert from degrees to radians, multiply by $\pi/180$, while to convert from radians to degrees, multiply by $180/\pi$.

$$\alpha_{\text{radians}} = \frac{\pi}{180°} \times \alpha_{\text{degrees}} \qquad\qquad \textbf{(15–9)}$$

$$\alpha_{\text{degrees}} = \frac{180°}{\pi} \times \alpha_{\text{radians}} \qquad\qquad \textbf{(15–10)}$$

Table 15–3 shows selected angles in both measures.

EXAMPLE 15–7

a. Convert 315° to radians.

b. Convert $5\pi/4$ radians to degrees.

Solution

a. $\alpha_{\text{radians}} = (\pi/180°)(315°) = 5.5 \text{ rad}$

b. $\alpha_{\text{degrees}} = (180°/\pi)(5\pi/4) = 225°$

Scientific calculators can perform these conversions directly. You will find this more convenient than using the above formulas.

Graphing Sine Waves

A sinusoidal waveform can be graphed with its horizontal axis scaled in degrees, radians, or time. When scaled in degrees or radians, one cycle is always 360° or 2π radians (Figure 15–27); when scaled in time, it is frequency dependent, since the length of a cycle depends on the coil's velocity of rotation as we saw in Figure 15–8. However, if scaled in terms of period T instead of in seconds, the waveform is also frequency independent, since one cycle is always T, as shown in Figure 15–27(c).

When graphing a sine wave, you don't need many points to get a good sketch: Values every 45° (one eighth of a cycle) are generally adequate, Table 15–4. Often, you can simply "eyeball" the curve in as illustrated in Example 15–8.

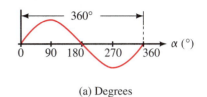

(a) Degrees

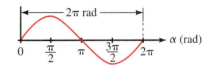

(b) Radians

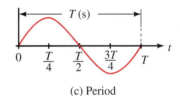

(c) Period

FIGURE 15–27 Comparison of various horizontal scales. Cycle length may be scaled in degrees, radians, or period. Each of these is independent of frequency.

TABLE 15–4	**Values for Rapid Sketching**		
α **(deg)**	α **(rad)**	t **(T)**	**Value of** $\sin \alpha$
0	0	0	0.0
45	$\pi/4$	$T/8$	0.707
90	$\pi/2$	$T/4$	1.0
135	$3\pi/4$	$3T/8$	0.707
180	π	$T/2$	0.0
225	$5\pi/4$	$5T/8$	−0.707
270	$3\pi/2$	$3T/4$	−1.0
315	$7\pi/4$	$7T/8$	−0.707
360	2π	T	0.0

Sketch the waveform for a 25-kHz sinusoidal current that has an amplitude of 4 mA. Scale the axis in seconds.

Solution For this waveform, $T = 1/25$ kHz = 40 μs. Thus,

1. Lay out the time axis with the end of the cycle marked as 40 μs, the half-cycle point as 20 μs, the quarter-cycle point as 10 μs, and so on (Figure 15–28).

2. The peak value (i.e., 4 mA) occurs at the quarter-cycle point, which is 10 μs on the waveform. Likewise, −4 mA occurs at 30 μs. Now sketch.

3. Values at other time points can be determined easily if needed. For example, the value at 5 μs can be calculated by noting that 5 μs is one eighth of a cycle, or 45°. Thus, $i = 4 \sin 45°$ mA = 2.83 mA. Alternately, from Table 15–4, at $T/8$, $i = (4$ mA$)(0.707) = 2.83$ mA. As many points as you need can be computed and plotted in this manner.

4. Values at particular angles can also be located easily. For instance, if you want a value at 30°, the required value is $i = 4 \sin 30°$ mA = 2.0 mA. To locate this point on the graph, note that 30° is one twelfth of a cycle or $T/12 = (40$ μs$)/12 = 3.33$ μs. The point is shown on Figure 15–28.

EXAMPLE 15–8

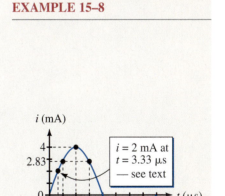

FIGURE 15–28

15.6 Voltages and Currents as Functions of Time

Relationship between ω, T, and f

Earlier you learned that one cycle of a sine wave may be represented as either $\alpha = 2\pi$ rads or $t = T$ s, Figure 15–27. Substituting these into $\alpha = \omega t$ (Equation 15–5), you get $2\pi = \omega T$. Transposing yields

$$\omega T = 2\pi \ (\text{rad}) \tag{15–11}$$

Thus,

$$\omega = \frac{2\pi}{T} \ (\text{rad/s}) \tag{15–12}$$

Recall, $f = 1/T$ Hz. Substituting this into Equation 15–12 you get

$$\omega = 2\pi f \ (\text{rad/s}) \tag{15–13}$$

EXAMPLE 15–9

In some parts of the world, the power system frequency is 60 Hz; in other parts, it is 50 Hz. Determine ω for each.

Solution For 60 Hz, $\omega = 2\pi f = 2\pi(60) = 377$ rad/s. For 50 Hz, $\omega = 2\pi f = 2\pi(50) = 314.2$ rad/s.

PRACTICE PROBLEMS 4

1. If $\omega = 240$ rad/s, what are T and f? How many cycles occur in 27 s?
2. If 56 000 cycles occur in 3.5 s, what is ω?

Answers
1. 26.18 ms, 38.2 Hz, 1031 cycles
2. 100.5×10^3 rad/s

Sinusoidal Voltages and Currents as Functions of Time

Recall from Equation 15–4, $e = E_m \sin \alpha$, and from Equation 15–5, $\alpha = \omega t$. Combining these equations yields

$$e = E_m \sin \omega t \tag{15–14a}$$

Similarly,

$$v = V_m \sin \omega t \tag{15–14b}$$

$$i = I_m \sin \omega t \tag{15–14c}$$

EXAMPLE 15–10

A 100-Hz sinusoidal voltage source has an amplitude of 150 volts. Write the equation for e as a function of time.

Solution $\omega = 2\pi f = 2\pi(100) = 628$ rad/s and $E_m = 150$ V. Thus, $e = E_m \sin \omega t = 150 \sin 628t$ V.

Equations 15–14 may be used to compute voltages or currents at any instant in time. Usually, ω is in radians per second, and thus ωt is in radians. You can work directly in radians or you can convert to degrees. For example, suppose you want to know the voltage at $t = 1.25$ ms for $e = 150 \sin 628t$ V.

Working in Rads. With your calculator in the RAD mode, $e = 150 \sin(628)(1.25 \times 10^{-3}) = 150 \sin 0.785 \text{ rad} = 106$ V.

Working in Degree. 0.785 rad = 45°. Thus, $e = 150 \sin 45° = 106$ V as before.

EXAMPLE 15–11

For $v = 170 \sin 2450t$, determine v at $t = 3.65$ ms and show the point on the v waveform.

Solution $\omega = 2450$ rad/s. Therefore $\omega t = (2450)(3.65 \times 10^{-3}) = 8.943$ rad = 512.4°. Thus, $v = 170 \sin 512.4° = 78.8$ V. Alternatively, $v = 170 \sin 8.943$ rad = 78.8 V. The point is plotted on the waveform in Figure 15–29.

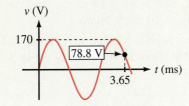

FIGURE 15–29

A sinusoidal current has a peak amplitude of 10 amps and a period of 120 ms.

PRACTICE PROBLEMS 5

a. Determine its equation as a function of time using Equation 15–14c.

b. Using this equation, compute a table of values at 10-ms intervals and plot one cycle of the waveform scaled in seconds.

c. Sketch one cycle of the waveform using the procedure of Example 15–8. (Note how much less work this is.)

Answers
a. $i = 10 \sin 52.36t$ A

c. Mark the end of the cycle as 120 ms, ½ cycle as 60 ms, ¼ cycle as 30 ms, etc. Draw the sine wave so that it is zero at $t = 0$, 10 A at 30 ms, 0 A at 60 ms, −10 A at 90 ms and ends at $t = 120$ ms. (See Figure 15–30.)

Determining When a Particular Value Occurs

Sometimes you need to know when a particular value of voltage or current occurs. Given $v = V_m \sin \alpha$. Rewrite this as $\sin \alpha = v/V_m$. Then,

$$\alpha = \sin^{-1}\frac{v}{V_m} \qquad (15\text{–}15)$$

Compute the angle α at which the desired value occurs using the inverse sine function of your calculator, then determine the time from

$$t = \alpha/\omega$$

EXAMPLE 15–12

A sinusoidal current has an amplitude of 10 A and a period of 0.120 s. Determine the times at which

a. $i = 5.0$ A,

b. $i = -5$ A.

Solution

a. Consider Figure 15–30. As you can see, there are two points on the waveform where $i = 5$ A. Let these be denoted t_1 and t_2 respectively. First, determine ω:

$$\omega = \frac{2\pi}{T} = \frac{2\pi}{0.120 \text{ s}} = 52.36 \text{ rad/s}$$

Let $i = 10 \sin \alpha$ A. Now, find the angle α_1 at which $i = 5$ A:

$$\alpha_1 = \sin^{-1}\frac{i}{I_m} = \sin^{-1}\frac{5 \text{ A}}{10 \text{ A}} = \sin^{-1}0.5 = 30° = 0.5236 \text{ rad}$$

Thus, $t_1 = \alpha_1/\omega = (0.5236 \text{ rad})/(52.36 \text{ rad/s}) = 0.01$ s $= 10$ ms. This is indicated in Figure 15–30. Now consider t_2. Because of symmetry, t_2 is the same distance back from the half-cycle point as t_1 is in from the beginning of the cycle. Thus, $t_2 = 60$ ms $- 10$ ms $= 50$ ms.

b. Similarly, t_3 (the first point at which $i = -5$ A occurs) is 10 ms past midpoint, while t_4 is 10 ms back from the end of the cycle. Thus, $t_3 = 70$ ms and $t_4 = 110$ ms.

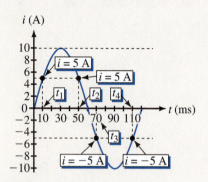

FIGURE 15–30

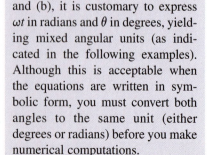

PRACTICE PROBLEMS 6

Given $v = 10 \sin 52.36t$, determine both occurrences of $v = -8.66$ V.

Answer
80 ms, 100 ms

Voltages and Currents with Phase Shifts

If a sine wave does not pass through zero at $t = 0$ s as in Figure 15–30, it has a **phase shift**. Waveforms may be shifted to the left or to the right (see Figure 15–31). For a waveform shifted left as in (a),

$$v = V_m\sin(\omega t + \theta) \qquad \qquad \textbf{(15–16a)}$$

while, for a waveform shifted right as in (b),

$$v = V_m\sin(\omega t - \theta) \qquad \qquad \textbf{(15–16b)}$$

NOTES . . .

With equations such as 15–16(a) and (b), it is customary to express ωt in radians and θ in degrees, yielding mixed angular units (as indicated in the following examples). Although this is acceptable when the equations are written in symbolic form, you must convert both angles to the same unit (either degrees or radians) before you make numerical computations.

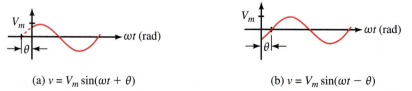

(a) $v = V_m \sin(\omega t + \theta)$ (b) $v = V_m \sin(\omega t - \theta)$

FIGURE 15–31   Waveforms with phase shifts. Angle θ is normally measured in degrees, yielding mixed angular units. (See Notes.)

EXAMPLE 15–13

Demonstrate that $v = 20 \sin(\omega t - 60°)$, where $\omega = \pi/6$ rad/s (i.e, $= 30°/$s), yields the shifted waveform shown in Figure 15–32.

Solution

1. Since ωt and $60°$ are both angles, $(\omega t - 60°)$ is also an angle. Let us define it as x. Then $v = 20 \sin x$, which means that the shifted wave is also sinusoidal.

2. Consider $v = \sin(\omega t - 60°)$. At $t = 0$ s, $v = 20 \sin(0 - 60°) = 20 \sin (-60°) = -17.3$ V as indicated in Figure 15–32.

3. Since $\omega = 30°/$s, it takes 2 s for ωt to reach $60°$. Thus, at $t = 2$ s, $v = 20 \sin(60° - 60°) = 0$ V, and the waveform passes through zero at $t = 2$ s as indicated.

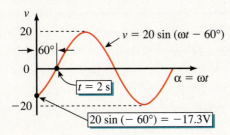

FIGURE 15–32

◀ MULTISIM

As you can see, this example confirms that $\sin (\omega t - \theta)$ describes the wave shape of Figure 15–31(b).

EXAMPLE 15–14

a. Determine the equation for the waveform of Figure 15–33(a), given $f = 60$ Hz. Compute current at $t = 4$ ms.

b. Repeat (a) for Figure 15–33(b).

Solution

a. $I_m = 2$ A and $\omega = 2\pi(60) = 377$ rad/s. This waveform corresponds to Figure 15–31(b) with $\theta = 120°$. Therefore,

$$i = I_m \sin(\omega t - \theta) = 2 \sin(377t - 120°) \text{ A}$$

At $t = 4$ ms, current is

$$i = 2 \sin(377 \times 4 \text{ ms} - 120°) = 2 \sin(1.508 \text{ rad} - 120°)$$
$$= 2 \sin(86.4° - 120°) = 2 \sin(-33.64°) = -1.11 \text{ A}.$$

b. This waveform matches Figure 15–31(a) if you extend the waveform back $90°$ from its peak as in (c). Note that $\theta = 40°$. Thus,

$$i = 2 \sin(377t + 40°) \text{ A}$$

At $t = 4$ ms, current is

$$i = 2 \sin(377 \times 4 \text{ ms} + 40°) = 2 \sin(126.4°)$$
$$= 1.61 \text{ A}.$$

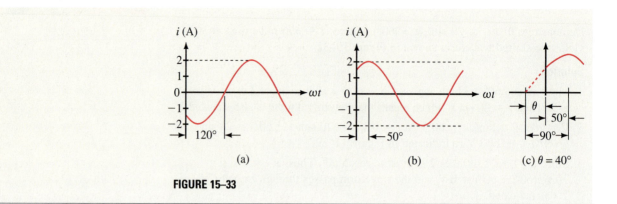

FIGURE 15–33

◀ **PRACTICE PROBLEMS 7**

1. Given $i = 2 \sin(377t + 60°)$, compute the current at $t = 3$ ms.
2. Sketch each of the following:
   a. $v = 10 \sin(\omega t + 20°)$ V.   b. $i = 80 \sin(\omega t - 50°)$ A.
   c. $i = 50 \sin(\omega t + 90°)$ A.   d. $v = 5 \sin(\omega t + 180°)$ V.
3. Given $i = 2 \sin(377t + 60°)$, determine at what time $i = 1.8$ A.

Answers
1. 1.64 A
2. a. Same as Figure 15–31(a) with $V_m = 10$ V, $\theta = 20°$.
    b. Same as Figure 15–31(b) with $I_m = 80$ A, $\theta = 50°$.
    c. Same as Figure 15–39(b) except use $I_m = 50$ A instead of V_m.
    d. A negative sine wave with magnitude of 5 V.
3. 0.193 ms

Probably the easiest way to deal with shifted waveforms is to use phasors. We introduce the idea next.

15.7 Introduction to Phasors

◀ **Online Companion**

A **phasor** is a rotating line whose projection on a vertical axis can be used to represent sinusoidally varying quantities. To get at the idea, consider the red line of length V_m shown in Figure 15–34(a). (It is the phasor.) The vertical projection of this line (indicated in dotted red) is $V_m \sin \alpha$. Now, assume that the phasor rotates at angular velocity of ω rad/s in the counterclockwise direction. Then, $\alpha = \omega t$, and its vertical projection is $V_m \sin \omega t$. If we designate this projection (height) as v, we get $v = V_m \sin \omega t$, which is the familiar sinusoidal voltage equation.

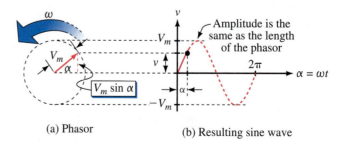

(a) Phasor   (b) Resulting sine wave

FIGURE 15–34   As the phasor rotates about the origin, its vertical projection creates a sine wave. (Figure 15–35 illustrates the process.)

If you plot a graph of this projection versus α, you get the sine wave of Figure 15–34(b). Figure 15–35 illustrates the graphing process. It shows snapshots of the phasor and the evolving waveform at various instants of time for a phasor of magnitude $V_m = 100$ V rotating at $\omega = 30°/s$. For example, consider $t = 0, 1, 2,$ and 3s:

1. At $t = 0$ s, $\alpha = 0$, the phasor is at its $0°$ position, and its vertical projection is $v = V_m\sin \omega t = 100 \sin 0° = 0$ V. The point is at the origin.

2. At $t = 1$ s, the phasor has rotated $30°$ and its vertical projection is $v = 100 \sin 30° = 50$ V. This point is plotted at $\alpha = 30°$ on the horizontal axis.

3. At $t = 2$ s, $\alpha = 60°$ and $v = 100 \sin 60° = 87$ V, which is plotted at $\alpha = 60°$ on the horizontal axis. Similarly, at $t = 3$ s, $\alpha = 90°$, and $v = 100$ V. Continuing in this manner, the complete waveform is evolved.

From the foregoing, we conclude that *a sinusoidal waveform can be created by plotting the vertical projection of a phasor that rotates in the counterclockwise direction at constant angular velocity ω. If the phasor has a length of V_m, the waveform represents voltage; if the phasor has a length of I_m, it represents current.* Note carefully: **Phasors apply only to sinusoidal waveforms.**

NOTES . . .

1. Although we have indicated phasor rotation in Figure 15–35 by a series of "snapshots," this is too cumbersome; in practice, we show only the phasor at its $t = 0$ s (reference) position and imply rotation rather than show it explicitly.

2. Although we are using maximum values (E_m and I_m) here, phasors are normally drawn in terms of effective (rms) values (considered in Section 15.9). For the moment, we will continue to use maximum values. We make the change to rms in Chapter 16.

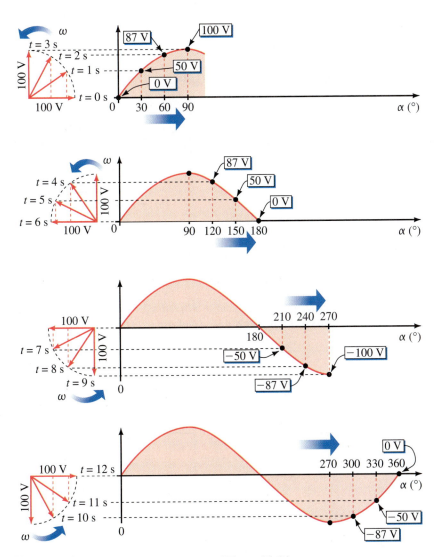

FIGURE 15–35   Evolution of the sine wave of Figure 15–34.

EXAMPLE 15–15

Draw the phasor and waveform for current $i = 25 \sin \omega t$ mA for $f = 100$ Hz.

Solution   The phasor has a length of 25 mA and is drawn at its $t = 0$ position, which is zero degrees as indicated in Figure 15–36. Since $f = 100$ Hz, the period is $T = 1/f = 10$ ms.

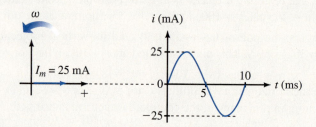

FIGURE 15–36   The reference position of the phasor is its $t = 0$ position.

Shifted Sine Waves

Phasors may be used to represent shifted waveforms, $v = V_m \sin(\omega t \pm \theta)$ or $i = I_m \sin(\omega t \pm \theta)$ as indicated in Figure 15–37. Angle θ is the position of the phasor at $t = 0$ s.

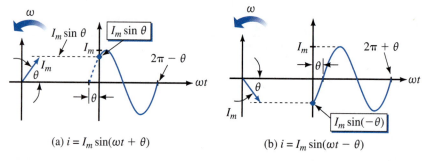

(a) $i = I_m \sin(\omega t + \theta)$ (b) $i = I_m \sin(\omega t - \theta)$

FIGURE 15–37   Phasors for shifted waveforms. Angle θ is the position of the phasor at $t = 0$ s.

EXAMPLE 15–16

Consider $v = 20 \sin(\omega t - 60°)$, where $\omega = \pi/6$ rad/s (i.e., 30°/s). Show that the phasor of Figure 15–38(a) represents this waveform.

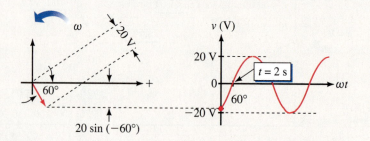

(a) Phasor   (b) $v = 20 \sin(\omega t - 60°)$, with $\omega = 30°/s$

FIGURE 15–38

Solution   The phasor has length 20 V and at time $t = 0$ is at –60° as indicated in (a). Now, as the phasor rotates, it generates a sinusoidal waveform, oscillating between ±20 V as indicated in (b). Note that the zero crossover point occurs at $t = 2$ s, since it takes 2 seconds for the phasor to rotate from –60° to 0° at 30 degrees per second. Now compare the waveform of (b) to the waveform of Figure 15–32, Example 15–13. They are identical. Thus, the phasor of (a) represents the shifted waveform $v = 20 \sin(\omega t - 60°)$.

With the aid of a phasor, sketch the waveform for $v = V_m \sin(\omega t + 90°)$.

EXAMPLE 15–17

Solution   Place the phasor at 90° as in Figure 15–39(a). Note that the resultant waveform (b) is a cosine waveform, i.e., $v = V_m \cos \omega t$. From this, we conclude that

$$\sin(\omega t + 90°) = \cos \omega t$$

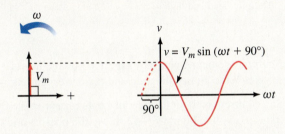

(a) Phasor at 90° position

(b) Waveform can also be described as a cosine wave

FIGURE 15–39   Demonstrating that $\sin(\omega t + 90°) = \cos \omega t$.

With the aid of phasors, show that

a. $\sin(\omega t - 90°) = -\cos \omega t$,

b. $\sin(\omega t \pm 180°) = -\sin \omega t$,

PRACTICE PROBLEMS 8

Phase Difference

Phase difference refers to the angular displacement between different waveforms of the same frequency. Consider Figure 15–40. If the angular displacement is 0° as in (a), the waveforms are said to be **in phase**; otherwise, they are **out of phase.** When describing a phase difference, select one waveform as reference. Other waveforms then lead, lag, or are in phase with this reference. For example, in (b), for reasons to be discussed in the next paragraph, the current waveform is said to lead the voltage waveform, while in (c) the current waveform is said to lag.

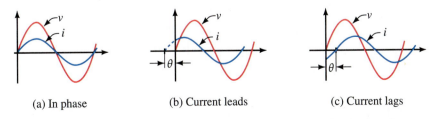

(a) In phase

(b) Current leads

(c) Current lags

FIGURE 15–40   Illustrating phase difference. In these examples, voltage is taken as reference.

The terms **lead** and **lag** can be understood in terms of phasors. If you observe phasors rotating as in Figure 15–41(a), the one that you see passing first is leading and the other is lagging. By definition, *the waveform generated by the leading phasor leads the waveform generated by the lagging phasor and vice versa.* In Figure 15–41, phasor I_m leads phasor V_m; thus current $i(t)$ leads voltage $v(t)$.

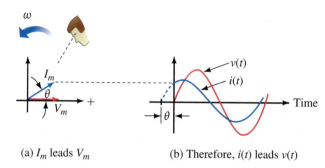

(a) I_m leads V_m (b) Therefore, $i(t)$ leads $v(t)$

FIGURE 15–41   Defining lead and lag.

EXAMPLE 15–18

Voltage and current are out of phase by 40°, and voltage lags. Using current as the reference, sketch the phasor diagram and the corresponding waveforms.

Solution   Since current is the reference, place its phasor in the 0° position and the voltage phasor at –40°. Figure 15–42 shows the phasors and corresponding waveforms.

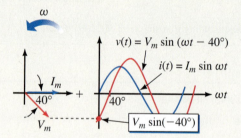

FIGURE 15–42

EXAMPLE 15–19

Given $v = 20 \sin(\omega t + 30°)$ and $i = 18 \sin(\omega t - 40°)$, draw the phasor diagram, determine phase relationships, and sketch the waveforms.

Solution   The phasors are shown in Figure 15–43(a). From these, you can see that v leads i by 70°. The waveforms are shown in (b).

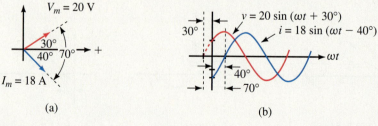

(a)   (b)

FIGURE 15–43

Figure 15–44 shows a pair of waveforms v_1 and v_2 on an oscilloscope. Each major vertical division represents 20 V and each major division on the horizontal (time) scale represents 20 μs. Voltage v_1 leads. Prepare a phasor diagram using v_1 as reference. Determine equations for both voltages.

EXAMPLE 15–20

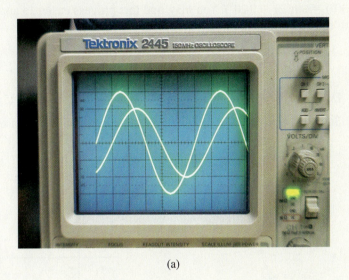

(a)

(b)

FIGURE 15–44

◀ **MULTISIM**

Solution   From the photograph, the magnitude of v_1 is V_{m_1} = 3 div × 20 V/div = 60 V. Similarly, V_{m_2} = 40 V. Cycle length is T = 6 × 20 μs = 120 μs, and the displacement between waveforms is 20 μs which is ⅙ of a cycle (i.e., 60°). Selecting v_1 as reference and noting that v_2 lags yields the phasors shown in (b). Angular frequency $\omega = 2\pi/T = 2\pi/(120 \times 10^{-6}$ s$) = 52.36 \times 10^3$ rad/s. Thus, $v_1 = V_{m_1} \sin \omega t = 60 \sin(52.36 \times 10^3 t)$ V and $v_2 = 40 \sin(52.36 \times 10^3 t - 60°)$ V.

Sometimes voltages and currents are expressed in terms of cos ωt rather than sin ωt. As Example 15–17 shows, a cosine wave is a sine wave shifted by $+90°$, or alternatively, a sine wave is a cosine wave shifted by $-90°$. For sines or cosines with an angle, the following formulas apply.

$$\cos(\omega t + \theta) = \sin(\omega t + \theta + 90°) \qquad \textbf{(15–17a)}$$

$$\sin(\omega t + \theta) = \cos(\omega t + \theta - 90°) \qquad \textbf{(15–17b)}$$

To illustrate, consider $\cos(\omega t + 30°)$. From Equation 15–17a, $\cos(\omega t + 30°) = \sin(\omega t + 30° + 90°) = \sin(\omega t + 120°)$. Figure 15–45 illustrates this relationship

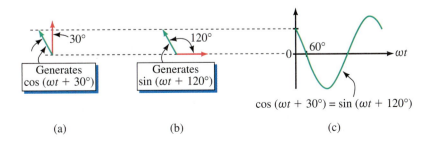

(a)   (b)   (c)

FIGURE 15–45   Using phasors to show that cos $(\omega t + 30°) =$ sin $(\omega t + 120°)$.

graphically. The red phasor in (a) generates cos ωt as was shown in Example 15–17. Therefore, the blue phasor generates a waveform that leads it by 30°, namely cos(ωt + 30°). For (b), the red phasor generates sin ωt, and the blue phasor generates a waveform that leads it by 120°, i.e., sin(ωt + 120°). Since the blue phasor is the same in both cases, you can see that cos(ωt + 30°) = sin(ωt + 120°). You may find this process easier to apply than trying to remember equations 15–17(a) and (b).

EXAMPLE 15–21

Determine the phase angle between $v = 30 \cos(\omega t + 20°)$ and $i = 25 \sin(\omega t + 70°)$.

Solution $i = 25 \sin(\omega t + 70°)$ may be represented by a phasor at 70°, and $v = 30 \cos(\omega t + 20°)$ by a phasor at (90° + 20°) = 110°, Figure 15–46(a). Thus, v leads i by 40°. Waveforms are shown in (b).

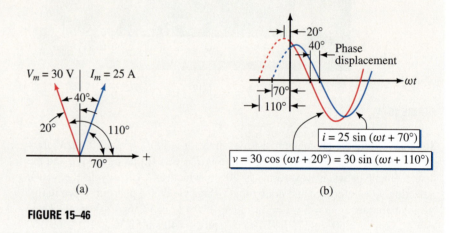

(a)

(b)

FIGURE 15–46

Sometimes you encounter negative waveforms such as $i = -I_m \sin \omega t$. To see how to handle these, refer back to Figure 15–36, which shows the waveform and phasor for $i = I_m \sin \omega t$. If you multiply this waveform by −1, you get the inverted waveform $-I_m \sin \omega t$ of Figure 15–47(a) with corresponding phasor (b). Note that the phasor is the same as the original phasor except that it is rotated by 180°. This is always true—thus, if you multiply a waveform by −1, the phasor for the new waveform is 180° rotated from the original phasor, regardless of the angle of the original phasor.

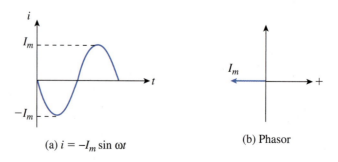

(a) $i = -I_m \sin \omega t$

(b) Phasor

FIGURE 15–47   For a negative sine wave, the phasor is at 180°.

Find the phase relationship between $i = -4 \sin(\omega t + 50°)$ and $v = 120 \sin (\omega t - 60°)$.

Solution $i = -4 \sin(\omega t + 50°)$ is represented by a phasor at $(50° - 180°) = -130°$ and $v = 120 \sin (\omega t - 60°)$ by a phasor at $-60°$, Figure 15–48. The phase difference is 70° and voltage leads. From this, you can see that i can also be written as $i = 4 \sin(\omega t - 130°)$.

EXAMPLE 15–22

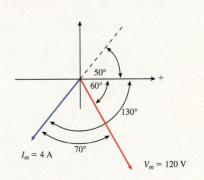

FIGURE 15–48

The importance of phasors to ac circuit analysis cannot be overstated—you will find that they are one of your main tools for representing ideas and for solving problems in later chapters. We will leave them for the moment, but pick them up again in Chapter 16.

IN-PROCESS
LEARNING CHECK 2

(Answers are at the end of the chapter.)

1. If $i = 15 \sin \alpha$ mA, compute the current at $\alpha = 0°, 45°, 90°, 135°, 180°, 225°, 270°, 315°,$ and $360°$.

2. Convert the following angles to radians:
   a. 20°   b. 50°
   c. 120°   d. 250°

3. If a coil rotates at $\omega = \pi/60$ radians per millisecond, how many degrees does it rotate through in 10 ms? In 40 ms? In 150 ms?

4. A current has an amplitude of 50 mA and $\omega = 0.2\pi$ rad/s. Sketch the waveform with the horizontal axis scaled in
   a. degrees   b. radians   c. seconds

5. If 2400 cycles of a waveform occur in 10 ms, what is ω in radians per second?

6. A sinusoidal current has a period of 40 ms and an amplitude of 8 A. Write its equation in the form of $i = I_m \sin \omega t$, with numerical values for I_m and ω.

7. A current $i = I_m \sin \omega t$ has a period of 90 ms. If $i = 3$ A at $t = 7.5$ ms, what is its equation?

8. Write equations for each of the waveforms in Figure 15–49 with the phase angle θ expressed in degrees and ω in rad/s.

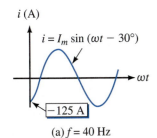

(a) $f = 40$ Hz

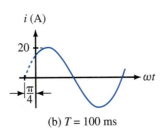

(b) $T = 100$ ms

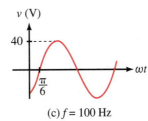

(c) $f = 100$ Hz

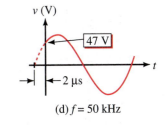

(d) $f = 50$ kHz

FIGURE 15–49

9. Given $i = 10 \sin \omega t$, where $f = 50$ Hz, find all occurrences of
   a. $i = 8$ A between $t = 0$ and $t = 40$ ms
   b. $i = -5$ A between $t = 0$ and $t = 40$ ms

10. Sketch the following waveforms with the horizontal axis scaled in degrees:
    a. $v_1 = 80 \sin(\omega t + 45°)$ V   b. $v_2 = 40 \sin(\omega t - 80°)$ V
    c. $i_1 = 10 \cos \omega t$ mA   d. $i_2 = 5 \cos(\omega t - 20°)$ mA

11. Given $\omega = \pi/3$ rad/s, determine when voltage first crosses through 0 for
    a. $v_1 = 80 \sin(\omega t + 45°)$ V
    b. $v_2 = 40 \sin(\omega t - 80°)$ V

12. Consider the voltages of Question 10:
    a. Sketch phasors for v_1 and v_2.
    b. What is the phase difference between v_1 and v_2?
    c. Determine which voltage leads and which lags.

13. Repeat Question 12 for the currents of Question 10.

15.8 AC Waveforms and Average Value

While we can describe ac quantities in terms of frequency, period, instantaneous value, etc., we do not yet have any way to give a meaningful value to an ac current or voltage in the same sense that we can say of a car battery that it has a voltage of 12 volts. This is because ac quantities constantly change and thus there is no one single numerical value that truly represents a waveform over its complete cycle. For this reason, ac quantities are generally described by a group of characteristics, including instantaneous, peak, average, and effective values. The first two of these we have already seen. In this section, we look at average values; in Section 15.9, we consider effective values.

Average Values

Many quantities are measured by their average, for instance, test and examination scores. To find the average of a set of marks for example, you add them, then divide by the number of items summed. For waveforms, the process is conceptually the same. For example, to find the average of a waveform, you can sum the instantaneous values over a full cycle, then divide by the number of points used. The trouble with this approach is that waveforms do not consist of discrete values.

Average in Terms of the Area Under a Curve

An approach more suitable for use with waveforms is to find the area under the curve, then divide by the baseline of the curve. To get at the idea, we can use an analogy. Consider again the technique of computing the average for a set of numbers. Assume that you earn marks of 80, 60, 60, 95, and 75 on a group of tests. Your average mark is therefore

$$\text{average} = (80 + 60 + 60 + 95 + 75)/5 = 74$$

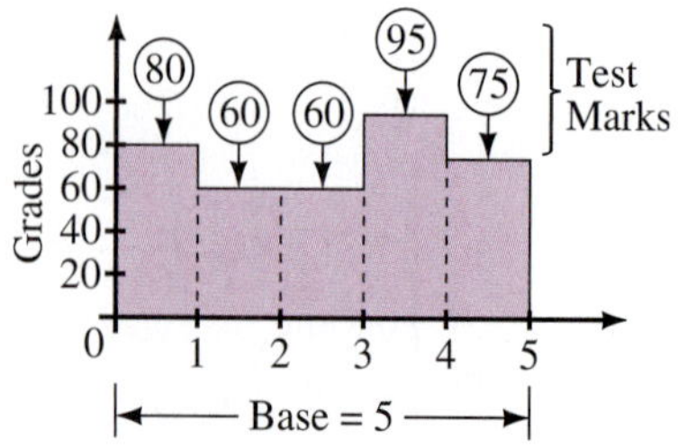

An alternate way to view these marks is graphically as in Figure 15–50. The area under this curve can be computed as

$$\text{area} = (80 \times 1) + (60 \times 2) + (95 \times 1) + (75 \times 1)$$

Now divide this by the length of the base, namely 5. Thus,

$$\frac{(80 \times 1) + (60 \times 2) + (95 \times 1) + (75 \times 1)}{5} = 74$$

FIGURE 15–50   Determining average by area.

which is exactly the answer obtained above. That is,

$$\text{average} = \frac{\text{area under curve}}{\text{length of base}} \qquad \textbf{(15–18)}$$

This result is true in general. Thus, *to find the average value of a waveform, divide the area under the waveform by the length of its base. Areas above the axis are counted as positive, while areas below the axis are counted as negative.* This approach is valid regardless of waveshape.

Average values are also called **dc values,** because dc meters indicate average values rather than instantaneous values. Thus, if you measure a non-dc quantity with a dc meter, the meter will read the average of the waveform, i.e., the value calculated according to Equation 15–18.

EXAMPLE 15–23

a. Compute the average for the current waveform of Figure 15–51.

b. If the negative portion of Figure 15–51 is −3 A instead of −1.5 A, what is the average?

c. If the current is measured by a dc ammeter, what will the ammeter indicate for each case?

Solution

a. The waveform repeats itself after 7 ms. Thus, $T = 7$ ms and the average is

$$I_{avg} = \frac{(2\text{ A} \times 3\text{ ms}) - (1.5\text{ A} \times 4\text{ ms})}{7\text{ ms}} = \frac{6 - 6}{7} = 0\text{ A}$$

b. $I_{avg} = \dfrac{(2\text{ A} \times 3\text{ ms}) - (3\text{ A} \times 4\text{ ms})}{7\text{ ms}} = \dfrac{-6\text{ A}}{7} = -0.857\text{ A}$

c. A dc ammeter measuring (a) will indicate zero, while for (b) it will indicate −0.857 A.

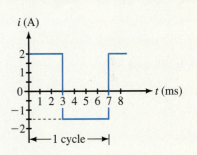

FIGURE 15–51

EXAMPLE 15–24

Compute the average value for the waveforms of Figures 15–52(a) and (c). Sketch the averages for each.

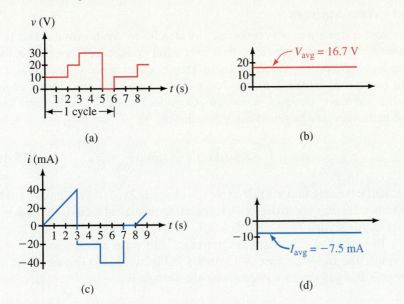

(a)

(b)

(c)

(d)

FIGURE 15–52

Solution For the waveform of (a), $T = 6$ s. Thus,

$$V_{avg} = \frac{(10\,V \times 2\,s) + (20\,V \times 1\,s) + (30\,V \times 2\,s) + (0\,V \times 1\,s)}{6\,s} = \frac{100\,V\text{-}s}{6\,s} = 16.7\,V$$

The average is shown as (b). A dc voltmeter would indicate 16.7 V. For the waveform of (c), $T = 8$ s and

$$I_{avg} = \frac{\frac{1}{2}(40\,mA \times 3\,s) - (20\,mA \times 2\,s) - (40\,mA \times 2\,s)}{8\,s} = \frac{-60}{8}\,mA = -7.5\,mA$$

In this case, a dc ammeter would indicate −7.5 mA.

PRACTICE PROBLEMS 9

Determine the averages for Figures 15–53(a) and (b).

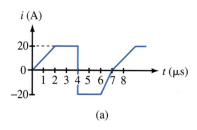

(a)

(b)

FIGURE 15–53

Answers
a. 1.43 A; b. 6.67 V

Sine Wave Averages

Because a sine wave is symmetrical, its area below the horizontal axis is the same as its area above the axis; thus, over a full cycle its net area is zero, independent of frequency and phase angle. Thus, the average of sin ωt, sin($\omega t \pm \theta$), sin $2\omega t$, cos ωt, cos($\omega t \pm \theta$), cos $2\omega t$, and so on are each zero. The average of half a sine wave, however, is not zero. Consider Figure 15–54. The area under the half-cycle may be found using calculus as

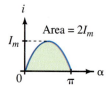

FIGURE 15–54 Area under a half-cycle.

$$\text{area} = \int_0^{\pi} I_m \sin\alpha \, d\alpha = \left[-I_m \cos \alpha \right]_0^{\pi} = 2I_m \qquad (15\text{–}19)$$

Similarly, the area under a half-cycle of voltage is $2V_m$. (If you haven't studied calculus, you can approximate this area using numerical methods as described later in this section.)

Two cases are important in electronics; full-wave average and half-wave average. The full-wave case is illustrated in Figure 15–55. The area from 0 to 2π is $2(2I_m)$ and the base is 2π. Thus, the average is

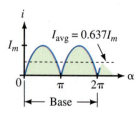

FIGURE 15–55 Full-wave average.

$$I_{avg} = \frac{2(2I_m)}{2\pi} = \frac{2I_m}{\pi} = 0.637 I_m$$

For the half-wave case (Figure 15–56),

$$I_{avg} = \frac{2I_m}{2\pi} = \frac{I_m}{\pi} = 0.318I_m$$

The corresponding expressions for voltage are

$$V_{avg} = 0.637V_m \text{ (full-wave)}$$
$$V_{avg} = 0.318V_m \text{ (half-wave)}$$

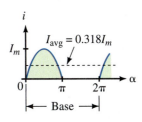

FIGURE 15–56   Half-wave average.

Numerical Methods

If the area under a curve cannot be computed exactly, it can be approximated. One method is to approximate the curve by straight line segments as in Figure 15–57. (If the straight lines closely fit the curve, the accuracy is very good.) Each element of area is a trapezoid (b) whose area is its average height times its base. Thus, $A_1 = \frac{1}{2}(y_0 + y_1)\Delta x$, $A_2 = \frac{1}{2}(y_1 + y_2)\Delta x$, etc. Summing areas and combining terms yields

$$\text{area} = \left(\frac{y_0}{2} + y_1 + y_2 + \cdots + y_{k-1} + \frac{y_k}{2}\right)\Delta x \qquad \textbf{(15–20)}$$

This result is known as the **trapezoidal rule.** Example 15–25 illustrates its use.

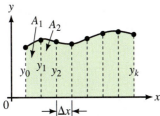

(a) Approximating the curve

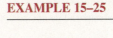

(b) Element of area

FIGURE 15–57   Determining area using the trapezoidal rule.

EXAMPLE 15–25

Approximate the area under $y = \sin(\omega t - 30°)$, Figure 15–58. Use an increment size of $\pi/6$ rad, i.e., 30°.

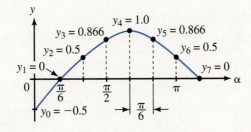

FIGURE 15–58

Solution   Points on the curve $\sin(\omega t - 30°)$ have been computed by calculator and plotted as Figure 15–58. Substituting these values into Equation 15–20 yields

$$\text{area} = \left(\frac{1}{2}(-0.5) + 0 + 0.5 + 0.866 + 1.0 + 0.866 + 0.5 + \frac{1}{2}(0)\right)\left(\frac{\pi}{6}\right) = 1.823$$

The exact area (found using calculus) is 1.866; thus, the approximation of Example 15–25 is in error by 2.3%.

1. Repeat Example 15–25 using an increment size of $\pi/12$ rad. What is the percent error?

2. Approximate the area under $v = 50 \sin(\omega t + 30°)$ from $\omega t = 0°$ to $\omega t = 210°$. Use an increment size of $\pi/12$ rad.

Answers
1. 1.855; 0.59%
2. 67.9 (exact 68.3; error = 0.6%)

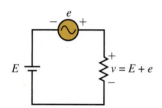

FIGURE 15–59

Superimposed AC and DC

Sometimes ac and dc are used in the same circuit. For example, amplifiers are powered by dc but the signals they amplify are ac. Figure 15–59 shows a simple circuit with combined ac and dc.

Figure 15–60(c) shows superimposed ac and dc. Since we know that the average of a sine wave is zero, the average value of the combined waveform will be its dc component, E. However, peak voltages depend on both components as illustrated in (c). Note for the case illustrated that although the waveform varies sinusoidally, it does not alternate in polarity since it never changes polarity to become negative.

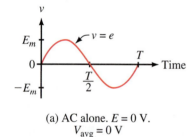

(a) AC alone. $E = 0$ V. $V_{avg} = 0$ V

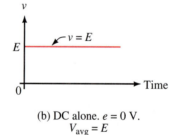

(b) DC alone. $e = 0$ V. $V_{avg} = E$

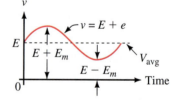

(c) Superimposed ac and dc. $V_{avg} = E$

FIGURE 15–60 Superimposed dc and ac.

EXAMPLE 15–26

Draw the waveform for voltage v for the circuit of Figure 15–61(a). Determine its average, peak, and minimum voltages.

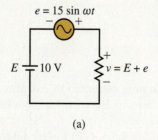

(a)

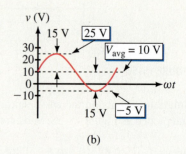

(b)

FIGURE 15–61 $v = 10 + 15 \sin \omega t$.

Solution   The waveform consists of a 10-V dc value with 15 V ac riding on top of it. The average is the dc value, $V_{avg} = 10$ V. The peak voltage is $10 + 15 = 25$ V, while the minimum voltage is $10 - 15 = -5$ V. This waveform alternates in polarity, although not symmetrically (as is the case when there is no dc component).

PRACTICE PROBLEMS 11

Repeat Example 15–26 if the dc source of Figure 15–61 is $E = -5$ V.

Answers
$V_{avg} = -5$ V; positive peak $= 10$ V; negative peak $= -20$ V

While instantaneous, peak, and average values provide useful information about a waveform, none of them truly represents the ability of the waveform to do useful work. In this section, we look at a representation that does. It is called the waveform's **effective value.** The concept of effective value is an important one; in practice, most ac voltages and currents are expressed as effective values. Effective values are also called **rms values** for reasons discussed shortly.

15.9 Effective (RMS) Values

What Is an Effective Value?

An effective value is an equivalent dc value: it tells you how many volts or amps of dc that a time-varying waveform is equal to in terms of its ability to produce average power. Effective values depend on the waveform. A familiar example of such a value is the value of the voltage at the wall outlet in your home. In North America its value is 120 Vac. This means that the sinusoidal voltage at the wall outlets of your home is capable of producing the same average power as 120 volts of steady dc.

Effective Values for Sine Waves

The effective value of a waveform can be determined using the circuits of Figure 15–62. Consider a sinusoidally varying current, *i(t)*. By definition, the effective value of *i* is that value of dc current that produces the same average power. Consider (b). Let the dc source be adjusted until its average power is the same as the average power in (a). The resulting dc current is then the effective value of the current of (a). To determine this value, determine the average power for both cases, then equate them.

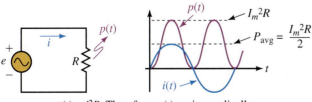

$p(t) = i^2R$. Therefore, $p(t)$ varies cyclically.

(a) AC circuit

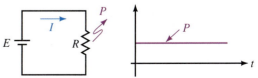

$P = I^2R$. Therefore, P is constant.

(b) DC circuit

FIGURE 15–62   Determining the effective value of sinusoidal ac.

First, consider the dc case. Since current is constant, power is constant, and average power is

$$P_{avg} = P = I^2R \tag{15–21}$$

Now consider the ac case. Power to the resistor at any value of time is $p(t) = i^2R$, where i is the instantaneous value of current. A sketch of $p(t)$ is shown in Figure 15–62(a), obtained by squaring values of current at various points along the axis, then multiplying by R. Average power is the average of $p(t)$. Since $i = I_m \sin \omega t$,

$$
\begin{aligned}
p(t) &= i^2R \\
&= (I_m \sin \omega t)^2 R = I_m^2 R \sin^2 \omega t \\
&= I_m^2 R \left[\frac{1}{2}(1 - \cos 2\omega t) \right]
\end{aligned} \tag{15–22}
$$

where we have used the trigonometric identity $\sin^2 \omega t = \frac{1}{2}(1 - \cos 2\omega t)$, from the mathematics tables inside the front cover of this book. Thus,

$$p(t) = \frac{I_m^2 R}{2} - \frac{I_m^2 R}{2} \cos 2\omega t \tag{15–23}$$

To get the average of $p(t)$, note that the average of $\cos 2\omega t$ is zero and thus the last term of Equation 15–23 drops off leaving

$$P_{avg} = \text{average of } p(t) = \frac{I_m^2 R}{2} \tag{15–24}$$

Now equate Equations 15–21 and 15–24, then cancel R.

$$I^2 = \frac{I_m^2}{2}$$

Now take the square root of both sides. Thus,

$$I = \sqrt{\frac{I_m^2}{2}} = \frac{I_m}{\sqrt{2}} = 0.707 I_m$$

Current I is the value that we are looking for; it is the effective value of current i. To emphasize that it is an effective value, we will initially use subscripted notation I_{eff}. Thus,

$$I_{eff} = \frac{I_m}{\sqrt{2}} = 0.707 I_m \tag{15–25}$$

Effective values for voltage are found in the same way:

$$E_{eff} = \frac{E_m}{\sqrt{2}} = 0.707 E_m \tag{15–26a}$$

$$V_{eff} = \frac{V_m}{\sqrt{2}} = 0.707 V_m \tag{15–26b}$$

As you can see, *effective values for sinusoidal waveforms depend only on magnitude.*

NOTES . . .

Because ac currents alternate in direction, you might expect average power to be zero, with power during the negative half-cycle being equal and opposite to power during the positive half-cycle and hence cancelling. However this is not true since, as Equation 15–22 shows, current is squared, and hence power is never negative. This is consistent with the idea that insofar as power dissipation is concerned, the direction of current through a resistor does not matter (Figure 15–63).

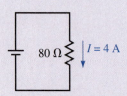

(a) $P = (4)^2(80) = 1280$ W

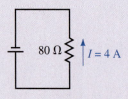

(b) $P = (4)^2(80) = 1280$ W

FIGURE 15–63 Since power depends only on current magnitude, it is the same for both current directions.

Determine the effective values of

a. $i = 10 \sin \omega t$ A

b. $i = 50 \sin(\omega t + 20°)$ mA

c. $v = 100 \cos 2\omega t$ V

EXAMPLE 15–27

Solution   Since effective values depend only on magnitude,

a. $I_{eff} = (0.707)(10 \text{ A}) = 7.07$ A

b. $I_{eff} = (0.707)(50 \text{ mA}) = 35.35$ mA

c. $V_{eff} = (0.707)(100 \text{ V}) = 70.7$ V

To obtain peak values from effective values, rewrite Equations 15–25 and 15–26. Thus,

$$I_m = \sqrt{2}I_{eff} = 1.414I_{eff} \qquad \textbf{(15–27)}$$

$$E_m = \sqrt{2}E_{eff} = 1.414V_{eff} \qquad \textbf{(15–28a)}$$

$$V_m = \sqrt{2}V_{eff} = 1.414V_{eff} \qquad \textbf{(15–28b)}$$

It is important to note that these relationships hold only for sinusoidal wave-forms. However, the concept of effective value applies to all waveforms, as we soon see.

Consider again the ac voltage at the wall outlet in your home. Since $E_{eff} = 120$ V, $E_m = (\sqrt{2})(120 \text{ V}) = 170$ V. This means that a sinusoidal voltage alternating between ±170 V produces the same average power in a resistive circuit as 120 V of steady dc (Figure 15–64).

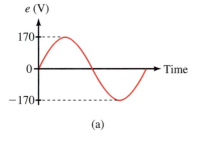
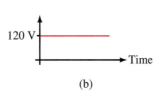

FIGURE 15–64   120 V of steady dc is capable of producing the same average power as sinusoidal ac with $E_m = 170$ V.

General Equation for Effective Values

The $\sqrt{2}$ relationship holds only for sinusoidal waveforms. For other waveforms, you need a more general formula. Using calculus, it can be shown that for any waveform

$$I_{eff} = \sqrt{\frac{1}{T} \int_0^T i^2 dt} \qquad \textbf{(15–29)}$$

with a similar equation for voltage. This equation can be used to compute effective values for any waveform, including sinusoidal. In addition, it leads to a graphic approach to finding effective values. In Equation 15–29, the integral of i^2 represents the area under the i^2 waveform. Thus,

$$I_{eff} = \sqrt{\frac{\text{area under the } i^2 \text{ curve}}{\text{base}}} \qquad (15\text{–}30)$$

To compute effective values using this equation, do the following:

Step 1: Square the current (or voltage) curve.

Step 2: Find the area under the squared curve.

Step 3: Divide the area by the length of the curve.

Step 4: Find the square root of the value from Step 3.

This process is easily carried out for rectangular-shaped waveforms since the area under their squared curves is easy to compute. For other waveforms, you have to use calculus or approximate the area using numerical methods. For the special case of superimposed ac and dc (Figure 15–60), Equation 15–29 leads to the following formula:

$$I_{eff} = \sqrt{I_{dc}^2 + I_{ac}^2} \qquad (15\text{–}31)$$

where I_{dc} is the dc current value, I_{ac} is the effective value of the ac component, and I_{eff} is the effective value of the combined ac and dc currents. Equations 15–30 and 15–31 also hold for voltage when V is substituted for I.

RMS Values

Consider again Equation 15–30. To use this equation, we compute the root of the mean square to obtain the effective value. For this reason, effective values are called **root mean square** or **rms** values and **the terms *effective* and *rms* are synonymous.** Since, in practice, sinusoidal ac quantities are almost always expressed as rms values, we shall assume from here on that, unless otherwise noted, *all sinusoidal ac voltages and currents are rms values.*

EXAMPLE 15–28

One cycle of a voltage waveform is shown in Figure 15–65(a). Determine its effective (rms) value.

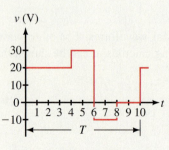

(a) Voltage waveform

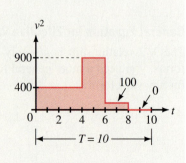

(b) Squared waveform

FIGURE 15–65

Solution Square the voltage waveform and plot it as in (b). Apply Equation 15–30:

$$V_{eff} = \sqrt{\frac{(400 \times 4) + (900 \times 2) + (100 \times 2) + (0 \times 2)}{10}}$$

$$= \sqrt{\frac{3600}{10}} = 19.0 \text{ V}$$

The waveform of Figure 15–65(a) has the same effective value as 19.0 V of steady dc.

Determine the effective (rms) value of the waveform of Figure 15–66(a).

EXAMPLE 15–29

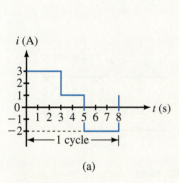

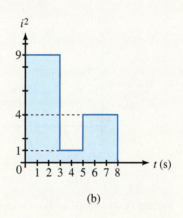

(a)	(b)

FIGURE 15–66

Solution Square the curve, then apply Equation 15–30. Thus,

$$I_{eff} = \sqrt{\frac{(9 \times 3) + (1 \times 2) + (4 \times 3)}{8}}$$

$$= \sqrt{\frac{41}{8}} = 2.26 \text{ A}$$

Compute the rms value of the waveform of Figure 15–61(b).

EXAMPLE 15–30

Solution Use Equation 15–31 (with I replaced by V). First, compute the rms value of the ac component. $V_{ac} = 0.707 \times 15 = 10.61$ V. Now substitute this into Equation 15–31. Thus,

$$V_{rms} = \sqrt{V_{dc}^2 + V_{ac}^2} = \sqrt{(10)^2 + (10.61)^2} = 14.6 \text{ V}$$

---

1. Determine the rms value of the current of Figure 15–51.

2. Repeat for the voltage graphed in Figure 15–52(a).

PRACTICE PROBLEMS 12

Answers
1. 1.73 A; 2. 20 V

One Final Note

The subscripts *eff* and *rms* are not used in practice. Once the concept is familiar, we drop them. They are thus implied rather than stated.

15.10 Rate of Change of a Sine Wave (Derivative)

As you will see later, several important circuit effects depend on the rate of change of sinusoidal quantities. The rate of change of a quantity is the slope (i.e., derivative) of its waveform versus time. Consider the waveform of Figure 15–67. As indicated, the slope is maximum positive at the beginning of the cycle, zero at both its peaks, maximum negative at the half-cycle crossover point, and maximum positive at the end of the cycle. This slope is plotted in Figure 15–68. Note that it is also sinusoidal, but it leads the original waveform by 90°. Thus, if *A* is a sine wave, *B* is a cosine wave. (This result is important to us in Chapter 16.)

NOTES . . .

The Derivative of a Sine Wave

The result developed intuitively here can be proven easily using calculus. To illustrate, consider the waveform $\sin \omega t$ shown in Figure 15–67. The slope of this function is its derivative. Thus,

$$\text{Slope} = \frac{d}{dt} \sin \omega t = \omega \cos \omega t$$

Therefore, the slope of a sine wave is a cosine wave as depicted in Figure 15–68.

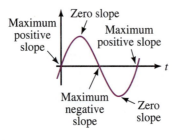

FIGURE 15–67 Slope at various places for a sine wave.

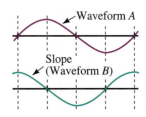

FIGURE 15–68 Showing the 90° phase shift.

15.11 AC Voltage and Current Measurement

Two of the most important instruments for measuring ac quantities are the multimeter and the oscilloscope. Multimeters read the magnitude of ac voltage and current, and sometimes frequency. Oscilloscopes show waveshape and period and permit determination of frequency, phase difference, and so on.

Meters for Voltage and Current Measurement

There are two basic classes of ac meters: one measures rms correctly for sinusoidal waveforms only (called "average responding" instruments); the other measures rms correctly regardless of waveform (called "true rms" meters). Most common meters are average responding meters.

Average Responding Meters

Average responding meters use a rectifier circuit to convert incoming ac to dc. They then respond to the average value of the rectified input, which, as shown in Figure 15–55, is $0.637V_m$ for a "full-wave" rectified sine wave. However, the rms value of a sine wave is $0.707V_m$. Thus, to make the meter display directly in rms, the scale is modified by the factor $0.707V_m/0.637V_m = 1.11$. Other meters use a "half-wave" circuit, which yields the waveform of Figure 15–56. In this case, its average is $0.318V_m$, yielding a scale factor of $0.707V_m/0.318V_m = 2.22$. Because these meters are calibrated for sinusoidal ac only, their readings are meaningless for all other waveforms. Figure 15–69 shows an average responding DMM.

FIGURE 15–69 An average responding DMM. While all DMMs can measure voltage, current, and resistance, this one can also measures frequency.

True RMS Measurement

To measure the rms value of a nonsinusoidal waveform, you need a true rms meter. A true rms meter indicates true rms voltages and currents regardless of waveform. For example, for the waveform of Figure 15–64(a), any ac meter will correctly read 120 V (since it is a sine wave). For the waveform of Figure 15–61(b), a true rms meter will correctly read 14.6 V (the rms value that we calculated earlier, in Example 15–30) but an average responding meter will yield only a meaningless value. True rms instruments are more expensive than standard meters.

Oscilloscopes

Oscilloscopes (frequently referred to as scopes, Figure 15–70) are used for time domain measurement, i.e., waveshape, frequency, period, phase difference, and so on. Usually, you scale values from the screen, although higher-priced models compute and display them for you on a digital readout.

Oscilloscopes measure voltage. To measure current, you need a current-to-voltage converter. One type of converter is a clip-on device, known as a **current gun** that clamps over the current-carrying conductor and monitors its magnetic field. (It works only with ac.) The varying magnetic field induces a voltage which is then displayed on the screen. With such a device, you can monitor current waveshapes and make current-related measurements. Alternately, you can place a small resistor in the current path, measure voltage across it with the oscilloscope, then use Ohm's law to determine the current.

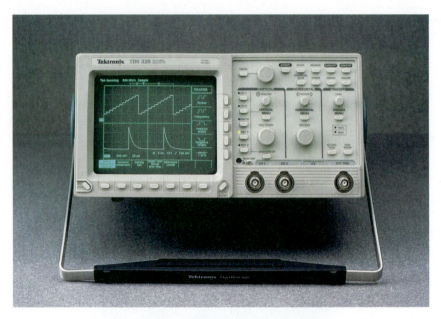

FIGURE 15–70 An oscilloscope may be used for waveform analysis.

A Final Note

AC meters measure voltage and current only over a limited frequency range, typically from 50 Hz to a few kHz, although some work up to the 100-kHz range. Note, however, that accuracy may be affected by frequency. (Check the manual.) Oscilloscopes, on the other hand, measure very high frequencies; even moderately priced oscilloscopes work at frequencies up to hundreds of MHz.

15.12 Circuit Analysis Using Computers

◀ MULTISIM

◀ CADENCE

MultiSIM and PSpice both provide a convenient way to study the phase relationships of this chapter, as they both incorporate easy-to-use graphing facilities. You simply set up sources with the desired magnitude and phase values and instruct the software to compute and plot the results. To illustrate, let us graph $e_1 = 100 \sin \omega t$ V and $e_2 = 80 \sin(\omega t + 60°)$ V. Use a frequency of 500 Hz.

MultiSIM

Create the circuit of Figure 15–71 on the screen. Double click Source 1 and when the dialog box opens, set Voltage Amplitude to 100 V, Phase to 0 deg, and Frequency to 500 Hz. Similarly, set Source 2 to 80 V, 60 deg (see Notes 1 and 2), and 500 Hz. Click the Simulate icon (or click *Simulate*) and select *Transient Analysis;* in the dialog box that opens, set TSTOP to 0.002 (to run the solution to 2 ms so that you see a full cycle) and *Minimum number of time points* to 1000 (see Note 3). Click the *Output Variables* tab and select nodes to plot during simulation, then click *Simulate.* Following simulation, graphs e_1 and e_2 (Figure 15–72) appear.

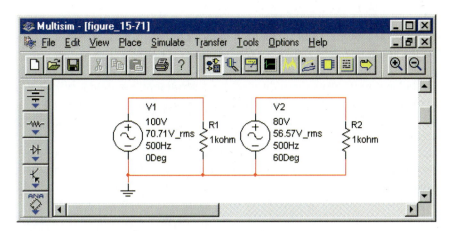

FIGURE 15–71   Studing phase relationships using MultiSIM.

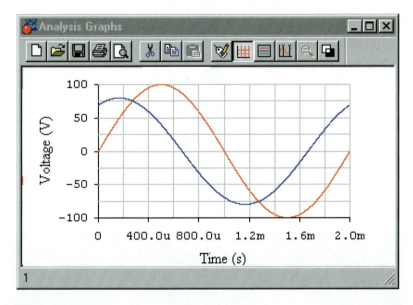

FIGURE 15–72

NOTES ...

1. MultiSIM accepts only positive angles. If you want to enter a negative value, you must subtract it from 360°. Thus, if you want −30°, you need to enter 330°.

2. For MultiSIM 2001 it is necessary to enter the negative of the angle that you actually want. For example, if you want 60°, you need to enter −60° (which means that you actually need to enter 300°). (Electronics Workbench does not have this problem; MultiSIM has indicated that they will change it in their new release.)

3. MultiSIM generates time steps automatically when it plots waveforms. Sometimes, however, it does not generate enough and you get a jagged curve. If this happens, click *Minimum number of time points* in the transient analysis dialog box (the dialog box in which you enter the value for TSTOP), and type in a larger number (say 1000). Experiment until you get a suitably smooth curve.

You can verify the angle between the waveforms using cursors. First, note that the period T = 2 ms = 2000 μs. (This corresponds to 360°.) Expand the graph to full screen, click the *Grid* icon, then the *Cursors* icon. Using the cursors, measure the time between crossover points, as indicated in Figure 15–73. You should get 333 μs. This yields an angular displacement of

$$\theta = \frac{333 \ \mu s}{2000 \ \mu s} \times 360° = 60°$$

as expected.

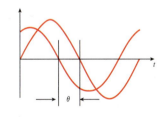

FIGURE 15–73

PSpice

For this problem, you need a sinusoidal time-varying ac voltage source. Use *VSIN* (it is found in the *SOURCE* library). For VSIN, you must specify the magnitude, phase, and frequency of the source, as well as its offset. Build the circuit on the screen as in Figure 15–74. Note the empty parameter boxes beside each source. Double click each in turn and enter 0V for the offset, 100V for the amplitude, and 500Hz for the frequency for Source 1 as shown in Figure 15–74. Similarly enter values for Source 2. Now double click the V2 source symbol and in its Properties editor window, scroll until you find a cell labeled PHASE, then enter 60deg. Click Apply, then close. (You don't need to do this for Source 1 because PSpice will automatically use the default value of zero degrees.) Click the *New Simulations* icon and enter the file name. When the dialog box opens, select *Time Domain,* set *TSTOP* to 2 ms (to display a full cycle), then OK. Select markers from the toolbar and place as shown. (This causes PSpice to automatically plot the traces.) Run the simulation. When the simulation is complete, the waveforms of Figure 15–75 should appear.

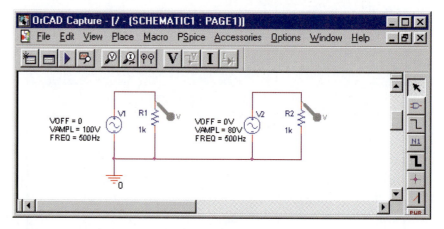

FIGURE 15–74   Studying phase relationships using PSpice.

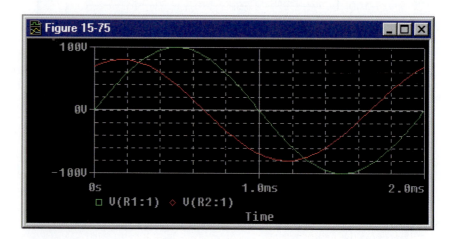

FIGURE 15–75

You can verify the angle between the waveforms using cursors. First, note that the period $T = 2$ ms $= 2000$ μs. (This corresponds to 360°.) Now using the cursors, measure the time between crossover points as indicated in Figure 15–73. You should get 333 μs. This yields an angular displacement of

$$\theta = \frac{333 \text{ μs}}{2000 \text{ μs}} \times 360° = 60°$$

which agrees with the given sources.

PROBLEMS

15.1 Introduction

1. What do we mean by "ac voltage"? By "ac current"?

15.2 Generating AC Voltages

2. The waveform of Figure 15–8 is created by a 600-rpm generator. If the speed of the generator changes so that its cycle time is 50 ms, what is its new speed?

3. a. What do we mean by instantaneous value?

   b. For Figure 15–76, determine instantaneous voltages at $t = 0, 1, 2, 3, 4, 5, 6, 7,$ and 8 ms.

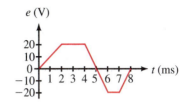

FIGURE 15–76

15.3 Voltage and Current Conventions for ac

4. For Figure 15–77, what is I when the switch is in position 1? When in position 2? Include sign.

5. The source of Figure 15–78 has the waveform of Figure 15–76. Determine the current at $t = 0, 1, 2, 3, 4, 5, 6, 7,$ and 8 ms. Include sign.

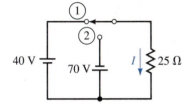

FIGURE 15–77   **FIGURE 15–78**

15.4 Frequency, Period, Amplitude, and Peak Value

6. For each of the following, determine the period:

   a. $f = 100$ Hz
   b. $f = 40$ kHz
   c. $f = 200$ MHz

7. For each of the following, determine the frequency:

   a. $T = 0.5$ s
   b. $T = 100$ ms
   c. $5T = 80$ μs

8. For a triangular wave, $f = 1.25$ MHz. What is its period? How long does it take to go through 8×10^7 cycles?

9. Determine the period and frequency for the waveform of Figure 15–79.

10. Determine the period and frequency for the waveform of Figure 15–80. How many cycles are shown?

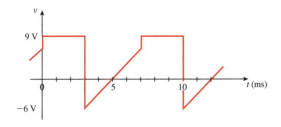

FIGURE 15–79

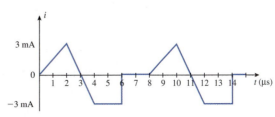

FIGURE 15–80

11. What is the peak-to-peak voltage for Figure 15–79? What is the peak-to-peak current of Figure 15–80?

12. For a certain waveform, $625T = 12.5$ ms. What is the waveform's period and frequency?

13. A square wave with a frequency of 847 Hz goes through how many cycles in 2 minutes and 57 seconds?

14. For the waveform of Figure 15–81, determine

    a. period   b. frequency   c. peak-to-peak value

15. Two waveforms have periods of T_1 and T_2 respectively. If $T_1 = 0.25\ T_2$ and $f_1 = 10$ kHz, what are T_1, T_2, and f_2?

16. Two waveforms have frequencies f_1 and f_2 respectively. If $T_1 = 4\ T_2$ and waveform 1 is as shown in Figure 15–79, what is f_2?

15.5   Angular and Graphic Relationships for Sine Waves

17. Given voltage $v = V_m \sin \alpha$. If $V_m = 240$ V, what is v at $\alpha = 37°$?

18. For the sinusoidal waveform of Figure 15–82,

    a. Determine the equation for i.

    b. Determine current at all points marked.

19. A sinusoidal voltage has a value of 50 V at $\alpha = 150°$. What is V_m?

20. Convert the following angles from radians to degrees:

    a. $\pi/12$ b. $\pi/1.5$

    c. $3\pi/2$ d. 1.43

    e. 17   f. 32π

21. Convert the following angles from degrees to radians:

    a. $10°$ b. $25°$

    c. $80°$ d. $150°$

    e. $350°$ f. $620°$

22. A 50-kHz sine wave has an amplitude of 150 V. Sketch the waveform with its axis scaled in microseconds.

23. If the period of the waveform in Figure 15–82 is 180 ms, compute current at $t = 30, 75, 140$, and 315 ms.

24. A sinusoidal waveform has a period of 60 μs and $V_m = 80$ V. Sketch the waveform. What is its voltage at 4 μs?

25. A 20-kHz sine wave has a value of 50 volts at $t = 5$ μs. Determine V_m and sketch the waveform.

26. For the waveform of Figure 15–83, determine v_2.

15.6   Voltages and Currents as Functions of Time

27. Calculate ω in radians per second for each of the following:

    a. $T = 100$ ns   b. $f = 30$ Hz

    c. 100 cycles in 4 s   d. period $= 20$ ms

    e. 5 periods in 20 ms

28. For each of the following values of ω, compute f and T:

    a. 100 rad/s   b. 40 rad in 20 ms   c. 34×10^3 rad/s

29. Determine equations for sine waves with the following:

    a. $V_m = 170$ V, $f = 60$ Hz   b. $I_m = 40$ μA, $T = 10$ ms

    c. $T = 120$ μs, $v = 10$ V at $t = 12$ μs

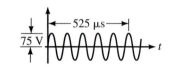

FIGURE 15–81

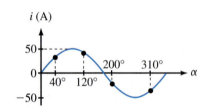

FIGURE 15–82

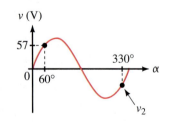

FIGURE 15–83

30. Determine f, T, and amplitude for each of the following:

    a. $v = 75 \sin 200\pi t$
    b. $i = 8 \sin 300t$

31. A sine wave has a peak-to-peak voltage of 40 V and $T = 50$ ms. Determine its equation.

32. Sketch the following waveforms with the horizontal axis scaled in degrees, radians, and seconds:

    a. $v = 100 \sin 200\pi t$ V

    b. $i = 90 \sin \omega t$ mA, $T = 80$ μs

33. Given $i = 47 \sin 8260t$ mA, determine current at $t = 0$ s, 80 μs, 410 μs, and 1200 μs.

34. Given $v = 100 \sin \alpha$. Sketch one cycle.

    a. Determine at which two angles $v = 86.6$ V.

    b. If $\omega = 100\pi/60$ rad/s, at which times do these occur?

35. Write equations for the waveforms of Figure 15–84. Express the phase angle in degrees.

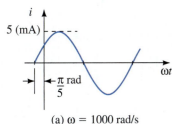

(a) $\omega = 1000$ rad/s

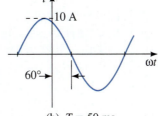

(b) $T = 50$ ms

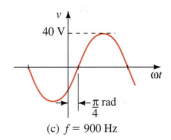

(c) $f = 900$ Hz

FIGURE 15–84

36. Sketch the following waveforms with the horizontal axis scaled in degrees and seconds:

    a. $v = 100 \sin(232.7t + 40°)$ V

    b. $i = 20 \sin(\omega t - 60°)$ mA, $f = 200$ Hz

37. Given $v = 5 \sin(\omega t + 45°)$. If $\omega = 20\pi$ rad/s, what is v at $t = 20$, 75, and 90 ms?

38. Repeat Problem 35 for the waveforms of Figure 15–85.

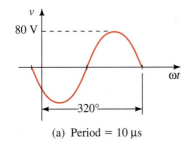

(a) Period = 10 μs

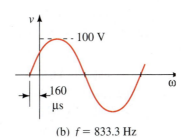

(b) $f = 833.3$ Hz

FIGURE 15–85

39. Determine the equation for the waveform shown in Figure 15–86.

40. For the waveform of Figure 15–87, determine i_2.

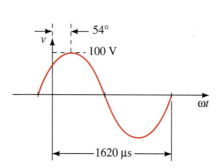

FIGURE 15–86

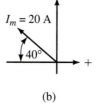 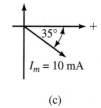

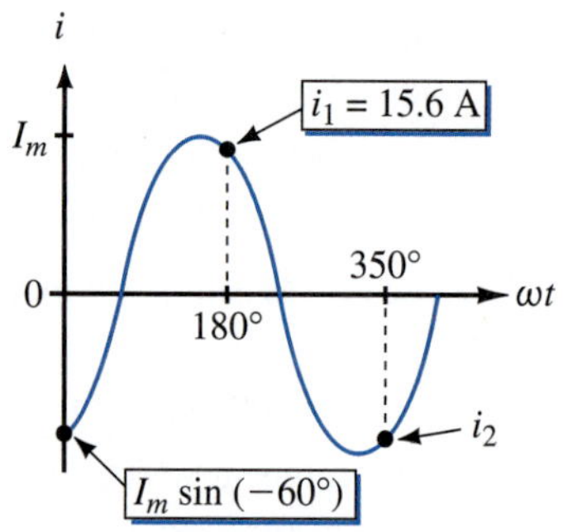

FIGURE 15–87

41. Given $v = 30 \sin(\omega t - 45°)$ where $\omega = 40\pi$ rad/s. Sketch the waveform. At what time does v reach 0 V? At what time does it reach 23 V and -23 V?

15.7   Introduction to Phasors

42. For each of the phasors of Figure 15–88, determine the equation for *v(t) or i(t) as applicable, and sketch the waveform.*

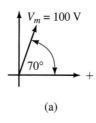

      (a)   (b)   (c)

FIGURE 15–88

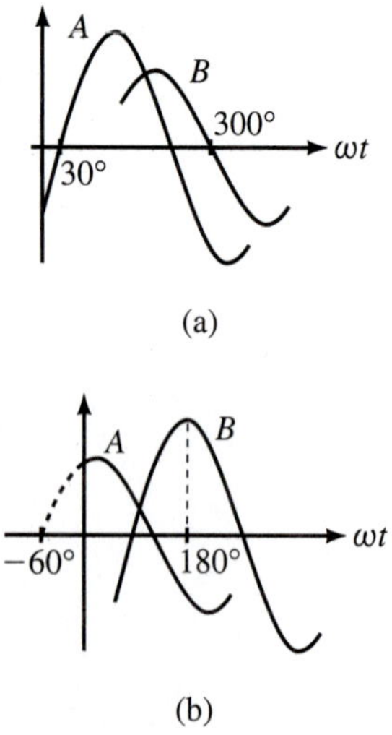

43. With the aid of phasors, sketch the waveforms for each of the following pairs and determine the phase difference and which waveform leads:

  a. $v = 100 \sin \omega t$ b. $v_1 = 200 \sin(\omega t - 30°)$
      $i = 80 \sin(\omega t + 20°)$ $v_2 = 150 \sin(\omega t - 30°)$

  c. $i_1 = 40 \sin(\omega t + 30°)$ d. $v = 100 \sin(\omega t + 140°)$
      $i_2 = 50 \sin(\omega t - 20°)$ $i = 80 \sin(\omega t - 160°)$

44. Repeat Problem 43 for the following.

  a. $i = 40 \sin(\omega t + 80°)$ b. $v = 20 \cos(\omega t + 10°)$
      $v = -30 \sin(\omega t - 70°)$ $i = 15 \sin(\omega t - 10°)$

  c. $v = 20 \cos(\omega t + 10°)$ d. $v = 80 \cos(\omega t + 30°)$
      $i = 15 \sin(\omega t + 120°)$ $i = 10 \cos(\omega t - 15°)$

45. For the waveforms in Figure 15–89, determine the phase differences. Which waveform leads?

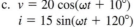

46. Draw phasors for the waveforms of Figure 15–89.

FIGURE 15–89

15.8   AC Waveforms and Average Value

47. What is the average value of each of the following over an integral number of cycles?

   a. $i = 5 \sin \omega t$ b. $i = 40 \cos \omega t$

   c. $v = 400 \sin(\omega t + 30°)$ d. $v = 20 \cos 2\omega t$

48. Using Equation 15-20, compute the area under the half-cycle of Figure 15–54 using increments of $\pi/12$ rad.

49. Compute I_{avg} or V_{avg} for the waveforms of Figure 15–90.

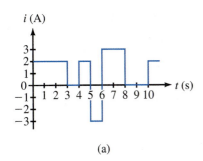

(a)

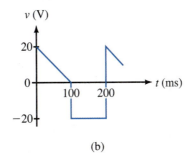

(b)

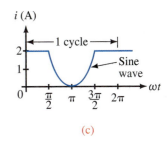

(c)

FIGURE 15–90

50. For the waveform of Figure 15–91, compute I_m.

51. For the circuit of Figure 15–92, $e = 25 \sin \omega t$ V and period $T = 120$ ms.

   a. Sketch voltage $v(t)$ with the axis scaled in milliseconds.

   b. Determine the peak and minimum voltages.

   c. Compute v at $t = 10, 20, 70,$ and 100 ms.

   d. Determine V_{avg}.

52. Using numerical methods for the curved part of the waveform (with increment size $\Delta t = 0.25$ s), determine the area and the average value for the waveform of Figure 15–93.

53. ⨍Using calculus, find the average value for Figure 15–93.

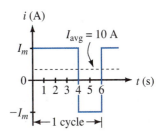

FIGURE 15–91

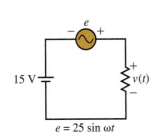

FIGURE 15–92

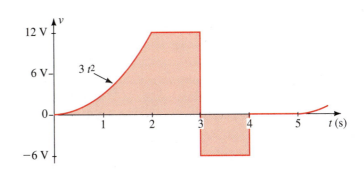

FIGURE 15–93

15.9   Effective Values

54. Determine the effective values of each of the following:

   a. $v = 100 \sin \omega t$ V   b. $i = 8 \sin 377t$ A

   c. $v = 40 \sin(\omega t + 40°)$ V   d. $i = 120 \cos \omega t$ mA

55. Determine the rms values of each for the following.
    a. A 12-V battery
    b. $-24 \sin(\omega t + 73°)$ mA
    c. $10 + 24 \sin \omega t$ V
    d. $45 - 27 \cos 2\, \omega t$ V

56. For a sine wave, $V_{\text{eff}} = 9$ V. What is its amplitude?

57. Determine the root mean square values for
    a. $i = 3 + \sqrt{2}(4) \sin(\omega t + 44°)$ mA
    b. Voltage v of Figure 15–92 with $e = 25 \sin \omega t$ V

58. Compute the rms values for Figures 15–90(a), and 15–91. For Figure 15–91, $I_m = 30$ A.

59. Compute the rms values for the waveforms of Figure 15–94.

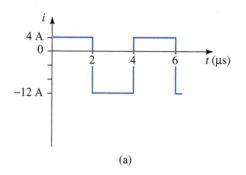

(a)

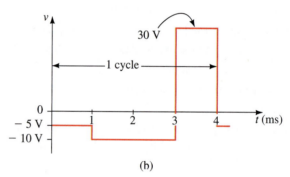

(b)

FIGURE 15–94

60. Compute the effective value for Figure 15–95.

61. Determine the rms value of the waveform of Figure 15–96. Why is it the same as that of a 24-V battery?

62. Compute the rms value of the waveform of Figure 15–52(c). To handle the triangular portion, use Equation 15–20. Use a time interval $\Delta t = 1$s.

63. ▌Repeat Problem 62, using calculus to handle the triangular portion.

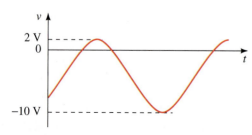

FIGURE 15–95

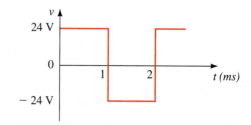

FIGURE 15–96

15.11   AC Voltage and Current Measurement

64. Determine the reading of an average responding AC meter for each of the following cases. (Note: Meaningless is a valid answer if applicable.) Assume the frequency is within the range of the instrument.

    a. $v = 153 \sin \omega t$ V

    b. $v = \sqrt{2}(120) \sin(\omega t + 30°)$ V

    c. The waveform of Figure 15–61

    d. $v = 597 \cos \omega t$ V

65. Repeat Problem 64 using a true rms meter.

15.12   Circuit Analysis Using Computers

Use MultiSIM or PSpice for the following.

66. Plot the waveform of Problem 37 and, using the cursor, determine voltage at the times indicated. Don't forget to convert the frequency to Hz.

 ◀ MULTISIM   ◀ CADENCE

67. Plot the waveform of Problem 41. Using the cursor, determine the time at which v reaches 0 V. Don't forget to convert the frequency to Hz.

◀ MULTISIM   ◀ CADENCE

68. Assume the equations of Problem 43 all represent voltages. For each case, plot the waveforms, then use the cursor to determine the phase difference between waveforms.

◀ MULTISIM   ◀ CADENCE

✓ ANSWERS TO IN-PROCESS LEARNING CHECKS

In-Process Learning Check 1

1. 16.7 ms

2. Frequency doubles, period halves

3. 50 Hz; 20 ms

4. 20 V; 0.5 ms and 2.5 ms; −35 V: 4 ms and 5 ms

5. (c) and (d); Since current is directly proportional to the voltage, it will have the same waveshape.

6. 250 Hz

7. $f_1 = 100$ Hz; $f_2 = 33.3$ Hz

8. 50 kHz and 1 MHz

9. 22.5 Hz

10. At 12 ms, direction →; at 37 ms, direction ←; at 60 ms, →

11. At 75 ms, $i = -5$ A

In-Process Learning Check 2

1.

α (deg)	0	45	90	135	180	225	270	315	360
i (mA)	0	10.6	15	10.6	0	−10.6	−15	−10.6	0

2. a. 0.349   b. 0.873

    c. 2.09   d. 4.36

3. 30°; 120°; 450°

4. Same as Figure 15–27 with $T = 10$ s and amplitude = 50 mA.

5. 1.508×10^6 rad/s

6. $i = 8 \sin 157t$ A

7. $i = 6 \sin 69.81t$ A

8. a. $i = 250 \sin(251t - 30°)$ A

   b. $i = 20 \sin(62.8t + 45°)$ A

   c. $v = 40 \sin(628t - 30°)$ V

   d. $v = 80 \sin(314 \times 10^3 t + 36°)$ V

9. a. 2.95 ms; 7.05 ms; 22.95 ms; 27.05 ms

   b. 11.67 ms; 18.33 ms; 31.67 ms; 38.33 ms

10.

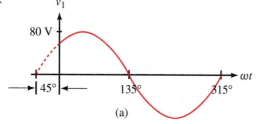

(a)

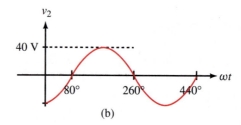

(b)

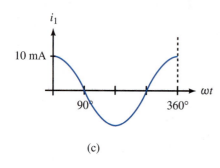

(c)

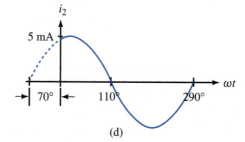

(d)

11. a. 2.25 s   b. 1.33 s

12. a.

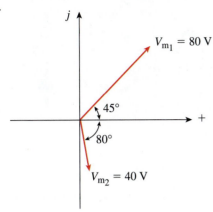

b. 125°

c. v_1 leads

13. a.

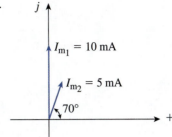

b. 20°

c. i_1 leads

■ OBJECTIVES

After studying this chapter, you will be able to

• express complex numbers in rectangular and polar forms,

• represent ac voltage and current phasors as complex numbers,

• represent ac sources in transformed form,

• add and subtract currents and voltages using phasors,

• compute inductive and capacitive reactance,

• determine voltages and currents in simple ac circuits,

• explain the impedance concept,

• determine impedance for $R, L,$ and C circuit elements,

• determine voltages and currents in simple ac circuits using the impedance concept,

• use MultiSIM and PSpice to solve simple ac circuit problems.

R, L, and C Elements and the Impedance Concept

16

In Chapter 15, you learned how to analyze a few simple ac circuits in the time domain using voltages and currents expressed as functions of time. However, this is not a very practical approach. A more practical approach is to represent ac voltages and currents as phasors, circuit elements as impedances, and analyze circuits in the phasor domain using complex algebra. With this approach, ac circuit analysis is handled much like dc circuit analysis, and all basic relationships and theorems—Ohm's law, Kirchhoff's laws, mesh and nodal analysis, superposition and so on—apply. The major difference is that ac quantities are complex rather than real as with dc. While this complicates computational details, it does not alter basic circuit principles. This is the approach used in practice. The basic ideas are developed in this chapter.

Since phasor analysis and the impedance concept require a familiarity with complex numbers, we begin with a short review. ■

Charles Proteus Steinmetz

CHARLES STEINMETZ WAS BORN IN BRESLAU, Germany in 1865 and emigrated to the United States in 1889. In 1892, he began working for the General Electric Company in Schenectady, New York, where he stayed until his death in 1923, and it was there that his work revolutionized ac circuit analysis. Prior to his time, this analysis had to be carried out using calculus, a difficult and time-consuming process. By 1893, however, Steinmetz had reduced the very complex alternating-current theory to, in his words, "a simple problem in algebra." The key concept in this simplification was the phasor—a representation based on complex numbers. By representing voltages and currents as phasors, Steinmetz was able to define a quantity called **impedance** and then use it to determine voltage and current magnitude and phase relationships in one algebraic operation.

Steinmetz wrote the seminal textbook on ac analysis based on his method, but at the time he introduced it he was practically the only person who understood it. Now, however, it is common knowledge and one of the basic tools of the electrical engineer and technologist. In this chapter, we learn the method and illustrate its application to the solution of basic ac circuit problems.

In addition to his work for GE, Charles Steinmetz was a professor of electrical engineering (1902–1913) and electrophysics (1913–1923) at Union University (now Union College) in Schenectady. ■

16.1 Complex Number Review

A **complex number** is a number of the form $\mathbf{C} = a + jb$, where a and b are real numbers and $j = \sqrt{-1}$. The number a is called the **real** part of $\mathbf{C}$ and b is called its **imaginary** part. (In circuit theory, j is used to denote the imaginary component rather than i to avoid confusion with current i.)

Geometrical Representation

Complex numbers may be represented geometrically, either in rectangular form or in polar form as points on a two-dimensional plane called the **complex plane** (Figure 16–1). The complex number $\mathbf{C} = 6 + j8$, for example, represents a point whose coordinate on the real axis is 6 and whose coordinate on the imaginary axis is 8. This form of representation is called the **rectangular form.**

Complex numbers may also be represented in **polar form** by magnitude and angle. Thus, $\mathbf{C} = 10 \angle 53.13°$ (Figure 16–2) is a complex number with magnitude 10 and angle 53.13°. This magnitude and angle representation is just an alternate way of specifying the location of the point represented by $\mathbf{C} = a + jb$.

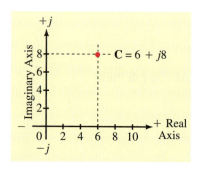

FIGURE 16–1 A complex number in rectangular form.

Conversion between Rectangular and Polar Forms

To convert between forms, note from Figure 16–3 that

$$\mathbf{C} = a + jb \quad \text{(rectangular form)} \tag{16–1}$$

$$\mathbf{C} = C \angle \theta \quad \text{(polar form)} \tag{16–2}$$

where C is the magnitude of $\mathbf{C}$. From the geometry of the triangle,

$$a = C \cos \theta \tag{16–3a}$$

$$b = C \sin \theta \tag{16–3b}$$

where

$$C = \sqrt{a^2 + b^2} \tag{16–4a}$$

and

$$\theta = \tan^{-1}\frac{b}{a} \tag{16–4b}$$

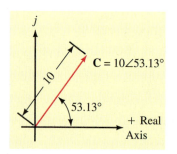

FIGURE 16–2 A complex number in polar form.

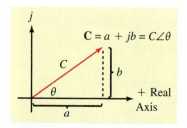

FIGURE 16–3 Polar and rectangular equivalence.

Equations 16–3 and 16–4 permit conversion between forms. When using Equation 16–4b, however, be careful when the number to be converted is in the second or third quadrant, as the angle obtained is the supplementary angle rather than the actual angle in these two quadrants. This is illustrated in Example 16–1 for the complex number **W.**

Determine rectangular and polar forms for the complex numbers **C, D, V,** and **W** of Figure 16–4(a)

EXAMPLE 16–1

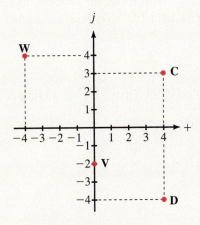

(a) Complex numbers

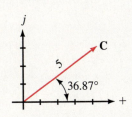

(b) In polar form, **C** = 5∠36.87°

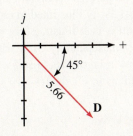

(c) In polar form, **D** = 5.66∠−45°

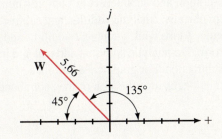

(d) In polar form, **W** = 5.66∠135°

FIGURE 16–4

Solution   *Point C:* Real part = 4; imaginary part = 3. Thus, **C** = 4 + j3. In polar form, $C = \sqrt{4^2 + 3^2}$ = 5 and θ_C = tan⁻¹ (3/4) = 36.87°. Thus, **C** = 5∠36.87° as indicated in (b).

Point D: In rectangular form, **D** = 4 − j4. Thus, $D = \sqrt{4^2 + 4^2}$ = 5.66 and θ_D = tan⁻¹ (−4/4) = −45°. Therefore, **D** = 5.66∠−45°, as shown in (c).

Point V: In rectangular form, **V** = −j2. In polar form, **V** = 2∠−90°.

Point W: In rectangular form, **W** = −4 + j4. Thus, $W = \sqrt{4^2 + 4^2}$ = 5.66 and tan⁻¹ (−4/4) = −45°. Inspection of Figure 16–4(d) shows, however, that this 45° angle is the supplementary angle. The actual angle (measured from the positive horizontal axis) is 135°. Thus, **W** = 5.66∠135°.

In practice (because of the large amount of complex number work that you will do), a more efficient conversion process is needed than that described previously. As discussed later in this section, inexpensive calculators are available that perform such conversions directly—you simply enter the complex number components and press the conversion key. With these, the problem of determining angles for numbers such as **W** in Example 16–1 does not occur; you just enter $-4 + j4$ and the calculator returns $5.66\angle135°$.

Powers of *j*

Powers of *j* are frequently required in calculations. Here are some useful powers:

$$j^2 = (\sqrt{-1})(\sqrt{-1}) = -1$$
$$j^3 = j^2 j = -j$$
$$j^4 = j^2 j^2 = (-1)(-1) = 1 \tag{16-5}$$
$$(-j)j = 1$$
$$\frac{1}{j} = \frac{1}{j} \times \frac{j}{j} = \frac{j}{j^2} = -j$$

Addition and Subtraction of Complex Numbers

Addition and subtraction of complex numbers can be performed analytically or graphically. Analytic addition and subtraction is most easily illustrated in rectangular form, while graphical addition and subtraction is best illustrated in polar form. For analytic addition, add real and imaginary parts separately. Similarly for subtraction. For graphical addition, add vectorially as in Figure 16–5(a); for subtraction, change the sign of the subtrahend, then add, as in Figure 16–5(b).

Given $\mathbf{A} = 2 + j1$ and $\mathbf{B} = 1 + j3$. Determine their sum and difference analytically and graphically.

Solution

$$\mathbf{A} + \mathbf{B} = (2 + j1) + (1 + j3) = (2 + 1) + j(1 + 3) = 3 + j4.$$
$$\mathbf{A} - \mathbf{B} = (2 + j1) - (1 + j3) = (2 - 1) + j(1 - 3) = 1 - j2.$$

Graphical addition and subtraction are shown in Figure 16–5.

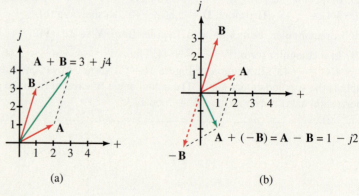

(a)   (b)

FIGURE 16–5

Multiplication and Division of Complex Numbers

These operations are usually performed in polar form. For multiplication, multiply magnitudes and add angles algebraically. For division, divide the magnitude of the denominator into the magnitude of the numerator, then subtract algebraically the angle of the denominator from that of the numerator. Thus, given $\mathbf{A} = A\angle\theta_A$ and $\mathbf{B} = B\angle\theta_B$,

$$\mathbf{A} \cdot \mathbf{B} = AB\underline{/\theta_A + \theta_B} \qquad\qquad (16\text{--}6)$$

$$\mathbf{A/B} = A/B\underline{/\theta_A - \theta_B} \qquad\qquad (16\text{--}7)$$

EXAMPLE 16–3

Given $\mathbf{A} = 3\angle 35°$ and $\mathbf{B} = 2\angle{-}20°$, determine the product $\mathbf{A} \cdot \mathbf{B}$ and the quotient $\mathbf{A/B}$.

Solution

$$\mathbf{A} \cdot \mathbf{B} = (3\angle 35°)(2\angle{-}20°) = (3)(2)\underline{/35° - 20°} = 6\angle 15°$$

$$\frac{\mathbf{A}}{\mathbf{B}} = \frac{(3\angle 35°)}{(2\angle{-}20°)} = \frac{3}{2}\underline{/35° - (-20°)} = 1.5\angle 55°$$

EXAMPLE 16–4

For computations involving purely real, purely imaginary, or small integer numbers, it is sometimes easier to multiply directly in rectangular form than it is to convert to polar. Compute the following directly:

a. $(-j3)(2 + j4)$.

b. $(2 + j3)(1 + j5)$.

Solution

a. $(-j3)(2 + j4) = (-j3)(2) + (-j3)(j4) = -j6 - j^2 12 = 12 - j6$

b. $(2 + j3)(1 + j5) = (2)(1) + (2)(j5) + (j3)(1) + (j3)(j5)$

$$= 2 + j10 + j3 + j^2 15 = 2 + j13 - 15 = -13 + j13$$

PRACTICE PROBLEMS 1

1. Polar numbers with the same angle can be added or subtracted directly without conversion to rectangular form. For example, the sum of $6\angle 36.87°$ and $4\angle 36.87°$ is $10\angle 36.87°$, while the difference is $6\angle 36.87° - 4\angle 36.87° = 2\angle 36.87°$. By means of sketches, indicate why this procedure is valid.

2. To compare methods of multiplication with small integer values, convert the numbers of Example 16–4 to polar form, multiply them, then convert the answers back to rectangular form.

Answers

1. Since the numbers have the same angle, their sum also has the same angle and thus, their magnitudes simply add (or subtract).

Reciprocals

In polar form, the reciprocal of a complex number $\mathbf{C} = C\angle\theta$ is

$$\frac{1}{C\angle\theta} = \frac{1}{C}\angle -\theta \qquad\qquad \textbf{(16–8)}$$

Thus,

$$\frac{1}{20\angle 30°} = 0.05\angle -30°$$

When you work in rectangular form, you must be somewhat careful—see the sidebar note.

Complex Conjugates

The **conjugate** of a complex number (denoted by an asterisk *) is a complex number with the same real part but the opposite imaginary part. Thus, the conjugate of $\mathbf{C} = C\angle\theta = a + jb$ is $\mathbf{C}^* = C\angle -\theta = a - jb$. For example, if $\mathbf{C} = 3 + j4 = 5\angle 53.13°$, then $\mathbf{C}^* = 3 - j4 = 5\angle -53.13°$.

Calculators for AC Analysis

The analysis of ac circuits involves a considerable amount of complex number arithmetic; thus, you will need a calculator that can work easily with complex numbers. There are several inexpensive calculators on the market that are suitable for this purpose in that they permit you to enter complex numbers and perform all required calculations in either rectangular or polar form without the need for conversion. This saves you a great deal of time and cuts down on errors. To illustrate, consider Example 16–5. Using a calculator with only basic complex number conversion capabilities requires that you do all the intermediate conversion shown. On the other hand, more capable calculators can work with mixed forms such as $6 + 30\angle 53.13°$ and thus require fewer intermediate conversion steps. You should consult your manual (if necessary) and learn how to use your calculator proficiently to handle problems such as this.

FIGURE 16–6   This calculator displays complex numbers in standard mathematical notation.

The following illustrates the type of calculations that you will encounter. Reduce the following.

EXAMPLE 16–5

$$(6 + j5) + \frac{(3 - j4)(10\angle 40°)}{6 + 30\angle 53.13°}$$

Solution   Using a calculator with basic capabilities requires a number of intermediate steps, some of which are shown below.

$$\text{answer} = (6 + j5) + \frac{(5\angle -53.13)(10\angle 40)}{6 + (18 + j24)}$$

$$= (6 + j5) + \frac{(5\angle -53.13)(10\angle 40)}{24 + j24}$$

$$= (6 + j5) + \frac{(5\angle -53.13)(10\angle 40)}{33.94\angle 45}$$

$$= (6 + j5) + 1.473\angle -58.13 = (6 + j5) + (0.778 - j1.251)$$

$$= 6.778 + j3.749 = 7.746\angle 28.95°$$

Representing AC Voltages and Currents by Complex Numbers

As you learned in Chapter 15, ac voltages and currents can be represented as phasors (see Notes). Since phasors have magnitude and angle, they can be viewed as complex numbers. To get at the idea, consider the voltage source of Figure 16–7(a). Its phasor equivalent (b) has magnitude E_m and angle θ. It therefore can be viewed as the complex number

$$\mathbf{E} = E_m\angle\theta \qquad\qquad (16\text{–}9)$$

16.2 Complex Numbers in AC Analysis

NOTES . . .

The basic definition of a phasor is a rotating radius vector with a length equal to the amplitude (E_m or I_m) of the voltage or current that it represents. However, for analytic purposes, the phasor is usually redefined in terms of its rms value. We will make this change on page 546. In the meantime, there are a few ideas that we need to explore using its fundamental definition.

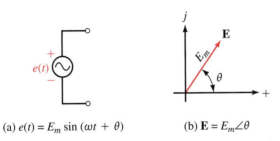

(a) $e(t) = E_m \sin(\omega t + \theta)$ (b) $\mathbf{E} = E_m\angle\theta$

FIGURE 16–7   Representation of a sinusoidal source voltage as a complex number.

From this point of view, the sinusoidal voltage $e(t) = 200 \sin(\omega t + 40°)$ of Figure 16–8(a) and (b) can be represented by its phasor equivalent, $\mathbf{E} = 200\,\text{V}\angle 40°$, as in (c).

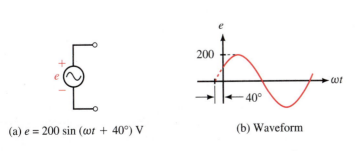
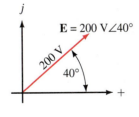

(a) $e = 200 \sin(\omega t + 40°)$ V   (b) Waveform   (c) Phasor equivalent

FIGURE 16–8   Transforming $e = 200 \sin(\omega t + 40°)$ V to $\mathbf{E} = 200\,\text{V}\angle 40°$

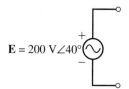

$\mathbf{E} = 200\ \text{V}\angle40°$

Transformed source

FIGURE 16–9 Direct transformation of the source.

We can take advantage of this equivalence. *Rather than show a source as a time-varying voltage e(t) that we subsequently convert to a phasor, we can represent the source by its phasor equivalent right from the start.* This viewpoint is illustrated in Figure 16–9. Since $\mathbf{E} = 200\ \text{V}\angle40°$, this representation retains all the original information of Figure 16–8 since the sinusoidal time variation and its associate angle as illustrated in Figure 16–8(b) is implicit in the definition of the phasor.

The idea illustrated in Figure 16–9 is of fundamental importance to circuit theory. *By replacing the time function e(t) with its phasor equivalent* **E,** *we have transformed the source from the time domain to the phasor domain.* The value of this approach is illustrated next.

Before we move on, we should note that both Kirchhoff's voltage law and Kirchhoff's current law apply in the time domain (i.e., when voltages and currents are expressed as functions of time) and in the phasor domain (i.e., when voltages and currents are represented as phasors). For example, $e = v_1 + v_2$ in the time domain can be transformed to $\mathbf{E} = \mathbf{V}_1 + \mathbf{V}_2$ in the phasor domain and vice versa. Similarly for currents.

Summing AC Voltages and Currents

Sinusoidal quantites must sometimes be added or subtracted as in Figure 16–10. Here, we want the sum of e_1 and e_2, where $e_1 = 10\sin\omega t$ and $e_2 = 15\sin(\omega t + 60°)$. The sum of e_1 and e_2 can be found by adding waveforms point by point as in (b). For example, at $\omega t = 0°$, $e_1 = 10\sin 0° = 0$ and $e_2 = 15\sin(0° + 60°) = 13$ V, and their sum is 13 V. Similarly, at $\omega t = 90°$, $e_1 = 10\sin 90° = 10$ V and $e_2 = 15\sin(90° + 60°) = 15\sin 150° = 7.5$, and their sum is 17.5 V. Continuing in this manner, the sum of $e_1 + e_2$ (the green waveform) is obtained.

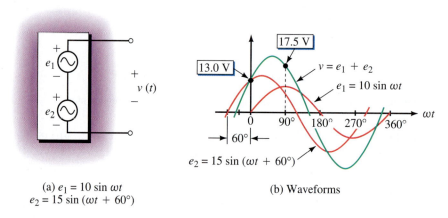

(a) $e_1 = 10\sin\omega t$
$e_2 = 15\sin(\omega t + 60°)$

(b) Waveforms

FIGURE 16–10 Summing waveforms point by point.

As you can see, the process is tedious and provides no analytic expression for the resulting voltage. A better way is to transform the sources and use complex numbers to perform the addition. This is shown in Figure 16–11. Here, we have replaced voltages e_1 and e_2 with their phasor equivalents, $\mathbf{E}_1$ and $\mathbf{E}_2$, and *v* with its phasor equivalent, **V**. Since $v = e_1 + e_2$, replacing *v*, e_1, and e_2 with their phasor equivalents yields $\mathbf{V} = \mathbf{E}_1 + \mathbf{E}_2$. Now **V** can be found by adding $\mathbf{E}_1$ and $\mathbf{E}_2$ as complex numbers. Once **V** is known, its corresponding time equation and companion waveform can be determined.

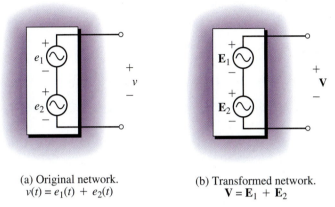

(a) Original network.
$v(t) = e_1(t) + e_2(t)$

(b) Transformed network.
$\mathbf{V} = \mathbf{E}_1 + \mathbf{E}_2$

FIGURE 16–11 Transformed circuit. This is one of the key ideas of sinusoidal circuit analysis.

Given $e_1 = 10 \sin \omega t$ V and $e_2 = 15 \sin(\omega t + 60°)$ V as before, determine *v* and sketch it.

EXAMPLE 16–6

Solution

$$e_1 = 10 \sin \omega t \text{ V. Thus, } \mathbf{E}_1 = 10 \text{ V}\angle 0°.$$
$$e_2 = 15 \sin(\omega t + 60°) \text{ V. Thus, } \mathbf{E}_2 = 15 \text{ V}\angle 60°.$$

Transformed sources are shown in Figure 16–12(a) and phasors in (b).

$$\mathbf{V} = \mathbf{E}_1 + \mathbf{E}_2 = 10\angle 0° + 15\angle 60° = (10 + j0) + (7.5 + j13)$$
$$= (17.5 + j13) = 21.8 \text{ V}\angle 36.6°$$

Thus, $v = 21.8 \sin(\omega t + 36.6°)$ V
Waveforms are shown in (c). (To verify that this produces the same result as adding the waveforms point by point, see Practice Problem 2.)

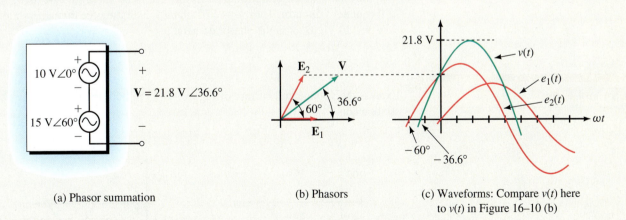

(a) Phasor summation

(b) Phasors

(c) Waveforms: Compare *v(t)* here to *v(t)* in Figure 16–10 (b)

FIGURE 16–12 Note that *v(t)*, determined from phasor **V** gives the same result as adding e_1 and e_2 point by point.

PRACTICE PROBLEMS 2

Verify by direct substitution that $v = 21.8 \sin(\omega t + 36.6°)$ V, as in Figure 16–12, is the sum of e_1 and e_2. To do this, compute e_1 and e_2 at a point, add them, then compare the sum to $21.8 \sin(\omega t + 36.6°)$ V computed at the same point. Perform this computation at $\omega t = 30°$ intervals over the complete cycle to satisfy yourself that the result is true everywhere. (For example, at $\omega t = 0°$, $v = 21.8 \sin(\omega t + 36.6°) = 21.8 \sin(36.6°) = 13$ V, as we saw earlier in Figure 16–10.)

Answers
Here are the points on the graph at 30° intervals:

ωt	0°	30°	60°	90°	120°	150°	180°	210°	240°	270°	300°	330°	360°
v	13	20	21.7	17.5	8.66	−2.5	−13	−20	−21.7	−17.5	−8.66	2.5	13

IMPORTANT NOTES . . .

1. To this point, we have used peak values such as V_m and I_m to represent the magnitudes of phasor voltages and currents, as this has been most convenient for our purposes. In practice, however, rms values are used instead. Accordingly, we will now change to rms. Thus, from here on, the Phasor $V = 120$ V $\angle 0°$ will be taken to mean a voltage of 120 volts rms at an angle of 0°. If you need to convert this to a time function, first multiply the rms value by $\sqrt{2}$, then follow the usual procedure. Thus, $v = \sqrt{2}\,(120) \sin \omega t = 170 \sin \omega t$.

2. To add or subtract sinusoidal voltages or currents, follow the three steps outlined in Example 16–6. That is,

   • convert sine waves to phasors and express them in complex number form,

   • add or subtract the complex numbers,

   • convert back to time functions if desired.

3. Although we use phasors to represent sinusoidal waveforms, it should be noted that sine waves and phasors are not the same thing. Sinusoidal voltages and currents are real—they are the actual quantities that you measure with meters and whose waveforms you see on oscilloscopes. *Phasors, on the other hand, are mathematical abstractions that we use to help visualize relationships and solve problems.*

4. Quantities expressed as time functions are said to be in the **time domain,** while quantities expressed as phasors are said to be in the **phasor (or frequency) domain.** Thus, $e = 170 \sin \omega t$ V is in the time domain, while $V = 120$ V $\angle 0°$ is in the phasor domain.

EXAMPLE 16–7

Express the voltage and current of Figure 16–13 in both the time and the phasor domains.

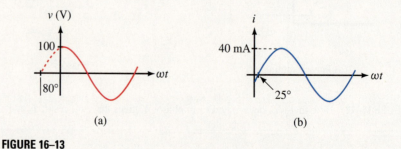

(a)   (b)

FIGURE 16–13

Solution

a. Time domain: $v = 100 \sin(\omega t + 80°)$ volts.

   Phasor domain: $\mathbf{V} = (0.707)(100 \text{ V} \angle 80°) = 70.7 \text{ V} \angle 80°$ (rms).

b. Time domain: $i = 40 \sin(\omega t - 25°)$ mA.

   Phasor domain: $\mathbf{I} = (0.707)(40 \text{ mA} \angle -25°) = 28.3 \text{ mA} \angle -25°$ (rms).

EXAMPLE 16–8

If $i_1 = 14.14 \sin(\omega t - 55°)$ A and $i_2 = 4 \sin(\omega t + 15°)$ A, determine their sum, i. Work with rms values.

Solution

$$\mathbf{I_1} = (0.707)(14.14 \text{ A}) \angle -55° = 10 \text{ A} \angle -55°$$
$$\mathbf{I_2} = (0.707)(4 \text{ A}) \angle 15° = 2.828 \text{ A} \angle 15°$$
$$\mathbf{I} = \mathbf{I_1} + \mathbf{I_2} = 10 \text{ A} \angle - 55° + 2.828 \text{ A} \angle 15°$$
$$= (5.74 \text{ A} - j8.19 \text{ A}) + (2.73 \text{ A} + j0.732 \text{ A})$$
$$= 8.47 \text{ A} - j7.46 \text{ A} = 11.3 \text{ A} \angle -41.4$$
$$i(t) = \sqrt{2}(11.3) \sin(\omega t - 41.4°) = 16 \sin(\omega t - 41.4°) \text{ A}$$

While it may seem silly to convert peak values to rms and then convert rms back to peak as we did here, we did it for a reason. The reason is that very soon, we will stop working in the time domain entirely and work only with phasors. At that point, the solution will be complete when we have the answer in the form $\mathbf{I} = 11.3 \angle -41.4°$. (To help focus on rms, voltages and currents in the next two examples (and in other examples to come) are expressed as an rms value times $\sqrt{2}$.)

EXAMPLE 16–9

For Figure 16–14, $v_1 = \sqrt{2}(16) \sin \omega t$ V, $v_2 = \sqrt{2}(24) \sin(\omega t + 90°)$ and $v_3 = \sqrt{2}(15) \sin(\omega t - 90°)$ V. Determine source voltage e.

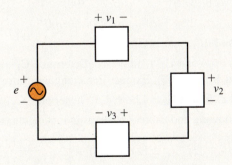

FIGURE 16–14

Solution   The answer can be obtained by KVL. First, convert to phasors. Thus, $\mathbf{V_1} = 16 \text{ V} \angle 0°$, $\mathbf{V_2} = 24 \text{ V} \angle 90°$, and $\mathbf{V_3} = 15 \text{ V} \angle -90°$. KVL yields $\mathbf{E} = \mathbf{V_1} + \mathbf{V_2} + \mathbf{V_3} = 16 \text{ V} \angle 0° + 24 \text{ V} \angle 90° + 15 \text{ V} \angle -90° = 18.4 \text{ V} \angle 29.4°$. Converting back to a function of time yields $e = \sqrt{2}(18.4) \sin(\omega t + 29.4°)$ V.

EXAMPLE 16–10

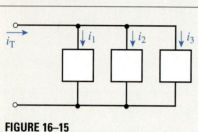

FIGURE 16–15

For Figure 16–15, $i_1 = \sqrt{2}(23) \sin \omega t$ mA, $i_2 = \sqrt{2}(0.29) \sin(\omega t + 63°)$ A and $i_3 = \sqrt{2}(127) \times 10^{-3} \sin(\omega t - 72°)$ A. Determine current i_T.

Solution   Convert to phasors. Thus, $\mathbf{I}_1 = 23$ mA∠0°, $\mathbf{I}_2 = 0.29$ A∠63°, and $\mathbf{I}_3 = 127 \times 10^{-3}$ A∠−72°. KCL yields $\mathbf{I}_T = \mathbf{I}_1 + \mathbf{I}_2 + \mathbf{I}_3 = 23$ mA∠0° + 290 mA∠63° + 127 mA∠−72° = 238 mA∠35.4°. Converting back to a function of time yields $i_T = \sqrt{2}(238) \sin(\omega t + 35.4°)$ mA.

PRACTICE PROBLEMS 3

1. Convert the following to time functions. Values are rms.
   a. $\mathbf{E} = 500$ mV∠−20°   b. $\mathbf{I} = 80$ A∠40°

2. For the circuit of Figure 16–16, determine voltage e_1.

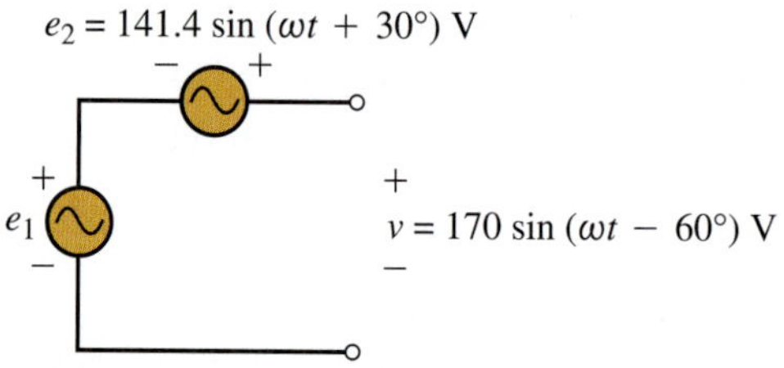

$$e_2 = 141.4 \sin (\omega t + 30°) \text{ V}$$

$v = 170 \sin (\omega t - 60°)$ V

FIGURE 16–16

Answers
1. a. $e = 707 \sin(\omega t - 20°)$ mV   b. $i = 113 \sin(\omega t + 40°)$ A
2. $e_1 = 221 \sin(\omega t - 99.8°)$ V

✓ IN-PROCESS
LEARNING CHECK 1

(Answers are at the end of the chapter.)

$i_T = i_1 + i_2 + i_3$

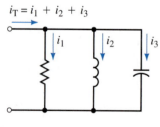

FIGURE 16–17

1. Convert the following to polar form:
   a. $j6$ b. $-j4$ c. $3 + j3$ d. $4 - j6$
   e. $-5 + j8$ f. $1 - j2$ g. $-2 - j3$

2. Convert the following to rectangular form:
   a. $4∠90°$ b. $3∠0°$ c. $2∠-90°$ d. $5∠40°$
   e. $6∠120°$ f. $2.5∠-20°$ g. $1.75∠-160°$

3. If $-\mathbf{C} = 12∠-140°$, what is $\mathbf{C}$?

4. Given: $\mathbf{C}_1 = 36 + j4$ and $\mathbf{C}_2 = 52 - j11$. Determine $\mathbf{C}_1 + \mathbf{C}_2$, $\mathbf{C}_1 - \mathbf{C}_2$, $1/(\mathbf{C}_1 + \mathbf{C}_2)$ and $1/(\mathbf{C}_1 - \mathbf{C}_2)$. Express in rectangular form.

5. Given: $\mathbf{C}_1 = 24∠25°$ and $\mathbf{C}_2 = 12∠-125°$. Determine $\mathbf{C}_1 \cdot \mathbf{C}_2$ and $\mathbf{C}_1/\mathbf{C}_2$.

6. Compute the following and express answers in rectangular form:
   a. $\dfrac{6 + j4}{10∠20°} + (14 + j2)$ b. $(1 + j6) + \left[2 + \dfrac{(12∠0°)(14 + j2)}{6 - (10∠20°)(2∠-10°)} \right]$

7. For Figure 16–17, determine i_T where $i_1 = 10 \sin \omega t$, $i_2 = 20 \sin(\omega t - 90°)$, and $i_3 = 5 \sin(\omega t + 90°)$.

R, L, and *C* circuit elements each have quite different electrical properties. Resistance, for example, opposes current, while inductance opposes changes in current, and capacitance opposes changes in voltage. These differences result in quite different voltage-current relationships as you saw earlier. We now investigate these relationships for the case of sinusoidal ac. Sine waves have several important characteristics that you will discover from this investigation:

1. When a circuit consisting of linear circuit elements *R, L,* and *C* is connected to a sinusoidal source, all currents and voltages in the circuit will be sinusoidal.

2. These sine waves have the same frequency as the source and differ from it only in terms of their magnitudes and phase angles.

16.3 *R, L,* and *C* Circuits with Sinusoidal Excitation

We begin with a purely resistive circuit. Here, Ohm's law applies and thus, current is directly proportional to voltage. Current variations therefore follow voltage variations, reaching their peak when voltage reaches its peak, changing direction when voltage changes polarity, and so on (Figure 16–18). From this, we conclude that *for a purely resistive circuit, current and voltage are in phase.* Since voltage and current waveforms coincide, their phasors also coincide (Figure 16–19).

16.4 Resistance and Sinusoidal AC

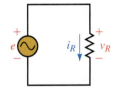
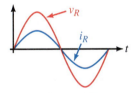

(a) Source voltage is a sine wave. Therefore, v_R is a sine wave

(b) $i_R = v_R/R$. Therefore i_R is a sine wave also

FIGURE 16–18   Ohm's law applies. Note that current and voltage are in phase.

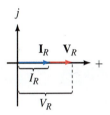

FIGURE 16–19   For a resistor, voltage and current phasors are in phase.

The relationship illustrated in Figure 16–18 may be stated mathematically as

$$i_R = \frac{v_R}{R} = \frac{V_m \sin \omega t}{R} = \frac{V_m}{R}\sin \omega t = I_m \sin \omega t \qquad \textbf{(16–10)}$$

where

$$I_m = V_m/R \qquad \textbf{(16–11)}$$

Transposing,

$$V_m = I_m R \qquad \textbf{(16–12)}$$

The in-phase relationship is true regardless of reference. Thus, if $v_R = V_m \sin(\omega t + \theta)$, then $i_R = I_m \sin(\omega t + \theta)$.

EXAMPLE 16–11

For the circuit of Figure 16–18(a), if $R = 5 \; \Omega$ and $i_R = 12 \sin(\omega t - 18°)$ A, determine v_R.

Solution $v_R = Ri_R = 5 \times 12 \sin(\omega t - 18°) = 60 \sin(\omega t - 18°)$ V. The waveforms are shown in Figure 16–20.

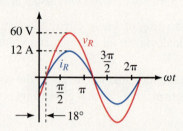

FIGURE 16–20

---

1. If $v_R = 150 \cos \omega t$ V and $R = 25 \; k\Omega$, determine i_R and sketch both waveforms.
2. If $v_R = 100 \sin(\omega t + 30°)$ V and $R = 0.2 \; M\Omega$, determine i_R and sketch both waveforms.

Answers
1. $i_R = 6 \cos \omega t$ mA. v_R and i_R are in phase.
2. $i_R = 0.5 \sin(\omega t + 30°)$ mA. v_R and i_R are in phase.

16.5 Inductance and Sinusoidal AC

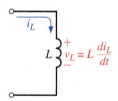

FIGURE 16–21 Voltage v_L is proportional to the rate of change of current i_L.

Phase Lag in an Inductive Circuit

As you saw in Chapter 13, for an ideal inductor, voltage v_L is proportional to the rate of change of current. Because of this, voltage and current are not in phase as they are for a resistive circuit. This can be shown with a bit of calculus. From Figure 16–21, $v_L = Ldi_L/dt$. For a sine wave of current, you get when you differentiate

$$v_L = L\frac{di_L}{dt} = L\frac{d}{dt}(I_m \sin \omega t) = \omega LI_m \cos \omega t = V_m \cos \omega t$$

Utilizing the trigonometric identity $\cos \omega t = \sin(\omega t + 90°)$, you can write this as

$$v_L = V_m \sin(\omega t + 90°) \qquad (16\text{–}13)$$

where

$$V_m = \omega LI_m \qquad (16\text{–}14)$$

Voltage and current waveforms are shown in Figure 16–22, and phasors in Figure 16–23. As you can see, *for a purely inductive circuit, current lags voltage by 90°* (i.e., ¼ cycle). Alternatively you can say that voltage leads current by 90°.

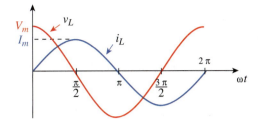

FIGURE 16–22 For inductance, current lags voltage by 90°. Here, i_L is reference.

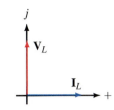

FIGURE 16–23 Phasors for the waveforms of Fig. 16–22 showing the 90° lag of current.

Although we have shown that current lags voltage by 90° for the case of Figure 16–22, this relationship is true in general, that is, current always lags voltage by 90° regardless of the choice of reference. This is illustrated in Figure 16–24. Here, $\mathbf{V}_L$ is at 0° and $\mathbf{I}_L$ at −90°. Thus, voltage v_L will be a sine wave and current i_L a negative cosine wave, i.e., $i_L = -I_m \cos \omega t$. Since i_L is a negative cosine wave, it can also be expressed as $i_L = I_m \sin(\omega t - 90°)$. The waveforms are shown in (b).

Since current always lags voltage by 90° for a pure inductance, you can, if you know the phase of the voltage, determine the phase of the current, and vice versa. Thus, if v_L is known, i_L must lag it by 90°, while if i_L is known, v_L must lead it by 90°.

Inductive Reactance

From Equation 16–14, we see that the ratio V_m to I_m is

$$\frac{V_m}{I_m} = \omega L \qquad (16\text{--}15)$$

This ratio is defined as **inductive reactance** and is given the symbol X_L. Since the ratio of volts to amps is ohms, reactance has units of ohms. Thus,

$$X_L = \frac{V_m}{I_m} \quad (\Omega) \qquad (16\text{--}16)$$

Combining Equations 16–15 and 16–16 yields

$$X_L = \omega L \ (\Omega) \qquad (16\text{--}17)$$

where ω is in radians per second and L is in henries. *Reactance X_L represents the opposition that inductance presents to current for the sinusoidal ac case.*

We now have everything that we need to solve simple inductive circuits with sinusoidal excitation, that is, we know that current lags voltage by 90° and that their amplitudes are related by

$$I_m = \frac{V_m}{X_L} \qquad (16\text{--}18)$$

and

$$V_m = I_m X_L \qquad (16\text{--}19)$$

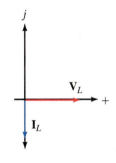

(a) Current $\mathbf{I}_L$ always lags voltage $\mathbf{V}_L$ by 90°

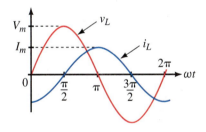

(b) Waveforms

FIGURE 16–24 Phasors and waveforms when $\mathbf{V}_L$ is used as reference.

EXAMPLE 16–12

The voltage across a 0.2-H inductance is $v_L = 100 \sin(400t + 70°)$ V. Determine i_L and sketch it.

Solution $\omega = 400$ rad/s. Therefore, $X_L = \omega L = (400)(0.2) = 80\ \Omega$.

$$I_m = \frac{V_m}{X_L} = \frac{100\ \text{V}}{80\ \Omega} = 1.25\ \text{A}$$

The current lags the voltage by 90°. Therefore $i_L = 1.25 \sin(400t - 20°)$ A as indicated in Figure 16–25.

NOTES . . .

Remember to show phasors as rms values from now on.

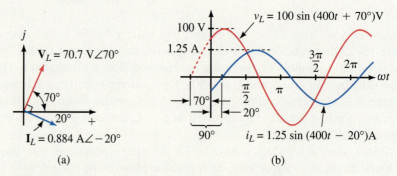

(a)   (b)

FIGURE 16–25   With voltage $\mathbf{V}_L$ at 70°, current $\mathbf{I}_L$ will be 90° later at −20°.

EXAMPLE 16–13

The current through a 0.01-H inductance is $i_L = 20 \sin(\omega t - 50°)$ A and $f = 60$ Hz. Determine v_L.

Solution

$$\omega = 2\pi f = 2\pi(60) = 377\ \text{rad/s}$$
$$X_L = \omega L = (377)(0.01) = 3.77\ \Omega$$
$$V_m = I_m X_L = (20\ \text{A})(3.77\ \Omega) = 75.4\ \text{V}$$

Voltage leads current by 90°. Thus, $v_L = 75.4 \sin(377t + 40°)$ V as shown in Figure 16–26.

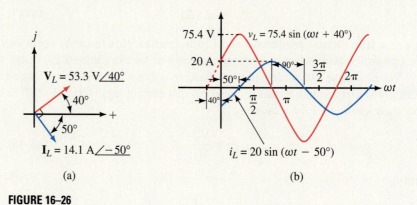

(a)   (b)

FIGURE 16–26

1. Two inductances are connected in series (Figure 16–27). If $e = 100 \sin \omega t$ and $f = 10$ kHz, determine the current. Sketch voltage and current waveforms.

2. The current through a 0.5-H inductance is $i_L = 100 \sin(2400t + 45°)$ mA. Determine v_L and sketch voltage and current phasors and waveforms.

Answers
1. $i_L = 1.99 \sin(\omega t - 90°)$ mA. Waveforms same as Figure 16–24.
2. $v_L = 120 \sin(2400t + 135° \text{ V})$. See following art for waveforms.

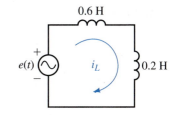

PRACTICE PROBLEMS 5

FIGURE 16–27

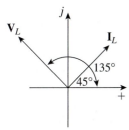

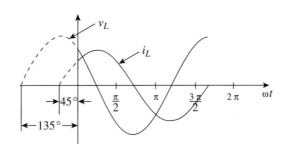

Variation of Inductive Reactance with Frequency

Since $X_L = \omega L = 2\pi f L$, inductive reactance is directly proportional to frequency (Figure 16–28). Thus, if frequency is doubled, reactance doubles, while if frequency is halved, reactance halves, and so on. In addition, X_L is directly proportional to inductance. Thus, if inductance is doubled, X_L is doubled, and so on. Note also that at $f = 0$, $X_L = 0 \ \Omega$. This means that inductance looks like a short circuit to dc. (We already concluded this earlier in Chapter 13.)

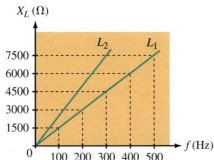

FIGURE 16–28   Variation of X_L with frequency. Note that $L_2 > L_1$.

PRACTICE PROBLEMS 6

A circuit has 50 ohms inductive reactance. If both the inductance and the frequency are doubled, what is the new X_L?

Answer
$200 \ \Omega$

16.6 Capacitance and Sinusoidal AC

Phase Lead in a Capacitive Circuit

For capacitance, current is proportional to the rate of change of voltage, i.e., $i_C = C\, dv_C/dt$ [Figure 16–29(a)]. Thus if v_C is a sine wave, you get upon substitution

$$i_C = C\frac{dv_C}{dt} = C\frac{d}{dt}(V_m \sin \omega t) = \omega C V_m \cos \omega t = I_m \cos \omega t$$

Using the appropriate trigonometric identity, this can be written as

$$i_C = I_m \sin(\omega t + 90°) \qquad (16\text{–}20)$$

where

$$I_m = \omega C V_m \qquad (16\text{–}21)$$

Waveforms are shown in Figure 16–29(b) and phasors in (c). As indicated, *for a purely capacitive circuit, current leads voltage by 90°,* or alternatively, voltage lags current by 90°. This relationship is true regardless of reference. Thus, if the voltage is known, the current must lead by 90° while if the current is known, the voltage must lag by 90°. For example, if $\mathbf{I}_C$ is at 60° as in (d), $\mathbf{V}_C$ must be at $-30°$.

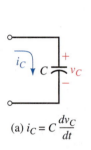

(a) $i_C = C\dfrac{dv_C}{dt}$

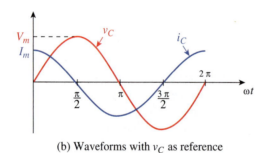

(b) Waveforms with v_C as reference

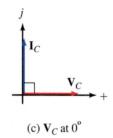

(c) $\mathbf{V}_C$ at $0°$

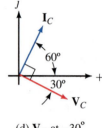

(d) $\mathbf{V}_C$ at $-30°$

FIGURE 16–29   For capacitance, current always leads voltage by 90°.

 PRACTICE PROBLEMS 7

1. The current source of Figure 16–30(a) is a sine wave. Sketch phasors and capacitor voltage v_C.
2. Refer to the circuit of Figure 16–31(a):
   a. Sketch the phasors.
   b. Sketch capacitor current i_C.

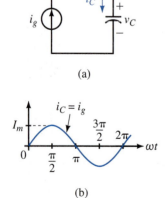

(a)

(b)

FIGURE 16–30

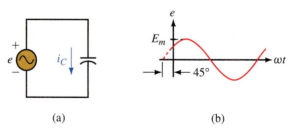

(a)

(b)

FIGURE 16–31

Answers
1. $\mathbf{I}_C$ is at 0°; $\mathbf{V}_C$ is at $-90°$; v_C is a negative cosine wave.
2. a. $\mathbf{V}_C$ is at 45° and $\mathbf{I}_C$ is at 135°.
   b. Waveforms are the same as for Problem 2, Practice Problem 5, except that voltage and current waveforms are interchanged.

Capacitive Reactance

Now consider the relationship between maximum capacitor voltage and current magnitudes. As we saw in Equation 16–21, they are related by $I_m = \omega C V_m$. Rearranging, we get $V_m/I_m = 1/\omega C$. The ratio of V_m to I_m is defined as **capacitive reactance** and is given the symbol X_C. That is,

$$X_C = \frac{V_m}{I_m} \quad (\Omega)$$

Since $V_m/I_m = 1/\omega C$, we also get

$$X_C = \frac{1}{\omega C} \quad (\Omega) \qquad \text{(16–22)}$$

where ω is in radians per second and C is in farads. *Reactance X_C represents the opposition that capacitance presents to current for the sinusoidal ac case.* It has units of ohms.

We now have everything that we need to solve simple capacitive circuits with sinusoidal excitation, i.e., we know that current leads voltage by 90° and that

$$I_m = \frac{V_m}{X_C} \qquad \text{(16–23)}$$

and

$$V_m = I_m X_C \qquad \text{(16–24)}$$

The voltage across a 10-μF capacitance is $v_C = 100 \sin(\omega t - 40°)$ V and $f = 1000$ Hz. Determine i_C and sketch its waveform.

EXAMPLE 16–14

Solution

$$\omega = 2\pi f = 2\pi(1000 \text{ Hz}) = 6283 \text{ rad/s}$$

$$X_C = \frac{1}{\omega C} = \frac{1}{(6283)(10 \times 10^{-6})} = 15.92 \ \Omega$$

$$I_m = \frac{V_m}{X_C} = \frac{100 \text{ V}}{15.92 \ \Omega} = 6.28 \text{ A}$$

Since current leads voltage by 90°, $i_C = 6.28 \sin(6283t + 50°)$ A as indicated in Figure 16–32.

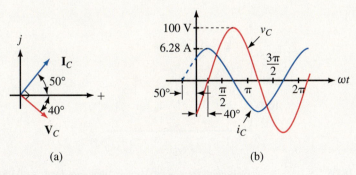

(a)   (b)

FIGURE 16–32   Phasors are not to scale with waveform.

EXAMPLE 16–15

The current through a 0.1-µF capacitance is $i_C = 5 \sin(1000t + 120°)$ mA. Determine v_C.

Solution

$$X_C = \frac{1}{\omega C} = \frac{1}{(1000 \text{ rad/s})(0.1 \times 10^{-6} \text{ F})} = 10 \text{ k}\Omega$$

Thus, $V_m = I_m X_C = (5 \text{ mA})(10 \text{ k}\Omega) = 50$ V. Since voltage lags current by 90°, $v_C = 50 \sin(1000t + 30°)$ V. Waveforms and phasors are shown in Figure 16–33.

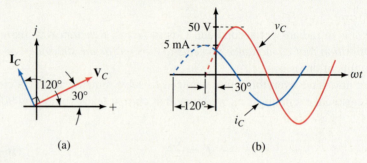

(a)   (b)

FIGURE 16–33   Phasors are not to scale with waveform.

PRACTICE PROBLEMS 8

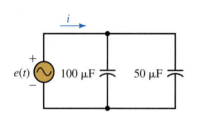

FIGURE 16–34

Two capacitances are connected in parallel (Figure 16–34). If $e = 100 \sin \omega t$ V and $f = 10$ Hz, determine the source current. Sketch current and voltage phasors and waveforms.

Answer: i = 0.942 sin(62.8*t* + 90°) = 0.942 cos 62.8*t* A

See Figure 16–29(b) and (c).

Variation of Capacitive Reactance with Frequency

Since $X_C = 1/\omega C = 1/2\pi f C$, the opposition that capacitance presents varies inversely with frequency. This means that the higher the frequency, the lower the reactance, and vice versa (Figure 16–35). At $f = 0$ (i.e., dc), capacitive reactance is infinite. This means that a capacitance looks like an open circuit to dc. (We already concluded this earlier in Chapter 10.) Note that X_C is also inversely proportional to capacitance. Thus, if capacitance is doubled, X_C is halved, and so on.

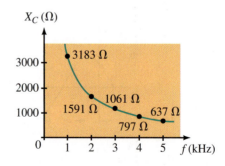

FIGURE 16–35 X_C varies inversely with frequency. Values shown are for $C = 0.05$ µF.

IN-PROCESS
LEARNING CHECK 2

(Answers are at the end of the chapter.)

1. For a pure resistance, $v_R = 100 \sin(\omega t + 30°)$ V. If $R = 2\ \Omega$, what is the expression for i_R?

2. For a pure inductance, $v_L = 100 \sin(\omega t + 30°)$ V. If $X_L = 2\ \Omega$, what is the expression for i_L?

3. For a pure capacitance, $v_C = 100 \sin(\omega t + 30°)$ V. If $X_C = 2\Omega$, what is the expression for i_C?

4. If $f = 100$ Hz and $X_L = 400\ \Omega$, what is L?

5. If $f = 100$ Hz and $X_C = 400\ \Omega$, what is C?

6. For each of the phasor sets of Figure 16–36, identify whether the circuit is resistive, inductive, or capacitive. Justify your answers.

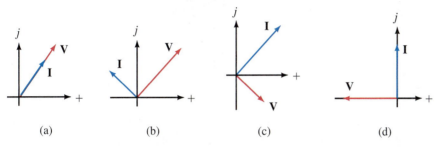

(a)　　　　　(b)　　　　　(c)　　　　　(d)

FIGURE 16–36

In Sections 16.5 and 16.6, we handled magnitude and phase analysis separately. However, this is not the way it is done in practice. In practice, we represent circuit elements by their impedance, and determine magnitude and phase relationships in one step. Before we do this, however, we need to learn how to represent circuit elements as impedances.

16.7 The Impedance Concept

◀ **Online Companion**

Impedance

The opposition that a circuit element presents to current in the phasor domain is defined as its **impedance.** The impedance of the element of Figure 16–37, for example, is the ratio of its voltage phasor to its current phasor. Impedance is denoted by the boldface, uppercase letter **Z.** Thus,

$$\mathbf{Z} = \frac{\mathbf{V}}{\mathbf{I}} \quad \text{(ohms)} \qquad (16\text{–}25)$$

(This equation is sometimes referred to as Ohm's law for ac circuits.)

Since phasor voltages and currents are complex, **Z** is also complex. That is,

$$\mathbf{Z} = \frac{\mathbf{V}}{\mathbf{I}} = \frac{V}{I}\angle\theta \qquad (16\text{–}26)$$

where V and I are the rms magnitudes of **V** and **I** respectively, and θ is the angle between them. From Equation 16–26,

$$\mathbf{Z} = Z\angle\theta \qquad (16\text{–}27)$$

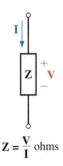

$$\mathbf{Z} = \frac{\mathbf{V}}{\mathbf{I}} \text{ ohms}$$

FIGURE 16–37　Impedance concept.

where $Z = V/I$. Since $V = 0.707V_m$ and $I = 0.707I_m$, Z can also be expressed as V_m/I_m. Once the impedance of a circuit is known, the current and voltage can be determined using

$$I = \frac{V}{Z} \qquad (16\text{–}28)$$

and

$$V = IZ \qquad (16\text{–}29)$$

Let us now determine impedance for the basic circuit elements R, L, and C.

Resistance

For a pure resistance (Figure 16–38), voltage and current are in phase. Thus, if voltage has an angle θ, current will have the same angle. For example, if $\mathbf{V}_R = V_R\angle\theta$, then $\mathbf{I} = I\angle\theta$. Substituting into Equation 16–25 yields:

$$\mathbf{Z}_R = \frac{\mathbf{V}_R}{\mathbf{I}} = \frac{V_R\angle\theta}{I\angle\theta} = \frac{V_R}{I}\angle 0° = R\angle 0° = R$$

Thus the impedance of a resistor is just its resistance. That is,

$$\mathbf{Z}_R = R \qquad (16\text{–}30)$$

This agrees with what we know about resistive circuits, i.e., that the ratio of voltage to current is R, and that the angle between them is $0°$.

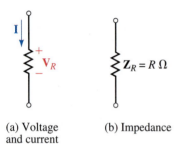

(a) Voltage and current

(b) Impedance

FIGURE 16–38   Impedance of a pure resistance.

Inductance

For a pure inductance, current lags voltage by $90°$. Assuming a $0°$ angle for voltage (we can assume any reference we want because we are interested only in the angle between $\mathbf{V}_L$ and $\mathbf{I}$), we can write $\mathbf{V}_L = V_L\angle 0°$ and $\mathbf{I} = I\angle -90°$. The impedance of a pure inductance (Figure 16–39) is therefore

$$\mathbf{Z}_L = \frac{\mathbf{V}_L}{\mathbf{I}} = \frac{V_L\angle 0°}{I\angle -90°} = \frac{V_L}{I}\angle 90° = \omega L\angle 90° = j\omega L$$

where we have used the fact that $V_L/I_L = \omega L$. Thus,

$$\mathbf{Z}_L = j\omega L = jX_L \qquad (16\text{–}31)$$

since ωL is equal to X_L.

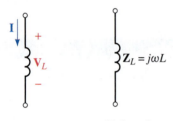

(a) Voltage and current

(b) Impedance

FIGURE 16–39   Impedance of a pure inductance.

EXAMPLE 16–16

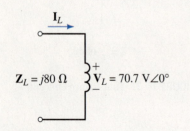

FIGURE 16–40

Consider again Example 16–12. Given $v_L = 100 \sin(400t + 70°)$ and $L = 0.2$ H, determine i_L using the impedance concept.

Solution   See Figure 16–40. Use rms values. Thus,

$$\mathbf{V}_L = 70.7 \text{ V}\angle 70° \quad \text{and} \quad \omega = 400 \text{ rad/s}$$
$$\mathbf{Z}_L = j\omega L = j(400)(0.2) = j80 \text{ }\Omega$$
$$\mathbf{I}_L = \frac{\mathbf{V}_L}{\mathbf{Z}_L} = \frac{70.7\angle 70°}{j80} = \frac{70.7\angle 70°}{80\angle 90°} = 0.884 \text{ A}\angle -20°$$

In the time domain, $i_L = \sqrt{2}(0.884) \sin(400t - 20°) = 1.25 \sin(400t - 20°)$ A, which agrees with our previous solution.

Capacitance

For a pure capacitance, current leads voltage by 90°. Its impedance (Figure 16–41) is therefore

$$\mathbf{Z}_C = \frac{\mathbf{V}_C}{\mathbf{I}} = \frac{V_C\angle 0°}{I\angle 90°} = \frac{V_C}{I}\angle -90° = \frac{1}{\omega C}\angle -90° = -j\frac{1}{\omega C} \quad \text{(ohms)}$$

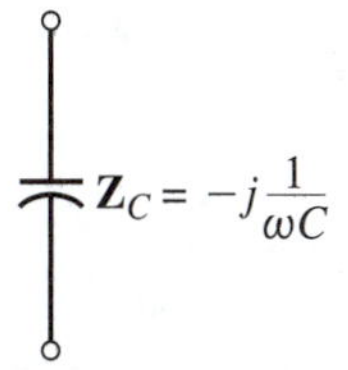

Thus,

$$\mathbf{Z}_C = -j\frac{1}{\omega C} = -jX_C \quad \text{(ohms)} \qquad \textbf{(16–32)}$$

since $1/\omega C$ is equal to X_C.

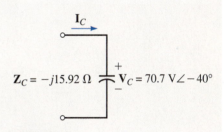

FIGURE 16–41 Impedance of a pure capacitance.

EXAMPLE 16–17

Given $v_C = 100 \sin(\omega t - 40°)$V, $f = 1000$ Hz, and $C = 10$ µF, determine i_C in Figure 16–42.

Solution

$$\omega = 2\pi f = 2\pi(1000 \text{ Hz}) = 6283 \text{ rads/s}$$
$$\mathbf{V}_C = 70.7 \text{ V}\angle -40°$$
$$\mathbf{Z}_C = -j\frac{1}{\omega C} = -j\left(\frac{1}{6283 \times 10 \times 10^{-6}}\right) = -j15.92 \text{ } \Omega.$$
$$\mathbf{I}_C = \frac{\mathbf{V}_C}{\mathbf{Z}_C} = \frac{70.7\angle -40°}{-j15.92} = \frac{70.7\angle -40°}{15.92\angle -90°} = 4.442 \text{ A}\angle 50°$$

In the time domain, $i_C = \sqrt{2}(4.442) \sin(6283t + 50°) = 6.28 \sin(6283t + 50°)$ A, which agrees with our previous solution, in Example 16–14.

FIGURE 16–42

(Figure: $\mathbf{Z}_C = -j15.92 \text{ } \Omega$, $\mathbf{V}_C = 70.7 \text{ V}\angle -40°$, current I_C)

PRACTICE PROBLEMS 9

1. If $\mathbf{I}_L = 5 \text{ mA}\angle -60°$, $L = 2$ mH, and $f = 10$ kHz, what is $\mathbf{V}_L$?
2. A capacitor has a reactance of 50 Ω at 1200 Hz. If $v_C = 80 \sin 800t$ V, what is i_C?

Answers
1. 628 mV$\angle 30°$
2. $0.170 \sin(800t + 90°)$ A

A Final Note

The real power of the impedance method becomes apparent when you consider complex circuits with elements in series, parallel, and so on. This we do later, beginning in Chapter 18. Before we do this, however, there are some ideas on power that you need to know. These are considered in Chapter 17.

In Chapter 15, you saw how to represent sinusoidal waveforms using PSpice and MultiSIM. Let us now apply these tools to the ideas of this chapter. To illustrate, recall that in Example 16–6, we summed voltages $e_1 = 10 \sin \omega t$ V and $e_2 = 15 \sin(\omega t + 60°)$ V using phasor methods to obtain their sum $v = 21.8 \sin(\omega t + 36.6°)$ V. We will now verify this summation numerically. Since the process is independent of frequency, let us choose $f = 500$ Hz. This yields a period of 2 ms; thus, 1/2 cycle is 1 ms, 1/4 cycle (90°) is 500 µs, 45° is 250 µs, etc.

16.8 Computer Analysis of AC Circuits

◀ MULTISIM

◀ CADENCE

NOTES . . .

MulltiSIM

As noted in Chapter 15, MultiSIM 2001 contains a bug that makes it necessary to enter the negative of the angle that you actually want. Thus, if you want an angle of 60°, you must enter −60°. However, since MultiSIM accepts only positive angles, you must enter 300° instead. (Because of this bug, your screen display for V2 will show 300Deg, rather than 60Deg as shown in Figure 16–43.) This bug is designated to be corrected in the new release.

MultiSIM

Procedure: Create the circuit of Figure 16–43 on the screen, then double click Source 1 and set Voltage Amplitude to **10 V,** Frequency to **500 Hz,** and Phase to **0 Deg.** Similarly set Source 2 to **15 V, 500 Hz,** and **60 Deg** (see Notes). Click the Analysis icon, select *Transient Analysis* and in the dialog box that opens, enter **0.002** for TSTOP (to display a full cycle), set *Minimum number of time points* to **1000** (to avoid getting a choppy waveform), select the nodes needed to display e_1 and v, then click Simulate. Following simulation, the graphs of Figure 16–43 appear. Expand the Analysis Graph window to full screen, then click the cursor icon and drag a cursor to 500 μs or as close as you can get it and read values. (You should get about 10 V for e_1 and 17.5 V for v.) Scale values from the graph at 200-μs increments and tabulate. Now replace the sources with a single source of $21.8 \sin(\omega t + 36.6°)$ V. Run a simulation. The resulting graph should be identical to v (the red curve) of Figure 16–43.

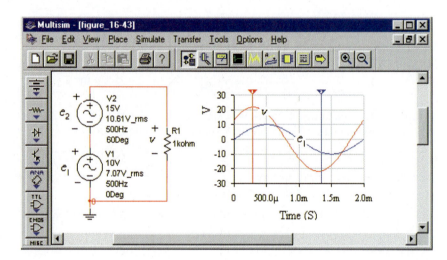

FIGURE 16–43   MultiSIM solution.

NOTES . . .

PSpice

1. Make sure that the polarities of the sources are as indicated. (You will have to rotate V2 three times to get it into the position shown.)

2. To display V2, use differential markers (indicated as + and − on the toolbar).

PSpice

With PSpice, you can plot all three waveforms of Figure 16–10 simultaneously. Procedure: Create the circuit of Figure 16–44 using source VSIN. Note the empty parameter boxes associated with each source. Double click each in turn and enter values as shown in Figure 16–44. Now double click the V2 source symbol and in the Properties editor window that opens, scroll until you find a cell labeled PHASE, enter **60deg,** click Apply, then close. (You do not need to enter a phase angle for Source 1, as it automatically defaults to zero if no value is entered.) Now place markers as shown (Note 2) so that PSpice will automatically create the plots. Click the New Simulation Profile icon, choose Transient, set TSTOP to **2ms** (to display a full cycle), set <u>M</u>aximum Step Size to **1us** (to yield a smooth plot), then click OK. Run the simulation and the waveforms of Figure 16–45 should appear. Using the cursor, scale voltages at 500 μs. You should get 10 V for e_1, 7.5 V for e_2 and 17.5 V for v. Read values at 200-μs intervals and tabulate. Now replace the sources of Figure 16–44 with a single source of 21.8 $\sin(\omega t + 36.6°)$ V, run a simulation, scale values, and compare results. They should agree.

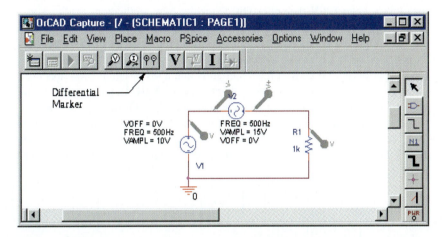

FIGURE 16–44   Use source VSIN.

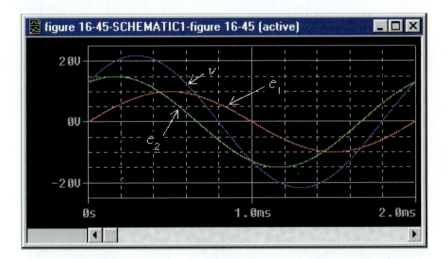

FIGURE 16–45   PSpice waveforms. Compare to Figure 16–10.

Another Example

PSpice makes it easy to study the response of circuits over a range of frequencies. This is illustrated in Example 16–18.

EXAMPLE 16–18

Compute and plot the reactance of a 12-μF capacitor over the range 10 Hz to 1000 Hz.

Solution   PSpice has no command to compute reactance; however, we can calculate voltage and current over the desired frequency range, then plot their ratio. This gives reactance. Procedure: Create the circuit of Figure 16–46 on the screen. (Use source VAC here as it is the source to use for phasor analyses.) Note its default of 0V. Double click the default value (not the symbol) and in the dialog box, enter **120V,** then click OK. Click the New Simulations Profile icon, enter a name and then in the dialog box that opens, select AC Sweep/Noise. For the S̲tart Frequency, key in **10Hz;** for the E̲nd Frequency, key **1kHz;** set AC Sweep type to L̲ogarithmic, select Decade and type **100** into the Pts/D̲ecade (points per decade) box. Run the simulation and a set of empty axes appears. Click T̲race, A̲dd Trace and in the dialog box, click **V1(C1),** press the / key on the keyboard, then click **I(C1)** to yield the ratio V1(C1)/I(C1) (which is the capacitor's reactance). Click OK and PSpice will compute and plot the capacitor's reactance versus frequency (Figure 16–47). Compare its shape to Figure 16–35. Use the cursor to scale some values off the screen and verify each point using $X_C = 1/\omega C$.

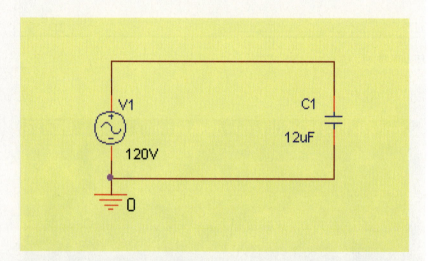

FIGURE 16–46   Circuit for plotting reactance. Use source VAC.

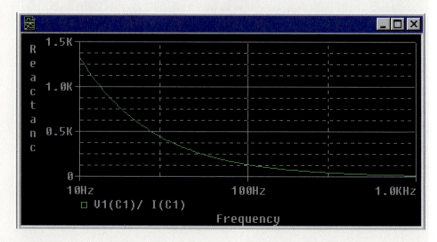

FIGURE 16–47   Reactance for a 12-μF capacitor versus frequency.

Phasor Analysis

As a last example, we will show how to use PSpice to perform phasor analysis—i.e., to solve problems with voltages and currents expressed in phasor form. To illustrate, consider again Example 16–17. Recall, $\mathbf{V}_C = 70.7\,V\angle-40°$, $C = 10\,\mu F$, and $f = 1000$ Hz. Procedure: Create the circuit on the screen (Figure 16–48) using source VAC and component IPRINT (Note 1). Double click the VAC symbol and in the Property Editor, set ACMAG to **70.7V** and ACPHASE to **−40deg.** (See Note 2). Double click IPRINT and in the Property Editor, type **yes** into cells AC, MAG, and PHASE. Click Apply and close the editor. Click the New Simulation Profile icon, select AC Sweep/Noise, set both S̲tart Frequency and E̲nd Frequency to **1000Hz** and T̲otal Points to **1.** Run the simulation. When the simulation window opens, click V̲iew, Output Fi̲le, then scroll until you find the answers (Figure 16–49 and Note 3). The first number is the frequency (1000 Hz), the second number (IM) is the magnitude of the current (4.442 A), and the third (IP) is its phase (50 degrees). Thus, $\mathbf{I}_C = 4.442\,A\angle50°$ as we determined earlier in Example 16–17.

N O T E S . . .

1. Component IPRINT is a software ammeter, found in the SPECIAL parts library. In this example, we configure it to display ac current in magnitude and phase angle format. Make sure that it is connected as shown in Figure 16–48, since if it is reversed, the phase angle of the measured current will be in error by 180°.

2. If you want to display the phase of the source voltage on the schematic as in Figure 16–48, double click the source symbol and in the Property Editor, click ACPHASE, Display, then select Value Only.

3. The results displayed by IPRINT are expressed in exponential format. Thus, frequency (Figure 16–49) is shown as 1.000E+03, which is $1.000 \times 10^3 = 1000$ Hz, etc.

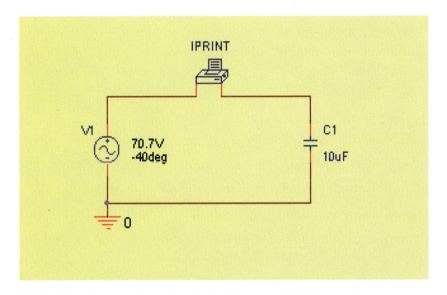

FIGURE 16–48 Phasor analysis using PSpice. Component IPRINT is a software ammeter.

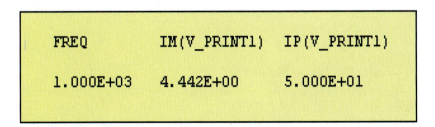

FREQ	IM(V_PRINT1)	IP(V_PRINT1)
1.000E+03	4.442E+00	5.000E+01

FIGURE 16–49 Result for the circuit of Figure 16–48. $I = 4.442\,A\angle50°$.

Modify Example 16–18 to plot both capacitor current and reactance on the same graph. You will need to add a second Y-axis for the capacitor current. (See Appendix A if you need help.)

PRACTICE PROBLEMS 10

PROBLEMS

16.1 Complex Number Review

1. Convert each of the following to polar form:
   a. $5 + j12$
   b. $9 - j6$
   c. $-8 + j15$
   d. $-10 - j4$

2. Convert each of the following to rectangular form:
   a. $6\angle 30°$
   b. $14\angle 90°$
   c. $16\angle 0°$
   d. $6\angle 150°$
   e. $20\angle -140°$
   f. $-12\angle 30°$
   g. $-15\angle -150°$

3. Plot each of the following on the complex plane:
   a. $4 + j6$
   b. $j4$
   c. $6\angle -90°$
   d. $10\angle 135°$

4. Simplify the following using powers of *j*:
   a. $j(1 - j1)$
   b. $(-j)(2 + j5)$
   c. $j[j(1 + j6)]$
   d. $(j4)(-j2 + 4)$
   e. $(2 + j3)(3 - j4)$

5. Express your answer in rectangular form.
   a. $(4 + j8) + (3 - j2)$
   b. $(4 + j8) - (3 - j2)$
   c. $(4.1 - j7.6) + 12\angle 20°$
   d. $2.9\angle 25° - 7.3\angle -5°$
   e. $9.2\angle -120° - (2.6 + j4.1)$
   f. $\dfrac{1}{3+j4} + \dfrac{1}{8-j6}$

6. Express your answer in polar form.
   a. $(37 + j9.8)(3.6 - j12.3)$
   b. $(41.9\angle -80°)(16 + j2)$
   c. $\dfrac{42 + j18.6}{19.1 - j4.8}$
   d. $\dfrac{42.6 + j187.5}{11.2\angle 38°}$

7. Reduce each of the following to polar form:
   a. $15 - j6 - \left[\dfrac{18\angle 40° + (12 + j8)}{11 + j11} \right]$
   b. $\dfrac{21\angle 20° - j41}{36\angle 0° + (1 + j12) - 11\angle 40°}$
   c. $\dfrac{18\angle 40° - 18\angle -40°}{7 + j12} - \dfrac{16 + j17 + 21\angle -60°}{4}$

16.2 Complex Numbers in AC Analysis

NOTES . . .

The answers given to Problems 8 to 11 assume that you are not using rms values since we did not start using rms until later in the chapter.

8. In the manner of Figure 16–9, represent each of the following as transformed sources.
   a. $e = 100 \sin(\omega t + 30°)$ V
   b. $e = 15 \sin(\omega t - 20°)$ V
   c. $e = 50 \sin(\omega t + 90°)$ V
   d. $e = 50 \cos \omega t$ V
   e. $e = 40 \sin(\omega t + 120°)$ V
   f. $e = 80 \sin(\omega t - 70°)$ V

9. Determine the sinusoidal equivalent for each of the transformed sources of Figure 16–50.

10. Given: $e_1 = 10 \sin(\omega t + 30°)$ V and $e_2 = 15 \sin(\omega t - 20°)$ V. Determine their sum $v = e_1 + e_2$ in the manner of Example 16–6, i.e.,

   a. Convert e_1 and e_2 to phasor form.

   b. Determine $\mathbf{V} = \mathbf{E}_1 + \mathbf{E}_2$.

   c. Convert $\mathbf{V}$ to the time domain.

   d. Sketch e_1, e_2, and v as per Figure 16–12.

11. Repeat Problem 10 for $v = e_1 - e_2$.

Note: For the remaining problems and throughout the remainder of the book, express phasor quantities as rms values rather than as peak values.

12. Express the voltages and currents of Figure 16–51 as time domain and phasor domain quantities.

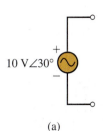

(a)

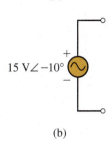

(b)

FIGURE 16–50

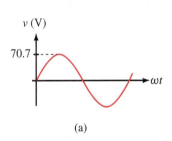

(a)

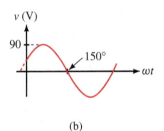

(b)

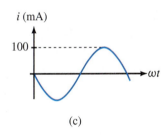

(c)

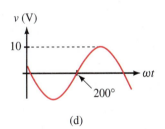

(d)

FIGURE 16–51

13. For Figure 16–52, $i_1 = 25 \sin(\omega t + 36°)$ mA and $i_2 = 40 \cos(\omega t - 10°)$ mA.

   a. Determine phasors $\mathbf{I}_1$, $\mathbf{I}_2$ and $\mathbf{I}_T$.

   b. Determine the equation for i_T in the time domain.

14. For Figure 16–52, $i_T = 50 \sin(\omega t + 60°)$ A and $i_2 = 20 \sin(\omega t - 30°)$ A.

   a. Determine phasors $\mathbf{I}_T$ and $\mathbf{I}_2$.

   b. Determine $\mathbf{I}_1$.

   c. From (b), determine the equation for i_1.

15. For Figure 16–17, $i_1 = 7 \sin \omega t$ mA, $i_2 = 4 \sin(\omega t - 90°)$ mA, and $i_3 = 6 \sin(\omega t + 90°)$ mA.

   a. Determine phasors $\mathbf{I}_1$, $\mathbf{I}_2$, $\mathbf{I}_3$ and $\mathbf{I}_T$.

   b. Determine the equation for i_T in the time domain.

16. For Figure 16–53, $i_T = 38.08 \sin(\omega t - 21.8°)$ A, $i_1 = 35.36 \sin \omega t$ A, and $i_3 = 28.28 \sin(\omega t - 90°)$ A. Determine the equation for i_2.

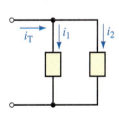

FIGURE 16–52

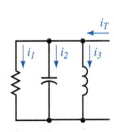

FIGURE 16–53

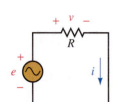

FIGURE 16–54

16.4 to 16.6

17. For Figure 16–54, $R = 12\ \Omega$. For each of the following, determine the current or voltage and sketch.

    a. $v = 120\ \sin \omega t$ V, $i =$ _____

    b. $v = 120\ \sin (\omega t + 27°)$ V, $i =$ _____

    c. $i = 17\ \sin (\omega t - 56°)$ mA, $v =$ _____

    d. $i = -17\ \cos(\omega t - 67°)$ μA, $v =$ _____

18. Given $v = 120\ \sin (\omega t + 52°)$ V and $i = 15\ \sin (\omega t + 52°)$ mA, what is R?

19. Two resistors $R_1 = 10\ \text{k}\Omega$ and $R_2 = 12.5\ \text{k}\Omega$ are in series. If $i = 14.7\ \sin (\omega t + 39°)$ mA,

    a. What are v_{R_1} and v_{R_2}?

    b. Compute $v_T = v_{R_1} + v_{R_2}$ and compare to v_T calculated from $v_T = i\ R_T$.

20. The voltage across a certain component is $v = 120\ \sin(\omega t + 55°)$ V and its current is $-18\ \cos(\omega t + 145°)$ mA. Show that the component is a resistor and determine its value.

21. For Figure 16–55, $V_m = 10$ V and $I_m = 5$ A. For each of the following, determine the missing quantity:

    a. $v_L = 10\ \sin(\omega t + 60°)$ V, $i_L =$ _____

    b. $v_L = 10\ \sin(\omega t - 15°)$ V, $i_L =$ _____

    c. $i_L = 5\ \cos(\omega t - 60°)$ A, $v_L =$ _____

    d. $i_L = 5\ \sin(\omega t + 10°)$ A, $v_L =$ _____

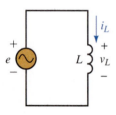

FIGURE 16–55

22. What is the reactance of a 0.5-H inductor at

    a. 60 Hz   b. 1000 Hz   c. 500 rad/s

23. For Figure 16–55, $e = 100\ \sin \omega t$ and $L = 0.5$ H. Determine i_L at

    a. 60 Hz   b. 1000 Hz   c. 500 rad/s

24. For Figure 16–55, let $L = 200$ mH.

    a. If $v_L = 100\ \sin 377t$ V, what is i_L?

    b. If $i_L = 10\ \sin(2\pi \times 400t - 60°)$ mA, what is v_L?

25. For Figure 16–55, if

    a. $v_L = 40\ \sin(\omega t + 30°)$ V, $i_L = 364\ \sin(\omega t - 60°)$ mA, and $L = 2$ mH, what is f?

    b. $i_L = 250\ \sin(\omega t + 40°)$ μA, $v_L = 40\ \sin(\omega t + \theta)$ V, and $f = 500$ kHz, what are L and θ?

26. Repeat Problem 21 if the given voltages and currents are for a capacitor instead of an inductor.

27. What is the reactance of a 5-μF capacitor at

    a. 60 Hz   b. 1000 Hz   c. 500 rad/s

28. For Figure 16–56, $e = 100\ \sin \omega t$ and $C = 5$ μF. Determine i_C at

    a. 60 Hz   b. 1000 Hz   c. 500 rad/s

29. For Figure 16–56, let $C = 50$ μF.

    a. If $v_C = 100\ \sin 377t$ V, what is i_C?

    b. If $i_C = 10\ \sin(2\pi \times 400t - 60°)$ mA, what is v_C?

30. For Figure 16–56, if

    a. $v_C = 362\ \sin(\omega t - 33°)$ V, $i_C = 94\ \sin(\omega t + 57°)$ mA, and $C = 2.2$ μF, what is f?

    b. $i_C = 350\ \sin(\omega t + 40°)$ mA, $v_C = 3.6\ \sin(\omega t + \theta)$ V, and $f = 12$ kHz, what are C and θ?

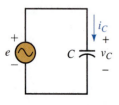

FIGURE 16–56

16.7 The Impedance Concept

31. Determine the impedance of each circuit element of Figure 16–57.

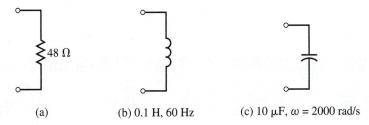

(a)   (b) 0.1 H, 60 Hz   (c) 10 μF, ω = 2000 rad/s

FIGURE 16–57

32. If $\mathbf{E} = 100\ V\angle 0°$ is applied across each of the circuit elements of Figure 16–58:
    a. Determine each current in phasor form.
    b. Express each current in time domain form.

33. If the current through each circuit element of Figure 16–58 is 0.5 A∠0°:
    a. Determine each voltage in phasor form.
    b. Express each voltage in time domain form.

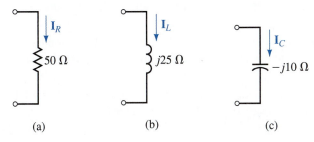

(a)   (b)   (c)

FIGURE 16–58

◀ **MULTISIM**

34. For each of the following, determine the impedance of the circuit element and state whether it is resistive, inductive, or capacitive.
    a. $\mathbf{V} = 240\ V\angle -30°$, $\mathbf{I} = 4\ A\angle -30°$.
    b. $\mathbf{V} = 40\ V\angle 30°$, $\mathbf{I} = 4\ A\angle -60°$.
    c. $\mathbf{V} = 60\ V\angle -30°$, $\mathbf{I} = 4\ A\angle 60°$.
    d. $\mathbf{V} = 140\ V\angle -30°$, $\mathbf{I} = 14\ mA\angle -120°$.

35. For each circuit of Figure 16–59, determine the unknown.

36. a. If $\mathbf{V}_L = 120\ V\ \angle 67°$, $L = 600\ \mu H$, and $f = 10\ kHz$, what is $\mathbf{I}_L$?
    b. If $\mathbf{I}_L = 48\ mA\ \angle -43°$, $L = 550\ mH$, and $f = 700\ Hz$, what is $\mathbf{V}_L$?
    c. If $\mathbf{V}_C = 50\ V\ \angle -36°$, $C = 390\ pF$, and $f = 470\ kHz$, what is $\mathbf{I}_C$?
    d. If $\mathbf{I}_C = 95\ mA\ \angle 87°$, $C = 6.5\ nF$, and $f = 1.2\ MHz$, what is $\mathbf{V}_C$?

16.8 Computer Analysis of AC Circuits

The version of MultiSIM current at the time of writing of this book is unable to measure phase angles. Thus, in the problems that follow, we ask only for magnitudes of voltages and currents.

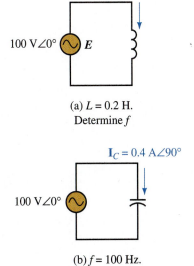

$I_L = 2\ A\angle -90°$

100 V∠0°   E

(a) $L = 0.2$ H.
Determine f

$I_C = 0.4\ A\angle 90°$

100 V∠0°

(b) $f = 100$ Hz.
Determine C

FIGURE 16–59

◀ MULTISIM

37. Create the circuit of Figure 16–60 on the screen. (Use the ac source from the Sources Parts bin and the ammeter from the Indicators Parts bin.) Double click the ammeter symbol and set Mode to AC. Click the ON/OFF switch at the top right hand corner of the screen to energize the circuit. Compare the measured reading against the theoretical value.

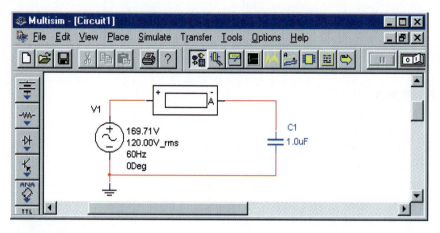

FIGURE 16–60

◀ MULTISIM

38. Replace the capacitor of Figure 16–60 with a 200-mH inductor and repeat Problem 37.

◀ CADENCE

39. Create the circuit of Figure 16–55 on the screen. Use a source of 100 V∠0°, $L = 0.2$ H, and $f = 50$ Hz. Solve for current $\mathbf{I}_L$ (magnitude and angle). See note below.

◀ CADENCE

40. Plot the reactance of a 2.39-H inductor versus frequency from 1 Hz to 500 Hz and compare to Figure 16–28. Change the x-axis scale to linear.

◀ CADENCE

41. For the circuit of Problem 39, plot current magnitude versus frequency from $f = 1$ Hz to $f = 20$ Hz. Measure the current at 10 Hz and verify with your calculator.

Note: PSpice does not permit source/inductor loops. To get around this, add a very small resistor in series, for example, $R = 0.00001$ Ω.

✓ **ANSWERS TO IN-PROCESS LEARNING CHECKS**

In-Process Learning Check 1

1. a. $6\angle 90°$
   b. $4\angle -90°$
   c. $4.24\angle 45°$
   d. $7.21\angle -56.3°$

   e. $9.43\angle 122.0°$
   f. $2.24\angle -63.4°$
   g. $3.61\angle -123.7°$

2. a. $j4$
   b. $3 + j0$
   c. $-j2$
   d. $3.83 + j3.21$

   e. $-3 + j5.20$
   f. $2.35 - j0.855$
   g. $-1.64 - j0.599$

3. $12\angle 40°$

4. $88 - j7; -16 + j15; 0.0113 + j\,0.0009;$
   $-0.0333 - j0.0312$

5. $288\angle -100°; 2\angle 150°$

6. a. $14.70 + j2.17$
   b. $-8.94 + j7.28$

7. $18.0 \sin(\omega t - 56.3°)$

In-Process Learning Check 2

1. $50 \sin(\omega t + 30°)$ A

2. $50 \sin(\omega t - 60°)$ A

3. $50 \sin(\omega t + 120°)$ A

4. 0.637 H

5. $3.98\ \mu F$

6. a. Voltage and current are in phase. Therefore, R
   b. Current leads by $90°$. Therefore, C
   c. Current leads by $90°$. Therefore, C
   d. Current lags by $90°$. Therefore, L

■ KEY TERMS

Active Power
Apparent Power
Average Power
Eddy Current Loss
Eddy Currents
Effective Resistance
F_p
Instantaneous Power
Power Factor (F_p)
Power Factor Correction
Power Triangle
Q
Radiation Resistance
Reactive Power
S
Skin Effect
Unity Power Factor
VAR
Volt-Amps (VA)
Wattmeter

■ OUTLINE

Introduction
Power to a Resistive Load
Power to an Inductive Load
Power to a Capacitive Load
Power in More Complex Circuits
Apparent Power
The Relationship Between P, Q, and S
Power Factor
AC Power Measurement
Effective Resistance
Energy Relationships for AC
Circuit Analysis Using Computers

■ OBJECTIVES

After studying this chapter, you will be able to

- explain what is meant by active, reactive, and apparent power,
- compute the active power to a load,
- compute the reactive power to a load,
- compute the apparent power to a load,
- construct and use the power triangle to analyze power to complex loads,
- compute power factor,
- explain why equipment is rated in VA instead of watts,
- measure power in single-phase circuits,
- describe why effective resistance differs from geometric resistance,
- describe energy relations in ac circuits,
- use PSpice to study instantaneous power.

Power in AC Circuits

17

CHAPTER PREVIEW

In Chapter 4, you studied power in dc circuits. In this chapter, we turn our attention to power in ac circuits. In ac circuits, there are additional considerations that are not present with dc. In dc circuits, for example, the only power relationship you encounter is $P = VI$ watts. This is referred to as *real power* or *active power* and is the power that does useful work such as light a lamp, power a heater, run an electric motor, and so on.

In ac circuits, you also encounter this type of power. For ac circuits that contain reactive elements however, (i.e., inductance or capacitance), a second component of power also exists. This component, termed *reactive power,* represents energy that oscillates back and forth throughout the system. For example, during the buildup of current in an inductance, energy flows from the power source to the inductance to create its magnetic field. When the magnetic field collapses, this energy is returned to the circuit. This movement of energy in and out of the inductance constitutes a flow of power. However, since it flows first in one direction, then in the other, it contributes nothing to the average flow of power from the source to the load. For this reason, reactive power is sometimes referred to as *wattless power.* (A similar situation exists regarding power flow to and from the electric field of a capacitor.)

For a circuit that contains resistive as well as reactive elements, some energy is dissipated while the remainder is shuttled back and forth as described above; thus, both active and reactive components of power are present. This combination of real and reactive power is termed *apparent power.*

In this chapter, we look at all three components of power. New ideas that emerge include the concept of power factor, the power triangle, the measurement of power in ac circuits, and the concept of effective resistance. ∎

PUTTING IT IN PERSPECTIVE

Henry Cavendish

CAVENDISH, AN ENGLISH CHEMIST and physicist born in 1731, is included here, not for what he did for the emerging electrical field, but for what he didn't do. A brilliant man, Cavendish was 50 years ahead of his time, and his experiments in electricity preceded and anticipated almost all the major discoveries that came about over the next half century (e.g., he discovered Coulomb's law before Coulomb did). However, Cavendish was interested in research and knowledge purely for its own sake and never bothered to publish most of what he learned, in effect depriving the world of his findings and holding back the development of the field of electricity by many years. Cavendish's work lay unknown for nearly a century before another great scientist, James Clerk Maxwell, had it published. Nowadays, Cavendish is better known for his work in the gravitational field than for his work in the electrical field. One of the amazing things he did was to determine the mass of the earth using the rather primitive technology of his day. ■

17.1 Introduction

◀ **Online Companion**

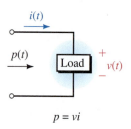

$$p = vi$$

FIGURE 17–1 Voltage, current and power references. When p is positive, power is in the direction of the reference arrow.

At any given instant, the power to a load is equal to the product of voltage times current (Figure 17–1). This means that if voltage and current vary with time, so will power. This time-varying power is referred to as **instantaneous power** and is given the symbol $p(t)$ or just p. Thus,

$$p = vi \quad \text{(watts)} \qquad \text{(17–1)}$$

Now consider the case of sinusoidal ac. Since voltage and current are positive at various times during their cycle and negative at others, instantaneous power may also be positive at some times and negative at others. This is illustrated in Figure 17–2, where we have multiplied voltage times current point by point to get the power waveform. For example, from $t = 0$ s to $t = t_1$, v and i are both positive; therefore, power is positive. At $t = t_1$, $v = 0$ V and thus $p = 0$ W. From t_1 to t_2, i is positive and v is negative; therefore, p is negative. From t_2 to t_3, both v and i are negative; therefore power is positive, and so on. As discussed in Chapter 4, a positive value for p means that power transfer is in the direction of the reference arrow, while a negative value means that it is in the opposite direction. Thus, during positive parts of the power cycle, power flows from the source to the load, while during negative parts, it flows out of the load back into the circuit.

The waveform of Figure 17–2 is the actual power waveform. We will now show that the key aspects of power flow embodied in this waveform can be described in terms of active power, reactive power, and apparent power.

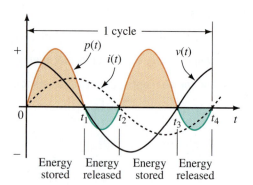

FIGURE 17–2 Instantaneous power in an ac circuit. Positive p represents power to the load; negative p represents power returned from the load.

Active Power

Since p represents the power flowing to the load, its average will be the **average power** to the load. Denote this average by the letter P. If P is positive, then, on average, more power flows to the load than is returned from it. (If P is zero, all power sent to the load is returned.) Thus, if P has a positive value, it represents the power that is really dissipated by the load. For this reason, P is called **real power.** In modern terminology, real power is also called **active power.** Thus, *active power is the average value of the instantaneous power, and the terms real power, active power, and average power mean the same thing.* (We usually refer to it simply as power.) In this book, we use the terms interchangeably.

Reactive Power

Consider again Figure 17–2. During the intervals that p is negative, power is being returned from the load. (This can only happen if the load contains reactive elements: L or C.) The portion of power that flows into the load then back out is called **reactive power.** Since it first flows one way then the other, *its average value is zero;* thus, reactive power contributes nothing to the average power to the load.

 Although reactive power does no useful work, it cannot be ignored. Extra current is required to create reactive power, and this current must be supplied by the source; this also means that conductors, circuit breakers, switches, transformers, and other equipment must be made physically larger to handle the extra current. This increases the cost of a system.

 At this point, it should be noted that real power and reactive power do not exist as separate entities. Rather, they are components of the power waveform shown in Figure 17–2. However, as you will see, we are able to conceptually separate them for purposes of analysis.

First consider power to a purely resistive load (Figure 17–3). Here, current is in phase with voltage. Assume $i = I_m\sin \omega t$ and $v = V_m\sin \omega t$. Then,

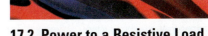

17.2 Power to a Resistive Load

$$p = vi = (V_m\sin \omega t)(I_m\sin \omega t) = V_mI_m\sin^2\omega t$$

Therefore,

$$p = \frac{V_mI_m}{2}(1 - \cos 2\ \omega t) \qquad (17\text{–}2)$$

where we have used the trigonometric relationship $\sin^2\omega t = \frac{1}{2}(1 - \cos 2\ \omega t)$ from inside the front cover of the book.

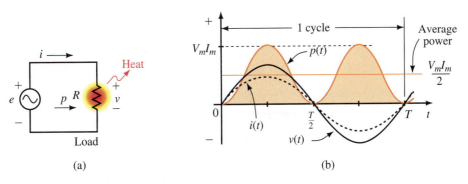

FIGURE 17–3   Power to a purely resistive load. The peak value of p is V_mI_m.

A sketch of p versus time is shown in (b). Note that p is always positive (except where it is momentarily zero). This means that power flows only from the source to the load. Since none is ever returned, all power delivered by the source is absorbed by the load. We therefore conclude that *power to a pure resistance consists of active power only*. Note also that the frequency of the power waveform is double that of the voltage and current waveforms. (This is confirmed by the 2ω in Equation 17–2.)

Average Power

Inspection of the power waveform of Figure 17–3 shows that its average value lies half way between zero and its peak value of $V_m I_m$. That is,

$$P = V_m I_m / 2$$

(You can also get the same result by averaging Equation 17–2 as we did in Chapter 15.) Since V (the magnitude of the rms value of voltage) is $V_m/\sqrt{2}$ and I (the magnitude of the rms value of current) is $I_m/\sqrt{2}$, this can be written as $P = VI$. Thus, average power to a purely resistive load is

$$P = VI \quad \text{(watts)} \tag{17–3}$$

Alternate forms are obtained by substituting $V = IR$ and $I = V/R$ into Equation 17–3. They are

$$P = I^2 R \quad \text{(watts)} \tag{17–4}$$

$$= V^2/R \quad \text{(watts)} \tag{17–5}$$

Thus the active power relationships for resistive circuits are the same for ac as for dc.

17.3 Power to an Inductive Load

For a purely inductive load as in Figure 17–4(a), current lags voltage by 90°. If we select current as reference, $i = I_m \sin \omega t$ and $v = V_m \sin(\omega t + 90°)$. A sketch of p versus time (obtained by multiplying v times i) then looks as shown in (b). Note that during the first quarter-cycle, p is positive and hence

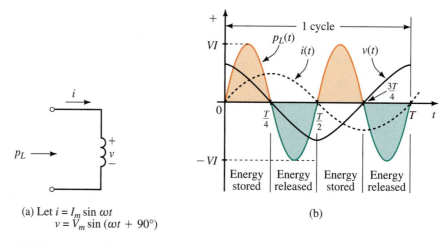

(a) Let $i = I_m \sin \omega t$ (b)
         $v = V_m \sin (\omega t + 90°)$

FIGURE 17–4   Power to a purely inductive load. Energy stored during each quarter-cycle is returned during the next quarter cycle. Average power is zero.

power flows to the inductance, while during the second quarter-cycle, p is negative and all power transferred to the inductance during the first quarter-cycle flows back out. Similarly for the third and fourth quarter-cycles. Thus, *the average power to an inductance over a full cycle is zero, i.e., there are no power losses associated with a pure inductance.* Consequently, $P_L = 0$ W and the only power flowing in the circuit is reactive power. This is true in general, that is, *the power that flows into and out of a pure inductance is reactive power only.*

To determine this power, consider again equation 17–1. With $v = V_m\sin(\omega t + 90°)$ and $i = I_m\sin \omega t$, $p_L = vi$ becomes

$$p_L = V_m I_m \sin(\omega t + 90°)\sin \omega t$$

After some trigonometric manipulation, this reduces to

$$p_L = VI \sin 2 \omega t \qquad \text{(17–6)}$$

where V and I are the magnitudes of the rms values of the voltage and current respectively.

The product VI in Equation 17–6 is defined as **reactive power** and is given the symbol Q_L. Because it represents "power" that alternately flows into, then out of the inductance, Q_L contributes nothing to the average power to the load and, as noted earlier, is sometimes referred to as wattless power. As you will soon see, however, reactive power is of major concern in the operation of electrical power systems.

Since Q_L is the product of voltage times current, its unit is the volt-amp (VA). To indicate that Q_L represents reactive volt-amps, an "R" is appended to yield a new unit, the **VAR** *(volt-amps reactive).* Thus,

$$Q_L = VI \quad \text{(VAR)} \qquad \text{(17–7)}$$

Substituting $V = IX_L$ and $I = V/X_L$ yields the following alternate forms:

$$Q_L = I^2 X_L = \frac{V^2}{X_L} \quad \text{(VAR)} \qquad \text{(17–8)}$$

By convention, Q_L is taken to be positive. Thus, if $I = 4$ A and $X_L = 2 \ \Omega$, $Q_L = (4 \text{ A})^2(2 \ \Omega) = +32$ VAR. Note that the VAR (like the watt) is a scalar quantity with magnitude only and no angle.

For a purely capacitive load, current leads voltage by 90°. Taking current as reference, $i = I_m\sin \omega t$ and $v = V_m\sin(\omega t - 90°)$. Multiplication of v times i yields the power curve of Figure 17–5. Note that negative and positive loops of the power wave are identical; thus, over a cycle, the power returned to the circuit by the capacitance is exactly equal to that delivered to it by the source. This means that *the average power to a capacitance over a full cycle is zero, i.e., there are no power losses associated with a pure capacitance.* Consequently, $P_C = 0$ W and the only power flowing in the circuit is reactive power. This is true in general, that is, *the power that flows into and out of a pure capacitance is reactive power only.* This reactive power is given by

$$p_C = vi = V_m I_m \sin \omega t \sin(\omega t - 90°)$$

which reduces to

$$p_C = -VI \sin 2 \omega t \qquad \text{(17–9)}$$

17.4 Power to a Capacitive Load

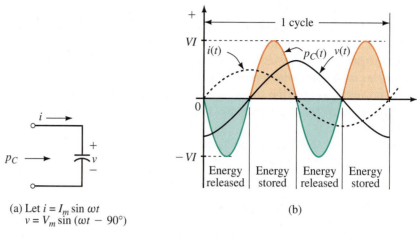

(a) Let $i = I_m \sin \omega t$
$v = V_m \sin (\omega t - 90°)$

(b)

FIGURE 17–5 Power to a purely capacitive load. Average power is zero.

where V and I are the magnitudes of the rms values of the voltage and current respectively. Now define the product VI as Q_C. This product represents reactive power. That is,

$$Q_C = VI \quad \text{(VAR)} \tag{17–10}$$

Since $V = IX_C$ and $I = V/X_C$, Q_C can also be expressed as

$$Q_C = I^2 X_C = \frac{V^2}{X_C} \quad \text{(VAR)} \tag{17–11}$$

By convention, reactive power to capacitance is defined as negative. Thus, if $I = 4$ A and $X_C = 2$ Ω, then $I^2 X_C = (4 \text{ A})^2(2 \text{ Ω}) = 32$ VAR. We can either explicitly show the minus sign as $Q_C = -32$ VAR or imply it by stating that Q represents capacitive vars, i.e., $Q_C = 32$ VAR (cap.).

EXAMPLE 17–1

For each circuit of Figure 17–6, determine real and reactive power.

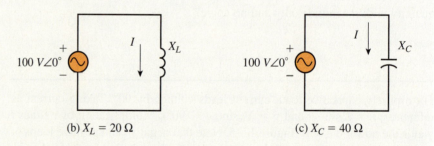

(a) $R = 25$ Ω

(b) $X_L = 20$ Ω

(c) $X_C = 40$ Ω

FIGURE 17–6

Solution Only voltage and current magnitudes are needed.

a. $I = 100 \text{ V}/25 \text{ Ω} = 4$ A. $P = VI = (100 \text{ V})(4 \text{ A}) = 400$ W. $Q = 0$ VAR

b. $I = 100 \text{ V}/20 \text{ Ω} = 5$ A. $Q = VI = (100 \text{ V})(5 \text{ A}) = 500$ VAR (ind.). $P = 0$ W

c. $I = 100 \text{ V}/40 \text{ Ω} = 2.5$ A. $Q = VI = (100 \text{ V})(2.5 \text{ A}) = 250$ VAR (cap.).
   $P = 0$ W

The answer for (c) can also be expressed as $Q = -250$ VAR.

PRACTICE PROBLEMS 1

1. If the power at some instant in Figure 17–1 is $p = -27$ W, in what direction is the power at that instant?

2. For a purely resistive load, v and i are in phase. Given $v = 10 \sin \omega t$ V and $i = 5 \sin \omega t$ A. Using graph paper, carefully plot v and i at 30° intervals. Now multiply the values of v and i at these points and plot the power. (The result should look like Figure 17–3(b).

   a. From the graph, determine the peak power and average power.

   b. Compute power using $P = VI$ and compare to the average value determined in (a).

3. Repeat Example 17–1 using equations 17–4, 17–5, 17–8, and 17–11.

Answers
1. From the load to the source.

2. a. 50 W; 25 W

   b. Same

17.5 Power in More Complex Circuits

The relationships described above were developed using the load of Figure 17–1. However, they hold true for every element in a circuit, no matter how complex the circuit or how its elements are interconnected. Further, in any circuit, total real power P_T is found by summing real power to all circuit elements, while total reactive power Q_T is found by summing reactive power, taking into account that inductive Q is positive and capacitive Q is negative.

It is sometimes convenient to show power to circuit elements symbolically as illustrated in the next example.

EXAMPLE 17–2

For the *RL* circuit of Figure 17–7(a), $I = 5$ A. Determine P and Q.

FIGURE 17–7   From the terminals, P and Q are the same for both (a) and (b).

Solution

$$P = I^2 R = (5 \text{ A})^2(3 \text{ } \Omega) = 75 \text{ W}$$

$$Q = Q_L = I^2 X_L = (5 \text{ A})^2(4 \text{ } \Omega) = 100 \text{ VAR (ind.)}$$

These can be represented symbolically as in Figure 17–7(b).

EXAMPLE 17–3

For the *RC* circuit of Figure 17–8(a), determine *P* and *Q*.

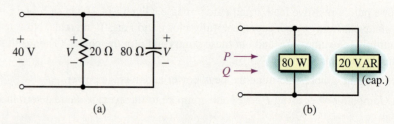

(a)   (b)

FIGURE 17–8   From the terminals, *P* and *Q* are the same for both (a) and (b).

Solution

$$P = V^2/R = (40 \text{ V})^2/(20 \text{ }\Omega) = 80 \text{ W}$$
$$Q = Q_C = V^2/X_C = (40 \text{ V})^2/(80 \text{ }\Omega) = 20 \text{ VAR (cap.)}$$

These can be represented symbolically as in Figure 17–8(b).

In terms of determining total *P* and *Q*, it does not matter how the circuit or system is connected or what electrical elements it contains. Elements can be connected in series, in parallel, or in series-parallel, for example, and the system can contain electric motors and the like, and total *P* is still found by summing the power to individual elements, while total *Q* is found by algebraically summing their reactive powers.

EXAMPLE 17–4

a. For Figure 17–9(a), compute P_T and Q_T.
b. Reduce the circuit to its simplest form.

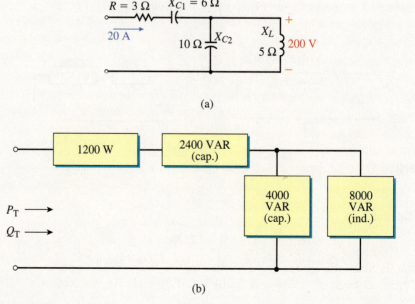

(a)

(b)

FIGURE 17–9

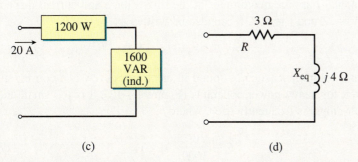

(c)   (d)

FIGURE 17–9   Continued.

Solution

a. $P = I^2R = (20 \text{ A})^2(3 \text{ }\Omega) = 1200$ W

   $Q_{C_1} = I^2X_{C_1} = (20 \text{ A})^2(6 \text{ }\Omega) = 2400$ VAR (cap.)

   $Q_{C_2} = \dfrac{V_2^2}{X_{C_2}} = \dfrac{(200 \text{ V})^2}{(10 \text{ }\Omega)} = 4000$ VAR (cap.)

   $Q_L = \dfrac{V_2^2}{X_L} = \dfrac{(200 \text{ V})^2}{5 \text{ }\Omega} = 8000$ VAR (ind.)

   These are represented symbolically in part (b). $P_T = 1200$ W and $Q_T = -2400$ VAR $-$ 4000 VAR $+$ 8000 VAR $= 1600$ VAR. Thus, the load is net inductive as shown in (c).

b. $Q_T = I^2X_{eq}$. Thus, $X_{eq} = Q_T/I^2 = (1600 \text{ VAR})/(20 \text{ A})^2 = 4 \text{ }\Omega$. Circuit resistance remains unchanged. Thus, the equivalent is as shown in (d).

PRACTICE PROBLEMS 2

For the circuit of Figure 17–10, $P_T = 1.9$ kW and $Q_T = 900$ VAR (ind.). Determine P_2 and Q_2.

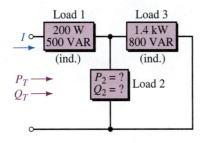

FIGURE 17–10

Answer
300 W, 400 VAR (cap.)

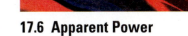

When a load has voltage V across it and current I through it as in Figure 17–11, the power that appears to flow to it is VI. However, if the load contains both resistance and reactance, this product represents neither real power nor reactive power. Since VI appears to represent power, it is called **apparent power.** Apparent power is given the symbol S and has units of **volt-amperes (VA).** Thus,

17.6 Apparent Power

$$S = VI \quad \text{(VA)} \qquad\qquad \textbf{(17–12)}$$

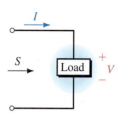

FIGURE 17–11   Apparent power $S = VI$.

where V and I are the magnitudes of the rms voltage and current respectively. Since $V = IZ$ and $I = V/Z$, S can also be written as

$$S = I^2Z = V^2/Z \quad \text{(VA)} \tag{17–13}$$

For small equipment (such as found in electronics), VA is a convenient unit. However, for heavy power apparatus (Figure 17–12), it is too small and kVA (kilovolt-amps) is frequently used, where

$$S = \frac{VI}{1000} \quad \text{(kVA)} \tag{17–14}$$

In addition to its VA rating, it is common practice to rate electrical apparatus in terms of its operating voltage. Once you know these two, it is easy to determine rated current. For example, a piece of equipment rated 250 kVA, 4.16 kV has a rated current of $I = S/V = (250 \times 10^3 \text{ VA})/(4.16 \times 10^3 \text{ V}) = 60.1$ A.

FIGURE 17–12   Power apparatus is rated in apparent power. The transformer shown is a 167-kVA unit. *(Courtesy Carte International Ltd.)*

17.7 The Relationship between *P, Q,* and *S*

Until now, we have treated real, reactive, and apparent power separately. However, they are related by a very simple relationship through the power triangle.

The Power Triangle

Consider the series circuit of Figure 17–13(a). Let the current through the circuit be $\mathbf{I} = I\angle 0°$, with phasor representation (b). The voltages across the resistor and inductance are $\mathbf{V}_R$ and $\mathbf{V}_L$ respectively. As noted in Chapter 16, $\mathbf{V}_R$ is in phase with $\mathbf{I}$, while $\mathbf{V}_L$ leads it by 90°. Kirchhoff's voltage law applies for ac voltages in phasor form. Thus, $\mathbf{V} = \mathbf{V}_R + \mathbf{V}_L$ as indicated in (c).

The voltage triangle of (c) may be redrawn as in Figure 17–14(a) with magnitudes of V_R and V_L replaced by IR and IX_L respectively. Now multiply all quantities by I. This yields sides of I^2R, I^2X_L, and hypotenuse VI as indicated in (b).

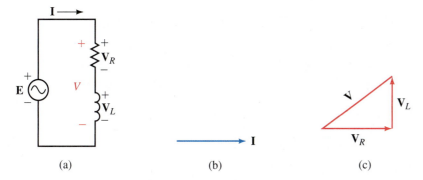

(a)   (b)   (c)

FIGURE 17–13   Steps in the development of the power triangle.

Note that these represent *P, Q,* and *S* respectively as indicated in (c). This is called the **power triangle.** From the geometry of this triangle, you can see that

$$S = \sqrt{P^2 + Q_L^2} \qquad \textbf{(17–15)}$$

Alternatively, the relationship between *P, Q,* and *S* may be expressed as a complex number:

$$\mathbf{S} = P + jQ_L \qquad \textbf{(17–16a)}$$

or

$$\mathbf{S} = S\angle\theta \qquad \textbf{(17–16b)}$$

If the circuit is capacitive instead of inductive, Equation 17–16a becomes

$$\mathbf{S} = P - jQ_C \qquad \textbf{(17–17)}$$

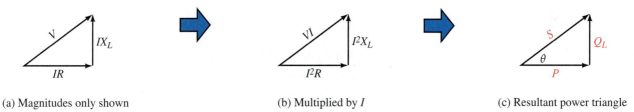

(a) Magnitudes only shown   (b) Multiplied by *I*   (c) Resultant power triangle

FIGURE 17–14   Steps in the development of the power triangle (continued).

The power triangle in this case has a negative imaginary part as indicated in Figure 17–15.

The power relationships may be written in generalized forms as

$$\mathbf{S} = \mathbf{P} + \mathbf{Q} \qquad \textbf{(17–18)}$$

and

$$\mathbf{S} = \mathbf{VI^*} \qquad \textbf{(17–19)}$$

where $\mathbf{P} = P\angle0°$, $\mathbf{Q}_L = jQ_L$, $\mathbf{Q}_C = -jQ_C$, and $\mathbf{I^*}$ is the conjugate of current **I.** These relationships hold true for all networks regardless of what they contain or how they are configured.

When solving problems involving power, remember that *P* values can be added to get P_T, and *Q* values to get Q_T (where *Q* is positive for inductive elements and negative for capacitive). However, apparent power values cannot be added to get S_T, i.e., $S_T \neq S_1 + S_2 + \cdots + S_N$. Instead, you must determine P_T and Q_T, then use the power triangle to obtain S_T.

FIGURE 17–15   Power triangle for capacitive case.

EXAMPLE 17–5

The P and Q values for a circuit are shown in Figure 17–16(a).

a. Determine the power triangle.

b. Determine the magnitude of the current supplied by the source.

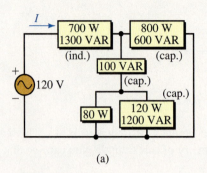

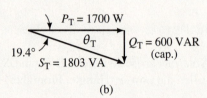

(a)   (b)

FIGURE 17–16

Solution

a. $P_T = 700 + 800 + 80 + 120 = 1700$ W

$Q_T = 1300 - 600 - 100 - 1200 = -600$ VAR $= 600$ VAR (cap.)

$\mathbf{S}_T = P_T + jQ_T = 1700 - j600 = 1803\angle -19.4°$ VA

The power triangle is as shown. The load is net capacitive.

b. $I = S_T/E = 1803$ VA/120 V $= 15.0$ A

EXAMPLE 17–6

A generator supplies power to an electric heater, an inductive element, and a capacitor as in Figure 17–17(a).

a. Find P and Q for each load.

b. Find total active and reactive power supplied by the generator.

c. Draw the power triangle for the combined loads and determine total apparent power.

d. Find the current supplied by the generator.

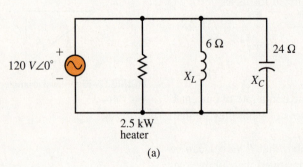

(a)

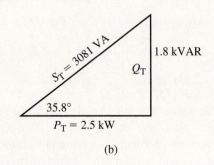

(b)

FIGURE 17–17

Solution

a. The components of power are as follows:

Heater: $P_H = 2.5$ kW $Q_H = 0$ VAR

Inductor: $P_L = 0$ W $Q_L = \dfrac{V^2}{X_L} = \dfrac{(120\text{ V})^2}{6\ \Omega} = 2.4$ kVAR (ind.)

Capacitor: $P_C = 0$ W $Q_C = \dfrac{V^2}{X_C} = \dfrac{(120\text{ V})^2}{24\ \Omega} = 600$ VAR (cap.)

b. $P_T = 2.5$ kW $+ 0$ W $+ 0$ W $= 2.5$ kW

$Q_T = 0$ VAR $+ 2.4$ kVAR $- 600$ VAR $= 1.8$ kVAR (ind.)

c. The power triangle is sketched as Figure 17–7(b). Both the hypotenuse and the angle can be obtained easily using rectangular to polar conversion. $\mathbf{S_T} = P_T + jQ_T = 2500 + j1800 = 3081\angle35.8°$. Thus, apparent power is $S_T = 3081$ VA.

d. $I = \dfrac{S_T}{E} = \dfrac{3081\text{ VA}}{120\text{ V}} = 25.7$ A

Active and Reactive Power Equations

An examination of the power triangle of Figures 17–14 and 17–15 shows that P and Q may be expressed respectively as

$$P = VI \cos \theta = S \cos \theta \quad \text{(W)} \qquad \textbf{(17–20)}$$

and

$$Q = VI \sin \theta = S \sin \theta \quad \text{(VAR)} \qquad \textbf{(17–21)}$$

where V and I are the magnitudes of the rms values of the voltage and current respectively and θ is the angle between them. P is always positive, while Q is positive for inductive circuits and negative for capacitive circuits. Thus, if $V = 120$ volts, $I = 50$ A, and $\theta = 30°$, $P = (120)(50)\cos 30° = 5196$ W and $Q = (120)(50)\sin 30° = 3000$ VAR.

PRACTICE PROBLEMS 3

A 208-V generator supplies power to a group of three loads. Load 1 has an apparent power of 500 VA with $\theta = 36.87°$ (i.e., it is net inductive). Load 2 has an apparent power of 1000 VA and is net capacitive with a power triangle angle of $-53.13°$. Load 3 is purely resistive with power $P_3 = 200$ W. Determine the power triangle for the combined loads and the generator current.

Answers
$S_T = 1300$ VA, $\theta_T = -22.6°$, $I = 6.25$ A

The quantity $\cos \theta$ in Equation 17–20 is defined as **power factor** and is given the symbol F_p. Thus,

$$F_p = \cos \theta \qquad \textbf{(17–22)}$$

From Equation 17–20, we see that F_p may be computed as the ratio of real power to apparent power. Thus,

$$\cos \theta = P/S \qquad \textbf{(17–23)}$$

Power factor is expressed as a number or as a percent. From Equation 17–23, it is apparent that power factor cannot exceed 1.0 (or 100% if expressed in percent).

17.8 Power Factor

The **power factor angle** θ is of interest. It can be found as

$$\theta = \cos^{-1}(P/S) \qquad (17–24)$$

Angle θ is the angle between voltage and current. For a pure resistance, therefore, $\theta = 0°$. For a pure inductance, $\theta = 90°$; for a pure capacitance, $\theta = -90°$. For a circuit containing both resistance and inductance, θ will be somewhere between $0°$ and $90°$; for a circuit containing both resistance and capacitance, θ will be somewhere between $0°$ and $-90°$.

Unity, Lagging, and Leading Power Factor

As indicated by Equation 17–23, a load's power factor shows how much of its apparent power is actually real power. For example, for a purely resistive circuit, $\theta = 0°$ and $F_p = \cos 0° = 1.0$. Therefore, $P = VI$ (watts) and all the load's apparent power is real power. This case ($F_p = 1$) is referred to as *unity* power factor.

For a load containing only resistance and inductance, the load current lags voltage. The power factor in this case is described as *lagging*. On the other hand, for a load containing only resistance and capacitance, current leads voltage and the power factor is described as *leading*. Thus, *an inductive circuit has a lagging power factor, while a capacitive circuit has a leading power factor.*

A load with a very poor power factor can draw excessive current. This is discussed next.

Why Equipment Is Rated in VA

As noted earlier, equipment is rated in terms of VA instead of watts. We now show why. Consider Figure 17–18. Assume that the generator is rated at 600 V, 120 kVA. This means that it is capable of supplying $I = 120$ kVA/600 V = 200 A. In (a), the generator is supplying a purely resistive load with 120 kW. Since $S = P$ for a purely resistive load, $S = 120$ kVA and the generator is supplying its rated current. In (b), the generator is supplying a load with $P = 120$ kW as before, but $Q = 160$ kVAR. Its apparent power is therefore $S = 200$ kVA, which means that the generator current is $I = 200$ kVA/600 V = 333.3 A. Even though it is supplying the same power as in (a), the generator is now greatly overloaded, and damage may result as indicated in (b).

This example illustrates clearly that rating a load or device in terms of power is a poor choice, as its current-carrying capability can be greatly exceeded (even though its power rating is not). Thus, *the size of electrical apparatus (generators, interconnecting wires, transformers, etc.) required to supply a load is governed, not by the load's power requirements, but rather by its VA requirements.*

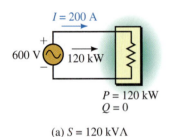

$P = 120$ kW
$Q = 0$

(a) $S = 120$ kVA

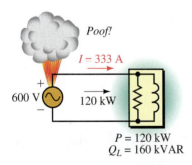

$P = 120$ kW
$Q_L = 160$ kVAR

(b) $S = \sqrt{(120)^2 + (160)^2} = 200$ kVA
The generator is overloaded

FIGURE 17–18 Illustrating why electrical apparatus is rated in VA instead of watts. Both loads dissipate 120 kW but the current rating of generator (b) is exceeded because of the power factor of its load.

Power Factor Correction

The problem shown in Figure 17–18 can be alleviated by cancelling some or all of the reactive component of power by adding reactance of the opposite type to the circuit. This is referred to as **power factor correction.** If you completely cancel the reactive component, the power factor angle is $0°$ and $F_p = 1$. This is referred to as **unity power factor correction.**

For the circuit of Figure 17–18(b), a capacitance with $Q_C = 160\,\text{kVAR}$ is added in parallel with the load as in Figure 17–19(a). Determine generator current I.

EXAMPLE 17–7

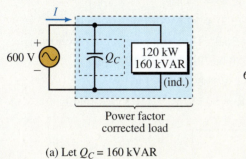

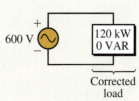

(a) Let $Q_C = 160\,\text{kVAR}$

(b) Load corrected to unity power factor

FIGURE 17–19   Power factor correction. The parallel capacitor greatly reduces source current.

Solution $Q_T = 160\,\text{kVAR} - 160\,\text{kVAR} = 0$. Therefore, $\mathbf{S}_T = 120\,\text{kW} + j0\,\text{kVAR}$. Thus, $S_T = 120\,\text{kVA}$, and current drops from 333 A to $I = 120\,\text{kVA}/600\,\text{V} = 200\,\text{A}$. Thus, the generator is no longer overloaded.

Residential customers are not charged directly for VARs—that is, they pay their electrical bills based solely on the number of the kilowatt-hours they use. This is because all residential customers have essentially the same power factor and the power factor effect is simply built into the rates that they pay. Industrial customers, on the other hand, often have widely differing power factors and the utility may have to monitor their VARs (or their power factor), as well as their watts, in order to determine a suitable charge.

To illustrate, assume that the loads of Figures 17–18(a) and (b) are two small industrial plants. If the utility based its charge solely on power, both customers would pay the same amount. However, it costs the utility more to supply customer (b) since larger conductors, larger transformers, larger switchgear, and so on are required to handle the larger current. For this reason, industrial customers may pay a penalty if their power factor drops below a prescribed value as set by the utility.

An industrial client is charged a penalty if the plant power factor drops below 0.85. The equivalent plant loads are as shown in Figure 17–20. The frequency is 60 Hz.

EXAMPLE 17–8

a. Determine P_T and Q_T.

b. Determine what value of capacitance (in microfarads) is required to bring the power factor up to 0.85.

c. Determine generator current before and after correction.

Solution

a. The components of power are as follows:

   Lights: $P = 12$ kW, $Q = 0$ kVAR

   Furnace: $P = I^2R = (150)^2(2.4) = 54$ kW

   $Q = I^2X = (150)^2(3.2) = 72$ kVAR (ind.)

   Motor: $\theta_m = \cos^{-1}(0.8) = 36.9°$. Thus, from the motor power triangle,

   $Q_m = P_m \tan \theta_m = 80 \tan 36.9° = 60$ kVAR (ind.)

   Total: $P_T = 12$ kW $+ 54$ kW $+ 80$ kW $= 146$ kW

   $Q_T = 0 + 72$ kVAR $+ 60$ kVAR $= 132$ kVAR (ind.)

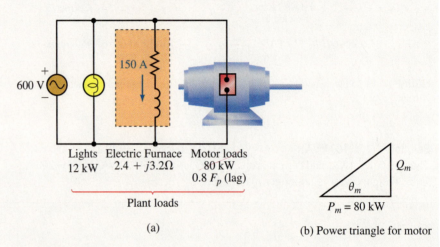

Lights Electric Furnace Motor loads
12 kW $2.4 + j3.2\Omega$ 80 kW
                               0.8 F_p (lag)

Plant loads

(a)

(b) Power triangle for motor

FIGURE 17–20

b. The power triangle for the plant is shown in Figure 17–21(a). However, we must correct the power factor to 0.85. Thus we need $\theta' = \cos^{-1}(0.85) = 31.8°$, where θ' is the power factor angle of the corrected load as indicated in Figure 17–21(b). The maximum reactive power that we can tolerate is thus $Q'_T = P_T \tan \theta' = 146 \tan 31.8° = 90.5$ kVAR.

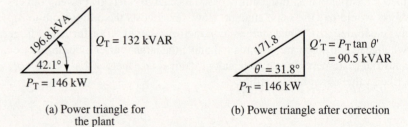

(a) Power triangle for
the plant

(b) Power triangle after correction

FIGURE 17–21   Initial and final power triangles. Note that P_T does not change when we correct the power factor, since for the capacitor, $P = 0$ W.

Now consider Figure 17–22. $Q'_T = Q_C + 132$ kVAR, where $Q'_T = 90.5$ kVAR. Therefore, $Q_C = -41.5$ kVAR $= 41.5$ kVAR (cap.). But $Q_C = V^2/X_C$. Therefore, $X_C = V^2/Q_C = (600)^2/41.5$ kVAR $= 8.67 \Omega$. But $X_C = 1/\omega C$. Thus a capacitor of

$$C = \frac{1}{\omega X_C} = \frac{1}{(2\pi)(60)(8.67)} = 306 \ \mu F$$

will provide the required correction.

c. For the original circuit Figure 17–21(a), S_T = 196.8 kVA. Thus,

$$I = \frac{S_T}{E} = \frac{196.8\ \text{kVA}}{600\ \text{V}} = 328\ \text{A}$$

For the corrected circuit 17–21(b), S'_T = 171.8 kVA and

$$I = \frac{171.8\ \text{kVA}}{600\ \text{V}} = 286\ \text{A}$$

Thus, power factor correction has dropped the source current by 42 A.

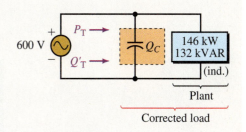

FIGURE 17–22

PRACTICE PROBLEMS 4

1. Repeat Example 17–8 except correct the power factor to unity.

2. Due to plant expansion, 102 kW of purely resistive load is added to the plant of Figure 17–20. Determine whether power factor correction is needed to correct the expanded plant to 0.85 F_p, or better.

Answers

1. 973 μF, 243 A. Other answers remain unchanged.

2. F_p = 0.88. No correction needed.

In practice, almost all loads (industrial, residential, and commercial) are inductive due to the presence of motors, fluorescent lamp ballasts, and the like. Consequently, you will likely never run into capacitive loads that need power factor correcting.

---

1. Sketch the power triangle for Figure 17–9(c). Using this triangle, determine the magnitude of the applied voltage.

2. For Figure 17–10, assume a source of E = 240 volts, P_2 = 300 W, and Q_2 = 400 VAR (cap.). What is the magnitude of the source current I?

3. What is the power factor of each of the circuits of Figure 17–7, 17–8, and 17–9? Indicate whether they are leading or lagging.

4. Consider the circuit of Figure 17–18(b). If P = 100 kW and Q_L = 80 kVAR, is the source overloaded, assuming it is capable of handling a 120-kVA load?

IN-PROCESS
LEARNING CHECK 1

(Answers are at the end of the chapter.)

To measure power in an ac circuit, you need a wattmeter. A wattmeter is a device that monitors current and voltage, and from these, determines power. Most modern units are digital instruments. For digital units, Figure 17–23, power is displayed on a numerical readout, while for analog instruments (see sidebar note), power is indicated by a pointer on a scale much like the analog meters of Chapter 2. Note however, although their details differ, their method of use and connection in a circuit are the same—thus, the measurement techniques described herein apply to both.

To help understand the concept of power measurement, consider Figure 17–1. Instantaneous load power is the product of load voltage times load current, and average power is the average of this product. One way to implement power measurement is therefore to create a meter with a current sensing circuit, a voltage sensing circuit, a multiplier circuit, and an averaging circuit. Figure 17–24 shows a simplified symbolic representation of such an instrument. Current is passed through its current coil (CC) to create a magnetic field proportional to the current, and a solid state sensor circuit connected across the load voltage reacts with this field to produce an output voltage proportional to the product of instantaneous

17.9 AC Power Measurement

NOTES . . .

Older wattmeters, (some are still in use), are electromechanical devices. They utilize a stationary coil connected in series with the load (the current coil or CC) and a pivoted coil (the potential coil or PC) connected across the load to monitor current and voltage. The interaction of the magnetic fields of these two coils creates a torque, and the pointer, attached to the pivoted coil, takes up a position on the scale corresponding to average power.

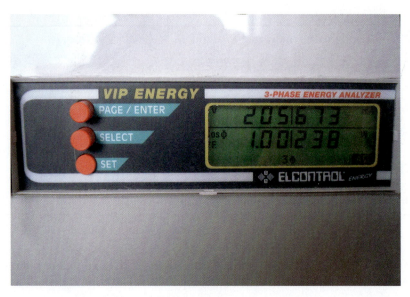

FIGURE 17–23 Multifunction power/energy meter. It can measure active power (W), reactive power (VARs), apparent power (VA), power factor, energy, and more.

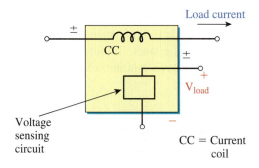

FIGURE 17–24 Conceptual representation of an electronic wattmeter.

voltage and current (i.e., proportional to instantaneous power). An averaging circuit averages this voltage and drives a display to indicate average power. (The scheme used by the meter of Figure 17–23 is actually considerably more sophisticated than this because it measures many things besides power—e.g., it measures, VARs, VA, energy, etc. However, the basic idea is conceptually correct.)

Figure 17–25 shows how to connect a wattmeter into a circuit. Load current passes through its current coil circuit, and load voltage is impressed across its voltage sensing circuit. With this connection, the wattmeter computes and displays the product of the magnitude of the load voltage, the magnitude of the load current, and the cosine of the angle between them, i.e., $V_{load} \cdot I_{load} \cdot \cos \theta_{load}$. Thus, it measures load power. Note the $\pm$ marking on the terminals. The meter is connected so that load current enters the $\pm$ current terminal and the higher

NOTES . . .

You need to be very clear on what angle to use in determining a wattmeter reading. The angle to use is the angle between the voltage across its voltage coil and the current through its current coil, using references as depicted in Figure 17–24.

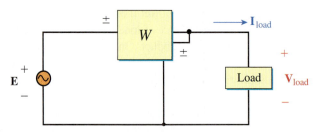

FIGURE 17–25 Connection of wattmeter.

potential end of the load is connected to the $\pm$ voltage terminal. (On many meters, the $\pm$ voltage terminal is internally connected so that only three terminals are brought out as in Figure 17–26.)

When power is to be measured in a low power factor circuit, a low power factor wattmeter must be used. This is because, for low power factor loads, currents can be very high, even though the power is low. Thus, you can easily exceed the current rating of a standard wattmeter and damage it, even though the power indication on the meter is small.

EXAMPLE 17–9

For the circuit of Figure 17–25, what does the wattmeter indicate if

a. $\mathbf{V}_{load} = 100\,\text{V}\angle 0°$ and $\mathbf{I}_{load} = 15\,\text{A}\angle 60°$,

b. $\mathbf{V}_{load} = 100\,\text{V}\angle 10°$ and $\mathbf{I}_{load} = 15\,\text{A}\angle 30°$?

Solution

a. $\theta_{load} = 60°$. Thus, $P = (100)(15)\cos 60° = 750\,\text{W}$,

b. $\theta_{load} = 10° - 30° = -20°$. Thus, $P = (100)(15)\cos(-20°) = 1410\,\text{W}$.

Note: For (b), since $\cos(-20°) = \cos(+20°)$, it does not matter whether we include the minus sign.

EXAMPLE 17–10

For Figure 17–26, determine the wattmeter reading.

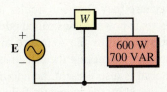

FIGURE 17–26   This wattmeter has its voltage side $\pm$ terminals connected internally.

Solution   A wattmeter reads only active power. Thus, it indicates 600 W.

It should be noted that the wattmeter reads power only for circuit elements on the load side of the meter. In addition, if the load consists of several elements, it reads the total power.

PRACTICE PROBLEMS 5

Determine the wattmeter reading for Figure 17–27.

FIGURE 17–27

Answer
750 W

17.10 Effective Resistance

Up to now, we have assumed that resistance is constant, independent of frequency. However, this is not entirely true. For a number of reasons, the resistance of a circuit to ac is greater than its resistance to dc. While this effect is small at low frequencies, it is very pronounced at high frequencies. AC resistance is known as **effective resistance.**

Before looking at why ac resistance is greater than dc resistance, we need to reexamine the concept of resistance itself. Recall from Chapter 3 that resistance was originally defined as opposition to current, that is, $R = V/I$. (This is ohmic resistance.) Building on this, you learned in Chapter 4 that $P = I^2R$. It is this latter viewpoint that allows us to give meaning to ac resistance. That is, we define ac or effective resistance as

$$R_{\text{eff}} = \frac{P}{I^2} \quad (\Omega) \qquad \qquad \text{(17–25)}$$

where P is dissipated power (as determined by a wattmeter). From this, you can see that anything that affects dissipated power affects resistance. For dc and low-frequency ac, both definitions for R, i.e., $R = V/I$ and $R = P/I^2$ yield the same value. However, as frequency increases, other factors cause an increase in resistance. We will now consider some of these.

Eddy Currents and Hysteresis

The magnetic field surrounding a coil or other circuit carrying ac current varies with time and thus induces voltages in nearby conductive material such as metal equipment cabinets, transformer cores, and so on. The resulting currents (called **eddy currents** because they flow in circular patterns like eddies in a brook) are unwanted and create power losses called **eddy current losses.** Since additional power must be supplied by the source to make up for these losses, P in Equation 17–25 increases, increasing the effective resistance of the coil.

If ferromagnetic material is also present, an additional power loss occurs due to hysteresis effects caused by the magnetic field alternately magnetizing the material in one direction, then the other. Hysteresis and eddy current losses are important even at low frequencies, such as the 60-Hz power system frequency. This is discussed in Chapter 23.

Skin Effect

AC currents create a time varying magnetic field about a conductor, Figure 17–28(a). This varying field in turn, induces voltage in the conductor. This voltage is of such a nature that it drives free electrons from the center of the wire to its periphery, Figure 17–28(b), resulting in a nonuniform distribution of

(a) Varying magnetic field induces voltage within the conductor

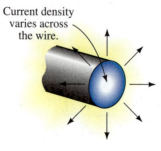

(b) This voltage drives free electrons toward the periphery, leaving few electrons in the center

(c) At high frequencies, the effect is so pronounced that hollow conductors may be used

FIGURE 17–28 Skin effect in ac circuits.

current, with current density greatest near the periphery and smallest in the center. This phenomenon is known as **skin effect.** Because the center of the wire carries little current, its cross-sectional area has effectively been reduced, thus increasing resistance. While skin effect is generally negligible at power line frequencies (except for conductors larger than several hundred thousand circular mils), it is so pronounced at microwave frequencies that the center of a wire carries almost no current. For this reason, hollow conductors are often used instead of solid wires, as shown in Figure 17–28(c).

Radiation Resistance

At high frequencies some of the energy supplied to a circuit escapes as radiated energy. For example, a radio transmitter supplies power to an antenna, where it is converted into radio waves and radiated into space. The resistance effect here is known as **radiation resistance.** This resistance is much higher than simple dc resistance. For example, a TV transmitting antenna may have a resistance of a fraction of an ohm to dc but several hundred ohms effective resistance at its operating frequency.

FINAL NOTES . . .

1. The resistance measured by an ohmmeter is dc resistance.

2. Many of the effects noted above will be treated in detail in your various electronics courses. We will not pursue them further here.

17.11 Energy Relationships for AC

Recall, power and energy are related by the equation $p = dw/dt$. Thus, energy can be found by integration as

$$W = \int p\, dt = \int vi\, dt \tag{17–26}$$

Inductance

For an inductance, $v = L\,di/dt$. Substituting this into Equation 17–26, cancelling dt, and rearranging terms yields

$$W_L = \int \left(L\frac{di}{dt} \right) i\, dt = L\int i\, di \tag{17–27}$$

Recall from Figure 17–4(b), energy flows into an inductor during time interval 0 to $T/4$ and is released during time interval $T/4$ to $T/2$. The process then repeats itself. The energy stored (and subsequently released) can thus be found by integrating power from $t = 0$ to $t = T/4$. Current at $t = 0$ is 0 and current at $t = T/4$ is I_m. Using these as our limits of integration, we find

$$W_L = L\int_0^{I_m} i\, di = \frac{1}{2}LI_m^2 = LI^2 \quad \text{(J)} \tag{17–28}$$

where we have used $I = I_m/\sqrt{2}$ to express energy in terms of effective current.

Capacitance

For a capacitance, $i = C\,dv/dt$. Substituting this into Equation 17–26 yields

$$W_C = \int v\left(C\frac{dv}{dt} \right) dt = C\int v\, dv \tag{17–29}$$

Consider Figure 17–5(b). Energy stored can be found by integrating power from $T/4$ to $T/2$. The corresponding limits for voltage are 0 to V_m. Thus,

$$W_C = C\int_0^{V_m} v\, dv = \frac{1}{2}CV_m^2 = CV^2 \quad \text{(J)} \tag{17–30}$$

where we have used $V = V_m/\sqrt{2}$. You will use these relationships later.

17.12 Circuit Analysis Using Computers

NOTES . . .

As of this writing, MultiSIM has no simple way to plot current and power; thus, no example is included here. If this changes when MSM is updated, we will add examples and problems to our web site.

The time-varying relationships between voltage, current and power described earlier in this chapter can be investigated easily using PSpice. To illustrate, consider the circuit of Figure 17–3 with $v = 1.2 \sin \omega t$, $R = 0.8\ \Omega$ and $f = 1000$ Hz. Create the circuit on the screen, including voltage and current markers as in Figure 17–29. Note the empty parameter boxes beside the source. Double click each in turn and enter values as shown. Click the New Simulation Profile icon, type a name, choose Transient, set TSTOP to **1ms,** then click OK. Run the simulation and the voltage and current waveforms of Figure 17–30 should appear. To plot power (i.e., the product of vi), click Trace, then Add Trace, and when the dialog box opens, use the asterisk to create the product V(R1:1)*I(R1), then click OK. The blue power curve should now appear. Compare to Figure 17–3. Note that all curves agree exactly.

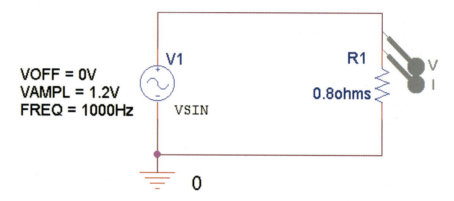

FIGURE 17–29 PSpice circuit.

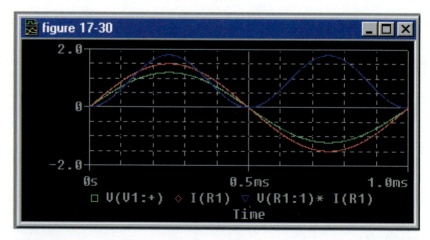

FIGURE 17–30 Voltage, current and power waveforms for Figure 17–29.

PROBLEMS

17.1–17.5

1. Note that the power curve of Figure 17–4 is sometimes positive and sometimes negative. What is the significance of this? Between $t = T/4$ and $t = T/2$, what is the direction of power flow?

2. What is real power? What is reactive power? Which power, real or reactive, has an average value of zero?

3. A pair of electric heating elements is shown in Figure 17–31.

   a. Determine the active and reactive power to each.

   b. Determine the active and reactive power delivered by the source.

4. For the circuit of Figure 17–32, determine the active and reactive power to the inductor.

5. If the inductor of Figure 17–32 is replaced by a 40-μF capacitor and source frequency is 60 Hz, what is Q_C?

6. Find R and X_L for Figure 17–33.

7. For the circuit of Figure 17–34, $f = 100$ Hz. Find

   a. R b. X_C c. C

8. For the circuit of Figure 17–35, $f = 10$ Hz. Find

   a. P b. X_L c. L

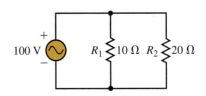

FIGURE 17–31

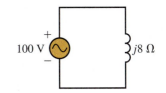

FIGURE 17–32

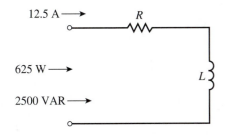

FIGURE 17–33

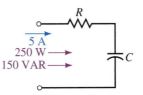

FIGURE 17–34

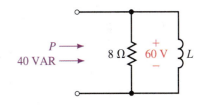

FIGURE 17–35

9. For Figure 17–36, find X_C.

10. For Figure 17–37, $X_C = 42.5$ Ω. Find R, P, and Q.

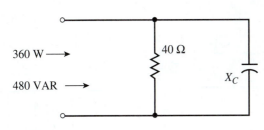

FIGURE 17–36

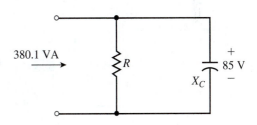

FIGURE 17–37

11. Find the total average power and the total reactive power supplied by the source for Figure 17–38.

12. If the source of Figure 17–38 is reversed, what is P_T and Q_T? What conclusion can you draw from this?

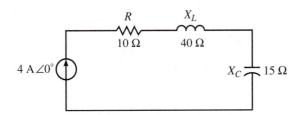

FIGURE 17–38

13. Refer to Figure 17–39. Find P_2 and Q_3. Is the element in Load 3 inductive or capacitive?

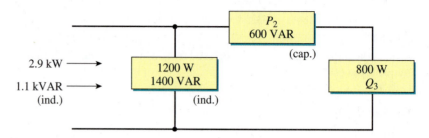

FIGURE 17–39

14. For Figure 17–40, determine P_T and Q_T.

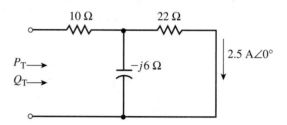

FIGURE 17–40

15. For Figure 17–41, $\omega = 10$ rad/s. Determine
   a. R_T b. R_2 c. X_C d. L_{eq}

16. For Figure 17–42, determine the total P_T and Q_T.

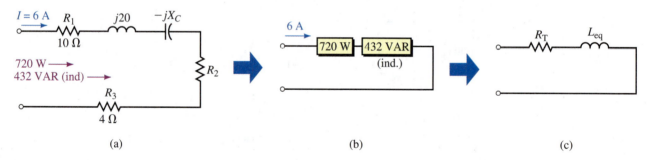

(a)   (b)   (c)

FIGURE 17–41

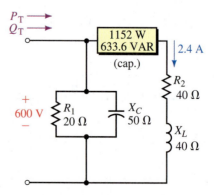

FIGURE 17–42

17.7   The Relationship between *P*, *Q*, and *S*

17. For the circuit of Figure 17–7, draw the power triangle and determine the apparent power.

18. Repeat Problem 17 for Figure 17–8.

19. Ignoring the wattmeter of Figure 17–27, determine the power triangle for the circuit as seen by the source.

20. For the circuit of Figure 17–43, what is the source current?

21. For Figure 17–44, the generator supplies 30 A. What is *R*?

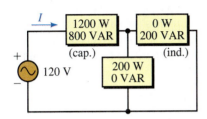

FIGURE 17–43

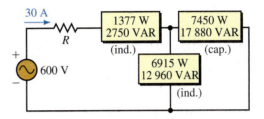

FIGURE 17–44

22. Suppose **V** = 100 V∠60° and **I** = 10A∠40°:

   a. What is θ, the angle between **V** and **I**?

   b. Determine *P* from *P* = *VI* cos θ.

   c. Determine *Q* from *Q* = *VI* sin θ.

   d. Sketch the power triangle and from it, determine **S**.

   e. Show that **S** = **VI*** gives the same answer as (d).

23. For Figure 17–45, S_{gen} = 4835 VA. What is *R?*

24. Refer to the circuit of Figure 17–16:

   a. Determine the apparent power for each box.

   b. Sum the apparent powers that you just computed. Why does the sum not equal S_T = 1803 VA as obtained in Example 17–5?

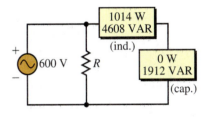

FIGURE 17–45

17.8   Power Factor

25. Refer to the circuit of Figure 17–46:

   a. Determine P_T, Q_T, and S_T.

   b. Determine whether the fuse will blow.

26. A motor with an efficiency of 87% supplies 10 hp to a load (Figure 17–47). Its power factor is 0.65 (lag).

   a. What is the power input to the motor?

   b. What is the reactive power to the motor?

   c. Draw the motor power triangle. What is the apparent power to the motor?

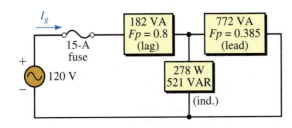

FIGURE 17–46

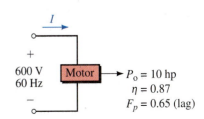

FIGURE 17–47

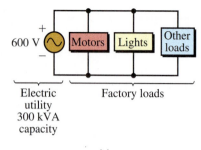

600 V

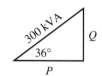

Electric
utility
300 kVA
capacity

Factory loads

(a)

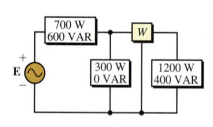

(b) Factory power triangle

FIGURE 17–48

27. To correct the circuit power factor of Figure 17–47 to unity, a power factor correction capacitor is added.

  a. Show where the capacitor is connected.

  b. Determine its value in microfarads.

28. Consider Figure 17–20. The motor is replaced with a new unit requiring $S_m = (120 + j35)$ kVA. Everything else remains the same. Find the following:

  a. P_T b. Q_T c. S_T

  d. Determine how much kVAR capacitive correction is needed to correct to unity F_p.

29. A small electrical utility has a 600-V, 300-kVA capacity. It supplies a factory (Figure 17–48) with the power triangle shown in (b). This fully loads the utility. If a power factor correcting capacitor corrects the load to unity power factor, how much more power (at unity power factor) can the utility sell to other customers?

17.9 AC Power Measurement

30. a. Why does the wattmeter of Figure 17–49 indicate only 1200 watts?

  b. Where would the wattmeter have to be placed to measure power delivered by the source? Sketch the modified circuit.

  c. What would the wattmeter indicate in (b)?

31. Determine the wattmeter reading for Figure 17–50.

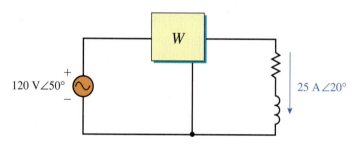

FIGURE 17–50

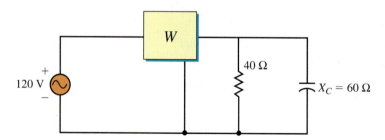

32. Determine the wattmeter reading for Figure 17–51.

FIGURE 17–49

FIGURE 17–51

17.10 Effective Resistance

33. Measurements on an iron-core solenoid coil yield the following values: $V = 80$ V, $I = 400$ mA, $P = 25.6$ W, and $R = 140$ Ω. (The last measurement was taken with an ohmmeter.) What is the ac resistance of the solenoid coil?

17.12 Circuit Analysis Using Computers

34. An inductance $L = 1$ mH has current $i = 4 \sin (2\pi \times 1000)t$. Use PSpice to investigate the power waveform and compare to Figure 17–4. Use current source ISIN (see Note).

◀ CADENCE

35. A 10-μF capacitor has voltage $v = 10 \sin(\omega t - 90°)$ V. Use PSpice to investigate the power waveform and compare to Figure 17–5. Use voltage source VSIN with $f = 1000$ Hz.

◀ CADENCE

36. The voltage waveform of Figure 17–52 is applied to a 200-μF capacitor.

◀ CADENCE

   a. Using your calculator and the principles of Chapter 10, determine the current through the capacitor and sketch. (Also sketch the voltage waveform on your graph.) Multiply the two waveforms to obtain a plot of $p(t)$. Compute power at its max and min points.

   b. Use PSpice to verify the results. Use voltage source VPWL. You have to describe the waveform to the source. It has a value of 0 V at $t = 0$, 10 V at $t = 1$ ms, -10 V at $t = 3$ ms, and 0 V at $t = 4$ ms. To set these, double click the source symbol and enter values via the Property Editor as follows: **0** for T1, **0V** for V1, **1ms** for T2, **10V** for V2, etc. Run the simulation and plot voltage, current, and power using the procedure we used to create Figure 17–30. Results should agree with those of (a).

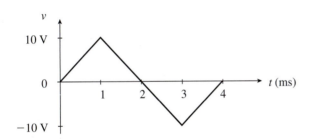

FIGURE 17–52

37. Repeat Question 36 for a current waveform identical to Figure 17–52 except that it oscillates between 2 A and -2 A applied to a 2-mH inductor. Use current source IPWL (see Note).

◀ CADENCE

✓ ANSWERS TO IN-PROCESS LEARNING CHECKS

In-Process Learning Check 1

1. 100 V

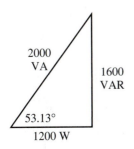

FIGURE 17–53

2. 8.76 A

3. Fig. 17–7: 0.6 (lag); Fig. 17–8: 0.97 (lead); Fig. 17–9: 0.6 (lag)

4. Yes. ($S = 128$ kVA)

Impedance Networks

V

As you have already observed, the impedance of an inductor or a capacitor is dependent upon the frequency of the signal applied to the element. When capacitors and inductors are combined with resistors and voltage or current sources, the circuit will behave in a predictable manner for all frequencies.

This part of the textbook examines how circuits consisting of various combinations of impedances and sources behave under specific conditions. In particular, we find that all the laws, rules, and theorems developed previously apply to even the most complicated impedance network.

Ohm's law and Kirchhoff's voltage and current laws are easily modified to give the framework for developing methods of network analysis. Just as in dc circuits, Thévenin's and Norton's theorems will allow us to simplify a complicated circuit to a single source and corresponding impedance.

The theorems and methods of analysis are applied to numerous types of circuits which are commonly encountered throughout electrical and electronics technology. Resonant circuits and filter circuits are commonly used to restrict the range of output frequencies for a given range of input frequencies.

The study of three-phase systems and transformers is particularly useful for anyone interested in commercial power distribution. These topics deal with practical applications and the drawbacks of using various types of circuits.

Finally, we examine how a circuit reacts to nonsinusoidal alternating voltages. This topic involves complex signals which are processed by impedance networks resulting in outputs which are often dramatically different from the input. ∎

■ OBJECTIVES

After studying this chapter, you will be able to

- apply Ohm's law to analyze simple series circuits,

- apply the voltage divider rule to determine the voltage across any element in a series circuit,

- apply Kirchhoff's voltage law to verify that the summation of voltages around a closed loop is equal to zero,

- apply Kirchhoff's current law to verify that the summation of currents entering a node is equal to the summation of currents leaving the same node,

- determine unknown voltage, current, and power for any series/ parallel circuit,

- determine the series or parallel equivalent of any network consisting of a combination of resistors, inductors, and capacitors.

AC Series-Parallel Circuits

18

In this chapter we examine how simple circuits containing resistors, inductors, and capacitors behave when subjected to sinusoidal voltages and currents. Principally, we find that the rules and laws which were developed for dc circuits will apply equally well for ac circuits. The major difference between solving dc and ac circuits is that analysis of ac circuits requires using vector algebra.

In order to proceed successfully, it is suggested that the student spend time reviewing the important topics covered in dc analysis. These include Ohm's law, the voltage divider rule, Kirchhoff's voltage law, Kirchhoff's current law, and the current divider rule.

You will also find that a brief review of vector algebra will make your understanding of this chapter more productive. In particular, you should be able to add and subtract any number of vector quantities. ∎

Heinrich Rudolph Hertz

HEINRICH HERTZ WAS BORN IN HAMBURG, Germany, on February 22, 1857. He is known mainly for his research into the transmission of electromagnetic waves.

Hertz began his career as an assistant to Hermann von Helmholtz in the Berlin Institute physics laboratory. In 1885, he was appointed Professor of Physics at Karlsruhe Polytechnic, where he did much to verify James Clerk Maxwell's theories of electromagnetic waves.

In one of his experiments, Hertz discharged an induction coil with a rectangular loop of wire having a very small gap. When the coil discharged, a spark jumped across the gap. He then placed a second, identical coil close to the first, but with no electrical connection. When the spark jumped across the gap of the first coil, a smaller spark was also induced across the second coil. Today, more elaborate antennas use similar principles to transmit radio signals over vast distances. Through further research, Hertz was able to prove that electromagnetic waves have many of the characteristics of light: they have the same speed as light; they travel in straight lines; they can be reflected and refracted; and they can be polarized.

Hertz's experiments ultimately led to the development of radio communication by such electrical engineers as Guglielmo Marconi and Reginald Fessenden.

Heinrich Hertz died at the age of 36 on January 1, 1894. ∎

18.1 Ohm's Law for AC Circuits

This section is a brief review of the relationship between voltage and current for resistors, inductors, and capacitors. Unlike Chapter 16, all phasors are given as rms rather than as peak values. As you saw in Chapter 17, this approach simplifies the calculation of power.

Resistors

In Chapter 16, we saw that when a resistor is subjected to a sinusoidal voltage as shown in Figure 18–1, the resulting current is also sinusoidal and in phase with the voltage.

The sinusoidal voltage $v = V_m\sin(\omega t + \theta)$ may be written in phasor form as $\mathbf{V} = V\angle\theta$. Whereas the sinusoidal expression gives the instantaneous value of voltage for a waveform having an amplitude of V_m (volts peak), the phasor form has a magnitude which is the effective (or rms) value. The relationship between the magnitude of the phasor and the peak of the sinusoidal voltage is given as

$$V = \frac{V_m}{\sqrt{2}}$$

Because the resistance vector may be expressed as $\mathbf{Z}_R = R\angle0°$, we evaluate the current phasor as follows:

$$\mathbf{I} = \frac{\mathbf{V}}{\mathbf{Z}_R} = \frac{V\angle\theta}{R\angle0°} = \frac{V}{R}\angle\theta = I\angle\theta$$

If we wish to convert the current from phasor form to its sinusoidal equivalent in the time domain, we would have $i = I_m\sin(\omega t + \theta)$. Again, the relationship between the magnitude of the phasor and the peak value of the sinusoidal equivalent is given as

$$I = \frac{I_m}{\sqrt{2}}$$

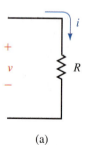

(a)

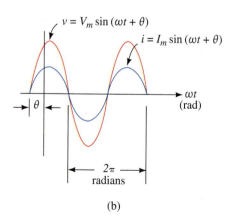

(b)

FIGURE 18–1   Sinusoidal voltage and current for a resistor.

The voltage and current phasors may be shown on a phasor diagram as in Figure 18–2.

Because one phasor is a current and the other is a voltage, the relative lengths of these phasors are purely arbitrary. Regardless of the angle θ, we see that the voltage across and the current through a resistor will always be in phase.

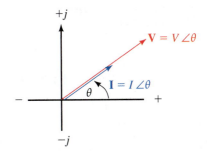

FIGURE 18–2   Voltage and current phasors for a resistor.

EXAMPLE 18–1

Refer to the resistor shown in Figure 18–3:

a. Find the sinusoidal current i using phasors.

b. Sketch the sinusoidal waveforms for v and i.

c. Sketch the phasor diagram of **V** and **I**.

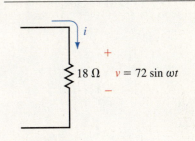

FIGURE 18–3

Solution

a. The phasor form of the voltage is determined as follows:

$$v = 72 \sin \omega t \Leftrightarrow \mathbf{V} = 50.9 \text{ V} \angle 0°$$

From Ohm's law, the current phasor is determined to be

$$\mathbf{I} = \frac{\mathbf{V}}{\mathbf{Z}_R} = \frac{50.9 \text{ V}\angle 0°}{18 \text{ }\Omega\angle 0°} = 2.83 \text{ A}\angle 0°$$

which results in the sinusoidal current waveform having an amplitude of

$$I_m = (\sqrt{2})(2.83 \text{ A}) = 4.0 \text{ A}$$

Therefore, the current i will be written as

$$i = 4 \sin \omega t$$

b. The voltage and current waveforms are shown in Figure 18–4.

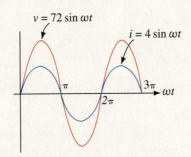

FIGURE 18–4

c. Figure 18–5 shows the voltage and current phasors.

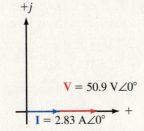

FIGURE 18–5

EXAMPLE 18–2

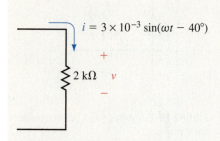

FIGURE 18–6

Refer to the resistor of Figure 18–6:

a. Use phasor algebra to find the sinusoidal voltage, v.

b. Sketch the sinusoidal waveforms for v and i.

c. Sketch a phasor diagram showing **V** and **I**.

Solution

a. The sinusoidal current has a phasor form as follows:

$$i = 3 \times 10^{-3} \sin(\omega t - 40°) \Leftrightarrow \mathbf{I} = 2.12 \text{ mA}\angle{-40°}$$

From Ohm's law, the voltage across the 2-kΩ resistor is determined as the phasor product

$$\mathbf{V} = \mathbf{IZ}_R$$
$$= (2.12 \text{ mA}\angle{-40°})(2 \text{ k}\Omega\angle{0°})$$
$$= 4.24 \text{ V}\angle{-40°}$$

The amplitude of the sinusoidal voltage is

$$V_m = (\sqrt{2})(4.24 \text{ V}) = 6.0 \text{ V}$$

The voltage may now be written as

$$v = 6.0 \sin(\omega t - 40°)$$

b. Figure 18–7 shows the sinusoidal waveforms for v and i.

c. The corresponding phasors for the voltage and current are shown in Figure 18–8.

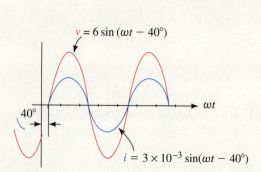

FIGURE 18–7

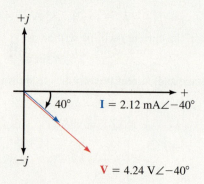

FIGURE 18–8

Inductors

When an inductor is subjected to a sinusoidal current, a sinusoidal voltage is induced across the inductor such that the voltage across the inductor leads the current waveform by exactly 90°. If we know the reactance of an inductor, then from Ohm's law the current in the inductor may be expressed in phasor form as

$$\mathbf{I} = \frac{\mathbf{V}}{\mathbf{Z}_L} = \frac{V\angle\theta}{X_L\angle 90°} = \frac{V}{X_L}\angle(\theta - 90°)$$

In vector form, the reactance of the inductor is given as

$$\mathbf{Z}_L = X_L\angle 90°$$

where $X_L = \omega L = 2\pi f L$.

Consider the inductor shown in Figure 18–9:

a. Determine the sinusoidal expression for the current i using phasors.

b. Sketch the sinusoidal waveforms for v and i.

c. Sketch the phasor diagram showing **V** and **I**.

Solution

a. The phasor form of the voltage is determined as follows:

$$v = 1.05 \sin(\omega t + 120°) \Leftrightarrow \mathbf{V} = 0.742 \text{ V } \angle 120°$$

From Ohm's law, the current phasor is determined to be

$$\mathbf{I} = \frac{\mathbf{V}}{\mathbf{Z}_L} = \frac{0.742 \text{ V} \angle 120°}{25 \text{ } \Omega \angle 90°} = 29.7 \text{ mA} \angle 30°$$

The amplitude of the sinusoidal current is

$$I_m = (\sqrt{2})(29.7 \text{ mA}) = 42 \text{ mA}$$

The current i is now written as

$$i = 0.042 \sin(\omega t + 30°)$$

b. Figure 18–10 shows the sinusoidal waveforms of the voltage and current.

c. The voltage and current phasors are shown in Figure 18–11.

EXAMPLE 18–3

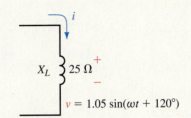

FIGURE 18–9

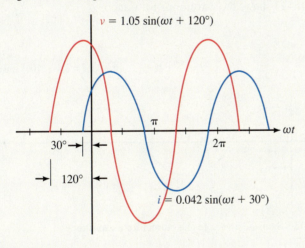

FIGURE 18–10   Sinusoidal voltage and current for an inductor.

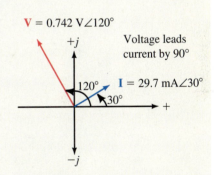

FIGURE 18–11   Voltage and current phasors for an inductor.

Capacitors

When a capacitor is subjected to a sinusoidal voltage, a sinusoidal current results. The current through the capacitor leads the voltage by exactly 90°. If we know the reactance of a capacitor, then from Ohm's law the current in the capacitor expressed in phasor form is

$$\mathbf{I} = \frac{\mathbf{V}}{\mathbf{Z}_C} = \frac{V \angle \theta}{X_C \angle -90°} = \frac{V}{X_C} \angle (\theta + 90°)$$

In vector form, the reactance of the capacitor is given as

$$\mathbf{Z}_C = X_C \angle -90°$$

where

$$X_C = \frac{1}{\omega C} = \frac{1}{2\pi f C}$$

EXAMPLE 18–4

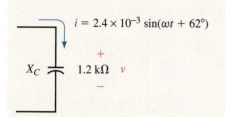

$i = 2.4 \times 10^{-3} \sin(\omega t + 62°)$

X_C ⊣⊢ $1.2\ \text{k}\Omega$ v

FIGURE 18–12

Consider the capacitor of Figure 18–12.

a. Find the voltage v across the capacitor.

b. Sketch the sinusoidal waveforms for v and i.

c. Sketch the phasor diagram showing **V** and **I**.

Solution

a. Converting the sinusoidal current into its equivalent phasor form gives

$$i = 2.4 \times 10^{-3} \sin(\omega t + 62°) \Leftrightarrow \mathbf{I} = 1.70\ \text{mA}\angle 62°$$

From Ohm's law, the phasor voltage across the capacitor must be

$$\mathbf{V} = \mathbf{I}\mathbf{Z}_C$$
$$= (1.70\ \text{mA}\angle 62°)(1.2\ \text{k}\Omega\angle -90°)$$
$$= 2.04\ \text{V}\angle -28°$$

The amplitude of the sinusoidal voltage is

$$V_m = (\sqrt{2})(2.04\ \text{V}) = 2.88\ \text{V}$$

The voltage v is now written as

$$v = 2.88 \sin(\omega t - 28°)$$

b. Figure 18–13 shows the waveforms for v and i.

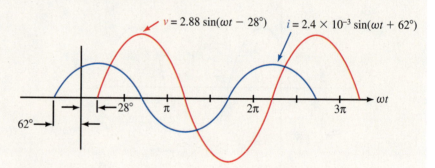

FIGURE 18–13   Sinusoidal voltage and current for a capacitor.

c. The corresponding phasor diagram for **V** and **I** is shown in Figure 18–14.

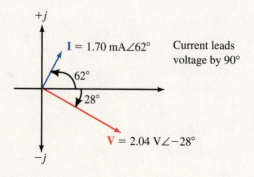

FIGURE 18–14   Voltage and current phasors for a capacitor.

The relationships between voltage and current, as illustrated in the previous three examples, will always hold for resistors, inductors, and capacitors.

IN-PROCESS
LEARNING CHECK 1

(Answers are at the end of the chapter.)

1. What is the phase relationship between current and voltage for a resistor?
2. What is the phase relationship between current and voltage for a capacitor?
3. What is the phase relationship between current and voltage for an inductor?

PRACTICE PROBLEMS 1

A voltage source, $\mathbf{E} = 10 \text{ V} \angle 30°$, is applied to an inductive impedance of 50 Ω.

a. Solve for the phasor current, $\mathbf{I}$.
b. Sketch the phasor diagram for $\mathbf{E}$ and $\mathbf{I}$.
c. Write the sinusoidal expressions for e and i.
d. Sketch the sinusoidal expressions for e and i.

Answers
  a. $\mathbf{I} = 0.2 \text{ A} \angle -60°$
  c. $e = 14.1 \sin(\omega t + 30°)$
    $i = 0.283 \sin(\omega t - 60°)$

PRACTICE PROBLEMS 2

A voltage source, $\mathbf{E} = 10 \text{ V} \angle 30°$, is applied to a capacitive impedance of 20 Ω.

a. Solve for the phasor current, $\mathbf{I}$.
b. Sketch the phasor diagram for $\mathbf{E}$ and $\mathbf{I}$.
c. Write the sinusoidal expressions for e and i.
d. Sketch the sinusoidal expressions for e and i.

Answers
  a. $\mathbf{I} = 0.5 \text{ A} \angle 120°$
  c. $e = 14.1 \sin(\omega t + 30°)$
    $i = 0.707 \sin(\omega t + 120°)$

18.2 AC Series Circuits

◀ **Online Companion**

When we examined dc circuits we saw that the current everywhere in a series circuit is always constant. This same applies when we have series elements with an ac source. Further, we had seen that the total resistance of a dc series circuit consisting of n resistors was determined as the summation

$$R_\text{T} = R_1 + R_2 + \cdots + R_n$$

When working with ac circuits we no longer work with only resistance but also with capacitive and inductive reactance. *Impedance is a term used to collectively determine how the resistance, capacitance, and inductance "impede" the current in a circuit.* The symbol for impedance is the letter Z and the unit is the ohm (Ω). Because impedance may be made up of any combination of resistances and reactances, it is written as a vector quantity $\mathbf{Z}$, where

$$\mathbf{Z} = Z \angle \theta \quad (\Omega)$$

Each impedance may be represented as a vector on the complex plane, such that the length of the vector is representative of the magnitude of the impedance. The diagram showing one or more impedances is referred to as an **impedance diagram.**

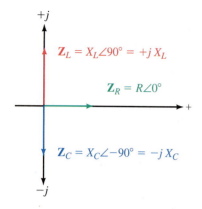

FIGURE 18–15

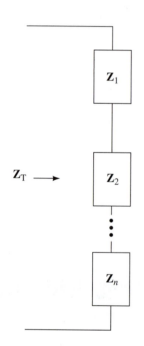

FIGURE 18–16

Resistive impedance $\mathbf{Z}_R$ is a vector having a magnitude of R along the positive real axis. Inductive reactance $\mathbf{Z}_L$ is a vector having a magnitude of X_L along the positive imaginary axis, while the capacitive reactance $\mathbf{Z}_C$ is a vector having a magnitude of X_C along the negative imaginary axis. Mathematically, each of the vector impedances is written as follows:

$$\mathbf{Z}_R = R\angle 0° = R + j0 = R$$
$$\mathbf{Z}_L = X_L\angle 90° = 0 + jX_L = jX_L$$
$$\mathbf{Z}_C = X_C\angle 90° = 0 - jX_C = -jX_C$$

An impedance diagram showing each of the above impedances is shown in Figure 18–15.

All impedance vectors will appear in either the first or the fourth quadrants, since the resistive impedance vector is always positive.

For a series ac circuit consisting of n impedances, as shown in Figure 18–16, the total impedance of the circuit is found as the vector sum

$$\mathbf{Z}_T = \mathbf{Z}_1 + \mathbf{Z}_2 + \cdots + \mathbf{Z}_n \qquad (18\text{–}1)$$

Consider the branch of Figure 18–17.

By applying Equation 18–1, we may determine the total impedance of the circuit as

$$\mathbf{Z}_T = (3\ \Omega + j0) + (0 + j4\ \Omega) = 3\ \Omega + j4\ \Omega$$
$$= 5\ \Omega\angle 53.13°$$

The above quantities are shown on an impedance diagram as in Figure 18–18.

From Figure 18–18 we see that the total impedance of the series elements consists of a real component and an imaginary component. The corresponding total impedance vector may be written in either polar or rectangular form.

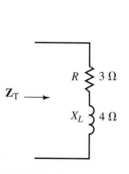

FIGURE 18–17

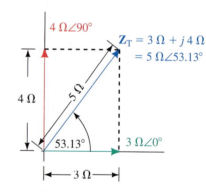

FIGURE 18–18

The rectangular form of an impedance is written as

$$\mathbf{Z} = R \pm jX$$

If we are given the polar form of the impedance, then we may determine the equivalent rectangular expression from

$$R = Z\cos\theta \qquad (18\text{–}2)$$

and

$$X = Z\sin\theta \qquad (18\text{–}3)$$

In the rectangular representation for impedance, the resistance term, R, is the total of all resistance looking into the network. The reactance term, X, is the difference between the total capacitive and inductive reactances. The sign for the imaginary term will be positive if the inductive reactance is greater than the capacitive reactance. In such a case, the impedance vector will appear in the first quadrant of the impedance diagram and is referred to as being an **inductive** impedance. If the capacitive reactance is larger, then the sign for the imaginary term will be negative. In such a case, the impedance vector will appear in the fourth quadrant of the impedance diagram and the impedance is said to be **capacitive.**

The polar form of any impedance will be written in the form

$$\mathbf{Z} = Z \angle \theta$$

The value Z is the magnitude (in ohms) of the impedance vector $\mathbf{Z}$ and is determined as follows:

$$Z = \sqrt{R^2 + X^2} \quad (\Omega) \qquad \textbf{(18–4)}$$

The corresponding angle of the impedance vector is determined as

$$\theta = \pm \tan^{-1}\left(\frac{X}{R}\right) \qquad \textbf{(18–5)}$$

Whenever a capacitor and an inductor having equal reactances are placed in series, as shown in Figure 18–19, the equivalent circuit of the two components is a short circuit since the inductive reactance will be exactly balanced by the capacitive reactance.

Any ac circuit having a total impedance with only a real component, is referred to as a **resistive** circuit. In such a case, the impedance vector $\mathbf{Z}_T$ will be located along the positive real axis of the impedance diagram and the angle of the vector will be 0°. The condition under which series reactances are equal is referred to as "series resonance" and is examined in greater detail in a later chapter.

If the impedance $\mathbf{Z}$ is written in polar form, then the angle θ will be positive for an inductive impedance and negative for a capacitive impedance. In the event that the circuit is purely reactive, the resulting angle θ will be either $+90°$ (inductive) or $-90°$ (capacitive). If we reexamine the impedance diagram of Figure 18–18, we conclude that the original circuit is inductive.

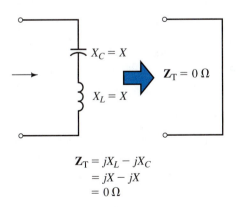

$$\mathbf{Z}_T = jX_L - jX_C$$
$$= jX - jX$$
$$= 0 \, \Omega$$

FIGURE 18–19

EXAMPLE 18–5

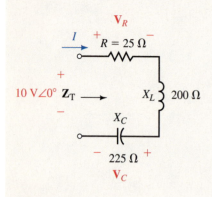

FIGURE 18–20

Consider the network of Figure 18–20.

a. Find $\mathbf{Z}_T$.

b. Sketch the impedance diagram for the network and indicate whether the total impedance of the circuit is inductive, capacitive, or resistive.

c. Use Ohm's law to determine $\mathbf{I}$, $\mathbf{V}_R$, and $\mathbf{V}_C$.

Solution

a. The total impedance is the vector sum

$$\mathbf{Z}_T = 25\ \Omega + j200\ \Omega + (-j225\ \Omega)$$
$$= 25\ \Omega - j25\ \Omega$$
$$= 35.36\ \Omega\angle{-45°}$$

b. The corresponding impedance diagram is shown in Figure 18–21.

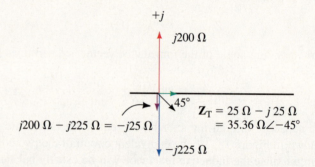

FIGURE 18–21

Because the total impedance has a negative reactance term ($-j25\ \Omega$), $\mathbf{Z}_T$ is capacitive.

c. $$\mathbf{I} = \frac{10\ \text{V}\angle0°}{35.36\ \Omega\angle{-45°}} = 0.283\ \text{A}\angle45°$$

$$\mathbf{V}_R = (282.8\ \text{mA}\angle45°)(25\ \Omega\angle0°) = 7.07\ \text{V}\angle45°$$

$$\mathbf{V}_C = (282.8\ \text{mA}\angle45°)(225\ \Omega\angle{-90°}) = 63.6\ \text{V}\angle{-45°}$$

Notice that the magnitude of the voltage across the capacitor is many times larger than the source voltage applied to the circuit. This example illustrates that the voltages across reactive elements must be calculated to ensure that maximum ratings for the components are not exceeded.

EXAMPLE 18–6

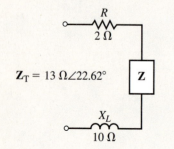

FIGURE 18–22

Determine the impedance $\mathbf{Z}$ which must be within the indicated block of Figure 18–22 if the total impedance of the network is $13\ \Omega\angle22.62°$.

Solution   Converting the total impedance from polar to rectangular form, we get

$$\mathbf{Z}_T = 13\ \Omega\angle22.62° \Leftrightarrow 12\ \Omega + j5\ \Omega$$

Now, we know that the total impedance is determined from the summation of the individual impedance vectors, namely

$$\mathbf{Z}_T = 2\ \Omega + j10\ \Omega + \mathbf{Z} = 12\ \Omega + j5\ \Omega$$

Therefore, the impedance **Z** is found as

$$\mathbf{Z} = 12\ \Omega + j5\ \Omega - (2\ \Omega + j10\ \Omega)$$
$$= 10\ \Omega - j5\ \Omega$$
$$= 11.18\ \Omega\angle-26.57°$$

In its most simple form, the impedance **Z** will consist of a series combination of a 10-Ω resistor and a capacitor having a reactance of 5 Ω. Figure 18–23 shows the elements that may be contained within **Z** to satisfy the given conditions.

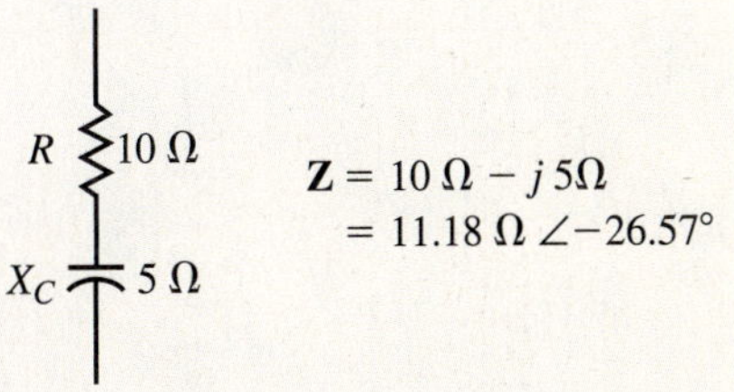

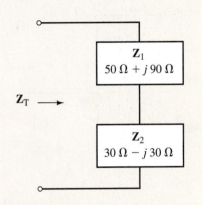

$$\mathbf{Z} = 10\ \Omega - j\,5\Omega$$
$$= 11.18\ \Omega\ \angle-26.57°$$

FIGURE 18–23

Find the total impedance for the network of Figure 18–24. Sketch the impedance diagram showing $\mathbf{Z}_1$, $\mathbf{Z}_2$, and $\mathbf{Z}_T$.

EXAMPLE 18–7

Solution

$$\mathbf{Z}_T = \mathbf{Z}_1 + \mathbf{Z}_2$$
$$= (50\ \Omega + j90\ \Omega) + (30\ \Omega - j30\ \Omega)$$
$$= (80\ \Omega + j60\ \Omega) = 100\ \Omega\angle36.87°$$

The polar forms of the vectors $\mathbf{Z}_1$ and $\mathbf{Z}_2$ are as follows:

$$\mathbf{Z}_1 = 50\ \Omega + j90\ \Omega = 102.96\ \Omega\angle60.95°$$
$$\mathbf{Z}_2 = 30\ \Omega - j30\ \Omega = 42.43\ \Omega\angle-45°$$

The resulting impedance diagram is shown in Figure 18–25.

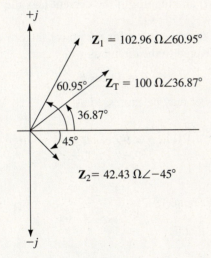

FIGURE 18–25

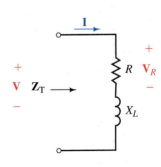

FIGURE 18–24

The phase angle θ for the impedance vector $\mathbf{Z} = Z\angle\theta$ provides the phase angle between the voltage **V** across **Z** and the current **I** through the impedance. For an inductive impedance the voltage will lead the current by θ. If the impedance is capacitive, then the voltage will lag the current by an amount equal to the magnitude of θ.

The phase angle θ is also useful for determining the average power dissipated by the circuit. In the simple series circuit shown in Figure 18–26, we know that only the resistor will dissipate power.

The average power dissipated by the resistor may be determined as follows:

$$P = V_R I = \frac{V_R^2}{R} = I^2R \qquad (18\text{–}6)$$

FIGURE 18–26

Notice that Equation 18–6 uses only the **magnitudes** of the voltage, current, and impedance vectors. *Power is never determined by using phasor products.* Ohm's law provides the magnitude of the current phasor as

$$I = \frac{V}{Z}$$

Substituting this expression into Equation 18–6, we obtain the expression for power as

$$P = \frac{V^2}{Z^2}R = \frac{V^2}{Z}\left(\frac{R}{Z}\right) \tag{18–7}$$

From the impedance diagram of Figure 18–27, we see that

$$\cos\theta = \frac{R}{Z}$$

The previous chapter had defined the power factor as $F_p = \cos\theta$, where θ is the angle between the voltage and current phasors. We now see that for a series circuit, the power factor of the circuit can be determined from the magnitudes of resistance and total impedance.

$$F_p = \cos\theta = \frac{R}{Z} \tag{18–8}$$

The power factor, F_p, is said to be **leading** if the current leads the voltage (capacitive circuit) and **lagging** if the current lags the voltage (inductive circuit).

Now substituting the expression for the power factor into Equation 18–7, we express power delivered to the circuit as

$$P = VI\cos\theta$$

Since $V = IZ$, power may be expressed as

$$P = VI\cos\theta = I^2 Z\cos\theta = \frac{V^2}{Z}\cos\theta \tag{18–9}$$

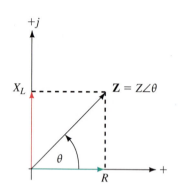

FIGURE 18–27

EXAMPLE 18–8

Refer to the circuit of Figure 18–28.

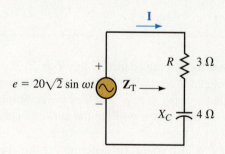

$e = 20\sqrt{2}\sin\omega t$

FIGURE 18–28

a. Find the impedance $\mathbf{Z}_T$.

b. Calculate the power factor of the circuit.

c. Determine **I**.

d. Sketch the phasor diagram for **E** and **I**.

e. Find the average power delivered to the circuit by the voltage source.

f. Calculate the average power dissipated by both the resistor and the capacitor.

Solution

a. $\mathbf{Z_T} = 3\ \Omega - j4\ \Omega = 5\ \Omega \angle -53.13°$

b. $F_p = \cos\theta = 3\ \Omega/5\ \Omega = 0.6$ (leading)

c. The phasor form of the applied voltage is

$$\mathbf{E} = \frac{(\sqrt{2})(20\text{ V})}{\sqrt{2}} \angle 0° = 20\text{ V}\angle 0°$$

which gives a current of

$$\mathbf{I} = \frac{20\text{ V}\angle 0°}{5\ \Omega\angle -53.13°} = 4.0\text{ A}\angle 53.13°$$

d. The phasor diagram is shown in Figure 18–29.

From this phasor diagram, we see that the current phasor for the capacitive circuit leads the voltage phasor by 53.13°.

e. The average power delivered to the circuit by the voltage source is

$$P = (20\text{ V})(4\text{ A})\cos 53.13° = 48.0\text{ W}$$

f. The average power dissipated by the resistor and capacitor will be

$$P_R = (4\text{ A})^2(3\ \Omega)\cos 0° = 48\text{ W}$$
$$P_C = (4\text{ A})^2(4\ \Omega)\cos 90° = 0\text{ W} \quad \text{(as expected!)}$$

Notice that the power factor used in determining the power dissipated by each of the elements is the power factor for that element and not the total power factor for the circuit.

As expected, the summation of powers dissipated by the resistor and capacitor is equal to the total power delivered by the voltage source.

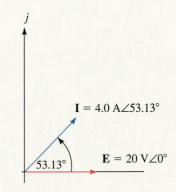

FIGURE 18–29

PRACTICE PROBLEMS 3

A circuit consists of a voltage source $\mathbf{E} = 50\text{ V}\angle 25°$ in series with $L = 20$ mH, $C = 50\ \mu\text{F}$, and $R = 25\ \Omega$. The circuit operates at an angular frequency of 2 krad/s.

a. Determine the current phasor, **I**.

b. Solve for the power factor of the circuit.

c. Calculate the average power dissipated by the circuit and verify that this is equal to the average power delivered by the source.

d. Use Ohm's law to find $\mathbf{V}_R$, $\mathbf{V}_L$, and $\mathbf{V}_C$.

Answers

a. $\mathbf{I} = 1.28\text{ A}\angle -25.19°$

b. $F_p = 0.6402$ lagging

c. $P = 41.0$ W

d. $\mathbf{V}_R = 32.0\text{ V}\angle -25.19°$

$\mathbf{V}_C = 12.8\text{ V}\angle -115.19°$

$\mathbf{V}_L = 51.2\text{ V}\angle 64.81°$

18.3 Kirchhoff's Voltage Law and the Voltage Divider Rule

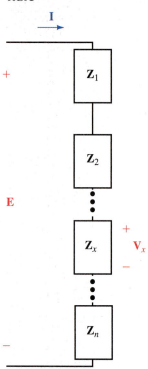

FIGURE 18–30

When a voltage is applied to impedances in series, as shown in Figure 18–30, Ohm's law may be used to determine the voltage across any impedance as

$$\mathbf{V}_x = \mathbf{I}\mathbf{Z}_x$$

The current in the circuit is

$$\mathbf{I} = \frac{\mathbf{E}}{\mathbf{Z}_T}$$

Now, by substitution we arrive at the voltage divider rule for any series combination of elements as

$$\mathbf{V}_x = \frac{\mathbf{Z}_x}{\mathbf{Z}_T}\mathbf{E} \qquad\qquad (18\text{–}10)$$

Equation 18–10 is very similar to the equation for the voltage divider rule in dc circuits. The fundamental differences in solving ac circuits are that we use impedances rather than resistances and that the voltages found are phasors. Because the voltage divider rule involves solving products and quotients of phasors, we generally use the polar form rather than the rectangular form of phasors.

Kirchhoff's voltage law must apply for all circuits whether they are dc or ac circuits. However, because ac circuits have voltages expressed in either sinusoidal or phasor form, Kirchhoff's voltage law for ac circuits may be stated as follows:

The phasor sum of voltage drops and voltage rises around a closed loop is equal to zero.

When adding phasor voltages, we find that the summation is generally done more easily in rectangular form rather than the polar form.

EXAMPLE 18–9

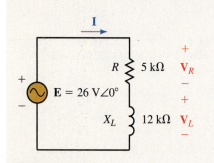

FIGURE 18–31

Consider the circuit of Figure 18–31.

a. Find $\mathbf{Z}_T$.

b. Determine the voltages $\mathbf{V}_R$ and $\mathbf{V}_L$ using the voltage divider rule.

c. Verify Kirchhoff's voltage law around the closed loop.

Solution

a. $\mathbf{Z}_T = 5\text{ k}\Omega + j12\text{ k}\Omega = 13\text{ k}\Omega\angle67.38°$

b. $\mathbf{V}_R = \left(\dfrac{5\text{ k}\Omega\angle0°}{13\text{ k}\Omega\angle67.38°}\right)(26\text{ V}\angle0°) = 10\text{ V}\angle-67.38°$

$\mathbf{V}_L = \left(\dfrac{12\text{ k}\Omega\angle90°}{13\text{ k}\Omega\angle67.38°}\right)(26\text{ V}\angle0°) = 24\text{ V}\angle22.62°$

c. Kirchhoff's voltage law around the closed loop will give

$$26\text{ V}\angle0° - 10\text{ V}\angle-67.38° - 24\text{ V}\angle22.62° = 0$$
$$(26 + j0) - (3.846 - j9.231) - (22.154 + j9.231) = 0$$
$$(26 - 3.846 - 22.154) + j(0 + 9.231 - 9.231) = 0$$
$$0 + j0 = 0$$

Consider the circuit of Figure 18–32:

a. Calculate the sinusoidal voltages v_1 and v_2 using phasors and the voltage divider rule.

b. Sketch the phasor diagram showing **E**, **V**$_1$, and **V**$_2$.

c. Sketch the sinusoidal waveforms of e, v_1, and v_2.

Solution

a. The phasor form of the voltage source is determined as

$$e = 100 \sin \omega t \Leftrightarrow \mathbf{E} = 70.71 \angle \text{V } 0°$$

Applying VDR, we get

$$\mathbf{V}_1 = \left(\frac{40\ \Omega - j80\ \Omega}{(40\ \Omega - j80\ \Omega) + (30\ \Omega + j40\ \Omega)} \right)(70.71\ \text{V}\angle 0°)$$

$$= \left(\frac{89.44\ \Omega \angle -63.43°}{80.62\ \Omega \angle -29.74°} \right)(70.71\ \text{V}\angle 0°)$$

$$= 78.4\ \text{V}\angle -33.69°$$

and

$$\mathbf{V}_2 = \left(\frac{30\ \Omega + j40\ \Omega}{(40\ \Omega - j80\ \Omega) + (30\ \Omega + j40\ \Omega)} \right)(70.71\ \text{V}\angle 0°)$$

$$= \left(\frac{50.00\ \Omega \angle 53.13°}{80.62\ \Omega \angle -29.74°} \right)(70.71\ \text{V}\angle 0°)$$

$$= 43.9\ \text{V}\angle 82.87°$$

The sinusoidal voltages are determined to be

$$v_1 = (\sqrt{2})(78.4)\sin(\omega t - 33.69°)$$

$$= 111 \sin(\omega t - 33.69°)$$

and

$$v_2 = (\sqrt{2})(43.9)\sin(\omega t + 82.87°)$$

$$= 62.0 \sin(\omega t + 82.87°)$$

b. The phasor diagram is shown in Figure 18–33.

c. The corresponding sinusoidal voltages are shown in Figure 18–34.

EXAMPLE 18–10

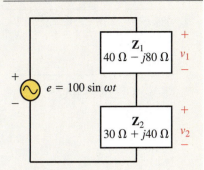

FIGURE 18–32

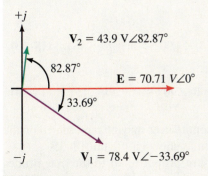

FIGURE 18–33

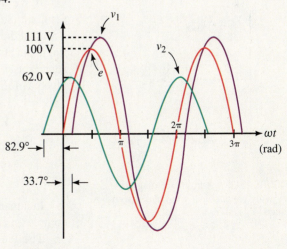

FIGURE 18–34

1. Express Kirchhoff's voltage law as it applies to ac circuits.

2. What is the fundamental difference between how Kirchhoff's voltage law is used in ac circuits as compared with dc circuits?

PRACTICE PROBLEMS 4

A circuit consists of a voltage source $\mathbf{E} = 50\text{ V}\angle 25°$ in series with $L = 20$ mH, $C = 50$ μF, and $R = 25\ \Omega$. The circuit operates at an angular frequency of 2 krad/s.

a. Use the voltage divider rule to determine the voltage across each element in the circuit.

b. Verify that Kirchhoff's voltage law applies for the circuit.

Answers

a. $\mathbf{V}_L = 51.2\text{ V}\angle 64.81°$, $\mathbf{V}_C = 12.8\text{ V}\angle -115.19°$

   $\mathbf{V}_R = 32.0\text{ V}\angle -25.19°$

b. $51.2\text{ V}\angle 64.81° + 12.8\text{ V}\angle -115.19° + 32.0\text{ V}\angle -25.19° = 50\text{ V}\angle 25°$

18.4 AC Parallel Circuits

The **admittance Y** of any impedance is defined as a vector quantity which is the reciprocal of the impedance **Z**.

Mathematically, admittance is expressed as

$$\mathbf{Y}_T = \frac{1}{\mathbf{Z}_T} = \frac{1}{Z_T\angle\theta} = \left(\frac{1}{Z_T}\right)\angle -\theta = Y_T\angle -\theta \quad \text{(S)} \qquad \text{(18-11)}$$

where the unit of admittance is the siemens (S).

In particular, we have seen that the admittance of a resistor R is called conductance and is given the symbol $\mathbf{Y}_R$. If we consider resistance as a vector quantity, then the corresponding vector form of the conductance is

$$\mathbf{Y}_R = \frac{1}{R\angle 0°} = \frac{1}{R}\angle 0° = G\angle 0° = G + j0 \quad \text{(S)} \qquad \text{(18-12)}$$

If we determine the admittance of a purely reactive component X, the resultant admittance is called the **susceptance** of the component and is assigned the symbol B. The unit for susceptance is siemens (S). In order to distinguish between inductive susceptance and capacitive susceptance, we use the subscripts L and C respectively. The vector forms of reactive admittance are given as follows:

$$\mathbf{Y}_L = \frac{1}{X_L\angle 90°} = \frac{1}{X_L}\angle -90° = B_L\angle -90° = 0 - jB_L \quad \text{(S)} \qquad \text{(18-13)}$$

$$\mathbf{Y}_C = \frac{1}{X_C\angle -90°} = \frac{1}{X_C}\angle 90° = B_C\angle 90° = 0 + jB_C \quad \text{(S)} \qquad \text{(18-14)}$$

In a manner similar to impedances, admittances may be represented on the complex plane in an **admittance diagram** as shown in Figure 18–35.

The lengths of the various vectors are proportional to the magnitudes of the corresponding admittances. The resistive admittance vector **G** is

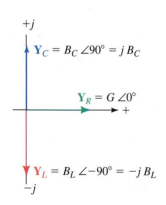

FIGURE 18–35 Admittance diagram showing conductance ($\mathbf{Y}_R$) and susceptance ($\mathbf{Y}_L$ and $\mathbf{Y}_C$).

shown on the positive real axis, whereas the inductive and capacitive admittance vectors $\mathbf{Y}_L$ and $\mathbf{Y}_C$ are shown on the negative and positive imaginary axes respectively.

Determine the admittances of the following impedances. Sketch the corresponding admittance diagram.

a. $R = 10\ \Omega$

b. $X_L = 20\ \Omega$

c. $X_C = 40\ \Omega$

EXAMPLE 18–11

Solutions

a. $\mathbf{Y}_R = \dfrac{1}{R} = \dfrac{1}{10\ \Omega\angle 0°} = 100\ \text{mS}\angle 0°$

b. $\mathbf{Y}_L = \dfrac{1}{X_L} = \dfrac{1}{20\ \Omega\angle 90°} = 50\ \text{mS}\angle -90°$

c. $\mathbf{Y}_C = \dfrac{1}{X_C} = \dfrac{1}{40\ \Omega\angle -90°} = 25\ \text{mS}\angle 90°$

The admittance diagram is shown in Figure 18–36.

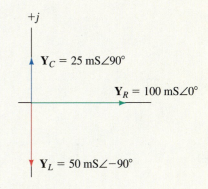

FIGURE 18–36

For any network of n admittances as shown in Figure 18–37, the total admittance is the vector sum of the admittances of the network. Mathematically, the total admittance of a network is given as

$$\mathbf{Y}_T = \mathbf{Y}_1 + \mathbf{Y}_2 + \cdots + \mathbf{Y}_n \quad (\text{S}) \qquad \textbf{(18–15)}$$

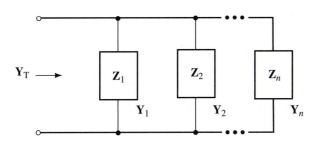

FIGURE 18–37

The resultant impedance of a parallel network of n impedances is determined to be

$$\mathbf{Z}_T = \frac{1}{\mathbf{Y}_T} = \frac{1}{\mathbf{Y}_1 + \mathbf{Y}_2 + \cdots + \mathbf{Y}_n}$$

$$\mathbf{Z}_T = \frac{1}{\dfrac{1}{\mathbf{Z}_1} + \dfrac{1}{\mathbf{Z}_2} + \cdots + \dfrac{1}{\mathbf{Z}_n}} \quad (\Omega) \qquad \textbf{(18–16)}$$

EXAMPLE 18–12

Find the equivalent admittance and impedance of the network of Figure 18–38. Sketch the admittance diagram.

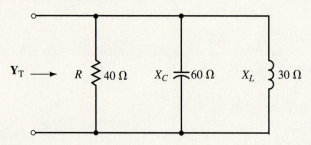

FIGURE 18–38

Solution The admittances of the various parallel elements are

$$\mathbf{Y}_1 = \frac{1}{40\ \Omega\angle 0°} = 25.0\ \text{mS}\angle 0° = 25.0\ \text{mS} + j0$$

$$\mathbf{Y}_2 = \frac{1}{60\ \Omega\angle -90°} = 16.\overline{6}\ \text{mS}\angle 90° = 0 + j16.\overline{6}\ \text{mS}$$

$$\mathbf{Y}_3 = \frac{1}{30\ \Omega\angle 90°} = 33.\overline{3}\ \text{mS}\angle -90° = 0 - j33.\overline{3}\ \text{mS}$$

The total admittance is determined as

$$\mathbf{Y}_T = \mathbf{Y}_1 + \mathbf{Y}_2 + \mathbf{Y}_3$$
$$= 25.0\ \text{mS} + j16.\overline{6}\ \text{mS} + (-j33.\overline{3}\ \text{mS})$$
$$= 25.0\ \text{mS} - j16.\overline{6}\ \text{mS}$$
$$= 30.0\ \text{mS}\angle -33.69°$$

This results in a total impedance for the network of

$$\mathbf{Z}_T = \frac{1}{\mathbf{Y}_T}$$
$$= \frac{1}{30.0\ \text{mS}\angle -33.69°}$$
$$= 33.3\ \Omega\angle 33.69°$$

The admittance diagram is shown in Figure 18–39.

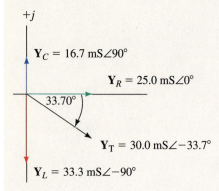

$+j$

$\mathbf{Y}_C = 16.7\ \text{mS}\angle 90°$

$\mathbf{Y}_R = 25.0\ \text{mS}\angle 0°$

$33.70°$

$\mathbf{Y}_T = 30.0\ \text{mS}\angle -33.7°$

$\mathbf{Y}_L = 33.3\ \text{mS}\angle -90°$

FIGURE 18–39

Two Impedances in Parallel

By applying Equation 18–14 for two impedances, we determine the equivalent impedance of two impedances as

$$\mathbf{Z}_T = \frac{\mathbf{Z}_1\mathbf{Z}_2}{\mathbf{Z}_1 + \mathbf{Z}_2} \quad (\Omega) \tag{18–17}$$

From the above expression, we see that for two impedances in parallel, the equivalent impedance is determined as the product of the impedances over the sum. Although the expression for two impedances is very similar to the expression for two resistors in parallel, the difference is that the calculation of impedance involves the use of complex algebra.

Find the total impedance for the network shown in Figure 18–40.

EXAMPLE 18–13

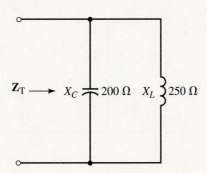

FIGURE 18–40

Solution

$$\mathbf{Z}_T = \frac{(200\ \Omega\angle{-90°})(250\ \Omega\angle90°)}{-j200\ \Omega + j250\ \Omega}$$

$$= \frac{50\ k\Omega\angle0°}{50\angle90°} = 1\ k\Omega\angle{-90°}$$

The previous example illustrates that unlike total parallel resistance, the total impedance of a combination of parallel reactances may be much larger that either of the individual impedances. Indeed, if we are given a parallel combination of equal inductive and capacitive reactances, the total impedance of the combination is equal to infinity (namely an open circuit). Consider the network of Figure 18–41.

The total impedance $\mathbf{Z}_T$ is found as

$$\mathbf{Z}_T = \frac{(X_L\angle90°)(X_C\angle{-90°})}{jX_L - jX_C} = \frac{X^2\angle0°}{0\angle0°} = \infty\angle0°$$

Because the denominator of the above expression is equal to zero, the magnitude of the total impedance will be undefined ($Z = \infty$). The magnitude is undefined and the algebra yields a phase angle $\theta = 0°$, which indicates that the vector lies on the positive real axis of the impedance diagram.

> *Whenever a capacitor and an inductor having equal reactances are placed in parallel, the equivalent circuit of the two components is an open circuit.*

The principle of equal parallel reactances will be studied in a later chapter dealing with "resonance."

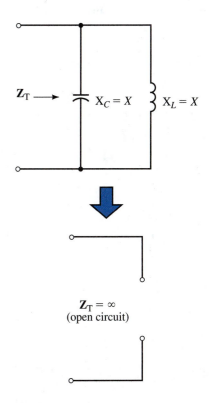

FIGURE 18–41

Three Impedances in Parallel

Equation 18–16 may be solved for three impedances to give the equivalent impedance as

$$\mathbf{Z}_T = \frac{\mathbf{Z}_1\mathbf{Z}_2\mathbf{Z}_3}{\mathbf{Z}_1\mathbf{Z}_2 + \mathbf{Z}_1\mathbf{Z}_3 + \mathbf{Z}_2\mathbf{Z}_3} \quad (\Omega) \qquad \textbf{(18–18)}$$

although this is less useful than the general equation.

EXAMPLE 18–14

Find the equivalent impedance of the network of Figure 18–42.

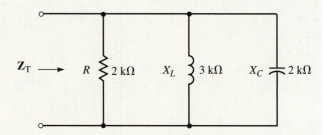

FIGURE 18–42

Solution

$$Z_T = \frac{(2\,\text{k}\Omega\angle 0°)(3\,\text{k}\Omega\angle 90°)(2\,\text{k}\Omega\angle -90°)}{(2\,\text{k}\Omega\angle 0°)(3\,\text{k}\Omega\angle 90°) + (2\,\text{k}\Omega\angle 0°)(2\,\text{k}\Omega\angle -90°) + (3\,\text{k}\Omega\angle 90°)(2\,\text{k}\Omega\angle -90°)}$$

$$= \frac{12 \times 10^9\,\Omega\angle 0°}{6 \times 10^6\angle 90° + 4 \times 10^6\angle -90° + 6 \times 10^6\angle 0°}$$

$$= \frac{12 \times 10^9\,\Omega\angle 0°}{6 \times 10^6 + j2 \times 10^6} = \frac{12 \times 10^9\,\Omega\angle 0°}{6.325 \times 10^6\angle 18.43°}$$

$$= 1.90\,\text{k}\Omega\angle -18.43°$$

And so the equivalent impedance of the network is

$$Z_T = 1.80\,\text{k}\Omega - j0.6\,\text{k}\Omega$$

PRACTICE PROBLEMS 5

A circuit consists of a current source, $i = 0.030 \sin 500t$, in parallel with $L = 20\,\text{mH}$, $C = 50\,\mu\text{F}$, and $R = 25\,\Omega$.
a. Determine the voltage V across the circuit.
b. Solve for the power factor of the circuit.
c. Calculate the average power dissipated by the circuit and verify that this is equal to the power delivered by the source.
d. Use Ohm's law to find the phasor quantities, I_R, I_L, and I_C.

Answers
a. $V = 0.250\,\text{V}\angle 61.93°$
b. $F_p = 0.4705$ lagging
c. $P_R = 41.0\,\text{W} = P_T$
d. $I_R = 9.98\,\text{mA}\angle 61.98°$
    $I_C = 6.24\,\text{mA}\angle 151.93°$
    $I_L = 25.0\,\text{mA}\angle -28.07°$

PRACTICE PROBLEMS 6

A circuit consists of a 2.5-A_{rms} current source connected in parallel with a resistor, an inductor, and a capacitor. The resistor has a value of $10\,\Omega$ and dissipates 40 W of power.
a. Calculate the values of X_L and X_C if $X_L = 3X_C$.
b. Determine the magnitudes of current through the inductor and the capacitor.

Answers
a. $X_L = 80\,\Omega$, $X_C = 26.7\,\Omega$
b. $I_L = 0.25\,\text{mA}$, $I_C = 0.75\,\text{mA}$

The current divider rule for ac circuits has the same form as for dc circuits with the notable exception that currents are expressed as phasors. For a parallel network as shown in Figure 18–43, the current in any branch of the network may be determined using either admittance or impedance.

$$\mathbf{I}_x = \frac{\mathbf{Y}_x}{\mathbf{Y}_T}\mathbf{I} \quad \text{or} \quad \mathbf{I}_x = \frac{\mathbf{Z}_T}{\mathbf{Z}_x}\mathbf{I} \qquad (18\text{–}19)$$

18.5 Kirchhoff's Current Law and the Current Divider Rule

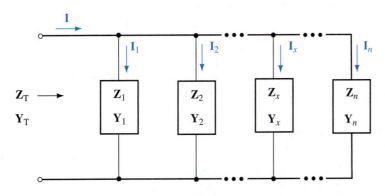

FIGURE 18–43

For two branches in parallel the current in either branch is determined from the impedances as

$$\mathbf{I}_1 = \frac{\mathbf{Z}_2}{\mathbf{Z}_1 + \mathbf{Z}_2}\mathbf{I} \qquad (18\text{–}20)$$

Also, as one would expect, Kirchhoff's current law must apply to any node within an ac circuit. For such circuits, KCL may be stated as follows:

The summation of current phasors entering and leaving a node is equal to zero.

Calculate the current in each of the branches in the network of Figure 18–44.

Solution

$$\mathbf{I}_1 = \left(\frac{250\ \Omega\angle{-90°}}{j200\ \Omega - j250\ \Omega}\right)(2\ A\angle{0°})$$

$$= \left(\frac{250\ \Omega\angle{-90°}}{50\ \Omega\angle{-90°}}\right)(2\ A\angle{0°}) = 10\ A\angle{0°}$$

and

$$\mathbf{I}_2 = \left(\frac{200\ \Omega\angle{90°}}{j200\ \Omega - j250\ \Omega}\right)(2\ A\angle{0°})$$

$$= \left(\frac{200\ \Omega\angle{90°}}{50\ \Omega\angle{-90°}}\right)(2\ A\angle{0°}) = 8\ A\angle{180°}$$

The above results illustrate that the currents in parallel reactive components may be significantly larger than the applied current. If the current through the component exceeds the maximum current rating of the element, severe damage may occur.

EXAMPLE 18–15

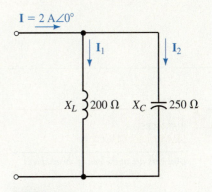

FIGURE 18–44

EXAMPLE 18–16

Refer to the circuit of Figure 18–45:

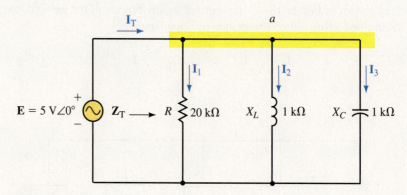

FIGURE 18–45

a. Find the total impedance, $\mathbf{Z}_T$.

b. Determine the supply current, $\mathbf{I}_T$.

c. Calculate $\mathbf{I}_1$, $\mathbf{I}_2$, and $\mathbf{I}_3$ using the current divider rule.

d. Verify Kirchhoff's current law at node a.

Solution

a. Because the inductive and capacitive reactances are in parallel and have the same value, we may replace the combination by an open circuit. Consequently, only the resistor R needs to be considered. As a result

$$\mathbf{Z}_T = 20 \text{ k}\Omega\angle 0°$$

b. $$\mathbf{I}_T = \frac{5 \text{ V}\angle 0°}{20 \text{ k}\Omega\angle 0°} = 250 \text{ }\mu\text{A}\angle 0°$$

c. $$\mathbf{I}_1 = \left(\frac{20 \text{ k}\Omega\angle 0°}{20 \text{ k}\Omega\angle 0°}\right)(250 \text{ }\mu\text{A}\angle 0°) = 250 \text{ }\mu\text{A}\angle 0°$$

$$\mathbf{I}_2 = \left(\frac{20 \text{ k}\Omega\angle 0°}{1 \text{ k}\Omega\angle 90°}\right)(250 \text{ }\mu\text{A}\angle 0°) = 5.0 \text{ mA}\angle -90°$$

$$\mathbf{I}_3 = \left(\frac{20 \text{ k}\Omega\angle 0°}{1 \text{ k}\Omega\angle -90°}\right)(250 \text{ }\mu\text{A}\angle 0°) = 5.0 \text{ mA}\angle 90°$$

d. Notice that the currents through the inductor and capacitor are 180° out of phase. By adding the current phasors in rectangular form, we have

$$\mathbf{I}_T = 250 \text{ }\mu\text{A} - j5.0 \text{ mA} + j5.0 \text{ mA} = 250 \text{ }\mu\text{A} + j0 = 250 \text{ }\mu\text{A}\angle 0°$$

The above result satisfies Kirchhoff's current law at the node.

✓ **IN-PROCESS**
LEARNING CHECK 3

(Answers are at the end of the chapter.)

1. Express Kirchhoff's current law as it applies to ac circuits.

2. What is the fundamental difference between how Kirchhoff's current law is applied to ac circuits as compared with dc circuits?

a. Use the current divider rule to determine current through each branch in the circuit of Figure 18–46.

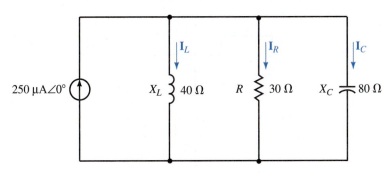

FIGURE 18–46

b. Verify that Kirchhoff's current law applies to the circuit of Figure 18–46.

Answers

a. $\mathbf{I}_L = 176 \ \mu A\angle -69.44°$

  $\mathbf{I}_R = 234 \ \mu A\angle 20.56°$

  $\mathbf{I}_C = 86.8 \ \mu A\angle 110.56°$

b. $\Sigma \mathbf{I}_{out} = \Sigma \mathbf{I}_{in} = 250 \ \mu A$

We may now apply the analysis techniques of series and parallel circuits in solving more complicated circuits. As in dc circuits, the analysis of such circuits is simplified by starting with easily recognized combinations. If necessary, the original circuit may be redrawn to make further simplification more apparent. Regardless of the complexity of the circuits, we find that the fundamental rules and laws of circuit analysis must apply in all cases.

Consider the network of Figure 18–47.

We see that the impedances $\mathbf{Z}_2$ and $\mathbf{Z}_3$ are in series. The branch containing this combination is then seen to be in parallel with the impedance $\mathbf{Z}_1$.

The total impedance of the network is expressed as

$$\mathbf{Z}_T = \mathbf{Z}_1 \parallel (\mathbf{Z}_2 + \mathbf{Z}_3)$$

Solving for $\mathbf{Z}_T$ gives the following:

$$\mathbf{Z}_T = (2 \ \Omega - j8 \ \Omega)\|(2 \ \Omega - j5 \ \Omega + 6 \ \Omega + j7 \ \Omega)$$

$$= (2 \ \Omega - j8 \ \Omega) \parallel (8 \ \Omega + j2 \ \Omega)$$

$$= \frac{(2 \ \Omega - j8 \ \Omega)(8 \ \Omega + j2 \ \Omega)}{2 \ \Omega - j8 \ \Omega + 8 \ \Omega + j2 \ \Omega}$$

$$= \frac{(8.246 \ \Omega\angle -75.96°)(8.246 \ \Omega\angle 14.04°)}{11.66 \ \Omega\angle -30.96°}$$

$$= 5.832 \ \Omega\angle -30.96° = 5.0 \ \Omega - j3.0 \ \Omega$$

18.6 Series-Parallel Circuits

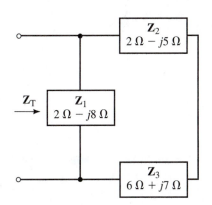

FIGURE 18–47

EXAMPLE 18–17

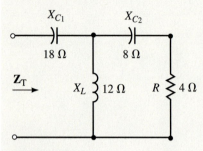

FIGURE 18–48

Determine the total impedance of the network of Figure 18–48. Express the impedance in both polar form and rectangular form.

Solution After redrawing and labelling the given circuit, we have the circuit shown in Figure 18–49.

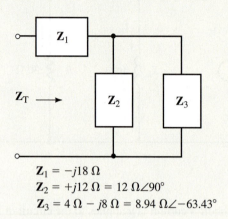

$\mathbf{Z}_1 = -j18\ \Omega$
$\mathbf{Z}_2 = +j12\ \Omega = 12\ \Omega\angle90°$
$\mathbf{Z}_3 = 4\ \Omega - j8\ \Omega = 8.94\ \Omega\angle-63.43°$

FIGURE 18–49

The total impedance is given as

$$\mathbf{Z}_T = \mathbf{Z}_1 + \mathbf{Z}_2 \parallel \mathbf{Z}_3$$

where

$$\mathbf{Z}_1 = -j18\ \Omega = 18\Omega\angle-90°$$
$$\mathbf{Z}_2 = j12\ \Omega = 12\Omega\angle90°$$
$$\mathbf{Z}_3 = 4\ \Omega - j8\ \Omega = 8.94\ \Omega\angle-63.43°$$

We determine the total impedance as

$$\mathbf{Z}_T = -j18\ \Omega + \left[\frac{(12\ \Omega\angle90°)(8.94\ \Omega\angle-63.43°)}{j12\ \Omega + 4\ \Omega - j8\ \Omega}\right]$$

$$= -j18\ \Omega + \left(\frac{107.3\ \Omega\angle26.57°}{5.66\angle45°}\right)$$

$$= -j18\ \Omega + 19.0\ \Omega\angle-18.43°$$

$$= -j18\ \Omega + 18\ \Omega - j6\ \Omega$$

$$= 18\ \Omega - j24\ \Omega = 30\ \Omega\angle-53.13°$$

EXAMPLE 18–18

Consider the circuit of Figure 18–50:

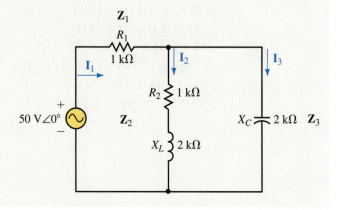

FIGURE 18–50

a. Find $\mathbf{Z}_T$.

b. Determine the currents $\mathbf{I}_1$, $\mathbf{I}_2$, and $\mathbf{I}_3$.

c. Calculate the total power provided by the voltage source.

d. Determine the average powers P_1, P_2, and P_3 dissipated by each of the impedances. Verify that the average power delivered to the circuit is the same as the power dissipated by the impedances.

Solution

a. The total impedance is determined by the combination

$$\mathbf{Z}_T = \mathbf{Z}_1 + \mathbf{Z}_2 \parallel \mathbf{Z}_3$$

For the parallel combination we have

$$\mathbf{Z}_2 \parallel \mathbf{Z}_3 = \frac{(1\text{ k}\Omega + j2\text{ k}\Omega)(-j2\text{ k}\Omega)}{1\text{ k}\Omega + j2\text{ k}\Omega - j2\text{ k}\Omega}$$

$$= \frac{(2.236\text{ k}\Omega\angle 63.43°)(2\text{ k}\Omega\angle -90°)}{1\text{ k}\Omega\angle 0°}$$

$$= 4.472\text{ k}\Omega\angle -26.57° = 4.0\text{ k}\Omega - j2.0\text{ k}\Omega$$

b. And so the total impedance is

$$\mathbf{Z}_T = 5\text{ k}\Omega - j2\text{ k}\Omega = 5.385\text{ k}\Omega\angle -21.80°$$

$$\mathbf{I}_1 = \frac{50\text{ V}\angle 0°}{5.385\text{ k}\Omega\angle -21.80°}$$

$$= 9.285\text{ mA}\angle 21.80°$$

Applying the current divider rule, we get

$$\mathbf{I}_2 = \frac{(2\text{ k}\Omega\angle -90°)(9.285\text{ mA}\angle 21.80°)}{1\text{ k}\Omega + j2\text{ k}\Omega - j2\text{ k}\Omega}$$

$$= 18.57\text{ mA}\angle -68.20°$$

and

$$\mathbf{I}_3 = \frac{(1\text{ k}\Omega + j2\text{ k}\Omega)(9.285\text{ mA}\angle 21.80°)}{1\text{ k}\Omega + j2\text{ k}\Omega - j2\text{ k}\Omega}$$

$$= \frac{(2.236\text{ k}\Omega\angle 63.43°)(9.285\text{ mA}\angle 21.80°)}{1\text{ k}\Omega\angle 0°}$$

$$= 20.761\text{ mA}\angle 85.23°$$

c.
$$P_T = (50\text{ V})(9.285\text{ mA})\cos 21.80°$$

$$= 431.0\text{ mW}$$

d. Because only the resistors will dissipate power, we may use $P = I^2R$:

$$P_1 = (9.285\text{ mA})^2(1\text{ k}\Omega) = 86.2\text{ mW}$$

$$P_2 = (18.57\text{ mA})^2(1\text{ k}\Omega) = 344.8\text{ mW}$$

Alternatively, the power dissipated by $\mathbf{Z}_2$ may have been determined as $P = I^2Z\cos\theta$:

$$P_2 = (18.57\text{ mA})^2(2.236\text{ k}\Omega)\cos 63.43° = 344.8\text{ mW}$$

Since $\mathbf{Z}_3$ is purely capacitive, it will not dissipate any power:

$$P_3 = 0$$

By combining these powers, the total power dissipated is found:

$$P_T = 86.2\text{ mW} + 344.8\text{ mW} + 0 = 431.0\text{ mW} \quad \text{(checks!)}$$

Refer to the circuit of Figure 18–51:

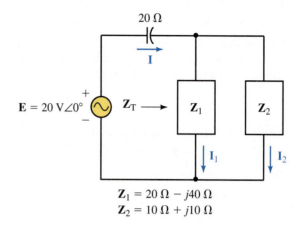

$$\mathbf{Z}_1 = 20\ \Omega - j40\ \Omega$$
$$\mathbf{Z}_2 = 10\ \Omega + j10\ \Omega$$

FIGURE 18–51

a. Calculate the total impedance, $\mathbf{Z}_T$.

b. Find the current $\mathbf{I}$.

c. Use the current divider rule to find $\mathbf{I}_1$ and $\mathbf{I}_2$.

d. Determine the power factor for each impedance, $\mathbf{Z}_1$ and $\mathbf{Z}_2$.

e. Determine the power factor for the circuit.

f. Verify that the total power dissipated by impedances $\mathbf{Z}_1$ and $\mathbf{Z}_2$ is equal to the power delivered by the voltage source.

Answers
a. $\mathbf{Z}_T = 18.9\ \Omega\angle{-45°}$

b. $\mathbf{I} = 1.06\ \text{A}\angle45°$

c. $\mathbf{I}_1 = 0.354\ \text{A}\angle135°$, $\mathbf{I}_2 = 1.12\ \text{A}\angle26.57°$

d. $F_{P(1)} = 0.4472$ leading, $F_{P(2)} = 0.7071$ lagging

e. $F_P = 0.7071$ leading

f. $P_T = 15.0\ \text{W}$, $P_1 = 2.50\ \text{W}$, $P_2 = 12.5\ \text{W}$

$P_1 + P_2 = 15.0\ \text{W} = P_T$

18.7 Frequency Effects

As we have already seen, the reactance of inductors and capacitors depends on frequency. Consequently, the total impedance of any network having reactive elements is also frequency dependent. Any such circuit would need to be analyzed separately at each frequency of interest. We will examine several fairly simple combinations of resistors, capacitors, and inductors to see how the various circuits operate at different frequencies. Some of the more important combinations will be examined in greater detail in later chapters that deal with resonance and filters.

RC Circuits

As the name implies, *RC* circuits consist of a resistor and a capacitor. The components of an *RC* circuit may be connected either in series or in parallel as shown in Figure 18–52.

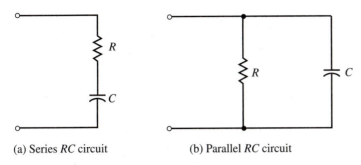

(a) Series *RC* circuit   (b) Parallel *RC* circuit

FIGURE 18–52

Consider the *RC* series circuit of Figure 18–53. Recall that the capacitive reactance, X_C, is given as

$$X_C = \frac{1}{\omega C} = \frac{1}{2\pi f C}$$

The total impedance of the circuit is a vector quantity expressed as

$$\mathbf{Z}_T = R - j\frac{1}{\omega C} = R + \frac{1}{j\omega C}$$

$$\mathbf{Z}_T = \frac{1 + j\omega RC}{j\omega C} \qquad \text{(18–21)}$$

If we define the **cutoff** or **corner frequency** for an *RC* circuit as

$$\omega_c = \frac{1}{RC} = \frac{1}{\tau} \quad \text{(rad/s)} \qquad \text{(18–22)}$$

or equivalently as

$$f_c = \frac{1}{2\pi RC} \quad \text{(Hz)} \qquad \text{(18–23)}$$

then several important points become evident.

For $\omega \le \omega_c/10$ (or $f \le f_c/10$) Equation 18–21 can be expressed as

$$\mathbf{Z}_T \simeq \frac{1 + j0}{j\omega C} = \frac{1}{j\omega C}$$

and for $\omega \ge 10\omega_c$, the expression of (18–21) can be simplified as

$$\mathbf{Z}_T \simeq \frac{0 + j\omega RC}{j\omega C} = R$$

Solving for the magnitude of the impedance at several angular frequencies, we have the results shown in Table 18–1.

If the magnitude of the impedance $\mathbf{Z}_T$ is plotted as a function of angular frequency ω, we get the graph of Figure 18–54. Notice that the abscissa and ordinate of the graph are not scaled linearly, but rather logarithmically. This allows for the the display of results over a wide range of frequencies.

The graph illustrates that the reactance of a capacitor is very high (effectively an open circuit) at low frequencies. Consequently, the total impedance of the series circuit will also be very high at low frequencies. Secondly, we notice that as the frequency increases, the reactance decreases. Therefore, as the frequency gets higher, the capacitive reactance has a diminished effect in the circuit. At very high frequencies (typically for $\omega \ge 10\omega_c$), the impedance of the circuit will effectively be $R = 1\ \text{k}\Omega$.

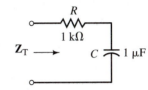

FIGURE 18–53

TABLE 18–1		
Angular Frequency, ω (Rad/s)	$X_C\ (\Omega)$	$\mathbf{Z}_T\ (\Omega)$
0	∞	∞
1	1 M	1 M
10	100 k	100 k
100	10 k	10.05 k
200	5 k	5.099 k
500	2 k	2.236 k
1000	1 k	1.414 k
2000	500	1118
5000	200	1019
10k	100	1005
100 k	10	1000

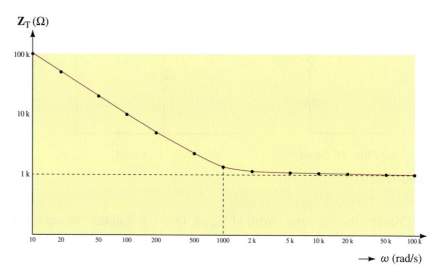

FIGURE 18–54 Impedance versus angular frequency for the network of Figure 18–53.

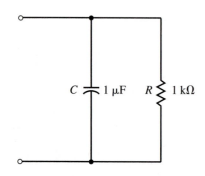

FIGURE 18–55

Consider the parallel RC circuit of Figure 18–55. The total impedance, $\mathbf{Z}_T$, of the circuit is determined as

$$\mathbf{Z}_T = \frac{\mathbf{Z}_R \mathbf{Z}_C}{\mathbf{Z}_R + \mathbf{Z}_C}$$

$$= \frac{R\left(\dfrac{1}{j\omega C}\right)}{R + \dfrac{1}{j\omega C}}$$

$$= \frac{\dfrac{R}{j\omega C}}{\dfrac{1 + j\omega RC}{j\omega C}}$$

which may be simplified as

$$\mathbf{Z}_T = \frac{R}{1 + j\omega RC} \qquad (18\text{--}24)$$

As before, the cutoff frequency is given by Equation 18–22. Now, by examining the expression of (18–24) for $\omega \le \omega_c/10$, we have the following result:

$$\mathbf{Z}_T \simeq \frac{R}{1 + j0} = R$$

For $\omega \ge 10\omega_c$, we have

$$\mathbf{Z}_T \simeq \frac{R}{0 + j\omega RC} = \frac{1}{j\omega C}$$

If we solve for the impedance of the circuit in Figure 18–55 at various angular frequencies, we obtain the results of Table 18–2.

Plotting the magnitude of the impedance $\mathbf{Z}_T$ as a function of angular frequency ω, we get the graph of Figure 18–56. Notice that the abscissa and ordinate of the graph are again scaled logarithmically, allowing for the display of results over a wide range of frequencies.

TABLE 18–2

Angular Frequency, ω (Rad/s)	X_C (Ω)	$\mathbf{Z}_T$ (Ω)
0	∞	1000
1	1 M	1000
10	100 k	1000
100	10 k	995
200	5 k	981
500	2 k	894
1 k	1 k	707
2 k	500	447
5 k	200	196
10 k	100	99.5
100 k	10	10

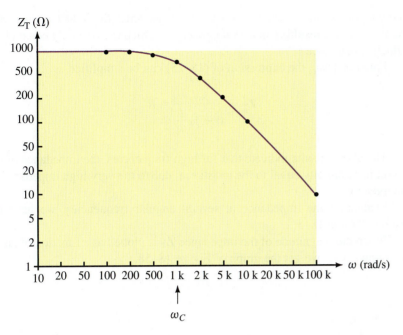

FIGURE 18–56   Impedance versus angular frequency for the network of Figure 18–55.

The results indicate that at dc ($f = 0$ Hz) the capacitor, which behaves as an open circuit, will result in a circuit impedance of $R = 1$ kΩ. As the frequency increases, the capacitor reactance approaches 0 Ω, resulting in a corresponding decrease in circuit impedance.

RL Circuits

RL circuits may be analyzed in a manner similar to the analysis of *RC* circuits. Consider the parallel *RL* circuit of Figure 18–57.

The total impedance of the parallel circuit is found as follows:

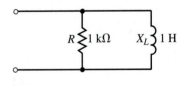

FIGURE 18–57

$$\mathbf{Z}_T = \frac{\mathbf{Z}_R \mathbf{Z}_L}{\mathbf{Z}_R + \mathbf{Z}_L}$$

$$= \frac{R(j\omega L)}{R + j\omega L} \qquad (18\text{–}25)$$

$$\mathbf{Z}_T = \frac{j\omega L}{1 + j\omega \dfrac{L}{R}}$$

If we define the *cutoff* or *corner frequency* for an *RL* circuit as

$$\omega_c = \frac{R}{L} = \frac{1}{\tau} \quad \text{(rad/s)} \qquad (18\text{–}26)$$

or equivalently as

$$f_c = \frac{R}{2\pi L} \quad \text{(Hz)} \qquad (18\text{–}27)$$

then several important points become evident.

For $\omega \leq \omega_c/10$ (or $f \leq f_c/10$) Equation 18–25 can be expressed as

$$\mathbf{Z}_T \simeq \frac{j\omega L}{1 + j0} = j\omega L$$

TABLE 18–3		
Angular Frequency, ω (Rad/s)	X_L (Ω)	Z_T (Ω)
0	0	0
1	1	1
10	10	10
100	100	99.5
200	200	196
500	500	447
1 k	1 k	707
2 k	2 k	894
5 k	5 k	981
10 k	10 k	995
100 k	100 k	1000

The previous result indicates that for low frequencies, the inductor has a very small reactance, resulting in a total impedance which is essentially equal to the inductive reactance.

For $\omega \geq 10\omega_c$, the expression of (18–25) can be simplified as

$$\mathbf{Z}_T \simeq \frac{j\omega L}{0 + j\omega \dfrac{L}{R}} = R$$

The above results indicate that for high frequencies, the impedance of the circuit is essentially equal to the resistance, due to the very high impedance of the inductor.

Evaluating the impedance at several angular frequencies, we have the results of Table 18–3.

When the magnitude of the impedance $\mathbf{Z}_T$ is plotted as a function of angular frequency ω, we get the graph of Figure 18–58.

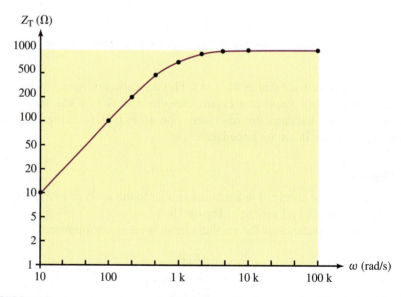

FIGURE 18–58 Impedance versus angular frequency for the network of Figure 18–57.

RLC Circuits

When numerous capacitive and inductive components are combined with resistors in series-parallel circuits, the total impedance $\mathbf{Z}_T$ of the circuit may rise and fall several times over the full range of frequencies. The analysis of such complex circuits is outside the scope of this textbook. However, for illustrative purposes we examine the simple series *RLC* circuit of Figure 18–59.

The impedance $\mathbf{Z}_T$ at any frequency will be determined as

$$\mathbf{Z}_T = R + jX_L - jX_C$$
$$= R + j(X_L - X_C)$$

At very low frequencies, the inductor will appear as a very low impedance (effectively a short circuit), while the capacitor will appear as a very high impedance (effectively an open circuit). Because the capacitive reactance will be much larger than the inductive reactance, the circuit will have a very large capacitive reactance. This results in a very high circuit impedance, $\mathbf{Z}_T$.

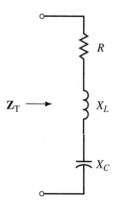

FIGURE 18–59

As the frequency increases, the inductive reactance increases, while the capacitive reactance decreases. At some frequency, f_0, the inductor and the capacitor will have the same magnitude of reactance. At this frequency, the reactances cancel, resulting in a circuit impedance which is equal to the resistance value.

As the frequency increases still further, the inductive reactance becomes larger than the capacitive reactance. The circuit becomes inductive and the magnitude of the total impedance of the circuit again rises. Figure 18-60 shows how the impedance of a series *RLC* circuit varies with frequency.

The complete analysis of the series *RLC* circuit and the parallel *RLC* circuit is left until we examine the principle of resonance in a later chapter.

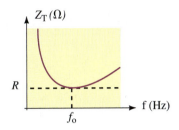

FIGURE 18–60

---

1. For a series network consisting of a resistor and a capacitor, what will be the impedance of the network at a frequency of 0 Hz (dc)? What will be the impedance of the network as the frequency approaches infinity?

2. For a parallel network consisting of a resistor and an inductor, what will be the impedance of the network at a frequency of 0 Hz (dc)? What will be the impedance of the network as the frequency approaches infinity?

IN-PROCESS
LEARNING CHECK 4

(Answers are at the end of the chapter.)

Given the series *RC* network of Figure 18–61, calculate the cutoff frequency in hertz and in radians per second. Sketch the frequency response of Z_T (magnitude) versus angular frequency ω for the network. Show the magnitude Z_T at $\omega_c/10$, ω_c, and $10\omega_c$.

Answers
$\omega_c = 96.7$ rad/s, $f_c = 15.4$ Hz
At 0.1 ω_c: $Z_T = 472$ kΩ, At ω_c: $Z_T = 66.5$ kΩ, At 10 ω_c: $Z_T = 47.2$ kΩ

PRACTICE PROBLEMS 9

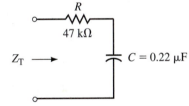

FIGURE 18–61

18.8 Applications

As we have seen, we may determine the impedance of any ac circuit as a vector $\mathbf{Z} = R \pm jX$. This means that any ac circuit may now be simplified as a series circuit having a resistance and a reactance, as shown in Figure 18–62.

Additionally, an ac circuit may be represented as an equivalent parallel circuit consisting of a single resistor and a single reactance as shown in Figure 18–63. *Any equivalent circuit will be valid only at the given frequency of operation.*

We will now examine the technique used to convert any series impedance into its parallel equivalent. Suppose that the two circuits of Figure 18–62 and

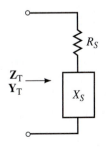

FIGURE 18–62

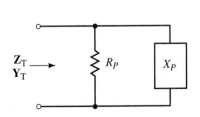

FIGURE 18–63

Figure 18–63 are exactly equivalent at some frequency. These circuits can be equivalent only if both circuits have the same total impedance, $\mathbf{Z}_T$, and the same total admittance, $\mathbf{Y}_T$.

From the circuit of Figure 18–62, the total impedance is written as

$$\mathbf{Z}_T = R_S \pm jX_S$$

Therefore, the total admittance of the circuit is

$$\mathbf{Y}_T = \frac{1}{\mathbf{Z}_T} = \frac{1}{R_S \pm jX_S}$$

Multiplying the numerator and denominator by the complex conjugate, we obtain the following:

$$\mathbf{Y}_T = \frac{R_S \mp jX_S}{(R_S \pm jX_S)(R_S \mp jX_S)}$$

$$= \frac{R_S \mp jX_S}{R_S^2 + X_S^2} \qquad (18\text{–}28)$$

$$\mathbf{Y}_T = \frac{R_S}{R_S^2 + X_S^2} \mp j\frac{X_S}{R_S^2 + X_S^2}$$

Now, from the circuit of Figure 18–63, the total admittance of the parallel circuit may be found from the parallel combination of R_P and X_P as

$$\mathbf{Y}_T = \frac{1}{R_P} + \frac{1}{\pm jX_P}$$

which gives

$$\mathbf{Y}_T = \frac{1}{R_P} \mp j\frac{1}{X_P} \qquad (18\text{–}29)$$

Two vectors can only be equal if both the real components are equal and the imaginary components are equal. Therefore the circuits of Figure 18–62 and Figure 18–63 can only be equivalent if the following conditions are met:

$$R_P = \frac{R_S^2 + X_S^2}{R_S} \qquad (18\text{–}30)$$

and

$$X_P = \frac{R_S^2 + X_S^2}{X_S} \qquad (18\text{–}31)$$

In a similar manner, we have the following conversion from a parallel circuit to an equivalent series circuit:

$$R_S = \frac{R_P X_P^2}{R_P^2 + X_P^2} \qquad (18\text{–}32)$$

and

$$X_S = \frac{R_P^2 X_P}{R_P^2 + X_P^2} \qquad (18\text{–}33)$$

A circuit has a total impedance of $\mathbf{Z}_T = 10 \ \Omega + j50 \ \Omega$. Sketch the equivalent series and parallel circuits.

EXAMPLE 18–19

Solution   The series circuit will be an inductive circuit having $R_S = 10 \ \Omega$ and $X_{LS} = 50 \ \Omega$.

The equivalent parallel circuit will also be an inductive circuit having the following values:

$$R_P = \frac{(10 \ \Omega)^2 + (50 \ \Omega)^2}{10 \ \Omega} = 260 \ \Omega$$

$$X_{LP} = \frac{(10 \ \Omega)^2 + (50 \ \Omega)^2}{50 \ \Omega} = 52 \ \Omega$$

The equivalent series and parallel circuits are shown in Figure 18–64.

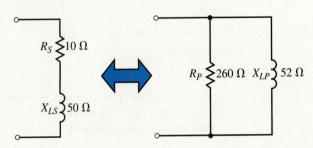

FIGURE 18–64

A circuit has a total admittance of $\mathbf{Y}_T = 0.559 \ \text{mS} \angle 63.43°$. Sketch the equivalent series and parallel circuits.

EXAMPLE 18–20

Solution   Because the admittance is written in polar form, we first convert to the rectangular form of the admittance.

$$G_P = (0.559 \ \text{mS}) \cos 63.43° = 0.250 \ \text{mS} \quad \Leftrightarrow \quad R_P = 4.0 \ \text{k}\Omega$$
$$B_{CP} = (0.559 \ \text{mS}) \sin 63.43° = 0.500 \ \text{mS} \quad \Leftrightarrow \quad X_{CP} = 2.0 \ \text{k}\Omega$$

The equivalent series circuit is found as

$$R_S = \frac{(4 \ \text{k}\Omega)(2 \ \text{k}\Omega)^2}{(4 \ \text{k}\Omega)^2 + (2 \ \text{k}\Omega)^2} = 0.8 \ \text{k}\Omega$$

and

$$X_{CS} = \frac{(4 \ \text{k}\Omega)^2(2 \ \text{k}\Omega)}{(4 \ \text{k}\Omega)^2 + (2 \ \text{k}\Omega)^2} = 1.6 \ \text{k}\Omega$$

The equivalent circuits are shown in Figure 18–65.

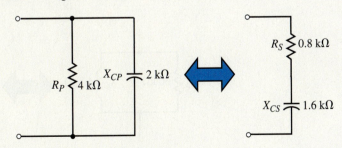

FIGURE 18–65

EXAMPLE 18–21

Refer to the circuit of Figure 18–66.

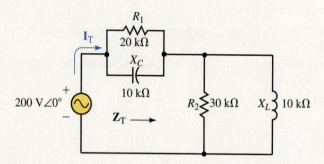

FIGURE 18–66

a. Find $\mathbf{Z}_T$.

b. Sketch the equivalent series circuit.

c. Determine $\mathbf{I}_T$.

Solution

a. The circuit consists of two parallel networks in series. We apply Equations 18–32 and 18–33 to arrive at equivalent series elements for each of the parallel networks as follows:

$$R_{S1} = \frac{(20 \text{ k}\Omega)(10 \text{ k}\Omega)^2}{(20 \text{ k}\Omega)^2 + (10 \text{ k}\Omega)^2} = 4 \text{ k}\Omega$$

$$X_{CS} = \frac{(20 \text{ k}\Omega)^2(10 \text{ k}\Omega)}{(20 \text{ k}\Omega)^2 + (10 \text{ k}\Omega)^2} = 8 \text{ k}\Omega$$

and

$$R_{S2} = \frac{(30 \text{ k}\Omega)(10 \text{ k}\Omega)^2}{(30 \text{ k}\Omega)^2 + (10 \text{ k}\Omega)^2} = 3 \text{ k}\Omega$$

$$X_{LS} = \frac{(30 \text{ k}\Omega)^2(10 \text{ k}\Omega)}{(30 \text{ k}\Omega)^2 + (10 \text{ k}\Omega)^2} = 9 \text{ k}\Omega$$

The equivalent circuits are shown in Figure 18–67.

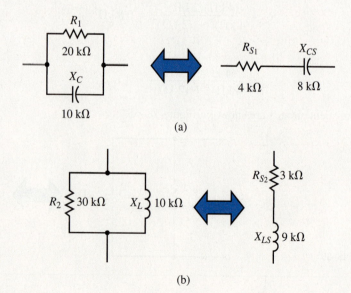

FIGURE 18–67

The total impedance of the circuit is found to be

$$Z_T = (4 \text{ k}\Omega - j8 \text{ k}\Omega) + (3 \text{ k}\Omega + j9 \text{ k}\Omega) = 7 \text{ k}\Omega + j1 \text{ k}\Omega = 7.071 \text{ k}\Omega\angle 8.13°$$

b. Figure 18–68 shows the equivalent series circuit.

c. $I_T = \dfrac{200 \text{ V}\angle 0°}{7.071 \text{ k}\Omega\angle 8.13°} = 28.3 \text{ mA}\angle -8.13°$

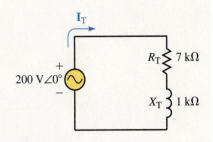

FIGURE 18–68

An inductor of 10 mH has a series resistance of 5 Ω.

a. Determine the parallel equivalent of the inductor at a frequency of 1 kHz. Sketch the equivalent showing the values of L_P (in henries) and R_P.

b. Determine the parallel equivalent of the inductor at a frequency of 1 MHz. Sketch the equivalent showing the values of L_P (in henries) and R_P.

c. If the frequency were increased still further, predict what would happen to the values of L_P and R_P.

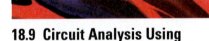

IN-PROCESS
LEARNING CHECK 5

(Answers are at the end of the chapter.)

A network has an impedance of $Z_T = 50 \text{ k}\Omega\angle 75°$ at a frequency of 5 kHz.

a. Determine the most simple equivalent series circuit (L and R).

b. Determine the most simple equivalent parallel circuit.

PRACTICE PROBLEMS 10

Answers
a. $R_S = 12.9 \text{ k}\Omega$, $L_S = 1.54$ H
b. $R_P = 193 \text{ k}\Omega$, $L_P = 1.65$ H

MultiSIM

In this section we will use MultiSIM to simulate how sinusoidal ac measurements are taken with an oscilloscope. The "measurements" are then interpreted to verify the ac operation of circuits. You will use some of the display features of the software to simplify your work. The following example provides a guide through each step of the procedure.

18.9 Circuit Analysis Using Computers

◀ MULTISIM   ◀ CADENCE

Given the circuit of Figure 18–69.

EXAMPLE 18–22

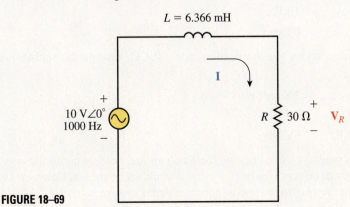

FIGURE 18–69

a. Determine the current **I** and the voltage $\mathbf{V}_R$.

b. Use MultiSIM to display the resistor voltage v_R and the source voltage e. Use the results to verify the results of part (a).

Solution

a. $X_L = 2\pi(1000 \text{ Hz})(6.366 \times 10^{-3} \text{ H}) = 40 \ \Omega$

$\mathbf{Z} = 30 \ \Omega + j40 \ \Omega = 50 \ \Omega\angle53.13°$

$\mathbf{I} = \dfrac{10 \text{ V}\angle0°}{50 \ \Omega\angle53.13°} = 0.200 \text{ A}\angle-53.13°$

$\mathbf{V}_R = (0.200 \text{ A}\angle-53.13°)(30 \ \Omega) = 6.00 \text{ V}\angle-53.13°$

b. Use the schematic editor to input the circuit shown in Figure 18–70.

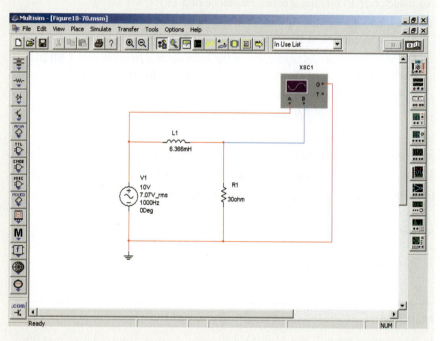

◀ MULTISIM   **FIGURE 18–70**

The ac voltage source is obtained from the Sources parts bin. The properties of the voltage source are changed by double clicking on the symbol and then selecting the Value tab. Change the values as follows:

Voltage RMS:	**10 V**
Frequency:	**1 kHz**
Phase:	**0 Deg**

After double-clicking on the oscilloscope (XSC1), change the settings as follows:

Time base:	**200μs/Div**
Channel A:	**5V/Div**
Channel B:	**5V/Div**

At this point, you can click on the Run tool and observe an oscilloscope display that will be very similar to what you would see in the lab. However, we can refine the display to provide information that is even more useful.

Click on the View menu and select Show Grapher (or simply strike Ctrl G). You will likely observe many cycles in the display window. In order to obtain a more useful display, it is necessary to change some of the display properties.

In the Analysis Graphs window, click on the Edit menu and select Properties. Click on the Bottom Axis tab and adjust the range to provide a display of one cycle ($T = 1.00$ ms). A suitable range will be from 0.200 to 0.201 s, although any other 1-ms display will be equally useful.

Since we would like to be able to measure the phase angle, it is necessary to provide a grid and cursors on the graph. This is done by clicking on both the Show/Hide Grid tool and on the Cursor tool. After moving the cursors, it is possible to determine the phase relationship between $\mathbf{V}_R$ and $\mathbf{E}$ directly from the table of data. The display will appear similar to that shown in Figure 18-71.

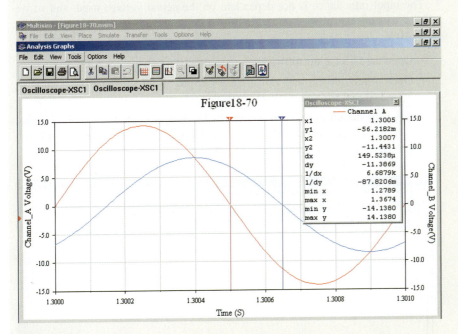

FIGURE 18–71

After positioning the cursors and using the display window, we are able to obtain various measurements for the circuit. The phase angle of the resistor voltage with respect to the source voltage is found by using the difference between the cursors in the bottom axis, $dx = 149.52$ μs. Now we have

$$\theta = \frac{149.52 \text{ μs}}{1000 \text{ μs}} \times 360° = 53.83°$$

The amplitude of the resistor voltage is 8.49 V, which results in an rms value of 6.00 V. As a result of the measurements we have

$$\mathbf{V}_R = 6.00 \text{ V}\angle{-53.83°}$$

and

$$\mathbf{I} = \frac{6 \text{ V}\angle{-53.83°}}{30 \text{ Ω}} = 0.200 \text{ A}\angle{-53.83°}$$

These values correspond very closely to the theoretical results calculated in part (a) of the example.

PSpice

In the following example we will use the Probe postprocessor of PSpice to show how the impedance of an *RC* circuit changes as a function of frequency. The Probe output will provide a graphical result that is very similar to the frequency responses determined in previous sections of this chapter.

EXAMPLE 18–23

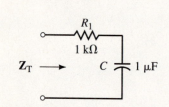

FIGURE 18–72

Refer to the network of Figure 18–72. Use the PSpice to input the circuit. Run the Probe postprocessor to provide a graphical display of network impedance as a function of frequency from 50 Hz to 500 Hz.

Solution Since PSpice is unable to analyze an incomplete circuit, it is necessary to provide a voltage source (and ground) for the circuit of Figure 18–72. The input impedance is not dependent on the actual voltage used, and so we may use any ac voltage source. In this example we arbitrarily select a voltage of 10 V.

- Open the CIS Demo software.
- Open a new project and call it Ch 18 PSpice 1. Ensure that the Analog or Mixed-Signal Circuit Wizard is activated.
- Enter the circuit as show in Figure 18–73. Simply click on the voltage value and change its value to **10V** from the default value of **0V**. (There must be no spaces between the magnitude and the units.)

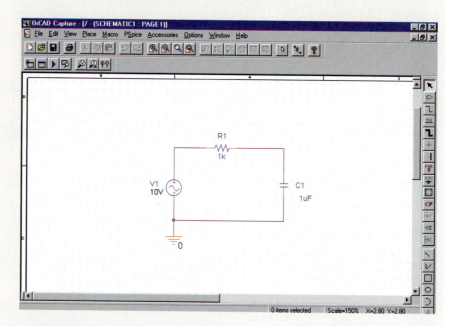

FIGURE 18–73

- Click on PSpice, New Simulation Profile and give a name such as Example 18–23 to the new simulation. The Simulation Settings dialog box will open.
- Click on the Analysis tab and select AC Sweep/Noise as the Analysis type. Select General Settings from the Options box.
- Select Linear AC Sweep Type (Quite often, logarithmic frequency sweeps such as Decade, are used). Enter the following values into the appropriate dialog boxes. Start Frequency; **50,** End Frequency: **500,** Total Points: **1001.** Click OK.
- Click on PSpice and Run. The Probe postprocessor will appear on the screen.

- Click on <u>T</u>race and <u>A</u>dd Trace. You may simply click on the appropriate values from the list of variables and use the division symbol to result in impedance. Enter the following expression into the Trace Expression box:

$$\textbf{V(R1:1)/I(R1)}$$

Notice that the above expression is nothing other than an application of Ohm's Law. The resulting display is shown in Figure 18–74.

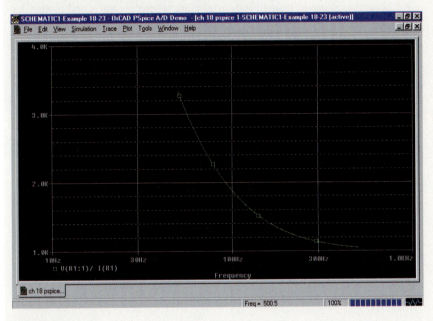

FIGURE 18–74

Given the circuit of Figure 18–75.

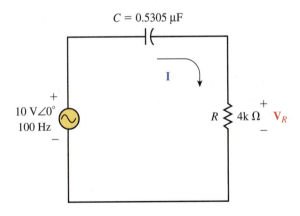

$C = 0.5305\ \mu F$

I

$10\ V\angle0°$
$100\ Hz$

$R\ \lessgtr\ 4k\ \Omega$ $+$ $\mathbf{V}_R$ $-$

FIGURE 18–75

MULTISIM

a. Determine the current $\mathbf{I}$ and the voltage $\mathbf{V}_R$.
b. Use MultiSIM to display the resistor voltage v_R and the source voltage, e. Use the results to verify the results of part (a).

Answers
a. $\mathbf{I} = 2.00\ mA\angle36.87°,\ \ \mathbf{V}_R = 8.00\ V\angle36.87°$
b. $v_R = 11.3\ \sin(\omega t + 36.87°),\ \ e = 14.1\ \sin \omega t$

PRACTICE PROBLEMS 12

◀ **CADENCE**

Use PSpice to input the circuit of Figure 18–55. Use the Probe postprocessor to obtain a graphical display of the network impedance as a function of frequency from 100 Hz to 2000 Hz.

PUTTING IT INTO PRACTICE

You are working in a small industrial plant where several small motors are powered by a 60-Hz ac line voltage of 120 Vac. Your supervisor tells you that one of the 2-Hp motors that was recently installed draws too much current when the motor is under full load. You take a current reading and find that the current is 14.4 A. After doing some calculations, you determine that even if the motor is under full load, it shouldn't require that much current.

However, you have an idea. You remember that a motor can be represented as a resistor in series with an inductor. If you could reduce the effect of the inductive reactance of the motor by placing a capacitor across the motor, you should be able to reduce the current since the capacitive reactance will cancel the inductive reactance.

While keeping the motor under load, you place a capacitor into the circuit. Just as you suspected, the current goes down. After using several different values, you observe that the current goes to a minimum of 12.4 A. It is at this value that you have determined that the reactive impedances are exactly balanced. Sketch the complete circuit and determine the value of capacitance that was added into the circuit. Use the information to determine the value of the motor's inductance. (Assume that the motor has an efficiency of 100%.)

PROBLEMS

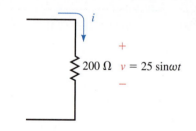

FIGURE 18–76

18.1   Ohm's Law for AC Circuits

1. For the resistor shown in Figure 18–76:
   a. Find the sinusoidal current i using phasors.
   b. Sketch the sinusoidal waveforms for v and i.
   c. Sketch the phasor diagram for **V** and **I**.

2. Repeat Problem 1 for the resistor of Figure 18–77.

3. Repeat Problem 1 for the resistor of Figure 18–78.

4. Repeat Problem 1 for the resistor of Figure 18–79.

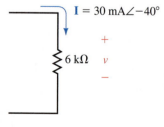

FIGURE 18–77

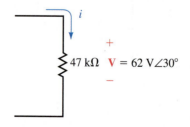

FIGURE 18–78

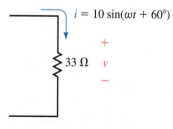

FIGURE 18–79

5. For the component shown in Figure 18–80:
   a. Find the sinusoidal voltage *v* using phasors.
   b. Sketch the sinusoidal waveforms for *v* and *i*.
   c. Sketch the phasor diagram for **V** and **I**.
6. For the component shown in Figure 18–81:
   a. Find the sinusoidal current *i* using phasors.
   b. Sketch the sinusoidal voltages for *v* and *i*.
   c. Sketch the phasor diagram for **V** and **I**.
7. For the component shown in Figure 18–82:
   a. Find the sinusoidal voltage *v* using phasors.
   b. Sketch the sinusoidal voltages for *v* and *i*.
   c. Sketch the phasor diagram for **V** and **I**.
8. For the component shown in Figure 18–83:
   a. Find the sinusoidal current *i* using phasors.
   b. Sketch the sinusoidal voltages for *v* and *i*.
   c. Sketch the phasor diagram for **V** and **I**.

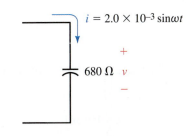

FIGURE 18–80

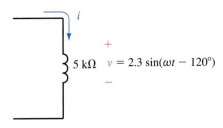

FIGURE 18–81

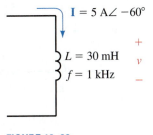

FIGURE 18–82

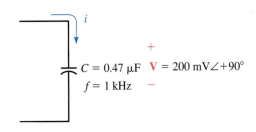

FIGURE 18–83

9. For the component shown in Figure 18–84:
   a. Find the sinusoidal voltage *v* using phasors.
   b. Sketch the sinusoidal voltages for *v* and *i*.
   c. Sketch the phasor diagram for **V** and **I**.
10. Repeat Problem 5 for the component shown in Figure 18–85.

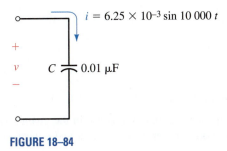

FIGURE 18–84

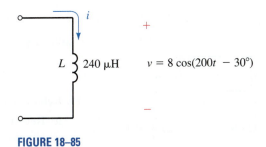

FIGURE 18–85

11. Repeat Problem 5 for the component shown in Figure 18–86.

12. Repeat Problem 5 for the component shown in Figure 18–87.

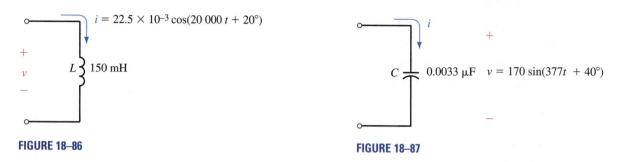

FIGURE 18–86   FIGURE 18–87

18.2   AC Series Circuits

13. Find the total impedance of each of the networks shown in Figure 18–88.

14. Repeat Problem 13 for the networks of Figure 18–89.

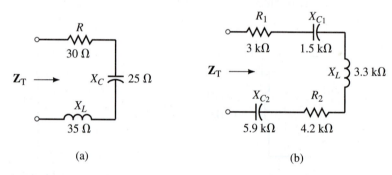

(a)   (b)

FIGURE 18–88

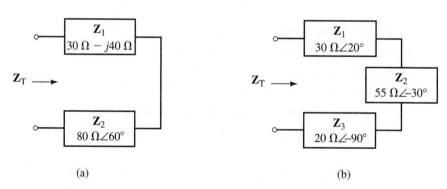

(a)   (b)

FIGURE 18–89

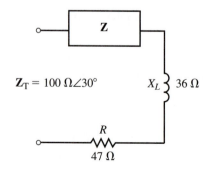

FIGURE 18–90

15. Refer to the network of Figure 18–90.

   a. Determine the series impedance $\mathbf{Z}$ that will result in the given total impedance, $\mathbf{Z}_T$. Express your answer in rectangular and polar form.

   b. Sketch an impedance diagram showing $\mathbf{Z}_T$ and $\mathbf{Z}$.

16. Repeat Problem 15 for the network of Figure 18–91.

17. A circuit consisting of two elements has a total impedance of $\mathbf{Z}_T = 2\ \text{k}\Omega\angle15°$ at a frequency of 18 kHz. Determine the values in ohms, henries, or farads of the unknown elements.

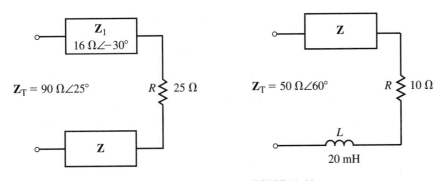

FIGURE 18–91   **FIGURE 18–92**

18. A network has a total impedance of $Z_T = 24.0\ k\Omega\angle-30°$ at a frequency of 2 kHz. If the network consists of two series elements, determine the values in ohms, henries, or farads of the unknown elements.

19. Given that the network of Figure 18–92 is to operate at a frequency of 1 kHz, what series components R and L (in henries) or C (in farads) must be in the indicated block to result in a total circuit impedance of $Z_T = 50\ \Omega\angle60°$?

20. Repeat Problem 19 for a frequency of 2 kHz.

21. Refer to the circuit of Figure 18–93.

    a. Find Z_T, I, V_R, V_L, and V_C.

    b. Sketch the phasor diagram showing I, V_R, V_L, and V_C.

    c. Determine the average power dissipated by the resistor.

    d. Calculate the average power delivered by the voltage source. Compare the result to (c).

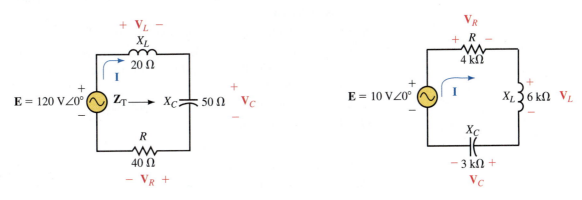

        FIGURE 18–93   **FIGURE 18–94**

22. Consider the circuit of Figure 18–94.

    a. Find Z_T, I, V_R, V_L, and V_C.

    b. Sketch the phasor diagram showing I, V_R, V_L, and V_C.

    c. Write the sinusoidal expressions for the current i and the voltages e, v_R, v_C, and v_L.

    d. Sketch the sinusoidal current and voltages found in (c).

    e. Determine the average power dissipated by the resistor.

    f. Calculate the average power delivered by the voltage source. Compare the result to (e).

23. Refer to the circuit of Figure 18–95.

    a. Determine the circuit impedance, Z_T.

    b. Use phasors to solve for i, v_R, v_C, and v_L.

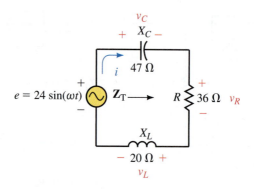

FIGURE 18–95

c. Sketch the phasor diagram showing $\mathbf{I}$, $\mathbf{V}_R$, $\mathbf{V}_L$, and $\mathbf{V}_C$.

d. Sketch the sinusoidal expressions for the current and voltages found in (b).

e. Determine the average power dissipated by the resistor.

f. Calculate the average power delivered by the voltage source. Compare the result to (e).

24. Refer to the circuit of Figure 18–96.

 a. Determine the value of the capacitor reactance, X_C, needed so that the resistor in the circuit dissipates a power of 200 mW.

 b. Using the value of X_C from (a), determine the sinusoidal expression for the current i in the circuit.

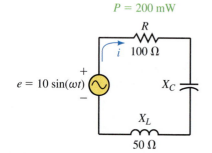

FIGURE 18–96

18.3 Kirchhoff's Voltage Law and the Voltage Divider Rule

25. a. Suppose a voltage of $10\ \mathrm{V}\angle0°$ is applied across the network in Figure 18–88a. Use the voltage divider rule to find the voltage appearing across each element.

 b. Verify Kirchhoff's voltage law for the network.

26. a. Suppose a voltage of $240\ \mathrm{V}\angle30°$ is applied across the network in Figure 18–89a. Use the voltage divider rule to find the voltage appearing across each impedance.

 b. Verify Kirchhoff's voltage law for each network.

27. Given the circuit of Figure 18–97:

 a. Find the voltages $\mathbf{V}_C$ and $\mathbf{V}_L$.

 b. Determine the value of R.

28. Refer to the circuit of Figure 18–98.

 a. Find the voltages $\mathbf{V}_R$ and $\mathbf{V}_L$.

 b. Determine the value of X_C.

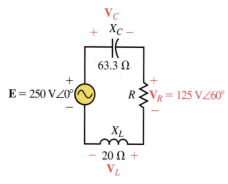

FIGURE 18–97

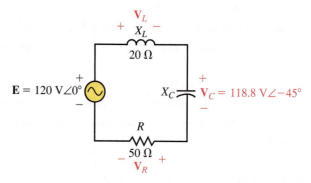

FIGURE 18–98

29. Refer to the circuit of Figure 18–99:

    a. Find the voltage across X_C.

    b. Use Kirchhoff's voltage law to find the voltage across the unknown impedance.

    c. Calculate the value of the unknown impedance **Z**.

    d. Determine the average power dissipated by the circuit.

30. Given that the circuit of Figure 18–100 has a current with a magnitude of 2.0 A and dissipates a total power of 500 W:

    a. Calculate the value of the unknown impedance **Z**. (Hint: Two solutions are possible.)

    b. Calculate the phase angle θ of the current **I**.

    c. Find the voltages $\mathbf{V}_R$, $\mathbf{V}_L$, and $\mathbf{V}_Z$.

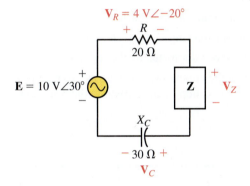

FIGURE 18–99

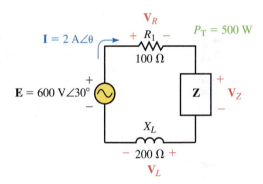

FIGURE 18–100

18.4 AC Parallel Circuits

31. Determine the input impedance, $\mathbf{Z}_T$, for each of the networks of Figure 18–101.

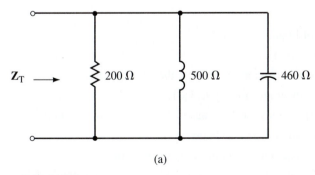

(a)

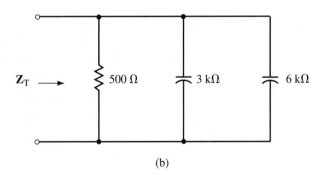

(b)

FIGURE 18–101

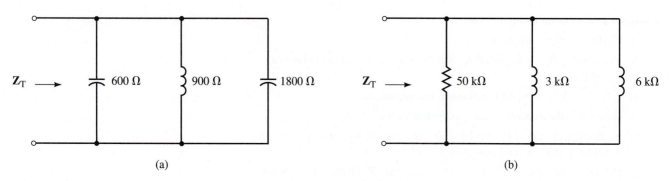

(a)   (b)

FIGURE 18–102

32. Repeat Problem 31 for Figure 18–102.
33. Given the circuit of Figure 18–103.
    a. Find $\mathbf{Z}_T$, $\mathbf{I}_T$, $\mathbf{I}_1$, $\mathbf{I}_2$, and $\mathbf{I}_3$.
    b. Sketch the admittance diagram showing each of the admittances.
    c. Sketch the phasor diagram showing $\mathbf{E}$, $\mathbf{I}_T$, $\mathbf{I}_1$, $\mathbf{I}_2$, and $\mathbf{I}_3$.
    d. Determine the average power dissipated by the resistor.
    e. Find the power factor of the circuit and calculate the average power delivered by the voltage source. Compare the answer to the result obtained in (d).

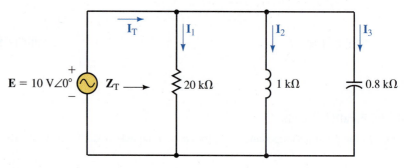

FIGURE 18–103

34. Refer to the circuit of Figure 18–104.
    a. Find $\mathbf{Z}_T$, $\mathbf{I}_T$, $\mathbf{I}_1$, $\mathbf{I}_2$, and $\mathbf{I}_3$.
    b. Sketch the admittance diagram for each of the admittances.
    c. Sketch the phasor diagram showing $\mathbf{E}$, $\mathbf{I}_T$, $\mathbf{I}_1$, $\mathbf{I}_2$, and $\mathbf{I}_3$.
    d. Determine the expressions for the sinusoidal currents i_T, i_1, i_2, and i_3.
    e. Sketch the sinusoidal voltage e and current i_T.
    f. Determine the average power dissipated by the resistor.
    g. Find the power factor of the circuit and calculate the average power delivered by the voltage source. Compare the answer to the result obtained in (f).

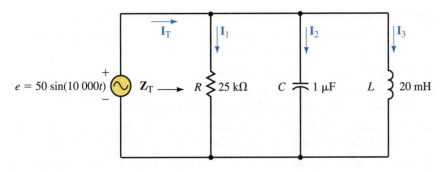

FIGURE 18–104

35. Refer to the network of Figure 18–105.
    a. Determine $\mathbf{Z}_T$.
    b. Given the indicated current, use Ohm's law to find the voltage, **V**, across the network.

36. Consider the network of Figure 18–106.
    a. Determine $\mathbf{Z}_T$.
    b. Given the indicated current, use Ohm's law to find the voltage, **V**, across the network.
    c. Solve for $\mathbf{I}_2$ and **I**.

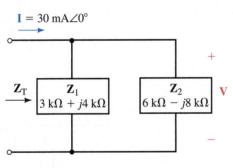

FIGURE 18–105

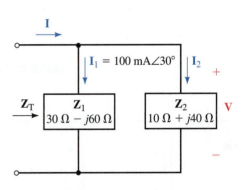

FIGURE 18–106

37. Determine the impedance, $\mathbf{Z}_2$, that will result in the total impedance shown in Figure 18–107.

38. Determine the impedance, $\mathbf{Z}_2$, that will result in the total impedance shown in Figure 18–108.

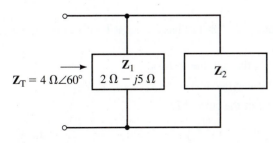

FIGURE 18–107

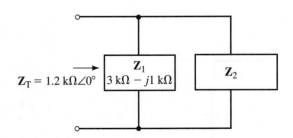

FIGURE 18–108

18.5 Kirchhoff's Current Law and the Current Divider Rule

39. Solve for the current in each element of the networks in Figure 18–101 if the current applied to each network is 10 mA$\angle-30°$.

40. Repeat Problem 39 for Figure 18–102.

41. Use the current divider rule to find the current in each of the elements in Figure 18–109. Verify that Kirchhoff's current law applies.

42. Given that $I_L = 4\ A\angle30°$ in the circuit of Figure 18–110, find the currents **I**, I_C, and I_R. Verify that Kirchhoff's current law appplies for this circuit.

43. Suppose that the circuit of Figure 18–111 has a current **I** with a magnitude of 8 A:

   a. Determine the current I_R through the resistor.

   b. Calculate the value of resistance, R.

   c. What is the phase angle of the current **I**?

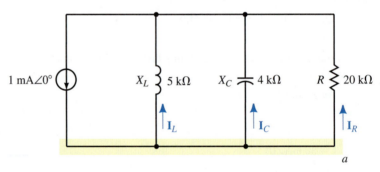

FIGURE 18–109

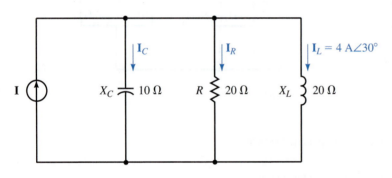

FIGURE 18–110

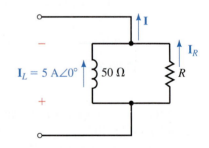

FIGURE 18–111

44. Assume that the circuit of Figure 18–112 has a current **I** with a magnitude of 3 A:

   a. Determine the current I_R through the resistor.

   b. Calculate the value of capacitive reactance X_C.

   c. What is the phase angle of the current **I**?

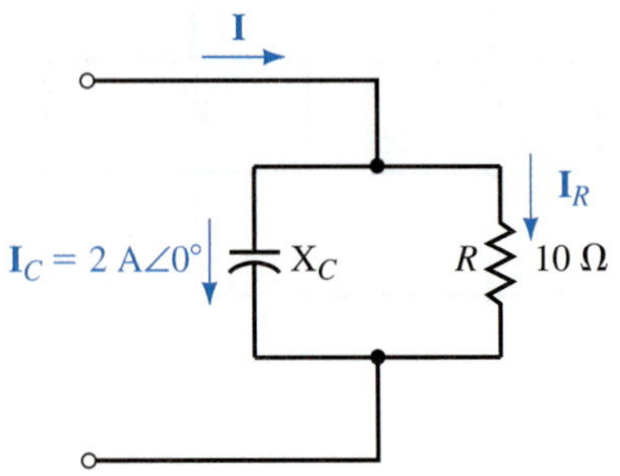

18.6 Series-Parallel Circuits

45. Refer to the circuit of Figure 18–113.

   a. Find Z_T, I_L, I_C, and I_R.

   b. Sketch the phasor diagram showing **E**, I_L, I_C, and I_R.

FIGURE 18–112

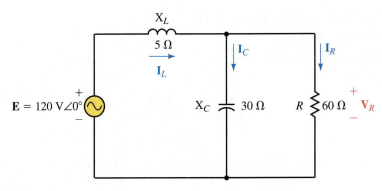

FIGURE 18–113

c. Calculate the average power dissipated by the resistor.

d. Use the circuit power factor to calculate the average power delivered by the voltage source. Compare the answer to the results obtained in (c).

46. Refer to the circuit of Figure 18–114.

    a. Find $\mathbf{Z_T}$, $\mathbf{I_1}$, $\mathbf{I_2}$, and $\mathbf{I_3}$.

    b. Sketch the phasor diagram showing $\mathbf{E}$, $\mathbf{I_1}$, $\mathbf{I_2}$, and $\mathbf{I_3}$.

    c. Calculate the average power dissipated by each of the resistors.

    d. Use the circuit power factor to calculate the average power delivered by the voltage source. Compare the answer to the results obtained in (c).

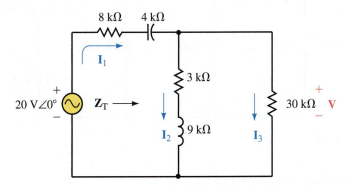

FIGURE 18–114

47. Refer to the circuit of Figure 18–115.

    a. Find $\mathbf{Z_T}$, $\mathbf{I_T}$, $\mathbf{I_1}$, and $\mathbf{I_2}$.

    b. Determine the voltage $\mathbf{V}_{ab}$.

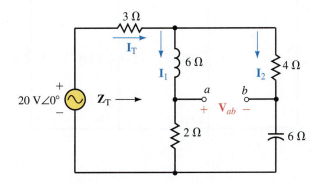

FIGURE 18–115

48. Consider the circuit of Figure 18–116.
   a. Find $\mathbf{Z}_T$, $\mathbf{I}_T$, $\mathbf{I}_1$, and $\mathbf{I}_2$.
   b. Determine the voltage **V**.

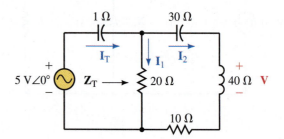

FIGURE 18–116

49. Refer to the circuit of Figure 18–117:
   a. Find $\mathbf{Z}_T$, $\mathbf{I}_1$, $\mathbf{I}_2$, and $\mathbf{I}_3$.
   b. Determine the voltage **V**.

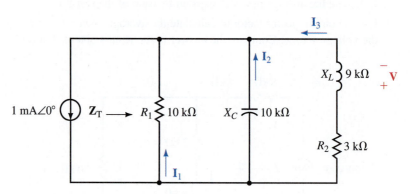

FIGURE 18–117

50. Refer to the circuit of Figure 18–118:
   a. Find $\mathbf{Z}_T$, $\mathbf{I}_1$, $\mathbf{I}_2$, and $\mathbf{I}_3$.
   b. Determine the voltage **V**.

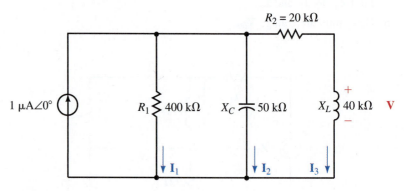

FIGURE 18–118

18.7   Frequency Effects

51. A 50-kΩ resistor is placed in series with a 0.01-μF capacitor. Determine the cutoff frequency ω_C (in rad/s) and sketch the frequency response (Z_T vs. ω) of the network.

52. A 2-mH inductor is placed in parallel with a 2-kΩ resistor. Determine the cutoff frequency ω_C (in rad/s) and sketch the frequency response (Z_T vs. ω) of the network.

53. A 100-kΩ resistor is placed in parallel with a 0.47-μF capacitor. Determine the cutoff frequency f_C (in Hz) and sketch the frequency response (Z_T vs. f) of the network.

54. A 2.7-kΩ resistor is placed in parallel with a 20-mH inductor. Determine the cutoff frequency f_C (in Hz) and sketch the frequency response (Z_T vs. f) of the network.

18.8   Applications

55. Convert each each of the networks of Figure 18–119 into an equivalent series network consisting of two elements.

56. Convert each each of the networks of Figure 18–119 into an equivalent parallel network consisting of two elements.

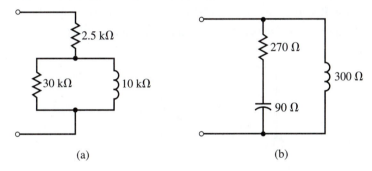

(a)   (b)

FIGURE 18–119

57. Show that the networks of Figure 18–120 have the same input impedance at frequencies of 1 krad/s and 10 krad/s. (It can be shown that these networks are equivalent at all frequencies.)

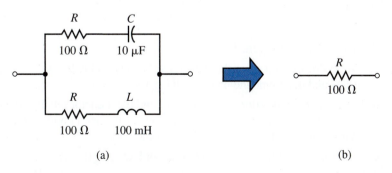

(a)   (b)

FIGURE 18–120

58. Show that the networks of Figure 18–121 have the same input impedance at frequencies of 5 rad/s and 10 rad/s.

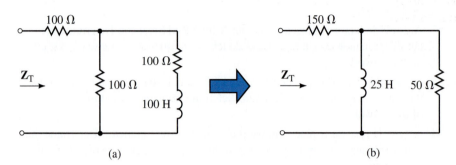

(a)   (b)

FIGURE 18–121

18.9 Circuit Analysis Using Computers

◀ MULTISIM

59. Given the circuit of Figure 18–122:

  a. Use MultiSIM to simultaneously display the capacitor voltage v_C and e. Record your results and determine the phasor voltage $\mathbf{V}_C$.

  b. Interchange the positions of the resistor and the capacitor relative to the ground. Use MultiSIM to simultaneously display the resistor voltage v_R and e. Record your results and determine the phasor voltage $\mathbf{V}_R$.

  c. Compare your results to those obtained in Example 18–8.

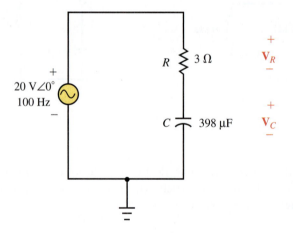

◀ MULTISIM

FIGURE 18–122

◀ MULTISIM

60. Given the circuit of Figure 18–123:

  a. Use MultiSIM to display the inductor voltage v_L and e. Record your results and determine the phasor voltage $\mathbf{V}_L$.

  b. Interchange the positions of the resistor and the inductor relative to the ground. Use MultiSIM to display the inductor voltage v_R and e. Record your results and determine the phasor voltage $\mathbf{V}_R$.

  c. Compare your results to those obtained in Example 18–9.

◀ CADENCE

61. A 50-kΩ resistor is placed in series with a 0.01-μF capacitor. Use PSpice to input these components into a circuit. Run the Probe postprocessor to provide a graphical display of the network impedance as a function of frequency from 50 Hz to 500 Hz. Let the frequency sweep logarithmically in octaves.

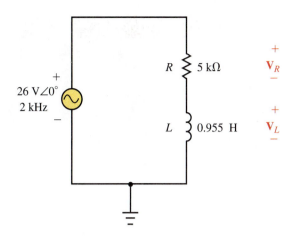

FIGURE 18–123

◀ MULTISIM

62. A 2-mH inductor is placed in parallel with a 2-kΩ resistor. Use PSpice to input these components into a circuit. Run the Probe postprocessor to provide a graphical display of the network impedance as a function of frequency from 50 kHz to 500 kHz. Let the frequency sweep logarithmically in octaves.

◀ CADENCE

63. A 100-kΩ resistor is placed in parallel with a 0.47-μF capacitor. Use PSpice to input these components into a circuit. Run the Probe postprocessor to provide a graphical display of the network impedance as a function of frequency from 0.1 Hz to 10 Hz. Let the frequency sweep logarithmically in octaves.

◀ CADENCE

64. A 2.7-kΩ resistor is placed in parallel with a 20-mH inductor. Use PSpice to input these components into a circuit. Run the Probe postprocessor to provide a graphical display of the network impedance as a function of frequency from 100 kHz to 1 MHz. Let the frequency sweep logarithmically in octaves.

◀ CADENCE

 ANSWERS TO IN-PROCESS LEARNING CHECKS

In-Process Learning Check 1

1. Current and voltage are in phase.

2. Current leads voltage by 90°.

3. Voltage leads current by 90°.

In-Process Learning Check 2

1. The phasor sum of voltage drops and rises around a closed loop is equal to zero.

2. All voltages must be expressed as phasors, i.e., $\mathbf{V} = V\angle\theta = V\cos\theta + jV\sin\theta$.

In-Process Learning Check 3

1. The phasor sum of currents entering a node is equal to the phasor sum of currents leaving the same node.

2. All currents must be expressed as phasors, i.e., $\mathbf{I} = I\angle\theta = I\cos\theta + jI\sin\theta$.

In-Process Learning Check 4

1. At $f = 0$ Hz, $Z = \infty$ (open circuit)

2. As $f\,\infty$, $Z = R$

In-Process Learning Check 5

1. $R_P = 795\ \Omega$, $L_P = 10.1$ mH

2. $R_P = 790\ M\Omega$, $L_P = 10.0$ mH

■ **OBJECTIVES**

After studying this chapter, you will be able to

- convert an ac voltage source into its equivalent current source, and conversely, convert a current source into an equivalent voltage source,

- solve for the current or voltage in a circuit having either a dependent current source or a dependent voltage source,

- set up simultaneous linear equations to solve an ac circuit using mesh analysis,

- use complex determinants to find the solutions for a given set of linear equations,

- set up simultaneous linear equations to solve an ac circuit using nodal analysis,

- perform delta-to-wye and wye-to-delta conversions for circuits having reactive elements,

- solve for the balanced condition in a given ac bridge circuit. In particular, you will examine the Maxwell, Hay, and Schering bridges,

- use MultiSIM to analyze bridge circuits,

- use PSpice to calculate current and voltage in an ac circuit.

Methods of AC Analysis

19

CHAPTER PREVIEW

To this point, we have examined only circuits having a single ac source. In this chapter, we continue our study by analyzing multisource circuits and bridge networks. You will find that most of the techniques used in analyzing ac circuits parallel those of dc circuit analysis. Consequently, a review of Chapter 8 will help you understand the topics of this chapter.

Near the end of the chapter, we examine how computer techniques are used to analyze even the most complex ac circuits. It must be emphasized that although computer techniques are much simpler than using a pencil and calculator, there is virtually no knowledge to be gained by mindlessly entering data into a computer. We use the computer merely as a tool to verify our results and to provide a greater dimension to the analysis of circuits. ■

Hermann Ludwig Ferdinand von Helmholtz

PUTTING IT IN PERSPECTIVE

HERMANN HELMHOLTZ WAS BORN IN POTSDAM (near Berlin, Germany) on August 31, 1821. Helmholtz was a leading scientist of the nineteenth century, whose legacy includes contributions in the fields of acoustics, chemistry, mathematics, magnetism, electricity, mechanics, optics, and physiology.

Helmholtz graduated from the Medical Institute in Berlin in 1843 and practiced medicine for five years as a surgeon in the Prussian army. From 1849 to 1871, he served as a professor of physiology at universities in Königsberg, Bonn, and Heidelberg. In 1871, Helmholtz was appointed Professor of Physics at the University of Berlin.

Helmholtz' greatest contributions were as a mathematical physicist, where his work in theoretical and practical physics led to the proof of the Law of Conservation of Energy in his paper "*Über die Erhaltung der Kraft,*" published in 1847. He showed that mechanics, heat, light, electricity, and magnetism were simply manifestations of the same force. His work led to the understanding of electrodynamics (the motion of charge in conductors), and his theory of the electromagnetic properties of light set the groundwork for later scientists to understand how radio waves are propagated.

For his work, the German emperor Kaiser Wilhelm I made Helmholtz a noble in 1883. This great scientist died on September 8, 1894, at the age of 73. ■

19.1 Dependent Sources

The voltage and current sources we have worked with up to now have been **independent sources,** meaning that the voltage or current of the supply was not in any way dependent upon any voltage or current elsewhere in the circuit. In many amplifier circuits, particularly those involving transistors, it is possible to explain the operation of the circuits by replacing the device with an equivalent electronic model. These models often use voltage and current sources which have values dependent upon some internal voltage or current. Such that are called **dependent sources.** Figure 19–1 compares the symbols for both independent and dependent sources.

Although the diamond is the accepted symbol for representing dependent sources, many articles and textbooks still use a circle. In this textbook we use both forms of the dependent source to familiarize the student with the various notations. The dependent source has a magnitude and phase angle determined by voltage or current at some internal element multiplied by a constant, k. The magnitude of the constant is determined by parameters within the particular model, and the units of the constant correspond to the required quantities in the equation.

(a) Independent sources

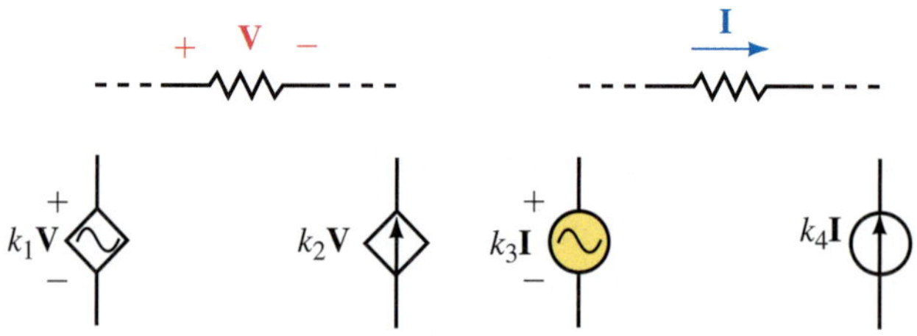

(b) Dependent sources

FIGURE 19–1

EXAMPLE 19–1

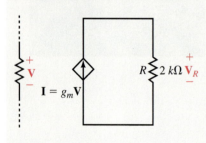

Given: $g_m = 4$ mS

FIGURE 19–2

Refer to the resistor shown in Figure 19–2. Determine the voltage $\mathbf{V}_R$ across the resistor given that the controlling voltage has the following values:

a. $\mathbf{V} = 0$ V.

b. $\mathbf{V} = 5$ V∠30°.

c. $\mathbf{V} = 3$ V∠−150°.

Solution   Notice that the dependent source of this example has a constant, g_m, called the transconductance. Here, $g_m = 4$ mS.

a. $\mathbf{I} = (4 \text{ mS})(0 \text{ V}) = 0$

$\mathbf{V}_R = 0$ V

b. $\mathbf{I} = (4 \text{ mS})(5 \text{ V}∠30°) = 20 \text{ mA}∠30°$

$\mathbf{V}_R = (20 \text{ mA}∠30°)(2 \text{ k}Ω) = 40 \text{ V}∠30°$

c. $\mathbf{I} = (4 \text{ mS})(3 \text{ V}∠−150°) = 12 \text{ mA}∠−150°$

$\mathbf{V}_R = (12 \text{ mA}∠−150°)(2 \text{ k}Ω) = 24 \text{ V}∠−150°$

The circuit of Figure 19–3 represents a simplified model of a transistor amplifier.

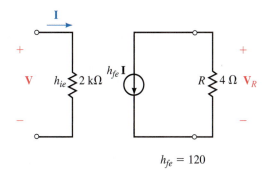

$h_{fe} = 120$

FIGURE 19–3

Determine the voltage $\mathbf{V}_R$ for each of the following applied voltages:

a. $\mathbf{V} = 10 \text{ mV}\angle 0°$.

b. $\mathbf{V} = 2 \text{ mV}\angle 180°$.

c. $\mathbf{V} = 0.03 \text{ V}\angle 90°$.

Answers
a. 2.4 mV∠180°; b. 0.48 mV∠0°; c. 7.2 mV∠−90°

When working with dc circuits, the analysis of a circuit is often simplified by replacing the source (whether a voltage source or a current source) with its equivalent. The conversion of any ac source is similar to the method used in dc circuit analysis.

19.2 Source Conversion

A voltage source **E** in series with an impedance **Z** is equivalent to a current source **I** having the same impedance **Z** in parallel. Figure 19–4 shows the equivalent sources.

From Ohm's law, we perform the source conversion as follows:

$$\mathbf{I} = \frac{\mathbf{E}}{\mathbf{Z}}$$

and

$$\mathbf{E} = \mathbf{IZ}$$

It is important to realize that the two circuits of Figure 19–4 are equivalent between points *a* and *b*. This means that any network connected to points *a* and *b* will behave exactly the same regardless of which type of source is used. However, the voltages or currents within the sources will seldom be the same.

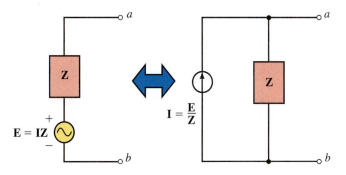

FIGURE 19–4

In order to determine the current through or the voltage across the source impedance, the circuit must be returned to its original state.

EXAMPLE 19–2

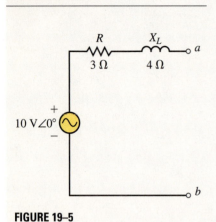

FIGURE 19–5

Convert the voltage source of Figure 19–5 into an equivalent current source.

Solution

$$\mathbf{Z}_T = 3\ \Omega + j4\ \Omega = 5\ \Omega\angle 53.13°$$

$$\mathbf{I} = \frac{10\ \text{V}\angle 0°}{5\ \Omega\angle 53.13°} = 2\ \text{A}\angle -53.13°$$

The equivalent current source is shown in Figure 19–6.

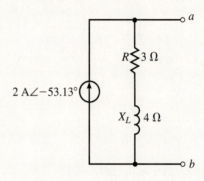

FIGURE 19–6

EXAMPLE 19–3

Convert the current source of Figure 19–7 into an equivalent voltage source.

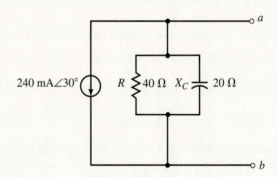

FIGURE 19–7

Solution   The impedance of the parallel combination is determined to be

$$\mathbf{Z} = \frac{(40\ \Omega\angle 0°)(20\ \Omega\ \angle -90°)}{40\ \Omega - j20\ \Omega}$$

$$= \frac{800\ \Omega\ \angle -90°}{44.72\angle -26.57°}$$

$$= 17.89\ \Omega\angle -63.43° = 8\ \Omega - j16\ \Omega$$

and so

$$\mathbf{E} = (240\ \text{mA}\angle 30°)(17.89\ \Omega\angle -63.43°)$$

$$= 4.29\ \text{V}\angle -33.43°$$

The resulting equivalent circuit is shown in Figure 19–8.

$\mathbf{Z} = 8\ \Omega - j16\ \Omega = 17.89\ \Omega\angle -63.43°$

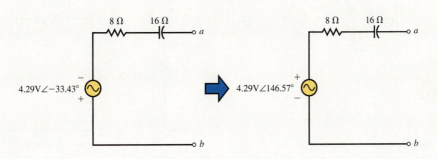

FIGURE 19–8

It is possible to use the same procedure to convert a dependent source into its equivalent provided that the controlling element is external to the circuit in which the source appears. *If the controlling element is in the same circuit as the dependent source, this procedure cannot be used.*

EXAMPLE 19–4

Convert the current source of Figure 19–9 into an equivalent voltage source.

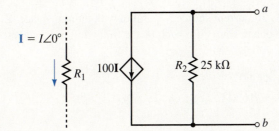

FIGURE 19–9

Solution In the circuit of Figure 19–9 the controlling element, R_1, is in a separate circuit. Therefore, the current source is converted into an equivalent voltage source as follows:

$$\begin{aligned}\mathbf{E} &= (100\mathbf{I}_1)(\mathbf{Z})\\ &= (100\mathbf{I}_1)(25\ \text{k}\Omega\angle 0°)\\ &= (2.5 \times 10^6\ \Omega)\mathbf{I}_1\end{aligned}$$

The resulting voltage source is shown in Figure 19–10. Notice that the equivalent voltage source is dependent on the current, **I,** just as the original current source.

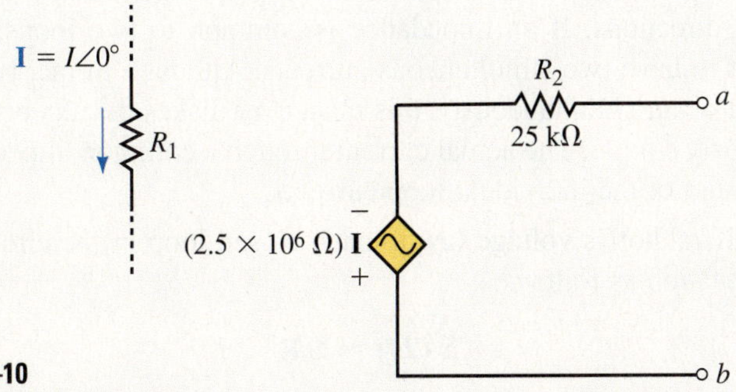

FIGURE 19–10

PRACTICE PROBLEMS 2

Convert the voltage sources of Figure 19–11 into equivalent current sources.

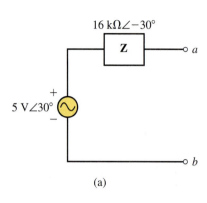

(a)

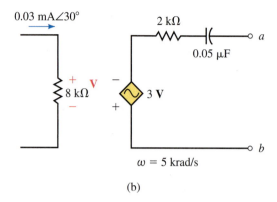

(b)

FIGURE 19–11

Answers

a. $\mathbf{I} = 0.3125$ mA$\angle 60°$ (from b to a) in parallel with $\mathbf{Z} = 16$ k$\Omega\angle -30°$

b. $\mathbf{I} = 0.161$ mA$\angle 93.43°$ (from a to b) in parallel with $\mathbf{Z} = 2$ k$\Omega - j4$ kΩ

IN-PROCESS
LEARNING CHECK 1

(Answers are at the end of the chapter.)

Given a 40-mA$\angle 0°$ current source in parallel with an impedance, $\mathbf{Z}$. Determine the equivalent voltage source for each of the following impedances:

a. $\mathbf{Z} = 25$ k$\Omega\angle 30°$.

b. $\mathbf{Z} = 100\ \Omega\ \angle -90°$.

c. $\mathbf{Z} = 20$ k$\Omega - j16$ kΩ.

19.3 Mesh (Loop) Analysis

◀ **Online Companion**

Mesh analysis allows us to determine each loop current within a circuit, regardless of the number of sources within the circuit. The following steps provide a format which simplifies the process of using mesh analysis:

1. Convert all sinusoidal expressions into equivalent phasor notation. Where necessary, convert current sources into equivalent voltage sources.

2. Redraw the given circuit, simplifying the given impedances wherever possible and labelling the impedances ($\mathbf{Z}_1$, $\mathbf{Z}_2$, etc.).

3. Arbitrarily assign clockwise loop currents to each interior closed loop within a circuit. Show the polarities of all impedances using the assumed current directions. If an impedance is common to two loops, it may be thought to have two simultaneous currents. Although in fact two currents will not occur simultaneously, this maneuver makes the algebraic calculations fairly simple. The actual current through a common impedance is the vector sum of the individual loop currents.

4. Apply Kirchhoff's voltage law to each closed loop in the circuit, writing each equation as follows:

$$\Sigma\ (\mathbf{ZI}) = \Sigma\ \mathbf{E}$$

If the current directions are originally assigned in a clockwise direction, then the resulting linear equations may be simplified to the following format:

Loop 1: $+(\Sigma \mathbf{Z}_1)\mathbf{I}_1 - (\Sigma \mathbf{Z}_{1-2})\mathbf{I}_2 - \cdots - (\Sigma \mathbf{Z}_{1-n})\mathbf{I}_n = (\Sigma \mathbf{E}_1)$

Loop 2: $-(\Sigma \mathbf{Z}_{2-1})\mathbf{I}_1 + (\Sigma \mathbf{Z}_2)\mathbf{I}_2 - \cdots - (\Sigma \mathbf{Z}_{2-n})\mathbf{I}_n = (\Sigma \mathbf{E}_2)$

.   .
.   .
.   .

Loop n: $-(\Sigma \mathbf{Z}_{n-1})\mathbf{I}_1 - (\Sigma \mathbf{Z}_{n-2})\mathbf{I}_2 - \cdots + (\Sigma \mathbf{Z}_n)\mathbf{I}_n = (\Sigma \mathbf{E}_n)$

In the above format, $\Sigma \mathbf{Z}_x$ is the summation of all impedances around loop x. The sign in front of all loop impedances will be positive.

$\Sigma \mathbf{Z}_{x-y}$ is the summation of impedances that are common between loop x and loop y. If there is no common impedance between two loops, this term is simply set to zero. All common impedance terms in the linear equations are given negative signs.

$\Sigma \mathbf{E}_x$ is the summation of voltage rises in the direction of the assumed current $\mathbf{I}_x$. If a voltage source has a polarity such that it appears as a voltage drop in the assumed current direction, then the voltage is given a negative sign.

5. Solve the resulting simultaneous linear equations using substitution or determinants. If necessary, refer to Appendix B for a review of solving simultaneous linear equations.

Solve for the loop equations in the circuit of Figure 19–12.

EXAMPLE 19–5

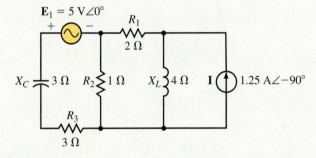

FIGURE 19–12

Solution
Step 1: The current source is first converted into an equivalent voltage source as shown in Figure 19–13.

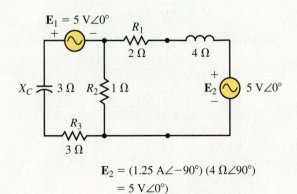

$\mathbf{E}_2 = (1.25 \text{ A}\angle-90°)(4 \text{ }\Omega\angle90°)$

$= 5 \text{ V}\angle0°)$

FIGURE 19–13

Steps 2 and 3: Next the circuit is redrawn as shown in Figure 19–14. The impedances have been simplified and the loop currents are drawn in a clockwise direction.

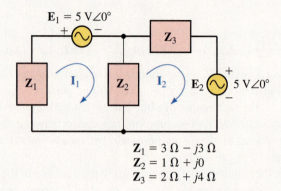

$$\mathbf{Z}_1 = 3 \, \Omega - j3 \, \Omega$$
$$\mathbf{Z}_2 = 1 \, \Omega + j0$$
$$\mathbf{Z}_3 = 2 \, \Omega + j4 \, \Omega$$

FIGURE 19–14

Step 4: The loop equations are written as

$$\text{Loop 1: } (\mathbf{Z}_1 + \mathbf{Z}_2)\mathbf{I}_1 - (\mathbf{Z}_2)\mathbf{I}_2 = -\mathbf{E}_1$$
$$\text{Loop 2: } -(\mathbf{Z}_2)\mathbf{I}_1 + (\mathbf{Z}_2 + \mathbf{Z}_3)\mathbf{I}_2 = -\mathbf{E}_2$$

The solution for the currents may be found by using determinants:

$$\mathbf{I}_1 = \frac{\begin{vmatrix} -\mathbf{E}_1 & -\mathbf{Z}_2 \\ -\mathbf{E}_2 & \mathbf{Z}_2 + \mathbf{Z}_3 \end{vmatrix}}{\begin{vmatrix} \mathbf{Z}_1 + \mathbf{Z}_2 & -\mathbf{Z}_2 \\ -\mathbf{Z}_2 & \mathbf{Z}_2 + \mathbf{Z}_3 \end{vmatrix}}$$

$$= \frac{(-\mathbf{E}_1)(\mathbf{Z}_2 + \mathbf{Z}_3) - \mathbf{E}_2\mathbf{Z}_2}{(\mathbf{Z}_1 + \mathbf{Z}_2)(\mathbf{Z}_2 + \mathbf{Z}_3) - \mathbf{Z}_2\mathbf{Z}_2}$$

$$= \frac{(-\mathbf{E}_1)(\mathbf{Z}_2 + \mathbf{Z}_3) - \mathbf{E}_2\mathbf{Z}_2}{\mathbf{Z}_1\mathbf{Z}_2 + \mathbf{Z}_1\mathbf{Z}_3 + \mathbf{Z}_2\mathbf{Z}_3}$$

and

$$\mathbf{I}_2 = \frac{\begin{vmatrix} \mathbf{Z}_1 + \mathbf{Z}_2 & -\mathbf{E}_1 \\ -\mathbf{Z}_2 & -\mathbf{E}_2 \end{vmatrix}}{\begin{vmatrix} \mathbf{Z}_1 + \mathbf{Z}_2 & -\mathbf{Z}_2 \\ -\mathbf{Z}_2 & \mathbf{Z}_2 + \mathbf{Z}_3 \end{vmatrix}}$$

$$= \frac{-\mathbf{E}_2(\mathbf{Z}_1 + \mathbf{Z}_2) - \mathbf{E}_1\mathbf{Z}_2}{(\mathbf{Z}_1 + \mathbf{Z}_2)(\mathbf{Z}_2 + \mathbf{Z}_3) - \mathbf{Z}_2\mathbf{Z}_2}$$

$$= \frac{-\mathbf{E}_2(\mathbf{Z}_1 + \mathbf{Z}_2) - \mathbf{E}_1\mathbf{Z}_2}{\mathbf{Z}_1\mathbf{Z}_2 + \mathbf{Z}_1\mathbf{Z}_3 + \mathbf{Z}_2\mathbf{Z}_3}$$

Solving these equations using the actual values of impedances and voltages, we have

$$\mathbf{I}_1 = \frac{-(5)(3 + j4) - (5)(1)}{(3 - j3)(1) + (3 - j3)(2 + j4) + (1)(2 + j4)}$$

$$= \frac{(-15 - j20) - 5}{(3 - j3) + (6 + j6 - j^2 12) + (2 + j4)}$$

$$= \frac{-20 - j20}{23 + j7}$$

$$= \frac{28.28\angle -135°}{24.04\angle 16.93°}$$

$$= 1.18 \text{ A}\angle -151.93°$$

and

$$I_2 = \frac{-(5)(4 - j3) - (5)(1)}{23 + j7}$$

$$= \frac{(-20 + j15) - (5)}{23 + j7}$$

$$= \frac{-25 + j15}{23 + j7}$$

$$= \frac{29.15\angle149.04°}{24.04\angle16.93°}$$

$$= 1.21\ A\angle132.11°$$

Given the circuit of Figure 19–15, write the loop equations and solve for the loop currents. Determine the voltage, **V.**

EXAMPLE 19–6

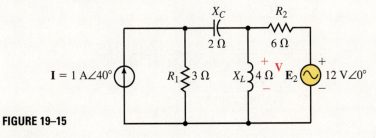

FIGURE 19–15

Solution

Step 1: Converting the current source into an equivalent voltage source gives us the circuit of of Figure 19–16.

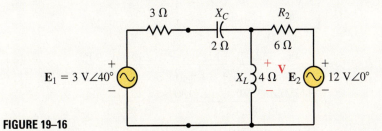

FIGURE 19–16

Steps 2 and 3: After simplifying the impedances and assigning clockwise loop currents, we have the circuit of Figure 19–17.

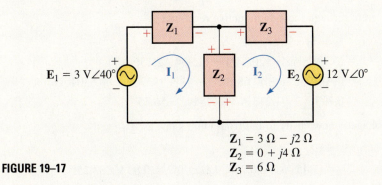

$$\mathbf{Z}_1 = 3\ \Omega - j2\ \Omega$$
$$\mathbf{Z}_2 = 0 + j4\ \Omega$$
$$\mathbf{Z}_3 = 6\ \Omega$$

FIGURE 19–17

Step 4: The loop equations for the circuit of Figure 19–17 are as follows:

$$\text{Loop 1: } (\mathbf{Z}_1 + \mathbf{Z}_2)\mathbf{I}_1 - (\mathbf{Z}_2)\mathbf{I}_2 = \mathbf{E}_1$$
$$\text{Loop 2: } -(\mathbf{Z}_2)\mathbf{I}_1 + (\mathbf{Z}_2 + \mathbf{Z}_3)\mathbf{I}_2 = -\mathbf{E}_2$$

which, after substituting the impedance values into the expressions, become

$$\text{Loop 1: } (3\,\Omega + j2\,\Omega)\mathbf{I}_1 - (j4\,\Omega)\mathbf{I}_2 = 3\text{ V}\angle 40°$$
$$\text{Loop 2: } -(j4\,\Omega)\mathbf{I}_2 + (6\,\Omega + j4\,\Omega)\mathbf{I}_2 = -12\text{ V}\angle 0°$$

Step 5: Solve for the currents using determinants, where the elements of the determinants are expressed as vectors.

Since the determinant of the denominator is common to both terms we find this value first.

$$\mathbf{D} = \begin{vmatrix} 3+j2 & -j4 \\ -j4 & 6+j4 \end{vmatrix}$$
$$= (3+j2)(6+j4) - (-j4)(-j4)$$
$$= 18 + j12 + j12 - 8 + 16$$
$$= 26 + j24 = 35.38\angle 42.71°$$

Now, the currents are found as

$$\mathbf{I}_1 = \frac{\begin{vmatrix} 3\angle 40° & -j4 \\ -12\angle 0° & 6+j4 \end{vmatrix}}{\mathbf{D}}$$

$$= \frac{(3\angle 40°)(7.211\angle 33.69°) - (12\angle 0°)(4\angle 90°)}{35.38\angle 42.71°}$$

$$= \frac{21.633\angle 73.69° - 48\angle 90°}{35.38\angle 42.71°}$$

$$= \frac{27.91\angle -77.43°}{35.38\angle 42.71°}$$

$$= 0.7887\text{ A}\angle -120.14°$$

and

$$\mathbf{I}_2 = \frac{\begin{vmatrix} 3+j2 & 3\angle 40° \\ -j4 & -12\angle 0° \end{vmatrix}}{\mathbf{D}}$$

$$\mathbf{I}_2 = \frac{(3.606\angle 33.69°)(-12\angle 0°) + (3\angle 40°)(4\angle 90°)}{35.38\angle 42.71°}$$

$$= \frac{-43.27\angle 33.69° + 12\angle 130°}{35.38\angle 42.71°}$$

$$= \frac{46.15\angle -161.29°}{35.38\angle 42.71°}$$

$$= 1.304\text{ A}\angle 156.00°$$

The current through the 4-Ω inductive reactance is

$$\mathbf{I} = \mathbf{I}_1 - \mathbf{I}_2$$
$$= (0.7887\text{ A}\angle -120.14°) - (1.304\text{ A}\angle 156.00°)$$
$$= (-0.3960\text{ A} - j0.6821\text{ A}) - (-1.1913\text{ A} + j0.5304\text{ A})$$
$$= 0.795\text{ A} - j1.213\text{ A} = 1.45\text{ A}\angle -56.75°$$

The voltage is now easily found from Ohm's law as

$$\mathbf{V} = \mathbf{I}\mathbf{Z}_L$$
$$= (1.45\text{ A}\angle -56.75°)(4\,\Omega\angle 90°) = 5.80\text{ V}\angle 33.25°$$

Given the circuit of of Figure 19–18, write the loop equations and show the determinant of the coefficients for the loop equations. Do not solve them.

EXAMPLE 19–7

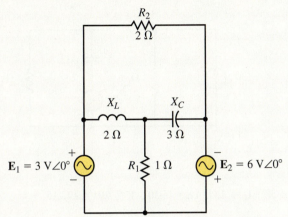

FIGURE 19–18

Solution   The circuit is redrawn in Figure 19–19, showing loop currents and impedances together with the appropriate voltage polarities.

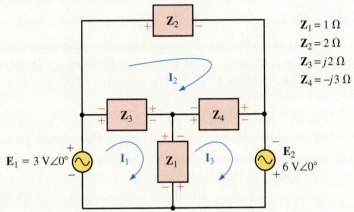

$Z_1 = 1\ \Omega$
$Z_2 = 2\ \Omega$
$Z_3 = j2\ \Omega$
$Z_4 = -j3\ \Omega$

FIGURE 19–19

The loop equations may now be written as

$$\text{Loop 1: } (\mathbf{Z}_1 + \mathbf{Z}_3)\mathbf{I}_1 - (\mathbf{Z}_3)\mathbf{I}_2 - (\mathbf{Z}_1)\mathbf{I}_3 = \mathbf{E}_1$$
$$\text{Loop 2: } -(\mathbf{Z}_3)\mathbf{I}_1 + (\mathbf{Z}_2 + \mathbf{Z}_3 + \mathbf{Z}_4)\mathbf{I}_2 - (\mathbf{Z}_4)\mathbf{I}_3 = 0$$
$$\text{Loop 3: } -(\mathbf{Z}_1)\mathbf{I}_1 - (\mathbf{Z}_4)\mathbf{I}_2 + (\mathbf{Z}_1 + \mathbf{Z}_4)\mathbf{I}_3 = \mathbf{E}_2$$

Using the given impedance values, we have

$$\text{Loop 1: } (1\ \Omega + j2\ \Omega)\mathbf{I}_1 - (j2\ \Omega)\mathbf{I}_2 - (1\ \Omega)\mathbf{I}_3 = 3\ \text{V}$$
$$\text{Loop 2: } -(j2\ \Omega)\mathbf{I}_1 + (2\ \Omega - j1\ \Omega)\mathbf{I}_2 - (-j3\ \Omega)\mathbf{I}_3 = 0$$
$$\text{Loop 3: } -(1\ \Omega)\mathbf{I}_1 - (-j3\ \Omega)\mathbf{I}_2 + (1\ \Omega - j3\ \Omega)\mathbf{I}_3 = 6\ \text{V}$$

Notice that in the above equations, the phase angles ($\theta = 0°$) for the voltages have been omitted. This is because $3\ \text{V}\angle 0° = 3\ \text{V} + j0\ \text{V} = 3\ \text{V}$.

The determinant for the coefficients of the loop equations is written as

$$\mathbf{D} = \begin{vmatrix} 1 + j2 & -j2 & -1 \\ -j2 & 2 - j1 & j3 \\ -1 & j3 & 1 - j3 \end{vmatrix}$$

Notice that the above determinant is symmetrical about the principle diagonal. The coefficients in the loop equations were given the required signs (positive for the loop impedance and negative for all common impedances). However, since the coefficients in the determinant may contain imaginary numbers, it is no longer possible to generalize about the signs ($+/-$) of the various coefficients.

Given the circuit of Figure 19–20, write the mesh equations and solve for the loop currents. Use the results to determine the current **I**.

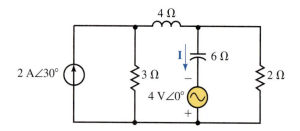

FIGURE 19–20

Answers
$\mathbf{I}_1 = 1.19\ \text{A}\angle 1.58°$, $\mathbf{I}_2 = 1.28\ \text{A}\angle{-46.50°}$, $\mathbf{I} = 1.01\ \text{A}\angle 72.15°$

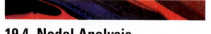

(Answers are at the end of the chapter.)

Briefly list the steps followed in using mesh analysis to solve for the loop currents of a circuit.

19.4 Nodal Analysis

Nodal analysis allows us to calculate all node voltages with respect to an arbitrary reference point in a circuit. The following steps provide a simple format to apply nodal analysis.

1. Convert all sinusoidal expressions into equivalent phasor notation. If necessary, convert voltage sources into equivalent current sources.

2. Redraw the given circuit, simplifying the given impedances wherever possible and relabelling the impedances as admittances ($\mathbf{Y}_1$, $\mathbf{Y}_2$, etc.).

3. Select and label an appropriate reference node. Arbitrarily assign subscripted voltages ($\mathbf{V}_1$, $\mathbf{V}_2$, etc.) to each of the remaining n nodes within the circuit.

4. Indicate assumed current directions through all admittances in the circuit. If an admittance is common to two nodes, it is considered in each of the two node equations.

5. Apply Kirchhoff's current law to each of the n nodes in the circuit, writing each equation as follows:

$$\Sigma(\mathbf{YV}) = \Sigma\mathbf{I}_{\text{sources}}$$

The resulting linear equations may be simplified to the following format:

$$\text{Node 1:} + (\Sigma\mathbf{Y}_1)\mathbf{V}_1 - (\Sigma\mathbf{Y}_{1-2})\mathbf{V}_2 - \cdots - (\Sigma\mathbf{Y}_{1-n})\mathbf{V}_n = (\Sigma\mathbf{I}_1)$$
$$\text{Node 2:} - (\Sigma\mathbf{Y}_{2-1})\mathbf{V}_1 + (\Sigma\mathbf{Y}_2)\mathbf{V}_2 - \cdots - (\Sigma\mathbf{Y}_{2-n})\mathbf{V}_n = (\Sigma\mathbf{I}_2)$$

$$\text{Node } n: - (\Sigma\mathbf{Y}_{n-1})\mathbf{V}_1 - (\Sigma\mathbf{Y}_{n-2})\mathbf{V}_2 - \cdots + (\Sigma\mathbf{Y}_n)\mathbf{V}_n = (\Sigma\mathbf{I}_n)$$

In the above format, $\Sigma\mathbf{Y}_x$ is the summation of all admittances connected to node x. The sign in front of all node admittances will be positive.

$\Sigma\mathbf{Y}_{x-y}$ is the summation of common admittances between node x and node y. If there are no common admittances between two nodes, this term is simply set to zero. All common admittance terms in the linear equations will have a negative sign.

$\Sigma \mathbf{I}_x$ is the summation of current sources entering node x. If a current source leaves the node, the current is given a negative sign.

6. Solve the resulting simultaneous linear equations using substitution or determinants.

Given the circuit of Figure 19–21, write the nodal equations and solve for the node voltages.

EXAMPLE 19–8

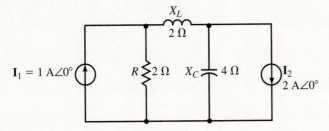

FIGURE 19–21

Solution   The circuit is redrawn in Figure 19–22, showing the nodes and a simplified representation of the admittances.

The nodal equations are written as

$$\text{Node 1: } (\mathbf{Y}_1 + \mathbf{Y}_2)\mathbf{V}_1 - (\mathbf{Y}_2)\mathbf{V}_2 = \mathbf{I}_1$$
$$\text{Node 2: } -(\mathbf{Y}_2)\mathbf{V}_1 + (\mathbf{Y}_2 + \mathbf{Y}_3)\mathbf{V}_2 = -\mathbf{I}_2$$

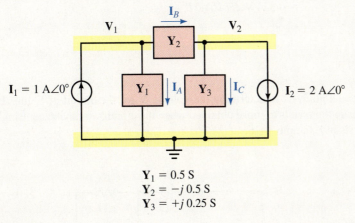

$$\mathbf{Y}_1 = 0.5 \text{ S}$$
$$\mathbf{Y}_2 = -j\,0.5 \text{ S}$$
$$\mathbf{Y}_3 = +j\,0.25 \text{ S}$$

FIGURE 19–22

Using determinants, the following expressions for nodal voltages are obtained:

$$\mathbf{V}_1 = \frac{\begin{vmatrix} \mathbf{I}_1 & -\mathbf{Y}_2 \\ -\mathbf{I}_2 & \mathbf{Y}_2 + \mathbf{Y}_3 \end{vmatrix}}{\begin{vmatrix} \mathbf{Y}_1 + \mathbf{Y}_2 & -\mathbf{Y}_2 \\ -\mathbf{Y}_2 & \mathbf{Y}_2 + \mathbf{Y}_3 \end{vmatrix}}$$
$$= \frac{\mathbf{I}_1(\mathbf{Y}_2 + \mathbf{Y}_3) - \mathbf{I}_2\mathbf{Y}_2}{\mathbf{Y}_1\mathbf{Y}_2 + \mathbf{Y}_1\mathbf{Y}_3 + \mathbf{Y}_2\mathbf{Y}_3}$$

and

$$V_2 = \frac{\begin{vmatrix} \mathbf{Y}_1 + \mathbf{Y}_2 & \mathbf{I}_1 \\ -\mathbf{Y}_2 & -\mathbf{I}_2 \end{vmatrix}}{\begin{vmatrix} \mathbf{Y}_1 + \mathbf{Y}_2 & -\mathbf{Y}_2 \\ -\mathbf{Y}_2 & \mathbf{Y}_2 + \mathbf{Y}_3 \end{vmatrix}}$$

$$= \frac{-\mathbf{I}_2(\mathbf{Y}_1 + \mathbf{Y}_2) + \mathbf{I}_1 \mathbf{Y}_2}{\mathbf{Y}_1 \mathbf{Y}_2 + \mathbf{Y}_1 \mathbf{Y}_3 + \mathbf{Y}_2 \mathbf{Y}_3}$$

Substituting the appropriate values into the above expressions we find the nodal voltages as

$$\mathbf{V}_1 = \frac{1(-j0.5 + j0.25) - (2)(-j0.5)}{(0.5)(-j0.5) + (0.5)(j0.25) + (-j0.5)(j0.25)}$$

$$= \frac{j0.75}{0.125 - j0.125}$$

$$= \frac{0.75\angle 90°}{0.1768\angle -45°}$$

$$= 4.243 \text{ V}\angle 135°$$

and

$$\mathbf{V}_2 = \frac{-2(0.5 - j0.5) + (1)(-j0.5)}{(0.5)(-j0.5) + (0.5)(j0.25) + (-j0.5)(j0.25)}$$

$$= \frac{-1 + j0.5}{0.125 - j0.125}$$

$$= \frac{1.118\angle 153.43°}{0.1768\angle -45°}$$

$$= 6.324 \text{ V}\angle 198.43°$$

$$= 6.324 \text{ V}\angle -161.57°$$

EXAMPLE 19–9

Use nodal analysis to determine the voltage **V** for the circuit of Figure 19–23. Compare the results to those obtained when the circuit was analyzed using mesh analysis in Example 19–6.

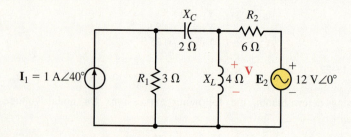

FIGURE 19–23

Solution
Step 1: Convert the voltage source into an equivalent current source as illustrated in Figure 19–24.

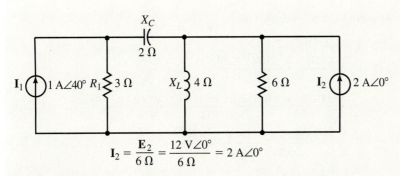

$$I_2 = \frac{E_2}{6\,\Omega} = \frac{12\text{ V}\angle0°}{6\,\Omega} = 2\text{ A}\angle0°$$

FIGURE 19–24

Steps 2, 3, and 4: The reference node is selected to be at the bottom of the circuit and the admittances are simplified as shown in Figure 19–25.

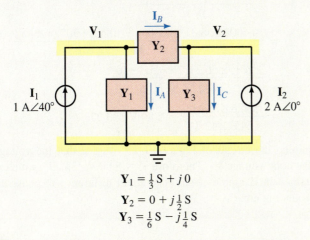

$$Y_1 = \tfrac{1}{3}\text{ S} + j\,0$$
$$Y_2 = 0 + j\tfrac{1}{2}\text{ S}$$
$$Y_3 = \tfrac{1}{6}\text{ S} - j\tfrac{1}{4}\text{ S}$$

FIGURE 19–25

Step 5: Applying Kirchhoff's current law to each node, we have the following:

$$\text{Node 1: } I_A + I_B = I_1$$
$$Y_1V_1 + Y_2(V_1 - V_2) = I_1$$
$$(Y_1 + Y_2)V_1 - Y_2V_2 = I_1$$

$$\text{Node 2: } I_C = I_B + I_2$$
$$Y_3V_2 = Y_2(V_1 - V_2) + I_2$$
$$-Y_2V_1 + (Y_2^2 + Y_3)V_2 = I_2$$

After substituting the appropriate values into the above equations, we have the following simultaneous linear equations:

$$(0.3333\text{ S} + j0.5\text{ S})V_1 - (j0.5\text{ S})V_2 = 1\text{ A}\angle40°$$
$$-(j0.5\text{ S})V_1 + (0.16667\text{ S} + j0.25\text{ S})V_2 = 2\text{ A}\angle0°$$

Step 6: We begin by solving for the determinant for the denominator:

$$\mathbf{D} = \begin{vmatrix} 0.3333 + j0.5 & -j0.5 \\ -j0.5 & 0.1667 + j0.25 \end{vmatrix}$$

$$= (0.3333 + j0.5)(0.1667 + j0.25) - (j0.5)(j0.5)$$
$$= 0.0556 + j0.0833 + j0.0833 - 0.125 + 0.25$$
$$= 0.1806 + j0.1667 = 0.2457\angle42.71°$$

Next we solve the node voltages as

$$V_1 = \frac{\begin{vmatrix} 1\angle40° & -j0.5 \\ 2\angle0° & 0.1667 + j0.25 \end{vmatrix}}{D}$$

$$= \frac{(1\angle40°)(0.3005\angle56.31°) - (2\angle0°)(0.5\angle-90°)}{D}$$

$$= \frac{0.3005\angle96.31° - 1.0\angle-90°}{0.2457\angle42.71°}$$

$$= \frac{1.299\angle91.46°}{0.2457\angle42.71°} = 5.29 \text{ V}\angle48.75°$$

and

$$V_2 = \frac{\begin{vmatrix} 0.3333 + j0.5 & 1\angle40° \\ -j0.5 & 2\angle0° \end{vmatrix}}{D}$$

$$= \frac{(0.6009\angle56.31°)(2\angle0°) - (0.5\angle-90°)(1\angle40°)}{D}$$

$$= \frac{1.2019\angle56.31° - 0.5\angle-50°}{0.2457\angle42.71°}$$

$$= \frac{1.426\angle75.98°}{0.2457\angle42.71°} = 5.80 \text{ V}\angle33.27°$$

Examining the circuit of Figure 19–23, we see that the voltage **V** is the same as the node voltage V_2. Therefore **V** = 5.80 V∠33.27°, which is the same result obtained in Example 19–6. (The slight difference in phase angle is the result of rounding error.)

EXAMPLE 19–10

Given the circuit of Figure 19–26, write the nodal equations expressing all coefficients in rectangular form. Do not solve the equations.

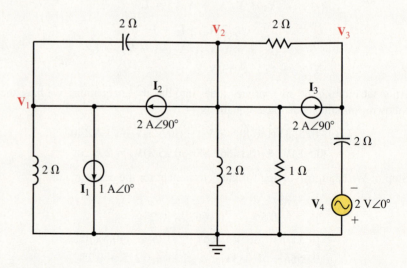

FIGURE 19–26

Solution As in the previous example, we need to first convert the voltage source into its equivalent current source. The current source will be a phasor $\mathbf{I}_4$, where

$$\mathbf{I}_4 = \frac{\mathbf{V}_4}{\mathbf{X}_C} = \frac{2\,\mathrm{V}\angle 0°}{2\,\Omega\angle -90°} = 1.0\,\mathrm{A}\angle 90°$$

Figure 19–27 shows the circuit as it appears after the source conversion. Notice that the direction of the current source is downward to correspond with the polarity of the voltage source $\mathbf{V}_4$.

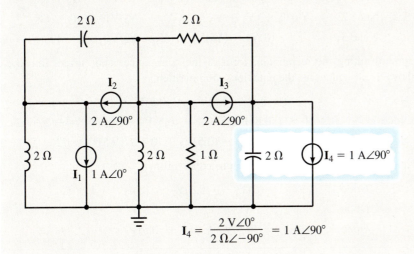

$$\mathbf{I}_4 = \frac{2\,\mathrm{V}\angle 0°}{2\,\Omega\angle -90°} = 1\,\mathrm{A}\angle 90°$$

FIGURE 19–27

Now, by labelling the nodes and admittances, the circuit may be simplified as shown in Figure 19–28.

$$\mathbf{Y}_1 = 0 - j\tfrac{1}{2}\,\mathrm{S} \qquad \mathbf{Y}_3 = \tfrac{1}{2}\,\mathrm{S} + j\,0 \qquad \mathbf{Y}_5 = 1\,\mathrm{S} - j\tfrac{1}{2}\,\mathrm{S}$$
$$\mathbf{Y}_2 = 0 + j\tfrac{1}{2}\,\mathrm{S} \qquad \mathbf{Y}_4 = 0 + j\tfrac{1}{2}\,\mathrm{S}$$

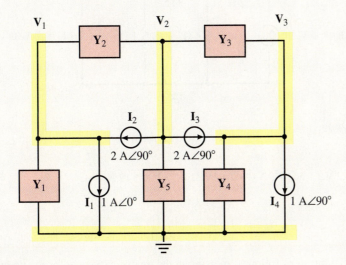

FIGURE 19–28

The admittances of Figure 19–28 are as follows:

$$\mathbf{Y}_1 = 0 - j0.5 \text{ S}$$
$$\mathbf{Y}_2 = 0 + j0.5 \text{ S}$$
$$\mathbf{Y}_3 = 0.5 \text{ S} + j0$$
$$\mathbf{Y}_4 = 0 + j0.5 \text{ S}$$
$$\mathbf{Y}_5 = 1.0 \text{ S} - j0.5 \text{ S}$$

Using the assigned admittances, the nodal equations are written as follows:

Node 1: $\quad (\mathbf{Y}_1 + \mathbf{Y}_2)\mathbf{V}_1 - (\mathbf{Y}_2)\mathbf{V}_2 - (0)\mathbf{V}_3 = -\mathbf{I}_1 + \mathbf{I}_2$

Node 2: $\quad -(\mathbf{Y}_2)\mathbf{V}_1 + (\mathbf{Y}_2 + \mathbf{Y}_3 + \mathbf{Y}_5)\mathbf{V}_2 - (\mathbf{Y}_3)\mathbf{V}_3 = -\mathbf{I}_2 - \mathbf{I}_3$

Node 3: $\quad -(0)\mathbf{V}_1 - (\mathbf{Y}_3)\mathbf{V}_2 + (\mathbf{Y}_3 + \mathbf{Y}_4)\mathbf{V}_3 = \mathbf{I}_3 - \mathbf{I}_4$

By substituting the rectangular form of the admittances and current into the above linear equations, the equations are rewritten as

Node 1: $\quad (-j0.5 + j0.5)\mathbf{V}_1 - (j0.5)\mathbf{V}_2 - (0)\mathbf{V}_3 = -1 + j2$

Node 2: $\quad -(j0.5)\mathbf{V}_1 + (j0.5 + 0.5 + 1 - j0.5)\mathbf{V}_2 - (0.5)\mathbf{V}_3 = -j2 - j2$

Node 3: $\quad -(0)\mathbf{V}_1 - (0.5)\mathbf{V}_2 + (0.5 + j0.5)\mathbf{V}_3 = j2 - j1$

Finally, the nodal equations are simplified as follows:

Node 1: $\quad (0)\mathbf{V}_1 - (j0.5)\mathbf{V}_2 - (0)\mathbf{V}_3 = -1 + j2$

Node 2: $\quad -(j0.5)\mathbf{V}_1 + (1.5)\mathbf{V}_2 - (0.5)\mathbf{V}_3 = -j4$

Node 3: $\quad (0)\mathbf{V}_1 - (0.5)\mathbf{V}_2 + (0.5 + j0.5)\mathbf{V}_3 = j1$

PRACTICE PROBLEMS 4

Given the circuit of Figure 19–29, use nodal analysis to find the voltages $\mathbf{V}_1$ and $\mathbf{V}_2$. Use your results to find the current $\mathbf{I}$.

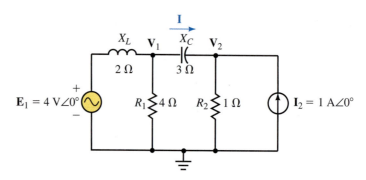

FIGURE 19–29

Answers
$\mathbf{V}_1 = 4.22 \text{ V}\angle{-56.89°}$, $\mathbf{V}_2 = 2.19 \text{ V}\angle 1.01°$, $\mathbf{I} = 1.19 \text{ A}\angle 1.85°$

**IN-PROCESS
LEARNING CHECK 3**

(Answers are at the end of the chapter.)

Briefly list the steps followed in using nodal analysis to solve for the node voltages of a circuit.

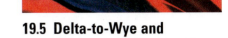

In Chapter 8, we derived the relationships showing the equivalence of "delta" (or "pi") connected resistance to a "wye" (or "tee") configuration.

In a similar manner, impedances connected in a Δ configuration are equivalent to a unique Y configuration. Figure 19–30 shows the equivalent circuits.

19.5 Delta-to-Wye and Wye-to-Delta Conversions

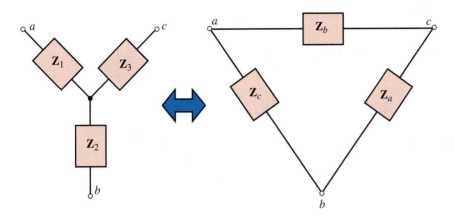

FIGURE 19–30 Delta-wye equivalence.

A Δ configuration is converted to a Y equivalent by using the following:

$$\mathbf{Z}_1 = \frac{\mathbf{Z}_b \mathbf{Z}_c}{\mathbf{Z}_a + \mathbf{Z}_b + \mathbf{Z}_c} \qquad (19\text{–}1)$$

$$\mathbf{Z}_2 = \frac{\mathbf{Z}_a \mathbf{Z}_c}{\mathbf{Z}_a + \mathbf{Z}_b + \mathbf{Z}_c} \qquad (19\text{–}2)$$

$$\mathbf{Z}_3 = \frac{\mathbf{Z}_a \mathbf{Z}_b}{\mathbf{Z}_a + \mathbf{Z}_b + \mathbf{Z}_c} \qquad (19\text{–}3)$$

The above conversion indicates that the impedance in any arm of a Y circuit is determined by taking the product of the two adjacent Δ impedances at this arm and dividing by the summation of the Δ impedances.

If the impedances of the sides of the Δ network are all equal (magnitude and phase angle), the equivalent Y network will have identical impedances, where each impedance is determined as

$$\mathbf{Z}_Y = \frac{\mathbf{Z}_\Delta}{3} \qquad (19\text{–}4)$$

A Y configuration is converted to a Δ equivalent by using the following:

$$\mathbf{Z}_a = \frac{\mathbf{Z}_1 \mathbf{Z}_2 + \mathbf{Z}_1 \mathbf{Z}_3 + \mathbf{Z}_2 \mathbf{Z}_3}{\mathbf{Z}_1} \qquad (19\text{–}5)$$

$$\mathbf{Z}_b = \frac{\mathbf{Z}_1 \mathbf{Z}_2 + \mathbf{Z}_1 \mathbf{Z}_3 + \mathbf{Z}_2 \mathbf{Z}_3}{\mathbf{Z}_2} \qquad (19\text{–}6)$$

$$\mathbf{Z}_c = \frac{\mathbf{Z}_1 \mathbf{Z}_2 + \mathbf{Z}_1 \mathbf{Z}_3 + \mathbf{Z}_2 \mathbf{Z}_3}{\mathbf{Z}_3} \qquad (19\text{–}7)$$

Any impedance in a "Δ" is determined by summing the possible two-impedance product combinations of the "Y" and then dividing by the impedance found in the opposite branch of the "Y."

If the arms of a "Y" have identical impedances, the equivalent "Δ" will have impedances given as

$$\mathbf{Z}_\Delta = 3\mathbf{Z}_Y \qquad\qquad (19\text{–}8)$$

EXAMPLE 19–11

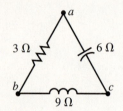

FIGURE 19–31

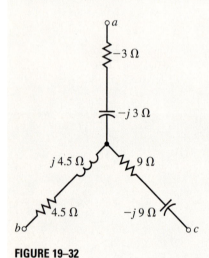

FIGURE 19–32

Determine the Y equivalent of the Δ network shown in Figure 19–31.

Solution

$$\mathbf{Z}_1 = \frac{(3\ \Omega)(-j6\ \Omega)}{3\ \Omega - j6\ \Omega + j9\ \Omega} = \frac{-j18\ \Omega}{3 + j3} = \frac{18\ \Omega\angle -90°}{4.242\angle 45°}$$
$$= 4.242\ \Omega\angle -135°$$
$$= -3.0\ \Omega - j3.0\ \Omega$$

$$\mathbf{Z}_2 = \frac{(3\ \Omega)(j9\ \Omega)}{3\ \Omega - j6\ \Omega + j9\ \Omega} = \frac{j27\ \Omega}{3 + j3} = \frac{27\ \Omega\angle 90°}{4.242\angle 45°}$$
$$= 6.364\ \Omega\angle 45°$$
$$= 4.5\ \Omega + j4.5\ \Omega$$

$$\mathbf{Z}_3 = \frac{(j9\ \Omega)(-j6\ \Omega)}{3\ \Omega - j6\ \Omega + j9\ \Omega} = \frac{54\ \Omega}{3 + j3} = \frac{54\ \Omega\angle 0°}{4.242\angle 45°}$$
$$= 12.73\ \Omega\angle -45°$$
$$= 9.0\ \Omega - j9.0\ \Omega$$

In the above solution, we see that the given Δ network has an equivalent Y network with one arm having a negative resistance. This result indicates that although the Δ circuit has an equivalent Y circuit, the Y circuit cannot actually be constructed from real components since *negative resistors* do not exist (although some active components may demonstrate negative resistance characteristics). If the given conversion is used to simplify a circuit we would treat the impedance $\mathbf{Z}_1 = -3\ \Omega - j3\ \Omega$ as if the resistance actually were a negative value. Figure 19–32 shows the equivalent Y circuit.

It is left to the student to show that the Y of Figure 19–32 is equivalent to the Δ of Figure 19–31.

EXAMPLE 19–12

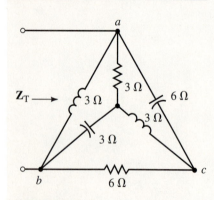

FIGURE 19–33

Find the total impedance of the network in Figure 19–33.

Solution If we take a moment to examine this network, we see that the circuit contains both a Δ and a Y. In calculating the total impedance, the solution is easier when we convert the Y to a Δ.

The conversion is shown in Figure 19–34.

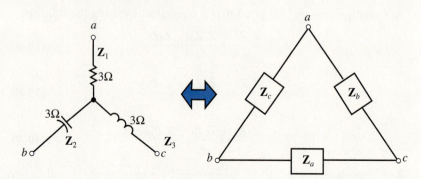

FIGURE 19–34

$$\mathbf{Z}_a = \frac{\mathbf{Z}_1\mathbf{Z}_2 + \mathbf{Z}_1\mathbf{Z}_3 + \mathbf{Z}_2\mathbf{Z}_3}{\mathbf{Z}_1}$$

$$= \frac{(3\ \Omega)(j3\ \Omega) + (3\ \Omega)(-j3\ \Omega) + (j3\ \Omega)(-j3\ \Omega)}{3\ \Omega}$$

$$= \frac{-j29\ \Omega}{3} = 3\ \Omega$$

$$\mathbf{Z}_b = \frac{\mathbf{Z}_1\mathbf{Z}_2 + \mathbf{Z}_1\mathbf{Z}_3 + \mathbf{Z}_2\mathbf{Z}_3}{\mathbf{Z}_2} = \frac{9\ \Omega}{-j3} = j3\ \Omega$$

$$\mathbf{Z}_c = \frac{\mathbf{Z}_1\mathbf{Z}_2 + \mathbf{Z}_1\mathbf{Z}_3 + \mathbf{Z}_2\mathbf{Z}_3}{\mathbf{Z}_3} = \frac{9\ \Omega}{j3} = -j3\ \Omega$$

Now, substituting the equivalent Δ into the original network, we have the revised network of Figure 19–35.

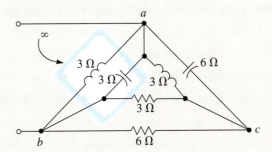

FIGURE 19–35

The network of Figure 19–35 shows that the corresponding sides of the Δ are parallel. Because the inductor and the capacitor in the left side of the Δ have the same values, we may replace the parallel combination of these two components with an open circuit. The resulting impedance of the network is now easily determined as

$$\mathbf{Z}_\mathrm{T} = 3\ \Omega \| 6\ \Omega + (j3\ \Omega)\|(-j6\ \Omega) = 2\ \Omega + j6\ \Omega$$

PRACTICE PROBLEMS 5

A Y network consists of a 60-Ω capacitor, a 180-Ω inductor, and a 540-V resistor. Determine the corresponding Δ network.

Answers
$\mathbf{Z}_a = -1080\ \Omega + j180\ \Omega$, $\mathbf{Z}_b = 20\ \Omega + j120\ \Omega$, $\mathbf{Z}_c = 360\ \Omega - j60\ \Omega$

A Δ network consists of a resistor, inductor, and capacitor, each having an impedance of 150 Ω. Determine the corresponding Y network.

IN-PROCESS
LEARNING CHECK 4

(Answers are at the end of the chapter.)

Bridge circuits, similar to the network of Figure 19–36, are used extensively in electronics to measure the values of unknown components.

Recall from Chapter 8 that any bridge circuit is said to be balanced when the current through the branch between the two arms is zero. In a practical circuit, component values of very precise resistors are adjusted until the current through the central element (usually a sensitive galvanometer) is exactly equal

19.6 Bridge Networks

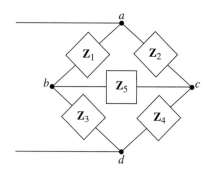

FIGURE 19–36

to zero. For ac circuits, the condition of a **balanced bridge** occurs when the impedance vectors of the various arms satisfy the following condition:

$$\frac{\mathbf{Z}_1}{\mathbf{Z}_3} = \frac{\mathbf{Z}_2}{\mathbf{Z}_4} \qquad (19\text{–}9)$$

When a balanced bridge occurs in a circuit, the equivalent impedance of the bridge network is easily determined by removing the central impedance and replacing it by either an open or a short circuit. The resulting impedance of the bridge circuit is then found as either of the following:

$$\mathbf{Z}_T = \mathbf{Z}_1\|\mathbf{Z}_2 + \mathbf{Z}_3\|\mathbf{Z}_4$$

or

$$\mathbf{Z}_T = (\mathbf{Z}_1 + \mathbf{Z}_3)\|(\mathbf{Z}_2 + \mathbf{Z}_4)$$

If, on the other hand, the bridge is not balanced, then the total impedance must be determined by performing a Δ-to-Y conversion. Alternatively, the circuit may be analyzed by using either mesh analysis or nodal analysis.

EXAMPLE 19–13

Given that the circuit of Figure 19–37 is a balanced bridge.

a. Calculate the unknown impedance, $\mathbf{Z}_x$.

b. Determine the values of L_x and R_x if the circuit operates at a frequency of 1 kHz.

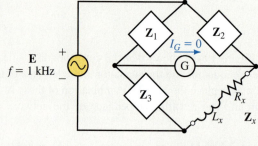

FIGURE 19–37

$$\mathbf{Z}_1 = 30 \text{ k}\Omega\angle{-20°}$$
$$\mathbf{Z}_2 = 10 \text{ k}\Omega\angle 0°$$
$$\mathbf{Z}_3 = 100 \ \Omega\angle 0°$$

Solution

a. The expression for the unknown impedance is determined from Equation 19–9 as

$$\mathbf{Z}_x = \frac{\mathbf{Z}_2\mathbf{Z}_3}{\mathbf{Z}_1}$$

$$= \frac{(10 \text{ k}\Omega)(100 \ \Omega)}{30 \text{ k}\Omega\angle{-20°}}$$

$$= 33.3 \ \Omega\angle 20°$$

$$= 31.3 + j11.4 \ \Omega$$

b. From the above result, we have

$$R_x = 31.3 \ \Omega$$

and

$$L_x = \frac{X_L}{2\pi f} = \frac{11.4 \ \Omega}{2\pi(1000 \text{ Hz})} = 1.81 \text{ mH}$$

We will now consider various forms of bridge circuits that are used in electronic circuits to determine the values of unknown inductors and capacitors. As in resistor bridges, the circuits use variable resistors together with very sensitive galvanometer movements to ensure a balanced condition for the bridge. However, rather than using a dc source to provide current in the circuit, the bridge circuits use ac sources operating at a known frequency (usually 1 kHz). Once the bridge is balanced, the value of unknown inductance or capacitance may be easily determined by obtaining the reading directly from the instrument. Most instruments using bridge circuitry will incorporate several different bridges to enable the measurement of various types of unknown impedances.

Maxwell Bridge

The **Maxwell bridge,** shown in Figure 19–38, is used to determine the inductance and series resistance of an inductor having a relatively large series resistance (in comparison to $X_L = \omega L$).

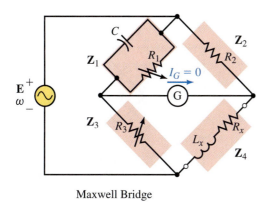

Maxwell Bridge

FIGURE 19–38 Maxwell bridge.

Resistors R_1 and R_3 are adjusted to provide the balanced condition (when the current through the galvanometer is zero: $I_G = 0$).

When the bridge is balanced, we know that the following condition must apply:

$$\frac{\mathbf{Z}_1}{\mathbf{Z}_2} = \frac{\mathbf{Z}_3}{\mathbf{Z}_4}$$

If we write the impedances using the rectangular forms, we obtain

$$\frac{\left[\dfrac{(R_1)\left(-j\dfrac{1}{\omega C}\right)}{R_1 - j\dfrac{1}{\omega C}}\right]}{R_2} = \frac{R_3}{R_x + j\omega L_x}$$

$$\frac{\left(-j\dfrac{R_1}{\omega C}\right)}{\left(\dfrac{\omega R_1 C - j1}{\omega C}\right)} = \frac{R_2 R_3}{R_x + j\omega L_x}$$

$$\frac{-jR_1}{\omega C R_1} - j = \frac{R_2 R_3}{R_x + j\omega L_x}$$

$$(-jR_1)(R_x + j\omega L_x) = R_2 R_3(\omega C R_1 - j)$$

$$\omega L_x R_1 - jR_1 R_x = \omega R_1 R_2 R_3 C - jR_2 R_3$$

Now, since two complex numbers can be equal only if their real parts are equal and if their imaginary parts are equal, we must have the following:

$$\omega L_x R_1 = \omega R_1 R_2 R_3 C$$

and

$$R_1 R_x = R_2 R_3$$

Simplifying these expressions, we get the following equations for a Maxwell bridge:

$$L_x = R_2 R_3 C \qquad\qquad (19\text{--}10)$$

and

$$R_x = \frac{R_2 R_3}{R_1} \qquad\qquad (19\text{--}11)$$

EXAMPLE 19–14

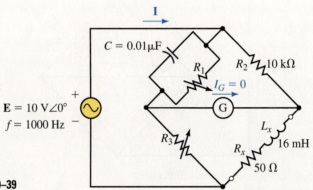

FIGURE 19–39

a. Determine the values of R_1 and R_3 so that the bridge of Figure 19–39 is balanced.

b. Calculate the current **I** when the bridge is balanced.

Solution

a. Rewriting Equations 19–10 and 19–11 and solving for the unknowns, we have

$$R_3 = \frac{L_x}{R_2 C} = \frac{16 \text{ mH}}{(10 \text{ k}\Omega)(0.01 \text{ }\mu\text{F})} = 160 \text{ }\Omega$$

and

$$R_1 = \frac{R_2 R_3}{R_x} = \frac{(10 \text{ k}\Omega)(160 \text{ }\Omega)}{50 \text{ }\Omega} = 32 \text{ k}\Omega$$

b. The total impedance is found as

$$\mathbf{Z_T} = (\mathbf{Z_C}\|\mathbf{R_1}\|\mathbf{R_2}) + [\mathbf{R_3}\|(\mathbf{R_x} + \mathbf{Z_{Lx}})]$$
$$\mathbf{Z_T} = (-j15.915 \text{ k}\Omega)\|32 \text{ k}\Omega\|10 \text{ k}\Omega + [160 \text{ }\Omega\|(50 \text{ }\Omega + j100.5 \text{ }\Omega)]$$
$$= 6.87 \text{ k}\Omega\angle{-25.6°} + 77.2 \text{ }\Omega\angle{38.0°}$$
$$= 6.91 \text{ k}\Omega\angle{-25.0°}$$

The resulting circuit current is

$$\mathbf{I} = \frac{10 \text{ V}\angle{0°}}{6.91 \text{ k}\Omega\angle{-25°}} = 1.45 \text{ mA}\angle{25.0°}$$

Hay Bridge

In order to measure the inductance and series resistance of an inductor having a small series resistance, a **Hay bridge** is generally used. The Hay bridge is shown in Figure 19–40.

By applying a method similar to that used to determine the values of the unknown inductance and resistance of the Maxwell bridge, it may be shown that the following equations for the Hay bridge apply:

$$L_x = \frac{R_2 R_3 C}{\omega^2 R_1^2 C^2 + 1} \qquad \textbf{(19–12)}$$

and

$$R_x = \frac{\omega^2 R_1 R_2 R_3 C^2}{\omega^2 R_1^2 C^2 + 1} \qquad \textbf{(19–13)}$$

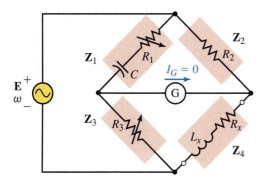

FIGURE 19–40 Hay bridge.

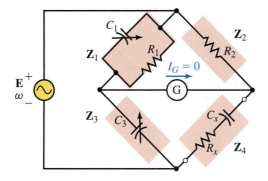

FIGURE 19–41 Schering bridge.

Schering Bridge

The **Schering bridge,** shown in Figure 19–41, is a circuit used to determine the value of unknown capacitance.

By solving for the balanced bridge condition, we have the following equations for the unknown quantities of the circuit:

$$C_x = \frac{R_1 C_3}{R_2} \qquad \textbf{(19–14)}$$

$$R_x = \frac{C_1 R_2}{C_3} \qquad \textbf{(19–15)}$$

EXAMPLE 19–15

Determine the values of C_1 and C_3 that will result in a balanced bridge for the circuit of Figure 19–42.

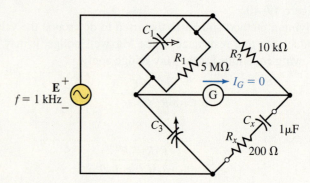

◀ MULTISIM

FIGURE 19–42

Solution   Rewriting Equations 19–14 and 19–15, we solve for the unknown capacitances as

$$C_3 = \frac{R_2 C_x}{R_1} = \frac{(10 \text{ k}\Omega)(1 \text{ }\mu\text{F})}{5 \text{ M}\Omega} = 0.002 \text{ }\mu\text{F}$$

and

$$C_1 = \frac{C_3 R_x}{R_2} = \frac{(0.002 \text{ }\mu\text{F})(200 \text{ }\Omega)}{10 \text{ k}\Omega} = 40 \text{ pF}$$

PRACTICE PROBLEMS 6

Determine the values of R_1 and R_3 so that the bridge of Figure 19–43 is balanced.

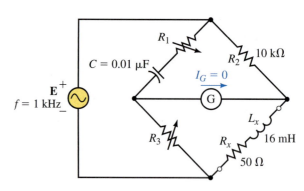

FIGURE 19–43

Answers
$R_1 = 7916 \text{ }\Omega$, $R_3 = 199.6 \text{ }\Omega$

19.7 Circuit Analysis Using Computers

◀ MULTISIM

◀ CADENCE

In some of the examples in this chapter, we analyzed circuits that resulted in as many as three simultaneous linear equations. You have no doubt wondered if there is a less complicated way to solve these circuits without the need for using complex algebra. Computer programs are particularly useful for solving such ac circuits. Both MultiSIM and PSpice have individual strengths in the solution of ac circuits. As in previous examples, MultiSIM provides an excellent simulation of how measurements are taken in a lab. PSpice, on the other hand, provides voltage and current readings, complete with magnitude and phase angle. The following examples show how these programs are useful for examining the circuits in this chapter.

Use MultiSIM to show that the bridge circuit of Figure 19–44 is balanced.

EXAMPLE 19–16

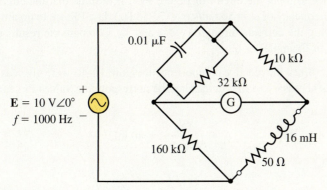

FIGURE 19–44

Solution   Recall that a bridge circuit is balanced when the current through the branch between the two arms of the bridge is equal to zero. In this example, we will use a multimeter set on its ac ammeter range to verify the condition of the circuit. The ammeter is selected by clicking on **A** and it is set to its ac range by clicking on the sinusoidal button. Figure 19–45 shows the circuit connections and the ammeter reading. The results correspond to the conditions that were previously analyzed in Example 19–14. (Note: When we use MultiSIM, the ammeter may not show exactly zero current in the balanced condition. This is due to the way the program does the calculations. Any current less than 5 µA is considered to be effectively zero.)

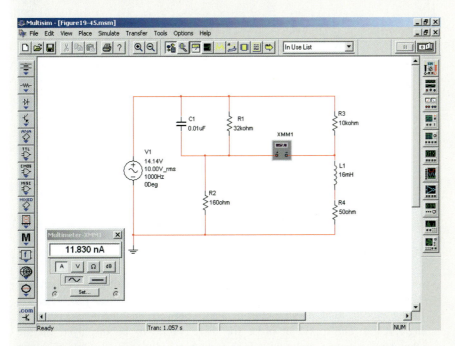

FIGURE 19–45

◀ MULTISIM

PRACTICE PROBLEMS 7

Use MultiSIM to verify that the results obtained in Example 19–15 result in a balanced bridge circuit. (Assume that the bridge is balanced if the galvanometer current is less than 5 μA.)

PSpice

EXAMPLE 19–17

Use PSpice to input the circuit of Figure 19–15. Assume that the circuit operates at a frequency of $\omega = 50$ rad/s ($f = 7.958$ Hz). Use PSpice to obtain a printout showing the currents through X_C, R_2, and X_L. Compare the results to those obtained in Example 19–6.

Solution   Since the reactive components in Figure 19–15 were given as impedance, it is necessary to first determine the corresponding values in henries and farads.

$$L = \frac{4\ \Omega}{50\ \text{rad/s}} = 80\ \text{mH}$$

and

$$C = \frac{1}{(2\ \Omega)(50\ \text{rad/s})} = 10\ \text{mF}$$

Now we are ready to use OrCAD Capture to input the circuit as shown in Figure 19–46. The basic steps are reviewed for you. Use the ac current source, ISRC from the SOURCE library and place one IPRINT part from the SPECIAL library. The resistor, inductor, and capacitor are selected from the ANALOG library and the ground symbol is selected by using the Place ground tool.

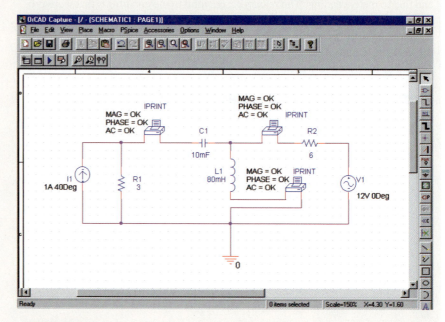

FIGURE 19–46

Change the value of the current source by double clicking on the part and moving the horizontal scroll bar until you find the field titled AC. Type **1A 40Deg** into this field. A space must be placed between the magnitude and phase angle. Click on Apply. In order for these values to be displayed on the schematic, you must click on the Display button and then <u>V</u>alue Only. Click on OK to return to the properties editor and then close the editor by clicking on X.

The IPRINT part is similar to an ammeter and provides a printout of the current magnitude and phase angle. The properties of the IPRINT part are changed by double clicking on the part and scrolling across to show the appropriate fields. Type **OK** in the AC, MAG, and PHASE fields. In order to display the selected fields on the schematic, you must click on the Display button and then select Name and Value after changing each field. Since we need to measure three currents in the circuit, we could follow this procedure two more times. However, an easier method is to click on the IPRINT part and copy the part by using <Ctrl><C> and <Ctrl><V>. Each IPRINT will then have the same properties.

Once the rest of the circuit is completed and wired, click on the New Simulation Profile tool. Give the simulation a name (such as **ac Branch Currents**). Click on the Analysis tab and select AC Sweep/ Noise as the analysis type. Type the following values:

Start Frequency: **7.958Hz**

End Frequency: **7.958Hz**

Total Points: **1**

Since we do not need the Probe postprocessor to run, it is disabled by selecting the Probe Window tab (from the Simulation Settings dialog box). Click on Display Probe window and exit the simulation settings by clicking on OK.

Click on the Run tool. Once PSpice has successfully run, click on the <u>V</u>iew menu and select the Output F<u>i</u>le menu item. Scroll through the file until the currents are shown as follows:

```
FREQ           IM(V_PRINT1) IP(V_PRINT1)
 7.958E+00      7.887E-01    -1.201E+02
FREQ           IM(V_PRINT2) IP(V_PRINT2)
 7.958E+00      1.304E+00     1.560E+02
FREQ           IM(V_PRINT3) IP(V_PRINT3)
 7.958E+00      1.450E+00    -5.673E+01
```

The above printout provides: $\mathbf{I}_1 = 0.7887$ A$\angle-120.1°$, $\mathbf{I}_2 = 1.304$ A$\angle156.0°$, and $\mathbf{I}_3 = 1.450$ A$\angle-56.73°$. These results are consistent with those calculated in Example 19–6.

PRACTICE PROBLEMS 8

Use PSpice to evaluate the node voltages for the circuit of Figure 19–23. Assume that the circuit operates at an angular frequency of $\omega = 1000$ rad/s ($f = 159.15$ Hz).

PUTTING IT INTO PRACTICE

The Schering bridge of Figure 19–74 (p. 779) is balanced. In this chapter, you have learned several methods that allow you to find the current anywhere in a circuit. Using any method, determine the current through the galvanometer if the value of $C_x = 0.07$ μF (All other values remain unchanged.) Repeat the calculations for a value of $C_x = 0.09$ μF. Can you make a general statement for current through the galvanometer if C_x is smaller that the value required to balance the bridge? What general statement can be made if the value of C_x is larger than the value in the balanced bridge?

19.1 Dependent Sources

1. Refer to the circuit of Figure 19–47.

   Find **V** when the controlling current **I** is the following:

   a. 20 μA∠0°

   b. 50 μA∠−180°

   c. 60 μA∠60°

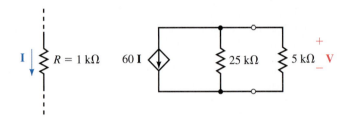

FIGURE 19–47

2. Refer to the circuit of Figure 19–48.

   Find **I** when the controlling voltage, **V,** is the following:

   a. 30 mV∠0°

   b. 60 mV∠−180°

   c. 100 mV∠−30°

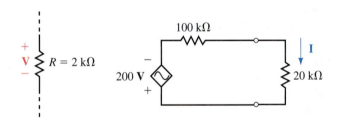

FIGURE 19–48

3. Repeat Problem 1 for the circuit of Figure 19–49.

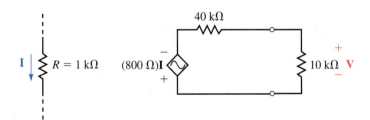

FIGURE 19–49

4. Repeat Problem 2 for the circuit of Figure 19–50.

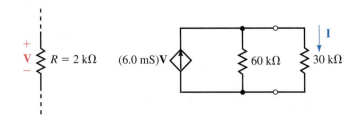

FIGURE 19–50

5. Find the output voltage, $\mathbf{V}_{out}$, for the circuit of Figure 19–51.

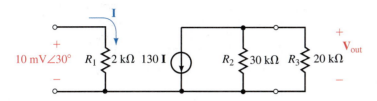

6. Repeat Problem 5 for the circuit of Figure 19–52.

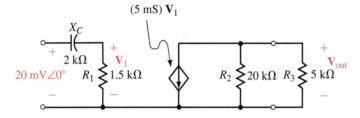

19.2 Source Conversion

7. Given the circuits of Figure 19–53, convert each of the current sources into an equivalent voltage source. Use the resulting circuit to find $\mathbf{V}_L$.

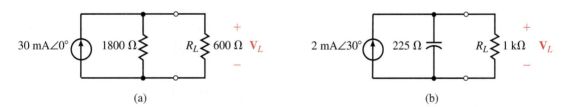

(a)   (b)

8. Convert each voltage source of Figure 19–54 into an equivalent current source.

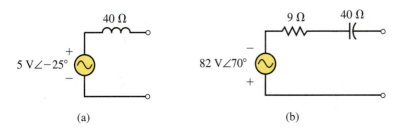

(a)   (b)

9. Refer to the circuit of Figure 19–55.

   a. Solve for the voltage, **V**.

   b. Convert the current source into an equivalent voltage source and again solve for **V**. Compare to the result obtained in (a).

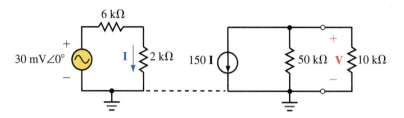

FIGURE 19–55

10. Refer to the circuit of Figure 19–56.

    a. Solve for the voltage, V_L.

    b. Convert the current source into an equivalent voltage source and again solve for V_L.

    c. If $I = 5\ \mu A\angle 90°$, what is V_L?

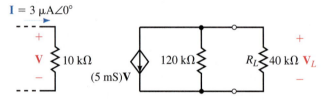

FIGURE 19–56

19.3   Mesh (Loop) Analysis

11. Consider the circuit of Figure 19–57.

    a. Write the mesh equations for the circuit.

    b. Solve for the loop currents.

    c. Determine the current **I** through the 4-Ω resistor.

12. Refer to the circuit of Figure 19–58.

    a. Write the mesh equations for the circuit.

    b. Solve for the loop currents.

    c. Determine the current through the 25-Ω inductor.

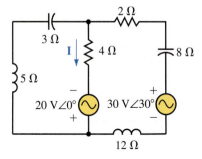

FIGURE 19–57

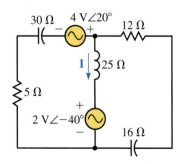

FIGURE 19–58

13. Refer to the circuit of Figure 19–59.

    a. Simplify the circuit and write the mesh equations.

    b. Solve for the loop currents.

    c. Determine the voltage **V** across the 15-Ω capacitor.

14. Consider the circuit of Figure 19–60.

    a. Simplify the circuit and write the mesh equations.

    b. Solve for the loop currents.

    c. Determine the voltage **V** across the 2-Ω resistor.

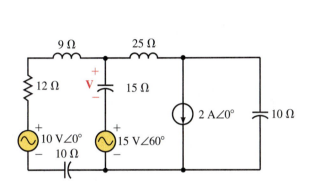

FIGURE 19–59

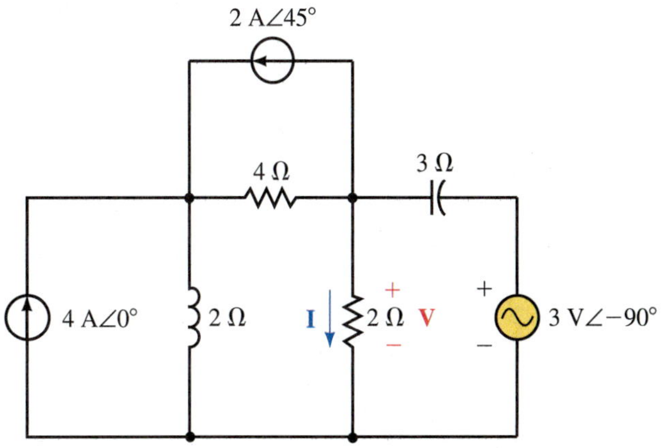

FIGURE 19–60

15. Use mesh analysis to find the current **I** and the voltage **V** in the circuit of Figure 19–61.

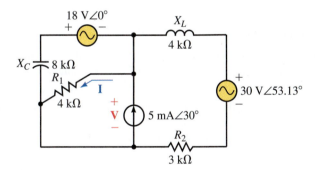

FIGURE 19–61

16. Repeat Problem 15 for the circuit of Figure 19–62.

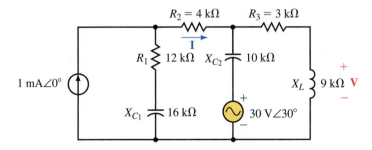

FIGURE 19–62

19.4 Nodal Analysis

17. Consider the circuit of Figure 19–63.

    a. Write the nodal equations.

    b. Solve for the node voltages.

    c. Determine the current **I** through the 4-Ω capacitor.

18. Refer to the circuit of Figure 19–64.

    a. Write the nodal equations.

    b. Solve for the node voltages.

    c. Determine the voltage **V** across the 3-Ω capacitor.

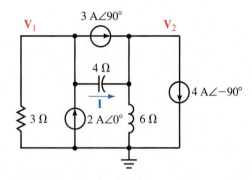

FIGURE 19–63

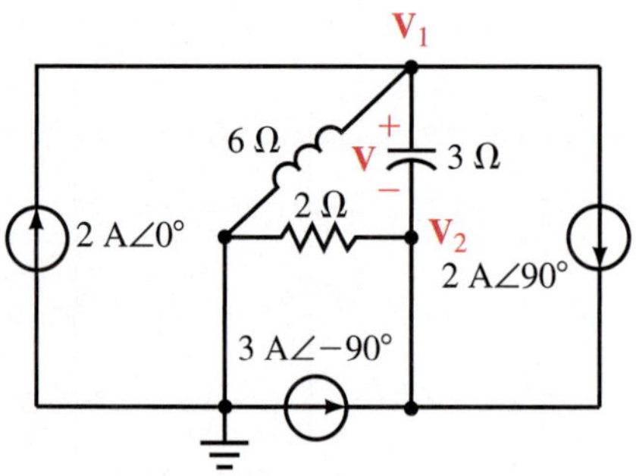

FIGURE 19–64

19. a. Simplify the circuit of Figure 19–59, and write the nodal equations.

    b. Solve for the node voltages.

    c. Determine the voltage across the 15-Ω capacitor.

20. a. Simplify the circuit of Figure 19–60, and write the nodal equations.

    b. Solve for the node voltages.

    c. Determine the current through the 2-Ω resistor.

21. Use nodal analysis to determine the node voltages in the circuit of Figure 19–61. Use the results to find the current **I** and the voltage **V.** Compare your answers to those obtained using mesh analysis in Problem 15.

22. Use nodal analysis to determine the node voltages in the circuit of Figure 19–62. Use the results to find the current **I** and the voltage **V.** Compare your answers to those obtained using mesh analysis in Problem 16.

19.5 Delta-to-Wye and Wye-to-Delta Conversions

23. Convert each of the Δ networks of Figure 19–65 into an equivalent Y network.

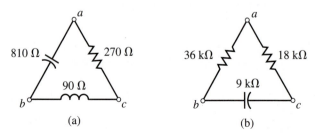

(a)

(b)

FIGURE 19–65

24. Convert each of the Y networks of Figure 19–66 into an equivalent Δ network.

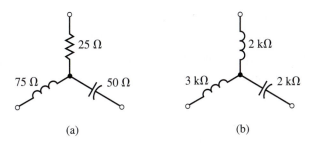

(a)   (b)

FIGURE 19–66

25. Using Δ→Y or Y→Δ conversion, calculate **I** for the circuit of Figure 19–67.

26. Using Δ→Y or Y→Δ conversion, calculate **I** for the circuit of Figure 19–68.

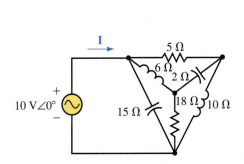

FIGURE 19–67

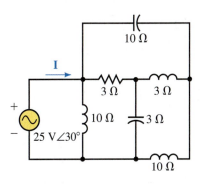

FIGURE 19–68

27. Refer to the circuit of Figure 19–69:
    a. Determine the equivalent impedance, Z_T, of the circuit.
    b. Find the currents **I** and I_1.

28. Refer to the circuit of Figure 19–70:
    a. Determine the equivalent impedance, Z_T, of the circuit.
    b. Find the voltages **V** and V_1.

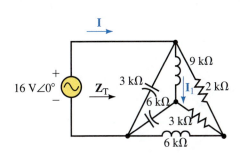

FIGURE 19–69

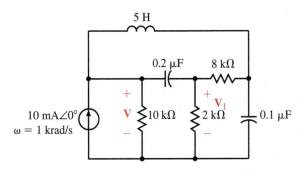

FIGURE 19–70

19.6   Bridge Networks

29. Given that the bridge circuit of Figure 19–71 is balanced:
    a. Determine the value of the unknown impedance.
    b. Solve for the current **I**.

30. Given that the bridge circuit of Figure 19–72 is balanced:
    a. Determine the value of the unknown impedance.
    b. Solve for the current **I**.

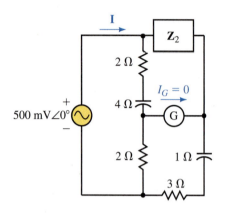

FIGURE 19–71

FIGURE 19–72

31. Show that the bridge circuit of Figure 19–73 is balanced.
32. Show that the bridge circuit of Figure 19–74 is balanced.

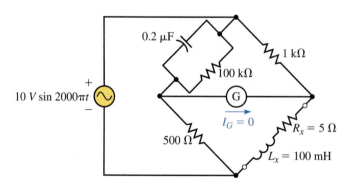

◀ MULTISIM   **FIGURE 19–73**

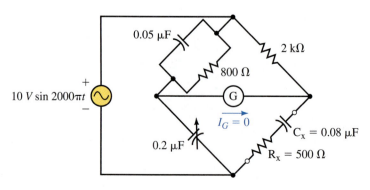

◀ MULTISIM   **FIGURE 19–74**

33. Derive Equations 19–14 and 19–15 for the balanced Schering bridge.

34. Derive Equations 19–12 and 19–13 for the balanced Hay bridge.

35. Determine the values of the unknown resistors that will result in a balanced bridge for the circuit of Figure 19–75.

36. Determine the values of the unknown capacitors which will result in a balanced bridge for the circuit of Figure 19–76.

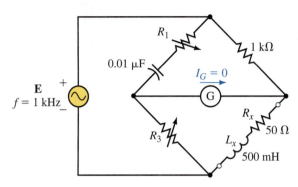

FIGURE 19–75   **FIGURE 19–76**

19.7   Circuit Analysis Using Computers

37. Use MultiSIM to show that the bridge circuit of Figure 19–73 is balanced. (Assume that the bridge is balanced if the galvanometer current is less than 5 μA.)

◀ MULTISIM

38. Repeat Problem 37 for the bridge circuit of Figure 19–74.

◀ MULTISIM

39. Use PSpice to input the file for the circuit of Figure 19–21. Assume that the circuit operates at a frequency $\omega = 2$ krad/s. Use IPRINT and VPRINT to obtain a printout of the node voltages and the current through each element of the circuit.

◀ CADENCE

40. Use PSpice to input the file for the circuit of Figure 19–29. Assume that the circuit operates at a frequency $\omega = 1$ krad/s. Use IPRINT and VPRINT to obtain a printout of the node voltages and the current through each element of the circuit.

◀ CADENCE

41. Use PSpice to input the file for the circuit of Figure 19–68. Assume that the circuit operates at a frequency $\omega = 20$ rad/s. Use IPRINT to obtain a printout of the current **I**.

◀ CADENCE

42. Use PSpice to input the file for the circuit of Figure 19–69. Assume that the circuit operates at a frequency $\omega = 3$ krad/s. Use IPRINT to obtain a printout of the current **I**.

◀ CADENCE

✔ ANSWERS TO IN-PROCESS LEARNING CHECKS

In-Process Learning Check 1

a. $\mathbf{E} = 1000\ \text{V}\angle 30°$ b. $\mathbf{E} = 4\ \text{V}\angle -90°$

c. $\mathbf{E} = 1024\ \text{V}\angle -38.66°$

In-Process Learning Check 2

1. Convert current sources to voltage sources.

2. Redraw the circuit.

3. Assign a clockwise current to each loop.

4. Write loop equations using Kirchhoff's voltage law.

5. Solve the resulting simultaneous linear equations to find the loop currents.

In-Process Learning Check 3

1. Convert voltage sources to current sources.

2. Redraw the circuit.

3. Label all nodes, including the reference node.

4. Write nodal equations using Kirchhoff's current law.

5. Solve the resulting simultaneous linear equations to find the node voltages.

In-Process Learning Check 4

$\mathbf{Z}_1 = 150\ \Omega\angle 90°$, $\mathbf{Z}_2 = 150\ \Omega\angle -90°$, $\mathbf{Z}_3 = 150\ \Omega\angle 0°$

■ **OBJECTIVES**

After studying this chapter you will be able to

- apply the superposition theorem to determine the voltage across or current through any component in a given circuit,

- determine the Thévenin equivalent of circuits having independent and/or dependent sources,

- determine the Norton equivalent of circuits having independent and/or dependent sources,

- apply the maximum power transfer theorem to determine the load impedance for which maximum power is transferred to the load from a given circuit,

- use PSpice to find the Thévenin and Norton equivalents of circuits having either independent or dependent sources,

- use MultiSIM to verify the operation of ac circuits.

AC Network Theorems

20

CHAPTER PREVIEW

In this chapter we apply the superposition, Thévenin, Norton, and maximum power transfer theorems in the analysis of ac circuits. Although the Millman and reciprocity theorems apply to ac circuits as well as to dc circuits, they are omitted since the applications are virtually identical with those used in analyzing dc circuits.

Many of the techniques used in this chapter are similar to those used in Chapter 9, and as a result, most students will find a brief review of dc theorems useful.

This chapter examines the application of the network theorems by considering both independent and dependent sources. In order to show the distinctions between the methods used in analyzing the various types of sources, the sections are labelled according to the types of sources involved.

An understanding of dependent sources is particularly useful when working with transistor circuits and operational amplifiers. Sections 20.2 and 20.5 are intended to provide the background for analyzing the operation of feedback amplifiers. Your instructor may find that these topics are best left until you cover this topic in a course dealing with such amplifiers. Consequently, the omission of sections 20.2 and 20.5 will not in any way detract from the continuity of the important ideas presented in this chapter. ■

William Bradford Shockley

PUTTING IT IN PERSPECTIVE

SHOCKLEY WAS BORN THE SON of a mining engineer in London, England on February 13, 1910. After graduating from the California Institute of Technology and Massachusetts Institute of Technology, Shockley joined Bell Telephone Laboratories.

With his co-workers, John Bardeen and Walter Brattain, Shockley developed an improved solid-state rectifier using a germanium crystal which had been injected with minute amounts of impurities. Unlike vacuum tubes, the resulting diodes were able to operate at much lower voltages without the need for inefficient heater elements.

In 1948, Shockley combined three layers of germanium to produce a device that was able to not only rectify a signal but to amplify it. Thus was developed the first transistor. Since its humble beginning, the transistor has been improved and decreased in size to the point where now a circuit containing thousands of transistors can easily fit into an area not much bigger than the head of a pin.

The advent of the transistor has permitted the construction of elaborate spacecraft, unprecedented communication, and new forms of energy generation.

Shockley, Bardeen, and Brattain received the 1956 Nobel Prize in Physics for their discovery of the transistor. ■

20.1 Superposition Theorem— Independent Sources

NOTES . . .

As in dc circuits, the superposition theorem can be applied only to voltage and current; it cannot be used to solve for the total power dissipated by an element. This is because power is not a linear quantity, but rather follows a square-law relationship ($P = V^2/R = I^2R$).

The superposition theorem states the following:

The voltage across (or current through) an element is determined by summing the voltage (or current) due to each independent source.

In order to apply this theorem, all sources other than the one being considered are eliminated. As in dc circuits, this is done by replacing current sources with open circuits and by replacing voltage sources with short circuits. The process is repeated until the effects due to all sources have been determined.

Although we generally work with circuits having all sources at the same frequency, occasionally a circuit may operate at more than one frequency at a time. This is particularly true in diode and transistor circuits which use a dc source to set a "bias" (or operating) point and an ac source to provide the signal to be conditioned or amplified. In such cases, the resulting voltages or currents are still determined by applying the superposition theorem. The topic of how to solve circuits operating at several different frequencies simultaneously is covered in Chapter 25.

EXAMPLE 20–1

Determine the current **I** in Figure 20–1 by using the superposition theorem.

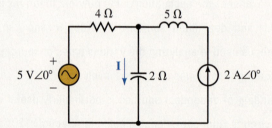

FIGURE 20–1

Solution
Current due to the 5 V∠0° voltage source: Eliminating the current source, we obtain the circuit shown in Figure 20–2.

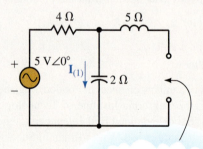

Current source is replaced with an open circuit.

FIGURE 20–2

Applying Ohm's law, we have

$$\mathbf{I}_{(1)} = \frac{5\text{ V}\angle 0°}{4 - j2\ \Omega} = \frac{5\text{ V}\angle 0°}{4.472\ \Omega\angle -26.57°}$$

$$= 1.118\text{ A}\angle 26.57°$$

Current due to the 2 A∠0° current source: Eliminating the voltage source, we obtain the circuit shown in Figure 20–3.

The current $\mathbf{I}_{(2)}$ due to this source is determined by applying the current divider rule:

$$\mathbf{I}_{(2)} = (2\text{ A}\angle 0°)\frac{4\text{ }\Omega\angle 0°}{4\text{ }\Omega - j2\text{ }\Omega}$$

$$= \frac{8\text{ V}\angle 0°}{4.472\text{ }\Omega\angle -26.57°}$$

$$= 1.789\text{ A}\angle 26.57°$$

The total current is determined as the summation of currents $\mathbf{I}_{(1)}$ and $\mathbf{I}_{(2)}$:

$$\mathbf{I} = \mathbf{I}_{(1)} + \mathbf{I}_{(2)}$$

$$= 1.118\text{ A}\angle 26.57° + 1.789\text{ A}\angle 26.57°$$

$$= (1.0\text{ A} + j0.5\text{ A}) + (1.6\text{ A} + j0.8\text{ A})$$

$$= 2.6 + j1.3\text{ A}$$

$$= 2.91\text{ A}\angle 26.57°$$

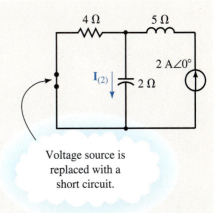

Voltage source is replaced with a short circuit.

FIGURE 20–3

Consider the circuit of Figure 20–4:

Find the following:

a. $\mathbf{V}_R$ and $\mathbf{V}_C$ using the superposition theorem.

b. Power dissipated by the circuit.

c. Power delivered to the circuit by each of the sources.

Solution

a. The superposition theorem may be employed as follows:

Voltages due to the current source: Eliminating the voltage source, we obtain the circuit shown in Figure 20–5.

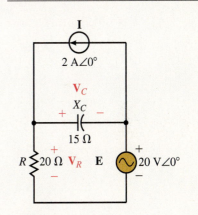

EXAMPLE 20–2

FIGURE 20–4

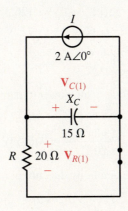

FIGURE 20–5

The impedance "seen" by the current source will be the parallel combination of $\mathbf{R}\|\mathbf{Z}_C$.

$$\mathbf{Z}_1 = \frac{(20\text{ }\Omega)(-j15\text{ }\Omega)}{20\text{ }\Omega - j15\text{ }\Omega} = \frac{300\text{ }\Omega\angle -90°}{25\text{ }\Omega\angle -36.87°} = 12\text{ }\Omega\angle -53.13°$$

The voltage $\mathbf{V}_{R(1)}$ is the same as the voltage across the capacitor, $\mathbf{V}_{C(1)}$. Hence,

$$\mathbf{V}_{R(1)} = \mathbf{V}_{C(1)}$$
$$= (2\text{ A}\angle 0°)(12\text{ }\Omega\angle -53.13°)$$
$$= 24\text{ V}\angle -53.13°$$

Voltages due to the voltage source: Eliminating the current source, we have the circuit shown in Figure 20–6.

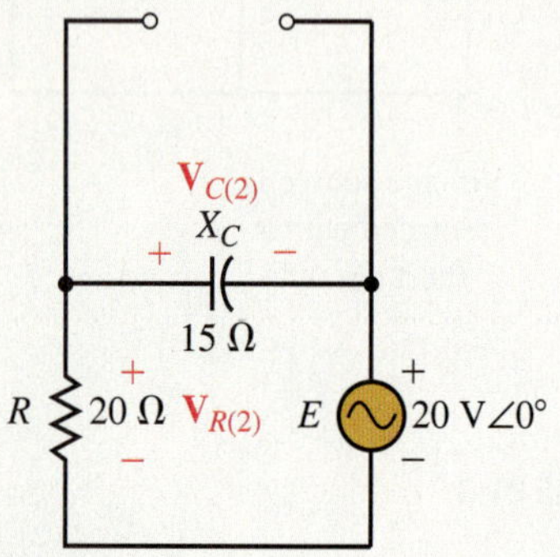

The voltages $\mathbf{V}_{R(2)}$ and $\mathbf{V}_{C(2)}$ are determined by applying the voltage divider rule,

$$\mathbf{V}_{R(2)} = \frac{20\text{ }\Omega\angle 0°}{20\text{ }\Omega - j15\text{ }\Omega}(20\text{ V}\angle 0°)$$
$$= \frac{400\text{ V}\angle 0°}{25\angle -36.87°} = 16\text{ V}\angle +36.87°$$

and

$$\mathbf{V}_{C(2)} = \frac{-15\text{ }\Omega\angle -90°}{20\text{ }\Omega - j15\text{ }\Omega}(20\text{ V}\angle 0°)$$
$$= \frac{300\text{ V}\angle 90°}{25\angle -36.87°} = 12\text{ V}\angle 126.87°$$

FIGURE 20–6

Notice that $\mathbf{V}_{C(2)}$ is assigned to be negative relative to the originally assumed polarity. The negative sign is eliminated from the calculation by adding (or subtracting) 180° from the corresponding calculation.

By applying superposition, we get

$$\mathbf{V}_R = \mathbf{V}_{R(1)} + \mathbf{V}_{R(2)}$$
$$= 24\text{ V}\angle -53.13° + 16\text{ V}\angle 36.87°$$
$$= (14.4\text{ V} - j19.2\text{ V}) + (12.8\text{ V} + j9.6\text{ V})$$
$$= 27.2\text{ V} - j9.6\text{ V}$$
$$= 28.84\text{ V}\angle -19.44°$$

and

$$\mathbf{V}_C = \mathbf{V}_{C(1)} + \mathbf{V}_{C(2)}$$
$$= 24\text{ V}\angle -53.13° + 12\text{ V}\angle 126.87°$$
$$= (14.4\text{ V} - j19.2\text{ V}) + (-7.2\text{ V} + j9.6\text{ V})$$
$$= 7.2\text{ V} - j9.6\text{ V}$$
$$= 12\text{ V}\angle -53.13°$$

b. Since only the resistor will dissipate power, the total power dissipated by the circuit is found as

$$P_\text{T} = \frac{(28.84\text{ V})^2}{20\text{ }\Omega} = 41.60\text{ W}$$

c. The power delivered to the circuit by the current source is

$$P_1 = V_1 I \cos\theta_1$$

where $\mathbf{V}_1 = \mathbf{V}_C = 12\text{ V}\angle -53.13°$ is the voltage across the current source and θ_1 is the phase angle between $\mathbf{V}_1$ and $\mathbf{I}$.

The power delivered by the current source is

$$P_1 = (12\text{ V})(2\text{ A})\cos 53.13° = 14.4\text{ W}$$

The power delivered to the circuit by the voltage source is similarly determined as

$$P_2 = EI_2 \cos\theta_2$$

where $\mathbf{I}_2$ is the current through the voltage source and θ_2 is the phase angle between $\mathbf{E}$ and $\mathbf{I}_2$.

$$P_2 = (20\ \text{V})\left(\frac{28.84\ \text{V}}{20\ \Omega}\right)\cos 19.44° = 27.2\ \text{W}$$

As expected, the total power delivered to the circuit must be the summation

$$P_T = P_1 + P_2 = 41.6\text{W}$$

Use superposition to find $\mathbf{V}$ and $\mathbf{I}$ for the circuit of Figure 20–7.

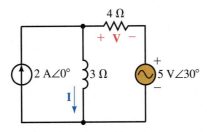

FIGURE 20–7

Answers
$\mathbf{I} = 2.52\ \text{A}\angle{-25.41°}$, $\mathbf{V} = 4.45\ \text{V}\angle 104.18°$

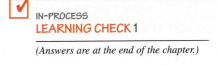
PRACTICE PROBLEMS 1

A 20-Ω resistor is in a circuit having three sinusoidal sources. After analyzing the circuit, it is found that the current through the resistor due to each of the sources is as follows:

$$I_1 = 1.5\ \text{A}\angle 20°$$
$$I_2 = 1.0\ \text{A}\angle 110°$$
$$I_3 = 2.0\ \text{A}\angle 0°$$

a. Use superposition to calculate the resultant current through the resistor.

b. Calculate the power dissipated by the resistor.

c. Show that the power dissipated by the resistor cannot be found by applying superposition, namely, $P_T \neq I_1^2 R + I_2^2 R + I_3^2 R$.

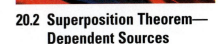
IN-PROCESS
LEARNING CHECK 1
(Answers are at the end of the chapter.)

Chapter 19 introduced the concept of dependent sources. We now examine ac circuits that contain dependent sources. In order to analyze such circuits, it is first necessary to determine whether the dependent source is conditional upon a controlling element in its own circuit or whether the controlling element is located in some other circuit.

If the controlling element is external to the circuit under consideration, the method of analysis is the same as for an independent source. However, if the controlling element is in the same circuit, the analysis follows a slightly different stategy. The next two examples show the techniques used to analyze circuits having dependent sources.

20.2 Superposition Theorem—Dependent Sources

EXAMPLE 20–3

Consider the circuit of Figure 20–8.

a. Determine the general expression for **V** in terms of **I**.

b. Calculate **V** if **I** = 1.0 A∠0°.

c. Calculate **V** if **I** = 0.3 A∠90°.

Solution

a. Since the current source in the circuit is dependent on current through an element that is located outside of the circuit of interest, the circuit may be analyzed in the same manner as for independent sources.

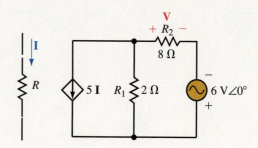

FIGURE 20–8

Voltage due to the voltage source: Eliminating the current source, we obtain the circuit shown in Figure 20–9.

$$\mathbf{V}_{(1)} = \frac{8\ \Omega}{10\ \Omega}\,(6\ \text{V}\angle 0°) = 4.8\ \text{V}\angle 0°$$

Voltage due to the current source: Eliminating the voltage source, we have the circuit shown in Figure 20–10.

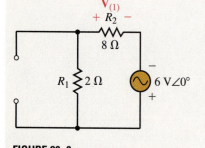

FIGURE 20–9

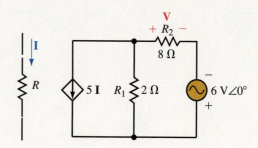

FIGURE 20–10

$$\mathbf{Z}_T = 2\ \Omega\|8\ \Omega = 1.6\ \Omega\angle 0°$$
$$\mathbf{V}_{(2)} = \mathbf{V}_{Z_T} = -(5\mathbf{I})(1.6\ \Omega\angle 0°) = -8.0\ \Omega\mathbf{I}$$

From superposition, the general expression for voltage is determined to be

$$\mathbf{V} = \mathbf{V}_{(1)} + \mathbf{V}_{(2)}$$
$$= 4.8\ \text{V}\angle 0° - 8.0\ \Omega\mathbf{I}$$

b.   If **I** = 1.0 A∠0°,

$$\mathbf{V} = 4.8\ \text{V}\angle 0° - (8.0\ \Omega)(1.0\ \text{A}\angle 0°) = -3.2\ \text{V}$$
$$= 3.2\ \text{V}\angle 180°$$

c.   If **I** = 0.3 A∠90°,

$$\mathbf{V} = 4.8\ \text{V}\angle 0° - (8.0\ \Omega)(0.3\ \text{A}\angle 90°) = 4.8\ \text{V} - j2.4\ \text{V}$$
$$= 5.367\ \text{V}\angle{-26.57°}$$

Given the circuit of Figure 20–11, calculate the voltage across the 40-Ω resistor.

Solution In the circuit of Figure 20–11, the dependent source is controlled by an element located in the circuit. Unlike the sources in the previous examples, the dependent source cannot be eliminated from the circuit since doing so would contradict Kirchhoff's voltage law and/or Kirchhoff's current law.

The circuit must be analyzed by considering all effects simultaneously.
Applying Kirchhoff's current law, we have

$$\mathbf{I}_1 + \mathbf{I}_2 = 2 \, A\angle 0°$$

From Kirchhoff's voltage law, we have

$$(10 \, \Omega) \, \mathbf{I}_1 = \mathbf{V} + 0.2 \, \mathbf{V} = 1.2 \, \mathbf{V}$$
$$\mathbf{I}_1 = 0.12 \, \mathbf{V}$$

and,

$$\mathbf{I}_2 = \frac{\mathbf{V}}{40 \, \Omega} = 0.025 \, \mathbf{V}$$

Combining the above expressions, we have

$$0.12 \, \mathbf{V} + 0.025 \, \mathbf{V} = 2.0 \, A\angle 0°$$
$$0.145 \, \mathbf{V} = 2.0 \, A\angle 0°$$
$$\mathbf{V} = 13.79 \, V\angle 0°°$$

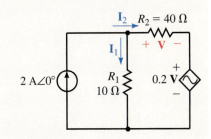

FIGURE 20–11

Determine the voltage **V** in the circuit of Figure 20–12.

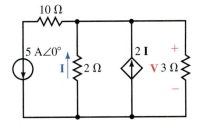

FIGURE 20–12

Answer
$\mathbf{V} = 2.73 \, V\angle 180°$

Thévenin's theorem is a method that converts any linear bilateral ac circuit into a single ac voltage source in series with an equivalent impedance as shown in Figure 20–13.

The resulting two-terminal network will be equivalent when it is connected to any external branch or component. If the original circuit contains reactive elements, the Thévenin equivalent circuit will be valid only at the frequency at which the reactances were determined. The following method may be used to determine the Thévenin equivalent of an ac circuit having either independent sources or sources which are dependent upon voltage or current in some other circuit. The outlined method may not be used in circuits having dependent sources controlled by voltage or current in the same circuit.

1. Remove the branch across which the Thévenin equivalent circuit is to be found. Label the resulting two terminals. Although any designation will do, we will use the notations *a* and *b*.

20.3 Thévenin's Theorem— Independent Sources

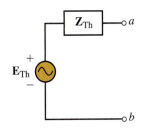

FIGURE 20–13 Thévenin equivalent circuit.

◀ **Online Companion**

2. Set all sources to zero. As in dc circuits, this is achieved by replacing voltage sources with short circuits and current sources with open circuits.

3. Determine the Thévenin equivalent impedance, $\mathbf{Z}_{Th}$ by calculating the impedance seen between the open terminals a and b. Occasionally it may be necessary to redraw the circuit to simplify this process.

4. Replace the sources removed in Step 3 and determine the open-circuit voltage across the terminals a and b. If any of the sources are expressed in sinusoidal form, it is first necessary to convert these sources into an equivalent phasor form. For circuits having more than one source, it may be necessary to apply the superposition theorem to calculate the open-circuit voltage. Since all voltages will be phasors, the resultant is found by using vector algebra. The open-circuit voltage is the Thévenin voltage, $\mathbf{E}_{Th}$.

5. Sketch the resulting Thévenin equivalent circuit by including that portion of the circuit removed in Step 1.

EXAMPLE 20–5

Find the Thévenin equivalent circuit external to $\mathbf{Z}_L$ for the circuit of Figure 20–14.

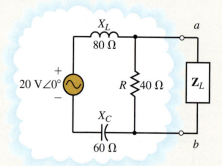

FIGURE 20–14

Solution
Steps 1 and 2: Removing the load impedance $\mathbf{Z}_L$ and setting the voltage source to zero, we have the circuit of Figure 20–15.

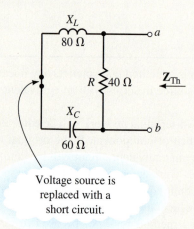

FIGURE 20–15

Step 3: The Thévenin impedance between terminals a and b is found as

$$\mathbf{Z}_{Th} = \mathbf{R}\|(\mathbf{Z}_L + \mathbf{Z}_C)$$
$$= \frac{(40\ \Omega\angle0°)(20\ \Omega\angle90°)}{40\ \Omega + j20\ \Omega}$$
$$= \frac{800\ \Omega\angle90°}{44.72\ \Omega\angle26.57°}$$
$$= 17.89\ \Omega\angle63.43°$$
$$= 8\ \Omega + j16\ \Omega$$

Step 4: The Thévenin voltage is found by using the voltage divider rule as shown in the circuit of Figure 20–16.

$$\mathbf{E}_{Th} = \mathbf{V}_{ab} = \frac{40\ \Omega\angle0°}{40\ \Omega + j80\ \Omega - j60\ \Omega}(20\ V\angle0°)$$
$$= \frac{800\ V\angle0°}{44.72\ \Omega\angle26.57°}$$
$$= 17.89\ V\angle-26.57°$$

Step 5: The resultant Thévenin equivalent circuit is shown in Figure 20–17.

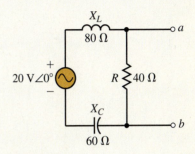

FIGURE 20–16

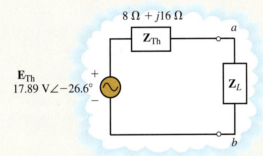

FIGURE 20–17

Determine the Thévenin equivalent circuit external to $\mathbf{Z}_L$ in the circuit in Figure 20–18.

EXAMPLE 20–6

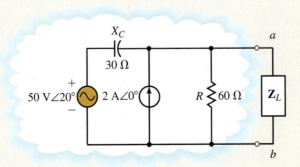

FIGURE 20–18

Solution

Step 1: Removing the branch containing $\mathbf{Z}_L$, we have the circuit of Figure 20–19.

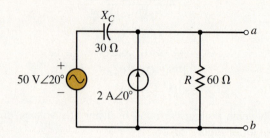

FIGURE 20–19

Step 2: After setting the voltage and current sources to zero, we have the circuit of Figure 20–20.

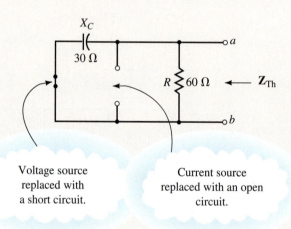

Voltage source replaced with a short circuit.

Current source replaced with an open circuit.

FIGURE 20–20

Step 3: The Thévenin impedance is determined as

$$\mathbf{Z}_{Th} = \mathbf{Z}_C \| \mathbf{Z}_R$$
$$= \frac{(30\ \Omega\angle-90°)(60\ \Omega\angle0°)}{60\ \Omega - j30\ \Omega}$$
$$= \frac{1800\ \Omega\angle-90°}{67.08\ \Omega\angle-26.57°}$$
$$= 26.83\ \Omega\angle-63.43°$$

Step 4: Because the given network consists of two independent sources, we consider the individual effects of each upon the open-circuit voltage. The total effect is then easily determined by applying the superposition theorem. Reinserting only the voltage source into the original circuit, as shown in Figure 20–21, allows us to find the open-circuit voltage, $\mathbf{V}_{ab(1)}$, by applying the voltage divider rule:

$$\mathbf{V}_{ab(1)} = \frac{60\ \Omega}{60\ \Omega - j30\ \Omega}(50\ V\angle20°)$$
$$= \frac{3000\ V\angle20°}{67.08\angle-26.57°}$$
$$= 44.72\ V\angle46.57°$$

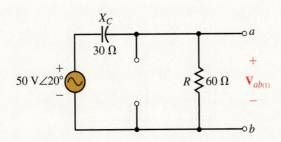

FIGURE 20–21

Now, considering only the current source as shown in Figure 20–22, we determine $\mathbf{V}_{ab(2)}$ by Ohm's law:

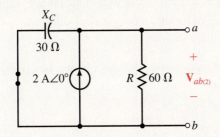

FIGURE 20–22

$$\mathbf{V}_{ab(2)} = \frac{(2\,\text{A}\angle 0°)(30\,\Omega\angle -90°)(60\,\Omega\angle 0°)}{60\,\Omega - j30\,\Omega}$$

$$= (2\,\text{A}\angle 0°)(26.83\,\Omega\angle -63.43°)$$

$$= 53.67\,\text{V}\angle -63.43°$$

From the superposition theorem, the Thévenin voltage is determined as

$$\mathbf{E}_{Th} = \mathbf{V}_{ab(1)} + \mathbf{V}_{ab(2)}$$

$$= 44.72\,\text{V}\angle 46.57° + 53.67\,\text{V}\angle -63.43°$$

$$= (30.74\,\text{V} + j32.48\,\text{V}) + (24.00\,\text{V} - j48.00\,\text{V})$$

$$= (54.74\,\text{V} - j15.52\,\text{V}) = 56.90\,\text{V}\angle -15.83°$$

Step 5: The resulting Thévenin equivalent circuit is shown in Figure 20–23.

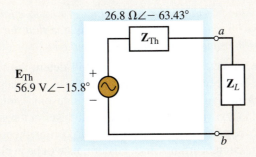

FIGURE 20–23

Refer to the circuit shown in Figure 20–24 of Practice Problem 3. List the steps that you would use to find the Thévenin equivalent circuit.

✓ IN-PROCESS
LEARNING CHECK 2

(Answers are at the end of the chapter.)

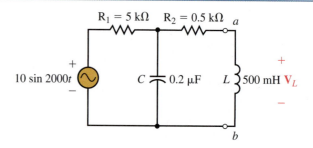

FIGURE 20–24

a. Find the Thévenin equivalent circuit external to the inductor in the circuit in Figure 20–24. (Notice that the voltage source is shown as sinusoidal.)

b. Use the Thévenin equivalent circuit to find the phasor output voltage, V_L.

c. Convert the answer of (b) into the equivalent sinusoidal voltage.

Answers

a. $Z_{Th} = 1.5 \text{ k}\Omega - j2.0 \text{ k}\Omega = 2.5 \text{ k}\Omega\angle{-53.13°}$, $E_{Th} = 3.16 \text{ V}\angle{-63.43°}$

b. $V_L = 1.75 \text{ V}\angle{60.26°}$

c. $v_L = 2.48 \sin(2000t + 60.26°)$

20.4 Norton's Theorem—Independent Sources

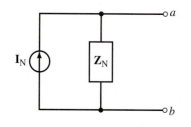

FIGURE 20–25   Norton equivalent circuit.

Norton's theorem converts any linear bilateral network into an equivalent circuit consisting of a single current source and a parallel impedance as shown in Figure 20–25.

Although Norton's equivalent circuit may be determined by first finding the Thévenin equivalent circuit and then performing a source conversion, we generally use the more direct method outlined below. The steps to find the Norton equivalent circuit are as follows:

1. Remove the branch across which the Norton equivalent circuit is to be found. Label the resulting two terminals *a* and *b*.

2. Set all sources to zero.

3. Determine the Norton equivalent impedance, Z_N, by calculating the impedance seen between the open terminals *a* and *b*.

   NOTE: Since the previous steps are identical with those followed for finding the Thévenin equivalent circuit, we conclude that the Norton impedance must be the same as the Thévenin impedance.

4. Replace the sources removed in Step 3 and determine the current that would occur between terminals *a* and *b* if these terminals were shorted. Any voltages and currents that are given in sinusoidal notation must first be expressed in equivalent phasor notation. If the circuit has more than one source it may be necessary to apply the superposition theorem to calculate the total short-circuit current. Since all currents will be in phasor form, any addition must be done using vector algebra. The resulting current is the Norton current, I_N.

5. Sketch the resulting Norton equivalent circuit by inserting that portion of the circuit removed in Step 1.

As mentioned previously, it is possible to find the Norton equivalent circuit from the Thévenin equivalent by simply performing a source conversion. We have already determined that both the Thévenin and Norton impedances are determined in the same way. Consequently, the impedances must be equivalent, and so we have

$$\mathbf{Z}_N = \mathbf{Z}_{Th} \qquad (20\text{–}1)$$

Now, applying Ohm's law, we determine the Norton current source from the Thévenin voltage and impedance, namely,

$$\mathbf{I}_N = \frac{\mathbf{E}_{Th}}{\mathbf{Z}_{Th}} \qquad (20\text{–}2)$$

Figure 20–26 shows the equivalent circuits.

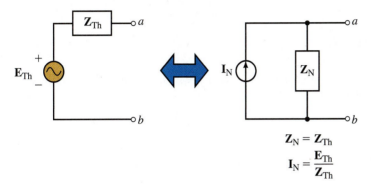

$$\mathbf{Z}_N = \mathbf{Z}_{Th}$$
$$\mathbf{I}_N = \frac{\mathbf{E}_{Th}}{\mathbf{Z}_{Th}}$$

FIGURE 20–26

Given the circuit of Figure 20–27, find the Norton equivalent.

EXAMPLE 20–7

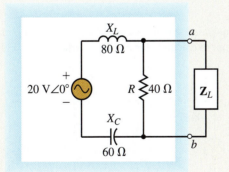

FIGURE 20–27

Solution
Steps 1 and 2: By removing the load impedance, $\mathbf{Z}_L$, and setting the voltage source to zero, we have the network of Figure 20–28.
Step 3: The Norton impedance may now be determined by evaluating the impedance between terminals a and b. Hence, we have

$$\mathbf{Z}_N = \frac{(40\ \Omega\angle 0°)(20\ \Omega\angle 90°)}{40\ \Omega + j20\ \Omega}$$

$$= \frac{800\ \Omega\angle 90°}{44.72\angle 26.57°}$$

$$= 17.89\ \Omega\angle 63.43°$$

$$= 8\ \Omega + j16\ \Omega$$

FIGURE 20–28

Step 4: Reinserting the voltage source, as in Figure 20–29, we find the Norton current by calculating the current between the shorted terminals, a and b.

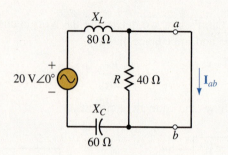

FIGURE 20–29

$$\mathbf{Z}_N = 8\ \Omega + j16\ \Omega = 17.89\ \Omega\angle63.43°$$

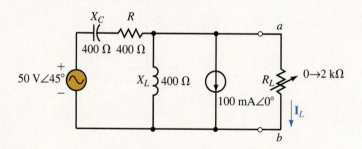

FIGURE 20–30

Because the resistor $R = 40\ \Omega$ is shorted, the current is determined by the impedances X_L and X_C as

$$\mathbf{I}_N = \mathbf{I}_{ab} = \frac{20\ \text{V}\angle0°}{j80\ \Omega - j60\ \Omega}$$

$$= \frac{20\ \text{V}\angle0°}{20\ \Omega\angle-90°}$$

$$= 1.00\ \text{A}\angle-90°$$

Step 5: The resultant Norton equivalent circuit is shown in Figure 20–30.

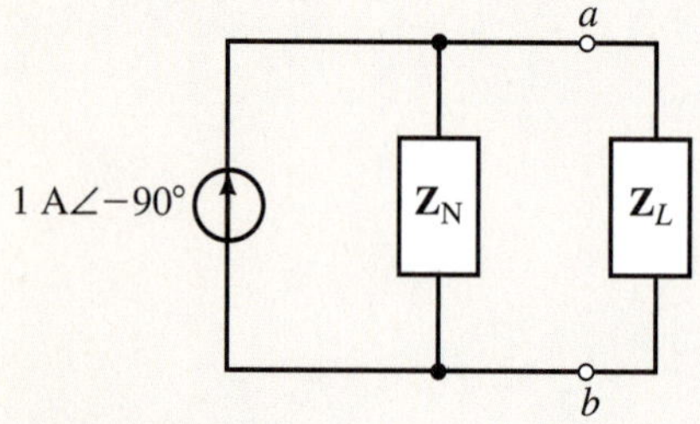

EXAMPLE 20–8

Find the Norton equivalent circuit external to R_L in the circuit of Figure 20–31. Use the equivalent circuit to calculate the current $\mathbf{I}_L$ when $R_L = 0\ \Omega$, $400\ \Omega$, and $2\ \text{k}\Omega$.

FIGURE 20–31

Solution
Steps 1 and 2: Removing the load resistor and setting the sources to zero, we obtain the network shown in Figure 20–32.
Step 3: The Norton impedance is determined as

$$\mathbf{Z}_N = \frac{(400\ \Omega\angle90°)(400\ \Omega - j400\ \Omega)}{j400\ \Omega + 400\ \Omega - j400\ \Omega}$$

$$= \frac{(400\ \Omega\angle90°)(565.69\ \Omega\angle-45°)}{400\ \Omega\angle0°}$$

$$= 565.69\ \Omega\angle+45°$$

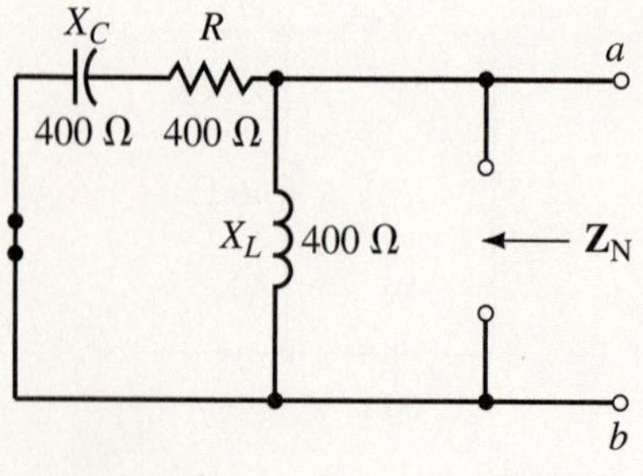

FIGURE 20–32

Step 4: Because the network consists of two sources, we determine the effects due to each source separately and then apply superposition to evaluate the Norton current source.

Reinserting the voltage source into the original network, we see from Figure 20–33 that the short-circuit current between the terminals a and b is easily found by using Ohm's law.

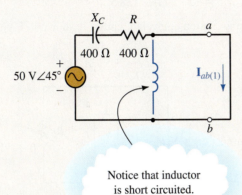

Notice that inductor is short circuited.

FIGURE 20–33

$$\mathbf{I}_{ab(1)} = \frac{50 \text{ V}\angle 45°}{400 \text{ }\Omega - j400 \text{ }\Omega}$$

$$= \frac{50 \text{ V}\angle 45°}{565.69 \text{ }\Omega\angle -45°}$$

$$= 88.4 \text{ mA}\angle 90°$$

Since short circuiting the current source effectively removes all impedances, as illustrated in Figure 20–34, the short-circuit current between the terminals a and b is given as follows:

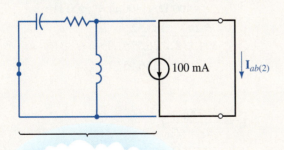

These components are short circuited.

FIGURE 20–34

$$\mathbf{I}_{ab(2)} = -100 \text{ mA}\angle 0°$$

$$= 100 \text{ mA}\angle 180°$$

Now applying the superposition theorem, the Norton current is determined as the summation

$$\mathbf{I}_N = \mathbf{I}_{ab(1)} + \mathbf{I}_{ab(2)}$$
$$= 88.4 \text{ mA}\angle 90° + 100 \text{ mA}\angle 180°$$
$$= -100 \text{ mA} + j88.4 \text{ mA}$$
$$= 133.5 \text{ mA}\angle 138.52°$$

Step 5: The resulting Norton equivalent circuit is shown in Figure 20–35.

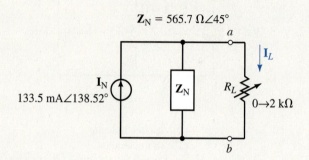

FIGURE 20–35

From the above circuit, we express the current through the load, $\mathbf{I}_L$, as

$$\mathbf{I}_L = \frac{\mathbf{Z}_N}{R_L + \mathbf{Z}_N}\mathbf{I}_N$$

$R_L = 0 \ \Omega$:

$$\mathbf{I}_L = \mathbf{I}_N = 133.5 \text{ mA}\angle 138.52°$$

$R_L = 400 \ \Omega$:

$$\mathbf{I}_L = \frac{\mathbf{Z}_N}{R_L + \mathbf{Z}_N}\mathbf{I}_N$$
$$= \frac{(565.7 \ \Omega\angle 45°)(133.5 \text{ mA}\angle 138.52°)}{400 \ \Omega + 400 \ \Omega + j400 \ \Omega}$$
$$= \frac{75.24 \text{ V}\angle 183.52°}{894.43 \ \Omega\angle 26.57°}$$
$$= 84.12 \text{ mA}\angle 156.95°$$

$R_L = 2 \ \text{k}\Omega$:

$$\mathbf{I}_L = \frac{\mathbf{Z}_N}{R_L + \mathbf{Z}_N}\mathbf{I}_N$$
$$= \frac{(565.7 \ \Omega\angle 45°)(133.5 \text{ mA}\angle 138.52°)}{2000 \ \Omega + 400 \ \Omega + j400 \ \Omega}$$
$$= \frac{75.24 \text{ V}\angle 183.52°}{2433.1 \ \Omega\angle 9.46°}$$
$$= 30.92 \text{ mA}\angle 174.06°$$

IN-PROCESS
LEARNING CHECK 3

(Answers are at the end of the chapter.)

Refer to the circuit shown in Figure 20–36 (Practice Problem 4). List the steps that you would use to find the Norton equivalent circuit.

Find the Norton equivalent circuit external to R_L in the circuit of Figure 20–36. Use the equivalent circuit to find the current I_L.

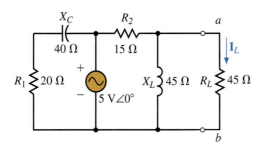

FIGURE 20–36

Answers
$Z_N = 13.5\ \Omega + j4.5\ \Omega = 14.23\ \Omega\angle18.43°$, $I_N = 0.333\ A\angle0°$
$I_L = 0.0808\ A\angle14.03°$

20.5 Thévenin's and Norton's Theorems for Dependent Sources

If a circuit contains a dependent source that is controlled by an element outside the circuit of interest, the methods outlined in Sections 20.2 and 20.3 are used to find either the Thévenin or Norton equivalent circuit.

EXAMPLE 20–9

Given the circuit of Figure 20–37, find the Thévenin equivalent circuit external to R_L. If the voltage applied to the resistor R_1 is 10 mV, use the Thévenin equivalent circuit to calculate the minimum and maximum voltage across R_L.

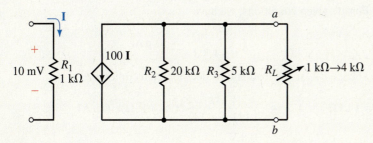

FIGURE 20–37

Solution
Step 1: Removing the load resistor from the circuit and labelling the remaining terminals *a* and *b*, we have the circuit shown in Figure 20–38.

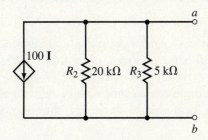

FIGURE 20–38

Steps 2 and 3: The Thévenin resistance is found by open circuiting the current source and calculating the impedance observed between the terminals a and b. Since the circuit is purely resistive, we have

$$R_{Th} = 20 \text{ k}\Omega \| 5 \text{ k}\Omega$$

$$= \frac{(20 \text{ k}\Omega)(5 \text{ k}\Omega)}{20 \text{ k}\Omega + 5 \text{ k}\Omega}$$

$$= 4 \text{ k}\Omega$$

Step 4: The open-circuit voltage between the terminals is found to be

$$\mathbf{V}_{ab} = -(100\mathbf{I})(4 \text{ k}\Omega)$$

$$= -(4 \times 10^5 \text{ }\Omega)\mathbf{I}$$

As expected, the Thévenin voltage source is dependent upon the current **I**.

Step 5: Because the Thévenin voltage is a dependent voltage source we use the appropriate symbol when sketching the equivalent circuit, as shown in Figure 20–39.

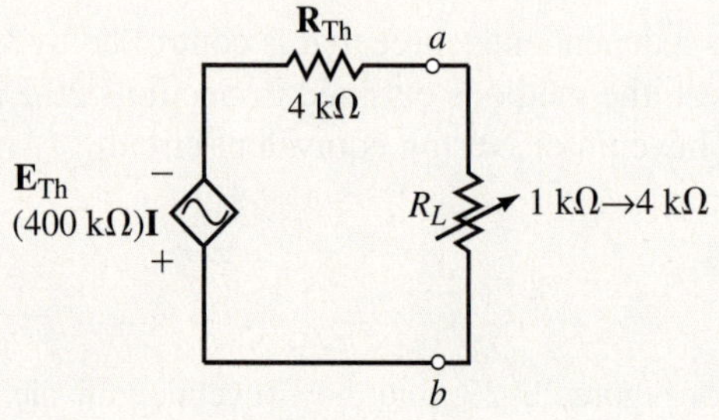

FIGURE 20–39

For the given conditions, we have

$$I = \frac{10 \text{ mV}}{1 \text{ k}\Omega} = 10 \text{ }\mu\text{A}$$

The voltage across the load is now determined as follows:

$$R_L = \mathbf{1} \text{ k}\Omega: \quad V_{ab} = -\frac{1 \text{ k}\Omega}{1 \text{ k}\Omega + 4 \text{ k}\Omega}(4 \times 10^5 \text{ }\Omega)(10 \text{ }\mu\text{A})$$

$$= -0.8 \text{ V}$$

$$R_L = \mathbf{4} \text{ k}\Omega: \quad V_{ab} = -\frac{4 \text{ k}\Omega}{4 \text{ k}\Omega + 4 \text{ k}\Omega}(4 \times 10^5 \text{ }\Omega)(10 \text{ }\mu\text{A})$$

$$= -2.0 \text{ V}$$

For an applied voltage of 10 mV, the voltage across the load resistance will vary between 0.8 V and 2.0 V as R_L is adjusted between $+$ kΩ and 4 kΩ.

If a circuit contains one or more dependent sources that are controlled by an element in the circuit being analyzed, all previous methods fail to provide equivalent circuits that correctly model the circuit's behavior. In order to determine the Thévenin or Norton equivalent circuit of a circuit having a dependent source controlled by a local voltage or current, the following steps must be taken:

1. Remove the branch across which the Norton equivalent circuit is to be found. Label the resulting two terminals a and b.

2. Calculate the open-circuit voltage (Thévenin voltage) across the two terminals *a* and *b*. Because the circuit contains a dependent source controlled by an element in the circuit, the dependent source may not be set to zero. Its effects must be considered together with the effects of any independent source(s).

3. Determine the short-circuit current (Norton current) that would occur between the terminals. Once again, the dependent source may not be set to zero, but rather must have its effects considered concurrently with the effects of any independent source(s).

4. Determine the Thévenin or Norton impedance by applying Equations 20–1 and 20–2 as follows:

$$\mathbf{Z_N} = \mathbf{Z_{Th}} = \frac{\mathbf{E_{Th}}}{\mathbf{I_N}} \qquad \qquad \textbf{(20–3)}$$

5. Sketch the Thévenin or Norton equivalent circuit, as shown previously in Figure 20–26. Ensure that the portion of the network that was removed in Step 1 is reinserted as part of the equivalent circuit.

For the circuit of Figure 20–40, find the Norton equivalent circuit external to the load resistor, R_L.

EXAMPLE 20–10

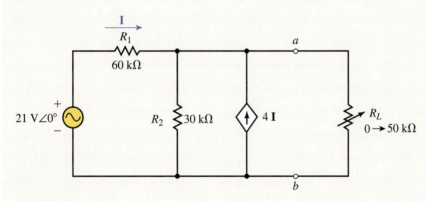

FIGURE 20–40

Solution
Step 1: After removing the load resistor from the circuit, we have the network shown in Figure 20–41.

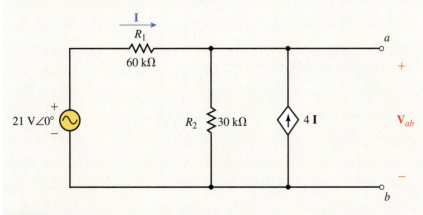

FIGURE 20–41

Step 2: At first glance, we might look into the open terminals and say that the Norton (or Thévenin) impedance appears to be 60 kΩ‖30 kΩ = 20 kΩ. However, we will find that this result is incorrect. The presence of the locally controlled dependent current source makes the analysis of this circuit slightly more complicated than a circuit that contains only an independent source. We know, however, that the basic laws of circuit analysis must apply to all circuits, regardless of the complexity. Applying Kirchhoff's current law at node a gives the current through R_2 as

$$\mathbf{I}_{R_2} = \mathbf{I} + 4\mathbf{I} = 5\mathbf{I}$$

Now, applying Kirchhoff's voltage law around the closed loop containing the voltage source and the two resistors, gives

$$21\ \text{V}\angle 0° = (60\ \text{k}\Omega)\mathbf{I} + (30\ \text{k}\Omega)(5\mathbf{I}) = 210\ \text{k}\Omega\mathbf{I}$$

which allows us to solve for the current $\mathbf{I}$ as

$$\mathbf{I} = \frac{21\ \text{V}\angle 0°}{210\ \text{k}\Omega} = 0.100\ \text{mA}\angle 0°$$

Since the open-circuit voltage, $\mathbf{V}_{ab}$ is the same as the voltage across R_2, we have

$$\mathbf{E}_{\text{Th}} = \mathbf{V}_{ab} = (30\ \text{k}\Omega)(5)(0.1\ \text{mA}\angle 0°) = 15\ \text{V}\angle 0°$$

Step 3: The Norton current source is determined by placing a short-circuit between terminals a and b as shown in Figure 20–42.

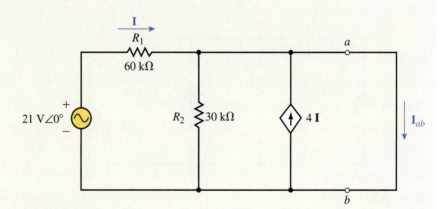

FIGURE 20–42

Upon further inspection of this circuit we see that resistor R_2 is short-circuited. The simplified circuit is shown in Figure 20–43.

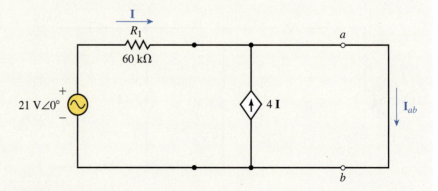

FIGURE 20–43

The short-circuit current $\mathbf{I}_{ab}$ is now easily determined by using Kirchhoff's current law at node a, and so we have

$$\mathbf{I}_N = \mathbf{I}_{ab} = 5\mathbf{I}$$

From Ohm's law, we have

$$\mathbf{I} = \frac{21\ \text{V}\angle 0°}{60\ \text{k}\Omega} = 0.35\ \text{mA}\angle 0°$$

and so

$$\mathbf{I}_N = 5(0.35\ \text{mA}\angle 0°) = 1.75\ \text{mA}\angle 0°$$

Step 4: The Norton (or Thévenin) impedance is now determined from Ohm's law as

$$\mathbf{Z}_N = \frac{\mathbf{E}_{Th}}{\mathbf{I}_N} = \frac{15\ \text{V}\angle 0°}{1.75\ \text{mA}\angle 0°} = 8.57\ \text{k}\Omega$$

Notice that this impedance is different from the originally assumed 20 kΩ. In general, this condition will occur for most circuits that contain a locally controlled voltage or current source.

Step 5: The Norton equivalent circuit is shown in Figure 20–44.

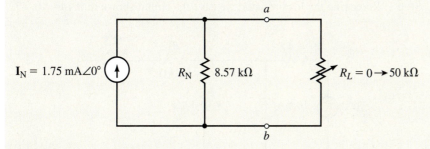

FIGURE 20–44

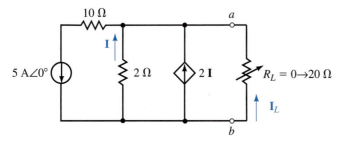

FIGURE 20–45

a. Find the Thévenin equivalent circuit external to R_L in the circuit of Figure 20–45.

b. Determine the current I_L when $R_L = 0$ and when $R_L = 20\ \Omega$.

Answers

a. $\mathbf{E}_{Th} = \mathbf{V}_{ab} = -3.33\ \text{V}, \ \mathbf{Z}_{Th} = 0.667\ \Omega$

b. For $R_L = 0$: $\mathbf{I}_L = 5.00\ \text{A}$ (upward); for $R_L = 20\ \Omega$: $\mathbf{I}_L = 0.161\ \text{A}$ (upward)

If a circuit has more than one independent source, it is necessary to determine the open-circuit voltage and short-circuit current due to each independent source while simultaneously considering the effects of the dependent source. The following example illustrates the principle.

EXAMPLE 20–11

Find the Thévenin and Norton equivalent circuits external to the load resistor in the circuit of Figure 20–46.

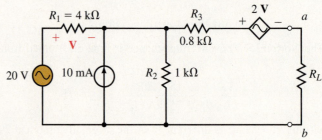

FIGURE 20–46

Solution   There are several methods of solving this circuit. The following approach uses the fewest number of steps.

Step 1: Removing the load resistor, we have the circuit shown in Figure 20–47.

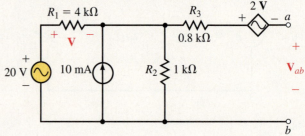

FIGURE 20–47

Step 2: In order to find the open-circuit voltage, V_{ab} of Figure 20–47, we may isolate the effects due to each independent source and then apply superposition to determine the combined result. However, by converting the current source into an equivalent voltage source, we can determine the open-circuit voltage in one step. Figure 20–48 shows the circuit that results when the current source is converted into an equivalent voltage source.

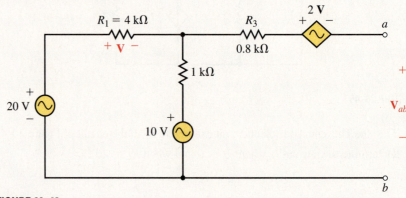

FIGURE 20–48

The controlling element (R_1) has a voltage, $\mathbf{V}$, determined as

$$\mathbf{V} = \left(\frac{4 \text{ k}\Omega}{4 \text{ k}\Omega + 1 \text{ k}\Omega}\right)(20 \text{ V} - 10 \text{ V})$$
$$= 8 \text{ V}$$

which gives a Thévenin (open-circuit) voltage of

$$\mathbf{E}_{Th} = \mathbf{V}_{ab} = -2(8 \text{ V}) + 0 \text{ V} - 8 \text{ V} + 20 \text{ V}$$
$$= -4.0 \text{ V}$$

Step 3: The short-circuit current is determined by examining the circuit shown in Figure 20–49.

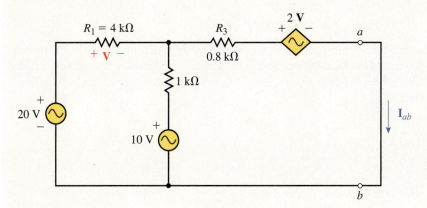

FIGURE 20–49

Once again, it is possible to determine the short-circuit current by using superposition. However, upon further reflection, we see that the circuit is easily analyzed using Mesh analysis. Loop currents $\mathbf{I}_1$ and $\mathbf{I}_2$ are assigned in clockwise directions as shown in Figure 20–50.

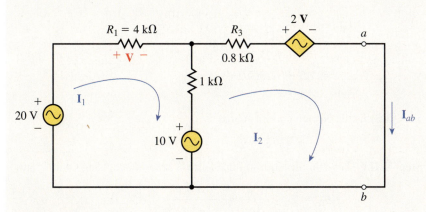

FIGURE 20–50

The loop equations are as follows:

Loop 1: $(5 \text{ k}\Omega) \, \mathbf{I}_1 - (1 \text{ k}\Omega) \, \mathbf{I}_2 = 10 \text{ V}$

Loop 2: $-(1 \text{ k}\Omega) \, \mathbf{I}_1 + (1.8 \text{ k}\Omega) \, \mathbf{I}_2 = 10 \text{ V} - 2 \text{ V}$

Notice that the voltage in the second loop equation is expressed in terms of the controlling voltage across R_1. We will not worry about this right now. We may simplify our calculations by ignoring the units in the above equations. It is obvious that if all impedances are expressed in $k\Omega$ and all voltages are in volts, then the currents I_1 and I_2 must be in mA. The determinant for the denominator is found to be

$$\mathbf{D} = \begin{vmatrix} 5 & -1 \\ -1 & 1.8 \end{vmatrix} = 9 - 1 = 8$$

The current $\mathbf{I}_1$ is solved by using determinants as follows:

$$\mathbf{I}_1 = \frac{\begin{vmatrix} 10 & -1 \\ 10 - 2\,\mathbf{V} & 1.8 \end{vmatrix}}{\mathbf{D}} = \frac{18 - (-1)(10 - 2\,\mathbf{V})}{8}$$

$$= 3.5 - 0.25\,\mathbf{V}$$

The above result illustrates that the current $\mathbf{I}_1$ is dependent on the controlling voltage. However, by examining the circuit of Figure 20–50, we see that the controlling voltage depends on the current $\mathbf{I}_1$, and is determined from Ohm's law as

$$\mathbf{V} = (4\ k\Omega)\mathbf{I}_1$$

or more simply as

$$\mathbf{V} = 4\mathbf{I}_1$$

Now, the current $\mathbf{I}_1$ is found as

$$\mathbf{I}_1 = 3.5 - 0.25(4\mathbf{I}_1)$$
$$2\mathbf{I}_1 = 3.5$$
$$\mathbf{I}_1 = 1.75\ \text{mA}$$

which gives $\mathbf{V} = 7.0\ \text{V}$.

Finally, the short circuit current (which in the circuit of Figure 20–50 is represented by $\mathbf{I}_2$) is found as

$$\mathbf{I}_2 = \frac{\begin{vmatrix} 5 & 10 \\ -1 & 10 - 2\,\mathbf{V} \end{vmatrix}}{\mathbf{D}} = \frac{50 - 10\,\mathbf{V} - (-10)}{8}$$

$$= 7.5 - 1.25\,\mathbf{V}$$
$$= 7.5 - 1.25(7.0\ \text{V})$$
$$= -1.25\ \text{mA}$$

This gives us the Norton current source as

$$I_N = I_{ab} = -1.25\ \text{mA}$$

Step 4: The Thévenin (or Norton) impedance is determined using Ohm's Law.

$$\mathbf{Z}_{Th} = \mathbf{Z}_N = \frac{\mathbf{E}_{Th}}{\mathbf{I}_N} = \frac{-4.0\ \text{V}}{-1.25\ \text{m A}} = 3.2\ k\Omega$$

The resulting Thévenin equivalent circuit is shown in Figure 20–51 and the Norton equivalent circuit is shown in Figure 20–52.

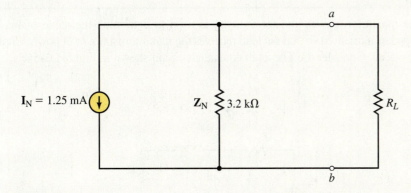

FIGURE 20–51

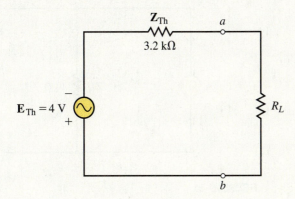

FIGURE 20–52

Find the Thévenin and Norton equivalent circuits external to the load resistor in the circuit of Figure 20–53.

PRACTICE PROBLEMS 6

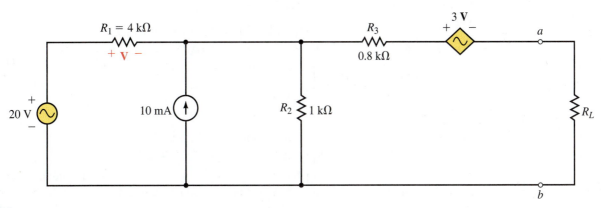

FIGURE 20–53

Answers

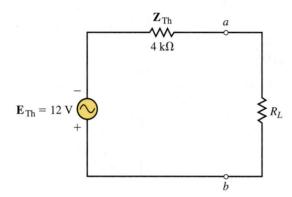

FIGURE 20–54

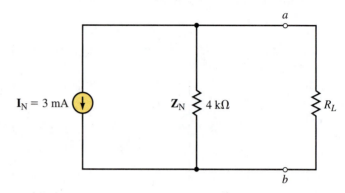

FIGURE 20–55

20.6 Maximum Power Transfer Theorem

The maximum power transfer theorem is used to determine the value of load impedance required so that the load receives the maximum amount of power from the circuit. Consider the Thévenin equivalent circuit shown in Figure 20–56.

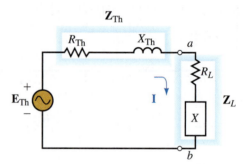

FIGURE 20–56

For any load impedance $\mathbf{Z}_L$ consisting of a resistance and a reactance such that $\mathbf{Z}_L = R_L \pm jX$, the power dissipated by the load will be determined as follows:

$$P_L = I^2 R_L$$

$$I = \frac{E_{Th}}{\sqrt{(R_{Th} + R_L)^2 + (X_{Th} \pm X)^2}}$$

$$P_L = \frac{E_{Th}^2 R_L}{(R_{Th} + R_L)^2 + (X_{Th} \pm X)^2}$$

Consider only the reactance portion, X, of the load impedance for the moment and neglect the effect of the load resistance. We see that the power dissipated by the load will be maximum when the denominator is kept to a minimum. If the load were to have an impedance such that $jX = -jX_{Th}$, then the power delivered to the load would be given as

$$P_L = \frac{E_{Th}^2 R_L}{(R_{Th} + R_L)^2} \tag{20–4}$$

We recognize that this is the same expression for power as that determined for the Thévenin equivalent of dc circuits in Chapter 9. Recall that maximum power was delivered to the load when

$$R_L = R_{Th}$$

For ac circuits, the maximum power transfer theorem states the following:

Maximum power will be delivered to a load whenever the load has an impedance which is equal to the complex conjugate of the Thévenin (or Norton) impedance of the equivalent circuit.

A detailed derivation of the maximum power transfer theorem is provided in Appendix C. The maximum power delivered to the load may be calculated by using Equation 20–4, which is simplified as follows:

$$P_{max} = \frac{E_{Th}^2}{4R_{Th}} \qquad (20\text{–}5)$$

For a Norton equivalent circuit, the maximum power delivered to a load is determined by substituting $E_{Th} = I_N Z_N$ into the above expression as follows:

$$P_{max} = \frac{I_N^2 Z_N^2}{4R_N} \qquad (20\text{–}6)$$

Determine the load impedance $\mathbf{Z}_L$ that will allow maximum power to be delivered to the load in the circuit of Figure 20–57. Find the maximum power.

Solution   Expressing the Thévenin impedance in its rectangular form, we have

$$\mathbf{Z}_{Th} = 500\ \Omega\angle 60° = 250\ \Omega + j433\ \Omega$$

In order to deliver maximum power to the load, the load impedance must be the complex conjugate of the Thévenin impedance. Hence,

$$\mathbf{Z}_L = 250\ \Omega - j433\ \Omega = 500\ \Omega\angle{-60°}$$

The power delivered to the load is now easily determined by applying Equation 20–5:

$$P_{max} = \frac{(20\ \text{V})^2}{4(250\ \Omega)} = 400\ \text{mW}$$

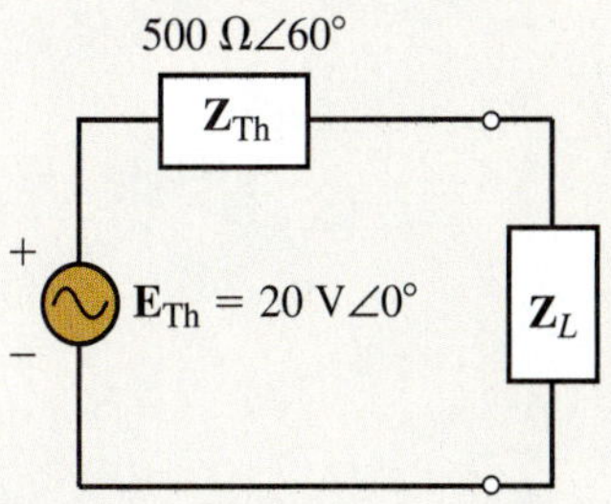

EXAMPLE 20–12

FIGURE 20–57

Given the circuit of Figure 20–57, determine the power dissipated by the load if the load impedance is equal to the Thévenin impedance, $\mathbf{Z}_L = 500\ \Omega\angle 60°$. Compare your answer to that obtained in Example 20–12.

PRACTICE PROBLEMS 7

Answer
$P = 100$ mW, which is less than P_{max}.

Occasionally it is not possible to adjust the reactance portion of a load. In such cases, a **relative maximum power** will be delivered to the load when the load resistance has a value determined as

$$R_L = \sqrt{R_{Th}^2 + (X \pm X_{Th})^2} \qquad (20\text{–}7)$$

If the reactance of the Thévenin impedance is of the same type (both capacitive or both inductive) as the reactance in the load, then the reactances are added.

If one reactance is capacitive and the other is inductive, however, then the reactances are subtracted.

To determine the power delivered to the load in such cases, the power will need to be calculated by finding either the voltage across the load or the current through the load. Equations 20–5 and 20–6 will no longer apply, since these equations were based on the premise that the load impedance is the complex conjugate of the Thévenin impedance.

EXAMPLE 20–13

For the circuit of Figure 20–58, determine the value of the load resistor, R_L, such that maximum power will be delivered to the load.

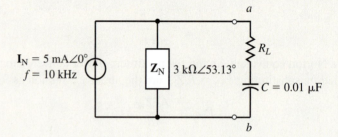

FIGURE 20–58

Solution   Notice that the load impedance consists of a resistor in series with a capacitance of 0.010 μF. Since the capacitive reactance is determined by the frequency, it is quite likely that the maximum power for this circuit may only be a relative maximum, rather than the absolute maximum. For the **absolute maximum power** to be delivered to the load, the load impedance would need to be

$$\mathbf{Z}_L = 3 \text{ k}\Omega\angle{-53.13°} = 1.80 \text{ k}\Omega - j2.40 \text{ k}\Omega$$

The reactance of the capacitor at a frequency of 10 kHz is determined to be

$$X_C = \frac{1}{2\pi(10 \text{ kHz})(0.010 \text{ μF})} = 1.592 \text{ k}\Omega$$

Because the capacitive reactance is not equal to the inductive reactance of the Norton impedance, the circuit will not deliver the absolute maximum power to the load. However, relative maximum power will be delivered to the load when

$$R_L = \sqrt{R_{\text{Th}}^2 + (X - X_{\text{Th}})^2}$$
$$= \sqrt{(1.800 \text{ k}\Omega)^2 + (1.592 \text{ k}\Omega - 2.4 \text{ k}\Omega)^2}$$
$$= 1.973 \text{ k}\Omega$$

Figure 20–59 shows the circuit with all impedance values.

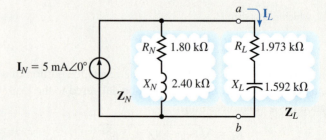

FIGURE 20–59

The load current will be

$$\mathbf{I}_L = \frac{\mathbf{Z}_N}{\mathbf{Z}_N + \mathbf{Z}_L}\mathbf{I}_N$$

$$= \frac{1.80 \text{ k}\Omega + j2.40 \text{ k}\Omega}{(1.80 \text{ k}\Omega + j2.40 \text{ k}\Omega) + (1.973 \text{ k}\Omega - j1.592 \text{ k}\Omega)}(5 \text{ mA}\angle0°)$$

$$= \frac{3 \text{ k}\Omega\angle53.13°}{3.773 \text{ k}\Omega + j0.808 \text{ k}\Omega}(5 \text{ mA}\angle0°)$$

$$= \frac{15.0 \text{ V}\angle53.13°}{3.859 \text{ k}\Omega\angle12.09°} = 3.887 \text{ mA}\angle41.04°$$

We now determine the power delivered to the load for the given conditions as

$$P_L = I_L^2 R_L$$
$$= (3.887 \text{ mA})^2(1.973 \text{ k}\Omega) = 29.82 \text{ mW}$$

If we had applied Equation 20–6, we would have found the absolute maximum power to be

$$P_{max} = \frac{(5 \text{ mA})^2(3.0 \text{ k}\Omega)^2}{4(1.8 \text{ k}\Omega)} = 31.25 \text{ mW}$$

PRACTICE PROBLEMS 8

Refer to the Norton equivalent circuit of Figure 20–60:

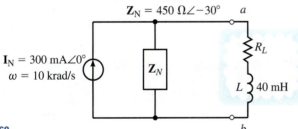

FIGURE 20–60

a. Find the value of load resistance, R_L, such that the load receives maximum power.

b. Determine the maximum power received by the load for the given conditions.

Answers
a. $R_L = 427 \ \Omega$; b. $P_L = 11.2$ W

IN-PROCESS
LEARNING CHECK 4

(Answers are at the end of the chapter.)

Refer to the Thévenin equivalent circuit of Figure 20–61.

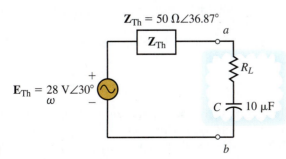

FIGURE 20–61

a. Determine the value of the unknown load resistance R_L that will result in a relative maximum power at an angular frequency of 1 krad/s.

b. Solve for the power dissipated by the load at $\omega_1 = 1$ krad/s.

c. Assuming that the Thévenin impedance remains constant at all frequencies, at what angular frequency, ω_2, will the circuit provide absolute maximum power?

d. Solve for the power dissipated by the load at ω_2.

20.7 Circuit Analysis Using Computers

◄ MULTISIM

◄ CADENCE

As demonstrated in Chapter 9, circuit analysis programs are very useful in determining the equivalent circuit between specified terminals of a dc circuit. We will use similar methods to obtain the Thévenin and Norton equivalents of ac circuits. Both MultiSIM and PSpice are useful in analyzing circuits with dependent sources. As we have already seen, the work in analyzing such a circuit manually is lengthy and very time consuming. In this section, you will learn how to use PSpice to find the Thévenin equivalent of a simple ac circuit. As well, we will use both programs to analyze circuits with dependent sources.

PSpice

The following example shows how PSpice is used to find the Thévenin or Norton equivalent of an ac circuit.

EXAMPLE 20–14

Use PSpice to determine the Thévenin equivalent of the circuit in Figure 20–18. Assume that the circuit operates at a frequency $\omega = 200$ rad/s ($f = 31.83$ Hz). Compare the result to the solution of Example 20–6.

Solution   We begin by using OrCAD Capture to input the circuit as shown in Figure 20–62.

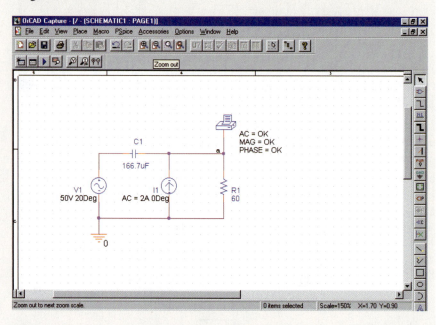

FIGURE 20–62

Notice that the load impedance has been removed and the value of capacitance is shown as

$$C = \frac{1}{\omega X_C} = \frac{1}{(200 \text{ rad/s})(30 \text{ }\Omega)} = 166.7 \text{ }\mu\text{F}$$

In order to adjust for the correct source voltage and current, we select VAC and ISRC from the source library SOURCE.slb of PSpice. The values are set for AC=**50V 20Deg** and AC=**2A 0Deg** respectively. The open-circuit output voltage is displayed using VPRINT1, which can be set to measure magnitude and phase of an ac voltage as follows. Change the properties of VPRINT1 by double clicking on the part. Use the horizontal scroll bar to find the AC, MAG, and PHASE cells. Once you have entered **OK** in each of the cells, click on Apply. Next, click on Display and select Nam*e* and Value from the display properties.

Once the circuit is entered, click on the New Simulation Profile tool and set the analysis for AC Sweep/Noise with a linear sweep beginning and ending at a frequency of **31.83Hz** (1 point). As before, it is convenient to disable the Probe postprocessor from the Probe Window tab in the simulation settings box.

As before, it is convenient to disable the Probe postprocessor prior to simulating the design. After simulating the design, the open-circuit voltage is determined by examining the output file of PSpice. The pertinent data from the output file is given as

```
FREQ          VM(a)          VP(a)
 3.183E+01    5.690E+01     -1.583E+01
```

The above result gives $\mathbf{E}_{Th} = 56.90 \text{ V}\angle -15.83°$. This is the same as the value determined in Example 20–6. Recall that one way of determining the Thévenin (or Norton) impedance is to use Ohm's law, namely

$$\mathbf{Z}_{Th} = \mathbf{Z}_N = \frac{\mathbf{E}_{Th}}{\mathbf{I}_N}$$

The Norton current is found by removing the VPRINT1 device from the circuit of Figure 20–62 and inserting an IPRINT device (ammeter) between terminal *a* and ground. The result is shown in Figure 20–63.

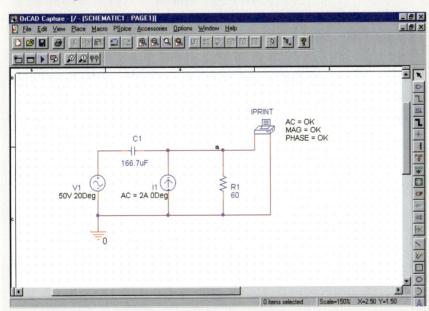

FIGURE 20–63

After simulating the design, the short-circuit current is determined by examining the output file of PSpice. The pertinent data from the output file is given as

```
FREQ          IM(V_PRINT2)IP(V_PRINT2)
3.183E+01    2.121E+00     4.761E+01
```

The above result gives $\mathbf{I}_N = 21.21 \text{ A}\angle 47.61°$ and so we calculate the Thévenin impedance as

$$\mathbf{Z}_{Th} = \frac{56.90 \text{ V}\angle -15.83°}{2.121 \text{ A}\angle 47.61°} = 26.83\Omega\angle -63.44°$$

which is the same value as that obtained in Example 20–6.

In the previous example, it was necessary to determine the Thévenin impedance in two steps, by first solving for the Thévenin voltage and then solving for the Norton current. It is possible to determine the value in one step by using an additional step in the analysis. The following example shows how to determine the Thévenin impedance for a circuit having a dependent source. The same step may also be used for a circuit having independent sources.

The following parts in OrCAD Capture are used to represent dependent sources:

Voltage-controlled voltage source:	**E**
Current-controlled current source:	**F**
Voltage-controlled current source:	**G**
Current-controlled voltage source:	**H**

When using dependent sources, it is necessary to ensure that any voltage-controlled source is placed *across* the controlling voltage and any current-controlled source is placed *in series with* the controlling current. Additionally, each dependent source must have a specified **gain.** This value simply provides the ratio between the output value and the controlling voltage or current. Although the following example shows how to use only one type of dependent source, you will find many similarities between the various sources.

EXAMPLE 20–15

Use PSpice to find the Thévenin equivalent of the circuit shown in Figure 20–46.

Solution   OrCAD Capture is used to input the circuit shown in Figure 20–64.

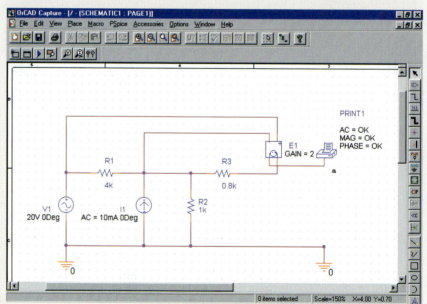

FIGURE 20–64

The voltage-controlled voltage source is obtained by clicking on the Place part tool and selecting E from the ANALOG library. Notice the placement of the source. Since R_1 is the controlling element, the controlling terminals are placed across this resistor. To adjust the gain of the voltage source, double click on the symbol. Use the scroll bar to find the GAIN cell and type **2.** In order to display this value on the schematic, you will need to click on Display and select Name and Value from the display properties.

As in the previous circuit, you will need to first use a VPRINT1 part to measure the open circuit voltage at terminal *a*. The properties of VPRINT1 are changed by typing **OK** in the AC, MAG, and PHASE cells. Click on Display and select Name and Value for each of the cells. Give the simulation profile a name and run the simulation. The pertinent data in the PSpice output file provides the open-circuit (Thévenin) voltage as follows.

```
FREQ          VM(N00431)   VP(N00431)
1.000E+03     4.000E+00    1.800E+02
```

The VPRINT1 part is then replaced with IPRINT, which is connected between terminal *a* and ground to provide the short-circuit current. Remember to change the appropriate cells using the properties editor. The PSpice output file gives the short-circuit (Norton) current as:

```
FREQ          IM(V_PRINT2) IP(V_PRINT2)
1.000E+03     1.250E-03    1.800E+02
```

The Thévenin impedance is now easily determined as

$$\mathbf{Z}_{Th} = \frac{4\ V}{1.25\ mA} = 3.2\ k\Omega$$

and the resulting circuit is shown in Figure 20–65.

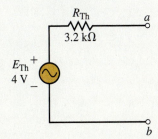

FIGURE 20–65

MultiSIM

MultiSIM has many similarities to PSpice in its analysis of ac circuits. The following example shows that the results obtained by MultiSIM are precisely the same as those obtained from PSpice.

EXAMPLE 20–16

Use MultiSIM to find the Thévenin equivalent of the circuit shown in Figure 20–47. Compare the results to those obtained in Example 20–15.

Solution   The circuit is entered as shown in Figure 20–66.

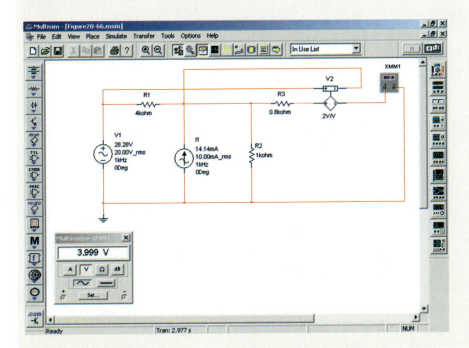

FIGURE 20–66

We use 1 kHz as the frequency of operation, although any frequency may be used. The Voltage-controlled voltage source is selected from the Sources parts bin and the gain is adjusted by double clicking on the symbol. The Value Tab is selected and the Voltage gain (E): is set to **2 V/V.** The multimeter must be set to measure ac volts. As expected, the reading on the multimeter is 4 V. Current is easily measured by setting the multimeter onto its ac ammeter range as shown in Figure 20–67. The current reading is 1.25 mA. A limitation of MultiSIM modeling is that the model does not indicate the phase angle of the voltage or current.

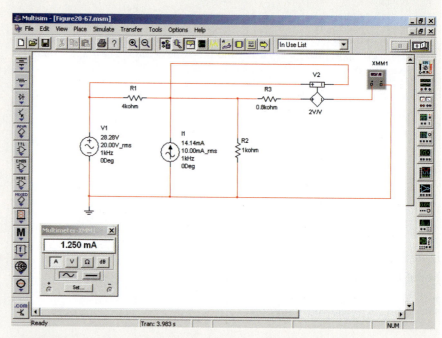

FIGURE 20–67

Now, the Thévenin impedance of the circuit is determined using Ohm's law, namely

$$\mathbf{Z}_{Th} = \frac{4V}{1.25\ mA} = 3.2\ k\Omega$$

These results are consistent with the calculations of Example 20–11 and the PSpice results of Example 20–15.

PRACTICE PROBLEMS 9

Use PSpice to find the Thévenin equivalent of the circuit shown in Figure 20–45. Compare your answer to that obtained in Practice Problems 5.

Answer
$\mathbf{E}_{Th} = \mathbf{V}_{ab} = -3.33$ V, $\mathbf{Z}_{Th} = 0.667\ \Omega$

PRACTICE PROBLEMS 10

Use MultiSIM to find the Thévenin equivalent of the circuit shown in Figure 20–45. Compare your answer to that obtained in Practice Problems 5 and Practice Problems 9.

Answer
$\mathbf{E}_{Th} = \mathbf{V}_{ab} = -3.33$ V, $\mathbf{Z}_{Th} = 0.667\ \Omega$

Use PSpice to find the Norton equivalent circuit external to the load resistor in the circuit of Figure 20–31. Assume that the circuit operates at a frequency of 20 kHz. Compare your results to those obtained in Example 20–8. Hint: You will need to place a small resistor (e.g., 1 mΩ) in series with the inductor.

Answers
$\mathbf{E}_{Th} = 7.75 \text{ V}\angle-176.5°$, $\mathbf{I}_N = 0.1335 \text{ A}\angle138.5°$, $\mathbf{Z}_N = 566\Omega\angle45°$

The results are consistent.

PUTTING IT INTO PRACTICE

In this chapter, you learned how to solve for the required load impedance to enable maximum power transfer to the load. In all cases, you worked with load impedances that were in series with the output terminals. This will not always be the case. The circuit shown in the accompanying figure shows a load that consists of a resistor in parallel with an inductor.

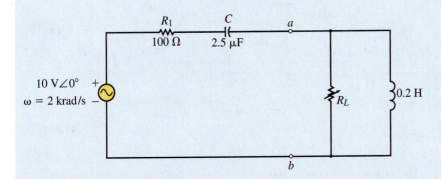

Determine the value of the resistor R_L needed to result in maximum power delivered to the load. Although several methods are possible, you may find that this example lends itself to being solved by using calculus.

PROBLEMS

20.1 Superposition Theorem—Independent Sources

1. Use superposition to determine the current in the indicated branch of the circuit in Figure 20–68.
2. Repeat Problem 1 for the circuit of Figure 20–69.
3. Use superposition to determine the voltage $\mathbf{V}_{ab}$ for the circuit of Figure 20–68.
4. Repeat Problem 3 for the circuit of Figure 20–69.
5. Consider the circuit of Figure 20–70.
   a. Use superposition to determine the indicated voltage, **V**.
   b. Show that the power dissipated by the indicated resistor cannot be determined by superposition.

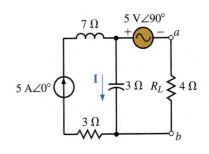

FIGURE 20–68

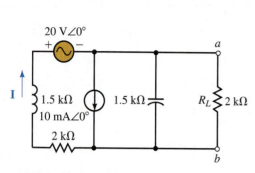

FIGURE 20–69

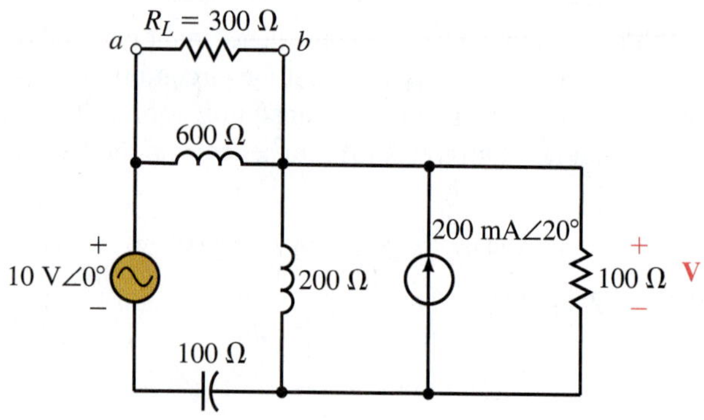

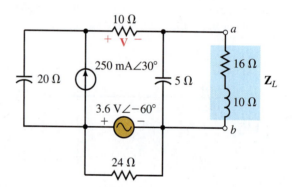

FIGURE 20–70

6. Repeat Problem 5 for the circuit of Figure 20–71.

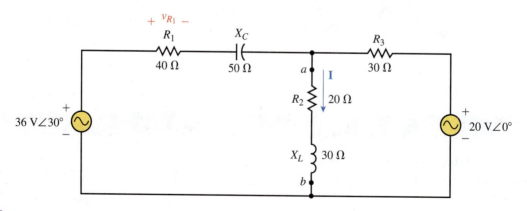

FIGURE 20–71

7. Use superposition to determine the current **I** in the circuit of Figure 20–72.

FIGURE 20–72

8. Repeat Problem 7 for the circuit of Figure 20–73.

9. Use superposition to determine the sinusoidal voltage, v_{R_1} for the circuit of Figure 20–72.

10. Repeat Problem 9 for the circuit of Figure 20–73.

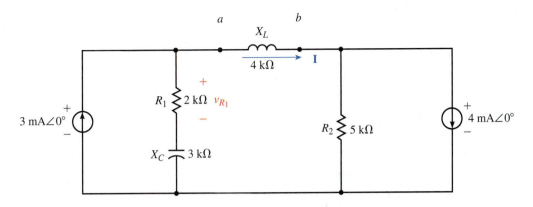

FIGURE 20–73

20.2 Superposition Theorem—Dependent Sources

11. Refer to the circuit of Figure 20–74.
    a. Use superposition to find V_L.
    b. If the magnitude of the applied voltage V is increased to 200 mV, solve for the resulting V_L.

12. Consider the circuit of Figure 20–75.
    a. Use superposition to find V_L.
    b. If the magnitude of the applied current I is decreased to 2 mA, solve for the resulting V_L.

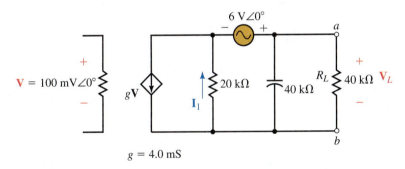

FIGURE 20–74   ◀ **MULTISIM**

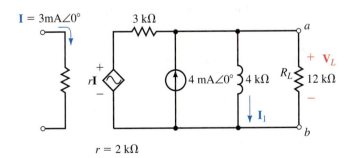

FIGURE 20–75

13. Use superposition to find the current I_1 in the circuit of Figure 20–74.

14. Repeat Problem 13 for the circuit of Figure 20–75.

15. Find $\mathbf{V}_L$ in the circuit of Figure 20–76.

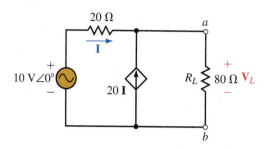

FIGURE 20–76

16. Find $\mathbf{V}_L$ in the circuit of Figure 20–77.

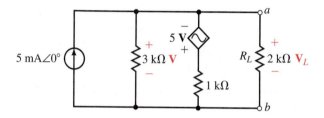

FIGURE 20–77

17. Determine the voltage $\mathbf{V}_{ab}$ for the circuit of Figure 20–78.

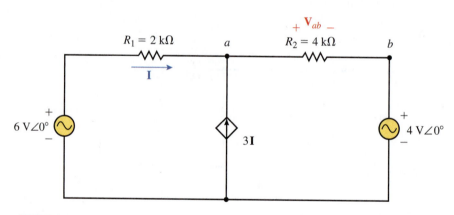

FIGURE 20–78

18. Determine the current $\mathbf{I}$ for the circuit of Figure 20–79.

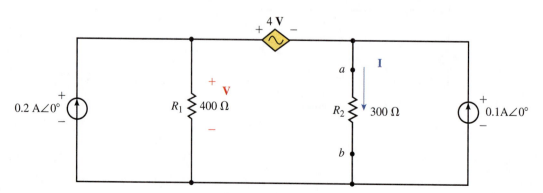

FIGURE 20–79

20.3 Thévenin's Theorem—Independent Sources

19. Find the Thévenin equivalent circuit external to the load impedance of Figure 20–68.

20. Refer to the circuit of Figure 20–80.

   a. Find the Thévenin equivalent circuit external to the indicated load.

   b. Determine the power dissipated by the load.

21. Refer to the circuit of Figure 20–81.

   a. Find the Thévenin equivalent circuit external to the indicated load at a frequency of 5 kHz.

   b. Determine the power dissipated by the load if $\mathbf{Z}_L = 100\ \Omega\angle30°$.

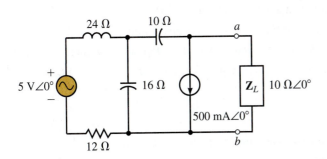

FIGURE 20–80

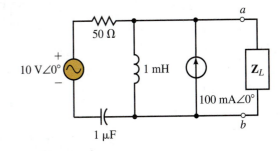

FIGURE 20–81

◀ MULTISIM

22. Repeat Problem 21 for a frequency of 1 kHz.

23. Find the Thévenin equivalent circuit external to R_L in the circuit of Figure 20–72.

24. Repeat Problem 23 for the circuit of Figure 20–69.

25. Repeat Problem 23 for the circuit of Figure 20–70.

26. Find the Thévenin equivalent circuit external to Z_L in the circuit of Figure 20–71.

27. Consider the circuit of Figure 20–82.

   a. Find the Thévenin equivalent circuit external to the indicated load.

   b. Determine the power dissipated by the load if $\mathbf{Z}_L = 20\ \Omega\angle-60°$.

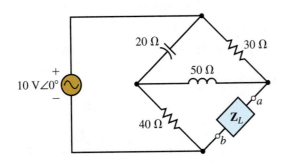

FIGURE 20–82

28. Repeat Problem 27 if a 10-Ω resistor is placed in series with the voltage source.

20.4 Norton's Theorem—Independent Sources

29. Find the Norton equivalent circuit external to the load impedance of Figure 20–68.

30. Repeat Problem 29 for the circuit of Figure 20–69.

31. a. Using the outlined procedure, find the Norton equivalent circuit external to terminals a and b in Figure 20–72.

    b. Determine the current through the indicated load.

    c. Find the power dissipated by the load.

32. Repeat Problem 31 for the circuit of Figure 20–73.

33. a. Using the outlined procedure, find the Norton equivalent circuit external to the indicated load impedance (located between terminals a and b) in Figure 20–70.

    b. Determine the current through the indicated load.

    c. Find the power dissipated by the load.

34. Repeat Problem 33 for the circuit of Figure 20–71.

35. Suppose that the circuit of Figure 20–81 operates at a frequency of 2 kHz.

    a. Find the Norton equivalent circuit external to the load impedance.

    b. If a 30-Ω load resistor is connected between terminals a and b, find the current through the load.

36. Repeat Problem 35 for a frequency of 8 kHz.

20.5   Thévenin's and Norton's Theorem for Dependent Sources

37. a. Find the Thévenin equivalent circuit external to the load impedance in Figure 20–74.

    b. Solve for the current through R_L.

    c. Determine the power dissipated by R_L.

38. a. Find the Norton equivalent circuit external to the load impedance in Figure 20–75.

    b. Solve for the current through R_L.

    c. Determine the power dissipated by R_L.

39. Find the Thévenin and Norton equivalent circuits external to the load impedance of Figure 20–76.

40. Find the Thévenin equivalent circuit external to the load impedance of Figure 20–77.

20.6   Maximum Power Transfer Theorem

41. Refer to the circuit of Figure 20–83.

    a. Determine the load impedance, $\mathbf{Z}_L$, needed to ensure that the load receives maximum power.

    b. Find the maximum power to the load.

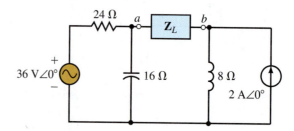

FIGURE 20–83

42. Repeat Problem 41 for the circuit of Figure 20–84.

43. Repeat Problem 41 for the circuit of Figure 20–85.

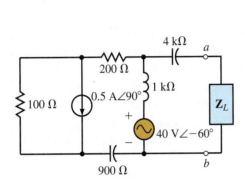

FIGURE 20–84

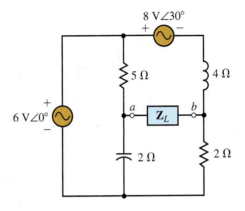

FIGURE 20–85

44. Repeat Problem 41 for the circuit of Figure 20–86.

45. What load impedance is required for the circuit of Figure 20–71 to ensure that the load receives maximum power from the circuit?

46. Determine the load impedance required for the circuit of Figure 20–82 to ensure that the load receives maximum power from the circuit.

47. a. Determine the required load impedance, $\mathbf{Z}_L$, for the circuit of Figure 20–81 to deliver maximum power to the load at a frequency of 5 kHz.

    b. If the load impedance contains a resistor and a 1-μF capacitor, determine the value of the resistor to result in a relative maximum power transfer.

    c. Solve for the power delivered to the load in (b).

48. a. Determine the required load impedance, $\mathbf{Z}_L$, for the circuit of Figure 20–81 to deliver maximum power to the load at a frequency of 1 kHz.

    b. If the load impedance contains a resistor and a 1-μF capacitor, determine the value of the resistor to result in a relative maximum power transfer.

    c. Solve for the power delivered to the load in (b).

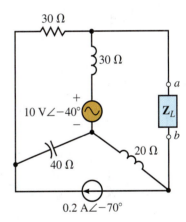

FIGURE 20–86

20.7 Circuit Analysis Using Computers

49. Use PSpice to find the Thévenin equivalent circuit external to R_L in the circuit of Figure 20–68. Assume that the circuit operates at a frequency of $\omega = 2000$ rad/s.

    Note: PSpice does not permit a voltage source to have floating terminals. Therefore, a large resistance (e.g., 10 GΩ) must be placed across the output.

50. Repeat Problem 49 for the circuit of Figure 20–69.

51. Use PSpice to find the Norton equivalent circuit external to R_L in the circuit of Figure 20–70. Assume that the circuit operates at a frequency of $\omega = 5000$ rad/s.

    Note: PSpice cannot analyze a circuit with a short-circuited inductor. Consequently, it is necessary to place a small resistance (e.g., 1 nΩ) in series with an inductor.

◀ CADENCE

◀ CADENCE
◀ CADENCE

52. Repeat Problem 51 for the circuit of Figure 20–71.

 Note: PSpice cannot analyze a circuit with an open-circuited capacitor. Consequently, it is necessary to place a large resistance (e.g., 10 GΩ) in parallel with a capacitor.

53. Use PSpice to find the Thévenin equivalent circuit external to R_L in the circuit of Figure 20–76. Assume that the circuit operates at a frequency of $f = 1000$ Hz.

54. Repeat Problem 53 for the circuit of Figure 20–77.

55. Use PSpice to find the Norton equivalent circuit external to $\mathbf{V}_{ab}$ in the circuit of Figure 20–78. Assume that the circuit operates at a frequency of $f = 1000$ Hz.

56. Repeat Problem 55 for the circuit of Figure 20–79.

57. Use MultiSIM to find the Thévenin equivalent circuit external to R_L in the circuit of Figure 20–76. Assume that the circuit operates at a frequency of $f = 1000$ Hz.

58. Repeat Problem 53 for the circuit of Figure 20–77.

59. Use MultiSIM to find the Norton equivalent circuit external to $\mathbf{V}_{ab}$ in the circuit of Figure 20–78. Assume that the circuit operates at a frequency of $f = 1000$ Hz.

60. Repeat Problem 55 for the circuit of Figure 20–79.

✓ **ANSWERS TO IN-PROCESS LEARNING CHECKS**

In-Process Learning Check 1

a. $\mathbf{I} = 3.39 \text{ A} \angle 25.34°$

b. $P_T = 230.4 \text{ W}$

c. $P_1 + P_2 + P_3 = 145 \text{ W} \neq P_T = 230.4 \text{ W}$
Superposition does not apply for power.

In-Process Learning Check 2

1. Remove the inductor from the circuit. Label the remaining terminals as a and b.

2. Set the voltage source to zero by removing it from the circuit and replacing it with a short circuit.

3. Determine the values of the impedance using the given frequency. Calculate the Thévenin impedance between terminals a and b.

4. Convert the voltage source into its equivalent phasor form. Solve for the open-circuit voltage between terminals a and b.

5. Sketch the resulting Thévenin equivalent circuit.

In-Process Learning Check 3

1. Remove the resistor from the circuit. Label the remaining terminals as a and b.

2. Set the voltage source to zero by removing it from the circuit and replacing it with a short circuit.

3. Calculate the Norton impedance between terminals a and b.

4. Solve for the short-circuit current between terminals a and b.

5. Sketch the resulting Norton equivalent circuit.

In-Process Learning Check 4

a. $R_L = 80.6 \ \Omega$

b. $P_L = 3.25 \text{ W}$

c. $\omega = 3333 \text{ rad/s}$

d. $P_L = 4.90 \text{ W}$

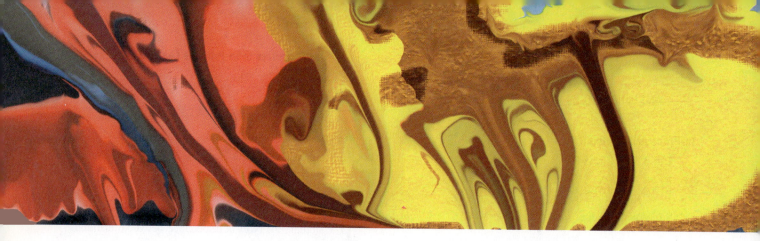

■ OBJECTIVES

After studying this chapter, you will be able to

- determine the resonant frequency and bandwidth of a simple series or parallel circuit,

- determine the voltages, currents, and power of elements in a resonant circuit,

- sketch the impedance, current, and power response curves of a series resonant circuit,

- find the quality factor, Q, of a resonant circuit and use Q to determine the bandwidth for a given set of conditions,

- explain the dependence of bandwidth on the L/C ratio and on R for both a series and a parallel resonant circuit,

- design a resonant circuit for a given set of parameters,

- convert a series RL network into an equivalent parallel network for a given frequency.

Resonance

21

CHAPTER PREVIEW

In this chapter, we build upon the knowledge obtained in previous chapters to observe how resonant circuits are able to pass a desired range of frequencies from a signal source to a load. In its most simple form, the **resonant circuit** consists of an inductor and a capacitor together with a voltage or current source. Although the circuit is simple, it is one of the most important circuits used in electronics. As an example, the resonant circuit, in one of its many forms, allows us to select a desired radio or television signal from the vast number of signals that are around us at any time. In order to obtain all the transmitted energy for a given radio station or television channel, we would like a circuit to have the frequency response shown in Figure 21–1(a). A circuit having an ideal frequency response would pass all frequency components in a band between f_1 and f_2, while rejecting all other frequencies. For a radio transmitter, the center frequency, f_r would correspond to the *carrier frequency* of the station. The difference between the upper and lower frequencies that we would like to pass is called the *bandwidth.*

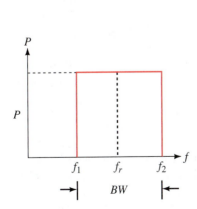

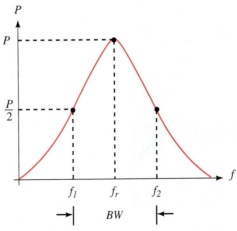

(b) Actual response curve of a resonant circuit

FIGURE 21–1

Whereas there are various configurations of resonant circuits, they all have several common characteristics. Resonant electronic circuits contain at least one inductor and one capacitor and have a bell-shaped response curve centered at some resonant frequency, f_r, as illustrated in Figure 21–1(b).

The response curve of Figure 21–1(b) indicates that power will be at a maximum at the resonant frequency, f_r. Varying the frequency in either direction results in a reduction of the power. The bandwidth of the resonant circuit is taken to be the difference between the half power points on the response curve of the filter.

If we were to apply variable-frequency sinusoidal signals to a circuit consisting of an inductor and capacitor, we would find that maximum energy will transfer back and forth between the two elements at the resonant frequency. In an ideal *LC* circuit (one containing no resistance), these oscillations would continue unabated even if the signal source were turned off. However, in the practical situation, all circuits have some resistance. As a result, the stored energy will eventually be dissipated by the resistance, resulting in **damped oscillations.** In a manner similar to pushing a child on a swing, the oscillations will continue indefinitely if a small amount of energy is applied to the circuit at exactly the right moment. This phenomenon illustrates the basis of how oscillator circuits operate and therefore provides us with another application of the resonant circuit.

In this chapter, we examine in detail the two main types of resonant circuits: the **series resonant circuit** and the **parallel resonant circuit.** ∎

PUTTING IT IN PERSPECTIVE

Edwin Howard Armstrong—Radio Reception

EDWIN ARMSTRONG WAS BORN in New York City on December 18, 1890. As a young man, he was keenly interested in experiments involving radio transmission and reception.

After earning a degree in electrical engineering at Columbia University, Armstrong used his theoretical background to explain and improve the operation of the triode vacuum tube, which had been invented by Lee de Forest. Edwin Armstrong was able to improve the sensitivity of receivers by using feedback to amplify a signal many times. By increasing the amount of signal feedback, Armstrong also designed and patented a circuit that used the vacuum tube as an oscillator.

Armstrong is best known for conceiving the concept of superheterodyning, in which a high frequency is lowered to a more usable intermediate frequency. Superheterodyning is still used in modern AM and FM receivers and in numerous other electronic circuits such as radar and communication equipment.

Edwin Armstrong was the inventor of FM transmission, which led to greatly improved fidelity in radio transmission.

Although Armstrong was a brilliant engineer, he was an uncompromising person who was involved in numerous lawsuits with Lee de Forest and the communications giant, RCA.

After spending nearly two million dollars in legal battles, Edwin Armstrong jumped to his death from his thirteenth-floor apartment window on January 31, 1954. ∎

A simple series resonant circuit is constructed by combining an ac source with an inductor, a capacitor, and optionally, a resistor as shown in Figure 21–2a. By combining the generator resistance, R_G, with the series resistance, R_S, and the resistance of the inductor coil, R_{coil}, the circuit may be simplified as illustrated in Figure 21–2b.

21.1 Series Resonance

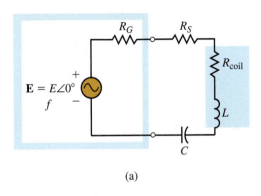 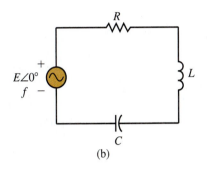

(a)   (b)

FIGURE 21–2

In this circuit, the total resistance is expressed as

$$R = R_G + R_S + R_{coil}$$

Because the circuit of Figure 21–2 is a series circuit, we calculate the total impedance as follows:

$$\mathbf{Z}_T = R + jX_L - jX_C$$
$$= R + j(X_L - X_C) \tag{21–1}$$

Resonance occurs when the reactance of the circuit is effectively eliminated, resulting in a total impedance that is purely resistive. We know that the reactances of the inductor and capacitor are given as follows:

$$X_L = \omega L = 2\pi f L \tag{21–2}$$

$$X_C = \frac{1}{\omega C} = \frac{1}{2\pi f C} \tag{21–3}$$

Examining Equation 21–1, we see that by setting the reactances of the capacitor and inductor equal to one another, the total impedance, $\mathbf{Z}_T$, is purely resistive since the inductive reactance which is on the positive j axis cancels the capacitive reactance on the negative j axis. The total impedance of the series circuit at resonance is equal to the total circuit resistance, R. Hence, at resonance,

$$Z_T = R \tag{21–4}$$

By letting the reactances be equal we are able to determine the series resonance frequency, ω_S (in radians per second) as follows:

$$\omega L = \frac{1}{\omega C}$$

$$\omega^2 = \frac{1}{LC} \tag{21–5}$$

$$\omega_S = \frac{1}{\sqrt{LC}} \quad \text{(rad/s)}$$

Since the calculation of the angular frequency, ω, in radians per second is easier than solving for frequency, f, in hertz, we generally express our resonant frequencies in the more simple form. Further calculations of voltage and current will usually be much easier by using ω rather than f. If, however, it becomes necessary to determine a frequency in hertz, recall that the relationship between ω and f is as follows:

$$\omega = 2\pi f \quad \text{(rad/s)} \tag{21–6}$$

Equation 21–6 is inserted into Equation 21–5 to give the resonant frequency as

$$f_S = \frac{1}{2\pi\sqrt{LC}} \quad \text{(Hz)} \tag{21–7}$$

The subscript s in the above equations indicates that the frequency determined is the series resonant frequency.

At resonance, the total current in the circuit is determined from Ohm's law as

$$\mathbf{I} = \frac{\mathbf{E}}{\mathbf{Z_T}} = \frac{E\angle 0°}{R\angle 0°} = \frac{E}{R}\angle 0° \tag{21–8}$$

By again applying Ohm's law, we find the voltage across each of the elements in the circuit as follows:

$$\mathbf{V}_R = IR\angle 0° \tag{21–9}$$

$$\mathbf{V}_L = IX_L\angle 90° \tag{21–10}$$

$$\mathbf{V}_C = IX_C\angle -90° \tag{21–11}$$

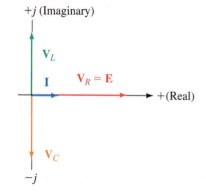

FIGURE 21–3

The phasor form of the voltages and current is shown in Figure 21–3.

Notice that since the inductive and capacitive reactances have the same magnitude, the voltages across the elements must have the same magnitude but be 180° out of phase.

We determine the average power dissipated by the resistor and the reactive powers of the inductor and capacitor as follows:

$$P_R = I^2 R \quad \text{(W)}$$
$$Q_L = I^2 X_L \quad \text{(VAR)}$$
$$Q_C = I^2 X_C \quad \text{(VAR)}$$

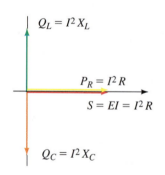

FIGURE 21–4

These powers are illustrated graphically in Figure 21–4.

21.2 Quality Factor, Q

For any resonant circuit, we define the **quality factor, Q,** as the ratio of reactive power to average power, namely,

$$Q = \frac{\text{reactive power}}{\text{average power}} \tag{21–12}$$

Because the reactive power of the inductor is equal to the reactive power of the capacitor at resonance, we may express Q in terms of either reactive power. Consequently, the above expression is written as follows:

$$Q_S = \frac{I^2 X_L}{I^2 R}$$

and so we have

$$Q_S = \frac{X_L}{R} = \frac{\omega L}{R} \qquad (21\text{--}13)$$

Quite often, the inductor of a given circuit will have a Q expressed in terms of its reactance and internal resistance, as follows:

$$Q_{\text{coil}} = \frac{X_L}{R_{\text{coil}}}$$

If an inductor with a specified Q_{coil} is included in a circuit, it is necessary to include its effects in the overall calculation of the total circuit Q.

We now examine how the Q of a circuit is used in determining other quantities of the circuit. By multiplying both the numerator and denominator of Equation 21–13 by the current, I, we have the following:

$$Q_S = \frac{IX_L}{IR} = \frac{V_L}{E} \qquad (21\text{--}14)$$

Now, since the magnitude of the voltage across the capacitor is equal to the magnitude of the voltage across the inductor at resonance, we see that the voltages across the inductor and capacitor are related to the Q by the following expression:

$$V_C = V_L = Q_S E \quad \text{at resonance} \qquad (21\text{--}15)$$

Note: Since the Q of a resonant circuit is generally significantly larger than 1, we see that the voltage across reactive elements can be many times greater than the applied source voltage. Therefore, it is always necessary to ensure that the reactive elements used in a resonant circuit are able to handle the expected voltages and currents.

Find the indicated quantities for the circuit of Figure 21–5.

EXAMPLE 21–1

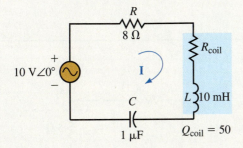

FIGURE 21–5

a. Resonant frequency expressed as ω(rad/s) and f(Hz).

b. Total impedance at resonance.

c. Current at resonance.

d. $\mathbf{V}_L$ and $\mathbf{V}_C$.

e. Reactive powers, Q_C and Q_L.

f. Quality factor of the circuit, Q_S.

Solution

a.
$$\omega_S = \frac{1}{\sqrt{LC}}$$

$$= \frac{1}{\sqrt{(10 \text{ mH})(1\mu\text{F})}}$$

$$= 10\,000 \text{ rad/s}$$

$$f_S = \frac{\omega}{2\pi} = 1592 \text{ Hz}$$

b.
$$X_L = \omega L = (10\,000 \text{ rad/s})(10 \text{ mH}) = 100 \text{ }\Omega$$

$$R_{\text{coil}} = \frac{X_L}{Q_{\text{coil}}} = \frac{100 \text{ }\Omega}{50} = 2.00 \text{ }\Omega$$

$$R_T = R + R_{\text{coil}} = 10.0 \text{ }\Omega$$

$$\mathbf{Z}_T = 10 \text{ }\Omega\angle 0°$$

c.
$$\mathbf{I} = \frac{\mathbf{E}}{\mathbf{Z}_T} = \frac{10 \text{ V}\angle 0°}{10 \text{ }\Omega\angle 0°} = 1.0 \text{ A}\angle 0°$$

d.
$$\mathbf{V}_L = (100 \text{ }\Omega\angle 90°)(1.0 \text{ A}\angle 0°) = 100 \text{ V}\angle 90°$$

$$\mathbf{V}_C = (100 \text{ }\Omega\angle -90°)(1.0 \text{ A}\angle 0°) = 100 \text{ V}\angle -90°$$

Notice that the voltage across the reactive elements is ten times greater than the applied signal voltage.

e. Although we use the symbol Q to designate both reactive power and the quality factor, the context of the question generally provides us with a clue as to which meaning to use.

$$Q_L = (1.0 \text{ A})^2(100 \text{ }\Omega) = 100 \text{ VAR}$$

$$Q_C = (1.0 \text{ A})^2(100 \text{ }\Omega) = 100 \text{ VAR}$$

f.
$$Q_S = \frac{Q_L}{P} = \frac{100 \text{ VAR}}{10 \text{ W}} = 10$$

PRACTICE PROBLEMS 1

Consider the circuit of Figure 21–6:

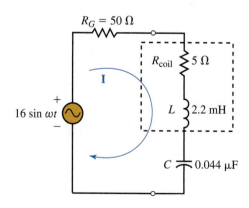

◄ MULTISIM

FIGURE 21–6

a. Find the resonant frequency expressed as ω(rad/s) and f(Hz).

b. Determine the total impedance at resonance.

c. Solve for $\mathbf{I}$, $\mathbf{V}_L$, and $\mathbf{V}_C$ at resonance.

d. Calculate reactive powers Q_C and Q_L at resonance.

e. Find the quality factor, Q_s, of the circuit.

Answers
a. 102 krad/s, 16.2 kHz; b. 55.0 $\Omega\angle0°$
c. 0.206 A$\angle0°$, 46.0 V$\angle90°$, 46.0 V$\angle-90°$; d. 9.46 VAR; e. 4.07

In this section, we examine how the impedance of a series resonant circuit varies as a function of frequency. Because the impedances of inductors and capacitors are dependent upon frequency, the total impedance of a series resonant circuit must similarly vary with frequency. For algebraic simplicity, we use frequency expressed as ω in radians per second. If it becomes necessary to express the frequency in hertz, the conversion of Equation 21–6 is used.

21.3 Impedance of a Series Resonant Circuit

The total impedance of a simple series resonant circuit is written as

$$\mathbf{Z}_T = R + j\omega L - j\frac{1}{\omega C}$$

$$= R + j\left(\frac{\omega^2 LC - 1}{\omega C}\right)$$

The magnitude and phase angle of the impedance vector, $\mathbf{Z}_T$, are expressed as follows:

$$Z_T = \sqrt{R^2 + \left(\frac{\omega^2 LC - 1}{\omega C}\right)^2} \qquad (21\text{–}16)$$

$$\theta = \tan^{-1}\left(\frac{\omega^2 LC - 1}{\omega RC}\right) \qquad (21\text{–}17)$$

Examining these equations for various values of frequency, we note that the following conditions will apply:

When $\omega = \omega_S$:

$$Z_T = R$$

and

$$\theta = \tan^{-1}0 = 0°$$

This result is consistent with the results obtained in the previous section.

When $\omega < \omega_S$:

As we decrease ω from resonance, Z_T will get larger until $\omega = 0$. At this point, the magnitude of the impedance will be undefined, corresponding to an open circuit. As one might expect, the large impedance occurs because the capacitor behaves like an open circuit at dc.

The angle θ will occur between $0°$ and $-90°$ since the numerator of the argument of the arctangent function will always be negative, corresponding to an angle in the fourth quadrant. Because the angle of the impedance has a negative sign, we conclude that the impedance must appear capacitive in this region.

When $\omega > \omega_S$:

As ω is made larger than resonance, the impedance Z_T will increase due to the increasing reactance of the inductor.

For these values of ω, the angle θ will always be within $0°$ and $+90°$ because both the numerator and the denominator of the arctangent function are positive. Because the angle of $\mathbf{Z}_T$ occurs in the first quadrant, the impedance must be inductive.

Sketching the magnitude and phase angle of the impedance $\mathbf{Z}_T$ as a function of angular frequency, we have the curves shown in Figure 21–7.

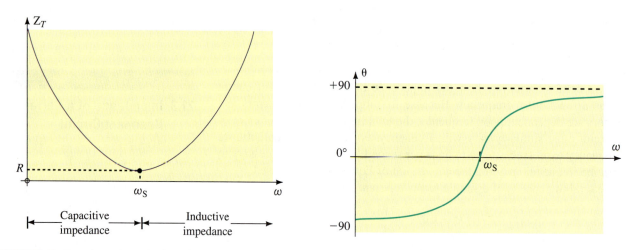

FIGURE 21–7   Impedance (magnitude and phase angle) versus angular frequency for a series resonant circuit.

21.4 Power, Bandwidth, and Selectivity of a Series Resonant Circuit

◀ **Online Companion**

Due to the changing impedance of the circuit, we conclude that if a constant-amplitude voltage is applied to the series resonant circuit, the current and power of the circuit will not be constant at all frequencies. In this section, we examine how current and power are affected by changing the frequency of the voltage source.

Applying Ohm's law gives the magnitude of the current at resonance as follows:

$$I_{max} = \frac{E}{R} \tag{21–18}$$

For all other frequencies, the magnitude of the current will be less than I_{max} because the impedance is greater than at resonance. Indeed, when the frequency is zero (dc), the current will be zero since the capacitor is effectively an open circuit. On the other hand, at increasingly higher frequencies, the inductor begins to approximate an open circuit, once again causing the current in the circuit to approach zero. The current response curve for a typical series resonant circuit is shown in Figure 21–8.

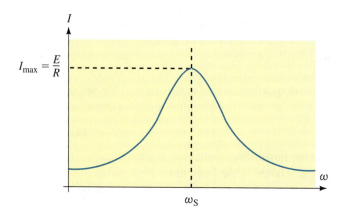

FIGURE 21–8   Current versus angular frequency for a series resonant circuit.

The total power dissipated by the circuit at any frequency is given as

$$P = I^2 R \qquad (21\text{–}19)$$

Since the current is maximum at resonance, it follows that the power must similarly be maximum at resonance. The maximum power dissipated by the series resonant circuit is therefore given as

$$P_{max} = I_{max}^2 R = \frac{E^2}{R} \qquad (21\text{–}20)$$

The power response of a series resonant circuit has a bell-shaped curve called the **selectivity curve,** which is similar to the current response. Figure 21–9 illustrates the typical selectivity curve.

Examining Figure 21–9, we see that only frequencies around ω_s will permit significant amounts of power to be dissipated by the circuit. We define the **bandwidth,** BW, of the resonant circuit to be the difference between the frequencies at which the circuit delivers half of the maximum power. The frequencies ω_1 and ω_2 are called the **half-power frequencies,** the **cutoff frequencies,** or the **band frequencies.**

> **NOTES . . .**
>
> We see from Figure 21–9 that the selectivity curve is not perfectly symmetrical on both sides of the resonant frequency. As a result, ω_s is not exactly centered between the half-power frequencies. However, as Q increases, we find that the resonant frequency approaches the midpoint between ω_1 and ω_2. *In general, if $Q > 10$, then we assume that the resonant frequency is at the midpoint of half-power frequencies.*

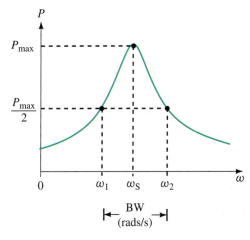

FIGURE 21–9   Selectivity curve.

If the bandwidth of a circuit is kept very narrow, the circuit is said to have a **high selectivity,** since it is highly selective to signals occurring within a very narrow range of frequencies. On the other hand, if the bandwidth of a circuit is large, the circuit is said to have a **low selectivity.**

The elements of a series resonant circuit determine not only the frequency at which the circuit is resonant, but also the shape (and hence the bandwidth) of the power response curve. Consider a circuit in which the resistance, R, and the resonant frequency, ω_s, are held constant. We find that by increasing the ratio of L/C, the sides of the power response curve become steeper. This in turn results in a decrease in the bandwidth. Inversely, decreasing the ratio of L/C causes the sides of the curve to become more gradual, resulting in an increased bandwidth. These characteristics are illustrated in Figure 21–10.

If, on the other hand, L and C are kept constant, we find that the bandwidth will decrease as R is decreased and will increase as R is increased. Figure 21–11 shows how the shape of the selectivity curve is dependent upon the value of resistance. A series circuit has the highest selectivity if the resistance of the circuit is kept to a minimum.

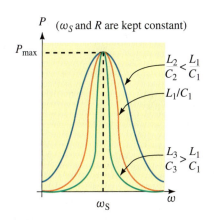

FIGURE 21–10

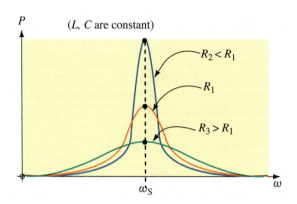

FIGURE 21–11

For the series resonant circuit the power at any frequency is determined as

$$P = I^2 R$$

$$= \left(\frac{E}{Z_T}\right)^2 R$$

By substituting Equation 21–16 into the above expression, we arrive at the general expression for power as a function of frequency, ω:

$$P = \frac{E^2 R}{R^2 + \left(\dfrac{\omega^2 LC - 1}{\omega C}\right)^2} \qquad (21\text{–}21)$$

At the half-power frequencies, the power must be

$$P_{hpf} = \frac{E^2}{2R} \qquad (21\text{–}22)$$

Since the maximum current in the circuit is given as $I_{max} = E/R$, we see that by manipulating the above expression, the magnitude of current at the half-power frequencies is

$$I_{hpf} = \sqrt{\frac{P_{hpf}}{R}} = \sqrt{\frac{E^2}{2R^2}} = \sqrt{\frac{I_{max}^2}{2}}$$

$$I_{hpf} = \frac{I_{max}}{\sqrt{2}} \qquad (21\text{–}23)$$

The cutoff frequencies are found by evaluating the frequencies at which the power dissipated by the circuit is half of the maximum power. Combining Equations 21–21 and 21–22, we have the following:

$$\frac{E^2}{2R} = \frac{E^2 R}{R^2 + \left(\dfrac{\omega^2 LC - 1}{\omega C}\right)^2}$$

$$2R^2 = R^2 + \left(\frac{\omega^2 LC - 1}{\omega C}\right)^2 \qquad (21\text{–}24)$$

$$\frac{\omega^2 LC - 1}{\omega C} = \pm R$$

$$\omega^2 LC - 1 = \pm \omega RC \quad \text{(at half-power)}$$

From the selectivity curve for a series circuit, we see that the two half-power points occur on both sides of the resonant angular frequency, ω_s.

When $\omega < \omega_S$, the term $\omega^2 LC$ must be less than 1. In this case the solution is determined as follows:

$$\omega^2 LC - 1 = -\omega RC$$

$$\omega^2 LC + \omega RC - 1 = 0$$

The solution of this quadratic equation gives the lower half-power frequency as

$$\omega_1 = \frac{-RC + \sqrt{(RC)^2 + 4LC}}{2LC}$$

or

$$\omega_1 = \frac{-R}{2L} + \sqrt{\frac{R^2}{4L^2} + \frac{1}{LC}} \qquad \textbf{(21–25)}$$

In a similar manner, for $\omega > \omega_s$, the upper half-power frequency is

$$\omega_2 = \frac{R}{2L} + \sqrt{\frac{R^2}{4L^2} + \frac{1}{LC}} \qquad \textbf{(21–26)}$$

Taking the difference between Equations 21–26 and 21–25, we find the bandwidth of the circuit as

$$\text{BW} = \omega_2 - \omega_1$$

$$= \frac{R}{2L} + \sqrt{\frac{R^2}{4L^2} + \frac{1}{LC}} - \left(-\frac{R}{2L} + \sqrt{\frac{R^2}{4L^2} + \frac{1}{LC}} \right)$$

which gives

$$\text{BW} = \frac{R}{L} \quad \text{(rad/s)} \qquad \textbf{(21–27)}$$

If the above expression is multiplied by ω_S/ω_S we obtain

$$\text{BW} = \frac{\omega_S R}{\omega_S L}$$

and since $Q_S = \omega_S L/R$ we further simplify the bandwidth as

$$\text{BW} = \frac{\omega_S}{Q_S} \quad \text{(rad/s)} \qquad \textbf{(21–28)}$$

Because the bandwidth may alternately be expressed in hertz, the above expression is equivalent to having

$$\text{BW} = \frac{f_S}{Q_S} \quad \text{(Hz)} \qquad \textbf{(21–29)}$$

Refer to the circuit of Figure 21–12.

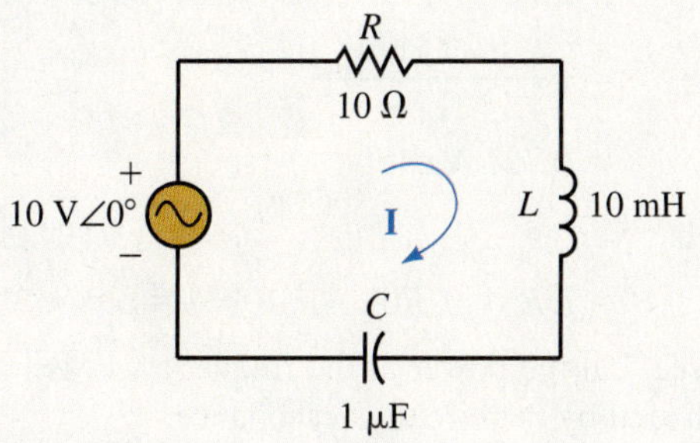

EXAMPLE 21–2

FIGURE 21–12

a. Determine the maximum power dissipated by the circuit.

b. Use the results obtained from Example 21–1 to determine the bandwidth of the resonant circuit and to arrive at the approximate half-power frequencies, ω_1 and ω_2.

c. Calculate the actual half-power frequencies, ω_1 and ω_2, from the given component values. Show two decimal places of precision.

d. Solve for the circuit current, **I**, and power dissipated at the lower half-power frequency, ω_1, found in Part (c).

Solution

a.
$$P_{max} = \frac{E^2}{R} = 10.0 \text{ W}$$

b. From Example 21–1, we had the following circuit characteristics:
$$Q_S = 10, \ \omega_S = 10 \text{ krad/s}$$

The bandwidth of the circuit is determined to be
$$\text{BW} = \omega_S Q_S = 1.0 \text{ krad/s}$$

If the resonant frequency were centered in the bandwidth, then the half-power frequencies occur at approximately
$$\omega_1 = 9.50 \text{ krad/s}$$

and
$$\omega_2 = 10.50 \text{ krad/s}$$

c.
$$\omega_1 = -\frac{R}{2L} + \sqrt{\frac{R^2}{4L^2} + \frac{1}{LC}}$$

$$= -\frac{10 \ \Omega}{(2)(10 \text{ mH})} + \sqrt{\frac{(10 \ \Omega)^2}{(4)(10 \text{ mH})^2} + \frac{1}{(10 \text{ mH})(1 \ \mu\text{F})}}$$

$$= -500 + 10 \ 012.49 = 9512.49 \text{ rad/s} \quad (f_1 = 1514.0 \text{ Hz})$$

$$\omega_2 = \frac{R}{2L} + \sqrt{\frac{R^2}{4L^2} + \frac{1}{LC}}$$

$$= 500 + 10 \ 012.49 = 10 \ 512.49 \text{ rad/s} \quad (f_2 = 1673.1 \text{ Hz})$$

Notice that the actual half-power frequencies are very nearly equal to the approximate values. For this reason, if $Q \geq 10$, it is often sufficient to calculate the cutoff frequencies by using the easier approach of Part (b).

d. At $\omega_1 = 9.51249$ krad/s, the reactances are as follows:
$$X_L = \omega L = (9.51249 \text{ krad/s})(10 \text{ mH}) = 95.12 \ \Omega$$

$$X_C = \frac{1}{\omega C} = \frac{1}{(9.51249 \text{ krad/s})(1 \ \mu\text{F})} = 105.12 \ \Omega$$

The current is now determined to be
$$\mathbf{I} = \frac{10 \text{ V}\angle 0°}{10 \ \Omega + j95.12 \ \Omega - j105.12 \ \Omega}$$

$$= \frac{10 \text{ V}\angle 0°}{14.14 \ \Omega\angle -45°}$$

$$= 0.707 \text{ A}\angle 45°$$

and the power is given as
$$P = I^2 R = (0.707 \text{ A})^2(10 \text{ V}) = 5.0 \text{ W}$$

As expected, we see that the power at the frequency ω_1 is indeed equal to half of the power dissipated by the circuit at resonance.

Refer to the circuit of Figure 21–13.

EXAMPLE 21–3

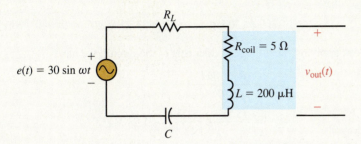

FIGURE 21–13

a. Calculate the values of R_L and C for the circuit to have a resonant frequency of 200 kHz and a bandwidth of 16 kHz.

b. Use the designed component values to determine the power dissipated by the circuit at resonance.

c. Solve for $v_{out}(t)$ at resonance.

Solution

a. Because the circuit is at resonance, we must have the following conditions:

$$Q_s = \frac{f_s}{BW}$$

$$= \frac{200 \text{ kHz}}{16 \text{ kHz}}$$

$$= 12.5$$

$$X_L = 2\pi f L$$

$$= 2\pi(200 \text{ kHz})(200 \text{ μH})$$

$$= 251.3 \text{ Ω}$$

$$R = R_L + R_{coil} = \frac{X_L}{Q_s}$$

$$= 20.1 \text{ Ω}$$

and so R_L must be

$$R_L = 20.1 \text{ Ω} - 5 \text{ Ω} = 15.1 \text{ Ω}$$

Since $X_C = X_L$, we determine the capacitance as

$$C = \frac{1}{2\pi f X_C}$$

$$= \frac{1}{2\pi(200 \text{ kHz})(251.3 \text{ Ω})}$$

$$= 3.17 \text{ nF} (\equiv 0.00317 \text{ μF})$$

b. The power at resonance is found from Equation 21–20 as

$$P_{max} = \frac{E^2}{R} = \frac{\left(\dfrac{30 \text{ V}}{\sqrt{2}}\right)^2}{20.1 \text{ Ω}}$$

$$= 22.4 \text{ W}$$

c. We see from the circuit of Figure 21–13 that the voltage $v_{out}(t)$ may be determined by applying the voltage divider rule to the circuit. However, we must first convert the source voltage from time domain into phasor domain as follows:

$$e(t) = 30 \sin \omega t \Leftrightarrow \mathbf{E} = 21.21 \text{ V}\angle0°$$

Now, applying the voltage divider rule to the circuit, we have

$$\mathbf{V}_{out} = \frac{(R_1 + j\omega L)}{R}\mathbf{E}$$

$$= \frac{(5 \text{ } \Omega + j251.3 \text{ } \Omega)}{20.1 \text{ } \Omega}21.21 \text{ V}\angle0°$$

$$= (251.4 \text{ } \Omega\angle88.86°)(1.056 \text{ A}\angle0°)$$

$$= 265.5 \text{ V}\angle88.86°$$

which in time domain is given as

$$v_{out}(t) = 375 \sin(\omega t + 88.86°)$$

PRACTICE PROBLEMS 2

Refer to the circuit of Figure 21–14.

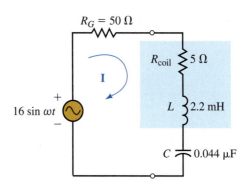

$R_G = 50 \text{ } \Omega$

$R_{coil} \gtrless 5 \text{ } \Omega$

I

16 sin ωt

$L \gtrless 2.2 \text{ mH}$

$C \rightleftharpoons 0.044 \text{ } \mu\text{F}$

MULTISIM

FIGURE 21–14

a. Determine the maximum power dissipated by the circuit.

b. Use the results obtained from Practice Problem 1 to determine the bandwidth of the resonant circuit. Solve for the approximate values of the half-power frequencies, ω_1 and ω_2.

c. Calculate the actual half-power frequencies, ω_1 and ω_2, from the given component values. Compare your results to those obtained in Part (b). Briefly explain why there is a discrepancy between the results.

d. Solve for the circuit current, $\mathbf{I}$, and power dissipated at the lower half-power frequency, ω_1, found in Part (c).

Answers

a. 2.33 W

b. BW = 25.0 krad/s (3.98 kHz), $\omega_1 \cong$ 89.1 krad/s, $\omega_2 \cong$ 114.1 krad/s

c. ω_1 = 89.9 krad/s, $\omega_2 \cong$ 114.9 krad/s. The approximation assumes that the power-frequency curve is symmetrical around ω_S, which is not quite true.

d. $\mathbf{I}$ = 0.145 A$\angle45°$, P = 1.16 W

Refer to the series resonant circuit of Figure 21–15.

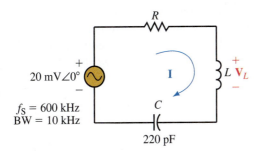

FIGURE 21–15

Suppose the circuit has a resonant frequency of 600 kHz and a bandwidth of 10 kHz:

a. Determine the value of inductor L in henries.

b. Calculate the value of resistor R in ohms.

c. Find $\mathbf{I}$, $\mathbf{V}_L$, and power, P, at resonance.

d. Find the appoximate values of the half-power frequencies, f_1 and f_2.

e. Using the results of Part (d), determine the current in the circuit at the lower half-power frequency, f_1, and show that the power dissipated by the resistor at this frequency is half the power dissipated at the resonant frequency.

Consider the series resonant circuit of Figure 21–16:

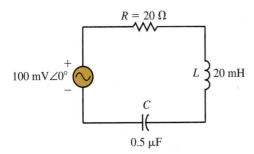

FIGURE 21–16

a. Solve for the resonant frequency of the circuit, ω_S, and calculate the power dissipated by the circuit at resonance.

b. Determine Q, BW, and the half-power frequencies, ω_1 and ω_2, in radians per second.

c. Sketch the selectivity curve of the circuit, showing P (in watts) versus ω (in radians per second).

d. Repeat Parts (a) through (c) if the value of resistance is reduced to 10 Ω.

e. Explain briefly how selectivity depends upon the value of resistance in a series resonant circuit.

21.5 Series-to-Parallel *RL* and *RC* Conversion

As we have already seen, an inductor will always have some series resistance due to the length of wire used in the coil winding. Even though the resistance of the wire is generally small in comparison with the reactances in the circuit, this resistance may occasionally contribute tremendously to the overall circuit response of a parallel resonant circuit. We begin by converting the series *RL* network as shown in Figure 21–17 into an equivalent parallel *RL* network. It must be emphasized, however, that *the equivalence is only valid at a single frequency, ω.*

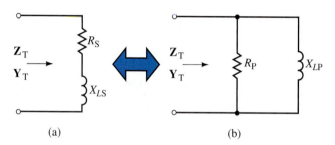

(a)   (b)

FIGURE 21–17

The networks of Figure 21–17 can be equivalent only if they each have the same input impedance, $\mathbf{Z}_T$ (and also the same input admittance, $\mathbf{Y}_T$).

The input impedance of the series network of Figure 21–17(a) is given as

$$\mathbf{Z}_T = R_S + jX_{LS}$$

which gives the input admittance as

$$\mathbf{Y}_T = \frac{1}{\mathbf{Z}_T} = \frac{1}{R_S + jX_{LS}}$$

Multiplying numerator and denominator by the complex conjugate, we have

$$\mathbf{Y}_T = \frac{R_S - jX_{LS}}{(R_S + jX_{LS})(R_S - jX_{LS})}$$
$$= \frac{R_S - jX_{LS}}{R_S{}^2 + X_{LS}{}^2} \tag{21–30}$$
$$= \frac{R_S}{R_S{}^2 + X_{LS}{}^2} - j\frac{X_{LS}}{R_S{}^2 + X_{LS}{}^2}$$

From Figure 21–17(b), we see that the input admittance of the parallel network must be

$$\mathbf{Y}_T = G_P - jB_{LP}$$

which may also be written as

$$\mathbf{Y}_T = \frac{1}{R_P} - j\frac{1}{X_{LP}} \tag{21–31}$$

The admittances of Equations 21–30 and 21–31 can only be equal if the real and the imaginary components are equal. As a result, we see that for a given frequency, the following equations enable us to convert a series *RL* network into its equivalent parallel network:

$$R_P = \frac{R_S^2 + X_{LS}^2}{R_S} \tag{21-32}$$

$$X_{LP} = \frac{R_S^2 + X_{LS}^2}{X_{LS}} \tag{21-33}$$

If we were given a parallel *RL* network, it is possible to show that the conversion to an equivalent series network is accomplished by applying the following equations:

$$R_S = \frac{R_P X_{LP}^2}{R_P^2 + X_{LP}^2} \tag{21-34}$$

$$X_{LS} = \frac{R_P^2 X_{LP}}{R_P^2 + X_{LP}^2} \tag{21-35}$$

The derivation of the above equations is left as an exercise for the student.

Equations 21–32 to 21–35 may be simplified by using the quality factor of the coil. Multiplying Equation 21–32 by R_S/R_S and then using Equation 21–13, we have

$$R_P = R_S \frac{R_S^2 + X_{LS}^2}{R_S^2}$$

$$R_P = R_S(1 + Q^2) \tag{21-36}$$

Similarly, Equation 21–33 is simplified as

$$X_{LP} = X_{LS} \frac{R_S^2 + X_{LS}^2}{X_{LS}^2}$$

$$X_{LP} = X_{LS}\left(1 + \frac{1}{Q^2}\right) \tag{21-37}$$

The quality factor of the resulting parallel network must be the same as for the original series network because the reactive and the average powers must be the same. Using the parallel elements, the quality factor is expressed as

$$Q = \frac{X_{LS}}{R_S}$$

$$= \frac{\left(\dfrac{R_P^2 X_{LP}}{R_P^2 + X_{LP}^2}\right)}{\left(\dfrac{R_P X_{LP}^2}{R_P^2 + X_{LP}^2}\right)}$$

$$= \frac{R_P^2 X_{LP}}{R_P X_{LP}^2}$$

$$Q = \frac{R_P}{X_{LP}} \tag{21-38}$$

EXAMPLE 21–4

For the series network of Figure 21–18, find the Q of the coil at $\omega = 1000$ rad/s and convert the series RL network into its equivalent parallel network. Repeat the above steps for $\omega = 10$ krad/s.

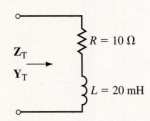

FIGURE 21–18

Solution
For $\omega = 1000$ rad/s,

$$X_L = \omega L = 20 \ \Omega$$

$$Q = \frac{X_L}{R} = 2.0$$

$$R_P = R(1 + Q^2) = 50 \ \Omega$$

$$X_{LP} = X_L\left(1 + \frac{1}{Q^2}\right) = 25 \ \Omega$$

The resulting parallel network for $\omega = 1000$ rad/s is shown in Figure 21–19.
For $\omega = 10$ krad/s,

$$X_L = \omega L = 200 \ \Omega$$

$$Q = \frac{X_L}{R} = 20$$

$$R_P = R(1 + Q^2) = 4010 \ \Omega$$

$$X_{LP} = X_L\left(1 + \frac{1}{Q^2}\right) = 200.5 \ \Omega$$

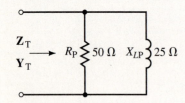

FIGURE 21–19

The resulting parallel network for $\omega = 10$ krad/s is shown in Figure 21–20.

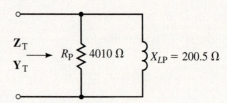

FIGURE 21–20

EXAMPLE 21–5

Find the Q of each of the networks of Figure 21–21 and determine the series equivalent for each.

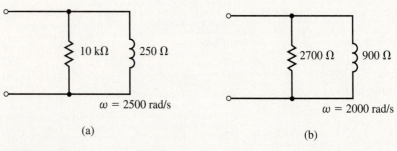

(a)

(b)

FIGURE 21–21

Solution   For the network of Figure 21–21(a),

$$Q = \frac{R_P}{X_{LP}} = \frac{10 \text{ k}\Omega}{250 \ \Omega} = 40$$

$$R_S = \frac{R_P}{1 + Q^2} = \frac{10 \text{ k}\Omega}{1 + 40^2} = 6.25 \ \Omega$$

$$X_{LS} = QR_S = (40)(6.25 \ \Omega) = 250 \ \Omega$$

$$L = \frac{X_L}{\omega} = \frac{250 \ \Omega}{2500 \text{ rad/s}} = 0.1 \text{ H}$$

For the network of Figure 21–21(b),

$$Q = \frac{R_P}{X_{LP}} = \frac{2700 \ \Omega}{900 \ \Omega} = 3$$

$$R_S = \frac{R_P}{1 + Q^2} = \frac{2700 \ \Omega}{1 + 3^2} = 270 \ \Omega$$

$$X_{LS} = QR_S = (3)(270 \ \Omega) = 810 \ \Omega$$

$$L = \frac{X_L}{\omega} = \frac{810 \ \Omega}{2000 \text{ rad/s}} = 0.405 \text{ H}$$

The resulting equivalent series networks are shown in Figure 21–22.

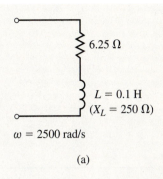

$\omega = 2500$ rad/s

(a)

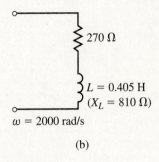

$\omega = 2000$ rad/s

(b)

FIGURE 21–22

Refer to the networks of Figure 21–23.

PRACTICE PROBLEMS 3

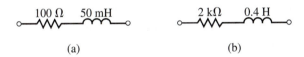

(a)   (b)

FIGURE 21–23

a. Find the quality factors, Q, of the networks at $\omega_1 = 5$ krad/s.

b. Use the Q to find the equivalent parallel networks (resistance and reactance) at an angular frequency of $\omega_1 = 5$ krad/s.

c. Repeat Parts (a) and (b) for an angular frequency of $\omega_2 = 25$ krad/s.

Answers
a. $Q_a = 2.5$ $Q_b = 1.0$

b. Network a: $R_P = 725 \ \Omega$ $X_{LP} = 290 \ \Omega$

    Network b: $R_P = 4 \text{ k}\Omega$ $X_{LP} = 4 \text{ k}\Omega$

c. Network a: $Q_a = 12.5$ $R_P = 15.725 \text{ k}\Omega$ $X_{LP} = 1.258 \text{ k}\Omega$

    Network b: $Q_b = 5$ $R_P = 52 \text{ k}\Omega$ $X_{LP} = 10.4 \text{ k}\Omega$

    The previous examples illustrate two important points which are valid if the Q of the network is large ($Q \geq 10$).

1. The resistance of the parallel network is approximately Q^2 larger than the resistance of the series network.

2. The inductive reactances of the series and parallel networks are approximately equal. Hence

$$R_P \cong Q^2 R_S \quad (Q \geq 10) \tag{21–39}$$

$$X_{LP} \cong X_{LS} \quad (Q \geq 10) \tag{21–40}$$

Although we have performed conversions between series and parallel *RL* circuits, it is easily shown that if the reactive element is a capacitor, the conversions apply equally well. In all cases, the equations are simply changed by replacing the terms X_{LS} and X_{LP} with X_{CS} and X_{CP} respectively. The Q of the network is determined by the ratios

$$Q = \frac{X_{CS}}{R_S} = \frac{R_P}{X_{CP}} \tag{21–41}$$

PRACTICE PROBLEMS 4

Consider the networks of Figure 21–24:

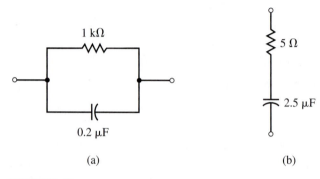

FIGURE 21–24

a. Find the Q of each network at a frequency of $f_1 = 1$ kHz.

b. Determine the series equivalent of the network in Figure 21–24(a) and the parallel equivalent of the network in Figure 21–24(b).

c. Repeat Parts (a) and (b) for a frequency of $f_2 = 200$ kHz.

Answers
a. $Q_a = 1.26$ $Q_b = 12.7$
b. Network a: $R_S = 388\ \Omega$ $X_{CS} = 487\ \Omega$
   Network b: $R_P = 816\ \Omega$ $X_{CP} = 64.1\ \Omega$
c. Network a: $Q_a = 251$ $R_S = 0.0158\ \Omega$ $X_{CS} = 3.98\ \Omega$
   Network b: $Q_b = 0.0637$ $R_P = 5.02\ \Omega$ $X_{CP} = 78.9\ \Omega$

IN-PROCESS LEARNING CHECK 3

(Answers are at the end of the chapter.)

Refer to the networks of Figure 21–25:

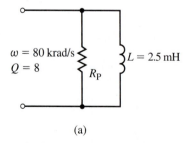
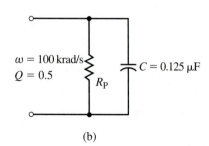

FIGURE 21–25

a. Determine the resistance, R_P, for each network.

b. Find the equivalent series network by using the quality factor for the given networks.

A simple parallel resonant circuit is illustrated in Figure 21–26. The parallel resonant circuit is best analyzed using a constant-current source, unlike the series resonant circuit which used a constant-voltage source.

21.6 Parallel Resonance

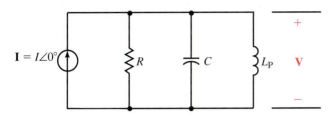

FIGURE 21–26 Simple parallel resonant circuit.

Consider the *LC* "tank" circuit shown in Figure 21–27. The tank circuit consists of a capacitor in parallel with an inductor. Due to its high *Q* and frequency response, the tank circuit is used extensively in communications equipment such as AM, FM, and television transmitters and receivers.

The circuit of Figure 21–27 is not exactly a parallel resonant circuit, since the resistance of the coil is in series with the inductance. In order to determine the frequency at which the circuit is purely resistive, we must first convert the series combination of resistance and inductance into an equivalent parallel network. The resulting circuit is shown in Figure 21–28.

At resonance, the capacitive and inductive reactances in the circuit of Figure 21–28 are equal. As we have observed previously, placing equal inductive and capacitive reactances in parallel effectively results in an open circuit at the given frequency. The input impedance of this network at resonance is therefore purely resistive and given as $Z_T = R_P$. We determine the resonant frequency of a tank circuit by first letting the reactances of the equivalent parallel circuit be equal:

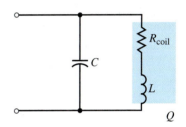

FIGURE 21–27

$$X_C = X_{LP}$$

Now, using the component values of the tank circuit, we have

$$X_C = \frac{(R_{coil})^2 + X_L^2}{X_L}$$

$$\frac{1}{\omega C} = \frac{(R_{coil})^2 + (\omega L)^2}{\omega L}$$

$$\frac{L}{C} = (R_{coil})^2 + (\omega L)^2$$

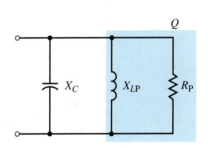

FIGURE 21–28

which may be further reduced to

$$\omega = \sqrt{\frac{1}{LC} - \frac{R_{coil}^2}{L^2}}$$

Factoring $\sqrt{LC}$ from the denominator, we express the parallel resonant frequency as

$$\omega_P = \frac{1}{\sqrt{LC}} \sqrt{1 - \frac{(R_{coil})^2 C}{L}} \qquad (21–42)$$

Notice that if $R_{coil}^2 \ll L/C$, then the term under the radical is approximately equal to 1.

Consequently, if $L/C \geq 100 R_{coil}$, the parallel resonant frequency may be simplified as

$$\omega_P \cong \frac{1}{\sqrt{LC}} \qquad \text{(for } L/C \geq 100 R_{coil}) \qquad (21–43)$$

NOTES . . .

For a high-*Q* circuit, ω_P can be approximated.

Recall that the quality factor, Q, of a circuit is defined as the ratio of reactive power to average power for a circuit at resonance. If we consider the parallel resonant circuit of Figure 21–29, we make several important observations.

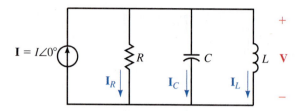

FIGURE 21–29

The inductor and capacitor reactances cancel, resulting in a circuit voltage simply determined by Ohm's law as

$$\mathbf{V} = \mathbf{I}R = IR\angle\, 0°$$

The frequency response of the impedance of the parallel circuit is shown in Figure 21–30.

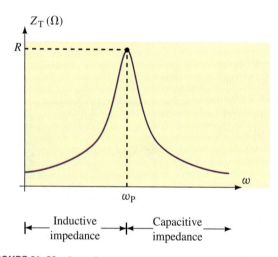

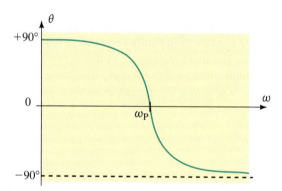

FIGURE 21–30   Impedance (magnitude and phase angle) versus angular frequency for a parallel resonant circuit.

Notice that the impedance of the entire circuit is maximum at resonance and minimum at the boundary conditions ($\omega = 0$ rad/s and $\omega \to \infty$). This result is exactly opposite to that observed in series resonant circuits which have minimum impedance at resonance. We also see that for parallel circuits, the impedance will appear inductive for frequencies less than the resonant frequency, ω_P. Inversely, the impedance is capacitive for frequencies greater than ω_P.

The Q of the parallel circuit is determined from the definition as

$$Q_P = \frac{\text{reactive power}}{\text{average power}}$$

$$= \frac{V^2/X_L}{V^2/R} \qquad\qquad \textbf{(21–44)}$$

$$Q_P = \frac{R}{X_{LP}} = \frac{R}{X_C}$$

This is precisely the same result as that obtained when we converted an *RL* series network into its equivalent parallel network. If the resistance of the coil is the only resistance within a circuit, then the circuit Q will be equal to the Q of the coil. However, if the circuit has other sources of resistance, then the additional resistance will reduce the circuit Q.

For a parallel *RLC* resonant circuit, the currents in the various elements are found from Ohm's law as follows:

$$\mathbf{I}_R = \frac{\mathbf{V}}{\mathbf{R}} = \mathbf{I} \qquad (21\text{-}45)$$

$$\begin{aligned} \mathbf{I}_L &= \frac{\mathbf{V}}{X_L \angle 90°} \\ &= \frac{\mathbf{V}}{R/Q_P} \angle -90° \\ &= Q_P I \angle -90° \end{aligned} \qquad (21\text{-}46)$$

$$\begin{aligned} \mathbf{I}_C &= \frac{\mathbf{V}}{X_C \angle -90°} \\ &= \frac{\mathbf{V}}{R/Q_P} \angle 90° \\ &= Q_P I \angle 90° \end{aligned} \qquad (21\text{-}47)$$

At resonance, the currents through the inductor and the capacitor have the same magnitudes but are 180° out of phase. Notice that the magnitude of current in the reactive elements at resonance is Q times greater than the applied source current. Because the Q of a parallel circuit may be very large, we see the importance of choosing elements that are able to handle the expected currents.

In a manner similar to that used in determining the bandwidth of a series resonant circuit, it may be shown that the half-power frequencies of a parallel resonant circuit are

$$\omega_1 = -\frac{1}{2RC} + \sqrt{\frac{1}{4R^2C^2} + \frac{1}{LC}} \quad (\text{rad/s}) \qquad (21\text{-}48)$$

$$\omega_2 = \frac{1}{2RC} + \sqrt{\frac{1}{4R^2C^2} + \frac{1}{LC}} \quad (\text{rad/s}) \qquad (21\text{-}49)$$

The bandwidth is therefore

$$\text{BW} = \omega_2 - \omega_1 = \frac{1}{RC} \quad (\text{rad/s}) \qquad (21\text{-}50)$$

If the quality factor of the circuit, $Q \geq 10$, then the selectivity curve is very nearly symmetrical around ω_P, resulting in half-power frequencies which are located at $\omega_P \pm \text{BW}/2$.

Multiplying Equation 21–50 by ω_P/ω_P results in the following:

$$\text{BW} = \frac{\omega_P}{R(\omega_P C)} = \frac{X_C}{R}\omega_P$$
$$\text{BW} = \frac{\omega_P}{Q_P} \quad (\text{rad/s}) \qquad (21\text{-}51)$$

Notice that Equation 21–51 is the same for both series and parallel resonant circuits.

EXAMPLE 21–6

Consider the circuit shown in Figure 21–31.

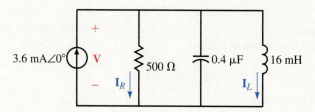

◀ MULTISIM

FIGURE 21–31

a. Determine the resonant frequencies, ω_P(rad/s) and f_P(Hz) of the tank circuit.
b. Find the Q of the circuit at resonance.
c. Calculate the voltage across the circuit at resonance.
d. Solve for currents through the inductor and the resistor at resonance.
e. Determine the bandwidth of the circuit in both radians per second and hertz.
f. Sketch the voltage response of the circuit, showing the voltage at the half-power frequencies.
g. Sketch the selectivity curve of the circuit showing P(watts) versus ω(rad/s).

Solution

a. $$\omega_P = \frac{1}{\sqrt{LC}} = \frac{1}{\sqrt{(16 \text{ mH}) (0.4 \text{ } \mu\text{F})}} = 12.5 \text{ krad/s}$$

$$f_P = \frac{\omega}{2\pi} = \frac{12.5 \text{ krad/s}}{2\pi} = 1989 \text{ Hz}$$

b. $$Q_P = \frac{R_P}{\omega L} = \frac{500 \text{ } \Omega}{(12.5 \text{ krad/s}) (16 \text{ mH})} = \frac{500 \text{ } \Omega}{200 \text{ } \Omega} = 2.5$$

c. At resonance, $\mathbf{V}_C = \mathbf{V}_L = \mathbf{V}_R$, and so

$$\mathbf{V} = \mathbf{I}R = (3.6 \text{ mA} \angle 0°) (500 \text{ } \Omega \angle 0°) = 1.8 \text{ V } \angle 0°$$

d. $$\mathbf{I}_L = \frac{\mathbf{V}_L}{\mathbf{Z}_L} = \frac{1.8 \text{ V} \angle 0°}{200 \text{ } \Omega \angle 90°} = 9.0 \text{ mA} \angle -90°$$

$$\mathbf{I}_R = \mathbf{I} = 3.6 \text{ mA} \angle 0°$$

e. $$\text{BW(rad/s)} = \frac{\omega_P}{Q_P} = \frac{12.5 \text{ krad/s}}{2.5} = 5 \text{ krad/s}$$

$$\text{BW(Hz)} = \frac{\text{BW(rad/s)}}{2\pi} = \frac{5 \text{ krad/s}}{2\pi} = 795.8 \text{ Hz}$$

f. The half-power frequencies are calculated from Equations 21–48 and 21–49 since the Q of the circuit is less than 10.

$$\omega_1 = -\frac{1}{2RC} + \sqrt{\frac{1}{4R^2C^2} + \frac{1}{LC}}$$

$$= -\frac{1}{0.0004} + \sqrt{\frac{1}{1.6 \times 10^{-7}} + \frac{1}{6.4 \times 10^{-9}}}$$

$$= -2500 + 12\,748$$

$$= 10\,248 \text{ rad/s}$$

$$\omega_2 = \frac{1}{2RC} + \sqrt{\frac{1}{4R^2C^2} + \frac{1}{LC}}$$

$$= \frac{1}{0.0004} + \sqrt{\frac{1}{1.6 \times 10^{-7}} + \frac{1}{6.4 \times 10^{-9}}}$$

$$= 2500 + 12\,748$$

$$= 15\,248 \text{ rad/s}$$

The resulting voltage response curve is illustrated in Figure 21–32.

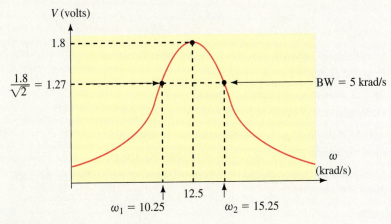

FIGURE 21–32

g. The power dissipated by the circuit at resonance is

$$P = \frac{V^2}{R} = \frac{(1.8 \text{ V})^2}{500 \text{ } \Omega} = 6.48 \text{ mW}$$

The selectivity curve is now easily sketched as shown in Figure 21–33.

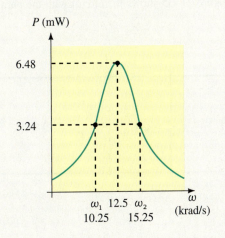

FIGURE 21–33

EXAMPLE 21–7

Consider the circuit of Figure 21–34.

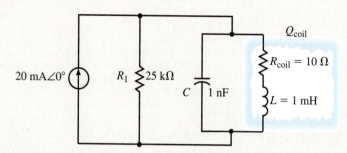

MULTISIM

FIGURE 21–34

a. Calculate the resonant frequency, ω_P, of the tank circuit.

b. Find the Q of the coil at resonance.

c. Sketch the equivalent parallel circuit.

d. Determine the Q of the entire circuit at resonance.

e. Solve for the voltage across the capacitor at resonance.

f. Find the bandwidth of the circuit in radians per second.

g. Sketch the voltage response of the circuit showing the voltage at the half-power frequencies.

Solution

a. Since the ratio $L/C = 1000 \geq 100R_{\text{coil}}$, we use the approximation:

$$\omega_p = \frac{1}{\sqrt{LC}} = \frac{1}{\sqrt{(1\text{ mH})(1\text{ nF})}} = 1\text{ Mrad/s}$$

b.

$$Q_{\text{coil}} = \frac{\omega L}{R_{\text{coil}}} = \frac{(1\text{ Mrad/s})(1\text{ mH})}{10\ \Omega} = 100$$

c.

$$R_P \cong Q_{\text{coil}}^2 R_{\text{coil}} = (100)^2(10\ \Omega) = 100\text{ k}\Omega$$

$$X_{LP} \cong X_L = \omega L = (1\text{ Mrad/s})(1\text{ mH}) = 1\text{ k}\Omega$$

The circuit of Figure 21–35 shows the circuit with the parallel equivalent of the inductor.

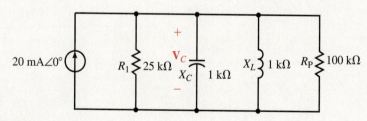

FIGURE 21–35

We see that the previous circuit may be further simplified by combining the parallel resistances:

$$R_{eq} = R_1 \| R_P = \frac{(25 \text{ k}\Omega)(100 \text{ k}\Omega)}{25 \text{ k}\Omega + 100 \text{ k}\Omega} = 20 \text{ k}\Omega$$

The simplified equivalent circuit is shown in Figure 21–36.

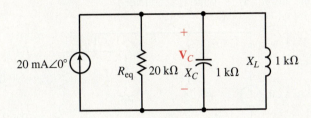

FIGURE 21–36

d.
$$Q_P = \frac{R_{eq}}{X_L} = \frac{20 \text{ k}\Omega}{1 \text{ k}\Omega} = 20$$

e. At resonance,

$$\mathbf{V}_C = \mathbf{I}R_{eq} = (20 \text{ mA}\angle 0°)(20 \text{ k}\Omega) = 400 \text{ V}\angle 0°$$

f.
$$\text{BW} = \frac{\omega_P}{Q} = \frac{1 \text{ Mrad/s}}{20} = 50 \text{ krad/s}$$

g. The voltage response curve is shown in Figure 21–37. Since the circuit $Q \geq 10$, the half-power frequencies will occur at the following angular frequencies:

$$\omega_1 \cong \omega_P - \frac{\text{BW}}{2} = 1.0 \text{ Mrad/s} - \frac{50 \text{ krad/s}}{2} = 0.975 \text{ Mrad/s}$$

and

$$\omega_2 \cong \omega_P + \frac{\text{BW}}{2} = 1.0 \text{ Mrad/s} + \frac{50 \text{ krad/s}}{2} = 1.025 \text{ Mrad/s}$$

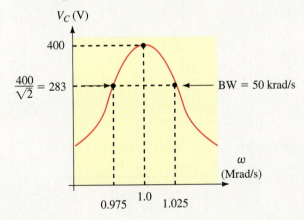

FIGURE 21–37

EXAMPLE 21–8

Determine the values of R_1 and C for the resonant tank circuit of Figure 21–38 so that the given conditions are met.

$L = 10$ mH, $R_{coil} = 30$ Ω

$f_P = 58$ kHz

BW $= 1$ kHz

Solve for the current, $\mathbf{I}_L$, through the inductor.

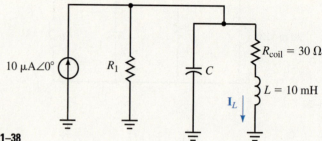

$10\ \mu A\angle0°$ R_1 C $R_{coil} = 30$ Ω $L = 10$ mH $\mathbf{I}_L$

FIGURE 21–38

Solution

$$Q_P = \frac{f_P}{BW(Hz)} = \frac{58 \text{ kHz}}{1 \text{ kHz}} = 58$$

Now, because the frequency expressed in radians per second is more useful than hertz, we convert f_P to ω_P:

$$\omega_P = 2\pi f_P = (2\pi)(58 \text{ kHz}) = 364.4 \text{ krad/s}$$

The capacitance is determined from Equation 21–43 as

$$C = \frac{1}{\omega_P^2 L} = \frac{1}{(364.4 \text{ krad/s})^2(10 \text{ mH})} = 753 \text{ pF}$$

Solving for the Q of the coil permits us to easily convert the series RL network into its equivalent parallel network.

$$Q_{coil} = \frac{\omega_P L}{R_{coil}}$$

$$= \frac{(364.4 \text{ krad/s})(10 \text{ mH})}{30 \text{ Ω}}$$

$$= \frac{3.644 \text{ kΩ}}{30 \text{ Ω}} = 121.5$$

$$R_P \cong Q_{coil}^2 R_{coil} = (121.5)^2(30 \text{ Ω}) = 443 \text{ kΩ}$$

$$X_{LP} \cong X_L = 3644 \text{ Ω}$$

The resulting equivalent parallel circuit is shown in Figure 21–39.

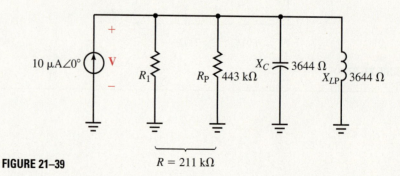

$10\ \mu A\angle0°$ $\mathbf{V}$ R_1 R_P 443 kΩ X_C 3644 Ω X_{LP} 3644 Ω

FIGURE 21–39 $R = 211$ kΩ

The quality factor, Q_P is used to determine the total resistance of the circuit as

$$R = Q_P X_C = (58)(3.644 \text{ k}\Omega) = 211 \text{ k}\Omega$$

But

$$\frac{1}{R} = \frac{1}{R_1} + \frac{1}{R_P}$$

$$\frac{1}{R_1} = \frac{1}{R} - \frac{1}{R_P} = \frac{1}{211 \text{ k}\Omega} - \frac{1}{443 \text{ k}\Omega} = 2.47 \ \mu S$$

And so

$$R_1 = 405 \text{ k}\Omega$$

The voltage across the circuit is determined to be

$$\mathbf{V} = \mathbf{I}R = (10 \ \mu A \angle 0°)(211 \text{ k}\Omega) = 2.11 \text{ V} \angle 0°$$

and the current through the inductor is

$$\mathbf{I}_L = \frac{\mathbf{V}}{R_{\text{coil}} + jX_L}$$

$$= \frac{2.11 \text{ V} \angle 0°}{30 + j3644 \ \Omega} = \frac{2.11 \text{ V} \angle 0°}{3644 \ \Omega \angle 89.95°} = 579 \ \mu A \angle -89.95°$$

Refer to the circuit of Figure 21–40:

PRACTICE PROBLEMS 5

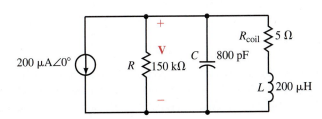

FIGURE 21–40

◀ MULTISIM

a. Determine the resonant frequency and express it in radians per second and in hertz.

b. Calculate the quality factor of the circuit.

c. Solve for the bandwidth.

d. Determine the voltage **V** at resonance.

Answers

a. 2.5 Mrad/s (398 kHz)

b. 75

c. 33.3 krad/s (5.31 kHz)

d. 7.5 V∠180°

IN-PROCESS
LEARNING CHECK 4

(Answers are at the end of the chapter.)

Refer to the parallel resonant circuit of Figure 21–41:

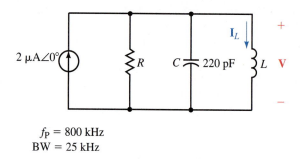

f_P = 800 kHz
BW = 25 kHz

FIGURE 21–41

Suppose the circuit has a resonant frequency of 800 kHz and a bandwidth of 25 kHz.

a. Determine the value of the inductor, L, in henries.

b. Calculate the value of the resistance, R, in ohms.

c. Find $\mathbf{V}$, $\mathbf{I}_L$, and power, P, at resonance.

d. Find the appoximate values of the half-power frequencies, f_1 and f_2.

e. Determine the voltage across the circuit at the lower half-power frequency, f_1, and show that the power dissipated by the resistor at this frequency is half the power dissipated at the resonant frequency.

21.7 Circuit Analysis Using Computers

PSpice is particularly useful in examining the operation of resonant circuits. The ability of the software to provide a visual display of the frequency response is used to evaluate the resonant frequency, maximum current, and bandwidth of a circuit. The Q of the given circuit is then easily determined.

PSpice

EXAMPLE 21–9

Use PSpice to obtain the frequency response for current in the circuit of Figure 21–12. Use cursors to find the resonant frequency and the bandwidth of the circuit from the observed response. Compare the results to those obtained in Example 21–2.

Solution
OrCAD Capture CIS is used to input the circuit as shown in Figure 21–42. For this example, the project is titled **EXAMPLE 21–9.** The voltage source used in this example is VAC and the value is changed to AC=**10V 0Deg.** In order to obtain a plot of the circuit current, use the Current Into Pin tool as shown.

Next, we change the simulation settings by clicking on the New Simulation Profile tool. Give the simulation a name such as **Series Resonance** and click on Create. Once you are in the simulation settings box, click on the Analysis tab and select AC Sweep/Noise as the analysis type. The frequency can be swept either linearly or logarithmically (decade or octave). In this example we select a logarithmic sweep through a decade. In the box titled AC Sweep Type, click on ⊙ Logarithmic and select Decade. Type the following values as the settings. *S*tart Frequency: **1kHz,** *E*nd Frequency: **10kHz,** and Points/*D*ecade: **10001.** Click OK.

Click on the Run tool. If there are no errors, the PROBE postprocessor will run automatically and display I(L1) as a function of frequency. You will notice that the selectivity curve is largely contained within a narrow range of frequencies. We may zoom into this region as follows. Select the *P*lot menu, and click

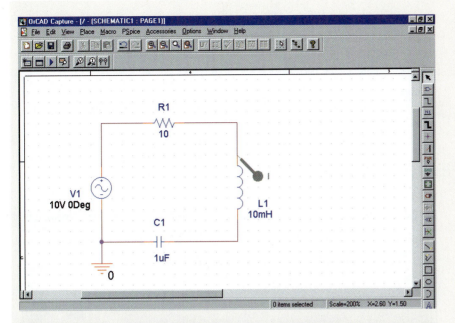

FIGURE 21–42

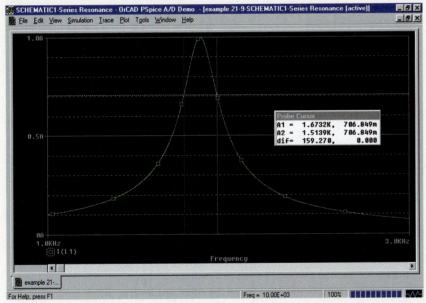

FIGURE 21–43

on Axis Settings menu item. Click on the X Axis tab and select Under Defined Data Range. Change the values to **1kHz** to **3kHz.** Click OK. The resulting display is shown in Figure 21–43.

Finally, we may use cursors to provide us with the actual resonant frequency, the maximum current and the half-power frequencies. Cursors are obtained as follows. Click on Trace, Cursor, and Display. The positions of the cursors are adjusted by using either the mouse or the arrow and <Shift> keys. The current at the maximum point of the curve is obtained by clicking on Trace, Cursor, and Max. The dialog box provides the values of both the frequency and the value of current. The bandwidth is determined by determining the frequencies at the half-power points (when the current is 0.707 of the maximum value). We obtain the following results using the cursors:

$I_{max} = 1.00$ A, $f_S = 1.591$ kHz, $f_1 = 1.514$ kHz, $f_2 = 1.673$ kHz, BW $= 0.159$ kHz.

These values correspond very closely to those calculated in Example 21–2.

EXAMPLE 21–10

Use PSpice to obtain the frequency response for the voltage across the parallel resonant circuit of Figure 21–34. Use the PROBE postprocessor to find the resonant frequency, the maximum voltage (at resonance), and the bandwidth of the circuit. Compare the results to those obtained in Example 21–7.

Solution

This example is similar to the previous example, with minor exceptions. The OrCAD Capture program is used to enter the circuit as shown in Figure 21–44. The ac current source is found in the SOURCE library as IAC. The value of the ac current source is changed AC = **20mA 0Deg.** The Voltage Level tool is used to provide the voltage simulation for the circuit.

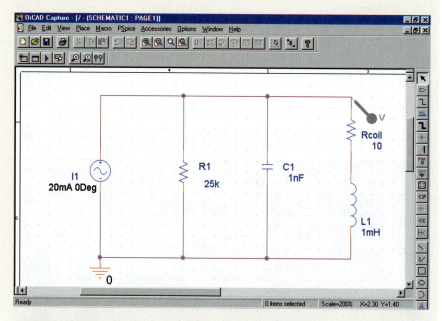

FIGURE 21–44

Use the New Simulation Profile tool to set the simulation for a logarithmic ac sweep from 100kHz to 300kHz with a total of 10001 points per decade. Click on the Plot menu to select the Axis Settings. Change the x-axis to indicate a User Defined range from **100kHz** to **300kHz** and change the y-axis to indicate a User Defined range from **0V** to **400V.** The resulting display is shown in Figure 21–45.

As in the previous example, we use cursors to observe that the maximum circuit voltage, $V_{max} = 400$ V, occurs at the resonance frequency as $f_P = 159.2$ kHz (1.00 Mrad/s). The half-power frequencies are determined when the output voltage is at 0.707 of the maximum value, namely at $f_1 = 155.26$ kHz (0.796 Mrad/s) and $f_2 = 163.22$ kHz (1.026 Mrad/s). These frequencies give a bandwidth of BW = 7.96 kHz (50.0 krad/s). The above results are the same as those found in Example 21–7.

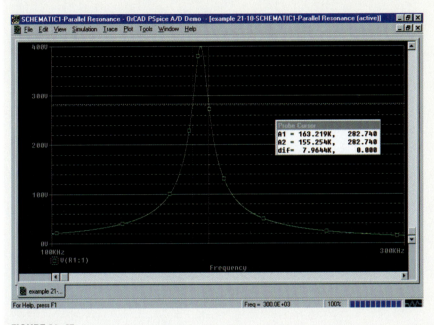

FIGURE 21–45

PRACTICE PROBLEMS 6

Use PSpice to obtain the frequency response of voltage, V versus f, for the circuit of Figure 21–40. Use cursors to determine the approximate values of the half-power frequencies and the bandwidth of the circuit. Compare the results to those obtained in Practice Problem 5.

Answers
$V_{max} = 7.50$ V, $f_P = 398$ kHz, $f_1 = 395.3$ kHz, $f_2 = 400.7$ kHz, BW = 5.31 kHz

PUTTING IT INTO PRACTICE

You are the transmitter specialist at an AM commercial radio station, which transmits at a frequency of 990 kHz and an average power of 10 kW. As is the case for all commercial AM stations, the bandwidth for your station is 10 kHz. Your transmitter will radiate the power using a 50-Ω antenna. The accompanying figure shows a simplified block diagram of the output stage of the transmitter. The antenna behaves exactly like a 50-Ω resistor connected between the output of the amplifier and ground.

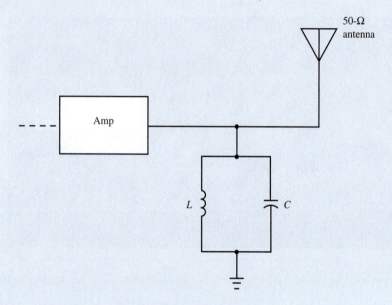

Transmitter stage of a commercial AM radio station.

You have been asked to determine the values of L and C so that the transmitter operates with the given specifications. As part of the calculations, determine the peak current that the inductor must handle and solve for the peak voltage across the capacitor. For your calculations, assume that the transmitted signal is a sinusoidal.

PROBLEMS

21.1 Series Resonance

1. Consider the circuit of Figure 21–46:
   a. Determine the resonant frequency of the circuit in both radians per second and hertz.
   b. Calculate the current, **I**, at resonance.
   c. Solve for the voltages $\mathbf{V}_R$, $\mathbf{V}_L$, and $\mathbf{V}_C$. (Notice that the voltage $\mathbf{V}_L$ includes the voltage dropped across the internal resistance of the coil.)
   d. Determine the power (in watts) dissipated by the inductor. (Hint: The power will not be zero.)

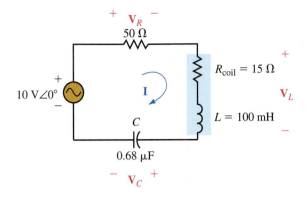

FIGURE 21–46   ◀ **MULTISIM**

2. Refer to the circuit of Figure 21–47:

   a. Determine the resonant frequency of the circuit in both radians per second and hertz.

   b. Calculate the phasor current, **I** at resonance.

   c. Determine the power dissipated by the circuit at resonance.

   d. Calculate the phasor voltages, $\mathbf{V}_L$ and $\mathbf{V}_R$.

   e. Write the sinusoidal form of the voltages v_L and v_R.

3. Consider the circuit of Figure 21–48:

   a. Determine the values of R and C such that the circuit has a resonant frequency of 25 kHz and an rms current of 25 mA at resonance.

   b. Calculate the power dissipated by the circuit at resonance.

   c. Determine the phasor voltages, $\mathbf{V}_C$, $\mathbf{V}_L$, and $\mathbf{V}_R$ at resonance.

   d. Write the sinusoidal expressions for the voltages v_C, v_L, and v_R.

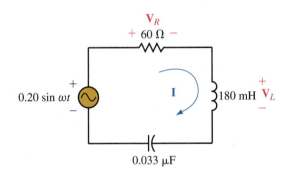

FIGURE 21–47   **FIGURE 21–48**

4. Refer to the circuit of Figure 21–49.

   a. Determine the capacitance required so that the circuit has a resonant frequency of 100 kHz.

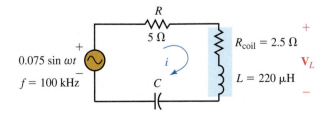

FIGURE 21–49

b. Solve for the phasor quantities $\mathbf{I}$, $\mathbf{V}_L$, and $\mathbf{V}_R$.

c. Find the sinusoidal expressions for i, v_L, and v_R.

d. Determine the power dissipated by each element in the circuit.

21.2 Quality Factor, Q

5. Refer to the circuit of Figure 21–50:

   a. Determine the resonant frequency expressed as ω(rad/s) and f(Hz).

   b. Calculate the total impedance, $\mathbf{Z}_T$, at resonance.

   c. Solve for current $\mathbf{I}$ at resonance.

   d. Solve for $\mathbf{V}_R$, $\mathbf{V}_L$, and $\mathbf{V}_C$ at resonance.

   e. Calculate the power dissipated by the circuit and evaluate the reactive powers, Q_C and Q_L.

   f. Find the quality factor, Q_S, of the circuit.

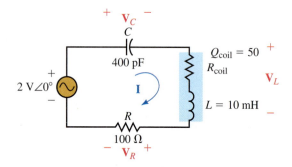

FIGURE 21–50

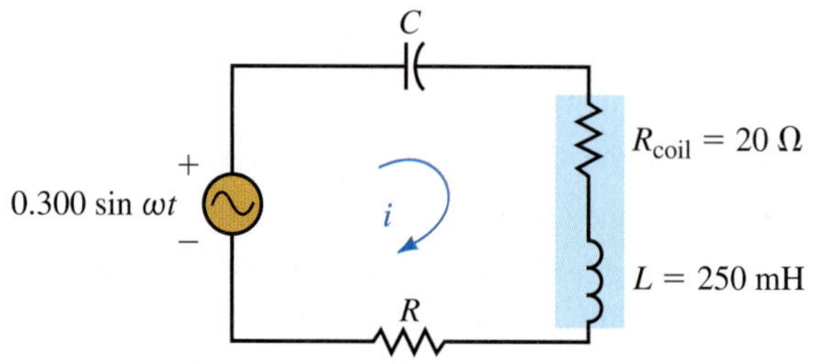

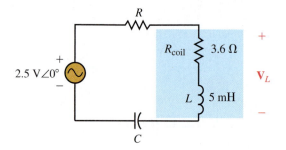

FIGURE 21–51

6. Suppose that the circuit of Figure 21–51 has a resonant frequency of $f_S = $ 2.5 kHz and a quality factor of $Q_S = 10$:

   a. Determine the values of R and C.

   b. Solve for the quality factor of inductor, Q_{coil}.

   c. Find $\mathbf{Z}_T$, $\mathbf{I}$, $\mathbf{V}_C$, and $\mathbf{V}_R$ at resonance.

   d. Solve for the sinusoidal expression of current i at resonance.

   e. Calcuate the sinusoidal expressions v_C and v_R at resonance.

   f. Calculate the power dissipated by the circuit and determine the reactive powers, Q_C and Q_L.

7. Refer to the circuit of Figure 21–52:

   a. Design the circuit to have a resonant frequency of $\omega = 50$ krad/s and a quality factor $Q_S = 25$.

   b. Calculate the power dissipated by the circuit at the resonant frequency.

   c. Determine the voltage, $\mathbf{V}_L$, across the inductor at resonance.

FIGURE 21–52

8. Consider the circuit of Figure 21–53:

a. Design the circuit to have a resonant frequency of $\omega = 400$ krad/s and a quality factor $Q_S = 10$.

b. Calculate the power dissipated by the circuit at the resonant frequency.

c. Determine the voltage, $\mathbf{V}_L$, across the inductor at resonance.

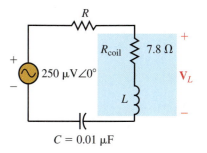

21.3 Impedance of a Series Resonant Circuit

9. Refer to the series resonant circuit of Figure 21–54.

a. Determine the resonant frequency, ω_S.

FIGURE 21–53

b. Solve for the input impedance, $\mathbf{Z}_T = Z\angle\theta$, of the circuit at frequencies of $0.1\omega_S, 0.2\omega_S, 0.5\omega_S, \omega_S, 2\omega_S, 5\omega_S,$ and $10\omega_S$.

c. Using the results from (b), sketch a graph of Z (magnitude in ohms) versus ω (in radians per second) and a graph of θ (in degrees) versus ω (in radians per second). If possible, use log-log graph paper for the former and semilog graph paper for the latter.

d. Using your results from (b), determine the magnitude of current at each of the given frequencies.

e. Use the results from (d) to plot a graph of I (magnitude in amps) versus ω (in radians per second) on log-log graph paper.

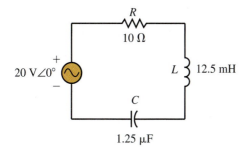

FIGURE 21–54

10. Repeat Problem 9 if the 10-Ω resistor is replaced with a 50-Ω resistor.

21.4 Power, Bandwidth, and Selectivity of a Series Resonant Circuit

11. Refer to the circuit of Figure 21–55.

a. Find ω_S, Q, and BW (in radians per second).

b. Calculate the maximum power dissipated by the circuit.

c. From the results obtained in (a) solve for the approximate half-power frequencies, ω_1 and ω_2.

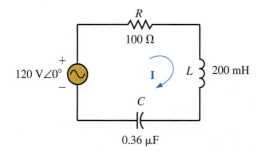

FIGURE 21–55

◀ MULTISIM

d. Calculate the actual half-power frequencies, ω_1 and ω_2 using the component values and the appropriate equations.

e. Are the results obtained in (c) and (d) comparable? Explain.

f. Solve for the circuit current, I, and power dissipated at the lower half-power frequency, ω_1, determined in (d).

12. Repeat Problem 11 for the circuit of Figure 21–56.

13. Consider the circuit of Figure 21–57.

a. Calculate the values of R and C for the circuit to have a resonant frequency of 200 kHz and a bandwidth of 16 kHz.

b. Use the designed component values to determine the power dissipated by the circuit at resonance.

c. Solve for v_{out} at resonance.

14. Repeat Problem 13 if the resonant frequency is to be 580 kHz and the bandwidth is 10 kHz.

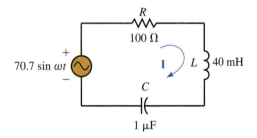

FIGURE 21–56

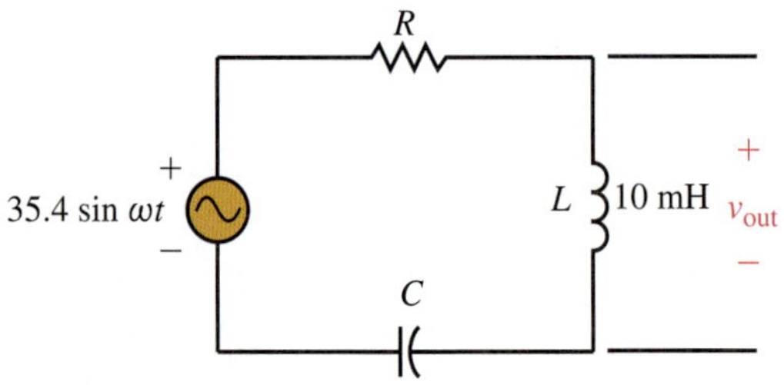

FIGURE 21–57

21.5 Series-to-Parallel *RL* and *RC* Conversion

15. Refer to the series networks of Figure 21–58.

a. Find the Q of each network at $\omega = 1000$ rad/s.

b. Convert each series *RL* network into an equivalent parallel network, having R_P and X_{LP} in ohms.

c. Repeat (a) and (b) for $\omega = 10$ krad/s.

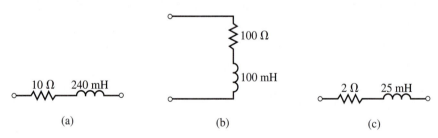

FIGURE 21–58

16. Consider the series networks of Figure 21–59.

a. Find the Q of each coil at $\omega = 20$ krad/s.

b. Convert each series *RL* network into an equivalent parallel network consisting of R_P and X_{LP} in ohms.

c. Repeat (a) and (b) for $\omega = 100$ krad/s.

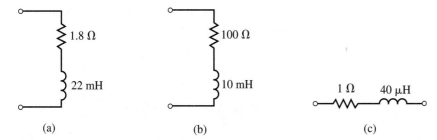

(a)   (b)   (c)

FIGURE 21–59

17. For the series networks of Figure 21–60, find the Q and convert each network into its parallel equivalent.

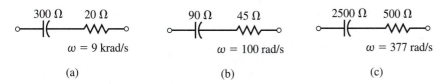

300 Ω   20 Ω   90 Ω   45 Ω   2500 Ω   500 Ω

$\omega = 9$ krad/s $\omega = 100$ rad/s $\omega = 377$ rad/s

(a)   (b)   (c)

FIGURE 21–60

18. Derive Equations 21–34 and 21–35 that enable us to convert a parallel RL network into its series equivalent. (Hint: Begin by determining the expression for the input impedance of the parallel network.)

19. Find the Q of each of the networks of Figure 21–61 and determine the series equivalent of each. Express all component values in ohms.

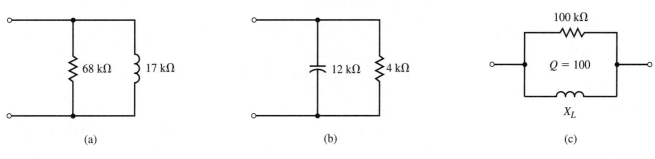

68 kΩ   17 kΩ   12 kΩ   4 kΩ   100 kΩ
                                                          $Q = 100$
                                                          X_L

(a)   (b)   (c)

FIGURE 21–61

20. Repeat Problem 19 for the networks of Figure 21–62.

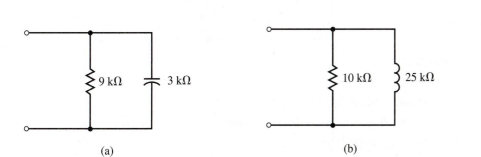
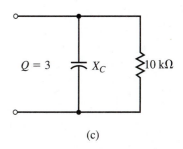

9 kΩ   3 kΩ   10 kΩ   25 kΩ $Q = 3$ X_C 10 kΩ

(a)   (b)   (c)

FIGURE 21–62

21. Determine the values of L_S and L_P in henries, given that the networks of Figure 21–63 are equivalent at a frequency of 250 krad/s.

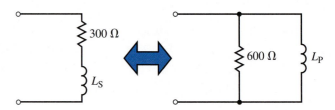

FIGURE 21–63

22. Determine the values of C_S and C_P in farads, given that the networks of Figure 21–64 are equivalent at a frequency of 48 krad/s.

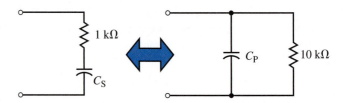

FIGURE 21–64

21.6 Parallel Resonance

23. Consider the circuit of Figure 21–65.

   a. Determine the resonant frequency, ω_P, in radians per second.

   b. Solve for the input impedance, $\mathbf{Z}_T = Z\angle\theta$, of the circuit at frequencies of $0.1\omega_P$, $0.2\omega_P$, $0.5\omega_P$, ω_P, $2\omega_P$, $5\omega_P$, and $10\omega_P$.

   c. Using the results obtained in (b), sketch graphs of Z (magnitude in ohms) versus ω (in radians per second) and θ (in degrees) versus ω. If possible, use log-log graph paper for the former and semilog for the latter.

   d. Using the results from (b), determine the voltage $\mathbf{V}$ at each of the indicated frequencies.

   e. Sketch a graph of magnitude V versus ω on log-log graph paper.

24. Repeat Problem 23 if the 20-kΩ resistor is replaced with a 40-kΩ resistor.

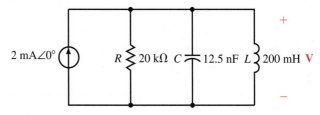

FIGURE 21–65

25. Refer to the circuit shown in Figure 21–66.

   a. Determine the resonant frequencies, ω_P(rad/s) and f_P(Hz).

   b. Find the Q of the circuit.

   c. Calculate $\mathbf{V}$, $\mathbf{I}_R$, $\mathbf{I}_L$, and $\mathbf{I}_C$ at resonance.

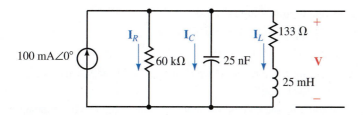

FIGURE 21–66

d. Determine the power dissipated by the circuit at resonance.

e. Solve for the bandwidth of the circuit in both radians per second and hertz.

f. Sketch the voltage response of the circuit, showing the voltage at the half-power frequencies.

26. Repeat Problem 25 for the circuit of Figure 21–67.

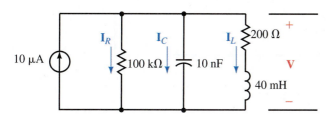

FIGURE 21–67

27. Determine the values of R_1 and C for the resonant tank circuit of Figure 21–68 so that the given conditions are met. Solve for current $\mathbf{I}_L$ through the inductor.

$$L = 25 \text{ mH}, R_{\text{coil}} = 100 \text{ V}$$
$$f_P = 50 \text{ kHz}$$
$$\text{BW} = 10 \text{ kHz}$$

28. Determine the values of R_1 and C for the resonant circuit of Figure 21–68 so that the given conditions are met. Solve for the voltage, **V,** across the circuit.

$$L = 50 \text{ mH}, R_{\text{coil}} = 50 \ \Omega$$
$$\omega_P = 100 \text{ krad/s}$$
$$\text{BW} = 10 \text{ krad/s}$$

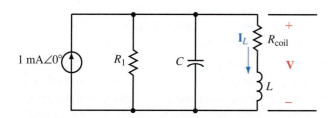

FIGURE 21–68

29. Refer to the circuit of Figure 21–69.

   a. Determine the value of X_L for resonance.

   b. Solve for the Q of the circuit.

   c. If the circuit has a resonant frequency of 2000 rad/s, what is the bandwidth of the circuit?

   d. What must be the values of C and L for the circuit to be resonant at 2000 rad/s?

   e. Calculate the voltage $\mathbf{V}_C$ at resonance.

30. Repeat Problem 29 for the circuit of Figure 21–70.

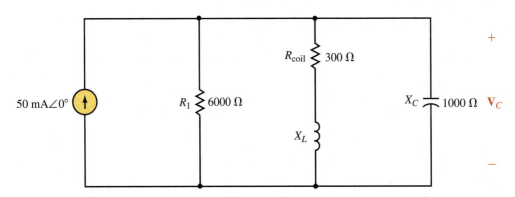

FIGURE 21–69

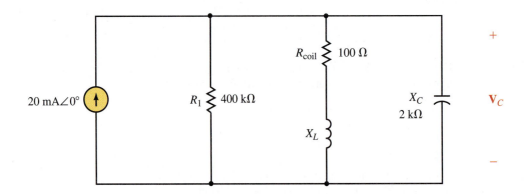

FIGURE 21–70

21.7 Circuit Analysis Using Computers

◀ CADENCE

31. Use PSpice to input the circuit of Figure 21–55. Use the Probe postprocessor to display the response of the inductor voltage as a function of frequency. From the display, determine the maximum rms voltage, the resonant frequency, the half-power frequencies, and the bandwidth. Use the results to determine the quality factor of the circuit.

◀ CADENCE

32. Repeat Problem 31 for the circuit of Figure 21–56.

◀ CADENCE

33. Use PSpice to input the circuit of Figure 21–66. Use the Probe postprocessor to display the response of the capacitor voltage as a function of frequency. From the display, determine the maximum rms voltage, the resonant frequency, the half-power frequencies, and the bandwidth. Use the results to determine the quality factor of the circuit.

34. Repeat Problem 33 for the circuit of Figure 21–67.

       ◀ CADENCE

35. Use the values of C and L determined in Problem 29 to input the circuit of Figure 21–69. Use the Probe postprocessor to display the response of the capacitor voltage as a function of frequency. From the display, determine the maximum rms voltage, the resonant frequency, the half-power frequencies, and the bandwidth. Use the results to determine the quality factor of the circuit.

       ◀ CADENCE

36. Use the values of C and L determined in Problem 30 to input the circuit of Figure 21–70. Repeat the measurements of Problem 35.

       ◀ CADENCE

✓ ANSWERS TO IN-PROCESS LEARNING CHECKS

In-Process Learning Check 1

   a. $L = 320 \; \mu H$

   b. $R = 20.1 \; \Omega$

   c. $\mathbf{I} = 0.995 \; mA\angle 0° \quad \mathbf{V}_L = 1.20 \; V\angle 90° \quad P = 20.0 \; \mu W$

   d. $f_1 = 595 \; kHz \quad f_2 = 605 \; kHz$

   e. $\mathbf{I} = 0.700 \; mA\angle 45.28° \quad P_1 = 9.85 \; \mu W$
     $P_1/P = 0.492 \cong 0.5$

In-Process Learning Check 2

   a. $\omega_S = 10 \; krad/s \quad P = 500 \; \mu W$

   b. $Q = 10 \quad BW = 1 \; krad/s \quad \omega_1 = 9.5 \; krad/s$
     $\omega_2 = 10.5 \; krad/s$

   d. $\omega_S = 10 \; krad/s \quad P = 1000 \; \mu W \quad Q = 20$
     $BW = 0.5 \; krad/s \quad \omega_1 = 9.75 \; krad/s$
     $\omega_2 = 10.25 \; krad/s$

   e. As resistance decreases, selectivity increases.

In-Process Learning Check 3

   a. $X_L = 200 \; \Omega \quad R_P = 1600 \; \Omega \quad X_C = 80 \; \Omega \quad R_P = 40 \; \Omega$

   b. $X_{LS} = 197 \; \Omega \quad R_S = 24.6 \; \Omega \quad X_{CS} = 16 \; \Omega \quad R_S = 32 \; \Omega$

In-Process Learning Check 4

   a. $L = 180 \; \mu H$

   b. $R_p = 28.9 \; k\Omega$

   c. $\mathbf{V} = 57.9 \; mV\angle 0° \quad \mathbf{I}_L = 64.0 \; \mu A\angle -90° \quad P = 115 \; nW$

   d. $f_1 = 788 \; kHz \quad f_2 = 813 \; kHz$

   e. $\mathbf{V} = 41.1 \; mV\angle 44.72° \quad P = 58 \; nW$

■ **OBJECTIVES**

After studying this chapter you will be able to

• evaluate the power gain and voltage gain of a given system,

• express power gain and voltage gain in decibels,

• express power levels in dBm and voltage levels in dBV and use these levels to determine power gain and voltage gain,

• identify and design simple (first-order) *RL* and *RC* low-pass and high-pass filters and explain the principles of operation of each type of filter,

• write the standard form of a transfer function for a given filter. The circuits that are studied will include band-pass and band-stop as well as low- and high-pass circuits,

• compute τ_c and use the time constant to determine the cutoff frequency(ies) in both radians per second and hertz for the transfer function of any first-order filter,

• sketch the Bode plots showing the frequency response of voltage gain and phase shift of any first-order filter,

• use PSpice to verify the operation of any first-order filter circuit.

Filters and the Bode Plot

22

CHAPTER PREVIEW

In the previous chapter we examined how *LRC* resonant circuits react to changes in frequency. In this chapter we will continue to study how changes in frequency affect the behavior of other simple circuits. We will analyze simple low-pass, high-pass, band-pass, and band-reject filter circuits. The analysis will compare the amplitude and phase shift of the output signal with respect to the input signal.

As their names imply, low-pass and high-pass filter circuits are able to pass low frequencies and high frequencies while blocking other frequency components. A good understanding of these filters provides a basis for understanding why circuits such as amplifiers and oscilloscopes are not able to pass all signals from their input to their output.

Band pass filters are designed to pass a range of frequencies from the input to the output. In the previous chapter, we saw that *L–C* circuits could be used to selectively pass a desired frequency range around a resonant frequency. In this chapter, we will observe that similar effects are possible by using only *R–C* or *R–L* components. Band reject filters, on the other hand, are used to selectively prevent certain frequencies from appearing at the output, while allowing both higher and lower frequencies to pass relatively unaffected.

The analysis of all filters may be simplified by plotting the output/input voltage relationship on a semilogarithmic graph called a *Bode plot*. ∎

Alexander Graham Bell

ALEXANDER GRAHAM BELL WAS BORN in Edinburgh, Scotland on March 3, 1847. As a young man, Bell followed his father and grandfather in research dealing with the deaf.

In 1873, Bell was appointed professor of vocal physiology at Boston University. His research was primarily involved in converting sound waves into electrical fluctuations. With the encouragement of Joseph Henry, who had done a great deal of work with inductors, Bell eventually developed the telephone.

In his now-famous accident in which Bell spilled acid on himself, Bell uttered the words "Watson, please come here. I want you." Watson, who was on another floor, ran to Bell's assistance.

PUTTING IT IN PERSPECTIVE

Although others had worked on the principle of the telephone, Alexander Graham Bell was awarded the patent for the telephone in 1876. The telephone he constructed was a simple device that passed a current through carbon powder. The density of the carbon powder was determined by air fluctuations due to the sound of a person's voice. When the carbon was compressed, resistivity would decrease, allowing more current.

Bell's name has been adopted for the decibel, which is the unit used to describe sound intensities and power gain.

Although the invention of the telephone made Bell wealthy, he continued experimenting in electronics, air conditioning, and animal breeding. Bell died at the age of 75 in Baddeck, Nova Scotia on August 2, 1922. ■

22.1 The Decibel

In electronics, we often wish to consider the effects of a circuit without examining the actual operation of the circuit itself. This black-box approach is a common technique used to simplify transistor circuits and for depicting integrated circuits that may contain hundreds or even thousands of elements. Consider the system shown in Figure 22–1.

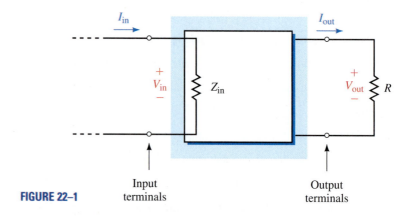

FIGURE 22–1

Input terminals   Output terminals

Although the circuit within the box may contain many elements, any source connected to the input terminals will effectively see only the input impedance, Z_{in}. Similarly, any load impedance, R_L, connected to the output terminals will have voltage and current determined by certain parameters in the circuit. These parameters usually result in an output at the load that is easily predicted for certain conditions.

We now define several terms that are used to analyze any system having two input terminals and two output terminals.

The **power gain,** A_P, is defined as the ratio of output signal power to the input signal power:

$$A_P = \frac{P_{out}}{P_{in}} \tag{22–1}$$

We must emphasize that the total output power delivered to any load can never exceed the total input power to a circuit. When we refer to the power gain of a system, we are interested in only the power contained in the ac signal, so we neglect any power due to dc. In many circuits, the ac power will be significantly less than the dc power. However, ac power gains in the order of tens of thousands are quite possible.

The **voltage gain,** A_v, is defined as the ratio of output signal voltage to the input signal voltage:

$$A_v = \frac{V_{out}}{V_{in}} \tag{22–2}$$

As mentioned, the power gain of a system may be a very large. For other applications the output power may be much smaller than the input power, resulting in a loss, or attenuation. Any circuit in which the output signal power is greater than the input signal power is referred to as an **amplifier.** Conversely, any circuit in which the output signal power is less than the input signal power is referred to as an **attenuator.**

The ratios expressing power gain or voltage gain may be either very large or very small, making it inconvenient to express the power gain as a simple ratio of two numbers. The **bel,** which is a logarithmic unit named after Alexander Graham Bell, was selected to represent a ten-fold increase or decrease in power. Stated mathematically, the power gain in bels is given as

$$A_{P(\text{bels})} = \log_{10}\frac{P_{\text{out}}}{P_{\text{in}}}$$

Because the bel is an awkwardly large unit, the **decibel** (dB), which is one tenth of a bel, has been adopted as a more acceptable unit for describing the logarithmic change in power levels. One bel contains 10 decibels and so the power gain in decibels is given as

$$A_{P(\text{dB})} = 10 \log_{10}\frac{P_{\text{out}}}{P_{\text{in}}} \qquad \textbf{(22–3)}$$

If the power level in the system increases from the input to the output, then the gain in dB will be positive. If the power at the output is less than the power at the input, then the power gain will be negative. Notice that if the input and output have the same power levels, then the power gain will be 0 dB, since log 1 = 0.

An amplifier has the indicated input and output power levels. Determine the power gain both as a ratio and in dB for each of the conditions:

a. $P_{\text{in}} = 1$ mW, $P_{\text{out}} = 100$ W.

b. $P_{\text{in}} = 4$ μW, $P_{\text{out}} = 2$ μW.

c. $P_{\text{in}} = 6$ mW, $P_{\text{out}} = 12$ mW.

d. $P_{\text{in}} = 25$ mW, $P_{\text{out}} = 2.5$ mW.

EXAMPLE 22–1

Solution

a. $A_P = \dfrac{P_{\text{out}}}{P_{\text{in}}} = \dfrac{100 \text{ W}}{1 \text{ mW}} = 100\,000$

     $A_{P(\text{dB})} = 10 \log_{10}(100\,000) = (10)(5) = 50$ dB

b. $A_P = \dfrac{P_{\text{out}}}{P_{\text{in}}} = \dfrac{2 \text{ μW}}{4 \text{ μW}} = 0.5$

     $A_{P(\text{dB})} = 10 \log_{10}(0.5) = (10)(-0.30) = -3.0$ dB

c. $A_P = \dfrac{P_{\text{out}}}{P_{\text{in}}} = \dfrac{12 \text{ mW}}{6 \text{ mW}} = 2$

     $A_{P(\text{dB})} = 10 \log_{10}2 = (10)(.30) = 3.0$ dB

d. $A_P = \dfrac{P_{\text{out}}}{P_{\text{in}}} = \dfrac{2.5 \text{ mW}}{25 \text{ mW}} = 0.10$

     $A_{P(\text{dB})} = 10 \log_{10}0.10 = (10)(-1) = -10$ dB

The previous example illustrates that if the power is increased or decreased by a factor of two, the resultant power gain will be either $+3$ dB or -3 dB, respectively. You will recall that in the previous chapter a similar reference was made when the half-power frequencies of a resonant circuit were referred to as the 3-dB down frequencies.

The voltage gain of a system may also be expressed in dB. In order to derive the expression for voltage gain, we first assume that the input resistance and the load resistance are the same value. Then, using the definition of power gain as given by Equation 22–3, we have the following:

$$A_{P(\text{dB})} = 10 \log_{10} \frac{P_{\text{out}}}{P_{\text{in}}}$$

$$= 10 \log_{10} \frac{V_{\text{out}}^2/R}{V_{\text{in}}^2/R}$$

$$= 10 \log_{10} \left(\frac{V_{\text{out}}}{V_{\text{in}}}\right)^2$$

which gives

$$A_{P(\text{dB})} = 20 \log_{10} \frac{V_{\text{out}}}{V_{\text{in}}}$$

Because the above expression represents a decibel equivalent of the voltage gain, we simply write the voltage gain in dB as follows:

$$A_{v(\text{dB})} = 20 \log_{10} \frac{V_{\text{out}}}{V_{\text{in}}} \tag{22–4}$$

EXAMPLE 22–2

The amplifier circuit of Figure 22–2 has the given conditions. Calculate the voltage gain and power gain in dB.

$Z_{\text{in}} = 10 \text{ k}\Omega$

$R_L = 600 \text{ }\Omega$

$V_{\text{in}} = 20 \text{ mV}_{\text{rms}}$

$V_{\text{out}} = 500 \text{ mV}_{\text{rms}}$

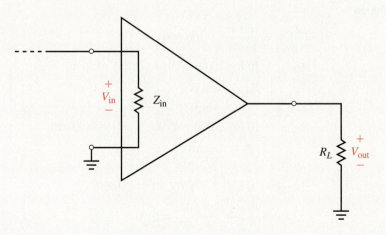

FIGURE 22–2

Solution The voltage gain of the amplifier is

$$A_v = 20 \log_{10} \frac{V_{out}}{V_{in}}$$

$$= 20 \log_{10} \frac{500 \text{ mV}}{20 \text{ mV}} = 20 \log_{10} 25 = 28.0 \text{ dB}$$

The signal power available at the input of the amplifier is

$$P_{in} = \frac{V_{in}^2}{Z_{in}} = \frac{(20 \text{ mV})^2}{10 \text{ k}\Omega} = 0.040 \text{ } \mu\text{W}$$

The signal at the output of the amplifier has a power of

$$P_{out} = \frac{V_{out}^2}{R_L} = \frac{(500 \text{ mV})^2}{600 \text{ }\Omega} = 416.7 \text{ } \mu\text{W}$$

The power gain of the amplifier is

$$A_P = 10 \log_{10} \frac{P_{out}}{P_{in}}$$

$$= 10 \log_{10} \frac{416.7 \text{ } \mu\text{W}}{0.040 \text{ } \mu\text{W}}$$

$$= 10 \log_{10}(10\ 417) = 40.2 \text{ dB}$$

PRACTICE PROBLEMS 1

Calculate the voltage gain and power gain (in dB) for the amplifier of Figure 22–2, given the following conditions:

$Z_{in} = 2 \text{ k}\Omega$

$R_L = 5 \text{ }\Omega$

$V_{in} = 16 \text{ } \mu\text{V}_{rms}$

$V_{out} = 32 \text{ } \mu\text{V}_{rms}$

Answers
66.0 dB, 92.0 dB

In order to convert a gain from decibels into a simple ratio of power or voltage, it is necessary to perform the inverse operation of the logarithm; namely, solve the unknown quantity by using the exponential. Recall that the following logarithmic and exponential operations are equivalent:

$$y = \log_b x$$
$$x = b^y$$

Using the above expressions, Equations 22–3 and 22–4 may be used to determine expressions for power and voltage gains as follows:

$$\frac{A_{P(dB)}}{10} = \log_{10} \frac{P_{out}}{P_{in}}$$

$$\frac{P_{out}}{P_{in}} = 10^{A_{P(dB)}/10} \qquad\qquad \textbf{(22–5)}$$

$$\frac{A_{v(dB)}}{20} = \log_{10} \frac{V_{out}}{V_{in}}$$

$$\frac{V_{out}}{V_{in}} = 10^{A_{v(dB)}/20} \qquad\qquad \textbf{(22–6)}$$

EXAMPLE 22–3

Convert the following from decibels to ratios:

a. $A_P = 25$ dB.

b. $A_P = -6$ dB.

c. $A_v = 10$ dB.

d. $A_v = -6$ dB.

Solution

a.
$$A_P = \frac{P_{out}}{P_{in}} = 10^{A_{P(dB)}/10}$$

$$= 10^{25/10} = 316$$

b.
$$A_P = 10^{A_{P(dB)}/10}$$

$$= 10^{-6/10} = 0.251$$

c.
$$A_v = \frac{V_{out}}{V_{in}} = 10^{A_{v(dB)}/20}$$

$$= 10^{10/20} = 3.16$$

d.
$$A_v = 10^{A_{v(dB)}/20}$$

$$= 10^{-6/20} = 0.501$$

Applications of Decibels

Decibels were originally intended as a measure of changes in acoustical levels. The human ear is not a linear instrument; rather, it responds to sounds in a logarithmic fashion. Because of this peculiar phenomenon, a ten-fold increase in sound intensity results in a perceived doubling of sound. This means that if we wish to double the sound heard from a 10-W power amplifier, we must increase the output power to 100 W.

The minimum sound level which may be detected by the human ear is called the threshold of hearing and is usually taken to be $I_0 = 1 \times 10^{-12}$ W/m². Table 22–1 lists approximate sound intensities of several common sounds. The decibel levels are determined from the expression

$$\beta_{(dB)} = 10 \log_{10} \frac{I}{I_0}$$

Some electronic circuits operate with very small power levels. These power levels may be referenced to some arbitrary level and then expressed in decibels in a manner similar to how sound intensities are represented. For instance, power levels may be referenced to a standard power of 1 mW. In such cases the power level is expressed in dBm and is determined as

$$P_{dBm} = 10 \log_{10} \frac{P}{1 \text{ mW}} \qquad (22\text{–}7)$$

If a power level is referenced to a standard of 1 W, then we have

$$P_{dBW} = 10 \log_{10} \frac{P}{1 \text{ W}} \qquad (22\text{–}8)$$

TABLE 22–1   Intensity Levels of Common Sounds

Sound	Intensity Level (dB)	Intensity (W/m²)
threshold of hearing, I_0	0	10^{-12}
virtual silence	10	10^{-11}
quiet room	20	10^{-10}
watch ticking at 1 m	30	10^{-9}
quiet street	40	10^{-8}
quiet conversation	50	10^{-7}
quiet motor at 1 m	60	10^{-6}
busy traffic	70	10^{-5}
door slamming	80	10^{-4}
busy office room	90	10^{-3}
jackhammer	100	10^{-2}
motorcycle	110	10^{-1}
loud indoor rock concert	120	1
threshold of pain	130	10

Express the following powers in dBm and in dBW.

EXAMPLE 22–4

a. $P_1 = 0.35\ \mu W$.

b. $P_2 = 20\ mW$.

c. $P_3 = 1000\ W$.

d. $P_4 = 1\ pW$.

Solution

a.
$$P_{1(dBm)} = 10\ \log_{10}\frac{0.35\ \mu W}{1\ mW} = -34.6\ dBm$$

$$P_{1(dBW)} = 10\ \log_{10}\frac{0.35\ \mu W}{1\ W} = -64.6\ dBW$$

b.
$$P_{2(dBm)} = 10\ \log_{10}\frac{20\ mW}{1\ mW} = 13.0\ dBm$$

$$P_{2(dBW)} = 10\ \log_{10}\frac{20\ mW}{1\ W} = -17.0\ dBW$$

c.
$$P_{3(dBm)} = 10\ \log_{10}\frac{1000\ W}{1\ mW} = 60\ dBm$$

$$P_{3(dBW)} = 10\ \log_{10}\frac{1000\ W}{1\ W} = 30\ dBW$$

d.
$$P_{4(dBm)} = 10\ \log_{10}\frac{1\ pW}{1\ mW} = -90\ dBm$$

$$P_{4(dBW)} = 10\ \log_{10}\frac{1\ pW}{1\ W} = -120\ dBW$$

Many voltmeters have a separate scale calibrated in decibels. In such cases, the voltage expressed in dBV uses 1 V_{rms} as the reference voltage. In general, any voltage reading may be expressed in dBV as follows:

$$V_{dBV} = 20\ \log_{10}\frac{V_{out}}{1\ V} \qquad (22\text{–}9)$$

Consider the resistors of Figure 22–3:

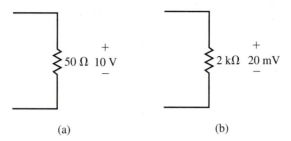

FIGURE 22–3   (a)   (b)

a. Determine the power levels in dBm and in dBW.

b. Express the voltages in dBV.

Answers

a. 33.0 dBm (3.0 dBW), −37.0 dBm (−67.0 dBW)

b. 20.0 dBV, −34.0 dBV

22.2 Multistage Systems

Quite often a system consists of several stages. In order to find the total voltage gain or power gain of the system, we would need to solve for the product of the individual gains. The use of decibels makes the solution of a multistage system easy to find. If the gain of each stage is given in decibels, then the resultant gain is simply determined as the summation of the individual gains.

Consider the system of Figure 22–4, which represents a three-stage system.

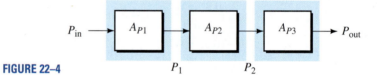

FIGURE 22–4

The power at the output of each stage is determined as follows:

$$P_1 = A_{P1} \, P_{in}$$
$$P_2 = A_{P2} \, P_1$$
$$P_{out} = A_{P3} \, P_2$$

The total power gain of the system is found as

$$A_{PT} = \frac{P_{out}}{P_{in}} = \frac{A_{P3} \, P_2}{P_{in}}$$
$$= \frac{A_{P3} \, (A_{P2} \, P_1)}{P_{in}}$$
$$= \frac{A_{P3} \, A_{P2} \, (A_{P1} P_{in})}{P_{in}}$$
$$= A_{P1} \, A_{P2} \, A_{P3}$$

In general, for n stages the total power gain is found as the product:

$$A_{PT} = A_{P1} \, A_{P2} \cdots A_{Pn} \tag{22–10}$$

However, if we use logarithms to solve for the gain in decibels, we have the following:

$$A_{PT}(dB) = 10 \, \log_{10} A_{PT}$$
$$= 10 \, \log_{10} (A_{P1} \, A_{P2} \cdots A_{Pn})$$
$$A_{PT}(dB) = 10 \, \log_{10} A_{P1} + 10 \, \log_{10} A_{P2} + \cdots + 10 \, \log_{10} A_{Pn}$$

The total power gain in decibels for n stages is determined as the summation of the individual decibel power gains:

$$A_{P(dB)} = A_{P1(dB)} + A_{P2(dB)} + \cdots + A_{Pn(dB)} \qquad (22\text{--}11)$$

The advantage of using decibels in solving for power gains and power levels is illustrated in the following example.

The circuit of Figure 22–5 represents the first three stages of a typical AM or FM receiver.

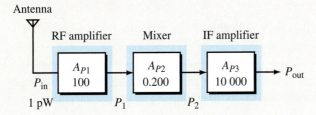

FIGURE 22–5

Find the following quantities:

a. $A_{P1(dB)}$, $A_{P2(dB)}$, and $A_{P3(dB)}$.

b. $A_{PT(dB)}$.

c. P_1, P_2, and P_{out}.

d. $P_{in(dBm)}$, $P_{1(dBm)}$, $P_{2(dBm)}$, and $P_{out(dBm)}$.

Solution

a. $A_{P1(dB)} = 10 \log_{10} A_{P1} = 10 \log_{10} 100 = 20$ dB

      $A_{P2(dB)} = \log_{10} A_{P2} = 10 \log_{10} 0.2 = -7.0$ dB

      $A_{P3(dB)} = 10 \log_{10} A_{P3} = 10 \log_{10}(10\ 000) = 40$ dB

b. $A_{PT(dB)} = A_{P1(dB)} + A_{P2(dB)} + A_{P3(dB)}$

             $= 20$ dB $- 7.0$ dB $+ 40$ dB

             $= 53.0$ dB

c. $P_1 = A_{P1} P_{in} = (100)\ (1 \text{ pW}) = 100$ pW

       $P_2 = A_{P2} P_1 = (100 \text{ pW})(0.2) = 20$ pW

     $P_{out} = A_{P3} P_2 = (10\ 000)(20 \text{ pW}) = 0.20$ μW

d. $P_{in(dBm)} = 10 \log_{10} \dfrac{P_{in}}{1 \text{ mW}} = 10 \log_{10} \dfrac{1 \text{ pW}}{1 \text{ mW}} = -90$ dBm

      $P_{1(dBm)} = 10 \log_{10} \dfrac{P_1}{1 \text{ mW}} = 10 \log_{10} \dfrac{100 \text{ pW}}{1 \text{ mW}} = -70$ dBm

      $P_{2(dBm)} = 10 \log_{10} \dfrac{P_2}{1 \text{ mW}} = 10 \log_{10} \dfrac{20 \text{ pW}}{1 \text{ mW}} = -77.0$ dBm

    $P_{out(dBm)} = 10 \log_{10} \dfrac{P_{out}}{1 \text{ mW}} = 10 \log_{10} \dfrac{0.20 \text{ μW}}{1 \text{ mW}} = -37.0$ dBm

Notice that the power level (in dBm) at the output of any stage is easily determined as the sum of the input power level (in dBm) and the gain of the stage (in dB). It is for this reason that many communication circuits express power levels in decibels rather than in watts.

Calculate the power level at the output of each of the stages in Figure 22–6.

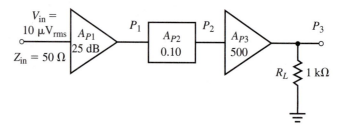

FIGURE 22–6

Answers
$P_1 = -62$ dBm, $P_2 = -72$ dBm, $P_3 = -45$ dBm

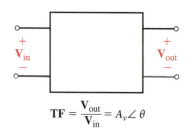

IN-PROCESS
LEARNING CHECK 1

(Answers are at the end of the chapter.)

1. Given that an amplifier has a power gain of 25 dB, calculate the output power level (in dBm) for the following input characteristics:

   a. $V_{in} = 10$ mV$_{rms}$, $Z_{in} = 50$ Ω.

   b. $V_{in} = 10$ mV$_{rms}$, $Z_{in} = 1$ kΩ.

   c. $V_{in} = 400$ μV$_{rms}$, $Z_{in} = 200$ Ω.

2. Given amplifiers with the following output characteristics, determine the output voltage (in volts rms).

   a. $P_{out} = 8.0$ dBm, $R_L = 50$ Ω.

   b. $P_{out} = -16.0$ dBm, $R_L = 2$ kΩ.

   c. $P_{out} = -16.0$ dBm, $R_L = 5$ kΩ.

22.3 Simple *RC* and *RL* Transfer Functions

Electronic circuits usually operate in a highly predictable fashion. If a certain signal is applied at the input of a system, the output will be determined by the physical characteristics of the circuit. The frequency of the incoming signal is one of the many physical conditions that determine the relationship between a given input signal and the resulting output. Although outside the scope of this text, other conditions that may determine the relationship between the input and output signals of a given circuit are temperature, light, radiation, etc.

For any system subjected to a sinusoidal input voltage, as shown in Figure 22–7, we define the **transfer function** as the ratio of the output voltage phasor to the input voltage phasor for any frequency ω (in radians per second).

$$\mathbf{TF}(\omega) = \frac{\mathbf{V}_{out}}{\mathbf{V}_{in}} = A_v \angle \theta \qquad (22\text{--}12)$$

Notice that the definition of transfer function is almost the same as voltage gain. The difference is that the transfer function takes into account both amplitude and phase shift of the voltages, whereas voltage gain is a comparison of only the amplitudes.

From Equation 22–12 we see that the amplitude of the transfer function is in fact the voltage gain, a scalar quantity. The phase angle, θ, represents the phase shift between the input and output voltage phasors. The angle θ will be positive if the output leads the input and negative if the output lags the input waveform.

If the elements within the block of Figure 22–7 are resistors, then the output and the input voltages will always be in phase. Also, since resistors have the same value at all frequencies, the voltage gain will remain constant at all

$$\mathbf{TF} = \frac{\mathbf{V}_{out}}{\mathbf{V}_{in}} = A_v \angle \theta$$

FIGURE 22–7

frequencies. (This text does not take into account resistance variations due to very high frequencies.) The resulting circuit is called an attenuator since resistance within the block will dissipate some power, thereby reducing (or attenuating) the signal as it passes through the circuit.

If the elements within the block are combinations of resistors, inductors, and capacitors, then the output voltage and phase will depend on frequency since the impedances of inductors and capacitors are frequency dependent. Figure 22–8 illustrates an example of how voltage gain and phase shift of a circuit may change as a function of frequency.

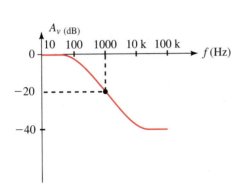

(a) Voltage gain as a function of frequency.

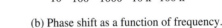

(b) Phase shift as a function of frequency.

FIGURE 22–8   Frequency response of a circuit.

In order to examine the operation of the circuit over a wide range of frequencies, the abscissa (horizontal axis) is usually shown as a logarithmic scale. The ordinate (vertical axis) is generally shown as a linear scale in decibels or degrees. The semilogarithmic graphs showing the frequency response of filters are referred to as **Bode plots** and are extremely useful in predicting and understanding the operation of filters. Similar graphs are used to predict the operation of amplifiers and many other electronic components.

By examining the frequency response of a circuit, we are able to determine at a glance the voltage gain (in dB) and phase shift (in degrees) for any sinusoidal input at a given frequency. For example, at a frequency of 1000 Hz, the voltage gain is −20 dB and the phase shift is 45°. This means that the output signal is one tenth as large as the input signal and the output leads the input by 45°.

From the frequency response of Figure 22–8, we see that the circuit that corresponds to this response is able to pass low-frequency signals, while at the same time partially attenuating high-frequency signals. Any circuit that passes a particular range of frequencies while blocking others is referred to as a **filter** circuit. Filter circuits are usually named according to their function, although certain filters are named after their inventors. The filter having the response of Figure 22–8 is referred to as a stepfilter, since the voltage gain occurs between two limits (steps). Other types of filters that are used regularly in electrical and electronic circuits include low-pass, high-pass, band-pass, and band-reject filters.

Although the design of filters is a topic in itself, we will examine Bode plots of a few common types of filters that are used in electronic circuits.

Sketching Bode Plots

In order to understand the operation of a filter (or any other frequency-dependent system) it is helpful to first sketch the Bode plots. All first order systems (namely those consisting of *R-C* or *R-L* combinations) have transfer functions that are constructed from only four possible forms (and perhaps a constant). If voltage gain is expressed in decibels, we find that the Bode plots of more complex expressions are simply determined as the arithmetic sum of these simple forms. We begin by examining the four possible forms, and then we will combine these forms to determine the frequency responses of more complicated transfer functions. In each of the forms, the value of tau (τ) is simply a time constant determined by the components of the circuit.

One important consideration of transfer functions is that the boundary conditions must always be satisfied. The boundary conditions of a transfer function or filter circuit are determined by examining the characteristics when $\omega = 0$ rad/s and as $\omega \rightarrow \infty$.

1. **TF** $= j\omega\tau = \omega\tau\angle 90°$

   The voltage gain of the above transfer function is $A_V = \omega\tau$. By examining the boundary conditions, we see that the voltage gain is $A_v = 0$ when $\omega = 0$ and $A_v = \infty$ as $\omega \rightarrow \infty$. (Although a transfer function consisting of only this term cannot occur for real systems, it is an important component when combined with other terms.) In decibels, the voltage gain becomes $[A_V]_{dB} = 20 \log \omega\tau$. Notice that the voltage gain of this expression increases as the frequency is increased.

   At a frequency of $\omega = 1/\tau$, the transfer function is evaluated as **TF** $= j1 = 1\angle 90°$. The voltage gain is $A_V = 1$ which is equivalent to $[A_V]_{dB} = 20 \log 1 = 0$ dB.

   If the frequency is increased by a factor of 10 (increasing by one **decade**) to $\omega = 10/\tau$, the transfer function becomes **TF** $= j10 = 10\angle 90°$. The voltage gain of $A_V = 10$ is equivalent to $[A_V]_{dB} = 20 \log 10 = 20$ dB. This illustrates that the gain of the transfer function increases at a rate of 20 dB/decade. Had the frequency been doubled (increasing by one **octave**), the transfer function would have been **TF** $= j2 = 2\angle 90°$. The voltage gain $A_V = 2$ is equivalent to $[A_V]_{dB} = 20 \log 2 = 6$ dB. This illustrates that the gain of the transfer function increases at a rate of 6 dB/octave. (Hence, a slope of 20 dB/decade is equivalent to 6 dB/octave.)

   For the given transfer function, the phase angle (of the output voltage relative to the input voltage) is a constant 90°. Consequently, a sinusoidal signal applied at the input will result in an output signal that always leads the input by 90°.

   The Bode plots for the transfer function appear as shown in Figure 22–9.

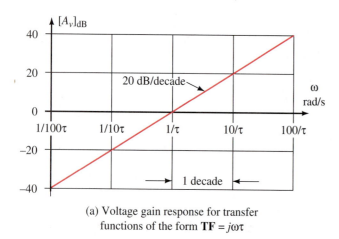

(a) Voltage gain response for transfer functions of the form **TF** $= j\omega\tau$

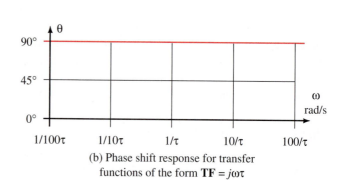

(b) Phase shift response for transfer functions of the form **TF** $= j\omega\tau$

FIGURE 22–9

2. **TF** $= 1 + j\omega\tau$

Examining the above expression, we determine that the magnitude of this transfer function also increases as ω is increased. However, unlike the previous transfer function form, the magnitude of this transfer function can never be less than 1. For small values of frequency, ω the real term "swamps" the imaginary term, resulting in **TF** $\approx 1 + j0 = 1\angle 0°$. For large values of frequency the opposite happens; the imaginary term "swamps" the real term, resulting in **TF** $\approx j\omega\tau = \omega\tau\angle 90°$.

Now, the question is "*what do we mean by a small frequency and large frequency?*" Let's examine what happens at a frequency of $\omega_c = 1/\tau$. At this frequency, the transfer function simply becomes **TF** $= 1 + j1 = \sqrt{2}\angle 45°$. The voltage gain is equivalent to $A_v = 20\log\sqrt{2} = 3.0$ dB. The frequency $\omega_c = 1/\tau$ is often referred to as the **cutoff frequency, the critical frequency,** or the **break frequency.** If we were to decrease the frequency so that it is 1/10 of the cutoff frequency we would have **TF** $= 1 + j0.1 = 1.005\angle 5.71°$. Two very important characteristics become evident for frequencies below the cutoff frequency:

i. The voltage gain is essentially constant at $A_v = 1 \equiv 0$ dB and

ii. The phase shift below $0.1\omega_c$ is essentially constant at $0°$.

Next, let's examine what happens at a frequency $\omega = 10\omega_c$. At this higher frequency, the transfer function becomes **TF** $= 1 + j10 = 10.05\angle 84.29°$. The voltage gain is equivalent to $A_v = 20\log 10.05 = 20.04$ dB. Further analysis would show that at $\omega = 100\omega_c$, the voltage gain is 40 dB. Two very important characteristics become evident for frequencies above the cutoff frequency:

i. The voltage gain rises at a rate of 20 dB/decade $\equiv 6$ dB/octave and

ii. The phase shift above $10\omega_c$ is essentially constant at $90°$.

When sketching the Bode plots for this transfer function, we approximate the voltage gain response as 0 dB for all frequencies less than $\omega_c = 1/\tau$ and rising at a rate of 20 dB/decade for all frequencies above $\omega_c = 1/\tau$. The phase shift response is a bit more complicated. The phase shift is approximated as $0°$ for $\omega < 0.1\omega_c$ and as $90°$ for $\omega > 10\omega_c$. In the region between $0.1\omega_c$ and $10\omega_c$, the phase shift is approximated as $45°$/decade. Figure 22–10 illustrates both the straight-line approximations and the actual curves for a transfer function of the form, **TF** $= 1 + j\omega\tau$.

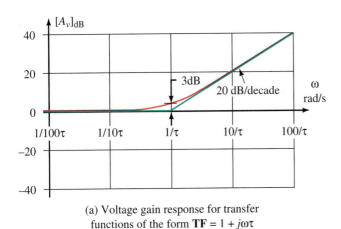

(a) Voltage gain response for transfer
functions of the form **TF** $= 1 + j\omega\tau$

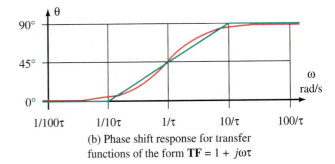

(b) Phase shift response for transfer
functions of the form **TF** $= 1 + j\omega\tau$

FIGURE 22–10

3. $\mathbf{TF} = \dfrac{1}{j\omega\tau}$

A transfer function of this form can also be written as $\mathbf{TF} = \dfrac{1}{\omega\tau\angle 90°} = \dfrac{1}{\omega\tau}\angle -90°$, which illustrates that the voltage gain is infinitely large when $\omega = 0$ and decreases as the frequency increases. Due to the j-operator in the denominator, it is also apparent that the phase shift is $-90°$ for all frequencies. (Once again, no real system of electronic components will have a transfer function consisting of only this term.) The Bode plots for transfer functions of the type $\mathbf{TF} = \dfrac{1}{j\omega\tau}$ are shown in Figure 22–11.

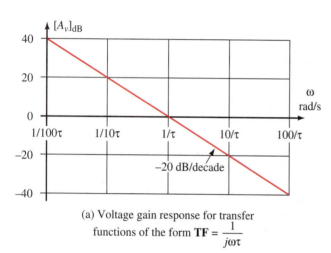

(a) Voltage gain response for transfer functions of the form $\mathbf{TF} = \dfrac{1}{j\omega\tau}$

(b) Phase shift response for transfer functions of the form $\mathbf{TF} = \dfrac{1}{j\omega\tau}$

FIGURE 22–11

4. $\mathbf{TF} = \dfrac{1}{1 + j\omega\tau}$

The magnitude of this transfer function decreases as ω is increased. However, unlike the previous transfer function form, the voltage gain of this transfer function can never be greater than 1. For small values of ω the real term "swamps" the imaginary term, resulting in $\mathbf{TF} \approx \dfrac{1}{1 + j0} = \dfrac{1}{1\angle 0°} = 1\angle 0°$.

For large values of frequency the opposite happens; the imaginary term "swamps" the real term, resulting in $\mathbf{TF} \approx \dfrac{1}{j\omega\tau} = \dfrac{1}{\omega\tau\angle 90°} = \dfrac{1}{\omega\tau}\angle -90°$.

Once again, the cutoff frequency occurs at $\omega_c = 1/\tau$. At this frequency, the transfer function results in a value of $\mathbf{TF} = \dfrac{1}{1 + j1} = \dfrac{1}{\sqrt{2}\angle 45°} = 0.7071\angle -45°$. At the cutoff frequency, the voltage gain (in decibels) is found to be $A_v = 20 \log 0.7071 = -3.0$ dB. For a filter having this transfer function, we see that the output power at the cutoff frequency is 3 dB down from its maximum value. You will recall that a "gain" of -3 dB results in an output power that is half of the maximum power.

For a filter having a transfer function of the form $\mathbf{TF} = \dfrac{1}{1 + j\omega\tau}$, two characteristics are evident for frequencies below the cutoff frequency:

i. The voltage gain is essentially constant at $A_v = 1 \equiv 0$ dB and

ii. The phase shift below $0.1\omega_c$ is essentially constant at $0°$.

For frequencies above the cutoff frequency:

i. The voltage gain drops at a rate of 20 dB/decade $\equiv 6$ dB/octave and

ii. The phase shift above $10\omega_c$ is essentially constant at $-90°$.

When sketching the Bode plots for this transfer function, we approximate the voltage gain response as 0 dB for all frequencies less than $\omega_c = 1/\tau$ and decreasing at a rate of 20 dB/decade for all frequencies above $\omega_c = 1/\tau$. The phase shift is approximated as $0°$ for $\omega < 0.1\omega_c$ and as $-90°$ for $\omega > 10\omega_c$. In the region between $0.1\omega_c$ and $10\omega_c$, the phase shift is approximated as having a slope of $-45°$/decade. Figure 22–12 shows the straight-line approximations and the actual curves for a transfer function of the form, $\mathbf{TF} = \dfrac{1}{1 + j\omega\tau}$.

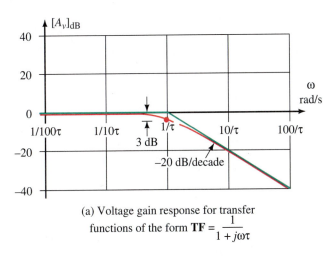

(a) Voltage gain response for transfer functions of the form $\mathbf{TF} = \dfrac{1}{1 + j\omega\tau}$

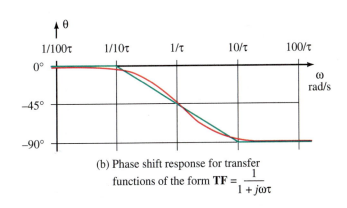

(b) Phase shift response for transfer functions of the form $\mathbf{TF} = \dfrac{1}{1 + j\omega\tau}$

FIGURE 22–12

Now we are ready to see how these various transfer function forms can be combined to sketch transfer functions of real circuits.

EXAMPLE 22–6

Given $\mathbf{TF} = \dfrac{10}{1 + j0.001\omega}$

a. Determine the boundary conditions.

b. Calculate the cutoff frequency in rad/s and in Hz.

c. Sketch the straight-line approximation of the voltage gain response.

d. Sketch the straight-line approximation of the phase shift response.

Solution

a. When $\omega = 0$, the transfer function simplifies to **TF** $= 10 = 10 \angle 0°$.

The voltage gain will be $A_v = 20 \log 10 = 20$ dB and the phase shift is 0°. As $\omega \to \infty$, the magnitude of the transfer function approaches zero and the phase shift will be $-90°$ (since the real term in the denominator will be swamped by the imaginary term).

b. The cutoff frequency is determined as

$$\omega_c = \frac{1}{\tau} = \frac{1}{0.001} = 1000 \text{ rad/s} \quad (f_c = \omega_c/2\pi = 159.15 \text{ Hz})$$

c. The voltage gain response is shown in Figure 22–13.

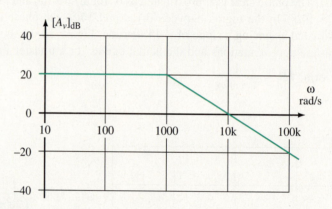

FIGURE 22–13

d. The phase shift response is shown in Figure 22–14.

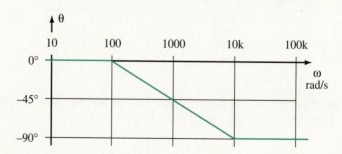

FIGURE 22–14

Note: By examining the response curves for the given transfer function, we see that this represents the characteristics of a low-pass filter. The notable exception is that a passive filter (consisting of only resistors, capacitors, or inductors) can not have a voltage gain larger than unity ($A_v = 1$). Consequently, we must conclude that the filter includes an active component such as a transistor or an operational amplifier.

EXAMPLE 22–7

Given $\mathbf{TF} = \dfrac{j0.001\omega}{1 + j0.002\omega}$

a. Determine the boundary conditions

b. Calculate the cutoff frequency in rad/s and in Hz.

c. Sketch the straight-line approximation of the voltage gain response.

d. Sketch the straight-line approximation of the phase shift response.

Solution

a. When $\omega = 0$, the numerator of the transfer function is j0 and the denominator is 1. Therefore the entire transfer function simplies to $\mathbf{TF} = j0 = 0\angle90°$. This indicates that for low frequencies, the voltage gain is very low, and that there is a 90° phase shift. As $\omega \to \infty$, the numerator will be j0.001ω, while the denominator simplifies to j0.002ω (since the real term is swamped). The reluctant transfer function will be simplified as $\mathbf{TF} = 0.5 = 0.5\angle0°$. This corresponds to a gain $A_v = 20 \log 0.5 = -6.0$ dB and a phase shift of $0°$.

b. The given transfer function has two primary frequencies of interest. The first is due to the numerator, which results in the straight-line component having a gain of 0 dB when $\omega = 1000$ rad/s. The second frequency is due to the denominator, resulting in a cutoff frequency of $\omega = 500$ rad/s (f = 79.6 Hz).

c. Figure 22–15 shows the straight-line approximations of the voltage gain due to each term in the transfer function as well as the resultant straight-line approximation of the combination. (Resultant in red.)

d. Figure 22–16 shows the phase response of the individual terms as well as the resultant combined effect. (Resultant in red.)

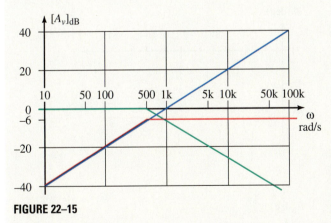

FIGURE 22–15

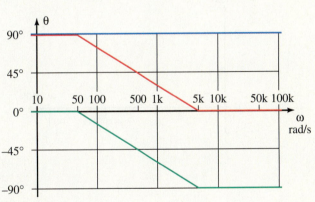

FIGURE 22–16

Note: The results of this example illustrate that frequency response plots of transfer functions are determined by arithmetically combining the results due to each component of the transfer function. Although the resultant plots may occasionally be complicated, it is important to realize that simple addition is used in all cases.

Another important consideration when sketching Bode plots is to remember that the boundary conditions of the frequency response curves must satisfy the calculated boundary conditions for the transfer function. Upon inspection of Figure 22–15, we see that for high frequencies (those above 500 rad/s) the voltage gain is −6 dB. This is precisely the value calculated from the transfer function. Similarly, we see that the phase response curve of Figure 22–16 satisfies the boundary conditions of 90° for $\omega = 0$ and 0° as $\omega \rightarrow \infty$ respectively.

PRACTICE PROBLEMS 4

Sketch the straight-line approximation for each of the transfer functions given.

a. **TF** $= 1 + j0.02\omega$

b. **TF** $= \dfrac{100}{1 + j0.005\omega}$

c. **TF** $= \dfrac{1 + j0.02\omega}{20(1 + j0.002\omega)}$

Answers

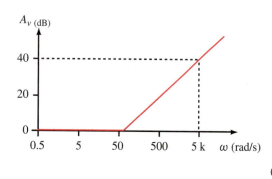
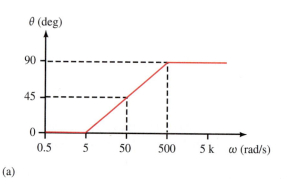

(a)

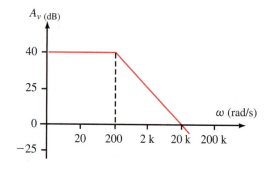
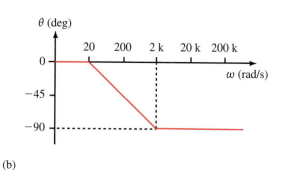

(b)

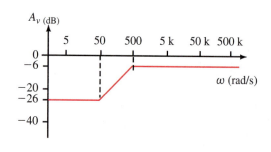
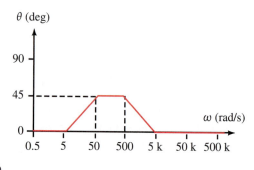

(c)

FIGURE 22–17

Writing Transfer Functions

The transfer function of any circuit is found by following a few simple steps. As we have already seen, a properly written transfer function allows us to easily calculate the cutoff frequencies and quickly sketch the corresponding Bode plot for the circuit. The steps are as follows:

1. Determine the boundary conditions for the given circuit by solving for the voltage gain when the frequency is zero (dc) and when the frequency approaches infinity. The boundary conditions are found by using the following approximations:

   At $\omega = 0$,   inductors are short circuits,
   capacitors are open circuits.

   At $\omega \rightarrow \infty$,   inductors are open circuits,
   capacitors are short circuits.

   By using the above approximations, all capacitors and inductors are easily removed from the circuit. The resulting voltage gain is determined for each boundary condition by simply applying the voltage divider rule.

2. Use the voltage divider rule to write the general expression for the transfer function in terms of the frequency, ω. In order to simplify the algebra, all capacitive and inductive reactance vectors are written as follows:

$$\mathbf{Z}_C = \frac{1}{j\omega C}$$

and

$$\mathbf{Z}_L = j\omega L$$

3. Simplify the resulting transfer function so that it is in the following format:

$$\mathbf{TF} = \frac{(j\omega\tau_{Z_1})(1 + j\omega\tau_{Z_2}) \cdots (1 + j\omega\tau_{Z_n})}{(j\omega\tau_{P_1})(1 + j\omega\tau_{P_2}) \cdots (1 + j\omega\tau_{P_m})}$$

   Once the function is in this format, it is good practice to verify the boundary conditions found in Step 1. The boundary conditions are determined algebraically by first letting $\omega = 0$ and then solving for the resulting dc voltage gain. Next, we let $\omega \rightarrow \infty$. The various $(1 + j\omega\tau)$ terms of the transfer function may now be approximated as simply $j\omega\tau$, since the imaginary terms will be much larger (≥ 10) than the real components. The resultant gain will give the high-frequency gain.

4. Determine the break frequency(ies) at $\omega = 1/\tau$ (in radians per second) where the time constants will be expressed as either $\tau = RC$ or $\tau = L/R$.

5. Sketch the straight-line approximation by separately considering the effects of each term in the transfer function.

6. Sketch the actual response of the circuit from the approximation. The actual voltage gain response will be a smooth, continuous curve which follows the asymptotic curve but that usually has a 3-dB difference at the cutoff frequency(ies). This approximation will not apply if two cutoff frequencies are separated by less than a decade. The actual phase shift response will have the same value as the straight-line approximation at the cutoff frequency. At frequencies one decade above and one decade below the cutoff frequency, the actual phase shift will be 5.71° from the straight-line approximation.

These steps will now be used in analyzing several important types of filters.

22.4 The Low-Pass Filter

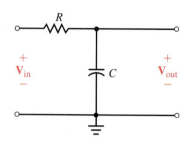

FIGURE 22–18   *RC* low-pass filter.

The *RC* Low-Pass Filter

The circuit of Figure 22–18 is referred to as a low-pass *RC* filter circuit since it permits low-frequency signals to pass from the input to the output while attenuating high frequency signals.

At low frequencies, the capacitor has a very large reactance. Consequently, at low frequencies the capacitor is essentially an open circuit resulting in the voltage across the capacitor, $\mathbf{V}_{out}$, to be essentially equal to the applied voltage, $\mathbf{V}_{in}$.

At high frequencies, the capacitor has a very small reactance, which essentially short circuits the output terminals. The voltage at the output will therefore approach zero as the frequency increases. Although we are able to easily predict what happens at the two extremes of frequency, called the boundary conditions, we do not yet know what occurs between the two extremes.

The circuit of Figure 22–18 is easily analyzed by applying the voltage divider rule. Namely,

$$\mathbf{V}_{out} = \frac{\mathbf{Z}_C}{\mathbf{R} + \mathbf{Z}_C}\mathbf{V}_{in}$$

In order to simplify the algebra, the reactance of a capacitor is expressed as follows:

$$\mathbf{Z}_C = -j\frac{1}{\omega C} = -j\frac{j}{j\omega C} = \frac{1}{j\omega C} \qquad (22\text{–}13)$$

The transfer function for the circuit of Figure 22–18 is now evaluated as follows:

$$\mathbf{TF}(\omega) = \frac{\mathbf{V}_{out}}{\mathbf{V}_{in}} = \frac{\dfrac{1}{j\omega C}}{R + \dfrac{1}{j\omega C}} = \frac{\dfrac{1}{j\omega C}}{\dfrac{1 + j\omega RC}{j\omega C}}$$

$$= \frac{1}{1 + j\omega RC}$$

We define the cutoff frequency, ω_c, as the frequency at which the output power is equal to half of the maximum output power (3 dB down from the maximum). This frequency occurs when the output voltage has an amplitude which is 0.7071 of the input voltage. For the *RC* circuit, the cutoff frequency occurs at

$$\omega_c = \frac{1}{\tau} = \frac{1}{RC} \qquad (22\text{–}14)$$

Then the transfer function is written as

$$\mathbf{TF}(\omega) = \frac{1}{1 + j\dfrac{\omega}{\omega_c}} \qquad (22\text{–}15)$$

The previous transfer function results in the Bode plot shown in Figure 22–19.

Notice that the abscissas (horizontal axes) of the graphs in Figure 22–19 are shown as a ratio of ω/ω_c. Such a graph is called a normalized plot and eliminates the need to determine the actual cutoff frequency, ω_c. The normalized plot will have the same values for all low-pass *RC* filters. The actual frequency

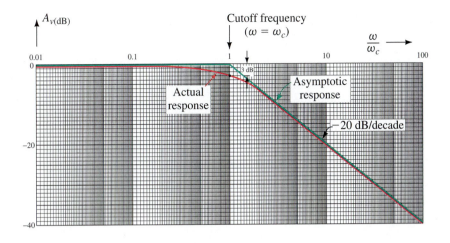

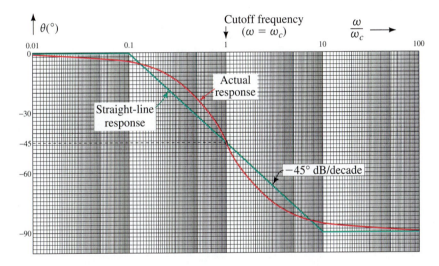

FIGURE 22–19   Normalized frequency response for an *RC* low-pass filter.

response of the *RC* low-pass filter can be approximated from the straight-line approximation by using the following guidelines:

1. At low frequencies ($\omega/\omega_c \leq 0.1$) the voltage gain is approximately 0 dB with a phase shift of about 0°. This means that the output signal of the filter is very nearly equal to the input signal. The phase shift at $\omega = 0.1\omega_c$ will be 5.71° less than the straight-line approximation.

2. At the cutoff frequency, $\omega_c = 1/RC$ ($f_c = 1/2\pi RC$), the gain of the filter is -3 dB. This means that at the cutoff frequency, the circuit will deliver half the power that it would deliver at very low frequencies. At the cutoff frequency, the output voltage will lag the input voltage by 45°.

3. As the frequency increases beyond the cutoff frequency, the amplitude of the output signal decreases by a factor of approximately ten for each tenfold increase in frequency; namely the voltage gain is -20 dB per decade. The phase shift at $\omega = 10\omega_c$ will be 5.71° greater than the straight-line approximation, namely at $\theta = -84.29°$. For high frequencies ($\omega/\omega_c \geq 10$), the phase shift between the input and output voltage approaches $-90°$.

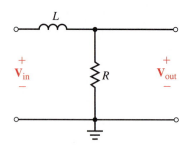

FIGURE 22–20 *RL* low-pass filter.

The *RL* Low-Pass Filter

A low-pass filter circuit may be made up of a resistor and an inductor as illustrated in Figure 22–20.

In a manner similar to that used for the *RC* low-pass filter, we may write the transfer function for the circuit of Figure 22–20 as follows:

$$\mathbf{TF} = \frac{\mathbf{V}_{\text{out}}}{\mathbf{V}_{\text{in}}}$$

$$= \frac{\mathbf{R}}{\mathbf{R} + \mathbf{Z}_L} = \frac{R}{R + j\omega L}$$

Now, dividing the numerator and the denominator by *R*, we have the transfer function expressed as

$$\mathbf{TF} = \frac{1}{1 + j\omega\dfrac{L}{R}}$$

Since the cutoff frequency is found as $\omega_c = 1/\tau$, we have

$$\omega_c = \frac{1}{\tau} = \frac{1}{\dfrac{L}{R}} = \frac{R}{L}$$

and so

$$\mathbf{TF} = \frac{1}{1 + j\dfrac{\omega}{\omega_c}} \qquad\qquad (22\text{–}16)$$

Notice that the transfer function for the *RL* low-pass circuit in Equation 22–16 is identical to the transfer function of an *RC* circuit in Equation 22–15. In each case, the cutoff frequency is determined as the reciprocal of the time constant.

EXAMPLE 22–8

◀ MULTISIM

Sketch the Bode plot showing both the straight-line approximation and the actual response curves for the circuit of Figure 22–21. Show frequencies in hertz.

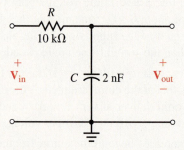

FIGURE 22–21

Solution   The cutoff frequency (in radians per second) for the circuit occurs at

$$\omega_c = \frac{1}{\tau} = \frac{1}{RC}$$

$$= \frac{1}{(10 \text{ k}\Omega)(2 \text{ nF})} = 50 \text{ krad/s}$$

which gives

$$f_c = \frac{\omega_c}{2\pi} = \frac{50 \text{ krad/s}}{2\pi} = 7.96 \text{ kHz}$$

In order to sketch the Bode plot, we begin with the asymptotes for the voltage gain response. The circuit will have a flat response until $f_c = 7.96$ kHz. Then the gain will drop at a rate of 20 dB for each decade increase in frequency. Therefore, the voltage gain at 79.6 kHz will be -20 dB, and at 796 kHz the voltage gain will be -40 dB. At the cutoff frequency for the filter, the actual voltage gain response will pass through a point which is 3 dB down from the intersection of the two asymptotes. The frequency response of the voltage gain is shown in Figure 22–22(a).

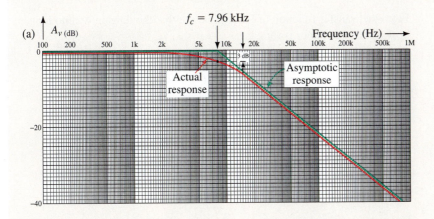

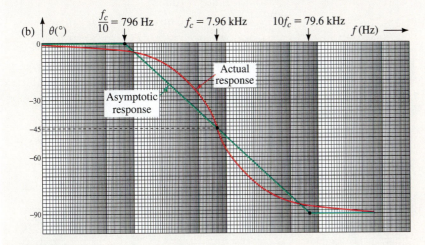

FIGURE 22–22

Next, we sketch the approximate phase shift response. The phase shift at 7.96 kHz will be $-45°$. At a frequency one decade below the cutoff frequency (at 796 Hz) the phase shift will be approximately equal to zero, while at a frequency one decade above the cutoff frequency (at 79.6 kHz) the phase shift will be near the maximum of $-90°$. The actual phase shift response will be a curve which varies slightly from the asymptotic response, as shown in Figure 22–22(b).

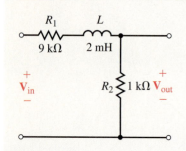

FIGURE 22–23

◀ MULTISIM

Consider the low-pass circuit of Figure 22–23:

a. Write the transfer function for the circuit.

b. Sketch the frequency response.

Solution

a. The transfer function of the circuit is found as

$$\mathbf{TF} = \frac{\mathbf{V}_{out}}{\mathbf{V}_{in}} = \frac{R_2}{R_2 + R_1 + j\omega L}$$

which becomes

$$\mathbf{TF} = \frac{R_2}{R_1 + R_2}\left(\frac{1}{1 + j\omega\dfrac{L}{R_1 + R_2}}\right)$$

b. From the transfer function of Part a), we see that the dc gain will no longer be 1 (0 dB) but rather is found as

$$A_{v(dc)} = 20\log\left(\frac{R_2}{R_1 + R_2}\right)$$

$$= 20\log\left(\frac{1}{10}\right)$$

$$= -20\ \text{dB}$$

The cutoff frequency occurs at

$$\omega_c = \frac{1}{\tau} = \frac{1}{\dfrac{L}{R_1 + R_2}}$$

$$\omega_c = \frac{R_1 + R_2}{L} = \frac{10\ \text{k}\Omega}{2\ \text{mH}}$$

$$= 5.0\ \text{Mrad/s}$$

The resulting Bode plot is shown in Figure 22–24. Notice that the frequency response of the phase shift is precisely the same as for other low-pass filters. However, the response of the voltage gain now starts at -20 dB and then drops at a rate of -20 dB/decade above the cutoff frequency, $\omega_c = 5$ Mrad/s.

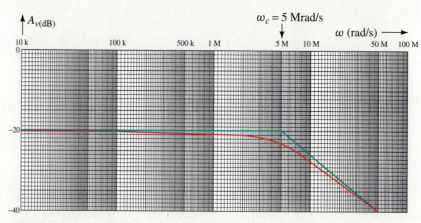

FIGURE 22–24   (Continued)

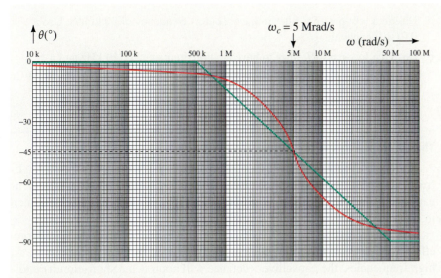

FIGURE 22–24   (Continued)

Refer to the low-pass circuit of Figure 22–25:

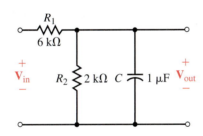

FIGURE 22–25

◖ MULTISIM

a. Write the transfer function for the circuit.

b. Sketch the frequency response.

Answers

a. $\mathbf{TF}(\omega) = \dfrac{0.25}{1 + j\omega 0.0015}$

b.

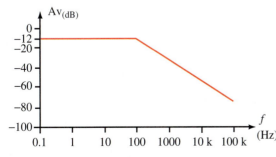

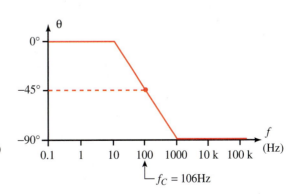

FIGURE 22–26

a. Design a low-pass *RC* filter to have a cutoff frequency of 30 krad/s. Use a 0.01-μF capacitor.

b. Design a low-pass *RL* filter to have a cutoff frequency of 20 kHz and a dc gain of −6 dB. Use a 10-mH inductor. (Assume that the inductor has no internal resistance.)

22.5 The High-Pass Filter

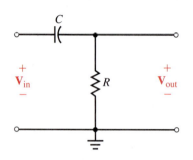

FIGURE 22–27   *RC* high-pass filter.

The *RC* High-Pass Filter

As the name implies, the high-pass filter is a circuit that allows high-frequency signals to pass from the input to the output of the circuit while attenuating low-frequency signals. A simple *RC* high-pass filter circuit is illustrated in Figure 22–27.

At low frequencies, the reactance of the capacitor will be very large, effectively preventing any input signal from passing through to the output. At high frequencies, the capacitive reactance will approach a short-circuit condition, providing a very low impedance path for the signal from the input to the output.

The transfer function of the low-pass filter is determined as follows:

$$\mathbf{TF} = \frac{\mathbf{V}_{out}}{\mathbf{V}_{in}} = \frac{\mathbf{R}}{\mathbf{R} + \mathbf{Z}_C}$$

$$= \frac{R}{R + \dfrac{1}{j\omega C}} = \frac{R}{\dfrac{j\omega RC + 1}{j\omega C}} = \frac{j\omega RC}{1 + j\omega RC}$$

Now, if we let $\omega_c = 1/\tau = 1/RC$, we have

$$\mathbf{TF} = \frac{j\dfrac{\omega}{\omega_c}}{1 + j\dfrac{\omega}{\omega_c}} \tag{22–17}$$

Notice that the expression of Equation 22–17 is very similar to the expression for a low-pass filter, with the exception that there is an additional term in the numerator. Since the transfer function is a complex number which is dependent upon frequency, we may once again find the general expressions for voltage gain and phase shift as functions of frequency, ω.

The voltage gain is found as

$$A_v = \frac{\dfrac{\omega}{\omega_c}}{\sqrt{1 + \left(\dfrac{\omega}{\omega_c}\right)^2}}$$

which, when expressed in decibels becomes

$$A_{v(dB)} = 20\log\frac{\omega}{\omega_c} - 10\log\left[1 + \left(\frac{\omega}{\omega_c}\right)^2\right] \tag{22–18}$$

The phase shift of the numerator will be a constant 90°, since the term has only an imaginary component. The overall phase shift of the transfer function is then found as follows:

$$\theta = 90° - \arctan\frac{\omega}{\omega_c} \qquad\qquad \textbf{(22–19)}$$

In order to sketch the asymptotic response of the voltage gain we need to examine the effect of Equation 22–18 on frequencies around the cutoff frequency, ω_c.

For frequencies $\omega \leq 0.1\omega_c$, the second term of the expression will be essentially equal to zero, and so the voltage gain at low frequencies is approximated as

$$A_{v(dB)} \cong 20 \log \frac{\omega}{\omega_c}$$

If we substitute some arbitrary values of ω into the above approximation, we arrive at a general statement. For example, by letting $\omega = 0.01\omega_c$, we have the voltage gain as

$$A_v = 20 \log(0.01) = -40 \text{ dB}$$

and by letting $\omega = 0.1\omega_c$, we have

$$A_v = 20 \log(0.1) = -20 \text{ dB}$$

In general, we see that the expression $A_v = 20 \log (\omega/\omega_c)$ may be represented as a straight line on a semilogarithmic graph. The straight line will intersect the 0-dB axis at the cutoff frequency, ω_c, and have a slope of $+20$ dB/decade.

For frequencies $\omega \gg \omega_c$, Equation 22–18 may be expressed as

$$A_{v(dB)} \cong 20 \log\frac{\omega}{\omega_c} - 10 \log\left[\left(\frac{\omega}{\omega_c}\right)^2\right]$$

$$= 20 \log\frac{\omega}{\omega_c} - 20 \log\frac{\omega}{\omega_c}$$

$$= 0 \text{ dB}$$

For the particular case when $\omega = \omega_c$, we have

$$A_{v(dB)} = 20 \log 1 - 10 \log 2 = -3.0 \text{ dB}$$

which is exactly the same result that we would expect, since the actual response will be 3 dB down from the asymptotic response.

Examining Equation 22–19 for frequencies $\omega \leq 0.1\omega_c$, we see that the phase shift for the transfer function will be essentially constant at 90°, while for frequencies $\omega \geq 10\omega_c$ the phase shift will be approximately constant at 0°. At $\omega = \omega_c$ we have $\theta = 90° - 45° = 45°$.

Figure 22–28 shows the normalized Bode plot of the high-pass circuit of Figure 22–27.

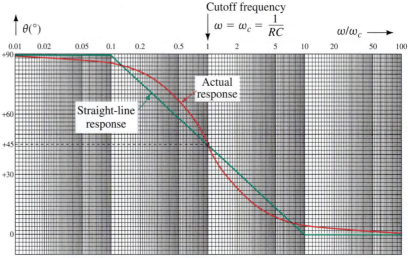

Cutoff frequency
$$\omega = \omega_c = \frac{1}{RC}$$

FIGURE 22–28   Normalized frequency response for an *RC* high-pass filter.

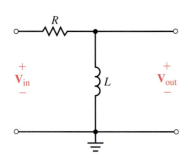

FIGURE 22–29   *RL* high-pass filter.

The *RL* High-Pass Filter

A typical *RL* high-pass filter circuit is shown in Figure 22–29.

At low frequencies, the inductor is effectively a short circuit, which means that the output of the circuit is essentially zero at low frequencies. Inversely, at high frequencies, the reactance of the inductor approaches infinity and greatly exceeds the resistance, effectively preventing current. The voltage across the inductor is therefore very nearly equal to the applied input voltage signal. The transfer function for the high-pass *RL* circuit is derived as follows:

$$\mathbf{TF} = \frac{\mathbf{Z}_L}{\mathbf{R} + \mathbf{X}_L}$$

$$= \frac{j\omega L}{R + j\omega L} = \frac{j\omega \dfrac{L}{R}}{1 + j\omega \dfrac{L}{R}}$$

Now, letting $\omega_c = 1/\tau = R/L$, we simplify the expression as

$$\mathbf{TF} = \frac{j\omega\tau}{1 + j\omega\tau}$$

The above expression is identical to the transfer function for a high-pass *RC* filter, with the exception that in this case we have $\tau = L/R$.

Design the *RL* high-pass filter circuit of Figure 22–30 to have a cutoff frequency of 40 kHz. (Assume that the inductor has no internal resistance.) Sketch the frequency response of the circuit expressing the frequencies in kilohertz.

EXAMPLE 22–10

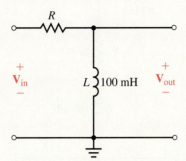

FIGURE 22–30

Solution   The cutoff frequency, ω_c, in radians per second is

$$\omega_c = 2\pi f_c = 2\pi(40 \text{ kHz}) = 251.33 \text{ krad/s}$$

Now, since $\omega_c = R/L$, we have

$$R = \omega_c L = (251.33 \text{ krad/s}(100 \text{ mH}) = 25.133 \text{ k}\Omega$$

The resulting Bode plot is shown in Figure 22–31.

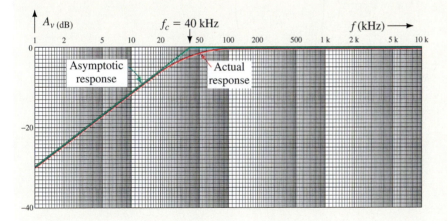

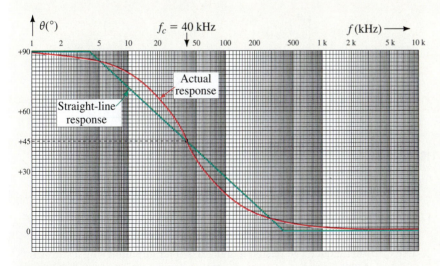

FIGURE 22–31

PRACTICE PROBLEMS 6

Consider the high-pass circuit of Figure 22–32:

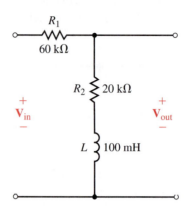

MULTISIM

FIGURE 22–32

a. Write the transfer function of the filter.
b. Sketch the frequency response of the filter. Show the frequency in radians per second. (Hint: The frequency response of the filter is a step response.)

Answers

a. $\text{TF}(\omega) = \left(\dfrac{R_2}{R_1 + R_2}\right)\left(\dfrac{1 + j\omega\dfrac{L}{R_2}}{1 + j\omega\dfrac{L}{R_1 + R_2}}\right)$

$f_1 = 31.8$ kHz (200 krad/s), $f_2 = 127$ kHz (800 krad/s)

b.

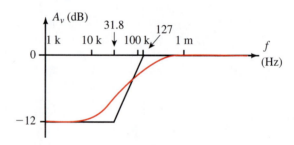

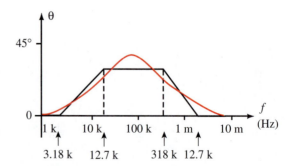

FIGURE 22–33

IN-PROCESS
LEARNING CHECK 3

(Answers are at the end of the chapter.)

1. Use a 0.05-μF capacitor to design a high-pass filter having a cutoff frequency of 25 kHz. Sketch the frequency response of the filter.

2. Use a 25-mH inductor to design a high-pass filter circuit having a cutoff frequency of 80 krad/s and a high-frequency gain of −12 dB. Sketch the frequency response of the filter.

A band-pass filter will permit frequencies within a certain range to pass from the input of a circuit to the output. All frequencies which fall outside the desired range will be attenuated and so will not appear with appreciable power at the output. Such a filter circuit is easily constructed by using a low-pass filter cascaded with a high-pass circuit as illustrated in Figure 22–34.

Although the low-pass and high-pass blocks may consist of various combinations of elements, one possibility is to construct the entire filter network from resistors and capacitors as shown in Figure 22–35.

The bandwidth of the resulting band-pass filter will be approximately equal to the difference between the two cutoff frequencies, namely,

$$BW \cong \omega_2 - \omega_1 \text{ (rad/s)} \qquad (22\text{–}20)$$

The above approximation will be most valid if the cutoff frequencies of the individual stages are separated by at least one decade.

22.6 The Band-Pass Filter

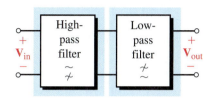

FIGURE 22–34 Block diagram of a band-pass filter.

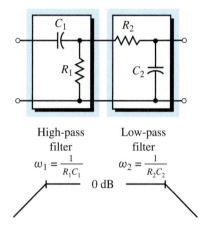

FIGURE 22–35

Write the transfer function for the circuit of Figure 22–36. Sketch the resulting Bode plot and determine the expected bandwidth for the band-pass filter.

EXAMPLE 22–11

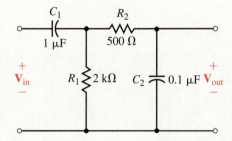

FIGURE 22–36

Solution While the transfer function for the circuit may be written by using circuit theory, it is easier to recognize that the circuit consists of two stages: one a low-pass stage and the other a high-pass stage. If the cutoff frequencies of each stage are separated by more than one decade, then we may assume that the impedance of one stage will not adversely affect the operation of the other stage. (If this is not the case, the analysis is complicated and is outside the scope of this textbook.) Based on the previous assumption, the transfer function of the first stage is determined as

$$\mathbf{TF_1} = \frac{\mathbf{V_1}}{\mathbf{V_{in}}} = \frac{j\omega R_1 C_1}{1 + j\omega R_1 C_1}$$

and for the second stage as

$$\mathbf{TF_2} = \frac{\mathbf{V_{out}}}{\mathbf{V_1}} = \frac{1}{1 + j\omega R_2 C_2}$$

Combining the above results, we have

$$\mathbf{TF} = \frac{\mathbf{V_{out}}}{\mathbf{V_{in}}} = \frac{(\mathbf{TF_2})(\mathbf{V_1})}{\dfrac{\mathbf{V_1}}{\mathbf{TF_1}}} = \mathbf{TF_1 TF_2}$$

which, when simplified, becomes

$$\mathbf{TF} = \frac{\mathbf{V_{out}}}{\mathbf{V_{in}}} = \frac{j\omega\tau_1}{(1 + j\omega\tau_1)(1 + j\omega\tau_2)} \qquad (22\text{--}21)$$

where $\tau_1 = R_1 C_1 = 2.0$ ms and $\tau_2 = R_2 C_2 = 50$ μs. The corresponding cutoff frequencies are $\omega_1 = 500$ rad/s and $\omega_2 = 20$ krad/s. The transfer function of Equation 22–21 has three separate terms which, when taken separately, result in the approximate responses illustrated in Figure 22–37.

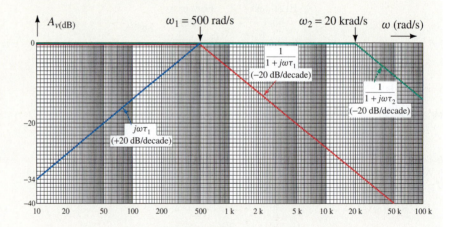

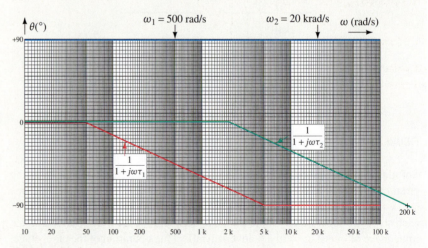

FIGURE 22–37

The resulting frequency response is determined by the summation of the individual responses as shown in Figure 22–38.

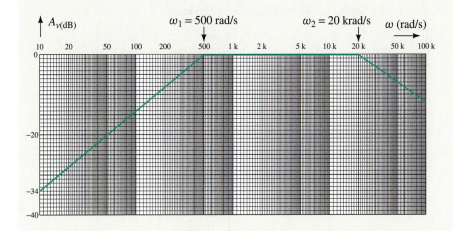

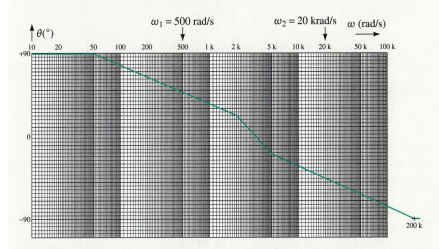

FIGURE 22–38

From the Bode plot, we determine that the bandwidth of the resulting filter is

$$BW(\text{rad/s}) = \omega_2 - \omega_1 = 20 \text{ krad/s} - 0.5 \text{ krad/s} = 19.5 \text{ krad/s}$$

Refer to the band-pass filter of Figure 22–39:

PRACTICE PROBLEMS 7

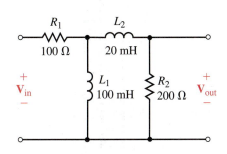

FIGURE 22–39

◀ MULTISIM

a. Calculate the cutoff frequencies in rad/s and the approximate bandwidth.

b. Sketch the frequency response of the filter.

Answers

a. $\omega_1 = 1.00$ krad/s, $\omega_2 = 10.0$ krad/s, BW = 9.00 krad/s

b.

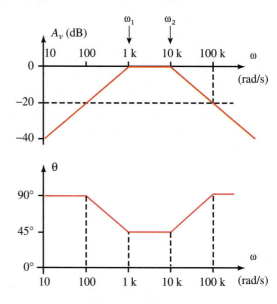

FIGURE 22–40

IN-PROCESS
LEARNING CHECK 4

(Answers are at the end of the chapter.)

Given a 0.1-μF capacitor and a 0.04-μF capacitor, design a band-pass filter having a bandwidth of 30 krad/s and a lower cutoff frequency of 5 krad/s. Sketch the frequency response of the voltage gain and the phase shift.

22.7 The Band-Reject Filter

The band-reject filter has a response that is opposite to that of the band-pass filter. This filter passes all frequencies with the exception of a narrow band which is greatly attenuated. A band-reject filter constructed of a resistor, inductor, and capacitor is shown in Figure 22–41.

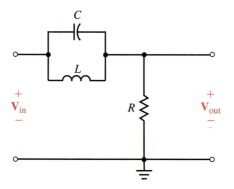

FIGURE 22–41   Notch filter.

Notice that the circuit uses a resonant tank circuit as part of the overall design. As we saw in the previous chapter, the combination of the inductor and the capacitor results in a very high tank impedance at the resonant frequency. Therefore, for any signals occurring at the resonant frequency, the output voltage is effectively zero. Because the filter circuit effectively removes any signal occurring at the resonant frequency, the circuit is often referred to as a **notch filter.** The voltage gain response of the notch filter is shown in Figure 22–42.

For low-frequency signals, the inductor provides a low-impedance path from the input to the output, allowing these signals to pass from the input and appear across the resistor with minimal attenuation. Conversely, at high frequencies the capacitor provides a low-impedance path from the input to the output. Although the complete analysis of the notch filter is outside the scope of this textbook, the transfer function of the filter is determined by employing the same techniques as those previously developed.

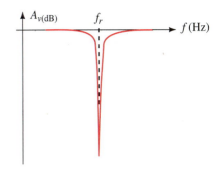

FIGURE 22–42 Voltage gain response for a notch filter.

$$\mathbf{TF} = \frac{\mathbf{R}}{\mathbf{R} + \mathbf{Z}_L \| \mathbf{Z}_C}$$

$$= \frac{R}{R + \dfrac{(j\omega L)\left(\dfrac{1}{j\omega C}\right)}{j\omega L + \dfrac{1}{j\omega C}}}$$

$$= \frac{R}{R + \dfrac{j\omega L}{1 - \omega^2 LC}}$$

$$= \frac{R(1 - \omega^2 LC)}{R - \omega^2 RLC + j\omega L}$$

$$\mathbf{TF} = \frac{1 - \omega^2 LC}{1 - \omega^2 LC + j\omega \dfrac{L}{R}} \qquad (22\text{–}22)$$

Notice that the transfer function for the notch filter is significantly more complicated than for the previous filter circuits. Due to the presence of the complex quadratic in the denominator of the transfer function, this type of filter circuit is called a second-order filter. The design of such filters is a separate field in electronics engineering.

Actual filter design often involves using operational amplifiers to provide significant voltage gain in the passband of the filter. In addition, such active filters have the advantage of providing very high input impedance to prevent loading effects. Many excellent textbooks are available for assistance in filter design.

Despite the complexity of designing a filter for a particular application, the analysis of the filter is a relatively simple process when done by computer. We have already seen the ease with which PSpice may be used to examine the frequency response of resonant circuits. In this chapter we will again use the Probe postprocessor of PSpice to plot the frequency characteristics of a particular circuit. We find that with some minor adjustments, the program is able to simultaneously plot both the voltage gain (in decibels) and the phase shift (in degrees) of any filter circuit.

MultiSIM provides displays that are similar to those obtained in PSpice. The method, though, is somewhat different. Since MultiSIM simulates actual lab measurements, an instrument called a Bode plotter is used by the software. As one might expect, the Bode plotter provides a graph of the frequency response (showing both the gain and phase shift) of a circuit even though no such instrument is found in a real electronics lab.

22.8 Circuit Analysis Using Computers

◀ MULTISIM

◀ CADENCE

PSpice

EXAMPLE 22–12

Use the PROBE postprocessor of PSpice to view the frequency response from 1 Hz to 100 kHz for the circuit of Figure 22–36. Determine the cutoff frequencies and the bandwidth of the circuit. Compare the results to those obtained in Example 22–11.

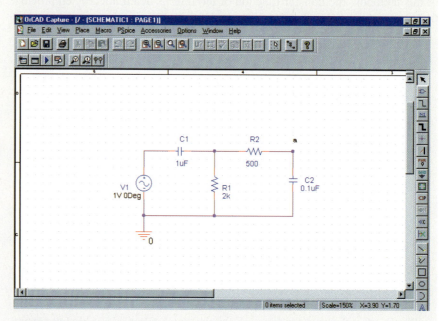

FIGURE 22–43

Solution   The OrCAD Capture program is used to enter the circuit as shown in Figure 22–43. The analysis is set up to perform an ac sweep from 1 Hz to 100 kHz using 1001 points per decade. The signal generator is set for a magnitude of 1 V and a phase shift of 0°.

In this example, we show both the voltage gain and the phase shift on the same display. Once the PROBE screen is activated, two simultaneous displays are obtained by clicking on Plot/Add Plot to Window. We will use the top display to show voltage gain and the bottom to show the phase shift.

PSpice does not actually calculate the voltage gain in decibels, but rather determines the output voltage level in dBV, referenced to 1 V_{rms}. (It is for this reason that we used a supply voltage of 1 V.) The voltage at point *a* of the circuit is obtained by clicking on Trace/Add Trace and then selecting **DB(V(C2:1))** as the Trace Expression. Cursors are obtained by clicking on Tools/Cursor/Display. The maximum of the function is found by clicking on Tools/Cursor/Max. The cursor indicates that the maximum gain for the circuit is -1.02 dB. We find the bandwidth of the circuit (at the -3 dB frequencies) by moving the cursors (arrow keys and <Ctrl> arrow keys) to the frequencies at which the output of the circuit is at -4.02 dB. We determine $f_1 = 0.069$ kHz, $f_2 = 3.67$ kHz, and BW $= 3.61$ kHz. These results are consistent with those found in Example 22–11.

Finally, we obtain a trace of the phase shift for the circuit as follows. Click anywhere on the bottom plot. Click on <u>T</u>race/<u>A</u>dd Trace and then select **P(V(C2:1))** as the <u>T</u>race Expression. The range of the ordinate is changed by clicking on the axis, selecting the Y Axis tab, and setting the <u>U</u>ser defined range for a value of −**90d** to **90d.** The resulting display is shown in Figure 22–44.

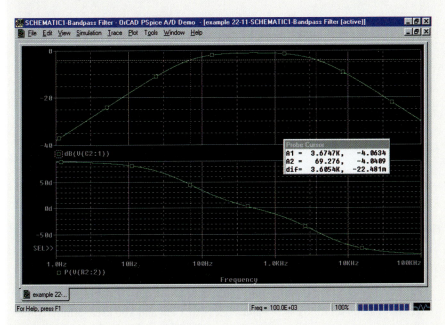

FIGURE 22–44

---

a. Use PSpice to input the circuit of Figure 22–39.

b. Use the Probe postprocessor to observe the frequency response from 1 Hz to 100 kHz.

c. From the display, determine the cutoff frequencies and use the cursors to determine the bandwidth.

d. Compare the results to those obtained in Practice Problem 7.

EXAMPLE 22–13

Use MultiSIM to obtain the frequency response for the circuit of Figure 22–36. Compare the results to those obtained in Example 22–12.

Solution   In order to perform the required measurements, we need to use the function generator and the Bode plotter, both located in the Instruments parts bin. The circuit is constructed as shown in Figure 22–45.

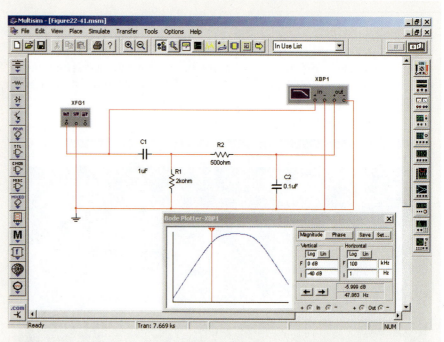

MULTISIM

FIGURE 22–45

The Bode plotter is adjusted to provide the desired frequency response by first double clicking on the instrument. Next, we click on the Magnitude button. The Vertical scale is set to log with values between −**40** dB and **0** dB. The Horizontal scale is set to log with values between **1** Hz and **100** kHz. Similarly, the Phase is set to have a Vertical range of −**90**° to **90**°. After clicking the run button, the Bode plotter provides a display of either the voltage gain response or the phase response. However, both displays are shown simultaneously by typing **Ctrl G.** By using the cursor feature, we obtain the same results as those found in Example 22–12. Figure 22–46 shows the frequency response as viewed using the Analysis Graphs window.

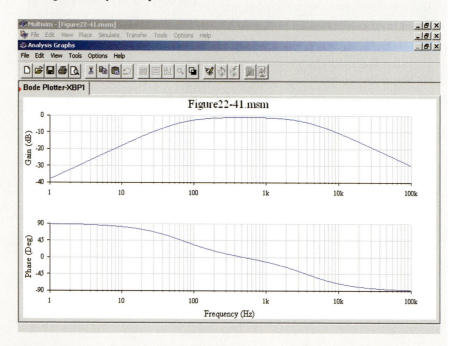

FIGURE 22–46

Use MultiSIM to obtain the frequency response for the circuit of Figure 22–25. Compare the results to those obtained in Practice Problem 5.

PUTTING IT INTO PRACTICE

As a designer for a sound studio, you have been asked to design band-pass filters for a color organ that will be used to provide lighting for a rock concert. The color organ will provide stage lighting that will correspond to the sound level and frequency of the music. You remember from your physics class that the human ear can perceive sounds from 20 Hz to 20 kHz.

The specifications say that the audio frequency spectrum is to be divided into three ranges. Passive *RC* filters will be used to isolate signals for each range. These signals will then be amplified and used to control lights of a particular color. The low-frequency components (20 to 200 Hz) will control blue lights, the mid-frequency components (200 Hz to 2 kHz) will control green lights, and the high-frequency components (2 kHz to 20 kHz) will control red lights.

Although the specifications call for 3 band-pass filters, you realize that you can simplify the design by using a low-pass filter with a break frequency of 200 Hz for the low frequencies and a high-pass filter with a break frequency of 2 kHz for the high frequencies. To simplify your work, you decide to use only 0.5-μF capacitors for all filters.

Show the design for each of the filters.

PROBLEMS

22.1 The Decibel

1. Refer to the amplifier shown in Figure 22–47. Determine the power gain both as a ratio and in decibels for the following power values.

   a. $P_{in} = 1.2$ mW, $P_{out} = 2.4$ W
   b. $P_{in} = 3.5$ μW, $P_{out} = 700$ mW
   c. $P_{in} = 6.0$ pW, $P_{out} = 12$ μW
   d. $P_{in} = 2.5$ mW, $P_{out} = 1.0$ W

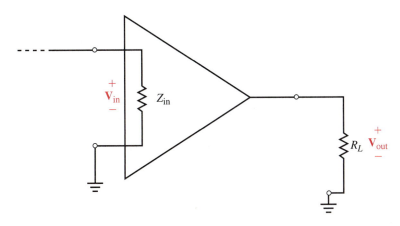

FIGURE 22–47

2. If the amplifier of Figure 22–47 has $Z_{in} = 600\ \Omega$ and $R_L = 2\ k\Omega$, find P_{in}, P_{out}, and $A_P(dB)$ for the following voltage levels:

   a. $V_{in} = 20$ mV, $V_{out} = 100$ mV

   b. $V_{in} = 100\ \mu$V, $V_{out} = 400\ \mu$V

   c. $V_{in} = 320$ mV, $V_{out} = 600$ mV

   d. $V_{in} = 2\ \mu$V, $V_{out} = 8$ V

3. The amplifier of Figure 22–47 has $Z_{in} = 2\ k\Omega$ and $R_L = 10\ \Omega$. Find the voltage gain and power gain both as a ratio and in dB for the following conditions:

   a. $V_{in} = 2$ mV, $P_{out} = 100$ mW

   b. $P_{in} = 16\ \mu$W, $V_{out} = 40$ mV

   c. $V_{in} = 3$ mV, $P_{out} = 60$ mW

   d. $P_{in} = 2$ pW, $V_{out} = 80$ mV

4. The amplifier of Figure 22–47 has an input voltage of $V_{in} = 2$ mV and an output power of $P_{out} = 200$ mW. Find the voltage gain and power gain both as a ratio and in dB for the following conditions:

   a. $Z_{in} = 5\ k\Omega$, $R_L = 2\ k\Omega$

   b. $Z_{in} = 2\ k\Omega$, $R_L = 10\ k\Omega$

   c. $Z_{in} = 300\ k\Omega$, $R_L = 1\ k\Omega$

   d. $Z_{in} = 1\ k\Omega$ $R_L = 1\ k\Omega$

5. The amplifier of Figure 22–47 has an input impedance of $5\ k\Omega$ and a load resistance of 250Ω. If the power gain of the amplifier is 35 dB, and the input voltage is 250 mV, find P_{in}, P_{out}, V_{out}, A_v, and $A_v(dB)$.

6. Repeat Problem 5 if the input impedance is increased to $10\ k\Omega$. (All other quantities remain unchanged.)

7. Express the following powers in dBm and in dBW:

   a. $P = 50$ mW

   b. $P = 1$ W

   c. $P = 400$ nW

   d. $P = 250$ pW

8. Express the following powers in dBm and in dBW.

   a. $P = 250$ W

   b. $P = 250$ kW

   c. $P = 540$ nW

   d. $P = 27$ mW

9. Convert the following power levels into watts:

   a. $P = 23.5$ dBm

   b. $P = -45.2$ dBW

   c. $P = -83$ dBm

   d. $P = 33$ dBW

10. Convert the following power levels into watts:

    a. $P = 16$ dBm

    b. $P = -43$ dBW

    c. $P = -47.3$ dBm

    d. $P = 29$ dBW

11. Express the following rms voltages as voltage levels (in dBV):

    a. 2.00 V

    b. 34.0 mV

    c. 24.0 V

    d. 58.2 μV

12. Express the following rms voltages as voltage levels (in dBV):

    a. 25 μV

    b. 90 V

    c. 72.5 mV

    d. 0.84 V

13. Convert the following voltage levels from dBV to rms voltages:

    a. −2.5 dBV

    b. 6.0 dBV

    c. −22.4 dBV

    d. 10.0 dBV

14. Convert the following voltage levels from dBV to rms voltages:

    a. 20.0 dBV

    b. −42.0 dBV

    c. −6.0 dBV

    d. 3.0 dBV

15. A sinusoidal waveform is measured as 30.0 $V_{p\text{-}p}$ with an oscilloscope. If this waveform were applied to a voltmeter calibrated to express readings in dBV, what would the voltmeter indicate?

16. A voltmeter shows a reading of 9.20 dBV. What peak-to-peak voltage would be observed on an oscilloscope?

22.2   Multistage Systems

17. Calculate the power levels (in dBm) at the output of each of the stages of the system shown in Figure 22–48. Solve for the output power (in watts).

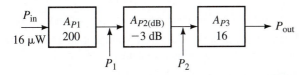

FIGURE 22–48

18. Calculate the power levels (in dBm) at the indicated locations of the system shown in Figure 22–49. Solve for the input and output powers (in watts).

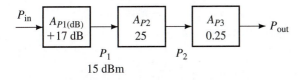

FIGURE 22–49

19. Given that power $P_2 = 140$ mW as shown in Figure 22–50. Calculate the power levels (in dBm) at each of the indicated locations. Solve for the voltage across the load resistor, R_L.

20. Suppose that the system of Figure 22–51 has an output voltage of 2 V:

    a. Determine the power (in watts) at each of the indicated locations.

    b. Solve for the voltage, V_{in}, if the input impedance of the first stage is 1.5 kΩ.

    c. Convert V_{in} and V_L into voltage levels (in dBV).

    d. Solve for the voltage gain, A_v (in dB).

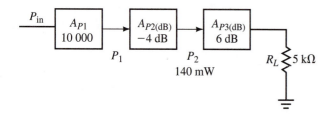

FIGURE 22–50

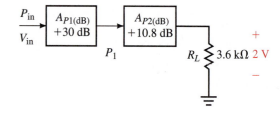

FIGURE 22–51

21. A power amplifier (P.A.) with a power gain of 250 has an input impedance of 2.0 kΩ and is used to drive a stereo speaker (output impedance of 8.0 Ω). If the output power is 100 W, determine the following:

    a. Output power level (dBm), input power level (dBm)

    b. Output voltage (rms), input voltage (rms)

    c. Output voltage level (dBV), input voltage level (dBV)

    d. Voltage gain in dB

22. Repeat Problem 21 if the amplifier has a power gain of 400 and $Z_{in} = 1.0$ kΩ. The power delivered to the 8.0-Ω speaker is 200 W.

22.3   Simple *RC* and *RL* Transfer Functions

23. Given the transfer function

$$\mathbf{TF} = \frac{200}{1 + j0.001\omega}$$

    a. Determine the cutoff frequency in radians per second and in hertz.

    b. Sketch the frequency response of the voltage gain and the phase shift responses. Label the abscissa in radians per second.

24. Repeat Problem 23 for the transfer function

$$\mathbf{TF} = \frac{1 + j0.001\omega}{200}$$

25. Repeat Problem 23 for the transfer function

$$\mathbf{TF} = \frac{1 + j0.02\omega}{1 + j0.001\omega}$$

26. Repeat Problem 23 for the transfer function

$$\mathbf{TF} = \frac{1 + j0.04\omega}{(1 + j0.004\omega)(1 + j0.001\omega)}$$

27. Repeat Problem 23 for the transfer function

$$\mathbf{TF}(\omega) = \frac{j0.02\omega}{1 + j0.02\omega}$$

28. Repeat Problem 23 for the transfer function

$$\mathbf{TF}(\omega) = \frac{j0.01\omega}{1 + j0.005\omega}$$

22.4   The Low-Pass Filter

29. Use a 4.0-µF capacitor to design a low-pass filter circuit having a cutoff frequency of 5 krad/s. Draw a schematic of your design and sketch the frequency response of the voltage gain and the phase shift.

30. Use a 1.0-µF capacitor to design a low-pass filter with a cutoff frequency of 2500 Hz. Draw a schematic of your design and sketch the frequency response of the voltage gain and the phase shift.

31. Use a 25-mH inductor to design a low-pass filter with a cutoff frequency of 50 krad/s. Draw a schematic of your design and sketch the frequency response of the voltage gain and the phase shift.

32. Use a 100-mH inductor (assume $R_{coil} = 0\ \Omega$) to design a low-pass filter circuit having a cutoff frequency of 15 kHz. Draw a schematic of your design and sketch the frequency response of the voltage gain and the phase shift.

33. Use a 36-mH inductor to design a low-pass filter having a cutoff frequency of 36 kHz. Draw a schematic of your design and sketch the frequency response of the voltage gain and the phase shift.

34. Use a 5-µF capacitor to design a low-pass filter circuit having a cutoff frequency of 100 krad/s. Draw a schematic of your design and sketch the frequency response of the voltage gain and the phase shift.

35. Refer to the low-pass circuit of Figure 22–52:

   a. Write the transfer function for the circuit.

   b. Sketch the frequency response of the voltage gain and phase shift.

36. Repeat Problem 35 for the circuit of Figure 22–53.

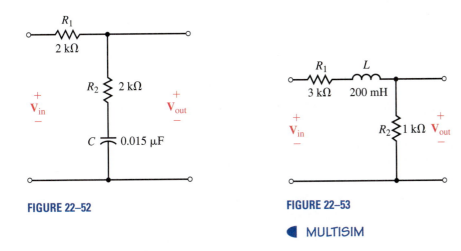

FIGURE 22–52

FIGURE 22–53

◀ MULTISIM

22.5   The High-Pass Filter

37. Use a 0.05-µF capacitor to design a high-pass filter to have cutoff frequency of 100 krad/s. Draw a schematic of your design and sketch the frequency response of the voltage gain and the phase shift.

38. Use a 2.2-nF capacitor to design a high-pass filter to have a cutoff frequency of 5 kHz. Draw a schematic of your design and sketch the frequency response of the voltage gain and phase shift.

39. Use a 2-mH inductor to design a high-pass filter to have a cutoff frequency of 36 krad/s. Draw a schematic of your design and sketch the frequency response of the voltage gain and phase shift.

40. Use a 16-mH inductor to design a high-pass filter circuit having a cutoff frequency of 250 kHz. Draw a schematic of your design and sketch the frequency response of the voltage gain and the phase shift.

41. Refer to the high-pass circuit of Figure 22–54.

    a. Write the transfer function for the circuit.

    b. Sketch the frequency response of the voltage gain and phase shift.

42. Repeat Problem 41 for the high-pass circuit of Figure 22–55.

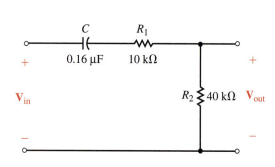

FIGURE 22–54

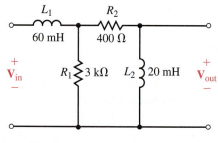

FIGURE 22–55

22.6   The Band-Pass Filter

43. Refer to the filter of Figure 22–56.

    a. Determine the approximate cutoff frequencies and bandwidth of the filter. (Assume that the two stages of the filter operate independently.)

    b. Sketch the frequency response of the voltage gain and the phase shift.

44. Repeat Problem 43 for the circuit of Figure 22–57.

45. a. Use two 0.01-μF capacitors to design a band-pass filter to have cutoff frequencies of 2 krad/s and 20 krad/s.

    b. Draw your schematic and sketch the frequency response of the voltage gain and the phase shift.

    c. Do you expect that the actual cutoff frequencies will occur at the designed cutoff frequencies? Explain.

46. a. Use two 10-mH inductors to design a band-pass filter to have cutoff frequencies of 25 krad/s and 40 krad/s.

    b. Draw your schematic and sketch the frequency response of the voltage gain and the phase shift.

    c. Do you expect that the actual cutoff frequencies will occur at the designed cutoff frequencies? Explain.

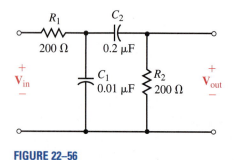

FIGURE 22–56

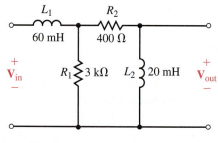

FIGURE 22–57

22.7   The Band-Reject Filter

47. Given the filter circuit of Figure 22–58:

   a. Determine the "notch" frequency.

   b. Calculate the Q of the circuit.

   c. Solve for the bandwidth and determine the half-power frequencies.

   d. Sketch the voltage gain response of the circuit, showing the level (in dB) at the "notch" frequency.

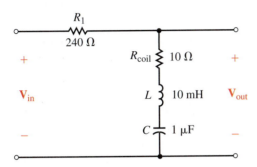

FIGURE 22–58   ◀ MULTISIM

48. Repeat Problem 47 for the circuit of Figure 22–59.

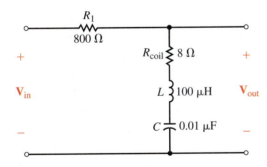

FIGURE 22–59   ◀ MULTISIM

49. Given the circuit of Figure 22–60:

   a. Determine the "notch" frequency in both rad/s and Hz.

   b. Solve for the Q of the circuit.

   c. Calculate the voltage gain (in dB) at the notch frequency.

   d. Determine the voltage gain at the boundary conditions

   e. Solve for the bandwidth of the notch filter and determine the frequencies at which the voltage gain is 3 dB higher than at the notch frequency.

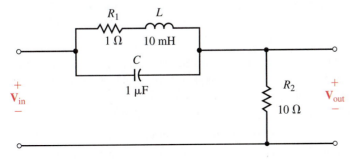

FIGURE 22–60   ◀ MULTISIM

50. Repeat Problem 49 for the circuit of Figure 22–61.

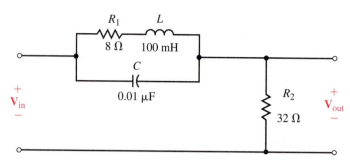

FIGURE 22–61

22.8 Circuit Analysis Using Computers

51. Use PSpice to input the circuit of Figure 22–62. Let the circuit sweep through frequencies of 100 Hz to 1 MHz. Use the Probe postprocessor to display the frequency response of voltage gain (in dBV) and phase shift of the circuit.

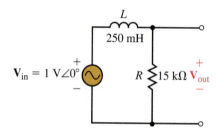

FIGURE 22–62

52. Repeat Problem 51 for the circuit shown in Figure 22–53.

53. Use PSpice to input the circuit of Figure 22–56. Use the Probe postprocessor to display the frequency response of voltage gain (in dBV) and phase shift of the circuit. Select a suitable range for the frequency sweep and use the cursors to determine the half-power frequencies and the bandwidth of the circuit.

54. Repeat Problem 53 for the circuit shown in Figure 22–57.

55. Repeat Problem 53 for the circuit shown in Figure 22–58.

56. Repeat Problem 53 for the circuit shown in Figure 22–59.

57. Repeat Problem 53 for the circuit shown in Figure 22–60.

58. Repeat Problem 53 for the circuit shown in Figure 22–61.

59. Use MultiSIM to obtain the frequency response for the circuit of Figure 22–62. Let the circuit sweep through frequencies of 100 Hz to 1 MHz.

60. Repeat Problem 59 for the circuit of Figure 22–53.

61. Use MultiSIM to obtain the frequency response for the circuit shown in Figure 22–58. Select a suitable frequency range and use cursors to determine the "notch" frequency and the bandwidth of the circuit.

62. Repeat Problem 61 for the circuit shown in Figure 22–59.

63. Repeat Problem 61 for the circuit shown in Figure 22–60.

64. Repeat Problem 61 for the circuit shown in Figure 22–61.

✓ ANSWERS TO IN-PROCESS LEARNING CHECKS

In-Process Learning Check 1

1. a. -1.99 dBm   b. -15.0 dBm   c. -66.0 dBm

2. a. 0.561 V$_{rms}$ b. 0.224 V$_{rms}$ c. 0.354 V$_{rms}$

In-Process Learning Check 2

a. $R = 3333$ Ω in series with $C = 0.01$ μF (output across C)

b. $R_1 = 630$ Ω in series with $L = 10$ mH and $R_2 = 630$ Ω (output across R_2)

In-Process Learning Check 3

1. $\mathbf{TF} = \dfrac{j\omega(6.36 \times 10^{-6})}{1 + j\omega(6.36 \times 10^{-6})}$

   $C = 0.05$ μF is in series with $R = 127.3$ Ω (output across R)

2. $\mathbf{TF} = \dfrac{j\omega(3.125 \times 10^{-6})}{1 + j\omega(12.5 \times 10^{-6})}$

   $R_1 = 8$ kΩ is in series with $L = 25$ mH$\|R_2 = 2.67$ kΩ (output across the parallel combination)

In-Process Learning Check 4

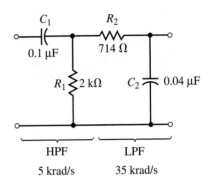

FIGURE 22–63

■ **OBJECTIVES**

After studying this chapter, you will be
able to

• describe how a transformer couples
  energy from its primary to its secondary via a changing magnetic
  field,

• describe basic transformer
  construction,

• use the dot convention to determine
  transformer phasing,

• determine voltage and current ratios
  from the turns ratio for iron-core
  transformers,

• compute voltage and currents in circuits containing iron-core and air-
  core transformers,

• use transformers to impedance
  match loads,

• describe some basic transformer
  applications,

• determine transformer equivalent
  circuits,

• compute iron-core transformer
  efficiency,

• use MultiSIM and PSpice to solve
  circuits with transformers and coupled circuits.

Transformers and Coupled Circuits

23

CHAPTER PREVIEW

In our study of induced voltage in Chapter 13 we found that the changing magnetic field produced by current in one coil induced a voltage in a second coil wound on the same core. A device built to utilize this effect is the **transformer.**

Transformers have many applications. They are used in electrical power systems to step up voltage for long-distance transmission and then to step it down again to a safe level for use in our homes and offices. They are used in electronic equipment power supplies to raise or lower voltages, in audio systems to match speaker loads to amplifiers, in telephone, radio, and TV systems to couple signals, and so on.

In this chapter, we look at transformer fundamentals and the analysis of circuits containing transformers. We discuss transformer action, types of transformers, voltage and current ratios, applications, and so on. Both iron-core and air-core transformers are covered. ■

PUTTING IT IN PERSPECTIVE

George Westinghouse

ONE OF THE DEVICES that made possible the commercial ac power system as we know it today is the transformer. Although Westinghouse did not invent the transformer, his acquisition of the transformer patent rights and his manufacturing business helped make him an important player in the battle of dc versus ac in the emerging electrical power industry (see Chapter 15). In concert with Tesla (see Chapter 24), Westinghouse fought vigorously for ac against Edison, who favored dc. In 1893, Westinghouse's company built the Niagara Falls power system using ac, and the battle was over with ac the clear winner. (Ironically, in recent years dc has been resurrected for use in commercial electrical power systems because at extremely high voltages, it is able to transmit power over longer distances than ac. However, this was not possible in Edison's day, and ac was and still is the correct choice for the commercial electrical power system.)

George Westinghouse was born in 1846 in Central Bridge, New York. He made his fortune with the invention of the railway air brake system. He died in 1914 and was elected to the Hall of Fame for Great Americans in 1955. ■

23.1 Introduction

A transformer is a magnetically **coupled circuit,** i.e., a circuit in which the magnetic field produced by time-varying current in one circuit induces voltage in another. To illustrate, a basic iron-core transformer is shown in Figure 23–1. It consists of two coils wound on a common core. Alternating current in one winding establishes a flux that links the other winding and induces a voltage in it. Power thus flows from one circuit to the other via the medium of the magnetic field, with no electrical connection between the two sides. The winding to which we supply power is called the **primary,** while the winding from which we take power is called the **secondary.** Power can flow in either direction, as either winding can be used as the primary or the secondary.

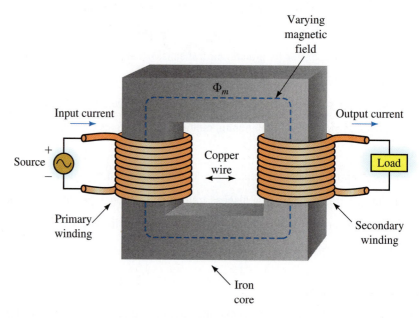

FIGURE 23–1   Basic iron-core transformer. Energy is transferred from the source to the load via the transformer's magnetic field with no electrical connection between the two sides.

Transformer Construction

Transformers fall into two broad categories, iron-core and air-core. We begin with **iron-core** types. Iron core transformers are generally used for low frequency applications such as audio- and power-frequency applications. Figures 23–2 and 23–3 show a few examples of iron-core transformers.

Iron (actually a special steel called transformer steel) is used for cores because it increases the coupling between coils by providing an easy path for magnetic flux. Two basic types of iron-core construction are used, the **core type** and the **shell type** (Figure 23–4). In both cases, cores are made from laminations of sheet steel, insulated from each other by thin coatings of ceramic or other material to help minimize eddy current losses (Section 23.6).

Iron, however, has considerable power loss due to hysteresis and eddy currents at high frequencies, and is thus not useful as a core material above about 50 kHz. For high-frequency applications (such as in radio circuits), **air-core** and **ferrite-core** types are used. Figure 23–5 shows a ferrite-core device. Ferrite (a magnetic material made from powdered iron oxide) greatly increases

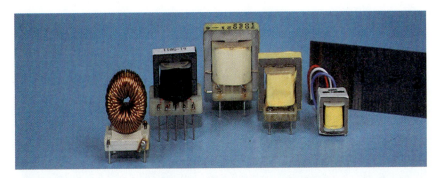

FIGURE 23–2   Iron-core transformers of the type used in electronic equipment. *(Courtesy Transformer Manufacturers Inc.)*

FIGURE 23–3   Distribution transformer (cutaway view) of the type used by electric utilities to distribute power to residential and commercial users. The tank is filled with oil to improve insulation and to remove heat from the core and windings. *(Courtesy Carte International Inc.)*

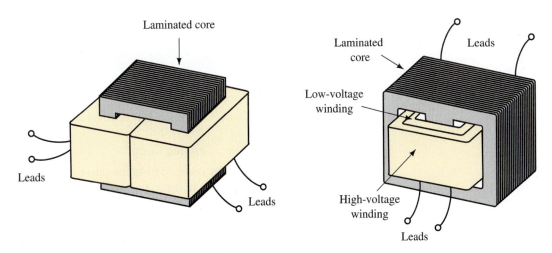

FIGURE 23–4   For the core type (left), windings are on separate legs, while for the shell type both windings are on the same leg. (Adapted with permission from Perozzo, *Practical Electronics Troubleshooting,* © 1985 Delmar Publishers Inc.)

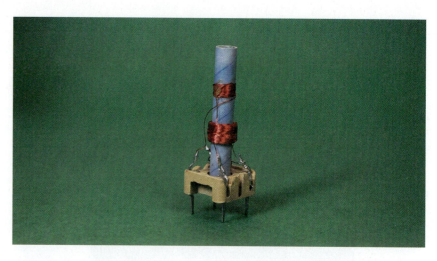

FIGURE 23–5   A slug-tuned ferrite-core transformer of the type used in radio circuits. Coupling between coils is varied by a ferrite slug inside the sleeve.

coupling between coils (compared with air) while maintaining low losses. Circuit symbols for transformers are shown in Figure 23–6.

Winding Directions

One of the advantages of a transformer is that it may be used to change the polarity of an ac voltage. This is illustrated in Figure 23–7 for a pair of iron-core transformers. For the transformer of (a) the primary and secondary voltages are in phase (for reasons to be discussed later), while for (b) they are 180° out of phase.

(a) Iron-core

(b) Air-core

(c) Ferrite-core

FIGURE 23–6   Transformer schematic symbols.

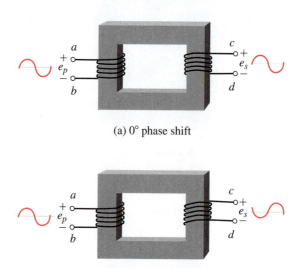

(a) 0° phase shift

(b) 180° phase shift

FIGURE 23–7   The relative direction of the windings determines the phase shift.

Tightly Coupled and Loosely Coupled Circuits

If most of the flux produced by one coil links the other, the coils are said to be **tightly coupled.** Thus, iron-core transformers are tightly coupled (since close to 100% of the flux is confined to the core and thus links both windings). For air- and ferrite-core transformers, however, much less than 100% of the flux links both windings. They are therefore **loosely coupled.** Since air-core and ferrite-core devices are loosely coupled, the same principles of analysis apply to both and we treat them together in Section 23.9.

Faraday's Law

All transformer operation is described by Faraday's law. Faraday's law (in SI units) states that the voltage induced in a circuit by a changing magnetic field is equal to the rate at which the flux linking the circuit is changing. When Faraday's law is applied to iron-core and air-core transformers, however, the results that emerge are quite different: Iron-core transformers are found to be characterized by their turns ratios, while air-core transformers are characterized by self- and mutual inductances. We begin with iron-core transformers.

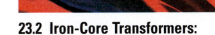

23.2 Iron-Core Transformers: The Ideal Model

At first glance, iron-core transformers appear quite difficult to analyze because they have several characteristics such as winding resistance, core loss, and leakage flux that appear difficult to handle. Fortunately, these effects are small and often can be neglected. The result is the **ideal transformer.** Once you know how to analyze an ideal transformer, however, it is relatively easy to go back and add in the nonideal effects. This is the approach we use here.

To idealize a transformer, (1) neglect the resistance of its coils, (2) neglect its core loss, (3) assume all flux is confined to its core, and (4) assume that negligible current is required to establish its core flux. (Well-designed iron-core power transformers are very close to this ideal.)

We now apply Faraday's law to the ideal transformer. Before we do this, however, we need to determine flux linkages. The flux linking a winding (as determined in Chapter 13) is the product of the flux that passes through the winding times the number of turns through which it passes. For flux Φ passing through N turns, flux linkage is $N\Phi$. Thus, for the ideal transformer (Figure 23–8), the primary flux linkage is $N_p\Phi_m$, while the secondary flux linkage is $N_s\Phi_m$, where the subscript "m" indicates mutual flux, i.e., flux that links both windings.

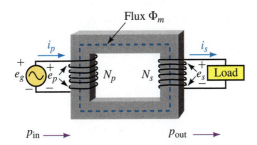

FIGURE 23–8   Ideal transformer. All flux is confined to the core and links both windings. This is a "tightly-coupled" transformer.

Voltage Ratio

Now apply Faraday's law. Since the flux linkage equals $N\Phi$ and since N is constant, the induced voltage is equal to N times the rate of change of Φ, i.e., $e = Nd\Phi/dt$. Thus, for the primary,

$$e_p = N_p\frac{d\Phi_m}{dt} \qquad (23\text{–}1)$$

while for the secondary

$$e_s = N_s\frac{d\Phi_m}{dt} \qquad (23\text{–}2)$$

Dividing Equation 23–1 by Equation 23–2 and cancelling $d\Phi_m/dt$ yields

$$\frac{e_p}{e_s} = \frac{N_p}{N_s} \tag{23–3}$$

Equation 23–3 states that *the ratio of primary voltage to secondary voltage is equal to the ratio of primary turns to secondary turns*. This ratio is called the **transformation ratio** (or **turns ratio**) and is given the symbol a. Thus,

$$a = N_p/N_s \tag{23–4}$$

and

$$e_p/e_s = a \tag{23–5}$$

For example, a transformer with 1000 turns on its primary and 250 turns on its secondary has a turns ratio of $1000/250 = 4$. This is referred to as a 4:1 ratio.

Since the ratio of two instantaneous sinusoidal voltages is the same as the ratio of their effective values, Equation 23–5 can also be written as

$$E_p/E_s = a \tag{23–6}$$

As noted earlier, e_p and e_s are either in phase or 180° out of phase, depending on the relative direction of coil windings. We can therefore also express the **voltage ratio** in terms of phasors as

$$\mathbf{E}_p/\mathbf{E}_s = a \tag{23–7}$$

where the relative polarity (in phase or 180° out of phase) is determined by the direction of the coil windings (Figure 23–7).

Step-Up and Step-Down Transformers

A **step-up** transformer is one in which the secondary voltage is higher than the primary voltage, while a **step-down** transformer is one in which the secondary voltage is lower. Since $a = E_p/E_s$, a step-up transformer has $a < 1$, while for a step-down transformer, $a > 1$. If $a = 1$, the transformer's turns ratio is **unity** and the secondary voltage is equal to the primary voltage.

EXAMPLE 23–1

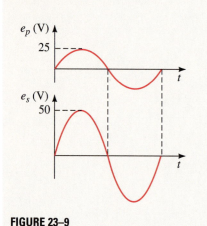

FIGURE 23–9

Suppose the transformer of Figure 23–7(a) has 500 turns on its primary and 1000 turns on its secondary.

a. Determine its turns ratio. Is it step-up or step-down?

b. If its primary voltage is $e_p = 25 \sin \omega t$ V, what is its secondary voltage?

c. Sketch the waveforms.

Solution

a. The turns ratio is $a = N_p/N_s = 500/1000 = 0.5$. This is a 1:2 step-up transformer.

b. From Equation 23–5, $e_s = e_p/a = (25 \sin \omega t)/0.5 = 50 \sin \omega t$ V.

c. Primary and secondary voltages are in phase as noted earlier. Figure 23–9 shows the waveforms.

If the transformers of Figure 23–7 have 600 turns on their primaries and 120 turns on their secondaries, and $\mathbf{E}_p = 120 \text{ V}\angle 0°$, what is $\mathbf{E}_s$ for each case?

EXAMPLE 23–2

Solution   The turns ratio is $a = 600/120 = 5$. For transformer (a), $\mathbf{E}_s$ is in phase with $\mathbf{E}_p$. Therefore, $\mathbf{E}_s = \mathbf{E}_p/5 = (120 \text{ V}\angle 0°)/5 = 24 \text{ V}\angle 0°$. For transformer (b), $\mathbf{E}_s$ is 180° out of phase with $\mathbf{E}_p$. Therefore, $\mathbf{E}_s = 24 \text{ V}\angle 180°$.

Repeat Example 23–1 for the circuit of Figure 23–7(b) if $N_p = 1200$ turns and $N_s = 200$ turns.

PRACTICE PROBLEMS 1

Answer
$e_s = 4.17 \sin(\omega t + 180°)$

Current Ratio

Because an ideal transformer has no power loss, its efficiency is 100% and thus power in equals power out. Consider again Figure 23–8. At any instant, $p_{\text{in}} = e_p i_p$ and $p_{\text{out}} = e_s i_s$. Thus,

$$e_p i_p = e_s i_s \tag{23–8}$$

and

$$\frac{i_p}{i_s} = \frac{e_s}{e_p} = \frac{1}{a} \tag{23–9}$$

since $e_s/e_p = 1/a$. (This means that if voltage is stepped up, current is stepped down, and vice versa.) In terms of current phasors and current magnitudes, Equation 23–9 can be written as

$$\frac{\mathbf{I}_p}{\mathbf{I}_s} = \frac{I_p}{I_s} = \frac{1}{a} \tag{23–10}$$

For example, for a transformer with $a = 4$, and $\mathbf{I}_p = 2 \text{ A}\angle -20°$, $\mathbf{I}_s = a\mathbf{I}_p = 4(2 \text{ A}\angle -20°) = 8 \text{ A}\angle -20°$, Figure 23–10.

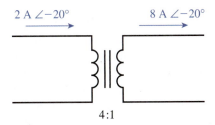

2 A ∠−20°   8 A ∠−20°

4:1

FIGURE 23–10   Currents are inverse to the turns ratio.

Polarity of Induced Voltage: The Dot Convention

As noted earlier, an iron core transformer's secondary voltage is either in phase with its primary voltage or 180° out of phase, depending on the relative direction of its windings. We will now demonstrate why.

A simple test, called the **kick test** (sometimes used by electrical workers to determine transformer polarity), can help establish the idea. The basic circuit is shown in Figure 23–11. A switch is used to make and break the circuit (since voltage is induced only while flux is changing).

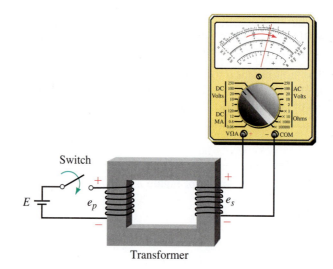

FIGURE 23–11   The "kick" test. For the winding directions shown, the meter kicks upscale at the instant the switch is closed. (This is the transformer of Figure 23–7a.)

For the winding directions shown, at the instant the switch is closed, the voltmeter needle "kicks" upscale, then settles back to zero. To understand why, we need to consider magnetic fields. Before we start, however, let us place a dot on one of the primary terminals; in this case, we arbitrarily choose the top terminal. Let us also replace the voltmeter by its equivalent resistance (Figure 23–12).

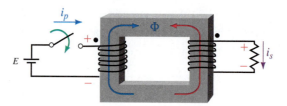

FIGURE 23–12   Determining dot positions.

At the instant the switch is closed, the polarity of the dotted primary terminal is positive with respect to the undotted primary terminal (because the + end of the source is directly connected to it). As current in the primary builds, it creates a flux in the upward direction as indicated by the blue arrow (recall the right-hand rule). According to Lenz's law, the effect that results must oppose the cause that produced it. The effect is a voltage induced in the secondary winding. The resulting current in the secondary produces a flux which, according to Lenz's law, must *oppose the buildup of the original flux,* i.e., it must be in the direction of the red arrow. Applying the right-hand rule, we see that secondary current must be in the direction indicated by i_s. Placing a plus sign at the tail of this arrow shows that the top end of the resistor is positive. This means that the top end of the secondary winding is also positive. Place a dot here. Dotted terminals are called **corresponding** terminals.

As you can see, corresponding terminals are positive (with respect to their companion undotted terminals) at the instant the switch is closed. If you perform a similar analysis at the instant the switch is opened, you will find that both dotted terminals are negative. Thus, *dotted terminals have the same polarity at all instants of time.* What we have developed here is known as the **dot convention for coupled circuits**—see sidebar note.

PRACTICAL NOTES . . .

1. While we developed the dot convention using a switched dc source, it is valid for ac as well. In fact, we will use it mostly for ac.

2. The resistor of Figure 23–12 is required only to work through the physics to establish the secondary voltage polarity. You can now remove it without affecting the resulting dot position.

3. In practice, corresponding terminals may be marked with dots, by color coded wires, or by special letter designations.

Determine the waveform for e_s in the circuit of Figure 23–13(a).

EXAMPLE 23–3

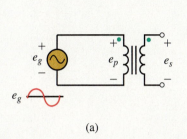
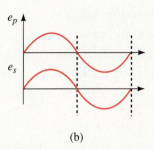

(a)   (b)

FIGURE 23–13

Solution   Dotted terminals have the same polarity (with respect to their undotted terminals) at all instants. During the first half-cycle, the dotted end of the primary coil is positive. Therefore, the dotted end of the secondary is also positive. During the second half-cycle, both are negative. The polarity marks on e_s mean that we are looking at the polarity of the top end of the secondary coil with respect to its bottom end. Thus, e_s is positive during the first half-cycle and negative during the second half cycle. It is therefore in phase with e_p as indicated in (b). Thus, if $e_p = E_{m_p} \sin \omega t$, then $e_s = E_{m_s} \sin \omega t$.

---

1. Determine the equation for e_s in the circuits of Figure 23–14(b), (c) and (d).

PRACTICE PROBLEMS 2

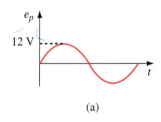
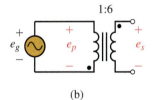
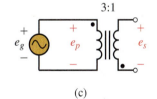

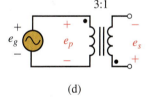

(a)   (b)   (c)   (d)

FIGURE 23–14

2. If $\mathbf{E}_g = 120\ \mathrm{V}\angle 30°$, determine $\mathbf{E}_s$ for each transformer of Figure 23–14.

3. Where do the dots go on the transformers of Figure 23–7?

Answers
1. $e_s = 72 \sin(\omega t + 180°)$ V; $e_s = 4 \sin(\omega t + 180°)$ V; $e_s = 4 \sin \omega t$
2. For (b), $\mathbf{E}_s = 720\ \mathrm{V}\angle{-150°}$; for (c), $\mathbf{E}_s = 40\ \mathrm{V}\angle{-150°}$; 40 V∠30°
3. For (a), place dots at a and c. For (b), place dots at a and d.

Analysis of Simple Transformer Circuits

Simple transformer circuits may be analyzed using the relationships described so far, namely $\mathbf{E}_p = a\mathbf{E}_s$, $\mathbf{I}_p = \mathbf{I}_s/a$, and $P_{\mathrm{in}} = P_{\mathrm{out}}$. This is illustrated in Example 23–4. More complex problems require some additional ideas.

EXAMPLE 23–4

For Figure 23–15(a), $\mathbf{E}_g = 120 \text{ V}\angle 0°$, the turns ratio is 5:1, and $\mathbf{Z}_L = 4 \text{ }\Omega\angle 30°$. Find

a. the load voltage,

b. the load current,

c. the generator current,

d. the power to the load,

e. the power output by the generator.

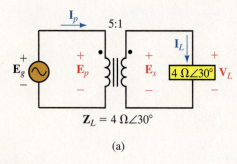

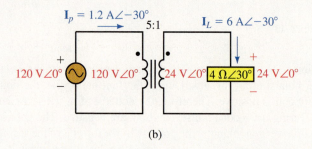

(a)   (b)

FIGURE 23–15

◀ **MULTISIM**

Solution

a. $\mathbf{E}_p = \mathbf{E}_g = 120 \text{ V}\angle 0°$

   $\mathbf{V}_L = \mathbf{E}_s = \mathbf{E}_p/a = (120 \text{ V}\angle 0°)/5 = 24 \text{ V}\angle 0°$

b. $\mathbf{I}_L = \mathbf{V}_L/\mathbf{Z}_L = (24 \text{ V }\angle 0°)/(4 \text{ }\Omega\angle 30°) = 6 \text{ A}\angle -30°$

c. $\mathbf{I}_g = \mathbf{I}_p$. But $\mathbf{I}_p = \mathbf{I}_s/a = \mathbf{I}_L/a$. Thus,

   $\mathbf{I}_g = (6 \text{ A}\angle -30°)/5 = 1.2 \text{ A}\angle -30°$

   Values are shown on Figure 23–15(b).

d. $P_L = V_L I_L \cos\theta_L = (24)(6)\cos 30° = 124.7 \text{ W}$.

e. $P_g = E_g I_g \cos\theta_g$, where θ_g is the angle between $\mathbf{E}_g$ and $\mathbf{I}_g$. $\theta_g = 30°$. Thus, $P_g = (120)(1.2)\cos 30° = 124.7 \text{ W}$, which agrees with (d) as it must since the transformer is lossless.

✔ IN-PROCESS
LEARNING CHECK 1

(Answers are at the end of the chapter.)

1. A transformer has a turns ratio of 1:8. Is it step-up or step-down? If $E_p = 25 \text{ V}$, what is E_s?

2. For the transformers of Figure 23–7, if $a = 0.2$ and $\mathbf{E}_s = 600 \text{ V}\angle -30°$, what is $\mathbf{E}_p$ for each case?

3. For each of the transformers of Figure 23–16, sketch the secondary voltage, showing both phase and amplitude.

4. For Figure 23–17, determine the position of the missing dot.

5. Figure 23–18 shows another way to determine dotted terminals. First, arbitrarily mark one of the primary terminals with a dot. Next, connect a jumper wire and voltmeter as indicated. From the voltmeter readings, you can determine which of the secondary terminals should be dotted. For the two cases indicated, where should the secondary dot be? (*Hint:* Use KVL.)

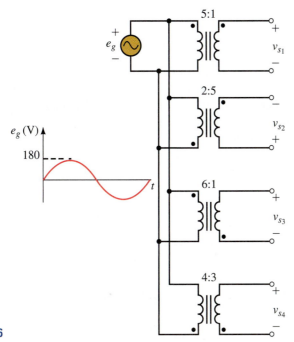

FIGURE 23–16

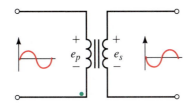

FIGURE 23–17

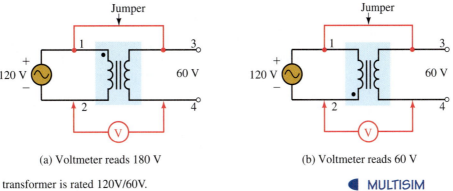

(a) Voltmeter reads 180 V   (b) Voltmeter reads 60 V

FIGURE 23–18   Each transformer is rated 120V/60V.

◀ MULTISIM

A transformer makes a load impedance look larger or smaller, depending on its turns ratio. To illustrate, consider Figure 23–19. When connected directly to the source, the load looks like impedance $\mathbf{Z}_L$, but when connected through a transformer, it looks like $a^2\mathbf{Z}_L$. This can be illustrated as follows. First, note that $\mathbf{Z}_p = \mathbf{E}_g/\mathbf{I}_p$. But $\mathbf{E}_g = \mathbf{E}_p$, $\mathbf{E}_p = a\mathbf{E}_s$, and $\mathbf{I}_p = \mathbf{I}_s/a$. Thus,

23.3 Reflected Impedance

$$\mathbf{Z}_p = \frac{\mathbf{E}_p}{\mathbf{I}_p} = \frac{a\mathbf{E}_s}{\left(\dfrac{\mathbf{I}_s}{a}\right)} = a^2\frac{\mathbf{E}_s}{\mathbf{I}_s} = a^2\frac{\mathbf{V}_L}{\mathbf{I}_L}$$

However $\mathbf{V}_L/\mathbf{I}_L = \mathbf{Z}_L$. Thus,

$$\mathbf{Z}_p = a^2\mathbf{Z}_L \qquad \text{(23–11)}$$

This means that $\mathbf{Z}_L$ now looks to the source like the transformer's turns ratio squared times the load impedance. The term $a^2\mathbf{Z}_L$ is referred to as the load's **reflected impedance.** Note that it retains the load's characteristics, that is, a capacitive load still looks capacitive, an inductive load still looks inductive, and so on. Note also from Figure 23–19(a) that the voltage looking into the transformer is $\mathbf{E}_g$. But $\mathbf{E}_g = a\mathbf{V}_L$. This means that the voltage across the reflected load is $a\mathbf{V}_L$ as indicated in (b).

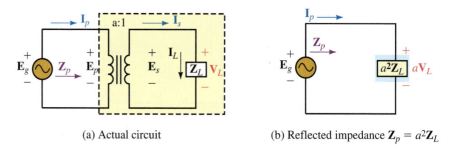

(a) Actual circuit   (b) Reflected impedance $\mathbf{Z}_p = a^2\mathbf{Z}_L$

FIGURE 23–19   Concept of reflected impedance. From the primary terminals, $\mathbf{Z}_L$ looks like an impedance of $a^2\mathbf{Z}_L$ with voltage $a\mathbf{V}_L$ across it and current $\mathbf{I}_L/a$ through it.

From Equation 23–11, we see that the load will look larger if $a > 1$ and smaller if $a < 1$. To illustrate, consider Figure 23–20. If a 1-Ω resistor were connected directly to the source, it would look like a 1-Ω resistor and the generator current would be $100\text{A}\angle 0°$. However, when connected to a 10:1 transformer it looks like a $(10)^2(1\ \Omega) = 100\text{-}\Omega$ resistor and the generator current is only 1 A$\angle 0°$.

The concept of reflected impedance is useful in a number of ways. It permits us to match loads to sources (such as amplifiers) as well as provides an alternate way to solve transformer problems.

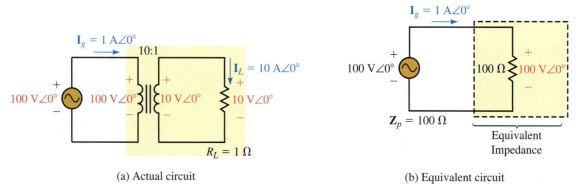

(a) Actual circuit   (b) Equivalent circuit

FIGURE 23–20   The equivalent impedance seen by the source is 100 Ω.

◀ **MULTISIM**

Use the reflected impedance idea to solve for primary and secondary currents and the load voltage for the circuit of Figure 23–21(a).

EXAMPLE 23–5

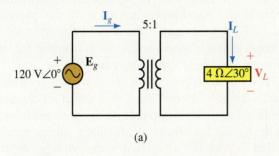

(a)

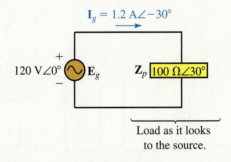

Load as it looks
to the source.

FIGURE 23–21   (b)

NOTES . . .

Example 23–5 is the same as Example 23–4. In Example 23–4, we analyzed the circuit in its original form while here we used the reflected impedance method. As can be seen, the amount of work required is about the same for both solutions, illustrating that there is no particular advantage of one method over the other for a circuit as simple as this. However, for complex problems, the advantages of the reflected impedance method are considerable—you will find that it reduces the amount of work involved and greatly simplifies analysis.

Solution

$$\mathbf{Z}_p = a^2\mathbf{Z}_L = (5)^2(4\ \Omega\angle 30°) = 100\ \Omega\angle 30°.$$

The equivalent circuit is shown in (b).

$$\mathbf{I}_g = \mathbf{E}_g/\mathbf{Z}_p = (120\ \text{V}\angle 0°) / (100\ \Omega\angle 30°) = 1.2\ \text{A}\angle -30°$$
$$\mathbf{I}_L = a\mathbf{I}_p = a\mathbf{I}_g = 5(1.2\ \text{A}\angle -30°) = 6\ \text{A}\angle -30$$
$$\mathbf{V}_L = \mathbf{I}_L\mathbf{Z}_L = (6\ \text{A}\angle -30°)(4\ \Omega\angle 30°) = 24\ \text{V}\angle 0°$$

The answers are the same as in Example 23–4—see sidebar note.

Power transformers are rated in terms of voltage and apparent power (for reasons discussed in Chapter 17). Rated current can be determined from these ratings. Thus a transformer rated 2400/120 volt, 48 kVA, has a current rating of 48 000 VA/2400 V = 20 A on its 2400-V side and 48 000 VA/120 V = 400 A on its 120-V side (Figure 23–22). This transformer can handle a 48-kVA load, regardless of power factor.

23.4 Power Transformer Ratings

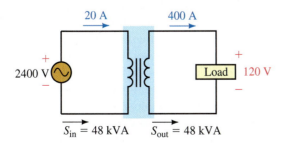

FIGURE 23–22 Transformers are rated by the amount of apparent power and the voltages that they are designed to handle.

23.5 Transformer Applications

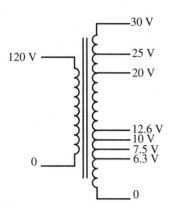

FIGURE 23–23   A multi-tapped power supply transformer. The secondary is tapped at various voltages.

Power Supply Transformers

On electronic equipment, **power supply transformers** are used to convert the incoming 120 Vac to the voltage levels required for internal circuit operation. A variety of commercial transformers are made for this purpose. The transformer of Figure 23–23, for example, has a multi-tapped secondary winding, each tap providing a different output voltage. It is intended for laboratory supplies, test equipment, or experimental power supplies.

Figure 23–24 illustrates a typical use of a power supply transformer. First, the incoming line voltage is stepped down, then a rectifier circuit (a circuit that uses diodes to convert ac to dc using a process called rectification) converts the ac to pulsating dc, a filter smooths it, and finally, a voltage regulator (an electronic device used to maintain a constant output voltage) regulates it to the required dc value.

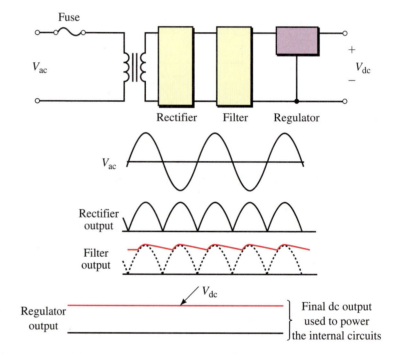

FIGURE 23–24   A transformer used in a power supply application.

Transformers in Power Systems

Transformers are one of the key elements that have made commercial ac power systems possible. Transformers are used at generating stations to raise voltage for long-distance transmission. This lowers the transmitted current and hence the I^2R power losses in the transmission line. At the user end, transformers reduce the voltage to a safe level for everyday use. A typical residential connection is shown in Figure 23–25. The taps on the primary permit the electric utility company to compensate for line voltage drops. Transformers located far from substations, for example, have lower input voltages (by a few percent) than those close to substations due to voltage drops on distribution lines. Taps permit the turns ratio to be changed to compensate. Note also the split secondary. It permits 120-V and 240-V loads to be supplied from the same transformer.

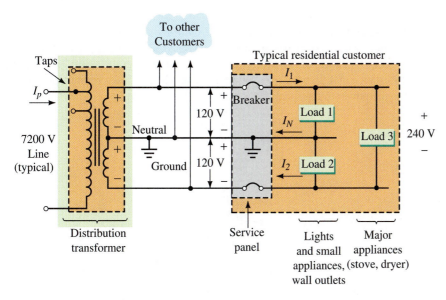

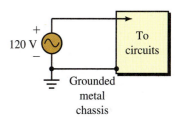

(a) Chassis is safe

FIGURE 23–25 Typical distribution transformer. This is how power is supplied to your home.

The transformer of Figure 23–25 is a single-phase unit (since residential customers require only single phase). By connecting its primary from line to neutral (or line to line), the required single-phase input is obtained from a three-phase line.

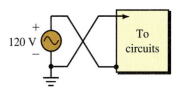

(b) Chassis accidentally connected to the "hot" side is at 120 V with respect to ground

Isolation Applications

Transformers are sometimes used to isolate equipment for safety or other reasons. If a piece of equipment has its frame or chassis connected to the grounded neutral of Figure 23–25, for example, the connection is perfectly safe as long as the connection is not changed. If, however, connections are inadvertently reversed as in Figure 23–26(b) (due to faulty installation for example), a dangerous situation results. A transformer used as in Figure 23–27 eliminates this danger by ensuring that the chassis is never directly connected to the "hot" wire. Isolation transformers are made for this purpose.

FIGURE 23–26 If the connections are inadvertently reversed as in (b), you will get a shock if you are grounded and you touch the chassis.

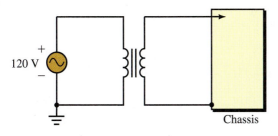

FIGURE 23–27 Using a transformer for isolation.

Impedance Matching

As you learned earlier, a transformer can be used to raise or lower the apparent impedance of a load by choice of turns ratio. This is referred to as **impedance matching.** Impedance matching is sometimes used to match loads to amplifiers to achieve maximum power transfer. If the load and source are not matched, a transformer can be inserted between them as illustrated next.

EXAMPLE 23–6

Figure 23–28(a) shows the schematic of a multi-tap sound distribution transformer with various taps to permit matching of speakers to amplifiers. Over their design range, speakers are basically resistive. If the speaker of Figure 23–29(a) has a resistance of 4 Ω, what transformer ratio should be chosen for max power? What is the power to the speaker?

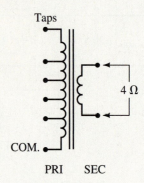

FIGURE 23–28   A tapped sound distribution transformer.

Solution   Make the reflected resistance of the speaker equal to the internal (Thévenin) resistance of the amplifier. Thus, $Z_p = 400\ \Omega = a^2Z_L = a^2(4\ \Omega)$. Solving for a yields

$$a = \sqrt{\frac{Z_p}{Z_L}} = \sqrt{\frac{400\ \Omega}{4\ \Omega}} = \sqrt{100} = 10$$

Now consider power. Since $Z_p = 400\ \Omega$, Figure 23–29(b), half the source voltage appears across it. Thus power to Z_p is $(40\ \text{V})^2/(400\ \Omega) = 4\ \text{W}$. Since the transformer is considered lossless, all power is transferred to the speaker. Thus, $P_{\text{speaker}} = 4\ \text{W}$.

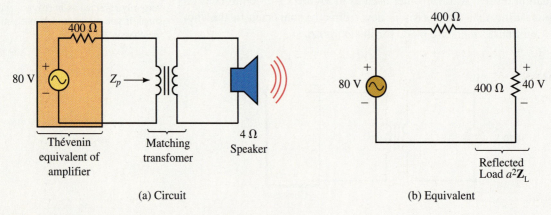

(a) Circuit

(b) Equivalent

FIGURE 23–29   Matching the 4-Ω speaker for maximum power transfer.

PRACTICE PROBLEMS 3

Determine the power to the speaker of Figure 23–29 if the transformer is not present (i.e., the speaker is directly connected to the amplifier). Compare to Example 23–6.

Answer
0.157 W (dramatically lower)

Transformers with Multiple Secondaries

For a transformer with multiple secondaries (Figure 23–30), each secondary voltage is governed by the appropriate turns ratio, that is, $\mathbf{E_1}/\mathbf{E_2} = N_1/N_2$ and $\mathbf{E_1}/\mathbf{E_3} = N_1/N_3$. Loads are reflected in parallel. That is, $\mathbf{Z'_2} = a_2^2\mathbf{Z_2}$ and $\mathbf{Z'_3} = a_3^2\mathbf{Z_3}$ appear in parallel in the equivalent circuit, (b).

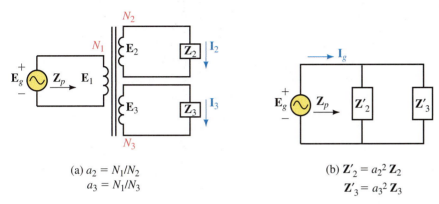

(a) $a_2 = N_1/N_2$
$a_3 = N_1/N_3$

(b) $\mathbf{Z'_2} = a_2^2\,\mathbf{Z_2}$
$\mathbf{Z'_3} = a_3^2\,\mathbf{Z_3}$

FIGURE 23–30 Loads are reflected in parallel.

For the circuit of Figure 23–31(a),

a. determine the equivalent circuit,

b. determine the generator current,

c. show that apparent power in equals apparent power out.

EXAMPLE 23–7

Solution

a. See Figure 23–31(b).

b. $\mathbf{I}_g = \dfrac{\mathbf{E}_g}{\mathbf{Z'_2}} + \dfrac{\mathbf{E}_g}{\mathbf{Z'_3}} = \dfrac{100\angle 0°}{10} + \dfrac{100\angle 0°}{-j10} = 10 + j10 = 14.14\ \text{A}\angle 45°$

c. Input: $S_{in} = E_g I_g = (100\ \text{V})(14.14\ \text{A}) = 1414$ VA
Output: From Figure 23–31(b), $P_{out} = (100\ \text{V})^2/(10\ \Omega) = 1000$ W and
$Q_{out} = (100\ \text{V})^2/(10\ \Omega) = 1000$ VAR. Thus $S_{out} = \sqrt{P_{out}^2 + Q_{out}^2} = 1414$ VA which is the same as S_{in}.

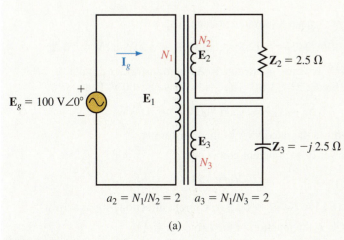

$a_2 = N_1/N_2 = 2$ $a_3 = N_1/N_3 = 2$

(a)

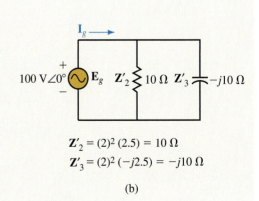

$\mathbf{Z'_2} = (2)^2\,(2.5) = 10\ \Omega$
$\mathbf{Z'_3} = (2)^2\,(-j2.5) = -j10\ \Omega$

(b)

FIGURE 23–31

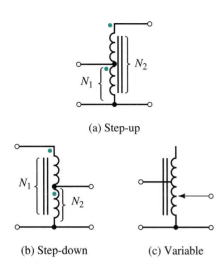

(a) Step-up

(b) Step-down   (c) Variable

FIGURE 23–32 Autotransformers.

Autotransformers

An important variation of the transformer is the **autotransformer** (Figure 23–32). Autotransformers are unusual in that their primary circuit is not electrically isolated from their secondary. However, they are smaller and cheaper than conventional transformers for the same load kVA since only part of the load power is transferred inductively. Figure 23–32 shows several variations. The transformer of (c) is variable by means of a slider, typically, from 0% to 110%.

For analysis, an autotransformer may be viewed as a standard two-winding transformer reconnected as in Figure 23–33(b). Voltage and current relationships between windings hold as they do for the standard connection. Thus, if you apply rated voltage to the primary winding, you will obtain rated voltage across the secondary winding. Finally, since we are assuming an ideal transformer, apparent power out equals apparent power in.

EXAMPLE 23–8

A 240/60-V, 3-kVA transformer [Figure 23–33(a)] is reconnected as an autotransformer to supply 300 volts to a load from a 240-V supply [Figure 23–33(b)].

a. Determine the transformer's rated primary and secondary currents.

b. Determine the maximum apparent power that can be delivered to the load.

c. Determine the supply current.

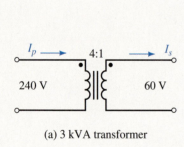

(a) 3 kVA transformer

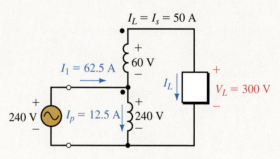

(b) Used as an autotransformer

FIGURE 23–33

◄ MULTISIM

Solution

a. Rated current = rated kVA/rated voltage. Thus,

$$I_p = 3 \text{ kVA}/240 \text{ V} = 12.5 \text{ A} \quad \text{and} \quad I_s = 3 \text{ kVA}/60 \text{ V} = 50 \text{ A}$$

b. Since the 60-V winding is rated at 50 A, the transformer can deliver 50 A to the load [Figure 23–33(b)]. The load voltage is 300 V. Thus,

$$S_L = V_L I_L = (300 \text{ V})(50 \text{ A}) = 15 \text{ kVA}$$

This is five times the rated kVA of the transformer.

c. Apparent power in = apparent power out:

$$240 I_1 = 15 \text{ kVA}.$$

Thus, $I_1 = 15 \text{ kVA}/240 \text{ V} = 62.5 \text{ A}$. Current directions are as shown. As a check, KCL at the junction of the two coils yields

$$I_1 = I_p + I_L = 12.5 + 50 = 62.5 \text{ A}$$

1. For each of the circuits of Figure 23–34, determine the required answer.

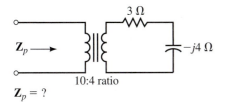

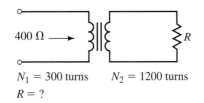

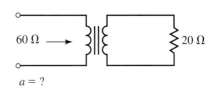

FIGURE 23–34

2. For Figure 23–35, if $a = 5$ and $\mathbf{I}_p = 5\ A\angle{-60°}$, what is $\mathbf{E}_g$?

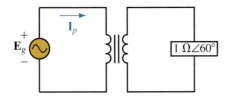

FIGURE 23–35

3. For Figure 23–35, if $\mathbf{I}_p = 30\ mA\angle{-40°}$ and $\mathbf{E}_g = 240\ V\angle20°$, what is a?

4. a. How many amps can a 24-kVA, 7200/120-V transformer supply to a 120-V unity power factor load? To a 0.75 power factor load?

   b. How many watts can it supply to each load?

5. For the transformer of Figure 23–36, between position 2 and 0 there are 2000 turns. Between taps 1 and 2 there are 200 turns, and between taps 1 and 3 there are 300 turns. What will the output voltage be when the supply is connected to tap 1? To tap 2? To tap 3?

6. For the circuit of Figure 23–37, what is the power delivered to a 4-ohm speaker? What is the power delivered if an 8-ohm speaker is used instead? Why is the power to the 4-ohm speaker larger?

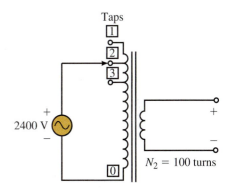

FIGURE 23–36

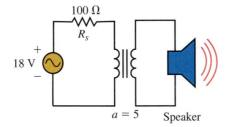

FIGURE 23–37

7. The autotransformer of Figure 23–38 has a 58% tap. The apparent power of the load is 7.2 kVA. Calculate the following:

   a. The load voltage and current.

   b. The source current.

   c. The current in each winding and its direction.

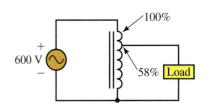

FIGURE 23–38

23.6 Practical Iron-Core Transformers

In Section 23.2, we idealized the transformer. We now add back the effects that we ignored.

Leakage Flux

While most flux is confined to the core, a small amount (called **leakage flux**) passes outside the core and through air at each winding as in Figure 23–39(a). The effect of this leakage can be modeled by inductances L_p and L_s as indicated in (b). The remaining flux, the **mutual flux** Φ_m, links both windings and is therefore accounted for by the ideal transformer as previously.

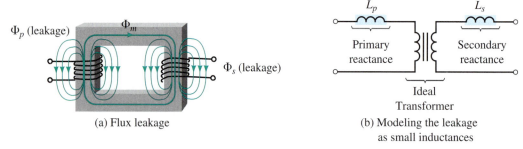

(a) Flux leakage

(b) Modeling the leakage as small inductances

FIGURE 23–39   Adding the effect of leakage flux to the model.

Winding Resistance

The effect of coil resistance can be approximated by adding resistances R_p and R_s as shown in Figure 23–40. The effect of these resistances is to cause a slight power loss and hence a reduction in efficiency as well as a small voltage drop. (The power loss associated with coil resistance is called **copper loss** (Sec. 23.7) and varies as the square of the load current.)

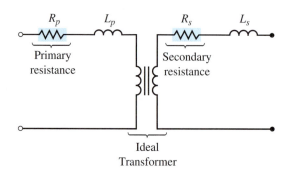

FIGURE 23–40   Adding the winding resistance to the model.

Core Loss

Losses occur in the core because of **eddy currents** and **hysteresis.** First, consider eddy currents. Since iron is a conductor, voltage is induced in the core as flux varies. This voltage creates currents that circulate as "eddies" within the core itself. One way to reduce these currents is to break their path of circulation by constructing the core from thin laminations of steel rather than using a solid block of iron. Laminations are insulated from each other by a coat of ceramic, varnish, or similar insulating material. (Although this does not eliminate eddy currents, it greatly reduces them.) Power and audio transformers are built this way (Figure 23–4). Another way to reduce eddy currents is to use powdered iron held together by an insulating binder. Ferrite cores are made like this.

Now consider hysteresis (Chapter 12, Section 12.14). Because flux constantly reverses, magnetic domains in the core steel constantly reverse as well.

This takes energy. In practice, this energy is minimized by using special grain-oriented transformer steel.

The sum of hysteresis and eddy current loss is called **core loss** or **iron loss.** In a well-designed transformer, it is small, typically one to two percent of the transformer rating. The effect of core loss can be modeled as a resistor, R_c in Figure 23–41. Core losses vary approximately as the square of applied voltage. As long as voltage is constant (which it normally is), core losses remain constant.

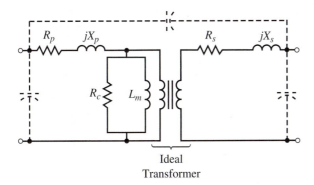

FIGURE 23–41   Final iron-core transformer equivalent circuit.

Other Effects

We have also neglected **magnetizing current.** In a real transformer, however, some current is required to magnetize the core. To account for this, add path L_m as shown in Figure 23–41. Stray capacitances also exist between various parts of the transformer. They can be approximated by lumped capacitances as indicated.

The Full Equivalent

Figure 23–41 shows the final equivalent with all effects incorporated. How good is it? Calculations based on this model agree extremely well with measurements made on real transformers—see sidebar note.

Voltage Regulation

Because of its internal impedance, voltage drops occur inside a transformer. Therefore, its output voltage under load is different than its output voltage under no load. This change in voltage (expressed as a percentage of full-load voltage) is termed **regulation.** For regulation analysis, parallel branches R_c and L_m and stray capacitance have negligible effect and can be neglected. This yields the simplified circuit of Figure 23–42(a). Even greater simplification is achieved by reflecting secondary impedances into the primary. This yields the circuit of (b). The reflected load voltage is $a\mathbf{V}_L$ and the reflected load current is $\mathbf{I}_L/a$. The simplified equivalent of (b) is the circuit that we use in practice to perform regulation analysis.

NOTES . . .

Although Figure 23–41 (which represents the full equivalent for iron core transformers) yields exceptionally good results, it is complex and awkward to use. Fortunately, we can simplify the model, since certain effects are negligible for specific applications. For example, at power frequencies, the effect of capacitance is negligible, while for regulation analysis (considered next), the core branch (R_c and L_m) also has negligible effect. Thus, in practice, they can usually be omitted, leading to the simplified model illustrated next (Figure 23–42).

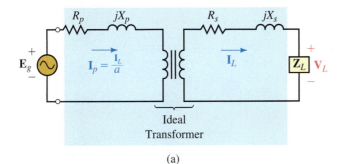

(a)

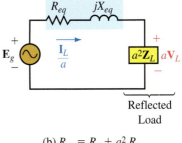

(b) $R_{eq} = R_p + a^2 R_s$
$X_{eq} = X_p + a^2 X_s$

FIGURE 23–42   Simplifying the equivalent.

EXAMPLE 23–9

A 10:1 transformer has primary and secondary resistance and reactance of 4 Ω + $j4$ Ω and 0.04 Ω + $j0.04$ Ω respectively as in Figure 23–43.

a. Determine its equivalent circuit.

b. If $\mathbf{V}_L = 120$ V∠0° and $\mathbf{I}_L = 20$ A∠−30°, what is the supply voltage, $\mathbf{E}_g$?

c. Determine the regulation.

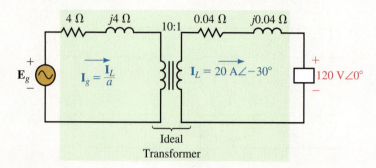

FIGURE 23–43

◀ MULTISIM

PRACTICAL NOTES . . .

1. From Figure 23–45, $a = E_g/V_{NL}$. This means that the turns ratio is the ratio of input voltage to output voltage at no load.

2. The voltage rating of a transformer (such as 1200/120 V) is referred to as its *nominal rating*. The ratio of nominal voltages is the same as the turns ratio. Thus, for a nonloaded transformer, if nominal voltage is applied to the primary, nominal voltage will appear at the secondary.

3. Power transformers are normally operated close to their nominal voltages. However, depending on operating conditions, they may be a few percent above or below rated voltage at any given time.

Solution

a. $R_{eq} = R_p + a^2 R_s = 4\ \Omega + (10)^2(0.04\ \Omega) = 8\ \Omega$

$X_{eq} = X_p + a^2 X_s = 4\ \Omega + (10)^2(0.04\ \Omega) = 8\ \Omega$

Thus, $\mathbf{Z}_{eq} = 8\ \Omega + j8\ \Omega$ as shown in Figure 23–44.

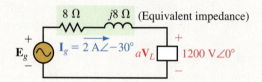

FIGURE 23–44

b. $a\mathbf{V}_L = (10)(120$ V∠0°$) = 1200$ V∠0° and $\mathbf{I}_L/a = (20$ A∠−30°$)/10 = 2$ A∠−30°. From KVL, $\mathbf{E}_g = (2$ A∠−30°$)(8\ \Omega + j8) + 1200$ V∠0° $= 1222$ V∠0.275°.

Thus, there is a phase shift of 0.275° across the transformer's internal impedance and a drop of 22 V, requiring that the primary be operated slightly above its rated voltage. (This is normal.)

c. Now consider the no-load condition (Figure 23–45). Let V_{NL} be the no-load voltage. As indicated, $aV_{NL} = 1222$ V. Thus, $V_{NL} = 1222/a = 1222/10 = 122.2$ volts and

$$\text{regulation} = \frac{V_{NL} - V_{FL}}{V_{FL}} \times 100 = \frac{122.2 - 120}{120} \times 100 = 1.83\%$$

Note that only magnitudes are used in determining regulation.

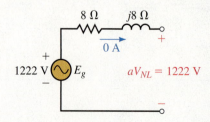

FIGURE 23–45   No-load equivalent: $aV_{NL} = E_g$.

A transformer used in an electronic power supply has a nominal rating of 120/12 volts and is connected to a 120-Vac source. Its equivalent impedance as seen from the primary is $10\ \Omega + j10\ \Omega$. What is the magnitude of the load voltage if the load is 5 ohms resistive? Determine the regulation.

PRACTICE PROBLEMS 4

Answer
11.8 V;   2.04%

Transformer Efficiency

Efficiency is the ratio of output power to input power.

$$\eta = \frac{P_{out}}{P_{in}} \times 100\% \qquad (23\text{--}12)$$

But $P_{in} = P_{out} + P_{loss}$. For a transformer, losses are due to I^2R losses in the windings (called copper losses) and losses in the core (called core losses). Thus,

$$\eta = \frac{P_{out}}{P_{out} + P_{loss}} \times 100\% = \frac{P_{out}}{P_{out} + P_{copper} + P_{core}} \times 100\% \quad (23\text{--}13)$$

Large power transformers are exceptionally efficient, of the order of 98 to 99 percent. The efficiencies of smaller transformers are around 95 percent.

Losses may be determined experimentally using the **short-circuit test** and the **open-circuit test.** (These tests are used mainly with power transformers.) They provide the data needed to determine a transformer's equivalent circuit and to compute its efficiency.

23.7 Transformer Tests

The Short-Circuit Test

Figure 23–46 shows the test setup for the short-circuit test. Starting at 0 V, gradually increase E_g until the ammeter indicates the rated current. (This occurs at about 5 percent rated input voltage.) Since core losses are proportional to the square of voltage, at 5 percent rated voltage, core losses are negligible. The losses that you measure are therefore only copper losses.

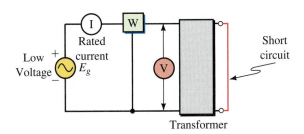

FIGURE 23–46   Short-circuit test. Measure from the HV side.

EXAMPLE 23–10

Measurements on the high side of a 240/120-volt, 4.8-kVA transformer yield $E_g = 11.5$ V and $W = 172$ W at the rated current of $I = 4.8$ kVA/240 = 20 A. Determine $\mathbf{Z}_{eq}$.

Solution   See Figure 23–47. Since $Z_L = 0$, the only impedance in the circuit is Z_{eq}. Thus, $Z_{eq} = E_g/I = 11.5$ V/20 A = 0.575 Ω. Further, $R_{eq} = W/I^2 = 172$ W/(20 A)^2 = 0.43 Ω. Therefore,

$$X_{eq} = \sqrt{Z_{eq}^2 - R_{eq}^2} = \sqrt{(0.575)^2 - (0.43)^2} = 0.382 \ \Omega$$

and $\mathbf{Z}_{eq} = R_{eq} + jX_{eq} = 0.43 \ \Omega + j0.382 \ \Omega$ as shown in (b).

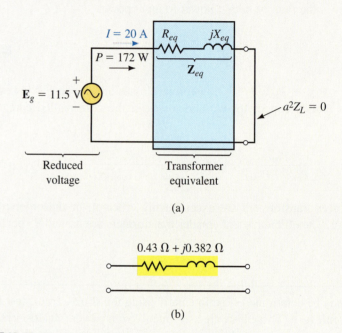

(a)

$$0.43 \ \Omega + j0.382 \ \Omega$$

(b)

FIGURE 23–47   Determining the equivalent circuit by test.

The Open-Circuit Test

The setup for the open-circuit test is shown in Figure 23–48. Apply the full rated voltage. Since the load current is zero, only exciting current results. Since the exciting current is small, power loss in the winding resistance is negligible, and the power that you measure is just core loss

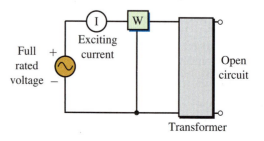

FIGURE 23–48   Open-circuit test. Measure from the LV side.

An open-circuit test on the transformer of Example 23–10 yields a core loss of 106 W. Determine this transformer's efficiency when supplying the full, rated VA to a load at unity power factor.

EXAMPLE 23–11

Solution   Since the transformer is supplying the rated VA, its current is the full, rated current. From the short-circuit test, the copper loss at full rated current is 172 W. Thus,

$$\text{copper loss} = 172 \text{ W}$$
$$\text{core loss} = 106 \text{ W (Measured above)}$$
$$\text{output} = 4800 \text{ W (Rated)}$$
$$\text{input} = \text{output} + \text{losses} = 5078 \text{ W}$$

Thus

$$\eta = P_{\text{out}}/P_{\text{in}} = (4800 \text{ W}/5078 \text{ W}) \times 100 = 94.5\%$$

Copper loss varies as the square of load current. Thus, at half the rated current, the copper loss is $(\frac{1}{2})^2 = \frac{1}{4}$ of its value at full rated current. Core loss remains constant since applied voltage remains constant.

For the transformer of Example 23–11, determine power input and efficiency at half the rated VA output to a unity power factor.

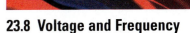

PRACTICE PROBLEMS 5

Answer
2549 W; 94.2%

Iron-core transformer characteristics vary with frequency and voltage. To determine why, we start with Faraday's law, $e = Nd\Phi/dt$. Specializing this to the sinusoidal ac case, it can be shown that

$$E_p = 4.44fN_p\Phi_m \qquad \text{(23–14)}$$

where Φ_m is the mutual core flux.

23.8 Voltage and Frequency Effects

Effect of Voltage

First, assume constant frequency. Since $\Phi_m = E_p/4.44fN_p$, the core flux is proportional to the applied voltage. Thus, if the applied voltage is increased, the core flux increases. Since magnetizing current is required to produce this flux, magnetizing current must increase as well. An examination of Figure 23–49 shows that magnetizing current increases dramatically when flux density rises above the knee of the curve; in fact, the effect is so pronounced that the primary current at no load may well exceed the full-load rated current of the transformer if the input voltage is large. For this reason, power transformers should be operated only at or near their rated voltage.

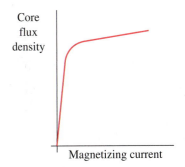

FIGURE 23–49

Effect of Frequency

Audio transformers must operate over a range of frequencies. Consider again $\Phi_m = E_p/4.44fN_p$. As this indicates, decreasing the frequency increases the core flux and hence the magnetizing current. At low frequencies, this larger current increases internal voltage drops and hence decreases

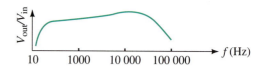

FIGURE 23–50 Frequency response curve, audio transformer.

the output voltage as indicated in Figure 23–50. Now consider increasing the frequency. As frequency increases, leakage inductance and shunt capacitance (recall Figure 23–41) cause the voltage to fall off. To compensate for this, audio transformers are sometimes designed so that their internal capacitances resonate with their inductances to extend the operating range. This is what causes the peaking at the high-frequency end of the curve.

IN-PROCESS
LEARNING CHECK 3

(Answers are at the end of the chapter.)

A transformer with a nominal rating of 240/120 V, 60 Hz, has its load on the 120-V side. Suppose $R_p = 0.4\ \Omega$, $L_p = 1.061$ mH, $R_s = 0.1\ \Omega$, and $L_s = 0.2653$ mH.

a. Determine its equivalent circuit as per Figure 23–42(b).

b. If $\mathbf{E}_g = 240\ \text{V}\angle 0°$ and $\mathbf{Z}_L = 3 + j4\ \Omega$, what is $\mathbf{V}_L$?

c. Compute the regulation.

23.9 Loosely Coupled Circuits

◀ **Online Companion**

We now turn our attention to coupled circuits that do not have iron cores. For such circuits, only a portion of the flux produced by one coil links another and the coils are said to be **loosely coupled.** Loosely coupled circuits cannot be characterized by turns ratios; rather, as you will see, they are characterized by self- and mutual inductances. Air-core transformers, ferrite-core transformers, and general inductive circuit coupling fall into this category. In this section, we develop the main ideas.

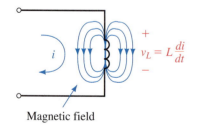

FIGURE 23–51 Place the plus sign for the self-induced voltage at the tail of the current arrow.

Voltages in Air-Core Coils

To begin, consider the isolated (noncoupled) coil of Figure 23–51. As shown in Chapter 13, the voltage across this coil is given by $v_L = L\,di/dt$, where i is the current through the coil and L is its inductance. Note carefully the polarity of the voltage; the plus sign goes at the tail of the current arrow. Because the coil's voltage is created by its own current, it is called a **self-induced voltage.**

Now consider a pair of coupled coils (Figure 23–52). When coil 1 alone is energized as in (a), it looks just like the isolated coil of Figure 23–51; thus its voltage is

$$v_{11} = L_1 di_1/dt \text{ (self-induced in coil 1)}$$

where L_1 is the self-inductance of coil 1 and the subscripts indicate that v_{11} is the voltage across coil 1 due to its own current. Similarly, when coil 2 alone is energized as in (b), its self-induced voltage is

$$v_{22} = L_2 di_2/dt \text{ (self-induced in coil 2)}$$

For both of these self-voltages, note that the plus sign goes at the tail of their respective current arrows.

Mutual Voltages

Consider again Figure 23–52(a). When coil 1 is energized, some of the flux that it produces links coil 2, inducing voltage v_{21} in coil 2. Since the flux here is due to i_1 alone, v_{21} is proportional to the rate of change of i_1. Let the constant of proportionality be M. Then,

$$v_{21} = M di_1/dt \text{ (mutually induced in coil 2)}$$

v_{21} is the **mutually induced voltage** in coil 2 and M is the **mutual inductance** between the coils. It has units of henries. Similarly, when coil 2 alone is energized as in (b), the voltage induced in coil 1 is

$$v_{12} = M di_2/dt \text{ (mutually induced in coil 1)}$$

When both coils are energized, the voltage of each coil can be found by superposition; *in each coil, the induced voltage is the sum of its self-voltage plus the voltage mutually induced due to the current in the other coil.* The sign of the self term for each coil is straightforward: It is determined by placing a plus sign at the tail of the current arrow for the coil as shown in Figures 23–52(a) and (b). The polarity of the mutual term, however, depends on whether the mutual voltage is additive or subtractive.

Additive and Subtractive Voltages

Whether self- and mutual voltages add or subtract depends on the direction of currents through the coils relative to their winding directions. This can be described in terms of the dot convention. Consider Figure 23–53(a). Comparing the coils here to Figure 23–12, you can see that their top ends correspond and thus can be marked with dots. Now let currents enter both coils at dotted ends. Using the right-hand rule, you can see that their fluxes add. The total flux linking coil 1 is therefore the *sum* of that produced by i_1 and i_2; therefore, the voltage across coil 1 is the sum of that produced by i_1 and i_2. That is $v_1 = v_{11} + v_{12}$. Expanded, this is

$$v_1 = L_1 \frac{di_1}{dt} + M \frac{di_2}{dt} \qquad \text{(23–15a)}$$

For coil 2, $v_2 = v_{21} + v_{22}$. Thus,

$$v_2 = M \frac{di_1}{dt} + L_2 \frac{di_2}{dt} \qquad \text{(23–15b)}$$

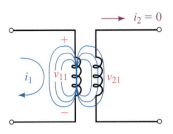

$i_2 = 0$

(a) v_{11} is the self-induced voltage in coil 1; v_{21} is the mutual voltage induced in coil 2

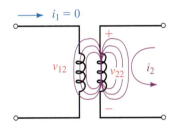

$i_1 = 0$

(b) v_{22} is the self-induced voltage in coil 2; v_{12} is the mutual voltage induced in coil 1

FIGURE 23–52   Self and mutual voltages.

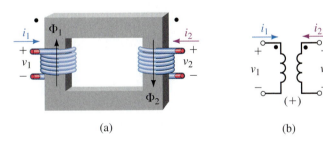

(a)   (b)

FIGURE 23–53   When both currents enter dotted terminals, use the $+$ sign for the mutual term in Equation 23–15.

Now consider Figure 23–54. Here, the fluxes oppose and the flux linking each coil is the *difference* between that produced by its own current and that produced by the current of the other coil. Thus, the sign in front of the mutual voltage terms will be negative.

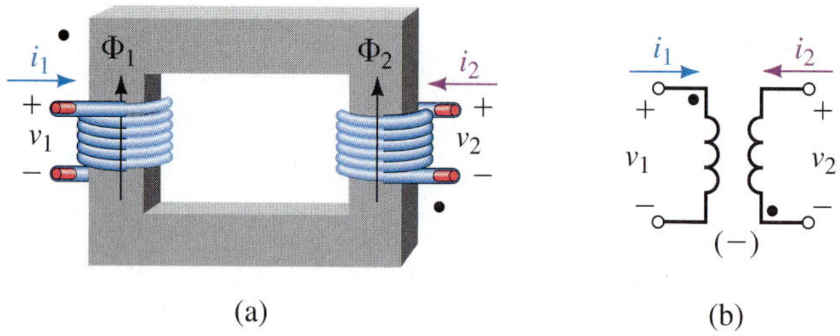

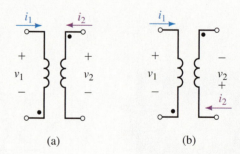

FIGURE 23–54 When one current enters a dotted terminal and the other enters an undotted terminal, use the − sign for the mutual term in Equation 23–15.

The Dot Rule

As you can see, the signs of the mutual voltage terms in Equations 23–15 are positive when both currents enter dotted terminals, but negative when one current enters a dotted terminal and the other enters an undotted terminal. Stated another way, *the sign of the mutual voltage is the same as the sign of its self-voltage when both currents enter dotted (or undotted) terminals, but is opposite when one current enters a dotted terminal and the other enters an undotted terminal.* This observation provides us with a procedure for determining voltage polarities in coupled circuits.

1. Assign a direction for currents i_1 and i_2.
2. Place a plus sign at the tail of the current arrow for each coil to denote the polarity of its self-induced voltage.
3. If both currents enter (or both leave) dotted terminals, make the sign of the mutually induced voltage the same as the sign of the self-induced voltage when you write the equation.
4. If one current enters a dotted terminal and the other leaves, make the sign of the mutually induced voltage opposite to the sign of the self-induced voltage.

EXAMPLE 23–12

Write equations for v_1 and v_2 of Figure 23–55(a).

FIGURE 23–55

Solution Since one current enters an undotted terminal and the other enters a dotted terminal, place a minus sign in front of M. Thus,

$$v_1 = L_1 \frac{di_1}{dt} - M \frac{di_2}{dt}$$

$$v_2 = -M \frac{di_1}{dt} + L_2 \frac{di_2}{dt}$$

Write equations for v_1 and v_2 of Figure 23–55(b).

PRACTICE PROBLEMS 6

Answer
Same as Equation 23–15.

Coefficient of Coupling

For loosely coupled coils, not all of the flux produced by one coil links the other. To describe the degree of coupling between coils, we introduce a **coefficient of coupling, k**. Mathematically, k is defined as the ratio of the flux that links the companion coil to the total flux produced by the energized coil. For iron-core transformers, almost all the flux is confined to the core and links both coils; thus, k is very close to 1. At the other extreme (i.e., isolated coils where no flux linkage occurs), $k = 0$. Thus, $0 \leq k < 1$. Mutual inductance depends on k. It can be shown that mutual inductance, self-inductances, and the coefficient of coupling are related by the equation

$$M = k\sqrt{L_1L_2} \qquad (23\text{–}16)$$

Thus, the larger the coefficient of coupling, the larger the mutual inductance.

Inductors with Mutual Coupling

If a pair of coils are in close proximity, the field of each coil couples the other, resulting in a change in the apparent inductance of each coil. To illustrate, consider Figure 23–56(a), which shows a pair of inductors with self-inductances L_1 and L_2. If coupling occurs, the effective coil inductances will no longer be L_1 and L_2. To see why, consider the voltage induced in each winding—it is the sum of the coil's own self-voltage plus the voltage mutually induced from the other coil. Since current is the same for both coils, $v_1 = L_1di/dt + Mdi/dt = (L_1 + M)di/dt$, which means that coil 1 has an effective inductance of $L_1' = L_1 + M$. Similarly, $v_2 = (L_2 + M)di/dt$, giving coil 2 an effective inductance of $L_2' = L_2 + M$. The effective inductance of the series combination [Figure 23–56(b)] is then

$$L_T^+ = L_1 + L_2 + 2M \quad \text{(henries)} \qquad (23\text{–}17)$$

If coupling is subtractive as in Figure 23–57, $L_1' = L_1 - M$, $L_2' = L_2 - M$, and

$$L_T^- = L_1 + L_2 - 2M \quad \text{(henries)} \qquad (23\text{–}18)$$

You can determine mutual inductance from Equations 23–17 and 23–18. It is

$$M = \frac{1}{4}(L_T^+ - L_T^-) \qquad (23\text{–}19)$$

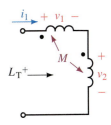

(a) $L_1' = L_1 + M$; $L_2' = L_2 + M$

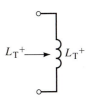

(b) $L_T^+ = L_1 + L_2 + 2M$

FIGURE 23–56 Coils in series with additive mutual coupling.

EXAMPLE 23–13

Three inductors are connected in series (Figure 23–57). Coils 1 and 2 interact, but coil 3 does not.

a. Determine the effective inductance of each coil.

b. Determine the total inductance of the series connection.

Solution

a. $L_1' = L_1 - M = 2 \text{ mH} - 0.4 \text{ mH} = 1.6 \text{ mH}$

$L_2' = L_2 - M = 3 \text{ mH} - 0.4 \text{ mH} = 2.6 \text{ mH}$

L_1' and L_2' are in series with L_3. Thus,

b. $L_T = 1.6 \text{ mH} + 2.6 \text{ mH} + 2.7 \text{ mH} = 6.9 \text{ mH}$

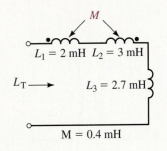

FIGURE 23–57

The same principles apply when more than two coils are coupled. Thus, for the circuit of Figure 23–58, $L_1' = L_1 - M_{12} - M_{31}$, etc.

For two parallel inductors with mutual coupling, the equivalent inductance is

$$L_{eq} = \frac{L_1 L_2 - M^2}{L_1 + L_2 \mp 2M} \tag{23–20}$$

If the dots are at the same ends of the coils, use the $-$ sign. For example, if $L_1 = 20$ mH, $L_2 = 5$ mH, and $M = 2$ mH, then $L_{eq} = 4.57$ mH if the dots are both at the same ends of the coils, and $L_{eq} = 3.31$ mH when the dots are at opposite ends.

PRACTICE PROBLEMS 7

For the circuit of Figure 23–58, different "dot" symbols are used to represent coupling between sets of coils.

a. Determine the effective inductance of each coil.

b. Determine the total inductance of the series connection.

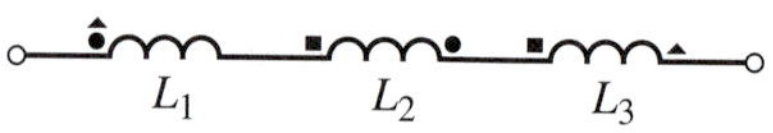

$$L_1 \qquad L_2 \qquad L_3$$

$L_1 = 10$ mH $M_{12} = 2$ mH (Mutual inductance between coils 1 and 2) (•)
$L_2 = 40$ mH $M_{23} = 1$ mH (Mutual inductance between coils 2 and 3) (■)
$L_3 = 20$ mH $M_{31} = 0.6$ mH (Mutual inductance between coils 3 and 1) (▲)

FIGURE 23–58

Answers
a. $L_1' = 7.4$ mH; $L_2' = 39$ mH; $L_3' = 20.4$ mH
b. 66.8 mH

The effect of unwanted mutual inductance can be minimized by physically separating coils or by orienting their axes at right angles. The latter technique is used where space is limited and coils cannot be spaced widely. While it does not eliminate coupling, it can help minimize its effects.

23.10 Magnetically Coupled Circuits with Sinusoidal Excitation

When coupling occurs between various parts of a circuit (whether wanted or not), the foregoing principles apply. However, since it is difficult to continue the analysis in general, we will change over to steady state ac. This will permit us to look for the main ideas. We will use the mesh approach. To use the mesh approach, (1) write mesh equations using KVL, (2) use the dot convention to determine the signs of the induced voltage components, and (3) solve the resulting equations in the usual manner using determinants, your calculator or a computer program such as Matlab.

To specialize to the sinusoidal ac case, convert voltages and currents to phasor form. To do this, recall from Chapter 16 that inductor voltage in phasor form is $\mathbf{V}_L = j\omega L\mathbf{I}$. (This is the phasor equivalent of $v_L = L\,di/dt$, Figure 23–51.) This means that $L\,di/dt$ becomes $j\omega L\mathbf{I}$; in a similar fashion, $M\,di_1/dt \Rightarrow j\omega M\mathbf{I}_1$ and $M\,di_2/dt \Rightarrow j\omega M\mathbf{I}_2$. Thus, in phasor form Equations 23–15 become

$$\mathbf{V}_1 = j\omega L_1\mathbf{I}_1 + j\omega M\mathbf{I}_2$$
$$\mathbf{V}_2 = j\omega M\mathbf{I}_1 + j\omega L_2\mathbf{I}_2$$

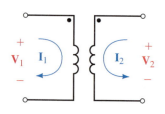

FIGURE 23–59 Coupled coils with sinusoidal ac excitation.

These equations describe the circuit of Figure 23–59 as you can see if you write KVL for each loop. (Verify this.)

EXAMPLE 23–14

For Figure 23–60, write the mesh equations and solve for $\mathbf{I}_1$ and $\mathbf{I}_2$. Let $\omega = 100$ rad/s, $L_1 = 0.1$ H, $L_2 = 0.2$ H, $M = 0.08$ H, $R_1 = 15\ \Omega$, and $R_2 = 20\ \Omega$.

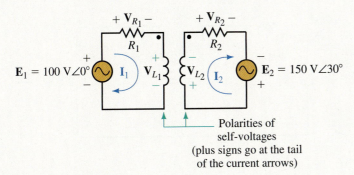

Polarities of
self-voltages
(plus signs go at the tail
of the current arrows)

FIGURE 23–60 Air-core transformer example.

Solution $\omega L_1 = (100)(0.1) = 10\ \Omega$, $\omega L_2 = (100)(0.2) = 20\ \Omega$, and $\omega M = (100)(0.08) = 8\ \Omega$. Since one current enters a dotted terminal and the other leaves, the sign of the mutual term is opposite to the sign of the self term. KVL yields

Loop 1: $\mathbf{E}_1 - R_1\mathbf{I}_1 - j\omega L_1\mathbf{I}_1 + j\omega M\mathbf{I}_2 = 0$

Loop 2: $\mathbf{E}_2 - j\omega L_2\mathbf{I}_2 + j\omega M\mathbf{I}_1 - R_2\mathbf{I}_2 = 0$

Thus,

$$(15 + j10)\mathbf{I}_1 - j8\mathbf{I}_2 = 100\angle 0°$$

$$-j8\mathbf{I}_1 + (20 + j20)\mathbf{I}_2 = 150\angle 30°$$

These can be solved by the usual methods such as determinants or by computer. The answers are $\mathbf{I}_1 = 6.36\angle -6.57°$ and $\mathbf{I}_2 = 6.54\angle -2.24°$.

EXAMPLE 23–15

For the circuit of Figure 23–61, determine $\mathbf{I}_1$ and $\mathbf{I}_2$.

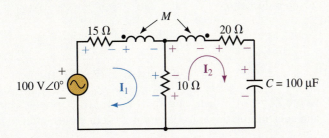

FIGURE 23–61 $L_1 = 0.1$ H, $L_2 = 0.2$ H, $M = 80$ mH, $\omega = 100$ rad/s

Solution $\omega L_1 = 10\ \Omega$, $\omega L_2 = 20\ \Omega$, $\omega M = 8\ \Omega$, and $X_C = 100\ \Omega$.

Loop 1: $100\angle 0° - 15\mathbf{I}_1 - j10\mathbf{I}_1 + j8\mathbf{I}_2 - 10\mathbf{I}_1 + 10\mathbf{I}_2 = 0$

Loop 2: $-10\mathbf{I}_2 + 10\mathbf{I}_1 - j20\mathbf{I}_2 + j8\mathbf{I}_1 - 20\mathbf{I}_2 - (-j100)\mathbf{I}_2 = 0$

Thus:

$$(25 + j10)\mathbf{I}_1 - (10 + j8)\mathbf{I}_2 = 100\angle 0°$$

$$-(10 + j8)\mathbf{I}_1 + (30 - j80)\mathbf{I}_2 = 0$$

Solution yields $\mathbf{I}_1 = 3.56$ A$\angle -18.6°$ and $\mathbf{I}_2 = 0.534$ A$\angle 89.5°$

Refer to the circuit of Figure 23–62.

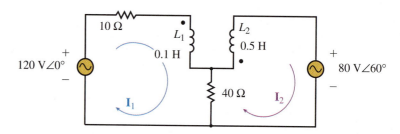

FIGURE 23–62 $M = 0.12$ H, $\omega = 100$ rad/s.

a. Determine the mesh equations.

b. Solve for currents I_1 and I_2.

Answer

a. $(50 + j10)I_1 - (40 - j12)I_2 = 120\angle 0°$

  $-(40 - j12)I_1 + (40 + j50)I_2 = -80\angle 60°$

b. $I_1 = 1.14$ A$\angle -31.9°$ and $I_2 = 1.65$ A$\angle -146°$

23.11 Coupled Impedance

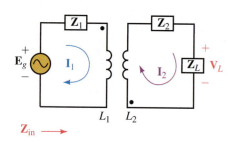

FIGURE 23–63

Earlier we found that an impedance Z_L on the secondary side of an iron-core transformer is reflected into the primary side as $a^2 Z_L$. A somewhat similar situation occurs in loosely coupled circuits. In this case however, the impedance that you see reflected to the primary side is referred to as **coupled impedance.** To get at the idea, consider Figure 23–63. Writing KVL for each loop yields

Loop 1: $E_g - Z_1 I_1 - j\omega L_1 I_1 - j\omega M I_2 = 0$

Loop 2: $-j\omega L_2 I_2 - j\omega M I_1 - Z_2 I_2 - Z_L L_2 = 0$

which reduces to

$$E_g = Z_p I_1 + j\omega M I_2 \qquad\qquad \text{(23–21a)}$$

$$0 = j\omega M I_1 + (Z_s + Z_L)I_2 \qquad\qquad \text{(23–21b)}$$

where $Z_p = Z_1 + j\omega L_1$ and $Z_s = Z_2 + j\omega L_2$. Solving Equation 23–21b for I_2 and substituting this into Equation 23–21a yields, after some manipulation,

$$E_g = Z_p I_1 + \frac{(\omega M)^2}{Z_s + Z_L}I_1$$

Now, divide both sides by I_1, and define $Z_{in} = E_g/I_1$. Thus,

$$Z_{in} = Z_p + \frac{(\omega M)^2}{Z_s + Z_L} \qquad\qquad \text{(23–22)}$$

The term $(\omega M)^2/(Z_s + Z_L)$, which reflects the secondary impedances into the primary, is the coupled impedance for the circuit. Note that since secondary impedances appear in the denominator, they reflect into the primary with reversed reactive parts. Thus, capacitance in the secondary circuit looks inductive to the source and vice versa for inductance.

EXAMPLE 23–16

For Figure 23–63, let $L_1 = L_2 = 10$ mH, $M = 9$ mH, $\omega = 1000$ rad/s, $\mathbf{Z}_1 = R_1 = 5\ \Omega$, $\mathbf{Z}_2 = 1\ \Omega - j5\ \Omega$, $\mathbf{Z}_L = 1\ \Omega + j20\ \Omega$ and $\mathbf{E}_g = 100$ V$\angle 0°$. Determine $\mathbf{Z}_{in}$ and $\mathbf{I}_1$.

Solution

$\omega L_1 = 10\ \Omega$. Thus, $\mathbf{Z}_p = R_1 + j\omega L_1 = 5\ \Omega + j10\ \Omega$.

$\omega L_2 = 10\ \Omega$. Thus, $\mathbf{Z}_s = \mathbf{Z}_2 + j\omega L_2 = (1\ \Omega - j5\ \Omega) + j10\ \Omega = 1\ \Omega + j5\ \Omega$.

$\omega M = 9\ \Omega$ and $\mathbf{Z}_L = 1\ \Omega + j20\ \Omega$. Thus,

$$\mathbf{Z}_{in} = \mathbf{Z}_p + \frac{(\omega M)^2}{\mathbf{Z}_s + \mathbf{Z}_L} = (5 + j10) + \frac{(9)^2}{(1 + j5) + (1 + j20)}$$

$$= 8.58\ \Omega\angle 52.2°$$

$$\mathbf{I}_1 = \mathbf{E}_g/\mathbf{Z}_{in} = (100\angle 0°)/(8.58\angle 52.2°) = 11.7\ \text{A}\angle -52.2°$$

The equivalent circuit is shown in Figure 23–64.

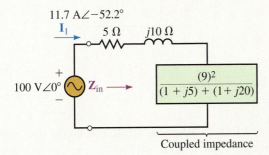

FIGURE 23–64

For Example 23–16, let $R_1 = 10\ \Omega$, $M = 8$ mH, and $\mathbf{Z}_L = (3 - j8)\ \Omega$. Determine $\mathbf{Z}_{in}$ and $\mathbf{I}_1$.

Answers
$28.9\ \Omega\angle 41.1°$; $3.72\ \text{A}\angle -41.1°$

MultiSIM and PSpice may be used to solve coupled circuits (see Notes). As a first example, let us solve for generator and load currents and the load voltage for the circuit of Figure 23–65. First, manually determine the answers to provide

23.12 Circuit Analysis Using Computers

NOTES...

PSpice handles both loosely coupled circuits and tightly coupled (iron-core) transformers, but as of this writing, MultiSIM handles only iron-core devices.

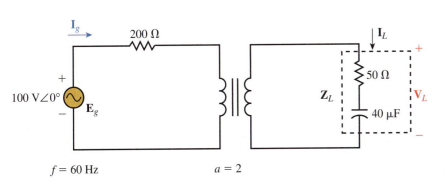

$f = 60$ Hz $a = 2$

FIGURE 23–65 Iron-core circuit for first MultiSIM and PSpice example.

a basis for comparison. Reflecting the load impedance using $a^2\mathbf{Z}_L$ yields the equivalent circuit of Figure 23–66. From this,

$$\mathbf{I}_g = \frac{100\ \text{V}\angle 0°}{200\ \Omega + (200\ \Omega - j265.3\ \Omega)} = 208.4\ \text{mA}\angle 33.5°$$

Thus,

$$\mathbf{I}_L = a\mathbf{I}_g = 416.8\ \text{mA}\angle 33.5°$$

and

$$\mathbf{V}_L = \mathbf{I}_L\mathbf{Z}_L = 34.6\ \text{V}\angle -19.4°.$$

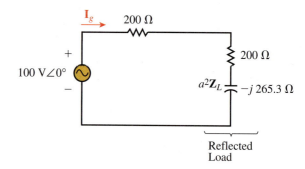

FIGURE 23–66   Equivalent of Figure 23–65.

MultiSIM

Read Notes 1 and 4 from "PSpice and MultiSIM Notes for Coupled Circuits," page 864, then create the circuit of Figure 23–67 on your screen. (Use TRANSFORMER_VIRTUAL. It is found in the basic parts bin. As indicated in Note 1, it models a transformer by means of its turns ratio.) Double click the transformer symbol and set its turns ratio to **2** and its primary and secondary winding resistances and leakage inductances to **0.000001.** (The values aren't critical; they just have to be small enough to be negligible.) Next, set the magnetizing inductance to **10000 H.** (This is L_m of Figure 23–41. Theoretically, for an ideal transformer it is infinity. All we need to do is make it very large.) Set all meters to AC, then click the ON/OFF button to activate the circuit. The answers of Figure 23–67 agree exactly with the analytic solution above.

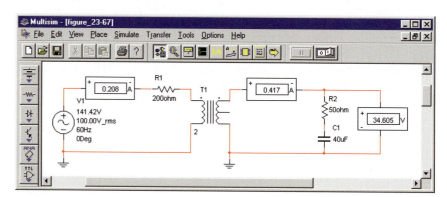

FIGURE 23–67   MultiSIM screen for Figure 23–65.

PSpice

Read Notes 2, 3, and 4 from "PSpice and MultiSIM Notes for Coupled Circuits" (see page 864). As indicated, transformer element XFRM_LINEAR may be used to model iron-core transformers based solely on their turns ratios. To do this, set coupling $k = 1$, choose an arbitrarily large value for L_1, then let $L_2 = L_1/a^2$ where a is the turns ratio. (The actual values for L_1 and L_2 are not critical; they simply must be very large.) For example, arbitrarily choose $L_1 = 100000$ H, then compute $L_2 = 100000/(2^2) = 25000$ H. This sets $a = 2$. Now proceed as follows. Create the circuit of Figure 23–65 on the screen as Figure 23–68. Use source VAC, set up as shown. Double click VPRINT1 and then set AC, MAG, and PHASE to **yes** in the Properties Editor. Repeat for the IPRINT devices. Double click the transformer and set COUPLING to **1,** L1 to **100000H,** and L2 to **25000H.** In the simulations settings dialog box, select AC Sweep and set $\underline{S}$tart and $\underline{E}$nd frequencies to **60Hz** and Points to **1.** Run the simulation, then scroll through the Output File. You should find $\mathbf{I}_g = 208.3$ mA$\angle 33.6°$, $\mathbf{I}_L = 416.7$ mA$\angle 33.6°$ and $\mathbf{V}_L = 34.6$ V$\angle -19.4°$. Note that these agree almost exactly with the computed results.

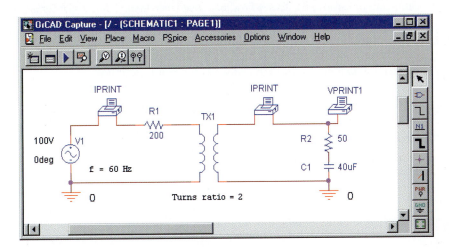

FIGURE 23–68   PSpice screen for Figure 23–65.

◀ MULTISIM

As a final PSpice example, consider the loosely coupled circuit of Figure 23–60, drawn on the screen as Figure 23–69. Use VAC for the sources and XFRM_LINEAR for the transformer. (Be sure to orient Source 2 as shown.) Compute $k = \dfrac{M}{\sqrt{L_1 L_2}} = 0.5657$. Now double click the transformer symbol and set $L_1 = \mathbf{0.1H}$, $L_2 = \mathbf{0.2H}$, and $k = \mathbf{0.5657.}$ Compute source frequency f (it is 15.9155 Hz). Select AC Sweep and set $\underline{S}$tart and $\underline{E}$nd frequencies to **15.9155Hz** and Points to **1.** Run the simulation. When you scroll the Output File, you will find $\mathbf{I}_1 = 6.36$ A$\angle -6.57°$ and $\mathbf{I}_2 = 6.54$ A$\angle -2.23°$ as determined earlier in Example 23–14.

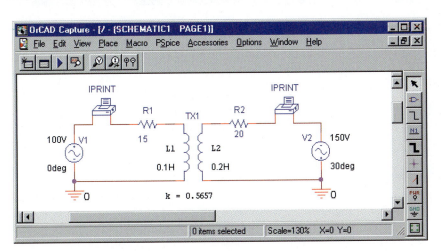

FIGURE 23–69   PSpice solution for Example 23–14.

PSPICE AND MULTISIM NOTES FOR COUPLED CIRCUITS

1. MultiSIM uses the ideal transformer equations $\mathbf{E}_p/\mathbf{E}_s = a$ and $\mathbf{I}_p/\mathbf{I}_s = 1/a$ to model transformers. It also permits you to set winding resistance, leakage flux, and exciting current effects as per Figure 23–41. As of this writing, however, MultiSIM cannot easily model loosely coupled circuits.

2. The PSpice transformer model XFRM_LINEAR is based on self-inductances and the coefficient of coupling and is thus able to handle loosely coupled circuits directly. It is also able to model tightly coupled circuits (such as iron-core transformers). To see how, note that basic theory shows that for an ideal iron-core transformer, $k = 1$ and L_1 and L_2 are infinite, but their ratio is $L_1/L_2 = a^2$. Thus, to approximate the transformer, just set L_1 to an arbitrary very large value, then compute $L_2 = L_1/a^2$. This fixes a, permitting you to model iron-core transformers based solely on their turns ratio.

3. The sign of the coefficient of coupling to use with PSpice depends on the dot locations. For example, if dots are on adjacent coil ends (as in Figure 23–60), make k positive; if dots are on opposite ends (Figure 23–63), make k negative.

4. PSpice and MultiSIM require grounds on both sides of a transformer.

PUTTING IT INTO PRACTICE

A circuit you are building calls for a 3.6-mH inductor. In your parts bin, you find a 1.2-mH and a 2.4-mH inductor. You reason that if you connect them in series, the total inductance will be 3.6 mH. After you build and test the circuit, you find that it is out of spec. After careful reasoning, you become suspicious that mutual coupling between the coils is upsetting operation. You therefore set out to measure this mutual inductance. However, you have a meter that measures only self-inductance. Then an idea hits you. You de-energize the circuit, unsolder the end of one of the inductors and measure total inductance. You get 6.32 mH. What is the mutual inductance?

PROBLEMS

23.1   Introduction

1. For the transformers of Figure 23–70, sketch the missing waveforms.

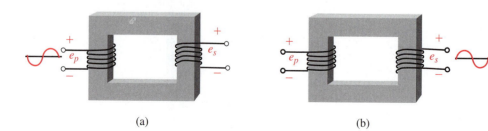

(a)   (b)

FIGURE 23–70

23.2 Iron-Core Transformers: The Ideal Model

2. List the four things that you neglect when you idealize an iron-core transformer.

3. An ideal transformer has $N_p = 1000$ turns and $N_s = 4000$ turns.

   a. Is it step-up or step-down voltage?

   b. If $e_s = 100 \sin \omega t$, what is e_p when wound as in Figure 23–7(a)?

   c. If $E_s = 24$ volts, what is E_p?

   d. If $\mathbf{E}_p = 24\ \text{V}\angle 0°$, what is $\mathbf{E}_s$ when wound as in Figure 23–7(a)?

   e. If $\mathbf{E}_p = 800\ \text{V}\angle 0°$, what is $\mathbf{E}_s$ when wound as in Figure 23–7(b)?

4. A 3:1 step-down voltage transformer has a secondary current of 6 A. What is its primary current?

5. For Figure 23–71, determine the expressions for v_1, v_2, and v_3.

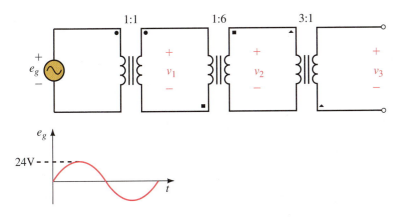

FIGURE 23–71

6. If, for Figure 23–72, $\mathbf{E}_g = 240\ \text{V}\angle 0°$, $a = 2$, and $\mathbf{Z}_L = 8\ \Omega - j6\ \Omega$, determine the following:

   a. $\mathbf{V}_L$ b. $\mathbf{I}_L$ c. $\mathbf{I}_g$

7. If, for Figure 23–72, $\mathbf{E}_g = 240\ \text{V}\angle 0°$, $a = 0.5$, and $\mathbf{I}_g = 2\ \text{A}\angle 20°$, determine the following:

   a. $\mathbf{I}_L$ b. $\mathbf{V}_L$ c. $\mathbf{Z}_L$

8. If, for Figure 23–72, $a = 2$, $\mathbf{V}_L = 40\ \text{V}\angle 0°$, and $\mathbf{I}_g = 0.5\ \text{A}\angle 10°$, determine $\mathbf{Z}_L$.

9. If, for Figure 23–72, $a = 4$, $\mathbf{I}_g = 4\ \text{A}\angle 30°$, and $\mathbf{Z}_L = 6\ \Omega - j8\ \Omega$, determine the following:

   a. $\mathbf{V}_L$ b. $\mathbf{E}_g$

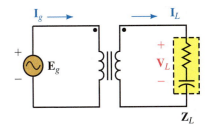

FIGURE 23–72

10. If, for the circuit of Figure 23–72, $a = 3$, $\mathbf{I}_L = 4\ \text{A}\angle 25°$, and $\mathbf{Z}_L = 10\ \Omega\angle -5°$, determine the following:

    a. Generator current and voltage.

    b. Power to the load.

    c. Power output by the generator.

    d. Does $P_{\text{out}} = P_{\text{in}}$?

23.3 Reflected Impedance

11. For each circuit of Figure 23–73, determine $\mathbf{Z}_p$.

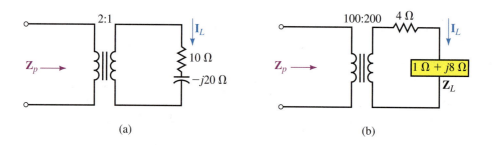

(a)   (b)

FIGURE 23–73

12. For each circuit of Figure 23–73, if $\mathbf{E}_g = 120\ \text{V}\angle 40°$ is applied, determine the following, using the reflected impedance of Problem 11.

   a. $\mathbf{I}_g$ b. $\mathbf{I}_L$ c. $\mathbf{V}_L$

13. For Figure 23–73(a), what turns ratio is required to make $\mathbf{Z}_p = (62.5 - j125)\ \Omega$?

14. For Figure 23–73(b), what turns ratio is required to make $\mathbf{Z}_p = 84.9\angle 58.0°\ \Omega$?

15. For each circuit of Figure 23–74, determine $\mathbf{Z}_T$.

16. For each circuit of Figure 23–74, if a generator with $\mathbf{E}_g = 120\ \text{V}\angle -40°$ is applied, determine the following:

   a. $\mathbf{I}_g$ b. $\mathbf{I}_L$ c. $\mathbf{V}_L$

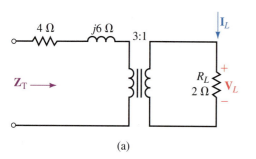

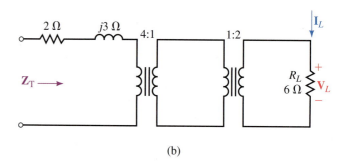

(a)   (b)

FIGURE 23–74

23.4 Power Transformer Ratings

17. A transformer has a rated primary voltage of 7.2 kV, $a = 0.2$, and a secondary rated current of 3 A. What is its kVA rating?

18. Consider a 48 kVA, 1200/120-V transformer.

   a. What is the maximum kVA load that it can handle at $F_p = 0.8$?

   b. What is the maximum power that it can supply to a 0.75 power factor load?

   c. If the transformer supplies 45 kW to a load at 0.6 power factor, is it overloaded? Justify your answer.

23.5 Transformer Applications

19. The transformer of Figure 23–25 has a 7200-V primary and center-tapped 240-V secondary. If Load 1 consists of twelve 100-W lamps, Load 2 is a 1500-W heater, and Load 3 is a 2400-W stove with $F_p = 1.0$, determine

   a. I_1 b. I_2 c. I_N d. I_p

20. An amplifier with a Thévenin voltage of 10 V and Thévenin resistance of 128 Ω is connected to an 8-Ω speaker through a 4:1 transformer. Is the load matched? How much power is delivered to the speaker?

21. An amplifier with a Thévenin equivalent of 10 V and R_{Th} of 25 Ω drives a 4-Ω speaker through a transformer with a turns ratio of $a = 5$. How much power is delivered to the speaker? What turns ratio yields 1 W?

22. For Figure 23–75, there are 100 turns between taps 1 and 2 and 120 between taps 2 and 3. What voltage at tap 1 yields 120 V out? At tap 3?

23. For Figure 23–30(a), $a_2 = 2$ and $a_3 = 5$, $\mathbf{Z}_2 = 20\ \Omega\angle 50°$, $\mathbf{Z}_3 = (12 + j4)\ \Omega$ and $\mathbf{E}_g = 120\ \text{V}\angle 0°$. Find each load current and the generator current.

24. It is required to connect a 5-kVA, 120/240-V transformer as an autotransformer to a 120-V source to supply 360 V to a load.

    a. Draw the circuit.

    b. What is the maximum current that the load can draw?

    c. What is the maximum load kVA that can be supplied?

    d. How much current is drawn from the source?

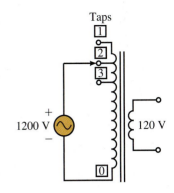

FIGURE 23–75 $N_s = 200$ turns.

23.6 Practical Iron-Core Transformers

25. For Figure 23–76, $\mathbf{E}_g = 1220\ \text{V}\angle 0°$.

    a. Draw the equivalent circuit,

    b. Determine $\mathbf{I}_g$, $\mathbf{I}_L$, and $\mathbf{V}_L$.

26. For Figure 23–76, if $\mathbf{V}_L = 118\ \text{V}\angle 0°$, draw the equivalent circuit and determine

    a. $\mathbf{I}_L$ b. $\mathbf{I}_g$ c. $\mathbf{E}_g$

    d. no-load voltage   e. regulation

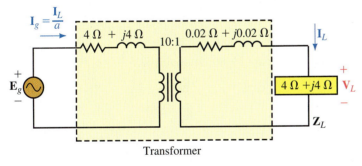

FIGURE 23–76

◀ MULTISIM

27. A transformer delivering $P_{out} = 48$ kW has a core loss of 280 W and a copper loss of 450 W. What is its efficiency at this load?

23.7 Transformer Tests

28. A short-circuit test (Figure 23–46) at rated current yields a wattmeter reading of 96 W, and an open-circuit test (Figure 23–48) yields a core loss of 24 W.

    a. What is the transformer's efficiency when delivering the full, rated output of 5 kVA at unity F_p?

    b. What is its efficiency when delivering one quarter the rated kVA at 0.8 F_p?

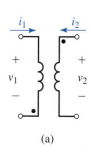

(a)

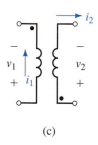

(b)

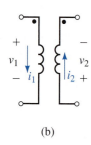

(c)

FIGURE 23–77

23.9 Loosely Coupled Circuits

29. For Figure 23–77,

$$v_1 = L_1\frac{di_1}{dt} \pm M\frac{di_2}{dt}, \qquad v_2 = \pm M\frac{di_1}{dt} + L_2\frac{di_2}{dt}$$

For each circuit, indicate whether the sign to use with M is plus or minus.

30. For a set of coils, $L_1 = 250$ mH, $L_2 = 0.4$ H, and $k = 0.85$. What is M?

31. For a set of coupled coils, $L_1 = 2$ H, $M = 0.8$ H and the coefficient of coupling is 0.6. Determine L_2.

32. For Figure 23–52(a), $L_1 = 25$ mH, $L_2 = 4$ mH, and $M = 0.8$ mH. If i_1 changes at a rate of 1200 A/s, what are the primary and secondary induced voltages?

33. Everything the same as Problem 32 except that $i_1 = 10\ e^{-500t}$ A. Find the equations for the primary and secondary voltages. Compute them at $t = 1$ ms.

34. For each circuit of Figure 23–78, determine L_T.

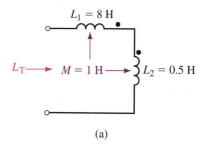

(a)

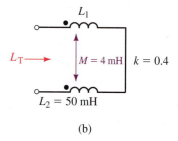

(b)

FIGURE 23–78

35. For Figure 23–79, determine L_T.

36. For the circuit of Figure 23–80, determine **I**.

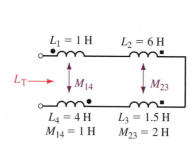

FIGURE 23–79

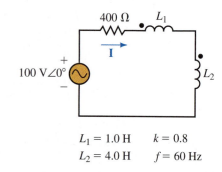

FIGURE 23–80 Coupled parallel inductors.

37. The inductors of Figure 23–81 are mutually coupled. What is their equivalent inductance? If $f = 60$ Hz, what is the source current?

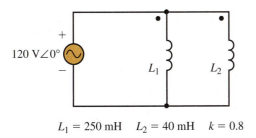

$L_1 = 250$ mH $\quad L_2 = 40$ mH $\quad k = 0.8$

FIGURE 23–81

23.10 Magnetically Coupled Circuits with Sinusoidal Excitation

38. For Figure 23–60, $R_1 = 10$ Ω, $R_2 = 30$ Ω, $L_1 = 100$ mH, $L_2 = 200$ mH, $M = 25$ mH, and $f = 31.83$ Hz. Write the mesh equations.

39. For the circuit of Figure 23–82, write mesh equations.

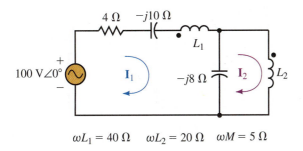

$\omega L_1 = 40$ Ω $\quad \omega L_2 = 20$ Ω $\quad \omega M = 5$ Ω

FIGURE 23–82

40. Write mesh equations for the circuit of Figure 23–83.

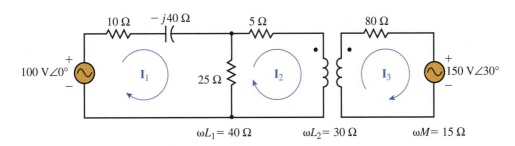

$\omega L_1 = 40$ Ω $\quad \omega L_2 = 30$ Ω $\quad \omega M = 15$ Ω

FIGURE 23–83

41. Write mesh equations for the circuit of Figure 23–84. (This is a very challenging problem.)

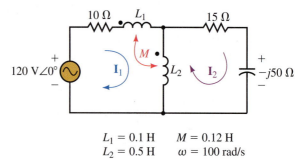

$$L_1 = 0.1 \text{ H} \quad M = 0.12 \text{ H}$$
$$L_2 = 0.5 \text{ H} \quad \omega = 100 \text{ rad/s}$$

FIGURE 23–84

23.11   Coupled Impedance

42. For the circuit of Figure 23–85,
   a. Determine $\mathbf{Z}_{in}$
   b. Determine $\mathbf{I}_g$.

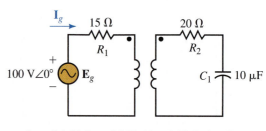

$$L_1 = 0.1 \text{ H}; L_2 = 0.2 \text{ H}; M = 0.08 \text{ H}; f = 60 \text{ Hz}$$

FIGURE 23–85

23.12   Circuit Analysis Using Computers

Notes: (1) At the time of writing, MultiSIM solves for magnitude only. (2) With PSpice, orient the IPRINT devices so that current enters the positive terminal. Otherwise the phase angles will be in error by 180°.

◀ MULTISIM   ◀ CADENCE

43. An iron-core transformer with a 4:1 turns ratio has a load consisting of a 12-Ω resistor in series with a 250-μF capacitor. The transformer is driven from a 120-V$\angle 0°$, 60-Hz source. Use MultiSIM or PSpice to determine the source and load currents. Verify the answers by manual computation.

◀ MULTISIM   ◀ CADENCE

44. Using MultiSIM or PSpice, solve for the primary and secondary currents and the load voltage for Figure 23–86.

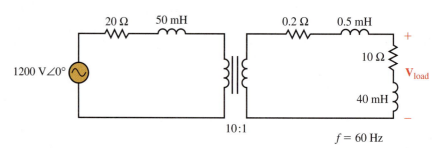

FIGURE 23–86

45. Using PSpice, solve for the source current for the coupled parallel inductors of Figure 23–81. Hint: Use XFRM_LINEAR to model the two inductors. You will need two very low-value resistors to avoid creating source-inductor loops.

46. Solve for the currents of Figure 23–61 using PSpice. Compare these to the answers of Example 23–15.

47. Solve for the currents of Figure 23–62 using PSpice. Compare these to the answers of Practice Problem 8.

48. Solve Example 23–16 for current $\mathbf{I}_1$ using PSpice. Compare answers. Hint: If values are given as X_L and X_C, you must convert them to L and C.

✓ ANSWERS TO IN-PROCESS LEARNING CHECKS

In-Process Learning Check 1

1. Step-up; 200 V

2. a. 120 V∠–30°; b. 120 V∠150°

3.

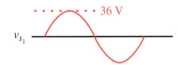

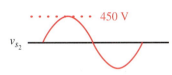

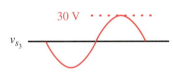

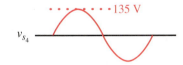

4. Secondary, upper terminal.

5. a. Terminal 4; b. Terminal 4

In-Process Learning Check 2

1. $\mathbf{Z}_p = 18.75\ \Omega - j25\ \Omega$; $R = 6400\ \Omega$; $a = 1.73$

2. 125 V∠0°

3. 89.4

4. a. 200 A; 200 A; b. 24 kW; 18 kW

5. Tap 1: 109.1 V; Tap 2: 120 V; Tap 3: 126.3 V

6. 0.81 W; 0.72 W; Maximum power is delivered when $R_s = a^2 R_L$.

7. a. 348 V; 20.7 A; b. 12 A; c. 12 A↓ 8.69 A ↑

In-Process Learning Check 3

1. a. $\mathbf{Z}_{eq} = 0.8\ \Omega + j0.8\ \Omega$

   b. 113.6 V ∠0.434°

   c. 5.63%

Foundation Electronic Concepts

VI

In previous chapters, we dealt with basic circuit analysis. In this section, we turn our attention to the second major topic area of the book—electronic principles and practices.

The electronic devices covered in this book are *semiconductor solid state* devices—that is they are made from solid crystalline semiconductor material. To understand how they operate requires an understanding of elementary atomic theory. Although we introduced the basics of this theory in Chapter 2, some additional ideas are needed. We begin with these. Following this, we examine various devices including diodes, transistors, operational amplifiers, etc. We then look at the application of these devices in various important circuits throughout the remainder of the book.

In keeping with the trend in teaching electronics today, less time is spent on the detailed analysis of transistor biasing techniques than in the past and more is devoted to the use of op-amps in small signal amplifiers. ■

■ OBJECTIVES

On completion of this chapter, you will
be able to

- sketch the symbolic representation
  of an atom, with particular attention
  to the valence shell,

- explain what is meant by energy
  levels, energy bands, the conduction band, the valence band, and
  energy gaps,

- compare the energy gaps of insulators, semiconductors, and conductors,

- describe how electron-hole pairs
  are created in semiconductors,

- explain what is meant by the terms
  trivalent, tetravalent, and *pentavalent*
  as they relate to atomic structure,

- describe the use of doping to foster
  the creation of conduction electrons
  and holes,

- describe electron current and how it
  accounts for conduction in *n*-material,

- describe how holes provide the
  mechanism by which electrons pass
  through *p*-material,

- describe what is meant by majority
  carriers and minority carriers,

- describe the important aspects of a
  p-n junction, including the depletion
  region and the barrier potential,

- identify the terminals of a diode and
  state the barrier potential for a typical diode,

- describe how conduction takes place
  through a forward-biased junction,

- describe why a reverse-biased
  junction blocks current,

- describe the basics of reverse
  junction breakdown and junction
  capacitance.

Introduction to Semiconductors

24

Semiconductor materials are the heart of solid-state electronics. From them, we fashion important devices such as diodes, transistors, and integrated circuits, and from these, we build computers, cell phones, home theatre systems, and the like. The most commonly used semiconductor material is silicon (Si), although germanium (Ge) is used in certain applications. Thus, most of our focus is on silicon.

Intrinsic (pure) semiconductors are materials that have properties between those of conductors and insulators—i.e. they conduct electricity better than insulators but not as well as conductors. However, by suitable "doping", we can alter their characteristics and fabricate the devices that we require from the modified materials. To understand how this is done and how such devices operate, you need an understanding of elementary atomic theory. Thus, we begin with a review of the atom. We then introduce energy levels, bonding, and the concept of electron and hole currents. Next, we describe doping and the *p-n* junction. The concepts of majority carriers and minority carriers are then described and the mechanism of charge flow in doped materials is studied. Finally, the simplest useful semiconductor device, the junction diode is described. We conclude with a look at forward and reverse biasing, thus setting the stage for our study of diodes in Chapter 25. ∎

CHAPTER PREVIEW

24.1 Semiconductor Basics

Consider Figure 24–1. As noted in Chapter 2, an atom consists of a positively charged nucleus of protons and neutrons surrounded by a cloud of orbiting negatively charged electrons. In its normal state, each atom has an equal number of protons and electrons and thus, is neutral, i.e., uncharged. Electrons are grouped into shells, designated *K, L, M, N,* etc. The last occupied shell is called the valence shell, and electrons in this shell are valence electrons. Materials are categorized electrically as conductors, insulators, or semiconductors, depending on how many electrons they have in their valence shell. Materials with 1 valence electron are conductors, those with 8 are insulators and those with 4 are semiconductors. Atoms with 4 valence electrons are called **tetravalent** atoms—thus, semiconductors are tetravalent materials.

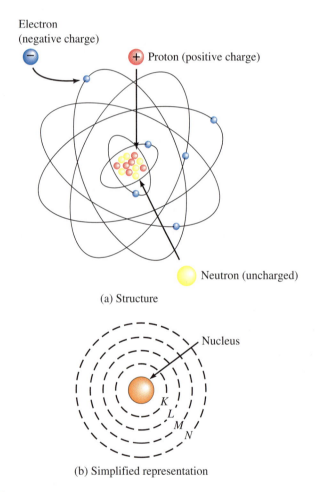

Electron
(negative charge)

+ Proton (positive charge)

Neutron (uncharged)

(a) Structure

Nucleus

K
L
M
N

(b) Simplified representation

FIGURE 24–1 Bohr model of the atom. Although this is a somewhat simplified picture, it is adequate for our purposes.

Two semiconductor materials, silicon (Si) and germanium (Ge), are of particular interest to the electronics industry. Of these, silicon is by far the most widely used because of its superior electrical properties. An important characteristic of both Si and Ge is that they tend to form crystals in their intrinsic state. This crystalline structure leads to some important electrical properties, as you will soon see.

Energy Levels of Orbiting Electrons

The electrons of Figure 24–1 travel about the nucleus at incredible speeds, making billions of trips per second. This means that they posses energy. A study of atomic physics shows that electrons with greater energy orbit at greater distances from the nucleus than those with less energy—see sidebar note. This means that valence electrons have the greatest energy of all. Studies also show that orbiting electrons can exist only at certain discrete energy levels—thus, only those orbits corresponding to these levels (the shells of Figure 24–1) are possible. Figure 24–2 symbolically illustrates the energy level associated with each shell for an isolated atom. Note the energy gaps between levels. These are called *forbidden* zones. No electron can orbit in a forbidden zone, although it may pass through such a zone if it changes orbit.

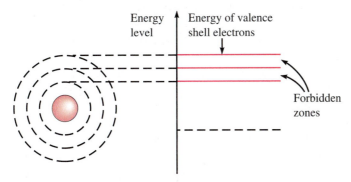

FIGURE 24–2 Energy levels for an isolated atom.

Energy Bands

Figure 24–2 suggests that an atom has discrete energy levels. However, single atoms do not exist in isolation. When atoms group together, the electrical forces of attraction and repulsion between adjacent atoms slightly alter the radiuses of orbiting electrons, causing the energy levels to widen into bands. Figure 24–3 illustrates bands for the three electrical classes of materials, insulators, semiconductors, and conductors. (Since we are only interested in valence electron energy levels, we have shown only their valence energy bands.) However, a new energy band, denoted the *conduction band*, has been added to each.

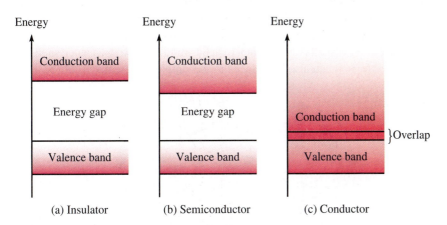

FIGURE 24–3 Conduction and valence bands for an insulator, a semiconductor, and a conductor.

N O T E S . . .

An Analogy

It may not be apparent why electrons in outer orbits have more energy than those in inner orbits. The following analogy may help. Consider a weight at the end of a rope. When you swing this weight in a circular motion, you experience a force pulling on the rope. If you twirl it faster (giving it more energy), the pull increases. If you let the rope slip through your hand (without letting go), the weight moves to a larger orbit. To keep the weight in this orbit, you must keep swinging it more energetically than when it was in the smaller orbit—in other words, you have to impart greater energy to the weight to keep it in the larger orbit. From this, you can see that the weight has greater energy in the larger orbit than it did in the smaller one.

The Conduction Band

As you saw in Chapter 2, electrons may be dislodged from atoms by the application of external energy. For example, for a metal like copper, huge numbers of valence electrons can gain sufficient energy from heat at room temperature to escape from their parent atoms and wander randomly from atom-to-atom throughout the material. Termed *free electrons,* they are (as you saw in Chapter 2) the electrons that take part in the conduction process, thus, they are also called *conduction electrons.* Since conduction electrons have more energy than those electrons that are still bound to their parent atoms, we have added an energy band (the conduction band) above the valence energy band to accommodate them.

The space between the valence band and the conduction band is called the *energy gap.* It represents the amount of energy, usually specified in electron volts (eV) (see sidebar note), that is required to break an electron free and move it into the conduction band. Now let us consider each class of material in turn.

Insulators For insulators (Figure 24–3(a)), the energy gap is quite large. This means that considerable energy is needed to move an electron from the valence band to the conduction band. Because of this, insulating materials have very few free electrons and thus, their conductivity is nearly zero.

Semiconductors The energy gap for semiconductors is much narrower than for insulators, typically around 1 eV. This is low enough that some valence electrons gain sufficient energy from heat alone (even at room temperature) to jump from the valence band to the conduction band (Figure 24–3(b)). Enough free electrons are created in this manner that semiconductors can actually conduct (although not well). As a result, semiconductors have electrical properties that are somewhere between those of a conductor and an insulator.

Conductors For conductors, the valence band and the conduction band overlap, Figure 24–3(c). In the overlap region, electrons in the valence band are also in the conduction band and thus, conductors have enough free electrons to permit easy conduction. Consequently, such materials have high conductivity.

Silicon and Germanium Atoms

Let us look specifically at semiconductors. Consider first a silicon atom. As shown in Figure 24–4(a), it has 14 orbiting electrons with 4 in the valence shell. Since we are interested only in valence electrons, we can simplify the representation to that of (b). The central group of charges contains the 14 positive charges ($14p^+$) of the nucleus plus the 10 electrons of the inner shells for a net positive charge of $4p^+$. Now consider a germanium atom (not shown). It has 32 protons and 32 electrons, its valence shell is shell N and all inner shells are full. Since its valence shell contains 4 electrons, it can also be reduced to the simple representation of (b).

NOTES . . .

The Electron Volt

Throughout this text, we generally use the SI unit of energy, the joule. However when dealing with energy levels of atoms, this is too large a unit. In atomic physics, energy is usually expressed in electron volts (eV). Recall from Chapter 2, Equation 2–3, energy is given by $W = QV$. Let $Q = e$ (the charge on an electron). If the electron is moved through a potential difference of V volts, then the energy change is $W = eV$ electron volts. Specifically, if $V = 1$ V, then $W = 1$ eV. Since $e = 1.602 \times 10^{-19}$ coulombs, 1 eV $= 1.602 \times 10^{-19}$ joules.

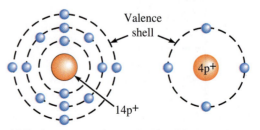

(a) Basic representation   (b) Simplified representation.
The net core change is $4p^+$

FIGURE 24–4 A silicon atom.

Covalent Bonding

In Chapter 2, we noted that a valence shell is full when it contains 8 electrons. Semiconductors contain 4. However, when crystals of silicon or germanium solidify, their atoms arrange themselves and bond together in such a manner as to fill all valence shells by sharing valence electrons with adjacent atoms. This is called *covalent bonding* and is illustrated symbolically in Figure 24–5. Note that each atom bonds with 4 neighboring atoms to gain the 4 required electrons. Since all valence shells are full, the material acts like an insulator. (Actually, thermal energy changes the situation somewhat so that this picture is true only at a temperature of 0 K as described in the next section.)

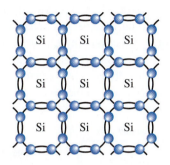

FIGURE 24–5 Covalent bonding. Each atom shares valence electrons with its four neighbors. Since all valence shells effectively posses 8 electrons, they are full.

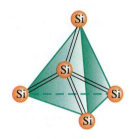

FIGURE 24–6 Si and Ge form three-dimensional tetrahedral shaped crystal. The double lines represent (symbolically) the shared valence electrons.

Electron-Hole Pairs

As noted above, at absolute zero temperature, all electrons are tied up in covalent bonds and there are no free electrons. However, as temperature is increased, some electrons in the valence band gain sufficient energy to break free and move into the conduction band. The vacancy left by a departing electron is called a *hole,* and each electron that moves into the conduction band leaves a corresponding hole in the valence band, creating an **electron-hole pair.** However, the life of a free electron is relatively short and soon after it jumps to the conduction band, it falls back into one of the available holes. This process is called *recombination.* Because of continual recombination and regeneration, at any given time, there are of the order of 10^{10} electron-hole pairs/cm^3 in silicon and 10^{12} in germanium at room temperature. This means that statistically (at room temperature) there are this many free electrons in their respective conduction bands. Note also that the atoms that have momentarily lost an electron are left with a net positive charge and are thus *positive ions.* However, the material still has no net charge because each positive ion bound in the lattice is balanced by a free electron elsewhere in the material.

Electron Current and Hole Current

Although the numbers of free electrons indicated above may seem large (see sidebar note), the numbers are actually quite small and thus, intrinsic silicon and germanium are very poor conductors. Nonetheless, a small amount of conduction does take place if an external source of emf is applied, although the details are somewhat different than for conductors since both an electron current and a hole current occur.

24.2 Conduction in Semiconductors

N O T E S . . .

Two Viewpoints on Holes

We now have two viewpoints on holes. On the one hand, we can view holes simply as a means whereby electrons move through a semiconductor material from the negative source terminal to the positive. On the other, we can view the holes as being positive charges that move through the material from the positive source terminal to the negative. Both viewpoints are useful and in what follows, we will use whichever is convenient.

Figure 24–7 illustrates the process. As indicated, electrons are attracted to the left (toward the positive terminal of the battery). However, only those electrons in the conduction band are free to move. This movement of electrons constitutes a current (referred to as *electron* current). Now consider the electrons that remain in the valence band. These cannot move in the same manner as conduction electrons since they are still bound to the parent atom and do not have enough energy to break free. However, as indicated in (a), the hole created by the displaced free electron exists in a valence band orbit and an electron from a neighboring atom may wander into it, filling the hole. When this happens, another hole is created as in (b). The process repeats itself, with the result that, as valence band electrons shift left into vacant holes, the holes effectively shift right as indicated by the progression in Figures 24–7(a), (b) and (c). Since a hole represents the absence of a negative charge in an otherwise neutral material, the moving holes look like moving positive charges. Since moving charges represent current, the movement of holes constitutes a current (referred to as *hole current*)—see sidebar note. Since there are equal numbers of free electrons and corresponding holes, the electron current is exactly equal to the hole current.

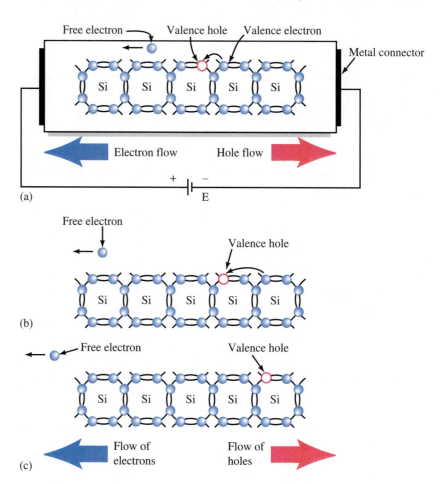

FIGURE 24–7   Illustrating the flow of electrons and holes in intrinsic silicon.

The Effect of Temperature

As temperature is increased, more electrons move into the conduction band. This increases the conductivity of the material and reduces its resistance. Consequently, semiconductors have a **negative temperature coefficient (NTC).** (In contrast, metals like copper have positive temperature coefficients since their resistance increases with temperature.)

A semiconductor in its intrinsic state is of little use. However, when impurities of the appropriate type and amount are added in the molten state, the properties of the material change dramatically. The addition of as little as 1 part per million (ppm) will foster the creation of a large number of holes or electrons, resulting in increased conductivity and greatly reduced resistivity. The process of adding impurities is referred to as *doping*. (Doped material is referred to as **extrinsic material.**) Doping can be used to create either **n-type** or **p-type** material.

24.3 Doping

n-Type Semiconductor

An *n*-type semiconductor is a semiconductor that has more free electrons than holes. It is created when intrinsic semiconductor is doped with atoms containing 5 electrons in their valence band. (Such atoms are called **pentavalent atoms** and include phosphorous (P), arsenic (As), and antimony (Sb)). To illustrate, consider again Figure 24–5. If the central atom of silicon (which has 4 valence electrons) is replaced by an atom of arsenic (which has 5 valence electrons), there will be 9 valence electrons available for bonding instead of the required 8. Thus, there is an extra electron not needed for the bonding process (Figure 24–8). (The material however, is still electrically neutral.) The arsenic atom, which provides the additional electron, is referred to as a **donor atom.** Because the extra electron is loosely bound, thermal energy may dislodge it, turning it into a free (conduction) electron. If you add 1 ppm of impurity, statistically every millionth atom will have an excess electron. For silicon (which has about 10^{23} atoms/cm^3), this adds about 10^{17} additional electrons/cm^3, greatly increasing the available free electrons and hence, the conductivity. Note that no new holes are created, i.e., the only holes present are the thermally-generated ones.

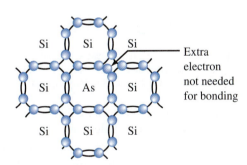

FIGURE 24–8   *n*-type material is created when intrinsic semiconductor is doped with pentavalent atoms. The extra electrons from the arsenic atoms are able to move freely throughout the material like free electrons do in a metal such as copper.

Majority Carriers and Minority Carriers

Recall intrinsic semiconductor has equal numbers of free electrons and holes. When we added the donor impurity, we got huge numbers of additional free electrons but no new holes. As a result in *n*-material, free-electrons vastly outnumber holes. Thus when a source is connected, electrons and holes move through the material as you saw in Figure 24–7, but since there are so many more electrons than holes, most of the conduction takes place as conduction electron flow (Figure 24–9). For this reason, in *n*-material, electrons are called **majority carriers** and holes are called **minority carriers.**

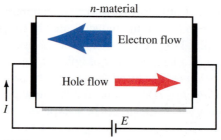

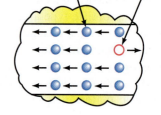

(a) Electrons are the majority carriers, and holes are the minority carriers

(b) Conceptual representation. Note that most of the carriers are electrons

FIGURE 24–9   Conduction in *n*-type material.

p-Type semiconductor

A *p*-type semiconductor is a semiconductor that has more holes than conduction electrons. It is created when intrinsic material is doped with **trivalent atoms** (atoms with three electrons in their valence band such as boron (B), aluminum (Al) and gallium (Ga)). To illustrate, consider Figure 24–10. Since boron has only three electrons in its valence shell, the covalent bond is missing an electron, i.e., a hole has been created. This valence hole will readily accept an available electron, thus permitting the hole to "move" through the material. The boron atom is referred to as an **acceptor atom.**

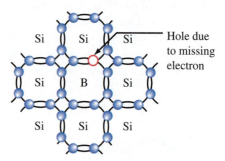

FIGURE 24–10   *p*-type material is created when intrinsic semiconductor is doped with trivalent atoms. The valence hole permits the acceptance of free electrons. Thus, the extra hole from the boron atom is free to move throughout the material similarly to what we saw in Figure 24–7.

Note that each acceptor atom creates a hole that is not counterbalanced by a new free electron. Thus, in *p*-material, holes vastly outnumber free electrons. Consequently, when an external voltage source is applied, there are many more holes that "move" through the material than there are electrons—thus, holes are the majority carriers and electrons the minority carriers in *p*-material (Figure 24–11).

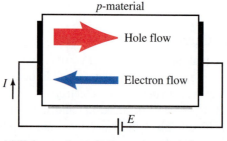

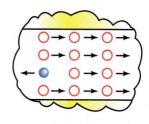

(a) Holes are the majority carriers, and electrons are the minority carriers

(b) Conceptual representation

FIGURE 24–11   Conduction in *p*-type material.

We now turn our attention to the ***p-n* junction.** A *p-n* junction is created when semiconductor material is fabricated with an abrupt transition from *p*-type to *n*-type material. However, junctions must be created by some sort of molten or diffusion process that maintains the continuous lattice structure of the material—you cannot simply push two pieces of material together to form a junction—see Notes.

24.4 The *p-n* Junction

NOTES . . .

Fabricating a *p-n* Junction

One way to create a *p-n* junction is by diffusion—you can diffuse donor impurities into *p*-material or acceptor impurities into *n*-material. The basic process is illustrated in Figure 24–12. Here, an *n*-type material is heated to a high temperature (about 950° C) in the presence of boron gas, causing boron impurities to diffuse into the *n*-material. The process is continued until sufficient boron atoms bond with the silicon atoms to create a region of *p*-material. Since the procedure is done at high temperature, the desired *p-n* junction is created while maintaining the required crystalline structure of the semiconductor.

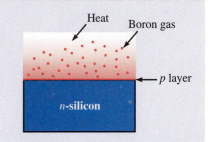

FIGURE 24–12 Fabricating a *p-n* junction by the diffusion process.

Although you cannot create a *p-n* junction by simply butting two pieces together, it is instructive to imagine that you can. Consider isolated *n*-type and *p*-type wafers (Figure 24–13). (Only majority carriers are depicted for each.) These carriers move about randomly in their respective materials under the influence of thermal energy.

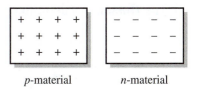

p-material   *n*-material

FIGURE 24–13 Isolated wafers with only majority carriers depicted. Both materials are electrical neutral.

Depletion Region and Barrier Potential

Now (conceptually) bring the two pieces together (Figure 24–14(a)). As they touch, some of the free electrons in the *n*-material diffuse across the junction into the holes in the *p*-material. Since the *p*-material was initially neutral, these additional electrons create a region of negative charge in the *p*-material near the junction. Note also that, since the electrons originate from an initially neutral *n*-material, their departure creates a region of positive charge in the *n*-material. The process continues until equilibrium is reached. Since the region straddling the junction has been depleted of majority carriers, it is called the *depletion region*. Figure 24–14(b) shows an expanded (conceptual) view. Since initially neutral atoms in the *n*-material have lost an electron, they are short one negative charge each and thus, are positive ions. Conversely, initially neutral atoms in the *p*-material have gained an electron and are thus negative ions. These separated charges produce an electric field in the depletion region in much the same manner as the charges on a capacitor's plates produce an electric field in the space between them. In fact, the separated charges create a voltage across the junction called the *junction* or **barrier** voltage (or **potential**) V_B. For silicon, V_B is approximately 0.7 V at 25° C, while for germanium it is approximately 0.3 V. (However, this barrier voltage cannot be used as a voltage source—i.e., if you connect a load across a *p-n* junction, it will not force current through the circuit as does a battery. It is best simply to think of V_B as a barrier voltage that must be overcome by an applied external source in order to force current through the device.) The polarity of V_B is as indicated in Figure 24–14.

Depletion region

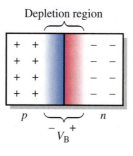

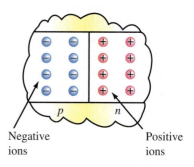

(a) *p-n* junction. Red represents a region of positive charge while blue represents a region of negative charge. (The junction width has been exaggerated for purposes of illustration.)

(b) Conceptual representation of the depletion region. Since atoms in the *p* material near the junction have gained an electron, they form negative ions, while atoms in the *n* material have lost an electron and form positive ions.

FIGURE 24–14   Conceptual joining of the two wafers to create a *p-n* junction. The migration of majority carriers across the junction results in a depletion region and a barrier potential V_B.

24.5 The Biased *p-n* Junction

The *p-n* junction is the basis for several important devices, the simplest of which is a two-terminal device called a **diode** (Figure 24–15). For reasons discussed below, a diode is a unidirectional device that passes current easily in one direction (in the direction of its arrow (c)), but blocks it in the other. Thus, when **forward biased** as in (d), the diode conducts easily, but when **reverse biased** as in (e), it doesn't. (Actually, there may be a small leakage current when reverse biased, but it is so small that it is usually negligible as described next.) The *p* end is called the **anode** and the *n* end is called the **cathode.**

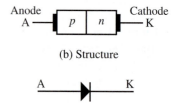

(b) Structure

(c) Symbol. The arrow indicates the direction that it passes current

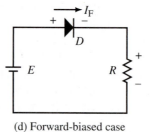

(a) Photo of a diode. The gray body band denotes the cathode end

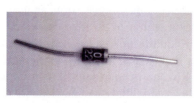

(d) Forward-biased case

(e) Reverse-biased case

FIGURE 24–15   The diode is a unidirectional device that passes current only in one direction.

The Reverse-Biased *p-n* Junction

For reverse biasing, the positive terminal of the source is connected to the *n*-material and the negative terminal to the *p*-material (Figure 24–16). Note that the depletion region has widened. This is because the positive terminal of the voltage source attracts electrons (the majority carriers) in the *n*-material toward its positive terminal, drawing them away from the junction, while at the same time, holes in the *p*-material (its majority carriers) are drawn to the negative terminal, again away from the junction. As you can see, no majority carriers are attracted toward and, hence, across the junction—thus, the majority current is zero.

Now consider minority carriers. Due to thermal energy, a small number of electron-hole pairs are created in each material. Because of their random motion, some of these will wander into the depletion region. For reasons that we won't go into (it requires a closer look at conduction band energy levels), all *p*-side minority electrons that enter the depletion region are swept through to the *n*-side. Similarly, all minority holes from the *n*-side that enter the depletion region are swept across into the *p*-side. This movement of charge constitutes a current. However, there are so few minority carriers in each material that the current, called **saturation current** I_s, is very small. (For example, saturation current for a typical signal diode is of the order of a few nanoamps to a few microamps.) Since minority carriers are generated by heat energy (not doping), saturation current is dependent on temperature (as we discuss more fully in Chapter 25).

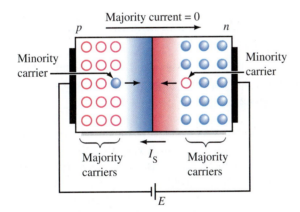

FIGURE 24–16 Reverse biased *p-n* junction. Since so few minority carriers are present, I_s is very small, typically in the nA range for signal diodes.

The Forward-Biased *p-n* Junction ($E > V_D$)

For forward-biasing, the positive terminal of the source is connected to the *p*-material and the negative terminal to the *n*-material, Figure 24–17. First consider the *n*-side. There are many free electrons (majority carriers) in the *n*-material, and these are repelled away from the negative terminal and attracted toward the positive terminal by the potential of the source—thus, they are propelled toward the junction. If the source voltage is greater than the barrier potential, many will acquire sufficient energy to cross where they recombine with holes to fill bonding sites in the *p*-material. (The battery provides a continuous supply of electrons at the negative terminal to replenish them.) Now consider those electrons that have crossed the junction. Because of the source potential, they drift from hole-to-hole in the valence band (as depicted earlier in Figure 24–7). From this, we see that current in the *n*-material is an electron current while the current in the *p*-material is a hole current—that is, conduction in each material is by means

PRACTICAL NOTES . . .

For the forward biased case Figure 24–18, the diode conducts when the source voltage E exceeds the barrier voltage V_B. We have not answered the question about what happens when E is less than V_B. However, this latter case is generally not of much practical interest and will not be considered here.

of its majority carriers. This current is referred to as $I_{majority}$. There is also a small amount of minority current (opposite in direction to the majority current in each material). Total current I_D (Figure 24–17) is thus composed of both majority and minority current. However, minority current is so small, it can usually be neglected. Thus, for all practical purposes, diode forward current consists of majority current only—that is, $I_D \approx I_{majority}$. The voltage across the diode is approximately the barrier voltage. This is illustrated in Figure 24–18 for a silicon diode: here, $V_D \approx 0.7$ V.

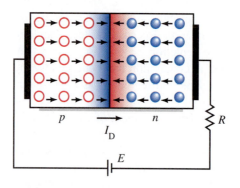

FIGURE 24–17 Forward-biased *p-n* junction. Although there is a small amount of minority current, for all practical purposes $I_D \approx I_{majority}$.

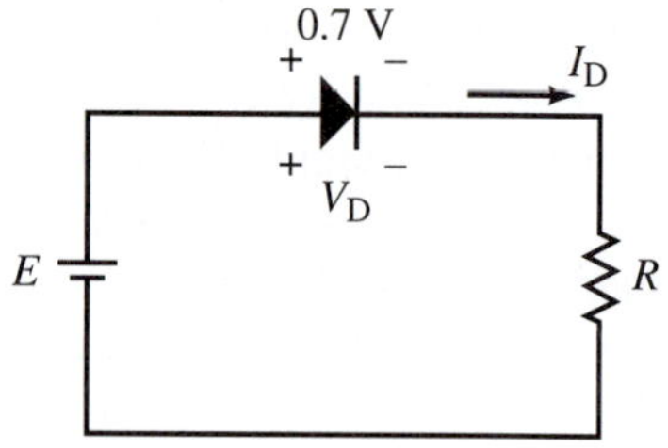

FIGURE 24–18 Forward-biased silicon diode.

24.6 Other Considerations

Junction Breakdown

When a reverse biased diode is subjected to too much voltage, its junction breaks down. When this happens, large reverse current results. There are two breakdown mechanisms.

Avalanche Breakdown As voltage is increased, minority carriers in the depletion region are accelerated to such velocities that when they collide with atoms in the crystal structure, electrons are knocked free, creating additional electron-hole pairs. These electrons are in turn accelerated and the process feeds upon itself until an "avalanche" of electrons results. Until breakdown occurs, the current is very small, but after breakdown, it is very high and, if not limited by the external circuit, will destroy the diode. The voltage at which this breakdown occurs is called *peak inverse voltage (PIV)* or *peak reverse voltage (PRV).* (We look at it in more detail in Chapter 25.)

Zener Breakdown If the semiconductor material is heavily doped, the depletion region narrows. This increases the electric gradient field at the junction, and at some applied voltage, the force on the electrons is so great that they are literally torn from orbit. This phenomenon is referred to as *Zener breakdown* and the voltage at which it occurs is called the *Zener voltage* V_Z. Diodes specifically developed to utilize this effect are called *Zener diodes.* We consider them also in Chapter 25.

Junction Capacitance

The structure of a reverse-biased *p-n* junction is similar to a capacitor in that its depletion region behaves as a dielectric (insulator), while its *n* and *p* regions because they are conductive, act like capacitor plates. As you have already seen, the thickness of the depletion region depends on the applied reverse voltage—

the larger the voltage, the wider the region. As you learned in Chapter 10, increasing the distance between capacitor plates decreases its capacitance—thus you can see that increasing reverse voltage lowers junction capacitance (and vice versa). The capacitance of a reverse-biased diode is very small, typically of the order of 5 to 100 pF. At low frequencies this represents a very large reactance $X_C = \dfrac{1}{2\pi f C}$ that has little effect on circuit operation. However, at high frequencies, X_C may become small enough that you have to take it into account. Some diodes (called *varistor diodes* or *EPICAP diodes*) are designed specifically to take advantage of this voltage controlled capacitance. We consider them in Chapter 25.

PROBLEMS

24.1   Semiconductor Basics

1. What is meant by the term *valence electron?*

2. Consider the copper atom of Figure 2–4 (Chapter 2). Which has the greater energy, an electron in the *M* shell or an electron in the *N* shell? Why?

3. Sketch an energy level diagram for gold.

4. If the energy gap for a certain material is 1.762×10^{-19} J, how many eV are needed to move an electron from its valence band to its conduction band?

5. How does the energy gap for a semiconductor compare to that of an insulator? Of a conductor? Use sketches to answer this question.

6. Sketch a germanium atom. Indicate the number of protons in its nucleus and the number of electrons in each shell.

7. A carbon atom nucleus contains 6 protons and 6 neutrons.

   a. How many electrons are in a neutral carbon atom?

   b. How many electrons are in its outermost shell?

   c. Based on the answer of part b, carbon would be classified as what type of atom, *trivalent, tetravalent,* or *pentavalent?* Explain your answer.

24.2   Conduction in Semiconductors

8. Copper has many more free electrons per unit volume than does intrinsic semiconductor. By what factor (approximately) are the numbers of the free electrons in copper larger than they are in intrinsic germanium at room temperature? Contrast this to silicon.

9. a. Explain what is meant by the term *positive temperature coefficient* and give an example of a material that has a PTC.

   b. Explain what is meant by the term *negative temperature coefficient* and give an example of a material that has an NTC.

10. The valence electrons of germanium are located at a greater distance from the nucleus than those of silicon. What effect does this have on the conductivity of the germanium (as compared to silicon) when temperature rises? Explain.

11. Describe hole current in intrinsic semiconductor, i.e., describe the means by which valence electrons move through the material.

24.3   Doping

12. What do the terms *intrinsic* and *extrinsic* mean as applied to semiconductors?

13. a. *p*-type impurity consists of _____ (tri, tetra, penta)-valent atoms.

    b. *n*-type impurity consists of _____ (tri, tetra, penta)-valent atoms.

14. How does the addition of an impurity affect the resistivity of a semiconductor such as silicon?

15. In *p*-type material, _____ (holes, electrons) are the majority carrier.

16. In *n*-type material, _____ (holes, electrons) are the minority carrier.

17. Charges moving in the external wires connected to a *p-n* junction are (holes, electrons).

18. If you add aluminum as an impurity to intrinsic silicon, what type of extrinsic material results?

19. Explain how the addition of antimony increases the number of free electrons in intrinsic silicon without increasing the number of holes.

24.4 The *p-n* Junction

20. Why can you not create a *p-n* junction by simply pressing wafers of *p*-type and *n*-type material together?

21. The barrier potential of a silicon *p-n* junction at 25° C is approximately _____ V.

22. When a *p-n* junction is constructed, it results in a device called a/an _____ .

23. If a voltmeter is placed across an unbiased silicon *p-n* junction at a temperature of 25° C, it will measure

    a. −0.7 V   b. +0.7 V

    c. 0 V   d. −0.3 V

    e. +0.3 V

24. For the *p-n* junction of Figure 24–19, what type of semiconductor material is used and which is the *p*-side. Explain your reasoning.

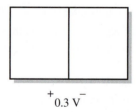

FIGURE 24–19

24.5 The Biased *p-n* Junction

25. Describe why forward biasing a junction narrows the depletion region.

26. Figure 24–20 shows an example of a _____ (forward/reverse) -biased diode. Its current will be _____ (large/small).

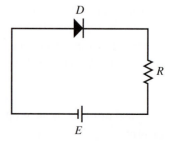

FIGURE 24–20

27. In a forward-biased diode, the current consists mainly of _____ (majority/minority) carriers.

28. In a reverse-biased diode, leakage current is referred to as _____ carrier current.

29. For the diode of Figure 24–21, sketch the schematic, label the anode and cathode, and indicate the direction of conventional current through the device when forward biased.

FIGURE 24–21

30. Redraw Figure 24–18 assuming a germanium diode.

■ OBJECTIVES

After studying this chapter, you will be able to

- analyze simple diode based circuits using the basic diode models,

- sketch the characteristic curve of a real diode and describe the important considerations relative to its forward- and reverse-biased regions,

- interpret the data presented on diode spec sheets,

- describe the operating characteristics of a zener diode,

- interpret the data presented on zener diode spec sheets,

- describe and analyze the basic zener voltage regulator circuit,

- describe the characteristics and the application of varactor diodes,

- explain and analyze the operation of half-wave and full-wave rectifier circuits,

- describe basic power supply filtering,

- analyze basic filter circuits to determine ripple and dc output voltage for full-wave and half-wave circuits,

- determine ripple factor for filtered power supply waveforms,

- use PSpice and MultiSIM to analyze diode based circuits.

Diode Theory and Application

25

There are many types of commercial diodes, including the basic junction diode, the zener diode, the varactor diode, the photodiode, and so on. There are also variations within families. For example, within the basic diode family, there are rectifier diodes (intended for heavy current, low frequency applications) and signal diodes, (intended for low current, high frequency applications). However, all share the basic *p-n* junction structure studied in Chapter 24, and thus have certain characteristics in common. In this chapter, we first look at these characteristics. We then examine some of the specialty diodes. Finally we look at the application of diodes in power supply circuits.

When studying diodes, it is customary (as in many other areas of electronics) to create models to help understand their operation and application. Often, devices can be modeled in several different ways. For example we sometimes start with a simple model to explore a device's main characteristics. Then, when we have a good understanding of these, we refine the model to take into account the things that we have ignored. This is how we approach the diode. We develop three different models, then, for any given problem, select the one that best suits the task with which we are faced. We begin with the ideal model. Here, we consider only the device's fundamental behavior and neglect everything else. ∎

TRADITIONALLY, AM RADIOS HAVE USED diode detection to separate the low frequency broadcast information (i.e., voice and music) from the high frequency radio wave carrier on which it rides. However, the path from early radio to modern radio contains some unusual devices. One of the most unusual was the "coherer." The coherer consisted of a pile of metal filings lying loosely in a glass tube between metallic electrodes. It was connected between the antenna and other circuitry and when a radio signal was received, its resistance dropped dramatically, allowing the passage of current. However, since the coherer did not resume its original high resistance, it was fitted with a tapping device that shook the filings loose again, returning it to its high resistance state. As crazy as it sounds, the device actually worked—it was the detector used by Marconi on his first successful transatlantic radio experiments in 1901. ∎

25.1 Diode Models

The Ideal Diode

As you saw in Chapter 24, the fundamental characteristic of a diode is that it passes current easily in one direction but blocks it in the other. This behavior can be modeled by means of a switch (Figure 25–1). When forward-biased as in (a), the diode looks like a closed switch and thus can effectively be replaced by a short circuit, while when reverse-biased (b), it looks like an off switch and can be replaced by an open circuit. This representation is called the **ideal model**. For the forward case, $V_D = 0$ V (since the diode looks like a short), while for the reverse case, $I_D = 0$ (since it looks like an open).

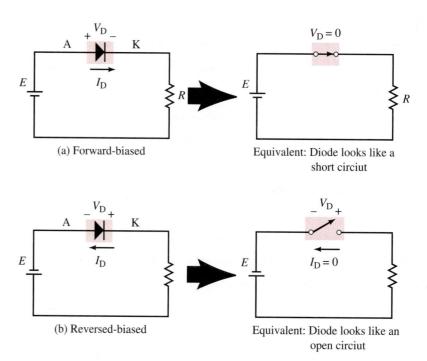

(a) Forward-biased   Equivalent: Diode looks like a short circiut

(b) Reversed-biased   Equivalent: Diode looks like an open circiut

FIGURE 25–1   The ideal diode model.

EXAMPLE 25–1

For Figure 25–2(a), determine diode voltages and circuit currents for the forward and reverse biased cases.

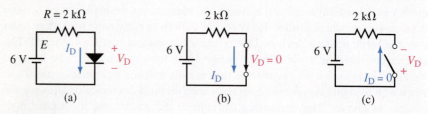

(a)   (b)   (c)

FIGURE 25–2

Solution

Forward-biased case (b): Since the diode looks like a short, $V_D = 0$ V and $I_D = E/R = (6 \text{ V})/(2\text{k}\Omega) = 3$ mA.

Reverse-biased case (c): Since the diode looks like an open circuit, current is zero and thus the voltage across R is also zero. KVL yields $V_D = 6$ V.

Ideal Diode Characteristic Curve

The behavior described above can be shown graphically. When forward-biased, the diode has zero volts across its terminals regardless of current. This plots as a vertical line (Figure 25–3). When reverse-biased, the diode has zero current, regardless of voltage. This plots as a horizontal line. The resulting graph is called a V-I **characteristic curve.**

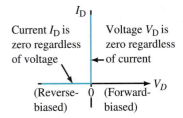

FIGURE 25–3 Characteristic curve for an ideal diode.

An Application

A simple application of a diode is the inexpensive two-level light dimmer circuit used in some table and floor lamps (Figure 25–4). When you switch the lamp on, contact 1 closes yielding the circuit of Figure 25–5(a). During the positive half cycle, the diode is forward-biased and conducts. During the negative half cycle, however, it is reverse-biased and does not. Since current results during only half the cycle, the light is dim. When you move the switch to its next position, you get circuit (b). With conduction over the full cycle, you get full rated current and the lamp is bright.

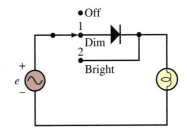

FIGURE 25–4 Dual level light dimmer circuit. Use a rectifier diode for this circuit.

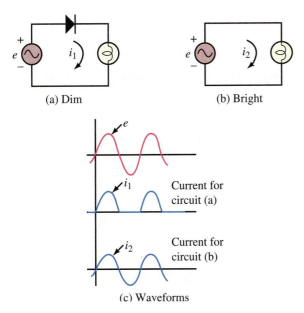

(a) Dim

(b) Bright

(c) Waveforms

FIGURE 25–5 Operation of the light dimmer circuit.

The Second Approximation

The ideal model is of limited value for numerical analysis. The second approximation yields an improvement by incorporating the barrier potential V_B. The forward-biased case is shown in Figure 25–6(b). Here the actual diode is represented as an ideal diode in series with a dc source equal to the barrier potential (approximately 0.7 V for silicon and 0.3 V for germanium)—see Note. Before the diode can conduct, the externally applied voltage must overcome this voltage. The characteristic curve is illustrated in (c). As you can see, current is zero when $E < V_B$ (i.e., below the knee of the curve), but when the applied voltage exceeds the barrier potential, the diode turns on and conduction takes place. Since the ideal diode part of the equivalent looks like a short circuit, voltage V_D remains constant at approximately 0.7 V (referred to as the *nominal voltage* of the diode), regardless of current. This yields a vertical line.

NOTES . . .

1. Voltage V_B shown in Figure 25–6 is not a source of emf. As noted in Chapter 24, it represents the junction (barrier) voltage of the diode and as we saw in Figure 24–18, it is the voltage that must be overcome by the external source to permit forward current.

2. Germanium diodes are seldom used in practice. Thus unless otherwise noted, all diodes used from here on are silicon.

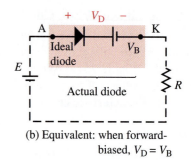

(a) Actual diode

(b) Equivalent: when forward-biased, $V_D = V_B$

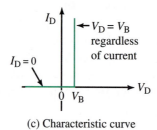

(c) Characteristic curve

FIGURE 25–6 Second diode approximation.

EXAMPLE 25–2

Using the second approximation, analyze the circuit of Figure 25–2(a).

Solution See Figure 25–7. For the forward-biased case, (a),

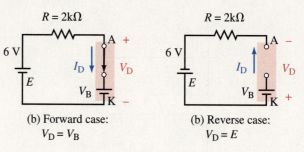

(b) Forward case:
$V_D = V_B$

(b) Reverse case:
$V_D = E$

FIGURE 25–7

$$I_D = \frac{E - V_D}{R} = \frac{6 \text{ V} - 0.7 \text{ V}}{2 \text{ k}\Omega} = 2.65 \text{ mA}$$

For reverse bias (b), $I_D = 0$ and the voltage across R is thus zero. KVL yields $V_D = 6$ V as in Example 25–1—thus, V_B has no effect here.

PRACTICE PROBLEMS 1

1. Repeat Example 25–2 for a germanium diode.
2. Add a 3-kΩ resistor across the diode of Figure 25–2 and solve for I_D for the forward-biased case. Hint: Use Thevenin's theorem.

Answers
1. Forward: 2.85 mA, 0.3 V; Reverse: 0 mA, 6 V
2. 2.42 mA

The Third Approximation

Semiconductor material has some resistance (recall Chapter 24), which can be incorporated as in Figure 25–8(b) to yield the third model. This resistance (called r_f) is quite small and ranges from a few ohms to several hundred ohms for commercially available diodes. When E exceeds V_B, the resultant current causes a drop across r_f and V_D increases slightly as current is increased, resulting in a sloped characteristic as in (c). For the reverse-biased case (not shown), the diode looks like an open circuit, current is zero, and thus, r_f has no effect. (We look at r_f in more detail in Section 25.2.)

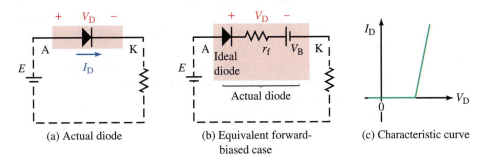

(a) Actual diode   (b) Equivalent forward-biased case   (c) Characteristic curve

FIGURE 25–8   Third diode approximation. The resistance r_f shown in (b) is studied in Section 25.2.

Real diodes differ somewhat from our models. Figure 25–9 shows the characteristic of a typical silicon diode. To help understand it, let us consider each region in detail.

25.2 Diode Characteristic Curve

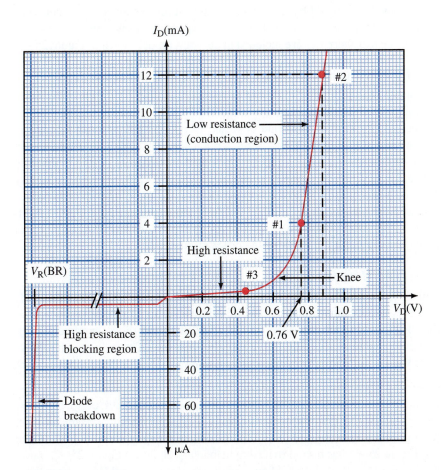

FIGURE 25–9   Characteristic curve of a real silicon diode.

The Forward Region

For low **forward voltage,** resistance is high since the depletion region at the *p-n* junction is relatively wide. Thus, current is low. As source voltage is increased, the depletion region narrows, and by the time that the *knee* of the curve is reached, it has been significantly reduced. As source voltage is increased further, more carriers become available until finally, the *conduction region* is reached. Note that voltage V_D is approximately 0.7 V here, but because of the internal voltage drop of the diode (due to its resistance r_f), V_D increases slightly as current increases. For a real diode, V_D may increase up to $\approx 1V$ under load.

Since resistance changes as you move along the curve, you must treat it as dynamic resistance. As noted in Chapter 4, Section 4.7, dynamic resistance is the inverse of the slope of the *V-I* curve. Thus, as shown in Figure 25–10,

$$r_f = \frac{\Delta V}{\Delta I} \qquad (25\text{–}1)$$

To illustrate, consider the region between points #1 and #2, i.e., the conduction region of Figure 25–9. Voltage V_D at 12 mA is 0.88 V and at 4 mA it is 0.76 V. Thus,

$$r_f = \frac{\Delta V}{\Delta I} = \frac{0.88\ V - 0.76\ V}{12\ mA - 4\ mA} = 15\ \Omega$$

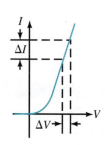

FIGURE 25–10 $r_f = \Delta V/\Delta I$.

Similarly, between the origin and point #3 you find $r_f = \dfrac{\Delta V}{\Delta I} = \dfrac{0.44\ V - 0\ V}{0.2\ mA - 0\ mA}$ = 2.2 kΩ

In practice, forward-biased diodes are almost always operated in their conducting regions, which means that their dynamic resistances will be relatively low. (This is the resistance that we included in our third model shown in Figure 25–8.)

The Reverse Region

For small reverse voltages, current is extremely small. (In Figure 25–9, we have expanded the current scale by a factor of 1000 to make it readable—thus, the current in the third quadrant is scaled in μA instead of mA.) As reverse voltage is further increased, you reach a point $V_{R(BR)}$ at which the diode breaks down and current increases rapidly. As noted in Chapter 24, this voltage is called the *peak inverse voltage* (PIV), *peak reverse voltage* (PRV), or simply $V_{R(max)}$. If this voltage is exceeded, the diode may be permanently damaged. (Some diodes, called zener diodes, are designed to operate beyond this point. We examine them in Section 25.4.) The resistance of the diode (before it breaks down) is very large, typically of the order of 10 MΩ.

Before We Move On

In reality, you will likely never need to compute resistance using Equation 25–1. However the concepts are important, as you will see in later sections. Note also that our third model describes the behavior of a real diode extremely well, except around the knee of the curve and at reverse breakdown. In practice, however, you do not normally operate a diode in either of these regions anyway.

25.3 Diode Data Sheets

Manufacturers generally create data sheets (also called *specs* or *specification sheets*) to describe their products. Typically they show electrical characteristics, recommended operating conditions, and maximum ratings. (Some specs are for ac and Figure 25–11 has been provided to help interpret these.) Figure 25–12 shows data for the 1N4001 to 1N4007 family of diodes—see Practical Notes (page 898).

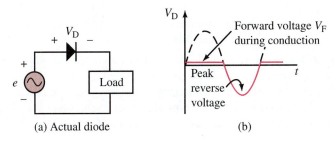

(a) Actual diode   (b)

FIGURE 25–11   Helping to understand diode ac characteristics.

1N4001, 1N4002, 1N4003, 1N4004, 1N4005, 1N4006, 1N4007

1N4004 and 1N4007 are Preferred Devices

Axial Lead Standard Recovery Rectifiers

This data sheet provides information on subminiature size, axial lead mounted rectifiers for general-purpose low-power applications.

Mechanical Characteristics
- Case: Epoxy, Molded
- Weight: 0.4 gram (approximately)
- Finish: All External Surfaces Corrosion Resistant and Terminal Leads are Readily Solderable
- Lead and Mounting Surface Temperature for Soldering Purposes: 220°C Max. for 10 Seconds, 1/16″ from case
- Shipped in plastic bags, 1000 per bag.
- Available Tape and Reeled, 5000 per reel, by adding a "RL" suffix to the part number
- Available in Fan-Fold Packaging, 3000 per box, by adding a "FF" suffix to the part number
- Polarity: Cathode Indicated by Polarity Band
- Marking: 1N4001, 1N4002, 1N4003, 1N4004, 1N4005, 1N4006, 1N4007

ON Semiconductor®

http://onsemi.com

**LEAD MOUNTED RECTIFIERS
50-1000 VOLTS
DIFFUSED JUNCTION**

**CASE 59-10
AXIAL LEAD
PLASTIC**

MARKING DIAGRAM

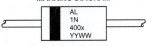

```
            AL
            1N
            400x
            YYWW
```

AL   = Assembly Location
1N400x = Device Number
x   = 1, 2, 3, 4, 5, 6 or 7
YY   = Year
WW   = Work Week

MAXIMUM RATINGS

Rating	Symbol	1N4001	1N4002	1N4003	1N4004	1N4005	1N4006	1N4007	Unit
*Peak Repetitive Reverse Voltage Working Peak Reverse Voltage DC Blocking Voltage	V_{RRM} V_{RWM} V_R	50	100	200	400	600	800	1000	Volts
*Non-Repetitive Peak Reverse Voltage (halfwave, single phase, 60 Hz)	V_{RSM}	60	120	240	480	720	1000	1200	Volts
*RMS Reverse Voltage	$V_{R(RMS)}$	35	70	140	280	420	560	700	Volts
*Average Rectified Forward Current (single phase, resistive load, 60 Hz, T_A = 75°C)	I_O	1.0							Amp
*Non-Repetitive Peak Surge Current (surge applied at rated load conditions)	I_{FSM}	30 (for 1 cycle)							Amp
Operating and Storage Junction Temperature Range	T_J T_{stg}	-65 to +175							°C

*Indicates JEDEC Registered Data

ORDERING INFORMATION

See detailed ordering and shipping information on page 2 of this data sheet.

Preferred devices are recommended choices for future use and best overall value.

FIGURE 25–12   Diode data sheet. (*Courtesy of ON Semiconductor*)

PRACTICAL NOTES . . .

1. Diodes are identified by part numbers, usually by a 1N prefix, for example, 1N4001, 1N4002, etc. as indicated in Figure 25–12.

2. Data sheets are typically divided into two main sections. The *Maximum Ratings* section shows limits that you must not exceed when you design your circuit, or else the device may be damaged or destroyed. The *Electrical Characteristics* section shows the typical and max values that you should expect during operation for key quantities, such as forward voltage drop and reverse current.

3. Although Figure 25–12 lists the complete family of 1N400X devices, the 1N4004 and 1N4007 are designated as *preferred* devices and are recommended for new designs.

4. When selecting a diode for a particular task it is important to include a safety margin, for example by choosing a component that is capable of handling at least 20% more current or voltage than the circuit is likely to require. (Some manufacturers of high quality equipment insist on a safety margin of 50%.) A safety factor of this magnitude ensures that the design will remain reliable for many years. In addition, you should use a preferred device if possible (as this reduces costs by cutting down on inventory). Thus, if the reverse voltage of a diode is determined to be 150 V, you might consider using a 1N4003, which has a $V_R = 200$ V. While this is acceptable, it is recommended instead that you use the 1N4004, which is a preferred device and is more likely to be stocked by your supplier.

5. You can get data sheets from the Internet.

Let us look at some of these specifications.

Reverse Voltage   All diodes in Figure 25–12 are identical except that each has a different maximum reverse voltage rating. For example the 1N4001 can handle a maximum dc voltage of $V_R = 50$ V, while the 1N4007 can handle 1000 V. With ac, the 1N4001 can handle a repetitive sinusoidal voltage with a peak reverse voltage of $V_{RRM} = 50$ V (Figure 25–11), a non-repetitive 60 Hz half sine wave maximum of $V_{RSM} = 60$ V or a continuous sine wave voltage with rms value of $V_{R(RMS)} = 35$ V, while the 1N4007 can handle corresponding values of 1000 V, 1200 V, and 700 V.

Forward Current   All members of the 1N400X family can handle a maximum **average forward current** of $I_0 = 1.0$ A (also called $I_{F(AVG)}$). However, when you first energize a circuit, you may get a brief current transient (called a *surge* current) that greatly exceeds the normal operating current. All devices can handle a surge current (denoted I_{FSM}) of 30 A for one cycle of a 60 Hz sinusoidal waveform.

Maximum Instantaneous Forward Voltage Drop v_F   This is the maximum instantaneous voltage across the forward-biased diode (Figure 25–11), and is guaranteed not to exceed 1.1 V at a current of $I_F = 1$ A.

Maximum Full-Cycle Average Voltage Drop $V_{F(AV)}$ When averaged over a full cycle, V_F is guaranteed not to exceed 0.8 V at a current of 1 A.

Temperature Derating

A diode generates heat and if the ambient temperature is high, this heat may not be dissipated fast enough and you may have to derate the device to protect it. For example, diode average forward current is typically specified at an ambient temperature (T_A) of 75° C, but if you wish to operate at a higher temperature, you must reduce the amount of current that you draw. Consider Figure 25–13. For the diode represented, if you maintain the ambient temperature at or below 75° C, you can draw 1 A; however if you wish to operate at an ambient of 125° C, you must design the circuit such that its average forward current does not exceed 0.4 A.

Parameter Shifts

Temperature also affects a diode's *V-I* characteristics, since an increase in temperature creates more thermally generated electron-hole pairs. For the forward case Figure 25–14, the greater number of carriers available results in a larger forward current for a given forward voltage, causing an effective shift of the characteristic to the left as temperature increases (and vice-versa). For both Si and Ge, this shift is about 0.25 mV for each °C change in junction temperature (T_J). To illustrate, consider a silicon diode with a barrier potential of $V_B = 0.7$ V at 25° C. If the junction temperature rises to 65° C (a rise of 40° C), V_B will drop by 40° C $\times$ 2.5 mV/°C = 0.10 V to 0.6 V.

For the reverse case, an increase in temperature also causes an increase in current, but here, the current is leakage (i.e., saturation) current. Experience has shown that each 10° C rise in junction temperature results in an approximate doubling of this current. For example, if $I_S = 10$ nA at 25° C, it will be 20 nA at 35° C, 40 nA at 45° C, etc. as illustrated in Figure 25–15.

25.4 Temperature Considerations and Other Effects

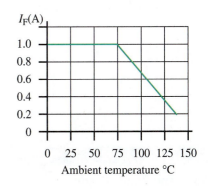

FIGURE 25–13   Typical derating curve.

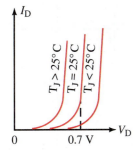

FIGURE 25–14   Effect of temperature on barrier voltage.

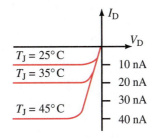

FIGURE 25–15   Effect of temperature on saturation current.

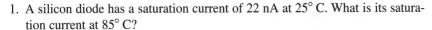

PRACTICE PROBLEMS 2

1. A silicon diode has a saturation current of 22 nA at 25° C. What is its saturation current at 85° C?

2. Repeat Problem 1 for a germanium diode with 15μA at 25° C

3. A silicon diode has a saturation current of 0.768 μA at 75° C. What is its saturation current at 25° C?

Answers
1. 1.41 μA; 2. 960 μA; 3. 24 nA

Reverse Recovery Time at Switching

When you switch a diode from its *on* state to its *off* state, the change does not occur instantaneously. The time that it takes (called **reverse recovery time**) is denoted t_{rr} and ranges from a few nanoseconds for switching diodes to a few microseconds for rectifier diodes. The larger t_{rr}, the slower is the diode.

Before We Move On

As you can see, temperature variations produce complex interactions in diode operation. Fortunately, for routine room temperature design you can generally ignore these. However if you wish to design for temperature extremes, you must take temperature effects into account. However, this is beyond the scope of this book and will not be considered here.

25.5 The Zener Diode

We look now at our first specialty diode, the **zener diode** (Figure 25–16). It is a special purpose diode designed to maintain a relatively constant voltage across its terminals when operated in its reverse-biased region. A typical characteristic is shown in Figure 25–17. The breakover voltage (the voltage at which the curve changes abruptly) is called the zener voltage V_Z. V_Z remains nearly constant from I_{ZK} (the knee current) to I_{ZM} (the maximum rated current). This means that a zener diode will maintain a nearly constant voltage across itself as long as its current is kept within these bounds.

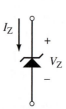

FIGURE 25–16 Zener diode symbol.

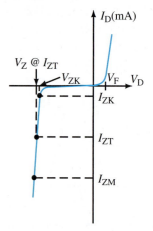

FIGURE 25–17 Characteristic curve of a typical silicon zener diode.

A Simple Application

To illustrate, let us look at a simple voltage regulator application (Figure 25–18). The job of the regulator is to hold the load voltage constant in spite of variations in input supply voltage and output load current. For this example, we have chosen a 12 V zener. Since the zener maintains (an almost) constant 12 V across itself, the voltage across the load will also remain at 12 V, even if the source voltage and/or the load current vary (within design limits as discussed later).

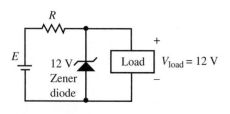

FIGURE 25–18 Using a zener diode for load voltage regulation.

The Zener Diode Spec Sheet

To understand how the circuit of Figure 25–18 works, you need to know more about zener characteristics. Such information can be found on data sheets such as that of Figure 25–19.

Zeners
1N4728A - 1N4752A

Absolute Maximum Ratings* $T_A = 25°C$ unless otherwise noted

Symbol	Parameter	Value	Units
P_D	Power Dissipation	1.0	W
	Derate above 50°C	6.67	mW/°C
T_{STG}	Storage Temperature Range	-65 to +200	°C
T_J	Operating Junction Temperature	+ 200	°C
$R_{\theta JL}$	Thermal resistance Junction to Lead	53.5	°C/W
$R_{\theta JA}$	Thermal resistance Junction to Ambient	100	°C/W
	Lead Temperature (1/16" from case for 10 seconds)	+ 230	°C
	Surge Power**	10	W

*These ratings are limiting values above which the serviceability of the diode may be impaired.
**Non-recurrent square wave PW = 8.3 ms, TA = 55 degrees C.

NOTES:
1) These ratings are based on a maximum junction temperature of 200 degrees C.
2) These are steady state limits. The factory should be consulted on applications involving pulsed or low duty cycle operations.

Tolerance: A = 5%

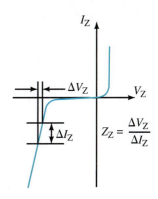

DO-41
COLOR BAND DENOTES CATHODE

Electrical Characteristics $T_A = 25°C$ unless otherwise noted

Device	V_Z (V)	Z_Z @ (Ω)	I_{ZT} (mA)	Z_{ZK} @ (Ω)	I_{ZK} (mA)	V_R @ (V)	I_R (μA)	I_{SURGE} (mA)	I_{ZM} (mA)
1N4728A	3.3	10	76	400	1.0	1.0	100	1380	276
1N4729A	3.6	10	69	400	1.0	1.0	100	1260	252
1N4730A	3.9	9.0	64	400	1.0	1.0	50	1190	234
1N4731A	4.3	9.0	58	400	1.0	1.0	10	1070	217
1N4732A	4.7	8.0	53	500	1.0	1.0	10	970	193
1N4733A	5.1	7.0	49	550	1.0	1.0	10	890	178
1N4734A	5.6	5.0	45	600	1.0	2.0	10	810	162
1N4735A	6.2	2.0	41	700	1.0	3.0	10	730	146
1N4736A	6.8	3.5	37	700	1.0	4.0	10	660	133
1N4737A	7.5	4.0	34	700	0.5	5.0	10	605	121
1N4738A	8.2	4.5	31	700	0.5	6.0	10	550	110
1N4739A	9.1	5.0	28	700	0.5	7.0	10	500	100
1N4740A	10	7.0	25	700	0.25	7.6	10	454	91
1N4741A	11	8.0	23	700	0.25	8.4	5.0	414	83
1N4742A	12	9.0	21	700	0.25	9.1	5.0	380	76
1N4743A	13	10	19	700	0.25	9.9	5.0	344	69
1N4744A	15	14	17	700	0.25	11.4	5.0	304	61
1N4745A	16	16	15.5	700	0.25	12.2	5.0	285	57
1N4746A	18	20	14	750	0.25	13.7	5.0	250	50
1N4747A	20	22	12.5	750	0.25	15.2	5.0	225	45
1N4748A	22	23	11.5	750	0.25	16.7	5.0	205	41
1N4749A	24	25	10.5	750	0.25	18.2	5.0	190	38
1N4750A	27	35	9.5	750	0.25	20.6	5.0	170	34
1N4751A	30	40	8.5	1000	0.25	22.8	5.0	150	30
1N4752A	33	45	7.5	1000	0.25	25.1	5.0	135	27

V_F Forward Voltage = 1.2 V Maximum @ I_F = 200 mA for all 1N4700 series

1N4700A Rev. C

FIGURE 25–19   A zener diode data sheet. *(Courtesy of Fairchild Semiconductor)*

Nominal Zener Voltage V_Z V_Z is the rated zener voltage and is specified at test current I_{ZT}.

Maximum Zener Current I_{ZM}   If this current is exceeded, the diode may be damaged.

Knee Current I_{ZK}   If I_Z drops below this value, regulation is lost.

Zener Impedance Z_Z @ I_{ZT}   This is dynamic impedance as defined by Equation 25–1 and determines how much V_z changes as I_z changes. As shown in Figure 25–20, if current changes by an amount ΔI_z, voltage changes by an amount

$$\Delta V_z = Z_Z \times \Delta I_z \qquad (25–2)$$

Z_Z, which is purely resistive and sometimes denoted as R_Z, is substantially constant between I_{ZK} and I_{ZM}. (Since Z_Z is specified at the test current I_{ZT}, it is sometimes denoted as Z_{ZT}.) For the devices of Figure 25–19, Z_Z ranges from 2 Ω to 45 Ω.

INTERPRETATION NOTES . . .

1. All voltages and currents on zener data sheets such as Figure 25–19 are given as positive values even though you know that operation is in the third quadrant where you would normally expect to see negative values.

2. I_{ZT}, I_{ZK} and I_{ZM} in these specs are defined by Figure 25–17.

FIGURE 25–20 $\Delta V_Z = Z_Z \times \Delta I_Z$. Z_Z is dynamic impedance.

Power Rating Maximum permissible *dc power dissipation* P_{Dmax} is given by

$$P_{Dmax} = V_Z I_{ZM} \text{ watts} \qquad (25\text{-}3)$$

Diodes are commercially available with maximum power ratings from 0.25 W to 50 W.

Power Derating Factor All diodes listed in Figure 25–19 are rated at 1W up to 50° C. Above 50° C you must derate them by 6.67 mW for each ° C rise above 50° C. For example, at 85° C, $P_{D \text{ derated}} = 1 \text{ W} - (6.67 \text{ mW/°C}) \times (85° \text{ C} - 50° \text{ C}) = 0.767 \text{ W}$. Sometimes derating information is given in the form of a curve as in Figure 25–21.

PRACTICE PROBLEMS 3

A 500 mW zener diode must be derated when operated above 75° C ambient as in Figure 25–21. What is its permissible power rating at 100° C?

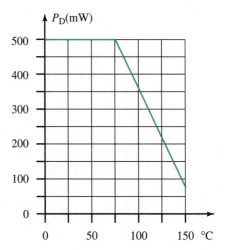

FIGURE 25–21

Answer
350 mW

NOTES . . .

Reminder

The voltage V_Z shown in Figure 25–22 represents the junction voltage of the diode. It is thus, the barrier voltage of the diode and cannot be used as a source of emf.

Modeling a Zener Diode

As a first approximation you can model a zener diode as a voltage source V_{ZT} (Figure 25–22(b)). This is called the *ideal* model. However, a real diode also includes internal impedance Z_Z. If you include this, you get the model of (c). It allows you to compute the diode's voltage at any current. To illustrate, note that the voltage at any point in the zener region may be found (relative to the test point) as $V_Z = V_{ZT} + \Delta V_Z$ where ΔV_Z is given by Equation 25–2. Thus,

$$V_Z = V_{ZT} + (Z_Z \times \Delta I_Z) = V_{ZT} + Z_T (I_Z - I_{ZT}) \qquad (25\text{-}4)$$

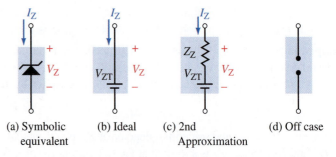

FIGURE 25–22 Zener diode equivalents. If I_Z drops below I_{ZK}, the zener looks like an open circuit as in (d).

For a 1N4742A zener, $V_{ZT} = 12$ V and $Z_Z = 9 \; \Omega$ @ $I_{ZT} = 21$ mA.

a. Use the second approximation to determine V_Z at $I_Z = 10$ mA and $I_Z = 50$ mA

b. Repeat using the ideal model. Compare results.

EXAMPLE 25–3

Solution

a. Since 10 mA and 50 mA are between I_{ZK} and I_{ZM}, Equation 25–4 applies. In each case, measure ΔI_Z from the test point. Thus, for $I_Z = 10$ mA, $\Delta I_Z = 10$ mA $-$ 21 mA $= -11$ mA. Substituting into Equation 25–4 yields $V_Z = 12$ V $+ (9 \; \Omega)(- 11$ mA$) = 11.9$ V. For $I_Z = 50$ mA, $V_Z = 12$ V $+ (9 \; \Omega)(50$ mA $-$ 21 mA$) = 12.3$ V. Thus, V_Z varies from 11.9 V to 12.3 V.

b. $V_Z = 12$ V regardless of current. (This yields an error of about 2.5% at 50 mA.)

Application: The Zener Voltage Regulator Revisited

You now have enough information to analyze the regulator circuit of Figure 25–18. First, determine whether the zener is in the *on* state or the *off* state. If it is on, replace it with the desired *on* equivalent from Figure 25–22. If it is *off*, replace it with an open circuit—see Practical Notes.

---

1. In most of what follows, we will use the ideal model. An obvious reason for this is simplicity—the ideal model is extremely easy to use. Another reason has to do with tolerances. Commercial diodes have tolerances typically of $\pm 5\%$ or $\pm 10\%$. With this level of uncertainty, it is generally unnecessary to worry about the small change in V_Z due to ΔI_Z

2. An easy way to design with zener diodes is to determine the limits imposed by I_{ZK} and I_{ZM} using the ideal model, then design your circuit to stay well within these limits to provide a safety margin. If you do this, even with the uncertainties, your circuit should work safely and satisfactorily.

PRACTICAL NOTES . . .

The zener diode of Figure 25–23(a) is a 12 V 1N4742A. Analyze the circuit using the ideal model.

EXAMPLE 25–4

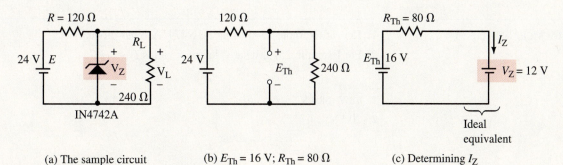

(a) The sample circuit   (b) $E_{Th} = 16$ V; $R_{Th} = 80 \; \Omega$ (c) Determining I_Z

FIGURE 25–23 Analyzing the regulator circuit.

Solution Remove the diode as in (b) and find the Thévenin's equivalent. Here, $E_{Th} = 16$ V and $R_{Th} = 80 \; \Omega$. Replace the diode as in (c). Since E_{Th} exceeds the rated zener voltage, the diode is on and $V_Z = 12$ V. Thus, $I_Z = (16 \text{ V} - 12 \text{ V})/80 \; \Omega = 50$ mA. Power dissipation is $P_D = V_Z I_Z = (12 \text{ V})(50 \text{ mA}) = 0.6$ W, well below the rated value of 1 W. From (a), we see that $V_L = V_Z = 12$ V. Thus $I_L = 12 \text{ V}/240 \; \Omega = 50$ mA. Source current is $I_R = I_L + I_Z = 100$ mA and the drop across R is $(120 \; \Omega)(100 \text{ mA}) = 12$ V.

In the previous example, all circuit elements are fixed. In practice, both input voltage and load current may vary. Let us look at each of these in turn.

Input (Line) Regulation

When input voltage E fluctuates, zener current I_Z fluctuates also. If I_Z drops below I_{ZK}, you lose regulation; if it rises above I_{ZM}, the diode may be damaged. Thus you must not let E drop below E_{min} where E_{min} is the voltage at which $I_Z = I_{ZK}$, and you must not let it rise above E_{max} where E_{max} is the voltage at which $I_Z = I_{ZM}$. To illustrate, consider Figure 25–24. Assuming an ideal zener and noting that $V_L = V_Z$, we see that

$$I_R = I_Z + I_L = I_Z + \frac{V_Z}{R_L} \tag{25–5}$$

and

$$E = I_R R + V_Z \tag{25–6}$$

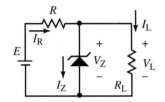

FIGURE 25–24

To find E_{min} use $I_Z = I_{ZK}$ in Equation 25–5, while to find E_{max}, use I_{ZM}.

EXAMPLE 25–5

If the zener of Figure 25–24 is a 1N4736A, $R = 330\ \Omega$ and $R_L = 1.5\ k\Omega$, what is the permissible range of E? Use the ideal model.

Solution   From the 1N4736A data sheet, $V_Z = 6.8$ V, $I_{ZK} = 1$ mA and $I_{ZM} = 133$ mA.

Min	From Equation 25–5:	$I_R = 1$ mA $+ 6.8$ V/1500 $\Omega = 5.53$ mA.
	From Equation 25–6:	$E_{min} = (5.53$ mA)(330 $\Omega) + 6.8$ V $= 8.6$ V
Max	From Equation 25–5:	$I_R = 133$ mA $+ 6.8$ V/1500 $\Omega = 137.5$ mA.
	From Equation 25–6:	$E_{max} = (137.5$ mA)(330 $\Omega) + 6.8$ V $= 52.2$ V

Thus you must keep E between 8.6 V and 52.2 V.

PRACTICE PROBLEMS 4

1. Repeat Example 25–5 using a 1N4744A zener diode.
2. For Example 25–5, what value of E yields $I_Z = I_{ZT}$?

Answers
1. 18 V; 38.4 V
2. 20.5 V

Load Regulation

If R_L becomes too small, the load will draw excessive current and the drop across R will cause V_Z to fall below the knee of the curve—thus, the minimum permitted value of R_L is the value at which $I_Z = I_{ZK}$. Alternatively, if R_L becomes too large, the resulting current diverted through the zener may exceed I_{ZM}. These considerations set the range for R_L. Problems can be solved using basic principles as in Example 25–6. Simply draw the circuit as it looks for each case, then solve.

If the zener of Figure 25–24 is a 1N4736A, $R = 330\ \Omega$ and $E = 24$ V. Determine the permissible range of R_L.

Solution From the data sheet, $V_Z = 6.8$ V, $I_{ZK} = 1$ mA and $I_{ZM} = 133$ mA. Consider Figure 25–25. For all cases, $I_R = \dfrac{E - V_Z}{R} = \dfrac{24\ \text{V} - 6.8\ \text{V}}{330\ \Omega} = 52.12$ mA.

Minimum load resistance is the resistance at which $I_Z = I_{ZK} = 1$ mA as indicated in (a). Here, $I_L = I_R - I_Z = 52.12$ mA – 1 mA = 51.12 mA. Thus $R_{L(min)} = 6.8$ V/51.12 mA = 133 Ω. To determine maximum resistance, note that for this particular problem, zener current will not exceed I_{ZM} no matter how large R_L is. To verify, assume an open circuit as in (b). Zener current is

$$I_Z = \frac{24\ \text{V} - 6.8\ \text{V}}{330\ \Omega} = 52.12\ \text{mA},$$ which is well below the maximum permitted I_{ZM} of 133 mA.

EXAMPLE 25–6

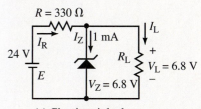

(a) Circuit as it looks for minimum R_L

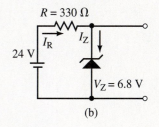

(b)

FIGURE 25–25

1. Repeat Example 25–6 using $E = 28$ V, $R = 120\ \Omega$ and a 1N4742A zener diode.

PRACTICE PROBLEMS 5

Answers
Min $R_L = 90.2\ \Omega$, Max $R_L = 209\ \Omega$

In practice, line variations and load variations may occur simultaneously. There are several end of chapter problems to help you understand this case.

Other Zener Applications

Zener diodes may also be used as *limiters* (also called *clippers*), Figure 25–26. (The job of a limiter is to limit the amplitude of a waveform.) During the positive half cycle of circuit (a), as long as the input voltage is below V_Z, zener current is extremely small, the drop across R is negligible and $v_o \approx e_{in}$. However, when e_{in} exceeds the zener voltage, the zener conducts and output voltage *clamps* at V_Z. During the negative half cycle, the zener is forward-biased and functions as a normal diode Thus, the output voltage will be approximately -0.7 V. Now consider circuit (b). Since the zener diode is reversed, it is forward-biased during the positive half cycle, and the output waveform is as shown.

PRACTICAL NOTES . . .

While the simple zener regulator described here is adequate for some applications (it is cheap and simple), such regulator circuits are not widely used anymore, as more efficient schemes are available. Some of these are described in Chapter 29.

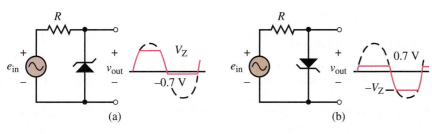

FIGURE 25–26

You can also place two zener diodes back-to-back to get clipping on both polarities as shown in Figure 25–27. This scheme is used to protect electronic equipment against transient voltage spikes. As indicated in (b), line voltage spikes larger than the zener voltage are clipped. Special diodes called *transient suppressors* are made for this purpose. They differ from standard zeners in their surge power handling capability—transient suppressors with over a kilowatt of short duration (a fraction of a cycle) surge power handling capability are available.

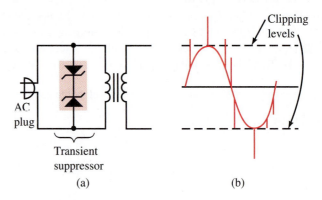

FIGURE 25–27   Using a transient suppresor to limit line voltage spikes.

25.6 The Varactor Diode

Another special-purpose diode is the **varactor** (also called *varicap, epicap,* or **tuning**) diode. It is, in essence, a voltage variable capacitor that takes advantage of the capacitance inherent in a reverse-biased *p-n* junction. This capacitance, as you saw in Chapter 24, varies with voltage—increasing the voltage across the junction lowers the capacitance and vice versa. Because their capacitance can be varied by means of bias voltages, varactors are used in electronically tuned systems, for example in TV receivers. Typical varactor characteristics are shown in Figure 25–28. Varactors are commercially available with capacitances (denoted C_T) that range from about 5 pF to 100 pF.

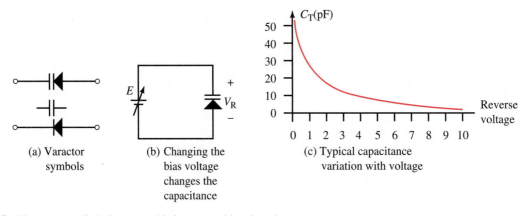

(a) Varactor symbols

(b) Changing the bias voltage changes the capacitance

(c) Typical capacitance variation with voltage

FIGURE 25–28   The varactor diode is operated in its reverse-biased mode.

Varactor Specifications

Figure 25–29 shows a data sheet. The device illustrated is intended for tuning applications and general frequency control. Let us look at some of the data sheet values.

Capacitance C_T   Nominal capacitance is 29 pF at $V_R = 3$ V. However, because of tolerance, the actual value may fall anywhere between 26 pF and 32 pF. Note the variation of C_T with voltage; at 3 V, $C_T = 29$ pF, but at 10 V, $C_T = 10$ pF. (The diagram looks different than Figure 25–28(c) since a semi-log scale is used here.)

ON Semiconductor ™

Silicon Epicap Diodes

Designed for general frequency control and tuning applications; providing solid–state reliability in replacement of mechanical tuning methods.

- High Q with Guaranteed Minimum Values at VHF Frequencies
- Controlled and Uniform Tuning Ratio
- Available in Surface Mount Package

MAXIMUM RATINGS

Rating	Symbol	MBV109T1	MMBV109LT1	MV209	Unit
Reverse Voltage	V_R		30		Vdc
Forward Current	I_F		200		mAdc
Forward Power Dissipation @ T_A = 25°C Derate above 25°C	P_D	280 2.8	200 2.0	200 1.6	mW mW/°C
Junction Temperature	T_J		+125		°C
Storage Temperature Range	T_{stg}		−55 to +150		°C

DEVICE MARKING

MBV109T1 = J4A, MMBV109LT1 = M4A, MV209 = MV209

ELECTRICAL CHARACTERISTICS (T_A = 25°C unless otherwise noted.)

Characteristic	Symbol	Min	Typ	Max	Unit
Reverse Breakdown Voltage (I_R = 10 μAdc)	$V_{(BR)R}$	30	—	—	Vdc
Reverse Voltage Leakage Current (V_R = 25 Vdc)	I_R	—	—	0.1	μAdc
Diode Capacitance Temperature Coefficient (V_R = 3.0 Vdc, f = 1.0 MHz)	TC_C	—	300	—	ppm/° C

	C_t, Diode Capacitance V_R = 3.0 Vdc, f = 1.0 MHz pF			Q, Figure of Merit V_R = 3.0 Vdc f = 50 MHz	C_R, Capacitance Ratio C_3/C_{25} f = 1.0 MHz (Note 1)	
Device	Min	Nom	Max	Min	Min	Max
MBV109T1, MMBV109LT1, MV209	26	29	32	200	5.0	6.5

1. C_R is the ratio of C_t measured at 3 Vdc divided by C_t measured at 25 Vdc.

MMBV109LT1 is also available in bulk packaging. Use **MMBV109L** as the device title to order this device in bulk.

© Semiconductor Components Industries, LLC, 2001
March, 2001 – Rev. 1

1

Publication Order Number:
MBV109T1/D

MBV109T1
MMBV109LT1 *
MV209 *

* ON Semiconductor Preferred Devices

26–32 pF
VOLTAGE VARIABLE CAPACITANCE DIODES

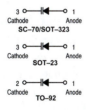

CASE 419–04, STYLE 3
SC–70/SOT–323

CASE 318–08, STYLE 6
SOT–23 (TO–236AB)

CASE 182–06, STYLE 1
TO–92 (TO–226AC)

3 o—▷|—o 1
Cathode   Anode
SC–70/SOT–323

3 o—▷|—o 1
Cathode   Anode
SOT–23

2 o—▷|—o 1
Cathode TO–92 Anode

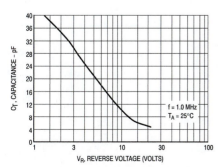

Figure 1. DIODE CAPACITANCE

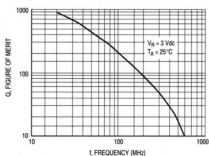

Figure 2. FIGURE OF MERIT

FIGURE 25–29 Data sheet. *(Courtesy of ON Semiconductor)*

Figure of Merit *(Q)*   An important consideration in tuned circuits is Q. To have a sharp tuning characteristic, you need a large Q. For this device, at $V_R = 3$ V and $f = 50$ MHz, Q has a minimum value of 200, which is comfortably high. However, Q is frequency dependent, and falls as frequency goes up.

Capacitance Ratio C_R The capacitance ratio (also called the *tuning ratio*) tells you by what factor C_T changes over its range. For example, the ratio C_3/C_{25} is the ratio of capacitance at 3 V to capacitance at 25 V. Thus, if, $C_3/C_{25} = 6$, the capacitance at 3 V is 6 times the capacitance at 25 V. For this particular device, C_R lies somewhere between 5.0 and 6.5. The larger the ratio, the more C_T changes with voltage, thus, the coarser is its tuning.

Capacitance Temperature Coefficient   The capacitance temperature coefficient (300 ppm/°C) tells you that capacitance changes by 300 millionth of its normal value per °C change. Thus, if $C_T = 29$ pF then C changes by $\Delta C = (300 \times 10^{-6})(29 \text{ pF}) = 0.0087$ pF/°C. Since $C_T = 29$ pF at 25° C, it will equal $C_T = 29$ pF + $(0.0087)(100° \text{ C} - 25° \text{ C}) = 29.7$ pF at 100° C. Thus, C_T is relatively insensitive to temperature.

PRACTICE PROBLEMS 6

For the varactor diode of Figure 25–29, at what temperature will nominal C_T equal 30 pF?

Answer
$\approx 140°$ C

A Varactor Tuning Application

As you saw in Chapter 21, the resonant frequency of a tank circuit may be controlled by means of a variable capacitor as illustrated in Figure 25–30(a). An alternate scheme using a varactor is shown in (b). By varying the potentiometer, you vary the bias voltage and thus the capacitance of the varactor. (Capacitor C_B is not part of the frequency determining circuitry. It is a blocking capacitor whose purpose is to prevent a dc path from the potentiometer wiper arm to ground through the inductor. C_B must be chosen to have negligible impedance to *ac* at the frequency of operation.) The control voltage must be very clean (i.e., free from voltage spikes or other disturbances) to prevent erratic changes in tuning. Because you can adjust the varactor's capacitance by means of its bias voltage, you can also create automatic tuning circuits as depicted symbolically in (c). In this scheme, the tuning voltage is developed electronically from elsewhere in the circuit. This is a commonly used scheme.

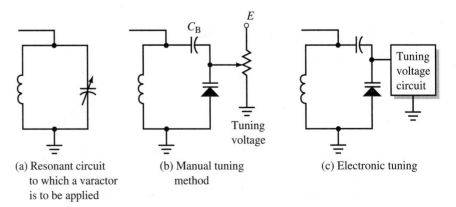

(a) Resonant circuit to which a varactor is to be applied

(b) Manual tuning method

(c) Electronic tuning

FIGURE 25–30   Using a varactor in a tuning circuit.

1. Using the second diode approximation, determine currents I_D and I_T (Figure 25–31).

2. A circuit consists of a source E, a 2-kΩ resistor and a forward biased diode with the characteristic of Figure 25–9. At what value of E will $I_D = 8$ mA?

3. A diode has the curve of Figure 25–13. How much current can this diode handle at 50° C? At 100° C?

4. Assuming nominal ratings for the 1N4737A zener, at what current does $V_z = 7.5$ V? At what current does $V_z = 7.54$ V?

5. Using a diode from Figure 25–19, you have designed a circuit where P_z does not exceed 750 mW. What is the maximum temperature at which you can operate this circuit?

6. You want to change the load voltage of Figure 25–18 to 18 V. What diode from Figure 25–19 should you choose? If $E = 30$ V and $R = 68$ Ω, over what range of load current can your circuit operate, assuming the ideal model?

7. For Question 6, if the load current is 50 mA, what is the permissible range of input voltage variation?

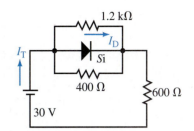

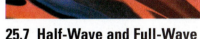

An important application of diodes is in power supply circuits. Here, they are used to convert ac to dc through a process called **rectification.** Following rectification, the dc voltage is filtered, then regulated to produce a constant voltage that is used to power electronic equipment such as computers, DVD players, and so on.

25.7 Half-Wave and Full-Wave Rectifier Circuits

Half-Wave Rectification

To begin, consider **half-wave rectification** in Figure 25–32. During the positive half-cycle the diode is forward-biased and conducts, but during the negative half cycle it is reverse-biased and does not. The result is a pulsating waveform (b) whose peak value V_m is one diode drop below the peak of the applied voltage. The average value (also called the dc value) of the rectified waveform is (as you learned in Chapter 15),

$$V_{dc} = \frac{V_m}{\pi} = 0.318\ V_m \qquad \textbf{(25–7)}$$

For current,

$$I_{dc} = \frac{I_m}{\pi} = 0.318\ I_m \qquad \textbf{(25–8)}$$

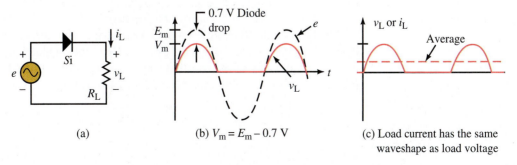

(a)

(b) $V_m = E_m - 0.7$ V

(c) Load current has the same waveshape as load voltage

FIGURE 25–32 Half-wave rectification. Maximum load voltage is one diode drop below maximum source voltage.

Required PIV

In practice, you need a diode that can withstand the maximum reverse voltage to which it is subjected. To determine this voltage, note that during the negative half cycle, current is zero and thus, the voltage across R_L is zero. Since there is no drop across R_L, full source voltage appears across the diode. You must therefore choose a diode with a PIV that can safely withstand E_m volts. To be on the safe side, allow a safety margin—see Practical Note 4, Section 25.3.

EXAMPLE 25–7

Determine the average load voltage and current and the peak voltage seen by the diode (Figure 25–33).

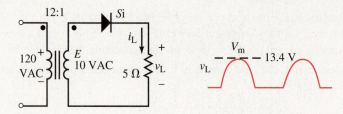

FIGURE 25–33

Solution   The 10 V rms secondary has a peak value of $E_m = \sqrt{2} \times 10$ V $=$ 14.1 V. Subtracting one diode drop yields $V_m = 14.1 - 0.7 = 13.4$ V. From Ohm's law, $I_m = 13.4$ V/5 $\Omega = 2.68$ A. From Equations 25–7 and 25–8, $V_{dc} = 0.318\ (13.4$ V$) = 4.26$ V and $I_{dc} = 0.318(2.68$ A$) = 0.852$ A. The peak reverse voltage seen by the diode is $E_m = 14.1$ V.

PRACTICE PROBLEMS 7

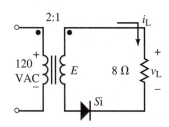

FIGURE 25–34

For Figure 25–34, determine V_{dc}, I_{dc} and the minimum required PIV for the diode. Sketch the load voltage waveform.

Answers
-26.8 V, -3.35 A, 84.9 V   Same shape as Figure 25–33 except inverted.

Full-Wave Rectification

Usually **full-wave rectification** is used instead of half-wave rectification because it provides a smoother dc output with twice the average voltage and current. There are two basic schemes, the center-tapped transformer type and the bridge type.

Center-Tapped Transformer Type Rectifier

Two diodes are needed as in Figure 25–35. During the positive half-cycle (a), transformer polarities are as shown and diode D_1 conducts while D_2 (which is reverse-biased) does not. During the negative half cycle (b), the situation is reversed and diode D_2 conducts—thus, load current is in the same direction for both half cycles. Since there is only one diode in the conduction path for each case, the peak value of the rectified waveform V_m is one diode drop below the peak of the applied waveform. Average (dc) values for these waveforms are

$$V_{dc} = \frac{2V_m}{\pi} = 0.637\ V_m \tag{25–9}$$

and

$$I_{dc} = \frac{2I_m}{\pi} = 0.637\,I_m \qquad\qquad \textbf{(25–10)}$$

To find the maximum reverse voltage that the off diode is subjected to, write KVL around the transformer secondary/D_1/D_2 loop. After some rearranging, this yields

$$PIV_{required} = 2\,E_m - 0.7\,V \approx 2\,E_m \qquad\qquad \textbf{(25–11)}$$

◀ MULTISIM

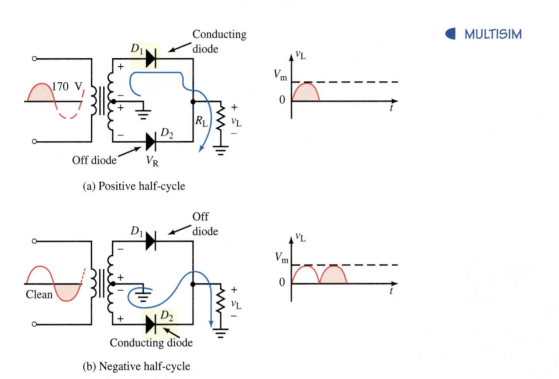

(a) Positive half-cycle

(b) Negative half-cycle

FIGURE 25–35   Full-wave rectification.

EXAMPLE 25–8

Assume the transformer of Figure 25–35 has a ratio of 5 and that a peak primary voltage of 170 V is applied. Draw the transformer secondary and load voltage waveforms and find the average load voltage and the peak reverse voltage seen by the diodes.

Solution   Peak secondary voltage is 170 V/5 = 34 V, yielding E_m = 17 V on each side of the center tap. Subtracting a diode drop of 0.7 V yields $V_m = E_m - 0.7\,V$ = 16.3 V. Waveforms are shown in Figure 25–36. Average load voltage is V_{dc} = 0.637 (16.3 V) = 10.4 V and peak inverse voltage seen by the diodes (Equation 25–11) is 2 × 17 − 0.7 = 33.3 V.

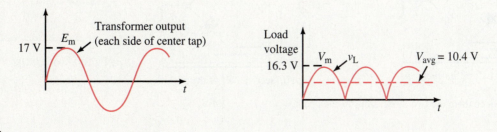

FIGURE 25–36

PRACTICE PROBLEMS 8

For Example 25–8, assume $R_L = 25\ \Omega$. Sketch the current waveform and determine I_m and I_{dc}.

Answer
The waveform looks like v_L in Figure 25–36, 652 mA, 415 mA

Full-Wave Bridge Rectifier

This configuration, shown in Figure 25–37, does not need a center-tapped transformer. Operation is depicted in Figure 25–38. (The transformer has been replaced by an equivalent voltage source for simplicity.) With input polarity as in (a), forward-biased diodes D_2 and D_3 conduct, while reverse-biased diodes D_1 and D_4 do not. This yields the current path indicated. When the polarity is reversed, diodes D_4 and D_1 conduct, yielding the path shown in (b). In both cases, current through R_L is in the same direction, resulting in the familiar waveform of (c). Since there are two diodes in each conduction path, V_m is two diode drops below the applied voltage E_m. The required peak inverse voltage (see end of chapter Problem 30) is

$$PIV_{required} \approx E_m \qquad \qquad (25\text{–}12)$$

(a) A typical bridge rectifier. All 4 diodes are encapsulated in a single package, and only 4 leads are brought out.

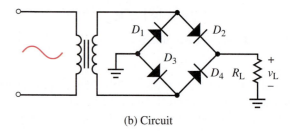

(b) Circuit

FIGURE 25–37   Bridge rectifier. Either a bridge as in (a) or 4 individual diodes may be used.

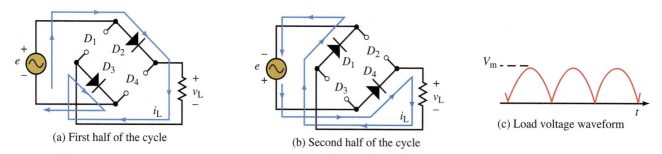

(a) First half of the cycle

(b) Second half of the cycle

(c) Load voltage waveform

FIGURE 25–38   Since there are two diode drops in each path, $V_m = E_m - 1.4\ \text{V}$.

PRACTICE PROBLEMS 9

For Figure 25–37 supply voltage is 120 V (rms), the transformer ratio is 8 and $R_L = 12\ \Omega$. Determine average load voltage and current.

Answers
12.6 V; 1.05 A

Waveforms produced by rectification are unsuited for most applications and must be *filtered*. The simplest filter consists of a large value capacitor placed across the load. Consider Figure 25–39. Assume that the capacitor is initially uncharged (and thus looks like a short). During the first quarter cycle (a), the diode is forward-biased and the capacitor charges to V_m volts. As the input voltage passes its peak and drops below that of the charged capacitor, the diode becomes reverse-biased and looks like an open circuit as in (b). The capacitor, which is now isolated from the supply, discharges through the load until the input voltage is again high enough to forward-bias the diode. If you make C large enough so that the time constant R_LC is considerably larger than the period of the waveform, the capacitor will not discharge appreciably between charging intervals. The final steady-state waveform is shown in (c). Since C charges through the small resistance of the diode circuit, but discharges through the much larger resistance of the load, the charging interval is much shorter than the discharging interval. A similar analysis for the full-wave case yields Figure 25–40.

25.8 Power Supply Filtering

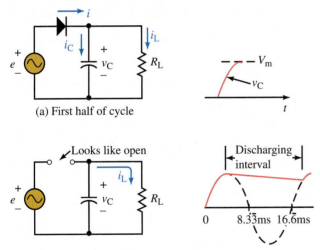

(a) First half of cycle

(b) Since v_C is greater than e, the diode is reverse-biased. The capacitor discharges through R_L until e again exceeds v_C.

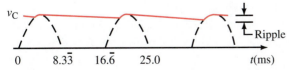

(c) Steady-state capacitor voltage. Times shown are for a 60 Hz input.

FIGURE 25–39 Evolving the capacitor-filtered waveform, half-wave case. Source = 60 Hz.

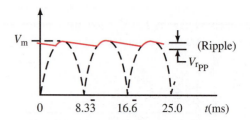

FIGURE 25–40 Steady state waveform for full-wave, 60 Hz case.

The small variation in output voltage is termed **ripple.** The size of the capacitor (for a given load) determines the size of this ripple. Ripple may be expressed as peak-to-peak volts or as rms volts. We will start with peak-to-peak.

Analysis of Filter Circuits

We will follow the usual practice of using an approximate analysis for filter circuits (since it is considerably simpler than a rigorous analysis and yields good results for small to moderate ripple). Let us begin with the full-wave case. First, note that $v_C = \dfrac{q}{C}$. Differentiating both sides yields $\dfrac{dv_C}{dt} = \dfrac{1}{C}\dfrac{dq}{dt}$. Now approximate the waveform of Figure 25–40 by straight lines as shown in Figure 25–41. For this waveform, we can write the above equation as $\dfrac{\Delta v_C}{\Delta t} = \dfrac{1}{C}\dfrac{\Delta q}{\Delta t}$. From Figure 25–41, we see that $\Delta v_C = -V_{r_{pp}}$ (since the slope of v_C is negative during discharge) and $\Delta t = T_2 \approx T$ (since $T_1 \ll T_2$). Furthermore, $\Delta q/\Delta t$ is the average capacitor discharge current I_C. Thus, $\dfrac{\Delta v_C}{\Delta t} = \dfrac{1}{C}\dfrac{\Delta q}{\Delta t}$ reduces to

$\dfrac{-V_{r_{pp}}}{T} = \dfrac{I_C}{C}$. From Figure 25–42, we see that load current is the negative of the capacitor current—that is, $I_C = -I_L$. However, average load current is just its dc value—that is, $I_L = I_{dc}$. Substituting and canceling the minus signs yields $\dfrac{V_{r_{pp}}}{T} = \dfrac{I_{dc}}{C}$. Finally,

$$V_{r_{pp}} = \frac{I_{dc}T}{C} \tag{25–13}$$

or

$$V_{r_{pp}} = \frac{V_{dc}T}{R_L C} \tag{25–14}$$

since $I_{dc} = V_{dc}/R_L$. For light loads, you can use I_m instead of I_{dc} in Equation 25–13 with negligible error where $I_m = V_m/R_L$.

DC load voltage is the average of capacitor voltage v_C. As indicated in Figure 25–41 the average lies half way between the min and max of the ripple. Thus,

$$V_{dc} = V_m - \frac{V_{r_{pp}}}{2} \tag{25–15}$$

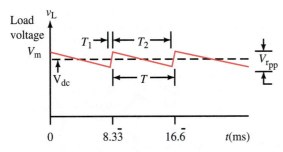

FIGURE 25–41 Approximating the filtered waveform. V_{dc} (the average value of the waveform) lies half way between the min and max of the ripple.

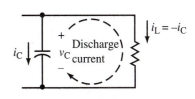

FIGURE 25–42 Discharge interval. From this, we can deduce that $I_L = -I_C$.

While Equations 25–13 and 25–14 permit you to determine ripple from load conditions, you sometimes need ripple in terms of input voltage. Combining Equations 25–14 and 25–15 yields

$$V_{r_{pp}} = \frac{V_{dc}T}{R_L C} = \left(V_m - \frac{V_{r_{pp}}}{2}\right)\left(\frac{T}{R_L C}\right)$$

Solving for ripple we get,

$$V_{r_{pp}} = \frac{2V_m}{1 + \dfrac{2R_L C}{T}} \qquad (25\text{–}16)$$

You can also combine Equations 25–15 and 25–16 (see end of chapter Problem 32) to get

$$V_{dc} = \frac{V_m}{1 + \dfrac{T}{2R_L C}} \qquad (25\text{–}17)$$

Note that all equations developed above apply to both full-wave and half-wave rectification by appropriate choice of T—see Notes.

NOTES . . .

1. For the half-wave case, T is the period of the supply voltage waveform and is thus equal to $1/f$. However, for the full-wave case (since the rectified waveform consists of two half sine waves), T is half the period of the supply voltage waveform. Thus, when using these equations at 60 Hz, set $T = 16.66$ ms for the half-wave case and $T = 8.33$ ms for the full-wave case.

2. Various authors use various approximate approaches to filter analysis. However, they ultimately all come down to the same thing. To illustrate, consider dc load voltage Equation 25–17. Noting that $\dfrac{1}{1 + x} \approx 1 - x$ for small x, we see that $V_{dc} = \dfrac{V_m}{1 + \dfrac{T}{2R_L C}} \approx V_m\left(1 - \dfrac{T}{2R_L C}\right)$, which is the equation (or some variation of it) that you sometimes encounter in other works.

EXAMPLE 25–9

If $C = 330\ \mu F$, what are the dc load voltage and load current for the circuit of Figure 25–43.

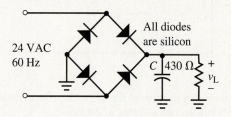

24 VAC
60 Hz

All diodes
are silicon

C 430 Ω $+$ v_L $-$

FIGURE 25–43

Solution   Peak input voltage is $\sqrt{2}(24) = 33.94$ V. Subtracting two diode drops yields $V_m = 32.54$ V. From Equation 25–16 (using $T = 8.33$ ms since this is full-wave),

$$V_{r_{pp}} = \frac{2(32.54\ \text{V})}{1 + \dfrac{2(430\Omega)(330\mu F)}{8.33\ \text{ms}}} = 1.86\ \text{V}$$

Load voltage (from Equation 25–15) is

$$V_{dc} = 32.54 - \frac{1.86}{2} = 31.6\ \text{V}$$

Load current is

$$I_{dc} = \frac{V_{dc}}{R_L} = \frac{31.6\ \text{V}}{430\ \Omega} = 73.5\ \text{mA}$$

Since there are a number of approximations in the above equations, just how good are they? Table 25–1 compares the results of Example 25–9 to computer solutions plus a more rigorous analysis (see end of chapter problem 42). Note that agreement is quite good. In fact, other sources contribute more uncertainty in practice than these approximations do. For example, filter capacitors typically have tolerances of $\pm 20\%$.

TABLE 25–1

	Approximate Solution	"Exact" Solution	PSpice	MultiSIM
$V_{r_{pp}}$	1.86 V	1.67 V	1.60 V	1.79 V
V_{dc}	31.6 V	31.7 V	31.7 V	31.8 V
I_{dc}	73.5 mA	73.7 mA	73.7 mA	74.0 mA

PRACTICE PROBLEMS 10

1. For Figure 25–43, determine the minimum value for C if $V_{r_{pp}}$ is to be no more than 1.2 V.

2. For Figure 25–39, $e = 18 \sin 377\ t$ volts, $C = 1000\ \mu F$, and $R_L = 680\ \Omega$. Determine ripple, V_{dc} and I_{dc}.

3. Consider Example 25–9. Assuming light loading, use Equations 25–13 and 25–15 to compute ripple and load voltage V_{dc}. Compare your answers to those obtained in Example 25–9.

4. Show that $V_{dc} = V_m - \dfrac{4.17 I_{dc}}{C}$ for full-wave by combining Equations 25–13

and 25–15 with $f = 60$ Hz and I_{dc} expressed in mA and C in μF.

Answers
1. 516 μF; 2. 0.419 V, 17.1 V, 25.1 mA; 3. 1.91 V, 31.6 V

Ripple Factor

Ripple is sometimes expressed in rms volts or as a ripple factor. Consider again Figure 25–41. Load voltage is approximately a triangular wave with a dc off-set. Using the techniques of Chapter 15, it can be shown that the rms value of

the triangular part is $V_{r(rms)} = \dfrac{V_{r_{pp}}}{2\sqrt{3}}$. The rms **ripple factor** is defined as

$$r = \text{ripple factor} = \frac{\text{rms ripple voltage}}{\text{dc voltage}} \qquad \textbf{(25–18)}$$

Combining Equations 25–16, 25–17 and 25–18, yields

$$r = \frac{T}{2\sqrt{3}R_L C} \qquad \textbf{(25–19)}$$

This formula applies to both the full-wave and half-wave cases. However (as usual), ripple for the half-wave case is double that for the full-wave case, since T is twice as large.

EXAMPLE 25–10

If $R_L = 180 \ \Omega$ for the circuit of Figure 25–44, determine the minimum capacitance needed if the rms ripple factor is not to exceed 2.5%. Determine load voltage, peak-to-peak ripple and rms ripple.

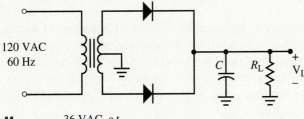

FIGURE 25–44   36 VAC, c.t.

120 VAC
60 Hz

$C \quad R_L \quad \overset{+}{\underset{-}{V_L}}$

Solution   From Equation 25–19, $C = \dfrac{T}{2\sqrt{3}rR_L} = \dfrac{8.33 \text{ ms}}{2\sqrt{3}(0.025)(180\Omega)} = 535 \ \mu\text{F}$

Since the transformer is center-tapped, the peak voltage applied to the diode circuit is $E_m = (\sqrt{2})(18 \text{ V}) = 25.46$ V. Subtracting a diode drop yields $V_m = 25.46 - 0.7 = 24.76$ V. From Equation 25–16 we get

$$V_{r_{pp}} = \frac{2V_m}{1 + \dfrac{2R_L C}{T}} = \frac{2(24.76)}{1 + \dfrac{2(180\Omega)(535 \ \mu\text{F})}{8.33 \text{ ms}}} = 2.06 \text{ V}$$

$$V_{dc} = V_m - \frac{V_{r_{p\text{-}p}}}{2} = 24.76 \text{ V} - \frac{2.06 \text{ V}}{2} = 23.7 \text{ V}$$

RMS ripple is 2.5% of V_{dc}. Thus, rms ripple $= (0.025)(23.7 \text{ V}) = 0.59$ V.

1. As you can see from Example 25–10, ripple expressed in volts rms is considerably smaller than ripple expressed in volts peak-to-peak. Knowing this is important when you purchase a power supply. When looking at specs, be consistent—that is, compare rms ripple to rms ripple or peak-to-peak ripple to peak-to-peak ripple.

2. When choosing a filter capacitor, there are two competing requirements, keeping the ripple voltage low versus keeping the charging current spikes (considered next) from getting too large. The problem is that for low ripple, you need large filter capacitors, but large capacitors result in large spikes. Thus, a compromise may be needed. In practice, filter capacitors typically range in size from a few hundred μF to a few thousand μF.

Diode Forward Current

Diode conduction occurs only for a brief interval each cycle (during the time that the output voltage decays below the voltage input to the rectifier), resulting in a stream of current pulses as illustrated in Figure 25–45(a). This current is called **diode repetitive forward current.**

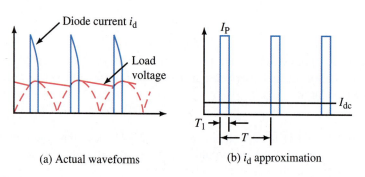

(a) Actual waveforms   (b) i_d approximation

FIGURE 25–45   Diode repetitive surge currents.

It is customary to approximate this current as a series of rectangular pulses as in (b). To estimate I_p, note that during charging, diode current must replenish the charge transferred from the capacitor to the load during discharging. But charge is equal to the area under the current curve (since $q = \int i_c \, dt$). Noting that $I_p >> I_{dc}$, we see $Q_{charge} \approx I_p \times T_1$, while $Q_{Load} \approx I_{dc} \times T$. Thus, $I_p \times T_1 \approx I_{dc} \times T_1$. Solving for I_p yields,

$$I_p \approx I_{dc} \frac{T}{T_1} \qquad (25\text{–}20)$$

From this, you can see that since $T_1 << T$, the diode is subjected to quite severe current spikes that are many times larger than the average load current, and that the shorter the charging interval, the larger will be these spikes. Many rectifier data sheets list the maximum permitted value for peak repetitive forward current as I_{FRM} (although unfortunately, the 1N400X family does not show it).

Effect of Filter on PIV

A filter capacitor doubles the diode PIV requirement for the half-wave circuit, but has no effect on the full-wave circuits. To illustrate, consider the half-wave case (Figures 25–39(b)). Capacitor voltage decays only slightly, so that $v_C \approx E_m$ when the source reaches its peak negative value of $-E_m$ volts. If you write KVL around the source/diode/capacitor loop at this instant of time, you find the voltage across the reverse-biased diode is approximately $2E_m$. Thus, you need a minimum PIV of 2 E_m here compared to E_m with no capacitor. A similar analysis for the full-wave case shows that the maximum reverse voltage is approximately E_m, the same as without the capacitor as we found in Equation 25–12.

Looking Ahead

The power supplies that we have described here are *unregulated supplies*—i.e., supplies whose output voltages vary if the input voltage or the load current varies. However, most power supplies used today are *regulated supplies*. In these, the unregulated dc output that we have described is fed to a regulator whose job it is to hold the output voltage constant. The simplest regulator is a zener diode, Figure 25–46. For some applications, it is adequate. Zener regulators are inefficient, however, because they waste a lot of power, and a more efficient scheme is considered in Chapter 29.

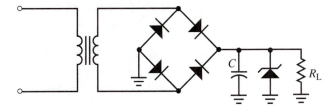

FIGURE 25–46 Simple regulated supply.

MultiSIM and PSpice may be used to study filter waveforms. Since the only time domain analysis they provide is a transient analysis, and since we are interested only in steady state analysis, we need some way to force the transient component of the solution to zero so that we arrive at the steady state immediately. We can do this by appropriately choosing the capacitor initial condition—see Operational Notes 4 and 5. Let us illustrate using Example 25–9.

25.9 Computer Analysis

Operational Notes for PSpice and MultiSIM

PRACTICAL NOTES . . .

1. Rotate C (3 times for PSpice, once for MultiSIM) so that its "1" end is up as described in Appendix A. (This is critical for setting initial conditions.) For PSpice, rotate R three times as well.

2. Since our objective here is to study capacitive filtering, we have simplified the circuits by replacing the transformers with equivalent voltage sources.

3. To permit comparison of the various solutions and, hence, to verify how good the approximate answers are, you need V_m to be the same for all cases. (For the computer solutions, suitable values for E_m had to be determined experimentally to achieve this since PSpice and MultiSIM use different values for diode-forward drops.)

4. To eliminate the transient part of the solution, set the initial condition on the capacitor equal to the voltage it would have in steady state at $t = 0$. From Figure 25–39, you can see that this is approximately V_{dc}. Thus, set $V_0 \approx V_{dc}$. (If you don't know V_{dc}, set V_0 to about 10% below V_m. For PSpice, don't set V_0 too high since it seems to fail if you set V_0 higher than its actual initial voltage.)

5. If some transient is still evident, make a closer estimate of V_0 from the plot then run the simulation again. Alternatively, if the first few ripples have some transient on them, step past it to where the waveform has stabilized then measure with your cursor.

MultiSIM

Bridge rectifiers are found in the bin marked by the diode icon. Select the MDA 2501, then build Figure 25–47(a) on the screen. Use a virtual capacitor so that you can easily set V_0. Double click C and set the Initial Condition to 31.5 V (the dc value that we determined earlier). Set the source to 60 Hz with Voltage RMS set to 24.

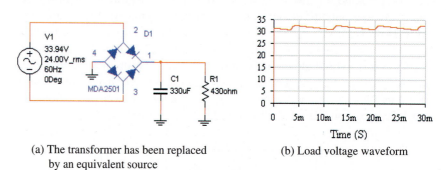

(a) The transformer has been replaced by an equivalent source

(b) Load voltage waveform

FIGURE 25–47   Waveform analysis using MultiSIM.

Next,

- Click *Options/Preferences/Show Node Names* to display node numbers. Determine the node number for the top end of the capacitor.

- Select the Analysis icon, choose Transient Analysis and in the dialog box under *Initial Conditions,* select *User Defined.* Increase the *Minimum Number of Time Points* to at least 1000 (to get a smooth plot). Set the transient run time (TSTOP) to 0.030. Click the *Output Variable* tab and specify the node number determined in the previous step.

- Click *Simulate.* When analysis is complete, the waveform of (b) should appear.

- Use the cursors to determine min and max voltages and hence, ripple.

- To find V_{dc}, connect a dc voltmeter (from the *Indicators* tool bin) across R_L. Click the simulation switch to *ON* to run the simulation. Read the meter when it stabilizes.

    Results are summarized in Table 25–1. As noted earlier, these reaffirm our contention that the approximate analysis yields quite good results.

PSpice

- Build the circuit on the screen as in Figure 25–48(a) using 1N4002 diodes and source VSIN.

- Double click the capacitor and set its initial voltage (IC) to 31.4 V. (When you run the simulation, if you get a dialog box that says "There are no data in section number 1", you have probably set IC too high. Try lowering it and run the simulation again.)

- Double click the source and set VAMPL to 34.1V (see Note 3), FREQ to 60 Hz and VOFF to 0.

- Click the *New Simulation Profile* icon and select *Time Domain (Transient)* analysis. Set *Run to time* to 30ms.

- Click the *Run* icon. You should get the voltage waveform shown in green (Figure 25–48(b)). (Your voltage scale may be different as the axis here has been re-scaled.) Using the cursor, measure the min and max voltage and, from this, compute ripple.

- Add a new Y-axis, then add the diode D_1 and D_2 current traces. Compare to Figure 25–45(a).

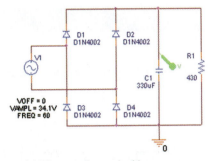

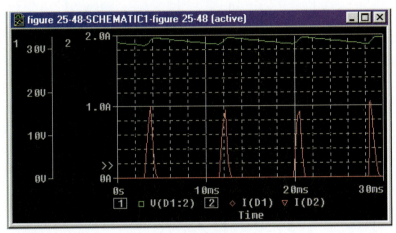

(a) The transformer had been
replaced by an equivalent source

(b) Load voltage (green) and
diode current (red)

FIGURE 25–48 Waveform analysis using PSpice.

Results are summarized in Table 25–1. As noted earlier, these reaffirm our contention that the approximate analysis yields quite good results.

PRACTICE PROBLEMS 11

Determine V_{dc}. Hints: Delete the voltage probe shown. Use a run time of about 120ms. When the empty display graph appears, use *Add Trace* to plot AVG(V1(R1)). PSpice computes and plots a running average of the waveform. However, you need to wait until the waveform stabilizes to determine its average—thus, measure with the cursor after the waveform has stabilized. Repeat for the load current.

Answer
About 31.7 V; 73.7 mA

PUTTING IT INTO PRACTICE

Some power supplies that you designed have experienced failures after several years of operation. Ultimately, the trouble was traced to inadequate ripple current rating of the filter capacitor. (You remember an experienced designer once telling you, "Ignore capacitor ripple currents at your peril.") Your design required a minimum of 4100 μF and you chose the next highest valued capacitor, which was 4700 μF. A suggested solution to your problem is to use two 2200 μF capacitors in parallel instead of the single 4700 μF unit. To investigate this suggestion, research ripple current ratings for capacitors, including the effects of excessive ripple current. Document your findings. (You will probably have to go to the Internet for this information.)

25.1   Diode Models

1. Assuming ideal diodes Figure 25–49, determine diode currents and voltages.

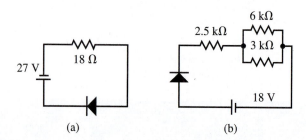

(a)   (b)

FIGURE 25–49

2. The circuit of Figure 25–50(a) has the waveform of (b) applied. Assume an ideal diode and sketch waveform v_R. Repeat if the diode is flipped to face the opposite direction.

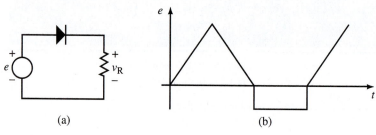

(a)   (b)

FIGURE 25–50

3. You have designed a control circuit with an electronic switch that turns current on and off to an inductor (Figure 25–51). However when the circuit turns the coil current off, a large voltage spike (recall Chapter 14, Section 14.3) occurs that "blows" your circuit. Show how you can connect a diode across the coil to protect your circuit. Sketch the coil voltage before and after adding the diode.

◀ MULTISIM

4. Box 1 of Figure 25–52 contains a rotary switch, while boxes 2 and 3 each contain a lamp. Each box also includes diodes. With the switch in position 1, Lamp 1 alone comes on; in position 2, Lamp 2 alone comes on; in position 3, both lamps come on. Sketch the circuit.

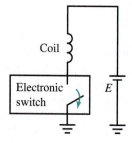

FIGURE 25–51

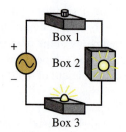

FIGURE 25–52   Only a single wire runs from box to box.

5. Using the second approximation, repeat Problem 1 assuming silicon diodes.

6. A second diode (in series with and pointed the same way as the existing diode) is added in Figure 25–49(b). If both are silicon, what is source current?

7. The diodes of Figure 25–53 are silicon. Using the second approximation, find diode currents. Hint: Use Thévenin's theorem.

8. Repeat Problem 7 if the diodes are germanium.

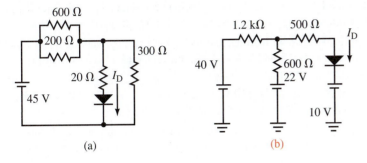

(a)   (b)

FIGURE 25–53

25.2 to 25.4   Diode Characteristic Curve, Data Sheet, Etc.

9. The diode of Figure 25–54 has the characteristic of Figure 25–9. If circuit current is 12 mA, what is V_D? What is E?

10. Repeat Problem 9 if circuit current is 4 mA.

11. For the circuit of Figure 25–55, average load current is 400 mA. You have a choice of a 1N4001 or a 1N4004. Which would you choose and why?

FIGURE 25–54   **FIGURE 25–55**

12. If you use a 1N4004 in Figure 25–55 and you want a safety margin of 20%, what is the maximum rms value sine wave that you can apply?

13. A 1 amp diode can deliver 1 A up to 75° C, but must be derated by 8.5 mA/° C above that. How much current can you draw at 110° C?

14. A circuit containing a diode with the derating curve of Figure 25–13 draws 0.6 A. What is the maximum ambient temperature in which you can operate the circuit?

15. A diode has a barrier potential of $V_B = 0.73$ V at 25° C. At approximately what temperature will V_B equal 0.76 V?

16. A diode has a reverse current of 420 nA at 45° C. What will be its reverse current at 25° C?

17. A diode has a reverse current of 125 nA at 25° C. At what temperature will it be 1 μA?

25.5   The Zener Diode

18. Over what range of current I_Z will the 1N4744A regulate? What is its maximum power rating at 40° C? At 60° C?

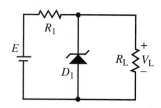

FIGURE 25–56

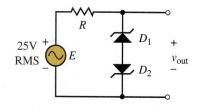

FIGURE 25–57

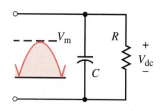

FIGURE 25–58 For all problems except #34, assume full-wave input.

19. If the diode in Figure 25–56 is a 1N4746A, $E = 24$ V, $R_1 = 82$ Ω and $R_L = 600$ Ω, use the ideal model to determine I_Z, P_D and load current I_L.

20. For Figure 25–56, $E = 30$ V, $R_1 = 150$ Ω, load current is 37.5 mA and the zener is 1N4746A. Use the ideal model and determine I_Z and P_D.

21. Using the component values of Problem 20 and the ideal model, determine the acceptable range for E.

22. For Figure 25–56, $E = 20$ V, $R_1 = 91$ Ω and the zener is 1N4742A. Using the ideal model, determine the range of load resistance that this circuit can handle while keeping operation within the limits specified by the data sheet.

23. For Figure 25–56, both load current and source voltage vary. If $R_1 = 120$ Ω, the zener is an ideal 1N4740A and load current varies from 20 mA to 50 mA, what are the min and max E?

24. For Figure 25–56, if $R_1 = 120$ Ω, the zener is an ideal 1N4740A and R_L varies from 150 to 800 Ω, what is the min and max E?

25. For the 1N4744A, what is V_Z at $I_Z = 17$ mA? Using the model of Figure 25–22(c), determine zener voltage at $I_Z = 1$ mA and at $I_Z = 55$ mA.

26. The voltage of a certain zener is 12.8 V at $I_{ZK} = 0.2$ mA and 13.4 V at $I_Z = 80$ mA. Using the model of Figure 25–22(c), determine its voltage at 22 mA.

27. In Figure 25–57, D_1 is a 1N4731A zener and D_2 is a 1N4740A.

    a. Assuming $v_F = 0$V for the each zener when forward-biased, sketch waveform v_{out}.

    b. Assuming $v_F = 0.7$ V for the each zener when forward-biased, sketch waveform v_{out}.

28. Redo Example 25–4 using the model of Figure 25–22(c). Hints: To find V_Z, solve iteratively using the results of Example 25–4 as a starting point. See also Example 25–3 for ideas.

25.7 Half-Wave and Full-Wave Rectifier Circuits

29. If the transformer of Figure 25–33 has a turns ratio of 5 and $R_L = 7$ Ω, determine average load current and the maximum reverse diode voltage.

30. Confirm Equation 25–12. Hint: Write KVL around the source, D_2, D_4 loop Figure 25–38(a).

25.8 Power Supply Filtering

31. For Example 25–9, solve for load voltage using Equation 25–17.

32. Show that Equations 25–15 and 16 yield Equation 25–17.

33. For Figure 25–58, $V_m = 20$ V, $C = 330$ μF and $R_L = 650$ Ω. What is peak-to-peak ripple?

34. Repeat Problem 33 assuming the waveform is half-wave and $C = 390$ μF.

35. If $C = 220$ μF, $I_{dc} = 20$ mA and $V_m = 20$ V for Figure 25–58, what are V_{rpp} and V_{dc}?

36. If $R_L = 1000$ Ω and $V_m = 20$ V for Figure 25–58 and ripple is not to exceed 0.6 V peak-to-peak, what is minimum required capacitance?

37. A 19.0 V dc load has current 500 mA dc. The load is supplied from a full-wave rectifier with $V_m = 20$ V. What is the peak-to-peak ripple and what is C?

38. For Figure 25–58, if $R_L = 300$ Ω, $V_m = 24$ V and $C = 470$ μF, what is the percent ripple factor?

39. For Figure 25–58, $R_L = 400\ \Omega$, $V_m = 20$ V and the ripple factor is not to exceed 3%. Determine peak-to-peak ripple, V_{dc} and the rms ripple voltage.

40. For Figure 25–58, $V_{r_{pp}} = 1.22$ V, $R_L = 600\ \Omega$ and $C = 220\ \mu F$. Find ripple factor and V_{dc}.

41. A dual voltage power supply with one side positive with respect to ground and the other negative can be created using a tapped transformer and a bridge rectifier. Add components (two resistors and two capacitors) to Figure 25–59 to create such a supply.

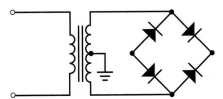

FIGURE 25–59

42. A more rigorous ("exact") analysis of the filtered waveform problem may be achieved using an iterative technique. To illustrate, consider Example 25–9. Capacitor discharge voltage is given by $V_m e^{-t/R_L C}$ where t is the discharge time. Discharge time to $v_{C(min)}$ Figure 25–40 looks to be about 90% of half a period, i.e, about 7.5 ms.

   a. Using this guess, compute $v_{C(min)}$.

   b. Using the technique of Chapter 15, Section 15.6, compute a new estimate for discharge time t. Repeat a. and b. until $v_{C(min)}$ does not change from trial-to-trial. Finally, determine $V_{r_{pp}} = V_m - V_{C(min)}$. Results are tabulated in Table 25–1.

25.9 Computer Analysis (MultiSIM and PSpice)

43. For Figure 25–60, $C_1 = C_2 = 470\ \mu F$.

   a. Estimate V_0 and set initial conditions for both capacitors. Run a simulation and refine your estimation of V_0 if necessary. When you have a satisfactory waveform, determine ripple and V_{dc}.

   b. Compute ripple and V_{dc} using the formulas of this chapter. (Use a value of 0.74 V for the diode drop in your analysis. This is fairly close to what the software packages use.) Compare your computed and computer results.

44. Replace both capacitors with 220 μF units, then repeat Problem 43 a and b.

45. Using a full-wave bridge rectifier, repeat Problem 43 a and b.

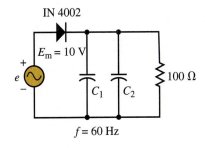

FIGURE 25–60

✓ **ANSWERS TO IN-PROCESS LEARNING CHECKS**

In-Process Learning Check 1

1. 46.5 mA, 48.8 mA

2. 16.82 V

3. 1 A, 0.7 A

4. 34 mA, 44 mA

5. 87.5° C

6. 1N4746A, 126.5 mA to 176.25 mA

7. 21.4 V to 24.8 V

■ OBJECTIVES

On completion of this chapter, you will be able to

- identify the three terminals of a junction transistor and explain the biasing of these terminals,

- use the transistor β and α to calculate the relationship between emitter, collector, and base currents,

- use manufacturers' specification sheets to determine the limits of operation of a transistor,

- calculate the power dissipation of a transistor,

- predict the *saturation* and *cutoff* points of a transistor circuit,

- sketch a dc load line on a graph of transistor collector characteristic curves and determine the operating point of the transistor circuit,

- analyze the operation of several bias arrangements,

- examine the operation of a transistor used as an electronic switch,

- use a multimeter to determine the pin allocation of pnp and npn transistors and measure the operating point of a transistor in a given amplifier circuit,

- analyze JFET amplifier circuits using Shockley's equation and the JFET transfer curve,

- analyze D-MOSFET and E-MOSFET amplifier circuits,

- use computer programs to predict the bias points of transistor circuits,

- properly handle static sensitive components.

Basic Transistor Theory

26

Since its invention in 1947, the transistor has revolutionized electrical, electronics, and computer engineering. From its crude beginning, the transistor has been miniaturized, modified, and combined with many other components onto complex integrated circuits that have revolutionized communications, travel, medicine, and most other industries.

Miniaturization of the transistor has permitted us to use cell phones and to directly receive television-transmitted signals from satellites. Without the transistor, flying the space shuttle or any other modern aircraft would be impossible. CT scanners and MRI (magnetic resonance imaging) scanners would not function without the circuitry that relies on transistors to process the signals. Even land surveying, which has for many decades used optical and mechanical methods, has been modernized by using GPS (global positioning systems) satellites and receivers.

This chapter examines the structure and basic operation of the transistor. Although the use of discrete transistors is disappearing, it is important to have a rudimentary understanding of transistor operation to appreciate the operation of more complex circuitry using hundreds (or perhaps thousands) of transistors. Students will be introduced to transistor biasing, the use of load lines, and troubleshooting techniques. ∎

PUTTING IT IN PERSPECTIVE

As so many other things that we take for granted, modern electronics began with Thomas Edison, who in 1883 discovered that when an electric current is applied to a conductor in a vacuum, electrons are given off in a process called *thermionic emission* (or Edison Effect).

In 1904, an Englishman by the name of John Fleming produced an "electronic valve" by heating a cathode with a small electric current. The heat of the cathode released a small cloud of electrons. Next, he connected a dc voltage between another terminal called the "plate" (anode) and the cathode. Since the plate was positive with respect to the cathode, the electrons were repelled from the cathode and attracted by the positive charge on the plate, resulting in current in one direction. This created the first diode.

In 1906, Lee De Forest inserted a fine metal grid between the plate and the cathode. By applying a small grid current, it was possible to control the current between the cathode and the plate. The triode (which De Forest called the "audion") revolutionized radio transmission, since it was now possible to amplify small electrical signals so that they could be easily transmitted over great distances.

The vacuum tube went through numerous changes between 1906 and 1960, with the introduction of various control grids resulting in tetrodes and pentodes. Miniature vacuum tubes were used in consumer electronics, while enormous vacuum tubes were used to power the radio and television stations that were transmitting the nation's AM, FM, and television stations.

In 1947, three American physicists working for Bell Laboratories made a discovery that would ultimately see the demise of the vacuum tube. William Shockley, John Bardeen, and Walter Brattain invented the transistor. ∎

26.1 Transistor Construction

The transistor is the basic building block used in complex integrated circuits. We will begin by examining the structure of the **bipolar junction transistor** (BJT), which is constructed of three layers of doped semiconductor material with two p-n junctions. The term *bipolar* refers to the fact that both holes and electrons are used as carriers in the device. Figure 26–1 shows the structure of a BJT, which is built on a substrate of pure silicon. The transistor is manufactured by progressively building and etching each layer over the previous layer. Very thin connections of aluminum are deposited onto the emitter, base, and collector regions. Finally, the entire transistor is covered with a very thin layer of silicon dioxide (SiO_2), which acts as a good insulator.

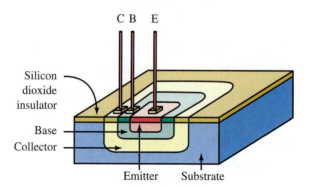

FIGURE 26–1   Construction of a bipolar junction transistor.

The three layers of the BJT are the emitter, the base, and the collector which can be constructed as either pnp or npn transistors as shown in Figure 26–2. Notice that the emitter and collector are always constructed of the same type of material (n-type material in the npn transistor) and are heavily doped to result in semiconductor material that has a low resistivity. The arrow of the BJT is always between the emitter and base, pointing from the p-type material towards the n-type material just like a diode. The base is constructed of the other type of material (p-type in the npn transistor) and is very lightly doped, resulting in a higher resistivity.

Each BJT shown in Figure 26–2 has two p-n junctions, the B-E (base-emitter) junction and the C-E (collector-emitter) junction. These junctions determine the operation of the transistor.

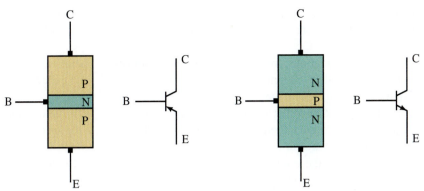

(a) Structure and electronic symbol of a pnp transistor

(b) Structure and electronic symbol of an npn transistor

FIGURE 26–2

26.2 Transistor Operation

In order for a transistor to operate as an amplifier, two conditions must be met:

- the B-E junction must be forward-biased, and
- the C-B junction must be reverse-biased.

Figure 26–3 shows that the forward-biased base-emitter junction results in a large majority carrier current from the heavily doped p-type material of the emitter into the lightly doped n-type base region.

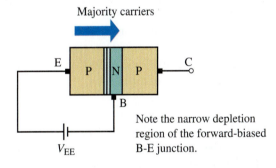

Note the narrow depletion region of the forward-biased B-E junction.

FIGURE 26–3   The B-E junction is forward-biased in a transistor amplifier circuit.

Figure 26–4 shows that the reverse-biased collector-base junction results in only a very small minority carrier current from the lightly doped n-type material of the base into the heavily doped collector region. The minority carrier current illustrated is often included in many specification sheets as I_{CBO}, which means that this is the leakage current between the base and the collector with

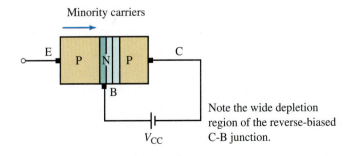

Note the wide depletion region of the reverse-biased C-B junction.

FIGURE 26–4   The C-B junction is reverse-biased in a transistor amplifier circuit.

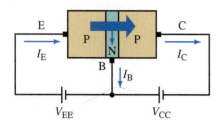

FIGURE 26–5   Biasing a pnp transistor.

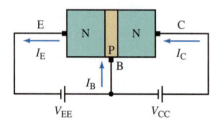

FIGURE 26–6   Biasing an npn transistor.

the emitter open. (The manufacturer will also provide the test voltage V_{CB} for which the specification is provided.) In most cases, I_{CBO} can be ignored since its effect is small.

When both junctions are correctly biased, the transistor operation is profoundly different than the simple operation that one expects from the diode-like behavior of the individual junctions.

Figure 26–5 shows that when the majority hole current from the emitter, I_E, enters the base region, there are few electrons in the thin base region to combine with the large injection of holes. Consequently, the base current, I_B is very small, typically in the order of microamps. The large number of holes in the narrow base region is able to drift to the reverse-biased C-B junction where the positive-charged holes are attracted to the negative terminal of the bias voltage, V_{CC}. This large movement of charge results in a correspondingly large collector current, I_C. Figure 26–5 illustrates another very important characteristic of the bipolar junction transistor, namely that the emitter current, collector current, and base current of the transistor are related by the expression.

$$I_E = I_C + I_B \qquad (26-1)$$

The above equation follows Kirchhoff's current law and is true regardless of whether the transistor is pnp as shown in Figure 26–5 or npn as illustrated in Figure 26–6.

Although Figures 26–5 and 26–6 show voltage sources connected directly to transistors, such an arrangement would never be used in an operational circuit. Rather, any transistor circuit will always contain resistors to limit the amount of current to prevent damaging the device. Another important characteristic of any properly biased BJT amplifier circuit is that, since the base-emitter junction is forward biased, the voltage across this p-n junction will be approximately 0.7 V (for a silicon transistor). Figure 26–7 shows properly biased pnp and npn transistor amplifier circuits. We will examine these and other circuits in greater detail later in the chapter.

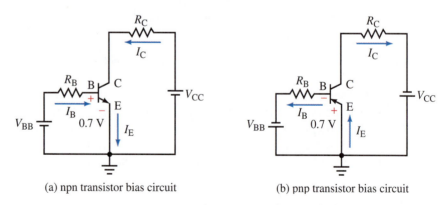

(a) npn transistor bias circuit   (b) pnp transistor bias circuit

FIGURE 26–7

Notice that for the npn transistor, the base-emitter voltage, $V_{BE} = +0.7$ V, while for the pnp transistor $V_{BE} = -0.7$ V. It is not surprising that these properties are exactly the same as for a forward-biased diode. Remember that the p-type material of a forward-biased junction will always be positive (p) with respect to the n-type material, which will be negative (n). Just as was the case with the diode, the arrow of the emitter points in the direction of conventional current (when the transistor is "on").

dc Beta (β_{dc})

We have already determined that due to the very light doping of the base material of any transistor, the base current, I_B will be very small when compared to both the emitter current, I_E and the collector current, I_C. Although small, the base current will not be zero. We define the **dc beta,** which is also called the **dc current gain,** β_{dc} (or h_{FE} in specification sheets) of a transistor as the ratio of collector current to base current.

$$\beta_{dc} = \frac{I_C}{I_B} \qquad \text{(26–2)}$$

Typical values of β_{dc} for a transistor range from about 40 for power transistors to about 400 for small-signal transistors. Even for a particular transistor type, the value of β_{dc} can range quite drastically. For example, the specification for the 2N3904 transistor (commonly used npn transistor) shows that β_{dc} ranges from 100 to 300. Although we often treat it as one, β_{dc} is not a constant for a given transistor. The value depends on the operating point and on the ambient temperature.

dc Alpha (α_{dc})

A less commonly used specification that is used to analyze transistor circuits is the dc alpha, α_{dc}, which is specified as the ratio of collector current to emitter current.

$$\alpha_{dc} = \frac{I_C}{I_E} \qquad \text{(26–3)}$$

Since I_C is always smaller than I_E, we see that α will always be less than 1, with typical values between 0.95 and 0.99.

EXAMPLE 26–1

A transistor circuit is measured to have $I_C = 2.00$ mA and $I_E = 2.05$ mA. Determine I_B, α, and β.

Solution
By applying Equations 26–1, 26–2, and 26–3, we have

$$I_B = I_E - I_C = 2.05 \text{ mA} - 2.00 \text{ mA} = 0.05 \text{ mA} = 50 \text{ }\mu\text{A}$$

$$\alpha_{dc} = \frac{I_C}{I_E} = \frac{2.00 \text{ mA}}{2.05 \text{ mA}} = 0.976$$

$$\beta_{dc} = \frac{I_C}{I_B} = \frac{2.00 \text{ mA}}{0.05 \text{ mA}} = 40$$

PRACTICE PROBLEMS 1

A transistor circuit is measured to have $\beta = 100$ and $I_C = 4.00$ mA. Determine I_B, I_E, and α.

Answers
40 μA, 4.04 mA, 0.9901

By manipulating Equations 26–1 and 26–2, it is possible to determine a relationship between I_E, β, and I_B, namely

$$I_E = (\beta + 1)I_B \qquad \text{(26–4)}$$

Now, by applying algebraic manipulation to Equations 26–2, 26–3, and 26–4 we arrive at two other expressions for α and β.

$$\alpha = \frac{\beta}{\beta + 1} \qquad (26\text{–}5)$$

$$\beta = \frac{\alpha}{1 - \alpha} \qquad (26\text{–}6)$$

It is left as end-of-chapter problems for the student to derive Equations 26–4 through 26–6.

PRACTICE PROBLEMS 2

A transistor has $\beta = 150$ and $I_B = 30\ \mu A$. Determine α, I_C, and I_E.

Answers
0.9934, 4.50 mA, 4.53 mA

26.3 Transistor Specifications

All semiconductor manufacturers publish comprehensive specifications for each component that is manufactured. These specifications provide the technologist or engineer with a complete guide to the performance of the particular component, allowing us to analyze and design electronic circuits. While a typical specification contains many pages of data, it is often necessary to consider only a few of the important characteristics. It is extremely important that the first-time user not be intimidated by the large amount of information that is contained in a manufacturer's specification.

Consider the specification for the 2N3904 npn transistor that is provided in Figure 26–8. Notice that the specification corresponds to three different packages, each with a different part number. The TO-92 case is a fairly large package that can be easily handled in the lab. The SOT-23 and SOT-223 are surface mount packages that are much smaller, requiring a much smaller "footprint" on a printed surface board. The disadvantage of the smaller cases is that they are more difficult to handle and are not practical for most lab measurements by students.

Absolute Maximum Ratings and Thermal Characteristics

$V_{CEO} = 40\ V$ is the maximum voltage (V) that can be applied between the collector and emitter (CE) with the base terminal left open (O). For the 2N3904 this value is 40 V, which means that this is the maximum supply voltage that would be used for any circuit. If this voltage is exceeded, there is a very good likelihood that the transistor would be permanently damaged.

$V_{CBO} = 60V$ is the maximum voltage that can be applied to the reverse-biased collector-base junction (with the emitter terminal left open). If the supply voltage is kept less than 40 V, we see that this rating will not be exceeded.

$V_{EBO} = 6.0V$ is the maximum voltage that can be applied to the reverse-biased emitter-base junction. (In this case the collector terminal is left open.) Recall that this junction will not normally be reverse-biased. However, you can see that it is extremely important to ensure that the B-E junction is not inadvertently connected to a supply with the wrong polarity.

$I_C = 200\ mA$ is the largest continuous collector current that can be handled by the transistor.

$P_D = 625\ mW$ is the maximum power that the transistor can safely dissipate at an ambient temperature, $T_A = 25°$. For any transistor, the power dissipation is determined as the product of collector current, I_C and the collector-emitter voltage, V_{CE}, namely

$$P_D = I_C V_{CE} \qquad (26\text{–}7)$$

We will examine an application of the power calculation later in the chapter.

NPN General Purpose Amplifier (continued)

Electrical Characteristics $T_A = 25°C$ unless otherwise noted

Symbol	Parameter	Test Conditions	Min	Max	Units
OFF CHARACTERISTICS					
$V_{(BR)CEO}$	Collector-Emitter Breakdown Voltage	$I_C = 1.0$ mA, $I_B = 0$	40		V
$V_{(BR)CBO}$	Collector-Base Breakdown Voltage	$I_C = 10$ µA, $I_E = 0$	60		V
$V_{(BR)EBO}$	Emitter-Base Breakdown Voltage	$I_E = 10$ µA, $I_C = 0$	6.0		V
I_{BL}	Base Cutoff Current	$V_{CE} = 30$ V, $V_{EB} = 3$V		50	nA
I_{CEX}	Collector Cutoff Current	$V_{CE} = 30$ V, $V_{EB} = 3$V		50	nA
ON CHARACTERISTICS*					
h_{FE}	DC Current Gain	$I_C = 0.1$ mA, $V_{CE} = 1.0$ V	40		
		$I_C = 1.0$ mA, $V_{CE} = 1.0$ V	70		
		$I_C = 10$ mA, $V_{CE} = 1.0$ V	100	300	
		$I_C = 50$ mA, $V_{CE} = 1.0$ V	60		
		$I_C = 100$ mA, $V_{CE} = 1.0$ V	30		
$V_{CE(sat)}$	Collector-Emitter Saturation Voltage	$I_C = 10$ mA, $I_B = 1.0$ mA		0.2	V
		$I_C = 50$ mA, $I_B = 5.0$ mA		0.3	V
$V_{BE(sat)}$	Base-Emitter Saturation Voltage	$I_C = 10$ mA, $I_B = 1.0$ mA	0.65	0.85	V
		$I_C = 50$ mA, $I_B = 5.0$ mA		0.95	V
SMALL SIGNAL CHARACTERISTICS					
f_T	Current Gain - Bandwidth Product	$I_C = 10$ mA, $V_{CE} = 20$ V, f = 100 MHz	300		MHz
C_{obo}	Output Capacitance	$V_{CB} = 5.0$ V, $I_E = 0$, f = 1.0 MHz		4.0	pF
C_{ibo}	Input Capacitance	$V_{EB} = 0.5$ V, $I_C = 0$, f = 1.0 MHz		8.0	pF
NF	Noise Figure	$I_C = 100$ µA, $V_{CE} = 5.0$ V, $R_S = 1.0$kΩ, f=10 Hz to 15.7kHz		5.0	dB
SWITCHING CHARACTERISTICS					
t_d	Delay Time	$V_{CC} = 3.0$ V, $V_{BE} = 0.5$ V,		35	ns
t_r	Rise Time	$I_C = 10$ mA, $I_{B1} = 1.0$ mA		35	ns
t_s	Storage Time	$V_{CC} = 3.0$ V, $I_C = 10$mA		200	ns
t_f	Fall Time	$I_{B1} = I_{B2} = 1.0$ mA		50	ns

*Pulse Test: Pulse Width ≤ 300 µs, Duty Cycle ≤ 2.0%

Spice Model

NPN (Is=6.734f Xti=3 Eg=1.11 Vaf=74.03 Bf=416.4 Ne=1.259 Ise=6.734 Ikf=66.78m Xtb=1.5 Br=.7371 Nc=2 Isc=0 Ikr=0 Rc=1 Cjc=3.638p Mjc=.3085 Vjc=.75 Fc=.5 Cje=4.493p Mje=.2593 Vje=.75 Tr=239.5n Tf=301.2p Itf=.4 Vtf=4 Xtf=2 Rb=10)

2N3904 / MMBT3904 / PZT3904

FAIRCHILD SEMICONDUCTOR™

2N3904 **MMBT3904** **PZT3904**

TO-92

SOT-23 Mark: 1A

SOT-223

NPN General Purpose Amplifier

This device is designed as a general purpose amplifier and switch.
The useful dynamic range extends to 100 mA as a switch and to 100 MHz as an amplifier.

Absolute Maximum Ratings* $T_A = 25°C$ unless otherwise noted

Symbol	Parameter	Value	Units
V_{CEO}	Collector-Emitter Voltage	40	V
V_{CBO}	Collector-Base Voltage	60	V
V_{EBO}	Emitter-Base Voltage	6.0	V
I_C	Collector Current - Continuous	200	mA
T_J, T_{stg}	Operating and Storage Junction Temperature Range	-55 to +150	°C

*These ratings are limiting values above which the serviceability of any semiconductor device may be impaired.

NOTES:
1) These ratings are based on a maximum junction temperature of 150 degrees C.
2) These are steady state limits. The factory should be consulted on applications involving pulsed or low duty cycle operations.

Thermal Characteristics $T_A = 25°C$ unless otherwise noted

Symbol	Characteristic	Max 2N3904	*MMBT3904	**PZT3904	Units
P_D	Total Device Dissipation	625	350	1,000	mW
	Derate above 25°C	5.0	2.8	8.0	mW/°C
$R_{θJC}$	Thermal Resistance, Junction to Case	83.3			°C/W
$R_{θJA}$	Thermal Resistance, Junction to Ambient	200	357	125	°C/W

* Device mounted on FR-4 PCB 1.6" X 1.6" X 0.06."
** Device mounted on FR-4 PCB 36 mm X 18 mm X 1.5 mm; mounting pad for the collector lead min. 6 cm².

2N3904/MMBT3904/PZT3904, Rev. A

FIGURE 26–8 Data sheets of a 2N3904 transistor. (*Courtesy of Fairchild Semiconductor Corporation*)

Electrical Characteristics

All electrical characteristics of transistors are specified for certain test conditions. In many instances, the manufacturer provides both minimum and maximum values that one can expect to measure in a lab.

At this point, we will examine only two additional characteristics. It is left as an exercise for the student to examine some of the other specifications.

$V_{(BR)CEO} = 40$ V implies that this is the minimum collector-emitter voltage at which one can expect the transistor to breakdown. As expected, this is precisely the same value that was specified under the heading **Absolute Maximum Ratings.** However, this specification also provides the test conditions, namely that $I_C = 1.0$ mA and $I_B = 0$.

h_{FE} is the dc current gain (or beta) of the transistor. As seen in the specification, this value is not constant, but rather is very dependent on the operating conditions of the transistor. As an illustration, we see that the transistor beta varies between 100 and 300 for a collector current of 10 mA. This illustrates clearly that two circuits having the same component values and device numbers can easily have different operating conditions.

26.4 Collector Characteristic Curves

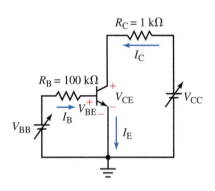

FIGURE 26–9

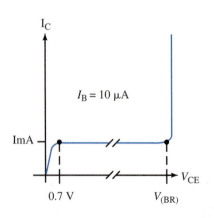

FIGURE 26–10

Consider the circuit shown in Figure 26–9. By changing the values of the supply voltages, it is possible to change the operating conditions of the transistor. The supply voltage, V_{BB}, determines the value of I_B, while the supply voltage, V_{CC}, determines both I_C and V_{CE}.

Imagine that we let $V_{CC} = 0$ and vary only V_{BB}. Once V_{BB} is greater than the barrier potential of the forward-biased B-E junction (0.7 V for silicon), base current results. For instance, if we were to set $V_{BB} = 1.7$ V, we can determine that the base current would remain constant at a value determined as

$$I_B = \frac{1.7 \text{ V} - 0.7 \text{ V}}{100 \text{ k}\Omega} = 10 \text{ }\mu\text{A}$$

With $V_{CC} = 0$, the collector will be at zero volts, which means that both the B-E and the C-B junctions are forward biased. Now, if we were to increase V_{CC}, collector current will increase linearly until the bias potential on the C-B junction is returned to 0 V. At this point, the voltage across the collector-emitter terminals can be determined from Kirchhoff's voltage law as $V_{CE} = 0.7$ V. Further increases in V_{CC} will simply increase the voltage across the now reverse-biased C-B junction, resulting in an increase of V_{CE}. Collector current will remain constant until the collector-emitter breakdown voltage, $V_{(BR)CEO}$, is exceeded. Figure 26–10 shows how of I_C varies as a function of V_{CE}.

If we were to plot the collector current as a function of the collector-emitter voltage, for numerous values of base current, we would arrive at a family of curves similar to those shown in Figure 26–11.

The collector characteristic curves of Figure 26–11 show several important points. The **cutoff region** is the portion of the graph that occurs below the line corresponding to $I_B = 0$. If no base current were applied to the transistor, one would expect that there would not be any collector current. This, however, is not true since there will always be a small leakage current (generally in the order of nanoamps) through the reverse-biased C-B junction. In the cutoff region, the transistor behaves very much like an open switch; current is very small and the voltage across the collector-emitter of the transis-

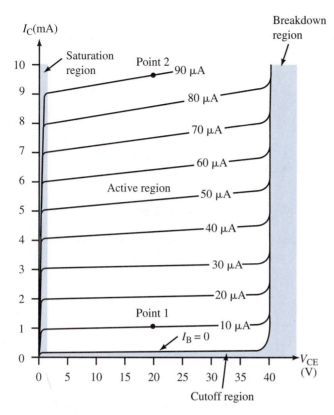

FIGURE 26–11 Collector characteristic curves.

tor is large. In the cutoff region, both the B-E and the C-B junctions are reverse-biased.

The **saturation region** is that portion of the graph where collector current increases rapidly for very small values of V_{CE}. In the saturation region, the transistor behaves very much like a closed switch; current is large and the voltage across the collector-emitter of the transistor is very small. In the saturation region, both the B-E and the C-B junctions are forward-biased. When transistors are used as switching devices, they will operate in either saturation (ON) or cutoff (OFF). We will examine the operation of the transistor switch in Section 26.7.

If the voltage across the collector-emitter of the transistor exceeds $V_{(BR)CE}$ of the transistor, a large amount of collector current will result in probable destruction of the device.

Finally, the **active region** is that portion of the collector characteristic curves in which the transistor can be used as an amplifier. Recall that in this region, the B-E junction is forward biased and the C-B junction is reverse-biased.

By examining the curves of Figure 26–11, we may conclude that the β_{dc} of a particular transistor is not constant, but rather is dependent on the **operating point.** The operating point (also called the quiescent point or simply the Q-point) of a transistor is a collection of dc bias conditions, which include I_B, I_C, and V_{CE}. The following example shows how we may use the collector characteristic curves to determine the β of a transistor at a given operating point.

EXAMPLE 26–2

Determine β_{dc} for the transistor having characteristic curves shown in Figure 26–11,

a. if the operating point of the transistor is $V_{CE} = 20$ V and $I_C = 1.0$ mA (point 1)

b. if the operating point of the transistor is $V_{CE} = 20$ V and $I_B = 90$ μA (point 2)

Solution

a. From the graph, we see that when $V_{CE} = 20$ V and $I_C = 1.0$ mA, the base current is $I_B = 10$ μA. Therefore,

$$\beta_1 = \frac{I_C}{I_B} = \frac{1.0 \text{ mA}}{0.01 \text{ mA}} = 100$$

b. At point 2, when $V_{CE} = 20$ V and $I_B = 90$ μA, the collector current is $I_C = 9.7$ mA, and so

$$\beta_2 = \frac{I_C}{I_B} = \frac{9.7 \text{ mA}}{0.09 \text{ mA}} = 108$$

Although Figure 26–11 does not illustrate it, the active region of a transistor is restricted by two other conditions that were mentioned previously; the maximum forward current $I_{C(MAX)}$ and the total device power dissipation, P_D. Figure 26–12 shows that the active region for the 2N3904 transistor is limited by $I_{C(MAX)} = 200$ mA, $V_{(BR)CEO} = 40$ V, and $P_D = 625$ mW.

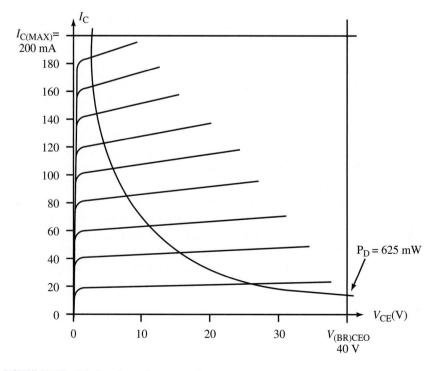

FIGURE 26–12 Limits of transistor operation.

Consider the transistor circuit shown in Figure 26–13.

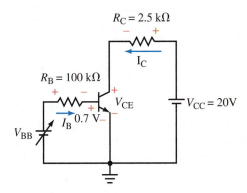

FIGURE 26–13

For any given set of component values, the transistor operation will occur somewhere between two extreme limits: *saturation* and *cutoff*. The components contained in the collector-emitter circuit determine these two points. When in saturation, the transistor behaves as a closed switch, resulting in maximum I_C through the transistor and a V_{CE} that is essentially zero. Conversely, when in cutoff the transistor behaves as an open switch where I_C is essentially zero and V_{CE} is at a maximum. Figure 26–14 shows equivalent circuits of the transistor in each of the two conditions described above.

By examining the circuit models that are shown in Figure 26–14, we see that the saturation current for the transistor is

$$I_{C(SAT)} = \frac{V_{CC}}{R_C} = \frac{20 \text{ V}}{2.5 \text{ k}\Omega} = 8.0 \text{ mA}$$

and the cutoff voltage is

$$V_{CE(OFF)} = V_{CC} = 20 \text{ V}$$

The cutoff and saturation points form the limits of the dc load line for a transistor as illustrated in Figure 26–15.

26.5 dc Load Line

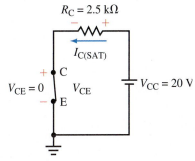

(a) A transistor in saturation behaves like a closed switch

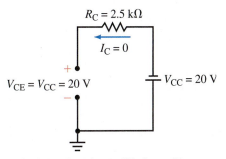

(b) A transistor in cutoff behaves like an open switch

FIGURE 26–14

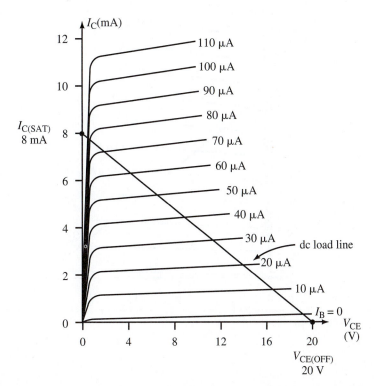

FIGURE 26–15

When the transistor is used as a linear amplifier, the actual operating point of the transistor should generally lie near the midpoint between cutoff and saturation and is determined by the base circuit. This important concept becomes evident if we vary V_{BB} in the circuit of Figure 26–13. For example, if we let $V_{BB} = 4.7$ V, we determine the base current to be

$$I_B = \frac{4.7 \text{ V} - 0.7 \text{ V}}{100 \text{ k}\Omega} = 40 \text{ }\mu\text{A}$$

The operating point for this base current is illustrated as Q_1 on the dc load line of Figure 26–16. Although a transistor operating point will generally be near the midpoint of the load line, this will not always be the case. Consider the condition where we let $V_{BB} = 1.7$ V, the base current would be determined as

$$I_B = \frac{1.7 \text{ V} - 0.7 \text{ V}}{100 \text{ k}\Omega} = 10 \text{ }\mu\text{A}$$

This operating point is illustrated as Q_2 on the dc load line, clearly indicating that by reducing the base current, we can move a transistor operating point towards cutoff. Alternatively if we let the base voltage, $V_{BB} = 8.7$ V, the base current is

$$I_B = \frac{8.7 \text{ V} - 0.7 \text{ V}}{100 \text{ k}\Omega} = 80 \text{ }\mu\text{A}$$

This last condition shows that by increasing the base current of the transistor circuit, we will cause the operating point to move towards saturation as shown by Q_3 in Figure 26–16. We may conclude, therefore, that for a given load line, the operating point is determined by the components in the base circuit.

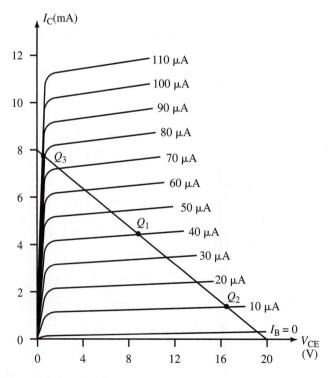

FIGURE 26–16

Refer to the circuit of Figure 26–13. Sketch the corresponding dc load lines for each of the following conditions.

a. $V_{CC} = 16$ V with $R_C = 2.5$ kΩ
b. $V_{CC} = 24$ V with $R_C = 2.5$ kΩ
c. $V_{CC} = 20$ V with $R_C = 2$ kΩ
d. $V_{CC} = 20$ V with $R_C = 4$ kΩ

Answers
a. $V_{CE(OFF)} = 16$ V, $I_{C(SAT)} = 6.4$ mA
b. $V_{CE(OFF)} = 24$ V, $I_{C(SAT)} = 9.6$ mA
c. $V_{CE(OFF)} = 20$ V, $I_{C(SAT)} = 10$ mA
d. $V_{CE(OFF)} = 20$ V $I_{C(SAT)} = 5$ mA

To this point, we have examined only the dc operation of a transistor. Although we will examine the ac operation in much greater detail in the next chapter, it is useful to see how a transistor can be used to amplify an ac signal. Consider the circuit shown in Figure 26–17.

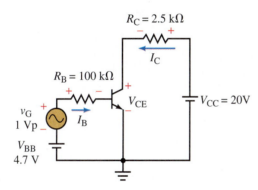

FIGURE 26–17

In Figure 26–17, we see that the ac voltage source is in series with the dc base bias voltage $V_{BB} = 4.7$ V. As in the previous illustration, we see that the dc load line has end points determined by $I_{SAT} = 8.0$ mA and $V_{CE(OFF)} = 20$ V. The operating point, which is determined by the base bias voltage, will be at the Q-point shown in Figure 26–18. ($I_{CQ} = 4.4$ mA and $V_{CEQ} = 9.0$ V).

As the alternating voltage source goes up, so too will the base current. When the generator voltage is at its maximum $v_G = +1$ V, the resulting base current will be

$$I_B = \frac{4.7 \text{ V} + 1 \text{ V} - 0.7 \text{ V}}{100 \text{ k}\Omega} = 50 \text{ μA}$$

At this point the collector-emitter voltage will be approximately $V_{CE} = 6.5$ V. Similarly, when the generator voltage is at its peak negative voltage, $v_G = -1$ V, the resulting base current will be

$$I_B = \frac{4.7 \text{ V} - 1 \text{ V} - 0.7 \text{ V}}{100 \text{ k}\Omega} = 30 \text{ μA}$$

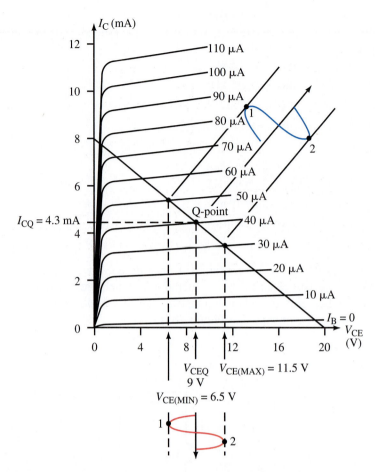

FIGURE 26–18

At this point on the dc load line we see that $V_{CE} = 11.5$ V. This illustrates that if the input voltage goes between $+1$ V and -1 V (total voltage swing of 2 $V_{p\text{-}p}$), the output voltage swings between 6.5 V and 11.5 V (5.0 $V_{p\text{-}p}$). We say that this circuit has a *voltage gain* of 2.5, since the output voltage, V_{CE}, is 2.5 times larger than the applied input ac voltage. We will examine the ac operation of transistor circuits in much greater detail in the following chapter.

26.6 Transistor Biasing

Although two separate voltage supplies may be used to bias a circuit as shown in Figure 26–13, it is generally not done. Rather, it is much more efficient to use a single supply to bias both the base bias and the collector. We will now examine several different types of biasing to determine how each type of circuit may be analyzed. You will also find that by making slight changes to the circuit, the circuits become increasingly less dependent on the transistor beta. This is particularly important when you consider that for a given transistor type, the beta may vary a large amount. Consider the specification for the 2N3904, which shows that h_{FE} (dc beta) can have values between 100 and 300. Clearly this can present quite a problem when designing a circuit for a given operating point.

The Fixed-Bias Circuit

The fixed-bias circuit of Figure 26–19 operates exactly the same as the dual-supply circuits examined previously. As in previous circuits, we determine the operating point of this circuit by starting at the input loop (the loop containing the base-emitter junction) and then transferring to the output loop (the loop containing the collector-emitter voltage).

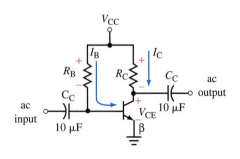

FIGURE 26–19

Although the circuit of Figure 26–19 does not initially look the same as the previous fixed-bias circuit, you will find that by redrawing the circuit, we do indeed have the same circuit. The circuit of Figure 26–19 shows capacitors, which are needed in the circuit to permit ac signals to pass. At this point, you simply need to remember that a capacitor operating at dc is effectively an open circuit and so may simply be removed when we are calculating the dc operating point. Figure 26–20 shows the equivalent redrawn circuit, with capacitors removed and the voltage supply, V_{CC}, shown as part of both the input loop and the output loop.

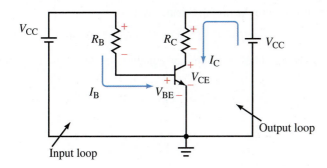

FIGURE 26–20

By examining the input loop, we see that we may easily determine the base current for the circuit as

$$I_B = \frac{V_{CC} - V_{BE}}{R_B} \qquad \text{(26–8)}$$

The base current allows us to transfer into the output loop by using the transistor beta, hence

$$I_C = \beta I_B \qquad \text{(26–9)}$$

Finally we use Kirchhoff's voltage law to write the output loop equation.

$$V_{CC} = I_C R_C + V_{CE}$$

and so we solve the collector-emitter voltage as

$$V_{CE} = V_{CC} - I_C R_C \qquad \text{(26–10)}$$

Determine I_B, I_C, and V_{CE} for the circuit of Figure 26–21 if the transistor has $\beta = 200$.

EXAMPLE 26–3

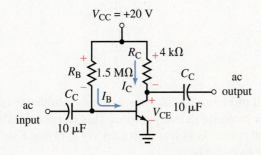

FIGURE 26–21

Solution

Writing the Kirchhoff Voltage Law equation for the input loop, we have

$$20 \text{ V} = 1.5 \text{ M}\Omega \, I_B + 0.7 \text{ V}$$

which allows us to calculate I_B as

$$I_B = \frac{20 \text{ V} - 0.7 \text{ V}}{1.5 \text{ M}\Omega} = 2.87 \text{ } \mu A$$

Now, transferring into the output circuit, we have

$$I_C = \beta I_B = (200)(12.87 \text{ } \mu A) = .57 \text{ mA}$$

and writing the Kirchhoff Voltage Law equation for the output loop, we have

$$20 \text{ V} = 4 \text{ k}\Omega \, I_C + V_{CE}$$
$$= (4 \text{ k}\Omega)(2.57 \text{ mA} + V_{CE}$$

and so

$$V_{CE} = 20 \text{ V} - 10.29 \text{ V} = 9.71 \text{ V}$$

PRACTICE PROBLEMS 4

Determine I_B, I_C, and V_{CE} for the circuit of Figure 26–21 if the transistor has
a. $\beta = 100$
b. $\beta = 300$

Answers
a. $I_B = 12.87 \text{ } \mu A$, $I_C = 1.287 \text{ mA}$, $V_{CE} = 14.85 \text{ V}$
b. $I_B = 12.87 \text{ } \mu A$, $I_C = 3.86 \text{ mA}$, $V_{CE} = 4.56 \text{ V}$

Emitter-Stabilized Bias Circuit

While the circuit of Figure 26–21 is relatively simple to analyze, the operating point is highly dependent upon the beta of the transistor. Introducing a "feedback resistor" into the emitter circuit of the transistor will stabilize the effects of changing beta. Figure 26–22 shows the schematic of an emitter-stabilized amplifier.

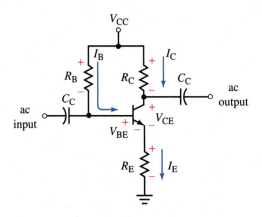

FIGURE 26–22

In the circuit of Figure 26–22, we see that the emitter resistor, R_E appears in both the input loop and the output loop. This resistor results in a feedback effect, which will reduce I_B as β is increased. By applying Kirchhoff's Voltage Law to the input loop, we have

$$V_{CC} = R_B I_B + V_{BE} + R_E I_E$$

However, from Equation 26–4 we have $I_E = (\beta + 1)I_B$ and so we have

$$V_{CC} = R_B I_B + V_{BE} + R_E (\beta + 1)I_B$$

which allows us to determine I_B as

$$I_B = \frac{V_{CC} - V_{BE}}{R_B + R_E (\beta + 1)} \qquad (26–11)$$

As before, $I_C = \beta I_B$. Writing the Kirchhoff Voltage Law equation for the output loop gives us

$$V_{CC} = R_C I_C + V_{CE} + R_E I_E$$

and since $I_C \cong I_E$, we may solve for V_{CE} as

$$V_{CE} \cong V_{CC} - (R_C + R_E)I_C \qquad (26–12)$$

The following example shows how the stabilizing resistor, R_E affects a transistor circuit.

Determine I_B, I_C, and V_{CE} for the circuit of Figure 26–23 if the transistor has $\beta = 200$.

EXAMPLE 26–4

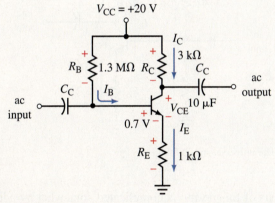

FIGURE 26–23

Solution
Using Kirchhoff's Voltage Law to write the loop equation for the input loop, we have

$$20\ V = (1.3\ M\Omega)I_B + 0.7\ V + (1\ k\Omega)I_E$$
$$= (1300\ k\Omega)I_B + 0.7\ V + (1\ k\Omega)(201 I_B)$$

Solving for I_B we get

$$I_B = \frac{20\ V - 0.7\ V}{1300\ k\Omega + 201\ k\Omega} = 12.86\ \mu A$$

Next, we transfer into the output loop to solve for the I_C as

$$I_C = \beta I_B = (200)(12.86\ \mu A) = 2.57\ mA.$$

Writing the output loop equation, we have

$$20\ V = 3\ k\Omega\ I_C + V_{CE} + 1\ k\Omega\ I_E$$
$$\cong (4\ k\Omega)(2.57\ mA) + V_{CE}$$

and so

$$V_{CE} \cong 20\ V - 10.29\ V = 9.71\ V$$

Determine I_B, I_C, and V_{CE} for the circuit of Figure 26–23 if the transistor has
a. $\beta = 100$
b. $\beta = 300$

Answers
a. $I_B = 13.78 \ \mu A$, $I_C = 1.38$ mA, $V_{CE} = 14.49$ V
b. $I_B = 12.05 \ \mu A$, $I_C = 3.62$ mA, $V_{CE} = 5.53$ V

Notice that the operating point for the circuit of Figure 26–23 does not change as dramatically with changes in beta as did the circuit of Figure 26–21. This effect is the result of the emitter-stabilizing resistor.

The Universal-Bias Circuit

Although the emitter-stabilized transistor circuit has an improved stability, it does not provide sufficient stability. The Universal-Bias circuit show in Figure 26–24 is a much-improved design over other bias arrangements.

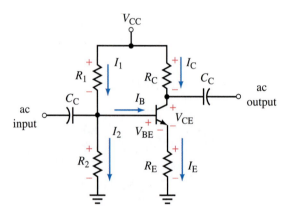

FIGURE 26–24   The universal-bias circuit.

The analysis of this design is complicated somewhat by adding the extra bias resistor into the circuit. We will discover that there are two methods that may be used to solve for the operating point of the transistor; the *exact method,* which always works, and the *approximate method,* which provides good results for most conditions.

Exact Method

First let's simplify the circuit by removing the two coupling capacitors that have no effect on the dc operation of the transistor. Next, we separate the circuit into two clearly identifiable circuits consisting of the input loop and the output loop as shown in Figure 26–25.

In Figure 26–25, we then further simplify the input loop by finding the Thévenin equivalent circuit as seen between the base and the reference point (ground). The Thévenin voltage is determined as

$$V_{BB} = E_{Th} = \frac{R_2}{R_1 + R_2} V_{CC} \tag{26–13}$$

and the Thévenin resistance is

$$R_{BB} = R_{Th} = R_1 \| R_2 = \frac{R_1 R_2}{R_1 + R_2} \tag{26–14}$$

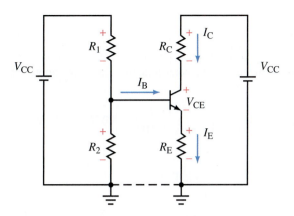

FIGURE 26–25

The resultant circuit is shown in Figure 26–26. It is now an easy matter to write the input loop equation

$$V_{BB} = R_{BB} I_B + V_{BE} + R_E I_E$$
$$= R_{BB} I_B + V_{BE} + R_E (\beta + 1)I_B$$

and solving for I_B, we have

$$I_B = \frac{V_{BB} - V_{BE}}{R_{BB} + (\beta + 1)R_E} \qquad (26\text{–}15)$$

Now, transferring into the output loop, we have the same results as those for the emitter stabilized bias circuit.

$$I_C = \beta I_B \qquad (26\text{–}16)$$

and

$$V_{CE} \cong V_{CC} - (R_C + R_E)I_C \qquad (26\text{–}17)$$

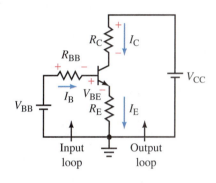

FIGURE 26–26

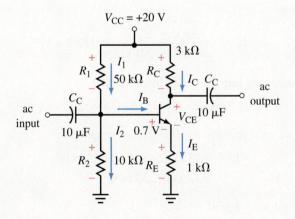

FIGURE 26–27

EXAMPLE 26–5

a. Determine I_B, I_C, and V_{CE} for the circuit of Figure 26–27 if the transistor has $\beta = 200$.

b. Use the above results to determine the current through each of the bias resistors, R_1 and R_2.

Solution

a. After removing the coupling capacitors, C_C, which have no effect on the dc operation of the circuit, we determine the Thévenin equivalent circuit of the input circuit as follows:

$$V_{BB} = \left(\frac{10 \text{ k}\Omega}{50 \text{ k}\Omega + 10 \text{ k}\Omega}\right)20 \text{ V} = 3.33 \text{ V}$$

and

$$R_{BB} = \frac{(10 \text{ k}\Omega)(50 \text{ k}\Omega)}{10 \text{ k}\Omega + 50 \text{ k}\Omega} = 8.33 \text{ k}\Omega$$

Now, by substituting the Thévenin equivalent of the input circuit, we have the equivalent amplifier shown in Figure 26–28.

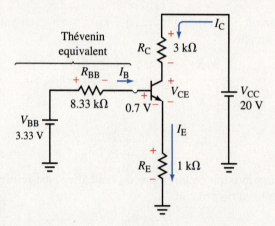

FIGURE 26–28

The loop equation of the input loop is

$$3.33 \text{ V} = (8.33 \text{ k}\Omega)I_B + 0.7 \text{ V} + (1 \text{ k}\Omega)I_E$$
$$= (8.33 \text{ k}\Omega)I_B + 0.7 \text{ V} + (1 \text{ k}\Omega)(\beta + 1)I_B$$
$$= (8.33 \text{ k}\Omega)I_B + 0.7 \text{ V} + (1 \text{ k}\Omega)(201I_B)$$

which gives

$$I_B = \frac{3.33 \text{ V} - 0.7 \text{ V}}{8.33 \text{ k}\Omega + 201 \text{ k}\Omega} = 12.58 \text{ }\mu\text{A}$$

Now, transferring into the output loop, we have

$$I_C = \beta I_B = (200)12.58 \text{ }\mu\text{A} = 2.52 \text{ mA} \cong I_E$$

Finally, we write the loop equation for the output loop.

$$20 \text{ V} = (3 \text{ k}\Omega)I_C + V_{CE} + (1 \text{ k}\Omega)I_E$$
$$\cong (4 \text{ k}\Omega)(2.52 \text{ mA}) + V_{CE}$$

Now we solve for V_{CE} as

$$V_{CE} \cong 20 \text{ V} - 10.06 \text{ V} = 9.94 \text{ V}$$

b. If we examine the circuit of Figure 26–28, we see that we can calculate the voltage at the base of the transistor as

$$V_B = 0.7 \text{ V} + R_E I_E$$
$$= 0.7 \text{ V} + (1 \text{ k}\Omega)(2.52 \text{ mA})$$
$$= 3.22 \text{ V}$$

Now, if we go back to the original circuit shown in Figure 26–27, we see that this is the same voltage that must appear across R_2. Therefore, we can calculate the current I_2 as

$$I_2 = \frac{3.22 \text{ V}}{10 \text{ k}\Omega} = 322 \text{ μA}$$

Applying Kirchhoff's Voltage Law, we determine the voltage across R_1 as

$$V_{R1} = 20 \text{ V} - 3.22 \text{ V} = 16.78 \text{ V}$$

which allows us to determine the current I_1 as

$$I_1 = \frac{16.78 \text{ V}}{50 \text{ k}\Omega} = 336 \text{ μA}$$

Determine I_B, I_C, and V_{CE} for the circuit of Figure 26–27 if the transistor has
a. $\beta = 100$
b. $\beta = 300$

Answers
a. $I_B = 24.09 \text{ μA}$, $I_C = 2.41 \text{ mA}$, $V_{CE} = 10.37 \text{ V}$
b. $I_B = 8.51 \text{ μA}$, $I_C = 2.55 \text{ mA}$, $V_{CE} = 9.78 \text{ V}$

 PRACTICE PROBLEMS 6

Notice that although the value of beta ranges from 100 to 300, the corresponding operating point (I_{CQ} and V_{CEQ}) changed only slightly. For this reason, the universal bias circuit is considered to be the most stable bias arrangement.

Approximate Method

In the previous example, I_B is very small in comparison to both I_1 and I_2. We may therefore assume that the base current is negligible in comparison to the current through the bias resistors, R_1 and R_2. However, this assumption is only valid if the bias current is relatively large. As a general rule of thumb, we say that the base current is negligible if

$$R_2 \leq \frac{1}{10} \beta R_E \qquad \text{(26–18)}$$

If the condition of Equation 26–18 is not met, it is necessary to analyze the circuit using the exact method that was shown previously.

Now, this assumption allows us to further conclude that the base voltage, V_B can be determined by applying the voltage divider rule. Therefore,

$$V_B = \left(\frac{R_2}{R_1 + R_2}\right) V_{CC} \qquad \text{(26–19)}$$

We can easily determine the voltage at the emitter (with respect to ground) as

$$V_E = V_B - V_{BE} \qquad \text{(26–20)}$$

Next, we determine the emitter current from Ohm's Law. Since $I_E \cong I_C$, we have the following expression.

$$I_E = \frac{V_E}{R_E} \cong I_C \qquad (26\text{--}21)$$

And finally

$$V_{CE} \cong V_{CC} - (R_C + R_E)I_C \qquad (26\text{--}22)$$

The following example shows how the approximate analysis method can be applied to the universal bias circuit.

EXAMPLE 26–6

Use the approximate method to solve for I_C and V_{CE} for the circuit of Figure 26–27 if the transistor has $\beta = 200$.

Solution

First, we must test to determine whether we are entitled to use the approximate method to analyze the circuit. In other words, is the condition of Equation 26–19 met? By inserting values into Equation 26–19, we have the following

$$10 \text{ k}\Omega \leq \frac{1}{10}(200)(1\text{k}\Omega) \text{ as required for the approximate method}$$

$$\leq 20\text{k}\Omega$$

$$V_B = \left(\frac{10 \text{ k}\Omega}{10 \text{ k}\Omega + 50 \text{ k}\Omega}\right)20 \text{ V} = 3.33 \text{ V}$$

$$V_E = 3.33 \text{ V} - 0.7 \text{ V} = 2.63 \text{ V}$$

$$I_E = \frac{2.63 \text{ V}}{1 \text{ k}\Omega} = 2.63 \text{ mA} \cong I_C$$

and finally

$$V_{CE} \cong V_{CC} - (R_C + R_E)I_C$$
$$= 20 \text{ V} - (3 \text{ k}\Omega + 1 \text{ k}\Omega)(2.63 \text{ mA})$$
$$= 9.47 \text{ V}$$

The above results are consistent with those found using the exact method, where we had $I_C = 2.52$ mA and $V_{CE} = 9.94$ V. The variation using the approximate method is less than 5% and certainly is a much easier approach than using the exact method.

The Common Collector Circuit

As the name implies, the common collector circuit is connected so that the collector of the transistor is connected to ground (rather than the emitter, as in previous circuits). Although the circuit may appear differently, the analysis follows the same technique as previous circuits. As before, we begin our analysis by examining the input loop, which contains the forward-biased base-emitter junction. Then we use beta to transfer into the output loop, which contains the collector-emitter of the transistor. The following example shows that the method is not dramatically different from that used previously.

Determine I_B, I_C, and V_{CE} for the circuit of Figure 26–29 if the transistor has $\beta = 200$.

EXAMPLE 26–7

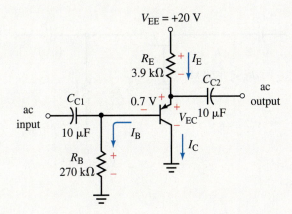

FIGURE 26–29

Solution

Notice that the transistor in the circuit of Figure 26–19 is a pnp transistor. Although this does not change the method of analysis, there are a few precautions. As one might expect, the voltage V_{BE} for the pnp transistor is -0.7 V rather than $+0.7$ V as in the npn transistor. The input loop equation begins at the supply voltage V_{EE} and is written as

$$V_{EE} = R_E I_E + V_{EB} + R_B I_B$$
$$20\ V = (3.9\ k\Omega)I_E + 0.7\ V + (270\ k\Omega)I_B$$
$$= (3.9\ k\Omega)(\beta + 1)I_B + 0.7\ V + (270\ k\Omega)I_B$$

and so we have

$$I_B = \frac{20\ V - 0.7\ V}{(3.9\ k\Omega)(201) + 270\ k\Omega}$$

$$= 18.3\ \mu A$$

Now, transferring into the output loop, we have

$$I_C = \beta I_B = (200)18.3\ \mu A = 3.66\ mA \cong I_E$$

Writing the output loop, we have

$$V_{EE} = R_E I_E + V_{EC}$$
$$20\ V = (3.9\ k\Omega)(3.66\ mA) + V_{EC}$$

We calculate V_{EC} as

$$V_{EC} = 20\ V - (3.9\ k\Omega)(3.66\ mA) = 5.72\ V$$

This results in a collector-emitter voltage of $V_{CE} = -5.72$ V, which is precisely the polarity that one expects in a transistor amplifier having a pnp transistor.

The Common Base Circuit

The common base amplifier is typically used for high-frequency applications and so is not a circuit that is as common as the common emitter or the common collector circuit. The following example shows the techniques used in determining the operating point of a common base amplifier.

EXAMPLE 26–8

Determine the V_{CE} and I_C of the common base amplifier circuit shown in Figure 26–30.

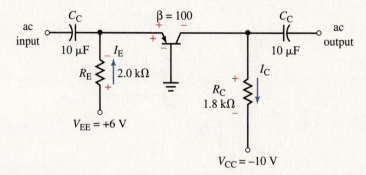

FIGURE 26–30

Solution

As always, we begin our analysis by examining the "input loop"—the loop containing the forward-biased base-emitter junction. The input loop equation is written as

$$R_E I_E + V_{EB} = +6\ V$$

$$(2.0\ k\Omega)I_E + 0.7\ V = +6\ V$$

which gives an emitter current (and the approximate value of collector current) as

$$I_E = \frac{6.0\ V - 0.7\ V}{2.0\ k\Omega} = 2.65\ mA \cong I_C$$

Now, we examine the output loop to arrive at the following equation:

$$6\ V + 10\ V = (2.0\ k\Omega)I_E + V_{EC} + (1.8\ k\Omega)I_C$$

$$= (3.8\ k\Omega)(2.65\ mA) + V_{EC}$$

which gives

$$V_{EC} = 16\ V - 10.07\ V$$

$$= 5.93\ V$$

and so the collector-emitter voltage is $V_{CE} = -5.93\ V$

26.7 The Transistor Switch

Although the transistor was once primarily used as an amplifier, its main application in modern electronic circuits is as a switch or as an inverter. In both cases, the operation of the transistor is either at saturation (analogous to a closed switch) or at cutoff (corresponding to an open switch). A low current signal applied at the base is translated into a high collector current that is capable of providing sufficient current to turn on LEDs (light emitting diodes) or operating relays. If a power transistor is used, the collector current may be sufficient to operate small motors.

When used as an inverter, a logic "1" ($\equiv 5\ V$) applied at the base input will result in a logic "0" ($\equiv 0\ V$) at the collector output. Conversely, a logic "0" applied at the base will result in a logic "1" at the collector.

The following example shows how a small base current can be translated into a fairly large collector current that can be used to operate an LED. A logic gate typically does not provide enough current to turn on an LED. In order to boost the current, a transistor circuit can be used as *buffer*.

EXAMPLE 26–9

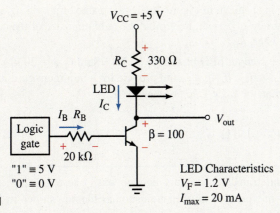

FIGURE 26–31

The transistor shown in Figure 26–31 is used as a buffer, providing sufficient current to turn on an LED when the output of the logic gate is "1".

a. Find the saturation current for the transistor.

b. Calculate the current through the LED when the output of the logic gate is "1".

c. Determine the output voltage at the collector (with respect to ground) for part a.

d. Calculate the current through the LED when the output of the logic gate is "0".

e. Determine the output voltage at the collector for part d.

Solution

a. In this example, the saturation current must take into account the forward voltage of the LED. Suppose V_F is the forward voltage drop across the LED when "on". Also recall that $V_{CE} = 0$ when the transistor is in saturation. Then

$$I_{C(SAT)} = \frac{V_{CC} - V_F}{R_C} = \frac{5\text{ V} - 1.2\text{ V}}{330\ \Omega} = 11.51\text{ mA}$$

b. When the output of the logic gate is a logic "1," the base current for the transistor is determined as

$$I_B = \frac{5\text{ V} - 0.7\text{ V}}{20\text{ k}\Omega} = 0.215\text{ mA}$$

Now, using the transistor beta, we evaluate the collector current as

$$I_C = \beta I_B = (100)(0.215\text{ mA}) = 21.5\text{ mA}$$

However, since the collector current cannot exceed the saturation current, we conclude that the collector current for the transistor is 11.5 mA.

c. With the transistor in saturation, the output voltage must be

$$V_{out} = 0\text{ V}$$

d. If the logic gate produces a logic "0" at its output, the transistor must go into cutoff since there can be no base current. Consequently, there can also be no collector current and so we have

$$I_C = 0$$

e. With the transistor in cutoff, the output voltage must be

$$V_{out} = 5\text{ V}$$

The results of part c. and e. illustrate that the transistor in this circuit behaves as both a buffer and a logic inverter. A logic "1" applied at the input results in a logic "0" at the output, while a logic "0" applied at the input results in a logic "1" at the output.

a. If the logic gate shown in Figure 26–31 is capable of providing a maximum "source" current of 50 µA, determine the minimum size of the current-limiting resistor, R_B.

b. If you were given an LED having $V_F = 1.8$ V and $I_{max} = 20$ mA calculate the range of resistance, R_C that would provide for a minimum current of 10 mA to the LED when the logic gate produces a "1."

Answers
a. 86 kΩ; b. $R_C = 160$ Ω → 320 Ω

26.8 Testing a Transistor with a Multimeter

Transistors and transistor circuits can be easily tested using a multimeter. Indeed, many inexpensive multimeters such as the one shown in Figure 26–32 include a provision for testing of both pnp and npn transistors. This type of multimeter allows us to determine the approximate dc beta of the transistor by simply inserting the transistor into the appropriate plugins. In this section you will find that there are several other methods that can be used to determine whether a transistor is operating correctly.

FIGURE 26–32 Multimeter with built-in transistor tester.

Since a transistor is essentially constructed as two back-to-back diodes, we may use an ohmmeter to measure the resistance of each of the p-n junctions. Recall that a forward-biased p-n junction has a small resistance, while a reverse-biased junction will essentially be an open circuit. By correctly using an ohmmeter, it is possible to determine whether an unknown transistor is pnp or npn and also to determine the correct pin configuration. If a transistor fails an ohmmeter test, we know that the transistor will not work properly when used in a circuit. Unfortunately, it is possible for a transistor to pass a simple ohmmeter test and yet fail in an operational circuit. We will examine this condition in the next chapter.

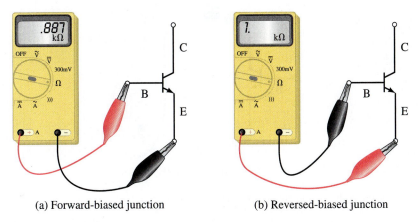

(a) Forward-biased junction   (b) Reversed-biased junction

FIGURE 26–33 Typical ohmmeter measurements of the BE junction of an npn transistor.

Consider the npn transistor shown in Figure 26–33. If we were to connect an ohmmeter with the positive (red) terminal at the base and the negative (black) terminal at the emitter, the BE junction would be forward-biased and we would expect to measure a relatively low resistance. Although the actual value is not important, here we see that we are measuring 0.887 kΩ. Remember that the ohmmeter uses a dc voltage source to generate a small current through the component under test. Now if we were to reverse the ohmmeter terminals and connect the negative terminal to the base and the positive terminal to the emitter, the BE junction would be reverse-biased. This time there would be no current through the junction, and the ohmmeter would indicate infinite resistance.

Figure 26–34 shows that a similar effect would occur if the ohmmeter were placed between the collector and base terminals. In this instance, we see that the forward-biased CB junction gives a reading of 0.849 kΩ with the ohmmeter in the same range as the first measurement.

PRACTICAL NOTES . . .

When using a digital ohmmeter to measure resistance of semiconductors, it is generally necessary to change the range of the meter to a location showing a diode symbol.

(a) Forward-biased junction   (b) Reversed-biased junction

FIGURE 26–34 Typical ohmmeter measurements of the CB junction of an npn transistor.

Finally, with the ohmmeter connected between the collector and emitter as shown in Figure 26–35, we see that both connections will indicate an open circuit. This is because each measurement results in one reverse-biased junction.

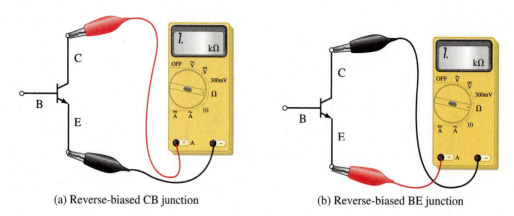

(a) Reverse-biased CB junction   (b) Reverse-biased BE junction

FIGURE 26–35   Typical ohmmeter measurements between the collector and emitter of an npn transistor.

Although the illustrations were for an npn transistor, we would notice similar effects for the pnp transistor. For all transistors, there will be two low resistance readings, and in all such cases, the terminal connected to the positive (red) terminal of the ohmmeter will be connected to p-type material, while the terminal connected to the negative (black) terminal of the ohmmeter will be connected to n-type material. Also, for all transistors (npn or pnp), the lower resistance reading will indicate that the measurement involves the collector of the transistor. The following steps may determine the type of transistor and the pin allocations.

1. Six different measurements must always be made.
2. If you measure an open circuit between any two terminals regardless of how the ohmmeter is connected, you may conclude that the remaining terminal is the base.
3. With any transistor, you will have exactly two low measurements if the transistor is functioning correctly. If you have a low reading when the positive terminal of the ohmmeter is connected to the base you have an npn transistor. Conversely, if the reading is low when the negative terminal is connected to the base, you have a pnp transistor.
4. The lower of the two low-resistance readings corresponds to the collector of the transistor, since the collector is the most heavily doped region. This, of course, means that the remaining terminal must be the emitter.

The following example shows how an ohmmeter can be used to determine the transistor type and pin configuration.

EXAMPLE 26–10

FIGURE 26–36

1 2 3

You are given the unknown transistor shown in Figure 26–36 and use an ohmmeter to take resistance measurements as tabulated. As always, the ohmmeter has two terminals; the positive (+ , red) terminal and the negative (− , black) terminal.

Pins 1 & 2		Pins 2 & 3		Pins 1 & 3	
Pin 1 (+) Pin 2 (−)	∞	Pin 2 (+) Pin 3 (−)	∞	Pin 1 (+) Pin 3 (−)	∞
Pin 2 (+) Pin 1 (−)	0.457 kΩ	Pin 3 (+) Pin 2 (−)	∞	Pin 3 (+) Pin 1 (−)	0.482 kΩ

a. What type of transistor were you given—pnp or npn?

b. What are the pin designations?

Solution

a. After examining the given results, we determine that pin 1 must be the base, since both measurements between pins 2 and 3 indicate open circuits. Since the base (pin 1) is connected to the negative terminal of the ohmmeter, we know that the base must be constructed of n-type material. Hence, your transistor must be pnp.

b. Finally, since the resistance measurement between pins 1 and 2 is smaller than the measurement between pins 1 and 3, we conclude that pin 2 must be the collector. Therefore, we have the following pin designations:

Pin 1: Base

Pin 2: Collector

Pin 3: Emitter

PRACTICE PROBLEMS 8

You are given the unknown transistor shown in Figure 26–36 and use an ohmmeter to take resistance measurements as tabulated below.

Pins 1 & 2		Pins 2 & 3		Pins 1 & 3	
Pin 1 (+) Pin 2 (−)	∞	Pin 2 (+) Pin 3 (−)	0.780 kΩ	Pin 1 (+) Pin 3 (−)	∞
Pin 2 (+) Pin 1 (−)	0.745 kΩ	Pin 3 (+) Pin 2	∞	Pin 3 (+) Pin 1 (−)	∞

a. What type of transistor were you given—pnp or npn?

b. What are the pin designations?

Answers

a. npn; b. pin 1 = collector, pin 2 = base, pin 3 = emitter

A multimeter can be used not only to test isolated transistors, but is extremely useful for assessing whether a transistor in a working circuit is operating as expected. Remember, when a transistor is used as an amplifier, the BE junction will always be forward-biased and one would expect to measure $V_{BE} \approx$ 0.7 V (for a silicon transistor). This means that the first test that one should perform on an amplifier is to determine whether this junction is biased correctly.

If the BE junction is forward-biased, we may then determine whether an amplifier is operating in its active region by measuring the operating point (V_{CEQ} and I_{CQ}). The voltage is very easy to measure since we would simply connect the voltmeter directly at the collector and emitter terminals of the transistor. In order to directly measure the current, we would need to break the circuit at the collector and insert the ammeter directly between collector resistor and the collector of the transistor. **This is not a practical method!** A much easier and preferred method for measuring current is to measure the voltage across a known resistance and then simply calculate the current using Ohm's Law.

Figure 26–37 illustrates typical measurements for an operational transistor circuit. The illustration shows that an additional voltage measurement can be taken to provide the current through the base bias resistor, R_B. This additional measurement allows us to calculate β for this transistor.

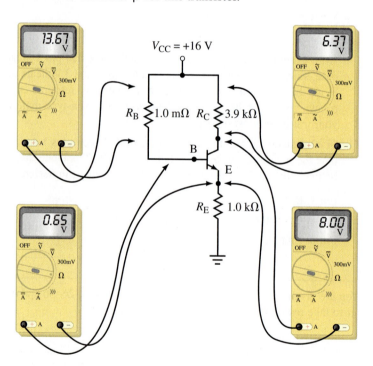

FIGURE 26–37

The measurements of Figure 26–37 provide much useful information on the operating point of the circuit. The BE voltage, $V_{BE} = 0.65$ V indicates that the base-emitter junction of the transistor is correctly biased. We solve for the collector current at the operating point as

$$I_{CQ} = \frac{6.37\ \text{V}}{3.9\ \text{k}\Omega} = 1.6\ \text{mA}$$

The collector-emitter voltage at the operating point is determined as $V_{CEQ} = 8.00$ V. The base current for the circuit is determined as

$$I_{BQ} = \frac{13.67\ \text{V}}{1.0\ \text{M}\Omega} = 14\ \mu\text{A}$$

which allows us to approximate β as

$$\beta_Q = \frac{1.6 \text{ mA}}{14 \text{ μA}} \approx 110$$

Given the circuit of Figure 26–38:

EXAMPLE 26–11

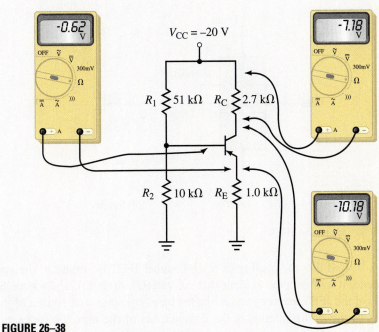

FIGURE 26–38

a. Is the BE junction biased correctly? Explain your answer.

b. What is the operating point of the transistor?

c. For the given information, can you calculate β of the circuit? If so, what is its value? If not, explain your answer.

Solution

a. The BE junction is biased correctly, since for a pnp transistor the base will be more negative than the emitter by a value of around 0.7 V. (In this case, a voltage $V_{BE} = -0.62$ V is within acceptable values.)

b. The collector current at the operating point is

$$I_{CQ} = \frac{-7.18 \text{ V}}{2.7 \text{ k}\Omega} = 2.7 \text{ mA}$$

(The negative in the above calculation is simply due to the voltage measurement being taken with the positive terminal of the voltmeter at the top of R_C.)

The collector-emitter voltage at the operating point is determined as $V_{CEQ} = -10.18$ V. As expected, for a pnp transistor, the collector-emitter voltage is negative.

c. β of the circuit cannot be found, since we are unable to calculate for base current by measuring across a known resistor. (A common mistake that is made is to measure the voltage across R_1 and then say that the base current is the same as the current through this resistor. It is not! The current entering the base is much smaller.)

26.9 Junction Field Effect Transistor Construction and Operation

Construction of the JFET

The JFET (**junction field effect transistor**) is similar to the BJT in that it has three terminals and can be used to amplify small ac signals. By embedding a p-type material around an n-type channel, as shown in Figure 26–39, we have an n-channel JFET. Notice that the arrow on the gate of the transistor points away from the p-type material towards the n-type material in the channel.

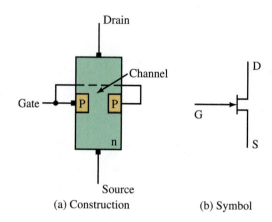

(a) Construction   (b) Symbol

FIGURE 26–39   n-channel JFET

We will find that the biasing of the n-channel JFET is similar to the npn BJT. Indeed, it will become evident that the gate(G), drain(D), and source(S) terminals of the JFET are very much like the base, collector, and emitter of the BJT. Just as the pnp transistor is the complement of the npn, the p-channel JFET is the complement to the n-channel. Operationally, the two FETs are the same, except that the n-channel is biased with a positive supply, while the p-channel is biased with a negative supply. Figure 26–40 shows both the structure and the symbol of the p-channel JFET. Once again notice that the arrow in the symbol points from the p-type material (channel) towards the n-type material (of the gate).

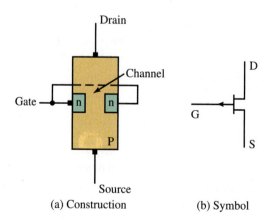

(a) Construction   (b) Symbol

FIGURE 26–40   p-channel JFET

While there are some similarities between the BJT and the JFET, there are several characteristics of the JFET that are quite different than the BJT. Whereas the BJT is a semiconductor device that provides current amplification, the JFET provides voltage amplification. The base-emitter junction of the BJT amplifier is always forward-biased, while in the JFET circuit, the gate-source is always reverse-biased. These conditions will become more evident as we examine the operation of the JFET. Other characteristics of the JFET are:

1. The input impedance of the device approaches infinity due to the reverse-biased gate-source junction.

2. The JFET amplifier is less sensitive to changes in input voltage than a junction transistor, since the JFET amplifier typically has a lower voltage gain (as we will find in the next chapter.)

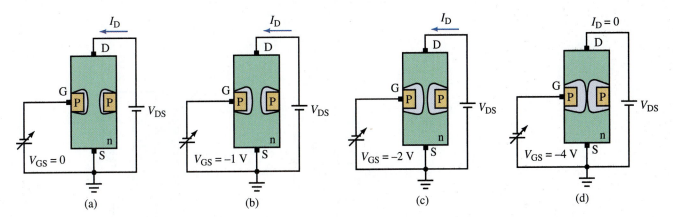

FIGURE 26–41   Channel width of a JFET decreases as the gate-source is reverse-biased.

Operation of the JFET

Although we will examine the operation of the n-channel JFET, the operation of the p-channel is precisely the same, except that currents and voltages will be the opposite directions and polarities. In some respects, the operation of the JFET is easier to understand than the operation of the BJT. The JFET has an operation that is similar to how a water hose works. Examine the structure and circuit of Figure 26–41.

If there is no control voltage applied at the gate ($V_{GS} = 0$) as shown in Figure 26–41(a), charge moves freely through the JFET, impeded only by the bulk resistance of the semiconductor material. The drain current, I_D, is relatively large. Figures 26–41(b) and (c) show that as the negative bias on the gate is increased, a depletion region is established in the channel region, effectively reducing the cross-sectional area of the channel. This is similar to squeezing a water hose that has water moving through. With light pressure, the flow will decrease somewhat; with increased pressure, the flow will reduce even further. Since the p-n junction in the channel region is reverse-biased, the gate current $I_G = 0$. Applying Kirchhoff's current law, we have a very important equation for FETs:

$$I_S = I_D \qquad (26\text{–}23)$$

In Figure 26–41(d), we see that when $V_{GS} = -4$ V, the depletion region is so large that there is effectively no channel between the drain and the source. Using the water hose analogy, this is the same as tightly pinching the hose. The value of the gate-source voltage at which the drain current just becomes zero (is $I_D = 0$) is expressed as $V_{GS(OFF)}$. Typical values for n-channel JFETs are between $V_{GS(OFF)} = -3$ V and $V_{GS(OFF)} = -8$ V. Making V_{GS} more negative than $V_{GS(OFF)}$ has no further effect. This is analogous to applying further pressure to a water hose that is already squeezed hard enough to stop flow.

Figure 26–42 shows that the depletion region can be increased even without applying an external bias voltage between the gate and the source. By

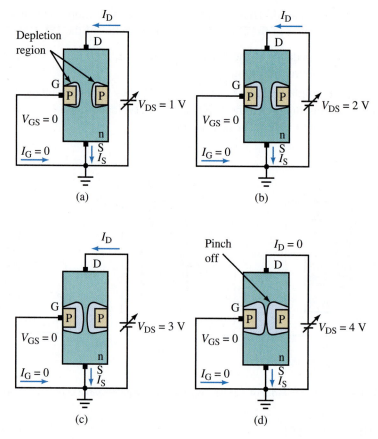

FIGURE 26–42   Channel width of a JFET decreases as V_{DS} increases.

increasing the drain-source voltage, the internal bulk resistance of the JFET sets up a voltage divider. Consequently, the channel in the region of the gate is at a higher potential than the gate (which is at ground). This results in the reverse-biased p-n junction at the gate and the corresponding depletion region. As V_{DS} is increased, I_D increases linearly and the depletion region at the gate becomes larger. Eventually, the depletion region becomes so large that additional drain current is no longer possible. At this point, $I_D = 0$ and the gate-source voltage at which the JFET is off is called the pinch-off voltage and is designated as V_P. For JFET transistors, the pinch-off voltage, V_P, and the gate shut-off voltage, $V_{GS(OFF)}$, are essentially the same magnitude and so we have

$$|V_P| = |V_{GS(OFF)}| \qquad \text{(26–24)}$$

The drain characteristics of the JFET are shown in Figure 26–43. Notice the similarity between these curves and the collector characteristic curves for a BJT. The notable difference is that each line corresponds to a different gate-to-source voltage, with the maximum drain current, I_{DSS}, occurring when $V_{GS} = 0$ V. When the forward breakdown voltage for the JFET is exceeded, the current through the device will generally destroy the component.

The output characteristic curves are used to generate the **transfer curve,** which relates the operation of the output portion of the circuit to the input. This curve is useful for determining the dc operating point (Q-point) of the JFET and will be used in the next chapter to determine the ac operation of a JFET amplifier. In all cases the transfer curve of any JFET is a parabolic segment between $V_{GS(OFF)}$ and I_{DSS} that provides the relationship between I_D and V_{GS}.

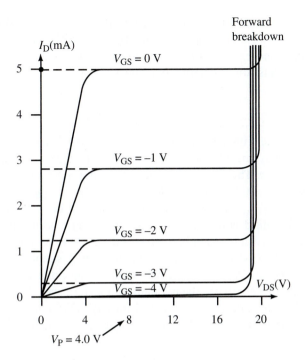

FIGURE 26–43   Drain curves of an n-channel JFET.

The equation that follows is called **Shockley's equation** after the Bell scientist who explained the *field-effect* theory for FETs.

$$I_D = I_{DSS}\left(1 - \frac{V_{GS}}{V_{GS(OFF)}}\right)^2 \qquad \textbf{Shockley's equation (26–25)}$$

The values for I_{DSS} and $V_{GS(OFF)}$ are constants that are generally found experimentally for a given transistor. Figure 26–44 shows the transfer curve as well as the drain curves that generate the transfer curve.

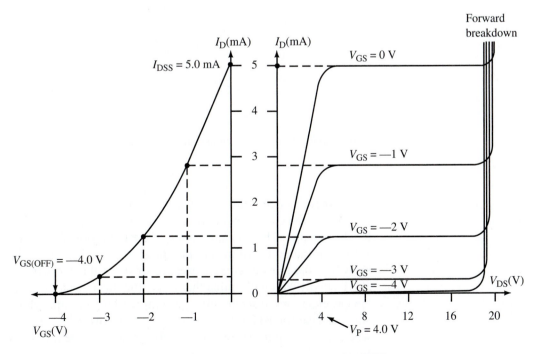

FIGURE 26–44   Transconductance curve and corresponding drain curves for an n-channel FET.

The transfer curve of Figure 26–44 shows that the drain current for values of V_{GS} less (more negative for an n-channel JFET) than $V_{GS(OFF)}$ is zero and the drain current can never exceed I_{DSS}. In the next section, we will use the transfer curve for a transistor to determine the operating point.

26.10 JFET Biasing

Just as was the case with the junction transistor, the JFET has several bias configurations. In analyzing a JFET circuit there are only three considerations that we must remember. The first is that the gate-source junction of the JFET is always reverse-biased. This means that we can expect V_{GS} to be any value between 0 V and $V_{GS(OFF)}$. The second point that must be remembered is that $I_S = I_D$, and the third is that the operating point must occur on the transfer curve (namely between I_{DSS} and $V_{GS(OFF)}$) for the given transistor. Although we could solve for the I_{DQ} and V_{GSQ} using algebra, a much simpler approach is to use graphical means. When solving for the operating point of a JFET using graphical method, we will use the normalized transfer curve illustrated in Figure 26–45. This diagram can be modified for any JFET transistor by simply inserting the correct values for I_{DSS} and $V_{GS(OFF)}$ and then re-scaling the divisions accordingly.

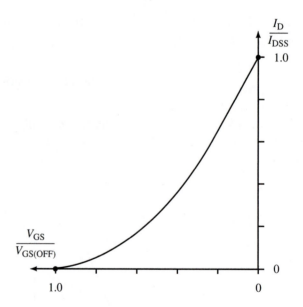

FIGURE 26–45 Normalized transconductance curve for an n-channel JFET.

In order to determine the operating point of the JFET, we use the output loop of the circuit (containing the drain and source terminals) to sketch the dc load line. The operating point of the transistor will occur at the intersection of the dc load line and the transfer curve. The following examples show how the transfer curve is used to solve for the operating point for any JFET circuit. You will notice that capacitors have been included in each of the circuits to more accurately represent working circuits. However, just as was the case for BJT circuits, we will ignore the effects of the capacitors when calculating the dc operating points since they behave as open circuits for dc.

The self-bias circuit shown in Figure 26–46 has a JFET with $I_{DSS} = 8.0$ mA and $V_{GS(OFF)} = -4.0$ V.

EXAMPLE 26–12

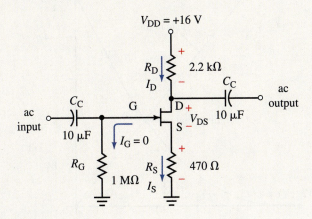

FIGURE 26–46

a. Find I_{DQ} and V_{GSQ}.

b. Solve for V_{DSQ}.

Solution

a. In order to determine the operating point we first sketch the dc load line on the transfer curve. To sketch this line, we need to arbitrarily select two points that will occur on the graph and yet be far enough apart to allow us to draw a suitable straight line.

Let $I_S = I_D = 0$ mA:

Examining the input loop shown in Figure 26–47, we write the input loop equation for the JFET as follows:

$$I_G R_G - V_{GS} - I_S R_S = 0$$

Now, since $I_G = 0$ due to the reverse-biased gate-source junction, we simplify the above expression for this circuit as

$$V_{GS} = -I_S R_S \qquad (26\text{–}26)$$

Now, because $I_S = I_D = 0$ mA, we have

$$V_{GS} = 0$$

For the second point on the dc load line, we again select a drain current value and use Equation 26–26 to solve for the corresponding value of V_{GS}.

Let $I_S = I_D = 8$ mA:

$$V_{GS} = -I_D R_S$$
$$= -(8 \text{ mA})(0.470 \text{ k}\Omega)$$
$$= -3.76 \text{ V}$$

The resulting straight line is found by connecting the two points and is shown in Figure 26–48. Finally, we approximate the intersection of the dc load line and the transfer curve as $I_{DQ} \approx 3.1$ mA and $V_{GSQ} \approx -1.5$ V. Clearly, the graphical method will not give precise results, since it is open to interpretation. However, it is a much simpler approach than

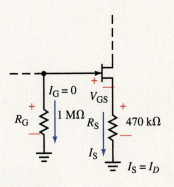

FIGURE 26–47

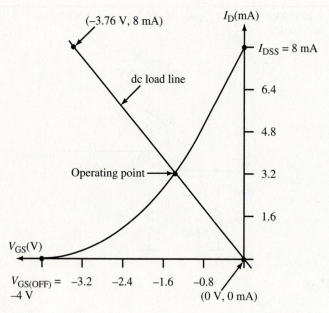

FIGURE 26–48

using the algebraic method, which involves the solution of two simultaneous equations, one being a quadratic equation. It is left as an exercise for the enthusiastic student to solve for the Q-point determined by the following two equations.

$$I_{DQ} = -\frac{V_{GSQ}}{0.47 \text{ k}\Omega}$$

$$I_{DQ} = 8 \text{ mA}\left(1 - \frac{V_{GSQ}}{-4 \text{ V}}\right)^2$$

(The algebraic results yield $I_{DQ} = 3.161$ mA and $V_{GSQ} = -1.486$ V.)

b. By examining the output loop of Figure 26–46, we write the output loop equation as

$$V_{DD} = I_D R_D + V_{DS} + I_S R_S \qquad (26\text{–}27)$$

Since $I_S = I_D$, we rewrite the above expression as

$$V_{DS} = V_{DD} - I_D (R_D + R_S) \qquad (26\text{–}28)$$

which gives

$$V_{DS} = 16 \text{ V} - 3.1 \text{ mA } (2.2 \text{ k}\Omega + 0.47 \text{ k}\Omega)$$
$$= 7.7 \text{ V}$$

Another bias arrangement uses a voltage divider similar to that used when working with the universal bias junction transistor. Just as in the previous example, we must again sketch the dc load line in order to determine the operating point on the transfer curve.

EXAMPLE 26–13

The circuit shown in Figure 26–49 has a p-channel JFET with $I_{DSS} = 10.0$ mA and $V_{GS(OFF)} = 5.0$ V.

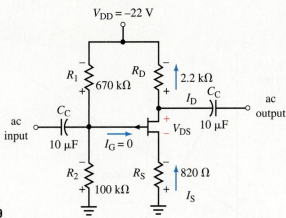

FIGURE 26–49

a. Find I_{DQ} and V_{GSQ}.

b. Solve for V_{DSQ}.

Solution

a. Due to the voltage divider of the resistors in the gate circuit, we see that the gate voltage will remain at a constant value determined as

$$V_G = \left(\frac{R_2}{R_1 + R_2}\right)V_{DD} \qquad (26\text{–}29)$$

and so we have

$$V_G = \left(\frac{100 \text{ k}\Omega}{670 \text{ k}\Omega + 100 \text{ k}\Omega}\right)(-22 \text{ V}) = -2.86 \text{ V}$$

Now, we examine the input loop of the JFET as shown in Figure 26–50.

Here we see that the input loop equation gives the following linear equation.

$$V_G = V_{GS} - I_S R_S \qquad (26\text{–}30)$$

Substituting the known values into the above expression we have

$$V_{GS} = -2.86 \text{ V} + (0.82 \text{ k}\Omega)I_D$$

In order to sketch the dc load line, we once again need to arbitrarily select two points. Remember that for the p-channel JFET, the operating point must occur between 0 and $V_{GS} = +5.0$ V.

Let $I_S = I_D = 4$ mA:

$$V_{GS} = -2.86 \text{ V} + (0.82 \text{ k}\Omega)(4 \text{ mA})$$
$$= 0.42 \text{ V}$$

and

Let $I_S = I_D = 8$ mA:

$$V_{GS} = -2.86 \text{ V} + (0.82 \text{ k}\Omega)(8 \text{ mA})$$
$$= 3.70 \text{ V}$$

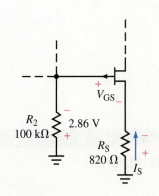

FIGURE 26–50

Now connecting the two points, we have the dc load line shown in Figure 26–51. At the intersection of the dc load line and the transfer curve, we obtain the operating point as $I_{DQ} \approx 5.2$ mA and $V_{GSQ} \approx 1.4$ V.

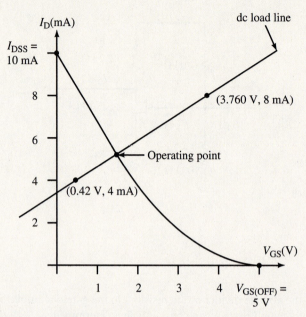

FIGURE 26–51

b. The output loop equation for the p-channel JFET of Figure 26–49 is written as

$$V_{DD} = -I_D R_D + V_{DS} - I_S R_S \qquad (26\text{--}31)$$

and so using the drain current found in part a., we find V_{DS} at the operating point as

$$\begin{aligned}
V_{DSQ} &= V_{DD} + I_D R_D + I_S R_S \\
&= -22 \text{ V} + 5.2 \text{ mA}(2.2 \text{ k}\Omega + 0.82 \text{ k}\Omega) \\
&= -6.30 \text{ V}
\end{aligned}$$

26.11 MOSFETs

Most integrated circuits today use metal oxide semiconductor field effect transistors to meet the size constraints and low power consumption requirements of modern electronic circuits. Although JFETs must be reverse-biased and are therefore restricted to working in the depletion region, you will discover that MOSFETs have no such restriction. Consequently, MOSFETs are typically able to handle more current than corresponding JFETs and are able to handle a broader operating range. There are two basic types of MOSFETs; depletion (or D-MOSFETs) and enhancement (or E-MOSFETs). Depletion MOSFETs can be reverse-biased to operate in the depletion region or they can be forward-biased to operate in the enhancement region. Figure 26–52 illustrates the structure and symbol of the n-channel depletion MOSFET, while Figure 26–53 shows the structure and symbol of the n-channel enhancement MOSFET.

Notice that for both the depletion and the enhancement MOSFET, the gate is isolated from the transistor by a layer of silicon dioxide insulator. Indeed, MOSFETs are often referred to as isolated gate FETs (or IGFETs). The origin of

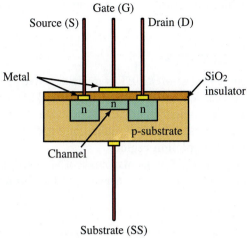

(a) Structure of an n-channel depletion MOSFET

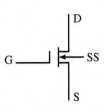

(b) Symbol of an n-channel depletion MOSFET

FIGURE 26–52

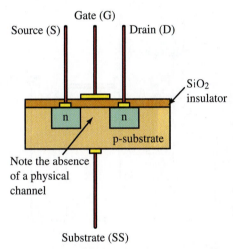

(a) Structure of an n-channel enhancement MOSFET

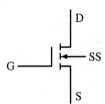

(b) Symbol of an n-channel enhancement MOSFET

FIGURE 26–53

the name MOSFETs for the devices shown in the previous figures is now fairly obvious. **Metal** is derived from the contacts at the drain, source, and particularly the gate. **Oxide** relates to the insertion of silicon dioxide insulator between the gate and the channel. The term **semiconductor** is due to the semiconductor materials that are used to construct the device and finally, the term **field effect** relates to the electric field that is generated by the gate to control the channel.

Since the gate of the MOSFET is isolated from the rest of the circuit, the input impedance of the MOSFET is much larger than the JFET (which already has an input impedance of many megohms). For this reason, the gate circuit of a MOSFET requires very little current (in the order of pA).

The principle difference between the D-MOSFET and the E-MOSFET is that for the former, there is a physical channel between the drain and the source, while for the latter a physical channel does not exist between the drain and the source. This means that an external voltage must be applied to the gate terminal to electrically generate a channel. In all circuits using MOSFETS, the substrate will be connected to a fixed potential, usually ground or the power supply. In many circuits, the source terminal will also be at ground. Therefore, a MOSFET will have either three or four terminals.

We will now examine the operation and biasing of both types of transistors in greater detail.

Depletion MOSFETs (D-MOSFETs)

When the gate to source of a D-MOSFET is made negative as shown in Figure 26–54, the majority carriers (electrons) of the channel are repelled into the p-substrate, effectively depleting the number of carriers available for the drain to source current. In this illustration we see that the MOSFET is very similar to a JFET operating in its **depletion region.**

As was the case with the JFET, if the magnitude of V_{GS} exceeds the $|V_{GS(OFF)}|$, there will be no available majority carriers available in the channel, and $I_D = 0$.

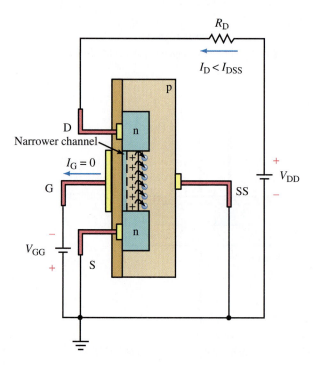

FIGURE 26–54 n-channel D-MOSFET operates in its depletion mode when V_{GS} is negative.

However, unlike the JFET, if the gate to source voltage of the D-MOSFET is made positive as shown in Figure 26–55, the majority carriers (holes) in the substrate will be repelled, effectively increasing (or enhancing) the channel width between the drain and the source. When the gate of an n-channel MOSFET is positive with respect to the source, we say that it is operating in its **enhancement region.**

With an increase in the size of the channel, additional current is now possible. The transfer curve for the n-channel D-MOSFET is illustrated in Figure 26–56.

Notice that the output current is no longer limited to a maximum value of I_{DSS}. However, the equation of the transfer curve for a D-MOSFET is precisely the same as that of a JFET, namely

$$I_D = I_{DSS}\left(1 - \frac{V_{GS}}{V_{GS(OFF)}}\right)^2 \qquad (26\text{–}32)$$

The operating point of a D-MOSFET amplifier circuit is found by determining the intersection of the dc load line and the transfer curve, exactly the same way as for a JFET. The following example illustrates the method.

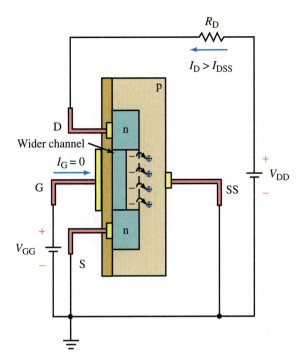

FIGURE 26–55 n-channel D-MOSFET operates in its enhancement mode when V_{GS} is positive.

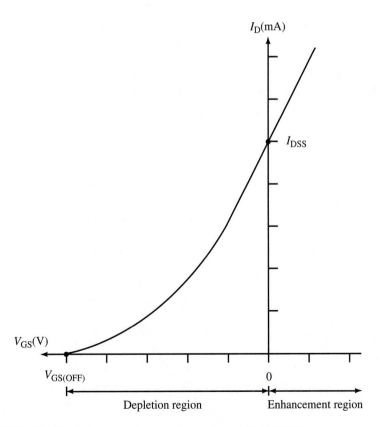

FIGURE 26–56 Transconductance curve of an n-channel D-MOSFET

EXAMPLE 26–14

The circuit shown in Figure 26–57 uses an n-channel D-MOSFET with $I_{DSS} = 10.0$ mA and $V_{GS(OFF)} = -5.0$ V.

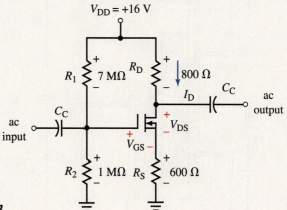

FIGURE 26–57

a. Find I_{DQ} and V_{GSQ}.

b. Is the MOSFET operating in its depletion mode or enhancement mode?

c. Solve for V_{DSQ}.

Solution

a. The gate voltage (with respect to ground) will remain constant due to the voltage divider and is calculated to be

$$V_G = \left(\frac{1 \text{ M}\Omega}{1 \text{ M}\Omega + 7 \text{ M}\Omega} \right)(16 \text{ V}) = 2.00 \text{ V}$$

As in previous examples, we arbitrarily select two points to help sketch the dc load line on the transfer curve shown in Figure 26–58.

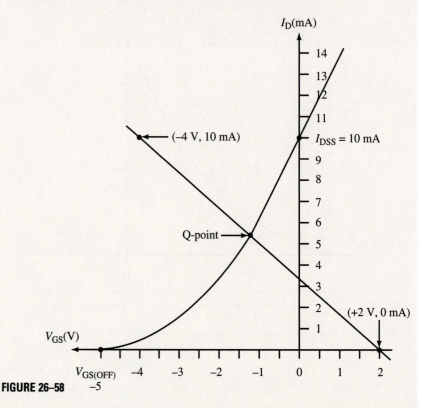

FIGURE 26–58

Let $I_S = I_D = 0$ mA:

$$V_{GS} = +2 \text{ V} - (0.6 \text{ k}\Omega)(0 \text{ mA})$$
$$= +2.00 \text{ V}$$

and

Let $I_S = I_D = 10$ mA:

$$V_{GS} = +2 \text{ V} - (0.6 \text{ k}\Omega)(10 \text{ mA})$$
$$= -4.00 \text{ V}$$

At the operating point, $I_{DQ} \approx 5.5$ mA and $V_{GSQ} \approx -1.28$ V.

b. Since I_{DQ} is less than I_{DSS}, the transistor is operating in its depletion region.

c. We find V_{DS} at the operating point as

$$V_{DSQ} = V_{DD} - I_D R_D - I_S R_S$$
$$= 16 \text{ V} - 5.5 \text{ mA}(0.8 \text{ k}\Omega + 0.6 \text{ k}\Omega)$$
$$= 8.3 \text{ V}$$

Repeat the work of Example 26–14, if the value of $R_S = 120$ Ω.

 PRACTICE PROBLEMS 9

Answers
a. $I_{DQ} \approx 12.2$ mA, $V_{GSQ} \approx +0.52$ V; b. Enhancement; c. $V_{DSQ} = 4.78$ V

Enhancement MOSFETs (E-MOSFETs)

Since E-MOSFETs do not have a channel, there can be no drain current until a channel is established by applying a positive gate to source voltage (for an n-channel E-MOSFET). This transistor can only operate in its enhancement mode, since the depletion mode is not possible. Although the transfer curve for an E-MOSFET is a parabolic curve, this curve is significantly different than for JFETs and D-MOSFETs. Refer to Figure 26–59.

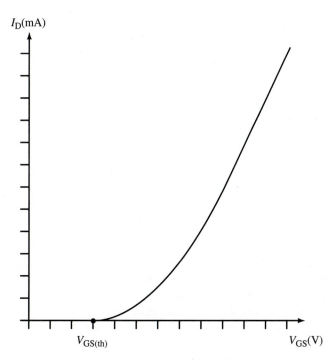

FIGURE 26–59   Transconductance curve of an n-channel E-MOSFET

There can be no current through the transistor until a minimum threshold voltage, $V_{GS(th)}$ is applied between the gate and the source. For the E-MOSFET, there is no value for I_{DSS}. Lastly, the relationship between the I_D and V_{GS} is given by the following expression.

$$I_D = k(V_{GS} - V_{GS(th)})^2 \qquad (26\text{-}33)$$

In the above expression, the value of the constant, k is dependent on the particular transistor and can be easily calculated from the specification sheets of the transistor. We can solve for k of any E-MOSFET as

$$k = \frac{I_{D(ON)}}{\left(V_{GS(ON)} - V_{GS(th)}\right)^2} \qquad (26\text{-}34)$$

For example, the 3N170 has a specified "ON" characteristic of $I_{D(ON)} = 10$ mA for $V_{GS(ON)} = 10$ V. The threshold voltage is $V_{GS(th)} = 1.5$ V. Equation 26-33 would be written as

$$I_{D(ON)} = k(V_{GS(ON)} - 1.5 \text{ V})^2$$

and so for the given transistor, we have

$$k = \frac{I_{D(ON)}}{\left(V_{GS(ON)} - 1.5 \text{ V}\right)^2}$$

$$= \frac{10 \text{ mA}}{(10 \text{ V} - 1.5 \text{ V})^2}$$

$$= 0.138 \text{ mA/V}^2$$

The biasing for E-MOSFETs must be such that V_{GS} is positive. The two best methods that provide correct biasing are the voltage divider circuit and the *drain-feedback circuit,* shown in Figure 26–60.

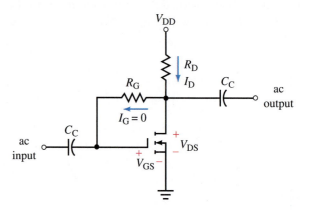

FIGURE 26–60

When we examine this circuit, several important characteristics become evident. First, due to the very high input impedance at the gate, we realize that there can be no gate current and so $I_G = 0$. Next, due to the zero gate current, we conclude that the voltage across R_G must also be zero. And finally, applying Kirchhoff's voltage law at both the input and output loops, we see that

$$V_{GS} = V_{DS} \qquad (26\text{-}35)$$

The E-MOSFET shown in Figure 26–61 has the following characteristics:

EXAMPLE 26–15

$$V_{GS(th)} = 2.00 \text{ V}$$
$$I_{D(ON)} = 200 \text{ mA}$$
$$V_{GS(ON)} = 4 \text{ V}$$

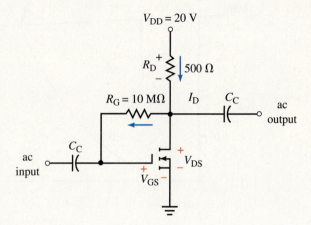

FIGURE 26–61

a. Write the expression for the I_D in terms of V_{GS}.

b. Solve for I_{DQ}, V_{GSQ}, V_{DSQ} at the quiescent (operating) point of the circuit.

Solution

a. We begin by using the given characteristics and Equation 26–34 to solve for the constant, k.

$$k = \frac{200 \text{ mA}}{(4 \text{ V} - 2 \text{ V})^2} = 50 \text{ mA/V}^2$$

Now, substituting this result into Equation 26–33, we get $I_D = (50 \text{ mA} / \text{V}^2)$ $(V_{GS} - 2 \text{ V})^2$. This expression can be plotted on a transfer curve. However, we need additional information to help scale the graph appropriately.

b. Examining the circuit of Figure 26–61, we see that the relationship between I_D and V_{DS} is determined as

$$I_D = \frac{V_{DD} - V_{DS}}{R_D} = \frac{20 \text{ V} - V_{DS}}{0.5 \text{ k}\Omega}$$

and for the drain-follower circuit, $V_{GS} = V_{DS}$. Therefore, the equation for the dc load line can now be written in terms of I_D and V_{GS} as

$$I_D = \frac{20 \text{ V} - V_{GS}}{0.5 \text{ k}\Omega}$$

Now, by inspection we notice that the absolute maximum current for the circuit will be $I_D = 40$ mA. We use this value to help scale the graph of the transfer curve with a maximum current of 50 mA as shown in Figure 26–62. (You may use the parabolic equation to show that at this value of I_D, $V_{GS} = 3.00$ V.) The rest of the graph is scaled accordingly.

Now, in order to sketch the dc load line we need to select suitable points on the graph of Figure 26–62.

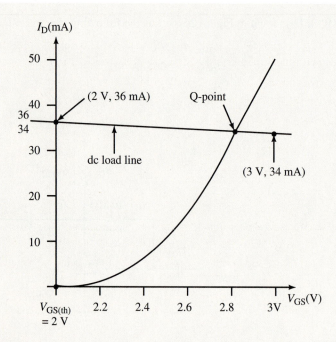

FIGURE 26–62

Let $V_{GS} = 2$ V:

And so we have $I_D = \dfrac{20\text{ V} - 2\text{ V}}{0.5\text{ k}\Omega} = 36$ mA

Let $V_{GS} = 3$ V:

And so we have $I_D = \dfrac{20\text{ V} - 3\text{ V}}{0.5\text{ k}\Omega} = 34$ mA

The intersection of the dc load line and the transfer curve reveals the operating point of the E-MOSFET circuit as $I_{DQ} \approx 34.2$ mA, $V_{GSQ} = V_{DSQ} \approx 2.82$ V.

MOSFET Handling Precautions

Because MOSFETs use very thin insulation between the gate and the channel, they are subject to damage by electrostatic discharge (ESD). Unfortunately, our bodies and many of the materials we use in daily activities generate substantial amounts of static electricity. Very often, the act of simply picking up a MOSFET transistor (or IC containing MOSFETs) will result in permanent damage or destruction of the component. In order to prevent damage to these sensitive components some simple, but effective procedures have been developed. They are outlined as follows:

1. When MOSFETs are shipped from a manufacturer, they will be packaged in static resistive bags or inserted into static conductive foam. Occasionally, a shorting wire is connected between all of the terminals of the MOSFET. Do not remove the components from these protective environments until you are at a workstation that is also protected. Never place a MOSFET circuit into ordinary plastic foam, since materials such as Styrofoam are extremely effective at generating dangerous amounts of static charge.

2. The single most important element when handling MOSFET circuits is to have a static-safe workstation. This means that you must use a wrist strap that is connected between you and ground. When properly connected, this wrist strap will effectively drain any static charge from your body through

a very large (usually 1–MΩ) resistor. All instruments at the workstation will be connected to earth ground through the ac plug. Additionally, in order to provide total protection, a workstation should have a static dissipative mat that is also connected to earth ground.

3. Never insert or remove a MOSFET component into or from an active circuit. Turn off all dc power supplies before working with the component.

4. Never apply an ac signal to MOSFET unless the circuit has been correctly biased.

26.12 Troubleshooting a Transistor Circuit

Now that you know how transistors are biased, you can use this knowledge to determine whether a given transistor circuit is operating as expected. Although troubleshooting a circuit is often more art than science, there are some basic rules that any successful technician or technologist will follow. If you have just built a circuit and it doesn't work properly, the chances are that you have either used a wrong component or you have made a faulty connection. Unless the components in the circuit were previously used, it is highly unlikely that you started with faulty components. In this case, before you go any further, check your resistor color codes and ensure that you have connected the transistor correctly. You may need to refer to the manufacturer's specifications to determine the correct pin allocations since each transistor type is different. Make sure that you have properly connected the dc voltage source; remember an npn transistor or n-channel FET will normally require a positive voltage supply while a pnp transistor or p-channel FET will need a negative supply. If all connections appear to be correct, you will need to take dc voltage measurements of the circuit.

Since we have not yet worked with transistor circuits using ac sources, it is assumed that your circuit does not use one. However, if there is an ac source, turn it off and remove it from the circuit. It is now a simple matter to use a dc voltmeter to determine whether the circuit is operating.

Consider the amplifier circuit shown in Figure 26–63. The operation of this circuit can be easily analyzed with the following steps:

1. The first step of any troubleshooting procedure is to plan a strategy of attack. Begin with the most obvious and progress to the less likely problems. The important thing is that you must know what to expect. Do the necessary calculations so that you have an idea of typical voltage measurements.

2. Your first measurement should be to measure the dc supply voltage (with respect to ground) to ensure that the correct bias voltage has been provided. Clearly the operation of the circuit is dependent upon the supply having both the correct magnitude and polarity.

3. Measure V_{BE}. Because this is an amplifier circuit, we know that the base-emitter junction must be forward-biased and since we have an npn transistor, we expect $V_{BE} \approx 0.7$ V. If the transistor is faulty, or if it is incorrectly connected, this voltage may be incorrect. The possible exception is if the collector and emitter terminals were reversed, since this would still result in a forward-biased junction. In this case however, the following step will generally indicate a problem.

4. Determine the operating point of the transistor under test by determining V_{CEQ} and I_{CQ} at the quiescent point. (Remember that we don't normally measure the current in an operating circuit, but rather measure the voltage across a known resistor—in this case $R_C = 2.2$ kΩ—and then calculating the current using Ohm's law.)

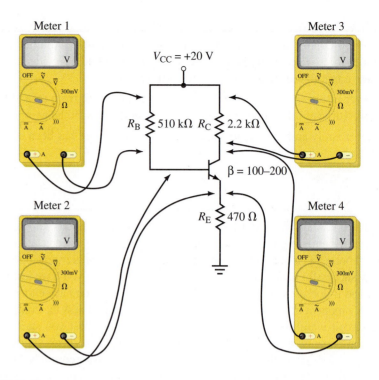

FIGURE 26–63

5. The circuit shown in Figure 26–63 is a beta-dependent circuit. If the transistor is not operating in its active region, it is possible that the value of the transistor beta is either much larger or much smaller than expected. The last measurement is to determine I_{BQ} and from this the value, calculate beta. Once again, remember that the preferred method of finding the base current is to measure the voltage across R_B and then solving for the current using Ohm's law.

Whenever you troubleshoot a circuit, it is imperative that you have a clear plan of action. You will not be an effective troubleshooter unless you know what readings to expect at each step of the way. In order to do this, you will need to know how the circuit should work.

PRACTICE PROBLEMS 10

For the circuit of Figure 26–63, the transistor specification sheet shows that the transistor type has $\beta = 100 \rightarrow 200$. What are the typical voltages that you would expect to measure on each of the voltmeters?

Answers
Meter 1: 16.3 V → 17.7 V; Meter 2: 0.7 V; Meter 3: 7.6 V → 14.1 V;
Meter 4: 3.0 V → 10.8 V

26.13 Computer Analysis of Transistor Circuits

MultiSIM and PSpice are extremely useful tools when analyzing the operation of complex circuits. In this section, we will enter several circuits and observe how the software predictions compare to the theoretical calculations. Needless to say, there will be some variation between the theoretical calculations and the values predicted by the software. In fact, you will find that if you were to build the circuits in the lab, you would have a third set of values. Although there will be differences between theoretical and measured values, it is important to observe that all values will normally have a discrepancy of less than 10%. When we take into account variations in the manufacturer's specifications, we find that the discrepancies are generally well within the expected limits.

MultiSIM

MultiSIM provides numerous semiconductor models and is extremely useful when simulating the operation of transistor circuits. The following example verifies that the theoretical calculations are very good approximations of transistor operation.

EXAMPLE 26–16

a. Use MultiSIM to find V_B, I_{CQ}, I_{EQ}, V_{CEQ}, and V_{BEQ} for the circuit of Figure 26–64.

b. Compare the values of I_{CQ} and V_{CEQ} to those predicted in Practice Problems 6.

Solution

a. Open the MultiSIM program and open a new file. Construct the circuit by selecting components and indicators from the Component toolbar. (You will need to select these components from the *Sources, Basic, Transistors,* and *Indicators* parts bins.) Remember to place ammeters into the circuit and voltmeters across the components to be measured. Change the component values to correspond to those shown in Figure 26–64. Select a 2N3904 transistor as follows: Click on the *Transistors* parts bin, whereupon you will be given a button bar showing a large number of transistor types. There are two buttons, one gray and one green, that allow you to select the npn transistor. The green button corresponds to a generic transistor, while the gray button allows you to select from a large number of actual component types. Click on the gray button. From the *Component Name List* scroll down until you come to the 2N3904. Highlight this transistor and click on *OK* to place the transistor into the circuit. Click on the *Run* button to simulate your design. You should have a display that appears as shown in Figure 26–65. The result of this simulation gives the following values:

$$V_B = 3.18 \text{ V}, I_{CQ} = 2.47 \text{ mA}, I_{EQ} = 2.49 \text{ mA}, V_{CEQ} = 10.09 \text{ V},$$
and $V_{BEQ} = 0.687 \text{ V}$

b. In Practice Problems 6, we assumed that the transistor had $\beta = 100 \rightarrow 300$, which is precisely the value specified for the 2N3904 transistor. The values $I_{CQ} = 2.47 \text{ mA}$ and $V_{CEQ} = 10.09 \text{ V}$ are well within the expected range of $I_C = 2.41 \text{ mA} \rightarrow 2.55 \text{ mA}$ and $V_{CEQ} = 9.78 \text{ V} \rightarrow 10.37 \text{ V}$.

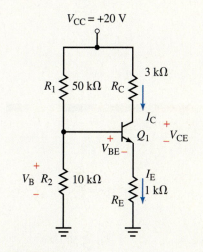

FIGURE 26–64

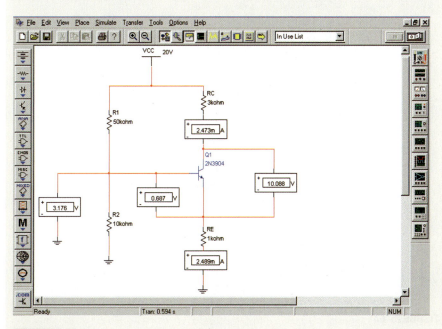

FIGURE 26–65

◀ **MULTISIM**

PRACTICE PROBLEMS 11

Remove the 2N3904 transistor and replace it with a 2N3906, its complement pnp transistor and change the supply to $V_{CC} = -20$ V. Simulate the design for the new transistor. What are the new values for V_B, I_{CQ}, I_{EQ}, V_{CEQ}, and V_{BEQ}?

Answers
$V_B = -3.22$ V, $I_{CQ} = -2.49$ mA, $I_{EQ} = -2.50$ mA, $V_{CEQ} = -10.01$ V, and $V_{BEQ} = -0.72$ V (As expected, the main difference between these values and those for the 2N3904 is that all voltages and currents are negative.)

PSpice

Although the student version of PSpice does not provide a comprehensive list of semiconductor components, there is a good selection allowing us to examine many different circuits. The following example analyzes the E-MOSFET circuit of Example 26–15.

EXAMPLE 26–17

a. Use PSpice to find V_{GSQ}, I_{DQ}, and V_{DSQ} for the circuit of Figure 26–66. (Since we are not applying an ac signal to the circuit, coupling capacitors are not shown.)

b. Compare the values to those found in Example 26–15.

Solution

a. Construct the PSpice circuit as illustrated in Figure 26–67. Obtain the MOSFET by holding down **Ctrl-G** (or by selecting the Draw menu item and selecting Get New Part). The semiconductor devices to which we have access are in the *EVAL* library. Once you are in this library, simply scroll down to the device labeled IRF150. In the description of this component, you will notice that this is an n-channel E-MOSFET as required. Place the transistor into the circuit and make all required connections. In order to show the bias voltages and currents, we can enable these features by clicking on the *Enable Bias Voltage Display* and *Enable Bias Current Display*. Save your file as Example 26–17 and click on the *Simulate* button. You should have an output that appears similar to that shown in Figure 26–67.

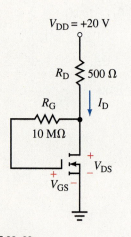

$V_{DD} = +20$ V

R_D 500 Ω

R_G I_D

10 MΩ

V_{DS}

V_{GS}

FIGURE 26–66

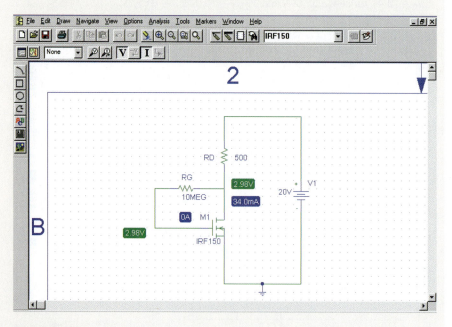

FIGURE 26–67

To disable the display of some of the bias voltages and currents, you may do so as follows:

To disable the display of a bias current, click on the part and then click on the button labeled as *Show/Hide Currents on Selected Part(s)*.

To disable the display of a bias voltage, click on the node and then click on the button labeled as *Show/Hide Voltage on Selected Net(s)*. The results of Figure 26–67 indicate that $V_{GSQ} = V_{DSQ} = 2.98$ V and $I_{DQ} = 34.0$ mA.

b. The above results are very close to those found in Example 26–15 where we had $V_{GSQ} = V_{DSQ} = 2.82$ V and $I_{DQ} = 34.2$ mA.

PROBLEMS

26.2   Transistor Operation

1. A transistor circuit is measured to have $I_C = 4.50$ mA and $I_E = 4.55$ mA. Determine I_B, α, and β.

2. A power transistor is measured to have $I_C = 24.5$ mA and $I_E = 25.1$ mA. Determine I_B, α, and β.

3. A transistor has $\beta = 120$ and $I_C = 5.00$ mA. Calculate the values of I_B, I_E, and α.

4. If a transistor has $\alpha = 0.92$ and $I_E = 3.50$ mA, determine the values of I_C, β, and I_B.

5. A transistor has $\beta = 100$ and $I_E = 4.94$ mA. Calculate the values of α, I_C, and I_B.

6. Use Equations 26–1 and 26–2 to derive Equation 26–4, $I_E = (\beta + 1)I_B$.

7. Use Equations 26–2, 26–3, and 26–4 to derive Equation 26–5, $\alpha = \dfrac{\beta}{\beta + 1}$.

8. Use Equations 26–2, 26–3, and 26–4 to derive Equation 26–6, $\beta = \dfrac{\alpha}{1 - \alpha}$.

26.3   Transistor Specifications

Refer to the specification sheets for the 2N3904 shown in Figure 26–8 for the problems in this section.

9. One of the electrical characteristics of the 2N3904 transistor is given as $V_{(BR)EBO}$. What is the value of this specification? What does the subscript of the symbol mean? What are the test conditions that are used to measure this characteristic?

10. One of the electrical characteristics of the 2N3904 transistor is given as $V_{(BR)CBO}$. What is the value of this specification? What does the subscript of the symbol mean? What are the test conditions that are used to measure this characteristic?

26.4   Collector Characteristic Curves

Refer to the collector characteristic curves shown in Figure 26–68 for the problems of this section.

11. Determine β_{dc} for a transistor having an operating point at $V_{CE} = 10$ V and $I_C = 5.5$ mA.

12. Determine β_{dc} for a transistor having an operating point at $V_{CE} = 6$ V and $I_C = 10.5$ mA.

13. Determine β_{dc} for a transistor having an operating point at $V_{CE} = 10$ V and $I_C = 8.0$ mA.

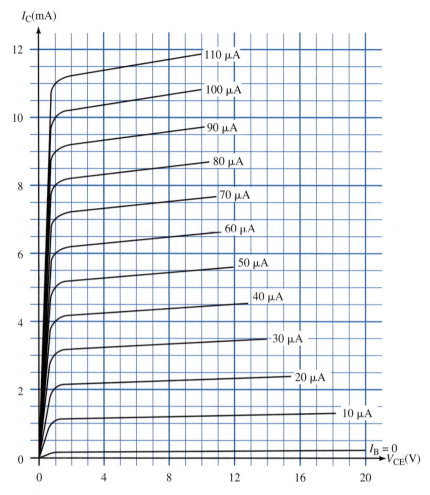

FIGURE 26–68

14. Determine β_{dc} for a transistor having an operating point at $V_{CE} = 8$ V and $I_C = 4.0$ mA.

15. A transistor is operating at a point having $I_C = 8.0$ mA and $I_B = 100$ µA. This operating point will place the transistor in which of the following regions:

   a. Active   b. Saturation   c. Cutoff

16. A transistor is operating at a point having $V_{CE} = 16$ V and $I_C = 200$ µA. This operating point will place the transistor in which of the following regions:

   a. Active   b. Saturation   c. Cutoff

26.5 dc Load Line

The transistors used in this section have the collector characteristics shown in Figure 26–68.

17. The circuit of Figure 26–69 has $V_{BB} = 5$ V, $V_{CC} = 16$ V, and $R_C = 2$ kΩ.

   a. Sketch the dc load line for the circuit.

   b. Determine the operating point and calculate β of the circuit if $R_B = 100$ kΩ.

   c. Determine the operating point and calculate β of the circuit if $R_B = 200$ kΩ.

d. Determine the operating point and calculate β of the circuit if $R_B = 20 \text{ k}\Omega$.

e. In general, what happens to the operating point as the value of R_B is increased?

18. The circuit of Figure 26–69 has $V_{BB} = 5$ V, $R_B = 100 \text{ k}\Omega$, and $R_C = 2 \text{ k}\Omega$.

a. Sketch the dc load line and indicate the operating point if $V_{CC} = 16$ V.

b. Sketch the dc load line and indicate the operating point if $V_{CC} = 20$ V.

c. Sketch the dc load line and indicate the operating point if $V_{CC} = 12$ V.

d. In general, what happens to the load line as the value of V_{CC} is increased?

19. The circuit of Figure 26–69 has $V_{BB} = 5$ V, $R_B = 100 \text{ k}\Omega$, and $V_{CC} = 16$ V.

a. Sketch the dc load line and indicate the operating point if $R_C = 2 \text{ k}\Omega$.

b. Sketch the dc load line and indicate the operating point if $R_C = 1.6 \text{ k}\Omega$.

c. Sketch the dc load line and indicate the operating point if $R_C = 4 \text{ k}\Omega$.

d. In general, what happens to the load line as the value of R_C is increased?

20. The circuit of Figure 26–69 has $V_{BB} = 5$ V, $V_{CC} = 16$ V, and $R_C = 2 \text{ k}\Omega$.

a. Calculate the value of R_B so that the transistor operates at the midpoint of the dc load line.

b. Determine the value of β at the operating point.

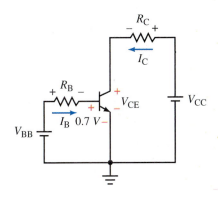

FIGURE 26–69

26.6 Transistor Biasing

21. For the circuit of Figure 26–70, let $V_{CC} = 16$ V, $R_B = 620 \text{ k}\Omega$, $R_C = 2 \text{ k}\Omega$, and $\beta = 120$.

a. Calculate the saturation current, $I_{C(SAT)}$.

b. Determine I_B, I_C, I_E, and V_{CE}.

c. Sketch the dc load line showing $I_{C(SAT)}$, $V_{CE(OFF)}$, and the values of the Q-point.

22. a. If the transistor beta of Problem 21 doubles to $\beta = 240$, determine the resulting values of I_B, I_C, I_E, and V_{CE}.

b. With the 100% increase in β, what is the percentage change in I_C?

c. Does the operating point of the transistor move towards saturation or cutoff?

23. For the circuit of Figure 26–70, let $V_{CC} = 24$ V, $R_B = 1.2 \text{ M}\Omega$, $R_C = 3.9 \text{ k}\Omega$, and $\beta = 150$.

a. Calculate the saturation current, $I_{C(SAT)}$.

b. Determine I_B, I_C, I_E, and V_{CE}.

c. Sketch the dc load line showing $I_{C(SAT)}$, $V_{CE(OFF)}$, and the values of the Q-point.

24. a. If the transistor beta of Problem 23 decreases to $\beta = 240$, determine the resulting values of I_B, I_C, I_E, and V_{CE}.

b. With the decrease in β, does the operating point of the transistor move towards saturation or cutoff?

25. For the circuit of Figure 26–71, let $V_{CC} = 22$ V, $R_B = 1.5 \text{ M}\Omega$, $R_C = 3.9 \text{ k}\Omega$, $R_E = 1.0 \text{ k}\Omega$, and $\beta = 150$.

a. Calculate the saturation current, $I_{C(SAT)}$.

b. Determine I_B, I_C, I_E, and V_{CE}.

c. Sketch the dc load line showing $I_{C(SAT)}$, $V_{CE(OFF)}$, and the values of the Q-point.

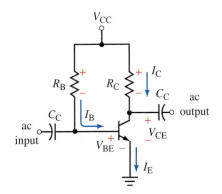

FIGURE 26–70   **MULTISIM**

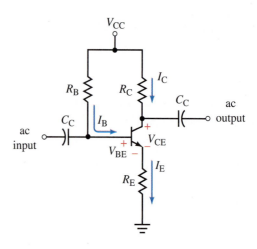

FIGURE 26–71

26. a. If the transistor beta of Problem 25 doubles to $\beta = 300$, determine the resulting values of I_B, I_C, I_E, and V_{CE}.

   b. With the 100% increase in β, what is the percentage change in I_C?

27. For the collector-feedback circuit of Figure 26–72, let $V_{CC} = 20$ V, $R_B = 270$ kΩ, $R_C = 2.0$ kΩ, and $\beta = 125$.

   a. Calculate the saturation current, $I_{C(SAT)}$.

   b. Determine I_B, I_C, I_E, I, and V_{CE}.

   c. Sketch the dc load line showing $I_{C(SAT)}$, $V_{CE(OFF)}$, and the values of the Q-point.

28. a. If the transistor beta of Problem 27 doubles to $\beta = 250$, determine the resulting values of I_B, I_C, I_E, I, and V_{CE}.

   b. With the 100% increase in β, what is the percentage change in I_C?

29. For the circuit of Figure 26–73, let $V_{EE} = 20$ V, $R_B = 1.0$ MΩ, $R_C = 2.0$ kΩ, $R_E = 2.0$ kΩ, and $\beta = 160$.

   a. Calculate the saturation current, $I_{C(SAT)}$.

   b. Determine I_B, I_C, I_E, and V_{CE}.

   c. Sketch the dc load line showing $I_{C(SAT)}$, $V_{CE(OFF)}$, and the values of the Q-point.

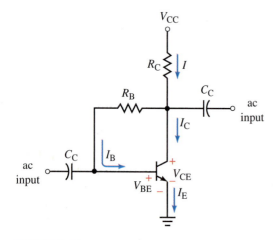

FIGURE 26–72   Collector-feedback circuit.

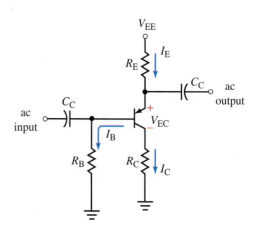

FIGURE 26–73

30. For the circuit of Figure 26–73, let $V_{EE} = 24V$, $R_B = 820$ kΩ, $R_C = 3.3$ kΩ, $R_E = 1.0$ kΩ, and $\beta = 120$.

a. Calculate the saturation current, $I_{C(SAT)}$.

b. Determine I_B, I_C, I_E, and V_{CE}.

c. Sketch the dc load line showing $I_{C(SAT)}$, $V_{CE(OFF)}$, and the values of the Q-point.

d. If you wanted a Q-point to have $I_C = 2.0$ mA, what value of base resistor, R_B would you need?

31. For the circuit of Figure 26–74, let $V_{CC} = 16$ V, $R_1 = 33$ kΩ, $R_2 = 5.1$ kΩ, $R_C = 1.5$ kΩ, $R_E = 510$ Ω, and $\beta = 120$.

a. Calculate the saturation current, $I_{C(SAT)}$.

b. Use the exact method to solve for I_B, I_C, I_E, and V_{CE}.

c. Sketch the dc load line showing $I_{C(SAT)}$, $V_{CE(OFF)}$, and the values of the Q-point.

32. a. If the transistor beta of Problem 31 doubles to $\beta = 240$, use the exact method to solve for the new values of I_B, I_C, I_E, and V_{CE}.

b. With the 100% increase in β, what is the percentage change in I_C.

33. For the circuit of Figure 26–74, let $V_{CC} = 16$ V, $R_1 = 330$ kΩ, $R_2 = 51$ kΩ, $R_C = 1.5$ kΩ, $R_E = 510$ Ω, and $\beta = 120$.

a. Calculate the saturation current, $I_{C(SAT)}$.

b. Use the exact method to solve for I_B, I_C, I_E, and V_{CE}.

c. Sketch the dc load line showing $I_{C(SAT)}$, $V_{CE(OFF)}$, and the values of the Q-point.

34. a. If the transistor beta of Problem 33 doubles to $\beta = 240$, use the exact method to solve for the new values of I_B, I_C, I_E, and V_{CE}.

b. With the 100% increase in β, what is the percentage change in I_C.

35. a. Use the approximate method and the component values of Problem 31 to solve for I_C and V_{CE}.

b. The approximate method often provides results that are close to the exact solution (using Thévenin's theorem) method. Determine the percent variation between the calculations of I_C and V_{CE} in part a. and those of Problem 31.

36. a. Use the approximate method and the component values of Problem 33 to solve for I_C and V_{CE}.

b. Determine the percent variation between the calculations of I_C and V_{CE} in part a. and those of Problem 33.

37. For the circuit of Figure 26–75, let $V_{CC} = -30$ V, $R_1 = 62$ kΩ, $R_2 = 8.2$ kΩ, $R_C = 3.9$ kΩ, $R_E = 1.0$ kΩ, and $\beta = 100$.

a. Calculate the saturation current, $I_{C(SAT)}$.

b. Use the approximate method to solve for I_C and V_{CE}.

c. Sketch the dc load line showing $I_{C(SAT)}$, $V_{CE(OFF)}$, and the values of the Q-point.

38. For the circuit of Figure 26–75, let $V_{CC} = -30$ V, $R_2 = 8.2$ kΩ, $R_C = 3.9$ kΩ, $R_E = 1.0$ kΩ, and $\beta = 100$. Calculate the value of R_1 needed to result in $I_C = 3.0$ mA.

39. The transistor of Figure 26–74 has $\beta = 100 \rightarrow 300$. Design the circuit for the following conditions:

$V_{CC} = +20$ V, $I_{C(SAT)} = 8.0$ mA, $I_{CQ} = 4.0$ mA, $R_E = R_C/4$, $R_2 \leq 0.1$ β R_E for all β.

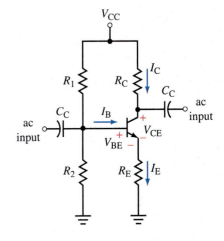

FIGURE 26–74

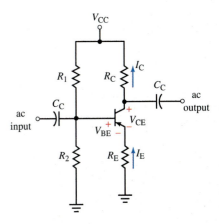

FIGURE 26–75

40. The transistor of Figure 26–74 has $\beta = 100 \rightarrow 300$. Design the circuit for the following conditions:

    $V_{CC} = +24$ V, $I_{C(SAT)} = 6.0$ mA, $I_{CQ} = 4.0$ mA, $R_E = R_C/5$, $R_2 \leq 0.1 \beta R_E$ for all β.

41. For the circuit of Figure 26–76, let $V_{EE} = +10$ V, $V_{CC} = -20$ V, $R_E = 2.0$ kΩ, $R_C = 2.0$ kΩ, and $\beta = 120$.

    a. Calculate the saturation current, $I_{C(SAT)}$.

    b. Solve for I_E, I_C, I_B, and V_{CE}.

    c. Sketch the dc load line showing $I_{C(SAT)}$, $V_{CE(OFF)}$, and the values of the Q-point.

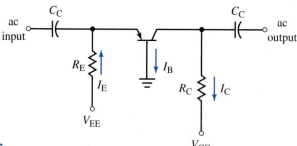

FIGURE 26–76

42. For the circuit of Figure 26–76, let $V_{EE} = +8$ V, $V_{CC} = -12$ V, $R_E = 2.2$ kΩ, $R_C = 1.8$ kΩ, and $\beta = 180$.

    a. Calculate the saturation current, $I_{C(SAT)}$.

    b. Solve for I_E, I_C, I_B, and V_{CE}.

    c. Sketch the dc load line showing $I_{C(SAT)}$, $V_{CE(OFF)}$, and the values of the Q-point.

26.7   The Transistor Switch

43. The transistor shown in Figure 26–77 is used as a buffer, providing sufficient current to turn on an LED when the output of the logic gate is "1." The circuit has $V_{CC} = 8$ V, $R_B = 33$ kΩ, $R_C = 470$ Ω, and $\beta = 120$.

    a. Find the saturation current for the transistor.

    b. Calculate the current through the LED when the output of the logic gate is "1."

    c. Determine the output voltage at the collector (with respect to ground) for part a.

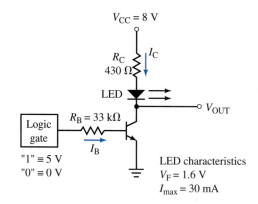

FIGURE 26–77

d. Calculate the current through the LED when the output of the logic gate is "0."

e. Determine the output voltage at the collector for part d.

44. For the circuit of Figure 26–77, which single component would you change to obtain a current of 25 mA through the diode? What would be the new value of this component? (Select the nearest commercial value of resistor having a tolerance of ±5%)

26.8 Testing a Transistor with a Multimeter

45. You are given the unknown transistor shown in Figure 26–36 and use an ohmmeter to take resistance measurements as tabulated. As always, the ohmmeter has two terminals: the positive (+, red) terminal and the negative (−, black) terminal.

a. What type of transistor were you given? pnp or npn?

b. What are the pin designations?

FIGURE 26–78

Pins 1 & 2		Pins 2 & 3		Pins 1 & 3	
Pin 1 (+) Pin 2 (−)	0.223 kΩ	Pin 2 (+) Pin 3 (−)	∞	Pin 1 (+) Pin 3 (−)	∞
Pin 2 (+) Pin 1 (−)	∞	Pin 3 (+) Pin 2 (−)	0.259 kΩ	Pin 3 (+) Pin 1 (−)	∞

46. Repeat Problem 45 if the measurements are as follows:

Pins 1 & 2		Pins 2 & 3		Pins 1 & 3	
Pin 1 (+) Pin 2 (−)	∞	Pin 2 (+) Pin 3 (−)	∞	Pin 1 (+) Pin 3 (−)	∞
Pin 2 (+) Pin 1 (−)	0.321 kΩ	Pin 3 (+) Pin 2 (−)	∞	Pin 3 (+) Pin 1 (−)	0.357 kΩ

47. Repeat Problem 45 if the measurements are as follows:

Pins 1 & 2		Pins 2 & 3		Pins 1 & 3	
Pin 1 (+) Pin 2 (−)	∞	Pin 2 (+) Pin 3 (−)	∞	Pin 1 (+) Pin 3 (−)	∞
Pin 2 (+) Pin 1 (−)	0.321 kΩ	Pin 3 (+) Pin 2 (−)	∞	Pin 3 (+) Pin 1 (−)	0.357 kΩ

48. You are given another transistor of the same type as the one in Problem 46. The measurements for this transistor are as follows:

Pins 1 & 2		Pins 2 & 3		Pins 1 & 3	
Pin 1 (+) Pin 2 (−)	∞	Pin 2 (+) Pin 3 (−)	∞	Pin 1 (+) Pin 3 (−)	0.354 kΩ
Pin 2 (+) Pin 1 (−)	0.344 kΩ	Pin 3 (+) Pin 2 (−)	∞	Pin 3 (+) Pin 1 (−)	0.362 kΩ

What conclusion can you make about this transistor? Explain your answer.

26.9 Junction Field Effect Transistor Construction and Operation

49. A JFET transistor has $V_{GS(OFF)} = -5$ V and $I_{DSS} = 8$ mA.

a. What type of JFET is this (n-channel or p-channel)?

b. Use Shockley's equation to solve for I_D when $V_{GS} = -3$ V.

c. Solve for I_D when $V_{GS} = -2$ V.

d. Solve for I_D when $V_{GS} = -6$ V.

50. A JFET transistor has $V_{GS(OFF)} = +4$ V and $I_{DSS} = 10$ mA.

    a. What type of JFET is this (n-channel or p-channel)?

    b. Use Shockley's equation to solve for I_D when $V_{GS} = +2$ V.

    c. Solve for I_D when $V_{GS} = +3$ V.

    d. Solve for I_D when $V_{GS} = +5$ V.

26.10   JFET Biasing

Use the following normalized JFET transfer curve to help you solve the problems of this section.

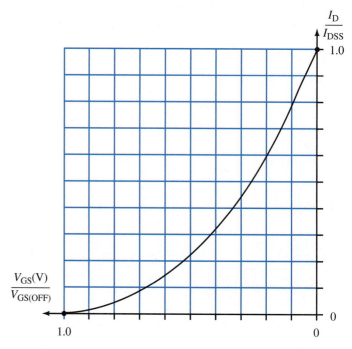

FIGURE 26–79   Normalized transconductance curve of an n-channel JFET.

51. For the circuit of Figure 26–80, let $V_{DD} = 16$ V, $R_D = 1.8$ kΩ, $R_S = 910$ Ω, $V_{GS(OFF)} = -4$ V, and $I_{DSS} = 8$ mA. At the operating point determine V_{GSQ}, I_{DQ}, and V_{DSQ}.

52. For the circuit of Figure 26–80, let $V_{DD} = 20$ V, $R_D = 2.4$ kΩ, $R_S = 1.0$ kΩ, $V_{GS(OFF)} = -5$ V, and $I_{DSS} = 10$ mA. At the operating point determine V_{GSQ}, I_{DQ}, and V_{DSQ}.

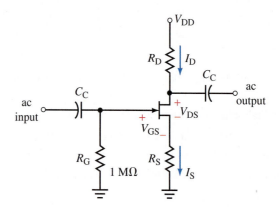

FIGURE 26–80

53. For the circuit of Figure 26–81, let $V_{DD} = 16$ V, $R_1 = 6.7$ MΩ, $R_2 = 1.0$ MΩ, $R_D = 3.9$ kΩ, $R_S = 2.0$ kΩ, $V_{GS(OFF)} = -4$ V, and $I_{DSS} = 8$ mA. At the operating point determine V_{GSQ}, I_{DQ}, and V_{DSQ}.

54. For the circuit of Figure 26–81, let $V_{DD} = 20$ V, $R_1 = 6.7$ MΩ, $R_2 = 1.0$ MΩ, $R_D = 3.9$ kΩ, $R_S = 2.0$ kΩ, $V_{GS(OFF)} = -5$ V, and $I_{DSS} = 10$ mA. At the operating point determine V_{GSQ}, I_{DQ}, and V_{DSQ}.

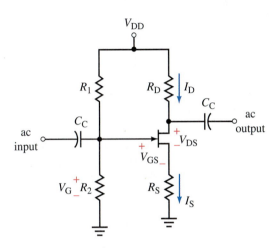

FIGURE 26–81   ◀ MULTISIM

26.11   MOSFETs

55. For the D-MOSFET circuit of Figure 26–82, let $V_{DD} = 20$ V, $R_1 = 1.8$ MΩ, $R_2 = 200$ kΩ, $R_D = 1.5$ kΩ, $R_S = 470$ Ω, $V_{GS(OFF)} = -5$ V, and $I_{DSS} = 10$ mA.

   a. Find I_{DQ} and V_{GSQ}.

   b. Is the MOSFET operating in its depletion mode or enhancement mode?

   c. Solve for V_{DSQ}.

56. For the D-MOSFET circuit of Figure 26–82, let $V_{DD} = 20$ V, $R_1 = 1.2$ MΩ, $R_2 = 680$ kΩ, $R_D = 680$ Ω, $R_S = 620$ Ω, $V_{GS(OFF)} = -5$ V, and $I_{DSS} = 10$ mA.

   a. Find I_{DQ} and V_{GSQ}.

   b. Is the MOSFET operating in its depletion mode or enhancement mode?

   c. Solve for V_{DSQ}.

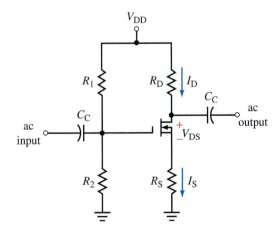

FIGURE 26–82

57. An E-MOSFET is specified as having $V_{GS(th)} = 2.0$ V and $I_{D(ON)} = 20$ mA for $V_{GS(ON)} = 12$ V. Determine the value of k for this MOSFET. Write the expression for I_D as a function of V_{GS}.

58. An E-MOSFET is specified as having $V_{GS(th)} = 3.0$ V and $I_{D(ON)} = 20$ mA for $V_{GS(ON)} = 15$ V. Determine the value of k for this MOSFET. Write the expression for I_D as a function of V_{GS}.

59. For the circuit of Figure 26–83, use the E-MOSFET of Problem 57 and let $V_{DD} = 20$ V. Find I_{DQ}, V_{GSQ}, and V_{DSQ}.

60. For the circuit of Figure 26–83, use the E-MOSFET of Problem 58 and let $V_{DD} = 30$ V. Find I_{DQ}, V_{GSQ}, and V_{DSQ}.

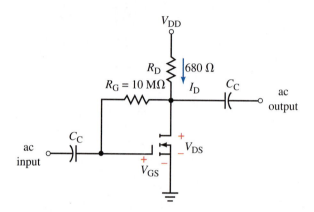

FIGURE 26–83

26.12   Troubleshooting a Transistor Circuit

61. Refer to the circuit shown in Figure 26–63. You obtain the following voltage readings on the indicated meters:

Meter 1: 0 V   Meter 2: 20.0 V   Meter 3: 0 V   Meter 4: 20.0 V

You have determined that the supply is operating correctly and that all components are the correct values and have been inserted correctly. What could be wrong with the transistor? Briefly explain your answer.

62. Refer to the circuit shown in Figure 26–63. You obtain the following voltage readings on the indicated meters:

Meter 1: 19.2 V   Meter 2: 0.7 V   Meter 3: 0.04 V   Meter 4: 19.94 V

You have determined that the supply is operating correctly and that all components are the correct values. What could be wrong with the transistor? Explain your answer.

26.13   Computer Analysis of Transistor Circuits

◀ MULTISIM

63. Use MultiSIM to find V_B, I_{CQ}, I_{EQ}, V_{CEQ}, and V_{BEQ} for the circuit of Figure 26–70. Use a 2N3904 transistor and the component values provided in Problem 21. Compare the values of I_{CQ} and V_{CEQ} to those predicted in Problem 21.

◀ MULTISIM

64. Use MultiSIM to find V_B, I_{CQ}, I_{EQ}, V_{CEQ}, and V_{BEQ} for the circuit of Figure 26–70. Use a 2N2222A transistor and the same component values of Problem 21. Compare the values of I_{CQ} and V_{CEQ} to those in Problem 63.

◀ MULTISIM

65. Use MultiSIM to find V_B, I_{CQ}, I_{EQ}, V_{CEQ}, and V_{BEQ} for the circuit of Figure 26–71. Use a 2N3904 transistor and the component values provided in Problem 25. Compare the values of I_{CQ} and V_{CEQ} to those predicted in Problem 25.

66. Use MultiSIM to find V_B, I_{CQ}, I_{EQ}, V_{CEQ}, and V_{BEQ} for the circuit of Figure 26–71. Use a 2N2222A transistor and the same component values of Problem 25. Compare the values of I_{CQ} and V_{CEQ} to those in Problem 65.

◀ MULTISIM

67. Use MultiSIM to find V_{GSQ}, I_{DQ}, and V_{DSQ} for the circuit of Figure 26–81. Use a 2N5045 JFET and the same component values of Problem 53. Compare the values of I_{DQ} and V_{DSQ} to those in Problem 53. Why is there a difference between the results?

◀ MULTISIM

68. Use MultiSIM to find V_{GSQ}, I_{DQ}, and V_{DSQ} for the circuit of Figure 26–81. Use a 2N5045 JFET and the same component values of Problem 54. Compare the values of I_{DQ} and V_{DSQ} to those in Problem 54. Why is there a difference between the results?

◀ MULTISIM

69. Use Capture to find V_B, I_{CQ}, I_{EQ}, V_{CEQ}, and V_{BEQ} for the circuit of Figure 26–71. Use a 2N3904 transistor and the component values provided in Problem 25. Compare the values of I_{CQ} and V_{CEQ} to those predicted in Problem 25.

◀ CADENCE

70. Use Capture to find V_B, I_{CQ}, I_{EQ}, V_{CEQ}, and V_{BEQ} for the circuit of Figure 26–71. Use a 2N2222A transistor and the same component values of Problem 25. Compare the values of I_{CQ} and V_{CEQ} to those in Problem 69.

◀ CADENCE

71. Use Capture to find V_{GSQ}, I_{DQ}, and V_{DSQ} for the circuit of Figure 26–81. Use a 2N3819 JFET and the same component values of Problem 53. Compare the values of I_{DQ} and V_{DSQ} to those in Problem 53. Why is there a difference between the results?

◀ CADENCE

72. Use Capture to find V_{GSQ}, I_{DQ}, and V_{DSQ} for the circuit of Figure 26–81. Use a 2N3819 JFET and the same component values of Problem 54. Compare the values of I_{DQ} and V_{DSQ} to those in Problem 54. Why is there a difference between the results?

◀ CADENCE

■ **OBJECTIVES**

On completion of this chapter, you will be able to

- use the frequency range of an amplifier to determine the size of coupling and bypass capacitors for amplifier circuits,

- determine the correct placement of electrolytic capacitors into a circuit and understand the importance of this placement,

- use the T-equivalent model and the h-parameter model to sketch the ac equivalent circuit of a BJT amplifier,

- calculate voltage gain, input impedance, output impedance, current gain, and power gain of a BJT amplifier,

- sketch the ac load line for a transistor circuit and use the ac load line to determine the maximum undistorted output signal for a given amplifier,

- determine the small-signal ac equivalent model of a FET transistor amplifier,

- calculate voltage gain, input impedance, and output impedance of JFET and MOSFET amplifiers,

- use troubleshooting skills to analyze ac operation of transistor amplifiers,

- use PSpice and MultiSIM to simulate the operation of amplifier circuits.

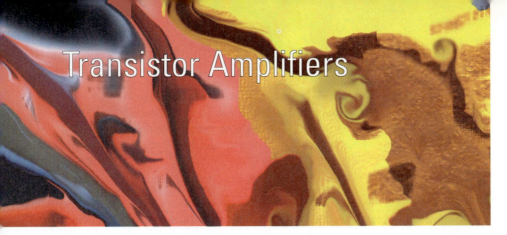

Transistor Amplifiers

27

In the previous chapter, you were introduced to the bipolar junction transistor (BJT) and several types of field effect transistors (FETs). Specifically, you learned that before a transistor can be used as an amplifier, it must be correctly biased so that it is ready to accept an ac signal. In this chapter you will examine how, once the transistor is biased, it is able to process an ac signal. To help understand transistor operation, several models have been developed. Each of these models uses dependent sources (refer to Chapter 19) to help explain how an ac signal at the input is ultimately amplified.

As mentioned previously, the use of discrete transistors in electronic circuits is less common. This chapter uses single transistors to develop concepts such as voltage gain, current gain, input impedance, and output impedance. Although we use a single transistor amplifier to define and analyze these characteristics, you will find that these same principles will be applied to IC amplifiers containing many hundreds or thousands of transistors. Most students of electronics find that it is much easier to start analyzing amplifiers that use a single three-terminal transistor than to immediately work with multi-terminal integrated circuits. ■

SINCE THE INVENTION OF THE transistor in 1947 by Bell physicists, William Shockley, John Bardeen, and Walter Brattain, the transistor has changed dramatically. The original transistor was a bipolar junction transistor (BJT) that consisted of a sandwich of p- and n-type materials, allowing both positive and negative charge carriers. By applying a small current variation current in the base circuit, a much larger variation in collector current is possible. Hence the BJT is a current amplifier.

Although the BJT is still used, many other types of transistors have been developed. For instance, the junction field effect transistor (JFET) and the metal-oxide-semiconductor field effect transistor (MOSFET) use a small variation in input voltage to control the size of an electric field, which in turn controls the current through the device. Consequently, FETs are referred to as voltage amplifiers.

Other developments in semiconductor have led to isolated gate FETs (IGFETs) and numerous other types of transistors. One of the greatest developments in semiconductor design occurred in the early 1960s with the invention of the integrated circuit (IC). Since then, the size of the transistor has shrunk to the point that several modern transistors can fit on the surface of a blood cell. Reducing the size of the transistor has resulted not only in more devices on a given IC wafer, but also improves the performance of the IC. This miniaturization has resulted in exponential improvements in the speed of computers and a corresponding reduction in the cost of all semiconductor devices. Indeed, this miniaturization led to Moore's Law, which states the speed of computers will double every eighteen months. Although many skeptics have predicted that this improvement cannot continue, Moore's law has been a surprisingly good predictor of computer speed and capacity for the past three decades. ■

27.1 The Use of Capacitors in Amplifier Circuits

In the previous chapter, you likely noticed that most of the transistor circuits included capacitors, even though they were ignored in calculation of the dc operating points. In this section, you will learn the importance of capacitors in a transistor circuits and be able to calculate the values of these capacitors. At this point, a brief review of capacitors is worthwhile. You have already seen that capacitors are electrical components capable of storing electrical charge. Capacitors have a reactance (or impedance) to ac that is inversely proportional to the frequency of the signal applied to the component. The reactance of a capacitor is given as

$$X_C = \frac{1}{2\pi f C} \quad [\Omega] \qquad (27\text{--}1)$$

From the above expression, we are able to observe two very important characteristics of capacitors. They are:

1. At dc ($f = 0$ Hz), a capacitor has infinite impedance, which means that a capacitor behaves as an open circuit.

2. As the frequency increases, the reactance of the capacitor decreases. If the frequency of a signal is sufficiently high, the capacitor will behave effectively as a short circuit.

The above characteristics are extremely useful in transistor amplifier circuits, since they allow capacitors to simultaneously prevent the dc operation of one stage from affecting other stages in the circuit, while at the same time allowing desired ac signals to pass readily between stages. Capacitors used in this manner are called **coupling capacitors,** since they couple the ac signal from one stage to the next.

Coupling Capacitors

Consider the simplified circuit shown in Figure 27–1. A voltage source, represented by v_S and a series resistance, R_S, is connected to a circuit having an input resistance, R_{in}.

The voltage, v_{in} applied to the input of the circuit is found by applying the voltage divider rule as

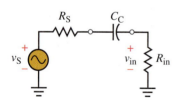

FIGURE 27–1

$$v_{in} = \left(\frac{R_{in}}{\sqrt{(R_{in} + R_S)^2 + (X_C)^2}} \right) v_S \qquad (27\text{--}2)$$

For frequencies above some **cutoff frequency,** the reactance of the capacitor will be very small relative to the size of the total resistance, $R = R_{in} + R_S$. Due to the negligible size of the capacitive reactance above the cutoff frequency, the expression for the input voltage can now be simply expressed as

$$v_{in} = \left(\frac{R_{in}}{R_{in} + R_S} \right) v_S \qquad (27\text{--}3)$$

The size of the coupling capacitor is selected to be such that $X_C \leq 0.1\, R$ at the lowest frequency of operation. This condition is referred to as **stiff coupling** and is used extensively throughout electronic design.

The simplified circuit shown in Figure 27–2, is intended to operate between a frequency range of 500 Hz and 10 kHz. Determine the size of the coupling capacitor that is required between the two stages if the output impedance of the first stage is 1200 Ω and the input impedance of the second stage is 800 Ω.

Solution

For stiff coupling, the value of the capacitive reactance must be kept to $X_C \leq 0.1\,R$ for the lowest frequency of 500 Hz. Letting $X_C = 0.1\,R$, we get

$$X_C = (0.1)(1200\ \Omega + 800\Omega) = 200\ \Omega$$

By applying Equation 27–1, we get

$$C_C = \frac{1}{2\pi f X_C}$$

$$= \frac{1}{2\pi(500\ \text{Hz})(200\ \Omega)}$$

$$= 1.59\ \mu\text{F}$$

EXAMPLE 27–1

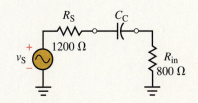

FIGURE 27–2

Bypass Capacitors

Although resistors are useful for establishing the bias point of a transistor, they may adversely affect the ac operation of the circuit. Once again, we use capacitors to improve operation of the circuit. Placing a capacitor across a resistor has no affect on the dc operation of the circuit, since it is effectively an open circuit at dc. However, we have seen that as the frequency of a signal increases, the reactance of the capacitor will effectively approach a short circuit. If this capacitor is placed across any resistance, the resistance is effectively shorted for frequencies above the cutoff frequency, f_c. As one might expect, a capacitor that is used to bypass ac signals around a component (or circuit) is called a **bypass capacitor.** In most cases, the unwanted ac signal is simply bypassed to the circuit ground.

The value of a bypass capacitor is found in much the same way as the value of a coupling capacitor. For stiff bypass, the value of capacitive reactance, $X_C \leq 0.1R$ where the value of R is the Thévenin resistance "seen" by the capacitor. The following example illustrates the principle.

The simplified circuit shown in Figure 27–3, is intended to operate between a frequency range of 500 Hz and 10 kHz. Determine the size of the bypass capacitor that is required across the resistor $R_E = 780\ \Omega$. The equivalent resistance of the rest of the circuit is 20 Ω.

Solution

The Thévenin resistance "seen" by the capacitor is

$$R = 20\ \Omega\|780\ \Omega$$

$$= 19.5\ \Omega$$

For stiff bypass, we let $X_C = 0.1R = 1.95\ \Omega$ and so the value of the bypass capacitor is now determined to be

$$C_C = \frac{1}{2\pi f X_C}$$

$$= \frac{1}{2\pi(500\text{Hz})(19.5\ \Omega)}$$

$$= 163\ \mu\text{F}$$

EXAMPLE 27–2

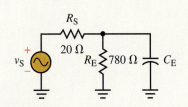

FIGURE 27–3

The previous examples clearly illustrate how capacitors are used to couple desired ac signals between various stages and to bypass unwanted ac signals to ground. Whenever coupling or bypass capacitors are used in transistor circuits, the circuit is analyzed using the following simplifications:

1. Capacitors are replaced by open circuits in order to determine dc voltages and currents and

2. Capacitors are replaced by short circuits when we determine ac voltages and currents. In any design, the value of a bypass or coupling capacitor is determined by using a stiff design, namely that $X_C \leq 0.1R$ for the lowest frequency of operation. The value of R is the Thévenin resistance external to the capacitor.

PRACTICE PROBLEMS 1

Given the circuit of Figure 27–4:

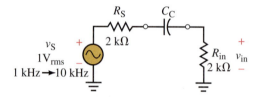

FIGURE 27–4

a. Determine the value of the coupling capacitor if the circuit is to operate in a frequency range from 1 kHz → 10 kHz.

b. Solve for the magnitude of the voltage, v_{in} at a frequency of 1 kHz. How does the result compare to the voltage if the coupling capacitor had been replaced by a short circuit?

c. Repeat part b. for a frequency of 10 kHz.

Answers
a. 0.398 μF; b. 0.498 V (as compared to 0.5 V); c. 0.500 V

PRACTICE PROBLEMS 2

Given the circuit of Figure 27–5:

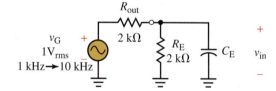

FIGURE 27–5

a. Determine the value of the bypass capacitor if the circuit is to operate in a frequency range from 1 kHz → 10 kHz.

b. Solve for the magnitude of the voltage, v_{in}, at a frequency of 1 kHz. How does the result compare to the voltage if the bypass capacitor had been replaced by a short circuit?

c. Repeat part b. for a frequency of 10 kHz.

Answers
a. 1.59 μF; b. 0.0498 V (as compared to 0V); c. 0.005 V

The operation of transistor amplifiers has been the source of much research and study. Numerous electrical models having been developed to explain the operation of the transistor; each of these models having advantages and disadvantages. For simplicity, our analysis examines only two models; the *T*-equivalent and the *h*-parameter small-signal models.

27.2 BJT Small-Signal Models

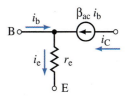

FIGURE 27–6 *T*-equivalent model of a BJT.

T-Equivalent Model

The *T*-equivalent model of a transistor shown in Figure 27–6 is the simplest representation of a transistor, having only one resistor and a single current source, each of which can be determined from dc characteristics of the circuit under test. There are several important points that need to be made with respect to this model. First, the *T*-equivalent model is exactly the same regardless of whether the transistor is npn or pnp. This is because both transistor types behave exactly the same when an ac signal is applied. All current directions are instantaneous values and could just as easily be shown in the opposite direction since, as the name implies, an ac signal must alternate. Lastly, because we are working with ac currents and voltages, we show the currents in lower case letters to keep the distinction between ac and dc. You will also notice that relationship between ac currents in transistors is exactly the same as for dc currents, namely:

$$i_e = i_b + i_c \tag{27–4}$$

and

$$i_e = i_b + \beta_{ac}i_b = (\beta_{ac} + 1)i_b \tag{27–5}$$

Unless we are given other information, we assume that $\beta_{ac} = \beta_{dc}$. Although not always correct, this is generally a fairly good approximation. Finally, the value of the ac resistance in the emitter of the model is determined from the dc operating point as

$$r_e = \frac{26 \text{ mV}}{I_{EQ}} \quad \text{at } 25° \text{ C} \tag{27–6}$$

The value, 26 mV, in the numerator of the above expression is due to the fact that a base-emitter junction behaves like a forward-biased diode. The resulting emitter resistance is typically very small (approximately 1 Ω to 50 Ω), as one might expect in a forward-biased diode. For example, if a transistor has an operating point of $I_{EQ} \approx I_{CQ} = 2.0$ mA, the emitter resistance would be

$$r_e = \frac{26 \text{ mV}}{20 \text{ mV}} = 1.3 \text{ Ω}$$

h-parameter Model

While the *T*-equivalent model circuit is very simple and generally provides a good model to explain most transistor operation, it is not accurate enough for other applications. For better representation of transistor ac operation, we use the more complex *h*-parameter model shown in Figure 27–7. This model consists of four variables, h_{fe} (the forward current transfer ratio—unitless), h_{ie} (the input impedance characteristic—in ohms, Ω), h_{re} (the reverse voltage transfer ratio—unitless), and h_{oe} (the output admittance characteristic—in siemens, S). Each of these parameters has a subscript *e*, indicating that the values are determined for the common emitter connection. Although less common, similar parameters may be found for common collector (h_{fc}, h_{ic}, h_{rc}, and h_{oc}) and for

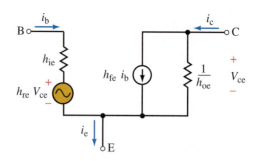

FIGURE 27–7   *h*-parameter model of a BJT.

common base (h_{fb}, h_{ib}, h_{rb}, and h_{ob}) connections. This textbook uses only the common emitter model.

Notice that the *h*-parameter model has a current source that uses h_{fe}, while the *T*-equivalent model uses β_{ac}. These values are, in fact, the same, and the terms are used interchangeably. As mentioned previously, the value of β_{ac} will often be close to the value of β_{dc}.

In order to find the values of the *h*-parameters, two methods are possible. The values can be measured by using an *h*-parameter tester or, more commonly, the values may be approximated from the manufacturer's specifications. Since the values of the *h*-parameters are dependent on the operating point, it is necessary to first determine the Q-point of the transistor. Figure 27–8 shows the *h*-parameter characteristics of a 2N3904 npn transistor.

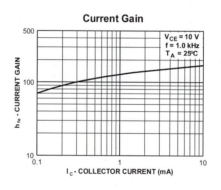

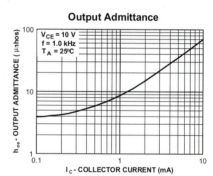

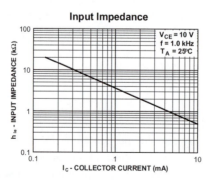

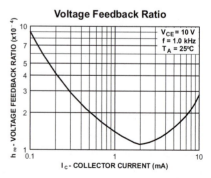

FIGURE 27–8   *h*-parameter curves of a 2N3904 transistor. (*Courtesy of Fairchild Semiconductor Corporation*)

NOTES . . .

Unless a transistor circuit is in its active region, it will not operate as a linear amplifier. In such a case it is not correct to use either of the ac models.

If we had a transistor with I_C = 2 mA (and V_{CE} = 10 V), we see that the values of the *h*-parameters are approximated as:

$$h_{fe} \approx 130,\ h_{oe} \approx 14\ \mu S,\ h_{ie} \approx 2.0\ k\Omega,\ \text{and}\ h_{re} \approx 1.1 \times 10^{-4}$$

The current source in the collector circuit of each of the models shows why the BJT is referred to as a current amplifier. A small current applied at the base of a transistor is amplified into a collector current that can be 100 to 200 times larger than the base current.

We are now ready to define some very important parameters of amplifier circuits; namely voltage gain (A_v), input impedance (z_{in}), current gain (A_i), output impedance (z_{out}), and power gain (A_p). Although we will analyze transistor circuits in this chapter, the above parameters apply equally well to any amplifier circuit. In the next chapter we will use the same parameters for circuits involving operational amplifiers. Imagine that any amplifier can be represented as a block diagram consisting of two ports as shown in Figure 27–9.

27.3 Calculating A_v, z_{in}, z_{out}, and A_i of a Transistor Amplifier

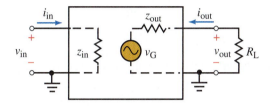

FIGURE 27–9   Two-port model of any amplifier circuit.

This model is called a two-port model since there are two distinct parts of the block diagram representing the amplifier; the input port and the output port. Although the model appears to show that these ports are isolated, the voltage source (which could just as easily have been the Norton equivalent with a current source and a parallel resistance) is dependent upon either voltage or current that occurs at the input. If we were to apply an input voltage, v_{in}, a current, i_{in} would result in the input loop. We see that this current is dependent on the input impedance, z_{in} of the amplifier and is easily calculated by using Ohm's law. The output circuit of the amplifier can be represented as a voltage source with a voltage, v_{out}, and an internal series impedance, z_{out}. Now, if some external load is applied to the output circuit, current would result in an output voltage across this load, R_L. The polarity of the output voltage, v_{out}, and the direction of the output current, i_{out}, are shown for reference purposes only. The actual voltage polarity and/or current direction may not necessarily be the same as those shown.

Whenever we analyze any amplifier circuit, we generally solve for five characteristics of the amplifier circuit: the voltage gain, the input impedance, the output impedance, the current gain, and the power gain. The use of the ac equivalent circuit is of fundamental importance in the analysis of all circuits.

Voltage Gain, A_v

Voltage gain is defined as the ratio of the output voltage to input voltage.

$$A_v = \frac{v_{out}}{v_{in}}$$

$$(27\text{–}7)$$

Input impedance, z_{in}

The **input impedance** of any circuit is defined as the ratio of input voltage to input current.

$$z_{in} = \frac{v_{in}}{i_{in}} \tag{27–8}$$

Output Impedance, z_{out}

The **output impedance** of an amplifier is determined by looking back into the transistor's output terminals with the load resistance, R_L, removed. The definition of output impedance of a circuit is very similar to the approach that was used to find the Thévenin impedance of a circuit; namely that the output impedance of any circuit is expressed as the ratio of open-circuit output voltage to short-circuit output current.

$$z_{out} = \frac{v_{out(OC)}}{i_{out(SC)}} \tag{27–9}$$

Current Gain, A_i

As expected, **current gain** is defined as the ratio of output current to input current.

$$A_i = \frac{i_{out}}{i_{in}} \tag{27–10}$$

Although the above expression is useful, there is an easier way to find the current gain of an amplifier once the voltage gain has been determined. The resulting expression can be used in any two-port circuit (i.e., any circuit consisting of two input terminals and two output terminals), regardless of the connection between the ports.

Remember that we defined voltage gain as

$$A_v = \frac{v_{out}}{v_{in}}$$

Applying Ohm's law on the circuit of Figure 27–9, results in

$$A_v = \frac{-i_{out}R_L}{i_{in}z_{in}}$$

Notice that the term in the numerator is negative. This is necessary to account for the reference direction of the output current, i_{out}. Finally, by using the definition of current gain we rewrite the equation as

$$A_i = -\frac{A_v z_{in}}{R_L} \tag{27–11}$$

Power Gain, A_p

As expected, it is a simple matter to find the power gain of the amplifier. **Power gain** is defined as the ratio of signal output power to signal input power, namely

$$A_p = \frac{P_{out}}{P_{in}} \tag{27–12}$$

However, since power is determined as $P = V \cdot I$, we rewrite the expression as

$$A_p = \frac{v_{out}i_{out}}{v_{in}i_{in}} = \left| A_v A_i \right| \tag{27–13}$$

The absolute value operation in the expression is necessary since power gain cannot have a negative value. Equation 27–13 is valid for all two-port systems regardless of the configuration. Although this expression seems to indicate that it is possible to have more output power than input power (a contradiction of the Law of Energy Conservation), what we have is an increase in signal (or ac) power. As we will observe later, the increase in signal power is the result of power conversion by the dc voltage source of the circuit.

In this section, we will examine several circuits and derive general equations that can be used to analyze circuits that have the same configuration. Although the circuit parameters can be easily found by simply plugging values into the appropriate equations, it is strongly recommended that you use applicable circuit theory in your analysis. The skills that you develop will help you to methodically analyze any circuit.

27.4 The Common-Emitter Amplifier

In order to determine how a transistor behaves with the application of an ac signal, we must first sketch the ac equivalent of a transistor circuit. To do this for any transistor circuit, we follow several important steps:

1. Find the operating point of the transistor to ensure that the transistor is in its active region.

2. Determine the ac parameters of the transistor model that you will use. If you use the T-equivalent model, you will need to calculate $r_e = 26 \text{ mV}/I_E$. If you use the h-parameter model you will need h_{fe} and h_{ie}. There are two other h-parameters that are seldom used. h_{re} is very small and has negligible effect on any circuit operation. The value of h_{oe} is only considered when $1/h_{oe}$ is within an order of magnitude of $R_C \| R_L$.

3. Remove all dc voltage sources from the given circuit and replace them with short circuits. (This step is similar to how we find the Thévenin equivalent circuit.)

4. Replace all coupling and bypass capacitors with short circuits. If the circuit was designed correctly, the values of these capacitors were designed to be stiff for the frequency range at which the transistor is to operate.

5. Replace the transistor with the appropriate model, whether the T-equivalent or the h-parameter.

6. Solve the resulting circuit for each of the required characteristics.

Let's examine how these steps can be applied to the common-emitter transistor amplifier circuit of Figure 27–10. If we assume that the operating point of the transistor has been found, we can apply Steps 3 and 4 to arrive at the equivalent circuit shown in Figure 27–11.

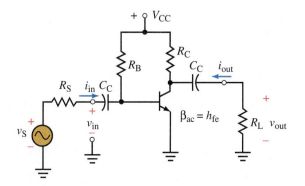

FIGURE 27–10   Fixed-bias CE amplifier.

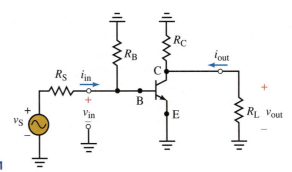

FIGURE 27–11

The circuit of Figure 27–11 is further redrawn in Figure 27–12 by substituting the transistor with the appropriate model (let's begin by using the *T*-equivalent) and by reconfiguring the resistors so that the grounds are at the bottom of the circuit. In order to help clarify the analysis, Figure 27–12 shows the circuit ground as a continuous connection. Notice also that the input and output terminals have been clearly identified. From the illustration we see that the output current is determined to be in the same direction as the ac collector current, i_c. However, since the output current is in the same direction as i_c, we conclude that the voltage across R_L must be opposite to the polarity that is shown in the diagram.

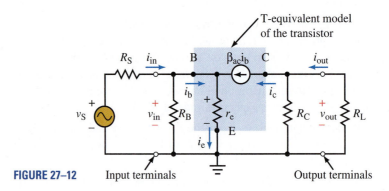

FIGURE 27–12   Input terminals   Output terminals

The polarities of all voltages are taken with respect to ground. Since the direction of the output current is into the collector (due to the current source in the transistor model), we see that the polarity of the output voltage must be negative. R_C and R_L are in a parallel connection and so we express the output voltage as $v_{out} = -i_c(R_C \| R_L)$.

Although the input voltage is across R_B, it is not useful to write the input voltage in terms of this component, since we know nothing of the current through R_B. Remember that Kirchhoff's voltage law states that the voltage at any point with respect to ground is the same, no matter which path is chosen. Consequently, we see that the input voltage may be written in terms of the current through r_e. This gives the input voltage as $v_{in} = i_e r_e$. The voltage gain is now easily written as

$$A_v = \frac{v_{out}}{v_{in}} = \frac{-i_c \left(R_C \| R_L \right)}{i_e r_e}$$

Now, since $i_c = \beta_{ac} i_b$ and $i_e = (\beta_{ac} + 1)i_b$ we express the voltage gain of the fixed-bias CE amplifier as

$$A_v = \frac{-\beta_{ac} i_b \left(R_C \| R_L \right)}{(\beta_{ac} + 1)i_b r_e} = -\frac{\beta_{ac} \left(R_C \| R_L \right)}{(\beta_{ac} + 1)r_e} \approx -\frac{\left(R_C \| R_L \right)}{r_e} \qquad \textbf{(27–14)}$$

Notice that the previous expression shows the voltage gain to be a negative value. This simply means that the output voltage is 180° out-of-phase (inverted) with respect to the input signal. This characteristic is true for all common-emitter circuits. In the final approximation, voltage gain is dependent on only the resistor values in the circuit and is not dependent on the ac beta (also specified as h_{fe}) of the transistor. This is because in most instances h_{fe} is quite large and so $\beta_{ac}/(\beta_{ac} + 1) \approx 1$. Typical magnitudes of voltage gain for the fixed-bias CE amplifier are in the order of 50 to 150.

In order to determine the input impedance of the circuit, we begin by first solving for the input impedance (looking into the base) of the transistor. By applying Equation 27–8 to the base terminal of the transistor, we determine the input impedance of the transistor, $z_{in(Q)}$ as

$$z_{in(Q)} = \frac{v_b}{i_b} = \frac{i_e r_e}{i_b} = \frac{(\beta_{ac} + 1)i_b r_e}{i_b} = (\beta_{ac} + 1)r_e \qquad \text{(27–15)}$$

The input impedance at the input terminals of the circuit can now be found by placing the base resistance, R_B in parallel with the input impedance of the transistor. The resultant input impedance of the fixed-bias CE transistor circuit is now found as

$$z_{in} = R_B \| (\beta_{ac} + 1)r_e \approx R_B \| (\beta_{ac} r_e) \qquad \text{(27–16)}$$

In general, the input impedance of the fixed-bias CE amplifier is moderately high and tends to be in a range of about 1 kΩ to 2 kΩ.

Although Equation 27–9 can be used to solve for the output impedance of this circuit, there is a much easier approach for determining the output impedance of most (though not all) transistor circuits. In a manner similar to that used in determining the input impedance of the circuit, we begin by looking into the collector terminal of the transistor. Recall that when we were finding the Thévenin impedance of a circuit, we first begin by removing all voltage and current sources. Removing the current source, $\beta_{ac} i_b$ means that the output impedance of this transistor circuit is an open circuit, namely $z_{out(Q)} = \infty$. Consequently, the impedance at the output terminals is simply found as

$$z_{out} = R_C \qquad \text{(27–17)}$$

The current gain of any amplifier can be found by using Equation 27–11. If we substitute Equations 27–14 and 27–16 into Equation 27–11, we arrive at an expression for the current gain of the fixed-bias CE amplifier as follows:

$$A_i = -\frac{A_v z_{in}}{R_L}$$

$$= -\left(\frac{\beta_{ac}(R_C \| R_L)}{(\beta_{ac} + 1)r_e} \right) \left(\frac{R_B \| [(\beta_{ac} + 1)r_e]}{R_L} \right)$$

which, when simplified using the approximation that $\beta_{ac} \approx \beta_{ac} + 1$, results in

$$A_i \approx \frac{\beta_{ac} R_B R_C}{(R_C + R_L)(R_B + \beta_{ac} r_e)} \qquad \text{(27–18)}$$

Typical current gain for the fixed-bias CE amplifier tends to be between 50 and 100.

The previous equations were derived using the T-equivalent model of the transistor. Similar derivations are possible if we were to use the h-parameter model. Figure 27–13 shows the ac equivalent of the fixed-bias CE amplifier using the *h*-parameter model of the transistor. As mentioned previously, we will not consider the effects of h_{re} for any ac model because the effects are minimal. In deriving the equations for the CE amplifier, we neglect the effects of the output impedance, $(1/h_{oe})$ of the transistor, since in most cases this impedance will be many times larger than the value of $R_C \| R_L$. However, as we will discover in Section 27.6 there are instances where the effects of h_{oe} cannot be ignored.

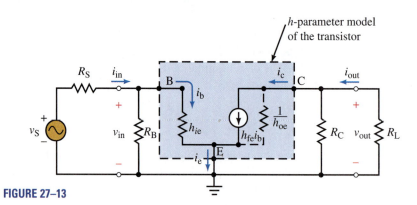

h-parameter model of the transistor

FIGURE 27–13

It is left as an exercise for the student to derive the following expressions using the *h*-parameter model of the transistor for the fixed-bias CE amplifier.

$$A_v = -\frac{h_{fe}\left(R_C \| R_L\right)}{h_{ie}} \tag{27-19}$$

$$z_{in} = R_B \| h_{ie} \tag{27-20}$$

$$z_{out} = R_C \tag{27-21}$$

$$A_i = \frac{h_{fe} R_B R_C}{\left(R_C + R_L\right)\left(R_B + h_{ie}\right)} \tag{27-22}$$

EXAMPLE 27–3

Given the fixed-bias CE amplifier circuit of Figure 27–14:

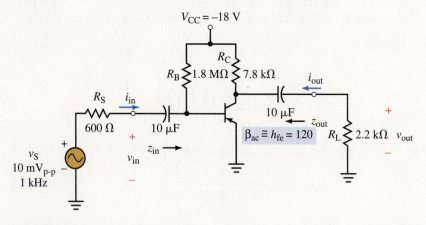

FIGURE 27–14

a. Find the dc operating point.

b. Sketch the ac equivalent circuit, using the *T*-equivalent model of the transistor.

c. Determine $A_v = v_{out}/v_{in}$, z_{in}, z_{out}, A_i, and A_p.

Solution

a.
$$(1.8 \text{M}\Omega)I_B + 0.7 \text{ V} = 18\text{V}$$

$$I_B = \frac{18 \text{ V} - 0.7 \text{ V}}{1.8 \text{ M}\Omega} = 9.61 \text{ } \mu\text{A}$$

$$I_C = (120)(9.61 \text{ } \mu\text{A}) = 1.15 \text{ mA}$$

$$V_{CE} = -18 \text{ V} + (7.8 \text{ k}\Omega)(1.15 \text{ mA}) = -9.00 \text{ V}$$

$$r_e = \frac{26 \text{ mV}}{1.15 \text{ mA}} = 22.5 \text{ } \Omega$$

b. The ac equivalent of the circuit is shown in Figure 27–15. It is important to note that the ac equivalent of the transistor is not dependent on whether the transistor is npn or pnp since either transistor type will respond to an ac signal in exactly the same way.

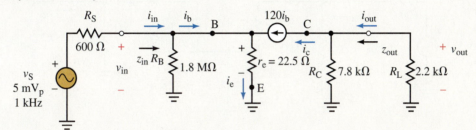

FIGURE 27–15

c. Voltage gain, $A_v = v_{out}/v_{in}$:

$$A_v = \frac{v_{out}}{v_{in}} = -\frac{i_c(R_C \| R_L)}{i_e r_e}$$

$$= -\frac{120i_b(7.8 \text{ k}\Omega \| 2.2 \text{ k}\Omega)}{121i_b(22.5 \text{ k}\Omega)}$$

$$= -75.6$$

Input impedance, z_{in}:

First, we find the input impedance of the transistor.

$$z_{in(Q)} = \frac{v_b}{i_b} = \frac{i_e r_e}{i_b} = \frac{(h_{fe} + 1)i_b r_e}{i_b} = 121(22.5 \text{ } \Omega) = 2.72 \text{ k}\Omega$$

The resultant input impedance is the parallel combination of R_B and $z_{in(Q)}$. Due to its much greater size, R_B has negligible effect on the input impedance of the amplifier circuit.

$$z_{in} = 2.72 \text{ k}\Omega \| 1.8 \text{ M}\Omega \approx 2.72 \text{ k}\Omega$$

Output impedance, z_{out}:

Due to the infinite output impedance of the T-equivalent transistor model, the output impedance of the amplifier is simply $z_{out} = 7.8 \text{ k}\Omega$.

Current gain, A_i:

Using Equation 27–11, the current gain of the amplifier circuit is

$$A_i = -\frac{A_v z_{in}}{R_L} = -\frac{(-75.6)(2.72 \text{ k}\Omega)}{2.2 \text{ k}\Omega} = 93.5$$

Power gain, A_p:

Using Equation 27–14, we determine the power gain to be

$$A_p = |(-75.6)(93.5)| = 7070$$

EXAMPLE 27–4

For the fixed-bias CE amplifier circuit of Figure 27–14, let $h_{ie} = 2600\ \Omega$ and $h_{oe} = 8.0\ \mu S$.

a. Sketch the ac equivalent circuit, using the h-parameter model of the transistor.

b. Determine A_v, z_{in}, z_{out}, A_i, and A_p using the h-parameter model.

c. If the signal generator has an unloaded output voltage of 10 mV$_{p-p}$, determine the expected output voltage. Sketch the waveforms of v_S, v_{in}, and v_{out}, showing amplitudes and relative phase angles.

Solution

a. The ac equivalent circuit is shown in Figure 27–16. Notice that the output impedance of the transistor is $1/h_{oe} = 125\ k\Omega$. Since this value is much larger than R_C and R_L its effect may be neglected.

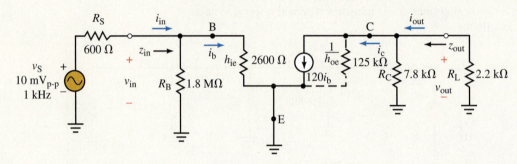

FIGURE 27–16

b.

$$A_v = \frac{v_{out}}{v_{in}} = \frac{-i_C(R_C\|R_L)}{i_B h_{ie}} = \frac{-(h_{fe}i_B)(R_C\|R_L)}{i_B h_{ie}} = \frac{-(120)(7.8\ k\Omega\|R2.2\ k\Omega)}{2.6\ k\Omega} = -79.2$$

$$Z_{in(Q)} = \frac{v_b}{i_b} = \frac{h_{ie}i_B}{i_B} = h_{ie} = 2.6\ k\Omega$$

$$Z_{in} = R_B\|Z_{in(Q)} = 1.8\ M\Omega\|2.6\ k\Omega \approx 2.6\ k\Omega$$

$$Z_{out} = R_C\left\|\frac{1}{h_{oe}}\right. \approx R_C = 7.8\ k\Omega$$

In general if $R_C\|R_L < 0.1(1/h_{oe})$ we may ignore the effect of h_{oe}

$$A_i = -\frac{A_v z_{in}}{R_L} = -\frac{(-79.2)(2.6\ k\Omega)}{2.2\ k\Omega} = 93.6$$

$$A_p = |A_v A_i| = (79.2)(93.6) = 7410$$

Notice that the values obtained by using the h-parameter model are somewhat different than those found using the T-equivalent model. For most circuits, the discrepancy is quite small. Because the h-parameter model is a closer representation of the actual behavior of a transistor, we find that this model provides a slightly better approximation than the T-equivalent model.

c. In order to solve for the output voltage, we must first determine the input voltage, v_{in}. A common error is to assume the input voltage is the same as the source voltage, v_S. This is not correct! As illustrated in Figure 27–17, we determine the input voltage by using the voltage divider rule between the input impedance, z_{in} of the amplifier and the source resistance, R_S.

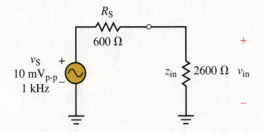

FIGURE 27–17

The input voltage is found as

$$v_{in} = \left(\frac{z_{in}}{z_{in} + R_S}\right)v_S = \left(\frac{2600\ \Omega}{2600\ \Omega + 600\ \Omega}\right)10\ mV_{p\text{-}p} = 8.12\ mV_{p\text{-}p}$$

Now since voltage gain is $A_v = v_{out}/v_{in} = -79.2$, we get

$$v_{out} = A_v v_{in} = (-79.2)(8.12\ mV_{p\text{-}p}) = 644\ mV_{p\text{-}p}$$

Notice that the output voltage is not written as $-644\ mV_{p\text{-}p}$ even though the voltage gain is preceded with a negative sign. By definition, peak-to-peak voltage is simply a magnitude. A negative sign would be meaningless in this context. Remember that the significance of the negative sign in the voltage gain simply means that the output voltage is $180°$ out-of-phase with respect to the input voltage. Figure 27–18 shows the waveforms as they would appear when observed with an oscilloscope.

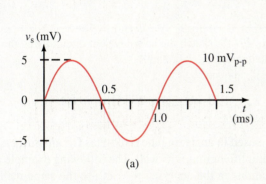

(a)

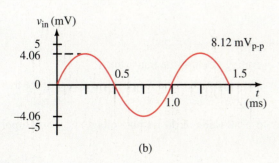

(b)

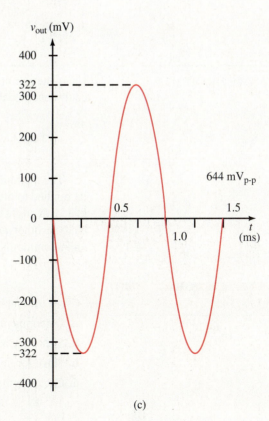

(c)

FIGURE 27–18

Given the amplifier circuit of Figure 27–19:

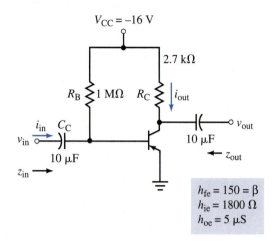

$V_{CC} = -16$ V

2.7 kΩ

R_B 1 MΩ R_C i_{out}

i_{in} C_C

v_{in}

10 μF

z_{in}

v_{out}

10 μF

z_{out}

$h_{fe} = 150 = \beta$
$h_{ie} = 1800$ Ω
$h_{oe} = 5$ μS

FIGURE 27–19

a. Determine the operating point.

b. Sketch the ac equivalent circuit, using the h-parameter model of the transistor.

c. Use the h-parameter model to calculate A_v, z_{in}, z_{out}, and A_i.

(Hint: When calculating the current gain, the load resistor is the component having the output current, i_{out}.)

Answers
a. $V_{CE} = -9.80$ V, $I_C = 2.30$ mA; c. $A_v = -225$, $z_{in} \approx 1.8$ kΩ, $z_{out} \approx 2.7$ kΩ, $A_i = 150$

Using the circuit of Figure 27–19, sketch the ac equivalent circuit using the T-equivalent model and calculate A_v, z_{in}, z_{out}, and A_i.

Answers
$r_e = 11.3$ Ω, $A_v = -237$, $z_{in} \approx 1.71$ κΩ, $z_{out} \approx 2.7$kΩ, $A_i = 150$

The methods that we used in analyzing the fixed-bias CE amplifier are easily applied to other CE amplifier circuits as well. The following example shows that with slight modifications in the analysis, we can use the same method to analyze other amplifier circuits such as the universal-bias CE amplifier, the most-commonly used transistor amplifier circuit.

EXAMPLE 27–5

Given the amplifier circuit of Figure 27–20:

a. Determine the dc operating point of the transistor.

b. Sketch the ac equivalent circuit using the T-equivalent model of the transistor.

c. Solve for A_v, z_{in}, z_{out}, and A_i.

d. Calculate the peak-to-peak value of the output voltage, v_{out} for a supply voltage of $v_S = 10$ mV$_{p-p}$.

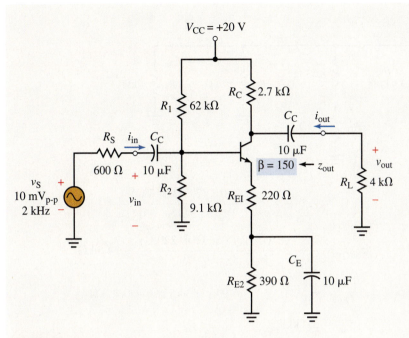

$V_{CC} = +20$ V

R_C 2.7 kΩ

R_1 62 kΩ

C_C i_{out}

R_S i_{in} C_C

10 μF

β = 150   ◄ z_{out}

600 Ω   10 μF

v_S

10 mV$_{p-p}$

2 kHz

v_{in}

R_2

9.1 kΩ

R_{EI} 220 Ω

R_L 4 kΩ

v_{out}

C_E

R_{E2} 390 Ω   10 μF

FIGURE 27–20

Solution

a. If we wish to use the voltage divider rule to determine the operating point, we must first examine the circuit to determine whether the following test is satisfied.

$$R_2 \le 0.1\, \beta_{dc}\, R_E$$

In this example, we see that the approximation is valid, since

$$9.1 \text{ k}\Omega \le 0.1\,(150)\,(0.61 \text{ k}\Omega) = 9.15 \text{ k}\Omega$$

(If the condition, $R_2 \le 0.1\, \beta_{dc}\, R_E$ had not been met then it would have been necessary find the Thévenin equivalent of the input of the circuit using β_{dc}). Now, using the approximation, we have the following:

$$V_B = \left(\frac{9.1\,\Omega}{9.1\,\Omega + 62 \text{ k}\Omega}\right)(20 \text{ V}) = 2.56 \text{ V}$$

$$V_E = 2.56 \text{ V} - 0.7 \text{ V} = 1.86 \text{ V}$$

$$I_E = \left(\frac{1.86 \text{ V}}{0.22\,\Omega + 0.39 \text{ k}\Omega}\right) = 3.05 \text{ mA} \approx I_C$$

$$V_{CE} \approx 20 \text{ V} - 3.05 \text{ mA}(2.7 \text{ k}\Omega + 0.22 \text{ k}\Omega + 0.39 \text{ k}\Omega) = 9.91 \text{ V}$$

b. $r_E = \dfrac{26 \text{ mV}}{3.05 \text{ mA}} = 8.53\,\Omega$

In the ac equivalent circuit, the bias resistors R_1 and R_2 are in a parallel connection from the transistor base to ground (Remember, the dc supply is replaced by a short circuit when doing ac analysis). The resulting ac equivalent circuit is shown in Figure 27–21.

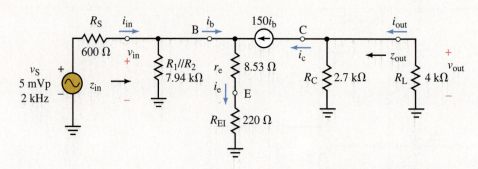

FIGURE 27–21

c.

$$A_v = \frac{v_{out}}{v_{in}} = -\frac{(150)i_b(2.7\ k\Omega\|4\ k\Omega)}{(151)i_b(8.53\ \Omega + 220\ \Omega)} = -7.00$$

$$z_{in(Q)} = \frac{v_b}{i_b} = \frac{(151)i_b(8.53\ \Omega + 220\ \Omega)}{i_b} = 34.5\ k\Omega$$

$$z_{in} = R_1\|R_2\|z_{in(Q)} = 7.94\ k\Omega\|34.5\ k\Omega = 6.45\ k\Omega$$

$$z_{out} \approx 2.7\ k\Omega$$

$$A_i = -\frac{A_v z_{in}}{R_L} = -\frac{(-7.00)(6.45\ k\Omega)}{4\ k\Omega} = 11.3$$

d. Using the voltage divider rule, we have

$$v_{in} = \left(\frac{6.45\ k\Omega}{6.45\ k\Omega + 0.6\ k\Omega}\right)(10\ mV_{p-p}) = 9.15\ mV_{p-p}$$

which results in

$$v_{out} = |(-7.00)(9.15\ mV_{p-p})| = 64.0\ mV_{p-p}$$

PRACTICE PROBLEMS 5

Given the circuit of Figure 27–22:

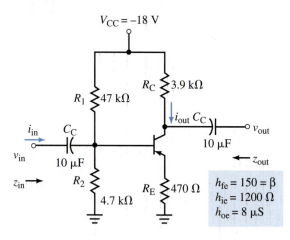

FIGURE 27–22

a. Determine the dc operating point of the transistor.

b. Sketch the ac equivalent circuit using the *h*-parameter model of the transistor.

c. Calculate A_v, z_{in}, z_{out}, and A_i.

Answers

a. $I_C \approx -1.99\ mA$, $V_{CE} \approx -9.29\ V$

b. $A_v = -8.11$, $z_{in} = 4.03\ k\Omega$, $z_{out} \approx 3.9\ k\Omega$, and $A_i = 8.38$.

Recall that the dc load line was used to determine the saturation current (maximum possible collector current) and the cutoff voltage (maximum possible collector-emitter voltage) that could occur in a transistor circuit. We found that the operating point or quiescent point (Q-point) of a transistor always occurred on the dc load line. In this section we will examine a similar line, called the **ac load line,** which is used to determine how the collector current and collector-emitter voltage behave when an ac signal is applied to the circuit.

The ac load line is used for determining the **maximum undistorted output signal** that can be expected for a given transistor amplifier. If we know the maximum undistorted output signal, we will be ready to calculate maximum power that we can expect to obtain from a particular transistor design.

Consider the CE amplifier circuit shown in Figure 27–23.

27.5 The ac Load Line

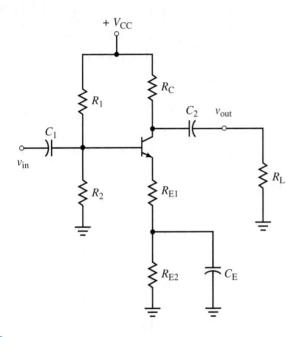

FIGURE 27–23

We know that the dc saturation current occurs when $V_{CE} = 0$ V, and is determined by considering the values of the resistors in the output loop of the transistor, namely

$$I_{C(DC-SAT)} = \frac{V_{CC}}{R_C + R_E}$$

Similarly, the dc cutoff voltage occurs when $I_C = 0$ mA, and in this circuit is simply given as

$$V_{CE(DC-OFF)} = V_{CC}$$

In determining the dc saturation and cutoff values, the capacitors are replaced by open circuits, since the value of capacitive reactance is infinite at dc ($f = 0$ Hz). As a result, we notice that the load resistor, R_L, has no effect on the dc operation of the circuit. This, however, is not the case when an ac signal is applied to the signal. When an ac signal is applied to the transistor, the bypass capacitor, C_E effectively shorts a portion of the emitter resistance, R_{E2} to ground and the coupling capacitor, C_2 introduces the load resistor, R_L into the collector portion of the circuit. Figure 27–24 illustrates a simplified ac version of the output of the transistor amplifier. For simplicity, the bias resistors, R_1 and R_2, are not included in the illustration.

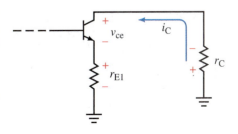

FIGURE 27–24

Now, using the direction and polarity of the ac current and voltage, we have the expressions

$$v_{CE} = -i_C(r_C + r_E) \tag{27-23}$$

and

$$i_C = \frac{v_{CE}}{r_C + r_E} \tag{27-24}$$

where r_C is the "ac resistance" connected to the collector and is determined as $r_C = R_C \| R_L$ and r_E is the "ac resistance" connected to the emitter and is given as $r_E = R_{E1}$. The resultant collector current and collector-emitter voltage in the circuit are determined (by superposition) as the summation of the dc and the ac components. Hence we have

$$i_C = I_{CQ} + i_c \tag{27-25}$$

and

$$v_{CE} = V_{CEQ} + v_{ce} \tag{27-26}$$

These results are more easily understood if we consider that the circuit has two load lines, the dc load line and the ac load line as shown in Figure 27–25.

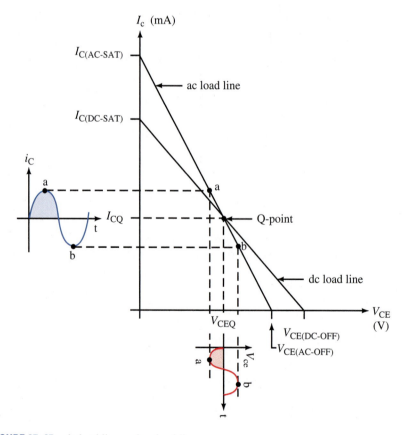

FIGURE 27–25 dc load line and ac load line.

Notice that both the ac load line and dc load line intersect at the Q-point. Any ac signal must follow the ac load line. If there were no ac signal, then the collector current in the circuit must be equal to the dc value, I_{CQ}. The illustration also clearly shows that as the collector current, i_c is increasing, the collector-emitter voltage, v_{ce} is decreasing. Clearly the two values are 180° out-of-phase. This important characteristic was demonstrated when we calculated the voltage gain for each of the CE amplifiers previously in this chapter. This characteristic is also observed as the negative sign in Equations 27–23 and 27–24.

Now we will derive simplified expressions for the ac saturation current and the ac cutoff voltage. Remember that saturation is defined as that point on the dc load line where the current is maximum and where the collector-emitter voltage is zero. At ac saturation, Equation 27–31 becomes

$$0 = V_{CEQ} + v_{ce}$$

which gives

$$v_{ce} = -V_{CEQ}$$

Substituting this result into Equation 27–24 gives

$$i_C = \frac{V_{CEQ}}{r_C + r_E}$$

Finally, inserting this expression into Equation 27–25, we get

$$I_{C(AC\text{-}SAT)} = I_{CQ} + \frac{V_{CEQ}}{r_C + r_E} \qquad (27\text{–}27)$$

Using a similar approach to solve for ac cutoff ($i_C = 0$), we get

$$V_{CE(AC\text{-}OFF)} = V_{CEQ} + I_{CQ}(r_C + r_E) \qquad (27\text{–}28)$$

Equations 27–27 and 27–28 indicate that it is possible to find ac saturation and cutoff simply by determining the combined "ac resistance" in the collector-emitter circuit and then using the values of current and voltage at the operating point (Q-point). Once these values have been determined it is possible to determine the magnitude of the largest undistorted signal that is possible for the given transistor circuit.

If we examine the plot of I_C vs. V_{CE} shown in Figure 27–25, we see that the Q-point is not in the center of the ac load line. Since a sinusoidal ac signal is symmetrical about the time-axis, we see that if the collector current, i_c were increased (due to an increase in base current caused by the input signal), point b will reach ac cutoff before point a reaches ac saturation. Therefore, we conclude that the maximum peak-to-peak undistorted collector current for this transistor is given as $i_{c(max)} = 2\,I_{CQ}$. If the Q-point were above the midpoint of the ac load line, the maximum peak-to-peak undistorted collector current for that condition would be given as $i_{c(max)} = 2\,(I_{C(AC\text{-}SAT)} - I_{CQ})$. From these results we concluded that in order to obtain the maximum undistorted signal at the output of a transistor amplifier, we would need to ensure that the amplifier is biased in the center of its ac load line.

EXAMPLE 27–6

Given the circuit of Figure 27–26:

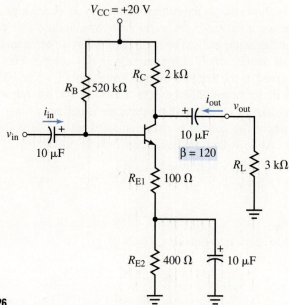

FIGURE 27–26

a. Determine the operating point of the transistor.

b. Sketch the simplified ac equivalent circuit of the output of the amplifier. (Do not model the transistor.)

c. Calculate the values of dc and ac saturation and cutoff.

d. Sketch a graph showing the dc load line and the ac load line.

e. Determine the maximum undistorted collector current and the maximum undistorted collector-emitter voltage. Solve for the maximum undistorted output voltage, v_{out}.

Solution:

a. The operating point of the transistor is found as

$$I_{BQ} = \frac{20\ V - 0.7\ V}{520\ k\Omega + 121(0.5\ k\Omega)} = 33.2\ \mu A$$

$$I_{CQ} = 120(33.2\ \mu A) = 3.99\ mA$$

$$V_{CEQ} = 20\ V - 3.99\ mA(2\ k\Omega + 0.5\ k\Omega) = 10.0\ V$$

b. Figure 27–27 shows the simplified ac circuit of the output of the transistor amplifier.

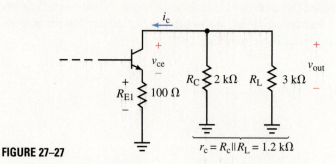

FIGURE 27–27

$$r_c = R_c \| R_L = 1.2\ k\Omega$$

c. The values of dc saturation and cutoff are determined as

$$I_{C(DC\text{-}SAT)} = \frac{20\text{ V}}{2\text{ k}\Omega + 0.1\text{ k}\Omega + 0.4\text{k}\Omega} = 8.0\text{ mA}$$

$$V_{CE(DC\text{-}OFF)} = 20\text{ V}$$

The values of ac saturation and cutoff are determined as

$$I_{C(AC\text{-}SAT)} = 3.99\text{ mA} + \frac{10\text{ V}}{1.2\text{ k}\Omega + 0.1\text{ k}\Omega} = 11.70\text{ mA}$$

$$V_{CE(AC\text{-}OFF)} = 10\text{ V} + 3.99\text{ mA}(1.2\text{ k}\Omega + 0.1\text{ k}\Omega) = 15.2\text{ V}$$

d. The corresponding load lines are shown in Figure 27–28.

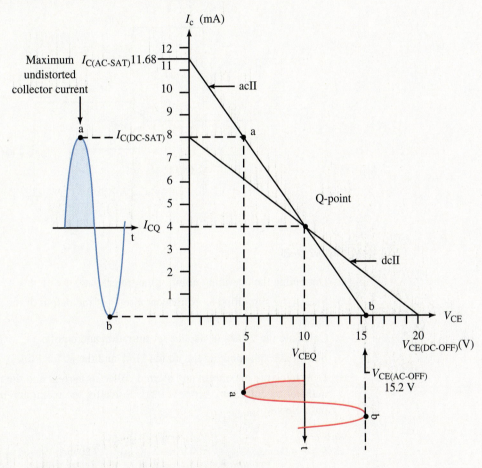

FIGURE 27–28

e. The maximum undistorted collector current is found to be

$$i_{c(max)} = 2(3.99\text{ mA}) = 7.98\text{ mA}_{p\text{-}p}$$

and the maximum undistorted collector-emitter voltage is found as

$$v_{ce(max)} = 2(15.2\text{ V} - 10.0\text{ V}) = 10.4\text{ V}_{p\text{-}p}$$

The maximum undistorted output voltage can be found either by using Ohm's law

$$v_{out(max)} = i_c(R_C \| R_L) = 7.98\text{ mA}(1.2\text{ k}\Omega) = 9.58\text{ V}_{p\text{-}p}$$

or by using the voltage divider rule

$$v_{out(max)} = v_{ce} + i_e r_E = 10.4\text{ V} - 7.98\text{ mA}(0.1\text{ k}\Omega) = 9.58\text{ V}_{p\text{-}p}$$

Note: The amplifier of Figure 27–26 has a voltage gain, of $A_v \approx -12$. Therefore, to ensure that the output is not distorted, the maximum signal that can be applied at the input is $v_{in(max)} = 0.80 \text{ V}_{p-p}$ (or 0.40 V_p). If the level of the input signal goes slightly above this value, the positive-going portion of the output will be clipped, since the signal causes the transistor to go into ac cutoff. If the input signal is increased still further, ac saturation will occur, resulting in distortion of both the positive- and negative-going portions of the output signal.

PRACTICE PROBLEMS 6

Given the circuit of Figure 27–29:

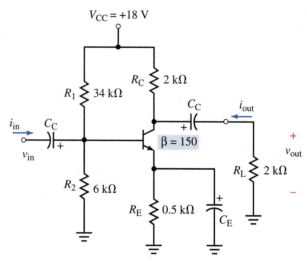

FIGURE 27–29

a. Determine the operating point of the transistor.

b. Sketch the simplified ac equivalent circuit of the output of the amplifier. (Do not model the transistor.)

c. Calculate the values of dc and ac saturation and cutoff.

d. Sketch a graph showing the dc load line and the ac load line.

e. Determine the maximum undistorted collector current and the maximum undistorted collector-emitter voltage. Solve for the maximum undistorted output voltage, v_{out}.

Answers
a. $I_{CQ} = 4.00 \text{ mA}$, $V_{CEQ} = 8.00 \text{ V}$; c. $I_{C(DC-SAT)} = 7.20 \text{ mA}$, $V_{CE(DC-OFF)} = 18.00 \text{ V}$,
$I_{C(AC-SAT)} = 12.0 \text{ mA}$, $V_{CE(DC-OFF)} = 12.00 \text{ V}$; e. $i_{c(max)} = 8.00 \text{ mA}_{p-p}$, $v_{ce(max)} = 8.00 \text{ V}_{p-p}$

Although Equations 27–27 and 27–28 were specifically derived by finding the ac load line for npn transistors, it is possible to use these same expressions for finding the ac load line of pnp transistors as well. In such circuits, the collector current, I_C and the collector-emitter voltage, V_{CE} will each be negative.

27.6 The Common-Collector Amplifier

The common-collector amplifier has several important characteristics that make it particularly useful in certain instances. The distinguishing characteristics of the CC amplifier are:

1. The input impedance tends to be very high, which means that this circuit will not adversely load the previous stage.

2. The output impedance of the CC amplifier is exceedingly low, which means that the following stage will not load the circuit.

3. The output voltage of the common-emitter is in-phase with the input voltage. For this reason this circuit is often referred to as a *voltage follower* circuit.

4. The common-collector circuit has an output voltage that is approximately equal to the input voltage. ($A_v \approx 1$). In actual fact, the output voltage is always slightly less than the input voltage.

5. The current gain of the CC amplifier is large, often just less than h_{fe}. Figure 27–30 shows a typical common-collector amplifier.

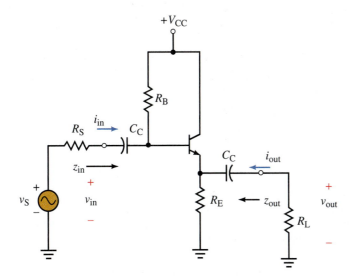

FIGURE 27–30

As in the common-emitter amplifier, the input signal is applied at the base of the circuit. However, the output is taken at the emitter. In the previous circuit, we see that any ac signal at the collector is connected to ground through the dc supply voltage, V_{CC}. As in previous circuits, we can use either the *T*-equivalent model or the *h*-parameter model of the transistor to assist in the ac analysis of the circuit. If the *T*-equivalent model of the transistor is used, the circuit will appear as shown in Figure 27–31.

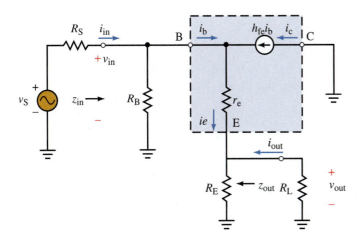

FIGURE 27–31 *T*-equivalent model of a CC amplifier.

For simplicity this circuit is redrawn in Figure 27–32. Notice that the only differences in the redrawn circuit are the locations of the collector and the emitter. All components are still in the same relative positions. Notice also that the output current, i_{out} has a reference direction opposite to the direction of emitter current, i_e. Let's not worry about this apparent conflict right now!

The ac analysis of the resulting ac circuit is done in a manner similar to that used in previous transistor circuits.

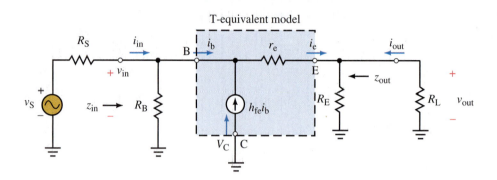

FIGURE 27–32

Voltage Gain, A_v

Applying the same procedure as that used for CE amplifiers, we determine the voltage gain of the circuit is as follows:

$$A_v = \frac{v_{out}}{v_{in}} = \frac{i_e\left(R_E \| R_L\right)}{i_e r_e + i_e\left(R_E \| R_L\right)}$$

or

$$A_v = \frac{\left(R_E \| R_L\right)}{r_e + \left(R_E \| R_L\right)} \tag{27–29}$$

Since r_e tends to be very small in comparison to typical values of R_E and R_L, its effect in the determination of the voltage gain is minimal and so we see that the denominator and the numerator cancel, allowing us to approximate the voltage gain as

$$A_v \approx 1 \tag{27–30}$$

Input Impedance, z_{in}

As in previous circuits, solve for the input impedance of the transistor (between the base terminal and the circuit ground) as

$$z_{in(Q)} = \frac{v_b}{i_b} = \frac{i_e r_e + i_e\left(R_E \| R_L\right)}{i_b} = \frac{\left(h_{fe} + 1\right)i_b\left(r_e + R_E \| R_L\right)}{i_b}$$

Simplifying, we get

$$z_{in(Q)} = \left(h_{fe} + 1\right)\left(r_e + R_E \| R_L\right) \tag{27–31}$$

The input impedance of the entire circuit is found as the parallel combination

$$z_{in} = R_B \| z_{in(Q)} = R_B \|\left[\left(h_{fe} + 1\right)\left(r_e + R_E \| R_L\right)\right] \tag{27–32}$$

which may be further approximated as

$$z_{in} \approx R_B \left\| \left[(h_{fe})(R_E \| R_L) \right] \right. \tag{27–33}$$

Current Gain, A_i

Recall that the current gain for any circuit is found by using the expression

$$A_i = \frac{i_{out}}{i_{in}} = -\frac{A_v z_{in}}{R_L}$$

For the circuit of Figure 27–32, we have

$$A_i = \frac{i_{out}}{i_{in}} = -\frac{\left[\dfrac{R_E R_L}{r_e + R_E \| R_L} \right] \left[R_B \left\| \left\{ (h_{fe} + 1)(r_e + R_E \| R_L) \right\} \right] \right.}{R_L} \tag{27–34}$$

The above expression shows that that the current gain is negative. Remember that this simply means that the actual output current is opposite to the reference direction, a condition that we noticed earlier.

Output Impedance, z_{out}

The output impedance of the common collector circuit is always quite low (typically between 5 Ω and 50 Ω). Solving for the output impedance for the common collector circuit uses a technique that is similar for finding the input impedance. **Unlike the common emitter circuit, the output impedance of the transistor cannot be assumed to be an open circuit.** In a manner similar to that used when finding the Thévenin impedance, we begin by sketching an equivalent circuit as seen from the open output terminals. Recall that when Thévenin's Theorem is used in finding the impedance, it is necessary to "zero" all sources.

Figure 27–33 shows that the voltage source, v_s is replaced by a short circuit, while the current source, $h_{fe}i_b$ is replaced by an open circuit. In order to find the output impedance of the transistor, it is necessary to use Ohm's law in combination with Kirchhoff's voltage law. The output impedance, as seen at the emitter of the transistor is

$$z_{out(Q)} = \frac{v_e}{i_e} = \frac{(R_S \| R_B) i_b + r_e i_e}{i_e} = \frac{(R_S \| R_B) i_b + r_e (h_{fe} + 1) i_b}{(h_{fe} + 1) i_b}$$

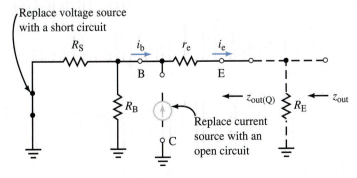

FIGURE 27–33 Thévenin equivalent circuit of a common-collector output. (*T*-equivalent model)

which, when simplified becomes

$$z_{out(Q)} = \frac{v_e}{i_e} = \frac{R_S \| R_B}{h_{fe} + 1} + r_e \qquad (27\text{–}35)$$

The output impedance of the circuit is simply the parallel combination of the emitter resistance, R_E and $z_{out(Q)}$

$$z_{out(Q)} = R_E \left\| \left(\frac{R_S \| R_B}{h_{fe} + 1} + r_e \right) \right. \qquad (27\text{–}36)$$

Since the value of R_E is much larger than the output impedance of the transistor, we may make the approximation

$$z_{out(Q)} \approx \frac{R_S \| R_B}{h_{fe} + 1} + r_e \qquad (27\text{–}37)$$

The following example demonstrates how a typical common collector circuit is analyzed using the T-equivalent model.

EXAMPLE 27–7

Given the circuit of Figure 27–34:

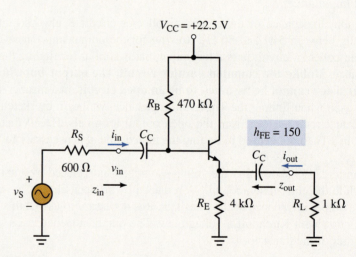

FIGURE 27–34

a. Find the dc operating point.

b. Sketch the ac equivalent circuit, using the T-equivalent model of the transistor.

c. Determine A_v, z_{in}, z_{out}, A_i, and A_p.

Solution

a.

$$I_B = \frac{22.5\,V - 0.7\,V}{470\,k\Omega + (151)(4\,k\Omega)} = 20.3\,\mu A$$

$$I_C = 150(20.3\,\mu A) = 3.04\,mA$$

$$V_{CE} = 22.5\,V - (4\,k\Omega)(3.04\,mA) = 10.3\,V$$

$$r_e \approx \frac{26\,mV}{3.04\,mV} = 8.51\,\Omega$$

b. The ac equivalent circuit is shown in Figure 27–35.

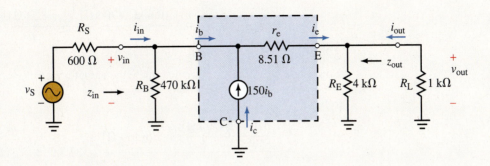

FIGURE 27–35

c. Although we have previously determined the voltage gain for the circuit as $A_v \approx 1$, the following calculation substantiates the result.

$$A_v = \frac{i_e(4 \text{ k}\Omega \| 1 \text{ k}\Omega)}{i_e(8.51 \Omega + 4 \text{ k}\Omega \| 1 \text{ k}\Omega)} = \frac{800 \ \Omega}{808.51 \ \Omega} = 0.989$$

The input impedance of the transistor is

$$z_{in(Q)} = \frac{v_b}{i_b} = \frac{i_e(8.51 \Omega + 4 \text{ k}\Omega \| 1 \text{ k}\Omega)}{i_b} = \frac{(151 i_b)(808.51 \ \Omega)}{i_b} = 122 \text{ k}\Omega$$

resulting in an input impedance for the circuit of

$$z_{in} = \frac{v_b}{i_b} = 470 \text{ k}\Omega \| 122 \text{ k}\Omega = 96.9 \text{ k}\Omega$$

The current gain is easily determined by using Equation 27–11.

$$A_i = -\frac{A_v z_{in}}{R_L} = -\frac{(0.989)(96.9 \text{ k}\Omega)}{1 \text{ k}\Omega} = -95.8$$

The power gain of the circuit is found to be

$$A_p = |A_v A_i| = (0.989)(95.8) = 94.7$$

In order to calculate the output impedance of the circuit, we first zero all sources and redraw the circuit as seen from the output terminals of the transistor. This circuit is shown in Figure 27–36.

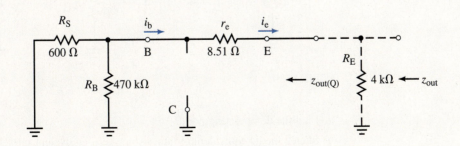

FIGURE 27–36

The output impedance of the transistor is found by using Kirchhoff's voltage law and Ohm's law.

$$z_{out(Q)} = \frac{v_e}{i_e} = \frac{i_b(0.6\ k\Omega\|470\ k\Omega) + i_e(8.51\ \Omega)}{i_e}$$

Now, since $i_b = \dfrac{i_e}{h_{fe} + 1}$, we determine the output impedance of the transistor to be

$$z_{out(Q)} = \frac{\dfrac{i_e(0.6\ k\Omega\|470\ k\Omega)}{151} + i_e(8.51\ k\Omega)}{i_e} = 3.97\ \Omega + 8.51\ \Omega = 12.5\ \Omega$$

Finally, the output impedance of the circuit becomes

$$z_{out} = 4\ k\Omega\|12.5\ \Omega = 12.4\ \Omega$$

The previous calculations clearly demonstrate that the common-collector amplifier has very high input impedance and very low output impedance. As expected, the voltage gain is very close to 1.0, which shows that the output voltage is essentially equal to the input voltage and at the same time is in-phase with the input voltage. The power gain of the ac signal is obtained purely as a result of the current gain.

PRACTICE PROBLEMS 7

Given the circuit of Figure 27–37.

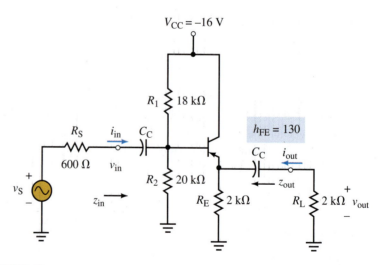

FIGURE 27–37

a. Find the dc operating point.

b. Sketch the ac equivalent circuit, using the *T*-equivalent model of the transistor.

c. Determine A_v, z_{in}, z_{out}, A_i, and A_p.

Answers

a. $I_E = 3.86\ mA \approx I_C$, $V_{CE} = -8.28\ V$; c. $A_v = 0.9933$, $z_{in} = 8.84\ k\Omega$, $z_{out} = 11.0\ \Omega$, $A_i = -4.39$, $A_p = 4.36$

Several techniques using the *h*-parameter model of a transistor are possible to analyze a common-collector amplifier. Although using the common collector model (h_{ic}, h_{rc}, h_{fc}, h_{oc}) of a transistor is possible, this technique is seldom used. Rather, the common emitter model (h_{ie}, h_{re}, h_{fe}, h_{oe}) is generally used, since the *h*-parameter values are readily available and the use of the CE model avoids a needless conversion between the model types. Substituting the CE *h*-parameter model for the transistor of Figure 27–30, we obtain the ac equivalent of Figure 27–38.

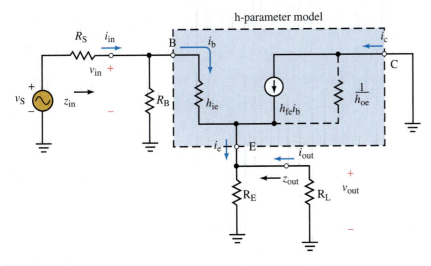

FIGURE 27–38

To further simplify the operation of the circuit, h_{oe} is removed (since its effect is negligible on the operation of the input). The redrawn circuit, showing the output on the right hand side is illustrated in Figure 27–39.

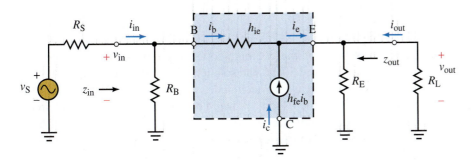

FIGURE 27–39

Voltage Gain, A_v

The voltage gain of the circuit is determined from the following expression:

$$A_v = \frac{v_{out}}{v_{in}} = \frac{i_e\left(R_E \| R_L\right)}{i_b h_{ie} + i_e\left(R_E \| R_L\right)}$$

Now since $i_e = i_b + i_c = (h_{fe} + 1)i_b$, we have

$$A_v = \frac{\left(h_{fe} + 1\right)\left(R_E \| R_L\right)}{h_{ie} + \left(h_{fe} + 1\right)\left(R_E \| R_L\right)} \qquad \textbf{(27–38)}$$

However, $h_{ie} << (h_{fe} + 1)(R_E \| R_L)$. The voltage gain may be simply stated as

$$A_v \approx 1 \tag{27-39}$$

Input Impedance, z_{in}

As always, we first determine the input impedance of the transistor (between the base terminal and the circuit ground).

$$z_{in(Q)} = \frac{v_b}{i_b} = \frac{i_b h_{ie} + i_e (R_E \| R_L)}{i_b} = \frac{i_b h_{ie} + (h_{ie} + 1)i_b (R_E \| R_L)}{i_b}$$

Simplifying, we get

$$z_{in(Q)} = h_{ie} + (h_{fe} + 1)\|(R_E R_L)$$

Now, the input impedance of the circuit is found as the parallel combination

$$z_{in} = R_B \| z_{in(Q)} = R_B \|[h_{ie} + (h_{fe} + 1)(R_E \| R_L)] \tag{27-40}$$

which may be approximated as

$$z_{in} \approx R_B \|[(h_{fe})(R_E \| R_L)] \tag{27-41}$$

Current Gain, A_i

Using Equation 27–11, we obtain the current gain for the circuit of Figure 27–44 as

$$A_i = \frac{i_{out}}{i_{in}} = -\frac{A_v z_{in}}{R_L}$$

$$= \frac{\left[\dfrac{(h_{fe} + 1)(R_E \| R_L)}{h_{ie} + (h_{fe} + 1)(R_E \| R_L)}\right]\left[R_B \|\{h_{ie} + (h_{fe} + 1)(R_E \| R_L)\}\right]}{R_L} \tag{27-42}$$

Output Impedance, z_{out}

The output impedance of the common-collector circuit is always quite low (typically between 5 Ω and 50Ω). The solution of the output impedance for the common-collector circuit uses a technique, which is similar to that used in finding the input impedance. First, we begin by sketching an equivalent circuit as seen from the output terminals. Recall that when Thévenin's theorem is used in finding the impedance, it is necessary to "zero" all sources. Figure 27–40

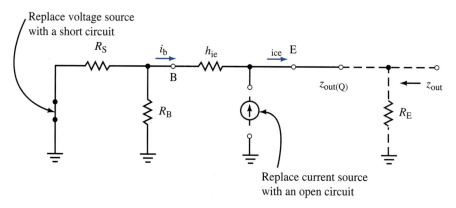

FIGURE 27–40　Thévenin equivalent circuit of a common-collector output.

shows that the voltage source, v_s is replaced by a short circuit, while the current source, $h_{fe}i_b$ is replaced by an open circuit.

In order to find the output impedance of the transistor, it is necessary to use Ohm's law and Kirchhoff's voltage law. The output impedance, as seen at the emitter of the transistor is found as

$$z_{out(Q)} = \frac{\left[\left(R_S \| R_B\right) + h_{ie}\right]i_b}{\left(h_{fe} + 1\right)i_b}$$

which is simplified as

$$z_{out(Q)} = \frac{\left(R_S \| R_B\right) + h_{ie}}{h_{fe} + 1} \qquad \textbf{(27–43)}$$

The output impedance of the circuit is found as the parallel combination of the emitter resistance, R_E and $z_{out(Q)}$, namely

$$z_{out} = R_E \left\| \left[\frac{\left(R_S \| R_B\right) + h_{ie}}{h_{fe} + 1}\right] \qquad \textbf{(27–44)}$$

Given the circuit of Figure 27–41.

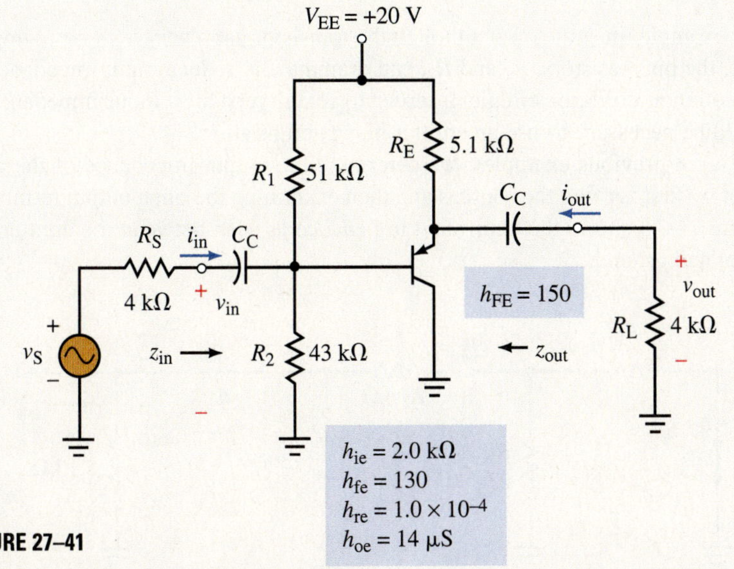

EXAMPLE 27–8

FIGURE 27–41

$h_{ie} = 2.0 \text{ k}\Omega$
$h_{fe} = 130$
$h_{re} = 1.0 \times 10^{-4}$
$h_{oe} = 14 \text{ }\mu\text{S}$

a. Find the dc operating point.

b. Sketch the ac equivalent circuit, using the *h*-parameter model of the transistor.

c. Determine A_v, z_{in}, z_{out}, A_i, and A_p.

Solution

a. The circuit of Figure 27–41 uses voltage divider bias, and so we have

$$V_B = \left(\frac{43 \text{ k}\Omega}{51 \text{ k}\Omega + 43 \text{ k}\Omega}\right) 20 \text{ V} = 9.15 \text{ V}$$

$$V_E = 9.15 \text{ V} + 0.7 \text{ V} = 9.85 \text{ V}$$

$$I_E = \frac{20 \text{ V} - 9.85 \text{ V}}{5.1 \text{ k}\Omega} = 1.99 \text{ mA} \approx I_C$$

$$V_{CE} = -(20 \text{ V} - 9.85 \text{ V}) = -10.15 \text{ V}$$

b. The simplified ac equivalent circuit is shown in Figure 27–42.

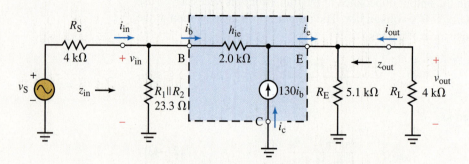

FIGURE 27–42

c. The voltage gain for the circuit is found as

$$A_v = \frac{(131)(5.1\ \text{k}\Omega \| 4\ \text{k}\Omega)}{2.0\ \text{k}\Omega + (131)(5.1\ \text{k}\Omega \| 4\ \text{k}\Omega)} = 0.993$$

The input impedance of the transistor (looking between the base and ground) is

$$z_{\text{in(Q)}} = 2.0\ \text{k}\Omega + (131)(5.1\ \text{k}\Omega \| 4\ \text{k}\Omega) = 296\ \text{k}\Omega$$

The input impedance of the circuit is found as

$$z_{\text{in}} = 51\ \text{k}\Omega \| 43\ \text{k}\Omega \| 296\ \text{k}\Omega = 21.6\ \text{k}\Omega$$

This example illustrates that although the transistor has a very high input impedance, the bias resistors, R_1 and R_2, can dramatically reduce input impedance of the common collector circuit. In order to retain very high input impedance, it would be necessary to use an emitter bias configuration.

As in previous examples, we determine the output impedance of the transistor by first zeroing the sources and then examining the open output terminals. Figure 27–43 shows the equivalent impedance as seen between the emitter terminal and ground.

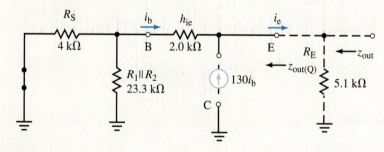

FIGURE 27–43

This results in output impedance for the circuit of

$$z_{\text{out}} = R_E \left\| \frac{\left(R_S \| R_1 \| R_2\right) + h_{\text{ie}}}{h_{\text{fe}} + 1} \right. = 5.1\ \text{k}\Omega \left\| \frac{\left(4\ \text{k}\Omega \| 51\ \text{k}\Omega \| 43\ \text{k}\Omega\right) + 2.0\ \text{k}\Omega}{131} \right. = 34.1\ \Omega$$

Using Equation 27–11, we solve for the current gain as

$$A_i = -\frac{(0.993)(21.6\ \text{k}\Omega)}{4\ \text{k}\Omega} = -5.36$$

Finally, the power gain is found as

$$A_p = |A_v A_i| = (0.993)(5.36) = 5.32$$

Given the circuit of Figure 27–44.

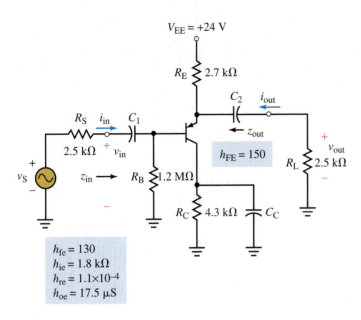

$h_{fe} = 130$
$h_{ie} = 1.8$ kΩ
$h_{re} = 1.1\times10^{-4}$
$h_{oe} = 17.5$ μS

FIGURE 27–44

a. Find the dc operating point.

b. Sketch the ac equivalent circuit, using the h-parameter model of the transistor.

c. Determine A_v, z_{in}, z_{out}, A_i, and A_p.

Answers

a. $I_B = 14.5$ μA, $I_C = 2.17$ mA, $V_{CE} = -8.78$ V

c. $A_v = 0.990$, $z_{in} = 150.$ kΩ, $z_{out} = 32.4$ Ω, $A_i = -59.5$, $A_p = 58.9$

When we examined the operation of field effect transistors you saw that when a voltage is applied to the gate of a FET, it affected the width of the channel between the drain and the source. It is for this reason that the FET is referred to as a voltage-controlled amplifier unlike the BJT, which is a current-controlled device. The small-signal ac model of JFETs and MOSFETs is precisely the same both types of transistors and is illustrated in Figure 27–45.

Notice that the impedance between the gate and the source is shown as an infinite resistance. Remember that in a JFET, the gate-source is always reverse-biased and so there can be no gate current. In both D-MOSFETs and E-MOSFETs, the gate is isolated from the channel by an insulating layer and, once again, there can be no gate current. Consequently, this model clearly shows that the ac drain current must be equal to the source current, namely

$$i_s = i_d \qquad (27\text{--}45)$$

The output impedance of the FET is r_d, which tends to be a fairly large value (around 20 kΩ to about 50 kΩ). As expected, the current source in the model shown in Figure 27–45 is dependent on the magnitude of the voltage applied between the gate and the source. Whereas in the h-parameter model of the BJT, the current source was determined on h_{fe} (the forward current transfer ratio), here we see that the magnitude of the current source is determined by using a conductance characteristic, g_m. This conductance is obtained from the

27.7 The FET Small-Signal Model

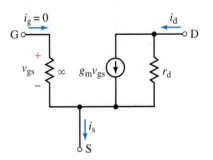

FIGURE 27–45   ac model of a field effect transistor.

transfer curve for the FET and for this reason, g_m is called the *transconductance* of the FET. Just as in the BJT, this term provides the relationship between the input and the output of the FET. The transconductance of a FET is not constant, but rather is dependent on the operating point of the FET. Consider an n-channel FET having the transfer curve shown in Figure 27–46.

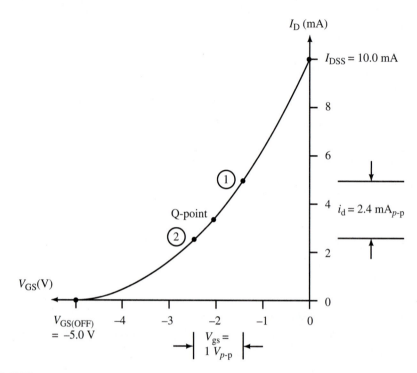

FIGURE 27–46

Let's assume that the operating point of the FET is at $V_{GSQ} = -2.0$ V and $I_{DQ} = 3.6$ mA and that an ac signal is applied between the gate and the source so that $v_{gs} = 1$ V_{p-p}. As the gate-source voltage alternates between -2.5 V and -1.5 V, the JFET will operate between points *1* and *2* on the transfer curve. The drain current will vary between 2.5 mA and 4.9 mA (2.4 mA_{p-p}). For small signals, the difference between points *1* and *2* will be relatively small and will be approximately linear. The slope of the line between the points *1* and *2* will be very close to the value of the transconductance, in this case

$$g_m \approx \frac{\Delta I_d}{\Delta V_{gs}} = \frac{4.9 \text{ mA} - 2.5 \text{ mA}}{-1.5 \text{ V} - (-2.5 \text{ V})} = \frac{2.4 \text{ mA}}{1.0 \text{ V}} = 2.5 \text{ mS}$$

The actual value of the transconductance at the Q-point is simply the slope of the tangent at the Q-point. This value can be determined graphically by sketching a line that is tangent at the Q-point and then calculating the slope of the tangent using

$$g_m \approx \frac{\Delta I_D}{\Delta V_{GS}} \tag{27–46}$$

Alternatively, we can use differential calculus to solve for the transconductance by solving for the derivative:

$$g_m = \frac{dI_D}{dV_{GS}} = \frac{d}{dV_{GS}}\left[I_{DSS}\left(1 - \frac{V_{GSQ}}{V_{GS(OFF)}} \right)^2 \right]$$

which results in

$$g_m = \left(-\frac{2I_{DSS}}{V_{GS(OFF)}} \right)\left(1 - \frac{V_{GSQ}}{V_{GS(OFF)}} \right) \qquad \text{(27–47)}$$

In the expression of Equation 27–47, the first term will always be positive and so this equation may be further simplified as

$$g_m = \left| \frac{2I_{DSS}}{V_{GS(OFF)}} \right|\left(1 - \frac{V_{GSQ}}{V_{GS(OFF)}} \right) \qquad \text{(27–48)}$$

In the case of a JFET, the term contained within the absolute value symbol represents the maximum slope (hence maximum transconductance that is possible for the given transistor). This value is simply referred to as

$$g_{mo} = \left| \frac{2I_{DSS}}{V_{GS(OFF)}} \right| \qquad \text{(27–49)}$$

and is a constant for any given transistor. Therefore, we may simply express Equation 27–53 as

$$g_m = g_{mo}\left(1 - \frac{V_{GSQ}}{V_{GS(OFF)}} \right) \qquad \text{(27–50)}$$

Given the transfer curve for an n-channel JFET illustrated in Figure 27–46.

a. Determine the transconductance of the JFET for an operating point having $V_{GSQ} = -2.0$ V,

b. Solve for g_m if $V_{GSQ} = -3.0$ V,

c. Solve for g_m if $V_{GSQ} = -1.0$ V.

EXAMPLE 27–9

Solution

a.

$$g_{mo} = \left| \frac{2I_{DSS}}{V_{GS(OFF)}} \right| = \left| \frac{2(10\text{ mA})}{-5.0\text{ V}} \right| = 4.0\text{ mS}$$

$$g_m = (4.0\text{ mS})\left(1 - \frac{-2.0\text{ V}}{-2.0\text{ V}} \right) = 2.4\text{ mS}$$

Note that this value is very close to the approximation that we determined previously.

b. $g_m = (4.0\text{ mS})\left(1 - \dfrac{-3.0\text{ V}}{-5.0\text{ V}} \right) = 1.6\text{ mS}$

c. $g_m = (4.0\text{ mS})\left(1 - \dfrac{-1.0\text{ V}}{-5.0\text{ V}} \right) = 3.2\text{ mS}$

The previous example clearly demonstrates that the transconductance, g_m of a FET is dependent on the operating point. In order for a JFET to have a larger transconductance, it is necessary for the bias point of the transistor to be closer towards the value of I_{DSS}. For a D-MOSFET, the value of g_{mo} and g_m is found the same way as for a JFET (using I_{DSS} and $V_{GS(OFF)}$). The transconductance of an E-MOSFET, however, is determined slightly differently. Recall from the previous chapter that the drain current of an E-MOSFET is found in terms of V_{GS} as

$$I_D = k(V_{GS} - V_{GS(th)})^2$$

where

$$k = \frac{I_{D(ON)}}{\left(V_{GS(ON)} - V_{GS(th)}\right)}$$

Using differential calculus to solve for the transconductance of an E-MOSFET, we have:

$$g_m = \frac{dI_D}{dV_{GS}} = \frac{d}{dV_{GS}}\left[k\left(V_{GS} - V_{GS(th)}\right)^2\right] \qquad \text{(27–51)}$$

which becomes

$$g_m = 2k(V_{GS} - V_{GS(th)}) \qquad \text{(27–52)}$$

EXAMPLE 27–10

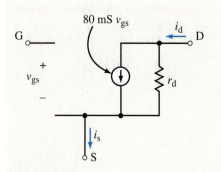

FIGURE 27–47

An E-MOSFET has the following characteristics:

$$V_{GS(th)} = 2.00 \text{ V}$$
$$I_{D(ON)} = 200 \text{ mA}$$
$$V_{GS(ON)} = 4 \text{ V}$$

If the operating point of the transistor occurs at $V_{GSQ} = 2.8$ V and $I_{DQ} = 32$ mA, find the value of the transconductance at the operating point. Sketch the small-signal ac equivalent circuit of the E-MOSFET.

$$k = \frac{I_{D(ON)}}{\left(V_{GS(ON)} - V_{GS(th)}\right)^2} = \frac{200 \text{ mA}}{(4 \text{ V} - 2 \text{ V})^2} = 50 \text{ mA / V}^2$$

$$g_m = 2k\left(V_{GS} - V_{GS(th)}\right) = 2(50 \text{ mA/V}^2)(2.8 \text{ V} - 2.0 \text{ V}) = 80 \text{ mS}$$

The resulting ac equivalent circuit of the E-MOSFET is shown in Figure 27–47.

27.8 The Common-Source Amplifier

The analysis of FET amplifiers is remarkably similar to the methods used when analyzing BJT amplifiers using the h-parameter model. As always, it is necessary to first determine whether the amplifier is correctly biased. Unless this is the case, the circuit cannot operate as an amplifier. In the case of all FET amplifiers, we need the dc operating point to calculate the transconductance, g_m of the transistor. The following examples show how the common source amplifier behaves very much like the common-emitter amplifier.

EXAMPLE 27–11

Given the JFET amplifier circuit of Figure 27–48:

a. Find I_{DSQ}, V_{GSQ}, and V_{DSQ} at the operating point.

b. Solve for g_{mo} and g_m of the amplifier.

c. Sketch the small-signal ac equivalent circuit.

d. Determine A_v, z_{in}, and z_{out} of the amplifier.

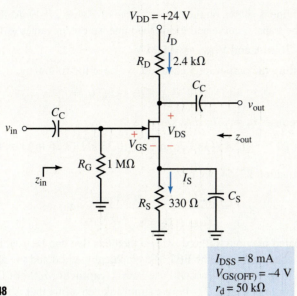

FIGURE 27–48

$I_{DSS} = 8$ mA
$V_{GS(OFF)} = -4$ V
$r_d = 50$ kΩ

Solution

a. In order to find the operating point, we first must determine the dc load line for the circuit. As before, we find two points on the dc load line by selecting suitable points.

Let $I_D = 0$: Since $I_G = 0$, $V_{GS} = 0$.

Let $I_D = 8.0$ mA: Now writing the loop equation for the input loop of the circuit, we have

$$V_{GS} = 0(1 \text{ M}\Omega) - (8.0 \text{ mA})(0.33 \text{ k}\Omega) = -2.64 \text{ V}$$

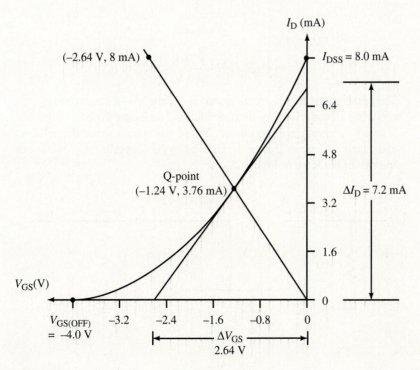

FIGURE 27–49

The load line is shown on the transfer curve of Figure 27–49. At the intersection of the transfer curve and the dc load line we obtain values as follows:

$I_{DQ} \approx 3.76$ mA and $V_{GSQ} \approx -1.24$ V

Now writing the output loop equation, we have the following:

$$(2.4 \text{ k}\Omega)I_D + V_{DS} + (0.33 \text{ k}\Omega)I_S = 24 \text{ V}$$

and since $I_S = I_D = 3.76$ mA, we have

$$V_{DSQ} = 24 \text{ V} - (2.4 \text{ k}\Omega)(3.76 \text{ mA}) - (0.33 \text{ k}\Omega)(3.76 \text{ mA}) = 13.74 \text{ V}$$

b. Using Equation 27–54, we have

$$g_{mo} = \left| \frac{2I_{DSS}}{V_{GS(OFF)}} \right| = \left| \frac{2(8.0 \text{ mA})}{-4.0 \text{ V}} \right| = 4.0 \text{ mS}$$

As indicated previously, there are two methods that can be used to solve for the transconductance of the FET: the graphical method and the algebraic method. In this example we will use both. **Graphical Method:** On the transfer curve of Figure 27–49, we simply sketch a line that is tangent at the Q-point. Recall that the value of the transconductance is simply expressed as the slope of the tangent line, namely

$$g_m = \left. \frac{\Delta I_d}{\Delta V_{gs}} \right|_{\text{at the point}}$$

and so we have

$$g_m = \frac{7.2 \text{ mA} - 0}{0 - (-2.64 \text{ V})} \approx 2.73 \text{ mS}$$

Algebraic Method: Using Equation 27–50, we have

$$g_m = g_{mo}\left(1 - \frac{V_{GSQ}}{V_{GS(OFF)}}\right)$$

$$= 4 \text{ mS}\left(1 - \frac{-1.24 \text{ V}}{-4.0 \text{ V}}\right) = 2.76 \text{ mS}$$

c. The small-signal ac equivalent circuit of Figure 27–48 is sketched by replacing all capacitors by short circuits and replacing the dc voltage source by a short circuit to ground. Notice that the source of the JFET is connected directly to ground through the bypass capacitor, C_S. The resulting circuit is shown in Figure 27–50.

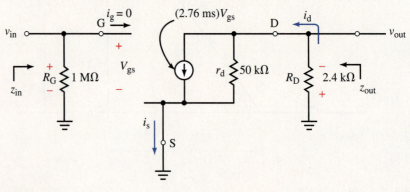

FIGURE 27–50

d. As in previous examples, we determine the voltage gain of the amplifier as follows:

$$A_v = \frac{v_{out}}{v_{in}} = \frac{-i_d R_D}{v_{gs}}$$

Now since r_d (the output impedance characteristic of the FET) is much larger than R_D, we neglect its effect on the gain of the circuit. (This is because there is very little current through the larger resistance.) Consequently, we let $i_d \approx g_m v_{gs}$. The voltage gain is now rewritten as

$$A_v \approx \frac{-g_m v_{gs} R_D}{v_{gs}}$$

which when simplified, becomes

$$A_v \approx - g_m R_D \qquad \text{(27–53)}$$

The voltage gain of our circuit is now easily calculated as

$$A_v \approx - (2.76 \text{ mS})(2.4 \text{ k}\Omega) = -6.62$$

Here we see that a FET amplifier having a bypassed source resistor has a relatively low voltage gain. A similar BJT amplifier has a much larger voltage gain (typically over 100), if the circuit has a bypassed emitter resistor. This demonstrates one of the major characteristic differences between BJT amplifiers and FET amplifiers.

The input impedance of all FET amplifiers is taken to be infinity, since the value will always be extremely large (10 MΩ to several thousand MΩ). Therefore, the input impedance of the circuit is determined simply by the value(s) of the bias resistors connected to the gate of the circuit. For the circuit of Figure 27–48, we have

$$z_{in} = R_G \qquad \text{(27–54)}$$

and so

$$z_{in} = 1 \text{ M}\Omega$$

Upon examining the ac equivalent circuit of Figure 27–50, we see that the output impedance characteristic, r_d is very large in comparison to the drain resistor, R_D, and so we conclude that for this circuit, the output impedance may simply be taken as

$$z_{out} \approx R_D \qquad \text{(27–55)}$$

giving

$$z_{out} \approx 2.4 \text{ k}\Omega$$

PRACTICE PROBLEMS 9

If the bypass capacitor, C_S is removed from the circuit of Figure 27–48:
a. Sketch the resulting small-signal ac equivalent circuit.

b. Determine A_v, z_{in}, and z_{out} of the amplifier. (Hint: The input voltage will now include the voltage across $R_S = 330 \text{ }\Omega$.)

Answers
b. $A_v = -3.47$, $z_{in} = 1 \text{ M}\Omega$, $z_{out} = 2.4 \text{ k}\Omega$

The analysis of D-MOSFETs is almost exactly the same as for JFETs, with the exception that D-MOSFETs can operate in the "enhancement" region. This means that it is possible to have the gate-to-source forward-biased. Remember that because the gate is isolated from the channel, there can be no current from the gate toward the source, once again resulting in an open circuit between the gate and the source in the ac small-signal model of the D-MOSFET.

E-MOSFET amplifiers follow an analysis process that is somewhat different than that used in JFET and D-MOSFET amplifiers. This is especially true if the circuit includes a source feedback resistor. The following example illustrates the point.

EXAMPLE 27–12

Given the E-MOSFET amplifier circuit of Figure 27–51:

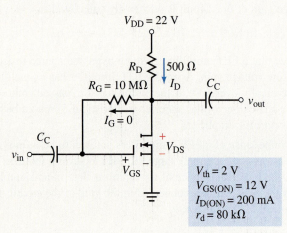

FIGURE 27–51

a. Find I_{DSQ}, V_{GSQ}, and V_{DSQ} at the operating point.

b. Solve for g_m of the amplifier.

c. Sketch the small-signal ac equivalent circuit.

d. Determine A_v, z_{in}, and z_{out} of the amplifier.

Solution

a. Solving for the constant, k of the E-MOSFET, gives

$$k = \frac{I_{D(ON)}}{\left(V_{GS(ON)} - V_{GS(th)}\right)^2} = \frac{200 \text{ mA}}{\left(12 \text{ V} - 2 \text{ V}\right)^2} = 2.0 \text{ mA/V}^2$$

For this circuit we see that $V_{GS} = V_{DS}$ for all values, since there can be no gate current. We select suitable points and sketch the dc load line as shown in Figure 27–52.

Let $V_{GS} = 10V$: $$I_D = \frac{22 \text{ V} - 10 \text{ V}}{500 \text{ }\Omega} = 24 \text{ mA}$$

Let $V_{GS} = 2V$: $$I_D = \frac{22 \text{ V} - 2 \text{ V}}{500 \text{ }\Omega} = 40 \text{ mA}$$

From the graph, we see that at the Q-point, $I_D = 32$ mA and $V_{GSQ} = 6$ V $= V_{DSQ}$.

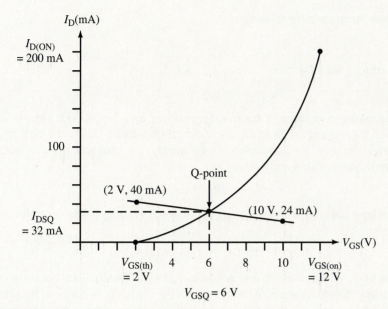

FIGURE 27–52

b. The transconductance at the operating point of the E-MOSFET is determined by applying Equation 27–57.

$$g_m = 2(2.0 \text{ mA} / \text{V}^2)(6.0 \text{ V} - 2.0 \text{ V}) = 16 \text{ mS}$$

c. Now the ac equivalent circuit of the E-MOSFET amplifier can be sketched as shown in Figure 27–53.

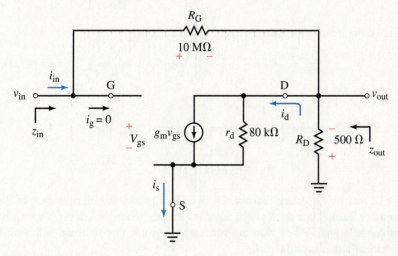

FIGURE 27–53

The ac equivalent circuit of Figure 27–53 shows that the resistance, R_G provides a feedback path from the output to the input. While the resistor has very little effect on the voltage gain and the output impedance of the circuit, we will find that it does have a significant effect on the input impedance.

d. Due to the very large resistance and the relatively large output impedance of the E-MOSFET, there will be very little current through these values. Consequently, the voltage gain is determined to be

$$A_v \approx \frac{-g_m v_{gs} R_D}{v_{gs}}$$

which when simplified, becomes

$$A_v \approx -g_m R_D$$

Therefore, we have

$$A_v \approx -(16 \text{ mS})(0.5 \text{ k}\Omega) = -8.0$$

The output impedance of the transistor will be $z_{\text{out}(Q)} = 80 \text{ k}\Omega$. The 10-M$\Omega$ resistor has a negligible effect on the output impedance, due to the very small current through the component. Consequently, we conclude that the output impedance may be simply solved as

$$z_{\text{out}} = r_d \| R_D \approx R_D = 500 \ \Omega$$

The input impedance however, must be calculated by using Ohm's law as follows:

$$z_{\text{in}} = \frac{v_{\text{in}}}{i_{\text{in}}}$$

The input voltage is easily determined as $v_{\text{in}} = v_{\text{gs}}$. The difficulty occurs in calculating the input current. Although $i_g = 0$, the current i_{in} is not zero. By examining the ac equivalent circuit, we see that the input current is the same as the current through the feedback resistor, R_G. Therefore, in order to solve for the current i_{in}, we simply need to find the voltage across R_G. If we go in a loop from the gate to source (ground) to drain, we obtain the voltage as

$$v_{RG} = v_{\text{gs}} + i_d R_D = v_{\text{gs}} + g_m v_{\text{gs}} R_D = v_{\text{gs}}(1 + g_m R_D)$$

Now we have

$$i_{\text{in}} = \frac{v_{\text{gs}}(1 + g_m R_D)}{R_G}$$

Finally, the input impedance is easily determined as

$$z_{\text{in}} = \frac{v_{\text{in}}}{i_{\text{in}}} = \frac{v_{\text{gs}}}{\dfrac{v_{\text{gs}}(1 + g_m R_D)}{R_G}} = \frac{R_G}{1 + g_m R_D} \tag{27-56}$$

which for this example results in input impedance of

$$z_{\text{in}} = \frac{10 \ \Omega}{1 + (16 \text{ mS})(0.5 \text{ k}\Omega)} = 1.11 \text{ M}\Omega$$

Notice that the feedback resistor, R_G, which improves the dc operation of the amplifier, results in a decrease of input impedance when the circuit is operating with an ac signal. When working with any electronic circuit, there will always be tradeoffs that must be made.

In the circuit of Figure 27–51, the FET had a very large value of output impedance, r_d in comparison to the drain resistance, R_D. If the value of r_d in the transistor circuit is not much (≥ 10 times) larger than R_D, we would need to use the following expressions to solve for voltage gain and input impedance.

$$A_v \approx -g_m\left(r_d \| R_D\right) \tag{27-57}$$

$$z_{\text{in}} = \frac{v_{\text{in}}}{i_{\text{in}}} = \frac{v_{gs}}{\dfrac{v_{gs}\left[1 + g_m\left(r_d \| R_D\right)\right]}{R_G}} = \frac{R_G}{1 + g_m\left(r_d \| R_D\right)} \tag{27-58}$$

If $R_D = 100\ \Omega$ in the circuit of Figure 27–51. (All other values remain unchanged.):

a. Find I_{DSQ}, V_{GSQ}, and V_{DSQ} at the operating point.

b. Solve for g_m of the amplifier.

c. Sketch the small-signal ac equivalent circuit.

d. Determine A_v, z_{in}, and z_{out} of the amplifier.

PRACTICE PROBLEMS 10

Answers

a. $I_{DSQ} = 122$ mA, $V_{GSQ} = V_{DSQ} = 9.81$ V; b. $g_m = 31.2$ mS; d. $A_v = -3.12$, $z_{in} = 2.43$ MΩ, $z_{out} = 100\ \Omega$

The common-drain amplifier is very similar to the common collector BJT amplifier that we examined previously. The characteristics of the common drain amplifier are:

Voltage gain is always less than unity ($A_v < 1$).

The output voltage is in-phase with the input voltage.

The input impedance is typically very high.

The output impedance is typically low.

The main application of common-drain FET amplifiers is as a buffer between a circuit having high output impedance and one having low input impedance. The following example illustrates the similarity between the common drain FET amplifier and the common collector BJT amplifier.

27.9 The Common-Drain (Source Follower) Amplifier

Given the JFET amplifier circuit of Figure 27–54:

EXAMPLE 27–13

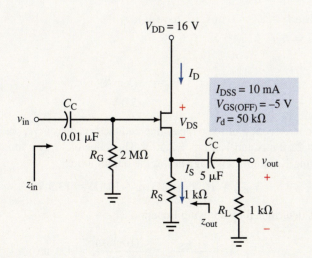

FIGURE 27–54

a. Find I_{DSQ}, V_{GSQ}, and V_{DSQ} at the operating point.

b. Solve for g_{mo} and g_m of the amplifier.

c. Sketch the small-signal ac equivalent circuit.

d. Determine A_v, z_{in}, and z_{out} of the amplifier.

Solution

a. As in previous problems, we determine the operating point by finding the point at which the dc load line intersects the JFET transfer curve.

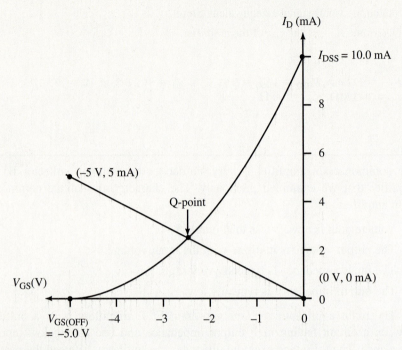

FIGURE 27-55

To find the dc load line, we arbitrarily select two points on the graph shown in Figure 27–55.

Let $I_S = I_D = 0$ mA:

Because $I_S = I_D = 0$ mA, we have $V_{GS} = 0$

Let $I_S = I_D = 5$ mA:

$$V_{GS} = -I_D R_S$$
$$= -(5 \text{ mA})(1 \text{ k}\Omega)$$
$$= -5.0 \text{ V}$$

At the intersection of the dc load line and the transfer curve, we have $I_{DQ} = 2.50$ mA and $V_{GSQ} = -2.50$ V. The value of V_{GSQ} is found by applying Kirchhoff's voltage law at the output loop.

$$V_{DSQ} = 16 \text{ V} - (2.5 \text{ mA})(1 \text{ k}\Omega) = 13.5 \text{ V}$$

b. g_{mo} is determined by applying Equation 27–54 as

$$V_{mo} = \left| \frac{2I_{DSS}}{V_{GS(OFF)}} \right| = \left| \frac{2(10 \text{ mA})}{-5 \text{ V}} \right| = 4.0 \text{ mS}$$

and the transconductance at the operating point is found from Equation 27–50 as

$$g_m = 4.0 \text{ mS} \left(1 - \frac{2.5 \text{ V}}{5.0 \text{ V}} \right) = 2.0 \text{ mS}$$

c. The small-signal ac equivalent circuit is shown in Figure 27–56.

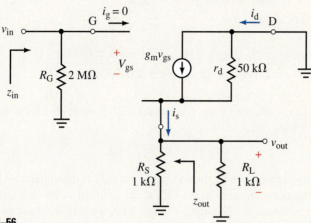

FIGURE 27–56

As was the case with the common collector circuit, we redraw the above equivalent circuit to make it easier to analyze. The redrawn circuit is shown in Figure 27–57.

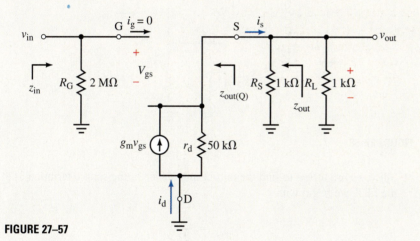

FIGURE 27–57

The direction of the current source in the drain circuit points away from the drain, just as it did in the original ac equivalent circuit. Notice also that $i_s = i_d = g_m v_{gs}$. Using this information, we write the expression for voltage gain as:

$$A_v = \frac{v_{out}}{v_{in}} = \frac{i_s \left(R_S \| R_L \right)}{v_{gs} + i_s \left(R_S \| R_L \right)}$$

$$= \frac{g_m v_{gs} \left(R_S \| R_L \right)}{v_{gs} + g_m v_{gs} \left(R_S \| R_L \right)}$$

which becomes

$$A_v = \frac{g_m \left(R_S \| R_L \right)}{1 + g_m \left(R_S \| R_L \right)} \qquad \text{(27–59)}$$

From Equation 27–59, we see that the voltage gain for a source follower amplifier must always be less than unity and that the output voltage will always be in-phase with the input voltage.

For our circuit, we have

$$A_v = \frac{g_m\left(R_S \| R_L\right)}{1 + g_m\left(R_S \| R_L\right)} = \frac{2.0 \text{ mS}\left(1 \text{ k}\Omega \| 1 \text{ k}\Omega\right)}{1 + 2.0 \text{ mS}\left(1 \text{ k}\Omega \| 1 \text{ k}\Omega\right)} = 0.5$$

The input impedance (at the gate) of the FET is an open circuit, and so we conclude that

$$z_{in} = R_G = 2 \text{ M}\Omega$$

The output impedance (at the source) of the FET is determined by first finding the output impedance of the FET alone and then combining this with the source resistance, R_S. To simplify things a bit, we zero the input voltage (by replacing it with a short circuit to ground) as shown in Figure 27–58.

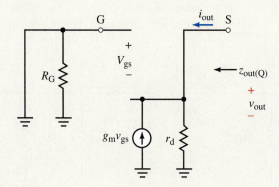

FIGURE 27–58

Since we are trying to find the output impedance at the source terminal of the FET, we begin with

$$z_{out(Q)} = \frac{v_{out}}{i_{out}}$$

But we see that

$$v_{out} = -v_{gs}$$

and

$$v_{out} = -v_{gs} = (g_m v_{gs} + i_{out})r_d$$

This last expression may be rewritten to solve for i_{out} as

$$i_{out} = \frac{-v_{gs}}{r_d} - g_m v_{gs}$$

The expression for output impedance may now be rewritten as

$$z_{out(Q)} = \frac{-v_{gs}}{i_{out}} = \frac{-v_{gs}}{\dfrac{-v_{gs}}{r_d} - g_m v_{gs}} = \frac{1}{\dfrac{1}{r_d} + g_m}$$

which is the same as

$$z_{out(Q)} = r_d \left\| \frac{1}{g_m} \right. \tag{27-60}$$

If r_d is large in comparison to $1/g_m$ then we approximate the output impedance of a common source FET amplifier as

$$z_{out(Q)} \approx \frac{1}{g_m} \tag{27-61}$$

And finally, the output impedance of the common source FET (without approximation) is taken as

$$z_{out} = z_{out(Q)} \| R_S = r_d \left\| \frac{1}{g_m} \right\| R_S \tag{27-62}$$

For the circuit of Figure 27–54, we have

$$z_{out} = 50 \text{ k}\Omega \left\| \frac{1}{2.0 \text{ mS}} \right\| 1 \text{ k}\Omega = 331 \text{ }\Omega$$

Given the circuit of Figure 27–59:

 **PRACTICE PROBLEMS 11**

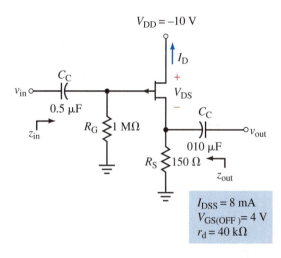

FIGURE 27–59

a. Find I_{DSQ}, V_{GSQ}, and V_{DSQ} at the operating point.
b. Solve for g_{mo} and g_m of the amplifier.
c. Sketch the small-signal ac equivalent circuit.
d. Determine A_v, z_{in}, and z_{out} of the amplifier.

Answers
a. $I_{DSQ} = 5.19$ mA, $V_{GSQ} = 0.79$ V, $V_{DSQ} = -9.22$ V; b. $g_{mo} = 4.0$ mS , $g_m = 3.22$ mS;
  d. $A_v = 0.76$, z_{in} 1 MΩ, and $z_{out} = 143$ Ω

27.10 Troubleshooting a Transistor Amplifier Circuit

NOTES . . .

It is extremely important to ensure that an electrolytic capacitor is correctly placed into a circuit. An incorrectly placed capacitor will tend to generate electrical noise, leading to poor circuit operation. It is also possible for an electrolytic capacitor to explode if the incorrect polarity is applied.

The most common problem encountered when a novice technologist builds a transistor amplifier circuit is the placement of electrolytic capacitors. If the capacitor is installed with incorrect polarity, it generally results in a "noisy" output signal. In such instances, the capacitor behaves as an antenna and picks up unwanted signals, usually 60 Hz. The signal may appear on an oscilloscope as shown in Figure 27–60.

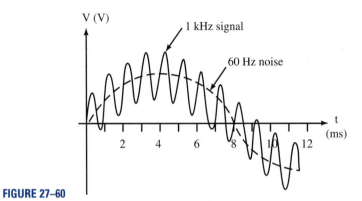

FIGURE 27–60

In the above illustration, we see that the desired 1-kHz component is superimposed on a fairly large 60 Hz component. Occasionally, the desired signal is entirely lost in the electrical "noise". The operation of the circuit is easily corrected. First, determine the proper polarity of the dc voltage across each capacitor and then place each electrolytic capacitor into the circuit accordingly. If an electrolytic capacitor has been inadvertently placed into a circuit with the wrong polarity, it is possible that the electrolyte in the capacitor may have been damaged. Therefore, if the problem persists, the capacitor will need to be replaced by a new one.

Another problem that may occur due to a faulty or incorrectly placed capacitor is that the measured voltage gain of a circuit is much different than the theoretical gain. A faulty capacitor will usually become an open circuit, although it is possible for it to develop an internal short. Consider the circuit shown in Figure 27–61.

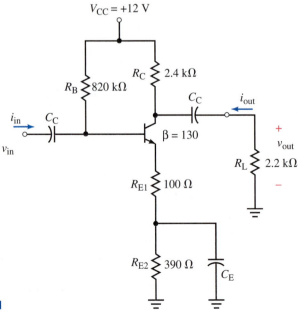

FIGURE 27–61

The theoretical gain of the above transistor (neglecting r_e) should be

$$A_v \approx - \frac{130 i_b \left(2.4 \text{ k}\Omega \| 2.2 \text{ k}\Omega\right)}{131 i_b \left(0.1 \text{ k}\Omega\right)} = -11$$

If the emitter bypass capacitor, C_E becomes an open circuit due to an internal fault, the voltage gain will decrease due to the added emitter resistance, R_{E2}. In such a case, we would measure the voltage gain to be

$$A_v \approx - \frac{130 i_b \left(2.4 \text{ k}\Omega \| 2.2 \text{ k}\Omega\right)}{131 i_b \left(0.1 \text{ k}\Omega + 0.39 \text{ k}\Omega\right)} = -2.2$$

In the unlikely event that the emitter bypass capacitor becomes a short circuit, the dc operating point can be significantly affected. In this instance the base current would now be

$$I_B = \frac{12 \text{ V} - 0.7 \text{ V}}{820 \text{ k}\Omega + 131(0.1 \text{ k}\Omega)} = 13.6 \, \mu\text{A}$$

rather than the actual value of

$$I_B = \frac{12 \text{ V} - 0.7 \text{ V}}{820 \text{ k}\Omega + 131(0.1 \text{ k}\Omega + 0.39 \text{ k}\Omega)} = 12.8 \, \mu\text{A}$$

As we have observed in previous chapters, it is a good practice to develop a strategy in troubleshooting a faulty circuit. In the case where a transistor amplifier circuit is not operating as expected, the following steps may help to isolate the problem:

1. Remove all ac signal sources from the circuit. If the dc bias point of the transistor is not in the active region, then it will certainly not work correctly with the application of an ac signal.

2. Calculate the theoretical operating point of the transistor.

3. Determine the actual operating point by measuring V_{CE} and solving for $I_C = V_{RC}/R_C$ if the circuit uses BJTs. If the amplifier circuit uses FETs, the operating point is found by measuring V_{DS} and solving for $I_D = V_{RD}/R_D$. If the transistor is not operating at the expected bias point, determine the reason. You may need replace a faulty resistor or transistor. If the transistor is properly biased, then the problem is due to a problem in the ac operation of the circuit.

4. Verify that the coupling and bypass capacitors have been correctly placed into the circuit. If any of these capacitors are electrolytic, it is necessary to determine the correct polarity and install the capacitor accordingly. (There is a possibility that the capacitor was permanently damaged if an incorrect dc voltage was applied for any length of time.)

5. Ensure that all connections, especially ground wires, are kept as short as possible. This is particularly important if a transistor amplifier operates at high frequencies, since it is possible for ground-loop currents to be generated. These currents can result in the appearance of all manners of unwanted signals.

6. If the output signal is distorted, as shown in Figure 27–62, then the result is likely due to an input signal that is too large for the transistor. In order to return the transistor amplifier to linear operation (where the output signal resembles the input signal), it is necessary to lower the amplitude of the applied ac signal.

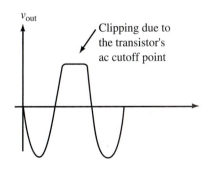

FIGURE 27–62

27.11 Computer Analysis of Transistor Amplifier Circuits

Computer analysis provides a powerful tool to analyze and predict the operation of complex circuits. Although the circuits that you are about to examine are not as complicated as circuits that appear in the "real world", you will find that the methods that are developed in entering and simulating the circuits will help when you work with much more complex circuits. Both PSpice and MultiSIM are excellent tools for simulating the operation of BJT and FET amplifiers.

MultiSIM

The following example shows that we can simulate a complete transistor amplifier circuit, using actual component values that are commonly available in the lab. While the software is more forgiving than using actual components in the lab, you will find that if you connect components incorrectly, there will still be problems with the circuit operation.

EXAMPLE 27–14

Use MultiSIM to construct the circuit shown in Figure 27–26. In the circuit, use a 2N2712 as the npn transistor and set the signal source for a voltage of $v_{in} = 0.1 \text{ V}_p$ and a frequency of 1000 Hz. Use 10-μF electrolytic capacitors and place them with the correct polarities.

a. Use the oscilloscope to measure both the input and output voltages, and use these values to calculate the voltage gain of the amplifier.

b. Observe the effect of increasing the input voltage to 0.55 V_p, a value that is greater than the theoretical maximum allowable input signal ($v_{in(max)} = 0.4 \text{ V}_p$).

c. Compare the results to those of Example 27–6.

Solution

a. Construct the circuit as you have done in previous chapters. Select an oscilloscope from the Instruments tool bar (to the right of the workspace). Place the oscilloscope into the circuit, so that Channel A displays the input voltage, v_{in} and Channel B displays the output voltage, v_{out}. Your circuit should appear as shown in Figure 27–63.

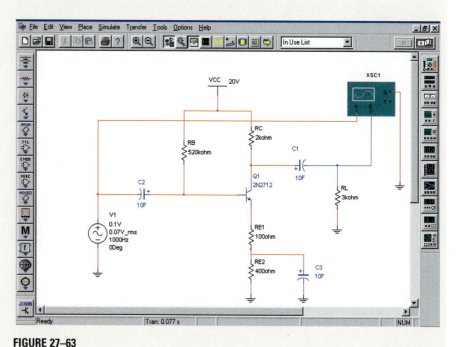

◀ MULTISIM   **FIGURE 27–63**

You may find that it is useful to display each of the measurements in a different color. In order to change the color of a particular display simply right-click on the wire, click *color,* and select the desired color from the *basic colors* that are provided. For instance, Figure 27–63 shows that the input voltage (Channel A) will appear in red, while the output voltage (Channel B) will appear in blue. You will find it much easier to interpret the results shown on the oscilloscope if you select different colors for each of the channels.

In order to observe the display on the oscilloscope, double click on the instrument. Now, click on the *Run* button to simulate the design. You should observe that a display appears on the oscilloscope. However, the display will not likely provide you with useful results. Just like in the lab, you will need to adjust the controls on the oscilloscope to provide you with useful observations. In the *Timebase* setting of the oscilloscope, set the scale for 200 μS/Div. In the *Channel A* and *Channel B* settings, set the volts per division as 50 mV/Div and 500 mV/Div, respectively. Your display should now be much more useful.

You will find, though, that the display is constantly refreshing and shifting its horizontal position. This is easily corrected by clicking on either the *Sing.* button or the *Nor.* button in the *Trigger* setting. Your display should now appear as shown in Figure 27–64, providing you with a display that is easy to interpret.

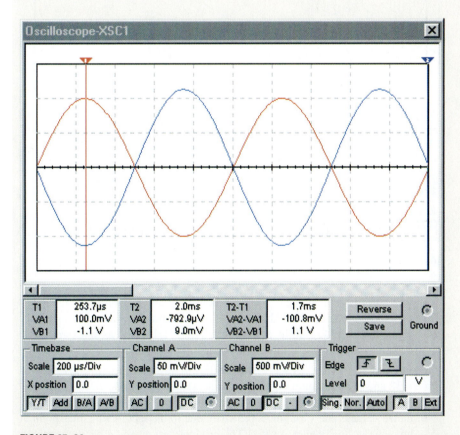

FIGURE 27–64

Let's use one more feature of the oscilloscope, the cursor. There are two cursors provided for this oscilloscope, one is red (labeled "1") and the other is blue (labeled "2"). Either cursor can be used to take measurements at any value of time by simply grabbing the cursor with the left mouse button and dragging it to the location. In this example, we see that when the input voltage is at its peak value, the output voltage will also be at a negative peak. From the results of the display, we observe the following:

$v_{in} = 100$ mV$_p$ and $v_{out} = 1.1$ V$_p$ and 180° out-of-phase with respect to the input.

Therefore, we conclude that this amplifier has a voltage gain of

$$A_v = \frac{v_{out}}{v_{out}} = -\frac{1.1 \text{ V}_p}{100 \text{ mV}_p} = -11$$

This result is consistent with the gain predicted in Example 27–6.

b. Increasing the input voltage to 0.55 V$_p$ results in the display of Figure 27–65.

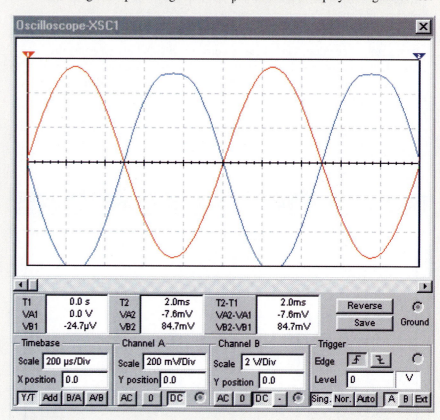

FIGURE 27–65

In the above display, we see that the positive-going portion of the output waveform is clipped due to the transistor going into ac cutoff. By adjusting the input voltage, it would be possible to determine the maximum input signal that would result in an undistorted output. It is left as an exercise for the student to show that this value occurs at around 0.52 V$_p$ (or 1.04 V$_{p-p}$).

c. In Example 27–6, the voltage gain of the circuit was approximately $A_v = -12$ and the maximum input signal that could be applied before distortion occurs, was 0.8 V$_{p-p}$. The observed results are relatively close to the predicted values. The variation is primarily due to the fact that the bias point is not equal to the predicted value. The original design would be more stable if we had used a universal-bias arrangement.

PSpice

The student version of PSpice does not have the full complement of transistors that a complete version has. However, you will find that there is a wide enough selection to observe how this software is applied to analyzing complete transistor circuits. The following example illustrates some of the features of the software. You are encouraged to experiment with the software to see some of the many other circuit applications.

EXAMPLE 27–15

a. Use PSpice to input the circuit of Figure 27–20. Use a 2N3904 npn transistor.

b. Use the Probe postprocessor to observe both the input voltage (at the base of the transistor) and the output voltage appearing across the load resistor.

c. Determine the voltage gain of the circuit and compare the value to the predicted gain of Example 27–5.

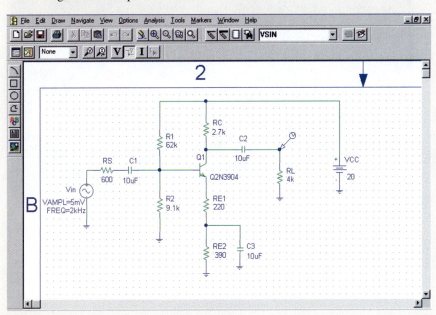

FIGURE 27–66

Solution

a. The circuit components are entered as shown in Figure 27–66. These components are obtained from the ANALOG.slb, EVAL.slb, and SOURCE.slb libraries. Notice that the component and voltage source labels have been changed to reflect those of Figure 27–20. While not important, this step helps to recognize the function of the component in the original circuit. Remember that you will need to change all component values default values. In order for the program to function correctly, the attributes of the VSIN voltage source (ac input signal) must be changed to the following values:

$$\text{VOFF=0 \quad VAMPL=5mV \quad FREQ=2kHz}$$

Remember that each of the attributes must be saved before closing the comment box. Notice that a voltage/level marker is placed at R_L, so that the Probe postprocessor will immediately display the output voltage, v_{out}.

Before simulating the design, it is important to instruct PSpice to perform the correct analysis. Click on the *Setup Analysis* tool and enable transient analysis. Click on the *Transient* button and set the analysis to provide *Print Step:* 0.2ns and *Final Time:* 1ms. (Note: PSpice does not permit spaces between the magnitude and the units for these values.)

b. If all values are correctly entered, the simulation will result in a sinusoidal voltage display of v_{out}. In order to display the input voltage, v_{in} as well as v_{out}, we need to have PSpice display an additional trace. This is done by first selecting *Add Plot* from the *Plot* menu. The previous step will provide us with a blank plot. Now, to display the input voltage, we now select *Add . . .* from the *Trace menu*. Since v_{in} is the same as the voltage across R_2, we simply select *V(R2:1)* from the list of measurements that were determined by PSpice. The resulting display is shown in Figure 27–67.

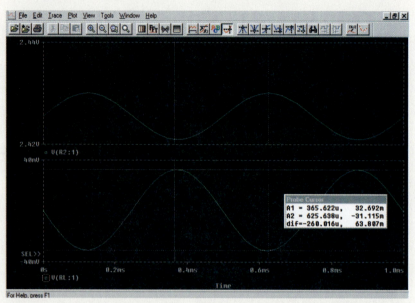

FIGURE 27–67

c. We use cursors to find the peak-to-peak voltages of both the input and output waveforms. The values are determined as $v_{in} = 9.138$ mV$_{p\text{-}p}$ and $v_{out} = 63.807$ mV$_{p\text{-}p}$. From the displays of Figure 27–67, we see that the output voltage is 180° out-of-phase with respect to the input. Consequently, the voltage gain for the circuit is determined to be

$$A_v = \frac{v_{out}}{v_{in}} = -\frac{63.807 \text{ mV}_{p\text{-}p}}{9.138 \text{ mV}_{p\text{-}p}} = -6.98$$

This value compares well to the value $A_v = -7.0$ that was found in Example 27–5.

PRACTICE PROBLEMS 12

a. Use PSpice to input the circuit of Figure 27–22. Make the following modifications to your circuit so that it will work correctly:

  • Use a 2N3906 pnp transistor.

  • Measure the output voltage at the collector of the transistor.

  • Do not include a coupling capacitor at the collector (since this will result in an error in a PSpice simulation).

  • Use a VSIN part having VOFF=0, VAMPL=5mV, FREQ=1kHz.

b. Use the Probe postprocessor to observe both the input voltage (at the base of the transistor) and the output voltage appearing at the collector of the transistor.

c. Determine the voltage gain of the circuit and compare the value to the predicted gain of Practice Problems 27–5.

Answers
b. $v_{in} = 9.967$ mV$_{p\text{-}p}$, $v_{out} = 79.256$ mV$_{p\text{-}p}$; c. $A_v = -7.95$ compares well with the theoretical gain of $A_v = -8.1$

27.1   The Use of Capacitors in Amplifier Circuits

1. The coupling capacitor, illustrated in Figure 27–68 is used to pass ac signals from a signal source having an output impedance (resistance), $z_{out} = 1200\ \Omega$ to a transistor circuit having an input impedance, $R_L = 800\ \Omega$. The signal varies within a frequency range of 100 Hz → 20 kHz.

   a. Calculate the minimum value of coupling capacitor required.

   b. Determine the amplitude of the input voltage, v_{in} if the coupling capacitor of part a. is used i) at 100 Hz and ii) at 20 kHz.

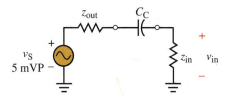

FIGURE 27–68

2. The coupling capacitor of Figure 27–68 is used to connect a circuit having $z_{out} = 600\ \Omega$ and $z_{in} = 20\ \Omega$ at a frequency range of 10 kHz → 200 kHz.

   a. Calculate the minimum value of coupling capacitor required.

   b. Determine the amplitude of the input voltage, v_{in} if the coupling capacitor of part a. is used i) at 10 kHz and ii) at 200 kHz.

3. A bypass capacitor is placed across an emitter resistance, $R_E = 470\ \Omega$, to increase the gain of a transistor amplifier. Determine the minimum value of bypass capacitor needed if the circuit is to operate between 500 Hz and 50 kHz.

4. A bypass capacitor is placed across an emitter resistance, $R_E = 1.2\ k\Omega$ of a transistor amplifier. Determine the minimum value of bypass capacitor needed if the circuit is to operate between 500 Hz and 50 kHz.

27.4   The Common-Emitter Amplifier

5. Given that $V_{CC} = +22$ V, $R_C = 4.0\ k\Omega$, $R_B = 1.1\ M\Omega$, $R_L = 4.0\ k\Omega$, and $\beta_{dc} = 150 = h_{fe} = \beta_{dc}$ in the circuit of Figure 27–69:

   a. Calculate the operating point of the transistor.

   b. Sketch the ac equivalent circuit using the T-equivalent model of the transistor.

   c. Use the ac equivalent circuit to find A_v, z_{in}, z_{out}, A_i, and A_p.

6. Given that $V_{CC} = +12$ V, $R_C = 2.2\ k\Omega$, $R_B = 560\ k\Omega$, $R_L = 3.3\ k\Omega$, and $\beta_{dc} = 130 = h_{fe} = \beta_{dc}$ in the circuit of Figure 27–69:

   a. Calculate the operating point of the transistor.

   b. Sketch the ac equivalent circuit using the T-equivalent model of the transistor.

   c. Use the ac equivalent circuit to find A_v, z_{in}, z_{out}, A_i, and A_p.

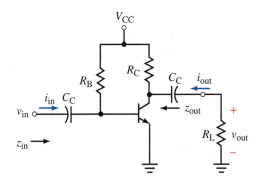

FIGURE 27–69

7. Consider the circuit of Figure 27–70.

   a. Given that $R_C = 3.3\ k\Omega$, $R_B = 910\ k\Omega$, and $\beta_{dc} = 150$, find the operating point of the transistor.

   b. Refer to the manufacturer's specifications of the 2N3906 transistor to obtain each of the h-parameters at the given operating point.

   c. Sketch the ac equivalent circuit using the appropriate values for the h-parameter model of the transistor.

   d. Use the ac equivalent circuit to find A_v, z_{in}, z_{out}, and A_i.

8. Refer to the circuit of Figure 27–70 and the values of part a. in Problem 7.

   a. Sketch the ac equivalent circuit using the T-equivalent model of the transistor.

   b. Use the ac equivalent circuit to find A_v, z_{in}, z_{out}, and A_i.

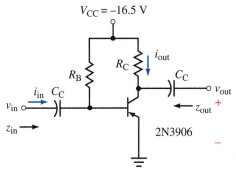

FIGURE 27–70

9. Refer to the circuit of Figure 27–71.

   a. Calculate the operating point of the transistor.

   b. Using the h-parameter values provided, sketch the ac equivalent circuit.

   c. Determine A_v, z_{in}, z_{out}, and A_i for the circuit.

   d. Calculate the amplitudes of the input voltage, v_{in} and the output voltage, v_{out}.

   e. Use graph paper to sketch two graphs, showing v_{in} and v_{out}. Correctly label each graph with the amplitudes and the phase relationship.

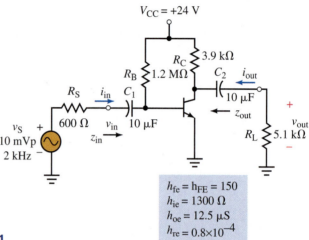

MULTISIM

FIGURE 27–71

$h_{fe} = h_{FE} = 150$
$h_{ie} = 1300\ \Omega$
$h_{oe} = 12.5\ \mu S$
$h_{re} = 0.8 \times 10^{-4}$

10. Refer to the circuit of Figure 27–72.

   a. Calculate the operating point of the transistor.

   b. Obtain the h-parameters of the transistor from the manufacturer's data sheets. Use the appropriate values to sketch the ac equivalent circuit.

   c. Determine A_v, z_{in}, z_{out}, and A_i for the circuit.

   d. Calculate the amplitudes of the input voltage, v_{in} and the output voltage, v_{out}.

   e. Use graph paper to sketch two graphs, showing v_{in} and v_{out}. Correctly label each graph with the amplitudes and the phase relationship.

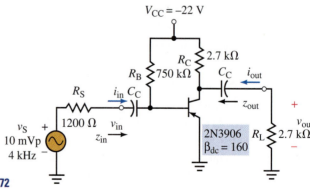

FIGURE 27–72

11. Given the circuit of Figure 27–73.

   a. Calculate the operating point of the transistor.

   b. Sketch the ac equivalent circuit using the T-equivalent model of the transistor.

   c. Use the ac equivalent circuit to find A_v, z_{in}, z_{out}, and A_i.

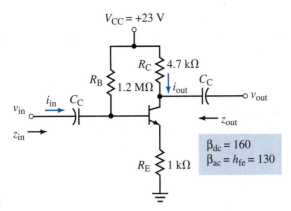

$V_{CC} = +23$ V

R_C 4.7 kΩ

R_B 1.2 MΩ

C_C

i_{out}

v_{out}

i_{in} C_C

v_{in}

z_{in}

z_{out}

R_E 1 kΩ

$\beta_{dc} = 160$
$\beta_{ac} = h_{fe} = 130$

FIGURE 27–73

12. Assume that a bypass capacitor is placed across the emitter resistance, R_E in the circuit of Figure 27–73.

a. Will the dc operation of the circuit change?

b. Sketch the ac equivalent circuit using the T-equivalent model of the transistor.

c. Use the ac equivalent circuit to find A_v, z_{in}, z_{out}, and A_i.

d. If a bypass capacitor is added across an emitter resistance of an emitter bias CE amplifier, what general statement can be made about the following?

   i. dc operating point

   ii. input impedance, z_{in} of the circuit

   iii. voltage gain, A_v of the circuit

13. Refer to the circuit of Figure 27–74. Given $V_{CC} = -18$ V, $R_S = 1200$ Ω, $R_B = 780$ kΩ, $R_C = 4.3$ kΩ, $R_{E1} = 180$ Ω, $R_{E2} = 820$ Ω, $R_L = 4.7$ kΩ

a. Calculate the operating point of the transistor.

b. Sketch the ac equivalent circuit using the h-parameter model of the transistor.

c. Use the ac equivalent circuit to find A_v, z_{in}, z_{out}, and A_i.

d. Given that $v_S = 5$ mV$_p$, determine the amplitudes of v_{in} and v_{out}.

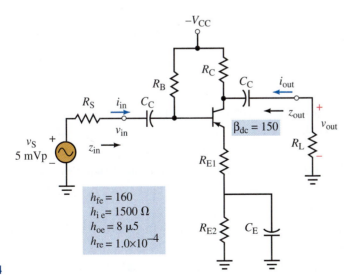

$-V_{CC}$

R_C

R_B

C_C

i_{out}

z_{out}

R_S i_{in} C_C

v_{in}

$\beta_{dc} = 150$

v_S +
5 mVp

z_{in}

R_{E1}

R_L

v_{out}

$h_{fe} = 160$
$h_{i\,e} = 1500$ Ω
$h_{oe} = 8$ μ5
$h_{re} = 1.0 \times 10^{-4}$

R_{E2} C_E

FIGURE 27–74

◀ MULTISIM

14. Refer to the circuit of Figure 27–74. Given $V_{CC} = -16$ V, $R_S = 2500$ Ω, $R_B = 680$ kΩ, $R_C = 2.7$ kΩ, $R_{E1} = 220$ Ω, $R_{E2} = 560$ Ω, $R_L = 3.3$ kΩ,

a. Calculate the operating point of the transistor.

b. Sketch the ac equivalent circuit using the h-parameter model of the transistor.

c. Use the ac equivalent circuit to find A_v, z_{in}, z_{out}, and A_i.

d. Given that $v_S = 5$ mV$_p$, determine the amplitudes of v_{in} and v_{out}.

15. Assume that the emitter bypass capacitor in the circuit of Problem 13 is removed.

a. Sketch the ac equivalent circuit using the h-parameter model of the transistor.

b. Use the ac equivalent circuit to find A_v, z_{in}, z_{out}, and A_i.

c. Given that $v_S = 5$ mV$_p$, determine the amplitudes of v_{in} and v_{out}.

16. Assume that the emitter bypass capacitor in the circuit of Problem 14 is removed.

a. Sketch the ac equivalent circuit using the h-parameter model of the transistor.

b. Use the ac equivalent circuit to find A_v, z_{in}, z_{out}, and A_i.

c. Given that $v_S = 5$ mV$_p$, determine the amplitudes of v_{in} and v_{out}.

17. Refer to the circuit of Figure 27–75.

Given that $V_{CC} = 16$ V, $R_1 = 47$ kΩ, $R_2 = 4.7$ kΩ, $R_E = 390$ Ω, $R_C = 3.9$ kΩ, $R_L = 3.9$ kΩ:

a. Calculate the operating point of the transistor.

b. Sketch the ac equivalent circuit using the T-equivalent model of the transistor.

c. Use the ac equivalent circuit to find A_v, z_{in}, z_{out}, and A_i.

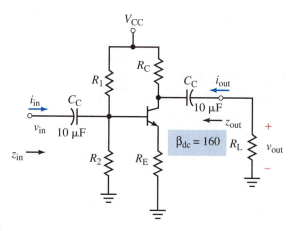

FIGURE 27–75

18. Refer to the circuit of Figure 27–75.

Given that $V_{CC} = 22$ V, $R_1 = 91$ kΩ, $R_2 = 10$ kΩ, $R_E = 220$ Ω, $R_C = 2.2$ kΩ, $R_L = 2.2$ kΩ:

a. Use Thévenin's theorem to calculate the operating point of the transistor.

b. Sketch the ac equivalent circuit using the T-equivalent model of the transistor.

c. Use the ac equivalent circuit to find A_v, z_{in}, z_{out}, and A_i.

19. Refer to the circuit of Figure 27–76.

Given that the circuit has $V_{CC} = -18.8$ V, $R_1 = 51$ kΩ, $R_2 = 8.2$ kΩ, $R_{E1} = 180$ Ω,

$R_{E2} = 670$ Ω, $R_C = 3.3$ kΩ, $R_L = 2.2$ kΩ, $R_S = 1.5$ kΩ:

a. Calculate the operating point of the transistor.

b. Obtain the h-parameters for the transistor. Sketch the ac equivalent circuit using the h-parameter model of the transistor.

c. Use the ac equivalent circuit to find A_v, z_{in}, z_{out}, and A_i.

d. Given that $v_S = 5$ mV$_p$, determine the amplitudes of v_{in} and v_{out}.

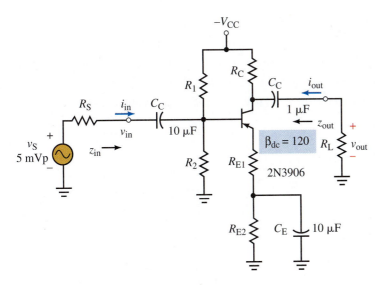

FIGURE 27–76

20. Refer to the circuit of Figure 27–76.

   Given that the circuit has $V_{CC} = -20$ V, $R_1 = 67$ kΩ, $R_2 = 9.1$ kΩ, $R_{E1} = 470$ Ω, $R_{E2} = 470$ Ω, $R_C = 4.3$ kΩ, $R_L = 1.8$ kΩ, $R_S = 2$ kΩ.

   a. Calculate the operating point of the transistor.

   b. Obtain the h-parameters for the transistor. Sketch the ac equivalent circuit using the h-parameter model of the transistor.

   c. Use the ac equivalent circuit to find A_v, z_{in}, z_{out}, and A_i.

   d. Given that $v_S = 5$ mV$_p$, determine the amplitudes of v_{in} and v_{out}.

21. Given the circuit of Figure 27–77:

   a. Calculate the operating point of the transistor.

   b. Sketch the ac equivalent circuit using the T-equivalent model of the transistor.

   c. Use the ac equivalent circuit to find A_v, z_{in}, z_{out}, and A_i.

   d. Given that $v_S = 5$ mV$_p$, determine the amplitudes of v_{in} and v_{out}.

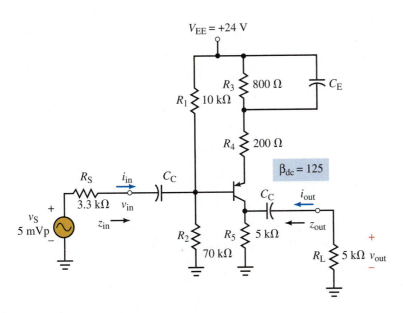

FIGURE 27–77

22. Given the circuit of Figure 27–78:
    a. Calculate the operating point of the transistor.
    b. Sketch the ac equivalent circuit using the T-equivalent model of the transistor.
    c. Use the ac equivalent circuit to find A_v, z_{in}, z_{out}, and A_i.
    d. Given that $v_S = 20$ mV$_p$, determine the amplitudes of v_{in} and v_{out}.

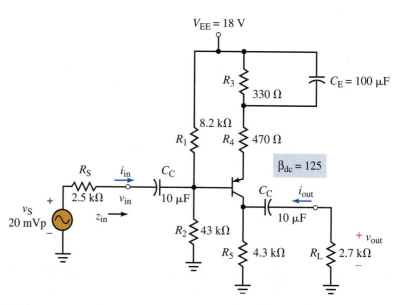

◀ MULTISIM

FIGURE 27–78

27.5   The ac Load Line

23. Given the circuit of Figure 27–79:
    a. Determine the operating point of the transistor.
    b. Sketch the simplified ac equivalent circuit of the output of the amplifier. (Do not model the transistor.)
    c. Calculate the values of dc and ac saturation and cutoff.
    d. Sketch a graph showing the dc load line and the ac load line.
    e. Determine the maximum undistorted collector current and the maximum undistorted collector-emitter voltage. Solve for the maximum undistorted output voltage, v_{out}.

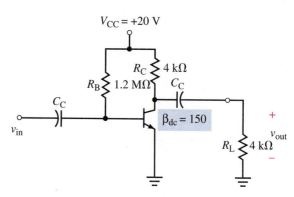

FIGURE 27–79

24. Repeat Problem 23 for the circuit of Figure 27–80.

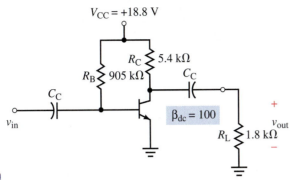

FIGURE 27–80

25. Repeat Problem 23 for the circuit of Figure 27–81.
26. Repeat Problem 23 for the circuit of Figure 27–82.

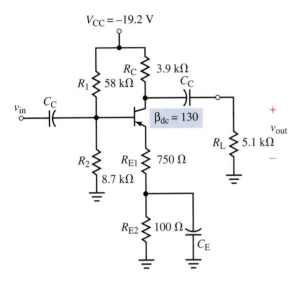

FIGURE 27–81

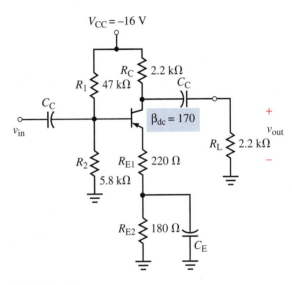

FIGURE 27–82

27. Refer to the circuit of Figure 27–74 and the component values of Problem 13.

   a. Determine the dc and ac cutoff and saturation values.

   b. Sketch a graph showing both the dc load line and the ac load line.

   c. Determine the maximum undistorted collector current and the maximum undistorted collector-emitter voltage. Solve for the maximum undistorted output voltage, v_{out}.

   d. Use the voltage gain calculated in Problem 13 to solve for the maximum input voltage, v_{in}, which can be applied to the circuit before the output signal is distorted.

28. Refer to the circuit of Figure 27–74 and the component values of Problem 14.

   a. Determine the dc and ac cutoff and saturation values.

   b. Sketch a graph showing both the dc load line and the ac load line.

   c. Determine the maximum undistorted collector current and the maximum undistorted collector-emitter voltage. Solve for the maximum undistorted output voltage, v_{out}.

   d. Use the voltage gain calculated in Problem 14 to solve for the maximum input voltage, v_{in} that can be applied to the circuit before the output signal is distorted.

27.6 The Common-Collector Amplifier

29. Refer to the circuit of Figure 27–83. Let $V_{CC} = 16$ V, $R_B = 67$ kΩ, $R_E = 470$ Ω

  a. Find the V_{CEQ} and I_{CQ} at the dc operating point.

  b. Sketch the ac equivalent circuit, using the T-equivalent model of the transistor.

  c. Determine A_v, z_{in}, z_{out}, A_i, and A_p of the amplifier.

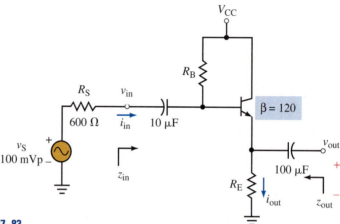

FIGURE 27–83

30. Refer to the circuit of Figure 27–83. Let $V_{CC} = 20$ V, $R_B = 150$ kΩ, $R_E = 1.0$ kΩ

  a. Find the V_{CEQ} and I_{CQ} at the dc operating point.

  b. Sketch the ac equivalent circuit, using the T-equivalent model of the transistor.

  c. Determine A_v, z_{in}, z_{out}, A_i, and A_p of the amplifier.

31. Refer to the circuit of Figure 27–84. Let $V_{CC} = 16$ V, $R_1 = 15$ kΩ, $R_2 = 15$ kΩ, $R_E = 1.5$ kΩ, $R_L = 1.0$ kΩ

  a. Find the V_{CEQ} and I_{CQ} at the dc operating point.

  b. Sketch the ac equivalent circuit, using the T-equivalent model of the transistor.

  c. Determine A_v, z_{in}, z_{out}, A_i, and A_p of the amplifier.

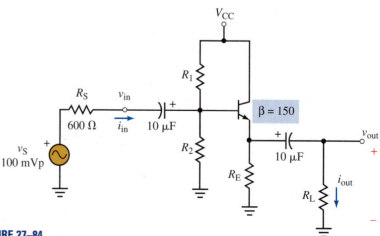

FIGURE 27–84

32. Refer to the circuit of Figure 27–84. Let $V_{CC} = 24$ V, $R_1 = 15$ kΩ, $R_2 = 20$ kΩ, $R_E = 2.0$ kΩ, $R_L = 2.4$ kΩ

   a. Find the V_{CEQ} and I_{CQ} at the dc operating point.

   b. Sketch the ac equivalent circuit, using the T-equivalent model of the transistor.

   c. Determine A_v, z_{in}, z_{out}, A_i, and A_p of the amplifier.

33. Refer to the circuit described in Problem 29.

   a. Use the manufacture's specifications to find h_{ie}, h_{fe}, h_{re}, and h_{oe}.

   b. Sketch the ac equivalent circuit using the h-parameter model for the 2N3904 transistor. (Use the simplified h-parameter model.)

   c. Determine A_v, z_{in}, z_{out}, A_i, and A_p of the amplifier.

34. Refer to the circuit described in Problem 30.

   a. Use the manufacture's specifications to find h_{ie}, h_{fe}, h_{re}, and h_{oe}.

   b. Sketch the ac equivalent circuit using the h-parameter model for the 2N3904 transistor. (Use the simplified h-parameter model.)

   c. Determine A_v, z_{in}, z_{out}, A_i, and A_p of the amplifier.

35. Refer to the circuit described in Problem 31.

   a. Use the manufacture's specifications to find h_{ie}, h_{fe}, h_{re}, and h_{oe}.

   b. Sketch the ac equivalent circuit using the h-parameter model for the 2N3904 transistor. (Use the simplified h-parameter model.)

   c. Determine A_v, z_{in}, z_{out}, A_i, and A_p of the amplifier.

36. Refer to the circuit described in Problem 32.

   a. Use the manufacture's specifications to find h_{ie}, h_{fe}, h_{re}, and h_{oe}.

   b. Sketch the ac equivalent circuit using the h-parameter model for the 2N3904 transistor. (Use the simplified h-parameter model.)

   c. Determine A_v, z_{in}, z_{out}, A_i, and A_p of the amplifier.

27.7 The FET Small-Signal Model

37. An n-channel JFET has $I_{DSS} = 7.5$ mA and $V_{GS(OFF)} = -4.0$ V:

   a. Determine the transconductance of the JFET for an operating point of $V_{GSQ} = -2.0$ V.

   b. Solve for g_m if $V_{GSQ} = -3.0$ V.

   c. Solve for g_m if $I_{DSQ} = 6.0$ mA.

38. A p-channel JFET has $I_{DSS} = 9.0$ mA and $V_{GS(OFF)} = +5.0$ V:

   a. Determine the transconductance of the JFET for an operating point of $V_{GSQ} = +2.0$ V.

   b. Solve for g_m if $V_{GSQ} = +1.5$ V.

   c. Solve for g_m if $I_{DSQ} = 6.0$ mA.

39. A p-channel E-MOSFET has $V_{th} = 2.5$ V and $I_{D(ON)} = 10$ mA at $V_{GS(ON)} = +4.0$ V:

   a. Determine the transconductance of the E-MOSFET for an operating point of $V_{GSQ} = +3.0$ V.

   b. Solve for g_m if $V_{GSQ} = +4.0$ V.

   c. Solve for g_m if $I_{DSQ} = 6.0$ mA.

   d. Solve for g_m if $I_{DSQ} = 12.0$ mA.

40. A p-channel E-MOSFET has $V_{th} = 3.0$ V and $I_{D(ON)} = 50$ mA at $V_{GS(ON)} = +5.0$ V:

    a. Determine the transconductance of the E-MOSFET for an operating point of $V_{GSQ} = +4.0$ V.

    b. Solve for g_m if $V_{GSQ} = +5.0$ V.

    c. Solve for g_m if $I_{DSQ} = 30$ mA.

    d. Solve for g_m if $I_{DSQ} = 60$ mA.

27.8   The Common-Source Amplifier

41. Refer to the circuit of Figure 27–85. Let $V_{DD} = 22$ V, $R_G = 1.0$ MΩ, $R_D = 2.2$ kΩ, $R_S = 750$ Ω, and $R_L = 1.0$ kΩ.

    a. Find I_{DSQ}, V_{GSQ}, and V_{DSQ} at the operating point.

    b. Solve for g_{mo} and g_m of the amplifier.

    c. Sketch the small-signal ac equivalent circuit.

    d. Determine A_v, z_{in}, and z_{out} of the amplifier.

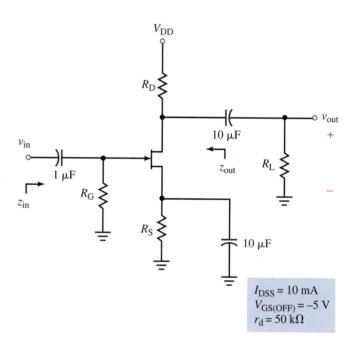

$I_{DSS} = 10$ mA
$V_{GS(OFF)} = -5$ V
$r_d = 50$ kΩ

FIGURE 27–85

42. Refer to the circuit of Figure 27–85. Let $V_{DD} = 16$ V, $R_G = 2.0$ MΩ, $R_D = 3.9$ kΩ, $R_S = 1.0$ kΩ, and $R_L = 2.2$ kΩ.

    a. Find I_{DSQ}, V_{GSQ}, and V_{DSQ} at the operating point.

    b. Solve for g_{mo} and g_m of the amplifier.

    c. Sketch the small-signal ac equivalent circuit.

    d. Determine A_v, z_{in}, and z_{out} of the amplifier.

43. Refer to the circuit of Figure 27–86. Let $V_{DD} = 25$ V, $R_1 = 390$ kΩ, $R_2 = 100$ kΩ, $R_D = 3.3$ kΩ, $R_S = 2.2$ kΩ, and $R_L = 3.3$ kΩ.

    a. Find I_{DSQ}, V_{GSQ}, and V_{DSQ} at the operating point.

    b. Solve for g_{mo} and g_m of the amplifier.

    c. Sketch the small-signal ac equivalent circuit.

    d. Determine A_v, z_{in}, and z_{out} of the amplifier.

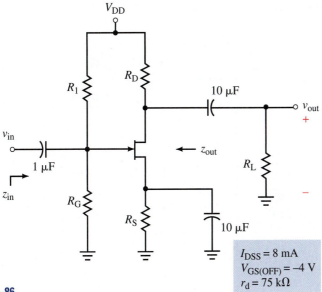

FIGURE 27–86

44. Refer to the circuit of Figure 27–86. Let $V_{DD} = 16$ V, $R_1 = 470$ kΩ, $R_2 = 220$ kΩ, $R_D = 1.1$ kΩ, $R_S = 1.1$ kΩ, and $R_L = 2.2$ kΩ.

   a. Find I_{DSQ}, V_{GSQ}, and V_{DSQ} at the operating point.

   b. Solve for g_{mo} and g_m of the amplifier.

   c. Sketch the small-signal ac equivalent circuit.

   d. Determine A_v, z_{in}, and z_{out} of the amplifier.

45. Refer to the circuit of Figure 27–87.

   a. Find I_{DSQ}, V_{GSQ}, and V_{DSQ} at the operating point.

   b. Solve for g_{mo} and g_m of the amplifier.

   c. Sketch the small-signal ac equivalent circuit.

   d. Determine A_v, z_{in}, and z_{out} of the amplifier.

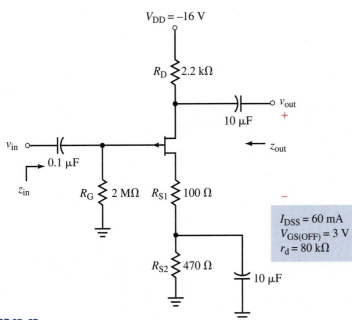

FIGURE 27–87

46. Repeat Problem 45 for the circuit of Figure 27–88.

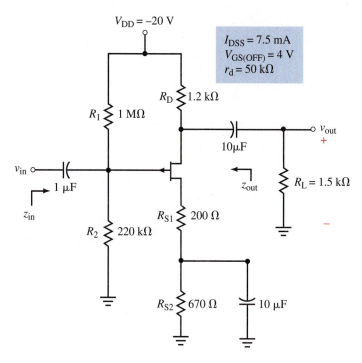

FIGURE 27–88

27.9   The Common-Drain (Source Follower) Amplifier

47. Refer to the circuit of Figure 27–89. Let $V_{DD} = 20$ V, $R_G = 1.0$ MΩ, $R_S = 1.0$ kΩ, and $R_L = 1.0$ kΩ.

   a. Find I_{DSQ}, V_{GSQ}, and V_{DSQ} at the operating point.

   b. Solve for g_{mo} and g_m of the amplifier.

   c. Sketch the small-signal ac equivalent circuit.

   d. Determine A_v, z_{in}, and z_{out} of the amplifier.

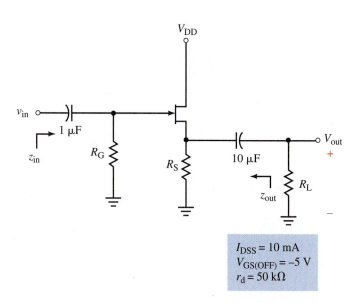

FIGURE 27–89

48. Repeat Problem 47 if V_{DD} = 16 V, R_G = 2.0 MΩ, R_S = 470 Ω, and R_L = 910 Ω.

49. Refer to the circuit of Figure 27–90. Let V_{DD} = −20 V, R_1 = 820 kΩ, R_2 = 220 kΩ, and R_S = 820 Ω.

    a. Find I_{DSQ}, V_{GSQ}, and V_{DSQ} at the operating point.

    b. Solve for g_{mo} and g_m of the amplifier.

    c. Sketch the small-signal ac equivalent circuit.

    d. Determine A_v, z_{in}, and z_{out} of the amplifier.

50. Repeat Problem 49 if V_{DD} = −16 V, R_1 = 1.0 MΩ, R_2 = 220 kΩ, and R_S = 670 Ω.

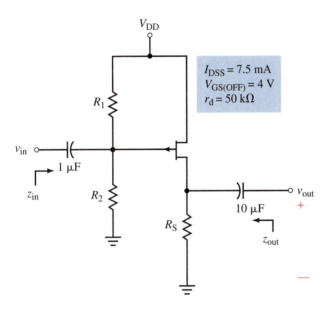

I_{DSS} = 7.5 mA
$V_{GS(OFF)}$ = 4 V
r_d = 50 kΩ

FIGURE 27–90

27.10 Troubleshooting a Transistor Amplifier Circuit

51. You have constructed the circuit of Figure 27–26. Upon measuring the voltage gain, you find that A_v = −2.4, rather than the expected value of A_v = −12. You have measured the operating point and found it to be close to the theoretical value. What is the most likely problem with the circuit?

52. The output of an amplifier circuit appears similar to that shown in Figure 27–60. List at least two problems that can cause such a problem.

53. a. If the coupling capacitor, C_2, in the circuit of Figure 27–71 were to develop a short circuit fault,

    i. what (if anything) would occur to the operating point of the transistor? Show calculations to support your conclusion.

    ii. what voltage gain would you expect the circuit to have? Show calculations.

    b. If the coupling capacitor, C_2, in the circuit of Figure 27–71 were to develop an open circuit fault,

    i. what (if anything) would occur to the operating point of the transistor? Show calculations to support your conclusion.

    ii. what voltage gain would you expect the circuit to have? Show calculations.

54. Refer to the circuit of Figure 27–74. Given $V_{CC} = -20$ V, $R_S = 600$ Ω, $R_B = 910$ kΩ, $R_C = 2.7$ kΩ, $R_{E1} = 100$ Ω, $R_{E2} = 560$ Ω, $R_L = 2.7$ kΩ.

   a. Calculate the operating point of the transistor.

   b. Determine the expected voltage gain of the amplifier circuit.

   c. If the emitter bypass capacitor, C_E, in the circuit were to develop a short circuit fault,

      i. what (if anything) would occur to the operating point of the transistor? Show calculations to support your conclusion.

      ii. what voltage gain would you expect the circuit to have? Show calculations.

   d. If the emitter bypass capacitor, C_E, in the circuit were to develop an open circuit fault,

      i. what (if anything) would occur to the operating point of the transistor? Show calculations to support your conclusion.

      ii. what voltage gain would you expect the circuit to have? Show calculations.

55. If the emitter bypass capacitor in the circuit of Problem 54 is an electrolytic capacitor, give two possible problems that could occur if it were inserted into the circuit with the wrong polarity.

56. Assume that the emitter bypass capacitor, C_E, in the circuit of Problem 19 develops an open-circuit fault.

   a. What effect would this have on the operating point? Give calculations to substantiate your answer.

   b. What effect would this have on the voltage gain and input impedance? Provide calculations to substantiate your answers.

27.11   Computer Analysis of Transistor Amplifier Circuits

◀ MULTISIM

57. Use MultiSIM to input the circuit of Figure 27–71. Use a 2N3904 transistor. Use the oscilloscope tool to simultaneously display the waveforms for v_{in} and v_{out}. Determine the voltage gain, $A_v = v_{out}/v_{in}$ of the circuit.

◀ MULTISIM

58. Use MultiSIM to input the circuit of Figure 27–75. Use a 2N3904 transistor and the component values of Problem 17. Use the oscilloscope tool to simultaneously display the waveforms for v_{in} and v_{out}. Determine the voltage gain, $A_v = v_{out}/v_{in}$ of the circuit.

◀ MULTISIM

59. Use MultiSIM to input the circuit of Figure 27–74. Use a 2N3906 transistor and the component values of Problem 13. Use the oscilloscope tool to simultaneously display the waveforms for v_{in} and v_{out}. Determine the voltage gain, $A_v = v_{out}/v_{in}$ of the circuit.

◀ MULTISIM

60. Use MultiSIM to input the circuit of Figure 27–78. Use a 2N3906 transistor. Use the oscilloscope tool to simultaneously display the waveforms for v_{in} and v_{out}. Determine the voltage gain, $A_v = v_{out}/v_{in}$ of the circuit.

◀ CADENCE

61. Use PSpice Capture to input the circuit of Figure 27–71. Use a 2N3904 transistor. Run the *Probe* postprocessor and obtain the waveforms for v_{in} and v_{out}. Use the results to determine the voltage gain, $A_v = v_{out}/v_{in}$ of the circuit.

62. Use PSpice Capture to input the circuit of Figure 27–75. Use a 2N3904 transistor and the component values of Problem 17. Run the *Probe* postprocessor and obtain the waveforms for v_{in} and v_{out}. Use the results to determine the voltage gain, $A_v = v_{out}/v_{in}$ of the circuit.

◀ CADENCE

63. Use PSpice Capture to input the circuit of Figure 27–74. Use a 2N3906 transistor and the component values of Problem 13. Run the *Probe* postprocessor and obtain the waveforms for v_{in} and v_{out}. Use the results to determine the voltage gain, $A_v = v_{out}/v_{in}$ of the circuit.

◀ CADENCE

64. Use PSpice Capture to input the circuit of Figure 27–78. Use a 2N3906 transistor. Run the *Probe* postprocessor and obtain the waveforms for v_{in} and v_{out}. Use the results to determine the voltage gain, $A_v = v_{out}/v_{in}$ of the circuit

◀ CADENCE

■ OBJECTIVES

On completion of this chapter, you will
be able to

- list the basic electrical characteristics of any operational amplifier
(op-amp),

- sketch the equivalent circuit of an
op-amp,

- correctly connect a dual power supply (V+ and V−) to an op-amp,

- explain the basic operation of a differential amplifier,

- calculate the common-mode voltage
gain of an amplifier,

- calculate the common-mode rejection ratio (CMRR) of an amplifier,

- explain the importance of using
negative feedback in an op-amp
circuit,

- analyze and design inverting amplifier circuits using op-amps,

- analyze and design non-inverting
amplifier circuits using op-amps,

- determine the effect of input offset
voltage, input offset current, and
input bias current on the output of
an op-amp circuit,

- use manufacturer's specifications to
determine the bandwidth of an op-
amp for a given voltage gain,

- use the specified slew rate of an
op-amp to calculate the maximum
amplitude for a sinusoidal output
waveform at a given frequency,

- modify an op-amp circuit to correctly compensate for variation due
to input offset,

- use PSpice and MultiSIM software
to observe the operation of an op-
amp circuit.

Operational Amplifiers

28

CHAPTER PREVIEW

This chapter examines the basic characteristics and operation of perhaps the most useful and versatile linear integrated circuit used in the electronics industry. In this chapter you will use your knowledge of impedance to analyze and design circuits using the operational amplifier. Although the actual circuit of the op-amp is complicated, we will treat this device as a simple block that uses quite simple concepts with which you are now familiar. This chapter examines the op-amp circuit using Ohm's law, Kirchhoff's current law, and Kirchhoff's voltage law. You will discover that we are no longer interested in exactly what goes on inside a given device, but rather we examine its operation as part of a complete system. ■

PUTTING IT IN PERSPECTIVE

ANALOG COMPUTERS, WHICH USED VACUUM tubes to perform arithmetic and feedback operations, were developed prior to World War II. These operational amplifiers were very slow and extremely inefficient, often requiring more than one voltage source. Due to the very slow reaction times, there were few practical uses for these amplifiers.

As so often is the case, war resulted in acceleration of research in feedback systems. The allies had made remarkable advances in radar, but needed some mechanism to aim antiaircraft guns using the information provided by radar. George A. Philbrick, a graduate of Massachusetts Institute of Technology and a researcher working with analog computers and operational amplifiers, was certain that the operational amplifier could be improved to provide improved response time necessary for real-time analysis of radar information. The operational amplifier design was greatly improved, thanks to the research of Loebe Julie, who worked for Professor Ragazzini at Columbia University, New York. Ultimately, the operational amplifier was successfully used in aiming antiaircraft guns.

After World War II, George Philbrick redesigned the operational amplifier into a modular package that was marketed in 1952 as the "K2-W" operational amplifier. The module consisted of two vacuum tubes that provided a gain of about 20,000 and a gain-bandwidth product of about 1 MHz, similar to many modern operational amplifiers.

With improvements in semiconductor technology, the modern operational amplifier is miniscule compared to the size of the original "K2-W". Modern op-amps need much less power to operate, requiring only 1/2 W of power compared to the almost 5 W for a K2-W. Perhaps the greatest improvement is that modern op-amps are a fraction of the cost of the older versions. ■

28.1 Introduction to the Operational Amplifier

The operational amplifier (op-amp) is one of the most versatile devices used in the electronics industry. George Philbrick of Huntington Engineering Labs originally designed the op-amp circuit in the late 1940s using vacuum tubes. Since digital computers were not yet available, the op-amp was used as an analog computer to perform mathematical operations (hence the name) such as addition, subtraction, multiplication, and solving differential equations. Modern op-amps are linear integrated circuits and are often included as part of even more complex ICs. A typical op-amp has several important characteristics:

- very high input impedance (generally several megohms)
- very low output impedance (generally less than 100 Ω)
- very high open-loop voltage gain (20,000 to 200,000)

Op-amps are used throughout electronics as comparators, voltage amplifiers, oscillators, active filters, and instrumentation amplifiers, to name just a few applications.

Figure 28–1 shows the symbol of a basic op-amp. The op-amp has two input terminals, the **inverting input** (−) and the **non-inverting input** (+) and a single output terminal. The impedance between the two input terminals and

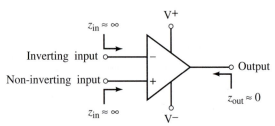

(a) Basic operational amplifier (op-amp)

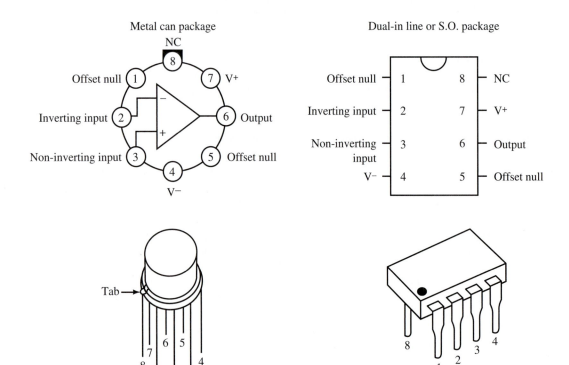

(b) LM741 connection diagrams
(*Courtesy of National Semiconductor Corporation*)

FIGURE 28–1

between each of the terminals and ground is very high and so we normally approximate $z_{in} \approx \infty$. The output impedance of the op-amp (looking from the output terminal to circuit ground) is very low and so $z_{out} \approx 0$. Figure 28–2 shows that the ideal op-amp may be represented as two separate parts, the input, which is effectively an open circuit, and the output that consists of an ideal voltage source in series with $z_{out} = 0 \ \Omega$.

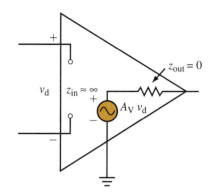

In addition to the characteristics shown in Figure 28–2, the ideal op-amp has infinite bandwidth and has unlimited output voltage. Clearly, it is not possible to have a real op-amp with these characteristics. In Section 28.6 we will examine the specifications of op-amps and see the limitations of the device.

Input signals can be applied at either of the two input terminals. Figure 28–3 shows the effect of applying a sinusoidal signal to each of the input terminals, with the other input grounded. When an op-amp is connected in this manner, it is said to be operating as a **single-ended amplifier,** since the input signal is applied at only one terminal, with the other grounded.

FIGURE 28–2

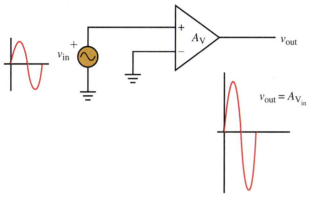

(a) Op-amp used as a non-inverting amplifier

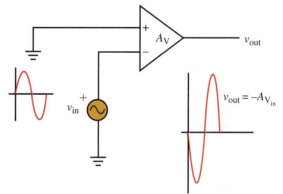

(b) Op-amp used as an inverting amplifier

FIGURE 28–3

Figure 28–4(a) shows that it is also possible to connect an op-amp so that the signal is applied directly between the two inputs, without using a ground connection. The voltage applied between the two input terminals is referred to as v_d, the difference voltage. In such instances, the op-amp is said to be operating as a **double-ended amplifier** or **differential amplifier** since it is amplifying the difference between the two inputs. When used as a differential amplifier, the op-amp may also have two different signals (relative to ground) applied to the two input terminals as illustrated in Figure 28–4(b).

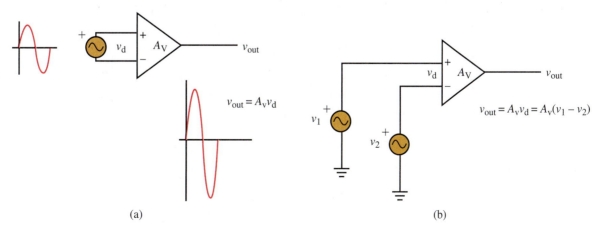

(a)

(b)

FIGURE 28–4

In order for an op-amp to operate, it is normally connected to two dc power supplies, one positive with respect to ground, the other negative. (Although less common, some op-amps require only a single power supply). Figure 28–5 shows the correct connection when using a dual power supply. In the lab, you will normally find that a single control changes voltage on both power supplies simultaneously. Notice that the ground terminal is connected between both supplies.

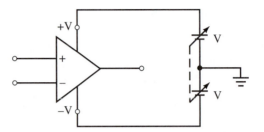

FIGURE 28–5

28.2 The Differential Amplifier and Common-Mode Signals

The Differential Amplifier

Figure 28–6 shows the schematic of the 741 operational amplifier. While the circuit is complex (21 transistors), the basic operation of the device can be understood by examining only two transistors, which together form the differential amplifier. In the circuit of Figure 28–6, the transistors Q_1 and Q_2 form

Schematic Diagram

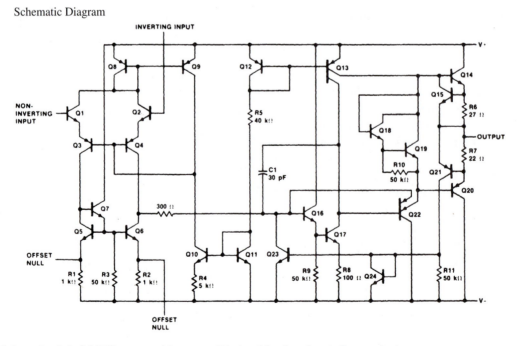

FIGURE 28–6   Schematic of the LM741 op-amp. *(Courtesy of National Semiconductor Corporation)*

part of the differential amplifier. Although we will not analyze the circuit in detail, it is worthwhile to briefly examine the operation of the differential amplifier stage. As the name implies, the differential amplifier amplifies only the difference between the signals appearing between the two inputs. The amplifier will reject any signal that is common to both inputs. Consider the transistor amplifier circuit shown in Figure 28–7.

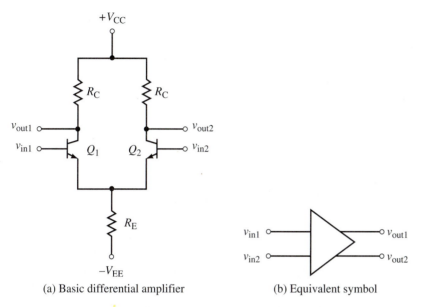

(a) Basic differential amplifier   (b) Equivalent symbol

FIGURE 28–7 The differential amplifier.

The components in the differential amplifier circuit are selected so that Q_1 and Q_2 have identical properties and both collector resistors are equal in value. When these components are constructed on a single IC, it is fairly easy to ensure that the required conditions are met. Consider that both inputs are grounded as illustrated in Figure 28–8.

We are able to easily determine the operating point of the transistors. Since both transistors are identical, the current through each emitter (and each collector) will be the same and so we may conclude that

$$I_{C1} = I_{C2} \approx I_{E1} = I_{E2} = \frac{I_E}{2} \qquad \textbf{(28–1)}$$

Also, we see that both emitters will be at the same potential, namely

$$V_E = -0.7 \text{ V}$$

Applying Ohm's law, gives

$$I_E = \frac{V_E - \left(-V_{EE}\right)}{R_E} = \frac{-0.7 \text{ V} - \left(-V_{EE}\right)}{R_E} = \frac{V_{EE} - 0.7 \text{ V}}{R_E} \qquad \textbf{(28–2)}$$

The voltage at the collector of each transistor is determined as

$$V_C = V_{CC} - I_C R_C \approx V_{CC} - \frac{I_E}{2} R_C \qquad \textbf{(28–3)}$$

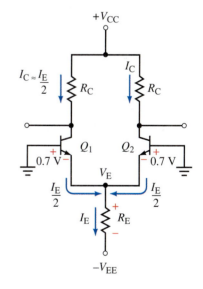

FIGURE 28–8

EXAMPLE 28–1

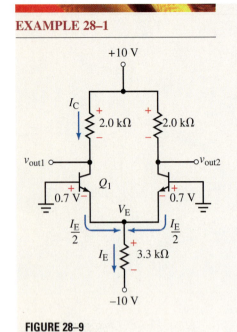

FIGURE 28–9

Given the circuit of Figure 28–9, find I_C and V_C.

Solution

The voltage at each emitter (relative to ground) is $V_E = -0.7\text{ V}$ and so we determine that

$$I_E = \frac{-0.7\text{ V} - (-10\text{ V})}{3.3\text{ k}\Omega} = 2.82\text{ mA}$$

and so

$$I_C \approx \frac{I_E}{2} = \frac{2.82\text{ mA}}{2} = 1.41\text{ mA}$$

Finally,

$$V_C = V_{CC} - I_C R_C = 10\text{ V} - (1.41\text{ mA})(2.0\text{ k}\Omega) = 7.18\text{ V}$$

In the above example, it is important to note that with the base of each transistor connected to ground, each collector will be at the same potential, namely 7.18 V. If the output is taken between the two output terminals we have the difference in potential between the terminals as $v_{out} = v_{out1} - v_{out2} = 0\text{ V}$.

Now, let's examine the effect of applying alternating voltages to each of the base terminals. Figure 28–10 illustrates what will happen to the voltages at various locations if a sinusoidal voltage is applied to the base of Q_1.

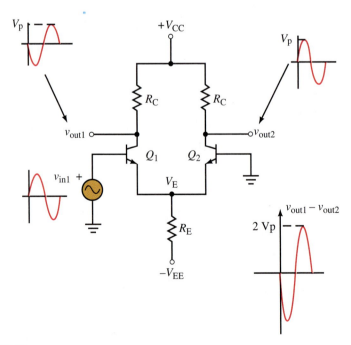

FIGURE 28–10

As v_{in1} increases, it results in a corresponding increase in collector current, I_{C1}. This increase in collector current results in a decrease in $v_{out1} = V_{C1} = V_{CC} - I_{C1}R_C$. Since the base of Q_2 is at ground, the voltage V_E will remain unchanged at $V_E = -0.7$ V. Consequently, there would be no change in I_E. In order for Kirchhoff's current law to be satisfied, the increase in collector current I_{C1}, would need to cause a decrease the collector current I_{C2} of the other transistor, Q_2. This decrease in I_{C2} results in an increased voltage at the collector of Q_2. Now, if we were to determine the output voltage as $v_{out} = v_{out1} - v_{out2}$, we would observe an output voltage that is $180°$ out-of-phase with respect to the input and which is significantly larger.

Using a similar approach, if a sinusoidal voltage were applied at the base of Q_2 (with the base of Q_1 grounded), the signal appearing at v_{out1} would be in phase with the input and v_{out2} would be $180°$ out-of-phase. In this case, the output voltage $v_{out} = v_{out1} - v_{out2}$, would be in phase with the input, v_{in2}.

What would happen if the input to each transistor's base were the same? In this case, both collector currents would increase simultaneously. The voltage V_E would also increase. More important though, is that the voltage at both collectors would always remain equal, since now there is no imbalance. This means that both outputs would change at exactly the same rate, and the difference, $v_{out} = v_{out1} - v_{out2} = 0$V. This means that if the same signal is applied to each input (referred to as a **common-mode signal**), the output will be zero.

Lastly, if the input signals were applied $180°$ out-of-phase, the output voltage, $v_{out} = v_{out1} - v_{out2}$ would have twice the magnitude as when only one signal was applied to the base of a transistor (with the other base grounded).

Figure 28–11, shows the effect of applying sinusoidal signals to one or both inputs of a differential amplifier. In the representation shown, the input, v_{in1}, is the inverting input and the input, v_{in2}, is the non-inverting input.

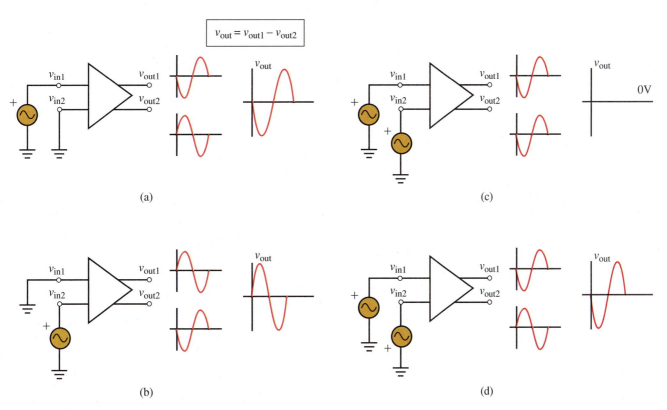

FIGURE 28–11

Common-Mode Signals

One of the characteristics of the ideal operational amplifier is that if the same (common-mode) signal is applied to both inputs, the output will be zero. When an op-amp is manufactured, it is impossible to make all transistors in the differential amplifiers exactly the same. This means that even when the same signal is applied to both inputs, there will be slight imbalances in the op-amp that result in a non-zero output.

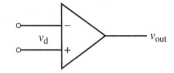

FIGURE 28–12

Consider the op-amp shown in Figure 28–12. We define the **differential voltage gain** of the amplifier as

$$A_{vd} = \frac{v_{out}}{v_d} \tag{28–4}$$

Another name for differential voltage gain is *open-loop voltage gain, A_{vol}*. The term *open-loop* refers to the fact that there is no feedback connection between the output and the input. As mentioned previously, the typical open-loop voltage gain for an op-amp is very large, normally between 20,000 and 200,000.

Now, if we were to connect a common-mode signal, v_c as shown in Figure 28–13, one expects that the output would be zero. However, due to slight unbalances within the differential amplifiers of the op-amp, there will normally be some detectable output signal. We define the *common-mode voltage gain* of the amplifier as

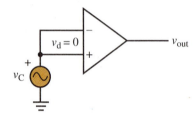

FIGURE 28–13

$$A_{vc} = \frac{v_{out}}{v_c} \tag{28–5}$$

The **common-mode rejection ratio (CMRR)** is the ability of an op-amp (or any other differential amplifier) to reject common-mode signals and is defined as

$$CMRR = \frac{A_{vd}}{A_{vc}} \tag{28–6}$$

Since the common-mode voltage gain is much smaller that the differential voltage gain, the CMRR of an op-amp will be very large. Therefore, the common-mode rejection ratio is normally expressed in decibels as

$$[CMRR]_{dB} = 20 \log CMRR \tag{28–7}$$

Typical values of common-mode rejection ratio for an op-amp are in the order of 70 to 90 dB.

EXAMPLE 28–2

An op-amp has a differential voltage gain of 200,000. Calculate its CMRR (in dB) if the output voltage is measured to be 2.0 $V_{p\text{-}p}$ when the input terminals are connected together and a 0.1 $V_{p\text{-}p}$ voltage is applied between the inputs and ground.

Solution
The common-mode voltage gain is

$$A_{vc} = \frac{v_{out}}{v_c} = \frac{2\ V_{p\text{-}p}}{0.1\ V_{p\text{-}p}} = 20$$

The common-mode rejection ratio is determined as

$$[CMMR]_{dB} = 20 \log \frac{A_{vd}}{A_{vc}} = 20 \log \frac{200,000}{20} = 80\ dB$$

One of the most important considerations in any electronic circuit is the amount of noise relative to the desired signal strength. We have all heard the effects of noise (or *static*) on an audio signal when we listen to far away radio stations. The static can be annoying at best and if large enough, can completely obliterate the desired signal. Imagine that the circuit shown in Figure 28–14 represents a signal being sent over a long distance.

As one might expect, the signal will decrease in amplitude as it moves from the source to the amplifier, since there will be losses in the line itself. Meanwhile, because the line behaves as an antenna, noise will be induced onto the line in series with the desired signal, with the result that the longer the line, the more noise. If we were to apply both the signal and noise to a single-ended amplifier as shown in Figure 28–14, both the signal and noise would be amplified equally. The desired signal would be lost in the noise.

Now, consider that we use an op-amp as a differential amplifier as shown in Figure 28–15. In this case, we see that neither input terminal is connected to ground. The desired signal is connected directly between the inverting and non-inverting terminals of the amplifier and so the signal provides a differential voltage. As in the first case, noise will once again be generated on the line. Since both lines are in the same electrical environment, and since neither line is connected to ground, each line will receive the same amount of induced noise induced (relative to ground). Because the noise is a common-mode signal, the differential amplifier will eliminate it and only the desired signal will appear at the output of the op-amp.

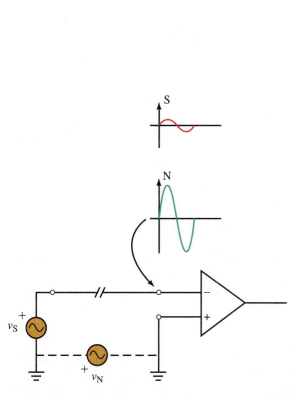

FIGURE 28–14 The single-ended amplifier amplifies both signal and noise equally.

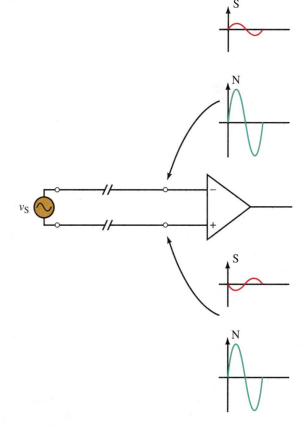

FIGURE 28–15 The differential amplifier amplifies only the signal, canceling the noise.

EXAMPLE 28–3

The input of an amplifier has a signal with amplitude of 1.0 mV and noise having amplitude of 100 mV.

a. If the above signal and noise are applied to the input of a single-ended amplifier having a differential voltage gain of 100:
   i) Calculate the signal voltage and the noise voltage at the output of the amplifier.
   ii) Determine the ratio of the signal voltage to the noise voltage at the output of the amplifier.
   iii) Will the signal be lost?

b. If the given signal and noise are applied to the input of a differential amplifier having a differential voltage gain of 100 and a CMMR of 10,000 (or 80 dB), repeat the calculations of part a.

Solution

a.   i) The signal and the noise will both be amplified by the same factor. The output signal voltage will be $v_s = 100(1.0 \text{ mV}) = 100 \text{ mV}$ and the output noise voltage will be $v_n = 100(100 \text{ mV}) = 10.0 \text{ V}$

   ii) The **signal-to-noise voltage ratio** will be

$$\text{S}/\text{N} = \frac{v_s}{v_n} = \frac{100 \text{ mV}}{10 \text{ V}} = 0.01$$

   iii) Since the signal voltage is $\frac{1}{100}$ of the noise, we can safely conclude that the signal will be lost in the noise.

b.   i) Recall that the CMRR was defined as $\text{CMRR} = \frac{A_{vd}}{A_{vc}}$. Therefore, the common-mode voltage gain is determined as

$$A_{vc} = \frac{A_{vd}}{\text{CMRR}} = \frac{100}{10,000} = 0.01$$

Now, since the desired signal is a differential voltage, we determine the output signal to have a value of

$$v_s = 100(1.0 \text{ mV}) = 100 \text{ mV}$$

The noise is a common-mode voltage, and so its level at the output of the op-amp will be

$$v_n = 0.01(100 \text{ mV}) = 1 \text{ mV}$$

   ii) The signal-to-noise voltage ratio for the differential amplifier will be

$$\text{S}/\text{N} = \frac{v_s}{v_n} = \frac{100 \text{ mV}}{1 \text{ mV}} = 100$$

   iii) Since the signal voltage is now 100 times larger than the noise, the signal will no longer be lost in the noise. In fact, the noise will be barely perceptible.

The previous example shows that if used correctly a differential amplifier can significantly reduce the amount of noise on a long line. This is especially important when working with audio or digital signals that must go through an electrically noisy environment, such as one having a large amount of 60 Hz wiring or in a location having many electric motors. The example also illustrates the importance of preventing ground loops (Figure 28–14), that can dramatically increase the noise in an amplifier.

One of the characteristics of an op-amp is that it has a very large differential voltage gain (in the order of 100,000). With a application of a very small input voltage, the output voltage will easily become saturated. Consider the op-amp circuit shown in Figure 28–16.

If the dc supply voltages for the op-amp are ±15 V as shown, the output voltage, v_{out}, could never exceed the values of the supply. (In fact the saturation voltage, V_{SAT} will normally be one or two volt less than the supply voltage.) Consequently, if we assume that the op-amp has a differential voltage gain of 100,000, the maximum signal that could be applied between the inputs would be 150 μV_p. If we were to apply 1 mV directly to the input of an op-amp, we expect an output of $v_{out} = (100,000)(1$ mV$) = 100$ V. Clearly this can't occur, and so the output voltage would have a magnitude of $\pm V_{SAT}$. Figure 28–17 shows what would happen if we were to connect a 1 mV dc source to each of the input terminals. Even if no input voltage were applied to the input terminals, the amount of electrical noise at the input will almost always be large enough to cause the op-amp to be saturated. This causes the output of the op–amp to have a value of either $+V_{SAT}$ or $-V_{SAT}$ even when no signal is applied to an input terminal.

28.3 Negative Feedback

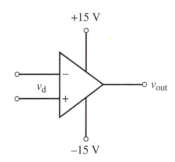

FIGURE 28–16

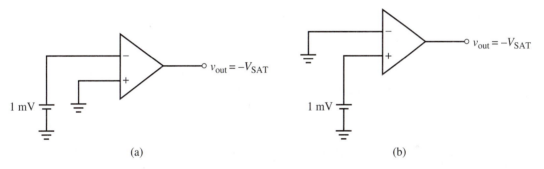

FIGURE 28–17   The differential voltage gain of an op-amp results in saturation with the application of a small input signal.

In order to make the op-amp more useful, *negative feedback* as shown in Figure 28–18, is used to reduce the voltage gain of the amplifier. Negative feedback takes a portion of the output signal and returns in to the inverting input of the op-amp. Because of the negative feedback, the voltage, v_d will be decreased, resulting in a corresponding decrease in the output voltage. The net result is that the voltage gain of the entire circuit is reduced.

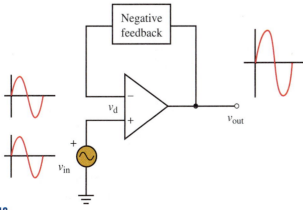

FIGURE 28–18

28.4 The Inverting Amplifier

Figure 28–19 shows an op-amp that uses negative feedback between the output and the inverting input terminal and has the input signal applied to the inverting input. This circuit is called the *inverting amplifier* since, as we will find, the output is 180° out-of-phase with respect to the input. By using basic circuit theory, several important characteristics of this amplifier become apparent. Using a feedback resistor results in a significant decrease in the voltage gain of the circuit. The voltage gain will no longer be equal to the open-loop voltage gain, but rather will be a much smaller value called the **closed-loop voltage gain,** A_{vcl}. You will find that even though the open-loop gain of the op-amp may vary from 20,000 to 200,000, the gain of the overall amplifier will not change at all. Additionally, you will find that by using feedback, the output impedance of the circuit will be reduced to almost zero.

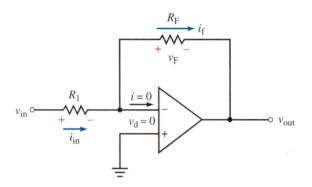

FIGURE 28–19   The inverting amplifier.

The op-amp of Figure 28–19 shows some very important characteristics of a typical op-amp circuit. The differential voltage, v_d will be about 100,000 times smaller than the output and so we say that $v_d \approx 0$. We therefore determine that the inverting input $(-)$ is at a **virtual ground,** since the potential at this point essentially zero. Additionally, there will be negligible current entering the op-amp due to the very high input impedance of the op-amp. By Kirchhoff's voltage law, we can solve for the input current of the circuit as

$$i_{in} = \frac{v_{in}}{R_1}$$

None of the input current enters the op-amp, and so the current must follow the path through R_F. We see therefore that $i_f = i_{in}$. Now applying Kirchhoff's voltage law at the output, the output voltage must be equal in magnitude to the voltage across the feedback resistor, namely

$$v_{out} = -v_f + v_d = -i_f R_F$$

Since $i_f = i_{in}$, the closed-loop voltage gain for the circuit is determined to be

$$A_{vcl} = \frac{v_{out}}{v_{in}} = -\frac{i_{in} R_F}{i_{in} R_1}$$

which, when simplified gives the voltage gain for the inverting amplifier as

$$A_{vcl} = -\frac{R_F}{R_1} \qquad \qquad \text{(28–8)}$$

The above result shows that the voltage gain of an amplifier is dependent only on the values of resistance that are used. The voltage gain of the circuit is not at all dependent on the open-loop voltage gain, thereby making the circuit operation much more stable.

Using the definition of input impedance allows us to solve for the input impedance of the inverting amplifier as

$$z_{in} = \frac{v_{in}}{i_{in}} = \frac{i_{in}R_1 + v_d}{i_{in}} \approx \frac{i_{in}R_1}{i_{in}}$$

which gives the input impedance for the inverting amplifier as

$$z_{in} \approx R_1 \qquad (28\text{--}9)$$

As you have already seen, the op-amp can be modeled as a circuit having a dependent source. Due to the dependent voltage source and the feedback resistor, we can no longer simply zero all sources to find the output impedance. Rather, it is necessary to apply the circuit theory that you learned in Chapter 19 to determine the output impedance of the circuit by solving for the ratio of the open-circuit voltage to short-circuit current.

$$z_{out} = \frac{v_{out(OC)}}{i_{out(SC)}} \qquad (28\text{--}10)$$

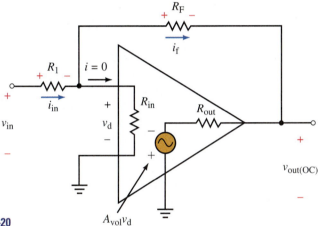

FIGURE 28–20

Figure 28–20 shows the inverting amplifier using the complete model of the op-amp. As determined previously, the open-circuit voltage of the circuit is simply found as

$$v_{out(OC)} = -\frac{R_F}{R_1}v_{in} \qquad (28\text{--}11)$$

In order to find the short circuit current, we examine the circuit shown in Figure 28–21, which shows the output terminal shorted to ground. Now, applying Kirchhoff's current law at the output terminal, we have

$$i_{out(SC)} = i_f - i_1 \qquad (28\text{--}12)$$

Applying Ohm's law in the output loop easily solves the current i_1 as

$$i_1 = \frac{A_{vol}v_d}{R_{out}} \qquad (28\text{--}13)$$

Further, we see that the voltage across R_F is equal to the differential voltage, v_d. Therefore we have

$$i_f = \frac{v_d}{R_F} \qquad (28\text{--}14)$$

Substituting Equations 28–13 and 28–14 into Equation 28–12 gives

$$i_{out(SC)} = \frac{v_d}{R_L} - \frac{A_{vol}v_d}{R_{out}}$$

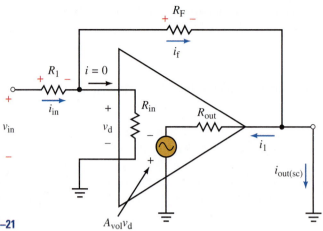

FIGURE 28–21

which can be rewritten as

$$i_{out(SC)} = v_d \left(\frac{1}{R_L} - \frac{A_{vol}}{R_{out}} \right)$$ (28–15)

In the circuit of Figure 28–21, we see that the differential voltage is found by applying the Ohm's law to the loop containing R_1 and R_F. (The input resistance of the op-amp is much larger than R_F, and so can be neglected.)

$$v_d = \frac{R_F}{R_1 + R_F} v_{in}$$ (28–16)

Substituting this result into Equation 28–15 gives

$$i_{out(SC)} = \left(\frac{R_F v_{in}}{R_1 + R_F} \right)\left(\frac{1}{R_L} - \frac{A_{vol}}{R_{out}} \right)$$

$$= -\left(\frac{R_F v_{in}}{R_1 + R_F} \right)\left(\frac{A_{vol}}{R_{out}} - \frac{1}{R_L} \right)$$

Now, substituting the above result and Equation 28–11 into Equation 28–10 we determine the output impedance as

$$z_{out} = \frac{v_{out(OC)}}{i_{out(SC)}}$$

$$= \frac{-\dfrac{R_F v_{in}}{R_1}}{-\left(\dfrac{R_F v_{in}}{R_1 + R_F} \right)\left(\dfrac{A_{vol}}{R_{out}} - \dfrac{1}{R_L} \right)}$$

$$= \frac{\dfrac{1}{R_1}}{\left(\dfrac{1}{R_1 + R_F} \right)\left(\dfrac{A_{vol}}{R_{out}} - \dfrac{1}{R_L} \right)}$$

$$= \frac{R_1 + R_F}{R_1}\left(\frac{1}{\left(\dfrac{A_{vol}}{R_{out}} - \dfrac{1}{R_L} \right)} \right)$$

Finally, since $\dfrac{A_{vol}}{R_{out}}$ is much larger than $\dfrac{1}{R_L}$ and since the closed-loop voltage

gain is $A_{vcl} = -\dfrac{R_F}{R_1}$, we simplify the above expression as

$$z_{out} = \left(1 - A_{vcl}\right)\left(\frac{R_{out}}{A_{vol}}\right) \qquad \text{(28–17)}$$

Given the circuit of Figure 28–22, find the voltage gain, input impedance, and output impedance. Sketch both the input and output voltage waveforms. The 741 op-amp has $R_{in} = 2.0$ MΩ, $R_{out} = 75$ Ω and open-loop gain of $A_{vol} = 200{,}000$.

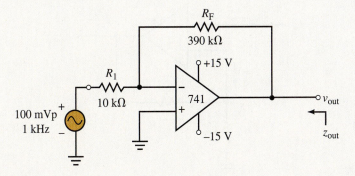

FIGURE 28–22

Solution

The voltage gain is

$$A_{vcl} = -\frac{R_F}{R_1} = -\frac{390 \text{ k}\Omega}{10 \text{ k}\Omega} = -39$$

The input impedance is

$$z_{in} = R_1 = 10 \text{ k}\Omega$$

The output impedance is

$$z_{out} = \left[1 - (-39)\right]\left(\frac{75 \text{ }\Omega}{200{,}000}\right) = 0.015 \text{ }\Omega$$

The output voltage will have amplitude of

$$v_{out} = \left|A_{vcl} v_{in}\right| = \left|(-39)\left(0.10 \text{ V}_p\right)\right| = 3.9 \text{ V}_p$$

Figure 28–23 shows the sketch of both the input and output voltages. This example clearly illustrates the simplicity of working with op-amps and also demonstrates that the output impedance is negligible.

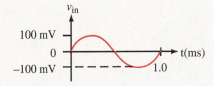

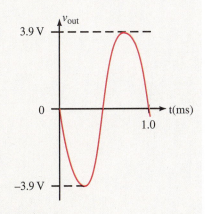

FIGURE 28–23

Using a 741 op-amp, design an inverting amplifier for a closed-loop gain of -5 and an input impedance of 15 kΩ. Use the characteristics of Example 28–4 to solve for the output impedance. What conclusion can you make about output impedance as it relates to closed-loop gain?

Answers
$R_1 = 15$ kΩ, $R_F = 75$ kΩ, $z_{out} = 2.25$ mΩ. As the voltage gain decreases, so too does the output impedance.

28.5 The Non-Inverting Amplifier

As the name implies, the non-inverting amplifier shown in Figure 28–24 has an output that is in phase with the input. In a manner similar to the previous section, we will determine expressions for the voltage gain, input impedance, and output impedance of this amplifier. Although the inverting amplifier had input impedance that was relatively low (determined by the value of the series resistance at the input terminal), you will discover that the input impedance of the non-inverting amplifier is many times larger than the already-large input resistance of the op-amp.

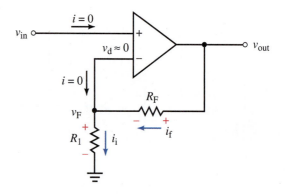

FIGURE 28–24   The non-inverting amplifier.

The voltage gain of the non-inverting amplifier is determined by starting with the definition of voltage gain, namely

$$A_v = \frac{v_{out}}{v_{in}}$$

By examining the circuit and applying Kirchhoff's voltage law, we are able to rewrite the voltage gain as

$$A_v = \frac{R_F i_F + R_1 i_1}{v_d + R_1 i_1}$$

Now, since the differential voltage is essentially zero and since $i = 0$ results in $i_f = i_1$, the expression is further simplified as

$$A_v = \frac{R_F i_F + R_1 i_F}{R_1 i_F}$$

which gives the closed-loop voltage gain of the non-inverting amplifier as

$$A_{vcl} = \frac{R_F}{R_1} + 1 \qquad (28–18)$$

The input impedance is determined by including the equivalent circuit of the op-amp as illustrated in Figure 28–25. When calculating the input impedance, we

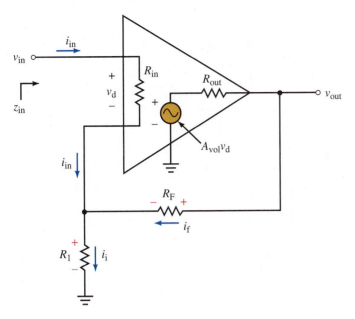

FIGURE 28–25

can no longer assume that $i_{in} = 0$ nor $v_d = 0$. Let's begin with the definition of input impedance and then simplify the resulting expression wherever possible.

$$z_{in} = \frac{v_{in}}{i_{in}} = \frac{v_d + R_1 i_1}{i_{in}} \approx \frac{R_1 i_1}{i_{in}} \qquad (28\text{--}19)$$

Now, since i_{in} is much smaller than i_1, we can simplify the numerator by letting

$$i_1 = i_f = \frac{A_{vol}v_d}{R_{out} + R_F + R_1} \approx \frac{A_{vol}v_d}{R_F + R_1}$$

Applying Ohm's law allows us to rewrite the denominator as

$$i_{in} = \frac{v_d}{R_{in}}$$

Substituting these values into Equation 28–19 gives

$$z_{in} = \frac{v_d + R_1\left(\dfrac{A_{vol}v_d}{R_F + R_1}\right)}{\dfrac{v_d}{R_{in}}} = \frac{1 + R_1\left(\dfrac{A_{vol}}{R_F + R_1}\right)}{\dfrac{1}{R_{in}}}$$

which, when simplified gives the input impedance as

$$z_{in} = \left(1 + \frac{A_{vol}}{A_{vcl}}\right)R_{in} \qquad (28\text{--}20)$$

Lastly, we solve for the output impedance of the non-inverting amplifier in much the same way as for the inverting amplifier. First, remember that we already have the open-circuit voltage for the non-inverting amplifier as

$$v_{out(OC)} = \left(\frac{R_F}{R_1} + 1\right)v_{in} \qquad (28\text{--}21)$$

The short-circuit current is found by placing a short between the output terminal and ground as illustrated in Figure 28–26.

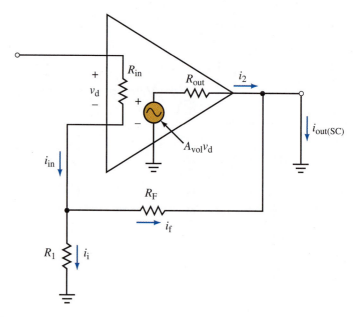

FIGURE 28–26

Applying Kirchhoff's current law gives the output short-circuit as

$$i_{out(SC)} = i_2 + i_f$$

However, i_f will be very small since $i_{in} \approx 0$. Also, by voltage divider, $v_d \approx v_{in}$ and so we write

$$i_{out(SC)} = i_2 = \frac{A_{vol}v_{in}}{R_{out}} \tag{28–22}$$

Now the output impedance is written as

$$z_{out} = \frac{v_{out(OC)}}{i_{out(SC)}} = \frac{A_{vcl}v_{in}}{\dfrac{A_{vol}v_{in}}{R_{out}}}$$

which becomes

$$z_{out} = \left(\frac{A_{vcl}}{A_{vol}}\right)R_{out} \tag{28–23}$$

EXAMPLE 28–5

Given the circuit of Figure 28–27, find the voltage gain, input impedance, and output impedance. The 741 op-amp has $R_{in} = 2.0$ MΩ, $R_{out} = 75$ Ω and open-loop gain of $A_{vol} = 200,000$.

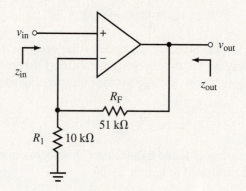

FIGURE 28–27

Solution

The voltage gain is

$$A_{vcl} = \frac{R_F}{R_1} + 1 = \frac{51 \text{ k}\Omega}{10 \text{ k}\Omega} + 1 = 6.1$$

The input impedance is

$$z_{in} = \left(1 + \frac{A_{vol}}{A_{vcl}}\right) R_{in} = \left(1 + \frac{200,000}{6.1}\right) 2.0 \text{ M}\Omega = 65.6 \text{ G}\Omega$$

The output impedance is

$$z_{out} = \left(\frac{A_{vcl}}{A_{vol}}\right) R_{out} = \left(\frac{6.1}{200,000}\right) 75 \text{ }\Omega = 0.0023 \text{ }\Omega = 2.3 \text{ m}\Omega$$

The previous example illustrates that the non-inverting amplifier has extremely high input impedance and very low output impedance. These characteristics make this design an ideal choice as a **buffer** circuit. The buffer circuit, also referred to as a **voltage follower,** has a unity voltage gain ($A_v = 1$) and provides an interface between two parts of a circuit that would normally be affected by loading. The following example illustrates how a buffer can be used between a high output impedance source and a low input impedance load.

Given the circuit of Figure 28–28:

EXAMPLE 28–6

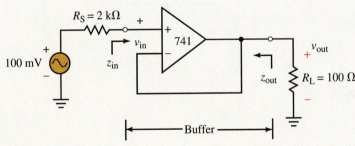

FIGURE 28–28

a. Solve for the voltage gain.

b. Determine the input impedance and output impedance of the buffer.

c. Calculate the output voltage.

d. Compare this to the output voltage that would have occurred if the buffer had not been used.

Solution

a. Notice that the feedback path results in $R_F = 0$, and $R_1 = \infty$. So the voltage gain of the amplifier will be

$$A_v = \frac{R_F}{R_1} + 1 = \frac{0}{\infty} + 1 = 1$$

b.

$$z_{in} = \left(1 + \frac{A_{vol}}{A_{vcl}}\right)R_{in} = \left(1 + \frac{200,000}{1}\right)2 \text{ M}\Omega = 400 \text{ G}\Omega$$

$$z_{out} = \left(\frac{1}{200,000}\right)75 \text{ }\Omega = 0.375 \text{ m}\Omega$$

c. The calculations of part b show that there will be no loading of the circuit. Now, since $v_{in} = 100$ mV, we have $v_{out} = A_v v_{in} = 1(100 \text{ mV}) = 100$ mV.

d. If the circuit had not used a buffer, the output voltage would have been

$$v_{out} = \left(\frac{R_L}{R_S + R_L}\right)v_{in} = \left(\frac{100 \text{ }\Omega}{2000 \text{ }\Omega + 100 \text{ }\Omega}\right)(100 \text{ mV}) = 4.76 \text{ mV}$$

This example clearly shows the importance of using a buffer circuit to prevent loading effects.

PRACTICE PROBLEMS 2

Given the circuit of Figure 28–29, find the voltage gain, output voltage, input impedance, and output impedance. The 741 op-amp has $R_{in} = 2.0$ MΩ, $R_{out} = 75$ Ω and open-loop gain of $A_{vol} = 200,000$.

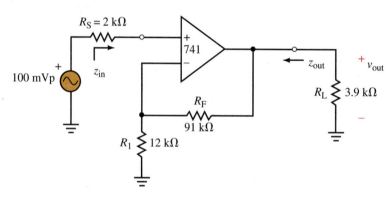

FIGURE 28–29

Answers
$A_v = 8.58$, $v_{out} = 858$ mV$_p$, $z_{in} = 46.6$ GΩ, $z_{out} = 3.22$ mΩ.

PRACTICE PROBLEMS 3

Use a 741 op-amp to design an inverting amplifier for a closed-loop gain of 10. Let $R_1 = 20$ kΩ. Use the characteristics of Example 28–5 to solve for the input and the output impedance of your design.

Answers
$R_F = 180$ kΩ, $z_{in} = 40.0$ GΩ, $z_{out} = 3.75$ mΩ.

28.6 Op-Amp Specifications

In this section, we examine typical electrical specifications of the op-amp. Although our examination uses the characteristics of the 741C op-amp summarized in Figure 28–30, you will find that all op-amps have similar characteristics.

Parameter	Conditions	LM741A			LM741			LM741C			Units
		Min	Typ	Max	Min	Typ	Max	Min	Typ	Max	
Input Offset Voltage	$T_A = 25°\,C$ $R_S \leq 10\,k\Omega$ $R_S \leq 50\,k\Omega$		0.8	3.0		1.0	5.0		2.0	6.0	mV mV
	$T_{AMIN} \leq T_A \leq T_{AMAX}$ $R_S \leq 50\,k\Omega$ $R_S \leq 10\,k\Omega$			4.0			6.0			7.5	mV mV
Average Input Offset Voltage Drift				15							$\mu V/°C$
Input Offset Voltage Adjustment Range	$T_A = 25°\,C, V_S = \pm 20V$	± 10				± 15			± 15		mV
Input Offset Current	$T_A = 25°\,C$ $T_{AMIN} \leq T_A \leq T_{AMAX}$		3.0	30 70		20 85	200 500		20	200 300	nA nA
Average Input Offset Current Drift				0.5							$nA/°C$
Input Bias Current	$T_A = 25°\,C$ $T_{AMIN} \leq T_A \leq T_{AMAX}$		30	80 0.210		80	500 1.5		80	500 0.8	nA μA
Input Resistance	$T_A = 25°\,C, V_S = \pm 20V$ $T_{AMIN} \leq T_A \leq T_{AMAX},$ $V_S = \pm 20V$	1.0 0.5	6.0		0.3	2.0		0.3	2.0		$M\Omega$ $M\Omega$
Input Voltage Range	$T_A = 25°\,C$ $T_{AMIN} \leq T_A \leq T_{AMAX}$				± 12	± 13		± 12	± 13		V V
Large Signal Voltage Gain	$T_A = 25°\,C, R_L \geq 2\,k\Omega$ $V_S = \pm 20V, V_O = \pm 15V$ $V_S = \pm 15V, V_O = \pm 10V$	50			50	200		20	200		V/mV V/mV
	$T_{AMIN} \leq T_A \leq T_{AMAX},$ $R_L \geq 2\,k\Omega,$ $V_S = \pm 20V, V_O = \pm 15V$ $V_S = \pm 15V, V_O = \pm 10V$ $V_S = \pm 5V, V_O = \pm 2V$	32 10			25			15			V/mV V/mV V/mV
Output Voltage Swing	$V_S = \pm 20V$ $R_L \geq 10\,k\Omega$ $R_L \geq 2\,k\Omega$	± 16 ± 15									V V
	$V_S = \pm 15V$ $R_L \geq 10\,k\Omega$ $R_L \geq 2\,k\Omega$				± 12 ± 10	± 14 ± 13		± 12 ± 10	± 14 ± 13		V V
Output Short Circuit Current	$T_A = 25°\,C$ $T_{AMIN} \leq T_A \leq T_{AMAX}$	10 10	25	35 40		25			25		mA mA
Common-Mode Rejection Ratio	$T_{AMIN} \leq T_A \leq T_{AMAX}$ $R_S \leq 10\,k\Omega, V_{CM} = \pm 12V$ $R_S \leq 50\,\Omega, V_{CM} = \pm 12V$	80	95		70	90		70	90		dB dB
Supply Voltage Rejection Ratio	$T_{AMIN} \leq T_A \leq T_{AMAX},$ $V_S = \pm 20V$ to $V_S = \pm 5V$ $R_S \leq 50\,\Omega$ $R_S \leq 10\,k\Omega$	86	96		77	96		77	96		dB dB
Transient Response Rise Time Overshoot	$T_A = 25°\,C$, Unity Gain		0.25 6.0	0.8 2.0		0.3 5			0.3 5		μs %
Bandwidth (Note 6)	$T_A = 25°\,C$	0.437	1.5								MHz
Slew Rate	$T_A = 25°\,C$, Unity Gain	0.3	0.7			0.5			0.5		V/μs
Supply Current	$T_A = 25°\,C$					1.7	2.8		1.7	2.8	mA

FIGURE 28–30   Electrical characteristics of the LM741 op-amp ($V_S = \pm 15$ V). *(Courtesy of National Semiconductor Corporation)*

Input Offset Voltage, V_{io}

There will always be imbalances in any differential amplifier. In the op-amp circuits that we have analyzed up to now, we have assumed that the op-amps are ideal, namely that there is no current entering the inverting or the non-inverting inputs. We have also assumed that the differential voltage between the input terminals is exactly zero. Although these assumptions are very good, they are not exactly correct. It is important to understand that real op-amp circuits may not work exactly as expected. If we examine the circuit shown in Figure 28–31, we see that the output voltage should ideally be zero. However, due to imbalances in the differential amplifier, there will be a slight error in the output voltage resulting in a value between a few microvolts and several millivolts. The specification sheet for the 741C op-amp indicates that a typical **input offset voltage** of 2 mV should be expected.

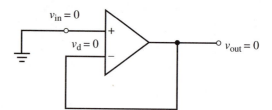

FIGURE 28–31

What possible effect could such small voltages have on the overall operation of the op-amp? Consider that the input offset voltage, V_{io} is modeled as a battery in series with the non-inverting (+) terminal of the op-amp as shown in Figure 28–32. We will discover that this insignificant value can have a dramatic effect on the operation of the amplifier. For example, if we were to connect an op-amp as illustrated in Figure 28–33(a), we see that the offset voltage will be multiplied by the open-loop gain of the op-amp, resulting in the output that will be equal to $\pm V_{SAT}$, depending on the polarity of the offset voltage.

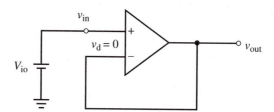

FIGURE 28–32

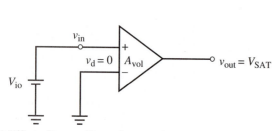

(a) Effect of input offset voltage on the output voltage of an op-amp having no feedback

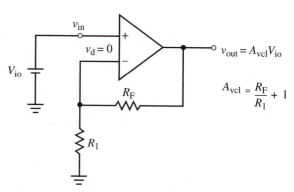

(b) Effect of input offset voltage on the output voltage of an op-amp using negative feedback

FIGURE 28–33

If the amplifier uses a feedback resistor as illustrated in Figure 28–33(b), the output will normally be less than the saturation voltage, but the offset voltage will nonetheless be amplified, in this case by the closed loop gain of the amplifier. The following example shows how to measure the input offset voltage of an op-amp.

EXAMPLE 28–7

The 741C op-amp shown in Figure 28–34 is measured to have output voltage of 180 mV. Determine the input offset voltage for this amplifier. Compare the result to the value of V_{io} specified by the manufacturer.

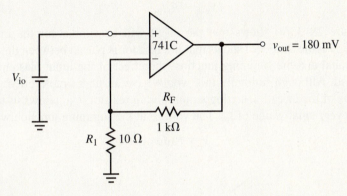

FIGURE 28–34

Solution
The voltage gain of the amplifier is

$$A_v = \frac{R_F}{R_1} + 1 = \frac{1000\ \Omega}{10\ \Omega} + 1 = 101$$

and so we have

$$V_{io} = \frac{V_{out}}{A_{vcl}} = \frac{180\ mV}{101} = 1.78\ mV$$

The manufacturer specifies that the input offset voltage has a typical value of 2 mV. The above result is consistent with the manufacturer's specification.

Input Bias Current, I_B

By examining the schematic of the op-amp, we see that the circuit consists of numerous transistors. Remember that in order for a transistor to operate correctly, it must be properly biased. The bias current for the base of the transistor connected to the inverting input terminal is referred to as $I_{B(-)}$, while the bias current for the base of the transistor connected to the non-inverting input terminal is referred to as $I_{B(+)}$. The base bias currents needed to operate the transistors of the differential amplifier will generally require slightly different bias currents for the two input terminals. We define the input bias current as the average of the two bias currents, namely

$$I_B = \frac{\left|I_{B(+)}\right| + \left|I_{B(-)}\right|}{2} \tag{28–24}$$

The 741C op-amp has a typical bias current of 80 nA. If the op-amp were designed using FET differential amplifiers, the bias current would have been in the order of pA (10^{-12} A).

Input Offset Current, I_{os}

The difference between the magnitude of the two bias currents is called the **input offset current** and is given as

$$I_{OS} = \left| I_{B(+)} \right| - \left| I_{B(-)} \right| \tag{28–25}$$

The 741C op-amp has a typical input offset current of 20 nA. Let's examine the effect of the input offset current on an amplifier.

Consider the voltage follower shown in Figure 28–35(a). If we imagine that the input offset current is applied to the inverting terminal as shown, we see that the output voltage of the amplifier will not be zero, but rather will be equal to the voltage across the feedback resistor.

$$V_{out} = R_F I_{B(-)} \tag{28–26}$$

Figure 28–35(b) shows that the result will be similar for the inverting amplifier. Figure 28–35(c) shows that if a resistor is placed between the inverting terminal and the summing junction, the effect of the input bias current is magnified. Although normally this would have an undesirable effect on the operation of the circuit, inserting a multiplying resistor, R_M, allows us to measure the very small value of I_{os}. The voltage at the summing junction will be

$$V = R_M I_{B(-)}$$

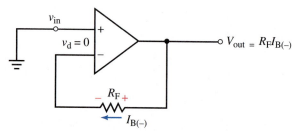

(a) Effect of input bias current on the voltage follower circuit

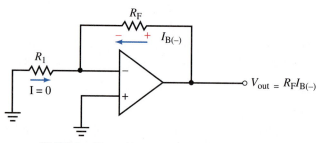

(b) Effect of input bias current on the inverting amplifier

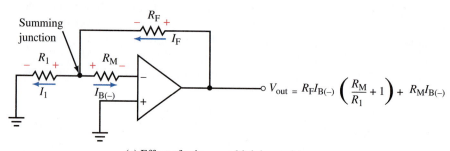

(c) Effect of using a multiplying resistor

FIGURE 28–35

and so the current through R_1 will be determined from Ohm's law as

$$I_1 = \frac{R_M I_{B(-)}}{R_1}$$

The current through the feedback resistor is determined by applying Kirchhoff's current law and is found to be

$$I_F = I_1 + I_{B(-)} = \frac{R_M I_{B(-)}}{R_1} + I_{B(-)} = I_{B(-)}\left(\frac{R_M}{R_1} + 1\right)$$

and so the output voltage is found by Kirchhoff's voltage law as

$$v_{out} = R_F I_F + R_M I_{B(-)}$$

which gives

$$V_{out} = R_F I_{B(-)}\left(\frac{R_M}{R_1} + 1\right) + R_M I_{B(-)} \qquad \text{(28–27)}$$

The following example shows that the multiplying resistor is easily used to find the value of the input offset current.

EXAMPLE 28–8

The circuit of Figure 28–36 is constructed to measure the value of the input offset current. The output voltage is measured to be $v_{out} = 0.6$ V. Determine the input offset current, $I_{io} = I_{B(-)}$.

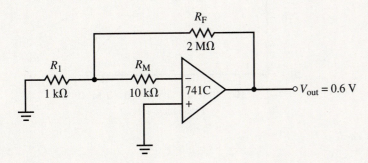

FIGURE 28–36

Solution

$$0.6 \text{ V} = (2 \text{ M}\Omega)I_{B(-)}\left(\frac{10 \text{ k}\Omega}{1 \text{ k}\Omega} + 1\right) + (10 \text{ k}\Omega)I_{B(-)}$$

$$I_{io} = I_{B(-)} \approx \frac{0.6 \text{ V}}{(2 \text{ M}\Omega)(11)} = 0.0273 \text{ }\mu\text{A} \equiv 27.3 \text{ nA}$$

Input Resistance

We have already done an in-depth examination of this characteristic. Although the input resistance of an op-amp has a value around 2 MΩ, we have observed that by using the inverting amplifier, the input impedance of a circuit can be significantly reduced. Conversely, using a non-inverting amplifier results in greatly increased input impedance for the circuit.

Open Loop Voltage Gain, A_{vol} and Bandwidth, *BW*

The open loop voltage gain and the bandwidth of the op-amp are closely related. Figure 28–37 shows that although the open loop voltage gain at low frequencies is very high (200,000) the gain very quickly decreases as the frequency of operation increases. For example, at a frequency of 10 Hz, the open loop gain of the 741C op-amp is 100,000, while at a frequency of 100 kHz, the gain has decreased to a value of 10. Notice that the product of gain and bandwidth is constant everywhere along the graph between 10 Hz and 1 Mz. We conclude, therefore, that the **gain-bandwidth product** for the 741 op-amp is a constant 10^6 Hz. This means that if we were to use negative feedback to set the closed loop gain of the amplifier, the bandwidth would be easily determined as

$$[BW]_{Hz} = \frac{10^6 \ Hz}{A_{vcl}} \qquad\qquad \textbf{(28–28)}$$

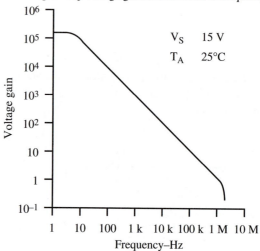

Open loop voltage gain as a function of frequency

FIGURE 28–37

PRACTICE PROBLEMS 4

A non-inverting amplifier has a gain of 20. Determine its bandwidth.

Answer
50 kHz

Slew Rate, $\Delta V/\Delta t$

The **slew rate** of an amplifier is a measure of how fast the output voltage can change. The lowest slew rate occurs for an amplifier having unity gain ($A_v = 1$). Imagine that a perfect square wave having amplitude of 10 V is applied to the input of a unity gain amplifier.

As illustrated in Figure 28–38, the output will not rise instantaneously, but rather have a slope determined by the slew rate of the op-amp. For the 741C op-amp, the value of slew rate is 0.5 V/μs. This means that it will take 20 μs for the output to go from 0 V to 10 V. The slew rate affects not only square wave inputs but results in distortion of any waveform that rises at a rate faster than 0.5 V/μs.

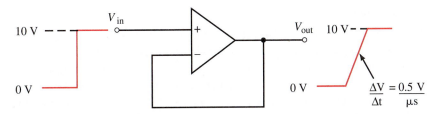

FIGURE 28–38

Consider that a sine wave is applied to the input of an amplifier. The input voltage is written as v(t) = A sin ωt. The instantaneous rate of change of the sinusoidal is determined by taking the derivative of this function, namely

$$\frac{dv}{dt} = \omega A \cos \omega t$$

Consequently, the maximum rate of change of the sinusoidal is simply

$$\left[\frac{dv}{dt}\right]_{max} = \omega A$$

which simplifies to

$$\text{slew rate} = 2\pi f A$$

This means that we can determine the maximum frequency or the maximum amplitude from the slew rate of an amplifier.

$$f_{max} = \frac{\text{slew rate}}{2\pi A} \qquad \textbf{(28–29)}$$

$$A_{max} = \frac{\text{slew rate}}{2\pi f} \qquad \textbf{(28–30)}$$

Determine the maximum peak undistorted output voltage that can be obtained from a non-inverting 741C op-amp having a gain of 5.

EXAMPLE 28–9

Solution
For a gain of 5, the bandwidth of the 741C op-amp is

$$[BW]_{Hz} = \frac{10^6}{A_{vcl}} = \frac{10^6}{5} = 200 \text{ kHz}$$

The maximum amplitude at the output of the amplifier is

$$A_{max} = \frac{\text{slew rate}}{2\pi f} = \frac{0.5 \text{ V/}\mu s}{2\pi (200 \text{ kHz})} = 0.398 \text{ V}_p$$

This means that the maximum sinusoidal signal that can be applied to the input of the amplifier is approximately 80 mV$_p$.

Determine the maximum sinusoidal frequency that can be applied to a unity gain non-inverting 741C op-amp if the amplitude of the input is 10 V$_p$.

PRACTICE PROBLEMS 5

Answer
8 kHz

Bias Compensation

We have observed that input bias current and input bias voltage can have a dramatic effect on the operation of an op-amp circuit. The effects of bias current can be minimized if the impedance seen between the non-inverting terminal and ground is the same as the impedance seen from the inverting terminal to ground. For example, placing a resistor, $R_C = R_1 \| R_F$ in series with the non-inverting input compensates for the effects of bias current in the circuit of Figure 28–39. If the signal source has a large internal resistance, its effect must also be included in the calculation.

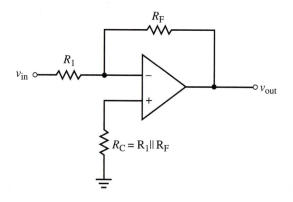

FIGURE 28–39

EXAMPLE 28–10

Given the circuit of Figure 28–40, determine the value of the compensating resistor, R_C.

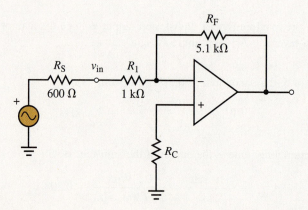

FIGURE 28–40

Solution

In this circuit, we must consider the effect of the source resistance, $R_S = 600\ \Omega$.

$$R_C = (R_S + R_1)\|R_F = (0.600\ \text{k}\Omega + 1\ \text{k}\Omega)\|5.1\ \text{k}\Omega = 1.22\ \text{k}\Omega$$

Manufacturers of op-amps have included null circuits that compensate for offset voltage. For example, the 741C op-amp has two external connections, −OFFSET NULL (pin 1) and +OFFSET NULL (pin 5), that permit very simple correction for offset voltage. Figure 28–41 shows the proper connections to minimize offset error for the 8-pin 741C DIP package.

The procedure used to null an op-amp is itemized below:

1. Remove all ac voltage sources and replace them with short circuits to ground.
2. Connect a 10-kΩ potentiometer between the offset null pins as illustrated in Figure 28–41.
3. Connect the wiper (center terminal) of the potentiometer to the negative supply.
4. With a dc coupled oscilloscope or a sensitive dc voltmeter connected between the output terminal and ground, adjust the wiper until the output voltage has dropped to zero as measured on the most sensitive scale.
5. Reinsert the ac voltage sources and do not readjust the potentiometer.

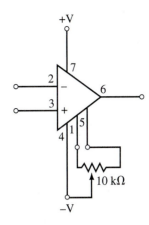

FIGURE 28–41 Connection of an offset voltage adjustment resistor.

Once an op-amp circuit is working, very little can go wrong unless there is a power surge resulting in a catastrophic failure. Most problems are encountered when the circuit is first built. The most important connection (and yet the most common source of errors) when building op-amp circuits occurs in connecting the dual power supply. Refer to the circuit of Figure 28–42, which shows all connections for an op-amp. The two supplies must be connected to the circuit reference point (ground) as illustrated.

28.7 Troubleshooting an Op-Amp Circuit

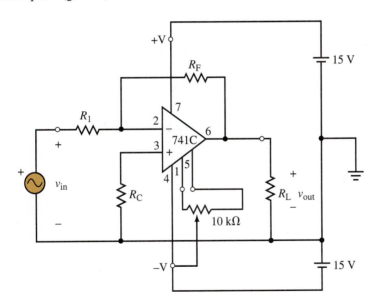

FIGURE 28–42

Notice that the input and the output of the amplifier use the same reference point as well. At this point it is worth noting that the reference point should only have one external connection to an earth ground, if required. Otherwise it is possible to generate a ground loop that acts as antenna, which picks up external noise, generally in the form of 60 Hz hum. For the same reason, it is good practice to keep all connecting wires short. If, after all other precautions, the amplifier is still picking up external noise, it may help to connect a small capacitor (e.g. 0.01 μF) across the feedback resistor, R_F.

In the unlikely event that a resistor in the op-amp circuit becomes faulty, it is possible to use voltage measurements to determine the origin of the fault. For example, if R_F of an inverting amplifier were to be open-circuited, there would be no feedback path. Consequently, the op-amp would have a gain equal to A_{vol}, resulting in an output voltage $v_{out} = \pm V_{SAT}$. On the other hand, if R_F were to be short-circuited, the entire output signal would be returned to the input, resulting in total cancellation, and so $v_{out} = 0$ V, regardless of the input value.

If R_1 of an inverting amplifier were to be short-circuited, the entire input voltage would be applied directly to the differential input of the op-amp, once again resulting in an output voltage $v_{out} = \pm V_{SAT}$. If R_1 were to be open-circuited, none of the input signal would appear at the op-amp, and so $v_{out} = 0$ V, regardless of the input value.

28.8 Computer Analysis of Op-Amp Circuits

In this section, we will use MultiSIM to show typical results that will be observed when measuring quantities in the lab. Although PSpice may be used to predict similar results, this software is used to create an op-amp model and allows us to measure both input impedance and output impedance of a non-inverting op-amp circuit. These values will be very difficult to observe in a lab, since the input impedance of the circuit will normally be in the order of many GΩ, while the output impedance will be in the order of a few mΩ. The PSpice simulation is used to demonstrate the validity of the equations used to solve for these values.

MultiSIM

EXAMPLE 28–11

Use MultiSIM to model the circuit of Figure 28–22. Apply a 1-kHz sinusoidal signal having amplitude of 100 mV$_p$ and use an oscilloscope to display both the input and the output waveforms. Calculate the voltage gain from the measurements.

Solution
The completed MultiSIM circuit is shown in Figure 28–43.

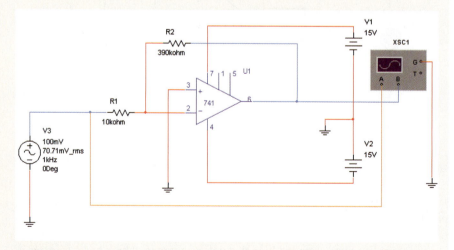

FIGURE 28–43

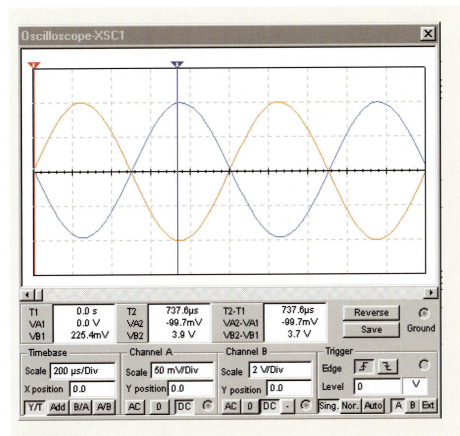

FIGURE 28–44

The waveforms appearing on the oscilloscope are shown in Figure 28–44. As expected, the output is 180° out-of-phase with respect to the input. The cursors allow us to calculate the gain of the amplifier as

$$A_v = \frac{v_{out}}{v_{in}} = -\frac{3.9 \text{ V}}{0.1 \text{ V}} = -39$$

The result is consistent with the gain determine in the calculations of Example 28–4.

PRACTICE PROBLEMS 6

Use MultiSIM to model the circuit of Figure 28–27. Apply a 1-kHz sinusoidal signal having amplitude of 100 mV$_p$ and use an oscilloscope to display both the input and the output waveforms. Calculate the voltage gain from the measurements.

Answer
$A_v = +6.1$ The output is in phase with the input.

PSpice

EXAMPLE 28–12

Use PSpice to create the model of a 741 op-amp. The model shall use $R_{in} = 2\ M\Omega$, $A_{vol} = 200{,}000$, and $R_{out} = 75\ \Omega$. Use the op-amp model to simulate the circuit shown in Figure 28–27. Apply an input voltage of 10 mV to the circuit and determine voltage gain, input impedance, and output impedance of the amplifier. Compare the results to those found in Example 28–5.

Solution

The circuit is constructed using the VSRC voltage source and changing the properties of the component (by pointing to the component and double-clicking) so that DC = 0V and AC = 1V. Figure 28–45 shows the PSpice capture file. Notice that the voltage controlled voltage source is used to model the open loop voltage gain (200,000) of the 741 op-amp.

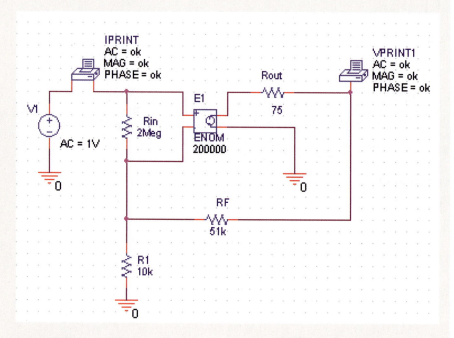

FIGURE 28–45

After entering the complete circuit and making the required changes to the circuit properties, the circuit can be simulated using an ac sweep. Once the Probe Postprocessor window appears, we obtain the desired voltage and current values by clicking on the <u>V</u>iew menu item and selecting Output F<u>i</u>le. The pertinent data from the output file is given as

```
FREQ            VM(N00083)       VP(N00083)
1.000E+03       6.100E+00        0.000E+00
```

and

```
FREQ            IM(V_PRINT1)     IP(V_PRINT1)
1.000E+03       1.527E-11        0.000E+00
```

The above data indicates that the output voltage is in phase with the input voltage and that $v_{out} = 6.1$ V. Therefore the gain of the amplifier is $A_v = 6.1$, as found in Example 28–5. The input impedance is determined by applying Ohm's law to the input of the circuit, namely

$$z_{in} = \frac{1 \text{ V}}{1.527 \times 10^{-11} \text{ A}} = 65.5 \text{ G}\Omega$$

The above result is almost identical to the calculated value of $z_{in} = 65.6$ GΩ. In order to find the output impedance, we need to determine the short circuit output current by replacing the VPRINT1 device with an IPRINT device. With an input voltage of 1 V, we can expect the short circuit current to be exceptionally large. Remember that we have modeled the 741 op-amp with its equivalent circuit, without regard to the actual circuit limits. Placing the IPRINT device between the output terminals and ground provides the following value for the short circuit current.

```
FREQ          IM(V_PRINT2)    IP(V_PRINT2)
1.000E+03     2.656E+03       0.000E+00
```

Clearly the short circuit current of 2656 A is much larger than the 741 op-amp is able to provide (25 mA). Although the current value is unrealistically large, we are nonetheless able to apply Thévenin's theorem for dependent sources to calculate the output impedance as follows:

$$z_{out} = \frac{V_{OC}}{I_{SC}} = \frac{6.1 \text{ V}}{2656 \text{ A}} = 2.30 \text{ m}\Omega$$

This value is precisely the value that was found in Example 28–5. We are therefore able to see that by using negative feedback, the input impedance of a non-inverting amplifier is increased to effectively an open circuit, while the output impedance is reduced to virtually zero.

PROBLEMS

28.1 Introduction to the Operational Amplifier

1. What are the three characteristics of a typical operational amplifier?

2. a. If a signal is applied to the inverting input, with the non-inverting input grounded, in what mode is the op-amp working?

   b. If a signal is applied between the inverting input and the non-inverting input, in what mode is the op-amp working?

28.2 The Differential Amplifier and Common-Mode Signals

3. Given that the circuit of Figure 28–46 has $V_{CC} = +15$ V, $V_{EE} = -15$ V, $R_E = 2.7$ kΩ, and $R_{C1} = R_{C2} = 2.2$ kΩ, find I_{C1} and V_{C1}.

4. Repeat Problem 1 if the circuit of Figure 28–46 has $V_{CC} = +12$ V, $V_{EE} = -12$ V, $R_E = 1.8$ kΩ, and $R_{C1} = R_{C2} = 1.2$ kΩ

5. An amplifier has a differential voltage gain of 200 and a common-mode voltage gain of 0.01. Calculate the common-mode rejection ratio in decibels.

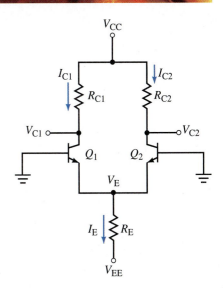

FIGURE 28–46

6. An op-amp is specified to have an open-loop voltage gain of 100,000 and a common-mode rejection ratio of at least 75 dB.

   a. If the amplifier is measured to have a common-mode gain of 0.02, calculate the actual $[\text{CMRR}]_{\text{dB}}$.

   b. Does the op-amp meet the manufacturer's specification?

7. An op-amp circuit has a differential voltage gain of 200 and CMRR of 80 dB. If the output signal voltage of the amplifier is 10 $V_{\text{p-p}}$, determine the maximum common-mode input voltage that can be applied to the amplifier to ensure that the output signal-to-noise ratio is no less than 1000 (60 dB).

8. Repeat Problem 5 if the op-amp were to have a differential voltage gain of 20. (All other values remain unchanged.)

28.4   The Inverting Amplifier

9. If $R_F = 100$ kΩ and $R_1 = 20$ kΩ in the circuit of Figure 28–47, determine the voltage gain, input impedance, and output impedance of the amplifier. Solve for v_{out} if $v_{\text{in}} = -2.0$ V.

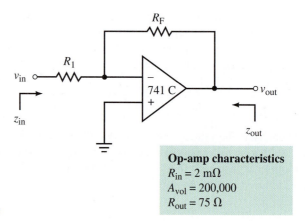

FIGURE 28–47

10. If $R_F = 68$ kΩ and $R_1 = 15$ kΩ in the circuit of Figure 28–47, determine the voltage gain, input impedance, and output impedance of the amplifier. Solve for v_{out} if $v_{\text{in}} = +0.5$ V.

11. Design the amplifier circuit of Figure 28–47 to have input impedance of 12 kΩ and a voltage gain of magnitude 5.

12. Design the amplifier circuit of Figure 28–47 to have input impedance of 7.5 kΩ and a voltage gain of magnitude 4.

28.5   Non-inverting Amplifier

13. If $R_F = 100$ kΩ and $R_1 = 20$ kΩ in the circuit of Figure 28–48, determine the voltage gain, input impedance, and output impedance of the amplifier. Solve for v_{out} if $v_{\text{in}} = -2.0$ V.

14. If $R_F = 68$ kΩ and $R_1 = 15$ kΩ in the circuit of Figure 28–48, determine the voltage gain, input impedance, and output impedance of the amplifier. Solve for v_{out} if $v_{\text{in}} = +0.5$ V.

15. Design the amplifier circuit of Figure 28–48 to have a voltage gain of magnitude 5.

16. Design the amplifier circuit of Figure 28–48 to have a voltage gain of magnitude 4.

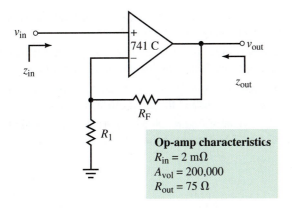

FIGURE 28–48

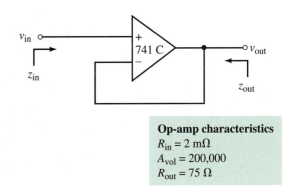

FIGURE 28–49

17. Determine the voltage gain, input impedance, and output impedance of the circuit shown in Figure 28–49.

18. The circuit of Figure 28–50(a) shows the loading effect when a low-resistance load is connected to a high-resistance source. Figure 28–50(b) shows how a voltage-follower (buffer) circuit is used to prevent loading effect.

    a. Calculate v_{out} for the circuit in Figure 28–50(a).

    b. Using the input impedance determined in Problem 17, find v_{in} for the circuit of Figure 28–50(b). Calculate v_{out} for the circuit.

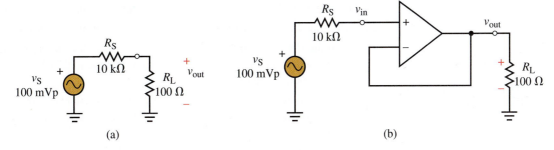

(a)   (b)

FIGURE 28–50

28.6 Op-Amp Specifications

19. Given that $R_F = 10\ k\Omega$ and $R_1 = 10\ \Omega$ in the circuit of Figure 28–51, calculate the value of output voltage that you would expect if the manufacturer of the op-amp gives a typical value of input offset voltage, $V_{io} = 2\ mV$.

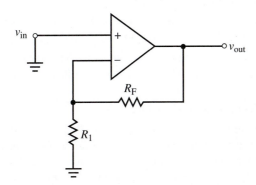

FIGURE 28–51

20. What value of output voltage would you expect for a voltage follower using the op-amp having input offset voltage, $V_{io} = 2$ mV?

21. An op-amp is measured to have $I_{B(+)} = 60$ nA and $I_{B(-)} = 40$ nA. Calculate the values of the input bias current, I_B and the input offset current, I_{os}.

22. An op-amp is measured to have $I_{B(+)} = 65$ nA and $I_{B(-)} = 45$ nA. Calculate the values of the input bias current, I_B and the input offset current, I_{os}.

23. If the output voltage in the circuit of Figure 28–52 is measured with a digital voltmeter and is found to be 0.5 V, determine the value of the input offset current, I_{os}.

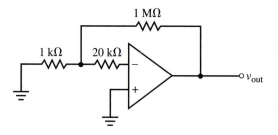

FIGURE 28–52

24. Repeat the calculation of Problem 23 if $v_{out} = 0.7$ V.

25. Determine the bandwidth for the inverting amplifier of Problem 9.

26. Determine the bandwidth for the inverting amplifier of Problem 10.

27. Determine the bandwidth for the non-inverting amplifier of Problem 15.

28. Determine the bandwidth for the non-inverting amplifier of Problem 16.

29. Using the slew rate for the 741C op-amp, calculate the maximum sinusoidal frequency that can be applied to the non-inverting amplifier of Problem 15 if the amplitude of the input is $1.0\ V_p$.

30. Using the slew rate for the 741C op-amp, calculate the maximum signal (amplitude) that can be applied to the input of the non-inverting amplifier of Problem 16.

31. Solve for the value of compensating resistor that would be required for the inverting amplifier of Problem 9. Sketch the resulting circuit, showing all resistor values.

32. Solve for the value of compensating resistor that would be required for the inverting amplifier of Problem 10. Sketch the resulting circuit, showing all resistor values.

28.7 Troubleshooting an Op-Amp Circuit

33. An inverting op-amp circuit has been operating properly for a long time. Given two failures that could result an output voltage of $v_{out} = 0$ V.

34. An inverting op-amp circuit has been operating properly for a long time. Given two failures that could result an output voltage of $v_{out} = \pm V_{SAT}$.

28.8 Computer Analysis of Op-Amp Circuits

◀ MULTISIM

35. Use MultiSIM to input the circuit of Problem 9 using a 741C op-amp. Apply a signal of 100 mV_p to the input of the amplifier and use the oscilloscope tool to simultaneously display the waveforms for v_{in} and v_{out}. Use the oscilloscope display to determine the voltage gain, $A_v = v_{out}/v_{in}$ of the circuit.

36. Use MultiSIM to input the circuit of Problem 10 using a 741C op-amp. Apply a signal of 100 mV$_p$ to the input of the amplifier and use the oscilloscope tool to simultaneously display the waveforms for v_{in} and v_{out}. Use the oscilloscope display to determine the voltage gain, $A_v = v_{out}/v_{in}$ of the circuit.

◀ MULTISIM

37. Use MultiSIM to input the circuit of Problem 15 using a 741C op-amp. Apply a signal of 100 mV$_p$ to the input of the amplifier and use the oscilloscope tool to simultaneously display the waveforms for v_{in} and v_{out}. Use the oscilloscope display to determine the voltage gain, $A_v = v_{out}/v_{in}$ of the circuit.

◀ MULTISIM

38. Use MultiSIM to input the circuit of Problem 16 using a 741C op-amp. Apply a signal of 100 mV$_p$ to the input of the amplifier and use the oscilloscope tool to simultaneously display the waveforms for v_{in} and v_{out}. Use the oscilloscope display to determine the voltage gain, $A_v = v_{out}/v_{in}$ of the circuit.

◀ MULTISIM

39. Use PSpice to input the equivalent circuit model of 741C op-amp. Using this model and the resistor values calculated in Problem 9, determine the input impedance and output impedance of circuit of the amplifier.

◀ CADENCE

40. Use PSpice to input the equivalent circuit model of 741C op-amp. Using this model and the resistor values calculated in Problem 10, determine the input impedance and output impedance of circuit of the amplifier.

◀ CADENCE

41. Use PSpice to input the equivalent circuit model of 741C op-amp. Using this model and the resistor values calculated in Problem 15, determine the input impedance and output impedance of circuit of the amplifier.

◀ CADENCE

42. Use PSpice to input the equivalent circuit model of 741C op-amp. Using this model and the resistor values calculated in Problem 16, determine the input impedance and output impedance of circuit of the amplifier.

◀ CADENCE

■ OBJECTIVES

On completion of this chapter, you will be able to

- analyze a comparator circuit and determine the circuit operation,

- analyze and design a voltage summing amplifier circuit,

- analyze the operation and determine the output waveform for an integrator circuit,

- analyze the operation and determine the output waveform for a differentiator circuit,

- analyze the operation of an instrumentation amplifier and be able to solve for the output of an unbalanced resistor bridge circuit,

- determine cutoff frequencies of low- and high-pass active filters and be able to sketch the frequency response of these filters,

- analyze and design wideband and narrowband bandpass filters for desired bandwidths,

- design an active notch filter circuit using a narrowband bandpass filter combined with a summing amplifier,

- analyze and design voltage regulator circuits using zener diodes, op-amps, and three-terminal voltage regulator ICs,

- calculate line regulation, load regulation, and ripple rejection ratio for given conditions in a regulator circuit,

- describe the operation of a voltage regulator circuit using an op-amp,

- use computer software to simulate the operation of op-amp circuits.

Applications of Op-Amps

29

CHAPTER PREVIEW

In the previous chapter, you learned that operational amplifiers can be used to provide inverting and non-inverting amplifiers. In this chapter we examine how op-amps can be used in both linear and non-linear applications. We begin by examining the comparator, which as the name implies, provides an output based on the result of voltage comparison at the input. We then examine the operation of op-amps in instrumentation amplifiers, where the signals are generally very small and often occur in environments that are electrically noisy. We then examine the operation of active filter circuits, which use the op-amp to provide additional gain and prevent loading effect that would occur if we used R-L-C components alone. Finally, the chapter is concluded by examining the operation of voltage regulators. ∎

PUTTING IT IN PERSPECTIVE

ALTHOUGH MOST PEOPLE THINK OF the computer as an invention that occurred in the late twentieth century, computers were in fact used much earlier. The first computers were not digital computers, as are modern computers, but rather were **analog computers,** constructed using mechanical components such as levers, gears, wheels, and springs to predict tides, positions of the planets and to measure and emulate the occurrence of other physical events.

William Thomson, 1824–1907 (who is better known as Lord Kelvin), was convinced that most mechanical events could be analyzed and predicted by using analogous mathematical models. Thomson invented and built devices called **planimeters** that used disk and wheel arrangements to solve mathematical integrals of harmonic motion such as tides. Similar analog computers are still in use today. For example, the mechanical odometer/speedometer converts the speed of tire rotation into an analogous distance traveled. (Remember from your physics course that the integral of speed is distance.)

The invention of the operational amplifier has made it possible to simplify analog computations without the necessity of using often-large, inefficient mechanical devices. By using operational amplifiers, it is possible to design and build electronic circuits that will add, subtract, and perform differentiation and integration.

Although the analog computer has been largely overshadowed by the much faster and smaller digital computer, it is highly unlikely that the operational amplifier will become obsolete anytime soon. ∎

29.1 Comparators

In the previous chapter you learned that the operational amplifier has a very large open-loop gain, which results in the device becoming saturated even when a very small differential input voltage is applied between the input terminals. In order to have an op-amp work as a linear device, where output is proportional to the input, you learned that it was necessary to implement negative feedback. Thus, the gain of the amplifier became much more stable with the addition of a few external resistors.

In this section, you will find that we will use the op-amp as a **comparator,** which is a non-linear device having an output voltage that is one of two values. If the voltage at the non-inverting (+) input is greater than the voltage at the inverting (−) input, the output will be $+V_{SAT}$, and if the opposite occurs the output will be $-V_{SAT}$. This occurs since even a very small difference will be amplified by the op-amp's very large open-loop voltage gain. Consider the circuit of Figure 29–1(a). Here we see that the inverting terminal of the op-amp is connected to ground, giving a reference voltage of $V_{REF} = 0$ V. Figure 29–1(b) shows that if a sine-wave is applied to the input, the output of the op-amp will be a square wave.

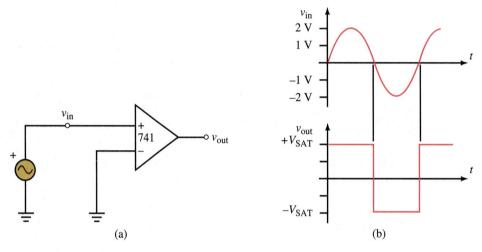

(a)   (b)

FIGURE 29–1 Comparator circuit.

Now, if we were to apply a dc reference voltage, $V_{REF} = 1$ V, as shown in Figure 29–2(a) to the inverting terminal, the output would be equal to $+V_{SAT}$ only when the input voltage exceeds $+1$ V. The resultant is shown in Figure 29–2(b).

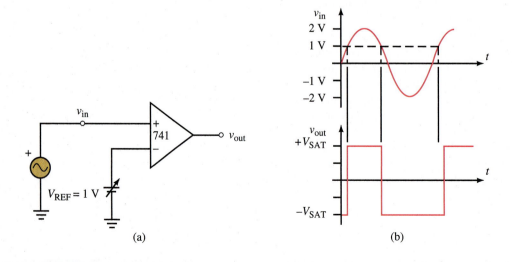

(a)   (b)

FIGURE 29–2

Although we have been connecting a reference voltage to the inverting terminal, it is often useful to connect the reference to the non-inverting terminal as illustrated in Figure 29–3.

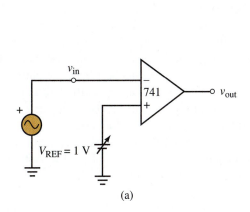

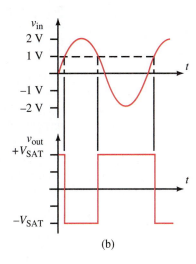

(a)

(b)

FIGURE 29–3

In this circuit, $v_{out} = -V_{SAT}$ whenever the input voltage is greater than V_{REF} as illustrated in Figure 29–3(b).

The comparator circuit has many applications in electronics. We will examine two possible applications. Remember, whenever you see a comparator used in any electronic circuit, the output will be either $-V_{SAT}$ or $+V_{SAT}$, depending on difference between the two input terminals of the op-amp.

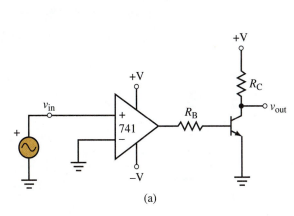

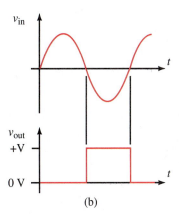

(a)

(b)

FIGURE 29–4   Zero-crossing detector.

Zero-Crossing Detector

In the zero-crossing detector circuit shown in Figure 29–4(a), the output of the comparator will go from $-V_{SAT}$ to $+V_{SAT}$ as the input signal goes from a negative value to a positive value (zero crossing). If the value of R_B is sufficiently small, the base current will saturate the transistor and result in $v_{out} = 0$ V. When the input voltage falls below 0 V, the output of the comparator goes to $-V_{SAT}$, reverse-biasing the base-emitter junction of the transistor and forcing the transistor into cutoff. The output will now be $+V$. Figure 29–4(b) shows the output of a zero-crossing detector for a sinusoidal input.

LED Voltage Level Indicator

The circuit of Figure 29–5 shows how comparators can be used to turn on progressively more LEDs as the input voltage increases. Such a circuit can be used in stereo amplifiers to show the relative signal strength at the input of a power amplifier. As the signal increases, more LEDs will turn on. (Normally, the top few LEDs in the stack will be red to indicate that the signal level is too large.) In the circuit of Figure 29–5, the four resistors, R_1, R_2, R_3, and R_4, establish a voltage divider network. By applying the voltage divider rule, we determine the reference voltage at the inverting input of the bottom comparator to be 1 V, while the reference voltages at the other two comparators will be 2 V and 3 V respectively. If the input voltage goes above 1 V, the output of the bottom comparator will go to $+V_{SAT} \approx 8$ V, causing current through LED1. If v_{in} is between 1 V and 2 V, the output of the top two comparators will be $-V_{SAT}$, resulting in no current through LED2 and LED3.

If v_{in} becomes greater than 2 V, the second comparator will have an output voltage of $+V_{SAT}$, resulting in current through LED2. Note that the output of comparator 1 remains at $+V_{SAT}$, and so diode LED1 will be on as well. Although the circuit of Figure 29–5 shows only three LEDs, it is possible to select any number of diodes by choosing appropriate resistors in the voltage divider network.

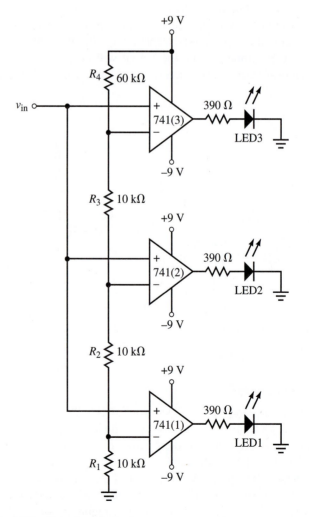

FIGURE 29–5 LED voltage level indicator.

As the name implies, the summing amplifier shown in Figure 29–6 has an output that is determined as the sum of the input voltages, each multiplied by some gain. If we examine the circuit of Figure 29–6 and assume that the input voltages, V_1, V_2, and V_3 are all positive with respect to ground, then we are able to determine the current through each of the input resistances as $I_1 = \dfrac{V_1}{R_1}$, $I_2 = \dfrac{V_2}{R_2}$, and $I_3 = \dfrac{V_3}{R_3}$. (Remember that the inverting terminal of the op-amp is at virtual ground.) Now, by applying Kirchhoff's current law at the summing junction, we determine that the current through the feedback resistor is

$$I_F = I_1 + I_2 + I_2$$

and the output voltage as

$$V_{out} = -I_F R_F$$

which gives

$$V_{out} = -\left(\frac{R_F}{R_1} V_1 + \frac{R_F}{R_2} V_2 + \frac{R_F}{R_3} V_3 \right) \qquad \textbf{(29–1)}$$

29.2 Voltage Summing Amplifier

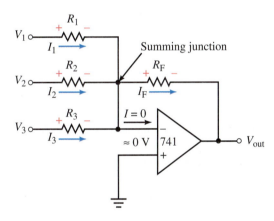

FIGURE 29–6   Op-amp used as a summing amplifier.

Given that the circuit of Figure 29–7 has the following resistor values:
$R_1 = 20 \text{ k}\Omega$, $R_2 = 10 \text{ k}\Omega$, $R_3 = 40 \text{ k}\Omega$, and $R_F = 100 \text{ k}\Omega$.
Solve for v_{out} if $V_1 = -0.8$ V, $V_2 = 1.2$ V, and $V_3 = -2.5$ V.

Solution

$$V_{out} = -\left[\frac{100 \text{ k}\Omega}{20 \text{ k}\Omega} (-0.8 \text{ V}) + \frac{100 \text{ k}\Omega}{10 \text{ k}\Omega} (1.2 \text{ V}) + \frac{100 \text{ k}\Omega}{40 \text{ k}\Omega} (-2.5 \text{ V}) \right]$$

$$= -\left[-4.0 \text{ V} + 12 \text{ V} + (-6.25 \text{ V}) \right]$$

$$= -1.75 \text{ V}$$

EXAMPLE 29–1

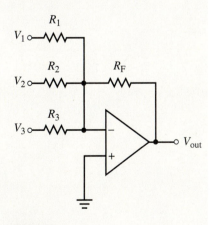

FIGURE 29–7

PRACTICE PROBLEMS 1

Solve for v_{out} in the circuit of Figure 29–7 if $R_1 = 75\ k\Omega$, $R_2 = 75\ k\Omega$, $R_3 = 75\ k\Omega$, and $R_F = 75\ k\Omega$. Let $V_1 = 1.5$ V, $V_2 = 2.2$ V, and $V_3 = -2.5$ V.

Answer
-1.2 V

IN-PROCESS
LEARNING CHECK 1

(Answers are at the end of the chapter.)

For the circuit shown in Figure 29–8, solve for the voltages at points *A*, *B*, and *C*. Determine the output voltage, v_{out}.

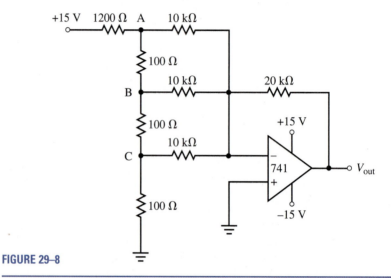

FIGURE 29–8

29.3 Integrators and Differentiators

Many of the op-amp circuits that we have examined so far have used negative feedback as shown in Figure 29–9, with all impedances being resistive. However, this is not always the case. Indeed, many op-amp circuits use combinations of resistors and capacitors in the blocks labeled Z_1 and Z_2 to provide interesting filter circuits. We will limit our analysis to two simple circuits: the integrator and the differentiator.

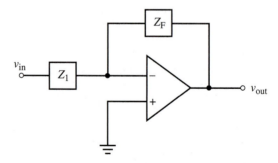

FIGURE 29–9

The Integrator

Figure 29–10 shows a simple integrator circuit. If we assume that v_{in} is positive with respect to ground, then the current through R_1 is determined as $i = \dfrac{v_{in}}{R_1}$. Due to the very high input impedance of the op-amp, there cannot be any current into the non-inverting input. Consequently, the current i must enter the feedback loop and charge the capacitor. Recall that the voltage across a capacitor is given by the expression:

$$v_C(t) = \frac{1}{C} \int_0^t i(t)\,dt + V_o \qquad \qquad \textbf{(29–2)}$$

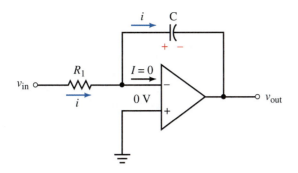

FIGURE 29–10   Op-amp integrator.

When we substitute the expression for current into this expression, we have

$$v_c(t) = \frac{1}{R_1 C} \int_0^t v_{in}(t)dt + V_o \qquad (29\text{–}3)$$

Now, if we assume that the capacitor is initially uncharged ($V_o = 0$ V), then we can write the expression for the output as

$$v_{out}(t) = -\frac{1}{R_1 C} \int_0^t v_{in}(t)dt \qquad (29\text{–}4)$$

The following example illustrates the operation of the integrator.

Given the circuit of Figure 29–11, sketch the output of the integrator for the given input waveform.

EXAMPLE 29–2

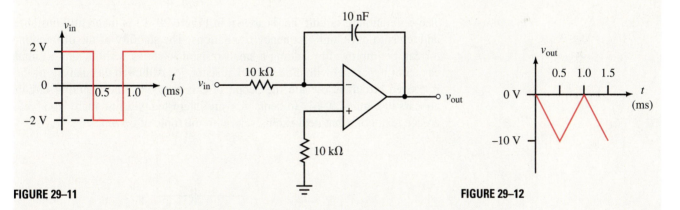

FIGURE 29–11

FIGURE 29–12

Solution
In the interval between $t = 0$ and $t = 0.5$ ms, the input voltage is a constant value of 2 V. Consequently, we determine that the voltage in this region is expressed as

$$v_{out}(t) = -\frac{1}{(10 \text{ k}\Omega)(10 \text{ nF})} \int 2 \text{ V} dt$$

$$= -20{,}000t \frac{\text{V}}{\text{s}}$$

If we were to assume that at $t = 0$, the capacitor is initially uncharged, then in 0.5 ms the voltage at the output will go from 0 V to -10 V. From $t = 0.5$ ms to $t = 1.0$ ms, the voltage across the capacitor will increase at a rate of 20,000 V/s. The resulting output waveform is shown in Figure 29–12.

PRACTICAL NOTES . . .

Due to offset voltage and current in the op-amp in Figure 29–11, the initial voltage across the capacitor will normally be $\pm V_{SAT}$. To cancel the output offset voltage, a large resistor ($R_F = 1$ MΩ) is generally placed across the capacitor in the feedback loop.

The Differentiator

The op-amp differentiator, shown in Figure 29–13 is less common than the integrator. Just as was the case with the integrator, the operation of the differentiator is easily explained by applying a few of the basic principles of capacitor operation. If we were to apply an input voltage to the circuit of Figure 29–13, this voltage appears across the capacitor. The current through the capacitor is therefore

$$i(t) = C \frac{dv_{in}}{dt} \tag{29–5}$$

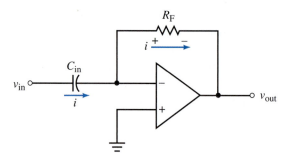

FIGURE 29–13   Op-amp differentiator.

Lastly, the voltage appearing at the output of the differentiator is easily determined to be

$$v_{out}(t) = -R_F C \frac{dv_{in}}{dt} \tag{29–6}$$

The differentiator circuit that is shown in Figure 29–13 is inherently unstable and tends to have high frequency oscillations. The stability of the differentiator can be improved by adding a small resistor in series with C_{in} and a small capacitor in parallel with the feedback resistor, R_F. Although the analysis of the improved circuit is outside the scope of this textbook, the resulting circuit appears in Figure 29–14. Students who are interested will find many good reference textbooks that deal exclusively with the topic of op-amps.

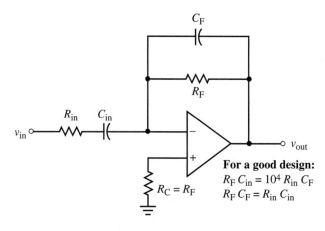

For a good design:
$R_F C_{in} = 10^4 R_{in} C_F$
$R_F C_F = R_{in} C_{in}$

FIGURE 29–14   Op-amp differentiator with oscillation suppression.

Instrumentation amplifiers are op-amp circuits that are designed to operate in environments that would normally result in excessive noise. By using the op-amp as a differential amplifier, any common-mode signal (noise) will be cancelled, while at the same time permitting the amplification of the desired difference voltage. The instrumentation amplifier shown in Figure 29–15 offers a very good common-mode rejection ratio, while at the same time providing a reasonable gain for a differential signal.

29.4 Instrumentation Amplifiers

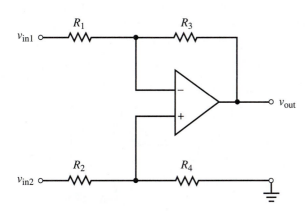

FIGURE 29–15   Instrumentation amplifier.

The following example shows the analysis of an instrumentation amplifier.

Given the instrumentation amplifier shown in Figure 29–16:

EXAMPLE 29–3

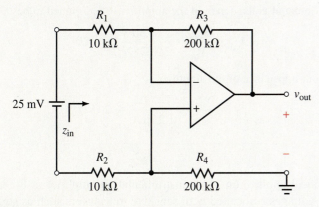

FIGURE 29–16

a. Determine the current through each resistor.

b. Solve for the output voltage, v_{out}.

c. Calculate the differential voltage gain of the amplifier.

d. Solve for the input impedance of the amplifier, z_{in}.

Solution

a. We begin by using the approximation that the voltage between the inverting and non-inverting terminals is zero. By examining the input loop, of the op-amp, we determine the currents

$$I_1 = I_2 = \frac{25 \text{ mV}}{10 \text{ k}\Omega + 10 \text{ k}\Omega} = 1.25 \text{ µA}$$

Now, due to the very high input impedance of the op-amp, there can be no current entering or leaving the input terminals of the op-amp and so applying Kirchhoff's current law, we have $I_3 = I_4 = 1.25$ µA. Figure 29–17 shows the currents and corresponding voltage drops.

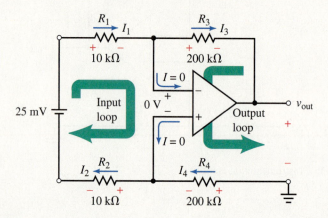

FIGURE 29–17

b. By applying Kirchhoff's voltage law to the output loop, we determine that the output voltage is

$$v_{\text{out}} = -I_3R_3 - I_4R_4 = -0.50 \text{ V}$$

c. The differential voltage gain of the amplifier is determined to be

$$A_{\text{vd}} = \frac{v_{\text{out}}}{v_{\text{in}}} = \frac{-0.50 \text{ V}}{0.025 \text{ V}} = -20$$

d. The input impedance of the amplifier is

$$z_{\text{in}} = \frac{v_{\text{in}}}{i_{\text{in}}} = \frac{24 \text{ mV}}{1.25 \text{ µA}} = 20 \text{ k}\Omega$$

A practical application of an instrumentation amplifier is for measuring very small voltages associated with sensitive transducers such as strain gages. In Chapter 3 you learned that a **transducer** is simply defined as any device that converts a physical change into an electrical change. A **strain gage** is a device that converts force into a change in electrical resistance. Figure 29–18 shows one application of a strain gage.

By applying an external force to a metal bar, the length of the bar (and hence the length of conductor of the strain gage) changes. Although the change in resistance is in the order of a few milliohms, this change can be measured by placing the strain gage into a bridge circuit and using an instrumentation amplifier to measure the resulting difference in potential. It is then possible to convert the voltage into a corresponding value of force (in newtons, pounds, or tons).

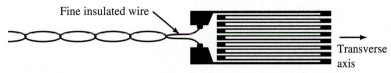

Fine insulated wire

(a) A strain gage is constructed of very thin metal foil mounted on a plastic backing.

(b) The strain gage is glued onto a metal bar, which is subjected to tension and compression. Here we have the strain gage in equilibrium. The resistance of the gage is equal to the equilibrium resistance, R.

$R+\Delta R$ F

(c) Strain gage under tension. The strain gage is stretched in the direction of its transverse axis resulting in an increase in resistance due to the extra length and decrease in cross-sectional area. The resistance is now $R+\Delta R$.

$R-\Delta R$

(d) Strain gage under compression. The strain gage is compressed in the direction of its transverse axis resulting in a decrease in resistance due to the reduction in length and increase in cross-sectional area. The resistance is now $R-\Delta R$.

FIGURE 29–18

An instrumentation amplifier and strain gage are used in the circuit shown in Figure 29–19. Determine the value of output voltage if the equilibrium resistance of the strain gage is 100 Ω, and the resistance, R_4 increases by 5 mΩ when a force is applied to the strain gage.

EXAMPLE 29–4

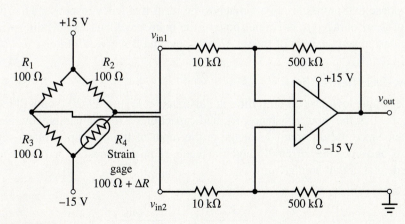

FIGURE 29–19

Solution

We begin by examining the bridge circuit. The voltage applied to v_{in2} is determined by voltage divider as

$$v_{in2} = \left(\frac{100\ \Omega}{100\ \Omega + 100\ \Omega} \right) 30\ \text{V} - 15.00\text{V} = 0.00\ \text{V}$$

and similarly v_{in1} is found as

$$v_{in1} = \left(\frac{100.005\ \Omega}{100\ \Omega + 100.005\ \Omega} \right) 30\ \text{V} - 15.0\ \text{V} = 0.000375\ \text{V}$$

Notice that we use all the significant digits in solving for the voltages. This is important, since there will normally be only a very small change in voltage in the bridge circuit. The difference in potential applied to the instrumentation amplifier is now easily seen to be $\Delta V = 375\ \mu\text{V}$. It is this value that will be amplified by the closed-loop gain, $A_v = -\dfrac{500\ \text{k}\Omega}{10\ \text{k}\Omega} = -50$ of the op-amp.

As a result, we have an output voltage of $V_{out} = (-50)(375\ \mu\text{V}) = -18.75\ \text{mV}$

This example clearly illustrates the need for having an amplifier with a high common-mode rejection ratio. With an output voltage in the order of a few millivolts, it would be very easy for the signal to be lost in a large noise voltage.

PRACTICE PROBLEMS 2

If the strain gage resistance in the circuit of Figure 29–19 was to decrease by 18 mΩ, calculate the corresponding output voltage of the instrumentation amplifier.

Answer
+0.135 V

29.5 Active Filters

In Chapter 22, we studied four basic types of passive filters: low-pass, high-pass, bandpass and band reject. The same types of filters are easily built with op-amps and may be designed to provide a gain. Another advantage of using an op-amp filter is that this circuit has minimal loading effect due to the load resistor. Although outside the scope of this textbook, other types of specialty filters can be designed and built with op-amps so that the resulting filter has a response that more closely approaches the ideal frequency response. The student is encouraged to examine some of the many excellent texts dealing with the design and analysis of Bessel, Chebyshev, and Butterworth filters.

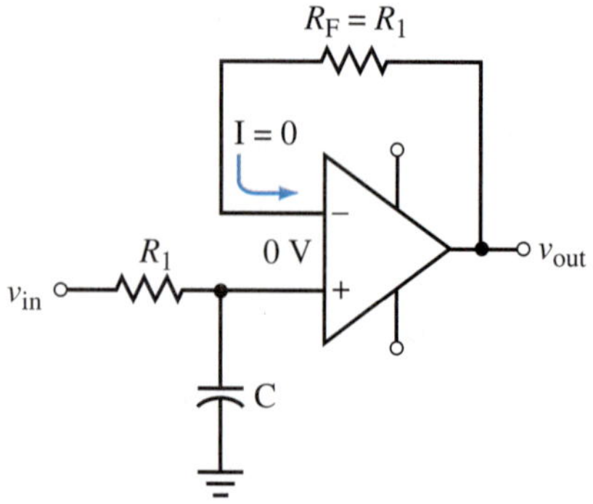

Low-Pass Active Filter

The simple low-pass op-amp filter is illustrated in Figure 29–20. You will recognize that the input resistance, R_1, and capacitor, C, form a passive low-pass filter. Since the differential voltage at the input of the op-amp is effectively zero, and since the current through the feedback resistor must be zero, we conclude that output voltage must be

FIGURE 29–20   Low-pass op-amp filter.

$$V_{out} = v_C = \frac{Z_C}{R_1 + Z_C} V_{in} = \frac{\dfrac{1}{j\omega C}}{R_1 + \dfrac{1}{j\omega C}} V_{in}$$

and so we see that the transfer function of the low-pass op-amp filter is the same as that for a passive low-pass RC filter, namely

$$\mathbf{TF}(j\omega) = \frac{1}{1 + j\omega R_1 C} \qquad (29\text{--}7)$$

Notice that the feedback resistor has the same value as R_1. Because the capacitor is an open circuit at dc, setting $R_F = R_1$ ensures that the dc resistance to ground at both the inverting and the non-inverting terminals is the same. This helps to prevent a dc offset at the output of the op-amp. Figure 29–21 shows that the amplifier may also be modified to provide low-frequency gain.

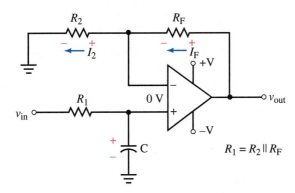

FIGURE 29–21   Low-pass op-amp filter with gain.

In Figure 29–21, we see that the voltage across the resistor R_2 must be the same as the voltage across the capacitor, C. Now since there is no current into or out of the non-inverting terminal, we have $I_F = I_2$. We see therefore that this amplifier has a dc gain of

$$A_v = 1 + \frac{R_F}{R_2} \qquad (29\text{--}8)$$

Given the filter circuit of Figure 29–22:

EXAMPLE 29–5

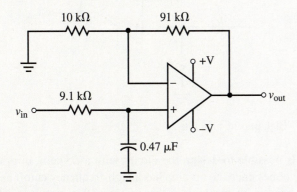

FIGURE 29–22

a. Determine the dc gain of the filter.

b. Calculate the cutoff frequency of the filter.

c. Sketch the straight-line approximation of the voltage gain (in dB) as a function of the frequency.

Solution

a. The dc voltage gain of the filter is

$$A_v = 1 + \frac{91\ k\Omega}{10\ k\Omega} = 10.1$$

which gives

$$[A_v]_{dB} = 20 \log(10.1) = 20.1\ dB$$

b. The cutoff frequency (in Hz) of the filter is

$$f_c = \frac{1}{2\pi R_1 C} = \frac{1}{2\pi(9.1\ k\Omega)(0.47\ \mu F)} = 37.2\ Hz$$

c. The straight-line approximation of the voltage gain (in dB) as a function of the frequency is shown in Figure 29–23.

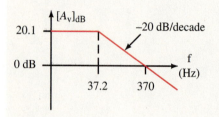

FIGURE 29–23

High-Pass Active Filter

As one might expect, the high-pass filter can be constructed as illustrated in Figure 29–24. In this circuit, the output voltage will be the same as the voltage across R_1, namely,

$$V_{out} = v_R = \frac{R_1}{R_1 + Z_C} V_{in} = \frac{R_1}{R_1 + \dfrac{1}{j\omega C}} V_{in}$$

and so the transfer function for the circuit is

$$\mathbf{TF}(j\omega) = \frac{R_1 C}{1 + j\omega R_1 C} \tag{29–9}$$

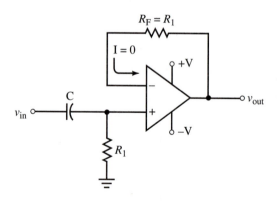

FIGURE 29–24   High-pass op-amp filter.

Although it is possible to design the circuit with a dc gain, in practice this is not generally done, since the op-amp has a high-frequency cutoff as determined by the gain-bandwidth product of the op-amp. The circuit would therefore result in a bandpass filter. For example, if an op-amp having a gain-bandwidth product of 1 MHz were designed to have a dc gain of 10, the resultant upper cutoff frequency of the filter circuit would be 100 kHz. If the gain were 100, the upper cutoff frequency would be 10 kHz, and so on.

You will notice that in this circuit, the feedback resistor once again has the same value as R_1, ensuring that the dc resistance to ground at both input terminals is the same.

Given that the filter circuit of Figure 29–24 has $R_1 = R_F = 9.1 \text{ k}\Omega$ and $C = 0.47 \text{ }\mu\text{F}$:

EXAMPLE 29–6

a. Determine the dc gain of the filter.

b. Calculate the cutoff frequency of the filter.

c. Sketch the straight-line approximation of the voltage gain (in dB) as a function of the frequency. Assume that the op-amp has a gain-bandwidth product of 1 MHz.

Solution

a. The dc voltage gain of the filter is $A_v = 1$, which is equivalent to $[A_v]_{dB} = 0$ db. (A gain of 1 results in an upper cutoff frequency for the circuit of 1 MHz.)

b. The cutoff frequency (in Hz) of the filter is

$$f_c = \frac{1}{2\pi R_1 C} = \frac{1}{2\pi (9.1 \text{ k}\Omega)(0.47 \text{ }\mu\text{F})} = 37.2 \text{ Hz}$$

c. The frequency response of the circuit is shown in Figure 29–25. Notice that the sketch shows the upper cutoff frequency due to the gain-bandwidth product of the op-amp. In any design, it is important to consider the effects of this quantity.

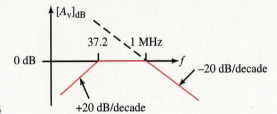

FIGURE 29–25

Active Wideband Bandpass Filter

It is possible to design a bandpass filter by cascading a low-pass op-amp filter with a high-pass filter. Each active filter is similar to the filters examined previously. Since the input impedance of the second filter does not affect the operation of the first, the filters can be cascaded in any order. Figure 29–26 shows

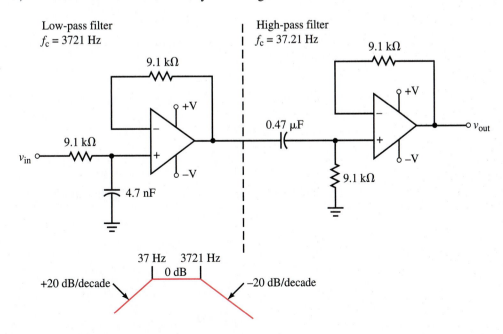

FIGURE 29–26   Wideband bandpass filter and frequency response.

a low-pass filter followed by a high-pass filter. The low-pass filter must have the higher cutoff frequency, in this case $f_c = 3721$ Hz. The high-pass filter must provide the lower cutoff frequency, in this case $f_c = 37$ Hz.

Active Narrowband Bandpass Filter

It is possible to construct a narrowband bandpass filter by using a single op-amp as illustrated in Figure 29–27.

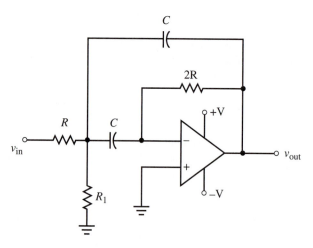

FIGURE 29–27   Narrowband bandpass filter.

This filter has a resonant frequency (in Hz) given as

$$f_0 = \frac{0.1125}{RC}\sqrt{1 + \frac{R}{R_1}} \tag{29-10}$$

and a bandwidth of

$$\text{BW} = \frac{0.1591}{RC} \tag{29-11}$$

EXAMPLE 29–7

Design a narrowband bandpass filter having a resonant frequency of 400 Hz and a bandwidth of 100 Hz.

Solution

We begin our design by selecting an appropriate value of capacitance. Let $C = 0.01$ μF. Next, we use the bandwidth value to determine the value of resistance, R from Equation 29–10.

$$R = \frac{0.1591}{\text{BW } C} = \frac{0.1591}{(100 \text{ Hz})(0.01 \, \mu\text{F})} = 159.1 \text{ k}\Omega$$

Now, we solve for the value of R_1 by applying Equation 29–10.

$$R_1 = \frac{(0.1125)^2 R}{f_0^2 R^2 C^2 - 0.1125^2}$$

$$= \frac{(0.1125)^2 (159.1 \text{ k}\Omega)}{(400 \text{ Hz})^2 (159.1 \text{ k}\Omega)^2 (0.01 \, \mu\text{F})^2 - 0.1125^2}$$

$$= 5.132 \text{ k}\Omega$$

The resulting design is shown in Figure 29–28. The results of this design are easily verified by using MultiSIM to simulate the operation. At the resonant frequency, the output of the bandpass filter is equal in magnitude and 180° out-of-phase with respect to the input signal.

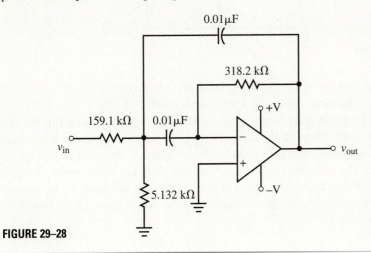

FIGURE 29–28

Active Notch Filter

The notch filter is easily constructed by cascading the narrowband bandpass filter with an adder circuit as illustrated in Figure 29–29. The output of the adder will have the same magnitude for all frequencies except at the resonant frequency of the bandpass filter. At the resonant frequency, both inputs to the adder will be equal but 180° out-of-phase. Consequently, the signal at the resonant frequency will be eliminated.

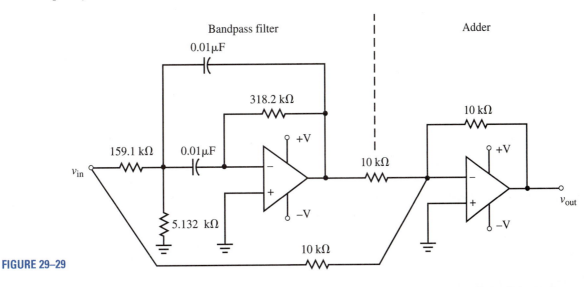

FIGURE 29–29

29.6 Voltage Regulation

As the name implies a **voltage regulator** is a circuit or component that regulates the voltage of a circuit. Normally as more current is demanded from a voltage source, the voltage of the source drops due to internal losses of the source. Also, if the input voltage were to increase, the voltage across the load will similarly increase. A voltage regulator ensures that the voltage delivered to a load remains constant for a specified range of load currents and within a specified input voltage. You were introduced to the zener diode regulator in Chapter 25, which is shown again in Figure 29–30.

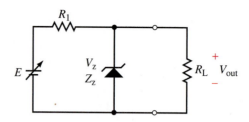

FIGURE 29–30

FIGURE 29–30

Recall that the zener diode regulator had the disadvantage of being inefficient, since the diode dissipates a significant amount of power, especially when the load resistor is quite large. Many types of regulators are used in the electronics industry. Some of the different types of voltage regulators that you are likely to encounter include the fixed voltage regulator, the variable voltage regulator, and the switching regulator. Although there are many types of regulators, they all have some common characteristics. In this section, we examine some of the characteristics of regulators and apply these characteristics to actual circuits.

Line Regulation

Consider the simplified regulator circuit shown in Figure 29–31. In the ideal regulator, there will be no change in output voltage even if the voltage at the input of the regulator were to change. This, however, will seldom be the case. We define **percent line regulation** as the following:

$$\% \text{ line regulation} = \frac{\Delta V_{out}}{\Delta V_{in}} \times 100\% \qquad (29\text{–}12)$$

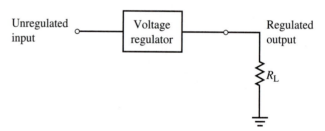

FIGURE 29–31

The voltage at the input of a voltage regulator can vary between 10 V and 30 V. For a given load resistor under these conditions, the output voltage varies between 5.20 V and 5.35 V. Determine the percent line regulation.

Solution

$$\% \text{ line regulation} = \frac{\Delta V_{out}}{\Delta V_{in}} \times 100\%$$

$$= \frac{5.35 \text{ V} - 5.20 \text{ V}}{30 \text{ V} - 10 \text{ V}} \times 100\% = 0.75\%$$

Occasionally, the line regulation is simply expressed as a ratio; in this case 7.5 mV/V. This means that for every volt that the input increases, there will be a corresponding 7.5 mV increase in output voltage.

The zener diode in the circuit shown in Figure 29–32a) has the characteristic shown in Figure 29–32b). Here we see that the voltage across the zener diode will not be constant, but rather will increase as the current through the diode increases.

EXAMPLE 29–9

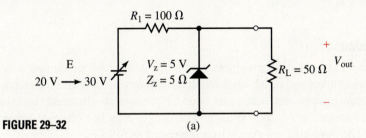

FIGURE 29–32   (a)

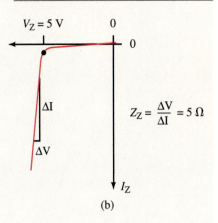

(b)

a. Determine the variation in output voltage as the unregulated input goes from 20 V to 30 V.

b. Solve for the line regulation of the circuit.

Solution

a. The analysis of this circuit is simplified when we determine the Thévenin equivalent circuit external to the zener diode. We begin by letting $E = 20$ V.

$$R_{Th} = 100 \ \Omega \parallel 50 \ \Omega = 33.33 \ \Omega$$

and

$$E_{Th} = \left(\frac{50 \ \Omega}{50 \ \Omega + 100 \ \Omega} \right) 20 \ V = 6.667 \ V$$

The resulting circuit is shown in Figure 29–33.

Here we see that the current through the zener diode is

$$I_Z = \frac{6.667 \ V - 5.0 \ V}{33.33 \ \Omega + 5 \ \Omega} = 0.04348 \ A$$

and the resulting output voltage is

$$V_{out} = 5 \ V + (0.043448 \ A)5 \ \Omega = 5.217 \ V$$

Next, if we let $E = 30$ V, we have $R_{Th} = 33.33 \ \Omega$

and

$$E_{Th} = \left(\frac{50 \ \Omega}{50 \ \Omega + 100 \ \Omega} \right) 30 \ V = 10.00 \ V$$

which results in

$$I_Z = \frac{10.0 \ V - 5.0 \ V}{33.33 \ \Omega + 5 \ \Omega} = 0.1304 \ A$$

and

$$V_{out} = 5 \ V + (0.1304 \ A)5 \ \Omega = 5.652 \ V$$

b. The line regulation is solved as

$$\% \text{ line regulation} = \frac{\Delta V_{out}}{\Delta V_{in}} \times 100\%$$

$$= \frac{5.652 \ V - 5.217 \ V}{30 \ V - 20 \ V} \times 100\% = 4.35\%$$

(or simply, line regulation = 43.5 mV/V)

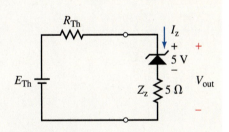

FIGUR.E 29–33

If the load resistance in the circuit of Figure 29–32 has a value of $R_L = 80 \, \Omega$,

a. Determine the variation in output voltage as the unregulated input goes from 20 V to 30 V.

b. Solve for the line regulation of the circuit.

Answers
a. 5.39 V to 5.84 V; b. 4.50 %

Load Regulation

Refer to the regulator diagram shown in Figure 29–31. As mentioned previously, the output voltage of the ideal regulator will remain constant for all conditions. However, in a real voltage regulator, if we were to decrease the value of load resistance, the output voltage will decrease as the current demand increases. The **percent load regulation** is defined as follows:

$$\% \text{ load regulation} = \frac{V_{NL} - V_{FL}}{V_{FL}} \times 100\% \qquad (29\text{–}13)$$

In the Equation of 29–12, V_{NL} is the "no-load" voltage and V_{FL} is the "full-load" voltage. All regulated voltage sources provide a relatively constant voltage until the load current reaches some maximum value. This maximum value will always be specified by a manufacturer and represents the full-load condition. The load regulation may be expressed as maximum variation of output voltage from the nominal output voltage of the regulator.

EXAMPLE 29–10

The manufacturer of a 12-V regulator specifies that the output regulation of the device has a typical value of 4 mV. Determine the percent load regulation.

Solution
Since the difference in the load voltage is 4 mV, we solve the load regulation as

$$\% \text{ load regulation} = \frac{V_{NL} - V_{FL}}{V_{FL}} \times 100\%$$

$$= \frac{4 \text{ mV}}{12 \text{ V}} \times 100\% = 0.033\%$$

EXAMPLE 29–11

Given the zener diode regulator circuit shown in Figure 29–34, determine the load regulation if the load resistance varies between 50 Ω and 100 Ω. Since the minimum load current occurs when $R_L = 100 \, \Omega$, let this be the no-load condition.

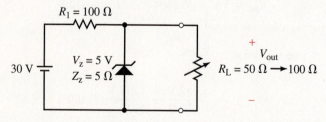

FIGURE 29–34

Solution

As in the previous example, we solve for the Thévenin equivalent circuit external to the zener diode as follows:

When $R_L = 100\ \Omega$, we have $R_{Th} = 50\ \Omega$
and

$$E_{Th} = \left(\frac{100\ \Omega}{100\ \Omega + 100\ \Omega}\right)30\text{ V} = 15.00\text{ V}$$

which results in

$$I_Z = \frac{15.0\text{ V} - 5.0\text{ V}}{50\ \Omega + 5\ \Omega} = 0.1818\text{ A}$$

and

$$V_{out} = 5\text{ V} + (0.1818\text{ A})5\ \Omega = 5.909\text{ V}$$

When $R_L = 50\ \Omega$, we have $R_{Th} = 33.33\ \Omega$
and

$$E_{Th} = \left(\frac{50\ \Omega}{50\ \Omega + 100\ \Omega}\right)30\text{ V} = 10.00\text{ V}$$

which results in

$$I_Z = \frac{10.0\text{ V} - 5.0\text{ V}}{33.33\ \Omega + 5\ \Omega} = 0.1304\text{ A}$$

and

$$V_{out} = 5\text{ V} + (0.1304\text{ A})5\ \Omega = 5.652\text{ V}$$

The load regulation is solved as

$$\% \text{ load regulation} = \frac{5.909\text{ V} - 5.652\text{ V}}{5.652\text{ V}} \times 100\% = 4.55\%$$

Besides being inefficient, the zener diode has the added disadvantage of allowing for regulation over only a narrow range of output load resistance. Both of these disadvantages can be largely overcome by using a series regulator as illustrated in Figure 29–35.

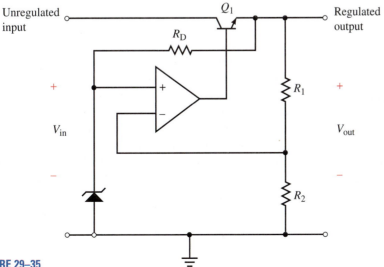

FIGURE 29–35

In the regulator circuit, we use a zener diode only as a means of providing a reference voltage rather than directly providing the voltage regulation. The current through the zener diode remains constant, since it is connected to the regulated output voltage. The op-amp comparator is used to detect feedback voltage and to provide a signal at the base of the power transistor, which in turn provides the necessary current for the load. The operation of the series regulator is really quite simple, and is outlined as follows:

1. When it is first turned on, power is provided to the op-amp by the unregulated input.

2. The output of the op-amp will go to some positive voltage, forward-biasing the B-E junction of transistor, Q_1.

3. As the transistor begins to conduct the emitter becomes increasingly positive.

4. The increase in output voltage causes the current through the zener diode to also increase, until the knee of the diode is reached. At this point the voltage across the zener diode (and at the non-inverting terminal of the op-amp) remains constant at V_Z.

5. The voltage divider that is set up by resistors R_1 and R_2 provides the feedback for the op-amp. If the voltage at the inverting input starts to increase, the differential voltage at input of the op-amp decreases, which in turn decreases the base bias current for Q_1, resulting in less output current, and correspondingly less output voltage.

6. At this point the circuit has reached equilibrium and the output voltage will remain constant.

The output voltage of the circuit is dependent on three quantities: the zener voltage and the values of the resistors R_1 and R_2. Since the voltage appearing at both the inverting and the non-inverting inputs of the op-amp are approximately the same, and since the current into the input terminals of the op-amp is essentially zero, we conclude that the voltage across R_2 must also be V_Z. Applying the voltage divider rule to R_1 and R_2 gives us the following:

$$V_{R_2} = V_Z = \frac{R_2}{R_1 + R_2} V_{out}$$

and so we are able to write the expression for output voltage as

$$V_{out} = \frac{R_1 + R_2}{R_2} V_Z \qquad \textbf{(29–14)}$$

When designing the regulator circuit of Figure 29–35, there are a few design considerations that will help to make the circuit more functional. Firstly, the transistor will generally dissipate a fair amount of power ($P_D \approx I_C V_{CE}$) and so it is generally a good design practice to used a power transistor. Secondly, the current through the zener should be sufficient to ensure that the zener diode is operating in its breakover region. Generally a current of $I_Z = 10$ mA is sufficient. Lastly, the values of the voltage divider resistors, R_1 and R_2, are designed to have a current of approximately 1 mA. The following circuit shows a typical design.

Refer to the circuit of the series regulator shown in Figure 29–35.

EXAMPLE 29–12

a. Determine all resistor values, given the following conditions:

Unregulated input voltage, $v_{in} = 40$ V

Regulated output voltage, $v_{out} = 20$ V

Zener diode has $V_Z = 5.0$ V

b. Solve for the approximate power dissipated by the power transistor if a load resistance of $R_L = 50$ Ω is place across the regulated output voltage.

Solution

a. For an output voltage of 20 V, the voltage across R_D must be $V_{RD} = 15$ V. If we let $I_Z = 10$ mA, we have

$$R_D = \frac{15 \text{ V}}{10 \text{ mA}} = 1.5 \text{ k}\Omega$$

Similarly, we solve for the resistance $R_1 + R_2$

$$R_1 + R_2 = \frac{20 \text{ V}}{1 \text{ mA}} = 20 \text{ k}\Omega$$

Now, using Equation 29–13, we determine R_2 as

$$R_2 = \frac{(R_1 + R_2)V_Z}{v_{out}}$$

$$= \frac{(20 \text{ k}\Omega)(5.0 \text{ V})}{20 \text{ V}} = 5 \text{ k}\Omega$$

And so $R_2 = 15$ kΩ.

The resulting design is shown in Figure 29–36.

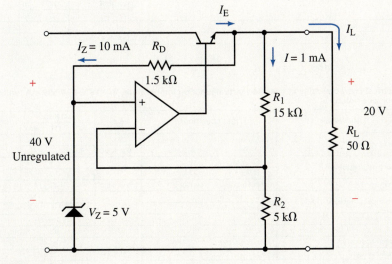

FIGURE 29–36

b. For a load resistor of $R_L = 50 \, \Omega$, the load current is determined to be

$$I_L = \frac{20 \text{ V}}{50 \, \Omega} = 400 \text{ mA}$$

The total emitter current is simply the sum

$$I_E = I_Z + I + I_L = 10 \text{ mA} + 1 \text{ mA} + 400 \text{ mA} = 411 \text{ mA}$$

The collector-emitter voltage of the transistor must be

$$V_{CE} = 40 \text{ V} - 20 \text{ V} = 20 \text{ V}$$

Finally, the power dissipated by the transistor is simply as $P_D = (0.411 \text{ A})$ $(20 \text{ V}) = 8.22 \text{ W}$. This example clearly demonstrates that the transistor must be capable of dissipating a large amount of power.

Although the previous circuit is easily built in a lab, it does involve using a substantial number of components. The circuit also has no short circuit protection, with the result that a shorted output will result in excessive current through the transistor. Although the circuit can be modified to provide additional protection, the extra effort is generally not practical. Many types of IC regulators are available to simplify and improve on the design of the series regulator. The μA7800 and μA 7900 series regulators allow the engineer, technologist, or technician to design and build voltage regulators for one of several possible output voltage values. The 7800 series are positive voltage regulators, while the μA 7900 series are negative regulators. Figure 29–37(a) shows the pinout of the μA 7800 series positive-voltage regulators, while Figure 29–37(b) gives typical

KC PACKAGE
(TOP VIEW)

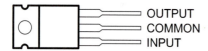

OUTPUT
COMMON
INPUT

The COMMON terminal is in electrical contact with the mounting base.

TO-220AB

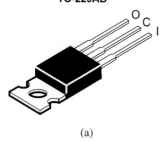

(a)

recommended operating conditions

			MIN	MAX	UNIT
V_I	Input voltage	μA7805C	7	25	V
		μA7808C	10.5	25	
		μA7810C	12.5	28	
		μA7812C	14.5	30	
		μA7815C	17.5	30	
		μA7824C	27	38	
I_O	Output current			1.5	A
T_J	Operating virtual junction temperature	μA7800C series	0	125	°C

(b)

electrical characteristics at specified virtual junction temperature, $V_I = 19$ V, $I_O = 500$ mA (unless otherwise noted)

PARAMETER	TEST CONDITIONS	T_J†	μA7808C			UNIT
			MIN	TYP	MAX	
Output voltage	$I_O = 5$ mA to 1 A, $V_I = 10.5$ V to 23 V, $P_D \leq 15$ W	25°C	7.7	8	8.3	V
		0°C to 125°C	7.6		8.4	
Input voltage regulation	$V_I = 10.5$ V to 25 V	25°C		6	160	mV
	$V_I = 11$ V to 17 V			2	80	
Ripple rejection	$V_I = 11.5$ V to 21.5 V, f = 120 Hz	0°C to 125°C	55	72		dB
Output voltage regulation	$I_O = 5$ mA to 1.5 A	25°C		12	160	mV
	$I_O = 250$ mA to 750 mA			4	80	
Output resistance	f = 1 kHz	0°C to 125°C		0.016		Ω
Temperature coefficient of output voltage	$I_O = 5$ mA	0°C to 125°C		−0.8		mV/°C
Output noise voltage	f = 10 Hz to 100 kHz	25°C		52		μV
Dropout voltage	$I_O = 1$ A	25°C		2		V
Bias current		25°C		4.3	8	mA
Bias current change	$V_I = 10.5$ V to 25 V	0°C to 125°C			1	mA
	$I_O = 5$ mA to 1 A				0.5	
Short-circuit output current		25°C		450		mA
Peak output current		25°C		2.2		A

† Pulse-testing techniques maintain the junction temperature as close to the ambient temperature as possible. Thermal effects must be taken into account separately. All characteristics are measured with a 0.33-μF capacitor across the input and a 0.1-μF capacitor across the output.

(c)

FIGURE 29–37   μA7800 Series Positive Voltage Regulators. *(Courtesy of Texas Instruments)*

input voltage limits. The output voltage for each regulator is determined by the last two numbers in the part number For example the μA7805 is a 5-V regulator, while the μA7812 is a 12-V regulator. Figure 29–37(c) shows the electrical characteristics of the μA7812 regulator. The complete set of specifications sheets for the regulators is available at the Texas Instruments web site, http://www.ti.com/.

Figure 29–38 shows the connection of a μA7812 regulator.

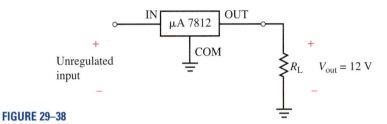

FIGURE 29–38

Refer to the manufacturer's specifications for the μA7800 series of voltage regulators. Answer the following questions relating to the specifications.

a. What is the range of input voltage that is permitted for the μA7805? The μA7824?

b. What is the typical ripple rejection ratio for the μA7812?

c. If the output terminals of the μA7812 were shorted, how much current would there be?

✔ IN-PROCESS
LEARNING CHECK **2**

(Answers are at the end of the chapter.)

Ripple Rejection

In Chapter 25 you examined the operation of rectifier circuits and were introduced to the concept of capacitor filtering. Recall that as the current in the load increased, two important changes occurred in the circuit. The ripple voltage increased since the capacitor discharged more quickly through the smaller resistance of the load. Secondly, the dc voltage delivered to the load was decreased. As you have already observed, a voltage regulator corrects for the variation in load voltage. Another characteristic of an IC regulator is its ability to greatly reduce the ripple voltage that appears across the load. The degree to which a voltage regulator is able to reject ripple is **ripple rejection,** which is calculated in decibels as

$$[\text{ripple rejection}]_{dB} = 20 \log \frac{V_{r(in)}}{V_{r(out)}} \qquad (29\text{--}15)$$

EXAMPLE 29–13

A manufacturer specifies that a three-terminal regulator has a ripple rejection greater than 75 dB. If the ripple at the output of a capacitor filtered full-wave rectifier is measured to be 400 mV$_{p-p}$, calculate the maximum ripple voltage that can be expected at the output of the three-terminal regulator.

Solution
Since we know that ripple rejection for a regulated power supply is expressed as

$$[\text{ripple rejection}]_{dB} = 20 \log \frac{V_{r(in)}}{V_{r(out)}}$$

we may rewrite the expression to solve for the output ripple voltage as

$$V_{r(out)} = \left(400 \text{ mV}_{p-p}\right)10^{-\frac{75}{20}} = 0.071 \text{ mV}_{p-p}$$

A ripple voltage of 0.071 mV$_{p-p}$ will be barely perceptible.

The input of a three-terminal regulator has a dc voltage of 24 V and a 2 $V_{p\text{-}p}$ ripple as illustrated in Figure 29–39. If the output of the regulator is 12 V_{dc} with a 0.5 $mV_{p\text{-}p}$ ripple, determine the following:

a. What is the ripple factor at the input of the regulator?

b. What is the ripple factor at the output of the regulator?

c. What is the ripple rejection (in dB) of the regulator?

d. If the manufacturer specifies that the regulator has a minimum ripple rejection of 55 dB, does this regulator meet the specification?

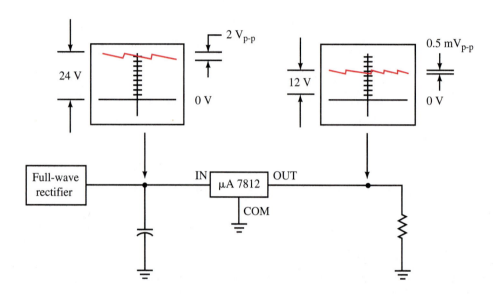

FIGURE 29–39

Solution

a. Recall (from Chapter 25) that ripple rejection is defined as

$$r = \frac{V_{r(rms)}}{V_{dc}} \times 100\%$$

and so at the input of the regulator, the ripple factor is found to be

$$r = \frac{\dfrac{2\,V_{p\text{-}p}}{2\sqrt{3}}}{24\ V} \times 100\% = 2.41\%$$

b. The ripple factor at the output of the regulator is

$$r = \frac{\dfrac{0.5\,mV_{p\text{-}p}}{2\sqrt{3}}}{12\ V} \times 100\% = 0.0012\%$$

c. The ripple rejection of the regulator is

$$[\text{ripple rejection}]_{dB} = 20 \log \frac{2\,V_{p\text{-}p}}{0.5\,mV_{p\text{-}p}} = 72.0\ dB$$

d. The ripple rejection of 72 dB is greater than the minimum specified and so the regulator meets the specification.

In Chapter 22, we used PSpice and MultiSIM software to predict frequency response of passive filters consisting of R-L-C elements. These software packages are equally adept at analyzing active filters. You will also find the software to be very useful in predicting other circuit operations involving op-amps used as comparators, voltage regulators, and amplifiers. The software calculations will easily predict whether a design is likely to work in a laboratory situation and allows us to quickly and easily adjust component values.

29.7 Computer Analysis

PSpice

In Example 29–7, we designed a narrowband bandpass filter to have a center frequency of 400 Hz and a bandwidth of 100 Hz. Use the PSpice Capture to input the circuit of Figure 29–28. Run the Probe postprocessor to view the frequency response (in dB) from 10 Hz to 10 kHz. Use the cursor tool to determine the actual center frequency of the design and solve for the bandwidth by evaluating the frequencies at which the output is 3 dB down from the value at the center frequency.

EXAMPLE 29–14

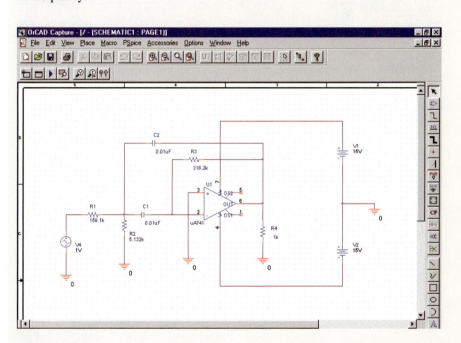

FIGURE 29–40

Solution

The circuit is constructed as shown in Figure 29–40. As in previous problems, we use the VAC part from the SOURCE library as the ac voltage source. After clicking on the part, we change the voltage to **1V.** The 741 op-amp is found in the EVAL library and is found by scrolling down or simple entering uA741 in the Part: box. The ac sweep settings are selected, and the scan is set to go from 10 Hz to 10 kHz. Refer to Chapter 22 if you need a review of how to use PSpice to provide a frequency scan of a filter. The desired display is obtained by selecting the Trace menu and clicking on Add Trace. . . . The output voltage is taken across the load resistor, $R_4 = 1k\Omega$. By using the cursor, we find that the center frequency is 400 Hz and that the bandwidth of the filter is 100 Hz as desired. The resulting voltage gain frequency response is shown in Figure 29–41.

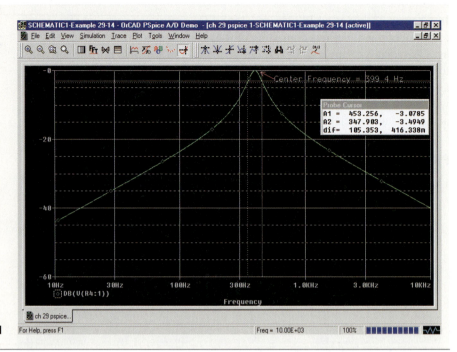

FIGURE 29–41

MultiSIM

EXAMPLE 29–15

a. Use MultiSIM to model the wideband bandpass circuit of Figure 29–26.

b. Use the spectrum analyzer tool to obtain the frequency response of the voltage gain from 1 Hz to 100 kHz.

c. Use the cursor to determine the 3-dB down frequencies. Compare the results to the theoretical values predicted by the component values.

Solution

a. Figure 29–42 shows the resultant circuit when MultiSIM is used to generate the circuit.

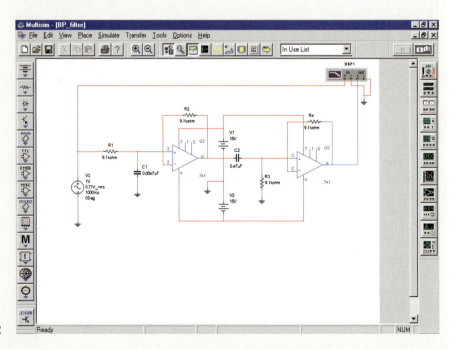

MULTISIM   **FIGURE 29–42**

b. By clicking on the Bode Plotter instrument, we obtain the display shown in Figure 29–43.

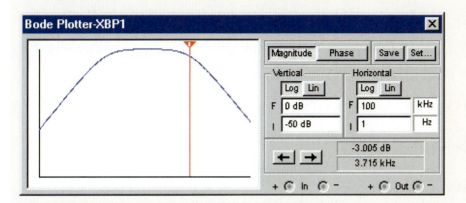

FIGURE 29–43

c. By dragging the cursor through the frequency axis, we determine the lower cutoff frequency of the filter (due to the high-pass section) to be 37 Hz, while the upper cutoff frequency (due to the low-pass section) is 3.7 kHz. These results are exactly the same as those predicted theoretically.

PUTTING IT INTO PRACTICE

Notch filters (also known as band reject filters) can be used to eliminate unwanted frequency components from a signal. As part of an engineering team for an aircraft manufacturer, you have been given the task of designing a notch filter to remove a 400–Hz signal from a monitoring system. You have decided to use two op-amps as shown in Section 29–5. Since you would like to remove only a very narrow band of frequencies, you select a fairly high $Q = 40$. Although you have many capacitors from which to choose, you select a fairly common value of $C = 0.022$ μF. Show the design of your filter and use Mult-SIM or PSpice to simulate the operation of your design.

PROBLEMS

29.1 Comparators

1. Refer to the circuit and input waveform shown in Figure 29–44. Sketch the corresponding waveform that will be observed at the output of the comparator. Label the resultant waveform, showing the correct amplitude and transition times if the reference voltage is $V_{\text{REF}} = +3$ V.

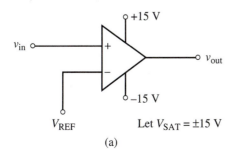

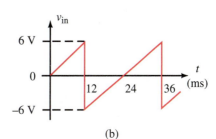

FIGURE 29–44   (a)   (b)

2. Repeat Problem 1 for $V_{REF} = +2$ V.

3. Repeat Problem 1 for $V_{REF} = -4$ V.

4. Repeat Problem 1 for $V_{REF} = +2$ V.

5. Refer to the circuit and input waveform shown in Figure 29–45. Sketch the corresponding waveform that will be observed at the output of the comparator. Label the resultant waveform, showing the correct amplitude and transition times if the reference voltage is $V_{REF} = +5$ V.

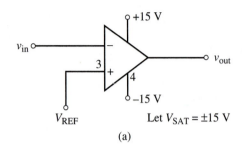

(a)

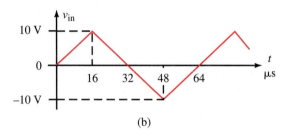

(b)

FIGURE 29–45

◀ MULTISIM

6. Repeat Problem 5 for $V_{REF} = +8$ V.

7. Repeat Problem 5 for $V_{REF} = -2$ V.

8. Repeat Problem 5 for $V_{REF} = -6$ V.

9. For the circuit of Figure 29–46:

   a. For what range of input voltage, v_{in}, will LED1 be on?

   b. For what range of input voltage, v_{in}, will LED2 be on?

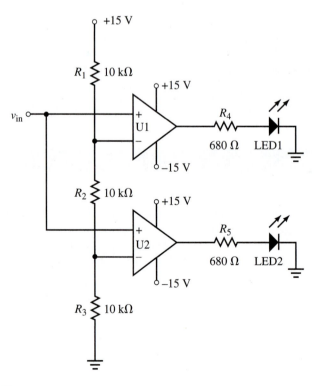

FIGURE 29–46

10. Refer to the circuit shown in Figure 29–47. If the indicated waveform is applied to the input of the op-amp comparators, sketch the output waveforms for v_{out1} and v_{out2}. Label the resultant waveforms, showing the correct amplitude and transition times.

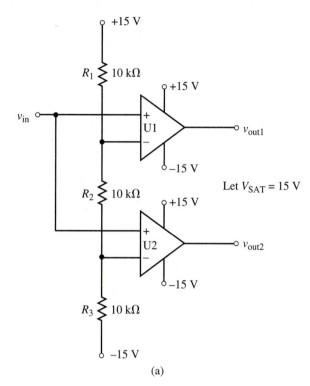

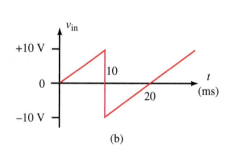

Let $V_{SAT} = 15$ V

(a)   (b)

FIGURE 29–47

29.2 Voltage Summing Amplifier

11. Refer to the circuit of Figure 29–48, letting $R_1 = 10$ kΩ, $R_2 = 10$ kΩ, $R_3 = 10$ kΩ, and $R_F = 20$ kΩ. Let $V_1 = -2$ V, $V_2 = +4.5$ V, and $V_3 = -1.5$ V.

a. Solve for the currents I_1, I_2, I_3, and I_F. (Note the reference directions.)

b. Determine the output voltage, V_{out}.

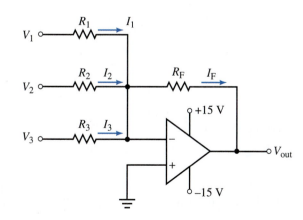

FIGURE 29–48   ◀ MULTISIM

12. Repeat Problem 11 by letting $R_1 = 10$ kΩ, $R_2 = 20$ kΩ, $R_3 = 30$ kΩ, and $R_F = 20$ kΩ. Let $V_1 = +2$ V, $V_2 = +1.5$ V, and $V_3 = -2.5$ V.

    a. Solve for the currents I_1, I_2, I_3, and I_F.

    b. Determine the output voltage, V_{out}.

13. Figure 29–49 shows a summing amplifier that can be used as a "mixer" to algebraically combine two input signals. Given the two waveforms shown in Figure 29–50, sketch the corresponding output waveform, showing the correct amplitude and transition times.

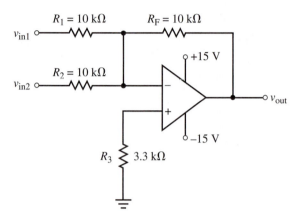

FIGURE 29–49

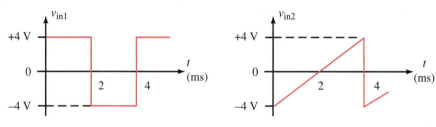

FIGURE 29–50

14. Repeat Problem 13 for the waveforms of Figure 29–51.

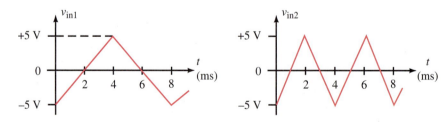

FIGURE 29–51

29.3   Integrators and Differentiators

15. Given that the integrator circuit of Figure 29–52 has $C = 0.05$ μF and a square wave input as shown.

    a. Determine the peak-to-peak value of the waveform that will appear at the output of the integrator if the applied square wave has amplitude of 2.0 V and frequency of 500 Hz.

    b. Sketch the corresponding output waveform. (Recall that the purpose of the feedback resistor, $R_F = 1$ MΩ is to remove the dc offset from the output voltage.)

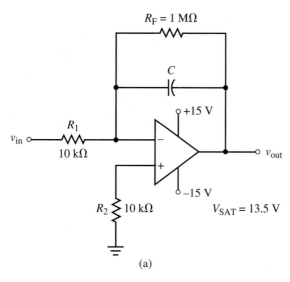
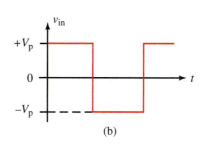

FIGURE 29–52

16. Repeat Problem 15 if the square wave has amplitude of 4.0 V and frequency of 500 Hz.

17. For the circuit of Figure 29–52, what is the maximum voltage peak voltage level that a 500 Hz signal can have before the output voltage goes into saturation?

18. If a square wave signal having amplitude of 2.0 V and frequency of 500 Hz is applied to the circuit of Figure 29–52, what is the minimum capacitor value that can be used before the output voltage goes into saturation?

19. Given that the differentiator circuit of Figure 29–53 has a sawtooth input wave as shown.

   a. Determine the expected waveform that will appear at the output of the integrator.

   b. Sketch the corresponding output waveform, showing the correct amplitude and transition times.

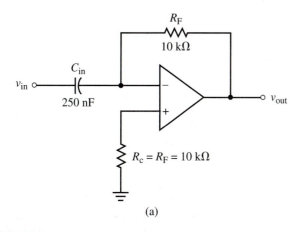
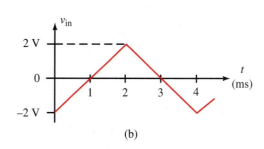

FIGURE 29–53

20. Repeat Problem 19 if the value of the capacitor were to be decreased to 100 nF.

21. The circuit of Figure 29–53 is prone to oscillation, due to bias currents. The operation of the circuit is improved by adding extra components into the circuit as illustrated previously in Figure 29–14. Redesign the circuit of Figure 29–53 by letting $R_F C_{in} = 10^4 R_{in} C_F$ and $R_F C_F = R_{in} C_{in}$. Sketch the new design.

22. Repeat Problem 20 using $C_{in} = 100$ nF.

29.4   Instrumentation Amplifiers

23. Determine the current (magnitude and direction) through each resistor in the circuit of Figure 29–54 and solve for the output voltage, v_{out}.

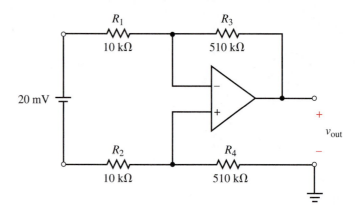

FIGURE 29–54

24. Determine the current (magnitude and direction) through each resistor in the circuit of Figure 29–55 and solve for the output voltage, v_{out}.

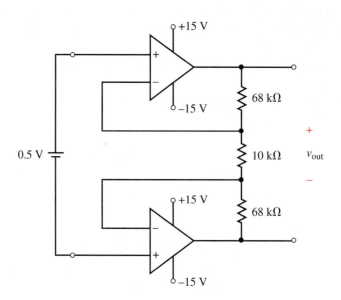

FIGURE 29–55

25. Two strain gages are bonded to a mounting beam that is subjected to an external force. When an external force is applied normal to the surface of the beam, the value of resistance in each strain gage increases by 2.0 mΩ. If the gages are connected as illustrated in Figure 29–56, determine the voltages, v_{in1} and v_{in2}. Calculate the corresponding output voltage, v_{out} of the instrumentation amplifier.

26. Repeat problem 25 if R_1 of the resistance bridge is replaced by a resistor having a constant value of 100 Ω.

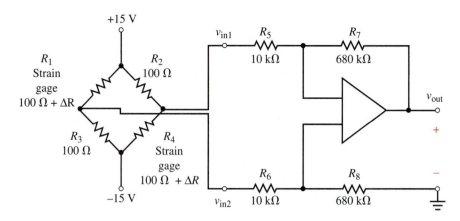

FIGURE 29–56

29.5   Active Filters

27. Given that $C = 0.01$ µF, design the low-pass filter of Figure 29–57 to have a low-frequency gain, $A_{v(dc)} = 10$ and a cutoff frequency of 10 kHz. Sketch the voltage gain frequency response (in dB) of the filter.

28. Repeat the design of Problem 27 if $C = 0.002$ µF, $A_{v(dc)} = 20$ and the cutoff frequency of the filter is 5 kHz.

29. Determine the low frequency cutoff of the op-amp circuit of Figure 29–58 due to the R-C network. Calculate the upper cutoff frequency of the op-amp due to the gain-bandwidth product of 10^6 Hz. Sketch the voltage gain frequency response (in dB) of the filter.

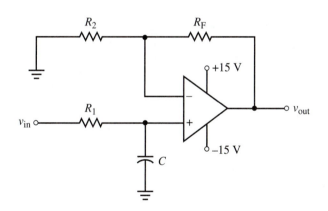

FIGURE 29–57

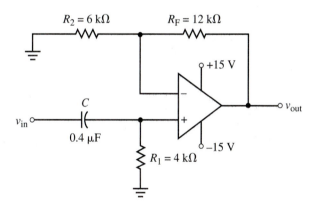

FIGURE 29–58

30. Using the same capacitor value, redesign the filter of Figure 29–58 to have a cutoff frequency of 250 Hz and a mid-frequency gain of 10. Calculate the upper cutoff frequency of the op-amp due to the gain-bandwidth product of 10^6 Hz. Sketch the voltage gain frequency response (in dB) of the filter.

31. Use a capacitor of 0.1 µF to design a narrowband bandpass filter to have a center frequency of 800 Hz and a bandwidth of 80 Hz.

32. Modify the design of Problem 31 to result in a notch filter at 800 Hz.

29.6   Voltage Regulation

33. If the voltage source in the circuit of Figure 29–59 varies between $E = 40$ V and $E = 60$ V, determine the corresponding variation in v_{out} for a load resistor, $R_L = 50$ Ω. Calculate the percent line regulation.

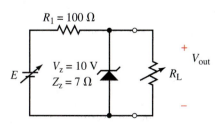

FIGURE 29–59

34. If the voltage source in the circuit of Figure 29–59 varies between $E = 40$ V and $E = 60$ V, determine the corresponding variation in V_{out} for a load resistor, $R_L = 100$ Ω. Calculate the percent line regulation.

35. If the load resistor in the circuit of Figure 29–59 varies between 50 Ω and 100 Ω, determine the corresponding variation in V_{out} for a voltage source, $E = 60$ V. Calculate the percent load regulation.

36. If the load resistor in the circuit of Figure 29–59 varies between 50 Ω and 100 Ω, determine the corresponding variation in V_{out} for a voltage source, $E = 40$ V. Calculate the percent load regulation.

37. Refer to the series regulator shown in Figure 29–60.

   a. Determine the range of output voltage over which the regulator provides a constant voltage. (Note: Assume that regulation is assured if V_{CE} of the transistor is maintained above 3 V.)

   b. If a 50-Ω load is connected across the output terminals, what is the maximum power that the transistor will dissipate?

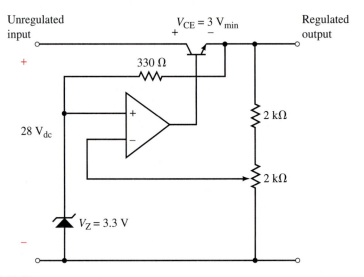

FIGURE 29–60

38. Repeat Problem 37 if the 3.3-V zener diode is replaced by a 6.2-V zener diode.

39. A three-terminal 20-V voltage regulator is measured to have a dc output voltage of 12 V with a 4.0 mV$_{p-p}$ ripple when it is under maximum load. If the input voltage of the regulator is measured to have a 2.0 V$_{p-p}$ ripple, calculate the ripple rejection ratio of the regulator (in dB).

40. A three-terminal 5-V voltage regulator is measured to have a dc input voltage of 26 V with a 4 V$_{p-p}$ ripple. If the manufacturer specifies that the voltage regulator has a ripple rejection ratio better than 58 dB, what is the maximum ripple voltage that you would expect to measure at the output of the regulator when it is under maximum load?

29.7 Computer Analysis

41. Use MultiSIM to input the circuit of Figure 29–45 using a 741C op-amp. Apply the indicated sawtooth signal to the input of the amplifier and a reference voltage of $V_{REF} = +5$ V. Use the oscilloscope tool to simultaneously display the waveforms for v_{in} and v_{out}.   ◀ MULTISIM

42. Use MultiSIM to input the circuit of Figure 29–45 using a 741C op-amp. Apply the indicated sawtooth signal to the input of the amplifier and a reference voltage of $V_{REF} = +8$ V. Use the oscilloscope tool to simultaneously display the waveforms for v_{in} and v_{out}.   ◀ MULTISIM

43. Use MultiSIM to input the circuit of Figure 29–45 using a 741C op-amp. Apply the voltages and resistor values of Problem 11 to the circuit and observe the output voltage with the multimeter tool.   ◀ MULTISIM

44. Use MultiSIM to input the circuit of Figure 29–45 using a 741C op-amp. Apply the voltages and resistor values of Problem 11 to the circuit and observe the output voltage with the multimeter tool.   ◀ MULTISIM

45. Use MutiSIM to input the circuit of Figure 29–49 using a 741C op-amp. Apply the voltages of Figure 29–50 to the circuit and observe the output voltage with the oscilloscope tool.   ◀ MULTISIM

46. Use MultiSIM to input the circuit of Figure 29–49 using a 741C op-amp. Apply the voltages of Figure 29–51 to the circuit and observe the output voltage with the oscilloscope tool.   ◀ MULTISIM

47. Use PSpice Capture to input the circuit of Figure 29–54 using a 741C op-amp. Apply a 20-mV input signal and determine the output voltage of the op-amp. Compare your result to the value determined in Problem 23.   ◀ CADENCE

48. Use PSpice to input the circuit of Figure 29–55 using a 741C op-amp. Apply a 20-mV input signal and determine the output voltage of the op-amp. Compare your result to the value determined in Problem 24.   ◀ CADENCE

49. Use PSpice to input the circuit of Figure 29–57 using a 741C op-amp and the component values of Problem 27. Let the input voltage equal 1 V. Use the Probe postprocessor to obtain a voltage gain frequency response (in dB) from 10 Hz to 10 MHz.   ◀ CADENCE

50. Repeat Problem 49 using the component values of Problem 28.   ◀ CADENCE

 ANSWERS TO IN-PROCESS LEARNING CHECKS

In-Process Learning Check 1

$V_A = +3.0$ V, $V_B = +2.0$ V, $V_C = +1.0$ V, $V_{out} = -12.0$ V

In-Process Learning Check 2

a. 7 V to 25 V, 27 V to 38 V

b. 71 dB

c. 350 mA

■ **OBJECTIVES**

On completion of this chapter, you will be able to

- use the control system equivalent of an amplifier to explain the principles of negative and positive feedback,
- calculate the closed loop of a non-inverting amplifier given the open-loop gain and the feedback ratio,
- analyze the operation of a Schmitt trigger,
- calculate the frequency of a relaxation oscillator that uses a Schmitt trigger as the active component,
- identify the two conditions required for oscillations to be sustained,
- analyze and design a Wien bridge oscillator,
- analyze and design a phase-shift oscillator,
- analyze a Colpitts oscillator,
- analyze a Hartley oscillator,
- explain the piezoelectric effect and sketch the electrical equivalent circuit of a quartz crystal,
- explain the operation of the 555 timer and be able to analyze and design circuits using monostable and astable operation,
- explain the basic operation of a voltage controlled oscillator and be able to design a voltage controlled oscillator circuit using an LM566 integrated circuit,
- use MutiSIM and PSpice software to simulate the operation of oscillator circuits.

Oscillators

30

A part from the filter circuit, the oscillator circuit is perhaps one of the most important circuits in electronics. Without the oscillator, it would be impossible to have any radio transmission. AM (amplitude modulation) and FM (frequency modulation) would be impossible since both modulation techniques rely on oscillator circuits to provide the *carrier frequency*. The oscillator circuit is also an integral part of the demodulation process in both AM and FM. Computers use crystal oscillators to ensure that data is sent the appropriate time and provide synchronization. A piezoelectric crystal is used in watches and cellular telephones to provide alarms and ringing. Despite the theoretical simplicity of the oscillator circuit, it is often one of the most difficult circuits to design. A common complaint of the designer is that if she designs an amplifier, the circuit will oscillate. Conversely, if she tries to design an oscillator, she often ends up with an amplifier. In fact, you will discover that there is very little difference between an amplifier and an oscillator circuit. In this chapter you will learn the principles of oscillator operation and examine several oscillator circuits. Some circuits use only resistors and capacitors to induce oscillations while others use the inherent characteristic of inductors and capacitors to oscillate. You will find that crystal oscillators, which are constructed from quartz (crystallized silicon dioxide), are used if a very stable oscillator frequency is desired. ∎

IN ORDER TO BE ABLE to transmit an audio signal (or any other type of signal) using radio waves, it is necessary to first **modulate** this signal. The modulation process takes the low-frequency signal and combines it with a high-frequency carrier signal in a circuit called a **mixer.** In order to be useful, the high-frequency carrier signal must be a very pure sinusoidal waveform that is generated in a circuit called an **oscillator.** All oscillators use the principle of positive feedback in order to generate an alternating waveform. Early oscillators used vacuum tubes, together with inductors and capacitors, to result in the required positive feedback needed to sustain oscillations. With tunable capacitors, the frequency of oscillation can be adjusted until the desired frequency is obtained. It was very difficult to control the frequency of these early oscillators, since the frequency would vary with aging of components, and with variation in the ambient temperature. By using **piezoelectric crystals** such as quartz, the frequency of oscillation can be controlled to within a few parts per million.

Once a signal is modulated and transmitted, it must be converted back into the original signal. Once again, a very stable oscillator is used to down-convert the original signal to a lower frequency called the **intermediate frequency.** This principle called **super heterodyning** was invented by Edwin Armstrong (see Chapter 21—Putting It in Perspective) and allows for better selectivity within the receiver. Improved selectivity means that any station within a band of frequencies can be isolated and amplified without interference from adjacent stations.

Besides being an integral part of communications circuits, oscillators provide the technician or technologist with an indispensable tool for measuring circuit parameters such as voltage gain and bandwidth. ■

30.1 Basics of Feedback

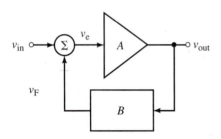

FIGURE 30–1

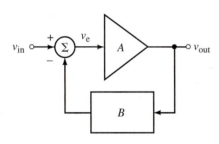

FIGURE 30–2

Figure 30–1 shows the control system equivalent of a feedback amplifier. This representation of the amplifier shows three main components: the forward gain element *(A)*, the feedback network *(B)*, and the summing junction (Σ).

The representation of Figure 30–1 can be used to explain the operation of oscillators as well as to derive the expressions for both the inverting and the non-inverting amplifier. For the amplifier circuits that we have studied to this point, the feedback voltage, v_f was always 180° out-of-phase with respect to the input voltage. This **negative feedback** gave rise to many of the desirable characteristics of the op-amp: high input impedance, low output impedance, wide bandwidth, and stable operation. We will briefly consider the feedback amplifiers using the principles of the control systems equivalent. These same principles will then be used to examine several oscillator circuits that use **positive feedback.**

Consider the diagram shown in Figure 30–2, which represents the non-inverting amplifier. The signal applied to the forward gain element, v_e (referred to as the error voltage in control systems) is determined by combining the signals entering the summing junction, as follows:

$$v_e = v_{in} - v_f \qquad (30\text{--}1)$$

This voltage is then amplified by the high gain of the forward gain element, and so we express the output voltage as

$$v_{out} = A(v_{in} - v_f) \qquad (30\text{--}2)$$

Now, since the feedback voltage represents that portion of the output voltage that is returned to the input, we have

$$v_f = Bv_{out} \qquad (30\text{--}3)$$

By combining Equation 30–3 with the expression of Equation 30–2, we get

$$v_{out} = A(v_{in} - Bv_{out})$$

and so,

$$v_{out} = Av_{in} - ABv_{out}$$

or

$$v_{out} + ABv_{out} = Av_{in}$$

Finally, we see that the closed-loop gain of the non-inverting amplifier is determined as

$$\frac{v_{out}}{v_{in}} = \frac{A}{1 + AB} \qquad (30\text{--}4)$$

Given the non-inverting amplifier shown in Figure 30–3, determine the feed-back ratio, B, and solve for the closed loop gain using Equation 30–4.

EXAMPLE 30–1

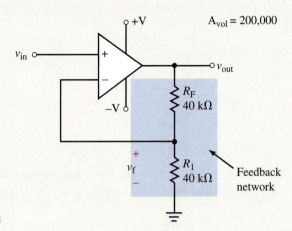

$A_{vol} = 200,000$

FIGURE 30–3

Solution
In the circuit of Figure 30–3, we see that the feedback voltage is calculated by applying the voltage divider rule to R_1 and R_F.

$$v_f = \frac{R_1}{R_1 + R_F} v_{out}$$

$$= \left(\frac{10\ k\Omega}{10\ k\Omega + 40\ k\Omega} \right) v_{out}$$

$$= 0.2 v_{out}$$

From the above calculation, the feedback ratio is simply determined from the resistor values, as $B = 0.2$. Now applying Equation 30–4, we solve for the closed-loop gain of the amplifier as

$$\frac{v_{out}}{v_{in}} = \frac{A}{1 + AB} = \frac{200,000}{1 + (0.2)(200,000)} = 5.00$$

As expected, the above result is consistent with the calculations that we used to find the gain of non-inverting amplifiers in previous chapters. Notice that if the open-loop gain of the op-amp is very large, the feedback ratio is simply the reciprocal of the closed loop gain; in other words, $B = \dfrac{1}{A_{vcl}}$.

The analysis of the inverting amplifier is equally easy. Consider the control system diagram of the inverting amplifier shown in Figure 30–4. The error voltage in this case is determined as

$$v_e = -v_{in} - v_f \qquad \text{(30–5)}$$

This voltage is then amplified by the high gain of the forward gain element, which gives the output voltage as

$$v_{out} = A(-v_{in} - v_f) \qquad \text{(30–6)}$$

As before, the feedback voltage represents that portion of the output voltage that is returned to the input. We have

$$v_f = Bv_{out} \qquad \text{(30–7)}$$

Now, by combining Equation 30–6 with the expression of Equation 30–5, we get

$$v_{out} = A(-v_{in} - Bv_{out})$$

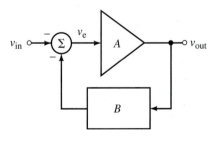

FIGURE 30–4

and so,

$$v_{out} = -Av_{in} - ABv_{out}$$

When simplified, the previous expression allows us to solve for closed-loop voltage of the inverting amplifier as

$$\frac{v_{out}}{v_{in}} = -\frac{A}{1 + AB} \qquad \text{(30–8)}$$

which can also be written as

$$\frac{v_{out}}{v_{in}} = -\frac{1}{\dfrac{1}{A} + B}$$

If the open-loop gain of the op-amp is very high, and the feedback ratio kept relatively low, we approximate the closed-loop gain as

$$\frac{v_{out}}{v_{in}} \approx -\frac{1}{B} \qquad \text{(30–9)}$$

From the result of Equation 30–9, we see that the feedback ratio of the inverting amplifier may be determined from the component values, as

$$B = \frac{R_1}{R_F} \qquad \text{(30–10)}$$

EXAMPLE 30–2

A non-inverting amplifier has $R_1 = 10$ kΩ and $R_F = 68$ kΩ. The op-amp has open–loop gain of 100.

a. Solve for the ideal closed-loop gain of the amplifier.

b. Determine the actual closed-loop gain.

c. Solve for the percent error between the two values.

Solution

a. The ideal closed-loop gain of the amplifier is

$$A_v = \frac{R_F}{R_1} + 1 = \frac{68 \text{ k}\Omega}{10 \text{ k}\Omega} + 1 = 7.8$$

b. The actual voltage gain of the amplifier is

$$A_v = \frac{v_{out}}{v_{in}} = \frac{A}{1 + B} = \frac{100}{1 + \left(\dfrac{1}{7.8}\right)(100)} = 7.24$$

c. The error is percent error $= \dfrac{7.24 - 7.8}{7.24} \times 100\% = -7.8\%$

PRACTICE PROBLEMS 1

A non-inverting amplifier uses an op-amp with an open-loop gain of 250 and has $R_1 = 10$ kΩ and $R_F = 68$ kΩ.

a. Solve for the ideal closed-loop gain of the amplifier.

b. Determine the actual closed loop gain.

c. Solve for the percent error between the two values.

Answers

a. 6.8; b. 6.62; c. −2.7%

In this section we will examine the operation of a simple circuit used to generate a square wave. By applying the knowledge learned in the previous chapter, we will integrate the square wave to obtain a triangular wave. Finally, we use a suitable filter to remove unwanted high-frequency components from the triangular wave to result in a relatively pure sine wave.

30.2 The Relaxation Oscillator

The **relaxation oscillator** uses a comparator to produce an output waveform that oscillates between the saturation voltages of the op-amp. In order to understand the operation of the relaxation oscillator, we begin by first considering the Schmitt trigger.

The Schmitt Trigger

The **Schmitt trigger** shown in Figure 30–5 is commonly used to generate a square wave from a periodic (or any other varying) waveform. A similar circuit is used to generate triggering pulses that provide synchronized scanning of the electron beam across the CRT of an analog oscilloscope.

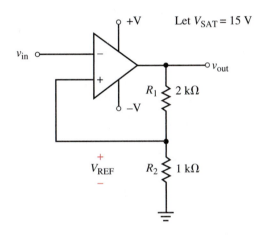

FIGURE 30–5 The Schmitt trigger.

The Schmitt trigger circuit is an op-amp circuit that uses positive feedback to generate a reference voltage on the non-inverting input terminal. (Recall that all amplifiers that we analyzed up to now used negative feedback.) As you learned in the previous chapter, if the voltage at the non-inverting (+) input terminal of a comparator is greater than the voltage at the inverting (−) terminal, the output will be forced to $+V_{SAT}$ due to the very high open-loop gain of the op-amp. Conversely, if the voltage at the (+) terminal is less that the voltage at the (−) terminal, then the output will be forced to $-V_{SAT}$.

If we examine the circuit of Figure 30–5, we see that the reference voltage (found by applying the voltage divider rule on R_1 and R_2) will be either +5 V or −5 V, depending on the input signal. No other values are possible.

Imagine that the input voltage, $v_{in} = +10$ V, in the circuit of Figure 30–5. Since $V_{REF} = \pm 5$ V, the voltage at the (+) terminal must be less than the input, and so the output will be forced to $v_{out} = -V_{SAT} = -15$ V. Now if we were to gradually decrease the input voltage from $v_{in} = +10$ V, the output will not change its value until v_{in} is just less than −5 V. At this point, the output of the comparator would change to $v_{out} = +V_{SAT} = +15$ V ($V_{REF} = +5$ V). If the input were decreased still further, the output would necessarily remain at the same value, namely $v_{out} = +V_{SAT} = +15$ V.

Now, if the input voltage were to start at $v_{in} = -10$ V, the opposite effect would occur. The transfer characteristic, showing the relationship between the input and the output, is illustrated in Figure 30–6. The overlap between the conditions is called *hysteresis*. Here we see that the output is not only dependent on the input, but also on the previous value of the input. The value of the reference voltage, which results in the output switching from $-V_{SAT}$ to $+V_{SAT}$ is called the **lower trip point** (LTP), while the reference voltage which results in the output switching from $+V_{SAT}$ to $-V_{SAT}$ is called the **upper trip point** (UTP).

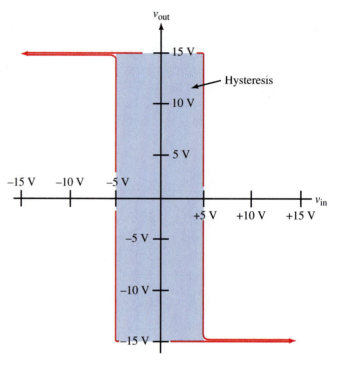

FIGURE 30–6

The following example shows the output of a Schmitt trigger for a given input signal.

EXAMPLE 30–3

If the signal shown in Figure 30–7 is applied to the input of Figure 30–5, determine the output of the Schmitt trigger circuit. Notice that the given signal shows the UTP and the LTP of the Schmitt trigger.

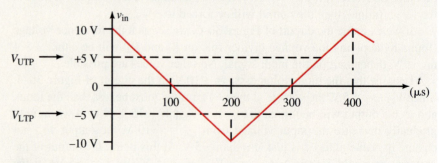

FIGURE 30–7

Solution

The Schmitt trigger will start at $v_{out} = -V_{SAT}$, since the input voltage is greater than the reference voltage. The output will change its state to $v_{out} = +V_{SAT}$ when the input signal goes to $v_{in} = V_{LTP} = -5$ V (at $t = 150$ μs). The next state change occurs at $t = 350$ μs, when the input signal goes above $v_{in} = V_{UTP} = +5$ V. At this point, the output will once again go to $-V_{SAT}$. The resulting output signal is shown in Figure 30–8.

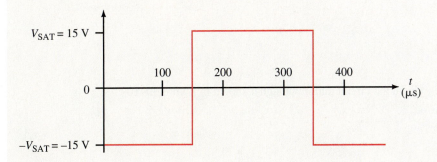

FIGURE 30–8

The Schmitt trigger circuit of Figure 30–5 has upper and lower trip points that have the same magnitude. The circuit shown in Figure 30–9 illustrates that the upper and lower trip points can be made adjustable with the addition of an op-amp buffer and through the selection of suitable resistor values.

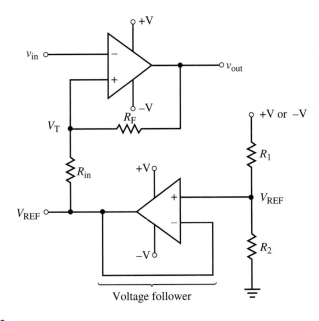

FIGURE 30–9

The reference voltage can be either positive or negative and has a magnitude determined by the values of resistors R_1 and R_2. In either case, the reference voltage will be

$$V_{REF} = \frac{R_2}{R_1 + R_2} (\pm V) \qquad\qquad (30\text{–}11)$$

The upper and lower trip points will be determined as

$$V_{UTP} = \frac{R_{in}(V_{SAT} - V_{REF})}{R_{in} + R_F} + V_{REF}$$ (30–12)

and

$$V_{LTP} = \frac{R_{in}(-V_{SAT} - V_{REF})}{R_{in} + R_F} + V_{REF}$$ (30–13)

EXAMPLE 30–4

Given the circuit and input signal shown in Figure 30–10:

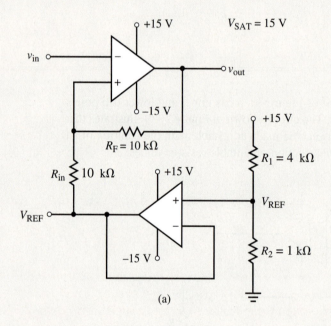

(a)

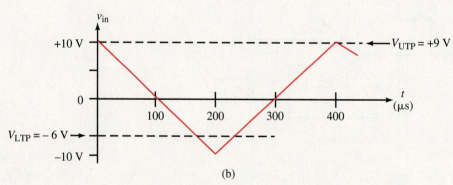

(b)

FIGURE 30–10

a. Determine the reference voltage and the values of the UTP and LTP voltages.

b. Sketch the corresponding output voltage of the Schmitt trigger circuit.

Solution

a. Applying the voltage divider rule gives us the reference voltage as

$$V_{\text{REF}} = \left(\frac{1\ k\Omega}{4\ k\Omega + 1\ k\Omega}\right)(+15\ \text{V}) = 3.0\ \text{V}$$

The upper trip point is determined as

$$V_{\text{UTP}} = \frac{10\ k\Omega(+15\ \text{V} - 3\ \text{V})}{10\ k\Omega + 10\ k\Omega} + 3.0\ \text{V} = 9.0\ \text{V}$$

and the lower trip point is

$$V_{\text{LTP}} = \frac{10\ k\Omega(-15\ \text{V} - 3\ \text{V})}{10\ k\Omega + 10\ k\Omega} + 3.0\ \text{V} = -6.0\ \text{V}$$

The Schmitt trigger will start at $v_{\text{out}} = -V_{\text{SAT}}$. Once the input voltage reaches the lower trip point, the output voltage will change to $v_{\text{out}} = +V_{\text{SAT}} = +15\ \text{V}$. The output remains at this level until the input voltage reaches the upper trip point. The corresponding output signal is shown in Figure 30–11.

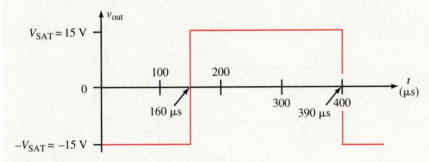

FIGURE 30–11

Schmitt Trigger Relaxation Oscillator

A relaxation oscillator is a circuit that has a frequency of oscillation determined by the alternate charging and discharging of a capacitor. Although there are many types of relaxation oscillators, the Schmitt trigger is perhaps the simplest to understand. Figure 30–12 shows that the Schmitt trigger circuit is easily converted into a relaxation oscillator by adding an extra capacitor and resistor into the circuit.

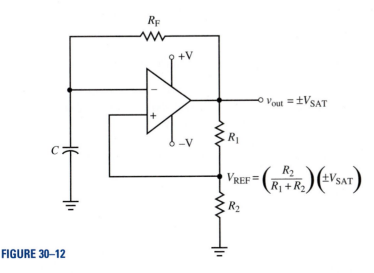

FIGURE 30–12

The circuit operation is explained as follows:

1. Resistors R_1 and R_2 are a voltage divider network that set the reference voltage of the Schmitt trigger. The reference voltage is determined as

$$V_{REF} = \frac{R_2}{R_1 + R_2}(\pm V_{SAT}) \tag{30–14}$$

2. If we assume for a moment that the voltage across the capacitor (voltage applied to the inverting input of the op-amp) is less than the reference voltage, then the capacitor will charge through R_F, attempting to reach a value of $+V_{SAT}$. Once the voltage across the capacitor reaches $V_C = V_{REF}$, the output of the Schmitt trigger will change states to $v_{out} = -V_{SAT}$. At this point the reference voltage will be negative.

3. The capacitor will now attempt to charge to $-V_{SAT}$.

4. Once again, when the voltage across the capacitor reaches the reference value, the output of the Schmitt trigger will change and the process repeats.

EXAMPLE 30–5

Given the circuit of Figure 30–13:

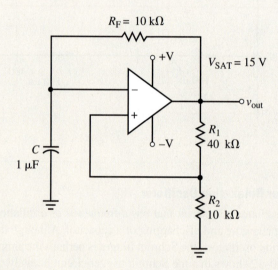

FIGURE 30–13

a. Determine the frequency of the relaxation oscillator.

b. Sketch the waveform that would appear at the inverting input terminal of the op-amp.

c. Sketch the waveform that would appear at the output of the Schmitt trigger.

Solution

a. We begin by finding the reference voltage determined by the voltage divider of R_1 and R_2. From Equation 30–14, we have

$$V_{REF} = \frac{10 \text{ k}\Omega}{40 \text{ k}\Omega + 10 \text{ k}\Omega}(\pm 15 \text{V}) = \pm 3.00 \text{ V}$$

Now, if we assume that the capacitor has an initial charge of -3.00 V, we can set up the charging equation of a capacitor to determine the length of time needed to charge to $+3.00$ V. Remember that the voltage to which the capacitor will attempt to charge is $+V_{SAT} = +15$ V.

$$v_C(t) = V_0 + \left(V_F - V_0\right)\left(1 - e^{-\frac{t}{RC}}\right)$$

$$+3.00 \text{ V} = -3.00 \text{ V} + \left(15 \text{ V} + 3 \text{ V}\right)\left(1 - e^{-\frac{t}{10 \text{ ms}}}\right)$$

Solving for the charging time, t, we have:

$$t = -(10 \text{ ms})(\ln 0.6667) = 4.055 \text{ ms}$$

Therefore, the period of oscillation will be $T = 2(4.055 \text{ ms}) = 8.11$ ms.

b. Figure 30–14 shows the voltage appearing at the input terminal of the op-amp, which is the same as the voltage across the capacitor.

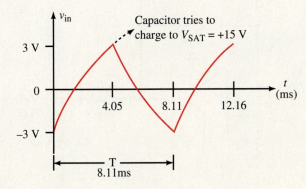

FIGURE 30–14

c. Figure 30–15 shows the corresponding output voltage.

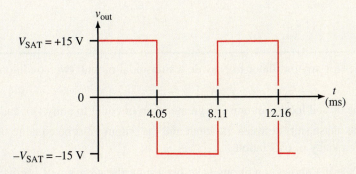

FIGURE 30–15

PRACTICE PROBLEMS 2

Solve for the period of the relaxation oscillator of Figure 30–13 if the voltage divider has resistor values of $R_1 = 10$ kΩ and $R_2 = 40$ kΩ.

Answer
$T = 43.94$ ms

For the relaxation oscillator of Figure 30–13, it can be shown that the general expression for the period of oscillation is determined as

$$T = 2R_FC \ln\left(1 + \frac{2R_2}{R_1}\right) \qquad \textbf{(30–15)}$$

As mentioned previously, adding an integrator and a filter circuit as shown in Figure 30–16, will result in further improvement in the operation of the relaxation oscillator. The modified circuit now has a square wave, a triangular wave, and a sinusoidal output.

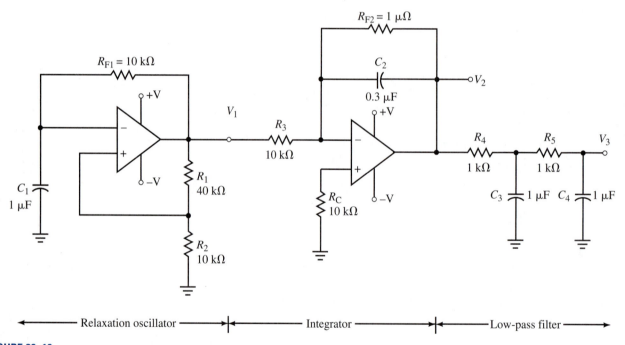

FIGURE 30–16

30.3 The Wien Bridge Oscillator

In order for any oscillator to provide a sinusoidal output, two conditions must be met:

1. the closed loop gain must be greater than or equal to unity (1), and
2. the phase shift between the input and the output must be equal to 0° at the frequency of oscillation.

Once the above conditions have been met, the circuit will oscillate even though no external signal is applied to the amplifier. Positive feedback (also called *regenerative feedback*) will return a portion of the output signal back to the input where this signal is further amplified together with other noise components at the same frequency. The cumulative effect of returning a portion of the output in phase with the input results in an oscillator very quickly reaching its steady state output at its resonant frequency. Figure 30–17 shows the output

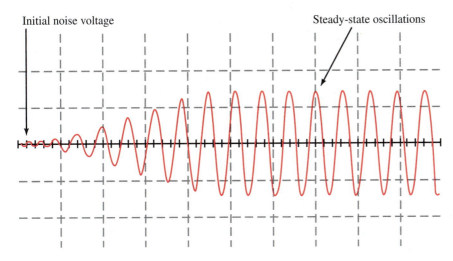

FIGURE 30–17

of an oscillator building from a small noise voltage to its steady state value. The origin of the noise that causes the oscillation to begin is primarily due to thermal effects within the amplifier as well as external random noise.

Figure 30–18 shows the Wien bridge oscillator, a circuit that uses positive feedback to provide sinusoidal oscillations. The bridge consists of two separate arms. The R-C network in the right arm of the bridge provides the positive feedback while the resistors in the left arm help to establish the required amplifier gain.

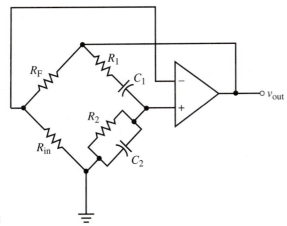

FIGURE 30–18

Figure 30–19 shows the isolated R-C network that provides the necessary positive feedback. In order to ensure that oscillations are maintained, it is necessary that the circuit provides 0° phase-shift at the resonant frequency. It can be shown that for the network of Figure 30–19, the frequency at which the output is in phase with the input occurs at

$$f_0 = \frac{1}{2\pi\sqrt{R_1 R_2 C_1 C_2}} \tag{30–16}$$

Further, it can also be shown that if the transfer function, **TF** $= v_{out}/v_{in}$ is evaluated at the frequency evaluated by Equation 30–16, the feedback gain of the R-C network is a real value (phase angle, $\theta = 0°$) determined as

$$B = \frac{R_2 C_1}{R_1 C_1 + R_2 C_2 + R_2 C_1} \tag{30–17}$$

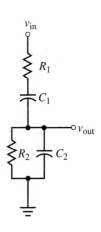

FIGURE 30–19

Since the components are generally selected so that $R_1 = R_2$ and $C_1 = C_2$, Equations 30–16 and 30–17 are simplified as follows:

$$f_0 = \frac{1}{2\pi RC} \tag{30–18}$$

and

$$B = \frac{1}{3} \tag{30–19}$$

In order to ensure that the oscillator has a closed-loop gain, $AB = 1$ (unity), the op-amp must provide the circuit with a gain of at least 3. Figure 30–20 shows the resultant circuit of the Wien bridge oscillator. Although the circuit does not show the bridge circuit in an easily recognized form, this circuit better illustrates both the positive feedback path ($A = 3$) and the negative feedback path ($B = \frac{1}{3}$).

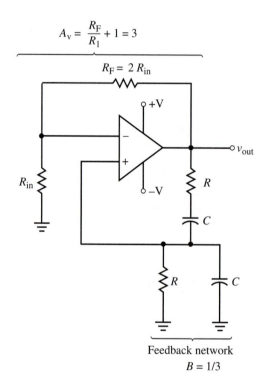

$$A_v = \frac{R_F}{R_1} + 1 = 3$$

$$R_F = 2\,R_{in}$$

Feedback network
$B = 1/3$

FIGURE 30–20

The operation of the circuit is further improved by separating the feedback resistor, R_F into two resistors, one variable and one fixed. The variable resistor is then adjusted to minimize the distortion on the output signal. Zener diodes are placed across the fixed resistor to form part of the **bounding circuit** to limit the range of the output voltage. When low-voltage zener diodes are used in the circuit, the variable resistor not only adjusts for distortion, but also provides better control of the output amplitude. The modified circuit is shown in Figure 30–21.

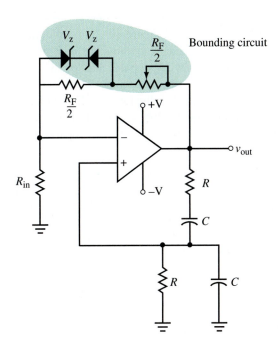

FIGURE 30–21

Design a Wien bridge oscillator having a resonant frequency of 1 kHz. Assume that you are given capacitors having values of $C = 0.01 \, \mu F$.

EXAMPLE 30–6

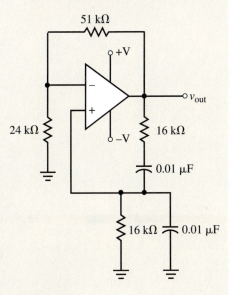

Solution
By applying Equation 30–18, we solve for the required resistance in the $R\text{-}C$ feedback network as

$$R = \frac{1}{2\pi f_0 C}$$

$$= \frac{1}{2\pi(1000 \text{ Hz})(0.01 \, \mu F)} = 15.9 \text{ k}\Omega$$

If we let $R = 16 \text{ k}\Omega$, our design would have a frequency variation of less than 1% from the desired value of 1 kHz. To minimize the effect of bias voltage and currents, we select $R_{in} = R = 24 \text{ k}\Omega$. Now the value of R_F is easily determined to be $R_F = 2R_{in} = 48 \text{ k}\Omega$. A 47-k$\Omega$ resistor would not be suitable for this circuit, since its value may not result in sufficient gain for the amplifier to oscillate. Therefore, in order to sustain the oscillations, it is better to use a larger resistor such as $R_F = 51 \text{ k}\Omega$, even though the output signal may have excessive distortion. Figure 30–22 shows the resulting design.

FIGURE 30–22

Refer to the $R\text{-}C$ feedback network of Figure 30–19. If $R_1 = R$, $R_2 = 2R$ and $C_1 = C_2 = C$, determine the feedback gain of this network. What must be the value of the amplifier gain in order for the oscillator to provide a sinusoidal output?

IN-PROCESS
LEARNING CHECK 1

(Answers are at the end of the chapter.)

30.4 The Phase-Shift Oscillator

The phase-shift oscillator, shown in Figure 30–23, uses a three-section R-C network to provide the required feedback to allow the amplifier to oscillate. At resonance, the R-C network provides a total phase shift of 180°. Although one might expect that each of the three sections contributes 60° of the total, the actual phase-shift of each section is determined by the loading effect of adjacent components. Since the oscillator circuit requires 360° ($\equiv$ 0°) of phase shift, the other 180° is obtained by applying the feedback to the inverting input of the op-amp.

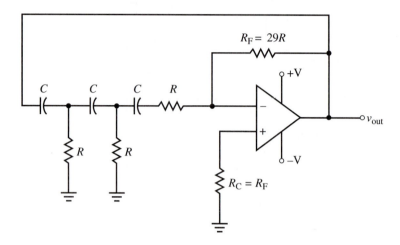

FIGURE 30–23

It can be shown that the output of the three-section R-C network provides exactly 180° of phase shift at a frequency determined as

$$f_0 = \frac{1}{2\pi\sqrt{6}RC} \qquad (30\text{--}20)$$

At this frequency, the phase-shift feedback network has a gain of $B = 1/29$, returning 1/29 of the output magnitude back to the input. Therefore, in order to ensure oscillations, it is necessary for the amplifier to provide a voltage gain of

$$A = 29 \qquad (30\text{--}21)$$

The following example illustrates the operation of the phase-shift oscillator.

Given the circuit shown in Figure 30–24:

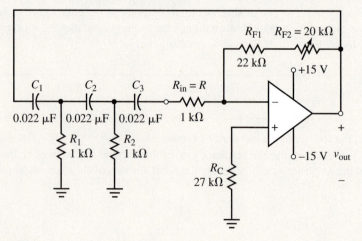

FIGURE 30–24

a) Determine the frequency of oscillation.

b) What is the minimum value of R_{F2} to ensure that oscillation will occur?

c) At the required amplifier gain, what is the bandwidth of the amplifier if the gain-bandwidth product for the op-amp is 10^6 Hz.

Solution

a. The frequency of oscillation is

$$f_0 = \frac{1}{2\pi\sqrt{6}RC} = \frac{1}{2\pi\sqrt{6}(1\text{ k}\Omega)(0.022\ \mu\text{F})} = 2953 \text{ Hz}$$

b. In order for oscillations to be sustained, the value of the feedback resistor must be

$$R_F = 29R = 29 \text{ k}\Omega \text{ and so } R_{F2} = 7 \text{ k}\Omega$$

c. At a gain of $A = 29$, the bandwidth of the amplifier will be

$$BW = \frac{10^6}{29} = 34.49 \text{ kHz}$$

The above calculation implies that the cutoff frequency of the amplifier is 34.49 kHz, illustrating that the amplifier will no longer be able to provide sufficient gain above this frequency to sustain oscillation. This example clearly illustrates that the gain-bandwidth product of an op-amp limits the oscillator frequency. If a high-frequency oscillator is desired, it is necessary to use an active component that is able to provide enough gain at the required frequency. Very often, high-frequency junction transistors or field-effect transistors are used for these applications.

PRACTICE PROBLEMS 3

Using the same capacitor values, redesign the phase-shift oscillator of Figure 20–24 to have a resonant frequency of 5.0 kHz.

Answers
$R_1 = R_2 = R_{in} = 591\ \Omega$; $R_F = R_C = 17.13 \text{ k}\Omega$

As we have already seen in previous chapters, inductor-capacitor circuits are inherently well-suited to produce sinusoidal oscillations. There are several types of *L-C* oscillators that have been used for many decades to provide desired output frequencies for applications including radio transmission and reception and test and measurement circuits. Many of these designs have been named for pioneer engineers who developed the designs, the most common being the Colpitts, Hartley, Clapp, and Armstrong oscillators. In this section we will examine the basic operation of the Colpitts and Hartley oscillators.

30.5 LC Oscillators

Colpitts Oscillator

The basic circuit of the Colpitts oscillator is shown in Figure 30–25. In this circuit, the feedback network consists of C_1, L, and C_2. It can be shown the impedance of this network (between the input and the output) is dependent on the frequency and is given as

$$\mathbf{Z}(j\omega) = \frac{1 + (j\omega)^2 LC_2}{j\omega(C_1 + C_2)\left(1 + \dfrac{(j\omega)^2 LC_1C_2}{C_1 + C_2}\right)} \qquad (30\text{–}22)$$

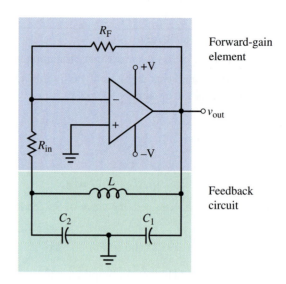

FIGURE 30–25   The Colpitts Oscillator.

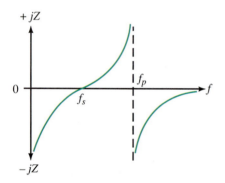

FIGURE 30–26   Impedance as a function of frequency for the feedback network of the Colpitts oscillator.

Figure 30–26 shows this impedance plotted as a function of frequency.

Notice that there are two frequencies of importance: f_s is the series resonant frequency determined when the numerator of Equation 30–22 is zero and f_p is the parallel resonant frequency (also referred to as *antiresonance*) found when the denominator is zero (called a *pole*). For the circuit shown, the L-C network provides 180° of phase shift at the parallel resonant frequency, which we see is determined by the expression,

$$f_0 = \frac{1}{2\pi\sqrt{L\dfrac{C_1 C_2}{C_1 + C_2}}} \qquad (30\text{–}23)$$

EXAMPLE 30–8

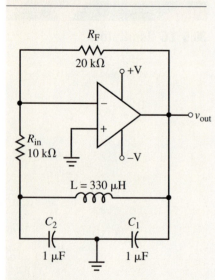

FIGURE 30–27

Determine the resonant frequency for the Colpitts oscillator shown in Figure 30–27.

Solution

By applying Equation 30–23, we determine the resonant frequency of this circuit to be

$$f_0 = \frac{1}{2\pi\sqrt{L\dfrac{C_1 C_2}{C_1 + C_2}}} = \frac{1}{2\pi\sqrt{(330\,\mu\text{H})(0.5\,\mu\text{F})}} = 12.39\ \text{kHz}$$

This circuit will be later examined using computer simulation.

Hartley Oscillator

Figure 30–28 shows the Hartley oscillator, which is similar to the Colpitts oscillator except that the locations of capacitors and inductors are interchanged. As expected, the impedance of this network is frequency dependent and can be shown to be

$$\mathbf{Z}(j\omega) = \frac{j\omega L_1 \left(1 + (j\omega)^2 L_2 C\right)}{1 + (j\omega)^2 \left(L_1 + L_2\right)C} \tag{30-24}$$

Figure 30–29 shows this impedance plotted as a function of frequency.

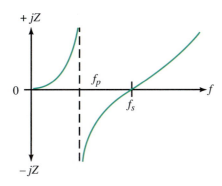

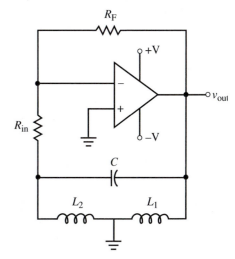

FIGURE 30–28 The Hartley oscillator.

FIGURE 30–29 Impedance as a function of frequency for the feedback network of the Hartley oscillator.

As before, there are two frequencies of importance: f_s, which is the series resonant frequency, and f_p, the parallel resonant frequency, which is found when the denominator is zero. Once again, the L-C network provides 180° of phase shift at the parallel resonant frequency, which from the expression for impedance (Equation 30–24) is determined as,

$$f_0 = \frac{1}{2\pi\sqrt{\left(L_1 + L_2\right)C}} \tag{30-25}$$

Determine the resonant frequency for the Hartley oscillator shown in Figure 30–30.

PRACTICE PROBLEMS 4

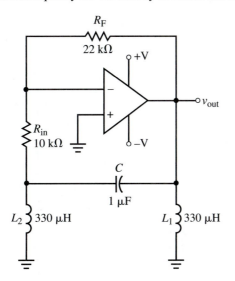

FIGURE 30–30

Answer
6195 Hz

As mentioned previously, the Colpitts and Hartley sinusoidal oscillators may also be constructed by using either bipolar junction transistors or junction field effect transistors. Figure 30–31 shows the transistor equivalent forms of the Colpitts oscillator, while Figure 30–32 shows the Hartley oscillator. The operation of these circuits is similar to the operation of op-amp oscillators. As is the case with the op-amp oscillators, each transistor must provide not only the required 180° of phase shift, but must also have sufficient forward gain to ensure oscillations can be sustained. The frequency of oscillation for each of

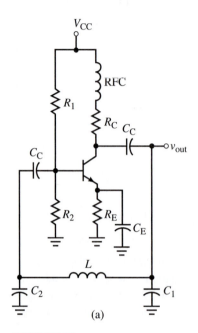
(a)

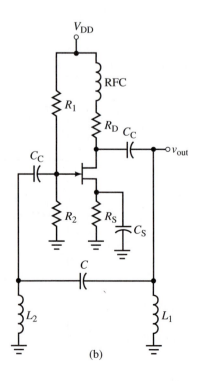
(b)

FIGURE 30–31

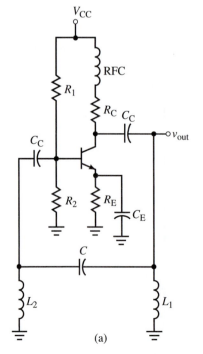
(a)

(b)

FIGURE 30–32

the circuits will be very nearly the same as those for op-amps, with the exception that stray capacitance due to the terminals of the transistors will result in slight variation. The RF choke (RFC) in each circuit helps to remove unwanted high-frequency (radio frequency) components from the output.

30.6 Crystal Oscillators

Although inductors, capacitors, and resistors are used to set oscillator frequencies for low-frequency requirements (less than 1 MHz), these components are not stable enough or accurate enough to provide the required feedback circuits for high-frequency oscillators. Quartz crystals offer improved stability, accuracy, and reliability for the high-frequency oscillators that are used in transmitters, receivers, computers, and watches. If a circuit requires lower (or higher) operating frequencies, the clock frequency can be applied to a divider or multiplier network.

The quartz crystal is a crystal of silicon dioxide, SiO_2, the same material that is used to provide insulation in an integrated circuit. The property that makes this crystal ideal for oscillators is its **piezoelectric effect.** When the crystal is subjected to mechanical pressure, an electric field is developed at right angles to the applied pressure. Conversely, if an electrical field is applied to the crystal, the material will deflect at right angles to the applied field. A quartz crystal that is used for oscillator circuits will be cut from very pure material and have a frequency that is dependent on the thickness of the cut and the angle at which the cut is taken within the base material. The most common crystal used in electronic circuits uses the "AT" cut, which consists of a slab cut so that one side of the crystal is along the x-axis and the main plane is at 35° with respect to the z-axis as illustrated in Figure 30–33.

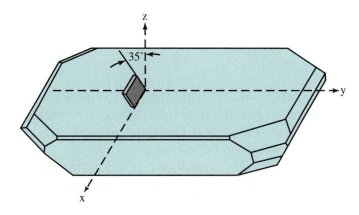

FIGURE 30–33   A quartz crystal showing the "AT" cut.

The crystal is mounted between two parallel plates, which are then connected to two external pins. Although the crystal is a mechanical device, the circuit shown in Figure 30–34 is an electrical model its operation. In the diagram shown, the capacitor, C_0, is the shunt capacitance between the crystal electrodes. The resistor, R_1, represents the bulk resistance between the terminals and gives rise to the losses within the crystal. The capacitance, C_1, is the motional capacitance representing the elasticity of the quartz; the inductance, L_1, is the motional inductance, which represents the vibrating mass of the crystal.

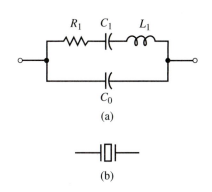

FIGURE 30–34   Electrical model and symbol of the quartz crystal.

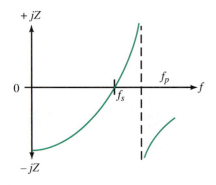

FIGURE 30–35
Impedance as a function of frequency for a quartz crystal.

The electrical properties shown in Figure 30–34 result in crystal impedance that varies as a function of frequency. This response is shown in Figure 30–35. You will notice that the response is very similar to that of the feedback circuit in the Colpitts oscillator, which we examined in the previous section. In fact, the crystal is normally connected into the feedback loop with two capacitors as illustrated in Figure 30–36. The active element in this circuit is a CMOS inverter. Since the inverter provides only two outputs ("1" ≈ +5 V and "0" ≈ 0 V), the circuit operates as a square oscillator.

The capacitors in the circuit of Figure 30–36 are selected to result in an equivalent circuit capacitance that will cause the crystal to oscillate at a frequency that falls between the series resonant frequency f_s and the parallel resonant frequency, f_p.

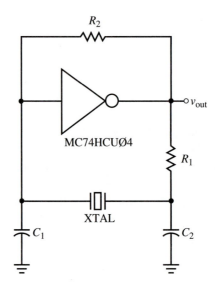

FIGURE 30–36

30.7 The 555 Timer

The 555 timer is an integrated circuit that allows accurate timing of single-occurrence events (monostable timing) or allows for oscillation in the astable mode. Figure 30–37 shows the simplified internal circuit of a 555 timer.

In the circuit, the 5-kΩ resistors form a voltage divider network. When connected into a circuit, each resistor will drop a voltage of $V_{CC}/3$. The bottom comparator will have an output of $+V_{SAT}$ when the *trigger* voltage is less than $V_{CC}/3$ and an output of 0 V when the trigger voltage is greater than $V_{CC}/3$. The top comparator will have an output of 0 V when the *threshold* voltage is less than the *control* voltage of $2V_{CC}/3$ and the output will be $+V_{SAT}$ when the *threshold* voltage goes above $2V_{CC}/3$. Notice that it is impossible for the outputs of both comparators to be $+V_{SAT}$ at the same time. (This characteristic becomes important when we consider the operation of the R-S flip-flop circuit.)

The outputs of both comparators are then applied to the input of a logic circuit called the **R-S flip-flop circuit.** This part of the timer circuit is instrumental in determining the output. The R-S flip-flop circuit is a sequential logic circuit, which means that output is not only dependent on the present values of the input, but rather may also be dependent on the previous output condition. (In other words, this device has memory.) Although the actual operation of the

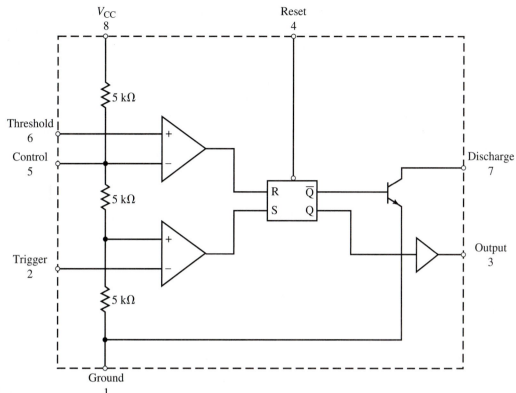

FIGURE 30–37

R-S flip-flop is outside the scope of this textbook, its operation can be summarized with a truth table as shown in Table 30–1.

The truth table shows us that the SET-action (S = 1, R = 0) causes the Q-output to go HIGH (≡1) with the complement, $\overline{Q}$ = 0. Conversely, the RESET-action (R = 1, S = 0) causes the $\overline{Q}$-output to go HIGH and the complement, Q = 0. In the case where S = R = 0, there will no change (N/C) occurring at the output. Consequently, the output of the R-S flip-flop will remain at whatever the previous condition was. Lastly, the case where S = R = 1 is referred to as the "forbidden condition" since this would result in Q = $\overline{Q}$, which is contradictory. If a 555-timer circuit is designed correctly, this condition will not occur.

S	R	Q	$\overline{Q}$
0	0	N/C	N/C
0	1	0	1
1	0	1	0
1	1	F	F

TABLE 30–1 Truth Table for an R-S Flip-Flop Circuit

Relaxation Oscillator (Astable Operation)

Figure 30–38 shows the 555 timer used as relaxation oscillator. If we assume that the capacitor is initially uncharged, the capacitor will go through a charge-discharge sequence as outlined below.

1. With the *trigger* voltage less than $V_{CC}/3$, the output of the bottom comparator will be HIGH, causing the R-S flip-flop to *set* (Q = 1 and $\overline{Q}$ = 0). The output of the 555 timer will be $+V_{CC}$ and the discharge transistor will be off.

2. As the voltage across the capacitor increases beyond $V_{CC}/3$, the output of the bottom comparator goes LOW. Since both inputs of the R-S flip-flop are LOW, the outputs will retain the same values as Step 1.

3. When the *threshold* voltage goes above $2\,V_{CC}/3$, then the output of the top comparator will go high, causing the R-S flip-flop to *reset* (Q = 0 and $\overline{Q}$ = 1). The output of the 555 timer will be zero volts. The B-E junction of the discharge transistor will be forward biased, allowing the transistor to provide a low-resistance path to discharge a capacitor through R_B.

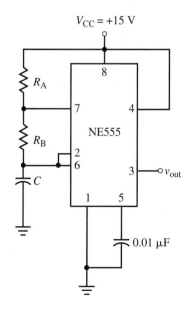

FIGURE 30–38

4. The outputs of the R-S flip-flop can also be reset ($Q = 0$ and $\overline{Q} = 1$) by applying a low voltage to the reset input. (The reset pin of the timer shown in Figure 30–38 is disabled by connecting it to V_{CC} to prevent noise pulses from inadvertently resetting the device.)

5. In the circuit of Figure 30–38, the capacitor, C, alternately charges from the trigger voltage level ($\approx V_{CC}/3$) to the threshold voltage level ($\approx 2V_{CC}/3$). The resulting waveforms observed across the capacitor and at the output of the 555 timer are shown in Figure 30–39. (In a lab you will not normally see the voltage buildup occurring from $t = 0$.)

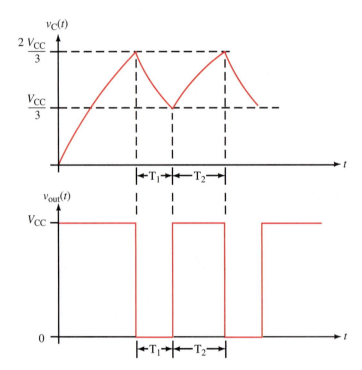

FIGURE 30–39

6. By applying discharge and charge equations to the circuit, it can be shown that the following expressions are derived.

$$T_1 = (\ln 2)(R_B C) \tag{30–26}$$

and

$$T_2 = (\ln 2)(R_A + R_B)C \tag{30–27}$$

Therefore, the period of oscillation for the relaxation oscillator is the sum of Equations 30–26 and 30–27, namely

$$T = (\ln 2)(R_A + 2R_B)C \tag{30–28}$$

Finally, the frequency of oscillation is determined as the reciprocal of Equation 30–28:

$$f = \frac{1}{(\ln 2)\left(R_A + 2R_B\right)C} \approx \frac{1.44}{\left(R_A + 2R_B\right)C} \tag{30–29}$$

EXAMPLE 30–9

If $R_A = R_B = 7.5\ k\Omega$ and $C = 0.1\ \mu F$ in the circuit of Figure 30–38, determine the frequency of oscillation.

Solution

$$f \approx \frac{1.44}{(R_A + 2R_B)C} = \frac{1.44}{(7.5\ k\Omega + 2\{7.5\ k\Omega\})0.1\ \mu F} = 640\ \text{Hz}$$

One-Shot (Monostable) Operation

As well as providing **astable operation** as a relaxation oscillator, the 555 timer can be used as a monostable circuit, which provides a single pulse each time the trigger voltage goes low. The resulting output pulse will have a minimum duration, regardless of how short the trigger pulse was. This condition is ideally suited for alarm circuits, which will continue to provide an alarm, even if the alarm trigger was reset. The duration of the output pulse is determined by the timing circuit and can be adjusted to be active from several seconds up to many minutes. The **monostable circuit** is shown in Figure 30–40.

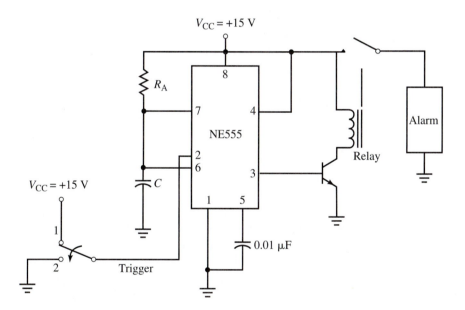

FIGURE 30–40

In Figure 30–40, we see that the voltage normally applied to the trigger input of the 555 timer circuit is +15 V. This high voltage will keep the output of the bottom comparator of the 555 timer low, and so the output voltage applied to the external transistor will also be low, keeping this transistor (and the alarm) off. At the same time, the internal discharge transistor will be turned on, preventing the capacitor from charging. When the switch in the trigger circuit is activated, the resulting low voltage at the trigger input will cause the bottom comparator to SET the R-S flip-flop, causing the output to go high. The high voltage forward biases the external transistor, closing the relay, and resulting in an alarm. Simultaneously, the internal transistor is turned off, allowing the capacitor to charge through R_A. The alarm will stay on until the capacitor, which is initially uncharged, reaches a voltage of 2 $V_{CC}/3$. Once the capacitor begins charging, the charging will continue even though the input switch may have been reset. The duration of the output pulse is given as

$$T_W = (\ln 3)R_A C \approx 1.10 R_A C \qquad \text{(30–30)}$$

The diagram of Figure 30–41 shows a typical trigger pulse and the corresponding waveforms that would be observed across the capacitor (threshold) and at the output terminal of the 555 timer.

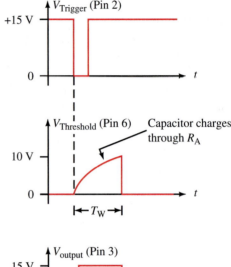

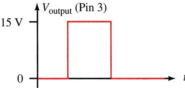

FIGURE 30–41

EXAMPLE 30–10

Determine component values for the circuit of Figure 30–40 so that the alarm will sound for a minimum of 30 seconds if the switch in the trigger circuit goes from position *1* to position *2*.

Solution
We begin by selecting a suitable value of resistance for R_A. The manufacturer suggests component values between $R_A = 2\ \text{k}\Omega$ to $100\ \text{k}\Omega$. We arbitrarily select a value of $R_A = 10\ \text{k}\Omega$. The capacitor value is then determined by applying Equation 30–30 as follows:

$$C \approx \frac{T_W}{1.10 R_A} = \frac{30\ \text{s}}{1.10(10\ \text{k}\Omega)} = 2730\ \mu\text{F}$$

If we were to use a larger capacitor, say $C = 3000\ \mu\text{F}$, the alarm would be activated for slightly longer than 30 s.

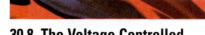

Voltage controlled oscillators (VCOs) are used in numerous electronic applications to provide an output frequency that changes as the magnitude of input voltage changes. In particular, VCOs are used as both **frequency modulator** (FM) and demodulator circuits. VCO circuits can be constructed using discrete components or by using 555 timers. However, the most common VCO circuits now use an integrated circuit specially designed for the application. The LM566C is a general purpose VCO that uses an external resistor and capacitor to produce both a square wave and a (sawtooth) triangular waveform. The output frequency varies linearly as a function of the input voltage. Figure 30–42 shows both the connection diagram and block diagram of the 8-pin package to help understand the basic circuit operation of the circuit. A 1-nF capacitor connected between pins 5 and 6 prevents unwanted parasitic oscillations that may occur during switching of the VCO.

30.8 The Voltage Controlled Oscillator—VCO

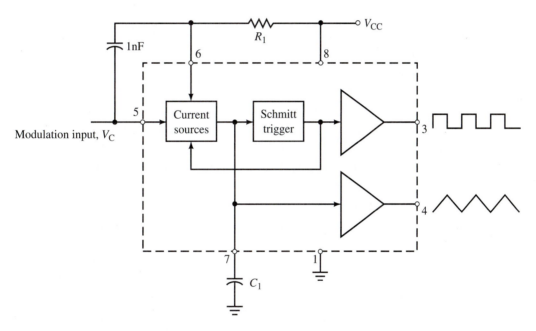

FIGURE 30–42

In order for the LM566 VCO to operate, the modulating voltage, V_C, must be applied between 0.75 V_{CC} and V_{CC}. The output frequency of the VCO is dependent on the supply voltage, R_1, C_1, and the modulating voltage, V_C, according to the relationship of Equation 30–31.

$$f_0 = \frac{2.4(V_{CC} - V_C)}{R_1 C_1 V_{CC}} \qquad (30\text{–}31)$$

The manufacturer recommends that the timing resistor have a value between $R_1 = 2 \text{ k}\Omega$ to 20 kΩ.

EXAMPLE 30–11

Refer to the circuit shown in Figure 30–43.

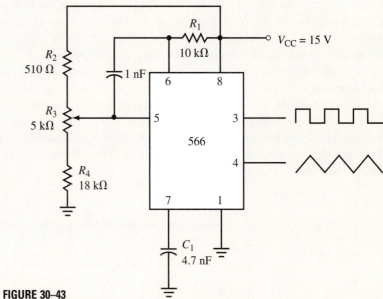

FIGURE 30–43

a. Determine the output frequency when the wiper (center terminal of R_3) is at 0% (bottom), 25%, 50%, 75%, and 100% (top).

b. Sketch the output frequency (in kHz) as a function of the modulating voltage (V).

Solution

a. When the wiper is at 0%, the modulating voltage is

$$V_C = \left(\frac{18 \text{ k}\Omega}{18 \text{ k}\Omega + 5 \text{ k}\Omega + 0.51 \text{ k}\Omega} \right) 15 \text{ V} = 11.5 \text{ V}$$

and so the output frequency is

$$f = \frac{2.4}{(10 \text{ k}\Omega)(4.7 \text{ nF})} \left(\frac{15 \text{ V} - 11.5 \text{ V}}{15 \text{ V}} \right)$$

$$= (51.06 \text{ kHz})(0.2344) = 12.0 \text{ kHz}$$

When the wiper is at 25%, we have

$$V_C = \left(\frac{19.25 \text{ k}\Omega}{23.51 \text{ k}\Omega} \right) 15 \text{ V} = 12.3 \text{ V}$$

and

$$f = (51.06 \text{ kHz}) \left(\frac{15 \text{ V} - 12.3 \text{ V}}{15 \text{ V}} \right) = 9.3 \text{ kHz}$$

When the wiper is at 50%, we have $V_C = 13.1$ V and $f = 6.5$ kHz; when the wiper is at 75%, we have $V_C = 13.9$ V and $f = 3.8$ kHz; and when the wiper is at 100%, we have $V_C = 14.7$ V and $f = 1.1$ kHz.

b. Figure 30–44 shows the plot of frequency as a function of modulating voltage. Notice the linearity of this graph. It is this linear relationship that makes the LM566 VCO an ideal circuit for low-frequency FM.

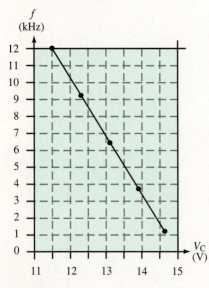

FIGURE 30–44

Figure 30–45 shows how the LM566 can be used to provide an FM output waveform. In this circuit, the resistors R_2 and R_3, establish the center frequency, f_C, of the oscillator. The modulating input will cause the VCO frequency to vary around the center frequency. This variation, called the **frequency deviation, δ** (delta) is dependent on the amplitude of the modulating signal.

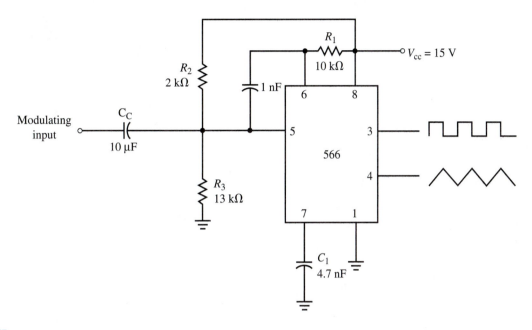

FIGURE 30–45

Determine the center frequency in the circuit of Figure 30–45. If the modulating input is 2.0 $V_{p\text{-}p}$, determine the frequency deviation at the output of the VCO.

✓ IN-PROCESS **LEARNING CHECK 2**

(Answers are at the end of the chapter.)

MultiSIM

MultiSIM and PSpice are extremely useful tools to analyze oscillator circuits. Oscillator circuits are difficult to design because the conditions required to sustain oscillations do not exist at all frequencies. Using a software package allows very simple modification of a circuit and easily shows the corresponding change in output. In this section, we examine some of the oscillators that were analyzed in previous sections of this chapter. You will find that the results very closely follow the theoretical calculations.

30.9 Computer Analysis

Use MultiSIM to observe the output waveform for the circuit of Figure 30–27. Use cursors to find the period of the waveform and compare the frequency to the theoretical frequency determined in Example 30–8.

EXAMPLE 30–12

Solution
After entering the circuit in MultiSIM, we have the schematic as illustrated in Figure 30–46.

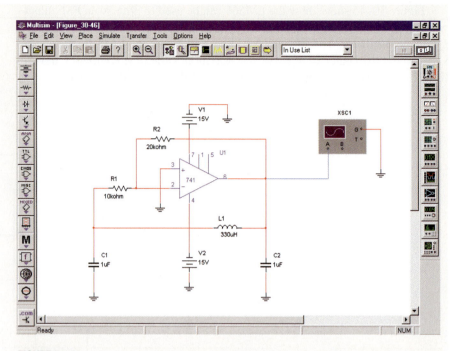

◀ MULTISIM

FIGURE 30–46

After the circuit has been entered, we click on the RUN/STOP button to obtain the oscilloscope display shown in Figure 30–47. Occasionally, it is necessary to wait a few seconds for the oscillations to begin.

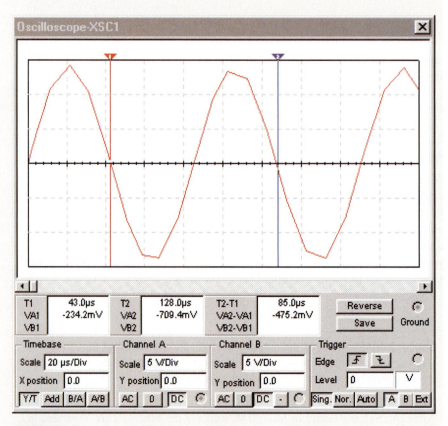

FIGURE 30–47

From the waveform observed in Figure 30–47, we see that the period of oscillation is $T = 85$ μs. The corresponding frequency of oscillation is determined as $f = 11.76$ kHz. These values correspond closely to the predicted value of 12.39 kHz. The discrepancy is likely due to the software accounting for parasitic capacitance of the component leads.

PRACTICE PROBLEMS 5

Use MultiSIM to construct the phase shift oscillator shown in Figure 30–24. Run the simulation to determine the frequency of oscillation and compare the result to the predicted frequency of oscillation in Example 30–7.

Answer
$f = 2878$ Hz compares well to the theoretical value of $f = 2953$ Hz.

PSpice

Use PSpice to construct the 555-relaxation oscillator circuit shown in Figure 30–38. Let $R_A = R_B = 7.5$ kΩ and $C = 0.1$ μF. Note: Since PSpice does not permit any open or unused terminals, it is necessary to connect a load resistor (say $R_L = 1$ kΩ) between the output terminal and ground. Display the output voltage (pin 3) and the capacitor voltage (pin 6) using the Probe postprocessor. Measure the period and calculate the frequency of oscillation. Compare these values to the theoretical calculations.

EXAMPLE 30–13

Solution
The resulting PSpice diagram is shown in Figure 30–48, and the display in the Probe postprocessor is shown in Figure 30–49.

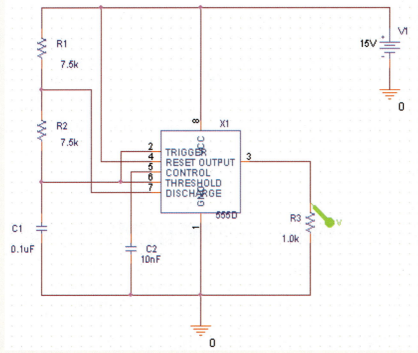

FIGURE 30–48

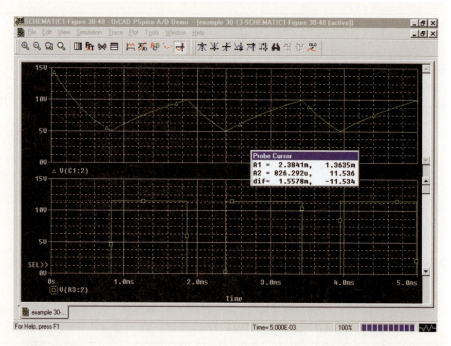

FIGURE 30–49

By using the cursors, we determine the period of oscillation to be $T = 1.56$ ms, which results in a frequency, $f = 641$ Hz. This result is same as that found in Example 30–10.

PUTTING IT INTO PRACTICE

As part of a home laboratory, you would like to have an oscillator that provides a square wave, a triangular wave, and a sine-wave output. You would like the frequency of each waveform to be within a range of 1 kHz to 2 kHz. Use the skills developed in this chapter to design a suitable circuit using commercially available component values. Simulate the operation of the circuit using either MultiSIM or PSpice.

PROBLEMS

30.1 Basics of Feedback

1. Given the non-inverting amplifier shown in Figure 30–50, determine the feedback ratio, B, and solve for the closed loop gain if the open-loop gain of the op-amp is 500.

2. Repeat Problem 1 if the open-loop gain of the op-amp is 1000.

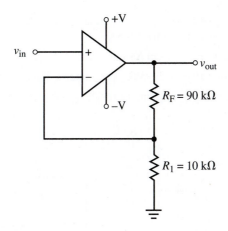

FIGURE 30–50

30.2 The Relaxation Oscillator

3. Given that $R_1 = 2$ kΩ and $R_2 = 3$ kΩ in the Schmitt trigger circuit shown in Figure 30–51:

   a. Sketch the hysteresis curve for the circuit.

   b. Sketch the corresponding output for the given input waveform.

4. If $R_1 = 4$ kΩ and $R_2 = 4$ kΩ in the Schmitt trigger circuit shown in Figure 30–51:

   a. Sketch the hysteresis curve for the circuit.

   b. Sketch the corresponding output for the given input waveform.

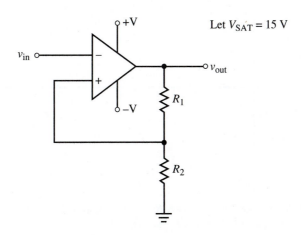

Let $V_{SAT} = 15$ V

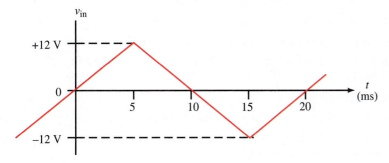

FIGURE 30–51

5. If $R_1 = 2\ k\Omega$, $R_2 = 3\ k\Omega$, $R_F = 10\ k\Omega$, and $R_{in} = 10\ k\Omega$ in the Schmitt trigger circuit shown in Figure 30–52.

a. Sketch the hysteresis curve for the circuit.

b. Determine the value of the reference voltage, V_{REF}.

c. Sketch the corresponding output for the given input waveform.

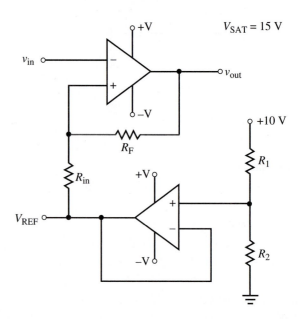

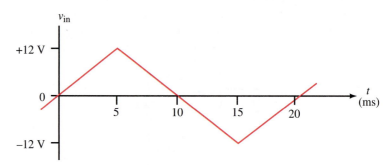

◀ MULTISIM   **FIGURE 30–52**

6. If $R_1 = 3\ k\Omega$, $R_2 = 2\ k\Omega$, $R_F = 20\ k\Omega$, and $R_{in} = 10\ k\Omega$ in the Schmitt trigger circuit shown in Figure 30–52.

a. Sketch the hysteresis curve for the circuit.

b. Determine the value of the reference voltage, V_{REF}.

c. Sketch the corresponding output for the given input waveform.

7. If $R_1 = 3\ k\Omega$, $R_2 = 2\ k\Omega$, $R_F = 20\ k\Omega$, and $C = 0.20\ \mu F$ in the circuit illustrated in Figure 30–53.

a. Sketch the capacitor voltage, v_C.

b. Sketch the output voltage, v_{out}. Label the times and amplitudes correctly.

8. If $R_1 = 10\ k\Omega$, $R_2 = 20\ k\Omega$, $R_F = 20\ k\Omega$, and $C = 0.10\ \mu F$ in the circuit illustrated in Figure 30–53.

a. Sketch the capacitor voltage, v_C.

b. Sketch the output voltage, v_{out}. Label the times and amplitudes correctly.

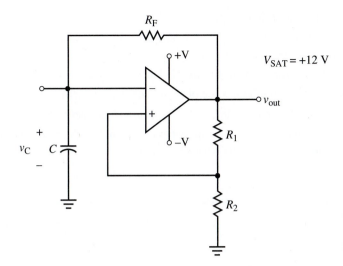

FIGURE 30–53

30.3   The Wien Bridge Oscillator

9. If $R_{in} = R_1 = R_2 = 10$ kΩ and $C_1 = C_2 = 0.22$ μF in the circuit shown in Figure 30–54:

   a. What is the minimum value of R_F to ensure that oscillations can begin?

   b. Determine the frequency of oscillation.

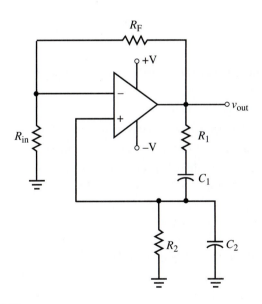

FIGURE 30–54   ◀ MULTISIM

10. What values of capacitor, C would be needed in the circuit of Problem 9 so that the output frequency is 1 kHz.

11. If $R_1 = 2R_2$ in the circuit of Figure 30–54, what would be the feedback ratio, B? How much forward gain would the amplifier need?

12. If $R_2 = 2R_1$ in the circuit of Figure 30–54, what would be the feedback ratio, B? How much forward gain would the amplifier need?

30.4   The Phase-Shift Oscillator

13. Given that the circuit of Figure 30–55 has component values $C = 0.047$ μF, $R = 1.5$ kΩ, and $R_{F1} = 33$ kΩ:

    a. Determine the frequency of oscillation.

    b. What is the minimum value of R_{F2} to ensure that oscillation will occur?

    c. What should be the value of the compensating resistor, R_C?

    d. At the required amplifier gain, what is the bandwidth of the amplifier if the gain-bandwidth product for the op-amp is 2×10^6 Hz.

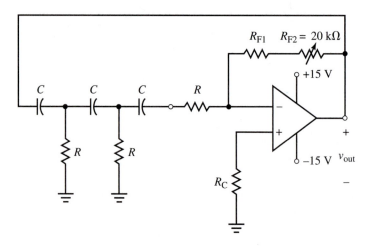

FIGURE 30–55

14. Given that the circuit of Figure 30–55 has component values $C = 0.01$ μF, $R = 2.0$ kΩ, and $R_{F1} = 47$ kΩ:

    a. Determine the frequency of oscillation.

    b. What is the minimum value of R_{F2} to ensure that oscillation will occur?

    c. What should be the value of the compensating resistor, R_C?

    d. At the required amplifier gain, what is the bandwidth of the amplifier if the gain-bandwidth product for the op-amp is 2×10^6 Hz.

15. Using the same capacitor values, redesign the phase-shift oscillator of Problem 13 to have a resonant frequency of 4.0 kHz.

16. Using the same capacitor values, redesign the phase-shift oscillator of Problem 14 to have a resonant frequency of 3.0 kHz.

30.5   LC Oscillators

17. Identify the type of oscillator that is illustrated in Figure 30–56. Calculate the frequency of oscillation.

18. Identify the type of oscillator that is illustrated in Figure 30–57. Calculate the frequency of oscillation.

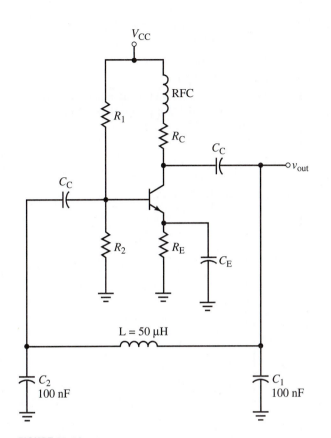

FIGURE 30–56

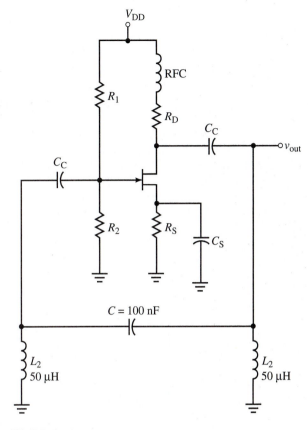

FIGURE 30–57

30.6 Crystal Oscillators

19. What is meant by the term *piezoelectric effect* as it relates to a quartz crystal?

20. Sketch the electrical equivalent circuit of a quartz crystal.

21. Sketch the impedance as a function of frequency for a quartz crystal.

22. When used with a CMOS inverter, a crystal oscillator will normally be tuned to operate in which of the following frequency ranges?

    a. Less than f_s?

    b. Between f_s and f_P?

    c. Greater than f_P?

30.7 The 555 Timer

23. Refer to the schematic representation of the 555 timer shown in Figure 30–37. Assume that V_{CC} and Ground have been connected to the timer circuit.

    a. If the trigger voltage is less than $V_{CC}/3$, what will be the value of the output voltage? Will the discharge transistor be ON or OFF?

    b. If the threshold voltage is greater than $2V_{CC}/3$, what will be the value of the output voltage? Will the discharge transistor be ON or OFF?

24. Explain the operation of the 555 timer when it is connected as an astable multivibrator.

25. Design a 555 timer as an astable multivibrator having $T_1 = 0.25$ s and $T_2 = 0.75$ s. As part of your design, let $C = 1.0$ μF.

26. Design a 555 timer as an astable multivibrator having $T_1 = 10$ ms and $T_2 = 90$ ms. As part of your design, let $C = 0.47$ μF.

27. The circuit shown in Figure 30–58 has the indicated input signal. Sketch the corresponding output signal showing expected voltage level and times.

28. Repeat Problem 27 if $R_A = 1.0$ MΩ.

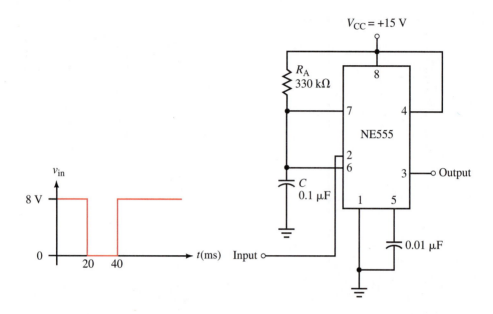

FIGURE 30–58

30.8 The Voltage Controlled Oscillator—VCO

29. a. Calculate the output frequencies for the VCO circuit shown in Figure 30–59 when the wiper of R_3 is at 0% (bottom), 50%, and 100% (top).

b. Sketch the output frequency (in kHz) as a function of the modulating voltage (V).

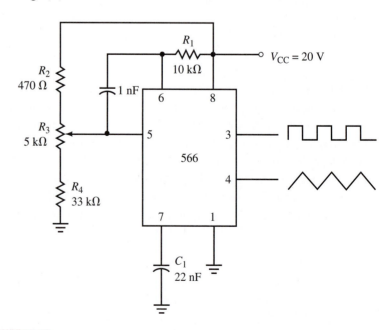

FIGURE 30–59

30. Repeat Problem 29 if the timing capacitor is replaced by $C_1 = 1$ nF.

30.9 Computer Analysis

31. Use MultiSIM to input the circuit of Figure 30–52, letting $R_1 = 2$ kΩ, $R_2 = 3$ kΩ, $R_F = 10$ kΩ, and $R_{in} = 10$ kΩ. Use 741 op-amps and supply voltages of ± 15 V for the op-amps. Set the signal generator to provide the indicated triangular waveform. Observe the output voltage and compare this result to the predicted output that you determined in Problem 5.   ◀ MULTISIM

32. Use MultiSIM and repeat Problem 31 letting $R_1 = 3$ kΩ, $R_2 = 2$ kΩ, $R_F = 20$ kΩ, and $R_{in} = 10$ kΩ.   ◀ MULTISIM

33. Use MultiSIM to input the circuit of Figure 30–54, letting $R_{in} = R_1 = R_2 = 10$ kΩ, $R_F = 33$ kΩ, and $C_1 = C_2 = 0.22$ μF. Use a 741 op-amp and supply voltages of ± 15 V. Observe the output voltage and compare this result to the predicted output that you determined in Problem 9.   ◀ MULTISIM

34. Use MultiSIM to input the circuit of Figure 30–54, letting $R_{in} = R_1 = R_2 = 10$ kΩ, $R_F = 33$ kΩ. Use the value of $C_1 = C_2 = C$ determined in Problem 10. Observe the output signal. You should observe that the frequency is 1 kHz.   ◀ MULTISIM

35. Use PSpice to input the circuit of Figure 30–53, letting $R_1 = 3$ kΩ, $R_2 = 2$ kΩ, $R_F = 20$ kΩ, and $C = 0.20$ μF. Use 741 op-amps and supply voltages of ± 15 V for the op-amps.   ◀ CADENCE

   a. Use the Probe postprocessor to obtain the capacitor voltage, v_C and the output voltage, v_{out}.

   b. Compare the predicted results to those obtained in Problem 7.

36. Use PSpice and repeat Problem 35, $R_1 = 10$ kΩ, $R_2 = 20$ kΩ, $R_F = 20$ kΩ, and $C = 0.10$ μF. Compare the observed results to the predicted results of Problem 8.   ◀ CADENCE

37. Use PSpice to input the astable multivibrator designed in Problem 25. Observe the output waveform at pin 3 of the 555 timer. Compare your results to the required characteristics.   ◀ CADENCE

38. Use PSpice to input the astable multivibrator designed in Problem 26. Observe the output waveform at pin 3 of the 555 timer. Compare your results to the required characteristics.   ◀ CADENCE

 ANSWERS TO IN-PROCESS LEARNING CHECKS

In-Process Learning Check 1

$B = 0.4$ $A = 2.5$

In-Process Learning Check 2

$f_C = 6810$Hz $\delta = 3404$ Hz

■ **OBJECTIVES**

On completion of this chapter, you will be able to

- relate the firing angle to the conduction angle of a thyristor,
- explain how various triggering devices are used to fire thyristors,
- determine the characteristics of a UJT (unijunction) transistor,
- calculate the frequency (or period) of oscillation of a UJT relaxation oscillator,
- explain the operation of an SCR (silicon controlled rectifier) circuit,
- explain the operation and be able to analyze a triac circuit,
- identify the modes of operation of a triac and indicate which modes are the most sensitive,
- use power control fundamentals to predict the rms voltage and power delivered to a load for half-wave and full-wave control,
- solve for the frequency or wavelength of a given E/M waveform,
- identify the advantages and disadvantages of LED displays,
- calculate the current-limiting resistor in an LED circuit,
- explain the basic operation of photodetectors, such as photodiodes and phototransistors,
- analyze and design circuits using optocouplers,
- describe the basic principle of operation of semiconductor lasers.

Thyristors and Optical Devices

31

CHAPTER PREVIEW

I n this chapter you will examine several electronic devices that use low-voltage signals to control high-voltage, high-current circuits. Almost everyone has used light dimmers to set the mood at a dinner table or used a potentiometer to control the speed of an electric motor. In this chapter you will learn about thyristors, four-layer devices that are used to perform these functions. You will also examine some of the ways that these devices can be triggered and, more importantly, how they are turned off.

We conclude the chapter by examining optical devices, from the common LED to less common photodiodes and phototransistors. You will discover that optical devices allow us to electrically isolate two circuits, functions that are extremely important in hospitals and anywhere an electrical fault can cause catastrophic results. You will also be introduced to fiber optic transmission that uses a laser as a light source to encode and transmit digital information. ■

PUTTING IT IN PERSPECTIVE

The LASER

LIGHT, WHICH IS SIMPLY MADE up of E/M (electromagnetic) waves, is normally generated when electrons that have been raised to a higher energy level (due to heat or some other external energy) drop from this excited level to their ground state. This process occurs randomly and is therefore called *spontaneous* emission.

In a LASER, electrons are similarly raised to a higher energy level. However, these electrons do not immediately drop to their ground state. Rather, they gather in a semi-stable level called the **metastable state.** When a photon ("particle") of light with exactly the right amount of energy strikes an electron in the metastable state, the electron drops to its ground state. This releases another photon with exactly the same energy (wavelength) and exactly in phase with the originating photon. This amplification process repeats until most of the electrons in the metastable state have dropped to the ground level. Consequently, the acronym LASER stands for *Light Amplification through Stimulated Emission of Radiation*.

The principle of laser operation was explained by two American physicists, Charles H. Townes and A. L Schawlow, in 1953. The first operational laser, consisting of a synthetic ruby cylinder, was built and patented in 1960 by Theodore H. Maiman. This laser used a xenon light source to provide the energy to raise the electrons from the ground state to a high energy state. One end of the ruby cylinder was totally silvered to reflect the light, while the other end was partially silvered to allow some of the intense beam to escape in very short duration pulses.

Modern lasers are used in numerous applications, including transmission of data over fiber optic cable, surgery, metal fabrication, surveying, and many other commercial and scientific uses. ■

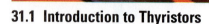

31.1 Introduction to Thyristors

Thyristors are electronic devices that behave essentially as switches with two stable states. In the ON-state a thyristor is conducting, while in the OFF-state the device blocks current. Although there are several types of devices that are categorized as thyristors, each of these devices is either a **unilateral (unidirectional) device,** meaning that it conducts in one direction or a **bilateral (bidirectional) device,** which conducts in either direction. All thyristors use internal feedback to provide *latching* (or switching) action. The silicon-controlled rectifier (SCR) and the triac, respectively, are examples of unilateral and bilateral thyristors. In this chapter we will examine the operation of thyristors which, when used in combination with triggering circuits, are used in ac circuits to control the amount of power delivered to a loads such as motors, heaters, or lights. Figure 31–1 illustrates a typical circuit that uses a trigger circuit to control the **firing angle** of a thyristor.

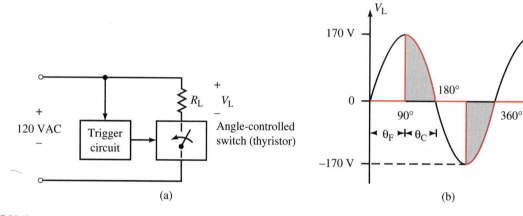

(a)   (b)

FIGURE 31–1

In the illustration of Figure 31–1, the trigger circuit causes the angle-controlled switch (generally a thyristor) to close at an angle of 90°. Since the load receives less than the full 180° of the supply voltage each half period, it stands to reason that the effective (rms) voltage delivered to the load will be significantly less than the rms voltage of the supply. We define the *firing angle,* θ_F of the circuit as the delay (in degrees) between the zero crossing and the closing of the angle-controlled switch. Conversely, the **conduction angle,** θ_C is defined as the number of degrees per half-cycle that the source voltage is applied to the load. Note that the following expression always holds:

$$\theta_C = 180° - \theta_F \qquad (31\text{–}1)$$

Although we could use calculus to determine the effective (rms) voltage and power applied to a load for a given firing angle, you will find that using graphs conveniently eliminates this tedious work.

31.2 Triggering Devices

In order to cause an angle-controlled switch to close at the correct instant, a trigger circuit must provide a suitable pulse to the switching device. Several types of triggering devices are available to provide such signals. Although the 555 timer that you examined in the previous chapter can be used to provide proper timing, the resulting circuit would be more complicated than necessary. In this section, we examine two fairly simple devices that will enable a thyristor to fire at the correct angle.

The Diac

The diac is a three-layer, bi-directional device, which behaves in a manner similar to having two back-to-back zener diodes. Whenever the voltage across the diac exceeds the breakover voltage, V_{BR} of the device, the diac will conduct. Figure 31–2(a) shows the schematic symbols that are used to represent the diac, and Figure 31–2(b) shows the characteristic curve of the device. In order for the diac to conduct, the voltage across the device must exceed the **breakover voltage**, V_{BR}. As long as the voltage across the diac is less than the breakover voltage, the diac is in its **blocking region,** meaning that there is very little current through the device. Once the voltage across the diac is greater than V_{BR}, there will be substantial current through the device, along with a reduction in voltage drop across the component. Since the diac operation is essentially the same regardless of the voltage polarity, the device can be connected into a circuit without regard to the terminal polarities.

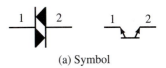

(a) Symbol

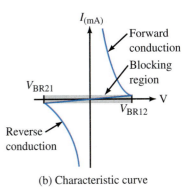

(b) Characteristic curve

FIGURE 31–2   The diac.

The Unijunction Transistor (UJT)

The UJT is a three-terminal device constructed as illustrated in Figure 31–3(a) and having the symbol as illustrated in Figure 31–3(b). The emitter region is heavily doped silicon while the inter-base region is lightly doped material. When the emitter terminal is open, the resistance between the bases, B_1 and B_2 is relatively high, typically having a total resistance in the order of $R_{BB} = 5$ kΩ to 10 kΩ. Figure 31–4 shows the electrical equivalent of the UJT and will be used to help explain the operation of this device. As you will discover shortly, the resistance between the emitter connection and the B_1 terminal is variable depending on the applied external voltage.

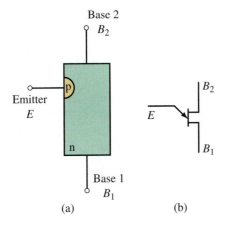

FIGURE 31–3   The n-channel unijunction transistor (UJT).

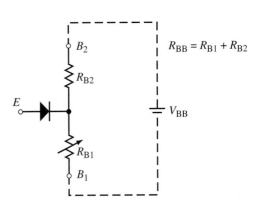

FIGURE 31–4   Equivalent circuit of an n-channel UJT.

When an external voltage source, V_{BB}, is applied between the base terminals (with the emitter open), an internal voltage will be developed across resistor, R_{B1}, in accordance with the voltage divider rule and is expressed as

$$V_1 = \frac{R_{B1}}{R_{B1} + R_{B2}} V_{BB} = \frac{R_{B1}}{R_{BB}} V_{BB} \qquad (31\text{–}2)$$

The **intrinsic standoff ratio** of the UJT is defined as the following resistance ratio:

$$\eta = \frac{R_{B1}}{R_{BB}} \qquad (31\text{–}3)$$

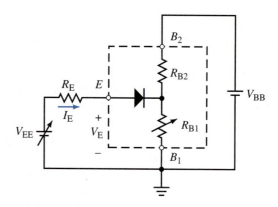

FIGURE 31–5   UJT test circuit.

Typical values for the intrinsic standoff ratio for UJTs are in the range between $\eta = 0.5$ to 0.9. The operation of a UJT can be examined by placing the device into a test circuit as shown in Figure 31–5.

The circuit operates as follows:

1. When $V_{EE} = 0V$, the p-n junction (represented by the diode) will be reverse-biased. The current in the emitter circuit will be the result of the leakage current through this junction. (Typically in the order of a few microamps.)

2. As the applied emitter voltage is increased to a point where $V_{EE} = V_1$, the diode will begin to conduct. The emitter voltage at this point will represent the *peak voltage* of the UJT and is expressed as

$$V_P = \eta V_{BB} + V_D \tag{31–4}$$

3. Since the emitter region of the UJT is heavily doped, the forward-biased p-n junction will allow holes to be injected into the base region of R_1, effectively reducing the resistance of this region. Consequently, the value of V_E decreases, while at the same time as the current through the emitter circuit increases.

4. The reduction in resistance causes a further increase in current. The increased current results in the injection of even more carriers into the base region of R_1, with a corresponding reduction in the value of R_1.

5. Eventually a point is reached where further increase in current does not inject any more holes the lower base region. The resistance, R_1, has reached its minimum value. This point is referred to as the **valley point** of the UJT.

6. Further increases in the emitter current, I_E, cause V_E to also increase. At this point the UJT is *saturated*.

Figure 31–6 illustrates the typical curve of V_E as a function of I_E. Notice that the peak voltage is not a constant for a given transistor, but rather depends on both the intrinsic standoff ratio and the supply voltage, V_{BB}.

If a circuit is designed with an emitter resistance, R_E, that ensures the UJT operates in the negative resistance region, we can use the UJT as a relaxation oscillator. The frequency of oscillation is dependent on the rate at which an emitter capacitor, C_E, charges. Figure 31–7 shows a simple UJT relaxation oscillator and the corresponding emitter voltage.

If we assume that the valley voltage of the UJT is $V_V = 0$ V, that the discharge time of the capacitor is $T_D = 0$, and that the voltage across the internal

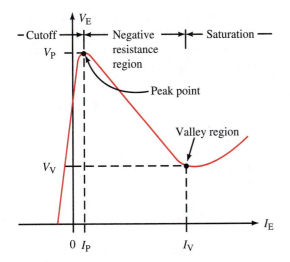

FIGURE 31–6   Emitter current-voltage characteristics of a UJT.

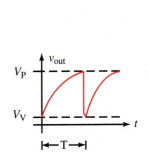

FIGURE 31–7   UJT relaxation oscillator.

p-n junction (forward-biased diode) is also 0 V, then it can be shown that the period of oscillation for the circuit of Figure 31–7 is approximated as:

$$T \approx R_E C_E \ln\left(\frac{1}{1 - \eta}\right) \tag{31–5}$$

Consequently, the frequency of oscillation for the UJT relaxation oscillator is approximately

$$f \approx \frac{1}{R_E C_E \ln\left(\dfrac{1}{1 - \eta}\right)} \tag{31–6}$$

Although the circuit of shown in Figure 31–7 is useful as a relaxation oscillator, it requires an additional resistor between B_1 and ground in order to be used as a trigger circuit. By adding a carbon resistor between the voltage supply and B_2, the frequency of oscillation will remain more constant despite changes in temperature. The resulting circuit is shown in Figure 31–8. The pulses appearing across R_1 can now be applied to trigger thyristor circuits.

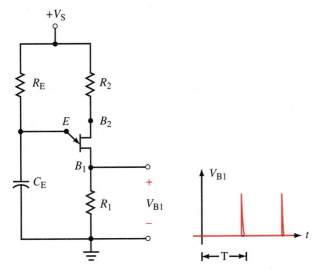

FIGURE 31–8

In order for the UJT oscillator to maintain its oscillations, it must operate in the negative resistance region. This means that the operating point must lie between the peak and valley points of the emitter characteristic curve. In order for the circuit to have enough current to cause the UJT to fire, the emitter resistance must have a value that is less than a value $R_{E(max)}$, determined from the peak point voltage and current as

$$R_{E(max)} = \frac{V_S - V_P}{I_P} \qquad (31\text{–}7)$$

Additionally, the value of R_E must be greater than the minimum value $R_{E(min)}$, determined from the valley point as

$$R_{E(min)} = \frac{V_S - V_V}{I_V} \qquad (31\text{–}8)$$

When designing a UJT relaxation oscillator circuit, a good rule of thumb is to select

$$R_E = 3R_{E(min)} \qquad (31\text{–}9)$$

EXAMPLE 31–1

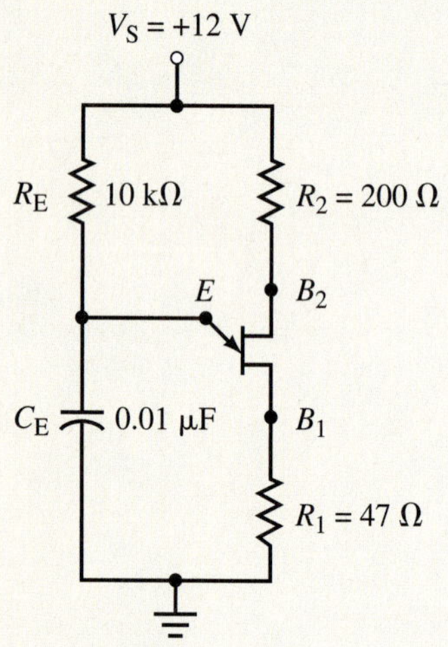

$V_S = +12$ V

R_E 10 kΩ $R_2 = 200$ Ω

E B_2

C_E ⎓ 0.01 μF B_1

$R_1 = 47$ Ω

FIGURE 31–9

Given that the UJT in the circuit of Figure 31–9 has the following characteristics:

Valley point: $V_V = 1.5$ V $I_V = 2.0$ mA
Peak point: $V_P = 8.0$ V $I_P = 2.0$ μA
$V_D = 0.6$ V
$I_{B2} = 1.8$ mA when $I_E = 0$

a. Solve for the frequency of oscillation,
b. Determine the value of R_{BB} of the UJT.
c. Sketch the waveforms that you would expect to observe at E, B_1, and B_2.

Solution

a. First we solve for the intrinsic standoff ratio. From Equation 31–4 we have

$$V_P = \eta V_{BB} + V_D$$

and η is determined as

$$\eta = \frac{V_P - V_D}{V_{BB}} = \frac{8.0 \text{ V} - 0.6 \text{ V}}{12.0 \text{ V}} = 0.617$$

The frequency of oscillation is now easily determined by applying Equation 31–6 as

$$f \approx \frac{1}{R_E C_E \ln\left(\dfrac{1}{1 - \eta}\right)}$$

$$= \frac{1}{(10 \text{ k}\Omega)(0.01 \text{ μF})\ln\left(\dfrac{1}{1 - 0.617}\right)} = 10.4 \text{ kHz}$$

b. The total resistance in the output loop of the UJT is determined from Ohm's law as

$$R_T = \frac{V_S}{I_{B2}} = \frac{12\ V}{1.8\ mA} = 6.67\ k\Omega$$

and so we solve for R_{BB} as

$$R_{BB} = R_T - R_2 - R_1$$
$$= 6.66\ K\Omega - 200\ \Omega - 47\ \Omega = 6.42\ k\Omega$$

c. The waveforms that are observed at the three terminals of the UJT appear as shown in Figure 31–10. The voltage spikes appearing across R_1 are the result of the capacitor rapidly discharging through the low-resistance path of the forward-biased p-n junction of the emitter.

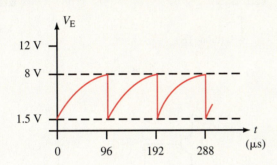

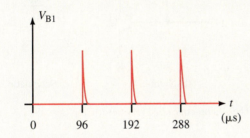

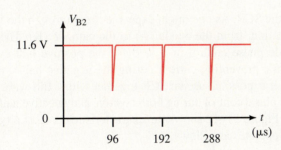

FIGURE 31–10

Determine the values of R_E and C_E so the circuit of Figure 31–9 oscillates at a frequency of 5 kHz. Let $R_E = 3R_{E(min)}$.

PRACTICE PROBLEMS 1

Answers
$R_E = 11.25\ k\Omega$, $C_E = 18.5\ nF$

31.3 Silicon-Controlled Rectifiers (SCRs)

The SCR is a four-layer device that is equivalent to two transistors connected so that once one of the transistors is turned on, the feedback due to the other transistor will keep both transistors on even though the triggering source is removed. Figure 31–11 shows the structure, the symbol, and the electronic

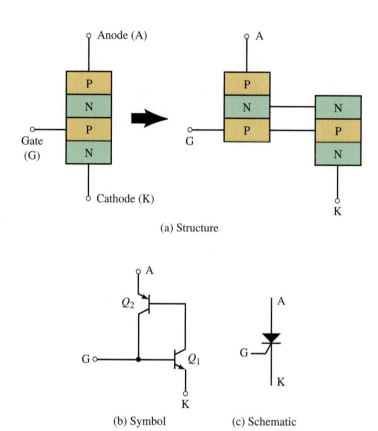

(a) Structure

(b) Symbol   (c) Schematic

FIGURE 31–11   The silicon-controlled rectifier (SCR).

equivalent of the SCR. As one might expect, current through the SCR can only be in one direction, from the anode (A) to the cathode (K). The SCR can be used to provide phase control of a load for the positive half-cycle of an ac waveform, while preventing current during the negative half-cycle. Although less common, it is possible use an SCR together with a full-wave bridge rectifier to provide phase control during both positive and negative half-cycles of an ac waveform. Figure 31–12 shows the characteristic curve of a typical SCR.

The SCR operates as follows:

1. When the anode to cathode is forward-biased (assuming there is no gate current, $I_G = 0$), there will be no anode current through the device until the forward breakover voltage, V_{DRM}, of the device is exceeded. In the *forward blocking region,* the C-B junction of Q_1 is reverse-biased, allowing very little leakage current.

2. If a positive voltage is applied to the gate (relative to the cathode), the base-emitter junction of Q_1 will be forward-biased, resulting in a large collector current.

3. The large collector current of Q_1 will cause an equal amount of base current to be applied to Q_2.

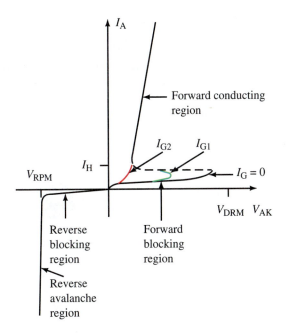

FIGURE 31–12   The SCR characteristic curve.

4. The resulting collector current of Q_2 will force Q_1 to be driven on even further. This regenerative feedback very quickly forces both transistors into saturation. Typical turn-on times for a low-power SCR is in the order of 0.1 to 1 μs. If the forward bias, which was initially applied between the gate and the cathode, is removed now, it will have no effect. The SCR will remain on.

5. The SCR will only turn off if the current through the device drops below the holding current, I_H, specified by the manufacturer. Typical holding current is in the order of $I_H = 10$ mA. The usual method to turn an SCR off is to either reverse the applied voltage (as in an ac voltage source) or to have a switch open the anode circuit.

6. Instead of applying a positive gate pulse, an SCR may also be latched on by providing a forward voltage that exceeds the forward breakover voltage, V_{FOM}. In this case, both transistors in the equivalent circuit go into saturation. Once again, the only way to turn the device off is by low-current dropout or by anode current interruption.

7. Other less common methods of turning an SCR on include:

   • high temperature, which would result in excessive leakage current, allowing the regenerative feedback to begin,

   • rate effect caused by high $\Delta V/\Delta t$ on the supply voltage due to a noise spike, and

   • radiation, which like an increased temperature will cause excessive charge movement.

The above methods of firing an SCR should be avoided, since they result in unpredictable and sometimes dangerous operation of high-voltage switching.

Figure 31–13 shows some of the specifications for the C106 Series SCR. There are many other SCRs available, each type having specifications that can be applied to a particular application. For example, some SCRs are capable of handling currents up to 2500 A and/or voltages up to 2500 V.

SCRs
4 AMPERES RMS
200 thru 600 VOLTS

MAXIMUM RATINGS (T$_J$ = 25°C unless otherwise noted)

Rating	Symbol	Value	Unit
Peak Repetitive Off–State Voltage (Note 1) (Sine Wave, 50–60 Hz, R$_{GK}$ = 1 kΩ, T$_C$ = –40° to 110°C)	V$_{DRM,}$ V$_{RRM}$		Volts
C106B		200	
C106D, C106D1		400	
C106M, C106M1		600	
On-State RMS Current (180° Conduction Angles, T$_C$ = 80°C)	I$_{T(RMS)}$	4.0	Amps
Average On–State Current (180° Conduction Angles, T$_C$ = 80°C)	I$_{T(AV)}$	2.55	Amps
Peak Non-Repetitive Surge Current (1/2 Cycle, Sine Wave, 60 Hz, T$_J$ = +110°C)	I$_{TSM}$	20	Amps

TO–225AA

CASE 077
STYLE 2

PIN ASSIGNMENT	
1	Cathode
2	Anode
3	Gate

ON CHARACTERISTICS

	Symbol				Unit
Peak Forward On–State Voltage (Note 3) (I$_{TM}$ = 4 A)	V$_{TM}$	–	–	2.2	Volts
Gate Trigger Current (Continuous dc) (Note 4) (V$_{AK}$ = 6 Vdc, R$_L$ = 100 Ohms) T$_J$ = 25°C T$_J$ = –40°C	I$_{GT}$	– –	15 35	200 500	μA
Peak Reverse Gate Voltage (I$_{GR}$ = 10 μA)	V$_{GRM}$	–	–	6.0	Volts
Gate Trigger Voltage (Continuous dc) (Note 4) (V$_{AK}$ = 6 Vdc, R$_L$ = 100 Ohms) T$_J$ = 25°C T$_J$ = –40°C	V$_{GT}$	0.4 0.5	0.60 0.75	0.8 1.0	Volts
Gate Non–Trigger Voltage (Continuous dc) (Note 4) (V$_{AK}$ = 12 V, R$_L$ = 100 Ohms, T$_J$ = 110°C)	V$_{GD}$	0.2	–	–	Volts
Latching Current (V$_{AK}$ = 12 V, I$_G$ = 20 mA) T$_J$ = 25°C T$_J$ = –40°C	I$_L$	– –	0.20 0.35	5.0 7.0	mA
Holding Current (V$_D$ = 12 Vdc) (Initiating Current = 20 mA, Gate Open) T$_J$ = 25°C T$_J$ = –40°C T$_J$ = +110°C	I$_H$	– – –	0.19 0.33 0.07	3.0 6.0 2.0	mA

FIGURE 31–13 The C106 series silicon-controlled rectifier. *(Courtesy of Semiconductor Components Industries, LLC 2002)*

The circuit shown in Figure 31–14 provides phase control of an SCR with firing angles of 0° to 90°.

As the sinusoidal voltage of the applied voltage source in the circuit of Figure 31–14 increases, a point is reached where there is sufficient forward current through the gate circuit to cause the SCR to fire. Once the SCR fires, it continues to conduct for the balance of the half-cycle, until the voltage of the ac source drops below the value needed to sustain the holding current of the SCR. The diode, D_1, protects the SCR gate circuit from excessive voltage during negative half-cycle because the SCR will not conduct during the negative half-cycle. The value of R_1 limits the gate current, and, consequently, determines the angle at which the SCR fires. The operation of an SCR can be simply explained by applying circuit theory.

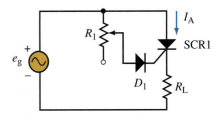

FIGURE 31–14 SCR providing $0° \rightarrow 90°$ phase control.

Refer to the circuit of Figure 31–14. Assume that you are given a voltage source having amplitude of 30 V, operating at a frequency of 60 Hz, and load resistance of $R_L = 15\ \Omega$. The SCR has a gate trigger current, $I_{GT} = 100\ \mu A$, and a gate trigger voltage of $V_{GT} = 0.6$ V. Determine the range of values for resistor R_1 so that the SCR will trigger between 10° and 90°.

EXAMPLE 31–2

Solution

At $\theta = 10°$ triggering: The instantaneous voltage is $v(10°) = 30 \sin 10° = 5.21$ V. Since the current through the SCR is $I_A = 0$, we write the Kirchhoff voltage law equation as

$$5.21\ V = I_G R_L + I_G R_1 + V_{D1} + V_{GT}$$
$$= (100\ \mu A)(15\ \Omega) + (100\ \mu A)R_1 + 0.7\ V + 0.6\ V$$

and so

$$R_1 = \frac{5.21\ V - 0.0015 - 0.7\ V - 0.6\ V}{100\ \mu A} = 39.1\ k\Omega$$

At $\theta = 90°$ triggering: The instantaneous voltage is $v(90°) = 30$ V. Using the same approach as before, we have

$$R_1 = \frac{30\ V - 0.0015 - 0.7\ V - 0.6\ V}{100\ \mu A} = 287\ k\Omega$$

In order for the SCR to have triggering control between 10° and 90°, the value of R_1 must be adjustable between 39 kΩ and 287 kΩ.

The circuit of Figure 31–15 shows an SCR circuit that provides phase angle control between 0° and 180°.

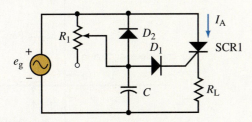

FIGURE 31–15 SCR providing $0° \rightarrow 180°$ phase control.

On the negative half-cycle of the input waveform, the SCR will not conduct. The capacitor, C, charges to the source voltage through the low forward resistance of diode D_2. On the positive half-cycle, the capacitor voltage will initially be negative, keeping the SCR off. As the capacitor charges through R_1, a point is reached at which the voltage is sufficiently positive to cause the SCR to fire. If the value of R_1 is small, the capacitor will quickly charge to the required positive voltage, allowing the SCR to fire at approximately 0°. Conversely, if the value of R_1 is large, the capacitor will slowly charge, allowing the SCR to fire as late as 180°. (If the value of R_1 is too large, the capacitor will charge too slowly, with the result that the SCR will not fire at all.)

Figure 31–16 shows that a UJT relaxation oscillator can be used as a source of positive pulses, enabling the SCR to fire at any angle between 0° and 180°. In this circuit, the 100-kΩ potentiometer is adjusted to provide the correct firing angle as observed with an oscilloscope connected across the load. Since the UJT will only operate on the positive half-cycle, full-wave control can be achieved if the circuit is modified with a full-wave rectifier bridge at the input.

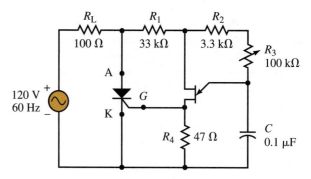

FIGURE 31–16 The UJT relaxation oscillator provides timing pulses to trigger the SCR.

IN-PROCESS
LEARNING CHECK 1

(Answers are at the end of the chapter.)

Describe the operation of the SCR circuit shown in Figure 31–16.

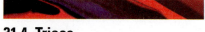

31.4 Triacs

The triac is a three-terminal semiconductor switch, which unlike the SCR, permits current in either direction. Whereas an SCR requires a positive pulse between the gate and the cathode in order to be triggered, a pulse of either polarity can trigger the triac. The symbol and corresponding voltage-current characteristic of a triac is illustrated in Figure 31–17.

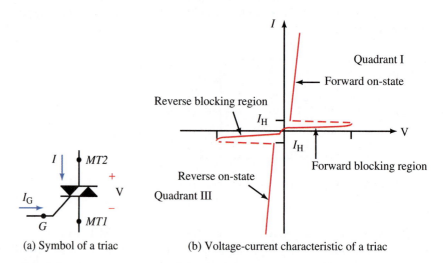

(a) Symbol of a triac   (b) Voltage-current characteristic of a triac

FIGURE 31–17

Since the triac is a bi-directional device, the terms *anode* and *cathode,* which are used for uni-directional devices, have no meaning. Rather, the terminals are labeled as main terminal 1, *MT1,* main terminal 2, *MT2,* and gate, *G.* (Some textbooks and data sheets use the term *anode* instead of *main terminal.*) Typical electrical specifications for a triac are shown in Figure 31–18.

TRIACS
4 AMPERES RMS
200 thru 600 VOLTS

TO–225AA

CASE 077
STYLE 5

PIN ASSIGNMENT	
1	Main Terminal 1
2	Main Terminal 2
3	Gate

MAXIMUM RATINGS (T_J = 25°C unless otherwise noted)

Rating	Symbol	Value	Unit
*Peak Repetitive Off-State Voltage(1) (T_J = −40 to 110°C, Sine Wave, 50 to 60 Hz, Gate Open) 2N6071A,B 2N6073A,B 2N6075A,B	V_{DRM}, V_{RRM}	200 400 600	Volts
*On-State RMS Current (T_C = 85°C) Full Cycle Sine Wave 50 to 60 Hz	$I_{T(RMS)}$	4.0	Amps
*Peak Non–repetitive Surge Current (One Full cycle, 60 Hz, T_J = +110°C)	I_{TSM}	30	Amps
Circuit Fusing Considerations (t = 8.3 ms)	I^2t	3.7	A^2s
*Peak Gate Power (Pulse Width ≤ 1.0 µs, T_C = 85°C)	P_{GM}	10	Watts
*Average Gate Power (t = 8.3 ms, T_C = 85°C)	$P_{G(AV)}$	0.5	Watt
*Peak Gate Voltage (Pulse Width ≤ 1.0 µs, T_C = 85°C)	V_{GM}	5.0	Volts

ON CHARACTERISTICS

	Symbol				Unit
*Peak On-State Voltage(1) (I_{TM} = ±6 A Peak)	V_{TM}	—	—	2	Volts
*Gate Trigger Voltage (Continuous dc) (Main Terminal Voltage = 12 Vdc, R_L = 100 Ohms, T_J = −40°C) All Quadrants	V_{GT}	—	1.4	2.5	Volts
Gate Non–Trigger Voltage (Main Terminal Voltage = 12 Vdc, R_L = 100 Ohms, T_J = 110°C) All Quadrants	V_{GD}	0.2	—	—	Volts
*Holding Current (Main Terminal Voltage = 12 Vdc, Gate Open, Initiating Current = ± 1 Adc) (T_J = −40°C) (T_J = 25°C)	I_H	— —	— —	30 15	mA
Turn-On Time (I_{TM} = 14 Adc, I_{GT} = 100 mAdc)	t_{gt}	—	1.5	—	µs

FIGURE 31–18   The 2N6071 series triac. *(Courtesy of Semiconductor Components Industries, LLC 2002)*

Characteristics of the Triac

Triacs may be used to provide direct replacement for ac mechanical relays or may be connected to a trigger circuit to provide full-wave control of an ac load such as a light, heater, or motor. The trigger circuit, which when connected to the gate, provides phase control by turning the triac on for a predetermined conduction angle for each half-cycle of an applied ac source. If no gate signal is applied to the triac, the device will remain in its *blocking region,* providing that the voltage across the main terminals does not exceed the *peak repetitive off-state voltage,* V_{DRM} of the device. In the blocking region, the current through the triac is very small—typically a few microamps.

There are four possible modes that will turn a triac on. When the *MT2* terminal is positive with respect to the *MT1* terminal, the triac is said to operate in Quadrant I as shown in Figure 31–17(b). If a positive gate pulse (with respect to the *MT1* terminal) is applied, the triac is said to operate in Mode I+, and if a negative gate pulse is applied, the triac is in Mode I−.

When the *MT2* terminal is negative with respect to the *MT1* terminal, the triac is said to operate in Quadrant III. If a positive gate pulse (once again, with respect to the *MT1* terminal) is applied, the triac is said to operate in Mode III+, and if a negative gate pulse is applied, the triac is in Mode III−.

A triac is most sensitive in Modes I+ and III−. It is less sensitive in Mode I− and least sensitive in Mode III+, and it should not be operated in this mode. Once a triac has been triggered and operates in its on-state, it will remain on until the current drops below the holding current, I_H.

Applications of Triacs

The circuit of Figure 31–19 shows a triac used as a phase-control light dimmer for an incandescent light bulb. The capacitor in the circuit charges through the load resistor (light bulb) and resistor, R_1. If the value of R_1 is small, the capacitor will quickly charge to a voltage large enough to cause the diac to breakover. When this happens, the corresponding positive voltage is applied to the gate of the triac, causing it to trigger. The triac, which is now effectively a short circuit, permits a large amount of current to keep the light bulb on for the balance of the half-cycle. The triac is now operating in Quadrant I. As the input sinusoidal voltage reaches the zero-crossing, the current in the circuit drops below the holding current, I_H, of the triac, and current ceases. During the negative half-cycle of the applied ac voltage, the capacitor again charges, this time to the opposite polarity. Since the diac is a bi-directional device, a negative gate current is now applied to the triac, causing the triac to operate in Quadrant III. The conduction angle will be essentially the same for

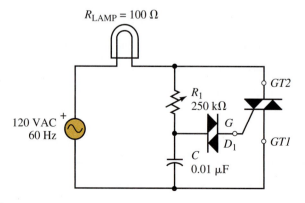

FIGURE 31–19

both the positive and negative half-cycles. Figure 31–20(a) shows the applied sinusoidal voltage, while Figure 31–20(b) and Figure 31–20(c) show the waveforms that will appear across the load and the triac respectively for a firing angle of $\theta_F = 90°$. (The conduction angle is $\theta_C = 90°$.) For a firing angle of 90°, the power delivered to the load will be exactly half the power that would be delivered had the full sinusoidal been applied.

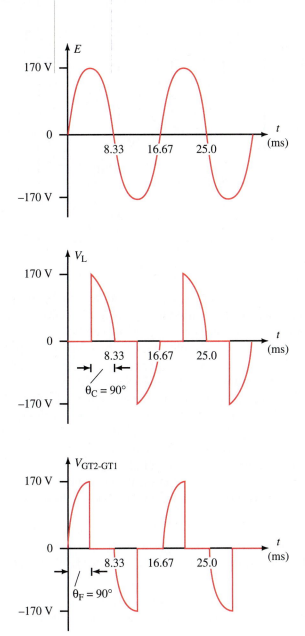

FIGURE 31–20

In what modes does the triac of Figure 31–19 operate? Explain your answers.

PRACTICE PROBLEMS 2

Answers
Mode I+. When the applied voltage is in the positive half-cycle (Quadrant I), the voltage across the capacitor will be positive. Mode III−. When the applied voltage is in the negative half-cycle (Quadrant III), the voltage across the capacitor will be negative.

Figure 31–21 shows how a triac is used to control temperature in a room. Notice that the ac voltage source is applied to the heater through the triac. In order for the heater to turn on, it is necessary to first trigger the triac. The required pulses for the gate of the triac are obtained as follows.

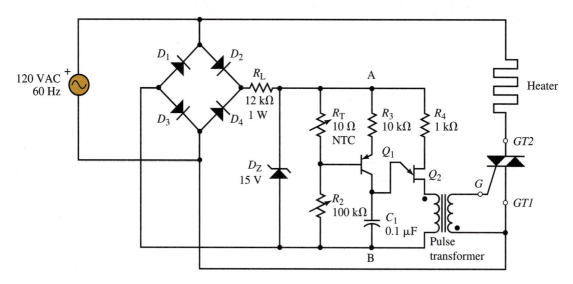

FIGURE 31–21   Temperature control circuit using a triac.

The full-wave rectifier bridge, together with R_1 and D_Z, form a dc supply for the two transistors. As the temperature in the room decreases below some set point determined by R_2, the negative temperature coefficient (NTC) of R_T causes an increase in resistance. This results in more voltage drop across R_T, which in turn increases the emitter current of Q_1. The capacitor charges quickly due to the additional current, causing the UJT transistor to fire early in the half-cycle. When Q_2 fires, the current pulse on the primary of the pulse transformer results in a negative-going pulse on the gate of the triac. Although Q_2 may produce additional pulses during the same half-cycle, these additional pulses will have no further effect on the triac, which remains on for the balance of the half-cycle.

At the end of each half-cycle, the voltage V_{AB} returns to zero, allowing the transistor circuits of Q_1 and Q_2 to reset. As the temperature increases, the value of R_T gets smaller, with the result that the feedback stabilizes the temperature of the room.

PRACTICE PROBLEMS 3

In what modes does the triac of Figure 31–21 operate? Explain your answers.

Answers
Mode I— and Mode III—. Regardless of whether the applied voltage is in the positive or the negative half-cycle, the voltage pulse between the gate and *MT1* of the triac will be negative.

31.5 Power Control Fundamentals

In the previous sections we observed that by delaying the firing time of thyristors, the power delivered to a load can be substantially less than the peak power that would be delivered if the load received the full cycle of ac voltage. By delaying the firing time of a thyristor, the resulting power can be a fraction of the peak value. The result is a circuit that can control the intensity of a lamp, the amount of heat from a resistive heater, or the speed of a motor. In this section, we examine the relationship between the firing angle and the resulting voltage and power delivered to the load. We will examine the effects of using both unidirectional (SCR) and bi-directional (triac) thyristors.

As a quick review, recall that the power delivered by an ac source to a resistive load is determined from the rms voltage, V, of a waveform and is given as

$$P = \frac{V_{rms}^2}{R} \tag{31-10}$$

In Chapter 15, you learned that the rms value of any periodic waveform can be calculated using calculus and is given as

$$V_{rms} = \sqrt{\frac{\int_0^T [V(t)]^2\, dt}{T}} \tag{31-11}$$

Specifically for a sinusoidal (or a full-wave rectified sinusoidal) having a peak voltage, V, the expression becomes

$$V_{rms(FW)} = \sqrt{\frac{\int_0^\pi [V\sin\theta]^2\, d\theta}{\pi}}$$

$$= V\sqrt{\frac{1}{\pi}\left[\frac{\theta}{2} - \frac{1}{4}\sin 2\theta\right]_0^\pi} \tag{31-12}$$

$$= \frac{V}{\sqrt{2}}$$

Similarly, the rms voltage of a half-wave rectified sinusoidal is reduced to

$$V_{rms(HW)} = \sqrt{\frac{\int_0^\pi [V\sin\theta]^2\, d\theta}{2\pi}}$$

$$= V\sqrt{\frac{1}{2\pi}\left[\frac{\theta}{2} - \frac{1}{4}\sin 2\theta\right]_0^\pi} \tag{31-13}$$

$$= \frac{V}{2}$$

Now, if the firing angle of the thyristor is delayed, the rms value of the resulting signal delivered to the load will be reduced. For a full-wave signal having a firing angle, θ_F we have

$$V_{rms(FW)} = \sqrt{\frac{\int_{\theta_F}^T [V\sin\theta]^2\, d\theta}{\pi}}$$

$$= V\sqrt{\frac{1}{\pi}\left[\frac{\theta}{2} - \frac{1}{4}\sin 2\theta\right]_{\theta_F}^\pi} \tag{31-14}$$

and for a half-wave rectified signal this becomes

$$V_{rms(HW)} = \sqrt{\frac{\int_{\theta_F}^\pi [V\sin\theta]^2\, d\theta}{2\pi}}$$

$$= V\sqrt{\frac{1}{2\pi}\left[\frac{\theta}{2} - \frac{1}{4}\sin 2\theta\right]_{\theta_F}^\pi} \tag{31-15}$$

The following examples show the difference between using an SCR and a triac as angle-controlled switches to control the intensity of an incandescent light bulb.

A trigger circuit delays firing of an SCR until the center of each half-cycle, as illustrated in Figure 31–22. Use calculus to determine the rms voltage and the power delivered to a 100-Ω load.

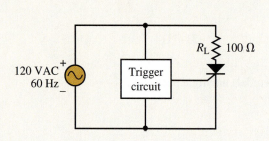

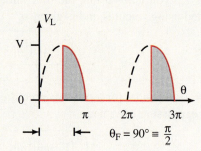

FIGURE 31–22

Solution

Since each half-cycle represents 180° or π radians, the trigger circuit results in a firing angle of $\theta_F = 90° \equiv \pi/2$ radians. Substituting this value into Equation 31–15 gives us an rms value of

$$V_{rms(HW)} = V\sqrt{\frac{1}{2\pi}\left[\frac{\theta}{2} - \frac{1}{4}\sin 2\theta\right]_{\frac{\pi}{2}}^{\pi}}$$

$$= 169.7\ V\sqrt{\frac{1}{2\pi}\left[\left(\frac{\pi}{2} - \frac{1}{4}\sin 2\pi\right) - \left(\frac{\left(\frac{\pi}{2}\right)}{2} - \frac{1}{4}\sin 2\left(\frac{\pi}{2}\right)\right)\right]}$$

$$= 169.7\ V\sqrt{\frac{1}{2\pi}\left[\frac{\pi}{2} - \frac{\pi}{4}\right]}$$

$$= 60.0\ V$$

The corresponding power delivered to the load is

$$P = \frac{(60.0\ V)^2}{100\ \Omega} = 36.0\ W$$

(Note that a power of 36.0 W represents exactly ¼ of the 144 W that would have been delivered to the load if the full sinusoidal wave had been applied.)

A trigger circuit delays firing of a triac until the center of each half-cycle, as illustrated in Figure 31–23. Use calculus to determine the rms voltage and the power delivered to a 100-Ω load.

EXAMPLE 31–4

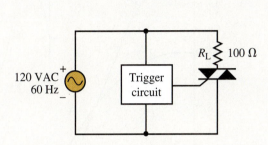

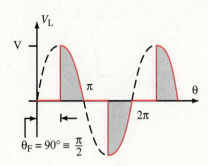

FIGURE 31–23

Solution

As in the previous example, we have a firing angle of $\theta_F = 90° \equiv \pi/2$ radians. Substituting this value into Equation 31–14 gives us an rms value of

$$V_{rms(FW)} = V\sqrt{\frac{1}{\pi}\left[\frac{\theta}{2} - \frac{1}{4}\sin 2\theta\right]_{\frac{\pi}{2}}^{\pi}}$$

$$= 169.7\ V\sqrt{\frac{1}{\pi}\left[\left(\frac{\pi}{2} - \frac{1}{4}\sin 2\pi\right) - \left(\frac{\left(\frac{\pi}{2}\right)}{2} - \frac{1}{4}\sin 2\left(\frac{\pi}{2}\right)\right)\right]}$$

$$= 169.7\ V\sqrt{\frac{1}{\pi}\left[\frac{\pi}{2} - \frac{\pi}{4}\right]}$$

$$= 84.85\ V$$

The corresponding power delivered to the load is

$$P = \frac{(84.85\ V)^2}{100\ \Omega} = 72.0\ W$$

As expected, since half the waveform appears across the load, we find that the load resistance receives exactly half of the 144 W that would have been delivered to the load if the full sinusoidal wave had been applied.

Although it is possible to use calculus to solve for the rms voltage for any firing angle, it is much more convenient to use a graphical approach when solving for the voltage and power. Figures 31–24 and 31–25 are normalized curves that allow us to solve for the rms voltage and power for any firing angle for half-wave or full-wave angle control.

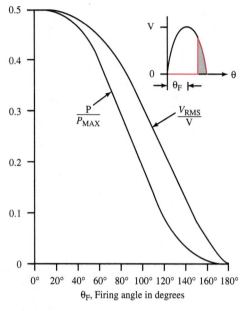

FIGURE 31–24   Voltage and power curves for half-wave control.

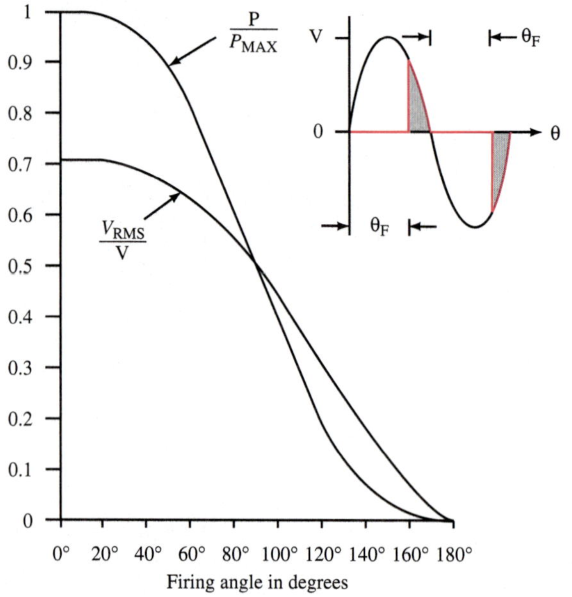

FIGURE 31–25   Voltage and power curves for full-wave control.

In order to use the power curves, we need to know three quantities for a given circuit; the peak voltage V (amplitude of the sinusoidal), the firing angle θ_F, and whether the device uses half-wave control (such as an SCR) or full-wave control (such as a triac or an SCR that is preceded by a full-wave rectifier bridge). The maximum power for any resistive load is based on the power that will be delivered to the load if the complete sinusoidal is applied. This value will be the same for both half-wave and full-wave control circuits. The following examples illustrate how the curves are used.

EXAMPLE 31–5

A trigger circuit delays firing of an SCR until $\theta_F = 120°$. If the supply voltage is a 120-VAC sinusoidal, determine the rms voltage and the power delivered to a 20-Ω load.

Solution
For a 20-Ω load, the maximum power that would be delivered is determined as

$$P_{max} = \frac{(120 \text{ V})^2}{20 \text{ }\Omega} = 720 \text{ W}$$

The peak voltage of the sinusoidal source is $V = (120 \text{ V})\sqrt{2} = 169.7$ V. For a firing angle of $\theta_F = 120°$, the normalized voltage ratio is approximately 0.222, and so we determine the rms voltage as

$$V_{rms} = (0.222)(169.7 \text{ V}) = 37.7 \text{ V}$$

The corresponding power can be determined either graphically as

$$P = (0.10)P_{max} = 72.0 \text{ W}$$

or mathematically from the rms voltage as

$$P = \frac{(37.7 \text{ V})^2}{20 \text{ }\Omega} = 71.0 \text{ W}$$

A trigger circuit fires a triac at $\theta_F = 130°$. If the supply voltage is a 240-VAC sinusoidal, determine the rms voltage and the power delivered to a 12–Ω load.

Answers
81.5 V, 576 W

Opto-electronic devices fall into two categories: devices that convert electric current into light and devices that convert light into electric current. Although we normally associate light with the visible spectrum, many optical devices operate in the infrared region of the electromagnetic (E/M spectrum). Figure 31–26 shows the electromagnetic frequency spectrum, together with various applications of the frequencies. Each horizontal division represents a decade (10-fold increase in frequency). Notice that the visible spectrum is a very narrow portion between 10^{14} Hz and 10^{15} Hz. Since all E/M waves propagate through free space (in a vacuum or in air) at a speed $c = 3.00 \times 10^8$ m/s, we relate wavelength and frequency as follows:

31.6 Introduction to Optical Devices

$$\lambda = \frac{c}{f} \qquad\qquad \textbf{(31–16)}$$

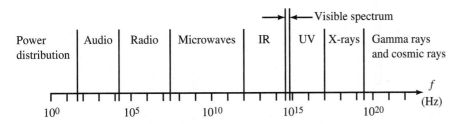

FIGURE 31–26 The electromagnetic frequency spectrum.

The wavelength of visible and near-visible light (which includes infrared—IR and ultraviolet—UV) is often measured in either nm (nanometers) or angstroms (Å), where

$$1 \text{ Å} \equiv 1 \times 10^{-10} \text{ m} \qquad\qquad \textbf{(31–17)}$$

Red light (at approximately 750 nm) and violet light (at approximately 380 nm) bound the visible light spectrum. All other colors have wavelengths between these values.

Determine the frequencies that correspond to red light and violet light.

EXAMPLE 31–6

Solution
From Equation 31–16, we have

$$f_{\text{red}} = \frac{c}{\lambda} = \frac{3.00 \times 10^8 \text{ m/s}}{750 \times 10^{-9} \text{ m}} = 4.0 \times 10^{14} \text{ Hz}$$

and

$$f_{\text{violet}} = \frac{3.00 \times 10^8 \text{ m/s}}{380 \times 10^{-9} \text{ m}} = 7.9 \times 10^{14} \text{ Hz}$$

Light Emitting Diodes (LEDs)

LEDs are similar to other diodes in that they consist of two layers of semiconductor material, one n-type and one p-type. The intrinsic material is doped with different types of impurities, such as gallium, arsenic, and phosphorous to result in the wide range of colors that are available. Typical current ratings of LEDs are between 10 mA and 20 mA. Since LEDs are constructed of semiconductor materials other than silicon, the voltage across a forward-biased LED will vary between 1.0 V and about 2.2 V. Figure 31–27(a) shows the construction of a typical surface-emitting LED, while Figure 21–27(b) shows the schematic symbol of an LED.

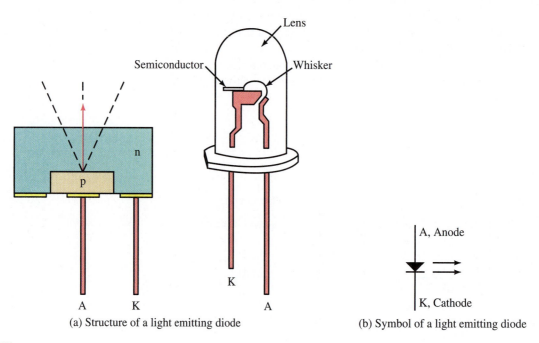

(a) Structure of a light emitting diode   (b) Symbol of a light emitting diode

FIGURE 31–27

When the LED is forward-biased, excess electrons from the n-type material will be injected into the p-type material, where there is an excess of *holes.* The *recombination* process between the holes and electrons causes the conduction-band electrons of the n-type material to drop from a high energy level to the lower energy level of the valence band in the p-type material. When the electron moves to a lower energy level, energy is conserved by the release of a *photon* of E/M radiation having an energy of

$$E = hf \qquad (31–18)$$

In Equation 31–18, E is the difference in energy between the conduction band and the valence band, h is Planck's constant ($h = 6.626 \times 10^{-34}$ J·s), and f is the frequency (Hz) of E/M radiation emitted. If the energy difference is large enough, the radiation will occur in the visible spectrum. LEDs have several advantages over other light sources:

- LEDs can operate from low-voltage sources. They are easily combined with logic ICs that use 5-V power supplies.

- Unlike incandescent light sources, which require filaments to heat up, LEDs react quickly to voltage changes, making them useful for digital transmission.

- LEDs have a very long life expectancy, provided that the maximum voltage and current ratings are not exceeded.
- The peak intensity of an LED can be matched to the peak sensitivity of a photodetector, making LEDs useful for data transmission.

Although LEDs have many advantages over other types of light sources, they have several disadvantages.

- LEDs are easily damaged by excessive voltage or current. Unlike other diodes, LEDs are unable to handle large reverse voltage. Typical reverse voltage for an LED is $V_R = 5$ V. If a circuit uses an alternating voltage source, it is necessary to include signal diodes across LEDs as protection.
- The brightness and peak response frequency of LEDs is dependent upon the ambient temperature.
- If an LED is used to transmit data, the pulses will spread due to the variation in the frequency of the emitted light (called *chromatic dispersion*).
- When used as displays (for calculators, or any battery operated commercial electronics), the LED is inefficient when compared to LCD (liquid crystal displays). A 7-segment display uses 7 LEDs to indicate numbers from zero to nine, and it could require as much as 7×20 mA $= 140$ mA to display a single decimal digit.

LEDs are very simple devices to design into a circuit. Manufacturers typically provide the forward current and voltage drop for a diode at a given light intensity. Light intensity for an LED operating in the visible spectrum is normally specified in the SI unit, the **candela** (cd). The candela is defined as the visible light intensity emitted from a $\frac{1}{60}$ cm^2 opening in a standard white-hot oven at a temperature of 2046 K. Since this amount of light is difficult to visualize, it is much easier to remember that 1 cd is approximately the amount of light that will be emitted from a candle (1 candle $= 0.981$ cd). When an LED is used in a circuit, it is generally necessary to insert a current-limiting resistor in series with the voltage source and the diode.

EXAMPLE 31–7

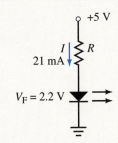

FIGURE 31–28

A high-intensity yellow LED providing 80 mcd of visible light is specified as having a rated voltage of 2.2 V at a current of 21 mA. Determine the current-limiting resistor that must be used to ensure that the diode does not exceed its specified operating point if the voltage source for the circuit is 5.0 Vdc.

Solution
The circuit appears as shown in Figure 31–28.
    We use Ohm's law to solve for the current limiting resistor as

$$R = \frac{5.0 \text{ V} - 2.2 \text{ V}}{21.0 \text{ mA}} = 133 \ \Omega$$

In this example, we would likely use a 150–Ω resistor to ensure that the current through the LED remains below the rated value.

Although LEDs are used extensively as indicators, many LEDs emit light in the infrared (IR) region. These IRLEDs are easily coupled to fiber optic cable that has optimum transmission characteristics at the same frequency as that at which the light is transmitted. Since IRLEDs can be switched on and off at relatively high rates, it is possible to use the resulting system to transmit binary data (in the form of 1s and 0s) at very high rates.

However, it is not sufficient to be able to simply transmit information as light; we must also be able to recover the data. There are several devices that allow light to be converted into electrical signals.

31.7 Photodetectors

There are many types of devices that convert light into electrical energy. In Chapter 3, you were introduced to the photoresistor, a device having a resistance that decreased as the amount of light increased. In this section, we examine semiconductor devices that produce a voltage or current variation that is dependent on the amount of light applied to a p-n junction. In each case, the principle of operation is the same. When a photon of light strikes the p-n junction, a valence electron absorbs the energy of the photon, moving into the higher energy level. The electron is now free to move and results in increased saturation or leakage current in the device. If necessary, the resulting current may be further amplified. The most common photodetectors are the photodiode, the phototransistor, and the LASCR (light activated silicon controlled rectifier).

Photodiodes

Figure 31–29 shows the symbol and typical current-voltage characteristics of a photodiode. Notice that the forward characteristic of the photodiode is similar to that of standard signal diode. The reverse characteristics are much different. When no light is applied to the p-n junction of the reverse-biased photodiode, there will be very little leakage current. However, as the light increases, we see a marked increase in the reverse current due to the increase in minority carriers.

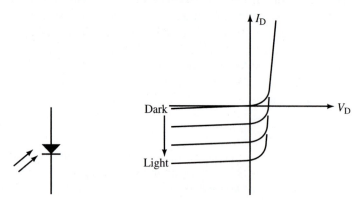

FIGURE 31–29　　(a) Symbol of a photodiode　　(b) Current-voltage characteristics of a photodiode

Phototransistors

The phototransistor shown in Figure 31–30 is an npn transistor, with the base normally left open. As light strikes the reverse-biased C-B junction, the number of minority carriers increases, thereby resulting in an increase in the collector current, I_C. Recall that the collector current of a transistor is given as

$$I_C = \beta I_B + (\beta + 1)I_{CBO} \qquad (31\text{–}19)$$

Since the base current is zero due to the open base, the collector current is entirely dependent on the contribution due to the minority carriers, namely

$$I_C = (\beta + 1)I_{CBO} \qquad (31\text{–}20)$$

Although the collector current of a phototransistor is dependent on the light intensity and could therefore be used as an amplifier, a phototransistor circuit is normally designed to operate as a switch. The transistor is cut off when there is no light (logic "0") or saturated (logic "1") when light is present. Fig-

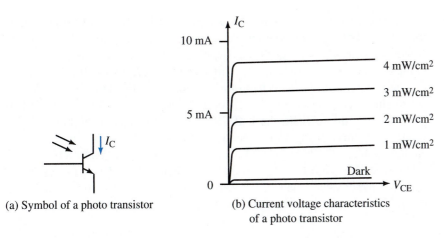

(a) Symbol of a photo transistor   (b) Current voltage characteristics
of a photo transistor

FIGURE 31–30

ure 31–31 shows how a photoemitter (such as an LED) can be combined with a photodetector (such as photodiode or phototransistor) as part of a fiber optic link. By using light transmission through a glass fiber, the attenuation of the signal will be very small over the length of glass fiber. Typical losses over high-quality glass fibers can be as low as 0.5 dB/km.

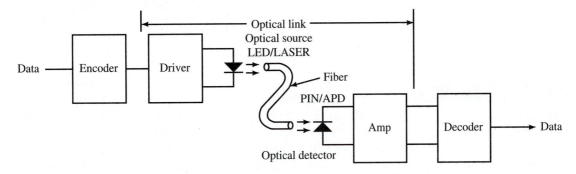

FIGURE 31–31   Fiber optic link uses an optical source and an optical detector to send digital signals.

Besides having very low attenuation characteristics, fiber optic transmission neither causes nor is susceptible to E/M interference. Although LEDs and photodiodes are used in data transmission, it is now much more common to use lasers as light sources and avalanche photodiodes as photodetectors. The main difference between avalanche photodiodes (APDs) and other optical detectors is that high electric fields within APDs cause freed carriers to accelerate and effectively knock other valence electrons out of the crystal lattice. This effect is called **impact ionization,** and it results in APDs providing a 30- to 100-fold increase in the number of carriers. While APDs have much faster response times that other photodetectors, they have several disadvantages:

- An avalanche photodiode is easily destroyed if the operating voltage exceeds the device breakdown voltage.

- Variation in operating temperature will adversely affect the operation of an APD.

- APDs are much more costly than other photodiodes.

Light-Activated SCRs-LASCRs

An LASCR uses light energy to cause an SCR to fire. Once the LASCR has fired, the device will remain on until the current drops below the holding current, I_H.

Figure 31–32 shows an LASCR used as a light-sensitive switch. When the ambient light level drops below a certain value, the LASCR will effectively be an open circuit, allowing C_1 to charge through R_1. Once the voltage across C_1 is sufficient to fire the diac D_5, the resulting gate current causes SCR_1 to also fire. This results in a large current through the lamp for the balance of the half-cycle. SCR_1 will not conduct on the negative half-cycle of the ac supply. If sufficient ambient light is available to the LASCR, the LASCR will effectively be a short circuit, preventing C_1 from charging. Consequently, SCR_1 will not be able to fire if the ambient light is high. The sensitivity of the circuit can be controlled by adjusting the value of R_2. If R_2 has low resistance, the LASCR will be less sensitive than if R_2 were a larger value. The LASCR is most sensitive when R_2 is left open. Notice that full-wave control can be achieved if the SCR_1 is replaced by a triac.

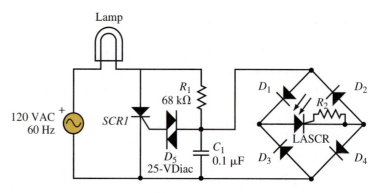

FIGURE 31–32 LASCR light control.

31.8 Optocouplers

As the name implies, an optocoupler is a photoelectric device used to couple two circuits using light as the common medium. An optocoupler generally uses an infrared emitting diode to convert electrical energy into light that is internally coupled to a photodiode, phototransistor, or other photodetector. Since the two circuits are electrically isolated, there will be no electrical interaction between the two circuits, making these circuits ideal for medical equipment and for interfacing high-voltage monitoring equipment to low-voltage microprocessor inputs. Figure 31–33 shows a typical 6-pin phototransistor optocoupler. Optocouplers can be used as either linear devices or as digital buffers.

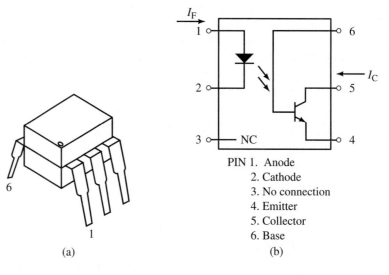

PIN 1. Anode
  2. Cathode
  3. No connection
  4. Emitter
  5. Collector
  6. Base

(a)   (b)

FIGURE 31–33 Phototransistor optocoupler. *(Courtesy of Fairchild Semiconductor Corporation)*

We define the current transfer, *CTR,* as the ratio of output collector current to input forward current, namely

$$CTR = \frac{I_C}{I_F} \qquad (31\text{--}21)$$

Typical values for CTR are between 0.1 and 1.0. The following examples show typical applications of optocouplers.

EXAMPLE 31–8

Given the optocoupler circuit shown in Figure 31–34. Assume that the forward diode voltage is $V_F = 1.2$ V and the current transfer ratio of the optocoupler is CTR = 0.3:

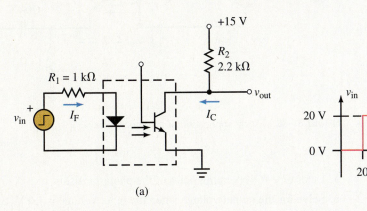

(a)   (b)

FIGURE 31–34 Digital application of an optocoupler.

a. Solve for the output voltage when the input voltage is 0 V.
b. Solve for the output voltage when the input voltage is 20 V.
c. Sketch the corresponding output voltage.

Solution

a. When the input voltage is 0 V, the current $I_F = 0$. Consequently, $I_C = 0$, and so we have $v_{out} = +15$ V.

b. When the input voltage is 20 V, the diode current will be

$$I_F = \frac{20 \text{ V} - 1.2 \text{ V}}{1 \text{ k}\Omega} = 18.8 \text{ mA}$$

The collector current will be

$$I_C = (0.3)(18.8 \text{ mA}) = 5.64 \text{ mA}$$

and so the output voltage will be

$$v_{out} = 15 \text{ V} - (2.2 \text{ k}\Omega)(5.64 \text{ mA}) = 2.59 \text{ V}$$

c. The output of the optocoupler is illustrated in Figure 31–35.

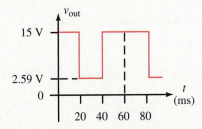

FIGURE 31–35

EXAMPLE 31–9

Given the optocoupler circuit shown in Figure 31–36. Assume that the forward diode voltage is $V_F = 1.2$ V, and the current transfer ratio of the optocoupler is CTR = 0.3:

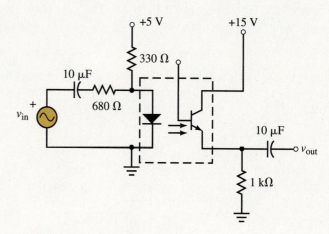

FIGURE 31–36   Linear application of an optocoupler.

a. Determine the operating point of the phototransistor.

b. Solve for the output voltage when the input voltage is 1.0 V_p and the frequency is 1 kHz.

Solution

a. The operating point of the phototransistor is determined by considering only the dc voltage sources. Using the same approach as the previous example, we have

$$I_F = \frac{5\ V - 1.2\ V}{330\ k\Omega} = 11.5\ mA$$

and so

$$I_{CQ} = (0.3)(11.5\ mA) = 3.45\ mA$$

The collector-emitter voltage at the operating point is now determined to be

$$V_{CEQ} = 15\ V - (1\ k\Omega)(3.45\ mA) = 11.55\ V$$

b. Since the diode ac resistance will be very small (normally $r_d \approx 7\ \Omega$) in comparison to the 680-Ω resistance, we may ignore its effect. Consequently, the variation in the diode forward current will be

$$i_f = \frac{1.0\ V_p}{680\ \Omega} = 1.47\ mA_p$$

The resulting variation in collector current will be

$$i_c = (0.3)(1.47\ mA) = 0.441\ mA_p$$

Since the coupling capacitor allows only the ac portion of the emitter voltage to pass to the output, we have

$$v_{out} = (0.441\ mA_p)(1.0\ k\Omega) = 0.441\ V_p$$

As we saw in the previous section, when an electron drops from a high energy level to a lower level, a photon of light is released. In the case of LEDs, an electron can drop to its lower level at any time, releasing a photon of light spontaneously and in any direction. In the case of lasers, electrons will drop to a lower energy level when they are stimulated to do so. Hence, the word LASER is an acronym for **L**ight **A**mplification through **S**timulated **E**mission of **R**adiation.

31.9 Semiconductor LASERs

The principle behind the operation of lasers is simple. As in the case of LEDs, an external voltage source raises the energy level of electrons within the semiconductor material. However, unlike LEDs, the number of electrons raised to this higher level is much greater. The electrons do not immediately drop to a lower level. An incoming photon with exactly the right energy level will stimulate an electron in this *metastable state* to drop to its lower energy level. When this happens, another photon having exactly the same wavelength as the original photon is created. The two photons will travel in the same direction, stimulating other electrons to also drop to a lower energy level. Hence the amplification. The external voltage source ensures that more electrons are raised to higher levels. A laser uses one totally reflective surface to ensure that light emerges from only one end. The opposite end is a partially reflective surface to sustain the stimulated emission of photons in a cavity between the p- and n-type materials. Part of the beam emerges from the partially reflective surface, as illustrated in Figure 31–37.

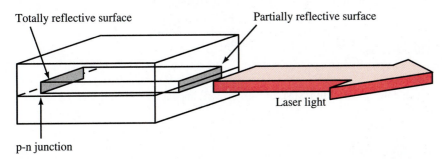

FIGURE 31–37   The semiconductor laser.

The beam of light that emerges from a laser will have much higher intensity that a similar beam from an LED. Additionally, the light will be **monochromatic** (having a single wavelength) and **coherent** (in phase and in the same direction). Figure 31–38 shows the major difference between an LED and a laser, each emitting light at 1330 nm. Besides having much greater intensity, a laser has a very small variation in wavelength, $\Delta\lambda$. When used to transmit digital pulses, this small variation results in a laser having less much less pulse-dispersion than an LED.

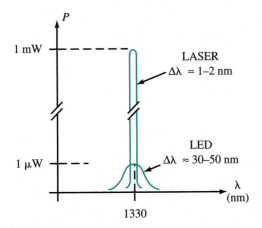

FIGURE 31–38   Comparison of the spectral purity and intensity of an LED and a laser.

31.10 Computer Analysis

In this section we examine how MultiSIM can be used to simulate the effects of triggering an SCR. We will use a variable resistor and observe how increasing the charging rate of a capacitor is used to delay the firing angle of the signal applied to the gate of the transistor.

EXAMPLE 31–10

Use MultiSIM to construct the circuit shown in Figure 31–39. Place an oscilloscope across the lamp and observe the voltage waveform as the variable resistor, R, is adjusted between 10% ($R = 5$ kΩ) and 50% ($R = 25$ kΩ). As the resistance value is increased, you will find that it will take longer for the capacitor to charge to a large enough voltage to cause the diac to reach its breakover voltage. This increase in time translates into an increased firing angle for the SCR. Determine the approximate firing angle for $R = 5$ kΩ and for $R = 25$ kΩ.

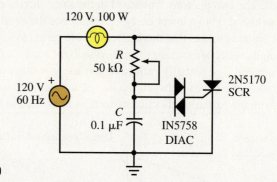

FIGURE 31–39

Solution
Once entered into MultiSIM, the circuit will appear as shown in Figure 31–40.

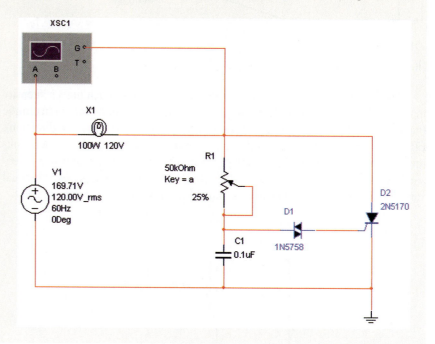

◄ MULTISIM   **FIGURE 31–40**

In order for the simulation to operate correctly, it is necessary to change the default instrument setting of EWB. The following settings will help to prevent simulation errors.

- Select the *Simulate* menu item.
- Click on *Default Instrument Settings . . .*
- In the tab labeled as *Defaults for Transient Analysis Instruments,* change the values as illustrated in Figure 31–41.

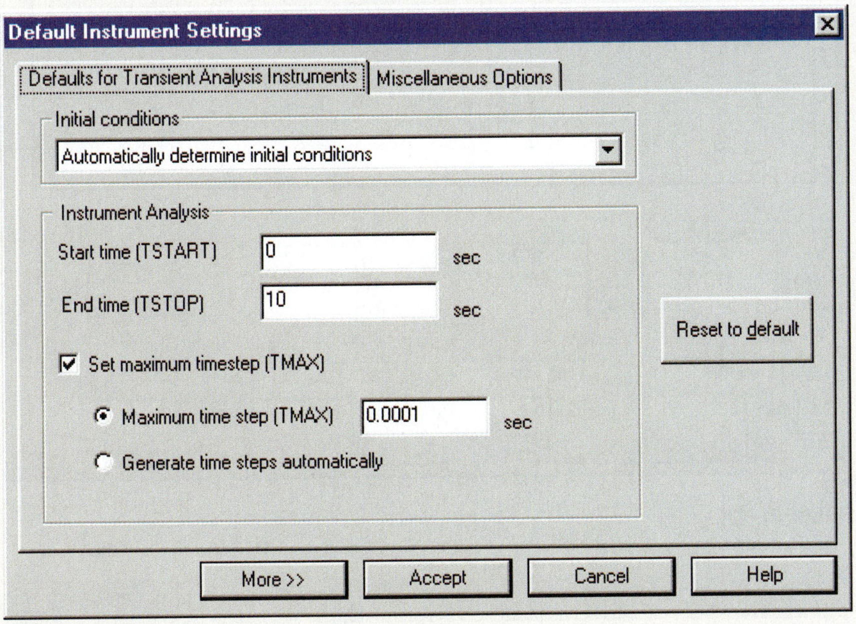

FIGURE 31–41

When the simulation is run, the displays for the resistor values are shown in Figure 31–42 and Figure 31–43. These results indicate that the conduction angles are

$$\theta_C = \frac{7.3 \text{ ms}}{8.33 \text{ ms}} \times 180° = 158° \quad \text{for } R = 5 \text{ k}\Omega \text{ and}$$

$$\theta_C = \frac{6.6 \text{ ms}}{8.33 \text{ ms}} \times 180° = 143° \quad \text{for } R = 25 \text{ k}\Omega$$

Therefore, the firing angles are $\theta_F = 22°$ for $R = 5 \text{ k}\Omega$ and $\theta_F = 37°$ for $R = 25 \text{ k}\Omega$.

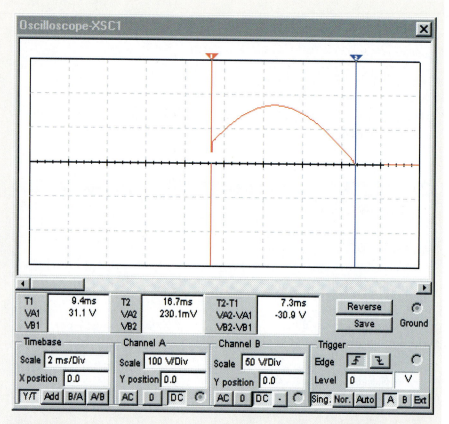

FIGURE 31–42

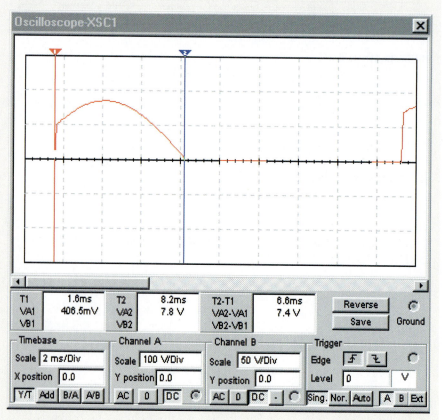

FIGURE 31–43

PUTTING IT INTO PRACTICE

Part of a monitoring system in a nuclear generating station requires that you observe the status of 24-V relays. The signals entering your computer must be either 0 V to indicate that the contact is closed or 5 V, to show the contact is open. Use an optocoupler that requires $I_F = 10.0$ mA and has CTR $= 0.4$ to produce the desired signal at your computer. Show the complete schematic of your design.

PROBLEMS

31.2   Triggering Devices

1. Given that the UJT in the circuit of Figure 31–44 has

    Valley point: $V_V = 1.8$ V $I_V = 2.0$ mA

    Peak point: $V_P = 10.0$ V $I_P = 4.0$ μA

    $V_D = 0.6$ V and $I_{B2} = 2.4$ mA when $I_E = 0$

    a. Solve for the frequency of oscillation if $R_E = 6.8$ kΩ and $C_E = 0.56$ μF.

    b. Determine the values of η and R_{BB} of the UJT

    c. Sketch the waveforms that appear at V_E, V_{out1}, and V_{out2} (relative to ground).

2. If the UJT in the circuit of Figure 31–44 has

    Valley point: $V_V = 1.0$ V $I_V = 1.5$ mA

    Peak point: $V_P = 10.0$ V $I_P = 5.0$ μA

    $V_D = 0.6$ V $R_{BB} = 6.0$ kΩ

    a. Solve for the frequency of oscillation if $R_E = 12$ kΩ and $C_E = 0.01$ μF.

    b. What is the purpose of the load resistor, R_L?

3. Using the UJT characteristics of Problem 1, design a UJT relaxation oscillator to oscillate at a frequency of 8.0 kHz.

4. Using the UJT characteristics of Problem 2, design a UJT relaxation oscillator to oscillate at a frequency of 6.25 kHz.

31.3   Silicon-Controlled Rectifiers (SCRs)

5. Refer to the illustration of Figure 31–45. Sketch the corresponding voltages V_{AK} and V_L if the triggering circuit provides positive-going gate pulses 5.0 ms after each zero-crossing of the applied ac voltage source.

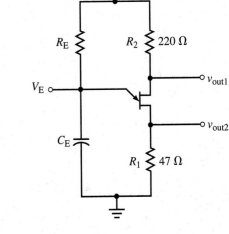

FIGURE 31–44

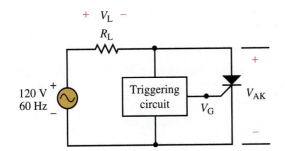

FIGURE 31–45

6. Repeat Problem 5 if the triggering circuit were adjusted to provide positive-going gate pulses 10.0 ms after each zero-crossing of the applied ac voltage source.

7. Refer to the illustration of Figure 31–46. Sketch the corresponding voltages V_{AK} and V_L if the triggering circuit provides positive-going gate pulses 5.0 ms after each zero-crossing of the applied ac voltage source.

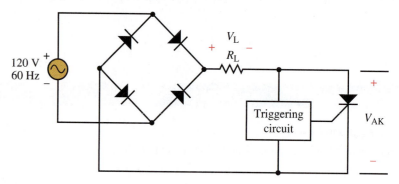

FIGURE 31–46

8. Repeat Problem 5 if the triggering circuit were adjusted to provide positive-going gate pulses 10.0 ms after each zero-crossing of the applied ac voltage source.

31.4   Triacs

9. Refer to the illustration of Figure 31–47. Sketch the corresponding voltages $V_{GT2\text{-}GT1}$ and V_L if the triggering circuit provides positive-going gate pulses 5.0 ms after each zero-crossing of the applied ac voltage source. In which modes of operation does the triac operate?

10. Repeat Problem 9 if the triggering circuit is adjusted to provide positive-going gate pulses 10.0 ms after each zero-crossing of the applied ac voltage source.

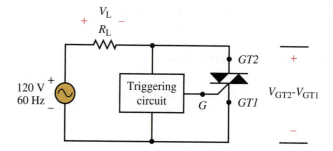

FIGURE 31–47

31.5   Power Control Fundamentals

11. A trigger circuit delays firing of an SCR until $\theta_F = 45°$. If the supply voltage is a 120-V (rms) sinusoidal, determine the rms voltage and power that is delivered to a 150-Ω load.

12. Repeat Problem 11 if the firing angle occurs at $\theta_F = 135°$.

13. A trigger circuit delays firing of a triac until $\theta_F = 60°$. If the supply voltage is a 240-V (rms) sinusoidal, determine the rms voltage and power that is delivered to a 20-Ω motor.

14. Repeat Problem 13 if the firing angle occurs at $\theta_F = 120°$.

15. An electric motor is rated as having 0.5 hp at 120 V. Assuming that the motor is 100% efficient, what must be the firing angle of a triac in order for the motor to provide 0.2 hp?

16. If an SCR circuit controls the motor of Problem 15, what must be the firing angle?

17. Use calculus to determine the rms voltage delivered to the load described in Problem 11.

18. Use calculus to determine the rms voltage delivered to the load described in Problem 14.

31.6   Introduction to Optical Devices

19. Green light has a wavelength of approximately 550 nm in air. Determine the frequency of this light.

20. Radiation emitted from a E/M source occurs at a frequency of 2.5×10^{14} Hz. Determine the wavelength of this radiation in air. In which region of the E/M spectrum is this radiation (infrared, visible, or ultraviolet)?

21. A GaAsP (gallium-arsenide-phoshide) amber LED requires 20 mA of current to provide its rated intensity. If the diode has a forward voltage of 2.0 V at this intensity, determine the current-limiting resistor needed for the LED to operate from a 9.0-V source.

22. White LEDs provide bright light that requires only 10% of the power of conventional incandescent light bulbs. A typical white LED providing 3200 mcd of light requires a forward current of $I_F = 20$ mA and has a forward voltage of 3.6 V. Determine the current-limiting resistor needed for the LED to operate from a 10-V source.

31.8   Optocouplers

23. Given the optocoupler circuit and corresponding input voltage shown in Figure 31–48, assume that the forward diode voltage is $V_F = 1.4$ V and the current transfer ratio of the optocoupler is CTR $= 0.4$:

   a. Solve for the output voltage when the input voltage is 0 V.

   b. Solve for the output voltage when the input voltage is 15 V.

   c. Sketch the corresponding output voltage.

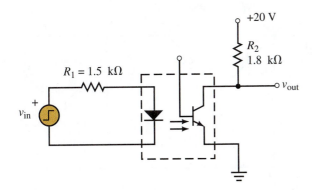 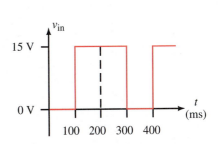

FIGURE 31–48

24. Given the optocoupler circuit and corresponding input voltage shown in Figure 31–49, assume that the forward diode voltage is $V_F = 1.5$ V, and the current transfer ratio of the optocoupler is CTR $= 0.4$:

   a. Determine the operating point of the phototransistor.

   b. Sketch the corresponding output voltage that will be observed.

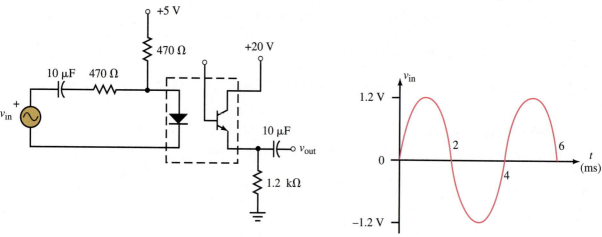

FIGURE 31–49

31.9 Semiconductor LASERs

25. What does the acronym LASER stand for?

26. Provide at least three characteristics that distinguish semiconductor lasers from LEDs.

31.10 Computer Analysis

27. Use MultiSIM to construct the circuit shown in Figure 31–50. Adjust the value of R_1 between 10% and 90%, while using the oscilloscope tool to observe the voltage across the lamp.

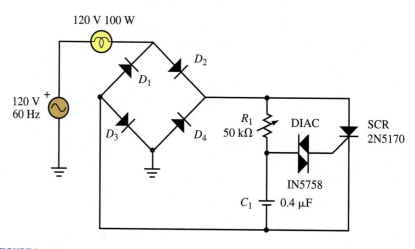

◀ MULTISIM   **FIGURE 31–50**

ANSWERS TO IN-PROCESS LEARNING CHECKS

In-Process Learning Check 1

1. Initially the anode current of the SCR is $I_A = 0$.

2. As the supply voltage increases, the voltage applied to B_1 of the UJT also increases. The capacitor begins to charge through resistors, R_2 and R_3. The rate of charge is determined by the value of R_3.

3. At some point in the positive half-cycle of the applied sinusoidal, the voltage across the capacitor will be sufficient to cause the UJT to fire. The resulting positive voltage spike across R_4 will trigger the SCR.

4. The SCR will conduct during the balance of the half-cycle. While it is conducting, the voltage across the SCR is $V_{AK} \approx 0V$ and so the UJT will no longer operate as a relaxation oscillator.

5. In the negative half-cycle of the applied sinusoidal, the SCR will remain off, since it can conduct in only one direction. At the same time, the UJT will not provide any pulses since $V_{B1\text{-}B2}$ must be positive in order for the UJT to oscillate.

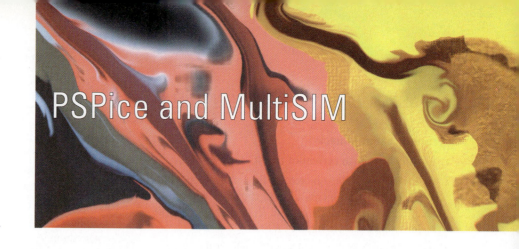

APPENDIX
A

PSPice and MultiSIM

A.1 PSpice

PSPICE HOUSEKEEPING NOTES

1. There are generally several ways to do things. The approaches shown here will get you started. As you gain experience, you will likely develop shortcuts and your own preferred ways to do things.

2. Capture provides a very structured way to organize projects. However, in this book we are concerned only with learning how to use PSpice to simulate circuits, not in how to manage project development—thus, we do not take advantage of its full organizational capability. Instead, for simplicity, we simply enter the figure number of the circuit that we are simulating each time "Name" appears in a dialog box. This works okay and makes life easier for us.

3. PSpice creates a great many intermediate files during simulation and these will add a lot of clutter to your hard drive unless you contain them. For this reason, we recommend that you use a separate folder for each problem as detailed in Chapter 4, Section 4.8, PSpice Operational Note 11.

The version of PSpice used in this book (the version current at the time of writing) is the Orcad Student Version 9.1 from Cadence Designs Systems Inc. (PSpice is a Cadence product, but when Cadence purchased the product line from Orcad, they retained the Orcad brand name.) An alternative choice is PSpice Lite, 9.2. Although both versions are free, the student version is easily downloaded and is intended for use in colleges and universities. It can be found at http://www.cadencepcb.com/products/downloads/PSpicestudent/default.asp

Orcad Capture versus Schematics

PSpice provides you with two choices of schematic capture editors, *Orcad Capture* or *Schematics*. (You make this choice after you click the starting icon.) Although Schematics is easier to use, Capture is Orcad's preferred editor because it permits better management of product development. This appendix, however, (except for the sections specifically labeled to the contrary) applies to both Capture and Schematic users.

Getting Started (Capture Users)

First, read the PSpice Housekeeping Notes. Then, assuming that you have the software loaded on your computer, click the Capture icon on your screen (or click Start, select Programs/PSpice Student/Capture Student). When the Capture session log opens, click the Create Document icon (or File/New/Project). The screen of Figure A–1 opens.

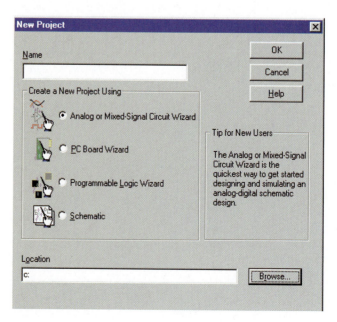

FIGURE A–1   New Project box.

1216

- Type in the name of your project and the path (in the Location box) to where you want to save your work as discussed in Notes 2 and 3. Ensure that the Analog or Mixed Signal Circuit Wizard button is selected, then click OK. Click Create a *b*lank project, then click OK.

- The Orcad Capture editor opens. Click anywhere on the screen to activate the tool palette. To familiarize yourself with the icons and menu items, place your cursor over each in turn, then view its function (Figure A–2).

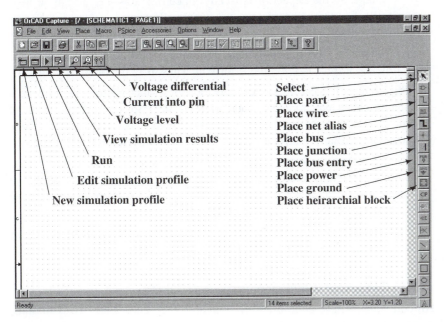

FIGURE A–2   Tool Palette.

The methodology for creating and simulating circuits is described in great detail, starting in Chapter 4, Section 4.8, and it will not be repeated here.

Getting Started (Schematic Users)

First, read PSpice Housekeeping Notes 1 and 3. Then, assuming that you have the software loaded on your computer, click the Schematics icon on your screen (or click Start, select Programs/PSPice Student/Schematics). This opens the schematics editor and you are ready to construct the circuit on your screen. The general process is similar to that for using Capture and is outlined in detail on our web site where we have provided duplicate coverage of the PSpice examples of this book using Schematics.

Some Key Points for PSpice Users

1. All PSpice circuits require a ground. If you don't use a ground, you will get an error message signifying "floating nodes."

2. All circuit components have default values. To change these, double click and type in the new values. You can include a unit if you wish, but if you don't, basic units (volts, amps, ohms) are assumed.

3. If you do include a unit, be sure that you do not leave a space. Thus, 25 and 25V are acceptable, but 25 V is not.

4. You can use either basic notation, scientific notation, or engineering prefixes. For example, you can specify a 12 000-Ω resistor as 12000, 1.2E04, 12k or 0.012Meg. Table A–1 shows PSpice prefixes.

TABLE A–1 Standard Prefixes Used by PSpice		
Symbol	**Scale**	**Name**
T or t	10^{12}	tera
G or g	10^{9}	giga
MEG or meg	10^{6}	mega
K or k	10^{3}	kilo
M or m	10^{-3}	milli
U or u	10^{-6}	micro
N or n	10^{-9}	nano
P or p	10^{-12}	pico

5. For transient analysis, capacitors and inductors always need initial conditions. In some cases, PSpice can determine these for you, while in others, you will have to enter them yourself—see Chapters 11 and 14 for examples. Note the following:

   a. PSpice can determine initial conditions only if it is possible to compute them by performing a pre-transient steady state analysis of the circuit. For circuits containing isolated capacitors, this is not possible, since the voltage on the capacitor is totally indeterminate—i.e., it can be anything. For cases like this, you must manually enter (i.e., type in) the values yourself.

   b. You can always manually enter an initial condition, even if it is possible for PSpice to compute it for you. Thus, if in doubt, do it yourself.

6. PSpice does not permit source/inductor loops—that is, you cannot have a source and an inductor alone in a loop. If you do, PSpice will display an error message. If you get such a message, add a small valued resistor to the loop. Make its resistance much smaller than any other impedance in the circuit.

7. PSpice requires a dc path from all nodes to the ground. If no path exists from a particular node to ground, add a large valued resistor from that node to ground. Make the value of this resistance much larger than the impedance of any other component in the circuit.

8. To activate a component that has already been placed on your screen, place your cursor over it and left click.

PSPice Component Orientation

All components have an implied "1" end and a "2" end. Whenever you place a component, it takes up a default position, for example, a resistor, capacitor, or inductor will take up a default position with its "1" end to the left as in Figure A–3(a). A component may be rotated by activating it, then, while holding down the Control key, typing R. (This is denoted Ctrl/R.) Each Ctrl/R rotates the component counterclockwise by 90°. To get the "1" end up, you must rotate the component 3 times from its default position as indicated in (c).

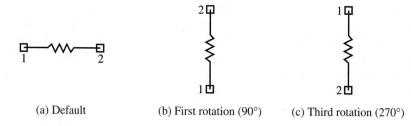

(a) Default   (b) First rotation (90°)   (c) Third rotation (270°)

FIGURE A–3

PSpice utilizes the implied "1" and "2" ends for its handling of current directions and voltage polarities—for example, it represents current as going through a device from its "1" end to its "2" end, and it represents voltage at its "1" end with respect to its "2" end. Knowing about orientation is important when you are setting up initial conditions. For example, if you set a capacitor's initial voltage to 10V, PSpice will place 10 volts across the capacitor with its "1" end positive. If you have inserted the capacitor in your circuit upside down, its polarity will be reversed from what you expect. If this happens, disconnect its wiring, rotate it to its desired orientation, rewire, then set its initial condition again.

Labeling Graph Traces and Naming Axes

To add text to a plot as in Figure A–4, select Plot, Label, and then Text. Type your text, then position it as desired. To add arrows select Plot, Label, and then Arrow, then position as desired. To add a y-axis title, select Plot, Axis Setting, click the Y-Axis tab, type in the desired title (e.g., Capacitor Voltage as in Figure A–4), then click OK.

Displaying Multiple Traces

Multiple traces may be displayed on the same vertical axis (see Figure 17–30, Chapter 17) or you may use independent axes as in Figure A–4 (which shows charging voltage and current for an *RC* circuit). Usually the first trace (assume the voltage trace here) is created using a marker probe. This automatically creates the first y-axis and scales it. To add the second trace (e.g., the current),

- If you require a second axis, click Plot, then Add Y Axis. You should now have two separate vertical axes.
- To add the current trace, click Trace, Add Trace, then click the variable whose trace you want to add [in the case of Figure A–4, this is I(C1)], then click OK.
- Label as desired.

Using the Cursors

Two cursors are used to read values from graphs such as that shown in Figure A–4. To activate the first cursor, click the cursor display icon (or click Trace, Cursor, Display), then click anywhere on the screen. To move the cursor, hold the left button down and drag to the desired position. (For fine control, you can also move it using the left and right arrow keys on your keyboard.) The x-y coordinates of the cursor appear in a display box. If you have several traces as in Figure A–4, you can move the cursor from one trace to another by clicking the small symbol immediately in front of the trace variable name that appears below the x-axis. To utilize the second cursor, use the right mouse button to activate and position it.

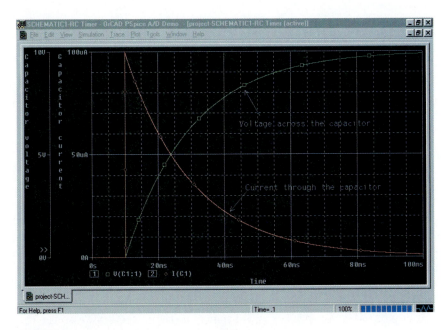

FIGURE A–4   Displaying and labeling multiple traces.

A.2 MultiSIM

The version of MultiSIM used in this book is MultiSIM 2001 Educational, the version current at the time of writing. For the most part, MultiSIM usage is straightforward, however, component orientation needs some consideration.

Component Orientation

All components have an implied "1" end and a "2" end. Whenever you place a component, it takes up a default position. For example, for resistors, capacitors, and inductors, the default position is with the "1" end to the left, while for sources, it is at the top. Components may be rotated by activating them, then while holding down the Control key, typing R. (This is denoted as Ctrl/R.) Each Ctrl/R rotates the component by 90°. To get the "1" end of a resistor up for example, rotate it once from its default position. Knowing about orientation is important when you are setting initial conditions. For example, if you set a capacitor's initial voltage to 10V, MultiSIM places 10 volts across the capacitor with its "1" end positive. If you have inserted the capacitor into your circuit upside down, its polarity will be reversed from what you expect. If this happens, disconnect its wiring, rotate it to its desired orientation, rewire, then set its initial condition again.

Using the Cursors

Cursors are used to read values from graphs created by MultiSIM (for example, Figure 11–43). To activate the cursors, click the Show/Hide Cursors icon (or click View on the Analysis Graph window, then Show/Hide Cursors). Two cursors appear. To move a cursor, position the mouse pointer over it, then press the left button and drag it to the desired position. The x-y coordinates of the cursors appear in a display box. To change the cursor from one curve to another, click anywhere on the curve to which you wish to change.

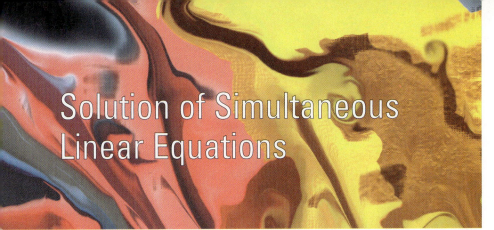

Solution of Simultaneous Linear Equations

$\mathbf{S}$imultaneous linear equations appear often in the solution of problems in electrical/electronics technology. The solutions of these equations has been simplified by a branch of mathematics called linear algebra. Although the actual theorems and proofs are well outside the scope of this textbook, we will use some of the principles of linear algebra to solve simple linear equations. The following is a set of n simultaneous linear equations in n unknowns:

$$a_{11}x_1 + a_{12}x_2 + \cdots + a_{1n}x_n = b_1$$
$$a_{21}x_1 + a_{22}x_2 + \cdots + a_{2n}x_n = b_2$$
$$\vdots$$
$$a_{n1}x_1 + a_{n2}x_2 + \cdots + a_{nn}x_n = b_n$$

The above equations may also be expressed in matrix form as

$$\mathbf{AX = B}$$

where

$$\mathbf{A} = \begin{bmatrix} a_{11}a_{12} \cdots a_{1n} \\ a_{21}a_{22} \cdots a_{2n} \\ \cdot \quad \cdot \quad \cdot \\ \cdot \quad \cdot \quad \cdot \\ \cdot \quad \cdot \quad \cdots \\ a_{n1}a_{n2} \cdots a_{nn} \end{bmatrix}, \quad \mathbf{X} = \begin{bmatrix} x_1 \\ x_2 \\ \cdot \\ \cdot \\ \cdot \\ x_n \end{bmatrix}, \quad \mathbf{B} = \begin{bmatrix} b_1 \\ b_2 \\ \cdot \\ \cdot \\ \cdot \\ b_n \end{bmatrix}$$

Substitution

Although simultaneous linear equations may be expressed in several unknowns, we begin with the most simple, namely two simultaneous linear equations in two unknowns. Consider the equations below:

$$a_{11}x_1 + a_{12}x_2 = b_1 \qquad \qquad \textbf{(B–1)}$$

$$a_{21}x_1 + a_{22}x_2 = b_2 \qquad \qquad \textbf{(B–2)}$$

If we multiply Equation B–1 by a_{22} and Equation B–2 by a_{12}, we have

$$a_{11}a_{22}x_1 + a_{12}a_{22}x_2 = a_{22}b_1$$
$$a_{12}a_{21}x_1 + a_{12}a_{22}x_2 = a_{12}b_2$$

Subtracting, we obtain

$$a_{11}a_{22}x_1 - a_{12}a_{21}x_1 = a_{22}b_1 - a_{12}b_2$$

which gives

$$x_1 = \frac{a_{22}b_1 - a_{12}b_2}{a_{11}a_{22} - a_{12}a_{21}}$$

Similarly, we solve for the unknown x_2 as

$$x_2 = \frac{a_{11}b_2 - a_{21}b_1}{a_{11}a_{22} - a_{12}a_{21}}$$

EXAMPLE B–1

Use substitution to find the solutions for the following following linear lequations:

$$2x_1 + 8x_2 = -2$$
$$x_1 + 2x_2 = 5$$

Solution   Rewriting the first equation, we have

$$2x_1 = -2 - 8x_2$$
$$x_1 = -1 - 4x_2$$

Now, substituting the above expression into the second equation, we have

$$(-1 - 4x_2) + 2x_2 = 5$$
$$-2x_2 = 6$$
$$x_2 = -3$$

Finally, we have

$$x_1 = -1 - 4(-3) = 11$$

Determinants

While substitution may be used for solving simultaneous linear equations in two variables, it is lengthy and particularly complicated when solving for more than two unknowns. An easier method used for solving simultaneous linear equations involves using *determinants*. We begin by expressing the simultaneous linear equations (B–1) and (B–2) as a product of matrices:

Column 1   Column 2   Column 3

$$\begin{bmatrix} a_{11}\, a_{12} \\ a_{21}\, a_{22} \end{bmatrix} \begin{bmatrix} x_1 \\ x_2 \end{bmatrix} = \begin{bmatrix} b_1 \\ b_2 \end{bmatrix} \qquad \textbf{(B–3)}$$

A determinant is a set of coefficients which has the same number of rows and columns and which may be expressed as a single value. The number of rows (or columns) defines the *order* of a determinant. The second-order determinant corresponding to the coefficients of the matrix equation (B–3) consists of the elements in columns 1 and 2 and is expressed as

$$D = \begin{vmatrix} a_{11}\, a_{12} \\ a_{21}\, a_{22} \end{vmatrix}$$

The value of the second-order determinant is found by taking the product of the upper left term and the lower right term (elements of the principal diagonal) and then subtracting the product of the lower left term and the upper right term (elements of the secondary diagonal). The result is given as

$$D = a_{11}a_{22} - a_{12}a_{21}$$

The unknowns of the simultaneous linear equations are found by using a technique called *Cramer's rule*. In applying this rule, we need to solve the following determinants:

$$x_1 = \frac{\begin{vmatrix} b_1 a_{12} \\ b_2 a_{22} \end{vmatrix}}{\begin{vmatrix} a_{11} a_{12} \\ a_{21} a_{22} \end{vmatrix}} \frac{a_{22}b_1 - a_{12}b_2}{a_{11}a_{22} - a_{21}a_{12}}$$

and

$$x_2 = \frac{\begin{vmatrix} a_{11} b_1 \\ a_{21} b_2 \end{vmatrix}}{\begin{vmatrix} a_{11} a_{12} \\ a_{21} a_{22} \end{vmatrix}} \frac{a_{11}b_2 - a_{21}b_1}{a_{11}a_{22} - a_{21}a_{12}}$$

The application of Cramer's rule gives the solution for each unknown by first placing the determinant of the coefficient matrix in the denominator. The numerator is then developed by using the same determinant with the exception that the coefficients of the variable to be found are replaced by the coefficients of the solution matrix. The resulting solutions are precisely those found when we used substitution.

EXAMPLE B–2

Use determinants to find solutions for the following linear equations:

$$2x_1 + 8x_2 = -2$$
$$x_1 + 2x_2 = 5$$

Solution   The determinant of the denominator is found as

$$D = \begin{vmatrix} 2 & 8 \\ 1 & 2 \end{vmatrix} = (2)(2) - (1)(8) = -4$$

The variables are now calculated as

$$x_1 = \frac{\begin{vmatrix} -2 & 8 \\ 5 & 2 \end{vmatrix}}{-4} = \frac{(-2)(2) - (5)(8)}{-4} = \frac{-44}{-4} = 11$$

and

$$x_2 = \frac{\begin{vmatrix} 2 & -2 \\ 1 & 5 \end{vmatrix}}{-4} = \frac{(2)(5) - (1)(-2)}{-4} = \frac{12}{-4} = -3$$

The solution of third-order simultaneous linear equations is similar to the method used for solving second-order equations. Consider the following third-order simultaneous linear equation:

$$a_{11}x_1 + a_{12}x_2 + a_{13}x_3 = b_1$$
$$a_{21}x_1 + a_{22}x_2 + a_{23}x_3 = b_2$$
$$a_{31}x_1 + a_{32}x_2 + a_{33}x_3 = b_3$$

The corresponding matrix equation is shown as follows:

$$\begin{bmatrix} a_{11} a_{12} a_{13} \\ a_{21} a_{22} a_{23} \\ a_{31} a_{32} a_{33} \end{bmatrix} \begin{bmatrix} x_1 \\ x_2 \\ x_3 \end{bmatrix} = \begin{bmatrix} b_1 \\ b_2 \\ b_3 \end{bmatrix}$$

The value of the third-order determinant may be found in one of several ways. The first method works for only third-order determinants, while the second method is a more general approach which evaluates any order of determinant.

Method I

This method works only for third-order determinants:

1. Begin by writing the original columns of the third-order determinant.

2. Copy the first two columns, placing them to the right of the original determinant.

3. Add the product of the elements of the principal diagonal to the products of the adjacent two parallel diagonals to the right of the principal diagonal.

4. Subtract the product of the elements of the secondary diagonal and also subtract the products of the elements along the two other parallel diagonals.

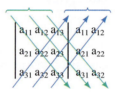

The resultant determinant is written as

$$D = a_{11}a_{22}a_{33} + a_{12}a_{23}a_{31} + a_{13}a_{21}a_{32} - a_{31}a_{22}a_{13} - a_{32}a_{23}a_{11} - a_{33}a_{21}a_{12}$$

EXAMPLE B–3

Evaluate the following determinant:

$$D = \begin{vmatrix} 3 & 1 & -2 \\ 1 & -2 & 3 \\ 2 & 3 & 2 \end{vmatrix}$$

Solution   We begin by rewriting the first two columns as follows:

$$D = \begin{vmatrix} 3 & 1 & -2 \\ 1 & -2 & 3 \\ 2 & 3 & 2 \end{vmatrix} \begin{matrix} 3 & 1 \\ 1 & -2 \\ 2 & 3 \end{matrix}$$

Now, adding the products of the principal diagonal and adjacent diagonals and subtracting the products of the secondary diagonal and adjacent diagonals, we have

$$D = (3)(-2)(2) + (1)(3)(2) + (-2)(1)(3)$$
$$- (2)(-2)(-2) - (3)(3)(3) - (2)(1)(1)$$
$$= -49$$

Method II

This evaluation of determinants is achieved by expansion by minors. The *minor* of an element is the determinant which remains after deleting the row and the column in which the element lies. The value of any nth-order determinant is found as follows:

1. For any row or column, find the product of each element and the determinant of its minor.

2. A product is given a positive sign if the sum of the row and the column of the element is even. The product is given a negative sign if the sum is odd.

3. The value of the determinant is the sum of the resulting terms.

As before, Cramer's rule is used to solve for the unknowns, x_1, x_2, and x_3, by using determinants and replacing the appropriate terms of the numerator with the terms of the solution matrix. The resulting determinants and solutions are given as follows:

$$x_1 = \frac{\begin{vmatrix} b_1 & a_{12} & a_{13} \\ b_2 & a_{22} & a_{23} \\ b_3 & a_{32} & a_{33} \end{vmatrix}}{\begin{vmatrix} a_{11} & a_{12} & a_{13} \\ a_{21} & a_{22} & a_{23} \\ a_{31} & a_{32} & a_{33} \end{vmatrix}}$$

By expansion of minors, the determinant of the denominator is found as

$$D = + a_{11}\begin{vmatrix} a_{22} & a_{23} \\ a_{32} & a_{33} \end{vmatrix} - a_{21}\begin{vmatrix} a_{12} & a_{13} \\ a_{32} & a_{33} \end{vmatrix} + a_{31}\begin{vmatrix} a_{12} & a_{13} \\ a_{22} & a_{23} \end{vmatrix}$$

$$= a_{11}(a_{22}a_{33} - a_{23}a_{32}) - a_{21}(a_{12}a_{33} - a_{13}a_{32}) + a_{31}(a_{12}a_{23} - a_{13}a_{22})$$

The solution for x_1 is now found to be

$$x_1 = \frac{+ b_1\begin{vmatrix} a_{22} & a_{23} \\ a_{32} & a_{33} \end{vmatrix} - b_2\begin{vmatrix} a_{12} & a_{13} \\ a_{32} & a_{33} \end{vmatrix} + b_3\begin{vmatrix} a_{12} & a_{13} \\ a_{22} & a_{23} \end{vmatrix}}{D}$$

$$= \frac{b_1(a_{22}a_{33} - a_{23}a_{32}) - b_2(a_{12}a_{33} - a_{13}a_{32}) + b_3(a_{12}a_{23} - a_{13}a_{22})}{D}$$

Similarly, for x_2, we get

$$x_2 = \frac{\begin{vmatrix} a_{11} & b_1 & a_{13} \\ a_{21} & b_2 & a_{23} \\ a_{31} & b_3 & a_{33} \end{vmatrix}}{\begin{vmatrix} a_{11} & a_{12} & a_{13} \\ a_{21} & a_{22} & a_{23} \\ a_{31} & a_{32} & a_{33} \end{vmatrix}}$$

$$= \frac{-b_1(a_{21}a_{33} - a_{23}a_{31}) + b_2(a_{11}a_{33} - a_{13}a_{31}) + b_3(a_{11}a_{23} - a_{13}a_{21})}{D}$$

and for x_3 we have

$$x_3 = \frac{\begin{vmatrix} a_{11} & a_{12} & b_1 \\ a_{21} & a_{22} & b_2 \\ a_{31} & a_{32} & b_3 \end{vmatrix}}{\begin{vmatrix} a_{11} & a_{12} & a_{13} \\ a_{21} & a_{22} & a_{23} \\ a_{31} & a_{32} & a_{33} \end{vmatrix}}$$

$$= \frac{b_1(a_{21}a_{32} - a_{22}a_{31}) - b_2(a_{11}a_{32} - a_{12}a_{31}) + b_3(a_{11}a_{22} - a_{12}a_{21})}{D}$$

EXAMPLE B–4

Solve for x_1 in the following system of linear equations using minors.

$$3x_1 + x_2 - 2x_3 = 1$$
$$x_1 - 2x_2 + 3x_3 = 11$$
$$2x_1 + 3x_2 + 2x_3 = -3$$

Solution The determinant of the denominator is evaluated as follows:

$$D = \begin{vmatrix} 3 & 1 & -2 \\ 1 & -2 & 3 \\ 2 & 3 & 2 \end{vmatrix}$$

$$= +(3) \begin{vmatrix} -2 & 3 \\ 3 & 2 \end{vmatrix} - (1) \begin{vmatrix} 1 & -2 \\ 3 & 2 \end{vmatrix} + (2) \begin{vmatrix} 1 & -2 \\ -2 & 3 \end{vmatrix}$$

$$= (3)(-4 - 9) - (2 + 6) + (2)(3 - 4)$$

$$= -49$$

and so the unknown x_1 is calculated to be

$$x_1 = \frac{\begin{vmatrix} 1 & 1 & -2 \\ 11 & -2 & 3 \\ -3 & 3 & 2 \end{vmatrix}}{-49}$$

$$= \frac{+(1) \begin{vmatrix} -2 & 3 \\ 3 & 2 \end{vmatrix} - (11) \begin{vmatrix} 1 & -2 \\ 3 & 2 \end{vmatrix} + (-3) \begin{vmatrix} 1 & -2 \\ -2 & 3 \end{vmatrix}}{-49}$$

$$= \frac{(-4 - 9) - (11)(2 + 6) - (3)(3 - 4)}{-49}$$

$$= 2$$

PRACTICE PROBLEM

Use expansion by minors to solve for x_2 and x_3 in Example B–4.

Answers
$x_2 = -3, \ x_3 = 1$

Maximum Power Transfer Theorem

Figure C–1 shows the Thévenin equivalent of a dc circuit.

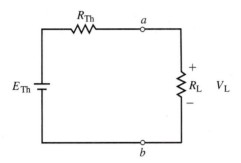

FIGURE C–1

For the above circuit, the values of E_{Th} and R_{Th} are constant. Therefore, power delivered to the load is determined as a function of the load resistance and is given as

$$P_L = \frac{V_L^2}{R_L} = \frac{\left(\dfrac{R_L E_{Th}}{R_L + R_{Th}}\right)^2}{R_L} = \frac{E_{Th}^2 R_L}{\left(R_L + R_{Th}\right)^2} \qquad \text{(C–1)}$$

Maximum power will be delivered to R_L when the first derivative, $\dfrac{dP_L}{dR_L} = 0$.

Applying the quotient rule, $\dfrac{d}{dx}\left(\dfrac{u}{v}\right) = \dfrac{v\dfrac{du}{dx} - u\dfrac{dv}{dx}}{v^2}$, we find the derivative of power with respect to load resistance as

$$\begin{aligned}
\frac{dP_L}{dR_L} &= \frac{(R_L + R_{Th})^2 (E_{Th})^2 - (E_{Th}^2 R_L)(2)(R_L + R_{Th})}{(R_L + R_{Th})^4} \\
&= \frac{E_{Th}^2[(R_L + R_{Th})^2 - 2R_L(R_L + R_{Th})]}{(R_L + R_{Th})^4}
\end{aligned} \qquad \text{(C–2)}$$

Now, since the first derivative can only be zero if the numerator of the above expression is zero, and since E_{Th} is a constant, we have

$$(R_L + R_{Th})^2 - 2R_L(R_L + R_{Th}) = 0 \qquad \text{(C–3)}$$

And so,

$$\begin{aligned}
R_L^2 + 2R_L R_{Th} + R_{Th}^2 - 2R_L R_L - 2R_L R_{Th} &= 0 \\
R_{Th}^2 - R_L^2 &= 0 \qquad \text{(C–4)} \\
R_L &= R_{Th}
\end{aligned}$$

Figure C–2 shows the Thévenin equivalent of an ac circuit.

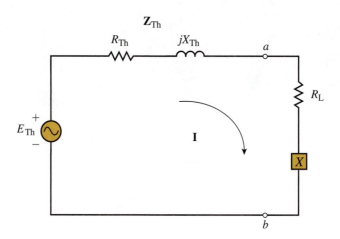

For the above circuit, the values of E_{Th}, R_{Th}, and X_{Th} are constant. Although X_{Th} is shown as an inductor, it could just as easily be a capacitor. The power delivered to the load is determined as a function of the load impedance as

$$P_L = I^2 R_L = \frac{E_{Th}^2 R_L}{(R_L + R_{Th})^2 + (X + X_{Th})^2} \tag{C–5}$$

Maximum power is dependent on two variables, R_L and X. Therefore, we will need to solve for partial derivatives. Maximum power will be transferred to the load when $\dfrac{\partial P_L}{\partial R_L} = 0$ and $\dfrac{\partial P_L}{\partial X} = 0$.

We begin by finding $\dfrac{\partial P_L}{\partial X}$. Using $\dfrac{d}{dx}\left(\dfrac{1}{v}\right) = -\dfrac{1}{v^2}\dfrac{dv}{dx}$, we get

$$\frac{\partial P_L}{\partial X} = -\frac{(E_{Th}^2 R_L)(2)(X + X_{Th})}{[(R_L + R_{Th})^2 + (X + X_{Th})^2]^2} \tag{C–6}$$

Now, since the partial derivative can only be zero if the numerator of the above expression is zero, and since E_{Th} and R_L are treated as constants, we have

$$X + X_{Th} = 0$$

or

$$X = -X_{Th} \tag{C–7}$$

This result implies that if the Thévenin impedance contains an inductor of magnitude X, the load must contain a capacitor with the same magnitude. (Conversely, if the Thévenin impedance were to contain a capacitor, then the load impedance would need to have an inductor of the same magnitude.)

Next, we determine the partial derivative, $\dfrac{\partial P_L}{\partial R_L}$ of equation (C–5). Applying the quotient rule,

$$\frac{d}{dx}\left(\frac{u}{v}\right) = \frac{v\dfrac{du}{dx} - u\dfrac{dv}{dx}}{v^2}$$

we get

$$\frac{\partial P_L}{\partial R_L} = \frac{[(R_L + R_{Th})^2 + (X + X_{Th})^2](E_{Th}^2) - (E_{Th}^2 R_L)[(2)(R_L + R_{Th})]}{[(R_L + R_{Th})^2 + (X + X_{Th})^2]^2} \quad \text{(C–8)}$$

Now, since the partial derivative can only be zero if the numerator of the above expression is zero, and since E_{Th} and X are treated as constants, we have

$$(R_L + R_{Th})^2 + (X + X_{Th})^2(E_{Th}^2) - (E_{Th}^2 R_L)(2)(R_L + R_{Th}) = 0$$
$$(R_L + R_{Th})^2 + (X + X_{Th})^2 - (R_L)(2)(R_L + R_{Th}) = 0$$
$$R_L^2 + 2R_L R_{Th} + R_{Th}^2 + (X + X_{Th})^2 - 2R_L^2 - 2R_L R_{Th} = 0$$
$$R_{Th}^2 - R_L^2 + (X + X_{Th})^2 = 0$$

In general, the above equation determines the value of load resistance regardless of the load reactance. Therefore, we have

$$R_L = \sqrt{R_{Th}^2 + (X + X_{Th})^2} \quad \text{(C–9)}$$

This result shows that the load resistance is dependent upon the load reactance, X. If the reactance of the load is the same type as the Thévenin reactive component (both inductive or both capacitive), then the reactances are added. If the load reactance is the opposite type of the Thévenin reactive component, then the reactances are subtracted. If the reactance of the load can be adjusted to result in maximum power transfer ($X = -X_{Th}$), then equation (C–9) is simplified to the expected result, namely

$$R_L = X_{Th} \quad \text{(C–10)}$$

CHAPTER 1

1. a. 1620 s   b. 2880 s   c. 7427 s

   d. 26 110 W   e. 2.45 hp   f. 8280°

3. a. 0.84 m^2 b. 0.0625 m^2

   c. 0.02 m^3 d. 0.0686 m^3

5. 4500 parts/h

7. 11.5 km/l

9. 150 rpm

11. 8.33 mi

13. 0.508 m/s

15. 7.45 km

17. 20.4 min

19. Machine 1: $25.80/h;   Machine 2: $25.00/h;   Machine 2

21. a. 8.675 × 10^3 b. 8.72 × 10^{-3} c. 1.24 × 10^3

   d. 3.72 × 10^{-1} e. 3.48 × 10^2 f. 2.15 × 10^{-7}

   g. 1.47 × 10^1

23. a. 1.25 × 10^{-1} b. 8 × 10^7

   c. 2.0 × 10^{-2} d. 2.05 × 10^4

25. a. 10   b. 10   c. 3.6 × 10^3

   d. 15 × 10^4 e. −12.0

27. 1.179;   4.450;   Direct computation is less work for these examples.

29. 6.24 × 10^{18}

31. 62.6 × 10^{21}

33. 1.16 s

35. 13.4 × 10^{10} l/h

37. a. kilo, k   b. mega, M   c. giga, G

   d. micro, μ   e. milli, m   f. pico, p

39. a. 1.5 ms   b. 27 μs   c. 350 ns

41. a. 150, 0.15   b. 0.33, 33

43. a. 680 V   b. 162.7 W

45. 1.5 kW

47. 187 A

49. Radio signal, 16.68 ms;   Telephone signal, 33.33 ms;   The radio signal by 16.65 ms.

CHAPTER 2

1. a. 10^{29} b. 10.4 × 10^{23}

3. Increases by a factor of 24.

5. a. Material with many free electrons (i.e., material with 1 electron in the valence shell).

   b. Inexpensive and easily formed into wires.

   c. Full valence shell. Therefore, no free electrons.

   d. The large electrical force tears electrons out of orbit.

7. It has an excess or deficiency of electrons.

9. 2 μC;   (Attraction)

11. 0.333 μC, 1.67 μC;   both (+) or both (−)

13. 30.4 μC

15. 27.7 μC (+)

17. 24 V

19. 2400 V

21. 4.25 mJ

23. 4.75 C

25. 50 mA

27. 334 μC

29. 3 mA

31. 80 A

33. 18 V, 0.966 A

35. a. 4.66 V   b. 1.50 V

37. 50 h

39. 11.7 h

41. 267 h

43. (c) Both

45. The voltmeter and ammeter are interchanged.

47. If you exceed a fuse's voltage rating, it may arc over when it "blows."

CHAPTER 3

1. a. 3.6 Ω   b. 0.90 Ω

   c. 36.0 kΩ   d. 36.0 mΩ

3. 0.407 inch

5. 300 m = 986 feet

7. 982 × 10^{-8} Ω · m (Resistivity is less than for carbon.)

9. 2.26 × 10^{-8} Ω · m (This alloy is not as good a conductor as copper.)

11. AWG 22: 4.86 Ω

   AWG 19: 2.42 Ω

   Diameter of AWG 19 is 1.42 times the diameter of AWG 22. The resistance of AWG 19 is half the resistance of an equal length of AWG 22.

13. AWG 19 should be able to handle 4 A.

AWG 30 can handle about 0.30 A.

15. 405 meters

17. a. 256 CM   b. 6200 CM   c. 1910 MCM

19. a. 16.2 Ω b. 0.668 Ω c. 2.17 $\times$ 10^{-3} Ω

21. a. 4148 CM = 3260 sq mil   b. 0.0644 inch

23. a. 1600 CM = 1260 sq mil   b. 1930 feet

25. $R_{-30°C}$ = 40.2 Ω $R_{0°C}$ = 46.1 Ω $R_{200°C}$ = 85.2 Ω

27. a. Positive temperature coefficient

   b. 0.00385 (°C)$^{-1}$

   c. $R_{0°C}$ = 18.5 Ω $R_{100°C}$ = 26.2 Ω

29. 16.8 Ω

31. $T = -260°C$

33. a. R_{ab} = 10 kΩ R_{bc} = 0 Ω

   b. R_{ab} = 8 kΩ R_{bc} = 2 kΩ

   c. R_{ab} = 2 kΩ R_{bc} = 8 kΩ

   d. R_{ab} = 0 kΩ R_{bc} = 10 kΩ

35. a. 150 kΩ ± 10%

   b. 2.8 Ω ± 5% with a reliability of 0.001%

   c. 47 MΩ ± 5%

   d. 39 Ω ± 5% with a reliability of 0.1%

37. Connect the ohmmeter between the two terminals of the light bulb. If the ohmmeter indicates an open circuit, the light bulb is burned out.

39. AWG 24 has a resistance of 25.7 Ω/1000 ft. Measure the resistance between the two ends and calculate the length as

$$\ell = \frac{R}{0.0257 \ \Omega/\text{ft}}$$

41. a. 380 Ω

   b. 180 Ω

   c. Negative temperature coefficient. Resistance decreases as temperature increases.

43. a. 4.0 S   b. 2.0 mS

   c. 4.0 μS   d. 0.08 μS

45 2.93 mS

CHAPTER 4

1. a. 2 A   b. 7.0 A   c. 5 mA

   d. 4 μA   e. 3 mA   f. 6 mA

3. a. 40 V   b. 0.3 V

   c. 400 V   d. 0.36 V

5. 96 Ω

7. 28 V

9. 6 A

11. Red, Red, Red

13. 22 V

15. a. 2.31 A   b. 2.14 A

17. 2.88 V

19. 4 Ω

21. 400 V

23. 3.78 mA

25. a. + 45 V −   b. 4 A ($\rightarrow$)

   c. − 90 V +   d. 7 A ($\leftarrow$)

27. 3.19 J/s;   3.19 W

29. 36 W

31. 14.1 A

33. 47.5 V

35. 50 V, 5 mA

37. 37.9 A

39. 2656 W

41. 23.2 V;   86.1 mA

43. 361 W $\rightarrow$ 441 W

45. a. 48 W ($\rightarrow$)   b. 30 W ($\leftarrow$)

   c. 128 W ($\leftarrow$)   d. 240 W ($\rightarrow$)

47. a. 1.296 $\times$ 10^6 J   b. 360 Wh   c. 2.88 cents

49. 26 cents

51. $5256

53. 5 cents

55. 51.5 kW

57. 82.7%

59. 2.15 hp

61. 8.8 hp

63. 1.97 hp

65. $137.45

67. a. 10 Ω b. 13.3 Ω

CHAPTER 5

1. a. +30 V   b. −90 V

3. a. +45 V   b. −60 V

   c. +90 V   d. −105 V

5. a. 7 V

   b. V_2 = 4 V V_1 = 4 V

7. V_3 = 12 V V_4 = 2 V

9. a. 10 kΩ b. 2.94 MΩ c. 23.4 kΩ

11. Circuit 1: 1650 Ω, 6.06 mA   Circuit 2: 18.15 kΩ, 16.5 mA

13. a. 10 mA   b. 13 kΩ c. 5 kΩ

   d. $V_{3\text{-k}\Omega}$ = 30 V $V_{4\text{-k}\Omega}$ = 40 V $V_{1\text{-k}\Omega}$ = 10 V V_R = 50 V

   e. $P_{1\text{-k}\Omega}$ = 100 mW $P_{3\text{-k}\Omega}$ = 300 mW $P_{4\text{-k}\Omega}$ = 400 mW P_R = 500 mW

15. a. 40 mA   b. V_{R_1} = 12 V V_{R_3} = 10 V   c. 26 V

17. a. V_{R_2} = 4.81 V V_{R_3} = 3.69 V

   b. 1.02 mA   c. 7.32 kΩ

19. a. 457 Ω b. 78.8 mA

   c. V_1 = 9.45 V V_2 = 3.07 V V_3 = 6.14 V V_4 = 17.33 V

   d. V_T = 36 V

   e. P_1 = 0.745 W P_2 = 0.242 W P_3 = 0.484 W P_4 = 1.365 W

   f. R_1: 1 W R_2: 1/4 W R_3: 1/2 W R_4: 2 W

   g. 2.836 W

21. a. 0.15 A   b. 0.115 mA

23. Circuit 1: $V_{6\text{-}\Omega} = 6$ V $V_{3\text{-}\Omega} = 3$ V $V_{5\text{-}\Omega} = 5$ V $V_{8\text{-}\Omega} = 8$ V
    $V_{2\text{-}\Omega} = 2$ V $V_T = 24$ V

    Circuit 2: $V_{4.3\text{-}k\Omega} = 21.6$ V $V_{2.7\text{-}k\Omega} = 13.6$ V
    $V_{7.8\text{-}k\Omega} = 39.2$ kΩ $V_{9.1\text{-}k\Omega} = 45.7$ V $V_T = 120$ V

25. Circuit 1:
    a. $R_1 = 0.104$ kΩ $R_2 = 0.365$ kΩ $R_3 = 0.730$ kΩ
    b. $V_1 = 2.09$ V $V_2 = 7.30$ V $V_3 = 14.61$ V
    c. $P_1 = 41.7$ mW $P_2 = 146.1$ mW $P_3 = 292.2$ mW

Circuit 2:
    a. $R_1 = 977$ Ω $R_2 = 244$ Ω $R_3 = 732$ Ω
    b. $V_1 = 25.0$ V $V_2 = 6.25$ V $V_3 = 18.75$ V
    c. $P_1 = 640$ mW $P_2 = 160$ mW $P_3 = 480$ mW

27. a. 0.2 A   b. 5.0 V   c. 1 W
    d. $R_T = 550$ Ω $I = 0.218$ A $V = 5.45$ V $P = 1.19$ W
    e. Life expectancy decreases.

29. Circuit 1: $V_{ab} = 9.39$ V $V_{bc} = 14.61$ V
    Circuit 2: $V_{ab} = +25.0$ V $V_{bc} = +6.25$ V

31. Circuit 1:
    $V_{3k\text{-}\Omega} = 9$ V $V_{9k\text{-}\Omega} = 27$ V $V_{6k\text{-}\Omega} = 18$ V $V_a = 45$ V
    Circuit 2: $V_{330\text{-}\Omega} = 2.97$ V $V_{670\text{-}\Omega} = 6.03$ V $V_a = 3.03$ V

33. a. 109 V   b. 9.20 Ω

35. Circuit 1:
    $I_{\text{actual}} = 0.375$ mA $I_{\text{measured}} = 0.3745$ mA
    loading error = 0.125%

    Circuit 2:
    $I_{\text{actual}} = 0.375$ mA $I_{\text{measured}} = 0.3333$ mA
    loading error = 11.1%

37. Circuit 1:
    a. 1 A
    b. $V_{6\text{-}\Omega} = 6$ V $V_{3\text{-}\Omega} = 3$ V $V_{5\text{-}\Omega} = 5$ V $V_{8\text{-}\Omega} = 8$ V
    $V_{2\text{-}\Omega} = 2$ V

    Circuit 2:
    a. 5.06 mA
    b. $V_{4.3\text{-}k\Omega} = 21.7$ V $V_{2.7\text{-}k\Omega} = 13.6$ V $V_{7.8\text{-}k\Omega} = 39.1$ V
    $V_{9.1\text{-}k\Omega} = 45.6$ V

39. a. 78.8 mA
    b. $V_1 = 9.45$ V $V_2 = 3.07$ V $V_3 = 6.14$ V $V_4 = 17.33$ V

CHAPTER 6

1. a. A and B are in series;   D and E are in series;
      C and F are parallel
   b. B, C, and D are parallel
   c. A and B are parallel;   D and F are parallel;
      C and E are in series
   d. A, B, C, and D are parallel

5. a. $I_1 = 3$ A $I_2 = -1$ A
   b. $I_1 = 7$ A $I_2 = 2$ A $I_3 = -7$ A
   c. $I_1 = 4$ mA $I_2 = 20$ mA

7. a. $I_1 = 1.25$ A $I_2 = 0.0833$ A $I_3 = 1.167$ A $I_4 = 1.25$ A
   b. $R_3 = 4.29$ Ω

9. a. $I_1 = 200$ mA $I_2 = 500$ mA $I_3 = 150$ mA
      $I_4 = 200$ mA
   b. 2.5 V   c. $R_1 = 12.5$ Ω $R_3 = 16.7$ Ω $R_4 = 50$ Ω

11. a. $R_T = 2.4$ Ω $G_T = 0.417$ S
    b. $R_T = 32$ kΩ $G_T = 31.25$ μS
    c. $R_T = 4.04$ kΩ $G_T = 247.6$ μS

13. a. 2.0 MΩ   b. 450 Ω

15. a. $R_1 = 1250$ Ω $R_2 = 5$ kΩ $R_3 = 250$ Ω
    b. $I_{R1} = 0.40$ A $I_{R2} = 0.10$ A
    c. 2.5 A

17. a. 900 mV   b. 4.5 mA

19. a. 240 Ω   b. 9.392 kΩ   c. 1.2 kΩ

21. $I_1 = 0.235$ mA $I_2 = 0.706$ mA $I_3 = 1.059$ mA
    $R_1 = 136$ kΩ $R_2 = 45.3$ kΩ $R_3 = 30.2$ kΩ

23. a. 12.5 kΩ   b. 0   c. 75 Ω

25. $R_T \cong 15$ Ω

27. $I = 0.2$ A $I_1 = 0.1$ A $= I_2$

29. a. $I_1 = 2$ A $I_2 = 8$ A   b. $I_1 = 4$ mA $I_2 = 12$ mA

31. a. $I_1 = 6.48$ mA $I_2 = 9.23$ mA $I_3 = 30.45$ mA
       $I_4 = 13.84$ mA
    b. $I_1 = 60$ mA $I_2 = 30$ mA $I_3 = 20$ mA $I_4 = 40$ mA
       $I_5 = 110$ mA

33. 12 Ω

35. a. 8 Ω
    b. 1.50 A
    c. $I_1 = 0.50$ A $I_2 = 0.25$ A $I_3 = 0.75$ A
    d. $\Sigma I_{\text{in}} = \Sigma I_{\text{out}} = 1.50$ A

37. a. 25 Ω $I = 9.60$ A
    b. $I_1 = 4.0$ A $I_2 = 2.40$ A $I_3 = 3.20$ A $I_4 = 5.60$ A
    c. $\Sigma I_{\text{in}} = \Sigma I_{\text{out}} = 9.60$ A
    d. $P_1 = 960$ W $P_2 = 576$ W $P_3 = 768$ W
       $P_T = 2304$ W $= P_1 + P_2 + P_3$

39. a. $I_1 = 1.00$ A $I_2 = 2.00$ A $I_3 = 5.00$ A $I_4 = 4.00$ A
    b. 12.00 A
    c. $P_1 = 20$ W $P_2 = 40$ W $P_3 = 100$ W $P_4 = 80$ W

41. a. $R_1 = 2$ kΩ $R_2 = 8$ kΩ $R_3 = 4$ kΩ $R_4 = 6$ kΩ
    b. $I_{R_1} = 24$ mA $I_{R_2} = 6$ mA $I_{R_4} = 8$ mA
    c. $I_1 = 20$ mA $I_2 = 50$ mA
    d. $P_2 = 288$ mW $P_3 = 576$ mW $P_4 = 384$ mW

43. $I_1 = 8.33$ A $I_2 = 5.00$ A $I_3 = 2.50$ A $I_4 = 7.50$ A
    $I_T = 15.83$ A

    The rated current of the fuse will be exceeded;   the fuse will
    "blow."

45. a. $V_{\text{measured}} = 20$ V   b. loading effect = 33.3%

47. 25.2 V

49. $I_1 = 4.0$ A $I_2 = 2.4$ A $I_3 = 3.2$ A

51. 20 V

53. $I_1 = 4.0$ A $I_2 = 2.4$ A $I_3 = 3.2$ A

CHAPTER 7

1. a. $R_T = R_1 + R_5 + [(R_2 + R_3)\|R_4]$

   b. $R_T = (R_1\|R_2) + (R_3\|R_4)$

3. a. $R_{T1} = R_1 + [(R_3 + R_4)\|R_2] + R_5$ $R_{T2} = R_5$

   b. $R_{T1} = R_1 + (R_2\|R_3\|R_5)$ $R_{T_2} = R_5\|R_3\|R_2$

7. a. $1500\ \Omega$ b. $2.33\ k\Omega$

9. $R_{ab} = 140\ \Omega$ $R_{cd} = 8.89\ \Omega$

11. a. $R_T = 314\ \Omega$

    b. $I_T = 63.7\ mA$ $I_1 = 19.2\ mA$ $I_2 = 44.5\ mA$
    $I_3 = 34.1\ mA$ $I_4 = 10.4\ mA$

    c. $V_{ab} = 13.6\ V$ $V_{bc} = -2.9\ V$

13. a. $I_1 = 5.19\ mA$ $I_2 = 2.70\ mA$ $I_3 = 1.081\ mA$
    $I_4 = 2.49\ mA$ $I_5 = 1.621\ mA$ $I_6 = 2.70\ mA$

    b. $V_{ab} = 12.43\ V$ $V_{cd} = 9.73\ V$

    c. $P_T = 145.3\ mW$ $P_1 = 26.9\ mW$ $P_2 = 7.3\ mW$
    $P_3 = 3.5\ mW$ $P_4 = 30.9\ mW$ $P_5 = 15.8\ mW$
    $P_6 = 7.0\ mW$ $P_7 = 53.9\ mW$

15. Circuit (a):

    a. $I_1 = 4.5\ mA$ $I_2 = 4.5\ mA$ $I_3 = 1.5\ mA$

    b. $V_{ab} = -9.0\ V$

    c. $P_T = 162\ mW$ $P_{6\text{-}k\Omega} = 13.5\ mW$ $P_{3\text{-}k\Omega} = 27.0\ mW$
    $P_{2\text{-}k\Omega} = 40.5\ mW$ $P_{4\text{-}k\Omega} = 81.0\ mW$

    Circuit (b):

    a. $I_1 = 0.571\ A$ $I_2 = 0.365\ A$ $I_3 = 0.122\ A$ $I_4 = 0.449\ A$

    b. $V_{ab} = -1.827\ V$

    c. $P_T = 5.14\ W$ $P_{10\text{-}\Omega} = 3.26\ W$ $P_{16\text{-}\Omega} = 0.68\ W$
    $P_{5\text{-}\Omega} = 0.67\ W$ $P_{6\text{-}\Omega} = 0.36\ W$ $P_{8\text{-}\Omega} = 0.12\ W$
    $P_{4\text{-}\Omega} = 0.06\ W$

17. $I_1 = 93.3\ mA$ $I_2 = 52.9\ mA$ $I_Z = 40.4\ mA$ $V_1 = 14\ V$
    $V_2 = 2.06\ V$ $V_3 = 7.94\ V$ $P_T = 2240\ mW$
    $P_1 = 1307\ mW$ $P_2 = 109\ mW$ $P_3 = 420\ mW$
    $P_Z = 404\ mW$

19. $R = 31.1\ \Omega \rightarrow 3900\ \Omega$

21. $I_C = 1.70\ mA$ $V_B = -1.97\ V$ $V_{CE} = -8.10\ V$

23. a. $I_D = 3.6\ mA$ b. $R_S = 556\ \Omega$ c. $V_{DS} = 7.6\ V$

25. $I_C \cong 3.25\ mA$ $V_{CE} \cong -8.90\ mV$

27. a. $V_L = 0 \rightarrow 7.2\ V$ b. $V_L = 2.44\ V$ c. $V_{ab} = 9.0\ V$

29. $V_{bc} = 7.45\ V$ $V_{ab} = 16.55\ V$

31. a. $V_{out(min)} = 0\ V$ $V_{out(max)} = 40\ V$

    b. $R_2 = 3.82\ k\Omega$

33. $0\ V, 8.33\ V, 9.09\ V$

35. a. $11.33\ V$ b. $8.95\ V$ c. 44.1% d. $1.333\ V$

37. a. Break the circuit between the 5.6-Ω resistor and the volt-
    age source. Insert the ammeter at the break, connecting
    the red (+) lead of the ammeter to the positive terminal
    of the voltage source and the black (−) lead to the 5.6-Ω
    resistor.

    b. $I_{1(loaded)} = 19.84\ mA$ $I_{2(loaded)} = 7.40\ mA$
    $I_{3(loaded)} = 12.22\ mA$

    c. loading effect (I_1) = 19.9%

    loading effect (I_2) = 18.0%

    loading effect (I_3) = 22.3%

39. $12.0\ V,\ 30.0\ V,\ 5.00\ A,\ 3.00\ A,\ 2.00\ A$

41. $14.1\ V$

43. $12.0\ V,\ 30.0\ V,\ 5.00\ A,\ 3.00\ A,\ 2.00\ A$

CHAPTER 8

1. $38\ V$

3. a. $12\ mA$ b. $V_S = 4.4\ V$ $V_1 = 2.0\ V$

5. $I_1 = 400\ \mu A$ $I_2 = 500\ \mu A$

7. $P_T = 7.5\ mW$ $P_{50\text{-}k\Omega} = 4.5\ mW$ $P_{150\text{-}k\Omega} = 1.5\ mW$

   $P_{current\ source} = 1.5\ mW$

   Note: The current source is absorbing energy from the circuit
   rather than providing energy.

9. Circuit (a):

   0.25 A-source in parallel with a 20-Ω resistor

   Circuit (b):

   12.5-mA source in parallel with a 2-kΩ resistor

11. a. $7.2\ A$ b. $E = 3600\ V$ $I_L = 7.2\ A$

13. a. $21.45\ V$ b. $6.06\ mA$ c. $0.606\ V$

15. $V_2 = -80\ V$ $I_1 = -26.7\ mA$

17. $V_{ab} = -7.52\ V$ $I_3 = 0.133\ mA$

19. $I_1 = 0.467\ A$ $I_2 = 0.167\ A$ $I_3 = 0.300\ A$

21. $I_2 = -0.931\ A$

23. a. $(8\ \Omega)I_1 + 0\ I_2 - (10\ \Omega)I_3 = 24\ V$
    $0\ I_1 + (4\ \Omega)I_2 + (10\ \Omega)I_3 = 16\ V$
    $I_1 - I_2 + I_3 = 0$

    b. $I = 3.26\ A$

    c. $V_{ab} = -13.89\ V$

25. $I_1 = 0.467\ A$ $I_2 = 0.300\ A$

27. $I_2 = -0.931\ A$

29. $I_1 = -19.23\ mA$ $V_{ab} = 2.77\ V$

31. $I_1 = 0.495\ A$ $I_2 = 1.879\ A$ $I_3 = 1.512\ A$

33. $V_1 = -6.73\ V$ $V_2 = 1.45\ V$

35. $V_1 = -6\ V$ $V_2 = 20\ V$

37. $V_{6\Omega} = 6.10\ V$

39. Network (a): $R_1 = 6.92\ \Omega$ $R_2 = 20.77\ \Omega$ $R_3 = 62.33\ \Omega$
    Network (b): $R_1 = 1.45\ k\Omega$ $R_2 = 2.41\ k\Omega$ $R_3 = 2.03\ k\Omega$

41. Network (a): $R_A = 110\ \Omega$ $R_B = 36.7\ \Omega$ $R_C = 55\ \Omega$
    Network (b): $R_A = 793\ k\Omega$ $R_B = 1693\ k\Omega$ $R_C = 955\ k\Omega$

43. $I = 6.67\ mA$

45. $I = 0.149\ A$

47. a. The bridge is not balanced.

    b. $(18\ \Omega)I_1 - (12\ \Omega)I_2 - (6\ \Omega)I_3 = 15\ V$
    $-(12\ \Omega)I_1 + (54\ \Omega)I_2 - (24\ \Omega)I_3 = 0$
    $-(6\ \Omega)I_1 - (24\ \Omega)I_2 + (36\ \Omega)I_3 = 0$

    c. $I = 38.5\ mA$

    d. $V_{R_5} = 0.923\ V$

49. $I_{R_5} = 0$ $I_{RS} = 60\ mA$ $I_{R_1} = I_{R_3} = 45\ mA$
    $I_{R_2} = I_{R_4} = 15\ mA$

51. $I_{R_1} = 0.495$ A $I_{R_2} = 1.384$ A $I_{R_3} = 1.879$ A
$I_{R_4} = 1.017$ A $I_{R_5} = 0.367$ A

53. $I_{R_1} = 6.67$ mA $I_{R_2} = 0$ $I_{R_3} = 6.67$ mA
$I_{R_4} = 6.67$ mA $I_{R_5} = 6.67$ mA $I_{R_6} = 13.33$ mA

CHAPTER 9

1. $I_{R_1} = 75$ mA (up) $I_{R_2} = 75$ mA (to the right)
$I_{R_3} = 87.5$ mA (down) $I_{R_4} = 12.5$ mA (to the right)

3. $V_a = -3.11$ V $I_1 = 0.1889$ A

5. $E = 30$ V $I_L(1) = 2.18$ mA $I_L(2) = 2.82$ mA

7. $R_{Th} = 20$ Ω $E_{Th} = 10$ V $V_{ab} = 6.0$ V

9. $R_{Th} = 2.02$ kΩ $E_{Th} = 1.20$ V $V_{ab} = -0.511$ V

11. a. $R_{Th} = 16$ Ω $E_{Th} = 5.6$ V
b. When $R_L = 20$ Ω: $V_{ab} = 3.11$ V
When $R_L = 50$ Ω: $V_{ab} = 4.24$ V

13. a. $E_{Th} = 75$ V $R_{Th} = 50$ Ω
b. $I = 0.75$ A

15. a. $E_{Th} = 50$ V $R_{Th} = 3.8$ kΩ
b. $I = 13.21$ mA

17. a. $R_{Th} = 60$ kΩ $E_{Th} = 25$ V
b. $R_L = 0$: $I = -0.417$ mA
$R_L = 10$ kΩ: $I = -0.357$ mA
$R_L = 50$ kΩ: $I = -0.227$ mA

19. a. $E_{Th} = 28.8$ V, $R_{Th} = 16$ kΩ
b. $R_L = 0$: $I = 1.800$ mA
$R_L = 10$ kΩ: $I = 1.108$ mA
$R_L = 50$ kΩ: $I = 0.436$ mA

21. $E_{Th} = 4.56$ V $R_{Th} = 7.2$ Ω

23. a. $E_{Th} = 8$ V $R_{Th} = 200$ Ω
b. $I = 22.2$ mA (upward)

25. $I_N = 0.5$ A, $R_N = 20$ Ω, $I_L = 0.2$ A

27. $I_N = 0.594$ mA, $R_N = 2.02$ kΩ, $I_L = 0.341$ mA

29. a. $I_N = 0.35$ A, $R_N = 16$ Ω
b. $R_L = 20$ Ω: $I_L = 0.156$ A
$R_L = 50$ Ω: $I_L = 0.085$ A

31. a. $I_N = 1.50$ A, $R_N = 50$ Ω
b. $I_N = 1.50$ A, $R_N = 50$ Ω

33. a. $I_N = 0.417$ mA, $R_N = 60$ kΩ
b. $I_N = 0.417$ mA, $R_N = 60$ kΩ

35. a. $I_N = 0.633$ A, $R_N = 7.2$ Ω
b. $I_N = 0.633$ A, $R_N = 7.2$ Ω

37. a. 60 kΩ   b. 2.60 mW

39. a. 31.58 Ω   b. 7.81 mW

41. a. $R_1 = 0$ Ω
b. 19.5 mW

43. $E = 1.5625$ V

45. $I = 0.054$ A, $P_L = 0.073$ W

47. $I = 0.284$ mA, $P_L = 0.807$ W

49. a. $I = 0.24$ A   b. $I = 0.24$ A
c. Reciprocity does apply.

51. a. $V = 22.5$ V   b. Reciprocity does apply.

53. $E_{Th} = 10$ V, $R_{Th} = 20$ Ω
$I_N = 0.5$ A, $R_N = 20$ Ω

55. $R_L = 2.02$ kΩ for maximum power.

CHAPTER 10

1. a. 800 μC   b. 2 μF   c. 100 μC
d. 30 V   e. 150 V   f. 1.5 μF

3. 200 V

5. 420 μC

7. 73 pF

9. 5.65×10^{-4} m^2

11. 117 V

13. a. 2.25×10^{12} N/C   b. 0.562×10^{12} N/C

15. 4.5 kV

17. 3.33 kV

19. a. points   b. spheres   c. points

21. 24.8 μF

23. 77 μF

25. 3.86 μF

27. a. 9.6 μF   b. 13 μF
c. 3.6 μF   d. 0.5 μF

29. 9 μF

31. 60 μF;   30 μF

33. 81.2 μF;   1.61 μF

35. The 10-μF capacitor is in parallel with the series combination of the 1-μF and 1.5-μF capacitors.

37. a. $V_1 = 60$ V; $V_2 = V_3 = 40$ V
b. $V_1 = 50$ V; $V_2 = V_3 = 25$ V; $V_4 = 25$ V;
$V_5 = 8.3$ V; $V_6 = 16.7$ V

39. 14.4 V;   36 V;   9.6 V

41. 800 μF

43. −50 mA from 0 to 1 ms;   50 mA from 1 ms to 4 ms;
0 mA from 4 ms to 6 ms;   50 mA from 6 ms to 7 ms;
−75 mA from 7 ms to 9 ms.

45. $-23.5 \, e^{-0.05t}$ μA

47. 0 mJ, 0.25 mJ, 1.0 mJ, 1.0 mJ, 2.25 mJ, 0 mJ

CHAPTER 11

1. a. 0 V; 5 A   b. 20 V; 0 A

3. a. Short circuit   b. Voltage source   c. Open circuit
d. $i(0^-) =$ current just before $t = 0$ s; $i(0^+) =$ current just after $t = 0$ s

5. 15.1 V

7. a. $45(1 - e^{-80t})$ V   b. $90e^{-80t}$ mA

c.
t (ms)	v_C (V)	i_C (mA)
0	0	90
20	35.9	18.2
40	43.2	3.67
60	44.6	0.741
80	44.93	0.150
100	44.98	0.030

9. $40(1 - e^{-t/39 \text{ ms}})$ V $10.3e^{-t/39 \text{ ms}}$ mA   28.9 V   2.86 mA

11. 40 μs;   200 μs

13. v_C: 0, 12.6, 17.3, 19.0, 19.6, 19.9 (all V)

i_C: 5, 1.84, 0.675, 0.249, 0.092, 0.034 (all A)

15. 25 kΩ;   8 μF

17. 45 V;   4.5 kΩ;   0.222 μF

19. 2.5 A

21. a. $20 + 10e^{-25\,000t}$ V   b. $-2.5e^{-25\,000t}$ A

c. v_C starts at 30 V and decays exponentially to 20 V in 200 μs. i_C is 0 A at $t = 0^-$, -2.5 A at $t = 0^+$, and decays exponentially to zero in 200 μs.

23. a. $50e^{-2t}$ V   b. $-2e^{-2t}$ mA   c. 0.5 s

d. v_C: 50 V, 18.4 V, 6.77 V, 2.49 V, 0.916 V, 0.337 V

i_C: -2 mA, -0.736 mA, -0.271 mA, -0.0996 mA, -0.0366 mA, -0.0135 mA

25. 14.4 V

27. a. 200 V; -12.5 mA   b. 8 ms

c. $200e^{-125t}$ V, $-12.5e^{-125t}$ mA

29. $45(1 - e^{-t/0.1857})$ V,   28.4 V (same)

31. a. $60(1 - e^{-500t})$ V   b. $1.5e^{-500t}$ A

33. 90 V;   15 kΩ;   100 μF

35. $V_{C1} = 65$ V; $V_{C2} = 10$ V; $V_{C3} = 55$ V; $I_T = 0.5$ A

37. 14.0 μF

39. a. 5 μs   b. 40%   c. 200 000 pulses/s

41. 0.8 μs;   0.8 μs;   4 μs

43. 6.6 ns

45. -17.8 V (theoretical)

47. a. 51.9 V   b. 203 mA

49. Verification point: At $t = 20$ ms, -17.8 V and -0.179 A

51. 29.3 V, 0.227 mA

CHAPTER 12

1. a. A_1 b. 1.4 T

3. 0.50 T

5. 1.23×10^{-3} Wb

7. 1 T;

9. 264 μWb;   738 μWb;   807 μWb

11. 1061 At/m

13. $N_1 I_1 = H_1 \ell_1 + H_2 \ell_2$; $N_2 I_2 = H_2 \ell_2 - H_3 \ell_3$

15. 0.47 A

17. 0.88 A

19. 0.58 A

21. 0.53 A

23. 0.86 A

25. 3.7 A

27. 4.4×10^{-4} Wb

29. 1.06×10^{-4} Wb

CHAPTER 13

1. 225 V

3. 6.0 V

5. 150 mV

7. 0.111 s

9. 79.0 μH

11. $L = \dfrac{N\Phi}{I} = \dfrac{N(B_g A_g)}{I} = \dfrac{N(\mu_0 H_g)A_g}{I}$

$= \dfrac{N\mu_0 \left(\dfrac{NI}{\ell_g}\right) A_g}{I} = \dfrac{\mu_0 N^2 A_g}{\ell_g}$

13. 4 H

15. a. 4 H   c. 5 A

17. 84 mH

19. 4.39 mH

21. a. 21 H   b. 2 H   c. 20 H

d. 4 H   e. 4 mH

23. 9 μH

25. Circuit (a): 6 H;   1.5 H

Circuit (b): 2 H;   8 H

27. 1.6 H, in series with 6 H‖4 H

29. 1.2 H in series with 8 H‖12 H

31. a. 1 H in series with 3 μF

b. 2 H in series with 10 μF

c. 10 Ω, 10 H, and 25 μF in series

d. 10 Ω in series with 40 Ω‖(50 H in series with 20 μF)

33. 0.32 J

35. The path containing L_1 and L_2 is open.

CHAPTER 14

1. a. open circuit

b. Circuit (a): 1.6 A

Circuit (b): 6 A;   60 V

Circuit (c): 0 A; E

Circuit (d): 2 A;   30 V

3. $v_{R_1} = 180$ V; $v_{R_2} = 120$ V; $v_{R_3} = 60$ V; $v_{R_4} = 32$ V

$v_{R_5} = 28$ V; $v_{R_6} = 0$ V; $i_T = 21$ A; $i_1 = 18$ A

$i_2 = 3$ A; $i_3 = 1$ A; $i_4 = i_5 = 2$A; $i_6 = 0$ A

5. a. 50 ms   b. 250 ms   c. $3(1 - e^{-20t})$ A; $180e^{-20t}$ V

d.

t	i_L (A)	v_L (V)
0	0	180
τ	1.90	66.2
2τ	2.59	24.4
3τ	2.85	8.96
4τ	2.95	3.30
5τ	2.98	1.21

7. a. 0.2 s   b. 1 s   c. $20 \, e^{-5t}$ V; $(1 - e^{-5t})$ A

d. v_L: 20, 7.36, 2.71, 0.996, 0.366, 0.135 (all V)

i_L: 0, 0.632, 0.865, 0.950, 0.982, 0.993 (all A)

9. $-182 (1 - e^{-393t})$ mA; $-40e^{-393t}$ V; -134 mA;
-10.5 V

11. 80 V; 20 Ω; 2 H

13. 40 V; 4 kΩ; 2 H

15. a. 2 ms   b. $5e^{-500t}$ A; $-1250e^{-500t}$ V

c.

t	i_L (A)	v_L (V)
0	5	-1250
τ	1.84	-460
2τ	0.677	-169
3τ	0.249	-62.2
4τ	0.092	-22.9
5τ	0.034	-8.42

17. -365 V

19. $R_1 = 20 \, \Omega$; $R_2 = 30 \, \Omega$

21. 5.19 A

23. a. 8.89 μs   b. $-203e^{-t/8.89\mu s}$ V; $5e^{-t/8.89\mu s}$ mA

c. -27.3 V; 0.675 mA

25. a. 10 ms   b. $90 (1 - e^{-t/10 \, ms})$ mA; $36e^{-t/10 \, ms}$ V

c. 2.96 V 82.6 mA

27. 103.3 mA; 3.69 V

29. 33.1 V

33. $i_L(25 \, ms) = 126$ mA; $i_L(50 \, ms) = 173$ mA

CHAPTER 15

1. AC voltage is voltage whose polarity cycles periodically between positive and negative. AC current is current whose direction cycles periodically.

3. a. The magnitude of a waveform (such as a voltage or current) at any instant of time.

b. 0, 10, 20, 20, 20, 0, -20, -20, 0 (all V)

5. 0 mA, 2.5 mA, 5 mA, 5 mA, 5 mA, 0 mA, -5 mA, -5 mA, 0 mA

7. a. 2 Hz   b. 10 Hz   c. 62.5 kHz

9. 7 ms; 142.9 Hz

11. 15 V; 6 mA

13. 149 919 cycles

15. 100 μs; 400 μs; 2500 Hz

17. 144.4 V

19. 100 V

21. a. 0.1745   b. 0.4363   c. 1.3963

d. 2.618   e. 6.1087   f. 10.821

23. 43.3 A; 25 A; -49.2 A; -50 A

25. $V_m = 85.1$ V. Waveform is like Figure 15–25 except $T = 50$ μs.

27. a. 62.83×10^6 rad/s b. 188.5 rad/s c. 157.1 rad/s

d. 314.2 rad/s   e. 1571 rad/s

29. a. $v = 170 \sin 377t$ V b. $i = 40 \sin 628t$ μA

c. $v = 17 \sin 52.4 \times 10^3 t$ V

31. $v = 20 \sin 125.7t$ V

33. 0, 28.8, -11.4, -22 (all mA)

35. a. $5 \sin(1000 \, t + 36°)$ mA

b. $10 \sin(40\pi \, t + 120°)$ A

c. $4 \sin(1800\pi \, t - 45°)$ V

37. 4.46 V; -3.54 V; 0.782 V

39. $v = 100 \sin(3491t + 36°)$ V

41. 6.25 ms; 13.2 ms; 38.2 ms

43. a. 20°; i leads b. in phase

c. 50°; i_1 leads d. 60°; i leads

45. a. A leads by 90°   b. A leads by 150°

47. Zero for each

49. a. 1.1 A   b. -5 V   c. 1.36 A

51. a. Similar to Figure 15–61(b), except positive peak is 40 V, negative peak is -10 V, and $V_{avg} = 15$ V. $T = 120$ ms.

b. 40 V; -10 V

c. 27.5 V; 36.7 V; 2.5 V; -6.65 V

d. 15 V

53. 2.80 V

55. a. 12 V   b. 17.0 mA

c. 19.7 V   d. 48.9 V

57. a. 5 mA   b. 23.2 V

59. a. 8.94 A   b. 16.8 A

61. 24 V; Its magnitude is always 24 V; therefore, it produces the same average power to a resistor as a 24 V battery.

63. 26.5 mA

65. a. 108 V   b. 120 V

c. 14.6 V   d. 422 V

67. 6.25 ms

CHAPTER 16

1. a. $13\angle67.4°$ b. $10.8\angle-33.7°$

c. $17\angle118.1°$ d. $10.8\angle-158.2°$

5. a. $7 + j6$ b. $1 + j10$

c. $15.4 - j3.50$ d. $-4.64 + j1.86$

e. $-7.2 - j12.1$ f. $0.2 - j0.1$

7. a. $14.2\angle-23.8°$ b. $1.35\angle-69.5°$ c. $5.31\angle167.7°$

9. a. $10 \sin(\omega t + 30°)$ V b. $15 \sin(\omega t - 10°)$ V

11. a. $10 \, V\angle30°$, $15 \, V\angle-20°$ b. $11.5 \, V\angle118.2°$

c. $11.5 \sin(\omega t + 118.2°)$ V

13. a. 17.7 mA ∠36°; 28.3 mA ∠80°; 42.8 mA ∠63.3°

  b. 60.5 sin(ωt + 63.3°) mA

15. a. 4.95 mA ∠0°; 2.83 mA ∠−90°; 4.2 mA ∠90°; 5.15 mA ∠15.9°

  b. 7.28 sin(ωt + 15.9°) mA

17. a. 10 sin ωt A   b. 10 sin(ωt + 27°) A

  c. 204 sin(ωt − 56°) mV   d. 204 sin(ωt − 157°) μV

19. a. 147 sin(ωt + 39°) V; 183.8 sin(ωt + 39°) V

  b. 330.8 sin(ωt + 39°) V; Identical

21. a. 5 sin(ωt − 30°) A

  b. 5 sin(ωt − 105°) A

  c. 10 sin(ωt + 120°) V

  d. 10 sin(ωt + 100°) V

23. a. 0.531 sin(377t − 90°) A

  b. 31.8 sin(6283t − 90°) mA

  c. 0.4 sin(500t − 90°) A

25. a. 8.74 kHz;   b. 50.9 mH;   130°

27. a. 530.5 Ω   b. 31.83 Ω   c. 400 Ω

29. a. 1.89 sin(377t + 90°) A

  b. 79.6 sin(2π × 400t − 150°) mV

31. a. 48 Ω∠0°   b. j37.7 Ω   c. −j50 Ω

33. a. V_R = 25 V∠0°; V_L = 12.5 V∠90°; V_C = 5 V∠−90°

  b. v_R = 35.4 sinωt V; v_L = 17.7 sin(ωt + 90°) V; v_C = 7.07 sin(ωt − 90°) V

35. a. 39.8 Hz   b. 6.37 μF

37. Theoretical: 45.2 mA

39. 1.59 A∠−90°

41. 7.96 A

CHAPTER 17

1. When p is +, power flows from source to load. When p is −, power flows out of load. Out of the load.

3. a. 1000 W and 0 VAR; 500 W and 0 VAR

  b. 1500 W and 0 VAR

5. 151 VAR (cap.)

7. a. 10 Ω   b. 6 Ω   c. 265 μF

9. 30 Ω

11. 160 W; 400 VAR (ind.)

13. 900 W; 300 VAR (ind.)

15. a. 20 Ω   b. 6 Ω

  c. 8 Ω   d. 1.2 H

17. 125 VA

19. 1150 W; 70 VAR (cap.); 1152 VA; θ = −3.48°

21. 2.36 Ω

23. 120 Ω

25. a. 721 W; 82.3 VAR (cap.); 726 VA

  b. I = 6.05 A; No

27. a. Across the load b. 73.9 μF

29. 57.3 kW

31. 2598 W

33. 160 Ω

35. Same as Figure 17–5 with peak value of p = 3.14 W

37. A sawtooth wave oscillating between 8 W and −8 W

CHAPTER 18

1. a. 0.125 sinωt

3. a. $1.87 × 10^{-3}$ sin(ωt + 30°)

5. a. 1.36 sin(ωt − 90°)

7. a. 1333 sin(2000πt + 30°)

9. a. 62.5 sin(10000t − 90°)

11. a. 67.5 sin(20000t − 160°)

13. Network (a): 31.6 Ω∠18.43°

  Network (b): 8.29 kΩ∠−29.66°

15. a. 42.0 Ω∠19.47° = 39.6 Ω + j14.0 Ω

17. R = 1.93 kΩ, L = 4.58 mH

19. R = 15 Ω, C = 1.93 μF

21. a. Z_T = 50 Ω∠−36.87°, I = 2.4 A∠36.87°, V_R = 96 V∠36.87°, V_L = 48 V∠126.87°, V_C = 120 V∠−53.13°

  c. 230.4 W

  d. 230.4 W

23. a. Z_T = 45 Ω∠−36.87°

  b. i = 0.533 sin(ωt + 36.87°), v_R = 19.20 sin(ωt + 36.87°), v_C = 25.1 sin(ωt − 53.13°), v_L = 10.7 sin(ωt + 126.87°)

  e. 5.12 W

  f. 5.12 W

25. a. V_R = 9.49 V∠−18.43°, V_L = 11.07 V∠71.57°, V_C = 7.91 V∠−108.43°

  b. ΣV = 10.00 V∠0°

27. a. V_C = 317 V∠−30°, V_L = 99.8 V∠150°

  b. 25 Ω

29. a. V_C = 6.0 V∠−110°

  b. V_Z = 13.87 V∠59.92°

  c. 69.4 Ω∠79.92°

  d. 1.286 W

31. Network (a): 199.9 Ω∠−1.99°,

  Network (b): 485 Ω∠−14.04°

33. a. Z_T = 3.92 kΩ∠−78.79°, I_T = 2.55 mA∠78.69°, I_1 = 0.5 mA∠0°, I_2 = 10.0 mA∠−90°, I_3 = 12.5 mA∠90°

  d. 5.00 mW

35. a. 5.92 kΩ∠17.4°

  b. 177.6 V∠17.4°

37. 2.55 Ω∠81.80°

39. Network (a): I_R = 10.00 mA∠−31.99°, I_L = 4.00 mA∠−121.99°, I_C = 4.35 mA∠58.01°

  Network (b): I_R = 9.70 mA∠−44.04°, I_{C1} = 1.62 mA∠45.96°, I_{C2} = 0.81 mA∠45.96°

41. $I_L = 2.83$ mA$\angle-135°$, $I_C = 3.54$ mA$\angle45°$,
    $I_R = 0.71$ mA$-45°$, $\Sigma I_{out} = \Sigma I_{in} = 1.00$ mA$\angle0°$

43. a. 6.245 A$\angle90°$ b. 40.0 Ω c. 8.00 A$\angle51.32°$

45. a. $Z_T = 22.5$ $\Omega\angle-57.72°$
    $I_L = 5.34$ A$\angle57.72°$
    $I_C = 4.78$ A$\angle84.29°$
    $I_R = 2.39$ A$\angle-5.71°$
    c. $P_R = 342$ W
    d. $P_T = 342$ W

47. a. $Z_T = 10.53$ $\Omega\angle10.95°$, $I_T = 1.90$ A$\angle-10.95°$,
    $I_1 = 2.28$ A$\angle-67.26°$, $I_2 = 2.00$ A$\angle60.61°$
    b. $V_{ab} = 8.87$ V$\angle169.06°$

49. a. $Z_T = 7.5$ k$\Omega\angle0°$, $I_1 = 0.75$ mA$\angle0°$,
    $I_2 = 0.75$ mA$\angle90°$, $I_3 = 0.79$A$\angle-71.57°$
    b. $V_{ab} = 7.12$ V$\angle18.43°$

51. $\omega_C = 2000$ rad/s

53. $f_C = 3.39$ Hz

55. Network (a): 5.5-kΩ resistor in series with a 9.0-kΩ inductive reactance

    Network (b): 207.7-Ω resistor in series with a 138.5-Ω inductive reactance

57. $\omega = 1$ krad/s: $Y_T = 0.01$ S $+ j0$, $Z_T = 100$ Ω
    $\omega = 10$ krad/s: $Y_T = 0.01$ S $+ j0$, $Z_T = 100$ Ω

CHAPTER 19

1. a. 5.00 V$\angle180°$ b. 12.50 V$\angle0°$ c. 15.00 V$\angle-120°$

3. a. 3.20 mV$\angle180°$ b. 8.00 mV$\angle0°$ c. 9.60 mV$\angle-120°$

5. 7.80 V$\angle-150°$

7. Circuit (a): $E = 54$ V$\angle0°$, $V_L = 13.5$ V$\angle0°$
   Circuit (b): $E = 450$ mV$\angle-60°$, $V_L = 439$ mV$\angle-47.32°$

9. a. 4.69 V$\angle180°$ b. $E = (7.5$ M$\Omega)I$, $V = 4.69$ V$\angle180°$

11. a. $(4\ \Omega + j2\ \Omega)I_1 - (4\ \Omega)I_2 = 20$ V$\angle0°$
    $-(4\ \Omega)I_1 + (6\ \Omega + j4\ \Omega)I_2 = 48.4$ V$\angle-161.93°$
    b. $I_1 = 2.39$ A$\angle72.63°$, $I_2 = 6.04$ A$\angle154.06°$
    c. $I = 6.15$ A$\angle-3.33°$

13. a. $(12\ \Omega - j16\ \Omega)I_1 + (j15\ \Omega)I_2 = 13.23$ V$\angle-79.11°$
    $(j15\ \Omega)I_1 + 0I_2 = 10.27$ V$\angle-43.06°$
    b. $I_1 = 0.684$ A$\angle-133.06°$, $I_2 = 1.443$ A$\angle-131.93°$
    c. $V = 11.39$ V$\angle-40.91°$

15. 27.8 V$\angle6.79°$ $I = 6.95$ mA$\angle6.79°$

17. a. $(0.417$ S$\angle36.87°)V_1 - (0.25$ S$\angle90°)V_2 = 3.61$ A$\angle-56.31°$
    $-(0.25$ S$\angle90°)V_1 + (0.083$ S$\angle90°)V_2 = 7.00$ A$\angle90°$
    b. $V_1 = 30.1$ V$\angle139.97°$, $V_2 = 60.0$ V$\angle75.75°$
    c. $I = 13.5$ V$\angle-44.31°$

19. a. $(0.0893$ S$\angle22.08°)V_1 + (0.04$ S$\angle90°)V_2 = 0.570$ A$\angle93.86°$
    $+(0.04$ S$\angle90°)V_1 + (0.06$ S$\angle90°)V_2 = 2.00$ A$\angle180°$
    b. $V_1 = 17.03$ V$\angle18.95°$, $V_2 = 31.5$ V$\angle109.91°$
    c. $V = 11.39$ V$\angle-40.91°$

21. $(0.372$ μS$\angle-5.40°)$, $V = 10.33$ mA$\angle1.39°$
    27.8 V$\angle6.79°$, $I = 6.95$ mA$\angle6.79°$

    As expected, the answers are the same as those in Problem 15.

23. Network (a):
    $Z_1 = 284.4$ $\Omega\angle-20.56°$, $Z_2 = 94.8$ $\Omega\angle69.44°$
    $Z_3 = 31.6$ $\Omega\angle159.44°$
    Network (b):
    $Z_1 = 11.84$ k$\Omega\angle9.46°$, $Z_2 = 5.92$ k$\Omega\angle-80.54°$
    $Z_3 = 2.96$ k$\Omega\angle-80.54°$

25. $I_T = 0.337$ A$\angle-2.82°$

27. a. $Z_T = 3.03$ $\Omega\angle-76.02°$
    b. $I = 5.28$ A$\angle76.02°$, $I_1 = 0.887$ A$\angle-15.42°$

29. a. $Z_2 = 1$ $\Omega - j7$ $\Omega = 7.07$ $\Omega\angle-81.87°$
    b. $I = 142.5$ mA$\angle52.13°$

31. $Z_1Z_4 = Z_2Z_3$ as required.

35. $R_3 = 50.01$ Ω, $R_1 = 253.3$ Ω

39. Same as Figure 19–21

41. Same as Problem 26

CHAPTER 20

1. $I = 4.12$ A$\angle50.91°$

3. 16 V$\angle-53.13°$

5. a. $V = 15.77$ V$\angle36.52°$
   b. $P_{(1)} + P_{(2)} = 1.826$ W $\neq P_{100\text{-}\Omega} = 2.49$ W

7. 0.436 A$\angle-9.27°$

9. $19.0\sin(\omega t + 68.96°)$

11. a. $V_L = 1.26$ V$\angle161.57°$ b. $V_L = 6.32$ V$\angle161.57°$

13. 0.361 mA$\angle-3.18°$

15. $V_L = 9.88$ V$\angle0°$

17. 1.78 V

19. $Z_{Th} = 3$ $\Omega\angle-90°$, $E_{Th} = 20$ V$\angle-90°$

21. a. $Z_{Th} = 37.2$ $\Omega\angle57.99°$ $E_{Th} = 9.63$ V$\angle78.49°$
    b. 0.447 W

23. $Z_{Th} = 22.3$ $\Omega\angle-15.80°$, $E_{Th} = 20.9$ V$\angle20.69°$

25. $Z_{Th} = 109.9$ $\Omega\angle-28.44°$, $E_{Th} = 14.5$ V$\angle-91.61°$

27. a. $Z_{Th} = 20.6$ $\Omega\angle34.94°$, $E_{Th} = 10.99$ V$\angle13.36°$
    b. $P_L = 1.61$ W

29. $Z_N = -j3$ Ω, $I_N = 6.67$ A$\angle0°$

31. a. $Z_N = 22.3$ $\Omega\angle-15.80°$, $I_N = 0.935$ A$\angle36.49°$
    b. 0.436 A$\angle-9.27°$
    c. 3.80 W

33. a. $Z_N = 109.9$ $\Omega\angle-28.44°$, $I_N = 0.131$ A$\angle-63.17°$
    b. 0.0362 A$\angle-84.09°$
    c. 0.394 W

35. a. $Z_N = 14.1$ $\Omega\angle85.41°$, $I_N = 0.181$ A$\angle29.91°$
    b. 0.0747 A$\angle90.99°$

37. a. $\mathbf{Z}_{Th} = 17.9 \ \Omega\angle-26.56°$, $\mathbf{E}_{Th} = 1.79 \ V\angle153.43°$
    b. $0.0316 \ \mu A\angle161.56°$
    c. $40.0 \ \mu W$
39. $\mathbf{E}_{Th} = 10 \ V\angle0°$, $\mathbf{I}_N = 10.5 \ A\angle0°$, $\mathbf{Z}_{Th} = 0.952 \ \Omega\angle0°$
41. a. $\mathbf{Z}_L = 8 \ \Omega\angle22.62°$ b. 40.2 W
43. a. $\mathbf{Z}_L = 2.47 \ \Omega\angle21.98°$ b. 1.04 W
45. $4.15 \ \Omega\angle85.24°$
47. a. $\mathbf{Z}_L = 37.2 \ \Omega\angle-57.99°$ b. $19.74 \ \Omega$
    c. 1.18 W
49. $\mathbf{Z}_{Th} = 3 \ \Omega\angle-90°$, $\mathbf{E}_{Th} = 20 \ \Omega\angle-90°$
51. $\mathbf{Z}_{Th} = 109.9 \ \Omega\angle-28.44°$, $\mathbf{E}_{Th} = 14.5 \ V \ \angle-91.61°$
53. $\mathbf{E}_{Th} = 10 \ V\angle0°$, $\mathbf{I}_N = 10.5 \ A\angle0°$, $\mathbf{Z}_{Th} = 0.952 \ \Omega\angle0°$
55. $\mathbf{Z}_N = 0.5 \ k\Omega\angle0°$, $\mathbf{I}_N = 4.0 \ mA\angle0°$
57. $\mathbf{E}_{Th} = 10 \ V\angle0°$, $\mathbf{I}_N = 10.5 \ A\angle0°$, $\mathbf{Z}_N = 0.952 \ \Omega\angle0°$
59. $\mathbf{Z}_N = 0.5 \ k\Omega\angle0°$, $\mathbf{I}_N = 4.0 \ mA\angle0°$

CHAPTER 21

1. a. $\omega_s = 3835$ rad/s $f_s = 610.3$ Hz
   b. $\mathbf{I} = 153.8 \ mA\angle0°$
   c. $\mathbf{V}_C = 59.0 \ V\angle-90°$ $\mathbf{V}_L = 59.03 \ V\angle87.76°$
      $\mathbf{V}_R = 7.69 \ V\angle0°$
   d. $P_L = 0.355$ W
3. a. $R = 25.0 \ \Omega$ $C = 4.05$ nF
   b. $P = 15.6$ mW
   c. $X_C = 1.57 \ k\Omega$ $\mathbf{V}_C = 39.3 \ V\angle-90°$
      $\mathbf{V}_L = 39.3 \ V\angle90°$
      $\mathbf{V}_R = \mathbf{E} = 0.625 \ V\angle-90°$
   d. $v_C = 55.5 \sin(50{,}000\pi t - 90°)$
      $v_L = 55.5 \sin(50{,}000\pi t + 90°)$
      $v_R = 0.884 \sin(50{,}000\pi t)$
5. a. $\omega_s = 500$ rad/s $f_s = 79.6$ kHz
   b. $\mathbf{Z}_T = 200 \ \Omega\angle0°$
   c. $\mathbf{I} = 10 \ mA\angle0°$
   d. $\mathbf{V}_R = 1 \ V\angle0°$ $\mathbf{V}_L = 50.01 \ V\angle88.85°$
      $\mathbf{V}_C = 50 \ V \ \angle-90°$
   e. $P_T = 20$ mW $Q_C = 0.5$ VAR (cap.)
      $Q_L = 0.5$ VAR (ind.)
   f. $Q_s = 25$
7. a. $C = 0.08 \ \mu F$ $R = 6.4 \ \Omega$
   b. $P_T = 0.625$ W
   c. $\mathbf{V}_L = 62.5 \ V \ \angle90°$
9. $\omega_s = 8000$ rad/s
11. a. $\omega_s = 3727$ rad/s, $Q = 7.45$, BW = 500 rad/s
    b. $P_{max} = 144$ W
    c. $\omega_1 \approx 3477$ rad/s $\omega_2 \approx 3977$ rad/s
    d. $\omega_1 = 3485.16$ rad/s $\omega_2 = 3985.16$ rad/s

e. The results are close, although the approximation will yield some error if used in further calculations. The error would be less if Q were larger.

13. a. $R = 1005 \ \Omega$, $C = 63.325$ pF
    b. $P = 0.625$ W, $\mathbf{V}_L = 312.5 \ V\angle90°$
    c. $v_{out} = 442 \sin(400\pi \times 10^3 t + 90°)$
15. Network (a):
    a. $Q = 24$
    b. $R_P = 5770 \ \Omega$, $X_{LP} = 240 \ \Omega$
    c. $Q = 240$, $R_P = 576 \ \Omega$
       $X_{LP} = 2400 \ \Omega$
    Network (b):
    a. $Q = 1$
    b. $R_P = 200 \ \Omega$, $X_{LP} = 200 \ \Omega$
    c. $Q = 10$, $R_P = 10.1 \ k\Omega$
       $X_{LP} = 1.01 \ k\Omega$
    Network (c):
    a. $Q = 12.5$
    b. $R_P = 314.5 \ \Omega$, $X_{LP} = 25.16 \ \Omega$
    c. $Q = 125$, $R_P = 31.25 \ \Omega$, $X_{LP} = 250 \ \Omega$
17. Network (a):
    $Q = 15$, $R_P = 4500 \ \Omega$, $X_{CP} = 300 \ \Omega$
    Network (b):
    $Q = 2$, $R_P = 225 \ \Omega$, $X_{CP} = 112.5 \ \Omega$
    Network (c):
    $Q = 5$, $R_P = 13 \ k\Omega$, $X_{CP} = 2600 \ \Omega$
19. Network (a):
    $Q = 4$, $R_s = 4 \ k\Omega$, $X_{LS} = 16 \ k\Omega$
    Network (b):
    $Q = 0.333$, $R_s = 3.6 \ k\Omega$, $X_{CS} = 1.2 \ k\Omega$
    Network (c):
    $Q = 100$, $R_s = 10 \ \Omega$, $X_{LS} = 1 \ k\Omega$
21. $L_s = 1.2$ mH, $L_P = 2.4$ mH
23. a. $\omega_P = 20$ krad/s
25. a. $\omega_P = 39.6$ krad/s, $f_P = 6310$ Hz
    b. $Q = 6.622$
    c. $\mathbf{V} = 668.2 \ V\angle0°$, $\mathbf{I}_R = 11.14 \ mA\angle0°$
       $\mathbf{I}_L = 668.2 \ mA\angle-82.36°$
       $\mathbf{I}_C = 662.2 \ mA\angle90°$
    d. $P_T = 66.82$ W
    e. BW = 5.98 krad/s, BW = 952 Hz
27. $R_1 = 4194 \ k\Omega$, $C = 405$ pF, $\mathbf{I}_L = 5.0 \ mA\angle-89.27°$
29. a. $900 \ \Omega$ b. 1.862   c. 1074 rad/s
    d. $C = 500$ nF, $L = 0.450$ H   e. $93.1 \ V\angle0°$
33. $V = 668.2$ V, $f_P = 6310$ Hz, BW = 952 Hz, $Q = 6.622$
35. $V = 93.1$ V, $f_P = 318.3$ Hz, BW = 170.9 Hz, $Q = 1.86$

CHAPTER 22

1. a. 2000 (33.0 dB)   b. 200,000 (53.0 dB)

   c. 2×10^6 (63.0 dB)   d. 400 (26.0 dB)

3. a. $A_P = 50 \times 10^6$ (77.0 dB), $A_V = 500$ (54.0 dB)

   b. $A_P = 10$ (10.0 dB), $A_V = 0.224$ (-13.0 dB)

   c. $A_P = 13.3 \times 10^6$ (71.3 dB), $A_V = 258$ (48.2 dB)

   d. $A_P = 320 \times 10^6$ (85.1dB), $A_V = 1265$ (62.0 dB)

5. $P_{in} = 12.5\mu W$, $P_{out} = 39.5$ mW

   $V_{out} = 3.14$ V, $A_V = 12.6$, $[A_V]_{dB} = 22.0$ dB

7. a. 17.0 dBm (-13.0 dBW)   b. 30.0 dBm (0 dBW)

   c. -34.0 dBm (-64.0 dBW)   d. -66.0 dBm (-96.0 dBW)

9. a. 0.224 W   b. 30.2 μW

   c. 5.01 pW   d. 1995 W

17. $P_1 = 5.05$ dBm, $P_2 = 2.05$ dBm, $P_3 = 14.09$ dBm
    $P_0 = 25.6$ dBm

19. $P_1 = 25.5$ dBm, $P_{in} = -14.5$ dBm, $V_{out} = 52.8$ V

23. a. $\omega_C = 1000$ rad/s, $f_C = 159.2$ Hz

25. a. $\omega_1 = 50$ rad/s, $f_C = 1000$ rad/s

35. a. $TF = \dfrac{1 + j0.00003}{1 + 0.00006}$

43. a. Low-pass filter: $\omega_C = 500$ rad/s

   High-pass filter: $\omega_C = 25$ krad/s

   BW = 475 krad/s

   c. The actual cutoff frequencies will be close to the designed values since the break frequencies are separated by more than one decade.

45. a. $R_1 = 5$ kΩ, $R_2 = 50$ kΩ

   c. The actual frequencies will not occur at the designed values since they are 1 decade apart.

47. a. 10 krad/s   b. 10

   c. BW = 1 krad/s, $\omega_1 = 9.5$ krad/s, $\omega_2 = 10.5$ krad/s

   d. At resonance, $[A_V]_{dB} = -28.0$ dB

49. a. 10 krad/s, 1592 Hz   b. 100   c. -60 dB

   d. -0.8 dB, 0 dB   e. 100 rad/s, 9.95 krad/s, 10.05 krad/s

CHAPTER 23

1. a. e_s is in phase with e_p.

   b. e_p is $180°$ out of phase with e_s.

3. a. Step-up   b. 25 sinωt V

   c. 6 V   d. 96 V$\angle 0°$

   e. 3200 V$\angle 180°$

5. $v_1 = 24$ sinωt V; $v_2 = 144$ sin($\omega t + 180°$) V;

   $v_3 = 48$ sinωt V

7. a. 1 A$\angle 20°$ b. 480 V$\angle 0°$ c. 480 $\Omega \angle -20°$

9. a. 160 V$\angle -23.1°$ b. 640 V$\angle -23.1°$

11. a. 40 $\Omega - j80\ \Omega$ b. 1.25 $\Omega + j2\ \Omega$

13. 2.5

15. a. 22 $\Omega + j6\ \Omega$ b. 26 $\Omega + j3\ \Omega$

17. 108 kVA

19. a. 20 A   b. 22.5 A   c. 2.5 A   d. 0.708 A

21. 0.64 W

23. 3 A$\angle -50°$;   1.90 A$\angle -18.4°$;   1.83 A$\angle -43.8°$

25. b. 2.12 A$\angle -45°$;   21.2 A$\angle -45°$;   120.2 V$\angle 0°$

27. 98.5%

29. All are minus.

31. 0.889 H

33. $-125\ e^{-500t}$ V; $-4\ e^{-500t}$ V; -75.8 V; -2.43 V

35. 10.5 H

37. 27.69 mH;   11.5 A$\angle -90°$

39. $(4 + j22)\ \mathbf{I}_1 + j13\ \mathbf{I}_2 = 100\angle 0°$

   $j13\ \mathbf{I}_1 + j12\ \mathbf{I}_2 = 0$

41. $(10 + j84)\ \mathbf{I}_1 - j62\ \mathbf{I}_2 = 120$ V$\angle 0°$

   $-j62\ \mathbf{I}_1 + 15\ \mathbf{I}_2 = 0$

43. 0.644 A$\angle -56.1°$;   6.44 A$\angle -56.1°$;   117 V$\angle 0.385°$

CHAPTER 24

1. An electron in the outer (i.e., last occupied) shell of an atom.

3. Gold is a conductor. Thus, same as Figure 24–3(c).

5. See Figure 24–3. It is smaller than that of an insulator but larger than that of a conductor.

7. a. 6   b. 4

   c. Tetravalent (has 4 electrons in valence shell).

9. a. Resistance increases as temperature increases. Copper.

   b. Resistance decreases as temperature increases. Silicon.

13. a. Trivalent   b. Pentavalent

15. Holes

17. Electrons

19. As shown in Figure 24–8, antimony has 5 valence electrons, which is one more than is needed for bonding. If the electron escapes, it will not leave behind a hole because it isn't needed for bonding anyway.

21. 0.7 V

23. 0 V

25. The depletion region is formed by free electrons from the n-material diffusing across the junction, leaving atoms of the n-material near the junction with a deficiency of electrons and atoms of the p-material with an excess as indicated in Figure 24–14. When a forward-biased source is connected, electrons injected into the n-material are propelled toward the junction, replacing those originally lost; similarly, holes injected into the p-material are propelled toward the junction, replacing those originally lost. As the ions near the junction have their lost charges replaced, the depletion region narrows.

27. Majority

29.

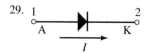

CHAPTER 25

1. a. 0 A: 27 V   b. 4 mA: 0 V

3. Connect the diode directly across the coil with its cathode at the source end. With the switch closed, the diode is reverse-biased and does not conduct. At the instant the switch is opened, the induced voltage across the coil changes polarity so that its bottom end is $+$ with respect to its top end $-$ thus, the diode is forward-biased and conducts. Now, instead of getting a huge voltage spike across the coil (recall Figure 14–16), the diode limits it to one diode drop.

5. a. 0 A: 27 V   b. 3.84 mA: 0.7 V

7. a. 0.244 A   b. 19.2 mA

9. 0.88 V: 22.48 V

11. 1N4004 to achieve a safety margin.

13. 703 mA

15. $13°$ C

17. $55°$ C

19. 43.2 mA: 777 mW: 30 mA

21. 23.6 V to 31.1 V

23. 16.0 V to 23.3 V

25. 14.77 V: 15.53 V

29. 1.51 A: 33.9 V

31. 31.6 V

33. 0.762 V

35. 0.758 V: 19.6 V

37. 2 V: 2083 μF

39. 1.98 V: 19.0 V 0.57 V

43. b. 1.51 V: 8.51 V

45. b. 0.723 V, 8.16 V

CHAPTER 26

1. $I_B = 0.05$ mA, $\alpha = 0.989$, $\beta = 90$

3. $I_B = 0.0417$ mA, $I_E = 5.04$ mA, $\alpha = 0.992$

5. $\alpha = 0.990$, $I_C = 4.89$ mA, $I_B = 0.0489$ mA

9. $V_{(BR)EBO} = 6.0$ V

    Emitter-base breakdown voltage (with open collector)

    $I_E = 10$ μA, $I_C = 0$

11. 110

13. 108

15. Saturation

17. b. $I_B = 43.0$ μA, $I_C = 4.5$ mA, $\beta = 105$

    c. $I_B = 21.5$ μA, $I_C = 2.5$ mA, $\beta = 116$

    d. $I_B = 0.215$ mA, $I_C = 8.0$ mA, β cannot be found. Transistor is in saturation.

    e. The transistor is in cutoff.

19. d. The slope becomes less negative. Also, the saturation current decreases in magnitude.

21. a. $I_{C(SAT)} = 8.0$ mA

    b. $I_B = 24.7$ μA, $I_C = 2.96$ mA, $V_{CE} = 10.08$ V

23. a. $I_{C(SAT)} = 6.15$ mA

    b. $I_B = 19.4$ μA, $I_C = 2.91$ mA, $V_{CE} = 12.64$ V

25. a. $I_{C(SAT)} = 4.49$ mA

    b. $I_B = 12.9$ μA, $I_C = 1.94$ mA, $V_{CE} = 12.5$ V

27. a. $I_{C(SAT)} = 10.0$ mA

    b. $I_B = 37.0$ μA, $I_C = 4.62$ mA, $I_E = I = 4.66$ mA, $V_{CE} = 10.68$ V

29. a. $I_{C(SAT)} = 5.00$ mA

    b. $I_B = 14.6$ μA, $I_C = 2.342$ mA, $I_E = 2.35$ mA, $I_{CE} = -10.60$ V

31. a. $I_{C(SAT)} = 7.96$ mA

    b. $I_B = 21.8$ μA, $I_C = 2.62$ mA $\approx I_E$, $V_{CE} = -10.60$ V

33. a. $I_{C(SAT)} = 7.96$ mA

    b. $I_B = 13.6$ μA, $I_C = 1.63$ mA $\approx I_E$, $I_{CE} = 12.7$ V

35. a. $I_{C(SAT)} = 7.96$ mA

    b. $I_C = 2.82$ mA $\approx I_E$, $V_{CE} = 10.3$ V

37. a. $I_{C(SAT)} = 6.12$ mA

    b. $I_C = 2.80$ mA $\approx I_E$, $I_{CE} = -16.3$ V

39. $R_C = 2$ kΩ, $R_E = 0.5$ kΩ, $R_2 = 5$ kΩ, $R_1 = 32$ kΩ

41. a. $I_{C(SAT)} = 7.5$ mA

    b. $I_C = 4.65$ mA $\approx I_E$, $V_{CE} = -11.4$ V

43. a. $I_{C(SAT)} = 14.9$ mA

    b. $I_C = 14.9$ mA (transistor is in saturation)

    c. $V_{out} \approx 0$ V

    d. $I_C = 0$ mA

    e. $V_{out} \approx 8$ V

45. Pin 2 is the base, Pin 3 is the emitter, and Pin 1 is the collector. The transistor is pnp.

47. Pin 1 is the base, Pin 3 is the emitter, and Pin 2 is the collector. The transistor is pnp.

49. a. p-channel   b. 1.28 mA

    c. 2.88 mA   d. 0 mA

51. $V_{GSQ} = -1.94$ V, $I_{DQ} = 2.15$ mA, $V_{DSQ} = 10.17$ V

53. $V_{GSQ} = -1.98$ V, $I_{DQ} = 2.03$ mA, $V_{DSQ} = 4.01$ V

55. a. $V_{GSQ} = -1.00$ V, $I_{DQ} = 6.40$ mA

    b. The MOSFET is in its depletion region.

    c. $V_{DSQ} = 7.41$ V

57. $k = 0.2$ mA/V^2, $I_D = (0.2$ mA/V$^2) (V_{GS} - 2.0$ V$)^2$

59. $V_{GSQ} = 10.4$ V, $I_{DQ} = 14.1$ mA, $V_{DSQ} = 10.4$ V

61. Meter 1 indicates there is no base current. This will be the result of the base-emitter being an open circuit. Meter 2 satisfies the same diagnosis. The absence of base current means there can be no collector current, as indicated by Meter 3. Consequently, the transistor is in cutoff, as indicated by Meter 4.

CHAPTER 27

1. a. 7.96 μF   b. 1.990 V, 2.000 V

3. 74.5 μF

5. a. $I_B = 17.4$ μA, $I_C = 2.60$ mA, $V_{CE} = -7.91$ V

    c. $A_v = -222$, $z_{in} = 1350$ Ω, $z_{out} = 4$ kΩ, $A_i = 74.9$

7. a. $I_B = 19.4$ μA, $I_C = 2.91$ mA, $V_{CE} = 10.4$ V

    b. $h_{fe} = 160$, $h_{ie} = 1600$ Ω

    d. $A_v = -330$, $z_{in} = 1600$ Ω, $z_{out} = 3.3$ kΩ, $A_i = 160$

9. a. $I_B = 19.4 \ \mu A$, $I_C = 2.91$ mA, $V_{CE} = 12.6$ V

    c. $A_v = 255$, $z_{in} = 1.30$ kΩ, $z_{out} = 3.9$ kΩ, $A_i = 64.9$

    d. $v_{in} = 6.84 \ mV_p$, $v_{out} = 1.744 \ V_p$

11. a. $I_B = 16.4 \ \mu A$, $I_C = 2.62$ mA, $V_{CE} = 10.7$ V

    b. $r_e = 9.91 \ \Omega$

    c. $A_v = -4.63$, $z_{in} = 143$ kΩ, $z_{out} = 4.7$ kΩ, $A_i = 141$

13. a. $I_B = 18.6 \ \mu A$, $I_C = 2.79$ mA, $V_{CE} = -3.23$ V

    c. $A_v = -11.8$, $z_{in} = 29.3$ kΩ, $z_{out} = 4.3$ kΩ, $A_i = 73.6$

    d. $v_{in} = 4.80 \ mV_p$, $v_{out} = 56.6 \ mV_p$

15. b. $A_v = -2.2$, $z_{in} = 134$ kΩ, $z_{out} = 4.3$ kΩ, $A_i = 63.3$

    c. $v_{in} = 4.96 \ mV_p$, $v_{out} = 11.0 \ mV_p$

17. a. $I_C = 1.93$ mA, $V_{CE} = 7.70$ V

    b. $r_e - 13.4 \ \Omega$

    d. $A_v = -4.8$, $z_{in} = 3990 \ \Omega$, $z_{out} = 3.9$ kΩ, $A_i = 4.91$

19. a. $I_C = 2.24$ mA, $V_{CE} = -9.50$ V

    b. $h_{ie} = 2.0$ kΩ, $h_{fe} = 150$

    c. $A_v = -6.79$, $z_{in} = 5.84$ kΩ, $z_{out} = 3.3$ kΩ, $A_i = 18.0$

21. a. $I_C = 2.3$ mA, $V_{CE} = -10.2$ V

    c. $A_v = -11.7$, $z_{in} = 6.87$ kΩ, $z_{out} = 5$ kΩ, $A_i = 16.1$

    d. $39.6 \ mV_p$

23. a. $I_C = 2.41$ mA, $V_{CE} = 10.35$ V

    c. $I_{C(AC-SAT)} = 7.59$ mA, $V_{CE(AC-OFF)} = 15.2$ V

    e. $9.65 \ V_{p-p}$, $4.83 \ mA_{p-p}$

25. a. $I_C = 2.12$ mA, $V_{CE} = -9.12$ V

    c. $I_{C(AC-SAT)} = 5.20$ mA, $V_{CE(AC-OFF)} = -15.4$ V

    e. $12.6 \ V_{p-p}$, $4.25 \ mA_{p-p}$

27. a. $I_{C(DC-SAT)} = 3.40$ mA, $V_{CE(DC-OFF)} = -18$ V

    b. $I_{C(AC-SAT)} = 4.12$ mA, $V_{CE(AC-OFF)} = -10.00$ V

    c. $6.46 \ V_{p-p}$, $2.66 \ mA_{p-p}$

    d. $0.548 \ V_{p-p}$

29. a. $I_C = 14.8$ mA, $V_{CE} = 9.03$ V

    c. $A_v = 0.996$, $z_{in} = 30.8$ kΩ, $z_{out} = 6.75 \ \Omega$, $A_i = 65.3$, $A_p = 65.1$

31. a. $I_C = 4.87$ mA, $V_{CE} = 8.7$ V

    c. $A_v = 0.991$, $z_{in} = 6.93$ kΩ, $z_{out} = 9.28$ V, $A_i = 6.87$, $A_p = 6.81$

33. a. $h_{ie} = 1600 \ \Omega$, $h_{re} = 1.1 \times 10^{-4}$, $h_{fe} = 150$, $h_{oe} = 20 \ \mu S$

    c. $A_v = 0.978$, $z_{in} = 34.8$ kΩ, $z_{out} = 14.1 \ \Omega$, $A_i = 72.5$, $A_p = 70.9$

35. a. $h_{ie} = 2200 \ \Omega$, $h_{re} = 1.1 \times 10^{-4}$, $h_{fe} = 140$, $h_{oe} = 12 \ \mu S$

    c. $A_v = 0.975$, $z_{in} = 6.90$ kΩ, $z_{out} = 19.6 \ \Omega$, $A_i = 6.73$, $A_p = 6.56$

37. a. 1.875 mS   b. 0.94 mS   c. 3.35 mS

39. a. 4.44 mS   b. 13.3 mS

    c. 10.3 mS   d. 14.6 mS

41. a. $I_{DQ} = 3.01$ mA, $V_{GSQ} = -2.26$ V, $V_{DSQ} = 13.1$ V

    b. $g_{m0} = 4.0$ mS, $g_m = 2.19$ mS

    d. $A_v = -1.51$, $z_{in} = 1$ MΩ, $z_{out} = 2.2$ kΩ

43. a. $I_{DQ} = 3.02$ mA, $V_{GSQ} = -1.54$ V, $V_{DSQ} = 8.39$ V

    b. $g_{m0} = 4.0$ mS, $g_m = 2.46$ mS

    d. $A_v = -4.06$, $z_{in} = 79.6$ kΩ, $z_{out} = 3.3$ kΩ

45. a. $I_{DQ} = 2.13$ mA, $V_{GSQ} = 1.21$ V, $V_{DSQ} = -10.10$ V

    b. $g_{m0} = 4.0$ mS, $g_m = 2.38$ mS

    d. $A_v = -4.23$, $z_{in} = 2$ MΩ, $z_{out} = 2.2$ kΩ

47. a. $I_{DQ} = 2.5$ mA, $V_{GSQ} = -2.5$ V, $V_{DSQ} = 17.5$ V

    b. $g_{m0} = 4.0$ mS, $g_m = 2.0$ mS

    d. $A_v = 0.667$, $z_{in} = 1$ MΩ, $z_{out} = 333 \ \Omega$

49. a. $I_{DQ} = 5.76$ mA, $V_{GSQ} = -4.23$ V, $V_{DSQ} = 4.72$ V

    b. $g_{m0} = 3.75$ mS, $g_m = 3.29$ mS

    d. $A_v = 0.730$, $z_{in} = 173$ kΩ, $z_{out} = 221 \ \Omega$

51. The emitter bypass capacitor is effectively an open circuit. This could be due to poor soldering or an internal fault.

53. a.   i. $I_C = 4.32$ mA, $V_{CE} = 7.16$ V

      ii. The voltage gain would remain unchanged.

    b.   i. The operating would not change.

      ii. -450

55. The capacitor may cause excessive noise.

    The capacitor may be damaged and possibly explode.

CHAPTER 28

1. Very high open-loop gain

   Very high input impedance.

   Very low output impedance.

3. $I_{C1} = 2.65$ mA, $V_{C1} = 9.17$ V

5. 86.0 dB

7. $0.5 \ V_{p-p}$

9. $A_v = -5.0$, $z_{in} = 20$ kΩ, $z_{out} = 2.25 \ m\Omega$

11. $R_1 = 12$ kΩ, $R_F = 60$ kΩ

13. $A_v = 6.0$, $z_{in} = 66.7$ GΩ, $z_{out} = 2.25 \ m\Omega$, $v_{out} = -12.0$ V

15. $R_1 = 12$ kΩ, $R_F = 40$ kΩ

17. $A_v = 1$, $z_{in} = 400$ GΩ, $z_{out} = 0.375 \ m\Omega$

19. 4 mV

21. 50 nA, 20 nA

23. 23.9 nA

25. 200 kHz

27. 1 MHz

29. 79.6 kHz

31. 16.67 kΩ

33. R_F might have failed.

    The power supply might have failed.

CHAPTER 29

1. $v_{out} = -15$ V, $0 < t < 6$ ms

   $v_{out} = +15$ V, 6 ms $< t < 12$ ms

3. $v_{out} = +15$ V, $0 < t < 12$ ms

   $v_{out} = -15$ V, 12 ms $< t < 16$ ms

5. $v_{out} = +15$ V, $0 < t < 8$ μs

   $v_{out} = -15$ V, 8 μs $< t < 24$ μs

7. $v_{out} = -15$ V, $0 < t < 35.2$ μs

   $v_{out} = +15$ V, 35.2 μs $< t < 60.8$ μs

9. LED1 will be on whenever $v_{in} > 10$ V

   LED2 will be on whenever $v_{in} > 5$ V

11. a. $I_1 = -0.20$ mA, $I_2 = 0.45$ mA, $I_3 = -0.15$ mA, $I_F = 0.10$ mA

    b. -2.0 V

15. a. 4.0 V_{p-p}

17. 15 V_p

19. a. $v_{out} = -0.5$ V, $0 < t < 1$ ms

    $v_{out} = 0.5$ V, 1 ms $< t < 2$ ms

21. $R_{in} = 100$ Ω, $C_F = 2.5$ nF

23. $I_1 = I_3 = 1.0$ μA (to the right), $I_2 = I_4 = 1.0$ μA (to the left), $v_{out} = 1.02$ V

25. -20.4 mV

27. $R_1 = 1590$ Ω, $R_F = 15.9$ kΩ, $R_2 = 1770$ Ω

29. $f_L = 99.5$ Hz, $f_H = 333$ kHz

31. $R = 19.89$ kΩ, $R_1 = 99.9$ Ω

33. $v_{out} = 12.3$ V $\rightarrow 13.5$ V, line regulation $= 5.79\%$

35. $v_{out} = 13.5$ V $\rightarrow 13.7$ V, load regulation $= 1.58\%$

37. a. $v_{out} = 6.6$ V $\rightarrow 25$ V

    b. 0.249 W

39. 54.0 dB

CHAPTER 30

1. $B = 0.1$, $A_{vcl} = 9.80$

3. UTP $= +9.0$ V, LTP $= -9.0$ V

5. UTP $= +10.5$ V, LTP $= -4.5$ V

7. $V_{REF} = \pm 4.8$ V, $T = 6.78$ mS

9. a. 20 kΩ   b. 72.3 Hz

11. $B = 0.25$, $A = 4$

13. a. 922 Hz   b. 10.5 kΩ

    c. 43.5 kΩ   d. 69.0 kHz

15. $R = 345$ Ω, $R_F = 10.0$ kΩ

17. 100.7 kHz

19. When the crystal is subjected to mechanical pressure, an electric field is developed at right angle to the applied pressure. Conversely, if an electric field is applied, the material will deflect at right angle to the electric field.

21. Refer to Figure 30–35.

23. a. $v_{out} = V_{CC}$ and the discharge transistor will be OFF.

    b. $v_{out} = 0$ and the discharge transistor will be ON.

25. $R_B = 361$ kΩ, $R_A = 721$ kΩ

27. $v_{out} = 0$ V $0 < t < 20$ ms

    $v_{out} = 15$ V $20 < t < 56.3$ ms

    $v_{out} = 0$ V 56.3 ms $< t$

29. $f_{0\%} = 34.1$ kHz, $f_{50\%} = 18.5$ kHz, $f_{100\%} = 2.93$ kHz

CHAPTER 31

1. a. 267 Hz

   b. $\eta = 0.627$ $R_{BB} = 5.98$ kΩ

3. $R_E = 12.7$ kΩ (Note: Any combination of $RC = 0.127$ ms will work.)

9. Triac operates in Modes I$^+$ and III$^+$.

11. $V = 81.1$ V $P = 43.9$ W

13. $V = 216$ V $P = 2320$ W

15. $\theta = 120°$

17. $V = 81.064$ V

19. 5.45×10^{14} Hz

21. 350 Ω

23. a. 20 V

    b. 13.5 V

    c. $v_{out} = +20$ V, $0 < t < 100$ ms

    $v_{out} = +13.5$ V, 100 ms $< t < 300$ ms

    $v_{out} = +20$ V, 300 ms $< t < 400$ ms

    $v_{out} = +13.5$ V, 400 ms $< t$

25. **L**ight **A**mplification through **S**timulated **E**mission of **R**adiation.

Glossary

ac Abbreviation for alternating current; used to denote periodically varying quantities such as ac current, ac voltage, and so on.

admittance (Y) A vector quantity (measured in siemens, S) that is the reciprocal of impedance. $Y = 1/Z$.

alternating current Current that periodically reverses in direction, commonly called an ac current.

alternating voltage Voltage that periodically changes in polarity, commonly called an ac voltage. The most common ac voltage is the sine wave.

American Wire Gauge (AWG) An American standard for classifying wire and cable.

ammeter An instrument that measures current.

ampere (A or amp) The SI unit of electrical current, equal to a rate of flow of one coulomb of charge per second.

ampere-hour (Ah) A measure of the storage capacity of a battery.

angular frequency (ω) Frequency of an ac waveform in radians/s. $\omega = 2\pi f$ where f is frequency in Hz.

apparent power (S) The power that apparently flows in an ac circuit. It has components of real power and reactive power, related by the power triangle. The magnitude of apparent power is equal to the product of effective voltage times effective current. Its unit is the VA (volt-amp).

atom The basic building block of matter. In the Bohr model, an atom consists of a nucleus of positively charged protons and uncharged neutrons, surrounded by negatively charged orbiting electrons. An atom normally consists of equal numbers of electrons and protons and is thus uncharged.

attenuation The amount that a signal decreases as it passes through a system. The attenuation is usually measured in decibels, dB.

audio frequency A frequency in the range of human hearing, which is typically from about 15 Hz. to 20 kHz.

autotransformer A type of transformer with a partially common primary and secondary winding. Part of its energy is transferred magnetically and part conductively.

average of a waveform The mean value of a waveform, obtained by algebraically summing the areas above and below the zero axis of the waveform, divided by the cycle length of the waveform. It is equal to the dc value of the waveform as measured by an ammeter or a voltmeter.

balanced (1) For a bridge circuit, the voltage between midpoints on its arms is zero. (2) In three-phase systems, a system (or a load) that is identical for all three phases.

band-pass filter A circuit that permits signals within a range of frequencies to pass through a circuit. Signals of all other frequencies are prevented from passing through the circuit.

band-stop filter (or notch filter) A circuit designed to prevent signals within a range of frequencies from passing through a circuit. Signals of all other frequencies freely pass through the circuit.

bandwidth (BW) The difference between the half-power frequencies for any resonant, band-pass, or band-stop filter. The bandwidth may be expressed in either hertz or radians per second.

Bode plot A straight line approximation that shows how the voltage gain of a circuit changes with frequency.

branch A portion of a circuit that occurs between two nodes (or terminals).

branch current The current through a branch of a circuit.

buffer An amplifier having a unity voltage gain ($A_v = 1$), very high input impedance, and very low output impedance. A buffer circuit is used to prevent loading effect.

capacitance A measure of charge storage capacity, for example, of a capacitor. A circuit with capacitance opposes a change in voltage. Unit is the farad (F).

capacitor A device that stores electrical charges on conductive "plates" separated by an insulating material called a dielectric.

cascade Two stages of a circuit are said to be in a cascade connection when the output of one stage is connected to the input of the next stage.

CGS system A system of units based on centimeters, grams, and seconds.

characteristic curve(s) A relationship between output current and output voltage of a semiconductor device. Characteristic curves may also show how output varies as a function of some other parameter such as input current, input voltage, and temperature.

charge (1) The electrical property of electrons and protons that causes a force to exist between them. Electrons are negatively charged while protons are positively charged. Charge is denoted by Q and is defined by Coulomb's law. (2) An excess or deficiency of electrons on a body. (3) To store electric charge as in to charge a capacitor or charge a battery.

choke Another name for an inductor.

circuit A system of interconnected components such as resistors, capacitors, inductors, voltages sources, and so on.

circuit breaker A resettable circuit protection device that trips a set of contacts to open the circuit when current reaches a preset value.

circuit common The reference point in a electrical circuit from which voltages are measured.

circular mil (CM) A unit used to specify the cross-section area of a cable or wire. The circular mil is defined as the area contained in a circle having a diameter of 1 mil (0.001 inch).

coefficient of coupling (k) A measure of the flux linkage between circuits such as coils. If $k = 0$, there is no linkage; if $k = 1$, all of the flux produced by one coil links another. The mutual inductance M between coils is related to k by the relationship $M = k\sqrt{L_1L_2}$, where L_1 and L_2 are the self-inductances of the coils.

coil A term commonly used to denote inductors or windings on transformers.

common-mode signal A signal that appears at both inputs of a differential amplifier.

conductance (G) The reciprocal of resistance. Unit is the siemens (S).

conductor A material through which charges move easily. Copper is the most common metallic conductor.

continuity of current This refers to the fact that current cannot change abruptly (i.e., in step-wise fashion) from one value to another in an isolated (i.e., non-coupled) inductance.

copper loss The I^2R power loss in a conductor due to its resistance, for example the power loss in the windings of a transformer.

core The form or structure around which an inductor or the coils of a transformer are wound. The core material affects the magnetic properties of the device.

core loss Power loss in the core of a transformer or inductor due to hysteresis and eddy currents.

coulomb (C) The SI unit of electrical charge, equal to the charge carried by 6.24×10^{18} electrons.

Coulomb's law An experimental law that states that the force (in Newtons) between charged particles is $F = Q_1Q_2/4\pi \in r^2$, where Q_1 and Q_2 are the charges (in coulombs), r is the distance between their centers in meters, and ϵ is the permittivity of the medium. For air, $\in = 8.854 \times 10^{-12} F/m$.

critical temperature The temperature below which a material becomes a superconductor.

current (I or i) The rate of flow of electrical charges in a circuit, measured in amperes.

current source A practical current source can be modeled as an ideal current source in parallel with an internal impedance.

cutoff frequency, f_c or ω_c The frequency at which the output power of a circuit is reduced to half of the maximum output power. The cutoff frequency may be measured in either hertz, (Hz) or radians per second, (rad/s).

cycle One complete variation of an ac waveform.

decade A tenfold change in frequency.

decibel (dB) A logarithmic unit used to represent an increase (or decrease) in power levels or sound intensity.

delta (Δ) A small change (increment or decrement) in a variable. For example, if current changes a small amount from i_1 to i_2, its increment is $\Delta i = i_2 - i_1$, while if time changes a small amount from t_1 to t_2, its increment is $\Delta t = t_2 - t_1$.

delta load A configuration of circuit components connected in the shape of a Δ (Greek letter delta). Sometimes called a pi (π) load.

derivative The instantaneous rate of change of a function. It is the slope of the tangent to the curve at the point of interest.

dielectric An insulating material. The term is commonly used with reference to the insulating material between the plates of a capacitor.

dielectric constant ($\in$) A common name for permittivity.

differentiator A circuit whose output is proportional to the derivative of its input.

diode A two-terminal component made of semiconductor material, which permits current in one direction while preventing current in the opposite direction.

direct current (dc) Unidirectional current such as that from a battery.

DMM A digital multimeter that displays results on a numeric readout. In addition to voltage, current, and resistance, some dmms measure other quantities such as frequency and capacitance.

duty cycle The ratio of on time to the duration of a pulse waveform, expressed in percent.

eddy current A small circulating current. Usually refers to the unwanted current that is induced in the core of an inductor or transformer by changing core flux.

effective resistance Resistance defined by $R = P/I^2$. For AC, effective resistance is larger than dc resistance due to skin effect and other effects such as power losses.

effective value An equivalent dc value of a time varying waveform, hence, that value of dc that has the same heating effect as the given waveform. Also called *rms* (root mean square) value. For sinusoidal current, $I_{eff} = 0.707 I_m$, where I_m is the amplitude of the ac waveform.

efficiency (η) The ratio of output power to input power, usually expressed as a percentage. $\eta = P_{out}/P_{in} \times 100\%$.

electron A negatively charged atomic particle. *See* atom.

energy (W) The ability to do work. Its SI unit is the joule; electrical energy is also measured in kilowatt-hours (kWh).

engineering notation A method of representing certain common powers of 10 via standard prefixes—for example, 0.125 A as 125 mA.

fall time (t_f) The time it takes for a pulse or step to change from its 90% value to its 10% value.

farad (F) The SI unit of capacitance, named in honor of Michael Faraday.

ferrite A magnetic material made from powdered iron oxide. Provides a good path for magnetic flux and has low enough eddy current losses that it is used as a core material for high frequency inductors and transformers.

field A region in space where a force is felt, hence a force field. For example, magnetic fields exist around magnets and electric fields exist around electric charges.

field intensity The strength of a field.

filter A circuit that passes certain frequencies while rejecting all other frequencies.

flux   A way of representing and visualizing force fields by drawing lines that show the strength and direction of a field at all points in space. Commonly used to depict electric or magnetic fields.

free electron   An electron that is weakly bound to its parent atom and is thus easily broken free. For materials like copper, there are billions of free electrons per cubic centimeter at room temperature. Since these electrons can break free and wander from atom to atom, they form the basis of an electric current.

frequency (f)   The number of times that a cycle repeats itself each second. Its SI unit is the hertz (Hz).

gain   The ratio of output voltage, current, or power to the input. Power gain for an amplifier is defined as the ratio of ac output power to ac input power, $A_p = P_{out}/P_{in}$. Gain may also be expressed in decibels. In the case of power gain, $A_p(dB) = 10 \log P_{out}/P_{in}$.

gauss   The unit of magnetic flux density in the CGS system of units.

giga (G)   A prefix with a value of 10^9.

ground   (1) An electrical connection to earth. (2) A circuit common. (*See* circuit common.) (3) A short to ground, such as a ground fault.

harmonics   Integer multiples of a frequency.

henry (H)   The SI unit of inductance, named in honor of Joseph Henry.

hertz (Hz)   The SI unit of frequency, named in honor of Heinrich Hertz. One Hz equals one cycle per second.

high-pass filter   A circuit that readily permits frequencies above the cutoff frequency to pass from the input to the output of the circuit, while attenuating frequencies below the cutoff frequency. (*See* cutoff frequency).

hysteresis loss   Power loss in a ferromagnetic material caused by the reversal of magnetic domains in a time varying magnetic field.

ideal current source   A current source having an infinite shunt (parallel) impedance. An ideal current source is able to provide the same current to all loads (except an open circuit). The voltage across the current source is determined by the value of the load impedance.

ideal transformer   A transformer having no losses and characterized by its turns ratio $a = N_p/N_s$. For voltage, $\mathbf{E}_p/\mathbf{E}_s = a$, while for current $\mathbf{I}_p/\mathbf{I}_s = 1/a$.

ideal voltage source   A voltage source having zero series impedance. An ideal voltage source is able to provide the same voltage across all loads (except a short circuit). The current through the voltage source is determined by the value of the load impedance.

impedance (Z)   Total opposition that a circuit element presents to sinusoidal ac in the phasor domain. $\mathbf{Z} = \mathbf{V}/\mathbf{I}$ ohms, where $\mathbf{V}$ and $\mathbf{I}$ are voltage and current phasors respectively. Impedance is a complex quantity with magnitude and angle.

induced voltage   Voltage produced by changing magnetic flux linkages.

inductance (L)   That property of a coil (or other current-carrying conductor) that opposes a change in current. The SI unit of inductance is the henry.

inductor   A circuit element designed to posses inductance, e.g., a coil of wire wound to increase its inductance.

initial-condition circuit   In transient analysis, this refers to a circuit drawn as it looks immediately after a disturbance (such as switching). In such a circuit, charged capacitors are represented by voltage sources, current-carrying inductors by current sources, uncharged capacitors by short circuits and non-current carrying inductors by open circuits.

instantaneous value   The value of a quantity (such as voltage or current) at some instant of time.

insulator   A material such as glass, rubber, bakelite, and so on, that does not conduct electricity.

integrator   A circuit whose output is proportional to the integral of its input.

internal impedance   The impedance that exists internally in a device such as a voltage source.

ion   An atom that has become charged. If it has an excess of electrons, it is a negative ion, while if it has a deficiency, it is a positive ion.

joule (J)   The SI unit of energy, equal to one newton-meter.

kilo   A prefix with the value of 10^3.

kilowatt-hour (kWh)   A unit of energy equal to 1000 W times one watt-hour and commonly used by electrical utilities.

Kirchhoff's current law   An experimental law that states that the sum of the currents entering a junction is equal to the sum leaving.

Kirchhoff's voltage law   An experimental law that states that the algebraic sum of voltages around a closed path in a circuit is zero.

lagging load   A load in which current lags voltage (e.g., an inductive load).

laser   A light source that emits very intense monochromatic (single color) coherent (in phase) light. The term is an acronym for Light Amplification through Stimulated Emission of Radiation.

leading load   A load in which current leads voltage (e.g., a capacitive load).

linear circuit   A circuit in which relationships are proportional. In a linear circuit, current is proportional to voltage.

load   (1) The device that is being driven by a circuit. Thus, the lamp in a flashlight is the load. (2) The current drawn by a load.

low-pass filter   A circuit that permits frequencies below the cutoff frequency to pass through from the input to the output of the circuit, while attenuating frequencies above the cutoff frequency. (*See* cutoff frequency.)

magnetic flux density (B)   The number of magnetic flux lines per unit area, measured in the SI system in tesla (T), where one T = one Wb/m^2.

magnetomotive force (mmf)   The flux producing ability of a coil. In the SI system, the mmf of a coil of N turns with current I is NI ampere-turns.

maxwell (Mx)   The CGS unit of magnetic flux Φ.

mega (M)   A prefix with the value of 10^6.

micro (μ)   A prefix with the value of 10^{-6}.

milli (m)   A prefix with the value of 10^{-3}.

multimeter   A multifunction meter used to measure a variety of electrical quantities such as voltage, current, and resistance. Its function and range is selected by a switch. (*See also* DMM.)

mutual inductance (*M*)   The inductance between circuits (such as coils) measured in henries. The voltage induced in one circuit by changing current in another circuit is equal to *M* times the rate of change of current in the first circuit.

nano (n)   A prefix with the value of 10^{-9}.

neutron   An atomic particle with no charge. (*See* atom.)

node   A junction where two or more components connect in an electric circuit.

octave   A two-fold increase (or decrease) in frequency.

ohm (Ω)   The SI unit of resistance. Also used as the unit for reactance and impedance.

ohmmeter   An instrument for measuring resistance.

open circuit   A discontinuous circuit, hence one that does not provide a complete path for current.

operational amplifier   An electronic amplifier characterized as having very high open-loop gain, very high input impedance, and very low output impedance.

oscilloscope   An instrument that electronically displays voltage waveforms on a screen. The screen is ruled with a scaled grid to permit measurement of the waveform's characteristics.

parallel   Elements or branches are said to be in a parallel connection when they have exactly two nodes in common. The voltage across all parallel elements or branches is exactly the same.

peak   The maximum instantaneous value (positive or negative) of a waveform.

peak-to-peak   The magnitude of the difference between a waveform's maximum and minimum values.

period (*T*)   The time for a waveform to go through one cycle. $T = 1/f$ where f is frequency in Hz.

periodic   Repeating at regular intervals.

permeability (μ)   A measure of how easy it is to magnetize a material. $B = \mu H$, where B is the resulting flux density and H is the magnetizing force that creates the flux.

permittivity ($\in$)   A measure of how easy it is to establish electric flux in a material. (*See also* relative dielectric constant and Coulomb's law.)

phase shift   The angular difference by which one waveform leads or lags another, hence the relative displacement between time varying waveforms.

phasor   A way of representing the magnitude and angle of a sine wave graphically or by a complex number. The magnitude of the phasor represents the rms value of the ac quantity and its angle represents the waveform's phase.

pico (p)   A prefix with the value of 10^{-12}.

potentiometer   A three-terminal resistor consisting of a fixed resistance between two end terminals and a third terminal that is connected to a movable wiper arm. When the end terminals are connected to a voltage source, the voltage between the wiper and either of the other terminals is adjustable.

power (*P*, *p*)   The rate of doing work, with units of watts, where one watt equals one joule per second. Also called real or active power.

power factor   The ratio of active power to apparent power, equal to cos θ, where θ is the angle between the voltage and the current.

power triangle   A way to represent the relationship between real power, reactive power, and apparent power using a triangle.

primary   The winding of a transformer to which we connect the source.

proton   A positively charged atomic particle. (*See* atom).

pulse   A short duration voltage or current that abruptly changes from one value to another, then back again.

pulse width   The duration of a pulse. For non-ideal pulses, it is measured at the 50% amplitude point.

quality factor (*Q*)   (1) A figure of merit. *Q* for a coil is the ratio of its reactive power to its real power. The higher the *Q*, the more closely the coil approaches the ideal. (2) A measure of the selectivity of a resonant circuit. The higher the *Q*, the narrower the bandwidth.

reactance (*X*)   The opposition that a reactive element (capacitance or inductance) presents to sinusoidal ac, measured in ohms.

reactive power   A component of power that alternately flows into then out of a reactive element, measured in VARs (volt-amps reactive). Reactive power has an average value of zero and is sometimes called "wattless" power.

reactor   Another name for an inductor.

rectifier   A circuit, generally consisting of a least one diode, which permits current in only one direction.

regulation   The change in voltage from no-load to full-load expressed as a percentage of full load voltage.

relative dielectric constant ($\in_r$)   The ratio of the dielectric constant of a material to that of a vacuum.

relay   A switching device that is opened or closed by an electrical signal. May be electromechanical or electronic.

reluctance   The opposition of a magnetic circuit to the establishment of flux.

resistance (*R*)   The opposition to current that results in power dissipation. Thus, $R = P/I^2$ ohms. For a dc circuit, $R = V/I$, while for an ac circuit containing reactive elements, $R = V_R/I$, where V_R is the component of voltage across the resistive part of the circuit.

resistor   A circuit component designed to posses resistance.

resonance, resonant frequency   The frequency at which the output power of an *L-R-C* circuit is at a maximum. $f = 1/(2\pi \sqrt{LC})$

rheostat   A variable resistor connected so that current through the circuit is controlled by the position of the wiper.

rise time (*t_r*)   The time that it takes for a pulse or step to change from its 10% value to its 90% value.

rms value   The root-mean-square value of a time varying waveform. (*See* effective value.)

saturation   The condition of a ferromagnetic material where it is fully magnetized. Thus, if the magnetizing force (current in a coil for example) is increased, no significant increase in flux results.

schematic diagram   A circuit diagram that uses symbols to represent physical components.

secondary winding   The output winding of a transformer.

selectivity   The ability of a filter circuit to pass a particular frequency, while rejecting all other frequency components.

semiconductor   A material such as silicon from which transistors, diodes, and the like are made.

series circuit   A closed loop of elements where two elements have no more than one common terminal. In a series circuit, there is only one current path and all series elements have the same current.

short circuit   A short circuit occurs when two terminals of an element or branch are connected together by a low-resistance conductor. When a short circuit occurs, very large currents may result in sparks or a fire, particularly when the circuit is not protected by a fuse or circuit breaker.

SI System   The international system of units used in science and engineering. It is a metric system and includes the standard units for length, mass, and time (e.g., meters, kilograms, and seconds), as well as the electrical units (e.g., volts, amperes, ohms, and so on).

siemens (S)   A unit of measure for conductance, admittance, and susceptance. The siemens is the reciprocal of ohm.

silicon controlled rectifier   A thyristor that permits current in only one direction once a suitable gate signal is present.

sine wave   A periodic waveform that is described by the trigonometric sine function. It is the principle waveform used in ac systems.

skin effect   At high frequencies, the tendency of current to travel in a thin layer near the surface of a conductor.

spectrum analyzer   An instrument that displays the amplitude of a signal as the function of frequency.

steady state   The condition of operation of a circuit after transients have subsided.

step   An abrupt change in voltage or current, as for example when a switch is closed to connect a battery to a resistor.

superconductor   A conductor that has no internal resistance. Current will continue unimpeded through a superconductor even though there is no externally applied voltage or current source.

susceptance   The reciprocal of reactance. Unit is the siemens.

tank circuit   A circuit consisting of an inductor and capacitor connected in parallel. Such an L-C circuit is used in oscillators and receivers to provide maximum signal at the resonant frequency. (See selectivity.)

temperature coefficient   (1) The rate at which resistance changes as the temperature changes. A material has a positive temperature coefficient if the resistance increases with an increase in temperature. Conversely, a negative temperature coefficient means that resistance decreases as temperature is increased. (2) Similarly for capacitance. The change in capacitance is due to changes in the characteristics of its dielectric with temperature.

tesla (T)   The SI unit of magnetic flux density. One T = one Wb/m^2.

time constant (τ)   A measure of how long a transient lasts. For example, during charging, capacitor voltage changes by 63.2% in

one time constant, and for all practical purposes, charges fully in five time constants, For an RC circuit, $\tau = RC$ seconds and for an RL circuit, $\tau = L/R$ seconds.

transformer   A device with two or more coils in which energy is transferred from one winding to the other by electromagnetic action.

transient   A temporary or transitional voltage or current.

transistor model   An electric circuit that simulates the operation of a transistor amplifier.

triac   A thyristor that permits current in either direction once a suitable gate signal is present.

turns ratio (a)   The ratio of primary turns to secondary turns; $a = N_p/N_s$.

valence shell   The outermost (last occupied) shell of an atom.

varactor diode (or varicap, epicap, and tuning diode)   A diode that behaves as a voltage-variable capacitor.

volt   The unit of voltage in the SI system.

voltage (V, v, E, e)   Potential difference created when charges are separated, as for example by chemical means in a battery. If one joule of work is required to move a charge of one coulomb from one point to another, the potential difference between the points is one volt.

voltage controlled oscillator   Provides an output frequency that is directly proportional to the magnitude of the applied input voltage.

voltage regulator   A device that maintains constant output voltage to a load regardless of the input voltage or the amount of output current.

voltage source   A practical voltage source can be modeled as an ideal voltage source in a series with an internal impedance.

watt (W)   The SI unit of active power. Power is the rate at which work is done; one watt equals one joule/s.

watthour (Wh)   A unit of energy, equal to one watt times one hour. One Wh = 3600 joules.

waveform   The variation versus time of a time varying signal, hence, the shape of a signal.

weber (Wb)   The SI unit of magnetic flux.

work (W)   The product of force times distance, measured in joules in the SI system, where one joule equals one newton-meter.

wye load   A configuration of circuit components connected in the shape of a Y. Sometimes called a star or T load.

zener diode   A diode that normally operates in its reverse region and is used to maintain a constant output voltage.

Index